Learning the Art of Electronics

This introduction to circuit design is unusual in several respects.

First, it offers not just explanations, but a full lab course. Each of the 25 daily sessions begins with a discussion of a particular sort of circuit followed by the chance to try it out and see how it actually behaves. Accordingly, students understand the circuit's operation in a way that is deeper and much more satisfying than the manipulation of formulas.

Second, it describes circuits that more traditional engineering introductions would postpone: thus, on the third day, we build a radio receiver; on the fifth day, we build an operational amplifier from an array of transistors. The digital half of the course centers on applying microcontrollers, but gives exposure to Verilog, a powerful Hardware Description Language.

Third, it proceeds at a rapid pace but requires no prior knowledge of electronics. Students gain intuitive understanding through immersion in good circuit design.

- Each session is divided into several parts, including Notes, Labs; many also have Worked Examples and Supplementary Notes
- An appendix introducing Verilog
- Further appendices giving background facts on oscilloscopes, Xilinx, transmission lines, pinouts, programs etc, plus advice on parts and equipment
- Very little math: focus is on intuition and practical skills
- A final chapter showcasing some projects built by students taking the course over the years

Thomas C. Hayes reached electronics via a circuitous route that started in law school and eventually found him teaching Laboratory Electronics at Harvard, which he has done for thirty-five years. He has also taught electronics for the Harvard Summer School, the Harvard Extension School, and for seventeen years in Boston University's Department of Physics. He shares authorship of one patent, for a device that logs exposure to therapeutic bright light. He and his colleagues are trying to launch this device with a startup company named Goodlux Technologies. Tom designs circuits as the need for them arises in the electronics course. One such design is a versatile display, serial interface and programmer for use with the microcomputer that students build in the course.

Paul Horowitz is a Research Professor of Physics and of Electrical Engineering at Harvard University, where in 1974 he originated the Laboratory Electronics course from which emerged *The Art of Electronics*.

Learning the Art of Electronics

A Hands-On Lab Course

Thomas C. Hayes

with the assistance of Paul Horowitz

CAMBRIDGE
UNIVERSITY PRESS

University Printing House, Cambridge CB2 8BS, United Kingdom

Cambridge University Press is part of the University of Cambridge.

It furthers the University's mission by disseminating knowledge in the pursuit of
education, learning, and research at the highest international levels of excellence.

www.cambridge.org
Information on this title: www.cambridge.org/9780521177238

© Cambridge University Press 2016

This publication is in copyright. Subject to statutory exception
and to the provisions of relevant collective licensing agreements,
no reproduction of any part may take place without the written
permission of Cambridge University Press.

First published 2016
3rd printing with corrections 2017

Printed in the United Kingdom by TJ International Ltd. Padstow Cornwall

A catalogue record for this publication is available from the British Library.

Library of Congress Cataloguing in Publication Data

ISBN 978-0-521-17723-8 Paperback

Cambridge University Press has no responsibility for the persistence or accuracy
of URLs for external or third-party Internet Web sites referred to in this publication
and does not guarantee that any content on such Web sites is, or will remain,
accurate or appropriate.

For Debbie, Tessa, Turner and Jamie

And in memory of my beloved friend, Jonathan

Contents

Preface		*page* xx
Overview, as the Course begins		xxv
Part I	**Analog: Passive Devices**	1
1N	**DC Circuits**	3
	1N.1 Overview	3
	1N.2 Three laws	5
	1N.3 First application: voltage divider	11
	1N.4 Loading, and "output impedance"	14
	1N.5 Readings in AoE	24
1L	**Lab: DC Circuits**	25
	1L.1 Ohm's law	25
	1L.2 Voltage divider	26
	1L.3 Converting a meter movement into a voltmeter and ammeter	27
	1L.4 The diode	29
	1L.5 I versus V for some mystery boxes	30
	1L.6 Oscilloscope and function generator	32
1S	**Supplementary Notes: Resistors, Voltage, Current**	35
	1S.1 Reading resistors	35
	1S.2 Voltage versus current	38
1W	**Worked Examples: DC circuits**	42
	1W.1 Design a voltmeter, current meter	42
	1W.2 Resistor power dissipation	44
	1W.3 Working around imperfections of instruments	45
	1W.4 Thevenin models	47
	1W.5 "Looking through" a circuit fragment, and R_{in}, R_{out}	48
	1W.6 Effects of loading	49
2N	**RC Circuits**	51
	2N.1 Capacitors	51
	2N.2 Time-domain view of RCs	53
	2N.3 Frequency domain view of RCs	58
	2N.4 Blocking and decoupling	74

	2N.5	A somewhat mathy view of *RC* filters	76
	2N.6	Readings in AoE	77
2L	**Labs: Capacitors**		**78**
	2L.1	Time-domain view	78
	2L.2	Frequency domain view	81
2S	**Supplementary Notes: RC Circuits**		**85**
	2S.1	Reading capacitors	85
	2S.2	*C* notes: trying for an intuitive grip on capacitors' behavior	90
	2S.3	Sweeping frequencies	93
2W	**Worked Examples: RC Circuits**		**100**
	2W.1	*RC* filters	100
	2W.2	*RC* step response	105
3N	**Diode Circuits**		**108**
	3N.1	Overloaded filter: another reason to follow our 10× loading rule	108
	3N.2	Scope probe	109
	3N.3	Inductors	112
	3N.4	*LC* resonant circuit	113
	3N.5	Diode Circuits	118
	3N.6	The most important diode application: DC from AC	119
	3N.7	The most important diode application: (unregulated-) power supply	123
	3N.8	Radio!	126
	3N.9	Readings in AoE	130
3L	**Lab: Diode Circuits**		**131**
	3L.1	*LC* resonant circuit	131
	3L.2	Half-wave rectifier	133
	3L.3	Full-wave bridge rectifier	134
	3L.4	Design exercise: AM radio receiver (fun!)	135
	3L.5	Signal diodes	136
3S	**Supplementary Notes and Jargon: Diode Circuits**		**138**
	3S.1	A puzzle: why *LC*'s ringing dies away despite Fourier	138
	3S.2	Jargon: passive devices	139
3W	**Worked Examples: Diode Circuits**		**141**
	3W.1	Power supply design	141
	3W.2	Z_{IN}	144
Part II	**Analog: Discrete Transistors**		**149**
4N	**Transistors I**		**151**
	4N.1	Overview of Days 4 and 5	151
	4N.2	Preliminary: introductory sketch	154

	4N.3	The simplest view: forgetting beta	155
	4N.4	Add quantitative detail: use beta explicitly	158
	4N.5	A strikingly different transistor circuit: the switch	166
	4N.6	Recapitulation: the important transistor circuits at a glance	167
	4N.7	AoE Reading	168
4L		**Lab: Transistors I**	169
	4L.1	Transistor preliminaries: look at devices out of circuit	169
	4L.2	Emitter follower	170
	4L.3	Current source	172
	4L.4	Common-emitter amplifier	172
	4L.5	Transistor switch	174
	4L.6	A note on power supply noise	176
4W		**Worked Examples: Transistors I**	178
	4W.1	Emitter follower	178
	4W.2	Phase splitter: input and output impedances of a transistor circuit	181
	4W.3	Transistor switch	185
5N		**Transistors II**	188
	5N.1	Some novelty, but the earlier view of transistors still holds	188
	5N.2	Reviewish: phase splitter	189
	5N.3	Another view of transistor behavior: Ebers–Moll	190
	5N.4	Complication: distortion in a high-gain amplifier	194
	5N.5	Complications: temperature instability	196
	5N.6	Reconciling the two views: Ebers–Moll meets $I_C = \beta \times I_B$	201
	5N.7	"Difference" or "differential" amplifier	201
	5N.8	Postscript: deriving r_e	207
	5N.9	AoE Reading	208
5L		**Lab: Transistors II**	209
	5L.1	Difference or differential amplifier	209
5S		**Supplementary Notes and Jargon: Transistors II**	220
	5S.1	Two surprises, perhaps, in behavior of differential amp	220
	5S.2	Current mirrors; Early effect	222
	5S.3	Transistor summary	230
	5S.4	Important circuits	232
	5S.5	Jargon: bipolar transistors	235
5W		**Worked Examples: Transistors II**	237
	5W.1	High-gain amplifiers	237
	5W.2	Differential amplifier	238
	5W.3	Op-amp innards: diff-amp within an IC operational amplifier	239

Part III Analog: Operational Amplifiers and their Applications — 243

6N Op-amps I — 245
- 6N.1 Overview of feedback — 245
- 6N.2 Preliminary: negative feedback as a general notion — 248
- 6N.3 Feedback in electronics — 249
- 6N.4 The op-amp golden rules — 251
- 6N.5 Applications — 252
- 6N.6 Two amplifiers — 252
- 6N.7 Inverting amplifier — 254
- 6N.8 When do the Golden Rules apply? — 256
- 6N.9 Strange things can be put into feedback loop — 259
- 6N.10 AoE Reading — 261

6L Lab: Op-Amps I — 262
- 6L.1 A few preliminaries — 262
- 6L.2 Open-loop test circuit — 263
- 6L.3 Close the loop: follower — 263
- 6L.4 Non-inverting amplifier — 265
- 6L.5 Inverting amplifier — 265
- 6L.6 Summing amplifier — 266
- 6L.7 Design exercise: unity-gain phase shifter — 266
- 6L.8 Push–pull buffer — 268
- 6L.9 Current to voltage converter — 269
- 6L.10 Current source — 271

6W Worked Examples: Op-Amps I — 273
- 6W.1 Basic difference amp made with an op-amp — 273
- 6W.2 A more exotic difference amp — 276
- 6W.3 Problem: odd summing circuit — 277

7N Op-amps II: Departures from Ideal — 280
- 7N.1 Old: subtler cases, for analysis — 281
- 7N.2 Op-amp departures from ideal — 284
- 7N.3 Four more applications — 294
- 7N.4 Differentiator — 300
- 7N.5 Op-amp Difference Amplifier — 301
- 7N.6 AC amplifier: an elegant way to minimize effects of op-amp DC errors — 301
- 7N.7 AoE Reading — 302

7L Labs: Op-Amps II — 303
- 7L.1 Integrator — 303
- 7L.2 Differentiator — 306
- 7L.3 Slew rate — 308
- 7L.4 AC amplifier: microphone amplifier — 308

7S Supplementary Notes: Op-Amp Jargon — 310

7W	**Worked Examples: Op-Amps II**	311
	7W.1 The problem	311
	7W.2 Op-amp millivoltmeter	314

8N	**Op-Amps III: Nice Positive Feedback**	319
	8N.1 Useful positive feedback	319
	8N.2 Comparators	320
	8N.3 RC relaxation oscillator	327
	8N.4 Sine oscillator: Wien bridge	331
	8N.5 AoE Reading	335

8L	**Lab. Op-Amps III**	336
	8L.1 Two comparators	336
	8L.2 Op-amp RC relaxation oscillator	338
	8L.3 Easiest RC oscillator, using IC Schmitt trigger	339
	8L.4 Apply the sawtooth: PWM motor drive	340
	8L.5 IC RC relaxation oscillator: '555	341
	8L.6 '555 for low-frequency frequency modulation ("FM")	342
	8L.7 Sinewave oscillator: Wien bridge	343

8W	**Worked Examples: Op-Amp III**	345
	8W.1 Schmitt trigger design tips	345
	8W.2 Problem: heater controller	348

9N	**Op-Amps IV: Parasitic Oscillations; Active Filter**	353
	9N.1 Introduction	353
	9N.2 Active filters	354
	9N.3 Nasty "parasitic" oscillations: the problem, generally	356
	9N.4 Parasitic oscillations in op-amp circuits	356
	9N.5 Op-amp remedies for keeping loops stable	361
	9N.6 A general criterion for stability	365
	9N.7 Parasitic oscillation without op-amps	367
	9N.8 Remedies for parasitic oscillation	370
	9N.9 Recapitulation: to keep circuits quiet…	372
	9N.10 AoE Reading	372

9L	**Labs. Op-Amps IV**	373
	9L.1 VCVS active filter	373
	9L.2 Discrete transistor follower	374
	9L.3 Op-amp instability: phase shift can make an op-amp oscillate	376
	9L.4 Op-amp with buffer in feedback loop	378

9S	**Supplementary Notes. Op-Amps IV**	380
	9S.1 Op-amp frequency compensation	380
	9S.2 Active filters: how to improve a simple RC filter	384
	9S.3 Noise: diagnosing fuzz	389
	9S.4 Annotated LF411 op-amp schematic	395
	9S.5 Quantitative effects of feedback	396

Contents

9W **Worked Examples: Op-Amps IV** — 401
- 9W.1 What all that op-amp gain does for us — 401
- 9W.2 Stability questions — 402

10N **Op-Amps V: PID Motor Control Loop** — 407
- 10N.1 Examples of real problems that call for this remedy — 408
- 10N.2 The PID motor control loop — 408
- 10N.3 Designing the controller (custom op-amp) — 410
- 10N.4 Proportional-only circuit: predicting how much gain the loop can tolerate — 412
- 10N.5 Derivative, D — 415
- 10N.6 AoE Reading — 420

10L **Lab. Op-Amps V** — 421
- 10L.1 Introduction: why bother with the PID loop? — 421
- 10L.2 PID motor control — 422
- 10L.3 Add derivative of the error — 428
- 10L.4 Add integral — 430
- 10L.5 Scope images: effect of increasing gain, in P-only loop — 432

11N **Voltage Regulators** — 433
- 11N.1 Evolving a regulated power supply — 434
- 11N.2 Easier: 3-terminal *IC* regulators — 439
- 11N.3 Thermal design — 441
- 11N.4 Current sources — 443
- 11N.5 Crowbar overvoltage protection — 444
- 11N.6 A different scheme: switching regulators — 445
- 11N.7 AoE Readings — 450

11L **Lab: Voltage Regulators** — 451
- 11L.1 Linear voltage regulators — 451
- 11L.2 A switching voltage regulator — 457

11W **Worked Examples: Voltage Regulators** — 462
- 11W.1 Choosing a heat sink — 462
- 11W.2 Applying a current-source IC — 463

12N **MOSFET Switches** — 465
- 12N.1 Why we treat FETs as we do — 465
- 12N.2 Power switching: turning something ON or OFF — 469
- 12N.3 A power switch application: audio amplifier — 471
- 12N.4 Logic gates — 473
- 12N.5 Analog switches — 474
- 12N.6 Applications — 475
- 12N.7 Testing a sample-and-hold circuit — 480
- 12N.8 AoE Reading — 485

12L	**Lab: MOSFET Switches**	486
	12L.1 Power MOSFET	486
	12L.2 Analog switches	489
	12L.3 Switching audio amplifier	495
12S	**Supplementary Notes: MOSFET Switches**	497
	12S.1 A physical picture	497
13N	**Group Audio Project**	503
	13N.1 Overview: a day of group effort	503
	13N.2 One concern for everyone: stability	506
	13N.3 Sketchy datasheets for LED and phototransistor	507
13L	**Lab: Group Audio Project**	508
	13L.1 Typical waveforms	508
	13L.2 Debugging strategies	509
Part IV	**Digital: Gates, Flip-Flops, Counters, PLD, Memory**	511
14N	**Logic Gates**	513
	14N.1 Analog versus digital	513
	14N.2 Number codes: Two's-complement	518
	14N.3 Combinational logic	520
	14N.4 The usual way to do digital logic: programmable arrays	526
	14N.5 Gate types: TTL and CMOS	528
	14N.6 Noise immunity	530
	14N.7 More on gate types	533
	14N.8 AoE Reading	535
14L	**Lab: Logic Gates**	537
	14L.1 Preliminary	537
	14L.2 Input and output characteristics of integrated gates: TTL and CMOS	540
	14L.3 Pathologies	541
	14L.4 Applying IC gates to generate particular logic functions	543
	14L.5 Gate innards; looking within the black box of CMOS logic	544
14S	**Supplementary Notes: Digital Jargon**	548
14W	**Worked Examples: Logic Gates**	550
	14W.1 Multiplexing: generic	550
	14W.2 Binary arithmetic	554
15N	**Flip-Flops**	567
	15N.1 Implementing a combinational function	568
	15N.2 Active-low, again	569
	15N.3 Considering gates as "Do this/do that" functions	573
	15N.4 XOR as Invert/Pass* function	574

	15N.5	OR as Set/Pass* function	575
	15N.6	Sequential circuits generally, and flip-flops	575
	15N.7	Applications: more debouncers	582
	15N.8	Counters	583
	15N.9	Synchronous counters	584
	15N.10	Another flop application: shift-register	586
	15N.11	AoE Reading	587
15L	**Lab: Flip-Flops**	588
	15L.1	A primitive flip-flop: *SR* latch	588
	15L.2	*D* type	588
	15L.3	Counters: ripple and synchronous	591
	15L.4	Switch bounce, and three debouncers	592
	15L.5	Shift register	594
15S	**Supplementary Note: Flip-Flops**	597
	15S.1	Programmable logic devices	597
	15S.2	Flip-flop tricks	599
16N	**Counters**	603
	16N.1	Old topics	603
	16N.2	Circuit dangers and anomalies	607
	16N.3	Designing a larger, more versatile counter	610
	16N.4	A recapitulation of useful counter functions	614
	16N.5	Lab 16L's divide-by-*N* counter	615
	16N.6	Counting as a digital design strategy	616
16L	**Lab: Counters**	617
	16L.1	A fork in the road: two paths into microcontrollers	617
	16L.2	Counter lab	619
	16L.3	16-bit counter	621
	16L.4	Make horrible music	629
	16L.5	Counter applications: stopwatch	631
16W	**Worked Examples: Applications of Counters**	634
	16W.1	Modifying count length: strange-modulus counters	634
	16W.2	Using a counter to measure period, thus many possible input quantities	636
	16W.3	Bullet timer	642
17N	**Memory**	648
	17N.1	Buses	648
	17N.2	Memory	651
	17N.3	State machine: new name for old notion	655
17L	**Lab: Memory**	661
	17L.1	RAM	661
	17L.2	State machines	663
	17L.3	State machine using a PAL programmed in Verilog	669

Contents — xv

17S   **Supplementary Notes: Digital Debugging and Address Decoding**   671
- 17S.1 Digital debugging tips   671
- 17S.2 Address decoding   675

17W   **Worked Examples: Memory**   678
- 17W.1 A sequential digital lock   678
- 17W.2 Solutions   681

Part V   **Digital: Analog–Digital, PLL, Digital Project Lab**   687

18N   **Analog ↔ Digital; PLL**   689
- 18N.1 Interfacing among logic families   689
- 18N.2 Digital ⇔ analog conversion, generally   693
- 18N.3 Digital to analog (DAC) methods   697
- 18N.4 Analog-to-digital conversion   701
- 18N.5 Sampling artifacts   712
- 18N.6 Dither   714
- 18N.7 Phase-locked loop   716
- 18N.8 AoE Reading   723

18L   **Lab: Analog ↔ Digital; PLL**   724
- 18L.1 Analog-to-digital converter   724
- 18L.2 Phase-locked loop: frequency multiplier   729

18S   **Supplementary Notes: Sampling Rules; Sampling Artifacts**   734
- 18S.1 What's in this chapter?   734
- 18S.2 General notion: sampling produces predictable artifacts in the sampled data   734
- 18S.3 Examples: sampling artifacts in time- and frequency-domains   735
- 18S.4 Explanation? The images, intuitively   739

18W   **Worked Examples: Analog ↔ Digital**   745
- 18W.1 ADC   745
- 18W.2 Level translator   748

19L   **Digital Project Lab**   749
- 19L.1 A digital project   749

Part VI   **Microcontrollers**   755

20N   **Microprocessors 1**   757
- 20N.1 Microcomputer basics   757
- 20N.2 Elements of a minimal machine   760
- 20N.3 Which controller to use?   762
- 20N.4 Some possible justifications for the hard work of the big-board path   764
- 20N.5 Rediscover the micro's control signals...   765
- 20N.6 Some specifics of our lab computer: big-board branch   771
- 20N.7 The first day on the SiLab branch   773
- 20N.8 AoE Reading   778

20L	**Lab: Microprocessors 1**	780
	20L.1 Big-board Dallas microcomputer	780
	20L.2 Install the GLUEPAL; wire it partially	781
	20L.3 SiLabs 1: startup	792
20S	**Supplementary Notes: Microprocessors 1**	803
	20S.1 PAL for microcomputers	803
	20S.2 Note on SiLabs IDE	805
20W	**Worked Examples: A Garden of Bugs**	809
21N	**Microprocessors 2. I/O, First Assembly Language**	813
	21N.1 What is assembly language? Why bother with it?	813
	21N.2 Decoding, again	818
	21N.3 Code to use the I/O hardware (big-board branch)	821
	21N.4 Comparing assembly language with C code: keypad-to-display	824
	21N.5 Subroutines: CALL	826
	21N.6 Stretching operations to 16 bits	830
	21N.7 AoE Reading	831
21L	**Lab: Microprocessors 2**	832
	21L.1 Big-board: I/O. Introduction	832
	21L.2 SiLabs 2: input; byte operations	844
21S	**Supplementary Notes: 8051 Addressing Modes**	857
	21S.1 Getting familiar with the 8051's addressing modes	857
	21S.2 Some 8051 addressing modes illustrated	867
22N	**Micro 3: Bit Operations**	869
	22N.1 BIT operations	869
	22N.2 Digression on conditional branching	874
22L	**Lab Micro 3. Bit Operations; Timers**	881
	22L.1 Big-board lab. Bit operations; interrupt	881
	22L.2 SiLabs 3: Timers; PWM; Comparator	886
22W	**Worked Examples. Bit Operations: An Orgy of Error**	901
	22W.1 The problem	901
	22W.2 Lots of poor, and one good, solutions	901
	22W.3 Another way to implement this "Ready" key	904
23N	**Micro 4: Interrupts; ADC and DAC**	905
	23N.1 Big ideas from last time	905
	23N.2 Interrupts	906
	23N.3 Interrupt handling in C	911
	23N.4 Interfacing ADC and DAC to the micro	912

	23N.5 Some details of the ADC/DAC labs	917
	23N.6 Some suggested lab exercises, playing with ADC and DAC	921
23L	**Lab Micro 4. Interrupts; ADC and DAC**	926
	23L.1 ADC → DAC	926
	23L.2 SiLabs 4: Interrupt; DAC and ADC	931
23S	**Supplementary Notes: Micro 4**	946
	23S.1 Using the RIDE assembler/compiler and simulator	946
	23S.2 Debugging	951
	23S.3 Waveform processing	955
24N	**Micro 5. Moving Pointers, Serial Buses**	959
	24N.1 Moving pointers	959
	24N.2 DPTR can be useful for SiLabs '410, too: tables	964
	24N.3 End tests in table eperations	964
	24N.4 Some serial buses	966
	24N.5 Readings	974
24L	**Lab Micro 5. Moving Pointers, Serial Buses**	975
	24L.1 Data table; SPI bus; timers	976
	24L.2 SiLabs 5: serial buses	982
24S	**Supplementary Note: Dallas Program Loader**	993
	24S.1 Dallas downloader	993
	24S.2 Hardware required	993
	24S.3 Procedure to try the loader: two versions	994
	24S.4 Debugging: LOADER420, in case you can't write to flash	999
	24S.5 Debugging in case of trouble with COM port assignments	1000
24W	**Worked Example: Table Copy, Four Ways**	1003
	24W.1 Several ways to copy a table	1003
25N	**Micro 6: Data Tables**	1006
	25N.1 Input and output devices for a microcontroller	1006
	25N.2 Task for big-board users: standalone micro	1008
	25N.3 Task for SiLabs users: off-chip RAM	1009
25L	**Lab: Micro 6: Standalone Microcontroller**	1012
	25L.1 Hardware alternatives: two ways to program the flash ROM	1012
	25L.2 SiLabs 6: SPI RAM	1018
	25L.3 Appendix: Program Listings	1021
26N	**Project Possibilities: Toys in the Attic**	1022
	26N.1 One more microcontroller that may interest you	1023
	26N.2 Projects: an invitation and a caution	1025
	26N.3 Some pretty projects	1025

	26N.4 Some other memorable projects	1030
	26N.5 Games	1041
	26N.6 Sensors, actuators, gadgets	1043
	26N.7 Stepper motor drive	1049
	26N.8 Project ideas	1051
	26N.9 Two programs that could be useful: LCD, Keypad	1052
	26N.10 And many examples are shown in AoE	1052
	26N.11 Now go forth	1052
A	**A Logic Compiler or HDL: Verilog**	**1053**
	A.1 The form of a Verilog file: design file	1053
	A.2 Schematics can help one to debug	1054
	A.3 The form of a Verilog file: simulation testbench	1055
	A.4 Self-checking testbench	1058
	A.5 Flip-flops in Verilog	1060
	A.6 Behavioral versus structural design description: easy versus hard	1064
	A.7 Verilog allows hierarchical designs	1065
	A.8 A BCD counter	1068
	A.9 Two alternative ways to instantiate a sub-module	1070
	A.10 State machines	1071
	A.11 An instance more appropriate to state form: a bus arbiter	1073
	A.12 Xilinx ISE offers to lead you by the hand	1076
	A.13 Blocking versus non-blocking assignments	1077
B	**Using the Xilinx Logic Compiler**	**1080**
	B.1 Xilinx, Verilog, and ABEL: an overview	1080
C	**Transmission Lines**	**1089**
	C.1 A topic we have dodged till now	1089
	C.2 A new case: transmission line	1090
	C.3 Reflections	1092
	C.4 But why do we care about reflections?	1094
	C.5 Transmission line effects for sinusoidal signals	1097
D	**Scope Advice**	**1099**
	D.1 What we don't intend to tell you	1099
	D.2 What we'd like to tell you	1099
E	**Parts List**	**1105**
F	**The Big Picture**	**1113**
G	**"Where Do I Go to Buy Electronic Goodies?"**	**1114**
H	**Programs Available on Website**	**1116**

I	**Equipment**		1119
	I.1	Uses for This List	1119
	I.2	Oscilloscope	1119
	I.3	Function generator	1120
	I.4	Powered breadboard	1120
	I.5	Meters, VOM and DVM	1121
	I.6	Power supply	1121
	I.7	Logic probe	1121
	I.8	Resistor substitution box	1121
	I.9	PLD/FPGA programming pod	1122
	I.10	Hand tools	1122
	I.11	Wire	1122
J	**Pinouts**		1123
	J.1	Analog	1123
	J.2	Digital	1125
Index			1128

Preface

A book and a course

This is a book for the impatient. It's for a person who's eager to get at the fun and fascination of putting electronics to work. The course squeezes what we facetiously call "all of electronics" into about twenty-five days of class. Of course, it is nowhere near *all*, but we hope it is enough to get an eager person launched and able to design circuits that do their tasks well.

Our title claims that this volume, which obviously is a *book*, is also a *course*. It is that, because it embodies a class that Paul Horowitz and I taught together at Harvard for more than 25 years. It embodies that course with great specificity, providing what are intended as day-at-a-time doses.

A day at a time: Notes, Lab, Problems, Supplements

Each day's dose includes not only the usual contents of a *book* on electronics – notes describing and explaining new circuits – but also a *lab* exercise, a chance to try out the day's new notions by building circuits that apply these ideas. We think that building the circuits will let you understand them in a way that reading about them cannot.

In addition, nearly every day includes a *worked example* and many days include what we call "supplementary notes." These – for example, early notes on how to read resistors and capacitors – are not for every reader. Some people don't need the note because they already understand the topic. Others will skip the note because they don't want to invest the time on a first pass through the book. That's fine. That's just what we mean by "supplementary:" it's something (like a supplementary vitamin) that may be useful, but that you can quite safely live without.

What's new?

If any reader is acquainted with the *Student Manual...*, published in 1989 to accompany the second edition of *The Art of Electronics*, it may be worth noting principal differences between this book and that one. First, this book means to be self-sufficient, whereas the earlier book was meant to be read alongside the larger work. Second, the most important changes in content are these:

- Analog:
    - we devote a day primarily to the intriguing and difficult topic of *parasitic* oscillations and their cures;
    - we give a day to building a "PID" circuit, stabilizing a feedback loop that controls a motor's position. We apply signals that form three functions of an *error* signal, the difference between target voltage and output voltage: "Proportional" (P), "Integral" (I), and "Derivative" (D) functions of that difference.

- Digital:
  - application of Programmable Logic Devices (PLDs or "PALs"), programmed with the high-level *hardware description language* (HDL), Verilog;
  - a shift from use of a microprocessor to a *microcontroller*, in the computer section that concludes the course. This microcontroller, unlike a microprocessor, can operate with little or no additional circuitry, so it is well-suited to the construction of useful devices rather than computers.
- Website: The book's website learningtheartofelectronics.com has a lot more things, in particular code in machine readable form. Appendix H lists these.

. . . And the style of this book

A reader will gather early on that this book, like the *Student Manual* is strikingly informal. Many figures are hand-drawn; notation may vary; explanations aim to help intuition rather than to offer a mathematical view of circuits. We emphasize *design* rather than *analysis*. And we try hard to devise applications for circuits that are fun: we like it when our designs make sounds (on a good day they emit *music*), and we like to see motors spin.

Who's likely to enjoy this book and course

You need not resemble the students who take our course at the university, but you may be interested to know who they are, since the course evolved with them in mind. We teach the course in three distinct forms. Most of our students take it during fall and spring daytime classes at the College. There, about half are undergraduates in the sciences and engineering; the other half are graduate students, including a few cross-registered from MIT who need an introduction quicker (and, admittedly, less deep) than electronics courses offered down there. (We don't get EE majors from there; we get people who want a less formal introduction to the subject.)

In the night version of the course, we get mostly older students, many of whom work with technology and who have become curious about what's in the "box" that they work with. Most often the mysterious "box" is simply a computer, and the student is a programmer. Sometimes the "box" is a lab setup (we get students from medical labs, across the river), or an industrial control apparatus that the student would like to demystify.

In the summer version of the course, about half our students are rising high school seniors – and the ablest of these prove a point we've seen repeatedly: to learn circuit design you don't need to know any substantial amount of physics or sophisticated math. We see this in the College course, too, where some of our outstanding students have been Freshmen (though most students are at least two or three years older).

And we can't help boasting, as we did in the preface to the 1989 *Student Manual*, that once in a great while a professor takes our course, or at least sits in. One of these buttonholed one of us recently in a hallway, on a visit to the University where he was to give a talk. "Well, Tom," he said, "one of your students finally made good." He was modestly referring to the fact that he'd recently won a Nobel Prize. We wish we could claim that we helped him get it. We can't. But we're happy to have him as an alumnus.[1]

We expect that some of these notes will strike you as elementary, some as excessively dense: your

---

[1] This was Frank Wilczek. He did sit quietly at the back of our class for a while, hoping for some insights into a simulation that he envisioned. If those insights came, they probably didn't come from us.

reaction naturally will reflect the uneven experience you have had with the topics we treat. Some of you are sophisticated programmers, and will sail through the assembly-language programming near the course's end; others will find it heavy going. That's all right. The course out of which this book grew has a reputation as fun, and not difficult in one sense, but difficult in another: the concepts are straightforward; abstractions are few. But we do pass a lot of information to our students in a short time; we do expect them to achieve literacy rather fast. This course is a lot like an introductory language course, and we hope to teach by the method sometimes called immersion. It is the laboratory exercises that do the best teaching; we hope this book will help to make those exercises instructive. I have to add though, in the spirit of modern jurisprudence, a reminder to read the legal notice appended to this Preface.

The mother ship: Horowitz & Hill's The Art of Electronics

Paul Horowitz launched this course, 40-odd years ago, and he and Winfield Hill wrote the book that, in its various editions, has served as textbook for this course. That book, now in its third edition and which we will refer to as "AoE," remains the reference work on which we rely. We no longer require that students buy it as they take our course. It is so rich and dense that it might cause intellectual indigestion in a student just beginning his study of electronics. But we know that some of our students and readers will want to look more deeply into topics treated in this book, and to help those people we provide cross-references to AoE throughout this book. The fortunate student who has access to AoE can get more than this book by itself can offer.

Analog and digital: a possible split

In our College course we go through all the book's material in one term of about thirteen weeks. In the night course, which meets just once each week, we do the same material in two terms. The first term treats analog (Days 1–13), the second treats digital (Days 14–26). We know that some other universities use the same split, analog versus digital. It is quite possible to do the digital half before the analog. Only on the first day of digital – when we ask that people build a logic gate from MOSFET switches – would a person without analog training need a little extra guidance. For the most part, the digital half treats its devices as black boxes that one need not crack open and understand. We do need to be aware of input and output properties, but these do not raise any subtle analog questions

It is also possible to pare the course somewhat, if necessary. We don't like to see any of our labs missed, but we know that the summer version of the class, which compresses it all into a bit more than six weeks, makes the tenth lab optional (Day 10 presents a "PID" motor controller). And the summer course omits the gratifying but not-essential digital project lab, 20L, in which students build a device of their own design.

Who helped especially with this book

First, and most obviously, comes Paul Horowitz, my teacher long ago, my co-teacher for so many years, and all along a demanding and invaluable critic of the book as it evolved. Most of the book's hand-drawn figures, as well, still are his handiwork. Without Paul and his support, this book would not exist.

Second, I want to acknowledge the several friends and colleagues who have looked closely at parts of the book and have improved and corrected these parts. Two are friends with whom I once taught, and who thus not only are expert in electronics but also know the course well. These are Steve Morss and Jason Gallicchio. Steve and I taught together nearly thirty years ago. Back then, he helped me to try out and to understand new circuits. He then went off to found a company, but we stayed in touch, and when we began to use a logic compiler in the course (Verilog) I took advantage of his experience. Steve was generous with his advice and then with a close reading of our notes on the subject. As I first met Verilog's daunting range of powers it was very good to be able to consult a patient and experienced practitioner.

Jason helped especially with the notes on sampling. He has the appealing but also intimidating quality of being unable to give half-power, light criticism. I was looking for pointers on details. The draft of my notes came back glowing red with his astute markups. I got more help than I'd hoped for – but, of course, that was good for the notes.

A happy benefit of working where I do is to be able to draw on the extremely knowledgeable people about me, when I'm stumped. Jim MacArthur runs the electronics shop, here, and is always overworked. I could count on finding him in his lab on most weekends, and, if I did, he would accept an interruption for questions either practical or deep. David Abrams is a similarly knowledgeable colleague who twice has helped me to explain to students results that I and the rest of us could not understand. With experience in industry as well as in teaching our course, David is another specially valuable resource.

Curtis Mead, one of Paul Horowitz's graduate students, gave generously of his skill in circuit layout, to help us make the LCD board that we use in the digital parts of this course. Jake Connors, who had served as our teaching assistant, also helped to produce the LCD boards that Curtis had laid out. Randall Briggs, another of our former TAs, helped by giving a keen, close reading.

It probably goes without saying, but let's say it: whatever is wrong in this book, despite the help I've had, is my own responsibility, my own contribution, not that of any wise advisor.

In the laborious process of producing readable versions of the book's thousand-odd diagrams, two people gave essential help. My son, Jamie Hayes, helped first by drawing, and then by improving the digital images of scanned drawings. Ray Craighead, a skilled illustrator whom we found online,[2] made up intelligently rendered computer images from our raggedy hand-drawn originals. He was able to do this in a style that does not jar too strikingly when placed alongside our many hand-drawn figures. We found no one else able to do what Ray did.

Then, when the pieces were approximately assembled, but still very ragged, the dreadfully hard job of putting the pieces together, finding inconsistencies and repetitions, cutting references to figures that had been cut, attempting to impose some consistency, ("Carry_Out" rather than "Carry$_{OUT}$" or "C$_{out}$" – at least on the same page – and so on), in 1000 pages or so, fell to my editor, David Tranah. He put up not only with the initial raggedness, but also with continual small changes, right to the end, and he did this soon after he had completed a similarly exhausting editing of AoE. For this unflagging effort I am both admiring and grateful.

And, finally, I should thank my wife, Debbie Mills, for tolerating the tiresome sight of me sitting, distracted, in many settings – on back porch, vacation terrace in Italy, fireside chair – poking away at revisions. She will be glad that the book, at last, is done.

---

[2] See craighead.com.

Legal notice

In this book we have attempted to teach the techniques of electronic design, using circuit examples and data that we believe to be accurate. However, the examples, data, and other information are intended solely as teaching aids and should not be used in any particular application without independent testing and verification by the person making the application. Independent testing and verification are especially important in any application in which incorrect functioning could result in personal injury or damage to property.

For these reasons, we make no warranties, express or implied, that the examples, data, or other information in this volume are free of error, that they are consistent with industry standards, or that they will meet the requirements for any particular application. THE AUTHORS AND PUBLISHER EXPRESSLY DISCLAIM THE IMPLIED WARRANTIES OF MERCHANTABILITY AND OF FITNESS FOR ANY PARTICULAR PURPOSE, even if the authors have been advised of a particular purpose, and even if a particular purpose is indicated in the book. The authors and publisher also disclaim all liability for direct, indirect, incidental, or consequential damages that result from any use of the examples, data, or other information in this book.

In addition, we make no representation regarding whether use of the examples, data, or other information in this volume might infringe others' intellectual property rights, including US and foreign patents. It is the reader's sole responsibility to ensure that he or she is not infringing any intellectual property rights, even for use which is considered to be experimental in nature. By using any of the examples, data, or other information in this volume, the reader has agreed to assume all liability for any damages arising from or relating to such use, regardless of whether such liability is based on intellectual property or any other cause of action, and regardless of whether the damages are direct, indirect, incidental, consequential, or any other type of damage. The authors and publisher disclaim any such liability.

Overview, as the Course begins

The circuits of the first three days in this course are humbler than what you will see later, and the devices you meet here are probably more familiar to you than, say, transistors, operational amplifiers – or microprocessors: Ohm's Law will surprise none of you; $I = CdV/dt$ probably sounds at least vaguely familiar.

But the circuit elements that this section treats – passive devices – appear over and over in later active circuits. So, if a student happens to tell us, "I'm going to be away on the day you're doing Lab 2," we tell her she will have to make up the lab somehow. We tell her that the second lab, on *RC* circuits, is the most important in the course. If you do not use that lab to cement your understanding of *RC* circuits – especially filters – then you will be haunted by muddled thinking for at least the remainder of the analog part of the course.

Resistors will give you no trouble; diodes will seem simple enough, at least in the view that we settle for: they are one-way conductors. Capacitors and inductors behave more strangely. We will see very few circuits that use inductors, but a great many that use capacitors. You are likely to need a good deal of practice before you get comfortable with the central facts of capacitors' behavior – easy to state, hard to get an intuitive grip on: they pass AC, block DC, and only *rarely* cause large phase shifts.

We should also restate a word of reassurance: you can manage this course perfectly even if the "$-j$" in the expression for the capacitor's impedance is completely unfamiliar to you. If you consult AoE, and after reading about complex impedances in AoE's spectacularly dense Math Review (Appendix A) you feel that you must be spectacularly dense, don't worry. That is the place in the course where the squeamish may begin to wonder if they ought to retreat to some slower-paced treatment of the subject. Do not give up at this point; hang on until you have seen transistors, at least. One of the most striking qualities of this book is its cheerful evasion of complexity whenever a simpler account can carry you to a good design. The treatment of transistors offers a good example, and you ought to stay with the course long enough to see that: the transistor chapter is difficult, but wonderfully simpler than most other treatments of the subject. You will begin designing useful transistor circuits on your first day with the subject.

It is also in the first three labs that you will get used to the lab instruments – and especially to the most important of these, the oscilloscope. It is a complex machine; only practice will teach you to use it well. Do not make the common mistake of thinking that the person next to you who is turning knobs so confidently, flipping switches and adjusting trigger level – all on the first or second day of the course – is smarter than you are. No, that person has done it before. In two weeks, you too will be making the scope do your bidding – assuming that you don't leave the work to that person next to you, who knew it all from the start.

The images on the scope screen make silent and invisible events visible, though strangely abstracted as well; these scope traces will become your mental images of what happens in your circuits. The scope will serve as a time microscope that will let you see events that last a handful of nanoseconds:

the length of time light takes to get from you to the person sitting a little way down the lab bench. You may even find yourself reacting emotionally to shapes on the screen, feeling good when you see a smooth, handsome sinewave, disturbed when you see the peaks of the sine clipped, or its shape warped; annoyed when fuzz grows on your waveforms.

Anticipating some of these experiences, and to get you in the mood to enjoy the coming weeks in which small events will paint their self-portraits on your screen, we offer you a view of some scope traces that never quite occurred, and that nevertheless seem just about right: just what a scope would show if it could. This drawing was posted on my door for years, and students who happened by would pause, peer, hesitate – evidently working a bit to put a mental frame around these not-quite-possible pictures. Sometimes a person would ask if these are scope traces. They are not, of course; the leap beyond what a scope can show was the artist's: Saul Steinberg's. Graciously, he has allowed us to show his drawing here. We hope you enjoy it. Perhaps it will help you to look on your less exotic scope displays with a little of the respect and wonder with which we have to look on the traces below.

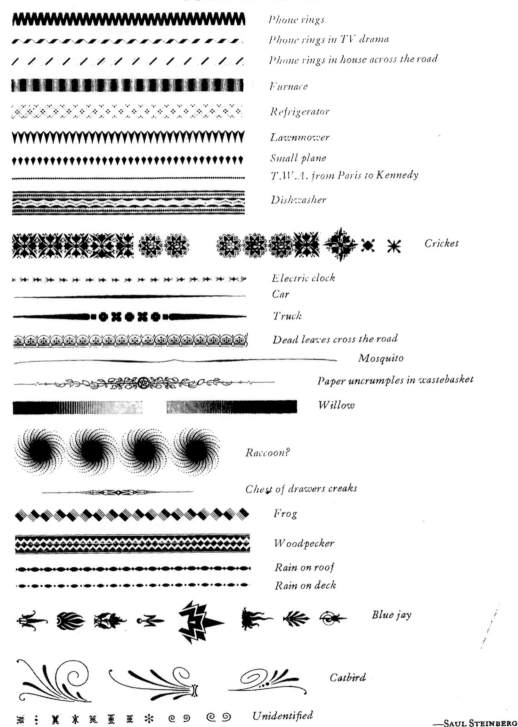

Drawing by Saul Steinberg, copyright Saul Steinberg Foundation; originally published in *The New Yorker Magazine*, 1979. reproduced with permission.

Part I

Analog: Passive Devices

1N DC Circuits

Contents

1N.1 Overview — 3
    1N.1.1 Why? — 3
    1N.1.2 What is "the art of electronics?" — 4
    1N.1.3 What the course is *not* about — 4
    1N.1.4 What the course *is* about: processing *information* — 5
1N.2 Three laws — 5
    1N.2.1 Ohm's law: $V = IR$ — 5
    1N.2.2 Kirchhoff's laws: V, I — 9
1N.3 First application: voltage divider — 11
    1N.3.1 A voltage divider to analyze — 13
1N.4 Loading, and "output impedance" — 14
    1N.4.1 Two possible methods — 14
    1N.4.2 Justifying the Thevenin shortcut — 15
    1N.4.3 Applying the Thevenin model — 17
    1N.4.4 VOM versus DVM: a conclusion? — 20
    1N.4.5 Digression on *ground* — 20
    1N.4.6 A rule of thumb for relating $R_{\text{OUT\_A}}$ to $R_{\text{IN\_B}}$ — 21
1N.5 Readings in AoE — 24

1N.1 Overview

We will start by looking at circuits made up entirely of

- DC voltage sources (things whose output voltage is constant over time; things like a battery, or a lab power supply); and ...
- resistors.

Sounds simple, and it is. We will try to point out quick ways to handle these familiar circuit elements. We will concentrate on one circuit fragment, the voltage divider.

1N.1.1 Why?

In each day's class notes we will sketch the sort of task that the day's material might let us accomplish. We do this to try to head off a challenge likely to occur to any skeptical reader: OK, this is a something-or-other circuit, but what's it for? Why do I need a something-or-other? This is an integrator – but why do I want an integrator? Here is our first try at providing such a sample application:

Problem Given a constant ("DC") voltage source, design a lower voltage source, strong enough to "drive" a particular "load" resistance.

Shorthand version of the problem: Make a voltage divider to deliver a specified voltage. Arrange things so that increasing load current to a maximum causes V_{out} to vary by no more than a specified percentage.

1N.1.2 What is "the art of electronics?"

Not art that you're likely to find in a museum,[1] but *art* in an older sense: a craft.[2] No doubt the title of *The Art of Electronics* (hereafter referred to as "AoE") was chosen with an awareness of the suggestion that there's something borderline-magical available here: perhaps a hint of "black art"?

AoE §1.1

Here is AoE's formulation of the subject of this course:

> *the laws, rules of thumb, and tricks that constitute the art of electronics as we see it.*

As you may have gathered, if you have looked at the text, this course differs from an engineering electronics course in concentrating on the "rules of thumb" and the bag of "tricks." You will learn to use rules of thumb and reliable "tricks" without apology. With their help you will be able to leave the calculator-bound novice engineer in the dust!

1N.1.3 What the course is *not* about

Wire my basement? Fix my TV?

Alumni of this course sometimes are asked for help that is beyond their capacities, and sometimes below – or beside – what they know. "So, now you can wire some outlets in my basement?" No. This course won't help much with that task, which is easy in a sense but difficult in another, in that it requires a detailed knowledge of electrical codes (required wire gauges; types of jacketing; where ground-fault-interrupters are required). And when your friend's TV quits, you're probably not going to want to fix it: much of the set's circuitry will be embodied in mysterious proprietary integrated circuits; an effective repair – if it were economically worthwhile – would likely amount to ordering a replacement for a substantial module, rather than replacing a burned-out resistor or transistor, as in the good old days of big and fixable devices.

Delivering power

A subtler point is worth making as well: only now and then, in this course, do we undertake to deliver *power* to something (the "something" is conventionally called a "load"). Occasionally, we are interested in doing that: when we want to make a loud sound from a speaker, or want to spin a motor. But much more often, we would like to minimize the flow of power; we are concerned, instead, with the flow of *information*.

On the wall of the lobby of MIT's Electrical Engineering building is a huge blowup of a photo of some MIT engineers standing among what look like large generators or motors, each about the size of a small cow. The photo in Fig. 1N.1 seems to date from the 1930s.

The "Electricals," back then, were concerned mostly with those big machines: with delivering power. It was the power companies that were hiring, when one of our uncles finished at MIT, around 1936. Hoover Dam, finished in 1935, was the engineering wonder of the day. Big was beautiful. (Even now, Hoover Dam's website boasts of the dam's *weight*! – 6.6 million tons, in case you were wondering.)

---

[1] But see, if you find yourself in Munich, a spectacular exception: the world's greatest museum of science and technology, the Deutsches Museum. There you will find wonderful machines demonstrating such arcana as the history of the manufacture of threaded fasteners.

[2] "An industrial pursuit... of a skilled nature; a craft, business, profession." Oxford English Dictionary (1989).

1N.2 Three laws

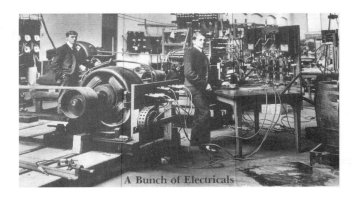

Figure 1N.1 Electronics ca. 1935 [used with permission of MIT.]

1N.1.4 What the course *is* about: processing *information*

Times have changed, as you may have noticed. Small is beautiful; nano is extra beautiful – and electronics, these days, is concerned mostly with processing information.³ So, we like circuits that pass and process signals while generating very little heat – using very little power. We like, for example, digital circuits made out of field-effect transistors that form switches; they offer low output impedance, gargantuan input impedance, and quiescent current of approximately zero. To a good approximation, they're not transferring power, not using it, not delivering it. They're dealing in information. That's almost always what we'll be doing in this course.

Obvious, perhaps? Perhaps.

We will postpone till next time – not to overload you, on the first day – discussion of a related topic: just what form the *information* is likely to take, in our circuits: *voltage* versus *current*. The answer may surprise you; or you may be inclined to reject the question as empty, since you know that long ago Ohm taught us that current and voltage in a device can be intimately related. Next time, we'll try to persuade you that you ought not to reject the question; that it's worth considering whether the signal is represented as a *voltage* or as a *current* (and see Note 1S on this topic).

Now on to less abstract topics, and our first useful circuit: a *voltage divider*.

1N.2 Three laws

AoE §1.2.1

A glance at three *laws*: Ohm's law, and Kirchhoff's laws (Voltage – "KVL," and current – "KCL").

We rely on these rules continually, in electronics. Nevertheless, we rarely will mention Kirchhoff again. We use his observations *implicitly*. By contrast, we will see and use Ohm's law a lot; no one has gotten around to doing what's demanded by the bumper sticker one sees around MIT: *Repeal Ohm's Law!*

1N.2.1 Ohm's law: $V = IR$

- V is the analog of water pressure or 'head' of water
- R describes the restriction of flow
- I is the rate of flow (volume/unit time)

The homely hydraulic analogy works pretty well, if you don't push it too far – and if you're not too proud to use such an aid to intuition.

---

³ We guess a potentially big exception is the continuing struggle to produce an efficient and economically-viable *electrically-powered* car. Some glory awaits the *electricals* who succeed at that task.

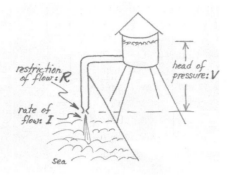

Figure 1N.2 Hydraulic analogy: voltage as head of water, etc. Use it if it helps your intuition.

What is "voltage," and other deeper questions

For the most part, we will evade such deep questions in this course. We're inclined to say, "Oh, a Volt is what pushes an Amp through an Ohm." But you don't have to be quite so glib (you don't have to sound so much like a Harvard student!). A less circular definition of *voltage* is the potential energy per unit charge. Or, equivalently, it can be defined as the work done to move a unit charge against an electric field (a word that we hope doesn't worry you; we suggest you try to get accustomed to use of the word, even if you have reservations about its usefulness[4]), from one electric potential (analogous to a position on a hillside) to a higher potential (higher on the hillside).

Figure 1N.3 Voltage is work to raise a unit charge from one level to a higher level (or "potential").

The voltage *difference* between two points on the hillside (or staircase, as in Fig. 1N.3) can be described as a difference in electric potential or voltage. The so-called "electric field" will tend to push that charge back down, just as gravity will tend to push the water down from the tank. You may or may not be interested to know that one volt is the work done as one adds one joule of potential energy to one coulomb of charge.[5] But we'll not again speak in these terms – which sound more like physics than like language for the "art of electronics."

"Ground"

Sometimes we speak of a voltage relative to some absolute reference – perhaps the planet earth (or, a bit more practically, the potential at the place where a copper spike has been driven into the ground, in the basement of the building where you are doing your electronics). In the hydraulic analogy, that absolute *zero*-reference might be sea-level. More often, as we will reiterate below, we are interested only in *relative* voltages: *differences* in potential, measured relative to an arbitrary reference point, not relative to planet earth.

Ohm's is a very useful rule; but it applies only to things that behave like *resistors*. What are these?

---

[4] You may be inclined to wonder, as Purcell suggests in his excellent book, "...what is a field? Is it something real, or is it merely a name for a factor in an equation which has to be multiplied by something else to give the numerical value of the force we measure in an experiment?" E.M. Purcell and D.J. Morin, *Electricity and Magnetism*, 3rd ed. (2013), §1.7. Purcell makes a persuasive argument that the concept of "field" is useful.

[5] See Purcell and Morin, §2.2.

1N.2 Three laws

They are things that obey Ohm's Law! (Sorry folks: that's as deeply as we'll look at this question, in this course.[6])

Why does Ohm's law hold?

The restriction of current flow that we call "resistance" – which we might contrast with the very-easy flow of current in a piece of wire[7] – occurs because the charge-carrying electrons, accelerated by an electric field, bump into obstacles (vibrations of the atomic lattice) after a short free flight, and then have to be re-accelerated in the direction of the field. Materials that are good conductors – metals – have a substantial population of electrons that are not tightly bound, and consequently are free to travel when pushed. The conductivity of a metal depends on the density of the population of charge carriers (usually, un-bound electrons), and it's kind of reassuring to find that conductivity degrades with rising temperature: the free flights become shorter, as electrons bump into the jumpier atoms of the hotter material. This effect you will see confirmed in Lab 1L if things come out right in your experiment (you'll have to do a little reasoning to see this effect confirmed; the notes to Lab 1L do not point out where this occurs). The stronger the field, the faster the drift of the electrons. Field strength goes with *voltage* difference between two points on the conductor; rate of drift of the electrons measures *current*. So, Ohm's Law is pretty plausible.

AoE §C.4

What determines the value of a resistor?

A "resistor" is also, of course, a "conductor"; it may seem a bit perverse to call this thing that is inserted in a circuit to permit current flow a "resistor." But the name comes from the assumption that the resistor is inserted where an excellent conductor – a piece of wire – might have stood, instead. To make a resistor, one can use either of two strategies: to make a "carbon composition" resistor (the sort that we'll use in lab because their values are relatively easy to read), one mixes up a batch of powdered insulator and powdered conductor (carbon), adjusting the proportions to give the material a particular resistivity. To make a "metal film" resistor (much the more common type, these days), one "deposits" a thin film of metal on a ceramic substrate, and then partially cuts away the thin conducting film.

How generally does Ohm's law apply? We begin almost at once to meet devices that do *not* obey Ohm's Law (see Lab 1L: a lamp; a diode). Ohm's Law describes *one* possible relation between V and I in a component; but there are others. As AoE says,

> Crudely speaking, the name of the game is to make and use gadgets that have interesting and useful I versus V characteristics.

In a resistor, current and voltage are proportional in a nice, linear way: double the voltage and you get double the current. Ohm's Law holds. Don't expect to use it where it doesn't fit. Even the lamp – whose filament is just a piece of metal that one might expect would behave like a resistor – doesn't follow Ohm's Law, as you'll see in Lab 1L. Why not?[8]

AoE §1.2.6

...But we can extend the reach of Ohm's law? Dynamic resistance: After today, we rarely will limit ourselves to devices that show simply *resistance* – and as we have said, even the resistor-like lamp that you meet in Lab 1L, along with the *diode*, defy Ohm's-Law treatment. But an extended version of Ohm's Law that we'll call *dynamic resistance* will allow us to apply the familiar rule in

---

[6] If this remark frustrates you, see an ordinary E&M book; for example, see the good discussion of the topic in E.M. Purcell and D.J. Morin, *Electricity & Magnetism*, 3rd ed. (2013), or in S. Burns and P. Bond, *Principles of Electronic Circuits* (1987).
[7] You may prefer the contrast with a superconductor, whose resistance is not just small but is *zero*.
[8] Here's a powerful clue: it *would* follow Ohm's Law if you could hold the filament's temperature constant.

settings where otherwise it would not work. The idea is just to define a *local* resistance – the tangent to the slope of the device's V–I curve:

$$R_{\text{dynamic}} \equiv \Delta V / \Delta I.$$

This redefinition allows us to talk about the effective resistance of a diode, a transistor, or a current source (a circuit that holds current constant). Here is a sketch of a diode's V–I curve – oriented so that V is the vertical axis. This orientation puts the curves' slopes into the familiar units, *Ohms* (rather than into 1/Ohms,[9] as in the more standard I–V plot).

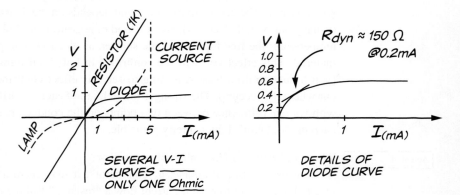

Figure 1N.4 Dynamic resistance illustrated: local slope can be defined for devices that are not Ohmic.

You may like the *resistor's* well-behaved straight line, because it is familiar. But the nice thing about the notion of R_{dynamic} is that is so broad-minded: it is happy to describe the V–I curve (or "I–V curve," as it is more often called) for any device. It will happily fit a transistor, an exotic current source – anything. The nearly-vertical plot of the *current source*, implying enormous R_{dynamic}, will become important to your understanding of transistors.

AoE §1.2.2C

Power in a resistor: Power is the rate of doing work, as you may recall from a course on mechanics. The concept comes up most often, in electronics, when one tries to specify a component that can safely handle the *power* that it is likely to have to dissipate. High power produces high heat, and calls for a component capable of unloading or dissipating that heat. The three resistors shown below illustrate the rough relation between power rating and size – because large size usually offers large area in contact with the surroundings or "ambient."

The indicated power ratings show the maximum that each can dissipate without damage. The tiny "surface-mount" on the left ("'0805 size," large by surface-mount standards) dissipates more than one might expect if one compares its size to that the of the 1/4W carbon-comp (the sort we use in the lab); it does better than one might expect because it is soldered directly to a circuit board, whose copper traces help to draw off and dissipate its heat.

In the coming labs you will sometimes run into the question whether your components can handle the power that is expected. Our usual resistors are rated at 1/4W. You can confirm that such a resistor can handle 15V (our usual maximum supply voltage) if the resistor's value is at least 1kΩ. Let's try that calculation: $P = I \times V$.

Thanks to Ohm's Law, the formula for power can be written in any of three ways:

- $P = I \times V$ (as we just said); but since $V = IR$,
- $P = I^2 R$; and since $I = V/R$,

[9] ...or *Siemens*, the official name for inverse Ohms.

1N.2 Three laws

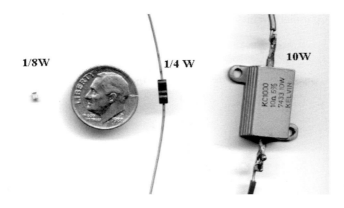

Figure 1N.5 Three resistors (plus a pretty good copper–nickel–alloy conductor).

- $P = V^2/R$

In the present case, it is the last form that is most useful. $1/4W = 15^2/R_{\min}$. So $R_{\min} = 225/(1/4) = 900$. So 1k is close to the minimum safe value, at 15V (910 would be safe, but let's not be so fussy; call it 1k).

So far, we have mentioned only power in a *resistor*. The notion is more general than that, and the formula,

$$P = V \times I$$

holds for any electronic component.

A closer look at what we mean by V and I makes this formula seem almost obvious:

- current measures charge/time
- voltage measures work/charge

So, the product, $V \times I$ = work/charge × charge/time = work/time, and this is *power*.

In this course, the exceptional cases where we do worry about power are those cases in which we either use large voltage swings (for example, the 30V output swing of the "comparator" in Lab 8L) or want to provide unusually large currents (for example, the speaker drive of Lab 6L, the light-emitting diode drive of the music-transmission lab, 13L, and the voltage regulators of Lab 11L).

1N.2.2 Kirchhoff's laws: *V, I*

These two 'laws' probably only codify what you think you know through common sense:

- Sum of voltages around the loop (or "circuit") is zero; see Fig. 1N.6, left.
- Sum of currents in and out of a node is zero (algebraic sum, of course); see Fig. 1N.6, right.

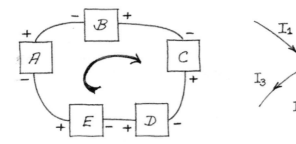

Figure 1N.6 Kirchhoff's two laws. Left: KVL – sum of voltages around a loop is zero; right: KCL – sum of currents in and out of a node is zero.

DC Circuits

Applications of these laws: series *and* parallel circuits:

Series $I_{total} = I_1 = I_2$
Parallel $I_{total} = I_1 + I_2$
Series $V_{total} = V_1 + V_2$
Parallel $V_{total} = V_1 = V_2$

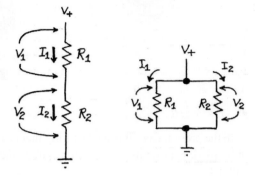

Figure 1N.7 Applications of Kirchhoff's laws: series and parallel circuits: a couple of truisms, probably familiar to you already

Query Incidentally, where is the "loop" that Kirchhoff's law refers to? *Answer:* the "loop" (or "circuit," a near-synonym) is apparent if one draws the voltage source as a circuit element, and ties its foot to the foot of the R: see Fig. 1N.8.

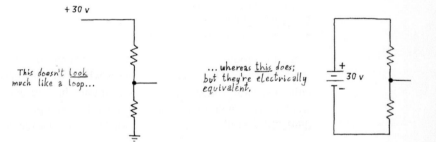

Figure 1N.8 Voltage divider redrawn to look more like a "loop" or "circuit".

Usually we don't bother to draw the voltage source that way; we label points with voltage values, and assume that you can picture the circuit path for yourself, if you choose to.

This is kind of *boring*. So, let's hurry on to less abstract circuits: to applications – and tricks. First, some labor-saving tricks.

Parallel resistances: calculating equivalent R

The *conductances* add:

$$\text{Conductance}_{total} = \text{Conductance}_1 + \text{Conductance}_2 = 1/R_1 + 1/R_2$$

Figure 1N.9 Parallel resistors: the *conductances* add; unfortunately, the *resistances* don't.

This is the easy notion to remember, but not usually convenient to apply, for one rarely speaks of conductances. The notion "resistance" is so generally used that you will sometimes want to use the formula for the effective resistance of two parallel resistors:

$$R_{tot} = (R_1 \cdot R_2)/(R_1 + R_2)$$

Believe it or not, even this formula is messier than what we like to ask you to work with in this course. So we proceed immediately to some tricks that let you do most work in your head. Consider the easy cases in Fig. 1N.10. The first two are especially important, because they help one to *estimate* the effect of a circuit one can liken to either case. Labor-saving tricks that give you an estimate are not to be scorned: if you see an easy way to an estimate, you're likely to make the estimate. If you have to work too hard to get the answer, you may find yourself simply *not* making the estimate. A deadly trap for the student doing a lab is the thought, "Oh, I'll calculate this later – some time this evening, when I'm comfortable in front of a spreadsheet." This student won't get to that calculation! The leftmost case in Fig. 1N.10 surely doesn't call for pulling out a formula: two equal Rs paralleled behave like $R/2$. The middle case is easier still, given that we're willing to ignore errors under 10%. (On the other hand, when you *do* want to trim an R value by 10% this is an easy way to do just that.) The rightmost calls for slightly more imagination: think of the lone R as a paralleling of two resistors of equal value: $1/R = 2/2R$. Then the whole looks like three paralleled resistors, each of value $2R$. The result then is $2R/3$.

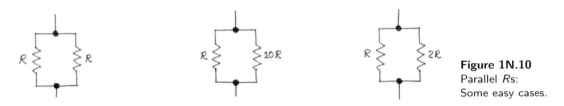

Figure 1N.10 Parallel Rs: Some easy cases.

In this course we usually are content with answers good to 10%. So, if two parallel resistors differ by a factor of ten or more, then we can ignore the larger of the two.

Let's elevate this observation to a **rule of thumb** (our first). While we're at it, we can state the equivalent rule for resistors in series.

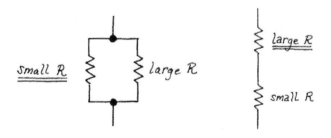

Figure 1N.11 Resistor calculation shortcut: parallel, series. In a parallel circuit, a resistor much *smaller* than others dominates. In a series circuit, the *larger* resistor dominates.

1N.3 First application: voltage divider

Why dividers? Are dividers necessary? Why not start with the *right* voltage? The answer, as you know, is just that a typical circuit needs several voltages, and building a "power supply" to deliver each voltage is impractical (meaning, mostly, *expensive*). You'll soon be designing power supplies, and certainly at that time will appreciate how much simpler a voltage divider is, compared to a full power supply.

To illustrate the point that voltage dividers are useful, and not just an academic device used to provide an easy introduction to circuitry, we offer here a piece of a fairly complex device, a "function generator" – the box that soon will be providing waveforms to the circuits that you build in Lab 2L. Fig. 1N.12 shows part of the circuitry that converts a triangular waveform into a sinusoidal shape.

12   DC Circuits

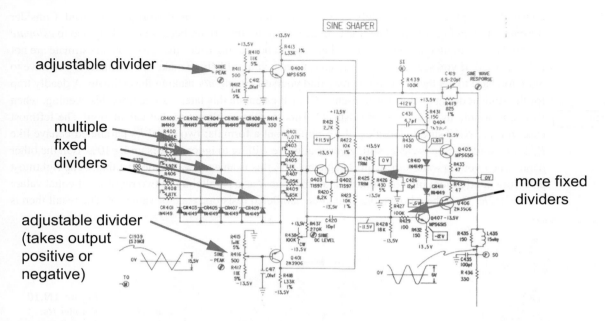

Figure 1N.12 Voltage dividers in a function generator: dividers are not just for beginners [Krohn–Hite 1400 function generator].

Aside: variable dividers or "potentiometers": Before we look closely at an ordinary divider, let's note a variation that's often useful: a voltage divider that is adjustable. This circuit, available as ready-made component, is called a "potentiometer."

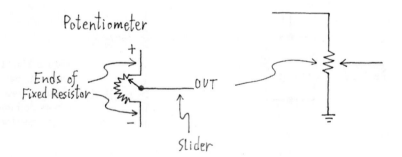

Figure 1N.13 Symbol for potentiometer, and its construction.

The name describes the circuit pretty well: the device "meters" or measures out "potential." Recalling this may help you to keep separate the two ways to use the device:

- as a potentiometer versus. . .
- as a variable resistor

"Variable resistor:" just one way to use a pot: The component is called a potentiometer (a 3-terminal device), but it can be used as a *variable resistor* (a 2-terminal device).

The pot becomes a variable resistor if one uses just one end of the fixed resistor and the slider, or if (somewhat better) one ties the slider to one end.[10]

---

[10] The difference between those two options is subtle. If the fixed resistance is, say 100k, the variable resistance range for

1N.3 First application: voltage divider

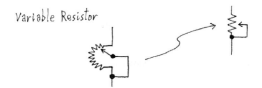

Figure 1N.14 A "pot" can be wired to operate as a variable resistor.

How the potentiometer is constructed: It helps to see how the thing is constructed. In Fig. 1N.15 are photos of two potentiometers. It is not hard to recognize how the large one on the left works. A (fixed) wire-wound resistor, not insulated, follows most of the way around a circle. A sliding contact presses against this wire-wound resistor. It can be rotated to either extreme.

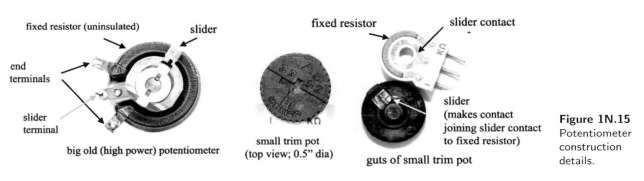

Figure 1N.15 Potentiometer construction details.

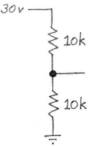

Figure 1N.16 Voltage divider.

At the position shown, the contact seems to be about 70% of the way between lower and upper terminals. If the upper terminal were at 10V and the lower one at ground, the output at the *slider* terminal would be about 7V.

The smaller pot shown in middle and right of Fig. 1N.15 (and shown enlarged, relative to the left-hand device) is fundamentally the same, but constructed in a way that makes it compact. Its fixed resistor – of value 1kΩ – is made not of wound wire but of "cermet."[11]

1N.3.1 A voltage divider to analyze

Figure 1N.16 is a simple example of the more common *fixed* voltage divider. At last we have reached a circuit that does something useful. It delivers a voltage of the designer's choice; a voltage less than the original or "source" voltage.

First, a note on labeling: we label the resistors "10k"; we omit "Ω." It goes without saying. The "k" means kilo- or 10^3, as you probably know.

One can calculate V_{out} in several ways. We will try to push you toward the way that makes it easy to get an answer in your head.

---

either arrangement is 0 to 100k. The difference – and the reason to prefer tying slider to an end – appears if the pot becomes dirty with age. If the slider should momentarily lose contact with the fixed resistor, the tied arrangement takes the effective R value to 100k. In contrast, the lazier arrangement takes R to *open* (call it "infinite resistance," if you prefer) when the slider loses contact.

The difference often is not important. But since it costs you nothing to use the "tied" configuration, you might as well make that your habit.

[11] "Cermet" is a composite material formed of ceramic and metal.

Three ways to analyze this circuit

First method: calculate the current through the series resistance...: Calculate the current (see Fig. 1N.17):
$$I = V_{IN}/(R_1 + R_2).$$
That's 30V /20kΩ = 1.5 mA. After calculating that current, use it to calculate the voltage in the lower leg of the divider:
$$V_{out} = I \cdot R_2.$$
Here that product is 1.5 mA · 10k = 15V. That takes *too long*.

Second method: rely on the fact that I is the same in top and bottom...: But rely on this equality only *implicitly*. If you want an algebraic argument, you might say,
$$V_2/(V_1+V_2) = I \cdot R_2/(I \cdot (R_1+R_2)) = R_2/(R_1+R_2),$$
or,
$$V_{out} = V_{in} \cdot R_2/(R_1+R_2). \quad (1N.1)$$
In this case that means
$$V_{out} = V_{in} \cdot (10k/20k) = V_{in}/2.$$

That's *much better*, and you will use formula (1N.1) fairly often. But we would like to push you not to memorize that equation, but instead to work less formally.

Third method: say to yourself in words how the divider works...: Something like

> Since the currents in top and bottom are equal, the voltage drops are proportional to the resistances (later, *impedances* – a more general notion that covers devices other than resistors).

So in this case, where the lower R makes up half the total resistance, it also will show half the total voltage.

For another example, if the lower leg is 10 times the upper leg, it will show about 90% of the input voltage (10/11, if you're fussy, but 90%, to our usual tolerances).

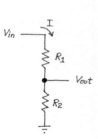

Figure 1N.17 Voltage divider: first method (too hard!): calculate current explicitly.

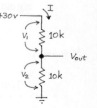

Figure 1N.18 Voltage divider: second method: (a little better): current implicit.

1N.4 Loading, and "output impedance"

Now – after you've calculated V_{out} for the divider – suppose someone comes along and puts in a third resistor, as in Fig. 1N.19. (*Query*: Are you entitled to be outraged? Is this not fair?[12])

Again there is more than one way to make the new calculation – but one way is tidier than the other.

1N.4.1 Two possible methods

Tedious method: Model the two lower Rs as one R; calculate V_{OUT} for this new voltage divider: see Fig. 1N.20. The new divider delivers 1/3 V_{IN}. That's reasonable, but it requires you to draw a new model to describe each possible loading.

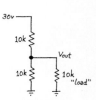

Figure 1N.19 Voltage divider loaded.

---

[12] We don't think you're entitled to be outraged – but perhaps you should be mildly offended. Certainly it's normal to see *something* attached to the output of your circuit (that "something" is called a "load"). You built the divider in order to provide current to something. But, as we'll soon be saying repeatedly, you are entitled to expect loads that are not too "heavy" (that is, don't draw too much current). By the standards of our course, this 10k load *is* too heavy. You'll see why in a moment.

1N.4 Loading, and "output impedance"

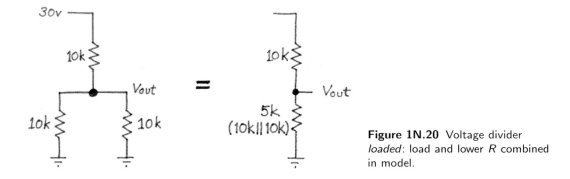

Figure 1N.20 Voltage divider *loaded*: load and lower *R* combined in model.

AoE §1.2.5

Better method: Thevenin's model: Here's how to calculate the two elements of the Thevenin model:

> **Thevenin's good idea.** Model the actual circuit (unloaded) with a simpler circuit – the *Thevenin model* – which is an idealized voltage source in series with a resistor. One can then see pretty readily how that simpler circuit will behave under various loads.

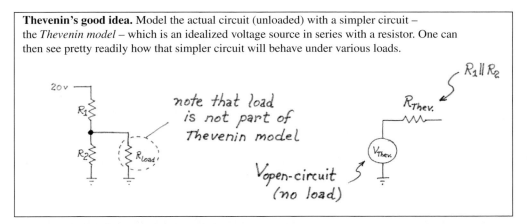

Figure 1N.21 Thevenin model: perfect voltage source in series with output resistance.

V_{Thevenin} Just $V_{\text{opencircuit}}$: the voltage out when nothing is attached ("no load")

R_{Thevenin} Often formulated as the quotient of $V_{\text{Thevenin}}/I_{\text{short-circuit}}$, which is the current that flows from the circuit output to ground if you simply *short* the output to ground.

In practice, you are not likely to discover R_{Thevenin} by so brutal an experiment. Often, shorting the output to ground is a very *bad* idea: bad for the circuit and sometimes dangerous to you. Imagine the result, for example, if you decided to try this with a 12V car battery! And if you have a diagram of the circuit to look at, a much faster shortcut is available: see Fig. 1N.22.

1N.4.2 Justifying the Thevenin shortcut

Our *shortcut*, shown in Fig. 1N.22, asserts that $R_{\text{Thevenin}} = R_{\text{parallel}}$, but the result still may strike you as a little odd: why should R_1, going up to the positive supply, be treated as *parallel* to R_2? Well, suppose the positive supply were set at zero volts. Then surely the two resistances would be in parallel, right?

And try another thought experiment: redefine the positive supply as 0V. It follows then that the voltage we had been calling "ground" or "zero volts" now is -30V (to use the numbers of Fig. 1N.16). Now it seems clear that the *upper* resistor, to ground, matters; the lower resistor, R_2, going to -30V

16 DC Circuits

> **Shortcut calculation of R_{Thevenin}.** Given a circuit diagram, the fastest way to calculate R_{Thevenin} is to see it as *the parallel resistance of the several resistances viewed from the output*. (This formulation assumes that the voltage sources are ideal, incidentally; when they are not, we need to include their output resistance. For the moment, let's ignore this complication.)

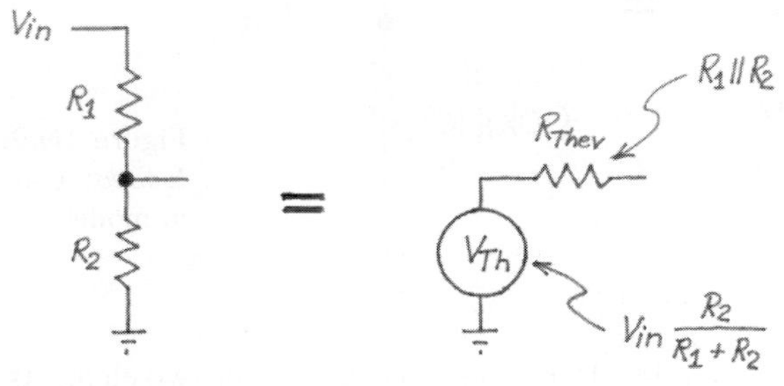

Figure 1N.22 $R_{\text{Thevenin}} = R_1$ parallel R_2.

now is the one that seems odd. But, of course, the circuit doesn't know or care how we humans have chosen to define our "zero volts" or "ground."

The point of this playing with "ground" definitions is just that the voltages at the far end of those resistors "seen" from the output do not matter. It matters only that those voltages are *fixed*.[13]

Or suppose a different divider (chosen to make the numbers easy): 20V divided by two 10k resistors. To discover the *impedance* at the output, do the usual experiment (one that we will speak of again and again):

> **A general definition and procedure for determining *impedance* at a point:**
> To discover the *impedance* at a point:
> apply a ΔV; find ΔI.
> The quotient is the impedance.

You will recognize the circuits in Fig. 1N.23 as just a "small signal" or "dynamic" version of Ohm's Law. In this case 1mA was flowing before the wiggle. After we force the output up by 1V, the currents in top and bottom resistors no longer match: upstairs: 0.9mA; downstairs, 1.1mA. The difference must come from you, the wiggler.

$$\text{Result: impedance} = \Delta V / \Delta I = 1\,\text{V}/0.2\,\text{mA} = 5\text{k}.$$

And – happily – that is the parallel resistance of the two Rs. Does that argument make the result easier to accept?

AoE §1.2.6

You may be wondering why this model is useful. Fig. 1N.23 shows one way to put the answer, though probably you will remain skeptical until you have seen the model at work in several examples.

---

[13] Later we will relax this requirement, too. Thevenin's model works as well to define the output resistance of a time-varying voltage source as it does for a DC source.

1N.4 Loading, and "output impedance"

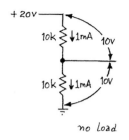
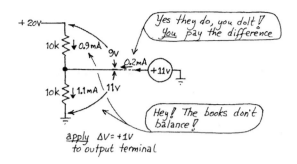

Figure 1N.23 Hypothetical divider: current = 1mA; apply a wiggle of voltage, ΔV; see what ΔI results.

Rationalizing Thevenin's result: parallel Rs

Any non-ideal voltage source "droops" when loaded. How much it droops depends on its "output impedance." The Thevenin equivalent model, with its R_{Thevenin}, describes this property neatly in a single number.

1N.4.3 Applying the Thevenin model

First, let's make sure Thevenin had it right: let's make sure his model behaves the way the original circuit does. We found that the 10k, 10k divider from 30V, which put out 15V when not loaded, drooped to 10V under a 10k load; see Fig. 1N.24. Does the model do the same?

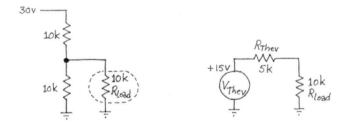

Figure 1N.24 Thevenin model and load: droops as original circuit drooped.

Yes, the model droops to the extent the original did: down to 10V. What the model provides that the original circuit lacked is that single value, R_{Thev}, expressing how droopy/stiff the output is.

If someone changed the value of the load, the Thevenin model would help you to see what droop to expect; if, instead, you didn't use the model and had to put the two lower resistors in parallel again and recalculate their parallel resistance, you'd take longer to get each answer, and still you might not get a feel for the circuit's *output impedance*.

Let's try using the model on a set of voltage sources that differ *only* in R_{Thev}. At the same time we can see the effect of an instrument's *input impedance*.

Suppose we have a set of voltage dividers, but dividing a 20V input by two; see Fig. 1N.25. Let's assume that we use 1% resistors (value good to ±1%): V_{Thev} is obvious, and is the same in all cases; but R_{Thevenin} evidently varies from divider to divider.

Suppose now that we try to *measure* V_{out} at the output of each divider. If we measured with a *perfect* voltmeter, the answer in all cases would be 10V. (*Query*: is it 10.000V? 10.0V?[14])

---

[14] It could be 10.0V – but it could be a bit less than 9.9V or a bit more than 10.1V, if the 20V source were good to 1% and we used 1% resistors. We should not expect perfection.

DC Circuits

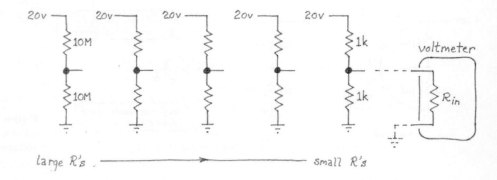

Figure 1N.25 Set of similar voltage dividers: same V_{Th}, differing $R_{Thevenin}$'s.

But if we actually perform the measurement, we will see the effect of loading by the R_{IN} of our imperfect lab voltmeters. Let's try it with a VOM ("volt–ohm-meter," the conventional name for the old-fashioned "analog" meter, which gives its answers by deflecting its needle to a degree that forms an analog to the quantity measured), and then with a DVM ("digital voltmeter," a more recent invention, which usually can measure current and resistance as well as voltage, despite its name; both types sometimes are called simply "multimeters").

Suppose you poke the several divider outputs, beginning from the right side, where the resistors are 1kΩ. Here's a table showing what we found, at three of the dividers:

R values	measured V_{OUT}	inference
1k	9.95	within R tolerance
10k	9.76	loading barely apparent
100k	8.05	loading obvious

The 8.05V reading shows such obvious loading – and such a nice round number, if we treat it as "8V" – that we can use this to calculate the meter's R_{IN} without much effort: see Fig. 1N.26.

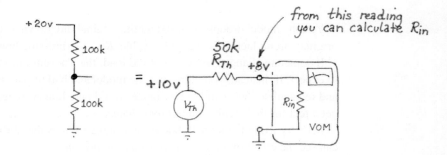

Figure 1N.26 VOM reading departs from ideal; we can infer R_{IN-VOM}.

As usual, one has a choice now, whether to pull out a formula and calculator, or whether to try, instead, to do the calculation "back-of-the-envelope" style. Let's try the latter method.

First, we know that R_{Thev} is 100k parallel 100k: 50k. Now let's talk our way to a solution (an *approximate* solution: we'll treat the measured V_{out} as just "8V"):

> The meter shows us 8 parts in 10; across the divider's R_{Thev} (or call it "R_{OUT}") we must be dropping the other 2 parts in 10. The relative sizes of the two resistances are in proportion to these two voltage drops: 8 to 2, so R_{IN-VOM} must be $4 \cdot R_{Thev}$: 200k.

1N.4 Loading, and "output impedance"

If we look closely at the front of the VOM, we'll find a little notation,

20,000 ohms/volt.

That specification means that if we used the meter on its 1V scale (that is, if we set things so that an input of 1V would deflect the needle fully), then the meter would show an input resistance of 20k. In fact, it's showing us 200k. Does that make sense? It will when you've figured out what must be inside a VOM to allow it to change voltage ranges: a set of big series resistors.[15] Our answer, 200k, is correct when we have the meter set to the 10V scale, as we do for this measurement.

This is probably a good time to take a quick look at what's inside a multimeter – VOM or DVM:

How a meter works: some meter types fundamentally sense *current*, others sense *voltage*; see Figs. 1N.27 and 1N.28. Both meter mechanisms, however, can be rigged to measure both quantities, operating as a "multimeter."

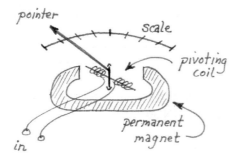

Figure 1N.27 An *analog* meter senses current in its guts.

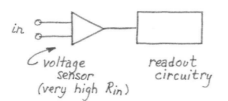

Figure 1N.28 A *digital* meter senses voltage in its innards.

The VOM specification, 20,000 ohms/volt, describes the sensitivity of the meter *movement* – the guts of the instrument. This movement puts a fairly low ceiling on the VOM's input resistance at a given range setting.

Let's try the same experiment with a DVM, and let's suppose we get the following readings:

R values	Measured V_{out}	Inference
100k	9.92	within R tolerance
1M	9.55	loading apparent
10M	6.69	loading obvious

Again let's use the case where the droop is obvious; again let's talk our way to an answer:

This time R_{Th} is 5M; we're dropping 2/3 of the voltage across R_{IN-DVM}, 1/3 across R_{Th}. So, R_{IN-DVM} must be $2 \times R_{Th}$, or 10M.

If we check the data sheet for this particular DVM we find that its R_{IN} is specified to be "10M, all ranges." Again our readings make sense.

---

[15] You'll understand this fully when you have done AoE Problem 1.8; for now, take our word for it.

DC Circuits

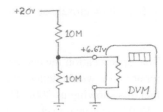

Figure 1N.29 DVM reading departs from ideal; we can infer $R_{\text{IN-DVM}}$.

1N.4.4 VOM versus DVM: a conclusion?

Evidently, the DVM is a better voltmeter, at least in its R_{IN} – as well as much easier to use. As a current meter, however, it is no better than the VOM: it drops 1/4V full scale, as the VOM does; it measures current simply by letting it flow through a small resistor; the meter then measures the voltage across that resistor.

1N.4.5 Digression on *ground*

The concept "ground" ("earth," in Britain) sounds solid enough. It turns out to be ambiguous. Try your understanding of the term by looking at some cases: see Fig. 1N.30.

Figure 1N.30 Ground in two senses.

Query What is the resistance between points A and B? (Easy, if you don't think about it too hard.[16]) We know that the ground symbol means, in any event, that the bottom ends of the two resistors are electrically joined. Does it matter whether that point is also tied to the pretty planet we live on? It turns out that it does not.

And where is "ground" in the circuit in Fig. 1N.31?

Two senses for "Ground"

"Local ground" Local ground is what we care about: the common point in our circuit that we arbitrarily choose to call zero volts. Only rarely do we care whether or not that local reference is tied to a spike driven into the earth.

---

[16] Well, not easy, unless you understand the "ground" indicated in Fig. 1N.30 to be the common zero-reference in the circuit at hand. In that case, the shared "ground" connection means that the 10k and 22k resistances are in series. If you took "ground" to mean "the world, with one connection in New York, the other in Chicago," the answer would not be so easy. Luckily, that would be a rare – and perverse – meaning to attribute to the circuit diagram.

1N.4 Loading, and "output impedance"

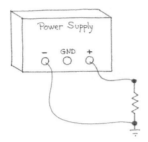

Figure 1N.31 Ground in two senses, revisited.

"Earth ground" But, *be warned*, sometimes you are confronted with lines that are tied to world ground – for example, the ground clip on a scope probe, and the "ground" of the breadboards that we use in the lab; then you must take care not to clip the scope ground to, say, +15 on the breadboard.

For a vivid illustration of "earth" grounding, see an image from Wikipedia:
`http://en.wikipedia.org/wiki/File:HomeEarthRodAustralia1.jpg`.

It might be useful to use different *symbols* for the two senses of ground, and such a convention does exist: see Fig. 1N.32.

Figure 1N.32 Symbols can distinguish the two senses of "ground".

Unfortunately, this distinction is not widely observed, and in this book as in AoE, we will not maintain this graphical distinction. Ordinarily, we will use the symbol that Fig. 1N.32 suggests for "local ground." In the rare case when we intend a connection to "world" ground, we will say so in words, not graphically.

1N.4.6 A rule of thumb for relating $R_{\text{OUT\_A}}$ to $R_{\text{IN\_B}}$

The voltage dividers whose outputs we tried to measure introduced us to a problem we will see over and over again: some circuit tries to "drive" a load. To some extent, the load changes the output. We need to be able to predict and control this change. To do that, we need to understand, first, the characteristic we call R_{IN} (this rarely troubles anyone) and, second, the one we have called R_{Thevenin} (this one takes longer to get used to). Next time, when we meet frequency-dependent circuits, we will generalize both characteristics to "Z_{IN}" and "Z_{OUT}."

Here we will work our way to another rule of thumb; one that will make your life as designer relatively easy. We start with a *Design goal*: When circuit A drives circuit B: arrange things so that B loads A lightly enough to cause only insignificant attenuation of the signal. And this goal leads to the rule of thumb in Fig. 1N.33.

How does this rule get us the desired result? Look at the problem as a familiar voltage divider question. If R_{OUTA} is much smaller than R_{INB}, then the divider delivers nearly all of the original signal. If the relation is 1:10, then the divider delivers 10/11 of the signal: attenuation is just under 10%, and that's good enough for our purposes.

> **Design rule of thumb**
> When circuit A drives circuit B.
> Let R_{OUT} for A be $\leq \frac{1}{10} R_{IN}$ for B.

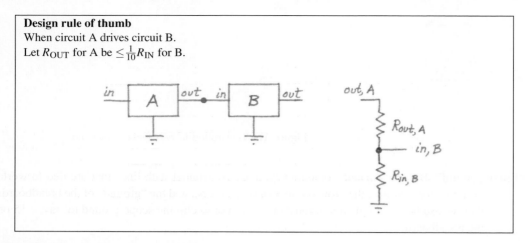

Figure 1N.33 Circuit A drives circuit B.

We like this arrangement not just because we like big signals. (If that were the only concern, we could always boost the output signal, simply amplifying it.) We like this arrangement above all because it *allows us to design circuit-fragments independently*: we can design A, then design B, and so on. We need not consider A,B as a large single circuit. That's good: makes our work of design and analysis lots easier than it would be if we had to treat every large circuit as a unit.

An example, with numbers as in Fig. 1N.34: what R_{Thev} for droop of $\leq 10\%$? What R's, therefore?

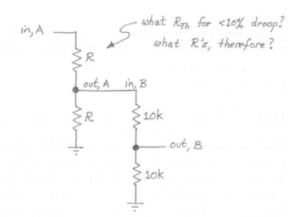

Figure 1N.34 One divider driving another: a chance to apply our rule of thumb.

The effects of this rule of thumb become more interesting if you extend this chain: from A and B, on to C, see Fig. 1N.35.

As we design C, what R_{Thev} should we use for B? Is it just 10k parallel 10k? That's the answer if we can consider B by itself, using the usual simplifying assumptions: source ideal ($R_{OUT} = 0$) and load ideal (R_{IN} infinitely large).

But should we be more precise? Should we admit that the upper branch really looks like 10K+2K: 12k? That's 20% different. Is our whole scheme crumbling? Are we going to have to look all the way back to the start of the chain, in order to design the next link? Must we, in other words, consider the whole circuit at once, not just the fragment B, as we had hoped?

No. Relax. That 20% error gets diluted to half its value: R_{Thev} for B is 10k *parallel* 12k, but that's a

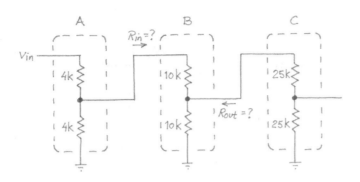

Figure 1N.35 Extending the divider: testing the claim that our rule of thumb lets us consider one circuit fragment at a time.

shade under 5.5k. So we need not look beyond B. We can, indeed, consider the circuit in the way we had hoped: fragment by fragment.

If this argument has not exhausted you, you might give our claim a further test by looking in the other direction: does C alter B's input resistance appreciably (>10%)? You know the answer, but confirming it would let you check your understanding of our rule of thumb and its effects.

Two important exceptions to our rule of thumb: signals that are currents, and transmission lines

The rule of §1N.4.6 for relating impedances is extremely useful and important. But we should acknowledge two important classes of circuits to which it does *not* apply:

- Signal sources that provide *current* signals rather than voltage signals. At the moment, this distinction may puzzle you. Yet the distinction between *voltage sources* – the sort familiar to most of us – versus *current sources* is substantial and important.[17]

- High-frequency circuits where the signal paths must be treated as "transmission lines" (see Appendix C).

  AoE §H.1

  Only at frequencies above what we will use in this course do such effects become apparent. These are frequencies (or *frequency components*, since "steep" waveform edges include high-frequency components, as Fourier teaches – see Chapter 3N) where the period of a signal is comparable to the time required for that signal to travel to the end of its path. For example, a six-foot cable's propagation time would be about 9ns, the period of a 110MHz sinusoid. Square waves are still more troublesome; one at even a few MHz would include quite a strong component at that frequency, and this component would be distorted by such a cable if we did not take care to "terminate" it properly. In fact, these calculations understate the difficulties; a round-trip path length greater than about 1/10th wavelength calls for termination.

We don't want you to worry, right now, about these two points. Signals as currents are unusual in this course, and in this course you are not likely to confront transmission-line problems. But be warned that you will meet these later, if you begin to work with signals at higher frequencies.

---

[17] You may be inclined to protest that you can convert a current into a voltage and vice versa. You will build your first circuit with an output that is a determined *current* in Lab 4L, and you will meet your first current-source transducer (a photodiode) in Lab 6L.

1N.5 Readings in AoE

This is the first instance in which we've tried to steer you toward particular sections of *The Art of Electronics*. That book is, of course a sibling (or is it the mother?) of the book you are reading. You should not consider the readings listed below to be *required*. But if you are proceeding by using the two books in parallel, these are the sections we consider most relevant.

Readings:
    Chapter 1, §§1.1–1.3.2.
    Appendix C on resistor types: resistor color code and precision resistors.
    Appendix H on transmission lines.
Problems:
    Problems in text.
    Exercises 1.37, 1.38.

1L Lab: DC Circuits

1L.1 Ohm's law

A preliminary note on *procedure*
The principal challenge here is simply to get used to the breadboard and the way to connect instruments to it. We do not expect you to find Ohm's law surprising. Try to build your circuit *on the breadboard*, not in the air. Novices often begin by suspending a resistor between the jaws of alligator clips that run to power supply and meters. Try to do better: plug the two leads of the DUT ("Device Under Test") into the plastic breadboard strip. Bring power supply and meters to the breadboard using banana-to-wire leads, if you have these, to go direct from banana jack on the source right into the breadboard. In Fig. 1L.1 is a sketch of the poor way and better way to build a circuit.

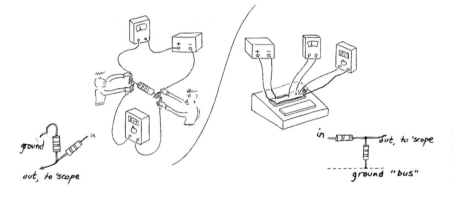

Figure 1L.1 Bad and good breadboarding technique: Left: labor intensive, mid-air method; Right: tidy method, circuit wired in place.

Color coding, and making the circuit look like its diagram
This is also the right time to begin to establish some conventions that will help you keep your circuits intelligible: in Fig. 1L.2 is a photo of one of our powered breadboard boxes, showing the usual power supply connections. The individual breadboard strips make connections by joining inserted wires in small spring-metal troughs: see Fig. 1L.3. Use a variable regulated DC supply, and the hookup shown in the Fig. 1L.9.

Caution! You must *not* connect the external supply to the colored banana jacks on the powered breadboard (Red, Yellow, Blue): these are the *outputs* of the internal supplies of the breadboard, and even when the breadboard power is **off**, these connectors tie to internal electronics that can be damaged by an attempt to drive it. The same warning applies to the three white horizontal power supply "buses" at the top of the board, strips that are internally tied to those three internal power supplies. (In case you are curious, Red is +5, while Yellow and Blue are adjustable, and normally are set to about ±15V.)

Note by the way, that voltages are measured *between* points in the circuit, while currents are measured *through* a part of a circuit. Therefore you usually have to "break" or interrupt the circuit in order to measure a current.

Lab: DC Circuits

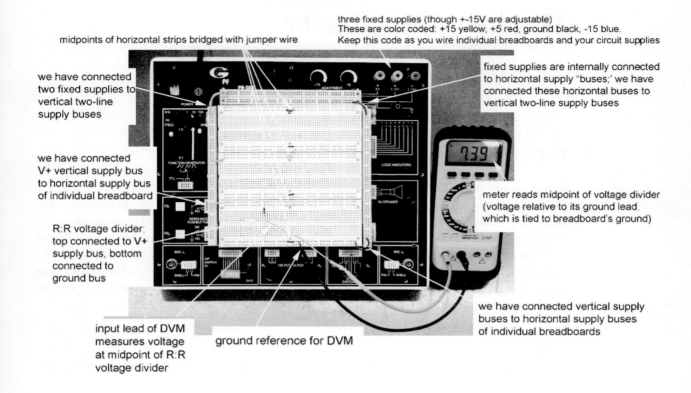

Figure 1L.2 Breadboard power bus connections, again.

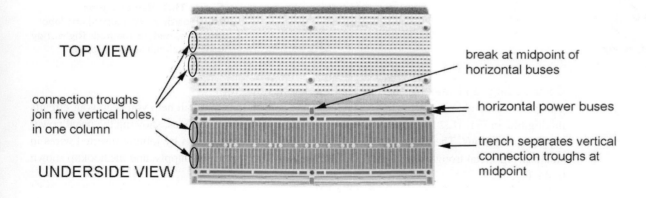

Figure 1L.3 Single breadboard strip: a look at underside reveals the pattern of connecting metal troughs.

1L.2 Voltage divider

Construct the voltage divider shown in Fig. 1L.4 Apply $V_{in}=15V$ (use the DC voltages on the breadboard). Measure the (open circuit) output voltage. Then attach a 7.5k load and see what happens.

1L.3 Converting a meter movement into a voltmeter and ammeter

Now measure the short circuit current. (That means "short the output to ground, but make the current flow through your current meter." Don't let the scary word "short" throw you: the current in this case will be very modest. You may have grown up thinking "a short blows a fuse." That's a good generalization around the house, but it often does not hold in electronics.)

From $I_{\text{ShortCircuit}}$ and $V_{\text{OpenCircuit}}$ you can calculate the Thevenin equivalent circuit.

Now build the Thevenin equivalent circuit, using the variable regulated DC supply as the voltage source, and check that its open circuit voltage and short circuit current match those of the circuit that it models. Then attach a 7.5k load, just as you did with the original voltage divider, to see if it behaves identically.

Figure 1L.4 Voltage divider.

> ### A Note on Practical Use of Thevenin Models
>
> You will rarely do again what you just did: *short* the output of a circuit to ground in order to discover its R_{Thevenin} (or "output impedance," as we soon will begin to call this characteristic). This method is too brutal in the lab, and too slow when you want to calculate R_{Th} on paper.
>
> In the lab, I_{SC} could be too large for the health of your circuit (as in your fuse-blowing experience). You will soon learn a gentler way to get the same information.
>
> On paper, if you are given the circuit diagram the fastest way to get R_{Th} for a divider is always to take the *parallel* resistance of the several resistances that make up the divider – again assuming R_{source} is ideal: zero ohms. (A hard point to get used to, here: look at all paths in parallel, going to any fixed voltage, not just to ground. See the attempt to rationalize this result in Fig. 1N.23 on p. 17.) The case you just examined is illustrated in Fig. 1L.5.

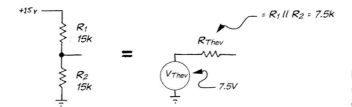

Figure 1L.5 R_{Th} = parallel resistances as seen from the circuit's output.

1L.3 Ohm's law applied to convert a meter movement into a voltmeter and ammeter

A Thevenin model is extremely useful as a *concept*; but you'll not again *build* such a model. The circuit you just put together is useful only as a device to help you get a grip on the concept. Now comes, instead, a chance to apply Ohm's law to make something almost useful! – a *voltmeter*, and then a *current meter* ("ammeter").

You will start with a bare-bones "meter movement" – a device that lets a current deflect a needle. This mechanism is basically a current-measuring device: the needle's deflection is proportional to the torque developed by a coil that sits in the field of a permanent magnet, and this torque is proportional to the current through the coil. See the sketch[1] in Fig. 1L.6.

An analog "multimeter" or VOM (volt–ohm-milliammeter) is just such a movement, with switchable resistor networks attached. We would now like you to re-invent the multimeter.

[1] After Google images file: "analog meter movement." Source unknown.

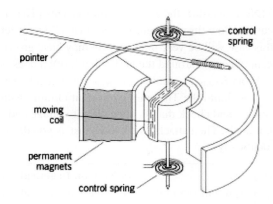

Figure 1L.6 Mechanism of analog current/volt-meter. From *The Free Dictionary*, see www.globalspec.com.

1L.3.1 Internal resistance of the movement

You could design a voltmeter – as you will in the next subsection – simply assuming that the movement's R_{INT} is negligibly small. The resistance of this *movement* is low – well under $1k\Omega$. But let's be more careful. Let's first measure the internal resistance. Here, *please note*, we want you to work with a simple *bare meter movement*, not with a multimeter. (It's more fun to start from scratch.)

Do this measuring any way you like, using the *variable power supply*, *DVM* (digital voltmeter – really a *multimeter*, despite its name), and *VOM* if you need it. If you use DVM as current meter, we should warn you that it is easy to blow the meter's internal fuse, because a moment's touch to a power supply will pass an essentially unlimited current. To make things worse, a DVM with blown fuse offers no display to indicate the problem. Instead, it simply persists in reading *zero* current. There are several good ways to do this task.

The meter movements are protected[2]. Don't worry about burning out the movement, even if you make some mistakes. Even "pinning" the needle against the stop will not damage the device, as we found by experiment – though in general you'll want to avoid doing this to analog meters. We have provided diodes that protect the movement against overdrive in both forward and reverse directions. You should *note*, as a result, that these diodes that protect the movement also make it behave very strangely if you overdrive it. (No doubt you can guess what device it behaves like.) So, don't use data gathered while driving the movement beyond full scale.

Sketch the arrangement you use to measure R_{INT} and note the values of R_{INT} and $I_{FULL-SCALE}$.

1L.3.2 10V voltmeter

Show how to use the movement, plus whatever else is needed, to form a 10V-full-scale voltmeter (that just means that 10V applied to the input of your circuit should deflect the needle fully). Draw your circuit.

Note: this is a good time to start getting used to our canny use of approximations, as we design. As you specify the resistor that you want to add to the bare meter movement, recall that you are limited to "5% values" – resistors whose true value is known only to lie within ±5% of the nominal value.

Does the movement's internal resistance cause a significant error? (What does "significant" mean? Well, you might compare the contribution of R_{INT} against the contribution of error from your uncertainty about the value of the resistor that (we hope!) you have included.)

---

[2] Note to instructors: please add two paralleled and oppositely-oriented silicon diodes such as 1N4004 in parallel with the meter movement. These will conduct at about 0.6V, protecting the movement during overdrive.

1L.3.3 10mA current meter ("ammeter")

Show how to use the movement, plus whatever else is needed, to form a 10mA-full-scale ammeter. (This is a bit trickier than the 10V voltmeter.) Sketch it, and test your circuit.

1L.4 The diode

Here is another device that does not obey Ohm's law: the **diode**. (We don't expect you to understand how the diode works yet; we just want you to meet it, to get some perspective on Ohm's Law devices: to see that they constitute an important but *special* case.

We need to modify the test setup here, because you can't just stick a voltage across a diode, as you did for the resistor and lamp above[3]. You'll see why after you've measured the diode's V versus I. Do that by wiring up the circuit shown in Fig. 1L.7.

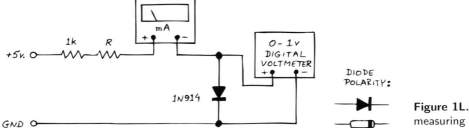

Figure 1L.7 Diode VI measuring circuit.

In this circuit you are applying a *current*, and noting the diode voltage that results; earlier, you applied a voltage and read resulting current. The 1k resistor limits the current to safe values. Vary R – use a 100k variable resistor (usually called a *potentiometer* or "pot" even when wired, as here, as a variable resistor), a resistor substitution box, or a selection of various fixed resistors) – and look at I versus V. First, get a feel for the behavior by sweeping the R value by hand, and noticing what happens to the diode current. Then sketch the plot in two forms: linear and lin–log (or semi-log), in Fig. 1L.8.

First, get an impression of the shape of the linear plot; just four or five points should define the shape of the curve. Then draw the same points on a *lin–log* plot, which compresses one of the axes. (Evidently, it is the fast-growing current axis that needs compressing, in this case.) If you have some lin–log paper use it. If you don't have such paper, you can use the small version laid out below. The point is to see the pattern.

See what happens if you reverse the direction of the diode. How would you summarize the V versus I behavior of a diode? Now explain what would happen if you were to put 5V across the diode (**don't try it!**). Look at a diode data sheet, if you're curious: see what the manufacturer thinks would happen. The data sheet won't say "Boom" or "Pfft," but that is what it will mean.

We'll do lots more with this important device.

---

[3] Well, you *can*; but you can't do it twice with the same diode!

Lab: DC Circuits

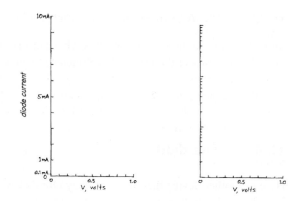

Figure 1L.8 Diode *I* versus *V*: linear plot; lin–log plot.

1L.5 I versus V for some mystery boxes

Find two *mystery* boxes which your instructor[4] will have set up for you (we will call these DUTs: Device Under Test). These are two-terminal devices, one of which is an ordinary resistor, the other an odder thing. The devices are hidden inside black plastic 35mm film cans. Apply voltages in the range zero to a couple of volts, using a *variable* power supply, and note voltage and current pairs. For the range between 0 and 1V, measure at increments of 0.1V (because this is the range where you need a detailed picture).

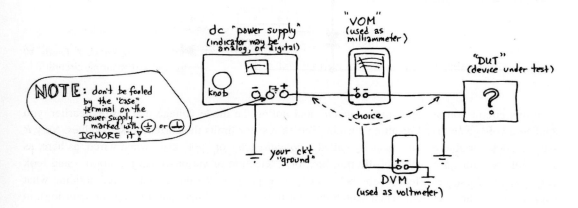

Figure 1L.9 Circuit for measurement of *I* versus *V*.

Sketch a graph of a few points to get the trend. Turning the power-supply voltage knob by hand, you may be able to get a sense of the shape of the curve, or see where more points are needed.

Decide which device is which: which ordinary, which odd. *Warning*: keep the applied voltage below 7V, or you may destroy one of the DUTs.

To make your task challenging, we ask that you measure *voltage* and *current simultaneously*, as you do this exercise. We ask this so that you will be obliged to consider the effects of the instruments on your measurements (more on this point, below).

---

[4] If you're working through this book on your own, no one will have set this up of course. But you and instructors can get more information at www.learningtheartofelectronics.com. For now, we reveal that one 'mystery' is a #47 lamp, the other a 33Ω power resistor.

Effects of the instruments on your readings: Consider a couple of practical questions that arise in even this simplest of "experiments."

A qualitative view: Is the voltmeter measuring the voltage at the place you want, namely across the object under test? Or does the voltage reading include the effect of the ammeter, in your arrangement? Does that matter? If you are measuring the object's *voltage* precisely, then are you reading its *current* or are you measuring the current in object plus the current passing through the DVM as well? If you can't have it both ways (as you can't), and must live with one of the two errors, which experimental setup gives you the smaller error?

Figure 1L.10 shows two sketches of the two possible placements of the voltmeter.

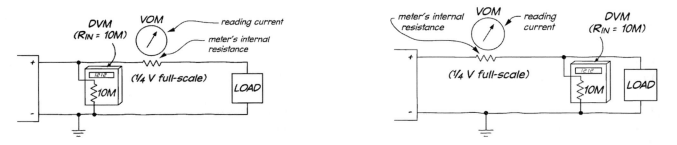

Figure 1L.10 Measuring *I* and *V* simultaneously: only one instrument gets a true reading at DUT.

If you know (and we'll now reveal this fact to you) that the *LOAD* resistance is under 1k ("resistance" or some near-equivalent; we're not promising you *Ohmic* behavior), then you can judge which of the two setups illustrated above gives the truer pair of *I* and *V* readings. Use that preferred setup to get a set of *I* and *V* readings, and try to infer an *R* value, if you can.

1L.5.1 Estimate % error caused by the instruments

Once you have your best estimate of *R* in hand, estimate the errors that the instruments cause. Specifically, *when the ammeter is on the 10mA full-scale range*, what percentage error in *inferred R* results from the presence of...

- ammeter, when you measure *V* where the ammeter causes an error?
- voltmeter, when you measure *V* where the ammeter causes no error?

After getting answers to these questions, probably you can say what an ideal voltmeter (or ammeter) should do to the circuit under test? What does that say about its "internal resistance"? (Perhaps you caught on to this theme, pages earlier.)

Now sketch the curves described by your data points. Don't work too hard: it's the shapes that we're after. The resistor isn't very interesting (what's its value?); the mystery device is a little more intriguing.

What do you suppose the mystery gadget is? We hope you saw its curve. In fact, it's made of material much like what forms the resistor. Why does its *I–V* curve look different? The bend can be useful: much later, in Lab 8L, we will exploit this curvature to make a circuit regulate its own *gain*. Now you have earned the right to open up the film canister and discover what's inside.

1L.5.2 Where should DVM and ammeter go, for large R_{LOAD}?

Now you may have "learned" in the previous exercise where to place the voltmeter in order to get the best simultaneous I and V readings. Just to make sure you don't go away thinking you've learned something[5], we'd like you to try *a pencil-and-paper* exercise, this time assuming that the DUT (R_{LOAD}) is a very large resistance: 10MΩ. Assume that the DVM's R_{IN} = 10MΩ. (Note that we're *not* asking you to carry out this experiment; only to predict what you would see if you did do the experiment.)

If you placed your DVM at the load, the DVM would appear in parallel with the load. What percentage error would this evoke in the simultaneous *current* measurement? In contrast, the *voltage* error caused by reading upstream of the ammeter in this case would be no larger than in the case you tried in the lab experiment.

Perhaps now you can begin to generalize about which cases oblige one to worry about the DVM's R_{IN}.

1L.6 Oscilloscope and function generator

This is just a first view of oscilloscopes to give you a head start (more about oscilloscopes can be found in Appendix D and AoE, Appendix O). You'll get a big (oscillo)scope workout later in Lab 2L.

We'll be using the oscilloscope and function generator in virtually every lab from now on. For today's lab, voltmeters were sufficient – and, in fact, more convenient than an oscilloscope, because our circuits' voltages and currents have been politely sitting still, giving us time to measure them. If you are familiar with the scope, go right on to the first real exercise in Lab 2L – or go home, having earned a rest.

The scope soon will become our favorite instrument as we begin to look at signals that are not static (called DC in electronics jargon): signals that, instead, vary with time. Lab 2L is concerned exclusively with such circuits, and with rare exceptions this will be true of all our work from now on. The scope draws a plot of voltage (on the vertical axis) versus time (on the horizontal axis).

Get familiar with scope and *function generator* (a box that puts out time-varying voltages, or "waveforms:" things like sinewaves, triangle waves and square waves) by generating a 1kHz sinewave with the function generator and displaying it on the scope. Connect the function generator directly to the scope, using a "BNC cable," not a *probe*.[6]

If it seems to you that both instruments present you with a bewildering array of switches and knobs, don't blame yourself. These front-panels just *are* complicated. You will need several lab sessions to get fully used to them – and at term's end you may still not know all: it may be a long time before you find any occasion for use of the *holdoff* control, for example, or of *single-shot* triggering, if your scope offers this.

Play with the scope's sweep and trigger controls. Most of the discussion, here, applies fully only to an *analog* scope. The digital scope automates some functions, such as triggering and gain setting, if you want it to. We suggest that you ought *not* to begin your scope career using such fancy features: *beware* the mind-stunting button labeled *AUTOSET*!

Specifically, try the following:

---

[5] Just kidding.
[6] BNC (Bayonet Neill–Concelman) names the connectors on the ends of this coaxial cable. (See AoE §1.8). We usually call the *cable* by this name, though strictly the term applies to its connectors.

1L.6 Oscilloscope and function generator

- The *vertical gain* switch. This controls "*volts/div*"; note that "div" or "division" refers to the *centimeter* marks, not to the tiny 0.2cm marks);
- The *horizontal sweep speed* selector: *time* per division.

    On this knob as on the vertical gain knob, make sure the switch is in its **CAL** position, not **VAR** or "variable." Usually that means that you should turn a small center knob clockwise till you feel the switch *detent* click into place. If you don't do this, you can't trust *any* reading you take.)

- The *trigger* controls. Don't feel dumb if you have a hard time getting the scope to trigger properly. Triggering is by far the subtlest part of scope operation. "Trigger" circuits tell the scope when to begin moving the trace across the screen: when to begin *drawing* a waveform. When you think you have triggering under control, invite your partner to prove to you that you *don't*: have your partner randomize some of the scope controls, then see if you can regain a sensible display (don't let your partner overdo it, here).

    Beware the tempting so-called "normal" settings (usually labeled "**NORM**"). They rarely help, and instead cause much misery when misapplied. Think of "normal" here as short for **abnormal**! Save it for the rare occasion when you know you need it. "**AUTO**" is almost always the better choice.

    The scope waits around till triggered. In AUTO mode, it sweeps either when triggered, or when it has waited so long that it loses patience. Thus you always get at least a trace, in AUTO mode.

    NORMAL, in contrast, makes the scope infinitely patient: it will wait forever, drawing nothing on the screen, until it sees a valid trigger signal. Meanwhile, you look at a dark screen; usually, that is not helpful!

    Trigger "source": here's another way to go wrong: trigger can be set to look at either of the scope's two input "channels" (CH1 or CH2), or at the external BNC connector, called EXT.

Switch the function generator to square waves and use the scope to measure the "risetime" of the square wave (defined as time to pass from 10% to 90% of its full amplitude).

At first you may be inclined to despair, saying "Risetime? The square wave rises instantaneously." The scope, properly applied, will show you this is not so.

> **A suggestion on triggering**
> It's a good idea to *watch* the waveform edge that *triggers* the scope, rather than trigger on one event and watch another. If you watch the *trigger event*, you will find that you can sweep the scope fast without losing the display off the right side of the screen.
> So, to measure *risetime* of a signal connected to Channel One, trigger on *Ch 1*, *rising-edge*.

What comes out of the function generator's **SYNC OUT** or **TTL**[7] connector?

Look at this on one channel while you watch a triangle or square wave on the other scope channel. To see how SYNC or TTL can be useful, try to trigger the scope on the peak of a sinewave *without* using these aids; then notice how entirely easy it is to trigger so when you *do* use SYNC or TTL to trigger the scope.

Triggering on a well-defined point in a waveform is especially useful when you become interested in measuring a difference in *phase* between two waveforms; this you will do several times in the next

---

[7] TTL stands for the equally cryptic phrase, Transistor–Transistor Logic, which is a kind of digital logic gate that you will find described in Chapter 14N. In the present context, TTL should be understood to mean simply Logic-Level square wave, a square wave that swings between ground and approximately four volts.

lab. You can use a *zero-crossing* (where the waveform crosses the midpoint of its excursion), or, when using SYNC OUT, you can use the peak or trough of the waveform.

How about the terminal marked **CALIBRATOR** (or CAL) on the scope's front panel? (We won't ask you to *use* this signal yet; not until Lab 3L do we explain how a scope probe works, and how you "calibrate" it with this signal. For now, just note that this signal is available to you). *Postpone* using scope probes until you understand what is within one of these gadgets. A "10×" scope probe is *not* just a piece of coaxial cable with a grabber on the end.

Put an "offset" onto the signal, if your function generator permits, then see what the **AC/DC** switch (located near the scope inputs) does.

> **Note on AC/DC switch**
> Common sense may seem to invite you to use the **AC** position most of the time: after all, are these time-varying signals that you're looking at not "AC" – alternating current (in some sense)? *Eschew* this plausible error. The *AC* setting on the scope puts a capacitor in series with the scope input, and this can produce startling distortions of waveforms if you forget it is there. (See what a 50Hz square wave looks like on AC, if you need convincing.) Furthermore, the AC setting washes away DC information, as you have seen: it hides from you the fact that a sinewave is sitting on a DC offset, for example. You don't want to wash away information except when you choose to do so knowingly and purposefully. Once in a while you *will* want to look at a little sine with its DC level stripped away; but always you will want to *know* that this DC information has been made invisible.

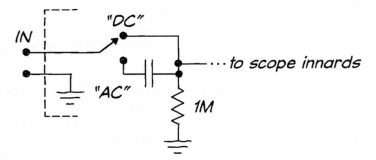

Figure 1L.11 AC/DC scope input select.

Set the function generator to some frequency in the middle of its range, then try to make an accurate frequency measurement with the scope. (Directly, you are obliged to measure *period*, of course, not frequency[8].) You will do this operation hundreds of times during this course. Soon you will be good at it.

Trust the scope period readings; distrust the analog function generator frequency markings; these are useful only for very *approximate* guidance.

---

[8] Well, if you're using a digital scope you could resort to the shabby trick of asking it to measure frequency for you. But don't start that way today.

1S Supplementary Notes: Resistors, Voltage, Current

1S.1   Reading resistors

In the lab exercises, from now and ever after, you will want to be able to read resistor values without pulling out a meter to *measure* the part's value (we do sometimes find desperate students resorting to such desperate means). The process will seem laborious, at first; but soon, as you get used to at least the common resistance values, you will be able to read many color codes at a glance. We will use resistors whose packages are big enough to be readable. The now-standard *surface-mount* parts are so tiny that the smaller ones normally are not labeled at all. This spares you the trouble (and opportunity) of reading them. If you mix a few surface-mount parts on the bench, you can only sweep them up and start over – unless you are willing to measure each.

1S.1.1   A menagerie of 10k resistors

Fig. 1S.1 shows a handsome collection of 10k resistors in a variety of packages.

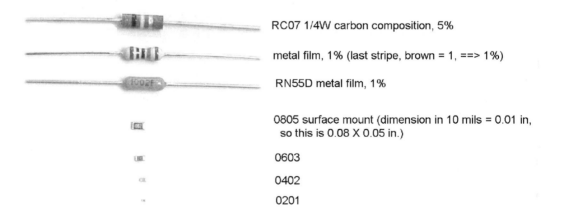

Figure 1S.1 10k resistors.

The type that we use – the carbon composition resistor at the top – is nearly obsolete, and relatively expensive, as a result. But we like them for lab work, because (on a good day) *we can read them*. The others are pretty nasty for breadboarding. The one with the value written out in numerals may appeal to you, if you don't want to learn the color codes – but it really isn't much fun to work with, because if it happens to be mounted with the value label *down*, you're out of luck.

1S.1.2 Resistor values and tolerances

Figure 1S.2 shows a resistor of the sort we use in this course's labs: it is a "5% carbon composition" part. "5%" "tolerance" means that its *actual* value is guaranteed to lie within 5% of its *nominal* value. If it is labeled "100k" (100,000Ω), its actual value can be expected to lie in the range ∼95−105kΩ.

The first problem one confronts on a first day with resistors, is *which way* to orient the part.

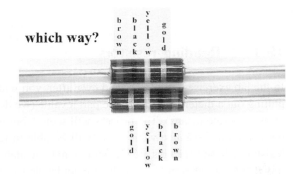

Figure 1S.2 Which way? Put the *tolerance* stripe to the right.

"*Tolerance stripe?*," you protest. "How do I know which that is?" In our labs, the answer is the fourth band, colored silver or gold. If you are using color-coded 1% resistors, the tolerance band will be the *fifth*. Fig. 1S.3 shows such a part.

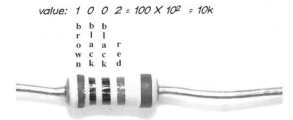

Figure 1S.3 Extra band for color-coded 1% resistor.

The use of four bands to define the value indicates (implicitly) that this is better than the usual 5% part; the color of the fifth band, brown represents "1," for this "1%" resistor. (A red band would indicate 2%, and so on.)

Tolerance: Only two tolerance values are common, for the 3-band carbon composition parts you are likely[1] to meet in the labs: 5% and 10%:

- silver: ±10%
- gold: ±5%

Value: Once you've oriented it properly – tolerance to the right – you can read off the value colors, and then can translate those to numbers. Finally, with the three numbers in hand, you will have enough information to discover the value. The resistor we just looked at is brown-black-yellow: see Fig. 1S.4.

Brown-black-yellow is one-zero-four, and the fourth band, gold, says "±5%."[2] The third band, the "four," is an exponent – a power of ten. So, this resistor's value is 100k.

---

[1] *No* tolerance band, in a 3-stripe part, indicates ±20%. Such resistors are rare.
[2] Sometimes the fourth band is *not* the last band; one more is added, for military components, to indicate the "failure rate."

1S.1 Reading resistors

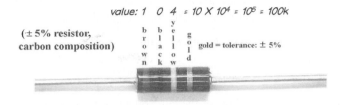

Figure 1S.4 Value stripes reveal what this resistor is.

The color code: The colors represent numbers, as set out below. A variety of mnemonics have arisen, most of them more-or-less offensive, in order to help engineers to memorize this code. One of the blander mnemonics is "Big Boys Race Our Young Girls, But Violet Generally Wins." This mnemonic is not very clever. It fails to distinguish the trio Black, Brown and Blue and also between the pair Green and Gray. *Black* is a plausible representation of the value zero, since black represents the absence of color; the color Brown is close to the color Black; perhaps one might remember Gray's position as "next to White." None of this is very satisfactory. We keep charts of the color codes up on the walls of our own teaching lab.

Here are the colors and the values they represent:

- black: zero
- brown: one
- red: two
- orange: three
- yellow: four
- green: five
- blue: six
- violet: seven
- gray: eight
- white: nine

  the next two are used as multipliers *only*, and are rarely seen:

- gold: 0.1
- silver: 0.01

1S.1.3 The set of "10% values"

It is hard to get used to the strange set of values that are "standard" in the lab. They are not the nice, round values that one might expect. They seem weird and arbitrary at first. But their strangeness does make sense. Because of our uncertainty about the *actual* value of a resistor, it doesn't make sense to specify distinct *nominal* values that are too close together; if we specify nominal values that are too close, their actual values are likely to overlap. To avoid this, the nominal values are placed far enough apart so as to make overlap slight.

A "10%" resistor of nominal value "10Ω," for example, could be as large as 11Ω. A "12Ω" resistor could be 10% smaller – a bit under 11Ω. So, 12 is about as close as it makes sense to place the next *10%* value after "10." And so on – the steps growing proportionally as the values rise, producing such unfamiliar numbers as *27, 39, 47,* and so on. Here is the 10% set (known as "E12," twelve values per decade. See Appendix C in AoE). Most of our laboratory circuits will use these *10%* values (with appropriate multipliers: we rarely use 10 ohm resistors, but often use 10kΩ parts, for example). This is so even though our lab resistors are better than 10% parts – normally they are good to ±5%:

$$10,\ 12,\ 15,\ 18,\ 22,\ 27,\ 33,\ 39,\ 47,\ 56,\ 68,\ 82,\ 100\ .$$

1S.1.4 Power

Only now and then are we obliged to consider *power* ratings of the components we use in the labs. That is true because our signal voltages are modest (under ±10V) and our currents are small (a few tens of milliamps). *Power* in our components, therefore, is modest as well, since power is the product of the two: Voltage *times* Current. 10V × 10mA, for example, dissipates 100mW – one tenth of a watt. But our standard resistors cannot handle much power: 1/4 watt is the most they can stand, sustained. So, recall this limitation – or you may be reminded by burned fingertips.

Incidentally, you may need to remind yourself, until your intuition catches up with your book knowledge, that it is *low-valued* resistors that are likely to overheat. 10V across *1k* is no problem; 10V across *100Ω* will hurt, if you touch that quarter-watt resistor.

1S.2 Voltage versus current

Two points to make here:

1. What is it that "flows" in most of our circuits? Current versus voltage.
2. It's usually information, not power, that interests us in this course.

1S.2.1 Is this note necessary?

I hope you don't feel insulted by this note – by its suggestion that you could get *current* and *voltage* muddled. Lest you should feel so insulted, let's adopt the device that we often find useful in this course: assume that we're talking not to you but to your slightly confused lab partner. We're trying to provide you with some arguments that will help you straighten out that intelligent but slightly-addled fellow.

Figure 1S.5 is a piece of evidence to support our claim that intelligent students can find it hard to keep I and V in their places. This is a drawing from a very good answer to a recent exam question. The answer was good – but includes a startling error that shows the student thinking *current* while handling *voltage*.

Do you see a place where the student seems to have interchanged the two concepts? (We don't think you need to understand this circuit fully to see the problem: you *do* need to understand that two circuits are included to provide a voltage gain of 2; we have indicated those.) We think this drawing shows that a smart student with quite a good grasp of electronics can still get this fundamental point wrong.

The evidence for the misunderstanding is just the presence of the two 2× amplifiers, with the slightly cryptic explanatory note at the left side, "1/2 of V_1 takes path 2× amplify." Notice that "V_1" appears again at the output of the upper 2× amplifier. Apparently, the author thought that V_1 coming from the 10k potentiometer would be cut in half because it goes two places – horizontally to the upper part of the circuit, and vertically to the lower part of the circuit.

This view is wrong: both destinations are very high-impedance inputs (input to triangle labeled "311", and input to lower triangle's "+" input). This *voltage* can go to many high-impedance inputs without suffering attenuation. The thought that "two similar destinations implies it's cut in half"

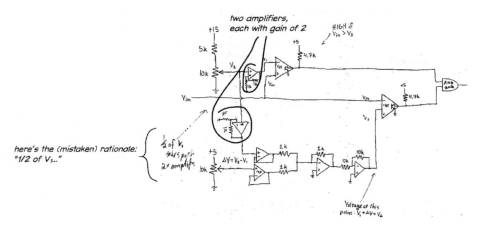

Figure 1S.5 Exam drawing: current and voltage confounded?

sounds like a correct view of how a *current* would behave. Fig. 1S.6 sketches what we guess the student imagined.

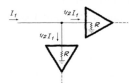

Figure 1S.6 *Currents*, in contrast to voltages, *do* split 50–50 when given two equal resistive paths.

Well, perhaps this example has convinced you that the question can puzzle a good student; perhaps we have not convinced you. Either way, let's carry on with the argument.

1S.2.2   What is it that "flows" in most of our circuits?

Your intuition may be inclined to say, "current flows," so the *signals* that we process and pass through our circuits probably are currents. That argument is plausible – but misleading at best, and usually just *wrong*.

It is *true* that *current* flows, whereas *voltage* emphatically does *not*. That's an important point – and many students need to correct themselves during the first weeks of the course as they find themselves referring to *voltages* flowing through a resistor.

But – and here is the subtler point – though *voltages* don't flow, we often speak of *signals* flowing through a circuit; indeed, all our circuits serve to process *signals* – and nearly always the *signals* that we process are *voltages*, not *currents*.

How can we reconcile these two truths: (1) currents, not voltages, flow; but (2) the signals that interest us as we design and analyze circuits nearly always are voltages, not currents? We can, as we hope some illustrations will persuade you.

And here is a preliminary question that we ought to get out of the way: *is this a question we need worry about?* Aren't voltages and current proportional, one to the other, in any case? So isn't it unimportant which we choose to describe – since one needs only to multiply by a constant in order to convert one into the other? Not a bad question, or challenge, but flawed: the deep flaw is not that some devices are not *ohmic* (though certainly this is true it is a minor point); the deep flaw is that circuit design strategy will head off in opposite directions, depending upon whether the *signals* that one intends to pass are *voltages* or *currents*.

Signal as *voltage*: Let's illustrate that point. Suppose one assumes – as we usually do – that our signal encodes its information as a *voltage*, not as a *current*. What follows? It follows that to keep this signal strong and healthy as it passes from one circuit to another – a passage that entails passing a voltage divider – we want $R_{IN_B} \gg R_{OUT_A}$. Ideally, we want R_{IN_B} infinite.

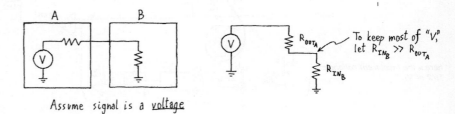

Figure 1S.7 Signal as voltage: preserving it implies impedance relation, A versus B.

This is the case that we normally assume. It is the case introduced in Chapter 1N, and then used to define a design "rule of thumb."

Signal as *current*: Notice how different our design rules would be if our signals were *currents*: we would want circuit B to interfere as little as possible with the output of A – and that means that we would want B to look like a *short circuit* to A. Ideally, we want R_{IN_B} zero!

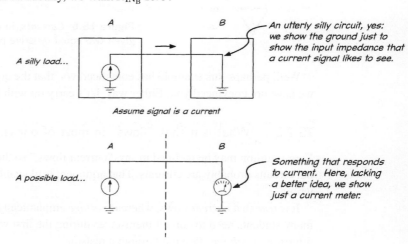

Figure 1S.8 Signal as current: preserving it implies a very different impedance relation, A versus B.

In our course, this case turns out to be *rare* (but you'll see it now and then: for example, see Lab 6L's photodiode circuit).

AoE Exercise 1.10.

A hybrid case? Signal as *power*: Sometimes our goal is to transfer, from A to B, *power*: say for the case where we want to drive a speaker and get a big sound out. In this case we want to deliver *both current and voltage* to the speaker. We will be frustrated if either one is small since the speaker's power will be the product of *voltage times current*. For a given R_{OUT}, the R_{IN} value that maximizes power in the load turns out to be neither large nor small relative to R_{OUT_A}; instead, we want to make the two values, R_{OUT_A} and R_{IN_B}, *equal*.[3]

---

[3] Note a subtlety here: we have stated a rule (first proposed by Jacobi) for choosing R_{LOAD} for a given R_{OUT}. The opposite problem, choosing R_{OUT} for a given R_{IN}, yields a different result. For that case, the maximum power to the load is delivered by R_{OUT} is zero.

But this, again, is a case very exceptional in our course. We are concerned almost always with processing and transferring *information*, not power; and that information we encode almost always as a voltage.

An illustrative example: low-pass filter (a preview): Next time you will meet *RC* frequency-dependent "filters." But we think you will be able to follow the argument presented in this example. A low-pass filter illustrates the point that it is a *voltage* signal that we want to *pass*: a wiggle of information. We do not want our filter to pass *current* from input to output. This example is particularly striking because the frequency range in which the *signal* is passed is the frequency range when input *current* is minimal – and vice versa.

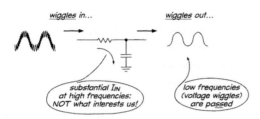

Figure 1S.9 Lowpass illustrates that it's voltage, not current, that we view as *signal*.

Incidental point: the decibel definition that interests us: Because we are concerned with processing *voltages*, we use a definition of the term *decibel* that reflects that interest:

$$dB = 20\log_{10} A_2/A_1 \ . \tag{1S.1}$$

We use this in preference to the definition of a ratio of *powers*:

$$dB = 10\log_{10} P_2/P_1 \tag{1S.2}$$

The difference between the two forms reflects the fact that power in a resistive load varies with the *square* of voltage.

1S.2.3 Punchline: it's information, not power that interests us in this course

So though the world still needs electrical power (and storage, delivery and control of electrical power for automobiles is a newly hot topic), our interests in this course concentrate rather on the use of electrical signals to convey and process *information*.

1W Worked Examples: DC circuits

1W.1 Design a voltmeter, current meter

> AoE 1.2.3, Multimeters, ex. 1.8

A 50μA meter movement has an internal resistance of 5k. What shunt resistance is needed to convert it to a 0–1A ammeter? What series resistance will convert it to a 0–10V voltmeter?

This exercise gives you an insight into the instrument, of course, but it also will give you some practice in judging when to use approximations: how precise to make your calculations, to say this another way.

1A meter: "50μA meter movement" means that the needle deflects fully when 50μA flows through the movement (a coil that deflects in the magnetic field of a permanent magnet: see Fig. 1L.6 for a sketch). The remaining current must bypass the movement; but the current through the movement must remain proportional to the whole current.

Such a long sentence makes the design sound complicated. In fact, as probably you have seen all along, the design is extremely simple: just add a resistance in parallel with the movement (this is the shunt mentioned in the problem): see Fig. 1W.1. What value?

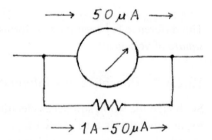

Figure 1W.1 Shunt resistance allows sensitive meter movement to measure a total current of 1A.

Well, what else do we know? We know the *resistance* of the meter movement. That characteristic plus the full-scale current tell us the *full-scale voltage drop* across the movement: that's

$$V_{\text{movement(full-scale)}} = I_{\text{full-scale}} \times R_{\text{movement}} = 50\mu\text{A} \times 5\text{k}\Omega = 250\,\text{mV}.$$

Now we can choose R_{shunt}, since we know current and voltage that we want to see across the parallel combination. At this point we have a chance to work too hard, or instead to use a sensible approximation. The occasion comes up as we try to answer the question, "How much current should pass through the shunt?"

One possible answer is "1A less 50μA, or 0.99995A." Another is "1A."

Which do you like? If you're fresh from a set of physics courses, you may be inclined toward the first answer. If we take that, then the resistance we need is

$$R = \frac{V_{\text{full-scale}}}{I_{\text{full-scale}}} = \frac{250\,\text{mV}}{0.99995\,\text{A}} = 0.2500125\,\Omega.$$

Now in some settings that might be a good answer. In this setting, it is not. It is a very silly answer. That resistor specification claims to be good to a few parts in a million. If that were possible at all, it would be a preposterous goal in an instrument that makes a needle move so we can peer at it.

So we should have chosen the second branch at the outset: seeing that the 50μA movement current is small relative to the the 1A total current, we should then ask ourselves, "About how small (in fractional or percentage terms)?" The answer would be 50 parts in a million. And that fraction is so small relative to reasonable resistor and meter tolerances that we should conclude at once that we should neglect the 50μA.

Neglecting the movement current, we find the shunt resistance is just 250mV/1A = 250mΩ. In short, the problem is very easy if we have the good sense to let it be easy. You will find this happening repeatedly in this course: if you find yourself churning through a lot of math, and especially if you are carrying lots of digits with you, you're probably overlooking an easy way to do the job. There is no sense carrying all those digits and then having to reach for a 5% resistor and 10% capacitor. In this case – designing a meter, we'll do a little better: we'll use 1% resistors – but approximations still make sense.

Voltmeter: Here we want to arrange things so that 10V applied to the circuit causes a full-scale deflection of the movement. Which way should we think of the cause of that deflection – "50μA flowing," or "250mV across the movement?"

Either is fine. Thinking in *voltage* terms probably helps one to see that most of the 10V must be dropped across some element we are to add, since only 0.25V will be dropped across the meter movement. That should help us sketch the solution: see Fig. 1W.2.

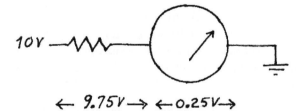

Figure 1W.2 Voltmeter: series resistance needed.

What series resistance should we add? There are two equivalent ways of answering:

1. The resistance must drop 9.75 volts out of 10, when 50μA flows; so $R = 9.75\text{V}/50\mu\text{A} = 195\text{k}\Omega$.
2. Total resistance must be 10V/50μA = 200kΩ. The meter movement looks like 5k, we were told; so we need to add the difference, 195kΩ.

If you got stung on the first part of this problem, giving an answer like "0.2500125Ω," then you might be inclined to say, "Oh, 50μA is very small; the meter is delicate, so I'll neglect it. I'll put in a 200k series resistor, and be just a little off."

Well, just to keep you off-balance, we must now push you the other way: this time, "50μA," though a small current, is not negligibly small because it is not to be compared with some much larger current. On the contrary, it is the crucial characteristic we need to work with: it determines the value of the series resistor. And we should *not* say "200k is close enough," though 195k is the exact answer. The difference is 2.5%: much less than what we ordinarily worry about in this course (because we need to get used to components with 5 and 10% tolerances); but in a meter it's surely worth a few pennies extra to get a 1% resistor: a 196k.

1W.2 Resistor power dissipation

Problem (Power dissipation in a lab circuit's resistors)   Power dissipation in a lab circuit's resistors. Please specify a *10%* value (of power) in each case. Here are the 10% values (of resistance), as a reminder:

$$10;\ 12;\ 15;\ 18;\ 22;\ 27;\ 33;\ 39;\ 47;\ 56;\ 68;\ 82;\ 100\ .$$

AoE §1.2.2C

Problem   What is the smallest 1/4W resistor one should put across a 5V supply?

Solution

$$P = V^2/R;\qquad R = 25V^2/0.25W = 100\Omega.$$

Problem   What is the lowest value surface-mount 1/8W resistor one should put between -15V and $+15$V?

Solution

$$P = V^2/R;\qquad R = 900V^2/0.125W = 7200\Omega \approx 7.5\text{k}.$$

8.2k is the nearest 10% value that would not overheat.

AoE §1.2.2C

Problem   What is the maximum voltage one ought to put across a 10W, 10Ω power resistor?

Solution

$$P = V^2/R;\qquad V = \sqrt{P \times R} = \sqrt{10\,\text{W} \times 10\,\Omega} = 10\text{V}.$$

Problem (Power transmission. Effect of voltage step-up.)   Electric power is transmitted at very high voltages to minimize power losses in the transmission cables. By what factor are long-distance *power line losses* reduced if the power company manages to raise its transmission voltage from 100,000 volts to 1 million volts? (We assume the power company is obliged to deliver a given amount of power to the customer, unchanged between the two cases.)

Solution   Power losses in the lines are proportional to $V \times I$ in the line, or, equivalently, to $I^2 \times R$, where R is the resistance of a unit length of the line (a fat cable, no doubt; but its resistance is not zero[1]).

Stepping *up* the voltage by a factor of 10 steps *down* the current by the same factor (while transmitting a given level of power). Since power loss is proportional to the *square* of the current, the power dissipated will fall by $(1/10)^2 = 1/100$. A big reward. The change sounds worthwhile – though extreme high voltages are troublesome to insulate; so, 1MV is about the practical limit.

Problem (Why AC?)   Why does high-voltage transmission militate strongly against Thomas Edison's program of DC power transmission?

Solution   AC transmission is favored, for most purposes, because stepping AC voltages up and down is easy, requiring only transformers. Stepping DC voltages up and down requires more complex processes. High-voltage DC transmission is, in fact, used for special cases, where AC losses caused by inductance and capacitance are unusually high, as in undersea cables. For the most part, however, AC transmission still prevails because of its simplicity and the low losses that are achieved in the stepping up/down process.

---

[1] ... not, at least, until all power lines are replaced by superconductors.

1W.3 Working around imperfections of instruments

§1L.5 asks you to go through the chore of confirming Ohm's Law. But it also confronts you at once with the difficulty that you cannot quite do what the experiment asks you to do: measure I and V in the resistor simultaneously. Two placements of the DVM are suggested in Fig. 1W.3 (one is drawn, the other hinted at).

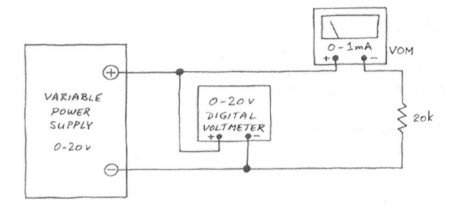

Figure 1W.3 §1L.5 setup: DVM and VOM cannot both measure the relevant quantity.

A qualitative view

Just a few minutes' reflection will tell you that the voltage reading is off, in the circuit as drawn; moving the DVM solves that problem (above), but now makes the current reading inaccurate.

A quantitative view

Here's the problem we want to spend a few minutes on.

Problem (Errors caused by the meters) If the analog meter movement is as described in §1W.1: what percentage error in the *voltage* reading results, if the voltage probe is connected as shown in the figure for the first Lab 1L experiment, when the measured resistor has the following values.

- $R = 20\text{k}\Omega$.
- $R = 200\Omega$.
- $R = 2\text{M}\Omega$.

Assume that you are applying 20V, and that you can find a meter setting that lets you get *full-scale deflection* in the current meter.

Solution This question is easier than it may appear. The error we get results from the voltage drop across the current meter; but we know what that drop is from earlier: full-scale: 0.25V. So the resistor values do not matter. Our voltage readings always are high by a quarter volt, if we can set the current meter to give full-scale deflection. The value of the resistor being measured does *not* matter.

When the DVM reads 20V, the true voltage (at the top of the resistor) is 19.75V. Our voltage reading is high by 0.25V/19.75V – about 0.25/20 or 1 part in 80: 1.25% (if we applied a lower voltage, the voltage error would be more important, assuming we still managed to get full-scale deflection from the current meter, as we might be able to by changing ranges).

Problem Same question, but concerning *current* measurement error, if the voltmeter probe is moved to connect directly to the top of the resistor, for the same resistor values. Assume the DVM has an input resistance of 20MΩ.

Worked Examples: DC circuits

Solution If we move the DVM to the top of the resistor, then the voltage reading becomes correct: we are measuring what we meant to measure. But now the current meter is reading a little high: it measures not only the resistor current but also the DVM current, which flows parallel to the current in R.

The size of *this* error depends directly on the size of R we are measuring. You don't even need a pencil and paper to calculate how large the errors are:

- If R is 20kΩ – and the DVM looks like 20M – then one part in a thousand of the total current flows through the DVM: the current reading will be high by 0.1%.
- If R is 200Ω, then the current error is minute: 1 part in 100,000: 0.001%.
- If R is 2MΩ, then the error is large: 1 part in ten.

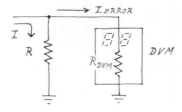

Figure 1W.4 DVM causes current-reading error: how large? % error same as ratio of R to R_{DVM}.

Conclusion? There is *no* general answer to the question "Which is the better way to place the DVM in this circuit?" The answer depends on R, on the applied voltage and on the consequent ammeter range setting.

And before we leave this question, let's notice the implication of that last phrase: the error depends on the VOM *range* setting. Why? Well, this is our first encounter with the concept we like to call Electronic Justice, or the principle that The Greedy Will Be Punished. No doubt these names mystify you, so we'll be specific: the thought is that if you want good resolution from the VOM, you will pay a price: the meter will alter results more than if you looked for less resolution, see Fig. 1W.5.

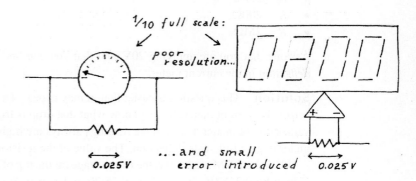

Figure 1W.5 Tradeoffs, or Electronic Justice I. VOM or DVM as *ammeter*: the larger the reading, the larger the voltage error introduced; VOM as *voltmeter*: the larger the deflection at a given V_{in}, the lower the input impedance.

If you want the current meter needle to swing nearly all the way across the dial (giving best resolution: small changes in current cause relatively large needle movement), then you'll get nearly the full-scale 1/4-volt drop across the ammeter. The same goes for the DVM as ammeter; if you understand that 'full scale' for the DVM means filling its digital range: "3 1/2 digits," as the jargon goes (the "half digit" is a character that takes only the values zero or one). So, if you set the DVM current range so that your reading looks like

$$0.093,$$

you have poor resolution: about 1%. If you are able to choose a setting that makes the same current look like 0.930, you've improved resolution tenfold. But you have also increased the voltage drop across the meter by the same factor; for the DVM, like the analog VOM drops 1/4 V full-scale, and proportionately less for smaller "deflection" (in the VOM) or smaller fractions of the full-scale range (for the DVM).

1W.4 Thevenin models

Problem Draw Thevenin Models for the circuits in Fig. 1W.6. Give answers to 10% and to 1%

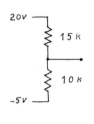
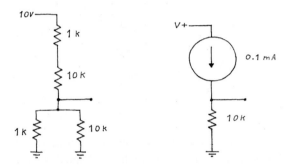

Figure 1W.6 Some circuits to be reduced to Thevenin models.

Some of these examples show typical difficulties that can slow you down until you have done a lot of Thevenin models.

The leftmost circuit is most easily done by temporarily redefining *ground*. That trick puts the circuit into the entirely familiar form in Fig 1W.7.

The only difficulty that the middle circuit presents comes when we try to approximate. The 1% answer is easy, here. The 10% answer is tricky. If you have been paying attention to our exhortations to use 10% approximations, then you may be tempted to model each of the resistor blocks with the dominant *R*: the small one, in the parallel case, the big one in the series case, see Fig. 1W.8.

Unfortunately, this is a rare case when the errors gang up on us; we are obliged to carry greater precision for the two elements that make up the divider.

This example is not meant to make you give up approximations. It makes the point that it's the *result* that should be good to the stated precision, not just the intermediate steps.

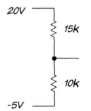
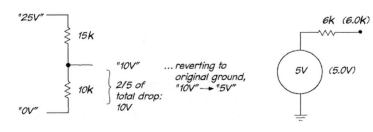

Figure 1W.7 A slightly novel problem reduced to a familiar one by temporary redefinition of ground.

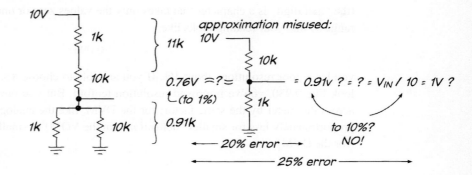

Figure 1W.8 10% approximations: errors can accumulate.

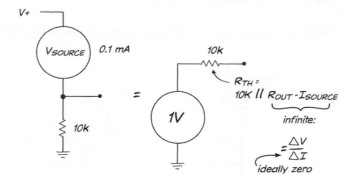

Figure 1W.9 Current source feeding resistor, and equivalent Thevenin model.

The *current source* shown in Fig. 1W.9 probably looks odd to you. But you needn't understand how to make one to see its effect; just take it on faith that it does what's claimed: sources (squirts) a fixed current, down toward the negative supply. The rest follows from Ohm's Law. (In Lab 4L you will learn to design these things – and you will discover that some devices just do behave like current sources without being coaxed into it: transistors behave this way – both bipolar and FET.)

The point that the current source shows a very high output impedance helps to remind us of the definition of impedance – always the same: $\Delta V / \Delta I$. It is better to carry that general notion with you than to memorize a truth like "Current sources show high output impedance." Recalling that definition of impedance, you can always figure out the current source's approximate output impedance (large versus small); soon you will know the particular result for a current source, just because you will have seen this case repeatedly.

1W.5 "Looking through" a circuit fragment, and R_{in}, R_{out}

Problem What are R_{in}, R_{out} at the indicated points in Fig. 1W.10?

Solution A. R_{in}. It's clear what R_{in} the divider should show: just $R_1 + R_2$. But when we say that are we answering the right question? Isn't the divider surely going to drive something down the line? If not, why was it constructed?

The answer is yes, it *is* going to drive something else – the *load*. But that something else should present an R_{in} high enough so that it does not appreciably alter the result you got when you ignored the load. If we follow our 10× rule of thumb (see §1N.4.6) you won't be far off this idealization: less

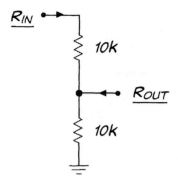

Figure 1W.10 Determining R_{in}, R_{out}; you need to decide what's beyond the circuit to which you're connecting.

than 10% off. To put this concisely, you might say simply that we assume an *ideal* load: a load with infinite input impedance.

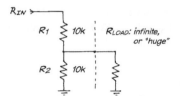

Figure 1W.11 R_{in}: we need an assumption about the load that the circuit drives, if we are to determine R_{in}.

Solution B. R_{out}. Here the same problem arises – and we settle it in the same way: by assuming an ideal source. The difficulty, again, is that we need to make some assumption about what is on the far side of the divider if we are to determine R_{out}: see Fig. 1W.12. The assumption we make in determining R_{out} is familiar to you from Thevenin models: we assume source impedance so low that we can neglect it, calling it zero. If each resistor is 10k, R_{out} is 5k.

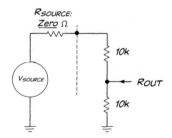

Figure 1W.12 R_{out}: we need an assumption about the source that drives the circuit, if we are to determine R_{out}.

1W.6   Effects of loading

Problem   In Fig. 1W.13, what is the voltage at X:

1. with no load attached?
2. when measured with a VOM labeled "10,000 ohms/volt?"
3. when measured with a scope whose input resistance is 1MΩ?

Worked Examples: DC circuits

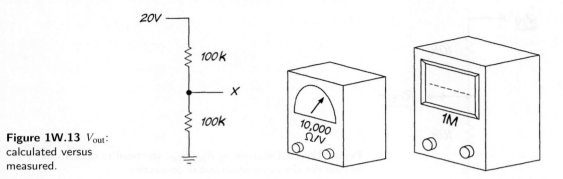

Figure 1W.13 V_{out}: calculated versus measured.

This example recapitulates a point made several times over in Chapter 1N, as you recognize. *Reminder*: The "... ohms/volt" rating implies that on the 1-volt scale (that is, when 1V causes full deflection of the meter) the meter will present that input resistance. What resistance would the meter present when set to the 10-volt scale? (Answer: 10 times the 1-volt R_{IN}: 100kΩ.)

We start, as usual, by trying to reduce the circuit to familiar form. The Thevenin model does that for us. Then we add meter or scope as *load*, forming a voltage divider, and see (Fig. 1W.14) what voltage results.

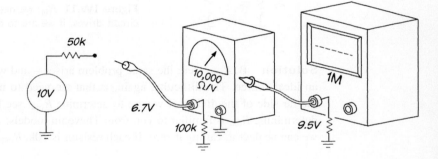

Figure 1W.14 Thevenin model of the circuit under test; and showing the "load" – this time, a meter or scope.

You will go through this general process again and again, in this course: reduce an unfamiliar circuit diagram to one that looks familiar. Sometimes you will do that by merely redrawing or rearranging the circuit; more often you will invoke a model, and often that model will be Thevenin's.

2N *RC* Circuits

Contents

2N.1	**Capacitors**		**51**
	2N.1.1	Why?	52
	2N.1.2	Capacitor structure	52
2N.2	**Time-domain view of *RC*s**		**53**
	2N.2.1	Integrators and differentiators	55
2N.3	**Frequency domain view of *RC*s**		**58**
	2N.3.1	The impedance or *reactance* of a cap	59
	2N.3.2	*RC* filters	61
	2N.3.3	Decibels	64
	2N.3.4	Estimating a filter's attenuation	66
	2N.3.5	Input and output impedance of an *RC* circuit	68
	2N.3.6	Phase shift	69
	2N.3.7	Phasor diagrams	70
2N.4	**Blocking and decoupling**		**74**
	2N.4.1	*Blocking* capacitor	74
	2N.4.2	Decoupling or bypass capacitor	75
2N.5	**A somewhat mathy view of *RC* filters**		**76**
2N.6	**Readings in AoE**		**77**

2N.1 Capacitors

Now things get a little more complicated, and more interesting, as we meet frequency-dependent circuits. We rely on the *capacitor* to implement this new trick, which depends on the capacitor's ability to "remember" its recent history.

That ability allows us to make timing circuits (circuits that let something happen a predetermined time after something else occurs); the most important of such circuits are *oscillators* – circuits that do this timing operation over and over, endlessly, in order to set the frequency of an output waveform. The capacitor's memory also lets us make circuits that respond mostly to changes (*differentiators*) or mostly to averages (*integrators*). And the capacitor's memory implements the *RC* circuit that is by far the most important to us: a circuit that favors one frequency range over another (a *filter*).[1]

All of these circuit fragments will be useful within later, more complicated circuits. The filters, above all others, will be with us constantly as we meet other analog circuits. They are nearly as ubiquitous as the (resistive-) *voltage divider* that we met in the first class.

---

[1] Incidentally, in case you need to be persuaded that remembering is the essence of the service that capacitors provide, note that much later in this course, partway into the digital material, we will meet large arrays of capacitors used simply and explicitly to remember: several sorts of digital memory (dynamic RAM, Flash, EEPROM and EPROM) use millions to billions of tiny capacitors to store their information, holding that data in some cases for many *years*.

2N.1.1 Why?

We have suggested a collection of applications for *RC* circuits. Here is a shorter answer to what we mean to do with the circuits we meet today:

- generate an output voltage transition that occurs a particular delay time after an input voltage transition;
- design a circuit that will treat inputs of different frequencies differently: it will pass more in one range of frequencies than in another (this circuit we call a "filter").

2N.1.2 Capacitor structure

The capacitor in Fig. 2N.1 is drawn to look like a ham sandwich: metal plates are the bread, some dielectric is the ham (*ceramic* capacitors really are about as simple as this). More often, capacitors achieve large area (thus large capacitance) by doing something tricky, such as putting the dielectric between two thin layers of metal foil, then rolling the whole thing up like a roll of paper towel (*mylar* capacitors are built this way).

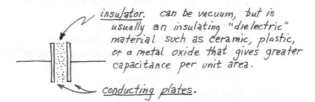

Figure 2N.1 The simplest capacitor configuration: sandwich.

A static description of cap behavior...

A *static* description of the way a capacitor behaves would say

$$Q = CV$$

where Q is total charge, C is the measure of how big the cap is (how much charge it can store at a given voltage: $C = Q/V$), and V is the voltage across the cap.

This statement just defines the notion of capacitance. It is the way a physicist might describe how a cap behaves, and rarely will we use it again.

AoE §1.4.1

...A dynamic description of cap behavior

Instead, we use a dynamic description – a statement of how things change with time:

$$I = C \frac{dV}{dt} \tag{2N.1}$$

This is just the time derivative of the "static" description. C is constant with time; I is defined as the rate at which charge flows. This equation isn't hard to grasp. It says "The bigger the current, the faster the cap's voltage changes."

A hydraulic analogy: Again, flowing water helps intuition: think of the cap (with one end grounded) as a tub that can hold charge: see Fig. 2N.2.

A tub of large diameter (cap) holds a lot of water (charge), for a given height or depth (V). If you fill the tub through a thin straw (small I), the water level – V – will rise slowly; if you fill or drain through a fire hose (big I) the tub will fill ("charge") or drain ("discharge") quickly. A tub of large diameter (large capacitor) takes longer to fill or drain than a small tub. Self-evident, isn't it?

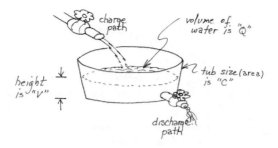

Figure 2N.2 A cap with one end grounded works a lot like a tub of water.

2N.2 Time-domain view of RCs

Now let's leave tubs of water, and anticipate what we will see when we watch the voltage on a cap change with time: when we look on a scope screen, as you will do in Lab 2L.

An easy case: constant I

This tidy waveform, called a *ramp*, is useful, and you will come to recognize it as the signature of this circuit fragment: capacitor driven by constant current (or "current source").

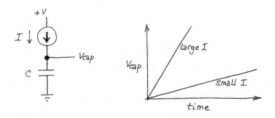

Figure 2N.3 Easy case: constant $I \longrightarrow$ constant dV/dt.

This arrangement is used to generate a triangle waveform, see Fig. 2N.4.

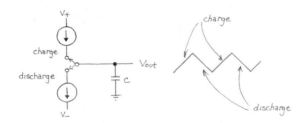

Figure 2N.4 How to use a cap to generate a triangle waveform: ramp up, ramp down.

But the ramp waveform is relatively rare, because current sources are relatively rare. Much more common is the next case.

A harder case but more common: constant voltage source in series with a resistor ("exponential" charging)

Here, the voltage on the cap approaches the applied voltage – but at a rate that diminishes toward zero as V_{cap} approaches its destination. Fig. 2N.5 shows V_{cap} starting out bravely, moving fast toward its V_{in} (charging at 10mA in the example above, thus at 10V/ms); but as it gets nearer to its goal, it loses its nerve. By the time it is 1V away it has slowed to 1/10 its starting rate.

(The cap behaves a lot like the hare in Xeno's paradox: remember him? Xeno teased his fellow Athenians by asking a question something like this: 'If a hare keeps going halfway to the wall, then again halfway to the wall, does he ever get there?' (Xeno really had a hare chase a tortoise; but the

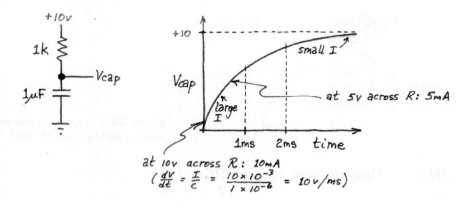

Figure 2N.5
The more usual case: cap charged and discharged from a *voltage* source, through a series resistor.

electronic analog to the tortoise escapes us, so we'll simplify his problem.) Hares do bump their noses; capacitors don't: V_{cap} never does reach $V_{applied}$, in an *RC* circuit. But it will come as close as you want.

Exponential charge and discharge

The behavior of *RC* charging and discharging is called "exponential" because it shows the quality common to members of that large class of functions. A function whose slope is proportional to its value is called exponential: the function e^x behaves as the charging *RC* circuit does. Its slope is equal to its value.

Familiar exponential functions are those describing population growth (the more bacteria, the faster the colony grows) and a draining bathtub (the shallower the water, the more slowly it drains). The capacitor's *discharge* curve follows this rule more obviously than the *charging* curve, so let's consider that case: as the capacitor discharges, the voltage across the *R* diminishes, reducing the rate at which it discharges. This is evident in the *discharge* curve of Fig. 2N.6. The *RC* discharges the way a bathtub drains.

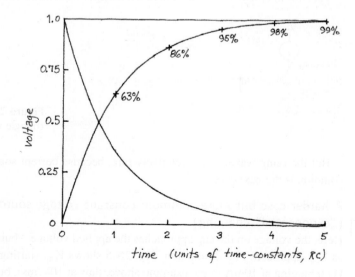

Figure 2N.6 *RC* charge, discharge curves.

Two numbers in the plot of Fig. 2N.6 are worth remembering:

- in *one RC* (called "one time-constant") V_{cap} goes 63% of the way toward its destination;
- in *five RCs*, 99% of the way

If you need an exact solution to such a timing problem:

$$V_{\text{cap}} = V_{\text{applied}} \cdot (1 - e^{-t/RC})$$

In case you can't see at a glance what this equation is trying to tell you, look at $e^{-t/RC}$ by itself:

- when $t = RC$, this expression is $e^{-1} = 1/e$, or 0.37.
- when $t \gg RC$, i.e., very large, this expression is tiny, and $V_{\text{cap}} \approx V_{\text{applied}}$

> A tip to help you calculate time-constants:
> MΩ and μF give time-constant in seconds
> kΩ and μF give time-constant in milliseconds

(The second case, kΩ and μF, is very common.) In the example above, RC is the product of 1k and 1μF: 1ms.

The "time constant," RC

You may be puzzled, at first, to see frequent reference to a circuit's "time constant," even in settings where no one plans to use the quantity RC directly, as one might, say, in an oscillator circuit. The *time constant* serves to give a ballpark measure of how quick or slow a circuit element may be. If you see fuzz on an oscilloscope screen at 1MHz, for example, you know at once not to blame an RC circuit whose time constant is in the millisecond range.

...and a trick for the calculator-less

The exponential expression, $e^{-t/RC}$, which describes the "error", may look pretty formidable to work with by hand. In fact, however, it is not hard to estimate. Just note the following pattern:

- in one time-constant the *error* decreases to about 40% of what it was at the start of that time-constant.

So, after a couple of time constants, the error is down to 0.4×0.4, or about 0.16. One more time constant takes the error down to about 0.06. And so on.

2N.2.1 Integrators and differentiators

The very useful formula, $I = C\,dV/dt$ will let us figure out when these two circuits perform pretty well as integrator and differentiator, respectively.

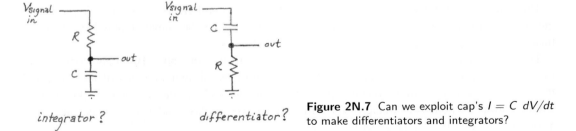

Figure 2N.7 Can we exploit cap's $I = C\,dV/dt$ to make differentiators and integrators?

Differentiator Let's first consider the simpler circuit in Fig. 2N.8. The current that flows in the cap is proportional to dV_{in}/dt; so, in once sense, the circuit differentiates the input signal perfectly: its *current* is proportional to the slope of V_{in}. But the circuit is pretty evidently useless (*perfectly* useless). It gives us no way to measure that current. If we could devise a way to measure the current, we would have a differentiator. Fig. 2N.9 shows our earlier proposal, again. Does it work?

Answer: Yes and No. Yes, to the extent that $V_{cap} = V_{in}$ (and thus $dV_{cap}/dt = dV_{in}/dt$), it works. But to the extent that V_{out} moves, it is imperfect, because that movement makes V_{cap} differ from V_{in}.

So the circuit errs to the extent that the output moves away from ground; but of course it must move away from ground to give us an output. This differentiator is compromised. So is the *RC* integrator, it turns out. When we meet operational amplifiers, we will manage to make nearly-ideal integrators, and pretty good differentiators.

Figure 2N.8
Useless "differentiator"?

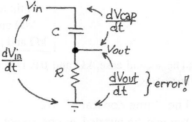

Figure 2N.9 Differentiator? – again.

AoE puts that point this way:

- for the differentiator: "...choose R and C small enough so that $dV_{out}/dt \ll dV_{in}/dt$..."
- for the integrator: "...[make sure that] $V_{out} \ll V_{in} \ldots \omega \cdot RC \gg 1$."[2]

AoE §1.4.3

AoE §1.4.4

We can put this simply – perhaps crudely. Assume a sinewave input. Then,

the *RC* differentiator (and integrator, too) works pretty well *if* it is *murdering* the signal (that is, attenuates it severely), so that V_{out} (and dV_{out}) is tiny: hardly moves away from ground.

It follows[3], along the way, that differentiator and integrator will impose a 90° phase shift on a sinusoidal input. This result, obvious here, should help you anticipate how *RC* circuits viewed as "filters" (below) will impose phase shifts.

Try out the differentiator: Fig. 2N.10 shows what a differentiator like the one you'll build in Lab 2L puts out, given a square wave input. The slow sweep on the left makes the output look pretty much like a spike, responding to each edge of the square wave. The detail on the right shows the decay of the *RC* that forms the differentiator[4] (the time constant, *RC*, is $100\Omega \times 100\text{pF} = 10^4 \times 10^{-12} = 10 \times 10^{-9} = 10\text{ns}$).

The square wave produces the most dramatic response from a differentiator. A sinusoid produces a quieter response: another sinusoid, simply phase shifted. Note that the output amplitude is much less than input: see Fig. 2N.11.

The output looks more or less correct: it measures the slope of the input. But it also looks a little funny: it is not a smooth sinusoid. That's the fault not of the differentiator but of the "sinusoid" that goes in – coming from the function generator. That sinusoid is a bit of a phoney, produced by whittling away the points of a triangle! The differentiator exposes this fraud: one can make out at least a constant-slope section as the underlying triangle crosses zero.

---

[2] This may be the first appearance of ω (*omega*) in these notes. ω describes frequency not in cycles-per-second, properly called "hertz," but in radians-per-second. Since a full cycle includes 2π radians, $\omega = 2\pi f$.

[3] When we say "it follows," we are assuming that you accept the proposition that the derivative or integral of a sinusoid is another sinusoid 90° phase-shifted relative to the original. If this point is new to you, hang on. We'll soon see this happening in today's lab.

[4] The amplitude is not the same as in the slower sweep, because the detail shows response to a positive 4V step (the edge of a "TTL" square wave), used because its edge is steeper than the edge of the square wave shown in the slower sweep. The steeper edge produces a more-nearly-ideal response pulse.

2N.2 Time-domain view of RCs

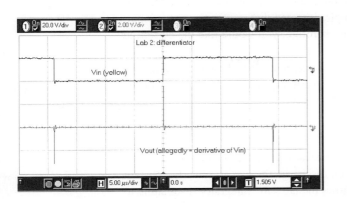
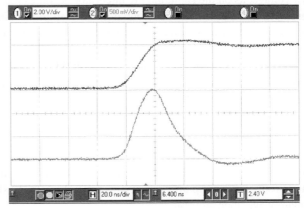

Figure 2N.10 Differentiator, fed a square wave as input; detail shows *RC*.

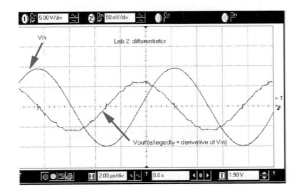

Figure 2N.11 Response of *RC* differentiator to sinusoid: shows the "sine" isn't clean (note scale change from IN to OUT: Ch1 is input).

AoE §1.4.4

Integrator

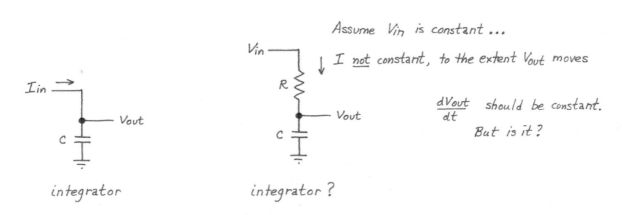

Figure 2N.12 Integrator? – again, only sort-of.

The circuit on the left in Fig. 2N.12 is a *perfect* integrator – but its input is a *current*. When the input is a voltage, as on the right, the *RC* integrator becomes imperfect. The limitations of the *RC* integrator resemble those of the *RC* differentiator. To keep things simple, imagine that you apply a *step* input; ask what waveform out you would like to see out, and what, therefore, you would like the current to do.[5]

The integrator works if $V_{OUT} \ll V_{IN}$ – or, less formally, it works if it's murdering the input. We achieve that by making sure that the cap doesn't have time to charge much between reversals of the input waveform: $RC \gg$ {half-period of a square wave input}, for example.

Try out the integrator: Again a square wave provides the easiest test of the integrator, since it's pretty plausible that a square wave input should charge, then discharge the cap at a nearly-constant rate, producing a ramp up, then a ramp down. Fig. 2N.13 shows the lab's *RC* integrator responding to square wave, then sinusoid.

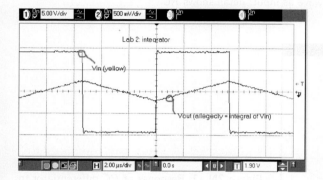 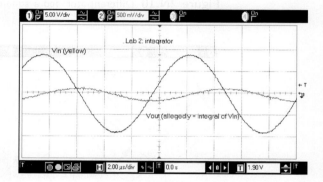

Figure 2N.13 Lab 2L's *RC* integrator responding to square wave and sinusoid (note scale change from IN to OUT: Ch1 is input).

Note that this integrator is severely attenuating the signal, as required for an *RC* integrator: the gain on the output channel is *ten times* the gain on the input channel, in Fig. 2N.13.

Incidentally, if you find it hard to get an intuitive grip on what the *integral* of a sinusoid should look like, try this trick: *reverse the process*. Check that the *input* waveform shows the *slope* of the output; in other words, that the input shows the derivative of the output – the *inverse* of the integral function.

2N.3 Frequency domain view of RCs

We'll now switch to speaking of the same old *RC* circuits in a different way, describing not what a scope image might show – the so-called "time-domain" view – but instead how the circuit behaves as frequency changes. This is the "frequency domain" view. Don't let the name scare you: the reference to "domain" (a word that evokes castles and lords and ladies) is just a high-fallutin' way to specify what's on the *x*-axis ("domain" describes the input to a mathematical function, as you may recall).

---

[5] What we'd like to see is an output *ramp*: constant dV_{out}/dt for constant V_{in}.

2N.3 Frequency domain view of RCs

2N.3.1 The impedance or *reactance* of a cap

A cap's impedance varies with frequency. Impedance is the generalized form of what we called "resistance" for "resistors;" "reactance" is the term reserved for capacitors and inductors. The latter usually are coils of wire, often wound around an iron core.

It's obvious that a cap cannot conduct a DC current: just recall what the cap's insides look like: an insulator separating two plates. That takes care of the cap's impedance at DC: clearly it's infinite (or *huge*, anyway).

A time-varying voltage across a cap can cause a *current* to pass through the capacitor. This is a point we hope we established in discussing differentiators, §2N.2.1.

A related but different issue concerns us when we consider filters, however: we would like to understand why a rapidly-varying *voltage* can pass through the capacitor. But that does happen, and we need to understand the process in order to understand a *high-pass* filter. If you're already happy with the result, skip this subsection.

When we say the AC signal passes through a filter, all we mean is that a wiggle on the left causes a wiggle of similar size on the right: see Fig. 2N.14.

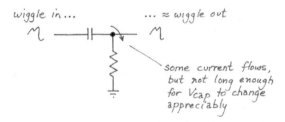

Figure 2N.14 How a cap "passes" a signal.

The wiggle makes it "across" the cap so long as there isn't time for the voltage on the cap to change much before the wiggle has ended – before the voltage heads in the other direction. In other words, quick wiggles pass; slow wiggles don't.

Please *note* that what we just described is the passing of a *voltage* signal through a filter. This is not the same as the passing of a *current* through a capacitor – an effect that cannot be continuous. The passing of current can occur only so long as the voltage difference between the capacitor's plates is *changing*, as it is for a continuous sinusoidal input, for example.

We can stop worrying about our intuition, if you like[6], and state the expression for the cap's *impedance*:

$$Z_C = -j/\omega C = -j/2\pi f C$$

We can see at once that the cap's impedance falls with frequency and is inversely proportional to C. But what's this "j"? Two answers: symbolically, it's just an electronics convention representing the mathematician's "i," i.e., $\sqrt{-1}$.[7] That answer may not help if you consider $\sqrt{-1}$ to be a pretty weird thing, as most right-minded people do! The second answer is that j happens to provide a mathematically convenient way to talk about phase shifts. But it is a way that we will not rely on in this course. Instead, we will usually sweep j and even phase-shift itself under the rug (see later in this section).

And once we have an expression for the impedance of the cap – an expression that shows it varying smoothly with frequency – we can see how capacitors will perform in voltage dividers.

---

[6] In §2S.2 we make another attempt to offer intuitive fingerholds on this idea: ways to grasp intuitively what the math is trying to tell us.

[7] Electrical engineers avoid saying "i" because it sounds like *current*.

Digression: deriving Z_C

Skip the following, if you're in a hurry. This is just for the person who feels uncomfortable when handed a formula that has not been justified.

Why does the formula $Z_C = -j/\omega C = -j/2\pi f C$, hold? We will sneak up on the answer, using some lazy approximations (very much in the style of this course).

What we mean by "impedance of the capacitor" is equivalent to what we mean by "impedance of the resistor:" the quotient of V/I. If we plug in expressions for the somewhat novel V and I in the capacitor, we should get the advertised formula for Z_C.

Since we're interested in how the capacitor responds to time-varying signals, and the sinusoid provides the simplest case,[8] we'll assume a V_{CAP} that is not DC, but, rather, sinusoidal:

$$V_{CAP} = A\sin(\omega t) .$$

Because the current through a capacitor is the time-derivative of the voltage across the cap (see §2N.1, above), the expression for the current through a capacitor across which one applies V_{CAP} will be

$$I_{CAP} = C\left(\frac{dV_{CAP}}{dt}\right) = CA\omega\cos\omega t .$$

and Z_{CAP} is the quotient, V/I, given by

$$\frac{V}{I} = \frac{V_{CAP}}{I_{CAP}} = \frac{A\sin(\omega t)}{A\omega\cos(\omega t)C} .$$

To evaluate this exactly one is obliged to confront the nasty fact that sine and cosine are out of phase, a fact that is handily expressed – but cryptically, to those of us not yet accustomed to complex quantities – in the "$-j$" that appears in the expression for Z_{CAP}. Just now, we choose *not* to confront that nasty fact, and choose instead to sweep $-j$ and phase shift under the rug.

Hiding from phase shift

If we neglect phase shift, then we can simplify this quotient, $\frac{A\sin(\omega t)}{A\omega\cos(\omega t)C}$. We can consider just the maximum values of the sine and cosine functions, i.e. 1, and then divide out the amplitude, A, along with the values of sin and cos which we take to be *one*, so as to get the very simple expression:

AoE §1.7.1A

$$X_{CAP} = |Z_{CAP}| = \frac{1}{\omega C} \quad \text{or} \quad \frac{1}{2\pi f C} .$$

The "|" indicates "magnitude, ignoring phase". *Reactance* thus describes the ratio of current and voltage *magnitudes* only. This is not quite the whole story, evidently, since this account ignores phase shift. But this expression provides a pretty good rough description of the way a capacitor's impedance behaves, and it is easy to understand. The expression tells us two important truths, as we noticed when we first met Z_C:

(1) the capacitor's impedance is *inversely* proportional to C; and
(2) the capacitor's impedance varies inversely with *frequency*.

The second point, of course, is the exciting one: here's a device that will let us build a new sort of voltage divider, more intriguing than the ones built with resistors. These dividers can be *frequency-dependent*. That sounds interesting.

Later, we will return to the topic of phase shift. For the moment, we won't worry about it.

---

[8] You'll recall that Fourier teaches that any waveform other than sinusoid – say, square wave, or triangle wave – can be formed as a sum of sinusoids. It follows that we need only consider how a circuit responds to a sinusoid, in order to gain a fully-general understanding of the circuit's behavior.

2N.3.2 RC filters

These are the most important applications of capacitors. These circuits are just voltage dividers, essentially like the resistive dividers you have met already. The *resistive* dividers treated DC and time-varying signals alike. To some of you that seemed obvious and inevitable (and maybe you felt insulted by the exercise at the end of Lab 1L that asked you to confirm that AC was like DC to the divider). Resistive dividers treat AC and DC alike because resistors can't remember what happened even an instant ago. They're little existentialists, living in the present. (We're talking about ideal resistors, of course.)

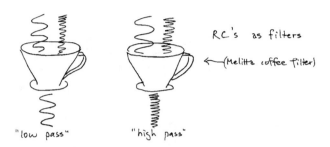

Figure 2N.15 *RC*s are most important as filters – more or less like coffee filters.

A rare but simple case: C:C divider

You know how a resistive divider works on a sine. How would you expect a divider made of capacitors to treat a time-varying signal? The C:C divider is not quite a realistic circuit, since we're normally stuck with substantial stray R values that complicate it. But it is useful to get us started in considering phase shifts.

Figure 2N.16 Two dividers that deliver $1/2$ of V_{in} – with no phase shift.

If this case worries you, good: you're probably worrying about phase shifts. Turns out they cause no trouble here: output is in phase with input. (If you can handle the complex notation, write $Z_C = -j/\omega C$, and you'll see the js wash out.)

But what happens in the combined case, where the divider is made up of both R and C? This turns out to be the case that usually concerns us. This problem is harder, but still fits the voltage-divider model. Let's generalize that model a bit in Fig. 2N.17.

A qualitative view of the filter's frequency response

The behavior of these voltage dividers – which we call *filters* when we speak of them in frequency terms, because each favors either high or low frequencies – is easy to analyze, at least roughly:

Figure 2N.17 Generalized voltage divider; and voltage dividers made up of R paired with C.

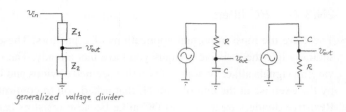

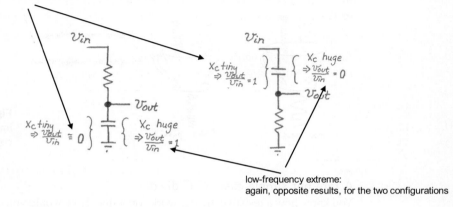

Figure 2N.18 Establish the endpoints of each filter's frequency response curve, by trying $f =$ tiny, $f=$ huge.

(1) See what the filter does at the two frequency extremes. (This looking at extremes is a useful trick; you will soon use it again to find the filters' worst-case Z_{in} and Z_{out}.)
   - At one end of the frequency range, one configuration of filter passes almost none of an input signal, while the other passes all, and vice versa:
      At $f = 0$, what fraction out?
      At very high f, what fraction out?
   As the annotations in Fig. 2N.18 say, each filter can deliver all of its input to the output at one frequency extreme, none at the other extreme; for the two filters those extremes occur, of course, at opposite ends of the frequency spectrum.
(2) Determine where the output "turns the corner" (corner defined arbitrarily[9]) as the frequency where the output is 3dB less than the input (always called just "the 3dB point"; the "minus" is understood).

Knowing the endpoints, which tell us whether the filter is *high-pass* or *low-pass*, and knowing the 3dB point, we can draw the full frequency-response curve; see Fig. 2N.19.

The "3dB point," the frequency where the filter "turns the corner" is

$$f_{3dB} = \frac{1}{2\pi RC}$$

Beware the more elegant formulation that uses ω:

$$\omega_{3dB} = \frac{1}{RC}.$$

---

[9] Well, not quite arbitrarily: a signal reduced by 3dB delivers half its original power.

2N.3 Frequency domain view of RCs

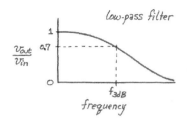
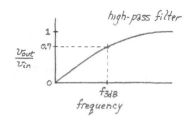

Figure 2N.19 *RC* filter's frequency response curve.

That is tidy, but is very likely to give you answers off by about a factor of 6, since you will be measuring *period* and its inverse in the lab: frequency in *hertz* (or "cycles-per-second," as it used to be called), *not* in radians. So avoid using ω as you evaluate frequency.

Two asides

Caution! **Do not confuse** these *frequency-domain* pictures with the earlier *RC* step-response picture, (which speaks in the *time-domain*). Both sorts of plots are sketched in Fig. 2N.20.

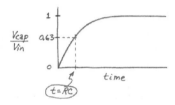
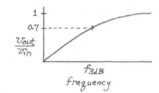

Figure 2N.20 Deceptive similarity between shapes of time- and frequency-plots of *RC* circuits.

Not only do the curves look vaguely similar but, to make things worse, details here seem tailor-made to deceive you:

Step response: In the *time RC* (time-constant), V_{cap} moves to about 0.6 of the applied step voltage (this fraction is $(1 - 1/e)$).
Frequency domain: At f_{3dB}, a frequency determined by *RC*, the filter's $V_{\text{out}}/V_{\text{in}}$ is about 0.7 (this is $1/\sqrt{2}$).

Don't fall into that trap. Don't confuse these two quite-different plots.

A note concerning log plots: You may wonder why the curves we have have sketched in Fig. 2N.19 the curves in AoE introducing filters (§1.7.8), and those you see on the scope screen (when you "sweep" the input frequency) don't look like the tidier curves shown in most books that treat frequency response, or like the curves in Chapter 6 of AoE treating filters. Our curves trickle off toward zero, in the low-pass configuration, whereas these other curves seem to fall smoothly and promptly to zero. This is an effect of the logarithmic compression of the axes on the usual graph. Our plots are linear; the usual plot ("Bode plot") is log–log: see Fig. 2N.21.

AoE §1.7.8

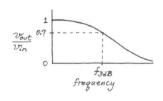
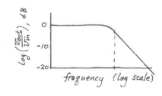

Figure 2N.21 Linear and log–log frequency-response plots contrasted.

We will illustrate the contrast in appearance between linear and log–log plots in §2N.3.3 after a brief introduction to *decibels*.

2N.3.3 Decibels

A "decibel"[10] is a unitless measure of a *ratio* between two values. Since it uses a logarithm of the ratio, it can help to describe values whose range is very large.

Since a decibel measures a *ratio*, it must always rely on a reference level. In the application most familiar to most of us, outside of electronics, *sound* level or loudness is measured relative to a level approximating the quietest audible sound. Sound measurement makes good use of the decibel because the range it needs to measure is enormous: the ratio of loudest sound to the quietest that is audible to a human ear can be as large as about three million to one (this is a ratio of sound pressure levels). This large quantity, awkward to express directly, shrinks to manageable form when expressed in decibels: 130dB. "0dB" is the reference level for audibility.

In electronics usage, decibel describes a *ratio*, as always – but sometimes the magnitudes that are compared represent *power*, sometimes *voltage* or *amplitude*. Since power varies as the square of voltage, the two definitions of decibel look different:

$$dB = 10\log_{10} P_2/P_1 = 20\log_{10} A_2/A_1 \tag{2N.2}$$

– where P represents power, A amplitude. It is the *amplitude* definition that will be useful to us.

Sometimes a label indicates the reference value to which a quantity is compared. *dBm*, for example, uses 1 milliwatt as its reference (with some assumed load impedance). In this course, however, we use dB most often to describe the attenuations achieved by a *filter*, and in this context a few ratios continually recur.

Some dB values we often encounter:

- -3dB = amplitude ratio of $1/\sqrt{2}$. This one comes up all the time, most often in the formula for the frequency where a filter attenuates enough so that we say it has "turned the corner" (but a "corner" that is so gentle that it hardly deserves the name, as you can see for example in Fig. 2N.27).
- -6dB = amplitude ratio of $1/2$. This comes up in the description of the slope of *RC* filters. "-6dB per octave" describes the rolloff of a low-pass filter, in jargon that could be translated to "halving amplitude for each doubling of frequency."
- $+3$dB and $+6$dB of course describe growth rather than attenuation: by amplitude factors of $\sqrt{2}$ or 2, respectively.
- 20dB = amplitude ratio of 10. This comes up in the alternative description of the slope of an *RC* filter: a low-pass, for example, falls at "-20dB per decade," meaning "cutting amplitude by a factor of ten for each 10-fold multiplication of frequency."

Rise and fall on linear and log plots

Linear plots: In lab, when we watch filter attenuations on a scope, we see and perhaps draw *linear* relations of amplitude to frequency. (Some function generators can be set to *sweep* frequencies using a logarithmic sweep that compresses the horizontal axis. But we don't ordinarily use such generators, and such a *semi-log* plot does not look like the usual textbook and datasheet plot, which is log–log.) We see the low-pass roll off rather gently with its $1/f$ shape. On a log–log plot (standard in electronics references that describe filters) this lazy and curvy rolloff is displayed as a straight line, as you can confirm with a look at Fig. 2N.24.

First, in Fig. 2N.22, is a reminder of what a plot rising or falling at ± 6dB/octave looks like. We have labeled the axes "Vo/Vi" and "frequency," as for a filter.

---

[10] A decibel is one tenth of a Bel – a unit defined by Bell Labs to measure signal attenuation in telephone cables, and named to honor Alexander Graham Bell. The Bel turned out (like Farad) to be impractically large, and is not used. You can find more lore on the decibel, if you like, in Watkinson's *The Art of Digital Audio* (3d ed., 2001), §2.18.

2N.3 Frequency domain view of RCs

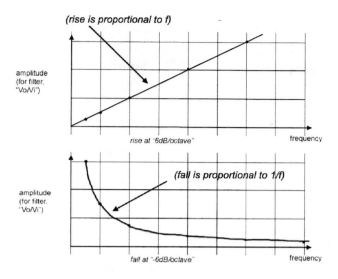

Figure 2N.22 Linear plot of rise and fall at ± "6dB/octave".

This slope can also be described, as we said above, as ±20dB/decade. We have not shown this in Fig. 2N.22 because we did not show a range sufficient to display a decade (a ten-fold) change in frequency.

Log–log plots: Log–log plots, like those in traditional "Bode plots," make the humble *RC* look more decisive. In particular, the lazy and curvy $1/f$ rolloff of a low-pass turns into a straight line that seems to dive resolutely and rapidly toward zero. The rise, too, looks tidier on a log–log plot, though not so drastically improved. In Fig. 2N.23 we have shown rises at "6dB/octave" and "12dB/octave." These are not measured filter results; they only show the behavior of curves on log–log paper.

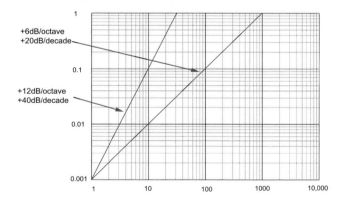

Figure 2N.23 Log–log plot of *RC* high-pass filter's *rising* response.

Fig. 2N.24 shows how a log–log plot can make an *RC* filter's response look more decisive, and can turn a curvy $1/f$ response into a tidy down ramp.

On these log–log plots, the behavior is easier to see if you describe it as "±20dB/decade" and "40dB/decade" rather than by *octave*.

RC Circuits

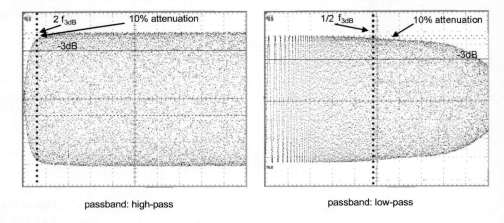

Figure 2N.24 Log–log plot of *RC* low-pass filter's *falling* response (often called "rolloff"). Low-pass rolloff figure on left side reproduced from AoE Fig. 1.104.

2N.3.4 Estimating a filter's attenuation

The shape of the frequency-response curves sketched in Fig. 2N.19 allows one to estimate the attenuation a filter will apply to a specified frequency range, given f_{3dB}. Here are some useful rules of thumb.

Attenuation in the **passband**

Attenuation will be under about 10%, if one keeps an octave (a factor of two in frequency) between *signal* and f_{3dB}. In a low-pass, for example, if you want to pass signals below 1kHz, put f_{3dB} at 2kHz. In a high-pass, if you want to pass signals above 10kHz, put f_{3dB} at 5kHz.

In Fig. 2N.25 are high- and low-pass filters, showing the passbands and the f_{3dB} frequencies.

Figure 2N.25 Passbands for high-and low-pass filters. Loss is about 10%, an octave from f_{3dB}. These are linear–linear plots; it is the (swept) waveforms that are shown, *not* (as in previous figures) a plot of response.

Details of passband: We rarely will need more detail about passband attenuation than the point mentioned above, that we lose 10% a factor of two away from f_{3dB}. (Specifically, we lose about 10% at $1/2\, f_{3dB}$ for a low-pass; at $2 \times f_{3dB}$ for a high-pass).

But, in case you are curious, Fig. 2N.26 shows passband attenuation in detail[11].

---

[11] This is modified from Fig. 4.8 of Tektronix 'ABC's of Probes' (2011), p. 35.

2N.3 Frequency domain view of RCs

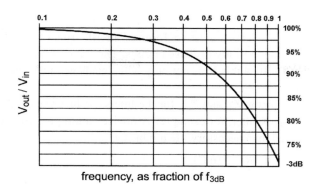

Figure 2N.26 Details of RC rolloff in the passband.

Attenuation in the **stopband**

Where the filter attenuates substantially, amplitude falls or rises in proportion to frequency. This simple pattern is accurate well away from f_{3dB} and is approximate close to f_{3dB}, as can be seen from Fig. 2N.27.

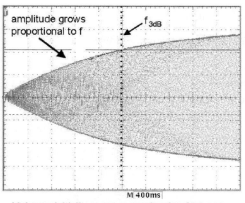
highpass: initially, grows proportional to frequency, approximation good up to about f_{3dB}

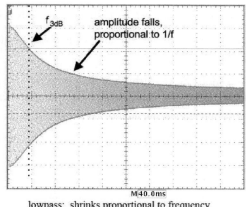
lowpass: shrinks proportional to frequency (amplitude proportional to 1/f)

Figure 2N.27 Stopbands for high- and low-pass filters. Growth or shrinkage with change of frequency.

In a low-pass, for example, if f_{3dB} is 2kHz, then noise at 20kHz ($10 \times f_{3dB}$) will be down to about 1/10 what its level would have been at f_{3dB}. By this reasoning, it will be down about $1/10 \times 0.7$, since the level is about 0.7 (by definition) at f_{3dB}. In fact, it will be attenuated somewhat less, because the $1/f$ shape is not achieved in the first octave or so above f_{3dB}.

In Fig. 2N.27, one can see the deviation of the high-pass response from a perfect straight line (that is, from strict proportionality to frequency) up to f_{3dB}. Far below f_{3dB}, the deviation from straight line would be slight.

It is harder to judge the low-pass deviation from a $1/f$ shape, but you may be able to see, if you look closely, that at $2 \times f_{3dB}$ amplitude is well above the 35% that one would predict if one expected a $1/f$ slope even in the first octave above f_{3dB}.

Still, the main point is not the imperfection of these approximations but their power. They let you give a quick and pretty-good estimate of what a filter will do to noise at a specified distance from f_{3dB}. That's valuable.

2N.3.5 Input and output impedance of an *RC* circuit

AoE §1.7.1

If filter A is to drive filter B – or anything else, in fact – it must conform to our $10\times$ rule of thumb, which we discussed earlier when we were considering only resistive circuits. The same reasons prevail, but here they are more urgent: if we don't follow this rule, not only will signals get attenuated; frequency response also is likely to be altered.

But to enforce our rule of thumb, we need to know Z_{in} and Z_{out} for the filters. At first glance, the problem looks nasty. What is Z_{out} for the low-pass filter, for example? A novice would give you an answer that's correct but much more complicated than necessary. He might say,

$$Z_{out} = Z_C \parallel R = \frac{R(-j/\omega C)}{R - j/\omega C}.$$

Yow! And then this expression doesn't really give an answer: it tells us that the answer is frequency-dependent.

"Worst case" impedances
We cheerfully sidestep this hard work, by considering only *worst case* values. We ask, "How bad can things get?"

We ask, "How bad can Z_{in} get?" And that means, "How *low* can it get?"

We ask, "How bad can Z_{out} get?" And that means, "How *high* can it get?"

This trick delivers a stunningly easy answer: the answer is always just R! Here's the argument for a *low-pass* for example (see Fig. 2N.28):

worst Z_{in}: cap looks like a short: $Z_{in} = R$ (this happens at highest frequencies).
worst Z_{out}: cap doesn't help at all; we look through to the source, and see only R: $Z_{out} = R$ (this happens at lowest frequencies).

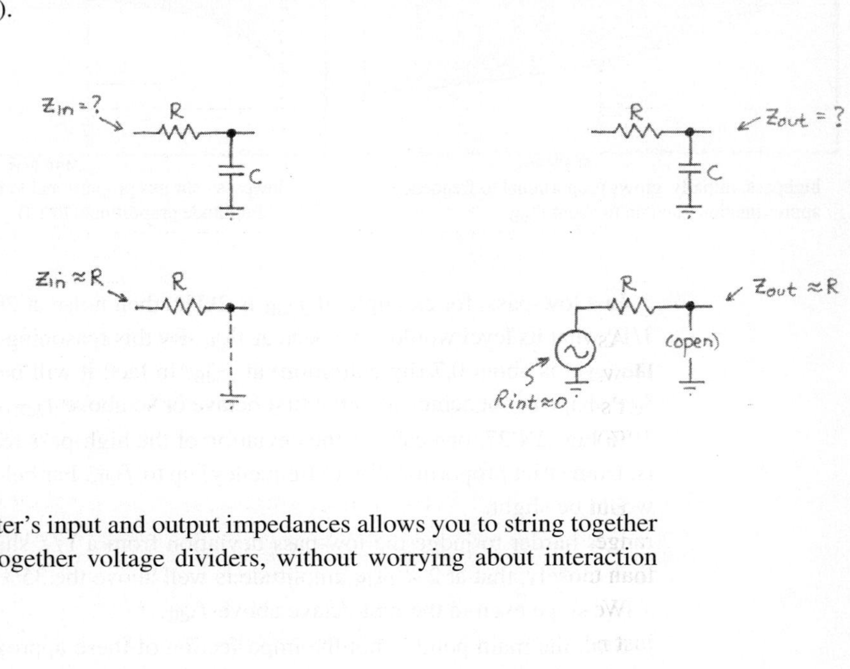

Figure 2N.28 Worst-case Z_{in} and Z_{out} for *RC* filter reduces to just *R*.

Having an easy way to handle the filter's input and output impedances allows you to string together *RC* circuits just as you could string together voltage dividers, without worrying about interaction among them.

2N.3.6 Phase shift

AoE §1.7.1

You already know roughly what to expect: the differentiator and integrator showed you phase shifts of 90°, and did this when they were severely attenuating a sinewave. But you need to *beware* the misconception that because a circuit has a cap in it, you should expect to see a 90° shift (or even just noticeable shift). You should *not* expect that. You need an intuitive sense of when phase shifting occurs, and of roughly its magnitude. You rarely will need to calculate the amount of shift. In fact, as we work with filters we usually get away with ignoring the issue of phase shift.

How we get away with largely ignoring phase shift in *RC* filters

Before we get into some details, let's try to explain why we can come close to ignoring phase shift. Here is a start: a rough account of phase shift in *RC* circuits:

> If the amplitude *out* is close to amplitude *in*, you will see little or no phase shift. If the output is much attenuated, you will see considerable shift (90° is maximum)

AoE §1.7.9

Fig. 2N.29 shows curves saying the same thing for the case of a low-pass filter.

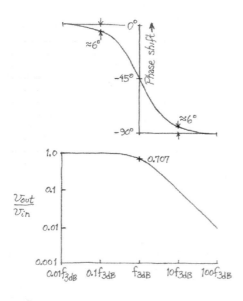

Figure 2N.29 Attenuation and phase shift (log–log plot).

As you can see, where the filter is *passing* a signal, the filter does not impose much phase shift. How much? If we assume that we can put f_{3dB} at twice the highest frequency that we want to pass (let's call that the highest "signal" frequency). Then the signal will be attenuated only slightly (about 10%) and will be shifted no more than about 25 degrees. Both attenuation and phase shift will be less at lower frequencies.

But, you may want to protest, more severe shifts occur over much of the possible input frequency range, shown in Fig. 2N.29. At many frequencies, phase shift is considerable: $-45°$ at f_{3dB}, and more than that as frequency climbs, approaching $-90°$. Why not worry about those shifts?

We don't worry because those shifts are applied to what we consider not signal but *noise*, and we're just not much interested in what nasty things we are doing to noise. We care about *attenuating* noise, it's true; we like the attenuation, and are interested in its degree. But that's only because we want to be rid of it. We don't care about the details of the noise's mutilation; we don't care about its phase shift.

Why phase shift occurs

Why does this happen? Here's an attempt to rationalize the result:

- in an *RC* series circuit, the voltages across the *R* and the *C* are 90° out of phase, as you know;
- the voltages across *R* and *C* must sum to V_{in}, at every instant;
- the voltage out of the filter, V_{out}, is the voltage across either *R* or *C* alone (*R* for a high-pass, *C* for a low-pass);
- as frequency changes, *R* and *C* share the total V_{in} differently, and V_{out} thus can look much like V_{in} or can look very different. In other words, the phase difference between V_{in} and V_{out} varies with frequency.

Consider, for example, a low-pass filter. If much of the total voltage V_{in} appears across the cap, then the phase of the input voltage (which appears across the *RC* series combination) will be close to the phase of the output voltage, V_{cap}. In other words, *R* plays a small part, and V_{out} is about the same as V_{in} in both amplitude and phase. Have we merely restated our earlier proposition? It almost seems so.

But let's try a drawing.

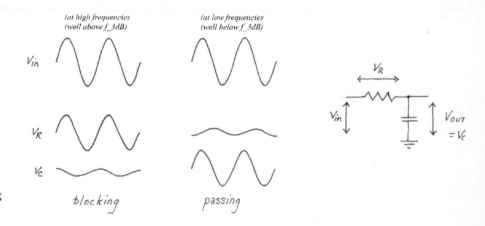

Figure 2N.30 *R* and *C* sharing input voltage differently at two different frequencies.

The sketch in Fig. 2N.30 shows what one would expect: little phase shift where the filter is *passing*, lots of phase shift where it is largely *attenuating*. Now let's try another aid to an intuitive understanding of phase shift: phasors.

2N.3.7 Phasor diagrams

AoE §1.7.12

These diagrams let you compare phase and amplitude of input and output of circuits that shift phases (circuits including *C*'s and *L*'s). They make the performance graphic, and allow you to get approximate results by use of geometry rather than by explicit manipulation of complex quantities.

The diagram uses axes that represent resistor-like ("real") impedances on the horizontal axis, and capacitor- or inductor-like impedances ("imaginary" – but don't let that strange name scare you; for our purposes it only means that voltages across such elements are 90° out of phase with voltages across the resistors). This plot is known by the extra-frightening name, "complex plane" (with nasty overtones, to the untrained ear, of 'too-complicated-for-you plane'!). But don't lose heart. It's all very easy to understand and use. Even better, *you don't need to understand phasors* if you don't want to. We use them rarely in the course, and always could use direct manipulation of the complex quantities instead. Phasors are meant to make you feel better. If they don't, forget them.

2N.3 Frequency domain view of RCs

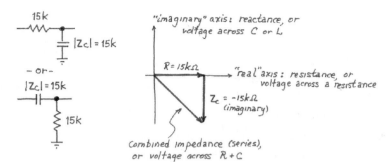

Figure 2N.31 Phasor diagram: "complex plane;" showing an *RC* at $f_{3\text{dB}}$.

Phasor diagram of an *RC* circuit

Fig. 2N.31 shows an *RC* filter at its 3dB point, where, as you can see, the *magnitude* of the impedance of *C* is the same as that of *R*. The arrows, or vectors, show phase as well as amplitude (notice that this is the *amplitude* of the waveform: the peak value, not a voltage varying with time); they point at right angles to indicate that the voltages in *R* and *C* are 90° out of phase.

Phase: We are to think of these vectors as rotating counterclockwise, during the period of an input signal. So, thinking of the input as aligned with the hypotenuse, we see the voltage across the resistor *leading* the input (that is true of such a circuit: a *high-pass*, which takes output across *R*). We see the voltage across the capacitor *lagging* the input (that is true of such a circuit: a *low-pass*, which takes output across *C*).

"Voltages?," you may be protesting, "but you said these arrows represent impedances." True. But in a series circuit, where the currents in the two elements are the same, the voltages are proportional to the impedances. So both interpretations of the figure are fair.

The total impedance that *R* and *C* present to the signal source is *not* 2*R*, but is a vector sum: it's the length of the hypotenuse, $\sqrt{2}R$. And from this diagram we now can read two familiar truths about how an *RC* filter behaves at its 3dB point:

- the amplitude of the output relative to input is down 3dB: down to $1/\sqrt{2}$: the length of either the *R* or the *C* vector, relative to the hypotenuse.
- the output is shifted 45° relative to the input: *R* or *C* vectors form an angle of 45° with the hypotenuse, which represents the phase of the input voltage. Whether output *leads* or *lags* input depends on where we take the circuit output, as we said just above.

So far, we're only restating what you know. But to get some additional information out of the diagram, try doubling the input frequency several times in succession, and watch what happens: each time, the length of the Z_C vector is cut to half what it was.

However, the first doubling also affects the length of the hypotenuse substantially; so the amplitude relative to input (let's assume a low-pass) is not cut quite so much as 50% (not quite so much as "6dB"). You can see that the output is attenuated a good deal more than at $f_{3\text{dB}}$ however, and also that phase shift between input and output has increased a good deal.

RC Circuits

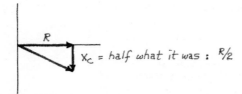

RC after a first doubling of frequency relative to Fig. 2N.31.

Again, on the second doubling of frequency, the length of the Z_C vector is halved, but this time the length of the hypotenuse is changed less than in the first doubling. So things are becoming simpler: the output shrinks nearly as much as the Z_C vector shrinks: nearly 50%. Here we are getting close to the ultimate slope of the filter's rolloff curve: -6dB/octave. Meanwhile, the phase shift between output and input is increasing too – approaching the limit of $-90°$.

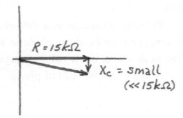

RC after a second doubling of frequency relative to previous diagram.

We've been assuming a *low-pass*. If you switch assumptions, and ask what these diagrams show happening to the output of a *high-pass*, you find all the information is there for you to extract. No surprise, there; but perhaps satisfying to notice.

Phasor diagram of an *LC* circuit

Finally, let's look (Fig. 2N.32) at an *LC trap* circuit on a phasor diagram. (This is a sneak preview of a circuit we have not yet talked about, and a circuit element that you'll encounter in the next lab. We can't resist including it here, where we're in the thick of *phasors*.)

> AoE §1.7.14

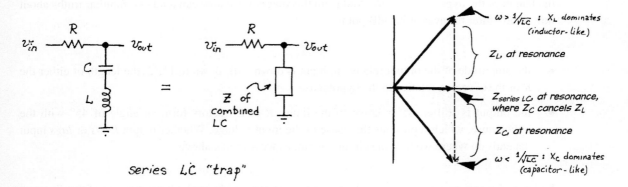

Figure 2N.32 *LC* trap circuit, and its phasor diagram.

This is less familiar, but pleasing because it reveals the curious fact – which you will see in Lab 3L when you watch a similar (parallel) *LC* circuit – that the *LC* combination looks sometimes like *L*, sometimes *C*, showing the phase shift characteristic of each – and at resonance, shows no phase

2N.3 Frequency domain view of RCs

shift at all. We'll talk about *LC*'s next time; but for the moment, see if you can enjoy how concisely this phasor diagram describes this behavior of the circuit (actually a *trio* of diagrams appears here, representing what happens at *three* sample frequencies).

To check that these *LC* diagrams make sense, you may want to take a look at what the old voltage-divider equation tells you ought to happen.

Here's the expression for the output voltage as a fraction of input:

$$\frac{V_{\text{out}}}{V_{\text{in}}} = \frac{Z_{\text{combination}}}{Z_{\text{combination}} + R}$$

But

$$Z_{\text{combination}} = \frac{-j}{\omega C} + j\omega L.$$

Figure 2N.33 *LC* trap: just another voltage divider.

And at some frequency – where the magnitudes of the expressions on the right side of that last equation are equal – the sum is zero, because of the opposite signs. Away from this magic frequency (the "resonant frequency"), either cap or inductor dominates. Can you see all this on the phasor diagram?

Better filters

Having looked hard at *RC* filters, maybe we should admit that you can make a filter better than the simple *RC* when you need one. You can do it by combining an inductor with a capacitor, or by using operational amplifiers (coming soon) to obviate use of the inductor. We will try that method in a later lab. Fig. 2N.34 shows a scope image showing three *RC* filters' frequency responses. Two of the three filters improve on the simple *RC* low-pass that we meet today.

AoE §6.2.2

By the way: "sweeping" frequencies for a scope display: The scope image in Fig. 2N.34 uses *frequency* for the horizontal axis – not *time*, as in the usual scope display. You may choose to use this technique in Lab 2L, and certainly will want to use it in Lab 3L (where it provides a pretty display of the dramatic response of the *LC resonant circuit*). The method is detailed in §2S.3 on p. 93. All we need say here is that at the left-hand edge of the scope display frequency is close to zero; the frequency climbs linearly to the right-hand extreme (where in this case the frequency reaches about 3.8kHz). f_{3dB} for all three filters is about 1kHz.

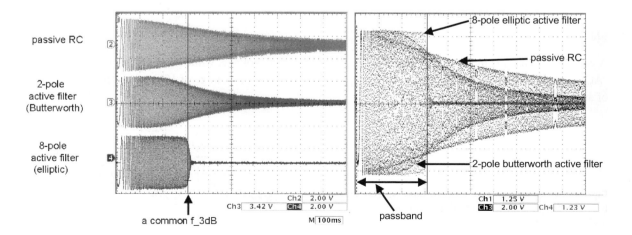

Figure 2N.34 Frequency response of three filters: improvements on the simple *RC*. Some things to look forward to. Left panel shows wider frequecy range; right panel shows more passband detail.

74 RC Circuits

1. The top trace in Fig. 2N.34 shows a passive *RC* filter. You can see that it's pretty droopy where it ought to be passing signals, and only very gradually "rolls off" (attenuates) signals at increasing frequencies.
2. The middle trace shows an "active" 2-pole[12] *RC* filter. Its frequency response improves on that of a passive filter (even a 2-pole passive filter) by some clever use of an amplifier and feedback (at the moment, the term "feedback" may sound pretty mysterious, to you. It will remain so until perhaps Lab 6L). This "active" filter improves on the passive *RC* in three senses: the "passband" is more nearly flat than for the passive filter, the rolloff (the way the output shrinks as frequency climbs beyond the "passband") is steeper, and the ultimate slope is steeper as well ($\propto 1/f^2$ rather than $\propto 1/f$). You will meet this filter in Lab 9L.
3. The bottom trace shows a fancier "active filter." This is an "8-pole" filter (like eight simple *RC* low-pass filters cascaded, but optimized so that it performs much better than a simple cascade of like filters would). You will meet this filter in Labs 12L and 23L. Its rolloff is, as you can see, quite spectacularly steep.

We don't want to make you a filter snob, though – don't want to make you look down on the simple *RC*. It would not be in the spirit of this course to make you pine after beautiful transfer functions. Nearly always, a simple *RC* filter is good enough for our purposes. It is nice to know, however, that in the exceptional case where you need a better filter, such a circuit is available.

2N.4 Two unglamorous but important cap applications: "blocking" and "decoupling"

In part, we mention these to introduce some jargon that otherwise might bother you: the names "blocking capacitor" and "decoupling capacitor" might, at first hearing, suggest that these are peculiar, specialized types of capacitor. The names do not mean that. The names simply refer to the application to which the capacitor is put (this usage is reminiscent of the "load" resistor that appeared in Chapter 1: not a special sort of resistor; just a resistor *considered as* nothing more than a load to the circuit under discussion).

2N.4.1 *Blocking* capacitor

A capacitor so used is included for the purpose of *blocking* a DC voltage, while passing a wiggle (an "AC" signal). In Fig. 2N.35, for example, the capacitor blocks the DC voltage of 10V, permitting use of an input signal centered on ground.

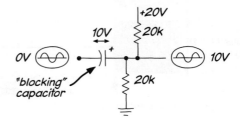

Figure 2N.35 A capacitor used to pass wiggles while blocking DC (permitting differing DC levels) is said to be a "blocking" cap.

You may be inclined to protest that the capacitor doing this job is also a *passing* capacitor. Yes, but

---
[12] "2-pole" is jargon meaning, approximately, "behaves like two ordinary *RC*'s cascaded (that is, placed in series)."

we're accustomed to such usage (a *resistor* is also a *conductor*) – and the blocking capacitor differs from a piece of wire not in its passing but in its blocking powers.

You may further protest that you recognize this combination of cap and resistors as nothing new: it is a *high-pass filter*, with the relevant R just R_{Thev} for the divider. Yes, you're right: it does behave like a high-pass.

But it makes sense to call the cap *blocking* if the goal is to permit the peaceful coexistence of differing DC levels – that is, to block DC. This motive is not the same as the usual for a high-pass, which is to attenuate low-frequency noise.

When blocking is the motive, the choice of f_{3dB} becomes not critical: just make f_{3dB} low enough so that all frequencies that you consider *signals* can pass. A huge C is OK, when the motive is only to block DC. No value is too large – though it may be foolish to choose a cap larger than needed, because large C values cost more and take more space than smaller ones, are often polarized, and are slower to settle.

A familiar case: scope input's "AC coupling"

When you switch the scope input to "AC" (as we suggest you ought to do only rarely), you put a blocking capacitor into the signal path. DC is blocked – and DC information is lost. That is why we advocate using DC coupling (which you can think of as "direct coupling") except in the exceptional case when you need to wash away a large DC offset in order to look at a small wiggle at high gain (a gain that would send a DC-coupled signal sailing off the screen).

2N.4.2 Decoupling or bypass capacitor

A very humble, but important application for a capacitor is just to minimize power-supply wiggles. Power supplies are not perfect, and the lines that link the supply to a circuit include some inductance (and some resistance, though the inductance is usually more troublesome).

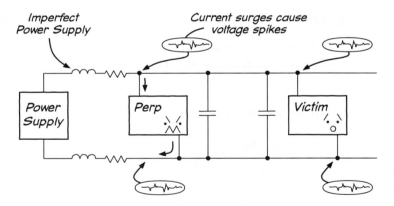

Figure 2N.36 A capacitor used to quiet a power supply is said to "decouple" one part of a circuit from another.

If one part of a circuit draws surges of current, those surges flowing through the stray inductance of the supply lines will cause voltage jumps. Those supply disturbances can "couple" between one part of the circuit (the "perp" or perpetrator, in Fig. 2N.36) and another (the "victim").

Decoupling capacitors tend to stabilize the power supply voltage. Such a capacitor provides a local source of charge for the *perp* – the circuit that needs a sudden surge of current – so not all of the current surge needs to flow in the longer path from the supply. A decoupling or "bypass" cap near the *victim* circuit simply tends to hold that supply voltage more nearly constant. In Fig. 2N.37, a digital oscillator performed the role of *perp*; about 6 inches distant, at the far end of a breadboard, the scope watched the power supply near the *victim*.

Small capacitors help a good deal (0.1µF). The cheapest, crummiest capacitors, the ceramics

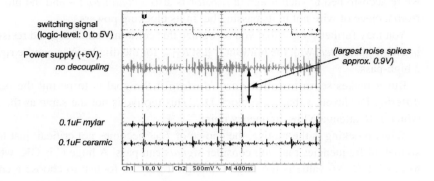

Figure 2N.37 Small capacitors between power supply and ground ("bypass") can stabilize the supply voltage.

scorned for other purposes, work fine; in fact, they work a little better than the more expensive mylars that we prefer for filters, as you can see in Fig. 2N.37. Often a large capacitor (say, a 4.7μF tantalum) is paralleled with a small ceramic (0.01 to 0.1μF).

Always the best general defense against such trouble is to provide heavy power supply lines or "traces" on a printed-circuit board, with their low inductance. On high-quality printed circuits, entire layers are dedicated to ground and supply "planes." But even after doing this, a wise designer invariably adds decoupling caps (sometimes as many as a few dozen).

The use of a capacitor to "bypass" a circuit element is more general than the *decoupling* application we have just looked at. You will see a resistor "bypassed" by a capacitor, for example, in the high-gain amplifier of §5N.5.2. You will see a similar application in the "split feedback" of Chapter 9N, for example in the amplifier of Fig. 9L.8.

2N.5 A somewhat mathy view of RC filters

In this course, we do what we can to dodge mathematical explanations for circuit behavior. Often, math doesn't help intuition. But here we propose to see if a little math may help to make sense of the frequency response of an *RC* circuit. If this does not help, ignore it.

A high-pass filter as voltage divider: Fig. 2N.38 shows the highpass and its response.

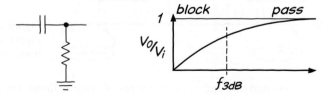

Figure 2N.38 Highpass filter.

And, treating it as a voltage divider, here is the fraction the output delivers:

$$\frac{V_{\text{out}}}{V_{\text{in}}} = \frac{R}{R+Z_C} = \frac{1}{1+(Z_c/R)} = \frac{1}{1-j(1/\omega RC)}.$$

Let's try looking at what this expression is trying (perhaps not very clearly) to tell us.

The interesting term, here, is $-j(1/\omega RC)$. Depending on frequency, this term can take a magnitude that is...

- tiny,

- huge or
- unity

These are not the only possibilities, but these three define three distinct regions in the filter's behavior – the ones labeled "block," "f_{3dB}," and "pass" in Fig. 2N.38. Let's consider these regions, one at a time.

- if $-j(1/\omega RC)$ is *tiny*, which means $\omega \gg 1/RC$, then we're well above f_{3dB} and $V_{out}/V_{in} \approx 1$. This is the "pass region." The near absence of j from the expression (because its coefficient is tiny) says, "No phase shift." This fits your experience in Lab 2L, experience that says phase shift in an RC filter goes with attenuation.
- if $-j(1/\omega RC)$ has value unity, which means $\omega = 1/RC$, then we're at f_{3dB} and $V_{out}/V_{in} = \frac{1}{1-j}$. This has magnitude $\frac{1}{\sqrt{2}}$, more usually described as "-3dB."[13] The presence of j in the expression promises phase shift. A phasor diagram like the one back in Fig. 2N.31 may be the best way to illustrate why that phase shift is $45°$. The phase shift is the angle between the hypotenuse (V_{in}) and the voltage across R. The latter voltage "leads" V_{in} by $45°$. Fig. 2N.39 shows a short form of that figure again.

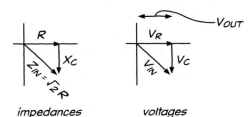

Figure 2N.39 Phasor diagram explains *geometrically* the RC's $45°$ phase shift at f_{3dB}.

- if $-j(1/\omega RC)$ is *huge*, which means $\omega \ll 1/RC$, then we're well below f_{3dB} and $V_{out}/V_{in} \approx \frac{1}{-j(1/\omega RC)} = j\omega RC$. This expression says a lot: "j" says "$+90°$ shift; "ω" says that as frequency grows amplitude grows proportionately.

2N.6 Readings in AoE

Readings:

AoE: Chapter 1.4–1.7.12 but *omit* §§1.5–1.6.8, and §1.7.2 about inductors, transformers, and diodes (we'll reach those next time).

Treat as *optional* §§1.7.3–1.7.9, where complex impedances are presented. This patch is heavy on math, and out of character for this course. These sections have scared many students, who later learned that they can manage perfectly well in this course without such rigor. This mathematical treatment is not characteristic even of AoE, and *very* unlike the treatments we use in *Learning the Art of Electronics* ....

Appendix on Math Review. (World's shortest calculus course, if you feel you want such a refresher. But in any event, you'll survive this course without it).

Exercises:

Exercises in text.
Additional Exercises 1.39–1.42.

---

[13] You can get this result by multiplying $\frac{1}{1-j}$ by its "complex conjugate," $\frac{1}{1+j}$. Their product gives a magnitude squared that is real and equals $1/2$. Taking its square root, we get $\frac{1}{\sqrt{2}}$, with the familiar value 0.7 or "-3dB."

2L Labs: Capacitors

2L.1 Time-domain view

2L.1.1 *RC* circuit: time-constant

Here's another try in our continuing effort to make your labs more exciting – more *suspenseful*: we ask you to *do the first exercise* (measuring the *RC* "time-constant") with *unknown R, C values*. To make this game possible, we engaged the great Christo to *wrap* an *R, C* pair. The *R* is the skinny object; the *C* is the chubbier device. See Fig. 2N.6, which illustrates exponential charging and discharging.

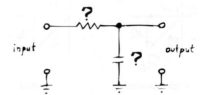

Figure 2L.1 *RC* circuit: step response.

Measure *RC*: Drive the circuit with a square wave at 500Hz or less, and look at the output. Adjust frequency so as to get a useful image: too high, and you won't allow time enough to see the waveform move far; too low, and you'll see the full waveform, but using just a small portion of the scope screen, and thus your time measurements will be only approximate. In Fig. 2L.2 is a scope image suggesting both possible errors:

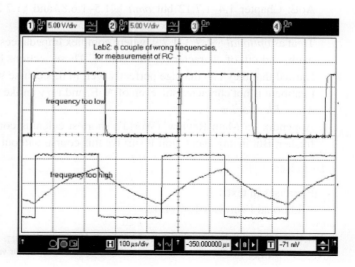

Figure 2L.2 A couple of wrong frequencies, for measurement of *RC*.

Be sure to use the scope's *DC* input setting, even though this is a time-varying waveform. (Remember the warning about the AC setting, last time?)

You will have no trouble determining *RC*. Measure the time constant by determining the time for the output to drop to 37% (= 1/e).

Suggestion The *percent* markings over at the left edge of the scope screen are made-to-order for this task: put the foot of the square wave on 0%, the top on 100%. Then crank up the sweep rate so that you use most of the screen for the fall from 100% to around 37%.

Measure the time to climb from 0% to 63%. Is it the same as the time to fall to 37%? (If not, something is amiss in your way of taking these readings!)

Try varying the frequency of the square wave.

Deduce *R* and *C* values: You would have no trouble determining *R* if we allowed you to use an ohmmeter, but we don't allow that. See if, instead, you can use what you know of the limiting values of the *RC* circuit's *input impedance* to discover *R*, experimentally. Then you can solve for *C*.

In case this advice seems a little cryptic, here are some hints.

Hints:

- First, try to determine *R* – despite the fact that *C* is also present. Form a voltage divider with a known resistor ahead of the *RC* circuit. We suggest you start with $R_{TEST} = 1k$. Apply a sinewave.

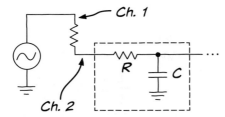

Figure 2L.3 Test setup to use R_{IN} to reveal *R*.

Queries:
- How will you know that it is the effect of *R* that you are observing, rather than some combination of *R* and X_C? (*Hint:* do you see a phase shift between the waveforms on Channels 1 and 2?) *Note:* keep f_{IN} under about 1MHz, so as not to complicate your search with the effect of the BNC cable's capacitance to ground. That capacitance – about 30pF/foot – becomes important at high frequencies; it forms a low-pass with $R_{Thevenin}$ at the point marked "Ch. 2" in Fig. 2L.3. So, as you push the frequency high, looking for the disappearance of phase shift, you will be frustrated if you go too far!
- Once you have eliminated the pesky phase shift, what should you assume has happened to the value of X_C?
- Once you have a value for *R*, you're about done.

2L.1.2 Differentiator

Construct the *RC* differentiator shown in Fig. 2L.4. Drive it with a square wave at 100kHz, using the function generator with its attenuator set to 20dB. Does the output make sense? Try a 100kHz triangle wave. Try a sine.

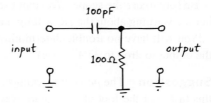

Figure 2L.4 *RC* differentiator.

Input impedance: Here's another chance to get used to quick *worst-case* impedance calculations, rather than exact and frequency-dependent calculations (which often are almost useless).

What is the impedance presented to the signal generator by the circuit (assume no load at the circuit's output)...

- ...at $f = 0$?
- ...at infinite frequency?

Questions like this become important when the signal source is less ideal than the function generators you are using.

2L.1.3 Integrator

Construct the integrator shown in Fig. 2L.5. Drive it with a 100kHz square wave at maximum output level (attenuator set at 0dB).

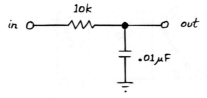

Figure 2L.5 *RC* integrator.

What is the input impedance at DC? At infinite frequency? Drive it with a triangle wave; what is the output waveform called? (Doesn't this circuit seem clever? Doesn't it remember its elementary calculus better than you do – or at least faster?)

To expose this as only an *approximate* or conditional integrator, try reducing the input frequency. Are we violating the stated condition (§2N.2.1):

$$V_{\text{OUT}} \ll V_{\text{IN}}?$$

The differentiator is similarly approximate, and fails unless (§2N.2.1):

$$dV_{\text{OUT}}/dt \ll dV_{\text{IN}}/dt?$$

Too large an *RC* tends to violate this restriction. If you are extra zealous you may want to look again at the differentiator of §2L.1.2 but this time increasing *RC* by a factor of, say, 1000. The "derivative" of the square wave gets ugly, and this will not surprise you; the derivative of the triangle looks odd in a less obvious way.

When we meet *operational amplifiers* in Chapter 3N, we will see how to make "perfect" differentiators and integrators – those that let us lift the restrictions we have imposed on these *RC* versions.

2L.2 Frequency domain view

2L.2.1 Low-pass filter

Construct the low-pass filter[1] shown in Fig. 2L.6.

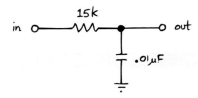

Figure 2L.6 *RC* low-pass filter.

What do you calculate to be the filter's -3dB frequency? Drive the circuit with a sinewave, sweeping over a large frequency range, to observe its low-pass property; the 1kHz and 10kHz ranges should be most useful.

Find $f_{3\text{dB}}$ *experimentally*: measure the frequency at which the filter attenuates by 3dB (V_{OUT} down to 70.7% of full amplitude).

Note Henceforth we will refer to "the 3dB point" and "$f_{3\text{dB}}$," not to the *minus* 3dB point, nor $f_{-3\text{dB}}$. This usage is confusing but conventional; you might as well start getting used to it.

What is the limiting phase shift, both at very low frequencies and at very high frequencies?

Suggestion As you measure phase shift, use the function generator's SYNC or TTL output to drive the scope's External Trigger. That will define the input phase cleanly. Then if you are using an analog oscilloscope, use its *continuously-variable* sweep rate[2] so as to make a full period of the input waveform use exactly eight major divisions (or eight centimeters). The output signal, viewed at the same time, should reveal its phase shift readily; see Fig. 2L.7.

Check to see if the low-pass filter attenuates 6dB/octave for frequencies well above the -3dB point; in particular, measure the output at 10 and 20 times $f_{3\text{dB}}$. While you're at it, look at phase shift versus frequency: what is the phase shift for

$$f \ll f_{3\text{dB}},$$
$$f = f_{3\text{dB}},$$
$$f \gg f_{3\text{dB}}?$$

Finally, measure the attenuation at $f = 2f_{3\text{dB}}$ and write down the attenuation figures at $f = 2f_{3\text{dB}}$, $f = 4f_{3\text{dB}}$ and $f = 10f_{3\text{dB}}$.

---

[1] Aside: *Integrator* versus *Low-pass Filter*. "Wait a minute!," you may be protesting, "Didn't I just build this circuit?" Yes, you did. Then why do it again? We expect that you will gradually divine the answer to that question as you work your way through this experiment. One of the two experiments might be called a special case of the other. When you finish, try to determine which is which.

[2] On most scopes you'll invoke this by turning a little red knob on the larger *sweep rate* knob: when the red knob is turned counter-clockwise, it comes out of a clicked "detent" position, usually labeled "CAL." Once you've done that, the scope screen no longer is usable to read *time*. So, don't leave it that way when you finish your *phase shift* measurement!

Labs: Capacitors

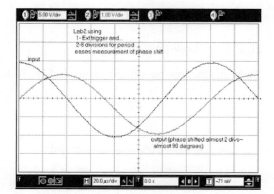

Figure 2L.7 It's easy to estimate phase shifts if you make a full *period* equal eight divisions.

> **Sweeping Frequencies** This circuit is a good one to look at with the function generator's **sweep** feature. This will let your scope draw you a plot of amplitude versus *frequency* instead of amplitude versus *time* as usual. If you have a little extra time, we recommend this exercise. If you feel pressed for time, save this task for next time, when the *LC* resonant circuit offers you another good target for sweeping.
>
> You may want to look at §2S.3, our more detailed note on *sweeping*, but here is the strategy in brief.
>
> In order to generate such a display of V_{OUT} versus frequency, let the generator's *ramp* output drive the scope's horizontal deflection, with the scope in "X–Y" mode: in X–Y, the scope ignores its internal horizontal deflection ramp (or "timebase") and instead lets the input labeled "X" determine the spot's horizontal position.
>
> The function generator's **ramp** time control now will determine sweep rate. Keep the ramp *slow*: a slow ramp produces a scope image that is annoyingly intermittent, but gives the truest, prettiest picture, since the slow ramp allows more cycles in a given frequency range than are permitted by a faster ramp.

2L.2.2 High-pass filter

Construct a high-pass filter with the components that you used for the low-pass. Where is this circuit's 3dB point? Check out how the circuit treats sinewaves: check to see if the output amplitude at low frequencies (well below the -3dB point) is proportional to frequency. What is the limiting phase shift, both at very low frequencies and at very high frequencies?

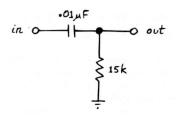

Figure 2L.8 *RC* high-pass filter.

2L.2.3 Filter application I: garbage detector

The circuit in Fig. 2L.9 will let you see the "garbage" on the 110V power line. First look at the output of the transformer, at **A**. It should look more or less like a classical sinewave. (The transformer, incidentally, serves two purposes – it reduces the 110V AC to a more reasonable 6.3V, and it "isolates" the circuit we're working on from the potentially lethal power line voltage)

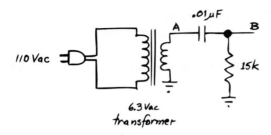

Figure 2L.9 High-pass filter applied to the 60Hz AC power.

To see glitches and wiggles, look at **B**, the output of the high-pass filter. All kinds of interesting stuff should appear, some of it curiously time-dependent. What is the filter's attenuation at 60Hz? (No complex arithmetic necessary. *Hint:* count octaves, or use the fact – which you confirmed just above – that amplitude grows linearly with frequency, well below f_{3dB}.)

2L.2.4 Design: filter application II: selecting signal from signal-plus-noise

Now we will try using high-pass and then low-pass filters to prefer one frequency range or the other in a composite signal, formed as shown in Fig. 2L.10. The transformer adds a large 60Hz sinewave (peak value about 10V) to the output of the function generator. Set the function generator frequency, initially, to around 10kHz.

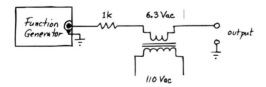

Figure 2L.10 Composite signal consisting of two sinewaves.

In order to choose the R value for your filter, you will need to determine the value of the output impedance[3] for the signal source you have constructed (function generator plus transformer). The function generator's R_{out} is 50Ω; the series impedance of the transformer winding is negligible at the frequencies of interest to us; the 1k resistor is included, incidentally, to protect the function generator in case the composite output accidentally is shorted to ground.

First design: high-pass: Design a high-pass filter that will keep most of the "signal" and get rid of most of the 60Hz "noise." Assume that the frequency of what you consider "signals" may range between about 2kHz and 20kHz. As you design, consider:

---
[3] We've been calling this Z_{out}, a name that certainly is not wrong – but this characteristic is not frequency-dependent, so it will be just as correct to call it R_{out}.

- what is an appropriate f_{3dB}?
- what Z_{in} is appropriate for your filter?

Run the composite waveform ("signal" plus "noise") through your high-pass filter.

High-pass filter (your design): Do you like the output of your filter? Is the attenuation of the 60Hz waveform about what you would expect? (As you will gather gradually, the 60Hz power lines are the most common and troublesome source of noise in the lab. This junk is often called "line noise.")

Second design: low-pass: Now let's change assumptions: let's suppose that we consider the 60Hz "signal," and the function generator's 10kHz "noise."

Design a low-pass filter that will keep most of the "signal" and get rid of most of the "noise."

Low-pass filter (your design): Now run the composite signal through your low-pass filter, and see if you like the result. If not, fix your design!

2S Supplementary Notes: *RC* Circuits

2S.1 Reading capacitors

2S.1.1 Why you may need this note

AoE §1.4.1

Most students learn pretty quickly to read resistor values. They tend to have more trouble finding, say, a 100pF capacitor.

That's not their fault. They have trouble, as you will agree when you have finished reading this note, because the cap[1] manufacturers don't want them to be able to read cap values. The cap markings have been designed by an international committee to be nearly unintelligible. With a few hints, however, you can learn to read cap markings, despite the manufacturers' efforts. Here are our hints.

2S.1.2 Big caps: electrolytics

These are easy to read, because there is room to write the value on the cap, including units. "16V" is the maximum voltage one can safely put across the capacitor without damaging the part. The *minus* sign and arrow indicate which is the negative terminal for this *polarized* capacitor.

Figure 2S.1 A big cap is labeled intelligibly.

Caps of 1µF and above usually are *polarized*

All of these big caps are *polarized* ("big," in our world, means 1µF or more). That means the capacitor's innards are not symmetrical, and that you may destroy the cap if you apply the wrong *DC* polarity to the terminals:[2] the terminal marked + (or the one *not* marked "−," as in the capacitor of Fig. 2S.1) must be at least as positive as the other terminal.[3] (Sometimes, violating this rule will generate gas that makes the cap blow up; more often, you will find the cap internally shorted, after a while. Often, you could get away with violating this rule, at low voltages. But don't try.)

One of our students brought a trophy to class after inadvertently wiring a polarized cap backwards.[4]

If caps wired backwards always reacted so spectacularly, they would never have a chance to subvert our circuits. Unfortunately, a reversed cap usually turns quietly into a short circuit, making your circuit behave weirdly while offering no clue to what is wrong. (The goo that looks like spilled guts of the cap in Fig. 2S.2, incidentally, is not guts, but only glue that was used to fix the cap's remains to a piece of paper, for display.)

---

[1] "Cap" is shorthand for "capacitor," as you probably know.
[2] This is a rule for *DC*, sustained levels; a short-term swing, not long enough to develop a substantial reverse voltage across the capacitor, is harmless.
[3] They are not symmetrical because the dielectric is a thin oxide layer on the surface of one metal electrode; the other terminal of the capacitor is immersed in goo that forms the second electrode – in case that primitive explanation is of any help.
[4] Image thanks to Rick Montesanti, who told us proudly that the cap announced the error with a *bang* and fired junk twenty feet.

Figure 2S.2 Polarized cap that was hooked up with reversed DC voltage.

2S.1.3 Pretty big caps: tantalum

As the caps get smaller, the difficulty in reading their markings begins. Tantalums, which can provide quite large capacitance, often are physically large enough to be easy to read.

The tantalums often are silver colored cylinders. They are polarized, like the big electrolytics. A + marks the positive end, on the tantalum cap pictured.[5] The tantalum cap in Fig. 2S.3 uses the relatively ample surface area of its packages to state its value twice.

The "4R7" means pretty much what it says, if you know that the "R" (radix point) marks the decimal place: it's a $4.7\mu F$ cap, and it can stand 50V without damage.

Figure 2S.3 Tantalum cap.

The second marking, "475M", under "CS13B" (which is the capacitor type), states the value in exponential form, as if this were a resistor: $47 \times 10^5 M$. What's "M"? Microfarads? (Surely we're not to take the capital seriously: Megafarads?!)

But we must resist the plausible assumption that "M" is a unit. It is not. It indicates *tolerance*, instead: ($\pm 20\%$). (Wasn't that nasty of the labelers to choose "M?" Guess what's another favorite letter for tolerance. That's right: K, which looks like "kilo," but instead means $\pm 10\%$. Pretty mean!)

What units? If "M" is not a unit but a tolerance marking, what are the units? 10^5 *what*? 10^5 of *something small*. You will meet this question repeatedly, and you must resolve it by relying on a few observations. The only units commonly used in the US in capacitor labeling are

- microfarads: 10^{-6} Farad,
- picofarads: 10^{-12} Farad

An intermediate unit, the nanofarad, nF: 10^{-9} Farad, is respectable in circuit diagrams, but fortunately not used to label capacitors. We say "fortunately" because we rely on the huge difference between "nF" and "pF" that allows us to guess which unit is intended, in a capacitor label.

"mF" (10^{-3} Farad) looks reasonable but is not used; instead, capacitors in this size range are measured in thousands of microfarads. For example, a cap will be described as "$4,000\mu F$" rather than "4mF." Strange, but dictated by tradition.

A Farad is a huge unit. The biggest cap you will use in this course is a few hundred μF. Such a cap is physically large. (We do keep a 1*Farad* cap around, but only for our freak show.)[6]

So, if you find a little cap labeled "680," you know it's 680 **pF**.[7]

---

[5] For this particular series of caps, a metal nipple also marks the positive end. The nipple is much easier to spot than the small "+" mark, which is quite invisible if the cap is not rotated just right – but unfortunately there seems to be no general consistency linking nipple and positive end.

[6] These giant caps are used as replacements for rechargeable batteries in calculators, portable computers, even flashing bike lights. Their advantage over batteries is long life: since their charge and discharge involve no chemical reactions, they should tolerate a great number of charge/discharge cycles.

[7] Except, of course, when it's *not* – that is, when the labeling scheme is exponential; see §2S.1.5.

2S.1 Reading capacitors

A picofarad is a tiny unit. You will not see a cap as small as 1pF in this course.[8] So, if you find a cap claiming that it is a fraction of some unstated unit – say ".01" – the unit is μFs: here ".01" means 0.01μF.

Beware the wrong assumption that a *picofarad* is only a bit smaller than a microfarad. A pF is *not* 10^{-9}F (i.e. $10^{-3}\mu$F); instead, it is 10^{-12}F: a *million* times smaller than a microfarad.

So, we conclude, this cap labeled "475" must be 4.7×10^6 picofarads. That, you will recognize, is a roundabout way to say

$$4.7\times 10^{-6}\text{F} .$$

We knew that was the answer, before we started this last decoding effort. This way of labeling is indeed roundabout, but at least it is unambiguous. It would be nice to see it used more widely. You will see another example of this *exponential* labeling in the case of the CK05 ceramics in §2S.1.5.

Figure 2S.4 Mylar: still big enough to be labeled clearly.

2S.1.4 A little smaller: polyester (mylar)

These are yellow cylinders, pretty clearly marked; see Fig. 2S.4.

" .01k" is just 0.01μF, of course; by now you are sophisticated enough to know that "k" does not mean "kilo." Instead, it is another *tolerance* label, like "M" on the $100\mu F$ cap above. "k" means $\pm 10\%$. These caps are not polarized; the black band marks the outer end of the foil winding. We don't worry about that fine point. Orient them at random in your circuits.

Ceramics (see next section) do better in this respect, though they are poor in other characteristics.

Figure 2S.5 Mylar ("polyester") cap, unrolled, reveals that it's a metal-foil coil, layers separated by mylar film.

2S.1.5 Small enough to be ambiguous: ceramic

Ceramics are the proletarians of the cap world: hard workers, not refined, not predictable, not precise, but good workers in difficult conditions.[9] They are used most often to "decouple" power supplies – that is, to stabilize the supplies with respect to ground. These ceramics are little rectangular boxes, or (the older style) little orange pancakes, and are used by the electronics industry in colossal numbers.[10] They act like capacitors even at high frequencies. The trick, in reading these, is to reject the markings that can't be units. As we'll see in a minute, we also need to be aware of the possibility of a non-standard labeling scheme.

CK05: These are little boxes, with their leads 0.2 inches apart. They are handy, therefore, for insertion into a printed circuit (or *were* handy, before the tiny *surface mount* parts took over). Fig. 2S.6 shows the two sides of a CK05-packaged part.

Let's find the value, throwing out what doesn't interest us, along the way:

- 200V: just what it sounds like: the maximum voltage one can safely put across the cap;
- K: this time not a tolerance but indicating the manufacturer (Kemet);
- 9902: not the value, but the week of the part's *birthday*: week 2 of [19]99.[11]

So far, no value information. That must lie on the other side. There we see:

---

[8] There's good reason for that: stray capacitance in our breadboarded circuits is on this same scale. The stray capacitance between two adjacent columns on our plastic breadboards is about 2pF. The input capacitance of a "10×" oscilloscope probe (a device you will meet in Lab 3L) is about 10pF. The capacitance of three feet of BNC cable is about 90pF.
[9] As AoE teaches, some ceramics perform well: the C0G (class-I) type.
[10] Wikipedia, http://en.wikipedia.org/wiki/Ceramic_capacitor, estimates annual production at around one trillion.
[11] Important for sentimental engineers who celebrate with all their components.

Figure 2S.6 CK05 capacitor markings.

- CK05, and on the next line...
- ...BX. These simply name the type of package
- 680k: here is the value, at last: 680 is in exponential notation; K is the tolerance, once more

So, this is a cap of 68pF (68×10^0 of something small, picofarads)

Possible ambiguity: old-style disc capacitors

Occasionally, you may run into a ceramic capacitor labeled acording to an older convention. Such *disk* capacitors come in relatively-large packages. In Fig. 2S.7 we show one of these, alongside a CK05 part.

Figure 2S.7 Some old disk caps are labeled using a non-exponential convention.

Having met and understood the CK05 labeling, one would assume that both these caps were 56pF parts. The one on the right *is* a 56pF cap. But the one on the left – in the older disc package – is a 560pF part. Nasty, but true. The comforting thought, though, is that this disc package and its labeling scheme are nearly obsolete.

2S.1.6 Caps too small to label intelligibly: the coming thing!

We have shown examples of cap markings you are likely to run into as you do labs in this course. But when you start building printed circuits, you are likely to use the smaller *surface mount* packages. Only the largest of these even attempt labeling. The smaller ones are blank; it's up to you to recall which bin you took each from. If you get confused, all you can do is reach for another, from a properly-labeled package or drawer.

Surface mount cap: large enough for labeling

The cap in Fig. 2S.8 is of large enough value ($4.7\mu F$) that it needs a package that is large by surface-mount standards. So, though it is small, it does show labeling.

The enlarged view in Fig. 2S.9 shows the "475K" that we would expect for a $4.7\mu F$ cap.

2S.1 Reading capacitors

Figure 2S.8 Surface mount cap large enough to show labeling.

Figure 2S.9 Closer view of labeled SMT cap: "475K" states value.

Figure 2S.10 Smaller surface mount packages altogether give up the attempt at labeling.

Surface mount caps too small for labeling

As the surface-mount packages get smaller, however, they cease even to try to announce their values. Fig. 2S.10 show some caps that dare not speak their names.

2S.1.7 Tolerance codes

Just to be thorough – and because this information is hard to come by – we've listed below all the tolerance codes. These apply both to capacitors and resistors; the tight tolerances are relevant only to resistors; the strangely-asymmetric tolerance is used only for capacitors.

Tolerance Code	Meaning	
Z	+80%, −20%	(for big filter capacitors, where you are assumed to have asymmetric worries: too small a cap allows excessive "ripple;" more on this in Lab 3L and Chapter 3N)
M	±20%	
K	±10%	
J	±5%	
G	±2%	
F	±1%	
D	±0.5%	
C	±0.25%	
B	±0.1%	
A	±0.05%	
Z	±0.025%	(precision resistors; context will show the asymmetric cap tolerance "Z" makes no sense here)
N	±0.02%	

2S.2 C notes: trying for an intuitive grip on capacitors' behavior

2S.2.1 Cs differ from Rs, of course

This note tries some very simple arguments, in the hope that they may help you to strengthen your intuitive grip on the behavior of capacitors: not the sort of grip provided by $Z_C = -j/\omega C$, but the sort of grip that might be provided by a mental picture of the way voltages change on the two plates of the cap, as an input is applied.

Resistors live in the present: capacitors, in contrast, are fundamentally conservative: see Fig. 2S.11.

The voltage across a cap can't change abruptly, and RC circuits may be much influenced by what happened a while ago.[12]

Let's look for some intuition that will let us *expect* caps to behave as they do.

2S.2.2 Low-pass filter

A low-pass attenuates high frequencies If the input changes fast, the voltage on the capacitor cannot keep up with changes at the input, as we can see from Fig. 2S.12.

The output (V_{cap}) moves toward the input voltage – but then the fickle input quickly changes its level, and V_{cap} never quite catches up. So, only a diminished sketch of the input waveform reaches the filter's output.

A low-pass passes low frequencies If you are patient (or, more exactly, if the waveform is patient) and give the capacitor time to charge or discharge almost to the level of the applied voltage, then the input signal does reach the output: the *low* frequency is *passed*; see Fig. 2S.13.

---

[12] As you'll soon see (if you haven't noticed it on your own), the characterization as "conservative" applies most clearly to the *low-pass* configuration.

Figure 2S.11 Top: Hippy Resistor: no sense of time. Bottom: Proustian Capacitor: ... believes that the past *does* matter.

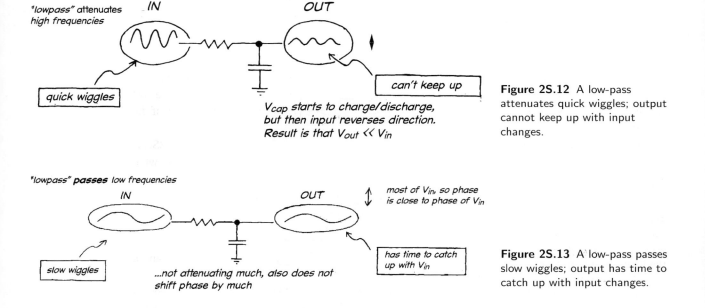

Figure 2S.12 A low-pass attenuates quick wiggles; output cannot keep up with input changes.

Figure 2S.13 A low-pass passes slow wiggles; output has time to catch up with input changes.

2S.2.3 High-pass

High-pass *does* transfer quick changes, from input to output

A quick edge passes: The high-pass behavior is a little bit harder to get a grip on. I find that it helps to begin by thinking about the circuit in *time domain* (that is, by considering what you might see on an oscilloscope screen, at input and output) rather than in the more abstract *frequency domain* view.

Set up the *RC* in the so-called high-pass configuration, and imagine applying a voltage *step* at the input, as in Fig. 2S.14.

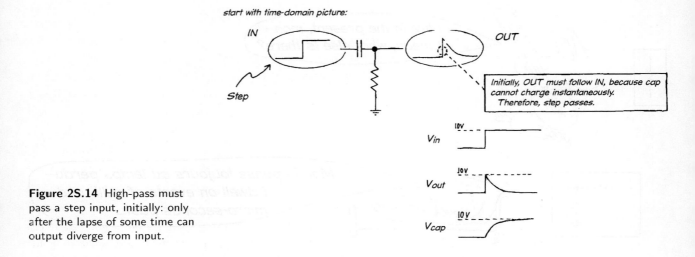

Figure 2S.14 High-pass must pass a step input, initially: only after the lapse of some time can output diverge from input.

A quick square wave passes: Repeat those quick edges, up, down, up ..., and you are applying a *square wave*; see Fig. 2S.15. Such a waveform, like the quick edge, passes, too – so long as you don't allow it to sit still long. If you do allow it to sit high or low for a while, then you'll see the output look droopy.

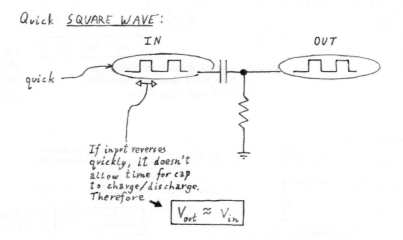

Figure 2S.15 High-pass can pass a quick square wave, too – if we don't let the input sit still for long.

A quick sinusoid also passes: A quick sinusoid[13] passes, too.[14], and for the same reasons: if we don't let the input sit anywhere for long, then the output voltage doesn't have time to move far from the input voltage. So, most of the input wiggle appears at the output (and, again, if the output *amplitude* looks like the input *amplitude*, then the output *phase* looks like the input *phase*). In other words, when you don't see attenuation, you'll not see phase shift (the same was true for the low-pass).

[13] Usually we don't speak so fancily – we call it a "sinewave."
[14] Fourier tells us that this isn't really a new point: what we called a "quick edge," says Fourier, can be described as a collection of sinusoids including some at *high-frequencies* So, to say, "the steep edge passes" is equivalent to saying, "the high-frequency sinusoids that form the edge pass." More on this in Lab 3L.

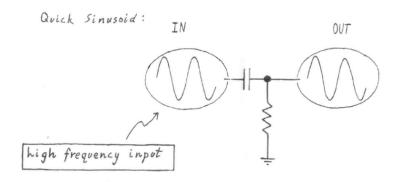

Figure 2S.16 High-pass will pass a "quick" (high-frequency) sinusoid.

...but the high-pass does not pass low frequencies

No surprise, here: if we wiggle slowly, the output voltage has time to come to differ a great deal from the input voltage. So, the input can wiggle without being able to make the output wiggle much. In this case, the phase of whatever wiggle gets through will differ a lot from the phase of the input (this we haven't proven to you; but it seems plausible, does it not?).

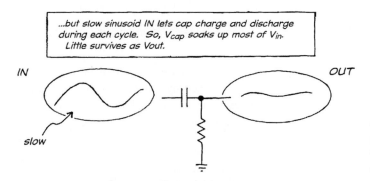

Figure 2S.17 High-pass does a poor job of passing slow wiggles; there's time for cap to charge and discharge.

2S.3 Sweeping frequencies

An oscilloscope can let one see at a glance what the frequency response of a circuit looks like. It can show amplitude versus *frequency* rather than amplitude versus *time*, as in the usual scope display.[15] All that's required is a function generator capable of "sweeping" frequencies – varying its output frequency in a regular manner. The function generators that we use in our labs can do this.

The best method for sweeping depends on whether the scope you are using is *analog* or *digital*. Our students begin with analog scopes, so our tips on sweeping frequencies also will begin by discussing these scopes.

2S.3.1 Function generator's "continuous sweep" operation

We usually set the generator frequency by hand. But the generator can be set to sweep its main output frequency from a start frequency to a stop frequency.

---

[15] Strictly, even in *X–Y* mode, the scope shows amplitude versus time when the *X*-axis is driven by a repeating waveform like the sweep *ramp*. But when frequency changes with time, and the scope is properly synchronized, the horizontal axis represents *frequency* as well as *time*.

Supplementary Notes: RC Circuits

Analog function generator

On our analog generators, this function is called "continuous sweep." When so used, the generator also puts out the *ramp* or *sawtooth* waveform that it uses to control the main generator's frequency. This waveform – from an output labeled "RAMP" on our generators – can be used to synchronize a scope display.

In Fig. 2S.18, one can see that the sinusoid frequency from the generator's output grows with the ramp voltage.

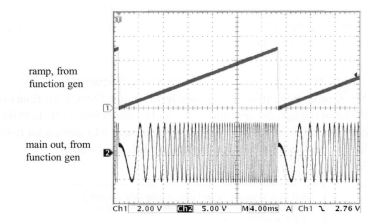

Figure 2S.18 Ramp and main outputs, with main frequency controlled by ramp voltage ("Continuous Sweep").

The *RAMP* amplitude is constant, but its repetition rate is adjustable. The frequency range at the generator's main output (a sinusoid, in Fig. 2S.18) is determined by START and STOP frequency settings. In our generators, those are set by two slider potentiometers (on the left edge of the generator front panel, in Fig. 2S.19), along with the usual *range* selection (three pushbuttons, on our generators).

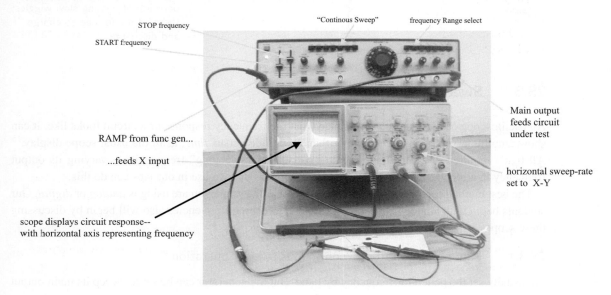

Figure 2S.19 Scope connections for a display of swept frequencies.

2S.3.2 Swept display on analog scope, analog generator

The best way to sweep differs for analog and digital scopes, in our experience. The analog sweep is best done in so-called "X–Y" mode. The digital is best done in the usual timed-sweep mode.

X–Y display

The best method for an analog scope takes advantage of the scope's ability to substitute the Channel One input for its usual *horizontal* sweep timebase. Channel One is called the "X" input, when this display mode is selected, because it controls the horizontal position of the scope's beam, and X is the traditional name for the horizontal axis of a 2-variable plot.

We illustrate the connections twice (perhaps overdoing it): first, Fig. 2S.19 showed what the setup looked like on our lab bench. Fig. 2S.20 gives it as a sketch.

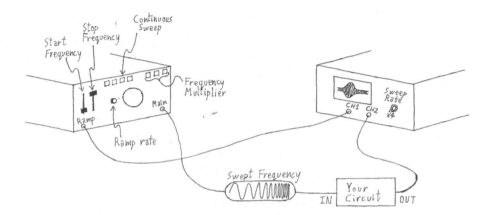

Figure 2S.20 Scope connections, for a display of swept frequencies.

The signal that one applies to X is the *sawtooth* waveform available at the function-generator terminal labeled "RAMP." The left edge of the screen shows response at the START frequency, the right edge shows response at the END frequency, and the screen plots the variation between these extremes, as was evident in Fig. 2S.18.

Scope settings First, set up the scope:

- Set the horizontal sweep to X–Y. On the analog scopes, this is the extreme counterclockwise position for the sweep-rate knob.
- Set the Y channel (channel 2, on a 2-channel scope) to GND – temporarily.
- While feeding RAMP into the X channel, note the width of the trace: adjust X gain so that the trace just fills the screen. You need to be careful not to overfill the screen: too much gain will try to send the beam off the right-hand edge of the screen, where it will disappear or even double back. So reduce the X gain till the trace fills just half the screen (as on left, in Fig. 2S.22):
- Then double the gain, just filling the screen (as on the right in Fig. 2S.22).
- Now you can take the Y channel off GND, to DC. The scope now should show you amplitude out versus frequency

2S.3.3 Function generator settings

- Adjust the START and STOP frequencies until you see a response that looks reasonable:

Supplementary Notes: RC Circuits

Figure 2S.21 For analog scope *X–Y* display, set horizontal sweep to select that display.

Figure 2S.22 Adjust *X* gain to use full screen width.

X gain (Ch 1) too low: just half screen...   ...X gain correct: fills screen

- For a low-pass or high-pass filter, START should be zero frequency (or the lowest that is available). STOP you can adjust to get the display you want. At least, STOP should be well above f_{3dB}.
- For the *RLC* resonant circuit of Lab 3L, you may want to START close to $f_{\text{resonance}}$ rather than at zero, to get a detailed view of the narrow passband. STOP must, of course, lie above $f_{\text{resonance}}$.
- Repetition rate:
  - in *X–Y* mode, you can freely vary the repetition rate (the ramp frequency) without disturbing the display's placement on the scope screen. That is why we recommend *X–Y* mode, rather than timed-sweep.
  - But the repetition rate does matter. If you set it too high, you will get a crude display, with too few cycles of stimulation reaching your circuit. In the case of the *RLC*, an excessive ramp frequency has even worse effects: it produces strange artifacts in the display. You will see bumps and depressions in the response, after $f_{\text{resonance}}$.

  - These occur because the resonant circuit continues to "ring" for a while, after it is stimulated. This ringing – persistent oscillation at $f_{\text{resonance}}$ – then interacts with the gradually changing input frequency.
    The two frequencies "beat:" sometimes the two add, reinforcing each other. In that case, a bump appears. At other times, the two frequencies subtract – oppose each other in phase. In that case, a depression appears in the response.
  - Therefore, you should use as low a ramp frequency as you can stand to watch: a low frequency repetition produces noticeable flicker on an analog scope.
    Low repetition rate does not produce flicker on a digital scope (which stores the waveform), but may try your patience by taking a long time to update the display. (At 1s/div, for example, a screen update takes 10 seconds.) Find a comfortable compromise for the ramp rate.

2S.3 Sweeping frequencies

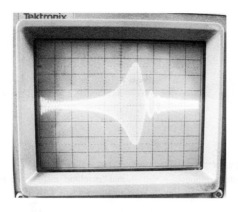

Figure 2S.23 If ramp repetition rate is too high, swept display goes bad.

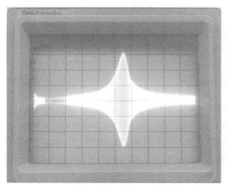

Figure 2S.24 Slow ramp repetition produces cleaner swept display.

Occasionally, timed scope sweep is better If you want to look at the response of *two* circuits, for comparison, on a two-channel scope, then X–Y is not available. You will need to use the scope's usual timed scope sweep (controlled by the "Horizontal" knob, as usual). This we ask you to do in Lab 9L, when you are to compare the frequency response of two filters: active versus passive. In the case of a digital scope, we recommend that you always use this familiar display method – as noted in §2S.3.4.

2S.3.4 Swept display on digital scope

A swept display is more straightforward, on a digital scope. There is no need to use the X–Y mode, though it is available. (We have found that it is somewhat buggy on our Tektronix TDS3014 scopes.) There is no need, because one can sweep very slowly without meeting the flicker problem that makes the adjustment of sweep on an analog scope a fussy operation, calling for continual readjustment.

So, use the usual *timed* scope sweep when using a digital scope. Use RAMP not to control the X axis but to *trigger* the sweep. Use the falling edge, which is steep (therefore a well-defined trigger signal) and which just precedes the ramp that controls the main generator's output frequency.

The only issue that persists, from those mentioned in the notes on sweeping for analog scope (§2S.3.2), is the problem of artifacts that one can introduce by setting the ramp repetion rate too high.

Figure 2S.25 is a contrast between such a too-fast repetition rate (leftmost image) and a clean sweep, done more slowly (right-hand images). The rightmost image shows a nice wrinkle: the scope can outline the "envelope" of the frequency response for you if you choose the display option "envelope" rather than the usual "sample."

Supplementary Notes: RC Circuits

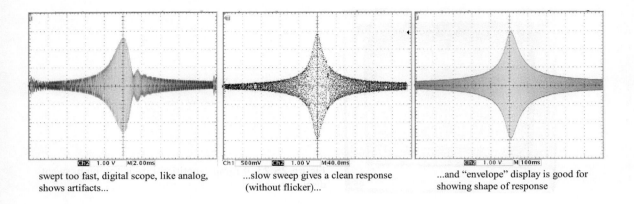

| swept too fast, digital scope, like analog, shows artifacts... | ...slow sweep gives a clean response (without flicker)... | ...and "envelope" display is good for showing shape of response |

Figure 2S.25 Repetition rate must be low, for clean digital display; "envelope" display suits sweep well.

2S.3.5 Swept display on digital scope, digital generator

No X–Y display, with digital function generator: *timed* sweep only

A *digital* function generator is in some ways better than analog: it allows great precision in its sweep. But it is in some ways worse: it offers no *ramp* output that can be used to drive an X–Y display. So, one is obliged to use a *timed* display. As a result, any change in function generator timing requires readjustment of the scope.

In Fig. 2S.26 we show the settings on an Agilent 33210A generator, and the resulting display, as we sweep the *RLC* circuit of Lab 3L. That circuit is resonant at about 16kHz, and we have set up the generator to sweep a narrow range of frequencies, between 15.5kHz and 16.5kHz.

The generator includes a feature it calls "marker," generating a falling edge at a designated frequency. We have used that feature to find (visually) the peak of the *RLC*'s response, in Fig. 2S.26. Once that falling edge is lined up with the response peak, which occurs at the resonant frequency, then we can read that resonance value as the marker frequency. This frequency – about 16.09kHz – is displayed in the lower-right corner of the figure.

This *marker* lets one find f_{resonant} approximately. But to get this value exactly one should use the method suggested in Lab 3L: watch input and output waveforms. Get close to resonance, then finely adjust f_{in} until *phases* of input and output match. When they do, you have found f_{resonant}.

Slow sweep works best

With a digital scope, since we don't need to worry about flicker, it is wise to use a slow sweep. The slow sweep prevents the ugly complication – visible in the right-hand image of Fig. 2S.27, a problem we saw back in §§2S.3.3 and 2S.3.4.

The one-second sweep provides a pretty clean display (there are hints of artifacts, even here). The 30ms sweep includes obvious beating between f_{in} and f_{resonant}, as we saw also in §2S.3.3 and Fig. 2S.23.

A stranger corruption appears as well: the seeming f_{resonant} is shifted when the circuit is swept fast. This is an illusion caused by the time required to drive the *RLC* into resonance. In the right-hand image of Fig. 2S.23 a delay of almost 1ms appears between the time when the resonant frequency is applied (indicated by the falling edge of the timing signal) and the time when the *RLC* shows its peak V_{out}. So, to get an honest display, be patient: wait a second or so for your swept image.

2S.3 Sweeping frequencies

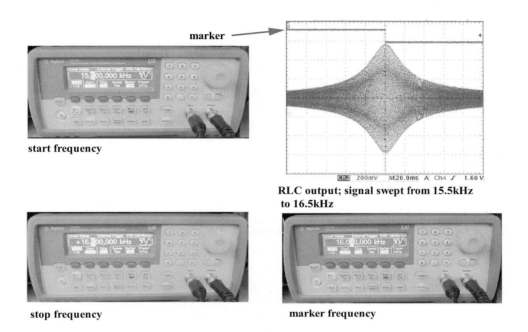

Figure 2S.26 Digital function generator permits precise sweep – but does not permit X–Y display.

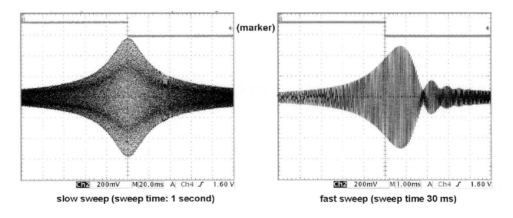

Figure 2S.27 Slow sweep provide an honest output; fast sweep corrupts the result with interactions between f_{in} and f_{resonant}.

2W Worked Examples: *RC* Circuits

2W.1 RC filters

2W.1.1 Filter to keep "signal" and reject "noise'

Problem: filter to remove fuzz

Suppose you are faced with a signal that looks like that in Fig. 2W.1: a signal of moderate frequency, polluted with some fuzz.

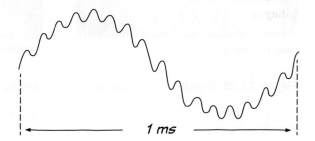

Figure 2W.1 Signal with fuzz added.

(1) Draw a *skeleton* circuit (no parts values, yet) that will keep most of the good signal, clearing away the fuzz.
(2) Now choose some values:
   (a) If the *load* has value $\geq 100k$, choose R for your circuit.
   (b) Choose f_{3dB}, explaining your choice briefly.
   (c) Choose C to achieve the f_{3dB} that you chose.
   (d) By about how much does your filter attenuate the noise "fuzz"?

   What is the circuit's input impedance:
   (a) at very low frequencies?
   (b) at very high frequencies?
   (c) at f_{3dB}?
(3) What happens to the circuit output if the load has resistance 10k rather than 100k?

A solution

Skeleton circuit: You need to decide whether you want a low-pass or high-pass, since the signal and noise are distinguishable by their frequencies (and are far enough apart so that you can hope to get one without the other, using the simple filters we have just learned about). Since we have called the lower frequency "good" or "signal," we need a *low-pass*: see Fig. 2W.2.

2W.1 RC filters

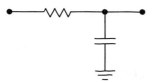

Figure 2W.2 Skeleton: just a low-pass filter.

Choose R, given the load: This dependence of R upon load follows from the observation that R of an RC filter defines the *worst-case* input and output impedance of the filter (see Chapter 2N). We want that output impedance low relative to the load's impedance; our rule of thumb says that 'low' means low by a factor of 10. So, we want $R \leq R_{\text{load}}/10$. In this case, that means R should be ≤ 10k. Let's use 10k.

Choose $f_{3\text{dB}}$: This is the only part of the problem that is not entirely straightforward. We know we want to pass the low and attenuate the high, but does that mean put $f_{3\text{dB}}$ halfway between good and bad? Does it mean put it close to good? ... Close to bad? Should both good and bad be on a steeply-falling slope of the filter's response curve?

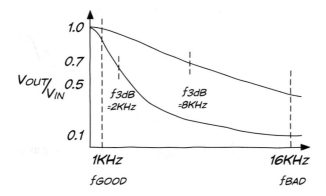

Figure 2W.3 Where should we put $f_{3\text{dB}}$? Some possibilities.

Assuming that our goal is to achieve a large ratio of good to bad signal, then we should not put $f_{3\text{dB}}$ close to the noise: if we did we would not do a good job of attenuating the bad. Halfway between is only a little better. Close to signal is the best idea: we will then attenuate the bad as much as possible while keeping the good, almost untouched.

An alert person might notice that the greatest relative preference for good over bad comes when both are on the steepest part of the curve showing frequency response: in other words, put $f_{3\text{dB}}$ so low that *both* good and bad are attenuated. This is a clever answer – but wrong, in most settings.

The trouble with that answer is that it assumes that the signal is a single frequency. Ordinarily the signal includes a range of frequencies, and it would be very bad to choose $f_{3\text{dB}}$ somewhere *within* that range: the filter would distort signals.

So, let's put $f_{3\text{dB}}$ at $2 \times f_{\text{signal(max)}}$: around 2kHz. This gives us 89% of the original signal amplitude (this we claimed in Chapter 2N; you can confirm this result, if you like, with a phasor diagram or by direct calculation). Incidentally, the phase shift at that frequency is also moderate – around 25° lag. Again phasor diagram or calculation can confirm this value. At the same time we should be able to attenuate the 16kHz noise a good deal (we'll see in a moment *how* much).

Worked Examples: RC Circuits

Choose C to achieve the f_{3dB} that you want: This calls for no more than plugging values into the formula for the 3dB point:

$$f_{3dB} = \frac{1}{(2\pi RC)} \Longrightarrow C = \frac{1}{(2\pi f_{3dB} R)} \approx \frac{1}{6(2 \times 10^3)(10 \times 10^3)} = \frac{1}{(120 \times 10^6)} \approx 0.008 \times 10^{-6} \text{F}.$$

We might as well use a $0.01\mu\text{F}$ cap. It will put our f_{3dB} about 25% low – 1.6kHz; but our choice was a rough estimate anyway.[1]

By about how much does your filter attenuate the noise ("fuzz")? There are two quick ways to get this answer:

- Look at the ratio of f_{noise} to f_{3dB}. Attenuation will be roughly proportional, since amplitude falls off as $1/f$ (or, equivalently "−6dB/octave").

  Here, $f_{noise}/f_{3dB} = 16\text{kHz}/2\text{kHz} = 8:1$. So, amplitude should be down to about 1/8 what it was at f_{3dB}. So the result is $\approx 0.125 \times 0.7 = 0.08$. Amplitude will be a bit more, though, since the slope does not reach the full $1/f$ shape till a few octaves above f_{3dB}.

- *Count octaves*: we could also say that the frequency is doubled three times ($= 2^3$) between f_{3dB} and the noise frequency. *Roughly*, that means that the fuzz amplitude is cut in half the same number of times: down to $(1/2)^3$: 1/8. Again, this is relative to the amplitude at f_{3dB}, where amplitude was already down to 70%.

What happens to the circuit output if the load has resistance 10k rather than 100k? Fig. 2W.4 is a picture of such loading.

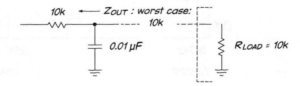

Figure 2W.4 Overloaded filter.

If you have gotten used to Thevenin models, then you can see in Fig. 2W.5 how to make this circuit look more familiar:

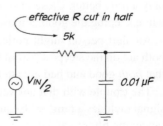

Figure 2W.5 Loaded circuit, redrawn.

The amplitude is down; but, worse, f_{3dB} has changed: it has doubled. Fig. 3N.2 is a plot showing this effect.

---

[1] To get exactly our target f_{3dB} we can readjust R. We have fewer choices of capacitor value, but can choose any R (though it doesn't make sense, given a "10%" cap, to do better than specify R to 10%).

What is the circuit's input impedance?

(1) at very low frequencies? Answer: *very large*: the cap shows a high impedance; the signal source sees only the load – which is assumed very high impedance (high enough so we can neglect it as we think about the filter's performance)
(2) at very high frequencies? Answer: *R*: The cap impedance falls toward zero – but *R* puts a lower limit on the input impedance.
(3) at f_{3dB}?

This is easy if you are willing to use phasors, a nuisance to calculate, otherwise. If you recall that the magnitude of $X_C = R$ at f_{3dB}, and if you accept the notion that the voltages across *R* and *C* are 90° out of phase so that they can be drawn at right angles to each other on a phasor diagram, then you get the phasor diagram of Fig. 2N.31, and can use a geometric argument to show that the hypotenuse – proportional to Z_{in} – is $R\sqrt{2}$.

2W.1.2 Bandpass

Problem: bandpass filter Design a bandpass filter to pass signals between about 1.5kHz and 8kHz. To do this, put the two f_{3dB} a factor of 2 from the nearest frequency. (Doing this will limit attenuation to about 10%, as you know from §2W.1.1.)

Assume that the *next* stage that your bandpass filter is to drive has an input impedance $\geq 1\text{M}\Omega$.

A solution Roughly, Fig. 2W.6 shows the shape of the frequency response that we want.

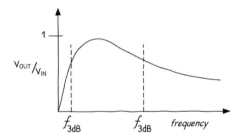

Figure 2W.6 Bandpass frequency response.

To get this frequency response from the filter we need to put *high-pass* and *low-pass* in series. f_{3dB} for the high-pass should be about half the lowest frequency of interest, so put it at about 750Hz. f_{3dB} for the low-pass should be double the highest frequency of interest, so put it at about 16kHz.

Viewed separately, the high-pass and low-pass responses are shown in Fig. 2W.7. The horizontal axis is frequency, and runs from close to zero up to 30kHz.

If the two filters are "cascaded" – placed in series, so that the poor old input has to pass through this double gauntlet – the result will be attenuation at the frequency extremes, and a bump in the middle: see Fig. 2W.8.

The linear frequency plot shows that the bandpass is pretty far from the ideal – a filter that blocks all of the bad and lets through all of the good. Instead of being shaped like a steep-walled mesa, it's more like a gentle foothill of a mountain range. The *log* frequency plot makes the response *look* a little more respectable – more as if there is a flat passband region; but of course the changed plot doesn't change the facts. Using simple *RC*s, we'll have to settle for this imperfect, slopey passband.

Worked Examples: RC Circuits

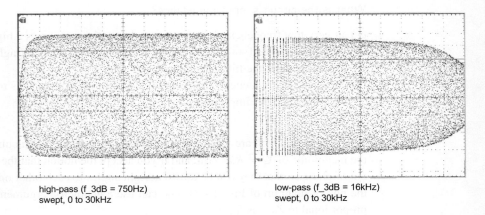

Figure 2W.7 High-pass and low-pass f_{3dB}'s at 750Hz and 16kHz, respectively.

high-pass (f_3dB = 750Hz) swept, 0 to 30kHz

low-pass (f_3dB = 16kHz) swept, 0 to 30kHz

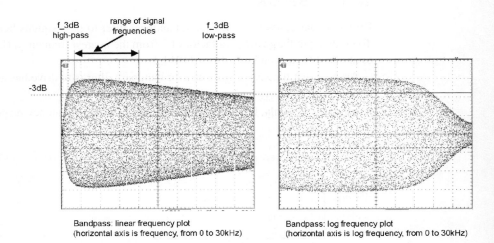

Figure 2W.8 Bandpass response that we want: linear and log frequency plots.

Bandpass: linear frequency plot (horizontal axis is frequency, from 0 to 30kHz)

Bandpass: log frequency plot (horizontal axis is log frequency, from 0 to 30kHz)

Choose Rs: Now we need to choose R values, because these will determine worst-case impedances for the two filter stages. The later filter must show Z_{out} low relative to the load, which is 1MΩ; the earlier filter must show Z_{out} low relative to Z_{in} of the second filter stage.

So, let the second-stage R be 100k; the first-stage R is 10k:

...then calculate Cs: Here the only hard part is to get the filters right: it's hard to say to oneself, "The high-pass filter has the lower f_{3dB};" but that is correct. Here are the calculations: notice that we try to keep things in *engineering notation* – writing "10×10^3" rather than "10^4." This form looks clumsy, but rewards you by delivering answers in standard units. It also helps you scan for nonsense in your formulation of the problem: it is easier to see that "10×10^3" is a good translation for "10k" than it is to see that, say, "10^5" is *not* a good translation.

Calculating the two C values is a purely mechanical process.

The low-pass:

$$f_{3dB-low} = 16\text{kHz} \Longrightarrow C = \frac{1}{2\pi \times 16 \times 10^3 \times 10 \times 10^3} \approx 1/10^9 = 0.001\mu\text{F}.$$

The high-pass:

$$f_{3dB-high} = 0.75\text{kHz} \Longrightarrow C = \frac{1}{2\pi \times 0.75 \times 10^3 \times 100 \times 10^3} \approx 1/(0.45 \times 10^9) \approx 0.002\mu\text{F}.$$

And the circuit looks as in Fig. 2W.9.

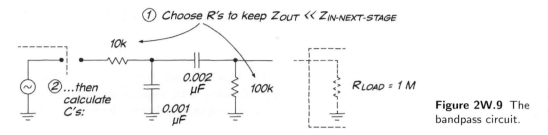

Figure 2W.9 The bandpass circuit.

2W.2 RC step response

We were struck recently by how difficult an "old" topic like *RC* behavior can be when it's a little different from the standard case. Several students convinced us we need more practice on such early learning. Here's a little workout on *RC*s.

2W.2.1 Problem: step response of *RC* circuit (time-domain)

Figure 2W.10 shows a capacitor feeding a pulse into several alternative circuit elements. Please draw the voltage waveform at the point where the capacitor meets "X" for each case. Show time and voltage scales, and label significant points on your waveform. As usual, we'll be happy with answers good to about 10%. Both R_{IN_X} and C_{IN_X} describe the effect of a *C* and an *R* in parallel; the far end of each is tied to a fixed voltage; let's assume that voltage is ground, just so all our drawings will look the same.

Note: We'd like you to draw V_{out}, not V_{cap}.

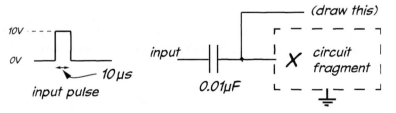

Figure 2W.10 Pulse, capacitively-coupled to various circuit fragments.

Here are the several alternatives for the circuit element labeled "X:"

- Idealized next stage: $R_{IN_X} = \infty$, $C_{IN_X} = 0$;
- Mixed next stage: $R_{IN_X} = \infty$, but $C_{IN_X} = 0.001\mu\text{F}$;
- Mixed next stage: $R_{IN_X} = 100k$, $C_{IN_X} = 0$;
- Mixed next stage: $R_{IN_X} = 1k$, $C_{IN_X} = 0$;
- Non-ideal next stage: $R_{IN_X} = 100k$, $C_{IN_X} = 0.01\mu\text{F}$.

2W.2.2 Solution

Idealized next stage: $R_{IN_X} = \infty$, $C_{IN_X} = 0$: Since there is no path permitting current to pass to or from the second plate of the cap, $I_{cap} = 0$ and V_{cap} remains zero. Therefore $V_{out} = V_{in}$.[2]

Figure 2W.11 No R, no C.

Mixed next stage: $R_{IN_X} = \infty$, but $C_{IN_X} = 0.001\mu F$: This is a C–C divider (rare or impossible in life). V_{out} *shape* is same as V_{in}'s, but *amplitude* is attenuated according to the impedances of the two caps. $V_{out} = 0.9 V_{in}$. (Again a DC path to ground is required on the output to make this circuit practical.)

Figure 2W.12 A capacitive divider.

Mixed next stage: $R_{IN_X} = 100k$, $C_{IN_X} = 0$: $RC = 1$ ms, $\gg$ pulse width. So V_{cap} will not change appreciably during pulse. Hence V_{out} looks like V_{in}. This case approximates the first case, of §2W.2.2. Again, $V_{out} = V_{in}$.

Figure 2W.13 Approximately the first case: all passes.

Mixed next stage: $R_{IN_X} = 1k$, $C_{IN_X} = 0$: $RC = 0.01$ ms, same as pulse width. So pulse output will decay to $V_{in} \times (1/e) \approx 37\% V_{in}$ during pulse. V_{out} steps to $+10V$, decays to about 4V, steps down to $-6V$, then decays toward zero.

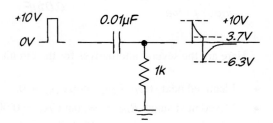

Figure 2W.14 Quick RC makes the exponential decays apparent.

---

[2] This is not a realistic case. A capacitor with no DC path to ground on the right side gradually would be charged by leakage currents. A practical circuit requires a resistive path to ground on the output to define the DC level at that terminal.

Non-ideal next stage: $R_{IN_X} = 100\text{k}$, $C_{IN_X} = 0.01\mu\text{F}$: Redraw as C–C divider driving R (this echoes our old Thevenin model, the R–R divider). As in the Thevenin model, the effective driving C is the two C values in parallel: $0.02\mu\text{F}$. So $RC = 2$ms. Again pulse width $\ll RC$, so pulse survives – but at half amplitude.

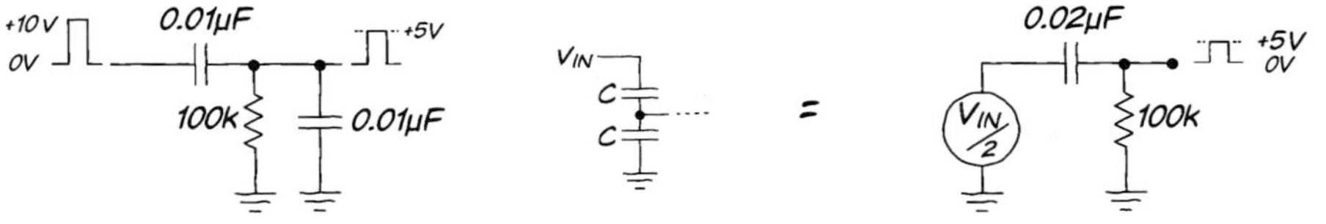

Figure 2W.15 A capacitive divider driving R: attenuates, but passes pulse.

3N Diode Circuits

Contents

3N.1	Overloaded filter: another reason to follow our $10\times$ loading rule	108
3N.2	Scope probe	109
	3N.2.1 Defective $10\times$ probe	110
3N.3	Inductors	112
3N.4	*LC* resonant circuit	113
	3N.4.1 Resonance	114
	3N.4.2 The meaning of Q	115
	3N.4.3 Testing whether Fourier was right	117
	3N.4.4 Response to a slow square wave: "ringing"	117
3N.5	Diode Circuits	118
3N.6	The most important diode application: DC from AC	119
	3N.6.1 Half-wave rectifier and clamp	119
	3N.6.2 Full-wave bridge rectifier	120
	3N.6.3 Zener diodes	122
3N.7	The most important diode application: (unregulated-) power supply	123
	3N.7.1 Transformer voltage	123
	3N.7.2 Capacitor	124
	3N.7.3 Transformer current rating (*rms* heating)	125
	3N.7.4 Fuse rating	125
3N.8	Radio!	126
	3N.8.1 Step 1: *LC* selects one "carrier" frequency	126
	3N.8.2 Step 2: detect "envelope" of AM waveform	127
	3N.8.3 What the result looks like	128
	3N.8.4 How to recover the AM information	129
3N.9	Readings in AoE	130

Why?

The sort of problem we mean to solve with the most important of today's circuits is the conversion of a sinusoidal power supply voltage – AC coming from the wall supply (often called "line" voltage) – to a constant DC level.

3N.1 Overloaded filter: another reason to follow our $10\times$ loading rule

Remember our claim that our *$10\times$ rule of thumb* would let us design circuit fragments? Let's confirm it by watching what happens to a filter when we violate the rule.

Suppose we have a low-pass filter, see Fig. 3N.1, designed to give us f_{3dB} a bit over 1kHz (this is a filter you built last time, you'll recall).

If R_{load} is around 150k or more, the load attenuates the signal only slightly, and f_{3dB} stays put. But

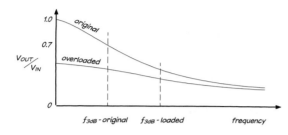

Figure 3N.1 *RC* low-pass: not loaded versus overloaded; redrawn to simplify.

what happens if we put $R_{\text{load}}=15\text{k}$? Attenuation is the lesser of the two bad effects. Look at what happens to $f_{3\text{dB}}$:

Figure 3N.2 Overloaded filter: we get something worse than attenuation: excessive loading shifts $f_{3\text{dB}}$.

The shift of $f_{3\text{dB}}$ from where it was designed to lie – in this case, a *doubling* of $f_{3\text{dB}}$ – is much more serious than the excessive attenuation at DC. The DC attenuation can be fixed by boosting circuit "gain" in a later stage; but the shifted $f_{3\text{dB}}$ cannot easily be fixed by doing something in a later stage. We hope this example will reinforce your faith in our "times-ten" rule – the rule that allows us to design circuit fragments as independent modules, confident that appending a new stage will not mess up performance of what he have designed.

3N.2 Scope probe

AoE Exercise 1.44

That mishap leads nicely into the problem of how to design a scope probe. We want to provide our scopes with higher input impedance.

You may be inclined to think we're awfully greedy, not to be content with the input impedance of a scope driven through a BNC cable: 1MΩ, in parallel with the capacitance of the cable. But the modest input impedance presented by cable and scope can be troublesome. It is particularly the stray capacitance that causes mischief.

"1MΩ" sounds nice enough – but consider what happens to the impedance of the cable as frequency climbs. Its capacitance to ground, along with the scope's input capacitance, adds up to about 100pF, if the cable is about three feet long. 100pF may still sound small – but at 1MHz, this is a heavy load:

$$X_C = |Z_C| = \frac{1}{(2\pi f C)} \approx \frac{1}{(6 \times 10^6 \times 0.1 \times 10^{-9})} = \frac{1}{(0.6 \times 10^{-3})} \approx 1.6\text{k}\Omega :$$

pretty low, and disastrously low when source impedance is high. We'll see this effect as we work our way to a good design for a probe. We'll first design the probe wrong.

The heavy capacitance of a bare BNC burdens the circuits you look at, and this burden can have effects even worse than just attenuation: it may make some of those circuits misbehave in strange ways. The most common misbehavior is a spontaneous oscillation.[1]

So, we nearly always use "10×" probes with a scope: that's a probe that makes the scope's input

---

[1] The problem of unintended *oscillation* – so-called "parasitic oscillation" – will concern us much, in later labs (it is the main topic of Lab 9L). It cannot occur until we add power-amplification to our circuitry. So, today, your circuits will not oscillate even when heavily loaded by a BNC cable. But in Lab 4L you may see exactly this misbehavior, since any of your

impedance 10× that of the bare scope. If the bare scope looks like 1MΩ in parallel with 100pF – cable and scope – the "10×" probe should look about 10 times better: 10MΩ in parallel with 10–12pF.

3N.2.1 Defective 10× probe

Figure 3N.3 shows a defective design for a 10× probe Do you see what's wrong? It works fine at DC. But try redrawing it as a Thevenin model driving a cap to ground, as in the example we did at the start of these notes. The flaw should appear. What is f_{3dB}? Quite low, since the effective R driving the stray capacitance ($R_{Thevenin}$) is large, almost 1M:

$$f_{3dB} = \frac{1}{2\pi RC} \approx \frac{1}{(2 \times 3 \times 10^6 \times 100 \times 10^{-12})}.$$

So

$$f_{3dB} \approx \frac{1}{0.6 \times 10^{-3}} \approx 1.6\text{kHz}.$$

That's a disaster for a scope designed to work up to many tens of MHz.

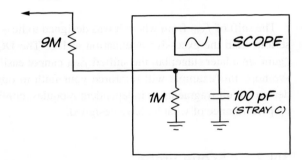

Figure 3N.3 Crummy 10× probe.

Remedy: We need to make sure our probe does *not* have this low-pass effect: scope and probe should treat alike all frequencies of interest (the upper limit is set by the scope's maximum frequency: for most in our lab that is up to 60 or even 100MHz).

The trick is just to build two voltage dividers in parallel: one resistive, the other capacitive: see Fig. 3N.4. At the two frequency extremes one or the other dominates (that is, passes most of the current); in between, they share. But if each delivers $V_{in}/10$, nothing complicated happens in this "in-between" range.

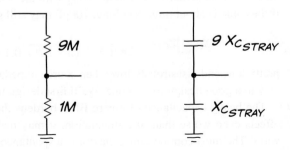

Figure 3N.4 Two dividers to deliver $V_{IN}/10$ to the scope.

What happens if we simply *join* the outputs of the two dividers? Do we have to analyze the resulting

transistor circuits can provide the amplification or "gain" necessary to sustain an oscillation: these parasitics are an interesting but troublesome novelty, coming soon.

composite circuit as one, fairly messy thing? No. No current flows along the line that joins the two dividers, so things remain utterly simple.

So, a good probe is just these two dividers joined, as in Fig. 3N.5 (where the "stray" C is assumed to sum scope and cable effects). Practical probes make the probe's added C adjustable. This adjustment raises a question: how do you know if the probe is properly adjusted, so that it treats all frequencies alike?

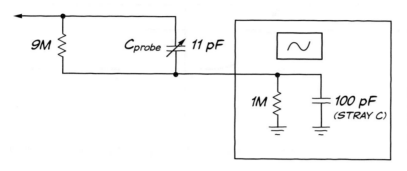

Figure 3N.5 A good 10× probe: one capacitor is trimmable, to allow use with scopes that differ in C_{IN}.

Probe "compensation": one way to check: sweep frequencies...

One way to check the frequency response of probe and scope, together, would be to sweep frequencies from DC to the top of the scope's range, and watch the amplitude the scope showed. In Fig. 3N.6 we've sketched what the response would look like for adjustments of $C_{\text{compensation}}$ that are wrong in each direction, and also correct.

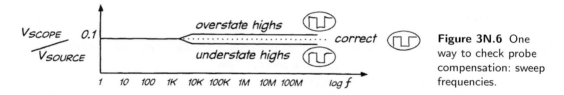

Figure 3N.6 One way to check probe compensation: sweep frequencies.

At low frequencies, the resistive divider dominates; at higher frequencies, above what we might call a "crossover frequency," the capacitive divider dominates. The two dividers ought each to deliver 1/10 amplitude. If the $C{:}C$ divider is wrongly adjusted, it delivers either more or less than that, as shown in Fig. 3N.6.[2]

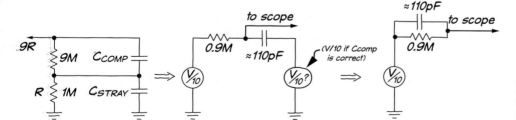

Figure 3N.7 Probe redrawn as equivalent RC network.

But this way of checking probe compensation is clumsy. It requires a good function generator, and would be a nuisance to set up each time you wanted to check a probe.

---

[2] The crossover frequency depends on the effective RC, where the circuit can be redrawn (modeled) as in Fig. 3N.7.

112 Diode Circuits

Probe "compensation": ... an easier way to check: Fourier, again

The easier way to do the same task is just to feed scope and probe a *square* wave, and then look to see whether the waveform looks square on the scope screen. If it does, good: all frequencies are treated alike. If it does not look square, just adjust the trimmable C in the probe until the waveform *does* look square.

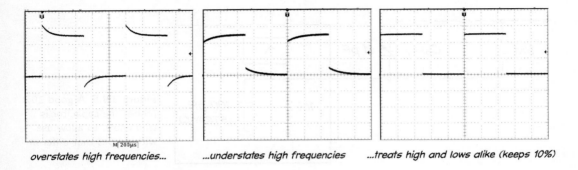

overstates high frequencies...   ...understates high frequencies   ...treats high and lows alike (keeps 10%)

Figure 3N.8 Using square wave to check frequency response of probe and scope.

Neat? This is so clearly the efficient way to check probe *compensation* (as the adjustment of the probe's C is called) that every respectable scope offers a square wave on its front panel. It's labeled something like *probe comp* or *probe adjust*. It's a small wave (of fixed amplitude) at around 1kHz.

3N.3 Inductors

In a physics course, inductors are treated with the same respect accorded capacitors, and the complementary behaviors of the two devices seem to support this view.

Here are the impedances of the two devices:

impedance of capacitor $Z_C = -j/(2\pi f C) = -j/(\omega C)$
impedance of inductor $Z_L = j(2\pi f L) = j\omega L$

The impedances of the two devices are complementary in two senses:

- the impedance of one device falls with frequency (true of Z_C), while the other rises with frequency (true of Z_L);
- the phase shifts between current and voltage are opposite for the two devices, as is indicated by the opposed signs of $-j$ and j. Current *leads* voltage in the capacitor; it *lags* voltage in the inductor.

It looks as if you could use L or C equally well to make a filter. Below, for example, are two versions of a low-pass.

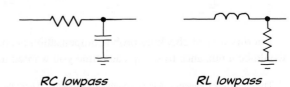

RC lowpass   RL lowpass

Figure 3N.9 In principle one can use *RL* or *RC* to form a filter....

Both versions of the low-pass filter work. But only the *RC* is practical, at all but very high frequencies.[3]

AoE §1.7.10

In this course – and in electronics generally – capacitors are used much more widely than inductors. The difference comes from the fact that inductors are relatively large and heavy (often including a core made of iron or another magnetically-permeable material), and that, owing to departures from ideal, they dissipate power.

In Fig. 3N.10 a self-righteous capacitor reminds us of this difference – and, like most of the self-righteous, he somewhat exaggerates his virtue.[4]

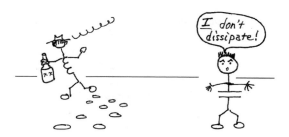

Figure 3N.10 Capacitors don't ordinarily dissipate much energy; inductors do.

That difference leads one to prefer capacitors and to avoid inductors altogether except at high frequencies (perhaps 1MHz, or more, where a small value of inductance is sufficient to do the job), or in power conversion circuits (where they are ubiquitous).

The resonant *RLC* circuit that we use in Lab 3L to select a radio broadcast frequency illustrates a typical case where inductors work well. The relatively high frequency permits use of an inductor of small value and small size.

AoE §6.2.4

In circuits running below a few megahertz, where ordinary *operational amplifiers* perform well, clever circuitry even permits capacitors to emulate the behavior of inductors, without bringing along their nasty properties.

3N.4 LC resonant circuit

AoE §1.7.14A

Figure 3N.11 shows Lab 3L's resonant *RLC* circuit. Before we acknowledge what's novel about this circuit – its *resonance* – let's take advantage of what we know of the impedances of capacitors and inductors to make a simple argument that this is, indeed, a *bandpass* filter: one that passes a range of intermediate frequencies, while attenuating both frequency extremes.

If we neglect the effect of the inductor – as in the lefthand circuit fragment in Fig. 3N.12 – we see a familiar *RC* low-pass. At high frequencies, it is *fair* to neglect the paralled inductor: its impedance is much larger than the impedance of the capacitor.

Toward the other end of the frequency range, the inductor's impedance is much less than that of the capacitor (recall that $Z_L = j\omega L = j2\pi f L$). In this frequency range, the *RL* forms a high-pass.

Over the full frequency range, then, the *RLC* forms a bandpass filter. So much, we can see without considering the *LC*'s resonance.

---

[3] "High" by the standards of this course: around 1MHz and up.
[4] In some circuits, such as switching power supplies, the capacitor's "equivalent series resistance" (ESR) can dissipate power and degrade performance. See AoE §§6.2.1A and 9.6.3A. In Lab 11L you will find that we needed to specify a low-ESR capacitor for use with the switching regulator.

114 Diode Circuits

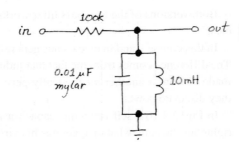

Figure 3N.11 Lab 3L's *RLC* circuit.

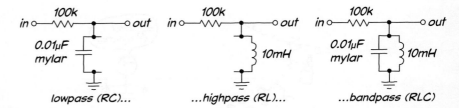

Figure 3N.12 A first approximation of the parallel-*RLC* circuit: highpass and lowpass.

3N.4.1 Resonance

AoE §1.7.14

But that rough analysis misses the interesting novelty of this circuit: the "resonance" to which we have referred. At some frequency – where the magnitude of the impedances of inductor and capacitor are equal – something startling occurs: the impedance of the parallel *LC* becomes very large.

It's not hard to persuade yourself that this happens, if you write out the formula for the impedance of the *LC* pair:

$$Z_{LC-\text{parallel}} = Z_C \parallel Z_L = \frac{Z_C \times Z_L}{Z_C + Z_L}.$$

But, because $Z_C = -j/\omega C$ while $Z_L = j\omega L$, the opposite signs of the "j" indicate that at some frequency, where the magnitudes are equal, the two impedances should sum to zero. When this occurs, taking the denominator of the parallel impedance to zero, the parallel impedance "blows up" – becomes very large.[5]

The *resonant frequency*, where $|Z_C| = |Z_L|$ occurs when $|-j/\omega C| = |j\omega L|$ and $1/\omega C = \omega L$:

$$\omega^2 = 1/LC \Rightarrow \omega = 1/\sqrt{LC} \quad \text{or} \quad f_{\text{resonance}} = 1/(2\pi\sqrt{LC}).$$

At this resonant frequency, where the impedance of the parallel *LC* becomes large, the circuit passes the largest-available fraction of its input. Ideally, that fraction would be 100%, but losses in the inductor mean that the maximum can be much less than 100% for large values of *R*. The *RLC* offers not just another way to make a bandpass; it permits making an extremely *narrow* passband, compared to what one can achieve with *RC*s.

The characteristic called "Q" describes just *how* narrow is the range of frequencies that are allowed to pass: $Q \equiv f_{\text{resonance}}/\Delta f$. Here Δf is the width at the amplitude that delivers half-power – the amplitude that is 3dB below the peak, as sketched in Fig. 3N.13.

The *RLC* does not work the way an *RC* filter does. It is more dynamic. It does more than form a frequency-selective voltage divider, although it does do that. It *stores energy*; it is more like a pendulum than like a coffee filter. The inductor–capacitor combination oscillates, once stimulated with a

---

[5] "Infinite," you may want to say. But *LC* imperfections – largely caused by resistance and core losses in the inductor – spoil this result. But it is enough that the impedance becomes very large, and is extremely sensitive to small frequency changes.

frequency close to its favorite – the frequency where it "resonates." In each cycle of oscillation, energy is transferred, back and forth, between inductor and capacitor.

When the *voltage* across the parallel pair is at a maximum, energy is stored in the electrostatic field between the plates of the capacitor; as the capacitor begins to discharge through the inductor, the current gradually grows, and when the *current* reaches a maximum, V_{CAP} is zero. At that point in the cycle, all the circuit's energy is stored in the *magnetic* field around the inductor. The energy sloshes back and forth between capacitor and inductor.[6]

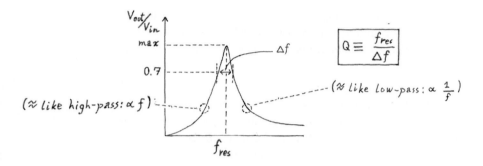

Figure 3N.13 Shape of *RLC* frequency response.

3N.4.2 The meaning of Q

Q measured using Δf: The lab asks you to estimate the circuit's *Q* ("quality factor," a term whose name apparently reflects the use of resonant circuits in radio tuners, where high selectivity is good).

Measured thus, *Q* is defined as follows, as we have said (Fig. 3N.13):

$$Q = \frac{f_{\text{resonance}}}{\Delta f},$$

where Δf" represents the width of the resonance peak between its "-3dB" points: see Fig. 3N.14.

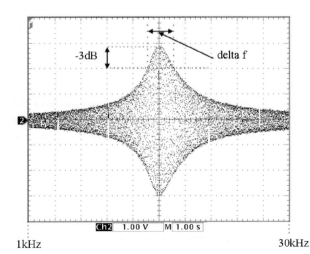

Figure 3N.14 *Q* measures the frequency-selectivity or *peakiness* of the resonant circuit.

---

[6] See Purcell & Morin, §8.1, and especially Fig. 8.3, showing a damped oscillation.

Diode Circuits

Q measured using decay time: Q can also be defined to describe how slowly energy leaks away from a resonant circuit, dissipated in its stray resistance:[7]

$$Q = \omega_0 \frac{\text{energy stored}}{\text{average power dissipated}}.$$

Or, less abstractly, Q can be defined as

$$Q = 2\pi \times \{\text{number of cycles required for the energy to diminish to } 1/e\}.$$

This latter definition isn't illuminating when you are looking at the frequency response of the *LC* circuit, as in the present exercise; this second definition will seem more helpful when you reach §3N.4.4, which looks at the *LC*'s response in the *time domain*. For the case of the *parallel LC*, one more rule is useful: $Q = \omega_0 RC$.

Modifying Q: Q describes how efficient the *RLC* is, and depends mostly upon the characteristics of the inductor. But you can modify Q with your choice of R. Large R produces higher Q, but lower amplitude out (the energy dissipated in the *LC* pair must be replaced; the larger the R feeding the *LC*, the larger the voltage drop across that R). So, one can trade off one virtue against another: high Q versus large amplitude.

If you're energetic and curious, you can look at the effect upon Q of substituting a 10k resistor for the 100k. Amplitude-out increases, while Q degrades. As usual, you're obliged to trade away one desirable trait to get another. But good Q is likely to be much more important than large amplitude; an amplifier can solve the problem of low amplitude.

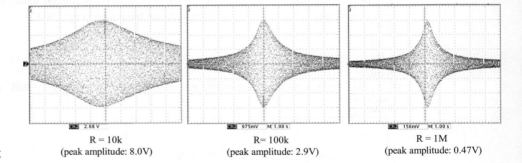

Figure 3N.15 Large R raises Q while diminishing output amplitude.

$R = 10k$ (peak amplitude: 8.0V) $R = 100k$ (peak amplitude: 2.9V) $R = 1M$ (peak amplitude: 0.47V)

Figure 3N.15 shows the effect of three R values in the *RLC* circuit: 10k, 100k and 1M. For the larger R values, Q does indeed improve – but at the expense of output amplitude. In Fig. 3N.15 we have adjusted scope gain so as to hold the *graphical* peak amplitude constant for the three images. We did this in order to make it easy to see the changes in the response *shape* (and Q). But note that the scope sensitivity increases from left to right (V/Div falls) – so that the amplitude out diminishes by more than a factor of 16. A smaller, secondary effect also appears, as R changes: the resonant frequency shifts slightly.[8]

In Lab 13L, where you are likely to use an *RLC* as part of an *FM demodulator*, you will take advantage of R's capacity to tailor the shape of the *RLC* response: and you will feel the tension between these two desirable characteristics, large amplitude versus high Q.

---

[7] See Purcell & Morin, equation 8.12.
[8] This is a "second order" for "light damping...," see Purcell, §8.1.

3N.4.3 Testing whether Fourier was right

The parallel *RLC* circuit allows you to test one of Fourier's claims – in case you doubted it. Fig. 3N.16 repeats a figure from today's lab, a sketch of how the Fourier series for a square wave begins.

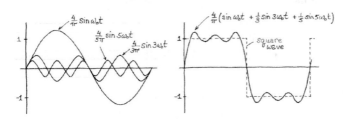

Figure 3N.16 Fourier series for square wave.

Less graphically, the Fourier series for a square wave of amplitude A is

$$\frac{4}{\pi}A(\sin(\omega t)) + \frac{1}{3}\sin(3\omega t)) + \frac{1}{5}\sin(5\omega t) + \cdots.$$

Your *LC* can pick out these frequency components for you: and the result, in the lab circuit, is a very pretty perspective-like display (the resemblance to receding telephone poles or fenceposts beside a highway is striking).

3N.4.4 Response to a slow square wave: "ringing"

A slow square wave whose frequency is not at all critical (Lab 3L suggests you might try 20Hz) stimulates the *LC*, even though 50Hz is very far from the *LC*'s resonant frequency of 16kHz. How come? Because, as you know, Fourier teaches that a square wave includes in its edges high-frequency "harmonics" close to the resonant frequency. These harmonics are small (since the Fourier series for a square wave falls off like $1/f$).

In Fig. 3N.17 is a figure showing the resonant circuit of Lab 3L driven by a square wave of a frequency far below $f_{\text{resonance}}$ (square wave frequency $\approx$50Hz).

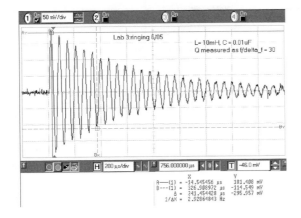

Figure 3N.17 Decay of *RLC*'s ringing can reveal Q.

Notice that it is the *energy*, not *amplitude* that should decay to $1/e$ at the rate thus related to Q. Since the energy stored in the capacitor is proportional to amplitude *squared* (energy$_{\text{cap}} = \frac{1}{2}CV^2$), we want to see V^2 – not V – fall to 37%, and V to about 60%.

In Fig. 3N.17 such a point is marked by a cursor, and it seems to occur after about 5 cycles. This would indicate a Q of about 30.

Why does it decay? A student recently asked "Why does it decay exponentially?" – a good question. The short answer is "because the rate at which it loses energy is proportional to its amplitude." A slightly longer version of that answer would note:

- energy in the LC circuit is proportional to V^2 (if you consider the time in a cycle when all the energy is in the capacitor, for example, at that time energy$_{\text{cap}} = \frac{1}{2}CV^2$: here V is V_{RMS}: we are not interested in the sign of V or in its instantaneous value);
- energy is dissipated in series resistance, R_{drive}, proportional to V^2/R_L (this is the power dissipated in the resistance).

Thus energy and rate of energy loss both are proportional to amplitude-squared. The *squared* factor only changes the time-constant, not the exponential shape of the decay.[9]

Poor grounding can evoke LC resonant waveforms

You will see such a response of an LC circuit to a step input whenever you happen to look at a square wave with an improperly grounded scope probe: when you fail to ground the probe close to the point you are probing, you force a ground current to flow through a long (inductive) path. Stray inductance and capacitance form a resonant circuit that produces ugly ringing. You might look for this effect now, if you are curious; or you might just wait for the day (almost sure to come) when you run into this effect inadvertently.

The scope images in Fig. 3N.18 show how shortening the ground return of a scope probe reduces ringing. The shorter ground return has less inductance, and pushes f_{resonant} higher while storing less energy, as well.

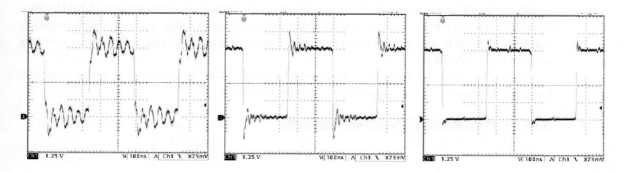

Figure 3N.18 Scope probe, poorly grounded, shows resonance of stray L and C. Left: no probe ground (except through scope power supply). Middle: probe grounded about 10″ from point probed. Right: probe grounded about 1/2″ from point probed.

3N.5 Diode Circuits

AoE §1.6.1

Diodes do a new and useful trick for us: they allow current to flow in one direction only: see Fig. 3N.19

Figure 3N.19 Diode as one-way current gate. $(\overrightarrow{yes})$ ——▶|—— $(\overleftarrow{no})$

---
[9] See R.E. Simpson, *Introductory Electronics*, (2d ed., 1984), pp. 98–100.

3N.6 The most important diode application: DC from AC

The symbol looks like a one-way street sign, and that's handy: it's telling conventional current which way to go. For many applications, it is enough to think of the diode as a one-way current valve, though you need to note also that when it conducts the diode doesn't behave quite like a wire. Instead, it shows a characteristic "diode drop" of about 0.6V. This you saw in Lab 1L; Fig. 3N.20 may remind you of the curve you drew on that first day (you did *not* see the breakdown at ≈ -100V, fortunately for your safety).

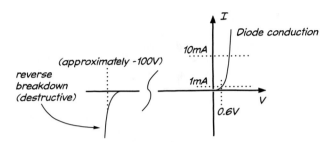

Figure 3N.20 Diode *I–V* curves: reverse current is in nanoamp range.

3N.6 The most important diode application: DC from AC

Electrical power is distributed in *AC* form rather than DC, in order to minimize power losses in transmission lines. Using AC permits transmission at high voltages, and correspondingly low currents. AC is easily "stepped" up or down using transformers – which are simply two windings placed close together so that they are "coupled" by their magnetic fields. In Worked Example 1W.2 you can find an example of the efficiency advantage that goes with high transmission voltage.

Because power is transmitted in AC form whereas every electronic device needs a DC power supply, making this transformation from AC to DC is an important mission of diode circuits.

3N.6.1 Half-wave rectifier and clamp

AoE §1.6.2

We often use diodes within voltage dividers, much the way we have used resistors, and then capacitors. Fig. 3N.21 shows a set of such dividers: what should the outputs look like?

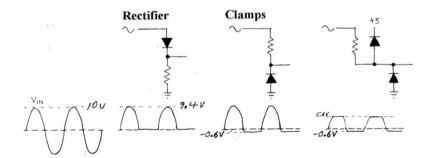

Figure 3N.21 Three dividers made with diodes: one rectifier, two clamps.

The outputs for the first two circuits look strikingly similar. Yet only the *rectifier* can used to generate the DC voltage needed in a power supply. Why? What important difference exists – invisible to the scope display – between the "rectifier" and the "clamp?"[10]

The bumpy output of the "rectifier" is still pretty far from what we need to make a power supply

---

[10] The answer is not the more obvious difference that one shows a 0.6V offset when passing the signal, whereas the other

– a circuit that converts the AC voltage that comes from the wall or "line" to a DC level. A capacitor will smooth the rectifier's output for us. But, first, let's look at a better version of the rectifier.

3N.6.2 Full-wave bridge rectifier

AoE §1.6.2

This clever circuit in Fig. 3N.22 gives a second bump out, on the *negative* swing of the input voltage.

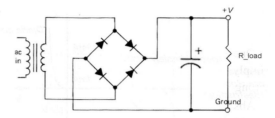

Figure 3N.22 Full-wave bridge.

Neat, isn't it? Once you have seen the circuit's output, you can see why the simpler rectifier is called *half-wave*.

Figure 3N.23 shows some details of the output of the full-wave rectifier, including the non-obvious fact that the bridge *input* drops *below* ground, on one half cycle.

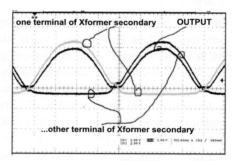

Figure 3N.23 Full-wave bridge rectifier.

Because of this wandering of the bridge input, the full-wave bridge can be used only with a "floating" source such as a transformer secondary. Neither of its inputs may be tied to ground, assuming that one of the two output terminals is to be defined as ground.

Rarely is there an excuse for using anything other than a full-wave bridge in a power supply, these days. Once upon a time, "diode" meant a $5 vacuum tube, and then designers tried to limit the number of these that they used. Now you just buy the bridge: a little epoxy package with four diodes and four legs; a big one may cost you a dollar.

Full-wave bridge in a "split supply"

A full-wave bridge can be used to generate a dual output: both positive and negative with respect to ground. Fig. 3N.24 shows such a circuit.

Evidently, the only difference from the single-supply of Fig. 3N.22 arises from use of the *center tap* on the transformer secondary, tying it to the point that we define as the *ground* of the split output. (We omit the load resistors from Fig. 3N.24; a load is, of course, assumed – if not, why build the supply?)

> does not; that difference would seem to favor the second circuit, the "clamp." No, what's important is the difference between the two *output impedances*. (Remember? We promised you that impedances would be a theme, in the analog part of this course.) The left-hand circuit – the rectifier – places only a conducting diode between source and load (the source usually is a power transformer). The clamp places a resistor between source and load. As a result, the rectifier provides the efficient way to build a power supply; the clamp does not.

3N.6 The most important diode application: DC from AC

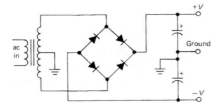

Figure 3N.24 Full-wave bridge used to generate both positive and negative ("split") outputs.

One might be tempted to think that the center tap is not necessary: that one could form a split supply simply by defining ground as the midpoint between the capacitors in Fig. 3N.24. That view would be wrong.

Absent use of the center-tap, the sharing of the total voltage, between positive and negative outputs would not be predictable; it would depend on the relative loading of the two outputs. The center-tap makes the two outputs independent, so that it is not just the total voltage, V_+ minus V_-, that the transformer determines. Instead, each of V_+ and V_- shows a voltage determined by the half-winding of the transformer.

If that proposition seems vague, and leaves you puzzled, you may want to trace the details of current flow on both half cycles of the line voltage, a flow sketched in Fig. 3N.25.

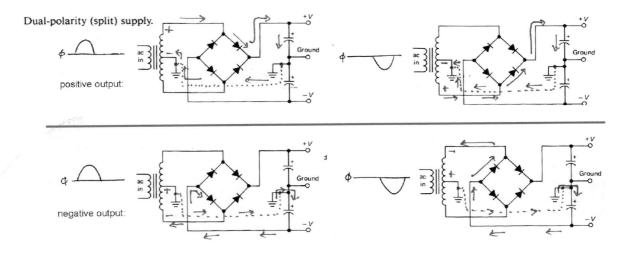

Figure 3N.25 Current flow in split-supply rectifier: positive and negative output flows shown separately.

Figure 3N.25 shows four cases: but there are, of course, only *two* half cycles to consider. We have done separate sketches for the positive and negative supplies, because we feared that thinking about both output polarities at once might be too much. You will notice, also, that we have drawn a dotted line between the two *ground* symbols, in order to make the return of current to the center-tap easier to envision.

In a minute, we'll proceed to looking at a full power supply circuit, which will include a full-wave bridge. But before we do, let's note one additional sort of diode: the zener.

3N.6.3 Zener diodes

You will not apply a simple zener in our labs (though you will meet a fancier *voltage reference* in Lab 11L). But we would like to mention the zener here alongside the other diodes that we'll meet in today's lab, 3L: the standard silicon diode (like the 1N914 we use repeatedly today) and the *Schottky diode*, a low-forward-drop device that you are likely to choose as you build the AM radio detector late in 3L.

AoE §1.2.6A

The zener diode conducts at some low *reverse* voltage, and *likes to*! If you put it into a voltage divider "backwards" – that is, "back-biased," with the current running the wrong way down the one-way street – you can form a circuit whose output voltage is pretty nearly constant despite variation at the input, and despite variation in loading.

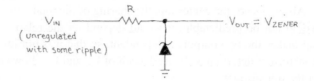

Figure 3N.26 Zener voltage source.

The plot of Fig. 3N.26 shows that the reverse conduction is not ideal: the curve is not vertical. The shape shows us why we need to follow AoE's rule of thumb that says "keep at least 10 mA flowing in the zener, even when the circuit is loaded."

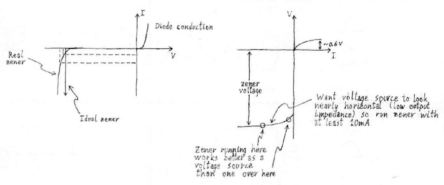

Figure 3N.27 Zener diode I–V curve: turned on its side, it reveals *impedances*.

Rotating the curve as in Fig. 3N.27 let's us see the curve's slope as a value in *Ohms*. The slope reveals $\Delta V/\Delta I$ – the device's *dynamic resistance*. That value describes how good a voltage reference the zener is: how much the output voltage will vary as current varies. (Here, current varies because of two effects: variation of *input* voltage, and variation in *output current*, or loading.) You can see how badly the diode would perform if you wandered up into the region of the curve where I_{zener} was very small.

A practical voltage source would never (well, hardly ever!) use a naked zener like the one we just showed you; it would always include a transistor or op-amp circuit after the zener, to limit the variation in *output current* (see AoE §2.2.4). The voltage regulators discussed on Day 11 (and in AoE Chapter 9) use such a scheme, though integrated references are used in preference to zeners. At least you should understand those circuits the better for having glimpsed a zener today. Now on to the diode application that you will see most often.

3N.7 The most important diode application: (unregulated-) power supply

Figure 3N.28 shows a standard unregulated power supply circuit. We'll learn later just what "unregulated" means: we can't understand fully until we meet regulated supplies, supplies that use negative feedback to hold V_{out} constant despite variations in both V_{in} and output current.

AoE §9.5

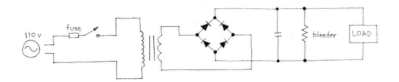

Figure 3N.28 AC-line-to-DC power supply circuit.

Now we'll look at a way to choose component values. In these notes, we will do the job incompletely, as if we were just sketching a supply. In Worked Example 3W you will find a similar case done more thoroughly. Assume we aim for the following specifications:

- V_{OUT}: about 12 volts
- Ripple: about 1 volt
- R_{load}: 120Ω

We must choose the following values:

- C size (μF)
- transformer voltage (V_{RMS})
- fuse rating

We will postpone until Worked Example 3W two less fundamental tasks (because often a ball-park guess will do). Those elements that we will omit here are:

- transformer current rating
- "bleeder" resistor value

Let's do this in stages. Surely you should begin by drawing the circuit without component values, as we did above.

3N.7.1 Transformer voltage

AoE §9.5.2

This is just V_{OUT} plus the voltage lost across the rectifier bridge. The bridge always puts two diodes in the path. Specify as *rms* voltage: for a sinewave, that means $V_{\text{PEAK}}/\sqrt{2}$. Here, that gives V_{peak} at the transformer of about $14V \approx 10V_{\text{RMS}}$: see Fig. 3N.29.

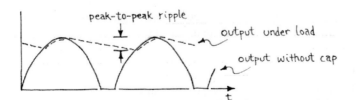

Figure 3N.29 Transformer voltage, given $V_{\text{OUT-Peak}}$.

Figure 3N.30 shows the surges of current that accompany the ripple. Current (labeled I_{xformer} in the figure) is measured on the ground line. (Note the high scope sensitivity on that trace: 50mV/div.)

Diode Circuits

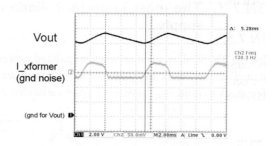

Figure 3N.30 Ripple on power supply: $I_{LOAD}=1A$, $C=3300\mu F$.

3N.7.2 Capacitor

This is the interesting part. This task could be hard, but we'll make it pretty easy by using two simplifying assumptions. Fig. 3N.31 shows what the "ripple" waveform will look like.

AoE §1.6.3A

AoE §9.5.3A

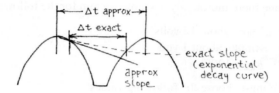

Figure 3N.31 Time details of ripple waveform.

We can estimate the ripple (or choose a C for a specified ripple, as in this problem) by using the general equation

$$I = C\frac{dV}{dt};$$

dV or ΔV is ripple; dt or ΔT is the time during which the cap discharges; I is the current taken out of the supply. To specify *exactly* the C that will allow a specified amount of ripple at the stated maximum load requires some thought. Specifically,

- What is I_{out}? If the load is resistive, the current out is not constant, but decays exponentially each half-cycle.
- What is ΔT? That is, for how long does the cap discharge (before the transformer voltage comes up again to charge it)?

We will, as usual, coolly sidestep these difficulties with some *worst-case* approximations:

- We assume I_{out} *is* constant at its maximum value;
- We assume the cap is discharged for the full time between peaks.

Both these approximations tend to overstate the ripple we will get. Since ripple is not a good thing, we don't mind building a circuit that delivers a bit less ripple than called for.

Try those approximations here:

- I_{out} is just $I_{out-max}$, i.e., 100mA
- ΔT is half the period of the 60Hz input waveform, i.e., $1/120Hz \approx 8ms$.

What cap size does this imply?

$$C = I \times \frac{\Delta T}{\Delta V} \approx 0.1 \times 8 \times \frac{10^{-3}}{1V} \approx 0.8 \times 10^{-3} F = 800\mu F.$$

That may sound big but it isn't for a power supply storage capacitor.

3N.7.3 Transformer current rating (*rms* heating)

Because the current that recharges the capacitor comes in surges, rather than continuously, the transformer heats more than one would assume if its current flowed steadily, as the *load* current does. (See Worked Example 3W on power supplies for more on this topic.) The extra heating can be calculated using the root-mean-square ("rms") value of the transformer current. Doing that is straightforward if you know the percentage of each cycle the transformer spends delivering current. That fraction is difficult to predict, but easy to observe.

Figure 3N.30 shows the ripple we saw on one of the lab's powered breadboards when we drew a steady 1A from the supply (here, we're digressing for a moment from the present design task, where the current is lower, at 100mA).

During the time when the transformer feeds the storage capacitor, all the charge that is removed during a cycle by the load must be replaced. So, the magnitude of the current that flows during the recharge portion of the full cycle – roughly 3/8 in Fig. 3N.30 – must be proportionally larger than the steady *load* current. In this case, it must be 8/3 the steady 1A output current. Such a current spike heats more than a steady current that would deliver the same charge, so we must use this rms current value to determine how large a transformer we need.

If, as in the scope image in Fig. 3N.30, the current flows during about 3/8 of the full cycle, the rms value is

$$I_{\text{rms}} = \sqrt{(\text{fraction of cycle during which current flows}) \times (\text{current})^2}$$
$$= \sqrt{3/8 \times (8I/3)^2} = \sqrt{3/8 \times 7.1I^2} = \sqrt{2.7I^2} \approx 1.6I.$$

So, for a load current of 100mA, you should specify a transformer that can handle at least 160mA. You'd not want to put your specification right at the edge, so you might specify 200mA.

3N.7.4 Fuse rating

The *current* in the primary is smaller than the current in the secondary, by about the same ratio as the primary *voltage* is larger. (This occurs because the transformer dissipates little power: $P_{\text{IN}} \approx P_{\text{OUT}}$).

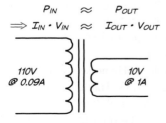

Figure 3N.32 Transformer preserves power, roughly; so, a 'step down' transformer draws less current than it puts out.

In the present case, where the *voltage* is stepped down about 11×, the primary current is lower than the secondary current by about the same factor. So, for 100mA out, about 9mA must flow in the primary.

In fact, the spikes of current that heated the transformer also heat the fuse, so the fuse feels its "9mA" as somewhat more (see §3N.7.3); call it 20mA. And we don't want the fuse to blow under the maximum current load. So, use a fuse that blows at perhaps four times the maximum I_{out}, adjusted for its heating effect: so 80mA. A 100mA, or 0.1A fuse is a pretty standard value, and we'd be content with that. The value is not critical, since the fuse is for emergencies, an event in which very large currents can be expected.

It's a good idea to use a *slow-blow* type, since on power-up a large initial current charges the filter capacitor, and we don't want the fuse to blow each time you turn on the supply. But note that it takes a pretty drastic overload to blow a fuse: a *fast-blow* fuse rated "1A" does not blow at 1A. The table below shows how long a small *fast-blow* glass fuse ("AGX"[11]) takes to blow, under several overloads:

% of Amp Rating	Time to Blow
110%	4 hours (min.)
135%	60 minutes (max)
200%	5 seconds (max)

3N.8 Radio!

In this section we'll meet the two elements essential to an AM ("amplitude modulation") radio receiver:

(1) a highly-selective bandpass filter that can pick out a single broadcast frequency; and
(2) a rectifier circuit that can *demodulate* the information encoded in the AM signal (with the help of an *RC* to knock out the high-frequency broadcast "carrier" signal).

3N.8.1 Step 1: *LC* selects one "carrier" frequency

Each broadcast station is alloted its peculiar frequency.[12] This is called a "carrier" frequency, chosen for its ability to propagate well, and conveying information by being varied or "modulated" in some way. The earliest and simplest of these modulation schemes, AM, does what the name suggests: varies the size of the carrier signal. It lets an audio signal in the range of a few kilohertz (talk or music) modulate a carrier of about 1MHz. (In Lab 13L you will use FM, a method that is more robust than AM but not quite so simple. The FM of Lab 13L will use not radio frequencies but a much lower-frequency carrier, easier to demodulate.)

An *LC* is well-adapted to the task of selecting one station's carrier: it is the most sensitively frequency-selective circuit we now know in the course, and can easily be set to resonate at standard AM frequencies, say 1MHz. One does not need to insert an *R* to feed the parallel *LC*; the antenna's impedance stands in for the *R* that we use in the first experiment of Lab 3L.

The *LC* not only rejects unwanted radio broadcasts, but also rejects the 60Hz noise that is a good deal larger than any of the radio signals. One can see this effect in Fig. 3N.33. The left-hand image shows the 60Hz sinusoid, strangely thick. The thickening is the ≈1MHz "carrier." It is hard to notice the variation in thickness – but this variation carries all the information. In the right-hand image, the 60Hz variation has disappeared, filtered out by the resonant circuit.

A second effect, more interesting and surprising, is visible in the right-hand image of Fig. 3N.33. The resonant *LC* is doing something intriguingly new. The output amplitude of the *LC* is *larger than the input*, for the selected carrier frequency (notice that the scope sensitivity is the same for the two screen shots of Fig. 3N.33: the *LC* really *does* enlarge the modulating signal).

---

[11] Source: Data sheet for Cooper–Bussmann AGX Series Fast-Acting Glass Tube Fuse, current ratings 1/16A to 2A.
[12] Strictly, each station is assigned not one frequency but a *range* of permitted frequencies: 10kHz wide, for AM broadcast, 200kHz wide for FM ("frequency modulation").

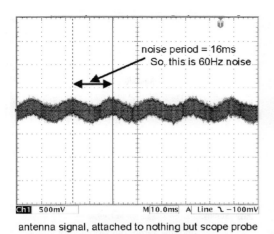 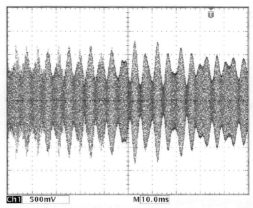

Figure 3N.33 Radio signal: raw, from antenna, and after *LC* frequency-selection.

That's a trick no *RC* could perform.[13] This is possible only when the input is more like a *current source* than *voltage source*, as is the case for the antenna. The *LC* stores energy, and – like a pendulum, or a child's swing pushed repeatedly by an attentive parent – the *LC* amplitude can keep growing beyond the voltage-amplitude of the driving signal. Similarly, the child's swing can move far beyond the limited range of the parent's push if the pushes are properly timed, adding a little energy each time the child passes.

Here's another way to say why output can grow larger than input (this account is somewhat less metaphorical): if Q is large – as it is, here – then a good deal more energy is put in than is lost in a cycle when amplitude out equals amplitude in. So the amplitude grows until the rate at which energy is dissipated reaches the rate at which it's coming in. That happens when the amplitude is a good deal larger than the original input level. Hence the enlargement of the signal by this *passive* circuit ("passive" meaning 'no borrowing of power from a supply, as in an amplifier;' all the energy comes from the signal source).

3N.8.2 Step 2: detect "envelope" of AM waveform

If we sweep the scope fast enough to resolve the carrier frequency, we see its amplitude modulation, which looks like strange uncertainty about amplitude: see Fig. 3N.34.

The left-hand image in Fig. 3N.34 shows varying amplitudes superimposed; the right-hand image shows a single-shot trace. Now, the task is to recover the information that lies in the amplitude variation, while ignoring the high-frequency carrier.

Remove the carrier: The detector should ignore or remove the "carrier" frequency (around 1MHz). The carrier can be stripped away, now that it has done its job, which was to deliver the audio information. The carrier is placed at a high frequency (around 1.5MHz in the images) because such signals travel much better than lower frequency signals.

---

[13] It's also a trick the *RLC* circuit of Lab 3L cannot perform. That circuit, driving the *LC* from a voltage source through a resistor, can never deliver amplitude greater than the input amplitude. This is because the "coupling" of source to *LC* through a resistor is "lossy" (in the jargon), whereas the coupling from antenna to *LC* in the radio circuit is not: it is capacitive, instead.

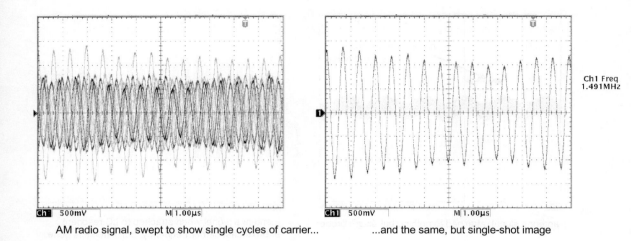

AM radio signal, swept to show single cycles of carrier...   ...and the same, but single-shot image

Figure 3N.34 Radio carrier resolved, showing its amplitude variation.

Two concerns call for use of a radio *carrier*. One strong reason to use a carrier, rather than try to transmit the audio itself as a radio signal, is the antenna size that audio frequency signals would require. The 1MHz AM signal is most efficiently received by an antenna a half-wavelength long: about 500 feet; we get by with a less-efficient antenna about 1/5 that long in our lab; if we were to transmit 1kHz audio as a radio signal, an optimal antenna (half-wavelength) would be about 150 *kilometers* long! Cellphones, incidentally, can use very small antennas, thanks to their high carrier frequencies: most GSM phones, for example (TMobile and AT&T) use about 2GHz, calling for an antenna about three inches long.[14]

Another powerful reason for using a carrier is the need to permit multiple broadcasts in a locality without mutual interference. That would not be possible if one simply transmitted audio frequencies without a carrier.

Removing the carrier sounds like a task for a lowpass filter. Well, it is and it isn't. A simple low-pass – meant to keep the "audio" signal (no more than 5kHz, under the AM standard) while knocking out the carrier – would unfortunately get rid of everything: baby would exit with bathwater. So, before we apply a lowpass, we need another of Lab 3L's circuits ....

Detect the envelope: A lowpass alone would fail: if applied to the signal arriving from the *LC*, a lowpass would get rid of the carrier – and with it, all its amplitude variations – since the voltage swings of the carrier are *symmetric*. The integrating effect of the lowpass delivers a nicely-centered *zero* as output! What's needed, to prevent this sad result, is a way to *spoil the symmetry* of the *LC*'s output. If we rectify the *LC* output before applying the lowpass, we find that the shape of the envelope survives.

3N.8.3 What the result looks like

Conceptually, the *rectification* of the modulated carrier, and then the *lowpass* elimination of the carrier are successive steps. The envelope, embodying the audio information, appears at last in the bottom

---

[14] A very good article on antennas (not very mathematical, published 2002), appears on the website of EDN Magazine: http://m.eet.com/media/1141963/21352-82250.pdf.

trace on the right in Fig. 3N.35. (Note that this *envelope* comes from the carrier shown above it, not from the left-hand scope image of Fig. 3N.35, which records a different moment in the radio broadcast.)

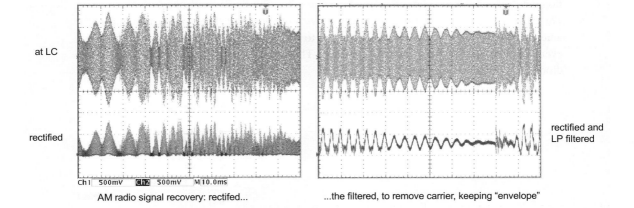

Figure 3N.35 AM signal recovery stages: rectification, then low-pass filtering.

3N.8.4 How to recover the AM information

Half-wave or full-wave?

- *Half-wave is enough . . .* Having met the *full-wave* rectifier as an improvement on the *half-wave*, for use in power supplies, you could quite reasonably suppose that the full-wave is needed for your radio. You might even fear that a half-wave could lose half the information conveyed by the AM signal – losing the excursions below ground, keeping only those above ground. But this turns out not to be the case. The symmetry of the carrier waveform means that we lose nothing by measuring its positive excursions only: we get the amplitude information. If you worry that the carrier may change amplitude between the positive and negative swings (indeed, *must* change amplitude, sometimes), don't worry: our detector will average the amplitudes of *hundreds* of cycles of the carrier, since one period of the highest audio signal (5kHz) will encompass about 200 cycles of a 1MHz carrier.
- *. . . half-wave is better than full-wave* In fact, there are good reasons to shun the full-wave rectifier, in this context:
  - *full-wave rectifier's input and output cannot share a single definition of ground*. If the output is referenced to ground (the usual scheme), then neither input terminal can be grounded. This truth, mentioned in §3N.6.2 and illustrated in Fig. 3N.22, was the basis for a warning that appears near the start of §3L.3.
    In a power supply derived from the output of a transformer, this restriction is not troublesome; the transformer winding needs no reference to ground. And if one looks closely at input and output waveforms (while heeding the warning not to ground the input), one sees each input going a diode drop *negative* on the half cycle when it does not squirt positive charge into the output. We saw this negative excursion in Fig. 3N.22.
    This property makes the bridge very inconvenient in the radio: either input or output must give up its connection to "ground." The antenna likes a connection to world ground; output doesn't absolutely need it, but the lack of it is very inconvenient. If you had to "float" your

output ground, you could not use the scope to watch V_{OUT} in the usual way, because the scope ground is internally tied to world ground. After grounding one of the input terminals of the bridge, you could take the trouble to use the scope in *differential* mode to watch V_{OUT}, by attaching *two* probes to the output: one to the normal V_{OUT} terminal, the other to your floating "ground," then subtracting one from the other. But that's a nuisance.

– *The full-wave's two-diode drops can lose too much of the signal*. The signal coming from the antenna is only about a volt, so it's better not to lose *two* diode drops from its amplitude – even when using low-drop Schottky diodes.[15] The simple half-wave rectifier loses just one diode drop voltage.

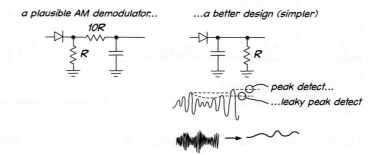

Figure 3N.36 Leaky peak detector is simpler than rectifier followed by lowpass.

Use a strange low-pass with the rectifier

The straightforward way to get rid of carrier and detect the envelope certainly seems to be to build a rectifier as usual (using a diode and an R to ground), then apply a low-pass (using a $10R$ resistor value, in order not to load the rectifier). That would work, but a simpler and sufficient circuit does the job with a single R. The diode feeds a cap, whose other end is tied to ground; diode and cap form a "peak detector," charging the cap to the highest input voltage (less a diode drop). Then, to allow the peak detector to droop when amplitude falls, we provide an R to ground (paralleling the cap), forming what we call a "leaky peak detector," as in Fig. 3N.36.

This circuit is hard to analyze in the frequency domain, but very easy in the time domain: we want the peak detector's decay-rate to be slow relative to the quick wiggles of the 1MHz carrier, but quick relative to the wiggles of the audio waveform. Those two frequencies – carrier versus audio – are so far apart that it is very easy to find an RC time-constant somewhere between. In fact, the RC value is not very critical. In lab, you may want to experiment, listening to determine how sensitive your radio's sound output is to your detector's RC value.

3N.9 Readings in AoE

Reading:
   Finish Chapter 1, including §§1.5–1.6.8; and §1.7.2 on inductors, transformers and diodes – sections omitted last time.
   Appendix on drawing schematics.

   Appendix I: *Television: A Compact Tutorial*.

Problems:
   Problems in text and Additional Exercises 1.43, 1.44.

---

[15] Forward voltage drop for the 1N5711 Schottky diode that we recommend for the lab radio is 0.4V @ 1mA.

3L Lab: Diode Circuits

3L.1 LC resonant circuit

3L.1.1 Response to sinusoid

Construct the parallel resonant circuit shown in Fig. 3L.1. Drive it with a sinewave, varying the frequency through a range that includes what you calculate to be the circuit's *resonant frequency*. Compare the resonant frequency that you observe with the one you calculated. (The circuit attenuates the signal considerably, even at its resonant frequency; the L is not perfectly efficient, but instead includes some series resistance.) As you watch for $f_{\text{resonance}}$ watch not for the frequency that delivers maximum amplitude out (this is difficult to determine), but instead watch for the other signature of resonance: the frequency where output is *in phase* with input.

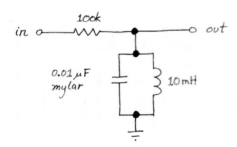

Figure 3L.1 *LC* parallel resonant circuit.

3L.1.2 Q: quality factor

Estimate the circuit's Q ("quality factor," a term whose name apparently reflects the use of resonant circuits in radio tuners, where high selectivity is good): Q is defined as

$$Q \equiv \frac{f_{\text{resonance}}}{\Delta f},$$

where "Δf" represents the width of the resonance peak between its "half-power" or -3dB points. The skinnier the peak, the higher the Q; see Fig. 3L.2.

You can make a very *good* measurement of Q if you use a *frequency counter*[1] to reveal the small change in frequency between the points, below and above $f_{\text{resonance}}$, where output amplitude is down 3dB. Note that this is "down 3dB" relative not to amplitude *in* but relative to the maximum amplitude *out*: amplitude at resonance. Amplitude out never equals amplitude in, which it would if our components were perfect.

---

[1] Your DVM may include such a frequency counter. This will be necessary for those using analog oscilloscopes. Or, if you are using a digital scope, it will measure the frequency for you.

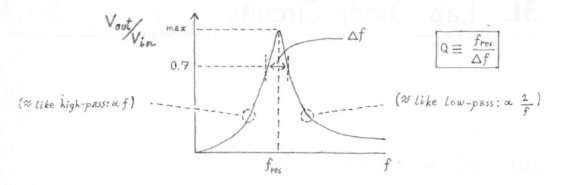

Figure 3L.2 You can measure Q, getting a precise Δf measure with a DVM.

If you're energetic and curious, you can look at the effect on Q of substituting a 10k resistor for the 100k. You'll notice that the amplitude out increases, with this reduced R. The fraction that survives, V_{OUT}/V_{IN}, grows. But as amplitude out grows, Q degrades. As usual, you're obliged to trade away one desirable trait to get another. Good Q, however, is likely to be much more important than large amplitude; an amplifier can solve the problem of low amplitude.

3L.1.3 A pretty sweep

Use the function-generator's *sweep* feature to show you a scope display of amplitude-out versus frequency. (See §2S.3 if you need some advice on how to do this trick.)

When you succeed in getting such a display of frequency response, try to explain why the display grows funny wiggles on one side of resonance as you increase the sweep rate. *Clue:* the funny wiggles appear on the side after the circuit has already been driven into resonant oscillation; the function generator there is driving an oscillating circuit.

3L.1.4 Finding Fourier components of a square wave

This resonant circuit can serve as a "Fourier Analyzer:" the circuit's response measures the amount of 16kHz (approx.) present in an input waveform.

Try driving the circuit with a *square* wave at the resonant frequency; note the amplitude of the (sinewave) response. Now gradually lower the driving frequency until you get another peak response (it should occur at 1/3 the resonant frequency) and check the amplitude (it should be 1/3 the amplitude of the fundamental response). With some care you can verify the amplitude and frequency of the first five or six terms of the Fourier series.

Figure 3L.3 shows the first few frequencies in the Fourier series for a square wave. You met this series earlier today as Fig. 3N.16.

3L.1.5 Classier: frequency spectrum display

If you *sweep* the *square wave* input to your 16kHz detector, you get a sort of inverse frequency spectrum: you should see a big bump at $f_{resonance}$, a smaller bump at $\frac{1}{3} f_{resonance}$, and so on.

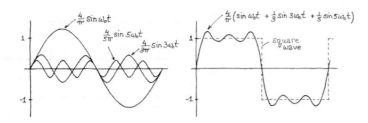

Figure 3L.3 Fourier series for square wave.

3L.1.6 Ringing

Now try driving the circuit with a low-frequency square wave: try 20Hz (but note that any low frequency will do; what matters is the steep edge of the waveform with its high-frequency components. Or, if you prefer, think of the edge as putting a jolt of energy into the resonant circuit, energy that sloshes back and forth between L and C until it has been dissipated). You should see a brief output in response to each edge of the input square wave. If you look closely at this output, you can see that it is a decaying sinewave. (If you find the display dim, increase the square wave frequency to around 100Hz.)

What is the frequency of this sinewave? (No surprise, here.)

Does it appear to decay exponentially? You may recall that we noted this behavior back in §3N.4.4.

Estimating Q from decay envelope: As we said in §3L.1.2, one can evaluate Q by noticing how fast the oscillation *envelope* decays away, after a stimulus to the *RLC*. Specifically (to quote ourselves),

$$Q = 2\pi \times (N_{\text{decay}}) \;,$$

where N_{decay} is the number of cycles required for energy to diminish to $1/e$ – where amplitude is down to about 60%, since energy is proportional to V^2 (see Chapter 3N).

When you see your own circuit's response to the slow square wave, count the cycles before decay to about 60%, and see if your Q, so calculated, matches the value you found by the method of §3L.1.2. See if you can confrm the claim we made in §3N.4.4, and illustrated with Fig. 3N.17, that poor probe grounding will produce ringing. While watching a square wave, remove the probe's ground clip. Satisfactorily ugly?

3L.2 Half-wave rectifier

Construct a half-wave rectifier circuit with a 6.3Vac (rms) transformer[2] and a 1N914 diode, as in Fig. 3L.4. Connect a 2.2k load, and look at the output on the scope. Is it what you expect? Polarity? Why is V_{peak} more than 6.3V? (Don't be troubled if V_{peak} is even a bit more than 6.3V$\times \sqrt{2}$: the transformer designers want to make sure your power supply gets at least what's advertised, even under heavy load; you're here loading it very lightly.)

---

[2] You might well wonder where such a weird voltage came from: "6.3V?" The answer comes from the history of vacuum tube radios. The tubes had filaments that needed to be heated in order to emit electrons. In early days they were heated by batteries: often by three lead–acid cells in series, each with open-circuit voltage of 2.1V. When AC supplies were adopted to take the place of the battery, their *rms* voltages were set to match the heating effect of the battery. Note that the 6.3V rms AC waveform has a peak value $\sqrt{2} \times V_{rms}$ and provides the same power as a DC supply at the sinusoid's *rms* value.

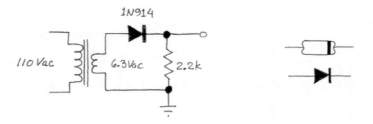

Figure 3L.4 Half-wave rectifier.

3L.3 Full-wave bridge rectifier

Now construct a full-wave bridge circuit, as in Fig. 3L.5. Be careful about polarities: the band on the diode indicates cathode, as in the figure. Look at the output waveform (but **don't** attempt to look at the input – the signal across the transformer's secondary – with the scope's other channel at the same time; this would require connecting the second "ground" lead of the scope to one side of the secondary. What disaster would that cause?[3]). Does it make sense? Why is the peak amplitude less than in the last circuit? How much should it be? What would happen if you were to reverse any one of the four diodes? (**Don't try it!**).

Don't be too gravely alarmed if you find yourself burning out diodes in this experiment. When a diode fails, does it usually fail *open* or *closed*? Do you see why diodes in this circuit usually fail in pairs – in a touching sort of suicide pact?[4]

Look at the region of the output waveform that is near zero volts. Why are there flat regions? Measure their duration, and explain.

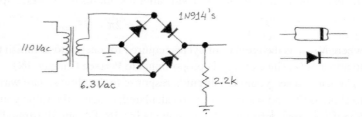

Figure 3L.5 Full-wave bridge.

3L.3.1 Ripple

Observe ripple, given C and load: Now connect a 15μF filter capacitor across the output (**Important**: observe polarity). Does the output make sense? Calculate what the "ripple" amplitude should be, then measure it. Does it agree? (If not, have you assumed the wrong discharge time, by a factor of 2?)

3L.3.2 Design exercise: choose C for acceptable ripple

Now suppose you want to let your power supply provide a current of up to 20mA with ripple of about 1V peak-to-peak. Your design task is to do the following:

---

[3] The disaster results from the consequent *shorting out* of one of the diodes. (Scrutinize Fig. 3L.5 to see why.) As a result, nothing limits the current in a second diode, which is guaranteed to burn out.

[4] This suicide pact makes troubleshooting the full-wave bridge interesting, in the case the output waveform looks wrong. Disconnect the transformer so that you can test the diodes (using a DVM's "diode test" function) and don't stop on detecting a single bad diode. If you find one dead diode, expect to find one more as well.

- choose R_{load} so as to draw about 20mA (peak)
- choose C so as to allow ripple of about 1V

Draw your design and try your circuit. Is the ripple about right? (Explain, to your own satisfaction, any deviation from what you expected.) This circuit is now a respectable voltage source, for loads of low current. To make a "power supply" of higher current capability, you'd use heftier diodes (e.g., 1N4002) and a larger capacitor. (In practice you would always follow the power supply with an active *regulator*, a circuit you will meet in Lab 11L.)

3L.4 Design exercise: AM radio receiver (fun!)

To make this exercise fun, you'll need a strong source of radio signals: that requires a pretty good antenna (or a poor antenna whose signal has been amplified for you by someone who knows how to make a high-frequency amplifier).

We get a strong signal in our teaching lab by running about 30 feet of wire from the window of the lab to a fire escape on the next building. The antenna is nothing fancy: just an old piece of wire, insulated from the fire escape by a piece of string. It gives us almost a volt in amplitude.

If you looked (with a scope) at the signal coming from the antenna, it would look something like the image on the left in Fig. 3N.33 on page 127. After selection with a resonant circuit, the signal would look like the scope image on the right side of that figure. The 60Hz noise is gone; that much is apparent. You cannot see a second benefit: the "carriers" of other radio stations also have been eliminated from the muddle of frequencies that came in on the antenna.

3L.4.1 A small but remarkable point

As we noticed in §3N.8.1, the resonant circuit not only selects a carrier frequency from among many; in addition, it makes the amplitude of the carrier that is selected much *larger* than in the original signal that came in the window.

3L.4.2 Detecting the AM signal

In order to detect an AM radio signal, you need to do two tasks – and you already know how to do them:

(1) rectify the signal (use a *Schottky* diode: 1N5711 or similar; its low forward-voltage will let it rectify a signal of just a few tenths of a volt;
(2) then low-pass filter this rectified signal.

The output of your circuit will be a small audio signal (much less than a volt). It may be audible on old-fashioned high-impedance earphones; it can be made audible on an ordinary 8-ohm speaker if someone provides you with an audio amplifier with a gain of 20 or so (an LM386 audio amplifier works fine). You will recognize that we have offered you only a strategy, not part values. We said "select the carrier," but didn't say how. We said "rectify," but did not suggest a value for the R to ground. We said "low-pass filter," but did not suggest $f_{3\text{dB}}$. So, we have left to you some hard – and interesting – parts of the job. Here are some suggestions:

- to detect the carrier, use an *LC* circuit like the one you built at the start of this lab, but showing the following differences:
  - the resonant frequency should be around 1MHz;

- you need no upper resistor in the "divider:" the antenna can drive the *LC* directly.
• The value of the resistor to ground is not critical; try 10k;
• the low-pass filter's job is to kill the carrier, keep the audio. Fortunately, these two frequencies are *very* far apart; so, you have a lot of freedom in placing f_{3dB}. The *form* of the low-pass you design may strike you as odd (though this depends on the way you chose to do the task: the "odd" configuration, described in §3N.8.4, Fig. 3N.36, uses the rectifier's resistor to ground as the *R* in the *RC* low-pass). Just make sure to put this in *time-domain* terms: that *RC* is very long relative to the period of the 1MHz carrier," but short relative to the "signal" or "audio" period.

The reward we hope you will get is, of course, to *hear* the radio signal. (You will probably need to experiment with your *LC* circuit by tacking in small additional caps in parallel, in order to select a particular station (or you can cheat by doing what everyone else who ever built a radio does: use a variable capacitor!) If you lack both high-impedance earphones and an audio amplifier, then at least *look* at the fruits of your labor on the scope.

You should see something more or less like the AM waveforms we showed in Chapter 3N. We hope you will look, with scope, first at the "raw" unfiltered mixture of signals coming off the antenna (before connecting this to the parallel *LC*), then at the selected carrier (selected by the resonant circuit), then at the rectified audio plus carrier, and finally the filtered and rectified output.

3L.5 Signal diodes

3L.5.1 Rectified differentiator

Use a diode to make a rectified differentiator, as in Fig. 3L.6. Drive it with a square wave at 10kHz or so, at the function generator's maximum output amplitude. Look at input and output, using both scope channels. Does it make sense? What does the 2.2k load resistor do? Try removing it.

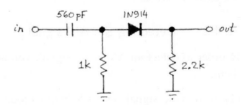

Figure 3L.6 Rectified differentiator.

Hint: You should see what appear to be *RC* discharge curves in both cases – with and without the 2.2k to ground. The challenge here is to figure out what determines the *R* and *C* that you are watching – and this problem is quite *subtle*![5]

3L.5.2 Diode clamps

Construct the simple diode clamp circuit shown in Fig. 3L.7. Drive it with a sinewave from your function generator, at maximum output amplitude, and observe the output. If you can see that the clamped voltage is not quite flat, then you can see the effect of the diode's non-zero impedance.

AoE 1.6.6C

---

[5] When you remove the 2.2k to ground, the scope probe itself becomes important. The discharge path now is extremely slow: the probe's capacitance (perhaps 12pF) discharges through the probe's 10M to ground. At a high input repetition rate, a square wave may produce what looks like a flat output.

3L.5 Signal diodes

Perhaps you can estimate a value for this *dynamic resistance* (see §3N.6.3); try a triangle waveform if you attempt this estimate.

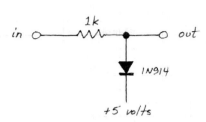

Figure 3L.7 Diode clamp.

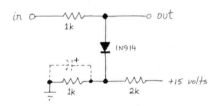

Figure 3L.8 Clamp with voltage divider reference.

AoE §1.6.6, Fig. 1.76

Now try using a voltage divider as the clamping voltage, as shown in Fig. 3L.8. Drive the circuit with a large sinewave, and examine the peak of the output waveform. Why is it rounded so much? (*Hint*: What is the impedance of the "voltage source" provided by the voltage divider? If you are puzzled, try drawing a Thevenin model for the whole circuit. Incidentally, this circuit is probably best analyzed in the *time* domain.) To check your explanation, drive the circuit with a triangle wave.

As a remedy, try adding a $15\mu F$ capacitor, as shown with dotted lines (note polarity). Try it out. Explain to your satisfaction why it works. (Here, you might use either a time- or frequency-domain argument.) This case illustrates well the concept of a bypass capacitor. What is it bypassing, and why?

3S Supplementary Notes and Jargon: Diode Circuits

3S.1 A puzzle: why LC's ringing dies away despite Fourier

A student asked a good, hard question, recently. I was stumped, till the answer struck me – more or less the way the apple is said to have bonked Newton on the head – next morning as I was pedalling to work.

3S.1.1 The puzzle

A low-frequency square wave (say, 50Hz) evokes a brief shiver from an *LC* circuit resonant at 16kHz (the circuit we built in Lab 3L). We explain this response by noticing that the square wave has a harmonic at 16kHz, though its amplitude is small. Our student put the puzzle this way: "Fourier says the 16kHz component is present not just at the edges, but throughout the waveform. Why, then, is the resonant circuit not stimulated continuously? Why only at the edges?"

A good question.

3S.1.2 The solution

The key point is this: the resonant circuit is responsive to *a range of frequencies*, not just to 16kHz. A first (and wrong-) inference might be, 'Oh: then we should see still more activity between the edges, stimulated by several harmonics.' That inference is wrong because *it's only at the edges* that the phases of all the several harmonics coincide, reinforcing one another. Away from the edges, they tend to cancel: and that is why (of course? well, none of this seems obvious to me) the square wave is *flat* between its edges: the several harmonics conspire to cancel one another, once the edge-step has occurred.

The resonant circuit passes a *set* of harmonics, not just one; the members of this set cancel one another, away from the edges. This happens because the circuit's Q is wide enough to let through this set of frequencies.

The limited-Q explanation finds confirmation in the *time domain*, as well. The ringing after a square-wave's edge decays at a rate described by Q: high Q implies slow decay (Q, in this view, measures energy loss per cycle of oscillation).

And you saw something like this earlier in the Lab 3L exercise When the square wave frequency is *not* very far below $f_{resonance}$ (say, at 1/3 or 1/5 of $f_{resonance}$), the "ringing" – the 16kHz harmonic – *does* persist all the way through, as the square wave sits high or low. In Fig. 3S.1 are images from Lab 3L showing the expected components of a square wave at 3, 7 and 9 times the frequency of the square wave. In all these cases, the fourier component persists across the "flat" section of the square wave. Why does it persist?

In the *time domain* the answer seems obvious: it hasn't had time to die away. But this persistence

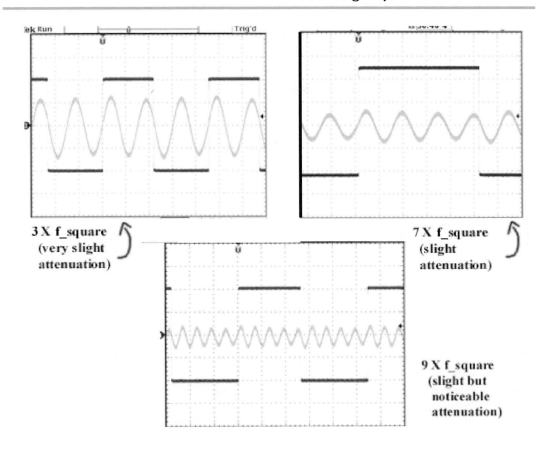

Figure 3S.1 Fourier components are sustained if f-square is not far from f-resonance.

also is (of course?) consistent with the frequency-domain view: the nearest *other* harmonics are far enough from $f_{\text{resonance}}$ so that they are knocked out.

In other words, we can explain this case (where the ringing *does* persist), as we could explain the case where the ringing died away, by referring to the effect of limited Q, in both time and frequency domains: in this case, the Q is high enough so that:

- in time domain: the ringing does not attenuate much, in the available time (the half-period of the square wave);
- in frequency domain: the circuit is selective enough to keep the 16kHz harmonic while killing any neighboring harmonics that would have cancelled the 16kHz, away from the edges.

And in Fig. 3S.2 is the case that evoked the original question, where f_{square} is far below $f_{\text{resonance}}$, and the ringing dies away.

3S.2   Jargon: passive devices

choke Inductor (noun).
droop Fall of voltage as effect of loading (loading implies drawing of current); also, waveform distortion from effect of high-pass filter, applied, say, to low-frequency square wave.

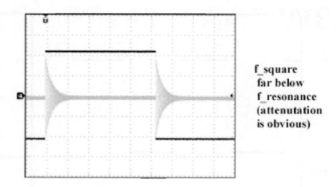

Figure 3S.2 Ringing dies away, for slow square wave, of course.

primary Input winding of transformer.

Q "quality factor" describes the sharpness of the peak of a frequency-selective *RLC* circuit:

$$Q = \frac{f_{\text{resonance}}}{\{\text{3dB width}\}},$$

or, equivalently, Q can be defined as time for resonant oscillation to decay:

$$Q = \text{number of radians for energy to decay to } 1/e \text{ of its peak value.}$$

ripple Variation of voltage resulting from partial discharge of power-supply filter capacitor between re-chargings by transformer.

risetime Time for waveform to rise from 10% of final value to 90%.

rms Root mean square (more literally, "root mean of squares"). Used to describe power delivered by time-varying waveform. For sine, $V_{\text{rms}} = V_{\text{peak}}/\sqrt{2}$. This is the DC voltage that would deliver the same *power* as the time-varying waveform.

secondary Output winding of transformer.

stiff When applied to a voltage source it means it "droops" little under load.

V_{peak} "Amplitude." For example in $v(t) = A \sin \omega t$, the term A is peak voltage (see AoE §1.3.1).

$V_{\text{peak-to-peak}}$ or $V_{\text{p-p}}$ Another way to characterize the size of a waveform; much less common than V_{peak}.

3W Worked Examples: Diode Circuits

3W.1 Power supply design

3W.1.1 Another power supply

Here we will do a problem much like the one we did more sketchily in Chapter 3N. If you are comfortable with the design process, skip to §§3W.1.6 and 3W.1.7, where we meet some new issues.

We are to design a standard *unregulated power supply* circuit. In this example we will look a little more closely than we did in class at the way to choose component values. Here's the particular problem.

Problem (Unregulated power supply)   Design a power supply to convert 110VAC 'line' voltage to DC. Aim for the following specifications:

- V_{out}: no less than 20V;
- Ripple: about 2V;
- I_{load}: 1A (maximum).

Choose:

- C size (μF);
- transformer voltage (V_{rms});
- fuse rating (I);
- "bleeder" resistor value;
- transformer current rating (current in secondary of transformer).

Questions: What difference would you see in the circuit output:

- If you took the circuit to Europe and plugged it into a wall outlet? There, the line voltage is 220V, 50Hz.
- If one diode in the bridge rectifier burned out (open, not shorted)?

3W.1.2 Skeleton circuit

First, as usual, we would draw the circuit without component values; see Fig. 3N.28. Fuse goes on primary side, to protect against as many mishaps as possible – including failures of the transformer and switch. Always use a *full-wave* rectifier (a bridge); most of the *half-wave rectifier* circuits you see in textbooks are relics of the days when "diode" meant an expensive vacuum tube. Now that diode means a little hunk of silicon, and they come as an integrated *bridge* package, there's rarely an excuse for anything but a bridge. A "bleeder" resistor is useful in a lab power supply, which might have no load: you want to be able to count on finding close to 0V a few seconds after you shut power off. The bleeder achieves that. Many power supplies are always loaded, at least by a regulator and perhaps by the circuit they were built to power; these supplies are sure to discharge their filter capacitors promptly and need no bleeder.

3W.1.3 Transformer voltage

This is just the peak value of V_{out} plus the two *diode drops* imposed by the bridge rectifier, as usual. If V_{out} is to be 20V *after* ripple, then $V_{\text{out(peak)}}$ should be two volts more: around 22V. The transformer voltage then ought to be about 23V.

When we specify the transformer we need to follow the convention that uses V_{rms}, not V_{peak} (remember, V_{rms} defines the *DC* voltage that would deliver the same power as the particular waveform). For a sinewave, V_{rms} is $V_{\text{peak}}/\sqrt{2}$, as you know.

In this case,
$$\frac{V_{\text{peak}}}{\sqrt{2}} \approx \frac{23\,\text{V}}{1.4} = 16 V_{\text{rms}} \;.$$

This happens to be a standard transformer voltage.

If it had not been standard, we would have needed to take the next higher standard value, or use a transformer with a 'tapped primary' that allows fine tuning of the step-down ratio.

3W.1.4 Capacitor

Figure 3W.1 is the "ripple" waveform, again. We have labeled the drawing with reminders that Δt depends on circumstances – in this particular, rather contrived, problem where we ask you to carry your supply to Europe. So Δt varies under the changed conditions suggested in the questions that conclude this problem.

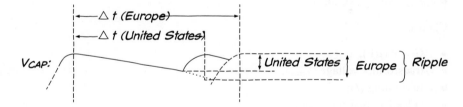

Figure 3W.1 Ripple.

Using
$$I = C\frac{dV}{dt} \;,$$
we plug in what we know, and solve for C. We know that:

- dV or ΔV, the ripple, is 2Vpp;
- dt or ΔT is the time between peaks of the input waveform: $\frac{1}{2\times 60\text{Hz}} \approx$ 8ms (in the United States);
- I is the peak output current: 1A.

  This specification of load *current* rather than load *resistance* may seem odd, at the moment. In fact, it is typical. The typical load for an unregulated supply is a *regulator* – a circuit that holds its output voltage constant; if the regulated supply drives a constant resistance, then it puts out a constant current despite the input ripple, and it thus draws a constant current from the filter capacitor at the same time).

Putting these numbers together, we get:
$$C = I \cdot \frac{\Delta T}{\Delta V} = 1\text{A} \cdot \frac{8 \times 10^{-3}\text{s}}{2\,\text{V}} = 4000\mu\text{F} \;.$$

Big, but not unreasonably so. You may want to call this "4mF." That would not be wrong – but would be unconventional and would mark you as an outsider.

3W.1.5 Fuse rating

This supply steps the voltage down from 110V to 16V; the current steps up proportionately: about sevenfold. So, the output (secondary) current of 1A implies an input (primary) current of about $\frac{1}{7}$A≈140mA.

But this calculation of *average* input current understates the heating effect of the primary current, and of the primary and secondary currents in the transformer. Because these currents come in surges, recharging the filter capacitor only during part of the full cycle, the currents during the surges are large. These surges heat a fuse more than a steady current delivering the same power, so we need to boost the fuse rating by perhaps a factor of 2, and then another factor of about 2 to prevent fuse from blowing at full load. (It's designed for emergencies.)

This set of rules of thumb carry us to something like:

fuse rating (current)=140mA×2 (for current surges)×2 (not to blow under normal full load) ≈560mA.

A 600mA slow-blow would do. Why slow-blow? Because on power-up (when the supply first is turned on) the filter capacitor is charged rapidly in a few cycles; large currents then flow. A fuse designed to blow during a brief overload could blow every time the supply was turned on. The slow-blow has larger thermal mass: needs overcurrent for a longer time than the normal fuse, before it will blow.

3W.1.6 Bleeder resistor

Polite power supplies include such a resistor, or some other fixed load, so as not to surprise their users. Again the value is not critical. Use an R that discharges the filter cap in no more than a few seconds; don't use a tiny R that substantially loads the supply.

Here, let's let RC equal a few seconds: $\implies R = \{\text{a few seconds}\}/C \approx 1\text{k}$

Before we go on to consider a couple of new issues, let's just draw the circuit with the values we have chosen, so far: see Fig. 3W.2.

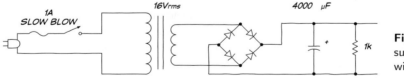

Figure 3W.2 Power supply: the usual circuit, with part values inserted.

3W.1.7 Transformer current rating

This is harder. The transformer provides brief surges of current into the cap. These heat the transformer more than a continuous flow of smaller current. We have made this point already today in Chapter 3N.

Figure 3W.3 is a sketch of current waveforms in relation to two possible ripple levels:

moderate ripple $I_{\text{rms}} = (\frac{1}{5}[5\,\text{A}]^2)^{1/2} = \sqrt{5}\,\text{A} = 2.2\,\text{A}.$
tiny ripple $I_{\text{rms}} = (\frac{1}{20}[20\,\text{A}]^2)^{1/2} = \sqrt{20}\,\text{A} = 4.4\,\text{A}$ *doubles* transformer heating

The left-hand figure shows current flowing for about 1/5 period, in pulses of 5A, to replace the charge drained at the steady 1A output rate. The right-hand figure shows current flowing for 1/20 period, in pulses of 20A.

Figure 3W.3 Transformer Current versus Ripple: Small Ripple ⟹ Brief High-Current Pulses, and excessive heating.

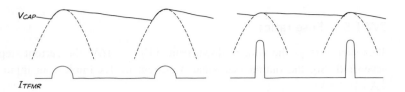

Moral: a little ripple is a good thing. You will see that this is so when you meet voltage regulators, which can reduce a volt of ripple out of the unregulated supply to less than a *millivolt* at the point where it goes to work (where the output of the regulated supply drives some load).

Questions

What difference would you see in the circuit output:

(1) If you took it to Europe, where the line voltage is 220V, 50Hz?
(2) If one diode in the bridge rectifier burned out (open, not shorted)?

Solutions

(1) In Europe, the obvious effect would be a doubling of output voltage. That's likely to cook something driven by this supply (that's why American travelers often carry small 2:1 step-down transformers – though contemporary power supplies usually are of the *switching* type, and smart enough to accept happily the doubling of *line* voltage). It might also cook transformer and filter capacitor, unless you had been very conservative in your design (it would be foolish, in fact, to specify a filter cap that could take double the anticipated voltage; caps grow substantially with the voltage they can tolerate).

So you probably would not have a chance to get interested in less obvious changes in the power supply. But let's look at them anyway: the output *ripple* would change: ΔT would be $1/50\text{Hz} = 10\text{ms}$, not 8ms.

Ripple should grow proportionately. If load current remained constant, ripple should grow to about 2.5V. If the load were resistive, then load current would double with the output voltage, and ripple would double relative to the value just estimated: to around 5V.

(2) The burned out diode would make the bridge behave like a *half-wave* rectifier. ΔT would double, so ripple amplitude would double, roughly. Ripple frequency would fall from 120Hz to 60Hz.

This information might someday tell you what's wrong with an old radio: if it begins to buzz at 60Hz, perhaps half of the bridge has failed; if it buzzes at 120Hz, probably the filter cap has failed. If you like such electronic detective work, many pleasures lie ahead of you.

3W.2 Z_{IN}

The problem: procedure to discover R_{IN} and C_{IN}

Describe a procedure for *measuring* R_{IN} and C_{IN} for the device in Fig. 3W.4. Note that your procedure should work even if the foot of the resistor shown to model R_{IN} is *not* tied to ground. (This last requirement makes the problem harder than it otherwise might be).

Solution

The first step is to add a known R in series with the mystery box (let's call this "Box"). We will use two scope channels to watch the original signal and the point loaded by Box; see Fig. 3W.5.

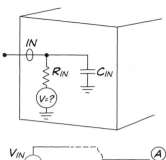

Figure 3W.4 Input impedance problem.

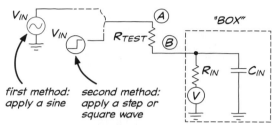

Figure 3W.5 Step 1: install known resistance in series with "Box".

Measure $R_{\rm IN}$

The challenge is essentially the same as the one presented by the exercise in §2L.1.1 that asks you to infer R and C once you have determined the RC product. You need to peel apart the effects of R versus C.

To measure $R_{\rm IN}$, our strategy will be to test the circuit at a frequency low enough so that the impedance of the capacitor is unimportant relative to that of the resistor: $X_C \gg R$. This will occur at a low frequency. But the point that calls for a little thought is *how* to determine when we have arrived at such a frequency. Here, we propose three alternative methods.

Measure $R_{\rm IN}$ using DC

We can observe the open-circuit input voltage. That reveals the internal V. Then use $R_{\rm TEST}$, as in Fig. 3W.5, to form the usual voltage divider. If the internal V is zero, then we need an external DC source as input.

Measure $R_{\rm IN}$ using a sinusoid

If we apply a sine to the input and watch points A and B, we need to make sure that whatever attenuation we see is caused by the R, not by X_C as well. An $R:R$ divider will show no phase shift; a divider with R on top and R parallel C will show phase shift, unless X_C is very large relative to the Box's R. So, we apply a low-frequency sinusoid, and watch for phase shift. If we see none, our measurement of $R_{\rm IN}$ is not being corrupted by the effect of $C_{\rm IN}$. The scope images in Fig. 3W.6 show results with the scope inputs *AC* coupled.

Two possible hazards:

- At a very low frequency, if you AC-couple only the channel watching B, you may see a phase shift that is a pure scope artifact. AC-coupling the scope input is a good idea, if you use the method of centering the scope display on the 0% mark: since the lower part of the waveform goes off-screen, a *DC offset* in the signal can deceive you. But when you use AC-coupling you can see the blocking-capacitor's effect, at extreme low frequencies: below about 10Hz.[1] This effect is visible in the leftmost image of Fig. 3W.6.

---

[1] $f_{\rm 3dB}$ for both our analog and digital scopes is about 7Hz (high-pass), when AC-coupled and fed by a BNC cable ("'1×'"). When fed by a 10× probe, the scope shows an $f_{\rm 3dB}$ so low it is not likely to trouble you: about 0.7Hz. But in the present case, where one is likely to watch both channels AC-coupled, the phase shift is harmless: it would affect both channels equally. Fig. 3W.6, in contrast, shows channel 1 DC-coupled, channel 2 AC-coupled.

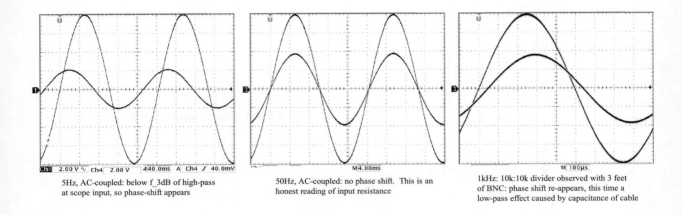

5Hz, AC-coupled: below f_3dB of high-pass at scope input, so phase-shift appears

50Hz, AC-coupled: no phase shift. This is an honest reading of input resistance

1kHz: 10k:10k divider observed with 3 feet of BNC: phase shift re-appears, this time a low-pass effect caused by capacitance of cable

Figure 3W.6 Using sinusoid, look for zero phase-shift to be sure you are reading R_{IN}.

- At higher frequencies, you may see phase shift (lag, this time) caused by a low-pass filter formed by the combination of the test resistor and the stray capacitance of the BNC cable and scope channel with which you are watching point B. This effect is visible in the rightmost image of Fig. 3W.6.

In order to be sure that neither effect is fooling you, make sure you see no phase shift between the two channels, when you watch points A and B – as in the center image of Fig. 3W.6. The attenuation that you see there is caused by the circuit's R_{IN} alone, not complicated by the effect of the capacitance to ground.

Measure R_{IN} using a square wave

A square wave can reveal R_{IN}, too. In Fig. 3W.7 the right-hand image shows the effect of a DC voltage at the foot of the unknown R of the box that we are measuring. We imposed this DC offset to remind ourselves of why we don't want to try to measure R_{IN} by applying a DC voltage to the test setup. The DC voltage at the foot of R would throw us off, if we tried to infer R_{IN} for this case. But the step input, like the sinusoid above, allows a measurement that is not thrown off by that DC voltage.

To infer R_{IN}, compare input step amplitude to the amplitude that appears at the input of Box.

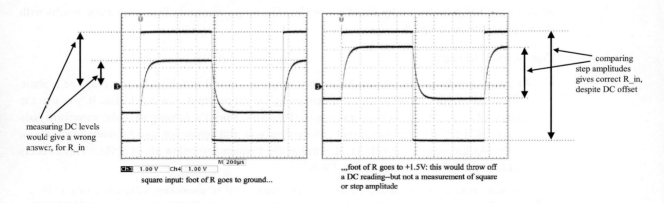

measuring DC levels would give a wrong answer, for R_in

square input: foot of R goes to ground...

,,,foot of R goes to +1.5V: this would throw off a DC reading–but not a measurement of square or step amplitude

comparing step amplitudes gives correct R_in, despite DC offset

Figure 3W.7 Square wave, like sinusoid, can allow one to measure R_{IN}.

As Fig. 3W.7 illustrates, whereas a single DC reading at points A and B (points defined in Fig. 3W.5) would be thrown off by such a DC offset, in contrast, a measurement of step- or square-wave amplitude is not so distorted, as the right-hand image of Fig. 3W.7 indicates.

Discover C_{IN}

Once one knows R_{IN}, the next step is to measure RC. Given that, we will be able to calculate C_{IN}. One can get RC in either of two ways, once more.

Use sinusoid: find RC, and then C, by measuring f_{3dB}

If one applies a sinusoid, as above, then one can gradually increase f_{IN}, watching amplitude at B fall, until it is about 30% below its low-frequency level: now one has found f_{3dB}. Given f_{3dB}, and knowing R one can easily calculate C_{IN}.

As you do this calculation, you need to recall that the effective R driving C_{IN} is $R_{IN} \parallel R_{TEST}$ (this is $R_{Thevenin}$).

Use square wave: find RC, and then C, by measuring RC directly

If you use a square wave, then you can measure RC directly, from the (time-domain) image on the scope screen. As usual, you need to choose an appropriate sweep rate: see Fig. 3W.8. This concern is not new to you, of course. You did this at the start of Lab 2L.

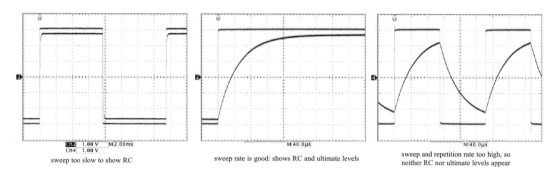

Figure 3W.8 To measure RC, choose an appropriate sweep rate.

sweep too slow to show RC

sweep rate is good: shows RC and ultimate levels

sweep and repetition rate too high, so neither RC nor ultimate levels appear

The middle image of Fig. 3W.8 would permit an RC reading – but one could do better, using still more of the scope screen to look at the time to rise from the lowest point to 63%.

Once you have RC, again you need to remember that the relevant R is R_{TEST} and R_{IN} in parallel, as we said for the case of the sinusoid.

Part II

Analog: Discrete Transistors

4N Transistors I

Contents

4N.1	**Overview of Days 4 and 5**	**151**
4N.2	**Preliminary: introductory sketch**	**154**
	4N.2.1 The name…	154
	4N.2.2 An intuitive model	154
	4N.2.3 The symbol	154
	4N.2.4 Assumptions	154
	4N.2.5 Starting simple: two views of transistor behavior	155
4N.3	**The simplest view: forgetting beta**	**155**
	4N.3.1 …that a current source provides a constant output current	155
	4N.3.2 …that a *common-emitter amplifier* shows voltage gain as advertised	157
	4N.3.3 …that a follower follows	157
	4N.3.4 …that a push–pull follower works, and also shows distortion	158
4N.4	**Add quantitative detail: use beta explicitly**	**158**
	4N.4.1 Digression on beta	158
	4N.4.2 Follower as impedance changer	159
	4N.4.3 Complication: *biasing*	161
	4N.4.4 Another complication: asymmetry can produce *clipping*	164
	4N.4.5 …A remedy for clipping: the push–pull	165
4N.5	**A strikingly different transistor circuit: the switch**	**166**
4N.6	**Recapitulation: the important transistor circuits at a glance**	**167**
	4N.6.1 Impedance at the collector, by the way	167
	4N.6.2 …and the switch stands apart	167
4N.7	**AoE Reading**	**168**

Why?

What problems do today's circuits solve? They can modify signals in any of the following ways:

- improve the apparent impedance of a circuit element (boost R_{in}, reduce R_{out});
- hold current constant as voltage varies (make a "current source");
- amplify a voltage signal (a ΔV);
- let a voltage signal switch current to a load On or Off.

4N.1 Overview of Days 4 and 5

A novel and powerful new sort of circuit performance appears today and tomorrow: a circuit that can *amplify*. Sometimes the circuit will amplify voltage; that's what most people think of as an amplifier's job. Sometimes the circuit will amplify only current; in that event one can describe its amplification as a transformation of *impedances*. As you know from your work in the earlier labs, that can be a valuable trick.

The transistors introduced in AoE Chapter 2 are called **bipolar** (because the charge carrying mechanism uses carriers of both polarities – but that is a story for another course). Only at the very end of the analog section of this course will you meet the other sort of transistor, which is called **field effect** (FET) rather than "unipolar" (though they were called "unipolar" at first). The FET type was developed after bipolar, though, as noted later, the discoverers of the bipolar transistor stumbled upon it while trying to develop a FET. But the FET, in a form called MOSFET[1] has turned out to be more important than bipolar in *digital* devices. You will see much of these FET logic circuits later in this course. In *analog* circuits, bipolar transistors still dominate, but even there FETs are gaining.

You will sometimes apply transistors singly – as *discrete* parts.[2] More often, you will pair the brawn of a discrete transistor with the brains of an operational amplifier (op-amp)[3] in order to get high-current or high-power that is beyond the capacity of an ordinary op-amp. But an understanding of transistor circuits is important in this course not so much for this reason as because an introduction to discrete transistor design will help you to understand the innards of the *integrated* circuits that you are certain to rely on.

After toiling through this section, you will find that you can recognize in the schematic of an otherwise mysterious IC a collection of familiar transistor circuits. In an effort to de-mystify the op-amp, we conclude Lab 5L by asking you to build yourself an op-amp from an array of transistors. Recognizing familiar circuits in the schematic of an op-amp, you will consequently recognize the op-amp's shortcomings as shortcomings of those familiar transistor circuit elements.

This section on transistors is difficult. It requires that you get used to a new device and at the same time apply techniques you learned in the section on passive devices. You will find yourself worrying about impedances once again: arranging things so that circuit-fragment A can drive circuit-fragment B without undue loading; you will design and build lots of *RC* circuits. Often you will need a Thevenin model to help you determine an effective *R*. We hope, of course, that you will find this chance to apply new skills gratifying; but you are likely to find it taxing as well.

We say this not to discourage you, but, on the contrary, to let you know that if you have trouble digesting the rules for design with transistors, that's not a sign that there's something wrong with you. This is a *rich* section of the course – perhaps the most difficult.

With Lab 6L, op-amps arrive, and suddenly your life as a circuit designer becomes radically easier. Operational amplifiers, used with *feedback*, will make the design of very good circuits very easy. At that point you may wonder why you ever labored with those difficult discrete-transistor problems. But we won't reveal to you now that op-amps are easy – because we don't want to sap your present enthusiasm for transistors. In return for your close attention to their demands, transistors will perform some pretty impressive work for you.

Instead of yearning for op-amps, let's try to put ourselves into a state of mind approaching that of the transistor's three inventors, who found to their delight, two days before Christmas 1947, that they had constructed a tiny amplifier on a chunk of germanium. Probably they could not yet envision a time when there would be no more vacuum tubes[4]; no more high-voltage supplies; no more power wasted

---

[1] MOSFET is Metal Oxide Field Effect Transistor: a strange way to say that its input terminal is *insulated* – by the "oxide," SiO_2. More on MOSFETs appears in Chapter 12N.
[2] "Discrete" doesn't mean "polite, and able to keep a secret." The origins of the words *discrete* and *discreet* happen to be the same. But *discrete*, in this context, means "used singly," rather than as one of many elements in an integrated circuit.
[3] Operational amplifiers are high-gain amplifiers used to implement circuits that exploit negative feedback. Op-amps become our principal analog circuit elements, beginning on Day 6.
[4] Well, almost no vacuum tubes. Rock'n roll guitarists and some mad audiophiles prefer the sound of tubes. The guitarists like the way the tubes sound when the amplifier is overdriven. The audiophiles like to spend money.

in heating filaments. The world could look forward to the microcomputer, the iPod, the smartphone, the – whatever comes next, made possible by amazingly tiny microcircuits.[5]

Here, in Fig. 4N.1 is a Nobel prize-winning device, looking wonderfully homemade. It's not *really* made from paper clips, Scotch tape and chewing gum.[6]

Figure 4N.1 The first transistor: point-contact type (1947).

---

[5] For some excellent history of the discovery of the bipolar transistor, see Michael Riordan and Lillian Hoddeson, *Crystal Fire*, Norton (1997) p. 138ff., and a short article by authors that include the former head of Bell Labs: William Brinkman et al., "A History of the Invention of the Transistor and Where It Will Lead Us," *IEEE Journal of Solid-state Circuits* vol. 32, no. 12, Dec. 1997.

[6] Photo used with permission of AT&T Bell Labs.

4N.2 Preliminary: introductory sketch

4N.2.1 The name...

A transistor is a "**trans**[fer]" "[res]**istor**:" a thing vaguely like a resistor, but one that can be controlled – that can transfer a signal from input to output. In having *three* terminals, it differs from the passive devices we have met to this point (with the exception of the 4- or 5-terminal *trans*former – whose name similarly signals its behavior). The three terminals will make your first days with transistors challenging.

But take heart, if you feel this difficulty: soon you will pop out above the clouds into to the bright and cheery world of operational amplifiers, as we promised in §4W.1 We made the point there that we'd like you to know what's within an op-amp, so that its operation is not *simply* magical.

4N.2.2 An intuitive model

A transistor is a *valve*; see Fig. 4N.2. Notice, particularly, that the transistor is not a *pump*: it does not force current to flow; it permits it to flow, to a controllable degree, when the power supply tries to force current through the device.

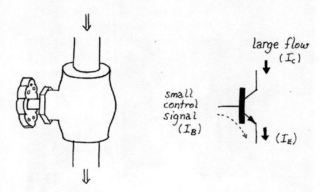

Figure 4N.2 A transistor is a valve (not a pump!).

4N.2.3 The symbol

The transistor symbol apparently amounts to a sketch of the earliest prototype.[7] In Fig. 4N.3 we have taken a fragment of the photo of the original Bell Labs transistor, rotating it to put emitter and collector above and below base.

The claim certainly looks plausible. The "base" was literally the base for the structure. Thomas Lee makes the intriguing observation that the bipolar transistor was *discovered* rather than invented by Shockley, Bardeen and Brattain – because the three investigators stumbled upon the structure. They were not trying to form a bipolar transistor. Instead, they were troubleshooting their defective field-effect device, probing the surface of the "base" piece to investigate disturbances in current flow.

4N.2.4 Assumptions

AoE §2.1.1

For NPN type:

(1) $V_C > V_E$ (by at least a couple of tenths of a volt);

---

[7] Thomas H. Lee, *Planar Microwave Engineering*, Cambridge University Press (2004), p. 342.

Bell labs transistor of 1947...

...turned on its side...

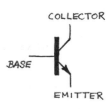
...and looking a good deal like its symbol

Figure 4N.3 Transistor symbol seems to sketch the original transistor.

(2) "things are arranged" so that $V_B - V_E$ is about 0.6V (V_{BE} is a diode junction, and must be forward biased)

4N.2.5 Starting simple: two views of transistor behavior

We begin with *two views of the transistor*: one simple, the other *very* simple. (Later, we will complicate things.)

- Pretty simple: current amplifier: $I_C = \beta \times I_B$; see Fig. 4N.4.
- Very simple: say nothing of beta (though assume it's at work): Say only:

$$V_{BE} = 0.6\text{V};$$
$$I_C \approx I_E.$$

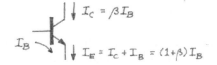

Figure 4N.4 Transistor as *current*-controlled valve or amplifier.

4N.3 The simplest view: forgetting beta

We can understand – and even design – many circuits without thinking explicitly about beta (also called "h_{FE}"), the ratio of current out to current in: $\beta = I_C/I_B$. The simplest view will let you understand the following circuits. You can understand...

4N.3.1 ...that a current source provides a constant output current

AoE §2.2.5B

See Fig. 4N.5: pretty charmingly simple, eh? It is nothing more than *Ohm's Law* that determines the magnitude of the output current, thanks to the predictability of the V_{BE} drop. Now turn to Fig. 4N.6 to see what happens if we vary the resistor on the collector – R_{LOAD}. We'll also tack in a base resistor, which will make no difference except in an extreme case that we'll consider in a minute, a case when the current source fails.

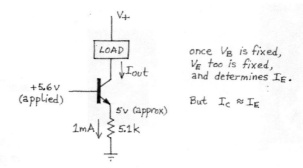

Figure 4N.5 Current source.

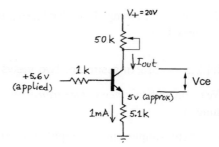

Figure 4N.6 Current source with variable load shown.

If we vary R_{LOAD}, how does I_{LOAD} behave? Suppose that R_{LOAD} initially is 10k and we drop it to 1k. Does the current increase?

If you're inclined to say, "Yes," then you are showing nostalgia for the *passive* section with which this book began, and for the resistors of your acquaintance. You are asking the transistor to behave like those old friends – like an ohmic device: more voltage across it, so more current. But, no, it doesn't behave that way.

As you lower the value of R_{LOAD}, it is true that V_C rises. But it does not follow that I_C grows. It does not, because $I_C \approx I_E$, and I_E is constrained – determined, in turn, by V_B. Unfamiliar, perhaps, but simple – and useful. If you take R_{LOAD} all the way down to zero, does the current stay constant? Yes.

What if you take R_{LOAD} to its maximum? Here, the answer is, "No. In this case, the current source fails." It fails because we have violated one of the assumptions that we mentioned in §4N.2.4.

In Fig. 4N.6, taking R_{LOAD} beyond about 15k forces V_C too low: we can no longer satisfy the requirement that V_C exceed V_E, and the circuit cannot hold the current constant. The transistor can no longer do its magic – adjusting the voltage, V_C, as needed to hold I_C constant. When V_{CE} falls below a couple of tenths of a volt, the transistor is said to be *saturated*.

When that happens, none of the usual rules that we have been advertising will hold. The circuit does become ohmic, the transistor looking like a small-valued resistor; further increases in R_{LOAD} simply reduce I_{LOAD}. Setting R_{LOAD} to 50k, for example, would take I_{LOAD} down to about 0.4mA (20V across about 55k): plain old Ohm's law behavior.

This limitation on the range of the current source – called its *output voltage compliance* – reminds us that although the transistor circuits are clever, they still require us to use our heads. They don't work if we don't play by the rules.

4N.3.2 ...that a *common-emitter amplifier* shows voltage gain as advertised

AoE §2.2.7

Adding a collector resistor to the current source (as in Fig. 4N.7) is about all that's required to turn the source into a voltage amplifier.

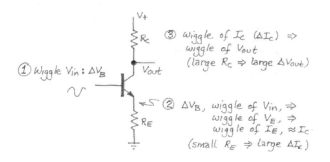

Figure 4N.7 Common-emitter amp.

One can regard the common-emitter amp as a two-stage circuit: see Fig. 4N.8. The first stage is a current source (with current controlled by V_{IN}); this is a *V*-to-*I* block. The second is a resistor that converts that output current, and its variations, into a voltage: this is an *I*-to-*V* block.

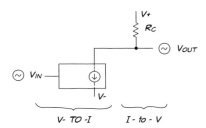

Figure 4N.8 Common-emitter amp, described as *V*-to-*I* block feeding an *I*-to-*V* block.

You may object that we're being a bit pompous in using the name *I*-to-*V* block for the humble resistor. Maybe so, but that is, indeed, the resistor's function in this circuit.[8]

4N.3.3 ...that a follower follows

AoE §2.2.3

The circuit in Fig. 4N.9 may look, at first glance, totally pointless: you start with a voltage wiggle; at the output you get the same voltage wiggle. Clearly it is not a voltage amplifier.

But it turns out that it amplifies *current* or, equivalently *changes impedances*. We'll appreciate how this works when we add one more element to our description of transistor behavior, the current-multiplication factor, β (see §4N.4). With the present, very simple view we can only confirm that the follower follows.

---

[8] And if you think we're giving a pretentious name to the resistor, consider a more spectacular instance: some bored engineers, apparently envious of their bosses' opportunities to go off and talk at conferences, got together to propose a paper. Their abstract described a new circuit element they had discovered: a very useful device that performs a bidirectional transformation between current and voltage, and one that – as an added benefit – kept the transformation *linear*! This novel device they named the *Linistor*. To their delight, they soon got the good news from the sponsors of the conference: they were invited to present the paper describing their invention. At that point, we are sad to report, they lost their nerve. The electronics world at large never got to hear about the linistor. The late Bob Pease tells this story in "What's All this Hoax Stuff?," *Electronic Design*, April 4, 1994.

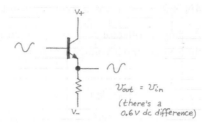

Figure 4N.9 Follower.

4N.3.4 ...that a push–pull follower works, and also shows distortion

Again, the circuit (Fig. 4N.10) is intelligible – but its virtues are not. We will see shortly why it is an extremely useful variation on the simpler *follower*.

AoE §2.4.1

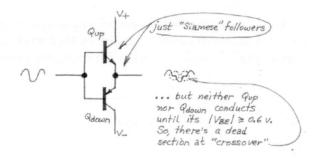

Figure 4N.10 Push–pull.

4N.4 Add quantitative detail: use beta explicitly

To understand why a *follower* is useful, we need to add a quantitative element that we omitted from the "simplest" view: we need to note that the *output current*, I_C (or the near-equivalent, I_E) is much larger than the *input current*, I_B. It is larger by the factor β. The notion of *current amplification* may be unfamiliar and uncongenial to you.

Though this is an accurate view of a follower, we more often describe the follower's behavior as *changing impedances*. That notion, too, takes getting used to. But the idea is simple, and builds on your familiarity with Ohm's law.[9] The point is simply this: voltage wiggles IN and OUT are the same; currents are very different. The contrasting quotients ($\Delta V/\Delta I$) give *high* R_{IN}, *low* R_{OUT}.

4N.4.1 Digression on beta

AoE §2.1.1

It can be hard to get used to our treatment of the transistor characteristic, β, a treatment that follows that in AoE.

On the one hand, we work hard to get across a piece of sound design advice: don't ever design a circuit on the assumption that you know the value of β. You should not, because the value is not predictable. It varies with individual transistors of a given type; it varies for an individual transistor as I_C varies; it varies with temperature.

---

[9] We're not renouncing our earlier point that the transistor's collector–emitter behavior is not ohmic (§4N.3.1, above). The impedances we speak of, here, are characteristics of input (at the base) and output (at the emitter).

On the other hand, we assume $\beta \approx 100$, for "small signal" transistors like the 2N3904. We don't know beta; beta is 100 (the king is dead, long live the king). This is just one more of Horowitz and Hill's permissions to "be wrong, but be wrong in the right direction."[10] We don't need to know beta as long as we *underestimate* its value, because beta is good.

4N.4.2 Follower as impedance changer

AoE §2.2.3

The follower amplifies current. The same truth makes it an impedance changer: a circuit that offers high R_{IN}, low R_{OUT}; see Fig. 4N.11.

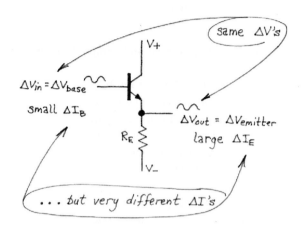

Figure 4N.11 How a follower changes impedances.

Figure 4N.12 shows a corny mnemonic device to describe this impedance-changing effect. Imagine an ill-matched couple gazing at each other in a dimly-lit cocktail lounge – and gazing through a rose-colored lens that happens to be a follower. Each sees what he or she wants to see!![11]

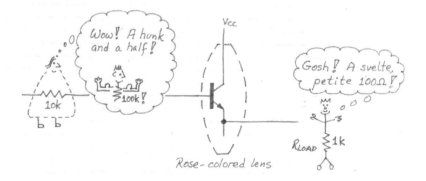

Figure 4N.12 Follower as rose-colored lens: it shows what one would *like* to see. If this figure seems very old-fashioned, feel free to interchange genders.

Estimating β by looking at R_{IN}

In Lab 4L, we ask you to infer beta from the observed value of R_{IN}. The experiment asks you to form a voltage divider using a known resistor (33k, in the example just below) to feed the follower's base, and then to compare amplitudes on the two sides of this known resistor. Fig. 4N.13 shows what we saw when we tried it (watching signal source and emitter).[12]

---

[10] See §4N.4.
[11] If Fig. 4N.12 seems very old-fashioned, feel free to interchange genders.
[12] Since we are watching signal swings (ΔV's) we may watch at either base or emitter.

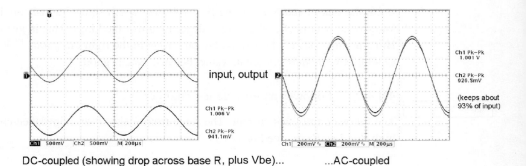

Figure 4N.13 Slight attenuation caused by follower's R_{IN} can reveal β at work.

DC-coupled (showing drop across base R, plus Vbe)... ...AC-coupled

The left-hand image shows the considerable DC offset between input and emitter – about 1.5V. Why does this appear?[13]

We get a clearer picture of R_{IN} when we strip away this DC difference by *AC-coupling* the two signals, as in the right-hand image of Fig. 4N.13. (Usually, we urge you *not* to use AC coupling in the lab, because it knocks out interesting information; here, we're happy to strip away the not-interesting DC difference.) AC-coupling helps one do a visual comparison – though the difference here is so slight that it would be hard to measure by eye. This oscilloscope helps us out by measuring the peak-to-peak amplitudes for us. We find we're keeping 93% to 94% of the original signal. Let's use that to get β.

If we take 93% as the truer reading (because it uses more of the scope screen, the AC image should give the better resolution), then we lose 7%, keep 93%. We keep about 13 parts while losing one. That means R_{IN} is about 13 times as large as the known resistance of 33k.

We could now calculate R_{IN}, and from this value infer β, in a two-step process. If you want to be lazier, though, try considering, as a sort of benchmark (see Fig. 4N.14), what you would see if β were 100, then compare what you see against that reference.

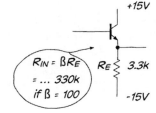

Figure 4N.14 A starting point or benchmark: estimate what R_{IN} would be if β were 100.

If β were a hundred, the R_{IN} would be 330k, just ten times R_{TEST} (we would see attenuation of just under 10%). But we see an R_{IN} that looks about 13 times R_{TEST}. So, β must be about 130 rather than 100. This is plausible. Recall that we do not predict that β will *equal* 100. Instead, we say, in effect, "Call it 100, and you will safely underestimate β."[14]

---

[13] Probably you're not puzzled by this effect; but it's worth recalling that we are interested, here, in passing a *wiggle*, not a DC level, so the DC difference doesn't interest us much. But let's see if the numbers make sense.
The 1.5V DC drop includes the substantial V_{BE} along with the drop caused by I_B flowing in the base resistor. Our base resistor is 33k, so the 0.9V drop attributable to that resistor would imply a current of $0.9V/33k\Omega \approx 27\mu A$. The emitter current is about $13V/3.3k\Omega \approx 4mA$. The ratio of I_E to I_B should be essentially β. Here, the ratio is $4mA/27\mu A \approx 145$. That's not far from what we estimate with the more careful reading of amplitude just below. The numbers seem to hang together.

[14] Here is yet another of Horowitz & Hill's licenses to be "right in the wrong direction." Compare our approximation, $Z_{IN_{RC-filter}} = R$, and the use of 8ms as Δt in the calculation of power-supply ripple as we mentioned in §4N.4.

4N.4.3 Complication: *biasing*

AoE §2.2.4

To this point – and early in Lab 4L – we have assumed the use of "split" power supplies (supplies of both polarities). Doing this allows our circuits to respond to signals symmetric about ground without *clipping* (that is, without the flattening that occurs when the waveform hits a limit).

The follower that you wire in the lab initially uses only a positive supply, and cannot follow an input voltage that swings below ground (or even below about one diode-drop positive). An input centered on ground evokes a half-wave-rectified output like the lower output trace in Fig. 4N.15.

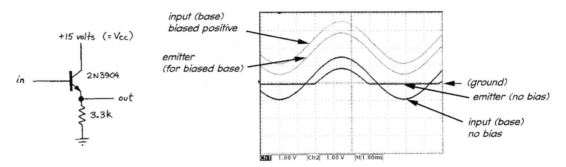

Figure 4N.15 A follower that uses only one supply must have its input biased positive.

This half-wave-rectified result reminds one how similar the *follower* may appear to be to a simple diode feeding a resistor. Wouldn't a diode and resistor (the first rectifier you met, back in Lab 3L), produce exactly the waveforms shown in Fig. 4N.15?

That is, wouldn't the two circuits of Fig. 4N.16 produce identical waveforms? Yes, to a very good approximation, they would.

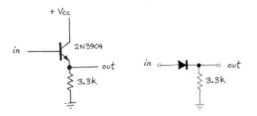

Figure 4N.16 Follower and half-wave rectifier: do they differ?

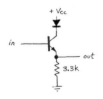

Figure 4N.17 Detail of "fraudulent" demo circuit.

A small fraud to make a point... In fact, we like to perpetrate a little fraud on our students during class, using the similarity of the two circuits. We talk about the follower, sketch the circuit, and then show a scope display of the follower's input and output. But – here's the fraud – we "forget" to turn on the power supply. The circuit doesn't seem to care.[15]

The very similar lower two traces of Fig. 4N.18 show how little difference the power supply makes to the voltage waveforms in this demonstration. How can this be? Is it possible that the power supply isn't necessary?

No. What this demonstration shows, instead, is that the oscilloscope, a voltage-sensing instrument, cannot directly show us what this circuit is doing.

---

[15] We should admit that the fraud requires us to insert a diode in series with the collector, so that the base–collector junction cannot conduct current into the positive supply, which would sink current even though switched off; see Fig. 4N.17.

Figure 4N.18 Emitter follower and a rectifier – implemented by failing to turn on the power supply: the voltage waveforms are hard to distinguish.

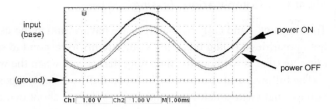

Making the follower's operation visible The follower is not a voltage amplifier. It is a *current* amplifier, instead – or, equivalently, it is an *impedance changer*. So, we need to insert a test *resistor* that will make the magnitude of *current* visible to the scope. At the input of each of the two competing circuits – follower versus rectifier, where the rectifier models the follower with power *off* – we'll put a 33k resistor. This will form a voltage divider that pairs it with the R_{in} of each circuit: see Fig. 4N.19.

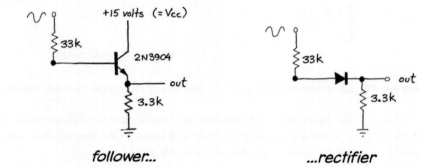

Figure 4N.19 Base resistor can make follower's *impedance change* or *current amplification* visible.

Now, what happens if we try the effect of power ON versus OFF as we watch signal source and emitter of the follower? It's a relief to see that power makes a giant difference to the follower's impedance: see Fig. 4N.20.

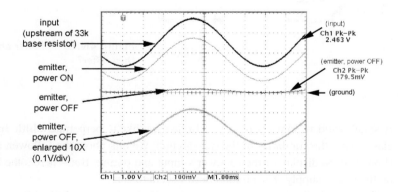

Figure 4N.20 Base resistor makes evident the follower's use of the power supply.

With power ON, the follower's input impedance is $\beta \times R_E$, as advertised: at least 330k. So the emitter waveform looks almost equal to the input swing (and shows the characteristic diode-drop of V_{BE}).

With power OFF, in contrast, the "follower" becomes just a diode rectifier, with V_{BE} serving as the diode. The input resistance is just 3.3k, and we see severe attenuation: less than 10% of the input swing survives.

This somewhat silly experiment reminds us that the transistor's output current – whether taken at emitter or collector – comes almost entirely from the power supply, not from the base. That's the key to the impedance-changing magic. The base current is just a small downpayment, needed to provide the big loan of power from the supply.[16] Incidentally, the experiment reminds us that the term "power supply" is named appropriately. It means what the phrase says.

AoE §2.2.4

A divider can provide needed bias In the lab exercises, we first provide this positive *bias* (pushing off-center) in the laziest way: simply by dialing up a positive offset from the function generator. But this is not a practical general solution.

The more general design scheme, as in Fig. 4N.21, allowing use of a *single* power supply, is to add a voltage divider that pulls the base positive. Thus the emitter can swing down as well as up, as the input swings negative and positive. A *blocking capacitor* allows base and signal source to adopt differing DC levels without a fight. The wiggles pass happily from source to base.

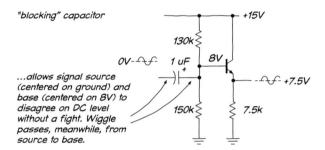

Figure 4N.21 Single-supply follower uses biasing.

The biasing divider must be stiff enough to hold the transistor where we want it (with V_{out} around the midpoint between V_{CC} and ground). It must not be too stiff: the *signal* source must be able to wiggle the transistor's base without much interference from the biasing divider.

The biasing problem is the familiar one: device A drives B; B drives C; see Fig. 4N.22. As usual, we want Z_{out} for each element to be low relative to Z_{in} for the next:

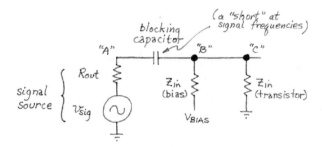

Figure 4N.22 Biasing arrangement.

You will notice that the biasing divider reduces the circuit's input impedance by a factor of ten, compared to what it would have been in a "split supply" design. This is regrettable; if you want to investigate complications, see the "bootstrap" circuit in AoE §2.4.3 for a way around this degradation.

You will have to get used to a funny convention: you will hear us talk about impedances not only

---

[16] Well, this metaphor isn't quite right, since we never repay the loan. But maybe, on second thought, that's about right. Isn't that the way our present debt-ridden culture works?

at points in a circuit, but also *looking* in a particular direction. For example: we will talk about the impedance "at the base" in two ways (illustrated in Fig. 4N.23):

AoE §2.2.4

- the impedance "looking into the base" (this is a characteristic of the transistor and its emitter load)
- the impedance at the base, looking back toward the input (this characteristic is *not* determined by the transistor; it depends on the *biasing* network, and (at signal frequencies) on the *source impedance*.

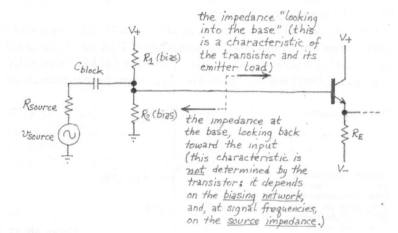

Figure 4N.23
Impedances "looking" in specified directions.

4N.4.4 Another complication: asymmetry can produce *clipping*

AoE §2.2.3D

We have asked you to admire the low R_{OUT} of the follower (and to measure this characteristic in Lab 4L. Now, candor compels us to admit[17] that the follower can misbehave, despite its low R_{OUT}. Fig. 4N.24 shows a hypothetical follower and a resistive "load."

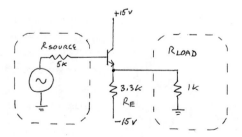

Figure 4N.24 Emitter follower, loaded. Can it drive this load?

Will this circuit work properly? Let's calculate the impedances IN and OUT.

- R_{IN} at the base: looks like $\beta \times (R_{\text{E}} \text{ parallel } R_{\text{LOAD}}) \approx 75\text{k}$, taking β to be at least 100. Since this $R_{\text{IN}} > 10 \times R_{\text{SOURCE}}$, we satisfy our "house rules" for impedances: OK.
- R_{OUT} looking back, at the emitter: R_{E} parallel $R_{\text{SOURCE}}/\beta \approx 50\Omega$ (the effect of R_{E} is negligible, as usual). This R_{OUT} is about 1/20 of R_{LOAD}. So, again, no problem.

---

[17] People our age – is there any such person in our readership? – may recognize this phrase as the one Lyndon Johnson liked to offer as preamble to a big fib! This is not a fib. We are not crooks.

4N.4 Add quantitative detail: use beta explicitly

This is correct – but only for an input signal of modest amplitude. If the signal amplitude grows to more than about 4V, then the circuit fails: the negative swings flatten – they "clip." You can see this occurring in Fig. 4N.25.

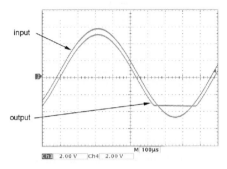

Figure 4N.25 Despite its low R_{OUT}, the follower can clip when loaded.

Why does this happen? And why only on the negative swings? The answer is simple, almost obvious, once you have heard it: the follower is radically asymmetric. It uses the *transistor* to drive current into the load; it uses a *resistor*, R_E, to pull current out of the load. The transistor is good at pushing current; the resistor is not so good at pulling current. At an extreme of negative swings, the transistor *shuts off*; see Fig. 4N.26.

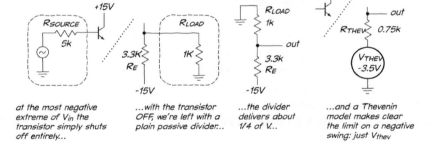

at the most negative extreme of V_{in} the transistor simply shuts off entirely...

...with the transistor OFF, we're left with a plain passive divider...

...the divider delivers about 1/4 of V...

...and a Thevenin model makes clear the limit on a negative swing: just V_{thev}

Figure 4N.26 Loaded follower clips when the transistor shuts off entirely, for an extreme negative V_{in}.

At that point, all the transistor magic that produced the nice low R_{OUT} – thanks to β and the "rose-colored lens" – dissipates like the magic that sustained Cinderella's beautiful carriage. The circuit turns back into a pumpkin: it becomes a simple resistive divider, and V_{OUT} gives up on its negative swings, sticking at a bit less than 1/4 of the negative supply ≈ -3.5V.

4N.4.5 . . . A remedy for clipping: the push–pull

AoE §2.4.1

What is to be done? We can offer a dumb answer and a smart one. The dumb answer is to decrease the value of R_E so as to push the clipping voltage as close to the negative supply as needed. To go to -10V, for example, we could cut R_E's value to 0.5k.

That doesn't sound bad. But it does sound bad if we change the numbers. Suppose our goal is to drive, as load, an 8Ω speaker. To clip at -10V calls for R_E of 4Ω. That works, but consider the power dissipation when V_{OUT} is zero volts – in other words, the power dissipated in order to achieve *silence*. The quiescent current is 15V/4Ω, almost 4 amps; this current flowing between the 30V power supply span dissipates almost 120W – and this for silence!

So, we need a better answer, and that is the *push–pull* follower, one that replaces the pull-down *resistor* with a pull-down *transistor*, as in Fig. 4N.10, repeated in Fig. 4N.27 with added scope image showing crossover distortion.

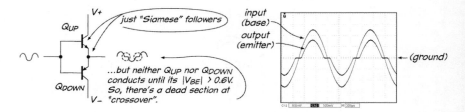

Figure 4N.27 Push–pull, again.

The push–pull is much more efficient than the simpler follower – dissipating little or no power with V_{OUT} at 0V.[18] The push–pull, with its good efficiency, dominates the output stages of amplifiers. Your lab function generator uses one; your audio amplifier probably uses one;[19] so does the output of the operational amplifiers that you soon will meet.

A respectable push–pull must include a remedy for the cross-over distortion that troubles the simple design shown above. A pair of diodes or a pair of transistor V_{BE}'s can bias the bases of the output transistors apart; or an operational amplifier and feedback can be used, as you will demonstrate in Lab 6L.

AoE §2.4.1A

4N.5 A strikingly different transistor circuit: the switch

AoE §2.2.1

AoE presents this circuit first, as it introduces transistors, because the switch is easy to understand. We present it last, because it is so exceptional. Either way, it is important to keep it apart from all the other transistor applications in your thinking, because the *switch* deliberately violates a fundamental rule governing the design of all the other circuits: the *switch* operates in only two conditions, both pathological for an ordinary transistor circuit: it is totally ON ("saturated," in the transistor jargon) or totally OFF. Since it is saturated when ON, it does not show the high-input impedance we think of as one of the virtues of a transistor circuit.

Nor does the usual rule, $I_{\text{C}} = \beta \times I_{\text{B}}$ hold. A switch should be purposefully overdriven, with about 10 times the minimum base current to pass the required I_{C}. That saturates the switch strongly, keeping V_{CE} low to minimize power lost in the switch. The best switch would disappear electrically when ON, putting all power into the load, none into itself.

Skeptical students sometimes ask, "Why bother? You start with a switched input level (either High or Low), and that simply generates a switched output level (either High or Low). Pointless." (Sounds rather like the follower, which takes an input wiggle and delivers the same wiggle as output.) It's a fair question.

The answer is that the input level usually comes not from a manual switch but from a circuit fragment not strong enough to switch the "load." So, you might think of the switch circuit as (once again) a current amplifier. It resembles the follower, but it is simpler. It is also true that one can build a system of switches driving switches, and of these one can make useful "binary digital" circuits. Those digital circuits are the subject of the second half of this course. We concede that almost none of those

Figure 4N.28 Switch: a useful circuit that may seem silly at first glance....

---

[18] "No power," you may insist, and you're right for the simple push–pull that we have pictured. The better version, however, which eliminates cross-over distortion by permitting a small current to flow through both transistors with V_{OUT} at 0V does dissipate a little power. See AoE §2.4.1B.

[19] Not your digital music player, though. It is likely to use a power switch instead, for its still greater efficiency. See AoE §2.4.1C and the switching amplifier of Lab 12L, the LM4667.

digital circuits use *bipolar* transistors like those we are discussing now (they use MOSFETs instead – see Lab 12L); but they do rely on switching just like what we have been looking at.

So, let's note that switches are important circuits – but let's also keep them in a compartment all their own, because the rules for their use are so different from those for followers, current sources, and amplifiers – the *linear* circuits that form the mainstream of our designing with transistors.

4N.6 Recapitulation: the important transistor circuits at a glance

AoE §2.2.9B

To start you on the process of getting used to what bipolar transistor circuits look like, and to the crucial differences that come from what *terminal* you treat as output, here is a family portrait, stripped of all detail:

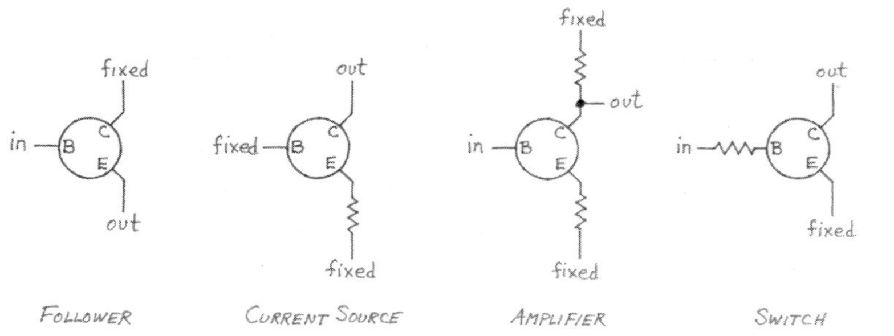

Figure 4N.29 The most important bipolar transistor circuits: sketch.

Note, among other differences in the circuits shown in Fig. 4N.29, that only the *follower* takes its output at the emitter. There, the impedance can be low.

4N.6.1 Impedance at the collector, by the way

An output at the collector, in contrast, provides *high* impedance, because at that terminal it is *current*, not *voltage* that is determined. So, the quotient $\Delta V_{CE}/\Delta I_C$ is very large (since ΔI_C is very small – ideally, zero). This is a case we foresaw back in §1N.2.1, when we promised you that the notion of *dynamic resistance* ($R_{dynamic}$) would become important to your understanding of transistor circuits. $R_{dynamic}$ at the collector is *very large*: ideally, infinite; practically, at least hundreds of kilohms. This fact is important not only for current sources, but also for the common-emitter amplifier, whose R_{out} is determined by R_C alone, since the effective R_{out} is R_C in parallel with the collector's huge $R_{dynamic}$.

4N.6.2 ... and the switch stands apart

The switch behaves entirely differently from all the other applications; it stands apart because it is run in the two conditions always *avoided* for the other circuits – as we said in §4N.5: OFF or fully ON ("saturated"). The uniqueness of the switch means you must use Fig. 4N.29 with care.

On Day 5 we will begin to use the more complicated *Ebers–Moll* model for the transistor. But the simplest model of the transistor, presented in this class, will remain important. We will always try to use the least complicated view that explains circuit performance, and often the very simplest will suffice.

4N.7 AoE Reading

Reading:
    Chapter 2, §§2.1–2.2.8 and §2.4.1 (push–pull).
    Take a quick, preliminary look at §2.2.9 (transconductance). Next time we will look more closely at the Ebers–Moll view of transistors. In §4L.4.1 the high-gain amp presents for the first time a circuit that requires analysis with the Ebers–Moll model.

Problems:
    Problems in text and Additional Exercise 2.25.

4L Lab: Transistors I

4L.1 Transistor preliminaries: look at devices out of circuit

4L.1.1 Transistor junctions are diodes

Here is a method for spot-checking a suspected bad transistor: the transistor must look like a pair of diodes when you test each junction separately. But, *caution*: do not take this as a description of the transistor's mechanism when it is operating: it does *not* behave like two back-to-back diodes when operating (the circuits of Fig. 4L.1, if made with a pair of ordinary diodes, would be a flop, indeed).

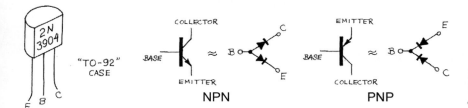

Figure 4L.1 Transistor junctions: (for testing, not to describe transistor operation).

4L.1.1.1 A game: discover transistor type and pinout, on a desert island

See if you can determine the type (*npn* or *pnp*) and *pinout* (base, emitter, collector leads) of a 2N3904 transistor by *experiment*. (On a desert island with DVM and a box of unknown transistors, you could sort and label the transistors; thus you could start the process of building a radio transmitter to summon your rescuers.) You can settle both *type* and *pinout* questions by flipping to the pinout section at the back of this book, of course.

But try using a DVM, instead, employing its *diode test* function. (Most meters use a diode symbol to indicate this function.) The diode test applies a small current (a few milliamps: current flows from Red to Black lead), and the meter reads the junction voltage. Using the diode test function, you can determine the directions in which pairs of leads conduct. That will let you answer the *npn* versus *pnp* question. You can also distinguish *BC from BE* junctions this way: the BC junction is the larger of the two, and the lower current density across that *larger* junction is revealed by a slightly *lower* voltage drop.

4L.1.2 'Decouple' power supplies: fuzz warning

This is the first lab in which you may see high-frequency oscillations on your circuit outputs. At a modest sweep rate, this fuzz will look like a thickening of the trace; at a high sweep rate it may reveal itself to be a more-or-less sinusoidal waveform in the range between a few hundred kHz and 100MHz. If you see such noise, you need to quiet your circuit with *decoupling* capacitors. See §4L.6.

4L.2 Emitter follower

Wire up an *npn* transistor as an emitter follower, as shown in Fig. 4L.2. Drive the follower with a sinewave that is symmetric about zero volts (be sure the dc "offset" of the function generator is set to zero), and look with a scope at the poor replica that comes out. Why does this happen?[1]

If you turn up the waveform amplitude you will begin to see bumps *below* ground. How do you explain these?[2]

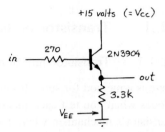

Figure 4L.2 Emitter follower. The small base resistor is often necessary to prevent oscillation.

Now try connecting the emitter return (the point marked V_{EE}[3]) to -15V instead of ground, and look at the output. Explain the improvement.

4L.2.1 Input and output impedance of follower

Measure Z_{in} and Z_{out} for the follower in Fig. 4L.3.[4]

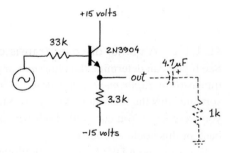

Figure 4L.3 Follower: circuit for measuring Z_{in} and Z_{out}. Disconnect the 1k *load* resistor while measuring Z_{in}.

Measure Z_{in}

In the circuit of Fig. 4L.2 replace the small base resistor with progressively larger resistors, in order to simulate a signal source of moderately high impedance, i.e., low current capability.

Start with 33k, as shown in Fig. 4L.3; you are likely to see attenuation of only 5% to 10%. With such small attenuation, it is possible but tedious to infer β.

---

[1] This is pretty simple: when V_{in} falls below about +0.6V we are violating one of our transistor "ground rules:" we are failing to permit forward-biasing of the V_{BE} junction. I_C and I_E fall to zero.
[2] Powerful hint: the data sheet for the 2N3904 shows V_{BE} breakdown can occur – for a *reverse* bias across the *BE* diode – at a voltage as low as 6V.
[3] This notation, repeating the subscript letter, is used to indicate power supply voltages: V_{EE} where the supply is negative (and applied to the emitter in an *npn* circuit), V_{CC} where the supply is positive (and applied to the collector.
[4] We are calling this Z, but what interests us here is really R; we want to avoid effects that are frequency dependent. Particularly, we don't want to find ourselves studying attenuations caused by stray capacitance.

If you install larger resistors in place of this 33k, eventually you will see attenuation of about 50%, and then your observation (and your arithmetic) will become easy. You'll know, if you get to 50% attenuation, that your installed R is about equal to R_{IN}. Use this data to make your inference of β.

Caution: Make sure that it is R_{IN} that you are observing, and not an effect of C_{IN} or C_{probe} that you are seeing. To make sure of this, adjust the input frequency until no appreciable phase shift appears between the original signal and what you see at the transistor's base.

With 1k "load" *detached*, measure Z_{in} for the circuit. In this case, that is the impedance looking into the transistor's base (the 33k series resistor, or the larger value you may have installed, is *not* a part of the follower. It is included to model a signal source with mediocre R_{out}). You can discover Z_{in} by using the scope's two channels to look at both sides of the base resistor. For this measurement the 3.3k emitter resistor is treated as the follower's "load". Use a small signal – less than a volt in amplitude.

The attenuation that you see may be *slight*, at least at first, and therefore hard to measure. If you are using a digital scope you can be lazy, using its amplitude-measurement functions. If you are using an analog scope, you need to work harder. The best way to measure a small attenuation using an analog scope is to take advantage of the *percent* markings on the scope screen.

Suggestions for measurement of Z_{in} (analog scope):

- Center the two waveforms on the 0% mark – these are the waveforms that appear on the two sides of the base resistor: let's call the waveforms "source" and "input";
- *AC couple* the signals to the scope, to ensure centering.
- Adjust the function-generator's amplitude to make the *source* peak just hit 100%;
- Now read the *input* waveform's amplitude in percent.

Does your result make sense? Is the follower transforming the impedance "seen through it," as promised? Once you have measured a value of Z_{IN} you can infer the transistor's β (or "h_{FE}:" see AoE §2.2.3B.) Make a note of this calculated β; in a few minutes you will want to compare it to the β you infer from your measurement of Z_{out}.

Measure Z_{out}

Replace any other base resistor you may have installed, *returning to the value of* 33k shown in Fig. 4L.3.[5]

Now measure Z_{out}, the output impedance of the follower, by connecting a 1k load to the output and observing the drop in output signal amplitude; again use a small input signal, less than a volt.

The procedure for measuring Z_{out} is slightly different from the one used to measure Z_{in}.

In order to measure Z_{out}, you need a two-step process:

- with *no load* attached, measure the amplitude of V_{out};
- then attach the 1k "load," and note the resulting attenuation;
- infer Z_{out}: you recall that you have built a voltage divider, where the upper "resistor" is the circuit's Z_{out} and the lower "resistor" in the divider is the load. The value of the *base resistor* is critical to your calculation, of course; this is the "source resistance" that the follower reduces by the factor β. So make note of what base resistor you are working with.

Infer β once again: in principle it should match the β you inferred from your measured Z_{in}.

---

[5] Larger R values make your work harder, not permitting you to ignore the effect of R_E, whose value strictly parallels the path through the transistor. $R_{\mathrm{source}}/\beta$ usually is so much smaller than R_E that can ignore the latter. A very large base resistor spoils that simplification. This exercise is hard enough without that complication.

Lab: Transistors I

The blocking capacitor used in measurement of Z_{out}: Why do we suggest that you use a blocking capacitor as you measure Z_{out}? The answer is rather subtle, in this case, because here you could get away with omitting the blocking cap; but in many other cases you could not get away with that.

We include the blocking capacitor to avoid disturbing *DC* levels in the circuit, while we watch what happens to time-varying (or *AC*) *signals*. In this case, the DC level at the emitter is approximately $-1V$; this is close enough to ground so that omitting the blocking cap would do no harm. Omitting the cap would alter that level only slightly. But in many other cases, where the emitter's DC voltage happened to be well away from ground, attaching a 1k *DC* "load" to ground would alter the DC or "bias" level appreciably. We do not want to do that; we mean only to see what the load does to *signal* amplitude (and we assume the signal is a voltage *wiggle*).

It's too bad that the cap in this case doesn't provide a more striking benefit. But we are trying to teach you a good habit: if you're interested in AC behavior, measure that behavior without disturbing DC conditions.

4L.3 Current source

Construct the current source shown in Fig. 4L.4 (sometimes called, more exactly, a current "sink").[6]

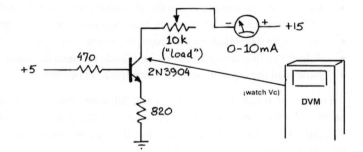

Figure 4L.4 Transistor current source.

Slowly vary the 10k variable load, and look for changes in current measured by the VOM. What happens at maximum resistance? Can you explain, in terms of *voltage compliance* of the current source ("compliance" is jargon for "range in which the circuit works properly")?[7]

Even within the compliance range, there are detectable variations in output current as the load is varied. What causes these variations?[8]

4L.4 Common-emitter amplifier

Wire up the common-emitter amplifier shown in Fig. 4L.5. What should its voltage gain be? Check it out. Is the signal's phase inverted?

---

[6] The term "current source" unfortunately can be ambiguous. It describes the entire class of circuits that hold current constant despite voltage variation. But the term also can be used to describe a particular *sub-class*, the sort of circuit that feeds current into a load whose other other terminal is more negative. Such a *source* then is contrasted against a current *sink*, one like the one in Fig. 4L.4, which draws current *from* a load. Thus current sources come in two flavors, awkwardly called "sinks" and "sources." Context usually makes clear whether a reference to a "current source" indicates sense of current flow as well as constancy of current.

[7] The current source fails when the transistor saturates. The DVM will let you measure or infer V_{CE}. Thus you can look for the relation between saturation and the fall of I_{out}.

[8] Early effect is the principal cause: varying V_{CE} slightly varies the effective base width. See §5S.2.

4L.4 Common-emitter amplifier

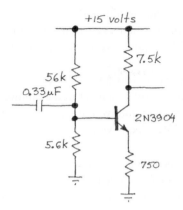

Figure 4L.5 Common-emitter amplifier.

Is the collector *quiescent* operating point correct (that is, its resting voltage)?[9] How about the amplifier's low frequency 3dB point?[10] What should the output impedance be?[11] Check it by connecting a resistive load – let's say 7.5k – using a blocking capacitor. (The blocking cap, again, lets you test impedance at signal frequencies without messing up the biasing scheme.)

4L.4.1 Maximizing gain: sneak preview of Ebers–Moll at work

Now let's try a case that our first, simple, view of the common-emitter amplifier does not describe accurately: check out Fig. 4L.6 to see what happens if we parallel the emitter resistor with a big capacitor, so that at "signal frequencies" R_E is shorted out.

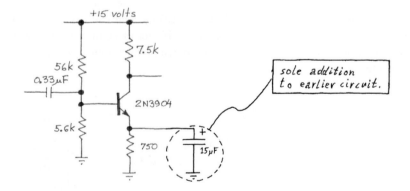

Figure 4L.6 Grounded emitter amplifier.

Modify your common-emitter amp to make the amplifier shown in Fig. 4L.6 (similar to Fig. 5N.17). What circuit properties does your addition of the *bypassing* capacitor affect? It affects gain, because R_E disappears from the gain equation. Does the C alter the biasing of the output (which was designed to be centered roughly at half the supply voltage)?[12]

---

[9] What level would be "correct?" The voltage that permits the widest swing without "clipping" – hitting either upper or lower limit. Roughly, that quiescent point would be half the supply voltage. A perfectionist would say, "Better to put it midway between the lower limit of about 1V (V_E) and the supply. But with a 15V supply we don't mind the lazier answer, "half the supply voltage." This will be our usual answer, for single-supply circuits.

[10] To calculate this you'll need to decide what is the effective *resistance* that should be paired with the blocking capacitor. C and that R form a high-pass, as you know.

[11] Yes, just R_C. You've seen this point explained in Chapter 4N.

[12] No: biasing is a simple *DC* effect.

Now drive this new circuit with a small *triangle* wave at 10kHz, at an amplitude that almost produces clipping (you'll need to use plenty of attenuation – 40dB or more – in the function generator). Does the output waveform look like Fig. 4L.7 (compare Fig. 5N.12)? Explain to yourself exactly why this "barn-roof" distortion occurs.

Installing the cap that bypasses R_E does *not* deliver infinite gain: sorry! In Day 5 you'll discover that we can predict the gain by adding to our transistor model a little resistor-like element in the emitter, a modelling device that we call (affectionately) "little r_e." In the present circuit, where we're running a quiescent current of about 1mA, little r_e takes a value of about 25Ω. See if your circuit's gain is consistent that value for r_e.

Figure 4L.7
Large-swing output of grounded emitter amplifier when driven by a triangle wave.

This measurement will be difficult. First, you will need to reduce the function generator output to a level close to the minimum possible. Then you will find the input is immeasurable small, viewed with a 10× probe. This, then, is one of the rare occasions when you need to revert to a 1× display: use a BNC rather than a 10× probe as you watch the *input* signal.[13]

Typically, the gain you observe is around 250 – lower than you predicted. Low I_C can help explain that; so can any curvature that remains in the output waveform: the gain revealed by such a waveform is sometimes higher than the quiescent value, sometimes lower, but the *net* effect of the curvature is to *reduce* gain.

4L.5 Transistor switch

The circuit in Fig. 4L.8 differs from all those you have built so far: the transistor, when ON, is *saturated*. In this regime you should not expect to see $I_C = \beta \times I_B$. Why not?[14] In addition, β for this big power transistor is much lower than what you have seen for the *small signal* transistor, 2N3904 –at least at high currents. Minimum β for the MJE3055T is a mere 20, at I_C=4A (see plot in Fig. 4L.9). But the transistor is big, and lives in a big package ("TO-220"), a package that can dissipate a lot of heat, especially if the metal package is thermally attached to a still bigger piece of metal (a "heat sink"). The large size of the transistor itself keeps current density down and saturation voltage low. You'll measure that $V_{CE(sat)}$ in a few minutes.

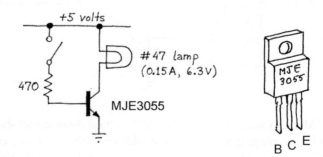

Figure 4L.8 Transistor switch: twist each transistor lead 90°, so that each fits easily into the breadboard's slots.

Turn the base current on and off by flipping the toggle switch – or, if you're too lazy to wire a switch, by pulling one end of the resistor out of the breadboard and touching it either to +5 or to ground. What is I_B, roughly? What is the minimum required β?[15]

---

[13] If your scope probe has a 10×/1× selector switch, just select 1×.
[14] One way to explain this is just to note that achieving a current as high as that would require either a smaller resistance on the collector or a larger power supply. When the transistor is saturated, the current is limited not by the transistor but by the load.
[15] You'll notice that we are *overdriving* the base, here, as is usual in a switch. Since base current here is about 10mA, even at

4L.5.1 Saturation or "On" voltage: $V_{CE(sat)}$

Now let's make the transistor work harder. Replace the lamp with a 10Ω *power* resistor (not an ordinary 1/4W resistor; consider how much power the resistor must dissipate: $V^2/R=25/10=2.5W$). A 1/4W resistor would cook in this circuit.

Measure the saturation voltage, $V_{CE(sat)}$, with DVM or scope. Then parallel the base resistor with 150Ω, and note the improved $V_{CE(sat)}$. Compare your results with the results promised by the data set out in Fig. 4L.9. Note that the curves assume heavy base drive: $I_B = I_C/10$, not I_C/β.

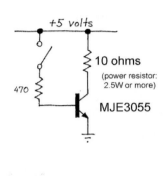
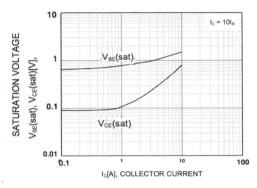

Figure 4L.9 Switch saturation: a heavier load current asks more of the switch here (data courtesy of ON Semiconductor).

We will return to transistor switches later, in Lab 12L, when we'll meet the leading competitor for the tasks a '3055 can do: a big power field-effect transistor ("MOSFET"). At that time, we'll set up a competition between a '3055 and a MOSFET and see which does a better job of delivering power to the load.

4L.5.2 Switch an inductive load (more exciting)

Now replace the resistor at the collector with a 10mH *inductor* as in Fig. 4L.10. Replace the 5V supply with 1.5V from a single AA cell (we don't want to overheat the inductor – and the low cell voltage makes all the more impressive the voltage spike that soon will appear). Drive the 1kΩ base resistor with a square wave from the function generator's "TTL" or "Sync" terminal (this is a "logic level" waveform: a square wave switching between ground and four or five volts).

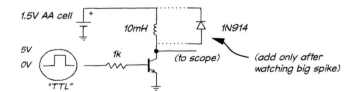

Figure 4L.10 Transistor switch with inductive load.

At the transistor's collector, you should find the inductor giving the transistor an alarming voltage spike (we saw about 100V). When the transistor tries to turn off, the inductor tries to keep the current flowing. Its method is to drive the collector voltage up, until the transistor "breaks down," permitting the inductor to have its way – keeping the current flowing despite the transistor's attempt to turn it off abruptly.

A diode from collector to V_+ tames this voltage spike; such a diode clamp is a standard protection

---

the specified minimum *beta* of 20, the transistor could pass more current than the load will permit. That's good: the switch will be well saturated, and its V_{CE} will therefore be low.

176 Lab: Transistors I

in circuits that switch inductive loads. Most transistors don't like to break down (this time we didn't mind since our goal was to show you this spike). Incidentally, we'll see a circuit that makes good use of this voltage spike when we look at switching power supplies later in Lab 12L.

4L.6 A note on power supply noise

Note §9S.3 is devoted to noise problems. Here, we offer just a few pointers related to one form of noise.

Fig. 4L.11 shows one of today's followers, swept rather slowly (at $20\mu s$/division). The follower has been fed a sine of about 15kHz from the function generator. A strange *thickening* of the trace appears.

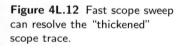

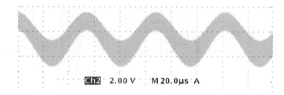

Figure 4L.11 Thickening of scope trace indicates a nasty "parasitic" oscillation.

Sweeping the scope faster (Fig. 4L.12, at 10ns/div) resolves the "thickening" into a very fast sinusoid – up in the *FM* radio broadcast range. Decoupling the power supplies should eliminate this problem (see below).

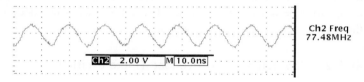

Figure 4L.12 Fast scope sweep can resolve the "thickened" scope trace.

Oscillations versus radio pickup

If you see this fuzz, try turning off your breadboard power supply. If the fuzz disappears, your circuit is guilty: it was causing the fuzz, by running as an unintended *oscillator* (we'll look closely at the question how this can happen, in Lab 9L). If the fuzz persists, your circuit is innocent and you're probably seeing radio broadcast stuff picked up by your wiring. There's no quick way to eliminate that; you see it whenever your scope gain is very high and the point you look at is not of low impedance at those high frequencies.

Remedies If your circuit is oscillating – not merely picking up radiated noise – there are three remedies you might try, in sequence:

(1) make sure that you are watching the circuit output with a scope *probe*, not with a *BNC* cable. The $10\times$ heavier capacitive load presented by the *BNC* cable often brings on oscillation.
(2) try shortening the power and ground leads that feed your circuit. Six inches of wire can show substantial inductive reactance at megahertz frequencies; shorter leads make the power supply and ground lines more nearly ideal – lower impedance – and harder for naughty circuits to wiggle.
(3) if the oscillation persists after those two attempted repairs, then add a "decoupling" or "bypass" capacitor between each supply and ground. Use a *ceramic* capacitor of value 0.01 or $0.1\mu F$; place it as close to your circuit as possible – again, inches can matter, at high frequencies.

Incidentally, we have listed this option last not because there's anything wrong with decoupling the supplies; soon you will be putting decoupling caps in routinely. We placed this remedy last just because it *is* so effective (and therefore mandatory): we wanted you to see first that the first two remedies also sometimes are sufficient.

4W Worked Examples: Transistors I

4W.1 Emitter follower

AoE §2.2.4A

AoE works a similar problem in detail: §2.2.4A. The example below differs in describing a follower for *AC* signals. That makes a difference, as you will see, but the problems are otherwise very similar.

4W.1.1 Problem: AC-coupled follower

Design a single-supply voltage follower that will allow this source to drive this load, without attenuating the signal more than 10%. Let $V_{CC} = 15V$, let the quiescent I_C be 0.5mA. Put the 3dB point around 100Hz.

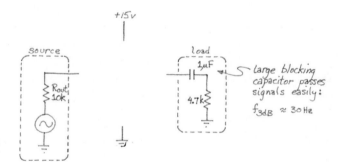

Figure 4W.1 Emitter follower (your design) to let given source drive given load.

4W.1.2 A solution

Before we begin, perhaps we should pause to recall why this circuit is useful. It does *not* amplify the signal voltage; in fact we concede in the design specification that we expect some *attenuation*; we want to limit that effect. But the circuit does something useful: before you met transistors, could you have let a 10k source drive a 4.7k load without having to settle for a good deal of attenuation? How much? We will try to explain our choices as we go along, in scrupulous – perhaps painful – detail.

Draw a skeleton circuit

Perhaps this is obvious; but start by drawing the circuit diagram without part values, as in Fig. 4W.2. Gradually we will fill those in.

Choose R_E to center V_{OUT}

To be a little more careful, we should say,

"we aim to center $V_{\text{out-quiescent}}$, given $I_{\text{C-quiescent}}$"

"Quiescent" means what it sounds like: it means conditions prevailing *with no input signal*. In effect, therefore, *quiescent* conditions mean *DC* conditions, in an *AC* amplifier like the present design.

4W.1 Emitter follower

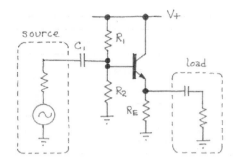

Figure 4W.2 Emitter follower skeleton circuit: load is AC coupled.

Figure 4W.3 Choose R_E to center V_{OUT}.

And anticipating some complications that we will meet in other transistor circuits, namely the *common-emitter* amplifier and the *differential amplifier*, we should acknowledge that, strictly, our goal is to center V_{out} *in the range available to it*. In the present case, that range does extend all the way from lowest voltage (ground) to most positive (V_+). It does not always behave so simply.

Center V_{base}

Here we'll be a little lazy: by centering the base voltage we will be sure to *miss* centering V_{OUT}. But we'll miss by only 0.6V, and that error won't matter if V_{CC} is big enough. The error is about 4% if we use a 15V supply, for example.

Centering the *base* voltage makes the divider resistors equal; that, in turn, makes their R_{Thevenin} very easy to calculate.

Choose bias divider Rs so as to make bias stiff enough

Stiff enough means, by our rule of thumb, $\leq 1/10$ of the R that it drives. What it drives is the base, and we need to know the value of this $R_{\text{in at base}}$ at DC (that is, not considering *signal* frequencies). If we follow that rule, loading by the base will not upset the bias voltage that we aimed for (to about 10%).

$R_{\text{in at base}}$ is just $\beta \times R_E$, as usual. That's straightforward. What is not so obvious is that we should *ignore* the AC-coupled *load*. That load is *invisible* to the bias divider, because the divider sets up *DC* conditions (steady state, quiescent conditions), whereas only *AC* signals pass through the blocking capacitor to the load. Fig. 4W.4 spells out the process – but note that the process begins on the right, with the impedance that the divider is to drive. That concludes the setting of DC conditions. Now we can finish by choosing the coupling capacitor (also called "blocking capacitor;" evidently both names fit: this cap couples one thing, blocks another).

Choose blocking capacitor

We choose C_1 to form a high-pass filter that passes any frequency of interest. Here we have been told to put f_{3dB} around 100Hz.

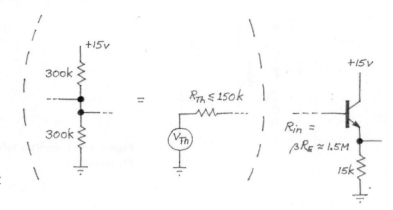

Figure 4W.4 Set $R_{\text{TH-bias}} \ll R_{\text{in at base}}$.

The only difficulty appears when we try to decide what the relevant R is in our high-pass filter: see Fig. 4W.5.

Figure 4W.5 What R for blocking cap as high-pass?

We need to look at the input impedance of the follower, seen from this point. The bias divider and transistor appear in *parallel*.

Digression on series versus parallel Stare at the circuit till you can convince yourself of that last proposition. If you have trouble, think of yourself as a little charge carrier – an electron, if you like – and note each place where you have a *choice* of routes: there, the circuit offers *parallel* paths; where the routes are obligatory, they are in *series*. Don't make the mistake of concluding that the bias divider and transistor are in series because they appear to come one after the other as you travel from left to right.

So, $Z_{\text{in-follower}} = R_{\text{THbias}}$ in parallel with $R_{\text{in at base}}$. The slightly subtle point appears as you try to decide what $R_{\text{in at base}}$ ought to be. Certainly it is $\beta \times$ something. But $\times$ what? Is it just R_E, which has been our usual answer? We did use R_E in choosing R_{TH} for the bias divider.

But this time the answer is, 'No, it's not just R_E,' because the *signal*, unlike the DC bias current, passes *through* the blocking capacitor that links the follower with its load. So we should put R_{load} in parallel with R_E, this time.

The impedance that gets magnified by the factor β, then, is not 15k but (15k || 4.7k), about 15k/4 or 3.75k. Even when increased by the factor β, this impedance cannot be neglected for a 10% answer:

$$R_{\text{in}} \approx 135\,\text{k} \,||\, 375\,\text{k}.$$

That's a little less than 3/4 of 135k (since 375k is a bit short of 3×135k, so we can think of the two resistors as one of value R, the other as three of those Rs in parallel – using a trick we mentioned in discussing the parallel circuits of Fig. 4N.11. This method is noted also in AoE Chapter 1). Result: we have four parallel resistors of 375k: roughly equivalent to 100k. (By unnatural good luck, we have landed within 1% of the exact answer, this time.)

4W.2 Phase splitter: input and output impedances of a transistor circuit

So, choose C_1 for f_{3dB} of 100Hz. Then

$$C_1 = \frac{1}{2\pi 100\text{Hz} \times 100\text{k}} \approx \frac{1}{6 \times 10^2 \times 100 \times 10^3} = \frac{1}{60} \times 10^{-6} \approx 0.016\mu\text{F};$$

$C_1 = 0.02\mu\text{F}$ would be generous.

Recapitulation

For people who hate to read through explanations in words, Fig. 4W.6 restates what we have just done.

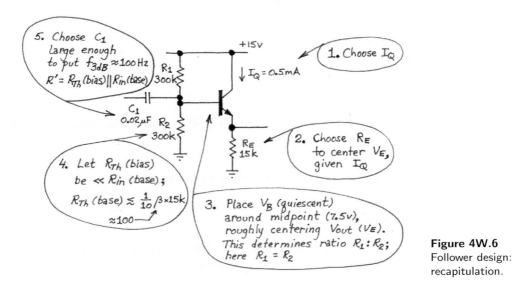

Figure 4W.6 Follower design: recapitulation.

4W.2 Phase splitter: input and output impedances of a transistor circuit

Problem Design a circuit that puts out waveforms inverted, with respect to each other

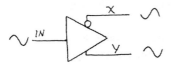

Figure 4W.7 Unity-gain phase splitter: generic.

Figure 4W.7 shows a circuit that you can make sense of with the tools from this, your first, day with transistors – our "simple" view of transistor operation. The circuit is called a "phase splitter," because it puts out two waveforms, 180° out of phase with each other. But it's useful to us, just now, because it gives a workout in calculating impedances – and that workout gives us a chance to review much of what we met in Lab 4L.

- Specifications:
  - Power supply: +20V only
  - Signal frequency range: 100Hz and up

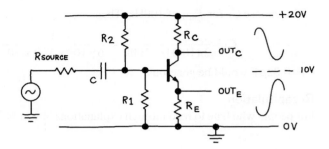

Figure 4W.8 Phase splitter, no component values.

- $R_{\text{OUT-signal-source}} \leq 5\text{k}\Omega$
- Quiescent current (I_C):
- Calculate input and output impedances

Solution

Skeleton circuit: First, let's start with the skeleton circuit of Fig. 4W.8 and calculate what component values we ought to choose.

Choose component values: We'll need the rules stated in the first transistor class:

- Simple: $I_C = I_B \times \beta$
- Simpler: $I_C \approx I_E$; $V_{BE} = 0.6\text{V}$

The second rule is enough to let us calculate the voltage gain of this circuit; the first rule is necessary to let us calculate most input and output impedances. Let's do this in stages:

Figure 4W.9 Choose value for collector resistor.

- Choose R_C, the collector resistor, see Fig. 4W.9

  In order to choose this value, we need to make an initial design choice – what the quiescent collector current should be. Let's be lazy and conventional and make it 1mA. Given that current, we can choose the R value by determining what *quiescent* voltage we would like to see at output Out_X.

  We don't want to reach, mechanically, for the answer, "Half the power supply." No. The basis for setting the quiescent voltage is to allow maximum swing without clipping – and this time the emitter voltage is not tamely sitting still (or nearly) as in the usual common-emitter amplifier. Instead, the emitter is swinging as widely as the collector swings – so, the range of voltage available to the collector is not the full power supply, but half of it.

  Therefore, we should put $V_{C\_quiescent}$ at the midpoint of the range available to it. That range is 10–20V, so the quiescent voltage should be 15V. To get this result, we want to drop 5V. So, $R_A = 5\text{k}$. The nearest 10% value is 4.7k.

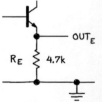

Figure 4W.10 Choose value for emitter resistor.

- Choose R_E, the emitter resistor, see Fig. 4W.10.

  Again the goal is to allow maximum swing, so we place $V_{Y\_quiescent}$ at the midpoint of its range: we place it at 5V. With 1mA flowing, we need 4.7k again.

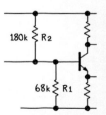

Figure 4W.11 Choose value for bias divider.

- The bias divider, see Fig. 4W.11.

  This is a little more involved.
  - Voltage: just a little above $V_{E\_quiescent}$. So, the exact target would be about 5.6V. Our standards aren't as high as that, in a 20V span. Let's be lazier and put the base voltage at 5V rather than 5.6V. That eases the arithmetic: we need 1/4 of the 20V supply, so the ratio R2:R1 is 3:1.
  - Values: we want the bias divider to be "stiff" relative to what it drives, the transistor base.

4W.2 Phase splitter: input and output impedances of a transistor circuit

- ○ $R_{\text{in\_base}}$ This, we learned last time, is $\beta \times R_E \geq 500\text{k}$.
- ○ $R_{\text{Thevenin\_divider}}$ should be small relative to what it drives: $\leq 50\text{k}$.
- ○ R_1, R_2 values: $R_{\text{Thevenin\_divider}} = R_1 \parallel R_2$. Since $R_2 = 3R_1$, $R_{\text{Thev}} = \frac{3}{4}R$. So the smaller R should be about $\frac{4}{3}R_{\text{Thev}} \approx 66\text{k}$: we're fine with 68k. The upper resistor should be about $3\times$ that. 180k is a little low – but we don't really mind since we're happy to put V_{base} a little above 5V, recalling the 0.6V V_{BE} drop.

- The blocking capacitor[1], see Fig 4W.12. (We've inserted the C value in the figure; we'll explain how we chose that value.)

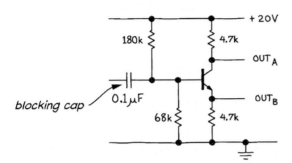

Figure 4W.12 Choose value for coupling/blocking capacitor.

We need an f_{3dB} as a design goal. Let's suppose we want to pass low-quality audio, so we can put f_{3dB} at 50Hz.

After that choice, we need only to determine what effective R this C is paired with. That effective resistance is the bias divider in parallel with the transistor's base.[2]

You could calculate this – but, as usual, we prefer to be intelligently lazy. We know that $R_{\text{in\_base}}$ is much larger than $R_{\text{Thevenin\_divider}}$. Why? Because we designed it that way, a moment ago. So, we can neglect $R_{\text{in\_base}}$, and treat the effective R as only $R_{\text{Thevenin\_divider}}$. We needn't calculate that value, because we know it. This is the value we started with as goal, when we designed the bias divider: 50k.

An exact C value would be about $0.06\mu\text{F}$. $0.1\mu\text{F}$ will do nicely. There is no harm at all in pushing f_{3dB} a little lower than the target, here: the cap's function is only to block DC. If it passes signals at a little lower frequency than originally planned, good.

Now that the values are in place, we can calculate the impedances in and out.

R_{out} at collector: This is the straightforward one. Put the two paths in parallel, as usual (by analogy to R_{Thevenin} – a model that is useful for impedance measurements far beyond its original scope, which covered only *resistors*).

Figure 4W.13 $R_{\text{out\_C}}$: turns out equal to collector resistor.

How should we treat the impedance seen "looking into" the collector? Step 1: don't fall for the error of thinking that the collector looks like the emitter, so that you'll see some *lens* effect using β. No. The collector behaves radically differently from the emitter.

Step 2: take advantage of the fact that, whereas the emitter is a *voltage* source, the collector is a *current* source: its *current* is determined. We have said this elsewhere, but let's say it again: when the current is determined, one can change voltage without changing current appreciably. So, the fraction

---

[1] As you know, this capacitor could just as well be called the "coupling capacitor." It *blocks* DC; it *couples* the AC signal.
[2] We know you wouldn't fall into the trap of thinking that these impedances are in *series*, just because you may see one drawn to the left of the other. A wire joins the two paths. We offered the same reminder earlier in §4W.1, the single-supply follower.

$\Delta V/\Delta I$ is very large (ideally, infinite). That fraction is just our definition of effective *resistance* at a point (we sometimes call this "dynamic resistance").

So, this very-large impedance looking into the collector has no effect when paralleled with R_C: the R_C dominates and in this circuit defines R_{out} at the collector. As for the common-emitter amplifier, R_{out} at the collector is the collector resistor, R_C.

R_{out} **at emitter:** This is not so simple as the impedance at the collector. Here, we do see the transistor's "lens" effect, and to determine R_{out} at the emitter we "look through" the transistor toward the signal source. You worked through such a problem in Lab 4L when you measured R_{out} for the follower.

R_{out} is R_E in parallel with the other path, through the emitter. Let's write this out, first, with painful rigor. Then we'll throw out what doesn't matter much:

$$R_{\text{out\_Emitter}} = R_E \parallel \left(r_e + \frac{R_{\text{Thev\_bias}} \parallel R_{\text{source}}}{(1+\beta)}\right).$$

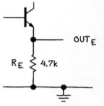

Figure 4W.14
$R_{\text{out\_E}}$: rather complicated, but approximately R-signal-source/beta.

This looks pretty formidable. So, let's start throwing out what has little effect.

- R_E normally can go: it should be much larger than what parallels it.
- $R_{\text{Thev\_bias}}$ can go, since it defines Z_{in} for the circuit, as we saw above, and this Z_{in} must be large relative to $R_{\text{OUT-signal-source}}$.
- We cannot neglect r_e,[3] since its value at $I_C=1\text{mA}=25\Omega$ is not negligible compared with the value of $R_{\text{OUT-signal-source}}/\beta$. That value is $5k/100 = 50\Omega$.

With the help of this intelligent laziness, then, we can reduce the formidable equation to the tame

$$R_{\text{out\_Emitter}} \approx r_e + \frac{R_{\text{source}}}{\beta} = 25 + \frac{5k}{100}.$$

The value for $R_{\text{out\_Emitter}}$ calculated in this lazy way is 75Ω; the exact value is 69Ω. The error is about 8%, quite tolerable by our standards.

Z_{in}: Again, we will make it simple for ourselves by neglecting what doesn't matter much.

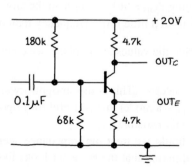

Figure 4W.15 Z_{in}: bias divider dominates.

- The capacitor: the key notion, here, is to recognize that what interests us is how this circuit behaves at *signal frequencies*. We don't care how it looks to what we consider *noise*: to frequencies below our input circuit's f_{3dB}. So, the C matters not at all: we sail right through it, as usual (if we didn't, we'd be using the wrong capacitor, and should go fetch a bigger one).

---

[3] We admit it's unfair to introduce this term before we reach its explanation in Day 5. But we feel obliged to mention it here to make this solution as general as possible. r_e will not puzzle you after you've been through Day 5.

- The bias divider: we know this impedance is low relative to what the base looks like – because we designed it that way. So. . .
- . . . we can neglect the input impedance at the base, which has been designed to be at least 10× larger than $R_\text{Thev\_bias}$

The short answer is that $Z_\text{in} \approx R_\text{Thev\_bias}$: i.e. 50k. This is within 10% of the exact answer, one that would take account of loading by $\beta \times (r_e + R_E)$; again, good enough, by our standards.

Perhaps we should admit that we've done something a little unrealistic here: we have neglected the *AC-coupled* load that may be attached to the emitter. That load should be included in the calculation, in a practical instance. In that case, we would amend the Z_in calculation:[4]

$$Z_\text{in} = R_\text{Thev\_bias} \parallel (\beta \times R_\text{Load\_AC\_coupled}).$$

That load, unlike R_E (which forms a *DC* load for the bias divider), cannot be assumed to be large relative to $R_\text{Thev\_bias}$. And if such a load is to be attached, then we should require that $R_\text{out\_signal\_source}$ be lower than what is shown in this example: lower than 5k (make the limit perhaps 1k; at this value, it would permit $R_\text{load} \geq 400\Omega$). With such a load attached, Z_in would drop to 22k.

Figure 4W.16 shows the whole circuit.

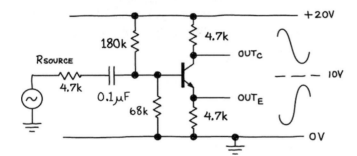

Figure 4W.16 Phase splitter circuit, with values.

Our main interest, in working through this example, was not in getting the refinements right. Instead, we hoped to remind you of a few basic notions, and to give you a chance to apply them:

- Output impedances are radically different at emitter and collector.
- AC and DC paths are to be distinguished – and "signal frequencies" to be distinguished from other frequencies (we do not design our circuit for the benefit of these other frequencies).

4W.3 Transistor switch

An ideal switch ought, when *ON*, to put maximum power into the load, while dissipating none itself. Fig. 4W.17 shows two possible *switch* circuits, and some questions about them.

Problem

1. **Collector current and switch power** What is approximate load current in the two cases? Assume that β is 100.
2. **Power in the switch** What power is dissipated in the transistor switch in the two cases, assuming that saturation voltage, $V_\text{CE-sat}$, is about 0.2V? (Please include power in the base resistor.)
3. **Why is one better than the other?** Explain briefly why one is the preferred switch configuration.

---

[4] We could, below, take account of R_E that parallels the load, to be a little more accurate. But we prefer the simplification that allowed us to neglect the effect of $\beta \times R_E$, much larger than the effect of the bias divider.

Worked Examples: Transistors I

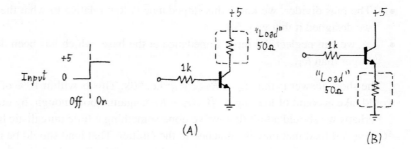

Figure 4W.17 Two possible "switch" circuits.

A solution

1. Collector current and switch power: Look at circuit **A** in Fig. 4W.17; it's a *switch* configuration:

$$I_B = 4.4\text{V}/1\text{k} = 4.4\text{mA}.$$

This current is ample to saturate the transistor. The transistor saturates at perhaps 0.2V, making $V_{\text{LOAD}} \approx 4.8\text{V}.$

So $I_{\text{LOAD}} = 4.8\text{V}/50\Omega = 100\text{mA}$. (Note that it is the *load* that sets the current.)

Now look at circuit **B**, which is a *follower* rather than a switch. Redraw as a voltage divider with R_E, V_{BE}, and R_B. Note R_B is transformed by Q's beta into what looks like $1\text{k}/\beta = 10\Omega$, see Fig. 4W.18:

$$I_C \approx I_E = \frac{4.4\text{V}}{60\Omega} \approx 70\text{mA}.$$

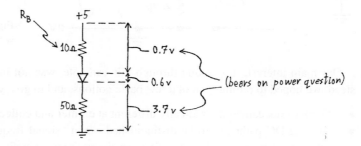

Figure 4W.18 Circuit B – a follower – redrawn as voltage divider to show where power is dissipated.

2. Power in the switch:

Circuit **A**, *switch*:

$$\begin{aligned} P = IV &= P_{\text{CE}} + P_{\text{BE}} + P_{\text{base-resistor}} \\ &= (0.1\text{A} \times 0.2\text{V}) + (4.4\text{mA} \times 0.6\text{V}) + [(4.4\text{mA})^2 \times 1\text{k}] \\ &= 0.02\text{W} + 2.6\text{mW} + 20\text{mW} = 43\text{mW}. \end{aligned}$$

Circuit **B**, *follower*:

$$\begin{aligned} P = IV &= P_{\text{CE}} + P_{\text{BE}} + P_{\text{base-resistor}} \\ &= (70\text{mA} \times 1.3\text{V}) + (0.7\text{mA} \times 0.6\text{V}) + (0.7\text{mA} \times 0.7\text{V}) \\ &= 91\text{mW} + 0.4\text{mW} + 0.5\text{mW} \approx 92\text{mW}. \end{aligned}$$

3. Why is one better than the other? The switch (**A**) puts more power into the *load*, less into the switch (less by a factor of about 1/2, as just above). The switch comes closer to putting the full power supply voltage across the load: 4.8V versus 3.7V for the follower.

If one puts the comparison into terms of *percentage* of total power that goes into the load, the contrast looks like this:

Switch (**A**):

$$P_{\text{LOAD}} = 0.1\text{A} \times 4.8\text{V} = 0.48\text{W}$$

$$P_{\text{switch}} = 0.043\text{W}$$

$$\implies P_{\text{load}}/P_{\text{total}} = 0.48/0.52 = 92\%$$

Follower (**B**):

$$P_{\text{LOAD}} = 0.07\text{A} \times 3.7\text{V} = 0.26\text{W}$$

$$P_{\text{follower}} = 0.09\text{W}$$

$$\implies P_{\text{load}}/P_{\text{total}} = 0.26/0.35 = 74\%$$

In short – as you have gathered – when ON versus OFF is all we need, we prefer the *switch* configuration because of its *efficiency*. We also like the switch configuration for permitting output swing to any rail, independent of the base-drive range.

5N Transistors II

Contents

5N.1	**Some novelty, but the earlier view of transistors still holds**		**188**
	5N.1.1	"Why transistors are hard"	189
5N.2	**Reviewish: phase splitter**		**189**
	5N.2.1	Output impedances	190
	5N.2.2	Input impedance	190
5N.3	**Another view of transistor behavior: Ebers–Moll**		**190**
	5N.3.1	A case that calls for Ebers–Moll treatment	190
	5N.3.2	Ebers–Moll equation describes transistor's V_{BE} versus I_C...	192
	5N.3.3	Little r_e	193
	5N.3.4	r_e solves impedance problems, too	194
5N.4	**Complication: distortion in a high-gain amplifier**		**194**
	5N.4.1	Distortion remedy: emitter resistor – at the price of gain	195
5N.5	**Complications: temperature instability**		**196**
	5N.5.1	Temperature stabilizing: feedback, using emitter resistor	197
	5N.5.2	Temperature stability *and* high gain	197
	5N.5.3	Another temperature-stabilizer: Explicit DC feedback	199
	5N.5.4	One more way to stabilize the circuit: add a compensating transistor	200
5N.6	**Reconciling the two views: Ebers–Moll meets $I_C = \beta \times I_B$**		**201**
5N.7	**"Difference" or "differential" amplifier**		**201**
	5N.7.1	Why a differential amp?	201
	5N.7.2	A differential amp circuit	204
	5N.7.3	Differential amp evolves into "op-amp"	206
5N.8	**Postscript: deriving r_e**		**207**
5N.9	**AoE Reading**		**208**

Why?

In this chapter we meet an amplifier sensitive to a *difference* between two inputs rather than to a difference from *ground*. This novelty permits implementation of the hugely important *operational amplifier*, which from the next class onward will be our principal analog building block.

5N.1 Some novelty, but the earlier view of transistors still holds

Today we encounter some familiar circuits that expose the limitations of our first view of transistors. High-gain amplifiers do this, with special clarity. But we will continue to use our first view whenever it suffices – because it is simpler.

The other important topic of this chapter is a circuit that does not require a new understanding of transistors, but builds a more complex circuit – and a very important one – out of circuit fragments that we met last time: the common-emitter amplifier and the follower.

The new circuit made up of these elements is a *difference amplifier* (often called a "differential amplifier;" we will use both terms), which when buffered with a *follower* output stage becomes an *operational amplifier*. We ask you to build such a circuit in Lab 5L, so that you can see that the op-amp is made up of modest and familiar elements. Armed with this comforting knowledge, you should be able to appreciate that the op-amp's wonderful performance – which you will witness next time – is not *just* the result of magic within the black box. The circuit's behavior is comprehensible.

The performance of op-amp circuits sometimes does seem magical – but we'll be pleased if you can enjoy that effect while also enjoying the comfortable sense that the device itself is just an aggregation of some old circuit friends of yours.

5N.1.1 "Why transistors are hard"

AoE §2.1

The basic difficulty in transistor circuits – or, at any rate, the basic *novelty* – arises from the fact that these devices have *three* terminals rather than the two that we are accustomed to in the passive devices we have met: resistors, capacitors, diodes, inductors.[1] We noted this novelty last time.

We have remarked before on the difficulty of discrete-transistor design, a difficulty that results perhaps from this three-terminal strangeness. On the other hand, two comforting thoughts: everyone else's treatment of transistors is much harder and more obscure than the one you will find in this book and in AoE. And here's a further comfort: soon you'll be using operational amplifiers, which will work very well while hiding from you the gritty difficulties of transistor circuit design.

5N.2 Reviewish: phase splitter

AoE §2.2.8

In Example §4W.2, we looked closely at the circuit in Fig. 5N.1, first putting together this design, and then calculating input and output impedances. Here, we'd like to use this circuit to bring out a few points that you saw in the first transistor lab, 4L.

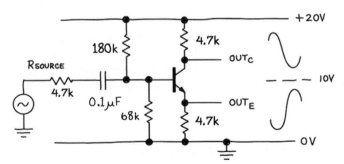

Figure 5N.1 Phase splitter.

We will use our two simple views of transistor operation from last time, and these will be almost sufficient for the whole analysis. We will bump up against one inadequacy – which may help motivate moving on, after these preliminaries, to the second view of transistors. That second view will form the central part of this chapter.

We can make sense of nearly all features of the splitter of Fig. 5N.1 with two descriptions of transistor behavior from last time:

- Simple: $I_C = I_B \times \beta$
- Simpler: $I_C \approx I_E$; $V_{BE} = 0.6\text{V}$

---

[1] Perhaps you will object that your old friend the transformer has four terminals, so three terminals don't daunt you. If so, then you can skip this part of the class note.

$I_E \ldots$: The second of the two propositions (labeled "simpler") lets us calculate emitter current. The base voltage in Fig. 5N.1 is 1/4 of the 20V supply, i.e. 5V. The emitter voltage, V_E, is a diode drop lower – still approximately 5V in a 20V span (the 0.6V error is fractionally small) – and the emitter current is just $V_E/R_E \approx 1$mA (our favorite current, as you will gather: it makes the arithmetic so easy!).

$\ldots$ hence, I_C: The collector current, I_C, is just about the same as I_E.

$\ldots$ and quiescent output voltages: So we can calculate the quiescent (resting) voltages at emitter and collector: 5V and 15V (20V less the 5V drop across R_C). Are these reasonable? Yes, because they allow maximum swing without bumping ("clipping") on upswing or downswing.

5N.2.1 Output impedances

At collector$\ldots$: Here, R_{out} is just the value of the collector resistor, R_C: 4.7k (if you wonder *why*, take a look at the phase-splitter example, §4W.2.

AoE §2.1.3B

$\ldots$ at emitter: Here, we see the "rose-colored lens" effect that you saw in the follower of Lab 4L: R_{source} is reduced by the transistor's current-gain, β. (This is the first question for which we have needed the first of the two propositions above – the one called "simple".) We would therefore predict $R_{out} \lesssim 50\Omega$.

AoE §2.2.7A

This is also the first point where our answer would be substantially wrong. The correct answer is about 75Ω, because of what we will soon call "little r_e," or "the intrinsic emitter resistance." (Intriguing new notion, coming later in §5N.3.3, and, we must admit, included in §4W.2, before you had heard the notion discussed.)

5N.2.2 Input impedance

This is dominated by the bias network – and thus is simpler than the case of the split-supply follower that you measured in Lab 4L. $R_{Thevenin}$ for the bias divider is much lower than the transistor's input impedance, so we do not meet that version of the "rose-colored lens" effect. (Of course, the *design* of the circuit, distinguished from the *analysis* that we are sketching here, did require paying close attention to the β transformation that determines R_{in} at the base. This process is spelled out in §4W.2 treating the splitter.)

5N.3 Another view of transistor behavior: Ebers–Moll

AoE §2.3

Some circuits that we worked with comfortably in the previous chapter will resist our first view, requiring use of the second, "Ebers–Moll" analysis. We just met such a circuit – one whose R_{out} is not as low as our simple view would predict. Our major example, however, will be another circuit, the common-emitter amplifier.

5N.3.1 A case that calls for Ebers–Moll treatment

The common-emitter amplifier of Lab 4L is easy to analyze: it's essentially a current source (the magnitude of the current varies with changes of V_{in}) and that current is fed to a resistor. That resistor converts the current back to a voltage – but scaled up to the extent that R_C exceeds R_E.

The amplifier's gain is $-R_C/R_E$, and we hoped, in Lab 4L, that you found this simplicity charming:

5N.3 Another view of transistor behavior: Ebers–Moll

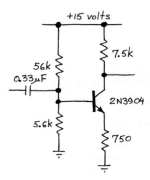

Figure 5N.2 Common emitter amplifier, gain of −10.

the expression for gain says nothing about the transistor, relying purely on the attached resistors. So far, good.

But what happens if we get greedy for gain? Could we set the ratio of R_C to R_E as high as 100? 1000? What would happen if we made R_E zero? Maybe you would shy away from that last as mathematically offensive; but the defect in this design is not mathematical. You would run into the limitations of the transistor long before reaching the impossibility of an infinite-gain amplifier.

Let's go right to the hardest case: let's take R_E to zero. How do we handle this? We seem to be applying a wiggle to V_{BE}, but we have said that V_{BE} is fixed at 0.6V.

AoE §2.3.4

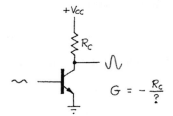

Figure 5N.3 Grounded-emitter amplifier. Infinite gain?

Well, we didn't mean *exactly* 0.6V. This is one of those circuits that obliges us to recognize that V_{BE} varies slightly as I_C varies. In fact, the I_C versus V_{BE} curve looks just like the diode curve, already familiar to you. It differs only in slope: the transistor's curve is steeper, see Fig. 5N.4:

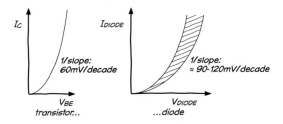

Figure 5N.4 Transistor I versus V looks like diode's.

If we omit R_E from the amplifier, then we are applying the input signal directly to V_{BE}. The current, therefore, changes a lot – given the exponential shape of the curve, see Fig. 5N.5. The gain for the amplifier is large, therefore – but not, of course, infinite.

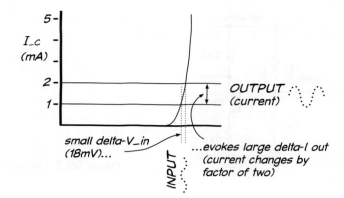

Figure 5N.5 An input applied to V_{BE} evokes large changes in I_C.

5N.3.2 Ebers–Moll equation describes transistor's V_{BE} versus I_C...

Ebers and Moll[2] describe the transistor as a voltage-to-current device – whereas our first view described it as a current-to-current converter (with β defining the multiplication factor). The two views are quite consistent, since the base current flows in a *diode*, the V_{BE} junction (see §5N.6).

We will write out the Ebers–Moll equation – but then we will whittle it down to make it more manageable. Then we rarely will refer to the full equation again. We prefer, instead, to model its consequences in forms that are more convenient. Here is the Ebers–Moll equation:

$$I_C = I_S(e^{\frac{V_{BE}}{kT/q}} - 1) . \tag{5N.1}$$

It turns out that kT/q is 25mV at room temperature, and the "-1" is unimportant once the transistor is operating.[3]

So we can simplify the equation somewhat:

$$I_C \approx I_S e^{V_{BE}/25\text{mV}} . \tag{5N.2}$$

Since I_S is a value that normally we don't know, Ebers–Moll usually boils down to the insight that the collector current is an exponential function of V_{BE}.

Only occasionally will we use the Ebers–Moll equation directly, most often to calculate relative values of I_C as V_{BE} changes. What happens, for example, if you increase V_{BE} by 18mV? Let's call the old collector current I_{C_1}, and the new one I_{C_2}:

$$\frac{I_{C_2}}{I_{C_1}} = \frac{I_S e^{V_{BE2}/25\text{mV}}}{I_S e^{V_{BE1}/25\text{mV}}}$$
$$= e^{(V_{BE2}-V_{BE1})/25} = e^{\Delta V_{BE}/25},$$

where $V_{BE2} - V_{BE1} = e^{\Delta V_{BE}}$ is measured in mV. But that is just

$$e^{18\text{mV}/25\text{mV}} \approx 2 .$$

This is a number perhaps worth remembering: 18mV ΔV_{BE} for a doubling of I_C; also sometimes handy: 60mV ΔV_{BE} per *decade* (that is, 10×) change in I_C.

---

[2] Their names make them sound like bearded German wise men living in Heidelberg. In fact, they're a pair of friendly-looking fellows from Ohio State.

[3] The "-1," delivering "$-I_S$", is important when the exponential term is negative: when the transistor is off. Then $I_C \approx -I_S$, which explains why I_S is called the "reverse leakage current."

5N.3.3 Little r_e

AoE §2.3.4

On our way to taming Ebers–Moll, we will take a strange preliminary step: we will turn the I_C versus V_{BE} plot on its side, as in Fig. 5N.6:

Now, we will move still farther from the scary original equation, heading for an application of Ebers–Moll that is pleasantly straightforward.

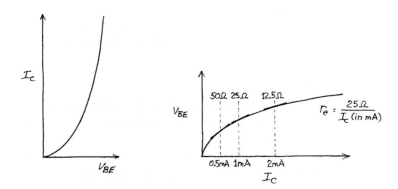

Figure 5N.6 Transistor transfer curve rotated: now slope is in *ohms*.

The reward for this odd redrawing comes immediately: we can model the transistor's *gain*[4] as a resistance. This "resistance" simply describes how much additional current you'd draw from the emitter if you tugged it down a bit (assuming the base voltage is fixed). (For a more elaborate exposition of how one might envision r_e, see §5N.8.) And, as you can see from the rotated plot in Fig. 5N.6, that effective resistance is wonderfully easy to calculate: simply divide 25mV by the collector current (in milliamps).

Grounded-emitter amplifier

If this notion still seems a little abstract, look at Fig. 5N.7 to see how wonderfully simple the problem of the grounded-emitter amplifier becomes when we draw in r_e as a little resistor.

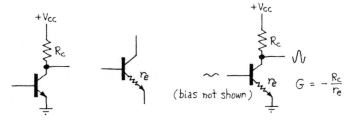

Figure 5N.7 Little r_e tames grounded emitter amplifier.

Let's try it with some values inserted, as in Fig. 5N.8. The gain, R_C/r_e, can be evaluated if we know I_C. Do we? Well, yes, if the circuit has been designed or set up properly: it should be set up so that $V_{\text{out\_quiescent}}$ is at the midpoint of its available range, in this case 5V. To get this result, I_C must be our favorite: 1mA.

Given this I_C, we can evaluate r_e: it is just 25Ω (*not* 25kΩ, please note: we divide by collector current *in milliamps* (10^{-3}A), not by the collector current). So, the gain is $-5k/25=-200$ (the minus

---

[4] This "gain" is peculiar – not the *change-in-voltage-out per change-in-voltage-in* that you used in Lab 4L. Instead, this is "transconductance" (g_m): *change-in*-current-*out per change-in-voltage-in*.

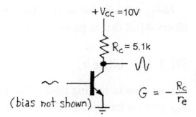

Figure 5N.8 Try r_e with some values inserted.

sign indicates phase inversion, not attenuation, incidentally: when V_{in} moves up, V_{out} moves down). We will soon learn that this is the answer only *at the quiescent point*. Still, this is a good start.

5N.3.4 r_e solves impedance problems, too

AoE §2.3.4

r_e puts a floor under R_{in}: This model solves another pair of problems that we had not mentioned: what is R_{in} for the grounded-emitter amplifier? It would be zero if we did not include r_e. It becomes $\beta \times r_e \geq 2.5k$ when we include r_e, and this is the correct answer.

AoE §2.3.3

r_e puts a floor under R_{out}: When our first view of the "lens effect" of a follower indicates an extremely low R_{out}, r_e brings things back down to earth.

For the circuit in Fig. 5N.9, our simple first view would predict R_{out} for this circuit might be under 0.5Ω. But r_e shows us this is a little too good to be true. Increasing I_C can reduce R_{out}, but not to the level predicted by the simple view of transistors, unless one pushes on to very large I_C. Again r_e tames the extreme case.

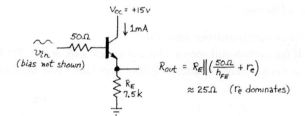

Figure 5N.9 r_e puts a realistic floor under R_{out}.

That concludes the good news about the grounded-emitter amplifier – a circuit that we soon will learn is not usable in the form we have presented. Now for some for some complications that oblige us to amend the circuit.

5N.4 Complication: distortion in a high-gain amplifier

AoE §2.3.4A

A gain of 200 is high. But evidently the gain is *not constant*, since I_C must vary as V_{OUT} moves (indeed, it is variation in I_C that *causes* V_{OUT} to move.)

Figures 5N.10–5N.12 illustrate the funny "barn-roof" distortion[5] that you see if you feed this circuit a small triangle. First, in Fig. 5N.10, is the variation in gain that we would predict, given the variation in I_C as V_{OUT} swings.

---

[5] We used this term back in §4L.4.1 without explanation. The name is not standard, incidentally; we just happen to live near Vermont's dairy barns with their gambrel roofs.

5N.4 Complication: distortion in a high-gain amplifier

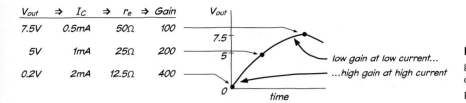

Figure 5N.10 Gain of grounded-emitter amp varies during output swing: gain evaluated at 3 points in output swing.

The plots in Fig. 5N.11 show how gain varies (continuously) during the output swing. This is bad distortion: -50% to $+100\%$.

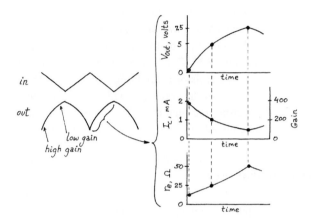

Figure 5N.11 During swing of V_{OUT}, I_C and thus r_e and gain *vary*.

And Fig. 5N.12 is a scope image confirming the predicted signal distortion. What is to be done?

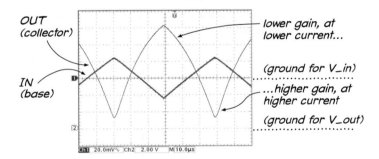

Figure 5N.12 Scope image of "barn roof" distortion. Note scope gain is 100× higher for input waveform.

5N.4.1 Distortion remedy: emitter resistor – at the price of gain

One cannot eliminate this variation in r_e, but one can make its effects negligible. Just add a constant resistance much larger than the varying r_e. That will hold the denominator of the gain equation nearly constant.

AoE §2.3.4A

With an emitter resistor added, the gain variation shrinks sharply, see Fig. 5N.13. r_e still varies as widely as before; but its variation is buried by the big constant in the denominator. Circuit gain now varies only from a low of -9.1 to a high of -9.75: a -4%, $+3\%$ variation about the midpoint gain of 9.5.

Punchline: an emitter resistor greatly reduces error (variation in gain, and consequent distortion).

196   Transistors II

Figure 5N.13 Emitter resistor cuts gain, but also cuts gain *variation*.

This we get at the price of giving up some gain. (This is one of many instances of Electronic Justice: here, *those greedy for gain will be punished: their output waveforms will be rendered grotesque*.)

We will see shortly that the emitter resistor helps solve other problems as well: the problem of temperature instability (see §5N.5), and even distortion caused by *Early effect*. How can a humble resistor do so much? It can because in the latter two cases the resistor is applying *negative feedback*, a design remedy of almost magical power. In the next segment of the course, in which we apply operational amplifiers, we will see negative feedback blossom from marginal remedy to central technique. Negative feedback is lovely to watch. Many such treats lie ahead.

5N.5   Complications: temperature instability

Semiconductor junctions respond so vigorously to temperature changes that they often are used as temperature sensors. If you hold V_{BE} fixed, for example, you can watch I_C vary exponentially with temperature.

But in any circuit not designed to measure temperature, the response of a transistor to temperature is a nuisance. Most of the time, the simple trick of adding an emitter resistor will let you forget about temperature effects. We will see below how this remedy works.

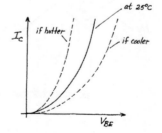

Figure 5N.14 Transconductance of bipolar transistor varies rapidly with temperature.

Preliminary warning. Do not look for a description of this temperature dependence in the Ebers–Moll equation. That equation (mis-read) will point you in exactly the wrong direction: increasing T should shrink the exponent:

$$I_C = I_S \left(e^{V_{BE}/(kT/q)} - 1\right).$$

AoE §2.3.1

Don't be fooled: Ebers–Moll equation seems to say I_C falls with temperature. Not so.

But actual results are quite contrary to what this point seems to suggest: in fact, rising T increases I_C, and does that *fast*. The solution to the riddle is that I_S grows *very fast* with temperature, overwhelming the effect of the shrinking exponent.

Here are two formulations for the way a transistor responds to temperature:

AoE §2.3.2C

5N.5 Complications: temperature instability

- I_C grows at about 9%/°C, if you hold V_{BE} constant;
- V_{BE} falls at 2mV/°C, if you hold I_C constant.

The first of these formulations is the easier to grasp intuitively: heat the device and it gets more vigorous, passes more current. The second formulation often makes your calculation easier. But if you use the second, just make sure that you don't get the feeling that the way to calm your circuits is to build small fires under them!

5N.5.1 Temperature stabilizing: feedback, using emitter resistor

The remedy described here is simple, and widely used. It is also quite subtle. The left-hand circuit in Fig. 5N.15 is so unstable that it is useless. An 8°C rise in temperature saturates the transistor.

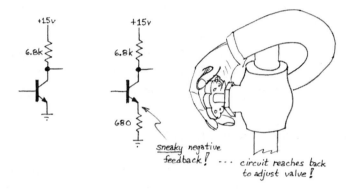

Figure 5N.15 An unstable circuit stabilized by emitter resistor. Note that we assume V_B is held constant.

Why does the right-hand circuit work better? How does the emitter resistor help, as I_C grows? Here is feedback at work. The circuit senses trouble as it begins:

- I_C begins to grow in response to increased temperature;
- V_E rises, as a result of increased I_C (this is just Ohm's law at work);

But this rise of V_E diminishes V_{BE}, since V_B is fixed. Squeezing V_{BE} tends to close the transistor "valve." Thus the circuit slows itself down (as the somewhat-grotesque hand in Fig. 5N.15 is meant to suggest).

We don't claim that I_C will change not at all. Some ΔI_C with temperature is necessary in order to generate the error signal. But the emitter resistor prevents wide movement of the quiescent point.

The plot in Fig. 5N.16 shows how the emitter resistor's behavior – plotted as a straight "load line" intersecting with the transistor's shifting curve – limits the variation in I_C with temperature change.

The larger the value of R_E, the stronger the feedback, and the less the variation in I_C with temperature.

5N.5.2 Temperature stability *and* high gain

You don't have to give up gain in order to get temperature stability – contrary to an impression we may have given back in §5N.5.1. You can get both stability *and* high gain, by arranging things so that R_E does its good stabilizing work while not reducing signal gain. The trick is to include R_E, as in a low-gain amplifier, but then to make it disappear at signal frequencies. R_E "disappears" at those frequencies because of the capacitor that parallels it, *bypassing* it, see Fig. 5N.17.

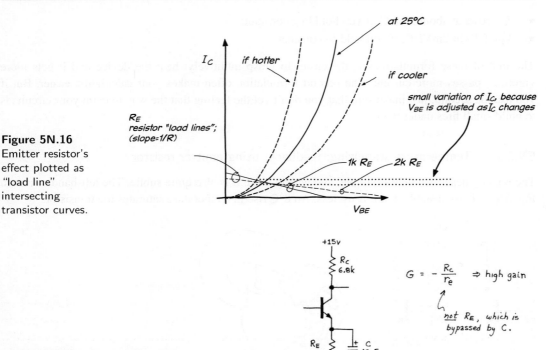

Figure 5N.16 Emitter resistor's effect plotted as "load line" intersecting transistor curves.

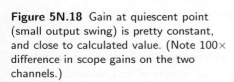

Figure 5N.17 Bypassed-emitter resistor: high gain plus temperature stability.

The circuit's gain, observed

The circuit in Fig. 5N.18 shows the same high gain as the unsatisfactory grounded-emitter amplifier. If we make the output swing very small, distortion is hardly noticeable:

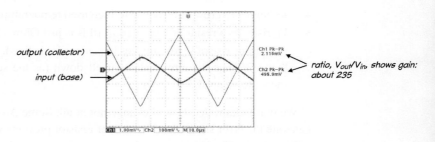

Figure 5N.18 Gain at quiescent point (small output swing) is pretty constant, and close to calculated value. (Note 100× difference in scope gains on the two channels.)

The gain at the quiescent point appears to be about -280 (the minus sign indicating only the signal's *inversion*). By calculation, we would predict gain of -300:

$$G \equiv \Delta V_{\text{out}}/\Delta V_{\text{in}} = -R_{\text{C}}/r_{\text{e}} = -7.5k/25 = -300 \ .$$

The difference between calculated gain and the gain observed in Fig. 5N.18 may be explained by the collector current being a little lower, in this circuit, than the target of 1mA.

The circuit's temperature stability, observed

In Fig. 5N.19 the stability of this circuit is contrasted with the nasty behavior of the grounded-emitter amplifier of Fig. 5N.8. We fed a small triangle to both circuits, stable and unstable, and then warmed both with a heat gun.

5N.5 Complications: temperature instability

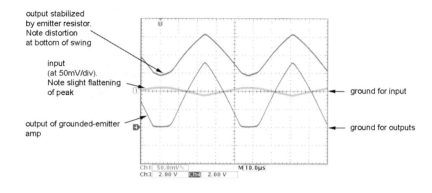

Figure 5N.19 Stable and unstable circuits contrasted: when heated, bad one clips.

The stable circuit's V_{out} stayed centered where it started; the unstable circuit's V_{out} drifted down close to ground, producing nasty clipping.

Curious detail: R_{in} degrades in saturation

A detail of Fig. 5N.19 is worth mentioning: the effect of saturation upon input impedance. The unstable output drifts down, and flattens ("clips") at ground. The transistor is *saturated* where the waveform clips: totally *on*. When that happens, the usual rules of transistor behavior do not apply (these are the rules that apply in what's called the "linear region" of transistor operation, where the device is neither totally *on* (saturated) nor totally *off* (cut off)).

When the transistor saturates its input resistance – usually $\beta \times R_{\text{E}}$ – becomes radically lower: no longer does any β multiplication occur. You can make out evidence of this effect in two details of the image in Fig. 5N.19:

- the input waveform itself is distorted – its peak flattened. That indicates that even the function generator, with its R_{out} of 50Ω is getting overloaded by the transistor's R_{in}.
- the output of the stable circuit – a circuit driven by the function generator that drives the unstable circuit – also shows distortion, distortion produced by overloading of the signal source shared by the two circuits.

The moral seems to be *don't hang around with bad company*: even the virtuous temperature-stable circuit ends up corrupted by sharing an input with the naughty grounded-emitter amp.

. . . but note that distortion persists: The emitter resistor neatly solves the temperature-stability problem. The distortion, however, remains; it appears also, for both circuits, back in Fig. 5N.12. There is no way around this, if you want so much gain in one stage.

5N.5.3 Another temperature-stabilizer: Explicit DC feedback

AoE §2.3.5C

Figure 5N.20 shows feedback in a form more obvious than through R_{E}, but used to similar effect.

When you have played with some operational amplifier circuits in the coming labs, you will recognize this feedback as very similar to what you apply in those op-amp circuits: 1/11 of V_{out} is fed back to the input.[6] But here a subtlety is at work that is not usual in the op-amp circuits: the feedback affects DC levels, but *not* circuit gain. It does not affect gain because the function generator, whose R_{out} is low, is able to overwhelm the relatively feeble feedback signal at signal frequencies (those that

---

[6] It is fed not quite to the *circuit* input but almost. The difference is only that it is fed to the base rather than to the far side of the blocking capacitor. As noted just below, this feedback affects DC levels, but not signals.

Transistors II

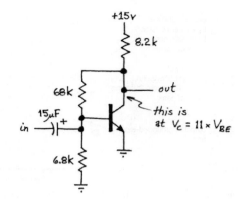

Figure 5N.20 DC feedback protects against temperature effects.

get through the blocking capacitor). Yet at DC the feedback is strong enough to handle the problem of drift.

5N.5.4 One more way to stabilize the circuit: add a compensating transistor

AoE §2.3.5B

The circuit in Fig. 5N.21 *compensates* for any change in one direction by planting a circuit element that tends to change at the same rate in the opposite direction.

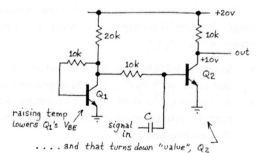

Figure 5N.21 Temperature stability through use of a second "compensating" transistor.

The circuit in Fig 5N.22 works if both transistors live on the same piece of silicon, so that their curves drift equally, as temperature changes.

If the circuit is heated, Q_1's V_{BE} shrinks. But this shrinkage squeezes down Q_2 as both transistors get hotter. So Q_2's current does not grow with temperature. (The 10k resistor on the base of Q_2 makes the biasing circuit not too stiff: the signal source (presumed to be of impedance $\ll$ 10k) can have its

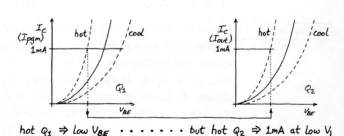

Figure 5N.22 Two matched transistors heated together can compensate to cancel drift.

way, as usual.) This circuit is a close kin to a "current mirror," a circuit discussed in §5S.2. AoE also treats mirrors in §2.3.7.

5N.6 Reconciling the two views: Ebers–Moll meets $I_C = \beta \times I_B$

At first, we promoted the current-to-current view: $I_C = \beta \times I_B$. A little later we offered Ebers–Moll's account, $I_C = I_S\, e^{V_{BE}/V_T}$.

In case you're troubled by the thought that our two views may not be consistent, here's an argument, and then a picture, meant to reassure you:

- An argument: since the BE junction is a diode, feeding it a current (I_B) generates a corresponding V_{BE}. Ebers–Moll teaches that this V_{BE} is the cause of the resulting I_C. If we use this account to justify the first, simplest view of transistor operation – which treated base *current* as input – we are acknowledging and using Ebers–Moll.
- If Fig. 5N.23 does not make you feel better, forget it; you don't *need* to reconcile the two views if you don't want to.

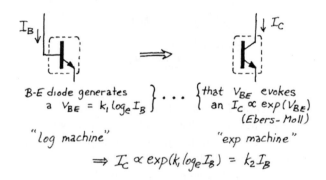

Figure 5N.23 Beta... and Ebers–Moll descriptions of transistor gain *reconciled*.

5N.7 "Difference" or "differential" amplifier

AoE §2.3.8

The differential amp is the last standard transistor circuit we will ask you to consider. It is especially important to us because it lets us understand the *operational amplifiers* that you soon will meet. These wonderful devices are in fact just very good differential amps, cleverly applied.[7]

5N.7.1 Why a differential amp?

A differential amplifier has an internal symmetry that allows it to cancel errors shared by its two sides, whatever the origin of those errors. Sometimes one takes advantage of that symmetry to cancel the effects of errors that arise within the amplifier itself: temperature effects, for example, which become harmless if they affect both sides of the amplifier equally. In other settings the shared error to be

---

[7] At the risk of complicating the thought too much, we should acknowledge that the *clever* application is of fundamental importance. The cleverness lies in the use of *negative feedback*. It is this, and not just the good diff-amp, that provides the really impressive behavior of the op-amp circuits that you soon will meet.

canceled is noise picked up by both of the amplifier's two inputs. Used this way, the amplifier picks out a signal that is mixed with noise of this particular sort: so-called "common-mode" noise.

You built a circuit back in Lab 2L that did something similar: passed a signal and attenuated noise. But that *RC* filter method works only if the noise and signal differ quite widely in *frequency*. The differential amp requires no such difference in frequency. It does require that the noise must be common to the two inputs, and that the *signal*, in contrast, must appear as a difference between the waveforms on the two lines. Such noise turns out to be rather common, and the differential amp faced with such noise can "reject" it (refuse to amplify it), while amplifying the signal: it can throw out the bad, keep the good.

An application: brain wave detector

Here is an example of a problem that might call for use of a differential amplifier. One can detect brain activity with skin contacts; the activity appears as small (microvolt range) voltage signals. The output impedance of these sources is high.

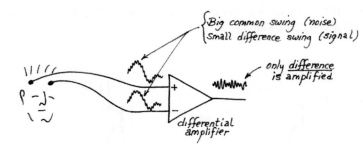

Figure 5N.24 An application for a differential amp: brain wave detection.

The feebleness of the signals makes their detection difficult. It is not hard to make a high-gain amplifier that can make the signals substantial. But the catch is that not only the signals but also *noise* will be amplified, if we are not careful. We can try to shield the circuit; that helps somewhat. But if the principal source of noise is something that affects both lines equally, we can use a differential amplifier instead – or as well; such a circuit ignores such "common mode" noise. The same difficulty applies to cardiac measurements – by the so-called EKG.

60Hz line noise will be coupled into both lines, and is likely to be much larger than the microvolt signal levels. A good differential amp can attenuate this noise by a factor of perhaps 1000 while amplifying the signal by, say, 100. An amp that could do that would show a "common-mode rejection ratio" – a preference for difference signals over common – of 10^5: 100dB.

...and a similar demo: music pulled out of nasty 60Hz noise

In a class demo, we made up a simple imitation of this sort of case. The differential signal source was an audio player (a CD player). Two unshielded wires ran from the CD player to a differential amplifier.[8]

A large 60Hz voltage swing appears on each of the two wires running three feet or so in Fig. 5N.25. That 60Hz noise is shown in Fig. 5N.26. A much smaller difference signal (music, in this case) looks like a barely noticeable tremor on the 60Hz swing. One can just make out a small difference, here and there in the left-hand image of Fig. 5N.26: it looks like a slight darkening, as one of the nearly-identical-twin waveforms pokes out from behind its sibling.

The difference between the two, however – shown in the right-hand scope image – is easy to make

---

[8] This amp we built with an operational amplifier, not with discrete transistors, because doing so is much the easiest method. You will soon share our preference for op-amps over discrete designs, we are confident.

5N.7 "Difference" or "differential" amplifier

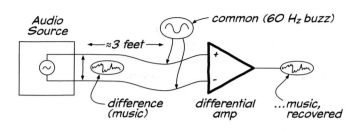

Figure 5N.25 Demonstration setup for pulling music signal from large common-mode noise.

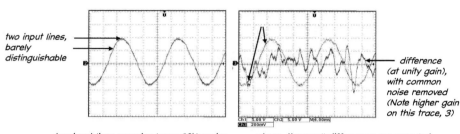

Figure 5N.26 Difference amplifier can pick out a small difference in presence of large common noise.

out once the large common-mode noise has been removed. Differential gain here is *unity*: so the "amplifier" here has done no more than remove the shared noise. (The "signal" looks as if it had been amplified in Fig. 5N.26; that's an illusion caused by the greater *scope* sensitivity for the output signal.) The effect, for a listener, is dramatic. The *original* signals sound like a loud 60Hz buzz, no music audible; the *difference* signal sounds like music, unpolluted by any audible 60Hz noise.

Differential signaling helps even digital circuits

When we reach the digital gates of Lab 14L we will encounter *differential* transmitters and receivers, applied for the purpose of achieving good noise immunity in the presence of daunting common-mode noise.

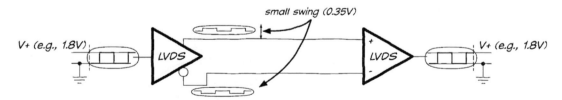

Figure 5N.27 Differential signalling applied in digital electronics: provides good noise immunity despite low supply voltage.

This method, used in the LVDS (Low Voltage Differential Signaling) circuits, can give good behavior even with the very low power supply voltages that are coming to dominate digital circuitry: supplies of 2.5V, 1.8V and lower. Fig. 5N.27 does not show common-mode noise, but when we tried this circuit (see Chapter 14N) the circuit easily rejected common noise five or six times larger than the *signal*. For a scope image showing rejection of drastic high-frequency noise by LVDS, see AoE Fig. 12.126.

5N.7.2 A differential amp circuit

> Compare AoE Fig. 2.63

Figure 5N.28 shows the lab's differential amp. This is the circuit you will build in Lab 5L. In order to achieve good matching of the two transistors, you will use an integrated array of transistors (a CA3096 array).

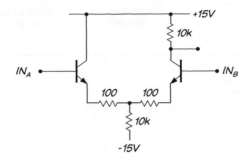

Figure 5N.28 Differential amp.

The circuit may look a bit daunting, but it is not hard to analyze. After we establish DC *quiescent* conditions, we will use a trick that AoE suggests: consider only pure cases – pure *common* signal, then pure *difference* signal.

Before we see what the circuit will do to input signals, let's first find the DC *quiescent* levels.

Quiescent points

V_{out}: Before you can predict V_{out} you need to determine currents.

Currents: If the bases are tied to ground, as we may assume for simplicity, the emitter voltages are close to ground, and it follows that point "A" is not far from ground (close to -1V.). From this observation you can estimate I_{tail} (in the lower 10k resistor): it is about 14V/10k≈1.5 mA.

Since the circuit's inputs are at the same voltage, symmetry[9] requires that the 1.5mA *tail* current must be shared equally by the two transistors. So I_C for each of them is about 0.75mA.

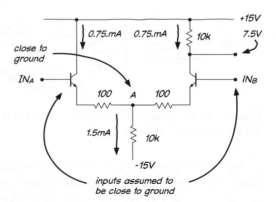

Figure 5N.29 Lab diff amp, showing quiescent currents and voltages.

From here $V_{out\_quiescent}$ is easy: centered as usual – but note that it is centered *not* between the supplies (that center would be 0V). Instead, it is centered *in the range through which it can swing*. That is always the deeper goal. Since the floor on the output swing is defined not by the negative

---

[9] "Symmetry?," you may want to protest. "There's a collector resistor on one side, and not on the other." True. But try to explain to yourself why that does not matter. The resistor does affect the *voltage* at each collector. But what influence does that voltage have on collector current? Not much. Ideally, none; Early effect says 'only a slight effect.'

supply but by the emitter voltage, the floor is close to ground, and the proper $V_{\text{out\_quiescent}}$ is around 7.5V.

Differential gain

Assume a pure *difference* signal: a wiggle up on one input, a wiggle down of the same size on the other input. It follows, you will be able to convince yourself after a few minutes' reflection, that the voltage at "A" of Fig. 5N.29 does not move.

That observation lets you treat the right-hand side of the amp as a familiar circuit: a common-emitter amp, see Fig. 5N.30.

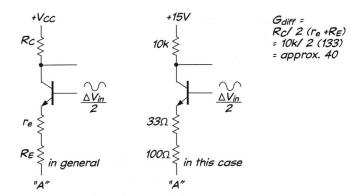

Figure 5N.30 Differential gain: just a common emitter amp again.

The gain might seem to be

$$G = -\frac{R_C}{r_e + R_E}.$$

That's almost correct. But you need to tack a factor of 2 into the denominator, just to reflect the way we stated the problem: the ΔV applied at the input to this "common emitter" amp is only *half* the difference signal we applied at the outset to the two inputs. You also might as well throw out the minus sign, since we have not defined what we might mean by positive or negative difference between the inputs.

So the expression for differential gain includes that factor of 2 in the denominator (*minus* sign discarded):

$$G_{\text{diff}} = \frac{R_C}{2(r_e + R_E)},$$

Common mode gain

Assume a pure *common* signal: tie the two inputs together and wiggle them. Now, point "A" is not fixed. Therefore, *this* common-emitter amp has much lower gain, because R_{tail} appears in the denominator of the gain equation. The denominator – which was just $2(r_e + R_E)$ for the differential case – now must include the much larger value, R_{tail}.

Again that is *almost* the whole story. But another odd factor of 2 appears, to reflect the fact that a twin common-emitter amp –the other side of the differential amp – is squirting a current of the same size into R_{tail}. The result is that the voltage at A jumps twice as far as one might otherwise expect: one can say this another way by calling the effective R_{tail} "$2R_{\text{tail}}$."

Let's redraw the circuit to show what's going on when we drive both inputs with the same signal ("common mode"). The redrawn circuit is Fig. 5N.31

To calculate gain, we can consider just the right-hand common-emitter amplifier (peeled away, in

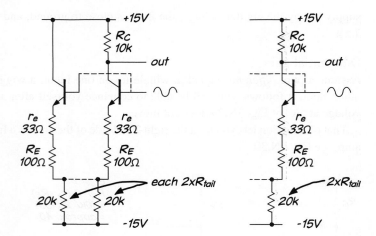

Figure 5N.31 Differential amp redrawn to show common-mode response.

Fig. 5N.31). And the common mode gain is

$$G_{\text{CM}} = -\frac{R_\text{C}}{r_\text{e} + R_\text{E} + 2R_\text{tail}}.$$

The bigger we can make R_tail, evidently, the better. A current source in the tail, therefore, provides best common-mode rejection. Any respectable differential amplifier normally includes a current source in the tail – as your lab circuit does, once you have moved on after first trying a tail *resistor*. You used the resistor to measure a predictable common-mode gain, and for contrast with the improvement effected by use of the current source.

AoE §2.3.8

5N.7.3 Differential amp evolves into "op-amp"

In Lab 5L we invite you to carry on, once you have tried out the differential amp, adding two more stages so as to convert your differential amplifier into a not-bad operational amplifier.

Fundamentally, an operational amplifier is only a high-gain differential (or "difference") amplifier. But a good op-amp also includes a buffer to provide reasonably-low output impedance. Fig. 5N.32 shows a block diagram of the lab 'op-amp,' with its gain of approximately 1000: pretty good.

Compare AoE Fig. 2.91 (a circuit that, however, takes its feedback internally).

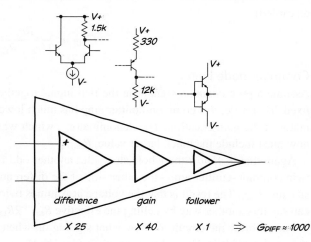

Figure 5N.32 3-stage op-amp of Lab 5L.

The IC op-amp you'll meet in Lab 6L provides gain a couple of hundred times better, along with better common-mode rejection. But this circuit offers a start – and one that we hope will demystify the triangles that you'll begin to invoke next time..

5N.8 Postscript: deriving r_e

Deriving the expression for r_e: The "intrinsic emitter resistance," r_e, is, as you know, the inverse of the slope of the transistor's gain curve: $\Delta I_{\text{out}}/\Delta V_{\text{in}}$ (see §5N.3.3). A little more formally, then, r_e is the derivative of the V as a function of I; that is, $dV_{\text{BE}}/dI_{\text{C}}$. Let's evaluate this, finding first the slope of the I–V curve.

If we write Ebers–Moll in the simplified form we use in Equation 5N.2

$$I_C \approx I_S \, e^{V_{\text{BE}}/25\text{mV}},$$

then

$$\frac{d(I_C)}{d(V_{\text{BE}})} = (1/25\text{mV})(I_S \, e^{V_{\text{BE}}/25\text{mV}}) \approx (1/25\text{mV})(I_C) = I_C/25\text{mV}.$$

And r_e, the reciprocal of the slope of the gain curve, is

$$\frac{25\,\text{mV}}{I_C \text{ (in amps)}},$$

which we prefer to write as

$$\frac{25\,\Omega}{I_C \text{ (in mA)}}.$$

Alternative 'derivation:' Tess o' Bipolarville: Here is an alternative argument to the same result: imagine a lovely milkmaid seated (in the summer twilight) on a stool, tugging dreamily at the emitter of a transistor whose base is fixed. She has pulled gently, until about 1mA flows. What *delta*-current falls into her milkpail, for an additional tug?

If the base is anchored, tugs on the emitter change V_{BE} a little; in response, I_C changes quite a lot. The squirts of ΔI_E (which we treat as equivalent to ΔI_C) reveal a relation between ΔV_{BE} and ΔI_C. The quotient

$$\Delta V_{\text{BE}}/\Delta I_C$$

Figure 5N.33 Dreamy milkmaid discovers the value of r_e, experimentally.

is r_e, which behaves just like a resistor whose far end is fixed.

"Yes," muses the charming milkmaid, her milkpail now filled (with *charge*, in fact; but she hasn't noticed). "Just as I thought: r_e, though only a model, does behave for all the world like a little resistor. So that's why we draw it that way." And with that she rises, little suspecting what her discovery portends, and carries her milkpail off into the gathering dusk.

5N.9 AoE Reading

AoE's Chapter 2 treats both bipolar transistors and elementary applications of operational amplifiers, so its material suits the topics of today's lab – though, as usual, it contains much that we do not require here. Here are some selected sections that are appropriate:

- Ebers–Moll: §2.3.
- §2.3 to end of chapter, but...
  - omit §2.3.7 (current mirrors) (we *like* current mirrors, but want to lighten your load a bit)
  - omit most details in introduction to negative feedback: §2.5; we will devote lots of attention to negative feedback soon – in the context of op-amp circuits. But highly relevant to us now is circuit B (non-inverting amp) of Fig. 2.85 in §2.5
- The most important sections of reading are:
  - The Ebers–Moll model: §2.3;
  - differential amplifiers: §2.3.8;
  - push-pull output stages: §2.4.1;
  - bootstrapping (a general notion of applications wider than just the emitter-follower that is described here, and shown in Fig.2.80): §2.4.3.
- Problems: Exercise 2.28.

In this book: you might glance at §9S.4 for the scary full circuit for the '411 op-amp. This note shows a circuit like the one you build today – but with a great many refinements added. Worry about the refinements later.

5L Lab: Transistors II

Overview of the lab

To do *all* of today's lab is a challenge: the circuit is the most complex that you've built so far, and if some stage holds you up, you're likely to run out of time. But that shouldn't worry you. Only the *differential amp* (§5L.1) – not its conversion into an *op-amp* – is fundamental. We hope you'll get at least partway into the op-amp construction, because that experience will make you feel more at home with the op-amps that enter next time; but you need not finish the exercise.

5L.1 Difference or differential amplifier

Predict differential and common-mode gains for the amplifier in Fig. 5L.1 (don't neglect r_e). Note that you will build this using an IC array of transistors, not our usual 2N3904s.

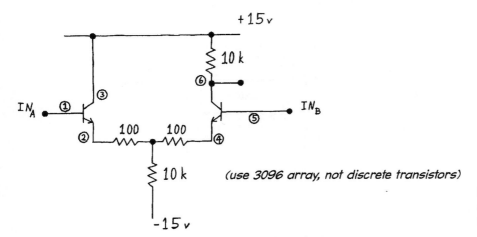

Figure 5L.1 First stage: differential amplifier, made from transistor array.

5L.1.1 Transistor array

We would like you to build the circuit on an array of bipolar transistors, the CA3096 or HFA3096. These transistors are fairly well matched,[1] and will track one another's temperatures; such tracking helps assure temperature stability, as you know.

Figure 5L.2 shows the pinout which applies to both packages: the CA3096 (DIP) and HFA3096 (SOIC). The "substrate," connected to pin 16, is the P-doped material on which the transistors are

---

[1] HFA3096 *npn* transistors show V_{BE} matching typically within 1.5mV, max 5mV (at $I_C = 10$mA).

built. Since we don't ever want that implicit set of diodes to conduct, we should connect *substrate* to the most negative point in the circuit. Here, that is −15V.

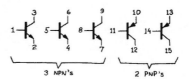

Figure 5L.2 CA3096 array of bipolar transistors.

If you have the HFA3096, which comes in a surface-mount package (SOIC), you (or *someone*) will need to solder it to an adapter, so that you can plug it into the breadboard.[2] Fig. 5L.3 shows the DIP and surface mount versions of the '3096.

Figure 5L.3 '3096 in two versions: CA3096 and HFA3096 mounted on DIP carrier.

Figure 5L.4 shows the way its pins are numbered, and the way it goes into the breadboard. It straddles the trench so that its 16 pins are independently accessible. (The same is true, of course, for the HFA3096 in its carrier.)

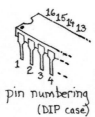
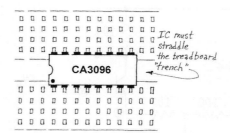

Figure 5L.4 DIP package pin numbering and insertion in breadboard.

5L.1.2 Setting up the test signals

Now you will use two function generators to generate a mixture of *common-mode* and *differential* signals.

5L.1.3 Setup I: using a function generator that can "float"

The signal setup is easy if you have a function generator that permits you to "float" its *common* lead – the one tied to the BNC cable's surrounding shield, a lead that until now we have always left tied to ground. Here, "float" simply means "disconnect from world ground." Not all function generators, however, permit such floating, and for these we suggest an alternative method in §5L.1.4.

---

[2] As you will see in this book's parts list, Appendix E, we use a company named Proto-Advantage to do this service for us.

5L.1 Difference or differential amplifier

Preliminaries

One generator will drive the other. In Fig. 5L.5 the scheme requires that you *"float"* the driven generator: find the switch or metal strap on the lab function generator that lets you *disconnect* the function generator's local ground from absolute or "world" ground. You will find such a switch or strap on the back of most generators.

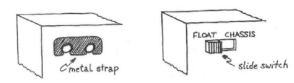

Figure 5L.5 Float the external function generator.

As you connect the two function generators to your amplifier, you will have to use care to avoid defeating the "floating" of the external function generator: recall that BNC cables and connectors can make *implicit* connections to absolute ground. You must avoid tying the external generator to ground through such inadvertent use of a cable and connector. So, you must avoid use of the BNC jacks on your breadboard; you must also avoid linking function generator to scope with a BNC for external trigger. You may find "BNC-to-mini-grabber" connectors useful to link function generator to circuit: these connectors do not oblige you to connect their *shield* lead to ground.

Composite signal to differential amplifier

Now let the breadboard's function generator (which cannot be "floated") drive the external function generator's *local ground* or *"common"* terminal. Use the output of the external function generator to feed your differential amplifier. That output can carry pure common-mode, pure differential, or a mixture of the two; see Fig. 5L.6. You will exploit this versatility later in §5L.1.5.

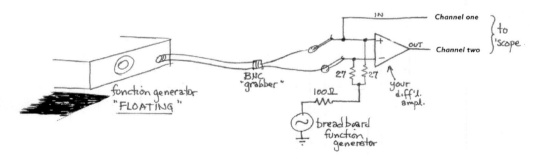

Figure 5L.6 Common-mode and differential signal summing circuit.

5L.1.4 Setup II: using a function generator that cannot "float"

If your main function generator lacks a switch that can float it, then you can use a transformer to provide the required separate control of *difference* and *shared* signals (usually called "differential" and "common-mode" signals).

Figure 5L.7 shows the arrangement, using a 6.3V transformer – the same type that you used in Lab 3L. Note that this transformer is *not* to be plugged into the 120V wall socket! We will, instead, use the external function generator to drive the transformer's power-input plug. Thus the external generator drives the *primary* of the transformer.

The breadboard function generator driving the center-tap of the transformer *secondary*, provides equal signals to the two inputs of the amplifier.

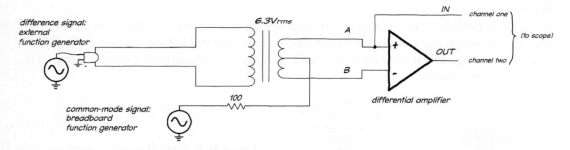

Figure 5L.7 Transformer allows two function generators to provide separate *differential* and *common-mode* signal sources.

Figure 5L.8 shows the signals that this arrangement can generate at the amplifier's inputs. These are the pure cases: only difference, or only common-mode.

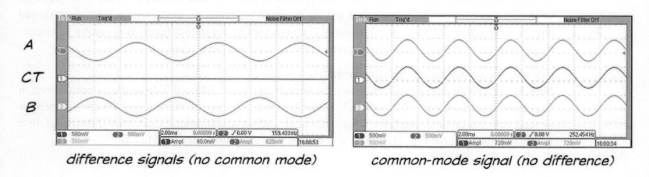

Figure 5L.8 Transformer use allows pure difference or pure common signals.

5L.1.5 A mediocre differential amp: resistor in "tail"

As a suggestion, try measuring common-mode and differential gain.

Common-mode gain (first try):

- **Measure common-mode gain:**
    - Shut off the differential signal (external function generator)[3] while driving the amplifier with a signal of a few volts amplitude. Does the common-mode gain match your prediction? If it is too high, the probable cause is the sneaking-in of a difference between the diff-amp's two inputs, so that the output you see includes some differential gain as well as common-mode. You can discover whether this is happening by simply shorting the diff-amp's two inputs together (parallel the series-pair of 27Ω resistors with a length of wire). That piece of wire assures that the applied signal is true common-mode. If you do insert this wire, be sure to remove it after you measure G_{CM}. The two inputs must be permitted to diverge in the next step, where you measure differential gain.

---

[3] You may prefer to cut *amplitude* to a minimum, rather than shut off power to the generator.

5L.1 Difference or differential amplifier

Differential gain:

- **Measure differential gain:**
    - Turn on the external function generator while cutting common-mode amplitude to a minimum (there is no Off switch on the breadboard function generator).
    - Apply a small differential signal. Does the differential gain match your prediction? If the differential gain appears to be high by about a factor of two, recall that when you watch a wiggle at a single input, you are looking at about one-half the difference signal you are applying to the amplifier. If you doubt this, try watching both inputs, on the scope's two channels. You should find approximately equal and opposite (180°-shifted) waveforms on the two inputs.
    - Now turn on *both* generators and compare the amplifier's output with the *composite* input. To help yourself distinguish the two signals, you may want to use two frequencies rather far apart; but do not let this experimental convenience obscure the point that this differential amp *needs* no such difference. The method you used in Lab 2 to pick out a signal while rejecting noise did, of course, require such a difference.

This experiment should give you a sense of what "common-mode rejection ratio" means: the small amplification of the common signal, and relatively large amplification of the difference signal.

Nevertheless, this circuit still lets a large common-mode signal produce noticeable effects at the output. The improvement in the next step should make common-mode effects much smaller.

Common-mode gain (second try): Apply a current source to improve common-mode rejection. Replace the 10k tail resistor with a 1.5mA current source. You may build this current source as you choose. The laziest way is to use a pair of field-effect transistors (JFETs: devices we will not discuss in this course) that serve as current-limiting diodes. These are a part called 1N5294, rated at 0.75mA ±10%. Two in parallel (see Fig. 5L.9) provide the desired 1.5mA.

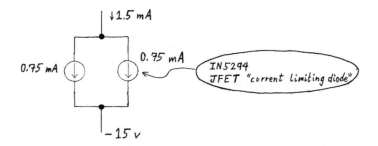

Figure 5L.9 JFET current-limiting diodes can provide the tail current source.

If this trick is too shabby and black-box for your taste, you know, of course, how to build current sources using bipolar transistors. Fig. 5L.10 shows two possible circuits.

Replacing the tail resistor with a current source should reduce the common-mode gain a great deal. (What is common-mode gain if the output impedance of the current source is around 1M?)[4] You should see very good CMRR at low frequencies: say, 100Hz. As frequency climbs, however, you'll see the output grow. This apparently results from capacitive coupling between input and output, as we argue in §5L.1.11.

Common-mode and difference signals mixed: Observe how this improved circuit treats a signal that combines common-mode and differential signals. Leave this circuit set up. You will be adding to it.

---
[4] Common mode gain is approximately $R_C/(2 \times R_{\mathrm{tail}}) \approx 10\mathrm{k}/2\mathrm{M} = 0.5 \times 10^{-2}$.

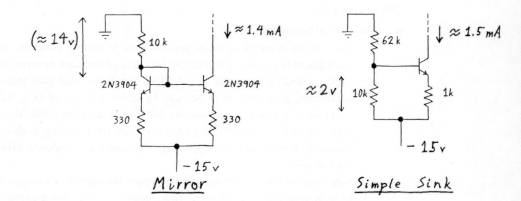

Figure 5L.10 ... or two alternative bipolar-transistor current sinks.

5L.1.6 A homemade operational amplifier

Here, we'll ask you to string together the three stages of the device that make up a standard operational amplifier. The op-amp is a just a good high-gain differential amplifier, so you can see that at this point in the lab you are partway to your destination.

An op-amp typically is a three-stage amplifier: a differential stage, a gain stage, and a push–pull output. Here, we ask you to add the two additional stages – a common emitter gain stage and a push–pull output – to the diff-amp you have built. These additions will convert this diff-amp into a modest op-amp. That device is, as you know, the building-block that you'll rely on in most of your analog designs, from Lab 6L onward. The op-amp you build today won't work as well as the IC version you meet in Lab 6L, but it should help you to gain some insight into what an op-amp is, and how it achieves its borderline-magical results.

Figure 5L.11 shows a block diagram that restates graphically the point we just made: an op-amp is a high-gain diff-amp with low output impedance.

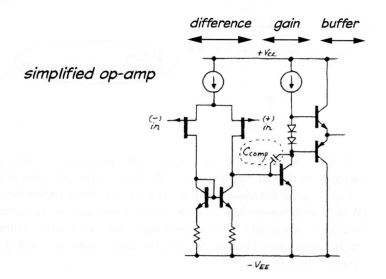

Figure 5L.11 Generic 3-stage operational amplifier.

We will ask you to modify your diff-amp somewhat to achieve higher gain, and to prepare it to drive the second stage conveniently. You'll test that; then the first two stages together; then the 3-stage amp. Finally, toward the end of this exercise, we'll ask you to apply overall feedback – a topic we have not yet discussed at any length. Perhaps you'll find the subject puzzling; we hope you won't mind this preview, even if the topic does come clear only later.

5L.1.7 Stage 1: increase the gain of the bipolar differential amplifier

First circuit change: maximize gain: Remove the 100Ω emitter resistors. Do you expect the circuit to lose *temperature stability*, with these gone? What happens to the *constancy of gain*?[5]
Test your views:

- to test temperature stability, watch $V_{\rm OUT}$ (with scope or DVM) as you try heating the CA3096 with your finger;
- to test constancy of gain, use a small *triangle* as input, and see whether you notice the "barn-roof" distortion that we saw in Lab 4L.

Second circuit change: move output quiescent point up: To get ready for addition of the next stage, change the collector resistor, $R_{\rm C}$ from 10k to 1.5k. This will violate our usual rule that calls for centering the output in the available range (here, 0–15V). This change will also lower the circuit gain. But your circuit's modest gain is not so sad as it may seem. We hope CMRR will remain respectable. Calculate your circuit's new differential and common-mode gains – or, if you are energetic, *measure* these gains.

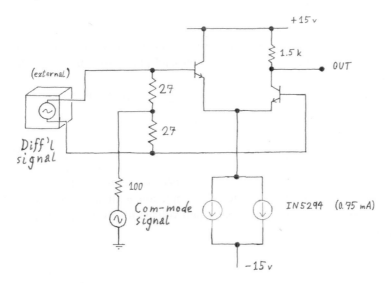

Figure 5L.12 Stage 1 diff-amp: preparing it to drive later stages.

5L.1.8 Stage 2: gain stage: common emitter amplifier

When you reduced the $R_{\rm C}$ to 1.5k, you placed the diff-amp's output quiescent voltage close to the positive supply, because we want this output to drive a common-emitter amplifier made with a *pnp*. This second stage will provide most of the voltage gain in the circuit.

---

[5] The circuit is temperature-stable without emitter resistors, because the two transistors run at equal temperatures. Though their $V_{\rm BE}$'s will change with temperature changes, their *sharing* of $I_{\rm tail}$ will not; so, the quiescent output voltage will not change as temperature changes. We noted this way of achieving stability back in §5N.5. See, especially, Fig. 5N.22. Constancy of gain, in contrast, will suffer: you will see distortion. Gain changes as $V_{\rm out}$ swings, for much the same reasons that cause "barn roof" distortion in a common-emitter amp, as noted in Chapter 5N. The diff-amp's distortion is shaped differently; §5S.1 explains this shape, in case you are curious.

Fig. 5L.13 shows the amplifier circuit that we propose. It's a conventional common-emitter amp (except that it probably looks annoyingly upside-down, to the *npn*-centric among us). The amplifier's input impedance is high enough not to load the preceding stage appreciably, as usual.

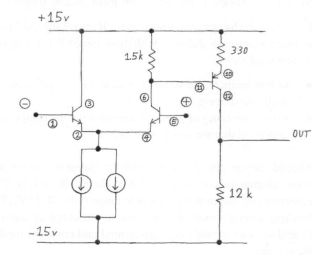

Figure 5L.13 Two-stage amplifier: differential and common-emitter.

Measure gain

Watch input and output of the CE amplifier stage, and measure this stage's gain. Then measure the overall differential gain, circuit input to circuit output (simply ground pin 1, applying a "pseudo-differential" input at pin 5).

You may need to tinker with the function generator's *DC offset* as you watch this high-gain amplifier, in order to make sure that neither first nor second stage clips.

5L.1.9 Stage 3: output stage: push–pull

In order to give the circuit low output impedance, we'll add a push–pull output stage. We won't bother to fix cross-over distortion, because we want to keep the circuit simple. In a minute, we'll let feedback try to undo this distortion. Fig. 5L.14 shows a push–pull voltage follower, made with two more transistors in the CA3096 array.

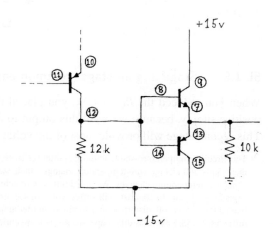

Figure 5L.14 Push–pull output stage (bipolar).

5L.1 Difference or differential amplifier

With this stage added, the baby op-amp is complete. The circuit – driven still with a pseudo-differential input, and still running "open-loop" rather than with overall feedback – is in Fig. 5L.15.

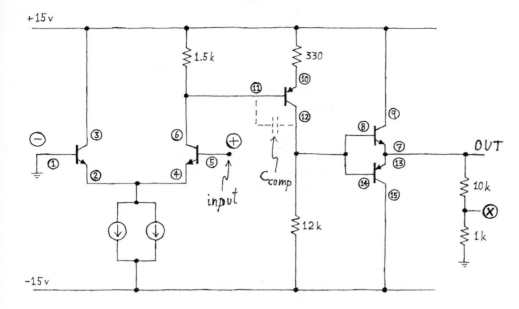

Figure 5L.15 Home-made op-amp: complete 3-stage circuit; still running open loop.

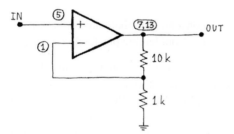

Figure 5L.16 X11 amp?

Feed a small sinewave differential signal to the input, at a low frequency: 1kHz or under; watch input and output of this push-pull stage. You should notice cross-over distortion: dead sections in the output, while the push–pull's input is too close to zero to turn on either the *pnp* or the *npn* transistor. To show this *crossover distortion*, the circuit output *must cross zero*. You may need to adjust the DC-offset of the input signal, in order to center the output waveform.

5L.1.10 Trying feedback

Op-amps almost always use overall feedback. Let's try it with your circuit.

A X11 amplifier?
Now we'll try the op-amp in the configuration that is normal for such devices: we feed back circuit output to circuit input (more precisely, we'll feed back a *fraction* of the circuit output). This arrangement is shown in Fig. 5L.16. We must keep the sense of feedback "negative:" output tending to diminish the input.

Connect the input at pin 1 to an attenuated version of the circuit output, marked X in Fig. 5L.15: 1/11 of the amplifier output that appears at pins 7 and 13. This connection, shown in Fig. 5L.16, will

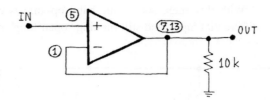

Figure 5L.18 Overall feedback imposed: a voltage follower.

force the amplifier to try to drive this input, (1), to the voltage applied at the other input, pin 5. As a consequence, we will trick your circuit into amplifying by about $11\times$ (this is the nominal gain, because we are feeding back one part in 11). Try it.

Gain is likely to be below the hoped-for X11, because our circuit gain is so modest.[6]

What's valuable and interesting about this *feedback* circuit is not, of course, that it delivers lower gain. As the British patent office reminded Mr. Black, *reduced* gain is not one's ultimate goal, in amplifier design.[7] Instead, one sacrifices gain for other desirable characteristics. In the present example, we hope you'll see two improvements in your amplifier's performance, now that you've applied negative feedback:

- Perhaps the most interesting difference from the open-loop case that you tried one stage back is the disappearance of cross-over distortion from the circuit output – at least at modest frequencies.
  Feed a small sine, and continue to watch input and output of the last stage. We hope you'll find the output of the op-amp looking sinusoidal, while the input to the push-pull (pins 8, 14) looks strange – because feedback is forcing that point to do something to cancel the cross-over distortion. Pretty magical?
- A less striking virtue: the amplifier should not show the inconstant gain that causes "barn-roof" distortion in a triangular waveform. Such distortion somewhat troubled the pre-feedback design.

If your circuit begins to oscillate on its own, you'll need to reduce its high-frequency gain. The best way to do this is to place a small capacitor (try 100pF) between collector and base of the common-emitter gain-stage transistor (pins 11, 12); see Fig 5L.17. This exploits Miller effect (see AoE §2.4.5), forming a low-pass filter whose apparent C is enlarged by the gain of this stage. Such reduction of high-frequency gain in order to achieve stability is called "compensation" and is routinely applied within op-amps.

A follower? (optional: risky business)

Oddly enough, the simpler circuit in Fig. 5L.18 – the voltage follower – is more difficult to stabilize than the $11\times$ amplifier; that is, the follower is more likely to show those nasty parasitic oscillations. If your circuit is very tidily built, you may be able to see a stable follower; some of these homemade op-amps, however, cannot be stabilized at unity gain, even with the compensation effort described above.

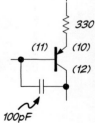

Figure 5L.17 Compensation capacitor can stabilize a feedback circuit by killing high-frequency gain.

5L.1.11 Appendix: CMRR degradation as frequency climbs

Here's a fine point we referred to in §5L.1.5: an explanation for the observation that CMRR degrades as frequency of the common-mode input grows.

The waveform's phase and frequency-response provides a clue that what we're seeing does not

---

[6] If you want to compare your circuit's gain against a theoretical estimate, see AoE §2.5.2. The circuit gain ought to be $A/(1+AB)$, where A is your circuit's "open loop gain" (the differential gain you just measured), and B is the "fraction fed back" (here, 1/11).

[7] See AoE §2.5.1 and §6N.1. Of course, the point of that story is that Black had the last laugh!

5L.1 Difference or differential amplifier

result from any failure of the current source in the tail. Here are a couple of scope images, showing output for common-mode inputs. The first, in Fig. 5L.19, shows outputs for sinusoids at two input frequencies: 1kHz and 10kHz. The amplitude is much larger at the higher frequency.

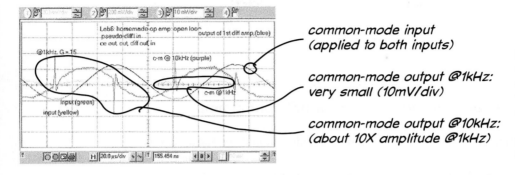

Figure 5L.19
Common-mode rejection diminishes at higher frequency: sinusoid inputs at 1kHz versus 10kHz. (Note that scope gain for output channels is 20× greater than for input.)

That sounds like *high-pass* behavior, certainly. The scope image in Fig. 5L.20 says the same thing in another way, showing the output looking like a *differentiation* of the input.

Does your circuit behave the same way? If so, what you're seeing is capacitive feedthrough, from input (base of the input transistor) to output (collector resistor). A cascode could minimize this effect (see Miller Effect discussion in AoE §2.4.5). But let's not pause for such perfectionism now.

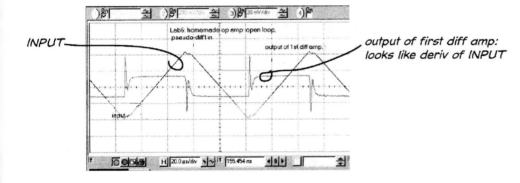

Figure 5L.20
Common-mode output acts like a differentiator: shows slope of triangle input.

5S Supplementary Notes and Jargon: Transistors II

5S.1 Two surprises, perhaps, in behavior of differential amp

We have advertised the differential amp as just a pair of common-emitter amplifiers, and have promised you that there's not much new to understand here: you can use what you know from the two earlier labs where you build C–E amps. But students have noticed some effects that are new: not what one might expect from experience with a single C–E amp. One of these effects may not puzzle you for long, if at all; the other is quite subtle.

5S.1.1 Clipping of first-stage diff amp in Lab 5L op-amp

In the common-emitter amp, we are accustomed to see clipping at the positive supply, and close to ground, where the transistor saturates. So, the clipping shown by this differential amp, when its R_C has been reduced to 1.5k, is unfamiliar. The scope image in Fig. 5S.1 shows *two* outputs, because we have inserted a collector resistor above the left-hand Q, making the circuit perfectly symmetric, except in its *drive*. The *drive* is what we call "pseudo-differential."[1]

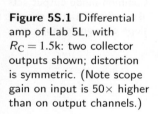

Figure 5S.1 Differential amp of Lab 5L, with $R_C = 1.5\text{k}$: two collector outputs shown; distortion is symmetric. (Note scope gain on input is 50× higher than on output channels.)

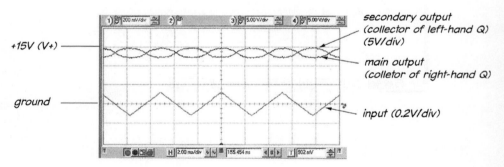

The flattening close the positive supply is nothing new. But two novelties appear:

- first, as the output swings *down*, the waveform flattens again, instead of growing steeper as in "barn-roof" distortion;
- second, this flattening occurs not where the transistor *saturates* (which would be at a voltage close to ground) but at about 12.5V *above* ground, where the transistor is very far from saturating.

This is unfamiliar, but we can make sense of this when we recall that the total current available to either transistor is limited – by the *tail* current source. So, the lowest collector voltage occurs not

---

[1] This sort of input mixes differential and common-mode signals, in fact – but harmlessly. The differential signal is the full amplitude of the applied signal; the common-mode is half that. As long as differential gain is much higher than common-mode, this sort of input is a very close equivalent to the true differential signal that we more laboriously applied earlier in Lab 5L.

when the transistor *saturates*, but when it is hogging the entire *tail* current, leaving none for the other side. Strange, perhaps; but explicable.

5S.1.2   A closer view of the distortion: barn-roof reflected in a puddle?

We saw in Fig. 5S.1 that the diff amp's distortion is symmetric, unlike the "barn roof" distortion of a common-emitter amp. If we look closely at that distortion, when driving the amplifier less strongly than in that Fig. 5S.1, we find what looks like barn roof distortion *mirrored* – offset to the side and perhaps reflected in a barnyard puddle.[2]

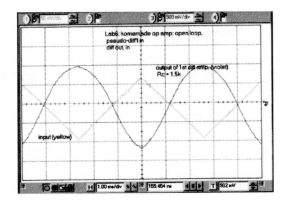

Figure 5S.2 Diff amp's distortion differs from C–E's: it's symmetric. (Note scope gain on input is 10× greater than on output channel.)

If you would like a mathematical argument – unusual in this course – to explain the symmetry in the distortion, we'll try one. The question we aim to answer is why gain should diminish on *both* both sides of the quiescent point. This is the barnyard–puddle symmetry. Here's the argument:

- Gain of the diff amp, with one input grounded:

$$G = -\frac{R_C}{r_{e_1} + r_{e_2}} \qquad (5S.1)$$

- ...and at the quiescent point this is simple enough; the gain is just $G = -R_C/2r_e$.
- But, as you recall from your experience with a common-emitter amplifier, the gain varies as the output swings, because I_C has to vary in order to cause the output swing; and the sum of $r_{e_1} + r_{e_2}$ is not constant as the diff amp output swings – even though one r_e grows as the other shrinks.
- More specifically, if we call the current in the output transistor I_1 while calling the total "tail" current I_T (both assumed measured in milliamps), then:
-

$$G = -\frac{R_C}{(r_{e_1} + r_{e_2})} = -\frac{R_C}{\left(\frac{25}{I_1} + \frac{25}{I_T - I_1}\right)} = -R_C \bigg/ \frac{25(I_T - I_1 + I_1)}{I_1(I_T - I_1)}$$
$$= -\frac{R_C}{25} \times \frac{I_1(I_T - I_1)}{I_T} = -\frac{R_C}{25} \times \left(I_1 - \frac{I_1^2}{I_T}\right). \qquad (5S.2)$$

If I_1 gets all or none of the tail current, gain is zero. If I_1 is half the tail current, gain is at a maximum – about 22, for R_C=1.5k and I_T=1.5mA, as in the lab circuit.

The gain curve looks like the plot of Fig. 5S.3. The figure makes evident that 50:50 sharing provides maximum gain.

---

[2] The output in Fig.5S.2, appears to be centered on zero volts. This is an illusion that results from AC-coupling of the scope input.

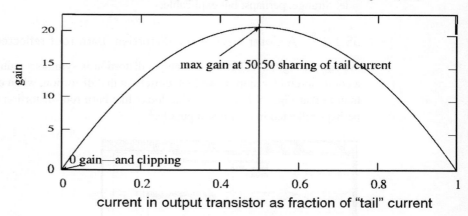

Figure 5S.3 Plot of diff amp's calculated gain as sharing of tail current diverges from 50:50

An expression for the derivative of gain: $dG/dI_1 = 1 - (2I_1/I_T)$ says the same thing: the maximum gain occurs when I_1 is 1/2 the tail current.

And the flattening that we saw in Fig. 5S.1 occurs when the circuit gain falls to zero. This occurs when either transistor hogs all the current.

5S.2 Current mirrors; Early effect

5S.2.1 Mirrors, a topic you can skip

We concede that there's something funny about opening a section with the remark, "you can skip this." We treat mirrors this way because of our evident ambivalence about these neat circuits. We chose to remove them from the lab exercises, judging them less important to a designer than the transistor circuits that we have asked you to build. But mirrors come up fairly often for a *reader* of circuits. They abound in op-amp ICs. So, it is useful to consider how they work, even if you are unlikely to adopt a mirror as an element in your own designs.

Mirrors also are pedagogically useful, to demonstrate *Early effect*, a topic that we similarly removed from labs, but would like to make available to the curious. Early effect lets us make a quantitative estimate of the output impedance of a current source. To this point we have had to content ourselves with saying that current sources show output impedance that is *high*.

We will begin by introducing the current mirror. Then we will adopt it to watch Early effect in action.

Applying the Ebers–Moll view to circuits: current mirror

A current mirror makes no sense without the help of the Ebers–Moll view of transistors, illustrated in Fig. 5S.4.

AoE §2.3.7

Why is a mirror useful, given that we know other ways to make current sources? A mirror makes it easy to link currents in a circuit, matching one to another. Such linking is useful in integrated circuits. A mirror also can show the widest possible *output voltage compliance*. You can see a zoo of mirrors, adopted for both these motives, in the schematic of the operational amplifier that we use in many of our analog labs, the LF411 (see §9S.4).

It's easy to make $I_{program} = I_{out}$, and only a little harder to scale I_{out} relative to $I_{program}$.

Early effect and *temperature* effects both could spoil the neat equality between $I_{program}$ and I_{out}. We

5S.2 Current mirrors; Early effect

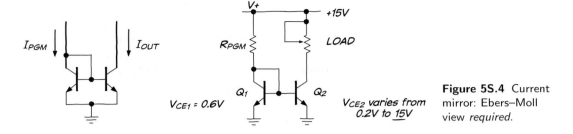

Figure 5S.4 Current mirror: Ebers–Moll view *required*.

will learn later (see §5S.2.7) how to fight these problems. For now, let's leave the mirror in its simplest form, as shown above, and notice how the circuit works.

5S.2.2 Feedback sets mirror programming current...

Look at Fig. 5S.5. The *program* side of the current mirror looks simple, if a bit weird (base and collector joined?!). It's quite neat, though: an application of negative feedback.

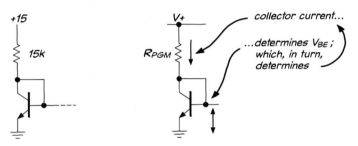

Figure 5S.5 Subtle negative feedback: programming side of the current mirror.

As you explain to yourself how this circuit fragment works – giving us our beloved 1mA – it's helpful to notice that nearly all of the current flows not in the base path, but from collector to emitter.[3]

5S.2.3 ...while equal V_{BE}'s assure equal I_C's: the "mirroring"

So, because of the equal V_{BE}'s of Q_1 and Q_2 (in Fig. 5S.4), $I_{out} = I_{program}$. So far, so neat.

The main virtue of this circuit is its ability to operate almost from "rail to rail" (jargon for "from one supply to the other"). The mirror will work until Q_2 saturates, so it shows wide "output voltage compliance." As R_{load} varies in the circuit of Fig. 5S.4, current stays pretty nearly constant at 1mA. The voltage range from V_+ (15V, in this case) down to about 0.2V (just above Q_2's saturation voltage). This wide compliance helps explain the widespread use of mirrors in operational amplifiers, as we have argued above (another reason is the preference for transistors over resistors, in IC fabrication).

5S.2.4 Complications

A difficulty easily solved: temperature effects

If the temperatures of Q_1 and Q_2 diverge, the claim of 1:1 mirroring fails. You saw in Chapter 5N how strongly temperature affects I_C.

But it is easy to hold Q_1 and Q_2 temperatures equal: just put the two transistors close together on

---

[3] Does it make sense? The base voltage – the same as collector voltage – is determined by the drop across $R_{program}$. But $I_C \times R_{program}$ sets a value for V_{BE}. Higher I_C drives down V_{BE}, so this is a negative feedback loop. It will stabilize, having found an I_C that generates a V_{BE} consistent with that I_C.

an integrated circuit. Any other way of building a mirror would be downright perverse. So, temperature matching takes care of itself. (You saw this stabilizing technique applied to a common emitter amplifier, as well, in Chapter 5N, Fig. 5N.21: the "compensating" transistor shown in that mirror-like amplifier circuit keeps the amplifier temperature-stable.) It is true that V_{BE} changes with temperature, at a given I_C: as temperature rises, V_{BE} falls, for example. But that altered V_{BE}, produced by Q_1, is applied to a heated Q_2, whose 1mA I_C requires just such a reduced V_{BE}.

5S.2.5 A harder problem, neatly illustrated by current mirrors: Early effect

AoE §2.3.7A

Another departure from the ideal simple *mirroring* is not so easily fixed: it is the slight degree to which collector current varies as voltage across the transistor varies.

Output impedance of a current source ideally is infinite (because the value $R_{dynamic}$, equivalent to $\Delta V_C / \Delta I_C$, is infinite if I_C does not change at all). But, in fact, I_C grows slightly as V_{CE} grows: so an actual current source behaves like a large resistor. In Fig. 5S.6 we model this imperfection.

Figure 5S.6 A non-ideal current source can be modelled as ideal parallel (large-) R.

You may have met this model as a Norton Equivalent Circuit. Like the Thevenin model, it permits us to give a value to a circuit's degree of imperfection.

Compare AoE Fig. 2.59

Figure 5S.7 shows another way to represent this behavior: as a non-zero slope in the I_C vs V_{CE} plot.

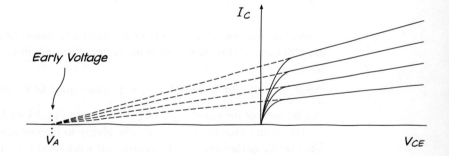

Figure 5S.7 Early voltage, where all the curves converge, provides a measure of departure from current constancy.

Strictly speaking, V_A is measured not relative to zero on the V_{CE} axis, but relative to the point at which the tangent is taken. The difference, however, between these two points is usually not substantial. V_{CE} ordinarily is much smaller than the value of V_A.

Figure 5S.8 gives the measured responses of three transistors as we increase V_{CE}:[4]

The leftmost transistor curves show the best current-source behavior, as you'll recognize. But in other respects (say, its impedance-changing virtues as a *follower*) that transistor, with its low β, might be considered the *worst* of the three transistors.

The mechanism of Early effect

The resistor-like response of the transistor to voltage across its C–E terminals is caused by *base width modulation*: an increase in V_{CE} widens a carrier *depletion region* between collector and base, shrinking the effective *base width* and thus increasing collector current. Early effect can be moderate

---

[4] Data taken from AoE Fig. 2.59. The middle plot is discussed in §2x.8 of *AoE – the x-Chapters*.

5S.2 Current mirrors; Early effect

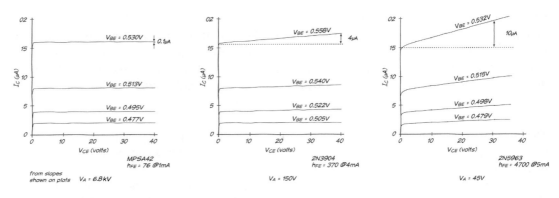

Figure 5S.8 Early effect: slopes of I_C versus V_{CE} reveal V_A. Taken from §2x.8 of *AoE – the X-Chapters*.

or extreme, depending on the intrinsic base width: a thin base region responds more dramatically to a given narrowing because that narrowing constitutes a larger fractional change in the effective width.

So transistors with narrow base regions, which are those with *high* β – are especially sensitive to V_{CE} variation. This appears in the pairings of values for V_A and β shown below, for the three transistors plotted in Fig. 5S.8.

Transistor	β (typical)	Early voltage, V_A (from measured curves)
MPSA42	25	6.8kV
2N3904	130	150V
2N5963	1200 (min.)	45V

V_A is useful – but also not ordinarily available as a transistor specification.[5] The curves of Fig. 5S.8 are difficult to measure, because temperature effects can muddle the attempt to see Early effect. If you have no V_A spec or curves to work with, you will have to settle for an estimate of V_A based on the transistors's β range. The estimate will be approximate, but helpful. For a small-signal transistor like the 2N3904, you can guess at a V_A of somewhere around 100V.

5S.2.6   Calculating the results of Early effect

A V_A specification gives a quick impression of the quality of a transistor as current source. But it is rather an abstraction. It would be useful to be able to convert V_A – or some other Early effect specification – into a measure of the behavior of a particular circuit.

Two results of Early effect would be especially useful...

- a prediction of the *output impedance* of a current source;
- a way to calculate a ceiling on the gain of a transistor amplifier, when its load (the impedance on the collector) is a current source

We can, indeed, calculate both results.

---

[5] This omission of V_A from transistor data sheets seems odd. We asked the late analog wise man Robert Pease why, and he answered (in an email), "Hardly anyone asks for it. And those that need this info, know how to get it," meaning that they can measure ΔV_{BE} versus ΔV_{CE}. We remain puzzled by the omission, though.

Calculating R_{out} for a current source

One way: use the slope of the I_C versus V_{CE} curves directly: If you are lucky enough to have a plot if I_C vs V_{CE}, like those in Fig. 5S.8, you can measure R_{out} from the plot.

For the 2N3904, for example, a simple current mirror (one with no emitter resistor) shows a slope that we have labeled for the particular curve that begins at about $17\mu A$:

$$\Delta I_C / \Delta V_{CE} = 4\mu A / 40V = 0.1 \times 10^{-6} \Omega^{-1} .$$

The units are inverse ohms: Ω^{-1} or siemens. The inverse of this slope is the output resistance at the collector, R_{out}, a value in ohms: $10M\Omega$. This is not a general answer, but an impedance at a particular output current (about $17\mu A$, in this case).

This is a start. We'd like to extend this to R_{out} at other currents. Let's recast the slope as a *percentage* change in current per volt. Fig. 5S.9 shows a hypothetical Early plot – similar to the 2N3904 curves, but showing a hypothetical Early voltage of 100V rather than our measured V_A. We chose this V_A to keep the arithmetic as simple as possible.

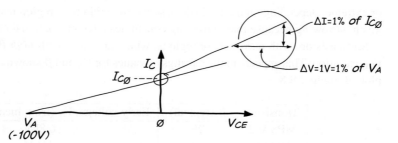

Figure 5S.9 Effect shown as fractional or percentage change per volt.

For $V_A = 100$, a 1V increase in V_{CE} – from an initial V_{CE} close to zero – is a 1% increase. The straightline ramp of the I_C plot, up from zero at V_A, enforces a resulting 1% increase in I_C.

This way to describe the Early effect slope is convenient, because it lets us calculate R_{out} at any initial current (I_{C_0}). At $I_{C_0} = 1mA$, for example, the 1% $\Delta I_C = 0.01mA = 10\mu A$. So, R_{out} at 1mA is the inverse of this slope: $0.1M = 100k\Omega$.

AoE §2.3.2D

Another way: Rather than use either V_A or the I_C versus V_{CE} slope directly, one can infer a useful relation, $\Delta V_{BE}/\Delta V_{CE}$. This relation – given the symbolic representation η – will help when we estimate R_{out} for current sources:

$$\eta \equiv -\frac{\Delta V_{BE}}{\Delta V_{CE}} \quad \text{at a } \textit{fixed value of } I_C .$$

This η is a little hard to get an intuitive grip on: it is the change in V_{BE} (a decrease) required to hold I_C constant as V_{CE} grows by a volt.

Using the hypothetical I_C versus V_{CE} curves again, from Fig. 5S.8, we can calculate the value of η. We saw a 1% increase in I_C for each 1V increase in V_{CE}. (Incidentally, here comes one of the rare instances in which we use the Ebers–Moll equation explicitly.)

Ebers–Moll tells us that $I_C \approx I_S(e^{V_{BE}/25mV})$. (We ignore the insignificant "-1" from the full expression, as usual. See AoE §2.3.1. The value 25mV is kT/q evaluated at room temperature.) The ratio of two collector currents is therefore

$$I_{C2}/I_{C1} = e^{(V_{BE2}-V_{BE1})/25mV} .$$

If I_{C2} is 1% more than I_{C1}, as in the plot of Fig. 5S.8, Ebers–Moll lets us calculate the associated change in V_{BE}:

5S.2 Current mirrors; Early effect

For this 1% change, then, $e^{\Delta V_{BE}/25mV} = 1.01$. We should admit the embarrassing fact that this describes a case *different* from the one that we mean to evaluate: it describes the case were V_{CE} is kept constant and an increase in V_{BE} causes the rise in I_C. Starting with this different case may strike you as a bit weird; it seems that way to us too. But hang on, the procedure does work.

$$\Delta V_{BE}/25mV = \log_e(1.01)$$
$$\Delta V_{BE} = \log_e(1.01) \times 25mV = 0.01 \times 25mV = 0.25mV.$$

Now we can pop out of this thought experiment and admit that what we're after is the *reduction* in V_{BE} that we would need in order to hold I_C fixed as V_{CE} grows: this is the same value for ΔV_{BE} but is opposite in sign:

$$\eta \equiv -\Delta V_{BE}/\Delta V_{CE} = 0.25mV/1V = 0.25 \times 10^{-3}.$$

For the moment, we will just note that we have found how to calculate this value. Soon, in §5S.2.7, we will put η to use.

5S.2.7 Improving R_{out} for a current source

AoE §2.3.7B

Cascode or "Wilson mirror"

One way to eliminate the effect of changes in V_{CE} is to add an extra transistor that prevents such V_{CE} changes. This trick – whose analogous design dates from vacuum-tube days – is called a "cascode" circuit.[6] The additional transistor simply passes the current fed it by the transistor that is to be protected from Early effect. The pass-through transistor does feel wide variation in V_{CE}, but this transistor does not set the output current; it merely passes the current determined by the protected transistor. This design, illustrated in Fig. 5S.10, is easily implemented in an integrated circuit, and is the standard form for mirrors within operational amplifiers, including the one that we rely on in most of our op-amp labs, the LF411.

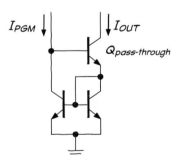

Figure 5S.10 Wilson mirror or "cascode" configuration: hides changes in V_{load}.

The circuit fragment shown at the left side of Fig. 5S.11, taken from the LF411 data sheet, looks at first like a bizarre new circuit component. Redrawn on the right, however, it reveals itself as the same circuit as that of Fig. 5S.10 (though done with PNPs rather than NPNs, to *source* rather than *sink* current). Q_{13}, you will notice – which looks pretty strange in the original drawing – is redrawn more conventionally on the right as two distinct transistors, sharing common emitter and base connections.

Current source improvement by use of *emitter resistors*:

The characteristic that we called η, in §5S.2.6 will help us to calculate the improvement in R_{out} that emitter resistors can provide. Fig. 5S.12 shows a simple mirror with emitter resistors added.

---

[6] The word, though now applied to transistor circuits, reflects the date of its birth, 1939: it merges the word "cascade" with "cathode," a terminal name familiar to you from its use for diodes but one that did not carry over from vacuum tubes to transistors.

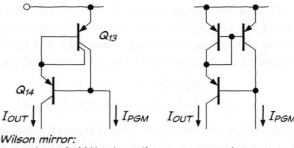

Figure 5S.11 Cascode or Wilson mirror in LF411 op-amp.

Wilson mirror:
...as shown in '411 schematic...   ...redrawn

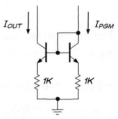

Figure 5S.12 Simple mirror improved by emitter resistors.

If you have accepted the argument (made in Chapter 5N) that installing an emitter resistor in a common-emitter amplifier stabilizes I_C, you will find it plausible that emitter resistors should tend to stabilize I_C in a mirror. In both cases, *negative feedback* is the key to the improvement.

We will try to do better than claim that the effect is plausible. We will use η to calculate the quantitative effect of these resistors.

Use η to calculate R_{out} for a mirror with R_E's: A rise of 1V at the collector causes V_{BE} to shrink by $\eta \times 1$V, thus causing a rise of emitter voltage by that amount: 0.25mV (for the η value of 0.25×10^{-3} that we found in §5S.2.6).

Ohm's law says that this change in V_E will increase I_E and I_C by

$$\Delta V_E / R_E = 0.25 \times 10^{-3} \text{V} / 10^3 \Omega = 0.25 \times 10^{-6} \text{A}.$$

At the collector, R_{out} will be the inverse of this result: 4MΩ. Generalizing this result, we can see that a rise of 1V at collector causes a rise η times smaller at the emitter. So the transistor, seen from its collector, behaves like a "lens": see Fig. 5S.13.

This lens metaphor is reminiscent of the follower that we likened to a rose-colored lens, but radically different in its result: this lens *increases* R_{out}, enlarging the emitter resistance by the factor $1/\eta$. In this example, that factor was 4000, converting the 1k R_E to an R_{out} of 4M.

Confirm η's utility, by calculating R_{out} for the original mirror: We earlier calculated R_{out} for the simple mirror using the slope of the I_C vs V_{CE} curves. Let's test our faith in η by recalculating that value using η instead, along with our usual Ebers–Moll sidestep, r_e. Let's see if we get the same result – 100k.[7]

Using the result for the case where we installed R_E's, we note that we'll get a ΔV_{BE} of 0.25mV for a 1V ΔV_{CE}, and (having faith in our "little r_e" model) we view this as raising the voltage across the model r_e by that amount.

---

[7] It would be strange if we did not – for we are simply running in reverse the calculation that led to our value for η.

5S.2 Current mirrors; Early effect

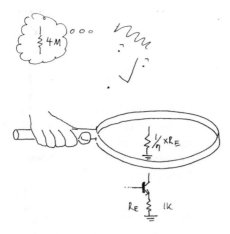

Figure 5S.13 Viewed from collector, transistor current source acts like a lens that magnifies R_E and r_e.

For the 1mA current source, the value of r_e is 25Ω at room temperature, so I_C grows…

$$\Delta I_C \approx \Delta I_E = \Delta V_{BE}/25\Omega \;.$$

For an assumed 1V change in V_{CE}, this quotient is $[(\eta \times \Delta V_{CE}) = 0.25\text{mV}]/25\Omega$:

R_{out} is the reciprocal, $1/\Delta I_C$:

$R_{\text{out}} = 25\Omega/0.25\text{mV} = 100\text{k}$, as we had hoped .

We could put it more simply (and more metaphorically), as we did in Fig. 5S.13, saying that once again the Early effect "lens" enlarges what's at the emitter by the factor $1/\eta$, which we found to be 4000 in the present case. Here, where the circuit includes no R_E, we must use r_e evaluated to 25Ω at 1mA. So $R_{\text{out}} = 4000 \times 25\Omega = 100\text{k}\Omega$. This *lens* method seems the most straightforward.

…and a mirror can be improved by use of both *both* cascode and emitter resistors
Wilson mirrors are available as ICs, and one from Texas Instruments, REF200, which includes two current limiting elements plus one Wilson mirror,[8] uses *both* of the improvements described above: *cascode* and *emitter resistors*. These are shown in Fig. 5S.14.

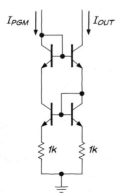

Figure 5S.14 REF200 IC mirror uses both cascode or Wilson configuration and additional stabilization by *emitter resistors*.

---

[8] See the discussion of this part in Chapter 11N. A TI application note shows many current source circuits, including applications of the REF200 cascoded to achieve enormous R_{out} (>10Gigohm). See http://www.ti.com/litv/pdf/sboa046, "Implementation and Applications of Current Sources and Current Receivers," 2000.

This schematic makes the Wilson configuration much easier to recognize than in the LF411 data sheet – and adds the new element of emitter resistors.[9] With this double protection, the mirror in REF200 achieves spectacular R_{out}: a *typical* value of 100MΩ.

5S.2.8 Another consequence of Early effect: a ceiling on amplifier gain

We will see, in operational amplifiers, that transistor current sources routinely replace collector resistors,[10] in order to maximize gain of differential and common-emitter stages. When this is done, the impedance of that current source matters – but so does the output impedance of the amplifier transistor. A naive view might see a current source load as providing gain that was *infinite*. As usual, we can bring such hyperbolic impressions down to earth, using Early effect to quantify the ceiling on amplifier gain.

Gain of ordinary common-emitter amp: determined by R_C and emitter resistance

> AoE §2.3.4A

In the cases familiar to us from this course, a transistor amplifier converts changing collector current to changing collector voltage using a resistor on the collector. We have said that gain, in a common-emitter amplifier, is determined by the ratio of R_C to either R_E or to the resistor-like model of transistor gain that we call "little r_e".

Equivalently, the maximum gain of our usual amplifier can be described as $g_m \times R_C$, where $g_m \equiv 1/r_e$.[11] The "transconductance," g_m, is current-change *out* per voltage-change *in*: $\Delta I_C / \Delta V_{BE}$.

This formulation ignores Early effect, and until now ignoring Early has worked. In order to take account of Early effect, one would acknowledge that the resistance driven by the transistor's collector, strictly, is $R_C \parallel r_{out-collector}$ (let's call it r_o), where r_o quantifies Early effect. In the cases we have seen to this point, R_C has been so much smaller than r_o that there was no need to mention more than R_C.

Gain of a common-emitter amp using a good current source on the collector

Now, if we consider the case where we maximize gain by using a very-good current source in place of R_C, we need to take account of Early effect: we need to include the effect of the transistor's own r_o. To keep things manageable, let's assume an *ideal* current source as load (or if "ideal" bothers you, use a source like the REF200 mirror, with its r_o of around 100MΩ).

> AoE x-Chapters, §2x.8

In this case, it will be the transistor's r_o, not the familiar R_C, that will define the circuit's voltage gain. Gain will be $g_m \times r_o$, or (equivalently) r_o/r_e. At 1mA, for the transistor we have been assuming in our discussion, with its V_A of 100V, this gain will be $100k/25 = 4000$. High, but a long way from infinite.

5S.3 Transistor summary

Why this note?

We're guessing that you're feeling overloaded with information, as we approach the end of the discrete-transistor section of the course. You've heard important points and details, and there's some danger that details will elbow out more fundamental learning. So, here's a sort of 3-minute-University version of what you've spent a week learning. Maybe it's a 6-minute-University.

---

[9] This circuit also adds a fourth transistor, a sort of *pass* element for the programming side, gilding the already impressive Wilson mirror's protections. This fourth transistor equalizes the two lowest transistors' V_{CE}'s at one V_{BE} drop, rather than one and two V_{BE}'s as in the standard Wilson mirror.

[10] Collector resistor equivalents are "drain" resistors for similar amplifiers made with field-effect transistors, as in the LF411 op-amp.

[11] This definition – suggesting that g_m is derived from r_e – makes sense in this course, since we have used r_e repeatedly. In the bigger world, however, the definition is a little odd, since g_m is the concept much the more generally used.

5S.3.1 Preliminary: why learn about discrete transistors?

Soon we'll be designing with op-amps, and the transistor designs we've studied (say, the common-emitter amp) will seem clumsy and defective and very complicated. Do you need to understand a C–E amp?

Maybe not, but the major argument *pro* is that you'll see such amplifiers within op-amps (second stage of '411, e.g.; and the differential amp that forms the first stage of every op-amp is essentially a pair of linked C–E amps). As we have said earlier, it's satisfying – though not necessary – to be able to understand how an op-amp works: to see that it's not *only* magic that lets an op-amp achieve its magical results.

Then there is a little room for discrete-transistor circuits: sometimes as high-frequency circuits, beyond the range of a garden-variety op-amp; more often, as higher-powered helpers that are controlled by an op-amp (say, a push–pull put within an op-amp's feedback loop, as in §6L.8).

5S.3.2 Bipolars

Main points, or truths worth recalling: bipolars

- Simple View (Day 4) (*note* that this simple view often is enough. Use it when it suffices.)
  - $V_{BE} \approx 0.6V$: this truth (*not* contradicted by the Ebers–Moll view) often allows quick analysis and design.
  - $I_C \approx I_E$: follows from notion that β is large (that is, that $I_C \gg I_B$).
  - I_C is β times larger than I_B: we need this notion in order to understand input impedance of follower or of C–E amp, and to understand the follower's ability to change impedances ("rose-colored lens" effect): the follower boosts R_{IN}, drops R_{OUT}.
- Fancier View: Ebers–Moll (Day 5)
  - I_C is determined by V_{BE} (rather than by I_B). More specifically, I_C is an exponential function of V_{BE}.[12]
  - Often we use this information – model it – with the curious device of "r_e," a model that shows a little resistor in the emitter lead of the transistor. This modelling device allows us to use our simpler (Lab 4L) view of transistors in order to calculate *gain* and *input impedance* of transistor circuits where R_E is very small or absent altogether.
- A few formulas.

  Input and output impedances of follower (rose-colored lens effect):
  $$R_{IN} = (1 + \beta) \times R_E\,^{13}.$$
  $$R_{OUT} = [r_e + R_{SOURCE}/\beta] \parallel R_E.$$
- Usually we can simplify this by ignoring R_E, because that value usually is much larger than the resistance seen through the transistor "lens," R_{SOURCE}/β.
- If the circuit is *biased*, then $R_{Thevenin-Bias}$ is parallel to R_{SOURCE}: $[R_{SOURCE}\|R_{Thevenin-Bias}]/\beta$. Again, R_{SOURCE} normally has been designed to be much lower than $R_{Thevenin-Bias}$, so the result usually boils down to R_{SOURCE}/β once more.

---

[12] We can reconcile this idea with Day 4's view (that I_C is determined by I_B) by recalling that the base-emitter junction is, after all, *a diode*, so that V_{BE} and I_B are linked. More specifically, I_B looks like an exponential function of V_{BE}. So, to say (as we do in Day 4) that I_C is a constant multiple of I_B is to say implicitly that I_C is an exponential function of V_{BE}, as Ebers–Moll teach us.

[13] Normally, we don't bother with the "1 + ···", since we don't begin to know β with that precision. We'll write β rather than the correct-but-silly "1 + β, in these formulas. Note, by the way, that "R_E" represents the whole impedance at the emitter (sometimes a *load* parallels the emitter resistor, for example)

- R_{OUT} for C–E amp (taken at collector): $\approx R_C$, because $R_{OUT} = R_C \parallel Z_{collector}$, and $Z_{collector}$ is huge.[14]
- Gain of C–E amp: $-R_C/(r_e + R_E)$.
- Gain of differential amp:
  Differential gain[15].
  Common-mode gain: $G_{CM} = -R_C/(r_e + R_E + 2R_{TAIL})$. This is approximately $-R_C/2R_{TAIL}$. A large value of R_{TAIL} permits low G_{CM}. A *current source* in the place of R_{TAIL} provides best – lowest – G_{CM}.
- …and *switches* are weird: different from all the other circuits we have seen:
  - we want switches either to *saturate*, or to turn *off* – whereas all the other circuits call for avoiding both of these extremes
  - none of the usual linear-region transistor rules apply to switches:
    No, $I_C \neq \beta \times I_B$, in a switch: we want, instead, to overdrive the base, by about a factor of 10. Load current is determined by the *load*, not by $\beta \times I_B$.
    No, input resistance at the base does *not* equal $\beta \times R_E$ or $\beta \times r_e$. Since the switch is saturated, R_{IN} is much lower than that. No helpful β limits the magnitude of base current.

5S.4 Important circuits

Most of us store circuits as graphic elements rather than as concepts. We invoke them like little linguistic idioms; we don't derive them each time from first principles or from formulas. Here, as a reminder, are some of your favorite transistor circuits.

- current sources:
  - single-transistor:

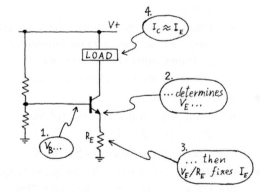

Current source (sink): single transistor

---

[14] That impedance is huge because that terminal behaves like a current source (or "sink"), and for any good current source, $\Delta V / \Delta I$ is very large. The transistor holds I_C nearly constant as V_C varies. *Early effect* describes the departure of the transistor from this ideal. But Early effect describes a small correction to the general truth that the transistor holds I_C constant.

[15] The factor of two in denominator reflects the way we set up the assumed input: a pure differential input looks like ΔV at one input, and $-\Delta V$ at the other input, so that the whole *difference* – or "differential" – signal is not ΔV but *two* ΔV: the differential gain is $R_C/2(r_e + R_E)$.

5S.4 Important circuits

- follower:
    - split-supply (the simpler circuit, but requires a second supply)

    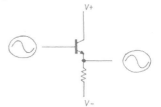

    Voltage follower ("emitter follower"): split-supply

    - single-supply (less elegant, but the cheaper form):

    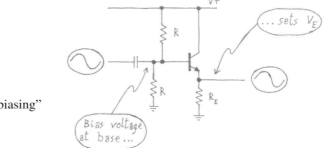

    Voltage follower "biasing"

    - ... and the input and output impedances:

    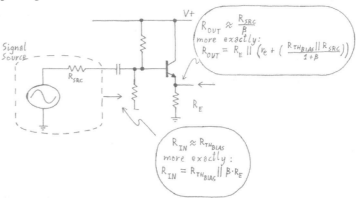

    Follower impedances

    - push–pull: solves the problem, in single-transistor follower, that the circuit's asymmetry makes it poor at *sinking* current from load (for the case of NPN).

    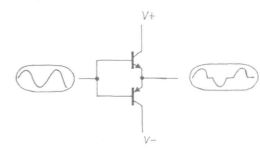

    Push–pull follower

- amplifiers (voltage amps):
    - common-emitter amplifiers

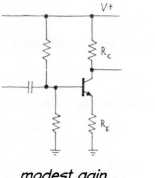 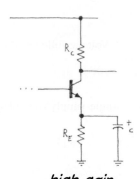

modest gain... ...high gain

$G = -R_C/r_e$, at "signal frequencies," where $X_C \ll R_E$

- common emitter impedances:

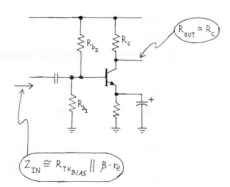

Impedances in and out, ce-amp

- differential amplifier:

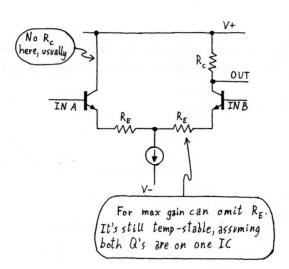

Impedances in and out, diff-amp

For max gain can omit R_E. It's still temp-stable, assuming both Q's are on one IC

- Switch (something completely different!):

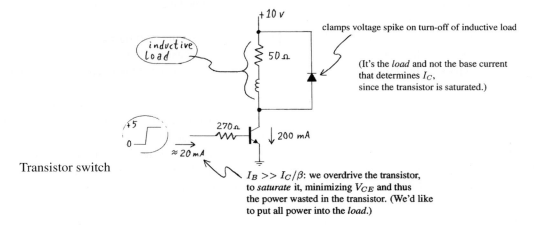

Transistor switch

5S.4.1 Fine points

Here are some refinements that we want you to be aware of, even though we don't expect you to be nimble with the details or with calculating the magnitude of these problems.

- three important ideas, but ideas you're not likely to treat quantitatively in this course:
  - temperature effects: the fact that transistors are very sensitive to temperature is a *fundamental* point; the formulas that predict response are not so important. Your designs should include protection against this sensitivity to temperature: most often, R_E does the job (providing *feedback*).
  - Early effect: describes the imperfection of a transistor current source. That is, describes the change in collector current with change in V_{CE}. Often negligible (usually, thanks to an R_E providing feedback); can cause substantial errors in a current mirror that lacks R_E's.
  - Miller Effect: a serious obstacle to design of high-frequency amplifiers. Miller effect results from capacitive feedback in any inverting amplifier, tending to oppose changes at the input. Because we don't build high-frequency amps in lab, we're not likely to feel the importance of Miller effect – but you will as soon as you try to design an amp that's to operate above, say, 1MHz. That will happen *after* this course.
- Interesting ideas, but less important.
  - bootstrapping (though it's a nifty idea: make a circuit element disappear electrically, by making its two ends move the same way);
  - Darlington connection.

5S.5 Jargon: bipolar transistors

biasing (see, for example, AoE §2.2.5). Setting *quiescent* conditions (see below) so that circuit elements work properly. To *bias* means, literally, to push off-center. We do that in transistor circuits to allow building with a single supply. The term is more general, as you know. (Compare AoE §1.6.6A, where a rectifying diode is *biased* into conduction.)

bootstrap (see, for example, AoE §2.3.5A). In general, any of several seemingly-impossible circuit tricks (source of term: "pull oneself up by the bootstraps:" impossible in life, possible in electronics!). In this chapter refers to the trick of making the impedance of a bias divider appear very large to improve the circuit's input impedance. Also *collector* bootstrap. See AoE §2.4.3A.

bypassed (-emitter resistor) (see, for example, AoE §2.3.5A). In common-emitter amp, a capacitor put in parallel with R_E is said to *bypass* the resistor because it allows *AC* current an easy path, bypassing the larger impedance of the resistor. Used to achieve high gain while keeping R_E large enough for good stability.

cascode (applied in Wilson Mirror: AoE §2.3.7). Circuit that uses one transistor to buffer or isolate another from voltage variation, so as to improve performance of the protected transistor. Used in cascode amplifier to beat Miller effect, in current source to beat Early effect.

clipping (illustrated in AoE §2.2.3D). Flattening of output waveform caused by hitting a limit on output swing. Example: single-supply follower will *clip* at ground and at the positive supply.

compliance (AoE §2.2.6D). Well defined in text: "The output voltage range over which a current source behaves well.... ."

Early effect (See, for example, AoE §2.3.7A). Variation of I_C with V_{CE} at a given value of V_{BE} or I_C. Thus it describes transistors's departure from true current-source behavior.

emitter degeneration (AoE §2.3.4). Placing of resistor between emitter and ground (or other negative supply) in common-emitter amp. It is done so as to stabilize the circuit despite variation in temperature. (Source of term: gain is reduced or "degenerated." General circuit *performance* is much *improved*, however!)

impedance "looking" in a direction . Impedance at a point considering only the circuit elements lying in one direction or another. Example: at transistor's base impedance *looking back* one "sees" bias divider and R_{source}; *looking into base* one "sees" only $\beta \times R_E$.

Miller effect (AoE §2.4.5). Exaggeration of actual capacitance between output and input of an inverting amplifier, tending to make a small capacitance behave like a much larger capacitance to ground: $(1+ \text{Gain})$ times as large as actual C.

quiescent (-current, -voltage) (for example, see AoE §2.2.5). Condition prevailing when *no* input signal is applied. So, describes DC conditions in an amplifier designed to amplify AC signals. Example: $V_{\text{out–quiescent}}$ should be midway between V_{CC} and ground in a single-supply follower to allow maximum output amplitude (or "swing") without clipping.

split supplies (AoE §2.2.5B). Power supplies of both polarities, negative as well as positive. Used in contrast to "single supply."

transconductance (AoE §2.2.9). Well defined in AoE. Briefly, $g_m \equiv \Delta I_{\text{out}}/\Delta V_{\text{in}}$.

Wilson mirror (AoE §2.3.7). Improved form of current mirror in which a third transistor protects the sensitive output transistor against effects of variation in voltage across the load (third transistor in cascode connection, incidentally).

5W Worked Examples: Transistors II

5W.1 High-gain amplifiers

Problem How do high-gain amplifiers, see Fig. 5W.1, compare with respect to "linearity" or constancy of gain over the output swing? Explain your conclusion, briefly. *Assume* that each amplifier is fed by a properly-biased input.

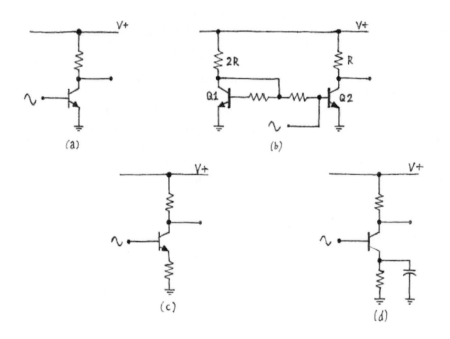

Figure 5W.1
High-gain amplifiers.

For case (b) we should provide some hints, because this circuit is very similar to a "current mirror," a circuit we didn't require you to learn. This circuit, mirror-like, will sink a current through Q1 of about $\frac{V+}{2R}$. That same current will flow in Q2, if we don't disturb things; Q2 is said to "mirror" the current passed by Q1, because the two V_{BE}'s are the same. This is the scheme that sets up the "biasing:" puts V_{out}, quiescent, at about $0.5V+$, as usual. Assume that the two transistors are well matched and live near each other on an integrated circuit. Enough said?

Solution All are the same except circuit (c). The others will show gain variation as the output voltage swings, because this output swing is caused by varying I_C, collector current, and the denominator of the gain equation for (a), (b), (d) is simply r_e, whose value varies with I_C. Circuit (c) differs because the constant value of R_E is added to the varying value of r_e. Circuit (c) gets its more constant gain at a price of course: the price is its diminished gain.

Another way to put the same point is to speak not – as we usually do – of r_e but instead of its inverse, g_m, the slope of the I_C versus V_{BE} curve. This slope, the transistor's *transconductance*, is proportional to I_C, and so is the circuit gain, when no R_E is present. Specifically, the circuit gain is $g_m \times R_C$. This formulation adds nothing new to our usual account which uses r_e, but perhaps makes the dependence on collector current more obvious.

Problem Which of the circuits in Fig 5W.1 are (is?) protected against temperature effects, and how? Explain the mechanism or mechanisms.

Solution Circuits (c) and (d) are temperature stable thanks to the emitter resistor, R_E, which provides the *negative feedback* described in Chapter 5N. Circuit (d) gives high gain along with stability (and also poor constancy of gain, as we have noted).

Circuit (a) is bias-unstable – and is, therefore a *bad* circuit.

Circuit (b) is a less familiar design, as the apologetic note in the problem statement suggests. This circuit is temperature-stable, but achieves it not by use of feedback but by taking advantage of the circuit's symmetry. This design is effective when the two transistors, Q1 and Q2 are well-matched and fabricated on one integrated circuit, as the problem statement declares they are.

Because the collector currents in the two transistors are equal, the two V_{BE}'s are equal. Imagine a change in temperature – let's imagine a rise of about 9°C. This drops the V_{BE} of Q1 by about 18mV (see Chapter 5N: at constant I_C, $\Delta V_{BE}/\Delta \text{temp} = -2.1\text{mV}/°C$). If this were an unprotected circuit, like circuit (a), that change would reduce Q2's output to one half what it was: in other words the output quiescent voltage would be radically upset.

But because Q2 is heated just as Q1 is, Q2's current is unaffected by the heating of both transistors. Heating reduced the V_{BE} of Q1; but that reduced value evokes the *original* current from *heated* Q2. This stabilizing scheme might be called stabilization by *compensation* rather than by feedback: a change in a component, otherwise disturbing, is *compensated* for by a matching change in a second component. We will see this effect at work again in the next section.

5W.2 Differential amplifier

Problem Lab 5L begins with a diff-amp of modest gain (collector resistor is 10k, emitter resistors are 100Ω, see Fig. 5W.2). Modify the circuit to maximize gain (without use of current sources). What is the modified circuit's differential gain? Common-mode gain?

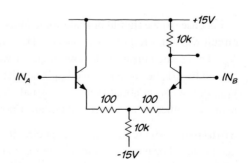

Figure 5W.2 Differential amplifier of Lab 5L.

Solution The change that maximizes gain (without using the fancy tricks of real op-amps – current mirror load, for example) is simply to remove the emitter resistors. Now the gain depends on r_e, whose quiescent value is about 33Ω at the 0.75mA collector current:

$$G_{\text{diff}} = R_C/2(R_E + r_e) = 10k/(2 \times 33\Omega) \approx 150$$
$$G_{\text{common-mode}} = -R_C/(R_E + r_e + 2R_{\text{tail}}) \approx -10k/20k = -1/2 .$$

Problem When you have maximized gain, does your circuit remain temperature stable? (Explain your answer.)

Solution Yes, as long as the two transistors run at the same temperatures, as they do when built on an integrated circuit, as in Lab 5L: the curves slide together, with temperature changes. If this were built with discrete transistors, it would not be temperature stable.

Problem (Tail of diff-amp) In Lab 5L, we suggest that you replace the 10k resistor in the tail with a current sink. Why? What is the argument that suggests a current sink will improve CMRR?

Solution CMRR is the ratio of good gain to bad gain, as you know: CMRR = $G_{\text{diff}}/G_{\text{CM}}$. The "tail" resistor influences common-mode gain but not differential, so boosting R_{tail} sounds like a good idea: the expression for common-mode gain (G_{CM}) shows $2R_{\text{tail}}$ in the denominator, whereas R_{tail} does not appear at all in the expression for $G_{\text{differential}}$. So a current source, with its large *dynamic resistance*, $\Delta V/\Delta I$, enlarges the effective R_{tail}, cutting G_{CM} while leaving differential gain untouched.

Problem Would increasing the value of R_{TAIL} by substituting a larger resistor – say, 100k – improve CMRR, relative to the circuit of Fig. 5W.2? Explain your answer.

Solution No, not at all. The larger *resistor*, unlike the current source, would cut the value of I_C by a factor of 10. To keep $V_{\text{out-quiescent}}$ properly placed (unchanged), we would need to boost the value of R_C by a factor of 10. Meanwhile, r_e would grow by the same factor, because of the reduction of I_C.

So, the sad result is that G_{CM} would be unchanged (still 1/2); G_{diff} would be unchanged: say 100k/(2×330)=150. CMRR, therefore, would be unchanged. We do indeed need a current source to improve CMRR.

5W.3 Op-amp innards: diff-amp within an IC operational amplifier

In Fig. 5W.3 is a much-simplified schematic of an op-amp, the '411 JFET-input device that you'll soon be using. We've included the gritty detail as well but we think you can answer by referring only to the left-hand, simplified figure. We will raise questions about each of the elements circled in Fig. 5W.3.

Observations
The *first stage* – differential input stage – uses JFETs rather than bipolars to achieve very low *bias current* (50pA, typical, for the '411). The *second stage* – common-emitter amp, gain stage – uses bipolars because the bipolar's *exponential I–V* curve is steeper than the FET's *quadratic* curve (thus providing higher gain[1]).

Problem Why all those current sources? Current source on "tail": (The sources in this design are current mirrors, favorites of IC designers.) In particular, why are current sources, rather than resistors, used on *tail* and *output* of first stage, and why on the *collector* of the second stage?

---

[1] To be more precise, the bipolar shows higher "transconductance" – $\Delta I_{\text{OUT}}/\Delta V_{\text{IN}}$.

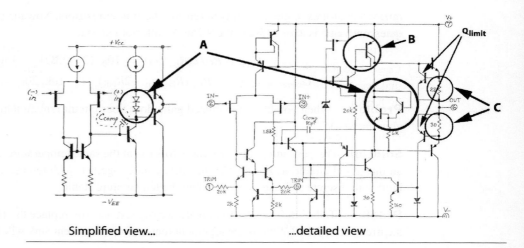

Figure 5W.3 Op-amp innards.   Simplified view...   ...detailed view

Solution

Why all those current sources? The current source on the tail of the input stage differential amplifier (the tail that's up in the air) is used to provide low common-mode gain, as usual (giving large CMRR: $G_{diff}/G_{com-mode}$). The common-mode gain when using bipolar transistors is, as you know, $G_{cm} = -R_C/(r_e + R_E + 2R_{tail})$: (for these FETs, one would change the labeling: $G_{CM} = -R_{drain}/(\frac{1}{g_m} + R_S + 2R_{tail})$). A large value for R_{tail} helps.

The current source makes R_{tail} huge because a good current source (as you well know) holds ΔI close to zero as voltage changes. In other words, a current source's $R_{dynamic} \equiv \Delta V/\Delta I = \Delta V/(\text{tiny})$ is huge. The huge effective R_{tail} makes common-mode gain tiny, and CMRR very large (100dB, typical, for the LF411).

Current mirror on drains of diff-amp Qs:

Short answer: A short answer, based on the view of current mirror as just a current source (here, a "sink"), is just that the very large $R_{dynamic}$ of the current source gives this differential stage high gain (the large $R_{dynamic}$ replacing the usual "R_C" – here, an "R_{drain}," because these are field-effect transistors).

A longer answer: If you have looked into current mirrors a bit, then you know that their mission in life is to hold their two currents equal. (Strictly, to hold one current – I_{out} – equal to the other – $I_{program}$.)

The differential amplifier's two JFET transistors' mission, by contrast, is to make the two currents *unequal* in response to any difference between the two input voltages. When these two circuit fragments with their opposed goals meet, the result is very high voltage gain.

Suppose the quiescent current for each of the two transistor is 1mA (a number that is not realistic here: much larger than the '411's – but easy to work with). Now suppose that an input-voltage difference makes the currents *unequal*: say, the input or "programming" current becomes 0.9mA (while the other Q's current grows to 1.1mA since the total, set by I_{Qtail}, is held constant). The mirror will accept only that 0.9mA, and the *difference* current – 0.2mA – must squirt out to the next stage (to the common-emitter amplifier). The mirror has *doubled* the output current from the first stage, relative to the current that would have issued to the second stage, if the first stage had fed an ordinary current sink.

Or, to put this in terms we normally apply to voltage amplifiers, the impedance the mirror presents is not just large (as any current source's dynamic impedance is). It is *extra large* because just as the current applied to the R_{drain} grows (to 1.1mA, in this case), the apparent impedance of this side of the mirror is boosted. The mirror not only tries to keep current constant (as any current source does); in this case it tries to shrink the current, further enhancing the ΔV response.

Probably the account in *current* terms is easier to follow than the one speaking of voltage gain.

Current source on collector of common-emitter stage ("B"): This current source, replacing the R_C that we are accustomed to putting on the collector of a common-emitter amplifier, again provides very high gain. If one did this with an ordinary amplifier (one that did not use overall feedback to make it well-behaved), this would be a "bad circuit." Used without feedback it would be useless because its output could only bang against one supply or the other. In an op-amp, using negative feedback as always, such extravagant gain is not a flaw but a virtue.

Problem What is the purpose of the elements labeled "A" in the two part of Fig. 5W.3? (Note that the two parts are equivalent; if the detailed schematic is a bit much, consider the "simplified" diagram instead!)

Solution These two transistors drop two V_{BE}'s. You may want to think of them as two diodes in series, as shown in the simplified diagram on the left side of Fig. 5W.3. Since the top of this V_{BE}-stack drives the base of the upper transistor and the bottom of the stack drives the base of the lower transistor, the stack biases the bases apart, eliminating the dead zone that otherwise occurs close to 0V in a push–pull. It is this dead zone that causes crossover distortion.

Having seen that feedback does a pretty good job of hiding the "dog" of crossover distortion (see Chapter 6N), you may wonder, "Why not let feedback solve this problem?" The answer is that feedback is not quick enough to do the job at high frequencies. In order to hide crossover distortion, the input to the push–pull must *slew* across about 1.2V, up or down, and doing this takes time. During that slewing, a glitch will appear on the circuit output. Fig. 5W.4 is a scope image of this effect in the home-made op-amp of Lab 5L. The circuit does hide most of the cross-over distortion, but a close look at the image – using the higher sweep rate of the right-hand image – shows that a glitch persists. Slewing past those two V_{BE}'s takes some time, and during that slewing, feedback, with its magic, fails.

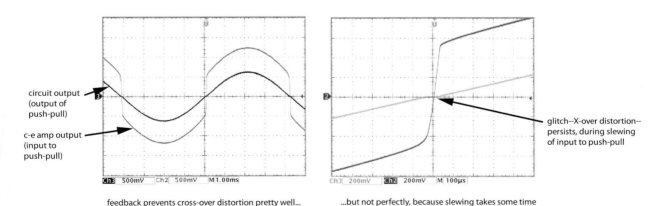

Figure 5W.4 Common-emitter stage slewing to hide crossover of output-stage push–pull: not perfect.

The time scale of Fig. 5W.4 is wrong for the '411, which slews much faster than our home-made op-amp of Lab 5L. But the problem for the '411 would be the same at some time scale, if the passive protections against crossover distortion had not been included.[2] The need for these passive protections

---
[2] Not all op-amps do include this passive protection against crossover distortion. The LM358, for example – the single-supply op-amp that we use in Labs 7L and 10L – does not include it.

reminds us of the limitations of feedback: it doesn't work well at high frequencies, when implemented with ordinary op-amps.

Problem   Are the resistors that are labeled "C" essential? Explain.

Solution   Yes they are. They prevent "thermal runaway" that otherwise would be very likely to destroy the push–pull transistors. The biasing-apart of the push–pull bases, described above, keeps both push–pull transistors just barely conducting, when input and output of the push–pull are at ground. This is the strategy that prevents a glitch when the circuit wakes up, in response to movement *away* from ground.

But that arrangement would be deadly without the emitter resistors. As a little current trickles through the push–pull transistors, it heats them slightly. But the heating increases their current, at a given V_{BE} (the V_{BE} imposed by the biasing network we met earlier). The increased current heats the transistors further. And so on. That's a positive-feedback loop, "thermal runaway." The emitter resistors provide the usual *negative* feedback to prevent runaway: if current grows, the emitter voltages move apart somewhat, reducing V_{BE}. You have seen this before, in the argument for including an emitter resistor in a high-gain common-emitter amplifier.

Those R_E's do another job in this circuit. They are part of a simple trick to limit the op-amp's output current to prevent damage by an overload. The scheme is so nifty that one can safely short a '411 output to its positive or negative supply.

Any current that flows in or out of the op-amp's output passes through one or the other of those resistors, developing a voltage drop. That drop is applied to one or other of the transistors labeled Q_{limit}, as its V_{BE}. When output current is sufficient to develop 0.5–0.6V across that resistor, the Q_{limit} transistor begins to steal current from the base of the push–pull transistor. This effect – combined with the limited current available from the common-emitter stage – puts a safe ceiling on the total current available from the op-amp output. (This is a trick you will see repeated in the voltage regulators of Chapter 11N and Lab 11L.)

Part III

Analog: Operational Amplifiers and their Applications

6N Op-amps I

Contents

6N.1	**Overview of feedback**	245
6N.2	**Preliminary: negative feedback as a general notion**	248
6N.3	**Feedback in electronics**	249
6N.4	**The op-amp golden rules**	251
6N.5	**Applications**	252
	6N.5.1 A follower	252
	6N.5.2 The effect of feedback on the follower's R_{OUT}	252
6N.6	**Two amplifiers**	252
	6N.6.1 Non-inverting amplifier	252
	6N.6.2 A skeptic's challenge to the Golden Rules	253
	6N.6.3 Some characteristics of the non-inverting amp	254
6N.7	**Inverting amplifier**	254
	6N.7.1 What virtual ground implies	255
6N.8	**When do the Golden Rules apply?**	256
	6N.8.1 More applications: improved versions of earlier circuits	257
6N.9	**Strange things can be put into feedback loop**	259
	6N.9.1 In general...	259
6N.10	**AoE Reading**	261

Why?

What problem do today's circuits address? The very general task of *improving* performance, through the application of *negative feedback*, of a great many of the circuits we have met to this point.

6N.1 Overview of feedback

Fig. 6N.1 shows a newspaper page on which Harold Black wrote one of his newly-conceived formalizations of feedback (this newspaper records not quite the basic notion, which he had sketched on a copy of the *New York Times* four days earlier, but rather an application of feedback in order to match impedances). He wrote this as he rode the ferry from Staten Island to work, one summer morning in 1927. (Was it chance, or caginess that led him to write it on a dated sheet?)[1]

We have been promising you the pleasures of feedback for some time. You probably know about the concept even if you haven't yet used it much in electronics. Now, at last, here it is.

Feedback is going to become more than just an item in your bag of tricks; it will be a central concept that you find yourself applying repeatedly, and in a variety of contexts, some far from operational

---

[1] (Copyright 1977 IEEE. Reprinted, with permission, from Harold S. Black, "Inventing the Negative Feedback Amplifier," *IEEE Spectrum*, Dec. 1977). You may protest that the dated newspaper page could hardly prove the date of his invention. Couldn't he have dug out an old newspaper page to backdate his invention? No, because he was prudent enough to have each newspaper sketch witnessed and signed, as soon as he arrived at work.

246 Op-amps I

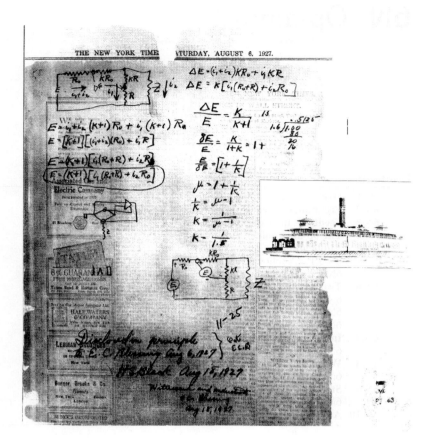

Figure 6N.1 Harold Black's notes on the feedback amplifier.

amplifiers. Already, you have seen feedback in odd corners of transistor circuits; you will see it constantly in the remaining analog labs; then you will see it again in a digital setting, when you build an analog-to-digital converter in Lab 18L, and then a phase-locked-loop in the same lab. It is a powerful idea.

Our treatment of negative feedback circuits begins, as did our treatment of bipolar transistors, with a simple, idealized view of the new circuits. At the conclusion of Lab 5L you built a simple "operational amplifier." From now on, we will rely on the integrated version of these little high-gain differential amplifiers. They make it easy to build good feedback circuits.

As this section of the course continues, we soon feel obliged once again to disillusion you – to tell you about the ways that op-amps are imperfect. But this is a minor theme. The major theme remains that op amps and feedback are wonderful. We keep trying additional applications for feedback, and we never lose our affection for these circuits. They work magically well.

Lab 8L, introduces the novelty of *positive* feedback: feedback of the sort that makes a circuit unstable – or makes its output slam to one power-supply limit or the other. Sometimes that is useful, and sometimes it is a nuisance. In Lab 8L we look at the benign effects of positive feedback, including its use in *oscillators*. Lab 9L opens with one more fruitful use of such feedback, in an *active filter*.

The remainder of Lab 9L is devoted to the more alarming effects of positive feedback: *parasitic* oscillations that occur when unwanted positive feedback sneaks up on you, destabilizing a circuit. We will learn some protections against such mischief. Knowing how to apply such defenses may make you a hero when your group is plagued by mysterious fuzz on its scope traces.

6N.1 Overview of feedback

Figure 6N.2 Righteous and deserving student, about to be rewarded for his travails with discrete transistors.

We then devote three labs to three particular applications of negative feedback:

- A motor-control loop using a technique called "PID" (describing the three modified error signals that are used in the feedback loop: "Proportional," "Derivative," and "Integral"). This is Lab 10L, the most complex circuit so far – and useful to let us see applications for many circuits that we have demonstrated as circuit fragments but not yet applied: the PID makes good use not only of differentiator and integrator but also depends upon a summing circuit, a diff-amp made with an op-amp, and a high-current motor-driver.
- Voltage regulators: these are specialized circuits designed for the narrow but important purpose of providing stable power supplies. We look at both linear and switching designs in Lab 11L.
- A group project, putting together a half dozen circuit fragments, some passive, most using feedback circuits. In this project, you students will design and combine the several circuit fragments to make a system that can transmit audio using frequency-modulated (FM) infrared light. This is Lab 13L (the number hops by one, to leave room for a final discrete-transistor lab, Lab 12L on MOSFETs, where feedback is not prominent).

With Lab 13L we conclude the analog half of the course, and in the very next lab you will find the rules of the game radically changed as you begin to build *digital* circuits. But we will save that story till later.

A piece of advice (unsolicited): How to get the greatest satisfaction out of the feedback circuits you are about to meet

Here are two thoughts that may help you to enjoy these circuits:

1. As you work with an op-amp circuit, recall the equivalent circuit made without feedback, and the

248   Op-amps I

difficulties it presented: for example, the transistor follower, or the transistor current sources. The op-amp versions in general will work better, to an extent that should astonish you.

Having labored through work with discrete transistors you have paid your dues. You are entitled to enjoy the ease of working with op-amps and feedback.

Figure 6N.2 shows you climbing – as you are about to do – out of the dark valleys through which you have toiled, up into that sunny region above the clouds where circuit performance comes close to the ideal. You reach the sunny alpine meadows where feedback blooms. Pat yourself on the back, and have fun with op-amps.

A second thought:

- Recall that negative feedback in electronics was not always used; was not always obvious – as Harold Black was able to persuade the patent office (Black comes as close as anyone to being the inventor of electronic feedback).

The faded and scribbled-on newspaper that is shown at the start of this chapter is meant to remind you of this second point – meant to help us feel some of the surprise and pleasure that the inventor must have felt as he jotted sketches and a few equations on his morning newspaper. A facsimile of this newspaper, recording the second of Black's basic inventions in the field, appeared in an article Black wrote years later to describe the way he came to conceive his invention. Next time you invent something of comparable value, don't forget to jot notes on a newspaper, preferably in a picturesque setting – and then keep the paper till you get a chance to write your memoirs (and get a witness to confirm the date).

6N.2   Preliminary: negative feedback as a general notion

This is the deepest, most powerful notion in this course. It is so useful that the phrase, at least, has passed into ordinary usage – and there it has been blurred. Let's start with some examples of such general use – one genuine cartoon (in the sense that it was not cooked up to illustrate our point), and three cartoons that we did cook up. Ask yourself whether you see feedback at work in the sense relevant to electronics, and if you see feedback, is its sense positive or negative?

Now comes an example of feedback misunderstood. Outside electronics, the word "negative" has

Figure 6N.3 Feedback: same sense as in electronics? Copyright 1985 Mark Stivers, first published in *Suttertown News*, reproduced with permission.

negative connotations: so, "negative feedback" sounds nasty, "positive feedback" sounds benign. A news story in the New York Times (October 8, 2008) concerning an economic slowdown began thus:

The technical term for it is "negative feedback loop." The rest of us just call it panic.

How else to explain yet another plunge in the stock market on Tuesday that sent the Standard & Poor's 500-stock index to its lowest level in five years – particularly in the absence of another nasty surprise.

What this reporter describes is, of course, *positive* feedback. But since the news is bad – "panic" – he can't resist his inclination to call it *negative*.

Sometimes a signal that usually would amount to negative feedback turns out to be positive (we will return to this topic, one that causes circuit instability, in Lab 9L). Is the observer applying positive or negative feedback to the two fellows on the left, in Fig. 6N.4? To decide, we need to judge whether his signal tends to increase or decrease the behavior that he senses.

Figure 6N.4 Freaks? Will he hurt their feelings?

The case in Fig. 6N.5 comes closer to fitting the electronic sense of *negative feedback*. In op-amp terms (not Hollywood's), who's playing what role?

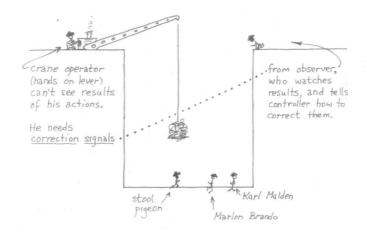

Figure 6N.5 Negative feedback: a case pretty much like op-amp feedback.

6N.3 Feedback in electronics

In conversation, people usually talk (as that reporter did above) as if "positive feedback" is nice, "negative feedback" is disagreeable. In electronics the truth is usually just the opposite. Generally speaking, negative feedback in electronics is wonderful stuff; positive feedback is nasty. Nevertheless the phrase means in electronics fundamentally what it should be used to mean in everyday speech.

AoE §2.5.1

Harold Black was the first to formalize the effects of negative feedback in electronic circuits. Here is the language of his patent:[2] explaining his idea:

```
     Applicant has discovered how to use larger
     amounts of 'negative feedback than were contem-
35   plated by prior art workers with a new and im-
     portant kind of improvement in tube operation.
     One improvement is in lowered distortion arising
     in the amplifier. Another improvement is
     greater consistency of operation, in particular a
40   more nearly constant gain despite variable fac-
     tors such as ordinarily would influence the gain.
     Various other operating characteristics of the
     circuit are likewise rendered more nearly con-
     stant. Applicant has discovered that these im-
45   provements are attained in proportion to the
     sacrifice that is made in amplifier gain, and that
     by constructing a circuit with excess gain and re-
     ducing the gain by negative feedback, any de-
     sired degree of linearity between output and
50   input and any desired degree of constancy or
     stability of operating characteristics can be real-
     ized, the limiting factor being in the amount of
     gain that can be attained, rather than any limi-
     tation in the method of improvement provided
55   by the Invention.
```

He summarized his idea in an article that he published many years afterward:

... by building an amplifier whose gain is made deliberately, say 40 decibels higher than necessary (10,000-fold excess on energy basis) and then feeding the output back to the input in such a way as to throw away the excess gain, it has been found possible to effect extraordinary improvement in constancy of amplification and freedom from nonlinearity.[3]

One proof of the originality of Black's idea was the fact that the British Patent Office rejected his application, seeing Black as a dull fellow who had not understood that an amplifier should *amplify*, so that "throwing away...gain" showed the inventor's confusion. Black had the last laugh.

Lest we oversell Black's invention, let's acknowledge that Black did not invent or discover feedback. Our bodies are replete with homeostatic systems, like the one that holds our body temperature close to 99°F, whether we are shivering in the snow or sweating on a beach (with shivering and sweating part of the stabilizing mechanism). And Newcomen used a speed "governour" on his mine-pumping steam pumps: the spin rate raised or lowered the weights, closing down or opening the steam valve appropriately, so as to hold the speed nearly constant despite variations in load.

AoE §2.5.1

Open-loop versus *feedback* circuits: Nearly *all* our circuits, so far, have operated open-loop – with some exceptions noted below. You may have gotten used to designing amplifiers to run open-loop (we will cure you of that); you would not consider driving a car open loop (we hope), and you probably know that it is almost impossible even to *speak* intelligibly *open-loop*.

Examples of feedback without op-amps

We know that feedback is not new to you, not only because you may have a pretty good idea of the notion from ordinary usage, but also because you have seen feedback at work in parts of some transistor circuits, see Fig. 6N.6.

AoE §2.3.4B

---

[2] Harold S. Black, "Wave Translation System," US Patent 2102671 (1937).
[3] *IEEE Spectrum*, Dec. 1977

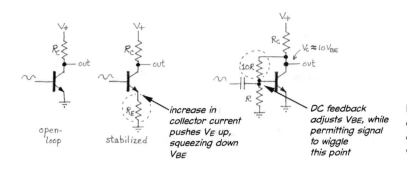

Figure 6N.6 Some examples of feedback in circuits we have built without op-amps.

Feedback with op-amps

Op-amp circuits make the feedback evident, and use a lot of it, so that they perform better than our improvised feedback fragments. The name "operational amplifier" derives from the assumption, early in their use, that they would be used primarily to do a sort of analog computing.[4]

Op-amps have enormous gain (that is, their *open-loop* gain is enormous: the chip itself, used without feedback, would show huge gain: ≈200,000 at DC, for the LF411, the chip you will use in most of our labs). As Black proposed, op-amp *circuits* deliberately throw away most of that gain, in order to improve circuit performance.

6N.4 The op-amp golden rules

Just as we began our treatment of bipolar transistors with a simple model of device behavior, and that model remained sufficient to let us analyze and design many circuits, so in this chapter we start with a simple, idealized view of the op-amp, and usually we will *continue* to use this view even when we meet a more refined model.

The *golden rules* (below) are approximations, but good ones.

(GR#1) The output tries to do whatever is necessary to make the voltage difference between the two inputs zero.

(GR#2) The inputs draw no current.

Three observations, before we start applying these rules:

- Rule (2): we're confident that you understand that the "inputs" that draw no current are the signal inputs, labeled "+" and "−", not the op-amp's power supply terminals(!)[5]
- Rule (1): the word "tries" is important. It remind us that it's up to us, the circuit designers, to make sure that the op-amp *can* hold its two inputs at equal voltages. If we blunder – say, by overdriving a circuit – we can make it impossible for the op-amp to do what it "tries" to do.
- More generally: these rules apply only to op-amp circuits that use *negative feedback*.

These simple rules will let you analyze a heap of clever circuits.

---

[4] They seem to have been named in a paper of 1947, which envisioned their important uses as follows: "The term 'operational amplifier' is a generic term applied to amplifiers whose gain functions are such as to enable them to perform certain useful operations such as summation, integration, differentiation, or a combination of such operations." Ragazzini, Randall, and Russell, "Analysis of Problems in Dynamics by Electronic Circuits," *Proceedings of the IRE*, **35**, May 1947, pp. 444–452, quoted in *op-amp Applications Handbook*, Walt Jung, editor emeritus, (Analog Devices Series, 2006), p. 779.

[5] You may think this is obvious, but every now and then some students, impressed by all our praise of op-amps and their magic, fail to connect the op-amps power supply pins. The students wait expectantly for truly miraculous circuit performance – and are disappointed.

6N.5 Applications

6N.5.1 A follower

Figure 6N.7 shows about the simplest circuit one can make with an op-amp. It may not seem very exciting – but it is the *best* follower you have seen.

The Golden Rules let you estimate...

- ...R_{IN} (try GR#2);
- ...V_{OFFSET} (try GR#1); and compare this to the DC-offset we are accustomed to in a bipolar transistor follower.
- ...R_{OUT}?...

Figure 6N.7 Follower.

6N.5.2 The effect of feedback on the follower's R_{OUT}

R_{OUT} raises a much subtler question. The Golden Rules don't answer it, but you can work your way through an argument that will show that R_{OUT} is very low. The novel point, here, is that it is low not because the IC itself has low R_{OUT}; this value, rarely specified, is moderately low – perhaps around 100Ω. What's new is that *feedback* accounts for the very low R_{OUT} of this circuit – and of nearly all other op-amp circuits.[6]

Let's redraw the op-amp to resemble a Thevenin model: perfect voltage source in series with a series resistance that we might call "r_{out}." Our drawing will include the op-amp's high *open-loop* gain, A. Then we can try the thought experiment of tugging the output by a ΔV, and seeing how the output responds, see Fig 6N.8.

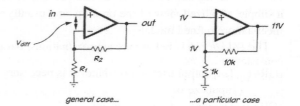

Figure 6N.8 Feedback lowers op-amp circuit's R_{OUT}.

If you try to change V_{OUT}, the circuit hears about it through the feedback wiring. It is *outraged*. Its high-gain amplifier squirts a large current at you, fighting your attempt to push the output away from where the op-amp wants it to stand. We'll leave the argument at this qualitative level, for now. We'll settle for calling the output impedance "very low." Later (mostly in Chapter 9N), we will be able to quantify the effects of feedback to arrive at a *value* for R_{OUT}.

6N.6 Two amplifiers

6N.6.1 Non-inverting amplifier

This circuit is a bit more exciting than the follower: it gives us voltage gain.[7]

---

[6] The one exception to this rule is an op-amp *current source*, whose output impedance should be *high* (the ideal current source, as you know, would show R_{OUT} infinite).

[7] Incidentally, it was this sort of circuit that interested Black. He was concerned about making "repeater" amplifiers for the telephone company, in order to transmit voice signals over long distances. It was this need to send a signal through a great

6N.6 Two amplifiers

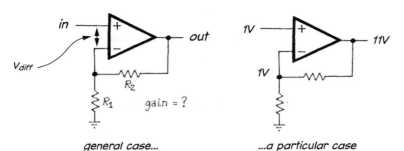

Figure 6N.9 Feedback lowers op-amp circuit's R_{OUT}.

Evidently, to satisfy the Golden Rules the op-amp has to take its output to 11V, if we apply 1V to the input. Generalizing, you may be able to persuade yourself that the gain is the inverse of the fraction fed back. 1/11 is fed back, so gain is 11. The gain usually is recited more simply as

$$G = 1 + R_2/R_1 \ .$$

Students are inclined to forget this pesky "1 +." We hope you won't forget it. Sometimes, of course, it is sensible to "forget it" in the sense that you neglect it: when the R_2/R_1 fraction is very large. But throw out the "+1" only after you have determined that its contribution is negligible.

6N.6.2 A skeptic's challenge to the Golden Rules

Suppose that a skeptical student wants to take advantage of his awareness that the op-amp is a differential amplifier (and a circuit familiar from the second transistor lab); such a student might present a plausible challenge to the Golden Rules. This skeptic doesn't fall for the notion that the op-amp is a weird little triangle that relies on pure magic.

This person might object that our Golden Rules' analysis must be wrong, because the analysis is not consistent with our understanding of the differential amp that is the core of any op-amp. Our Golden Rules predict $V_{\text{OUT}} = 11$. But if GR#2 says that the op-amp has driven the input voltages to equality, then – Aha! We have caught a self-contradiction in the analysis! Equal voltages at the input of a diff amp cannot produce 11V out; the Golden Rule must be wrong!

But, no; you won't fall for this argument. The way to resolve the seeming contradiction is only to acknowledge that the Golden Rules are approximate. The op-amp does not drive the inputs to equality, but to near-equality. How close? Just close enough so that your differential-amplifier view can work. In order to drive V_{OUT} to 11, the difference between the inputs – V_{diff}, in Fig. 6N.9 – must be not 0 but $V_{\text{OUT}}/G_{\text{OL}}$, where G_{OL} is the op-amp's open-loop gain. For a '411 op-amp at DC, that indicates $V_{\text{diff}} \approx \frac{11V}{200,000} = 50\mu V$. Not zero, but pretty close.

Doing that calculation can help one to appreciate why it is useful to have all that excess gain to throw away (as Black put it): the Golden Rule approximations come closer to the truth as open-loop gain rises. $A_{\text{OPEN-LOOP}} = 1,000,000$, for example (a value available on many premium op-amps) would reduce V_{diff}, in the example above, to just $10\mu V$.

many stages – hundreds, in order to cross the country – that made the phone company more concerned than anyone else with the problem of distortion in amplifiers. Distortion that is minor in one or two stages can be disastrous when superadded hundreds of times.

It's a curious historical fact that the telephone company, which we may be inclined to view as a stodgy old utility, repeatedly stood at the forefront of innovation: not only Black's invention, but also bipolar transistors, field-effect transistors, and fiber-optic signal transmission were developed largely at "the phone company," if one includes within this company the famous Bell Laboratories.

6N.6.3 Some characteristics of the non-inverting amp

A couple of other characteristics are worth noting – perhaps they seem obvious to you:

- R_{IN} is enormous, as for the follower;
- R_{OUT} is tiny, as for the follower[8]

Incidentally, you might note that the *follower* can be described as a special case of the non-inverting amplifier. What is its peculiar R_1 value?[9]

6N.7 Inverting amplifier

AoE §4.2.1

The circuit in Fig. 6N.10 also amplifies well. It differs from the non-inverting amplifier in one respect that is obvious – it *inverts* the signal – and in some respects that are not so obvious.

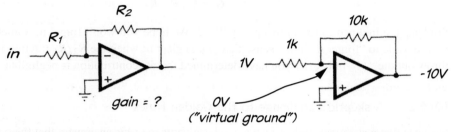

Figure 6N.10 Inverting amplifier.

GR#1 tells us that the inverting terminal voltage should be zero. From this observation, we can get at the circuit's properties

Gain: Let's look at the particular case: 1V causes 1mA to flow toward the inverting terminal. Where does this current go? Not into the op-amp (that is forbidden by GR#2). So, it must go around the corner, through the 10k feedback resistor. Doing that, it drops 10V. So, $V_{OUT} = -10$V.

Generalizing this result, you'll be pleased to see no nasty "$1 + \cdots$" in the gain expression: just $G = -R_2/R_1$. The *minus* sign means – as in the common-emitter amp – only that the circuit inverts: that output and input are 180° out of phase.

If you're not too proud to use a primitive way of looking at the circuit – a way that's based on its appearance, along with the fixed *virtual ground* at the inverting terminal – try thinking of the circuit as a child's see-saw. It pivots on virtual ground. If the two arms are of equal length, it is balanced with equal voltages: its gain is -1. If one arm is 10 times longer, as in Fig. 6N.10, the long arm swings 10 times as far as the short arm. And so on.

Impedances, in and out: Now, let's consider the input impedance of the inverting amps of Fig. 6N.10. From the observation that the inverting input is fixed at 0V also flow some of the circuit's peculiarities:

---

[8] We'll see later, when we look at the effects of feedback *quantitatively*, that the amplifier trades away a very little of its impedance virtues in exchange for gain. Still, this correction to the claim that amplifier and follower share good input and output impedances is only a minor change. The dependence of circuit impedances upon B, the fraction of output that is fed back, is treated in AoE §§2.5.3B and 2.5.3C. In this book we return to the topic in Day 9.
[9] For the follower, R_1 is infinite; so, gain goes to 1.

- $R_{\rm IN}$: let's see if we can get you to fall for either of two *wrong answers*:
    - Wrong answer #1: "$R_{\rm IN}$ is huge, because input goes to an op-amp terminal, and GR#2 says the inputs draw no current."
    - Wrong answer #2: "Oh, no – what I forgot is that it's the output of the circuit that is at low impedance (we saw that back in §6N.6.1). So $R_{\rm IN} = R_1 + R_2$."

  What's wrong with these answers? To give you a chance to see through our falsehoods on your own, we'll hide our explanations in a footnote.[10]

- The inverting amp's inverting terminal (the one marked "-") often is called "virtual ground." Why "virtual"?[11]) This point, often called by the suggestive name, "summing junction," turns out to be useful in several important circuits, and helps to characterize the circuit.

6N.7.1 What virtual ground implies

$R_{\rm in}$: First, recognizing "virtual ground" – the point locked by feedback at 0V – makes the circuit's $R_{\rm in}$ very simple: it is just R_1. So, compared to the non-inverting amp, the inverting amp's input resistance is only mediocre.

"Mediocre?," you may protest. "I can make it as large as I like, just by specifying R_1. I'll make mine 10M or 100M, if I'm in the mood." Well, no. Your defiant answer makes sense on this first day with op-amps, when we are pretending that they are ideal. But as soon as we come down to earth (splashdown is scheduled for the next chapter), we're obliged to admit that things begin to go awry when the R values become very large. So, if we treat approximately 10M as a practical limit on R values, we must concede that the inverting amp cannot match the non-inverting amp's $R_{\rm in}$.

What happens to the circuit's $R_{\rm in}$ if you omit R_1 entirely, replacing it with a wire? The resulting $R_{\rm in} \approx 0$, alarming though it looks, is exactly what pleases one class of signal sources: it pleases a *current source*. In Lab 6L you will exploit this characteristic when you convert *photodiode* current to a voltage.

AoE §4.3.1D

Summing circuit

The fact that virtual ground is fixed allows one to inject several *currents* into that point without causing any interaction among the signals; so, the several signals sum cleanly. The left-hand circuit in Fig. 6N.11 shows a *passive* summing circuit. It is inferior, because the contribution of any one input slightly alters the effect of any other input.

In the lab you will build a variation on this op-amp circuit: a potentiometer lets you vary the *DC offset* of the op-amp output. The circuits shown in Fig. 6N.11 form a *binary-weighted* sum.[12] In principle, this is one way to build a digital-to-analog converter (DAC), though this method is used only rarely.

---

[10] The first wrong answer confounds two distinct questions: *circuit-* and *op-amp-*input impedances. These are not the same, here. It's true that a Golden Rule says the inverting input will not pass current; but no rule says that the *circuit* should not draw current, through R_1, a current that can pass *around* the op-amp, through R_2 to the op-amp output terminal. That is, of course, how the input current does flow. The second wrong answer neglects the fact that the voltage at the inverting terminal is locked at 0V (by feedback), so that R_2 is quite invisible from the circuit input.

[11] Because it doesn't behave like ordinary (*real*) ground: current flowing to virtual ground bounces; it doesn't disappear, but runs off somewhere else (around the op-amp, using the op-amp's feedback path), where it remains available for measurement as it flows around the op-amp, using the op-amp's feedback path.

[12] Binary-weighted means that one can think of the three inputs as digits of a binary value, with each given double the weight of its "less significant" neighbor (you may find this notion easier to remember if you recall the weights of Homer Simpson and his less significant neighbor, Ned Flanders). So, the output voltage in the right-hand circuit of Fig. 6N.11 is $V_{\rm out} = -(4C + 2B + A) \times V_{\rm in}/4$.

256 Op-amps I

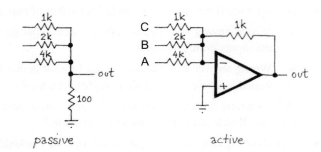

Figure 6N.11 Summing circuits.

One more summing circuit: Does a summing circuit have to use the *inverting* configuration? No, but the non-inverting form is not so simple, and is difficult to extend beyond two inputs. Does the circuit in Fig. 6N.12 form the sum of V_A and V_B?

This is *almost* a summing circuit. The output is the *average*, rather than sum. An op-amp circuit could easily convert it to a sum (what gain is required?[13]). But unlike the inverting summer, this one doesn't easily extend to three, four and more inputs. So, the usual summing circuit uses the inverting configuration.

Integrator, differentiator, and others

Virtual ground allows one to make a nearly-ideal integrator, as we'll see later. Again, the key is that the point where R meets C, a point that must be kept *close to ground* in the passive version, sits anchored *at ground*[14] in the op-amp version, see Fig. 6N.13.

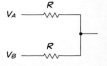

Figure 6N.12 Summing circuit?

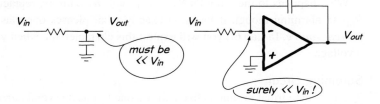

Figure 6N.13 Op-amp integrator does nearly perfectly what the passive version can only approximate.

The same sort of improvement is available for the op-amp differentiator, though there it is less impressive, compromised by a subtlety that we are happy to postpone: concern for circuit *stability*. We will meet differentiator and integrator in Day 7.

6N.8 When do the Golden Rules apply?

AoE §4.2.7

Now that we have applied the Golden Rules a couple of times, we are ready to understand that they sometimes do not apply.

Try some cases: do the rules apply at all?

The question may seem to you silly: Fig. 6N.14(a) is pretty clearly not a golden-rule circuit; it's just a high-gain amplifier – either a mistake or an anticipation of the circuit we will meet as a "comparator" in Lab 8L. It uses *no feedback*. Fig. 6N.14(b) works; Fig. 6N.14(c) does not, and the reason is that it uses the wrong flavor of feedback: positive.

---
[13] That's right: 2.
[14] Well, OK: *very close* to ground; we'll see, next time, that some departure from the ideal is inevitable.

6N.8 When do the Golden Rules apply?

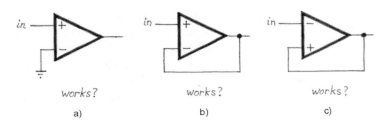

Figure 6N.14 Do the Golden Rules apply to these circuits?

So, generalizing a bit from these cases, one can see that the Golden Rules do not automatically apply to all op-amp circuits. Instead, they apply to op-amp circuits that...

- use feedback, and
- use feedback that is *negative*, and
- stay in the active region (don't let the op-amp output hit a limit, "saturating").

...Subtler cases: Golden Rules apply *for some inputs*

Will the output's *"attempt..."* to hold the voltages at its two inputs equal succeed in the two cases in Fig. 6N.15? No. But for the other polarity of input fed to each cicuit the answer would be "Yes." We will return to such conditional-feedback circuits next time.

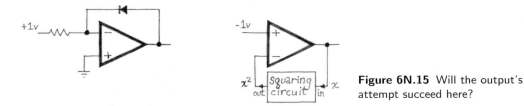

Figure 6N.15 Will the output's attempt succeed here?

6N.8.1 More applications: improved versions of earlier circuits

Nearly all the op-amp circuits that you meet will do what some earlier (open-loop) circuit did – but they will do it better. This is true of all the op-amp circuits you will see today in the lab. Let's consider a few of these: *current source, summing circuit, follower, and current-to-voltage converter.*

AoE §4.2.5

Current source: The left-hand circuit of Fig. 6N.16 is straightforward – but not usually satisfactory. For one thing, the *load* hangs in limbo, neither end tied to either ground or a power supply. For another, the entire load current must come from the op-amp itself – so, it is limited to about $\pm$ 25mA, for an ordinary op-amp.

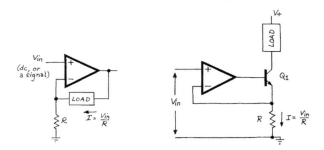

Figure 6N.16 Op-amp current sources.

The right-hand circuit is better. It gives you a chance to marvel at the op-amp's ability to make a device that's brought within the feedback loop behave as if it were *perfect*. Here, the op-amp will hide V_{BE} variations, with temperature and current, and also the slope of the I_C versus V_{CE} curve (a slope that reveals that the transistor is not a perfect current source; this imperfection is described by "Early effect," a topic discussed in §5S.2).

Do you begin to see *how* the op-amp can do this magic? It takes some time to get used to these wonders. At first it seems too good to be true.

AoE §4.3.1E

Followers: Figure 6N.17 shows three op-amp-assisted followers, alongside a simple bipolar follower.

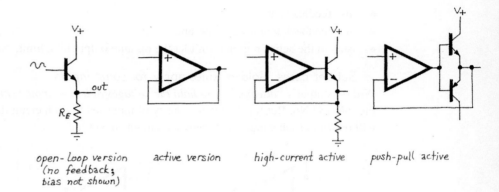

Figure 6N.17 Op-amp followers.

How are the op-amp versions better than the bare-transistor version? The obvious difference is that all the op-amp circuits hide the annoying 0.6V diode drop. A subtler difference – not obvious, by any means, is the much better output impedance of the op-amp circuits. How about input impedance?[15]

Current-to-voltage converter: Figure 6N.18 shows two applications for an op-amp "transresistance amplifier." *A Puzzle*: if you and I can design an "ideal" current meter so easily, why do our lab multimeters **not** work that way? Are we that much smarter than everyone else?[16]

AoE §4.3.1C

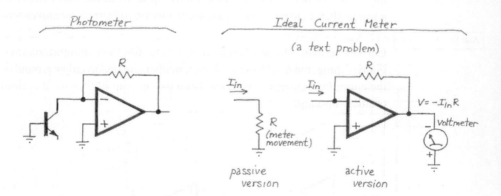

Figure 6N.18 Two applications for *I*-to-*V* converter: photometer; "ideal" current meter.

---

[15] Yes, the op-amp circuits win, hands down. Recall GR#1.
[16] Here's a thought or two on that question: as in the first current source (Fig. 6N.16), nearly the entire input current passes through the op-amp, so maximum current is modest, and the batteries powering the circuit will not last long.

6N.9 Strange things can be put into feedback loop

The push–pull follower within the feedback loop, shown in Figs. 6N.17 and 6N.20, begins to illustrate how neatly the op-amp can take care of and hide the eccentricities of circuit elements – like bipolar followers, or diodes.

6N.9.1 In general...

Figure 6N.19 illustrates the cheerful scheme, showing two distinct cases that fit all the op-amp circuits we will meet.

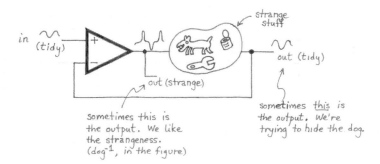

Figure 6N.19 Op-amps can tidy up after strange stuff within the loop.

Hiding the "dog"...
Sometimes, instead of treating the op-amp output as circuit output – and thus showing the inverse of the signal fed back, as we do in our op-amp voltage amplifiers for example – we use the op-amp to "hide the dog." The push–pull follower is such a case. The weirdness of the transfer function of the bare push–pull is not interesting or useful; it is a defect that we are happy to hide. Active rectifiers do a similar trick to hide diode drops.

To see the op-amp dutifully generating, at the input to the push–pull, whatever strange waveform is needed in order to hide the push–pull is quite dazzling the first time one sees it. See if you find it so. Figs. 6N.21 and 6N.22 show some scope images showing the process.

Crossover distortion in two forms...: We rigged a demonstration of the op-amp's cleverness (see Fig. 6N.20) by switching the feedback path between two points. The "silly" point, the op-amp output, produces a clean sinusoid in the wrong place while delivering nasty crossover distortion at the circuit output. The "smart" feedback point, the circuit's output, delivers a clean sinusoid where we want it.

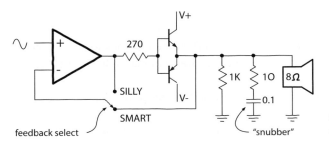

Figure 6N.20 Details of circuit set up to show op-amp cleaning up crossover distortion.

First, here is cross-over distortion, shown in two forms. The left-hand image of Fig. 6N.21 shows what this distortion looks like when the load is resistive. This might be called "classic" crossover distortion. The image on the right shows what the distortion looks like when the load is not a resistor but an 8Ω speaker, whose inductance produces stranger shapes.

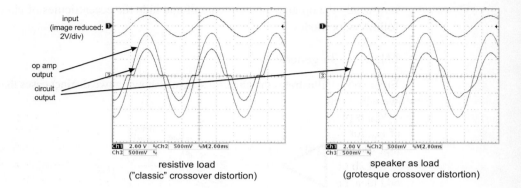

Figure 6N.21 Crossover distortion in two forms: resistive load, and speaker as load (stranger).

...Op-amp to the rescue: crossover fixed: Could the op-amp be clever enough to undo the strange distortion introduced by the push–pull – and rendered even stranger (asymmetric, too) by the speaker as load? Surely not!

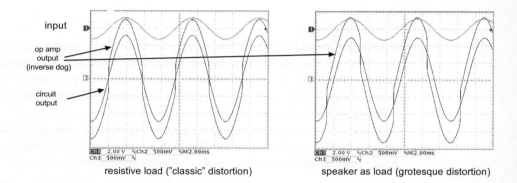

Figure 6N.22 Crossover distortion fixed by op-amp – which cleverly generates the inverse of the crossover transfer function.

Well, yes: the op-amp is that clever (at moderate frequencies): it generates just what's needed (dog^{-1}) to produce a clean output. Fig. 6N.22 shows the cleanup that the op-amp provides. (Note that the input waveform is shown with scope gain reduced $4\times$ relative to the output waveforms.)

AoE §4.3.1E

...Admiring the "dog": In other cases, as the comments in Fig. 6N.19 say, we want – and treat as circuit output – the *op-amp output*. In such cases, the "strange stuff" in the feedback loop is something we have planted in order to tease the op-amp into generating the inverse of that strange stuff ("inverse dog"[17]).

The "strange stuff" need not be strange, of course; if we insert a voltage divider that feeds back $1/10$ of V_{out}, then V_{out} will shows us $10 \times V_{in}$.

But here are two more exotic cases: two examples where the "strange signal" (dog^{-1}) evoked by the "strange stuff" in the feedback is useful. The left-hand example in Fig. 6N.23, with the diode in

---

[17] We would like to have included a visual representation, in the figure, of *inverse dog* – but we lost our nerve. What does *inverse dog* look like? Is it a dog with his paws in the air? Is it a cat? Is it a god? You can see why we abstained.

the feedback loop, may look at first like a rectifier. But consider what it does for an always-negative input voltage.

Figure 6N.23 Two cases where we plant strange stuff in loop to get "strange" and interesting op-amp output.

V_{out} is proportional to the *log* of V_{in}, for V_{in} negative; in the right-hand circuit, V_{out} is the square-root of V_{in}, for V_{in} positive. Both circuits fail for inputs of the wrong sign, as we meant to suggest when we first showed these two circuits, in Fig. 6N.15.

In Lab 6L you will be so bold as to put the oscilloscope itself inside one feedback loop. This is an odd idea, but one that produces entertaining results, see Fig. 6N.24.

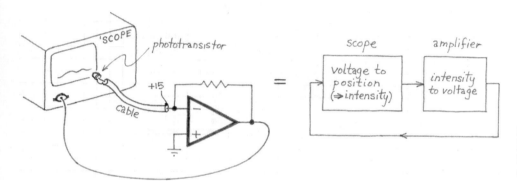

Figure 6N.24 Scope brought within feedback loop: adjusts location of CRT beam.

You'll see lots of nifty circuits in this chapter. Soon you may find yourself inventing nifty circuits. Op-amps give you wonderful powers. IC manufacturers show collections of application circuits for their devices. TI, for example, shows about 30 pages of circuits (reprinting National Semiconductor's application note 31): http://www.ti.com/ww/en/bobpease/assets/AN-31.pdf. Or for a sampling of application wisdom from the late analog wizard Jim Williams at Linear Technology, see http://cds.linear.com/docs/en/application-note/an6.pdf.

6N.10 AoE Reading

- Chapter 2: Once again, introduction to negative feedback: §2.5 (this was assigned for Lab 5L, but remains a relevant introduction).
- Chapter 4: §§4.1–4.2.

6L Lab: Op-Amps I

6L.1   A few preliminaries

First a few reminders.

Mini-DIP package: You saw a DIP[1] in the previous lab. Fig. 6L.1 shows another, this time an 8-pin mini-DIP, housing the operational amplifiers that we will meet in this and later labs.

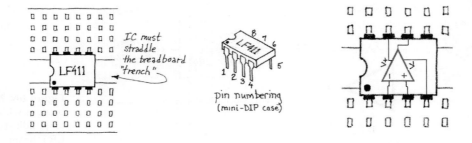

Figure 6L.1 Op-amp mini-DIP package.

The pinout (best represented by the rightmost image of Fig. 6L.1) was established by op-amps even earlier than the classic LM741[2] and is standard for single op-amps in this package. You will meet this pinout again in Lab 7L when you use the '741 and also the much more recent LT1150. Such standardization is kind to all of us users. Unfortunately, as parts get smaller, DIP parts are becoming scarce. Many new designs are issued in surface-mount packages only.

Power: Second, a point that may seem to go without saying, but sometimes needs a mention: the op-amp *always* needs power, applied at two pins; nearly always that means ± 15V in this course. We remind you of this because circuit diagrams ordinarily *omit* the power connections. On the other hand, many op-amp circuits make *no* direct connection between the chip and *ground*. Don't let that rattle you: the *circuit* always includes a ground in the important sense: a common reference treated as 0V.

Decoupling: You should always "decouple" the power supplies with a small ceramic capacitor (0.01–0.1μF), as suggested in Fig. 6L.2 and as we said in Lab 4L. If you begin to see fuzz on your circuit outputs, check whether you have forgotten to decouple. Most students don't believe in decoupling until they see that fuzz for the first time. Op-amp circuits, using feedback in all cases, are peculiarly vulnerable to such "parasitic oscillations." Op-amps are even more vulnerable to parasitic oscillations than the transistor circuits that evoked our warning on Day 4.

---

[1] "Dual in-line package".
[2] The pinout appears at least as early as the μ709, a Fairchild device introduced in 1965.

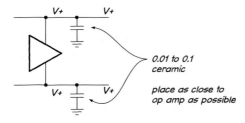

Figure 6L.2 Decouple the power supplies. Otherwise your circuits may show nasty fuzz (unwanted oscillations).

6L.2 Open-loop test circuit

Astound yourself by watching the output voltage as you slowly twiddle the pot in the circuit of Fig. 6L.3, trying to apply 0V. Is the behavior consistent with the 411 specification that claims "Gain (typical) = 200V/mV?" Don't spend long "astounding yourself" however; this is a most *abnormal* way to use an op-amp. Hurry on to the useful circuits!

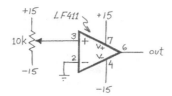

Figure 6L.3 Open-loop test circuit

6L.3 Close the loop: follower

Build the *follower* shown in Fig. 6L.4, using a 411. Check out its performance. In particular, measure (if possible) Z_{in} and Z_{out}, using processes to be described in later in this section.

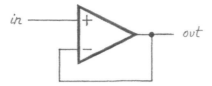

Figure 6L.4 Op-amp follower.

6L.3.1 Input impedance

Try to measure the circuit's input impedance at 1kHz, by putting a 1M resistor in series with the input. Here, watch out for two difficulties:

- *Beware* the finding "10MΩ".[3]
- R_{in} is so huge that C_{in} dominates. You can calculate what C_{in} must be, from the observed value of f_{3dB}. Again make sure that your result is not corrupted by the scope probe's impedance (this time, it's the probe's capacitance that could throw you off).

[3] It's very easy to be so deceived, because this is the first circuit we have met that shows an input resistance much larger than R_{in} for the oscilloscope-with-probe. That R_{in} is 10MΩ.

6L.3.2 Output impedance

The short answer to the value of the follower's output impedance is "low." It's so low that it is difficult to measure. Rather than ask you to measure it, we propose to let you show yourself that it is *feedback* that is producing the low output impedance.

Here's the scheme: add a 1k resistor in series with the output of the follower; treat this (perversely) as a part of the follower, and look at V_{out} with and without load attached. This is our usual procedure for testing output impedance. No surprises here: R_{out} had better be 1k.

Now move the feedback point from the op-amp's output to the point beyond the 1k series resistor – to apply "feedback #2," in Fig. 6L.5. What's the new R_{out}? How does this work? (Chapter 6N sketches an explanation of this result.)

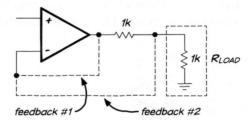

Figure 6L.5 Measuring R_{out} – and effect of feedback on this value.

If you're really determined, you can try to measure the output impedance of the bare follower (without series R). Note that no blocking capacitor is needed (why?[4]). You should expect to fail, here: you probably can do no more than confirm that Z_{out} is very low.

Do not mistake the effect of the op-amp's limited *current* output for high Z_{out}. You will have to keep the signal quite small here, to avoid running into this current limit. The curves in Fig. 6L.6 say this graphically. They say, concisely, that the current is limited to ±25mA over an output voltage range of ±10V, and you'll get less current if you push the output to swing close to either rail ("rail" is jargon for "the supply voltages"). This current limit is a useful self-protection feature designed into the op-amp. But we need to be aware of it.

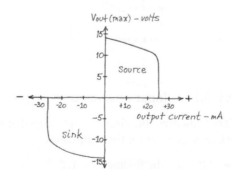

Figure 6L.6 Effects of limit on op-amp output current (LF411).

---

[4] No blocking cap because there's no *DC* voltage to block.

6L.4 Non-inverting amplifier

Wire up the non-inverting amplifier shown in Fig. 6L.7. You will recognize this as nearly the *follower* – except that we are tricking the op-amp into giving us an output bigger than the input. How much bigger?[5]

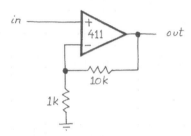

Figure 6L.7 Non-inverting amplifier.

What is the maximum output swing? How about linearity (try a triangle wave)? Try sinewaves of different frequencies. Note that at some fairly high frequency the amplifier ceases to work well: *sine in* does not produce *sine out*.

We are not asking you to quantify these effects today, only to get an impression of the fact that op-amp virtues fade at higher frequencies. In Lab 7L you will take time to *measure* the *slew rate* that imposes this limit on output swing at a given frequency. We are still on our honeymoon with the op-amp, after all; it is still *ideal*. "Yes, sweetheart, your slewing is flawless".

No need to measure input and output impedances again. Later, you will learn that you have traded away a little of some other virtues in exchange for the increased voltage gain; still, this amp's performance is pretty spectacular.

6L.5 Inverting amplifier

Construct the inverting amplifier drawn in Fig. 6L.8. If you are sly, you will notice that you *don't need to start fresh*: you can use the non-inverting amplifier, simply redefining which terminal is *input*, which is *grounded*. *Note*: Keep this circuit set up: you will use it again in §§6L.6 and 6L.7.

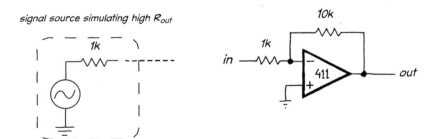

Figure 6L.8 Inverting amplifier.

Drive the amplifier with a 1kHz sinewave. What is the gain? Is it the same as for the *non-inverting* amp you built a few minutes ago?

Now drive the circuit with a sinewave at 1kHz again. Measure the input impedance of this amplifier

---

[5] Feeding back 1/11, we get $V_{out} = 11 \times V_{in}$. Looks familiar, from the final circuit of Lab 5L, does it not?

circuit by adding 1k in series with the signal source (simulating a source of crummy R_{out}). This time, you should have no trouble making the measurement. If you suppose that the 1k in series with your signal source represents R_{out} for your source, then what is the inverting amp's gain for such a source?[6]

Take advantage of the follower you built earlier, to solve the problem that we have created for you, the signal source's poor R_{out}. With the follower's help, your inverting amp's overall gain should jump back up to its original value (-10).

6L.6 Summing amplifier

Modify the inverting amplifier slightly, to form the circuit shown in Fig. 6L.9. This circuit sums a DC level with the input signal. Thus it lets you add a DC offset to a signal. (Could you devise other op-amp circuits to do the same task?[7])

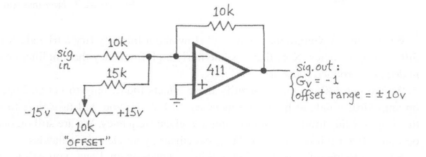

Figure 6L.9 Summing circuit: DC offset added to signal.

6L.7 Design exercise: unity-gain phase shifter

6L.7.1 Phase shifter I: using V_{in} and its complement

AoE §2.2.8

The circuit in Fig. 6L.10 applies a signal and its inverse to an R and C in series. By varying R, one can make V_{out} more similar to V_{in} or similar to its complement, $V_{in-inverted}$. Thus phase can be adjusted over a range of 180°. So far, so simple.

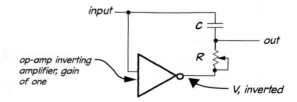

Figure 6L.10 Phase shifter, basic design.

The subtlety and cleverness of this circuit lies in the fact that the *amplitude* of the output always equals V_{in}, regardless of phase shift. (This is radically different from the behavior of, say, a low-pass filter, whose phase shift also can vary quite widely: between zero and almost $-90°$. But as the filter's phase shift varies, with changes of frequency, so does amplitude out.) Fig. 6L.11 helps to explain this happy and surprising result.

---

[6] That's right: gain falls to half of what it was, because the effective "R_1" is doubled, in the gain equation $G=-R_2/R_1$.
[7] Probably; but we doubt they'd be as simple as this one.

6L.7 Design exercise: unity-gain phase shifter

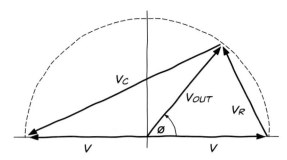

Figure 6L.11 Phasor diagram of phase shifter in Fig. 6L.10 (borrowed from AoE).

The input amplitude, fed to the series RC, is $2{\times}V$, and this is shown on the horizontal ("real") axis of Fig. 6L.11. The voltages V_R and V_C are $90°$ out of phase; this behavior is familiar to us from our experience with RC filters. The voltages V_R and V_C sum to $2{\times}V$. This relation is shown in the figure as the triangle whose hypotenuse – the sum – is the horizontal $2{\times}V$.

The really nifty element in this result is the fact that V_{out}, represented as the distance from ground to the junction of R and C, is always equal to V_{in} or "V." (The point we describe as "ground" is simply the midpoint on the horizontal axis, midway between V_{in} and its complement.)

To put this another way, the R–C junction lives on a semicircle of radius V. That radius is shown, in Fig. 6L.11, as a phasor *arrow* joining origin to that R–C junction.

Now design your own op-amp adjustable phase shifter and try it out. As you choose your values for R and C, note that you want your maximum R (the pot value) to be much larger than X_C at the driving frequency. Don't feel disappointed when you notice that at a given pot setting, phase shift varies with input frequency. That's just the way this circuit works.

6L.7.2 Phase shifter II: phase shifter with voltage control

Figure 6L.12 is a less obvious route to the same goal. The advantage of this circuit over the one you designed in §6L.7.1 is the fact that the phase shift can be adjusted by varying a resistance *to ground*. This feature renders straightforward the use of an electronic signal (rather than your hand adjusting a pot) to control phase shift. You could, for example, use a JFET in its resistive range to serve as adjustable resistor;[8] then let a repeating waveform drive the FET to sweep the phase shift. If you were to mix that signal with the non-shifted original signal, the result might sound like a "flanger." You could try this today, if you have some time on your hands. (Not likely!)

Compare AoE §6.3.5

How does it work? This is a subtler version of the phase shifter that you designed in §6L.7.1. To see how it works, imagine taking the variable R value to extremes:

- when $R = 0$, this is just an inverting amp, unity gain;
- when $R \gg X_C$, $V_{\text{out}} = V_{\text{in}}$.

These are the phase extremes, $180°$ apart.[9] Between these extremes, the phase at the non-inverting input is adjustable. The surprising virtue of *constant amplitude* results from essentially the argument that explains the simpler shifter (AoE §2.2.8A). We leave our explanation at that – a bit incomplete, "leaving the exercise to the reader" – because we don't want to stop you, at lab time, with a long digression on this topic.

---

[8] We don't build any JFET circuits in this course, but you will find such voltage-controlled-resistance circuits discussed in AoE in §3.2.7.
[9] The circuit is extremely similar to the "follower-to-inverter" circuit of AoE §4.3, Fig. 4.20.

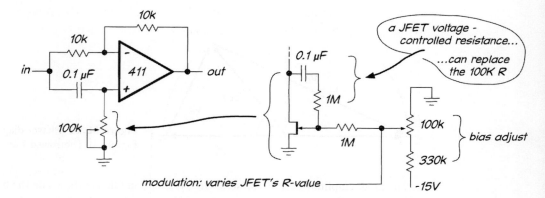

Figure 6L.12 Phase-shifter: second design.

6L.8 Push–pull buffer

AoE §4.3.1E, Fig. 4.26

Build the circuit shown in Fig. 6L.13. Drive it with a sinewave of 100–500Hz. Look at the output of the op-amp, and then at the output of the push–pull stage (make sure you have at least a few volts of output, and that the function generator is set for no DC offset). You should see classic crossover distortion.

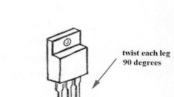

Figure 6L.13 Amplifier with push–pull buffer.

Listen to this waveform on the breadboard's 8Ω speaker – or on a classier speaker, if you can find one. Your ears (and those of people near you) should protect you from overdriving the speaker. But it would be prudent, before driving the speaker, to determine the maximum safe amplitude given the speaker's modest power rating. The transistors are tough guys, but you can check whether you need to lower the power-supply levels on your breadboard, to keep the transistors cool, given the following power ratings:

- transistors: 75W – if very well heat-sunk, so that case remains at 25°C. Much less (0.6W) if no heat-sink is used, as is likely in your setup.
- speaker: 250mW.

Now reconnect the right side of the feedback resistor to the push–pull output (as proposed near the end of Chapter 6N), and once again look at the push–pull output. The crossover distortion should be eliminated now. If that is so, what should the signal at the output of the op-amp look like? Take a look. (Doesn't the op-amp seem to be *clever*?)

Listen to this improved waveform: does it sound smoother (more flute-like) than the earlier waveform? Why did the crossover distortion sound harsh and metallic, more like a clarinet than like a flute – as if a higher frequency were mixed with the sine?[10]

---

[10] Well, because a higher frequency *is* mixed with the sine. The abrupt edges visible in the step from below to above the input

6L.9 Current to voltage converter

In earlier exercises we met and solved a "problem" presented by the inverting amplifier: its R_{in} is relatively low. Once in a while, this defect of the inverting amp becomes a virtue. This happens when a signal source is a *current*-source rather than the much more common *voltage*-sources that we are accustomed to. A photodiode is such a signal source, and is delighted to find the surprising R_{in} at the inverting terminal of the op-amp: what is that R_{in} value?[12]

Photodiode: Use a BPV11 or LPT100 phototransistor as a photo*diode* in the circuit of Fig. 6L.14. These devices are most sensitive in near-infrared (around wavelength of 850nm) but show about 80% of this sensitivity to visible *red* light. Look at the output signal. If the DC level is more than 10V, reduce the feedback resistor to 4.7M or even to 1M.

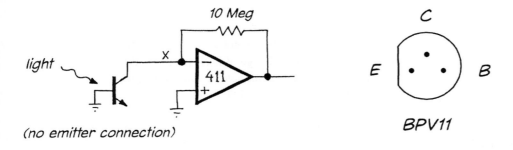

Figure 6L.14 Photodiode photometer circuit.

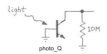

Figure 6L.15 A less good photodiode circuit.

If you see fuzz on the output – oscillations – put a small capacitor in parallel with the feedback resistor. The R is so big that a tiny cap should do: even 22pF causes circuit gain to fall off at an f_{3dB} ($1/2\pi RC$) of less than a kilohertz. In the *phototransistor* circuit in Fig. 6L.16, with its smaller $R_{feedback}$, you would need a proportionately (100×) larger C. Why does this capacitor douse the oscillation? Because the oscillation can't persist if the circuit gain is gone at the (high) frequency where the circuit "wants" to oscillate. It wants to oscillate up there because there it finds large, troublesome phase shifts; in Lab 9L we'll see much more of this problem.

What is the average DC output level, and what is the percentage "modulation?" (The latter will be relatively large if the laboratory has old-fashioned fluorescent lights, which flicker at 120Hz; much smaller with contemporary 40kHz fluorescents, or with incandescent lamps.) What input photocurrent

---

waveform, as the latter crosses zero volts, include high-frequency components. Your ears recognize this, even if you have for a moment forgotten the teachings of Fourier.

[11] The circuit requires the op-amp to snap its output from a diode drop *below* V_{in} to a diode drop above, as the output crosses zero volts. That excursion of about a volt takes time. The op-amp can "slew" its output only so fast – and here that rate is much lower than the maximum "slew rate" that you will measure next time, because the op-amp input is not strongly overdriven during this brief transition. While this slewing is occurring, the op-amp is not able to make the output do what it ought to do. The glitch becomes visually noticeable in the scope display when the output waveform is steep and scope sweep rate is high.

[12] Yes, ideally this R_{in} is zero. In fact, it is very small (later, in Chapter 9N, you'll learn to calculate it as $R_{feedback}/A$, where A is the op-amp's "open-loop gain").

does the output level correspond to? Try covering the phototransistor with your hand. Look at the "summing junction" (point **X**) with the scope, as V_{out} varies. What should you see?[13]

Make sure you understand how this circuit is preferable to a simpler "current-to-voltage converter," a resistor, used as in Fig. 6L.15[14]

Phototransistor: Now connect the BPV11 as a photo*transistor*, as shown in Fig. 6L.16 (the base is to be left *open*, as shown). Look again at the summing junction. Keep this circuit set up for the next stage.

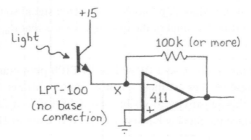

Figure 6L.16 Phototransistor photometer circuit.

Applying the photometer circuit: For some fun and frivolity, if you put the phototransistor at the end of a cable connected to your circuit, you can let the transistor look at an image of itself (so to speak) on the scope screen. (A BNC cable with grabbers on *both* ends is convenient; note that in this circuit *neither* terminal is to be grounded, so do not use one of the breadboard's fixed BNC connectors.) Such a setup is shown in Fig. 6L.17.

The image appears to be shy: it doesn't like to be looked at by the transistor. Notice that this scheme brings the scope itself within a feedback loop.

Figure 6L.17 Photosensor sees its own image.

You can make entertaining use of this curious behavior if you cut out a shadow mask, using heavy paper, and arrange things so that the CRT beam just peers over the edge of the mask. In this way you can generate arbitrary waveforms. If you try this, keep the "amplitude" of your cut-out waveform down to an inch or so. Have fun: this will be your last chance, for a while, to generate really silly

---

[13] The "summing junction" is a *virtual ground* and should sit at 0V. If it does not, then something has gone wrong. Most likely, the op-amp output has hit a limit ("saturated"), making it impossible for feedback to drive that terminal to match the grounded non-inverting input.

[14] The difference results from the photodiode's "disliking" large voltage variation. The passive circuit puts a changing voltage across the photodiode, and the diode varies its current somewhat in response. This variation distorts the relation between light intensity and diode current.

6L.10 Current source

Try the op-amp current source shown in Fig. 6L.18. What should the current be? Vary the *load* pot and watch the current, using a digital multimeter. This current source should be so good that it's boring. Now substitute a 10k pot for the 1k and use a second meter or a scope to watch the op-amp's output *voltage* as you vary R_{LOAD} (in this case, just the 1k variable R). This second meter should reveal to you why the current source fails when it does fail.

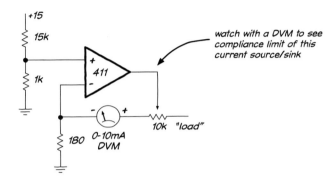

Figure 6L.18 Current source.

Note that this current source, although far more precise and stable than our simple transistor current source, has the disadvantage of requiring a "floating" load (neither side connected to ground); in addition, it has significant speed limitations, leading to problems in a situation where either the output current or load impedance varies at microsecond speeds.

The circuit of Fig. 6L.19 begins to solve the first of these two problems: this circuit sources a current into a load connected to ground.

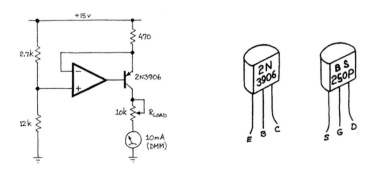

Figure 6L.19 Current source for load returned to ground.

Watch the variation in I_{out} as you vary R_{LOAD}. Again, performance should be so good that it bores you. The 10K resistor will cause the current source to fail when you dial R up to about 2k. You can see why if you use a second meter to watch V_{CE}.

You may not see any current variation using the bipolar transistor: a 2N3906. If you manage to discover a tiny variation, then you should try replacing the '3906 with a BS250P MOSFET (the pinout is equivalent, so you can plug either in exactly where you removed the 2N3906).

Should the circuit perform better with FET or with a bipolar transistor?[15] Do you find a difference that confirms your prediction (that difference will be extremely small)? With either kind of transistor the current source is so good that you will have to strain to see a difference between FET and bipolar versions. Note that you have no hope of seeing this difference if you try to use a VOM to measure the current; use a DVM.

---

[15] The FET version performs a little better. The feedback circuit monitors I_E, whereas I_C is the output; the difference is I_B – a small error. The FET shows no equivalent to I_B: its input current is zero. So, in the FET version the *output* current is the same as the current *monitored*.

6W Worked Examples: Op-Amps I

The problem – just analysis this time: This is a rare departure from our practice of asking you to *design*, not to *analyze*. Inventing a difference amp[1] seemed a tall order, and, on the other hand, the difference amplifier's behavior seems far from obvious. So, here's a little workout in seeing how the circuit operates.

6W.1 Basic difference amp made with an op-amp

AoE §4.2.4

The difference amplifier made with an op-amp looks like Fig. 6W.1, which shows the general configuration and the particularly straightforward case of *unity* gain, the case that this note treats first.

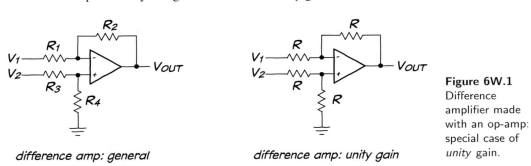

Figure 6W.1 Difference amplifier made with an op-amp: special case of *unity* gain.

In general, as you know, the amplifier requires that the ratios of the resistors in the two paths, from V_1 and from V_2, must be well matched. Common-mode rejection depends on this matching.[2] The amplifier's differential gain is $R_2/R_1 = R_4/R_3$.

Seeing the amp's gain by trying some cases

It is possible to derive the circuit's gain using a little algebra (see §6W.1.5). But that sort of description often lacks persuasive power. So, first let's use another approach: we will try particular cases of input voltages, to satisfy ourselves that the circuit does what it ought to do. This technique is one more application of a common technique that we use when confronted by a novel circuit: try to redraw it so that it looks like a familiar circuit that we understand well. In §6W.1.1, for example, the diff-amp becomes a good old inverting amp.

---

[1] This circuit is more often called a "differential amplifier." But we prefer the less grandiloquent term here, used also in AoE §4.2.4. The term "differential" seems to suggest that this rather straightforward circuit is doing some challenging calculus operations for you. The word "difference" better describes what the circuit measures. Nevertheless, we should admit, we often revert to the more traditional term and find ourselves calling the circuit a differential amp. A simple sidestep is to call it a "diff-amp," as many of us often do.

[2] Because close matching is hard to achieve, the best approach may be to buy an integrated array of resistors tightly matched: the best, to 0.001% (see AoE §4.2.4) or to buy an integrated difference amp like TI/Burr–Brown's INA105 with gain within 0.05% of unity, or the INA149, described below in §6W.2, with 100dB of common-mode rejection (CMR).

Recall that in order to keep things simple we have made all R values equal, giving the difference amp *unity* gain.

6W.1.1 First case: inverting amp

Ground V_2, apply a signal to V_1; see Fig. 6W.2. We're forcing the non-inverting input to 0V, so the inverting input goes to *virtual* ground, and the circuit forms a traditional inverting amplifier, gain of -1.

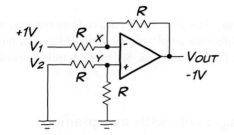

Figure 6W.2
Diff-amp: gain is -1.

Case	Inputs		Internal Points		Output	Notes
	V_1	V_2	X	Y	V_{OUT}	
A	1	0	0	0	-1	inverting amp

6W.1.2 Second Case: non-inverting amp

Ground V_1, apply a signal to V_2; see Fig. 6W.3. Now the circuit forms a non-inverting amplifier of gain 2 (recall that $G = 1 + R_2/R_1$).

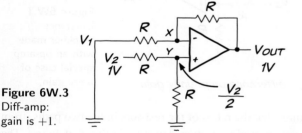

Figure 6W.3
Diff-amp: gain is $+1$.

Case	Inputs		Internal Points		Output	Notes
	V_1	V_2	X	Y	V_{OUT}	
B	0	1	0.5	0.5	1	non-inverting amp gain of $2 \times (1/2)V_1$

At first glance gain of *two* may sound wrong – but not when you recall that the non-inverting signal is attenuated to one-half before it reaches **Y**. So, the non-inverting gain is $+1$, as one would hope.

6W.1.3 Third case: equal inputs

Equal inputs should produce zero out. Try $+1V$ to both inputs, see Fig. 6W.4. Equal inputs send equal voltages to the two op-amp inputs, forcing equal voltages at the end of the two divider-chains. So V_{OUT} must be ground. In other words *common-mode* gain is zero. (This will be true to the extent that the four Rs are well matched.)

6W.1.4 Fourth case: non-zero inputs that differ by a volt

The output should be indifferent to any shared input, but should amplify the difference (1V) by 1.
Yes. Here we might say there is a *common-mode* signal of 1.5V, and a *difference* signal of 1V. The amplifier ignores the common signal, and amplifies the difference by 1.

6W.1 Basic difference amp made with an op-amp

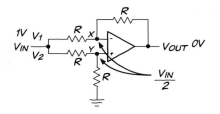

Case	Inputs		Internal Points		Output	Notes
	V_1	V_2	X	Y	V_{OUT}	
C	1	1	0.5	0.5	0	dividers match foot of each at gnd

Figure 6W.4 Diff-amp: equal inputs, output zero.

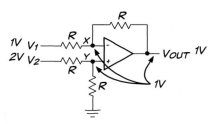

Case	Inputs		Internal Points		Output	Notes
	V_1	V_2	X	Y	V_{OUT}	
D	1	2	1	1	1	inv. amp but amp'ed with respect to V_Y and added to V_Y

Figure 6W.5 Diff-amp: two positive inputs, difference 1.

This case is exactly like the case that we called "non-inverting amp", §6W.1.2, except that both inputs are higher by 1V. This common difference the amplifier ignores.

Alternatively, one might see this as an inverting amp that amplifies V_1 with respect to V_X and V_Y. In this case there is no difference to amplify, so the output is just the baseline voltage, V_Y.

6W.1.5 Gain, formally derived

If looking at particular cases and trying to liken them to familiar amplifier configurations seems laborious, maybe you'll prefer a more traditional analysis. First, let's redraw the circuit to add a few labels; see Fig 6W.6.

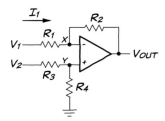

Figure 6W.6 Difference amplifier: a more general diagram.

Although we have generalized the diagram to permit gains other than unity, let us stick to unity gain, in order to make the arithmetic as easy as possible. So, again, let's assume all Rs are equal. They need not be for good differential performance; as we said at the outset, we only need $R_2/R_1 = R_4/R_3$.

The input current, I_1, is $(V_1 - V_X)/R$, and $V_X = V_Y = V_2/2$. So

$$I_1 = \frac{V_1 - (V_2/2)}{R}; \quad V_O = V_X - (I_1 R) = V_2/2 - \left(\frac{V_1 - (V_2/2)}{R}\right) R .$$

The Rs conveniently divide out, leaving $V_O = V_2 - V_1$

Though we looked, once more, at the unity gain case, the algebraic argument would work just as well for any matched set of resistor ratios: one where $R_2/R_1 = R_4/R_3$. If the resistor ratio is not unity, then of course V_O is not *equal to* the difference between the two circuit inputs but is a multiple of that difference.

Is this algebraic argument persuasive? Or do you prefer to look at the particular cases with which we began? Your answer will depend only on your tastes, of course.

6W.2 A more exotic difference amp: IC with wide common-mode input range: INA149

TI makes several integrated versions of the standard difference amp, ICs that include well-matched resistors delivering good common-mode rejection (100dB for the INA149). In addition, one of these ICs, the INA149, adds a 5th resistor that makes sure that the op-amp is not overdriven even for an input as large as ± 275V: the fifth resistor divides such a large input voltage down to 13.5V at the op-amp inputs. If you wonder why anyone would want such common-mode range, TI suggests, for example, that this range could be useful to monitoring current on a high-voltage line: the diff-amp could watch the voltage drop across a small resistor that carries that current.

Figure 6W.7 shows the circuit.[3] On the right side we have added labels in order to facilitate a discussion of the circuit.

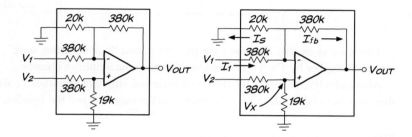

Figure 6W.7 Integrated difference amplifier, INA149.

This amplifier may look – if one recalls the general gain expression for a diff-amp (§6W.1) – as if it would deliver an attenuated output, with differential gain 1/20, reflecting the ratio of the 19k to the 380k input resistor on the V_2 side. But the circuit's differential gain turns out to be 1. We'll investigate how that result arises. We find the circuit's behavior far from obvious.

Again, we propose two ways to consider the circuit: first, using a couple of special cases that are easy to analyze; then chugging through some laborious algebra to get a general result.

6W.2.1 A couple of special input cases

As we found for the generic difference amp of §6W.1, some input cases are easy to analyze, and we'll try those to convince ourselves that, contrary to appearances, the circuit can indeed deliver a gain of 1.

Ground V_1: Grounding V_1 puts 380k and 20k to ground, in parallel: equivalent to 19k to ground. Thus we find a curious symmetry: V_2 is divided by 21 (=19k/(380k + 19k)). The inverting path divides its output voltage by the same fraction. So, $V_{\text{out}} = V_{\text{in}}$. You can put this argument differently, if you prefer: say V_2 is divided by 21, and the gain of the non-inverting amp driven by this value is $1 + 380\text{k}/(20\text{k}\|\|380\text{k}) = 1 + 380\text{k}/19\text{k} = 21$. Gain thus is $+1$.

Ground V_2: This case is easier. Grounding V_2 puts both op-amp input terminals ("Vx" on diagram) at ground. So, the amplifier looks like a simple inverting amp (the 20k has no effect, conducting no current with 0V across it). Gain is -1.

---

[3] The 20k and 19k resistors are shown grounded. They can be tied to other voltage references if an output with offset is wanted.

6W.2.2 A general solution

The algebra that generalizes these results is less easy and less fun. But here goes. We'll use the signals named on the right-hand side of Fig. 6W.7: V_1, V_2, V_X and the currents I_1, I_S and I_{fb}.

We'd like to find $I_{feedback}$. Knowing that value, and V_X, we will be able to calculate V_{out}:

$$V_X = V_2/21 \;.$$

Now let's calculate currents:

$$I_1 = (V_1 - V_X)/380k;$$

$$I_{feedback} = I_1 - I_S = \frac{V_1 - V_X}{380k} - \frac{V_x}{20k}$$

$$= \frac{V_1 - V_2/21}{380k} - \frac{V_2/21}{20k} \;.$$

Now we can use that current to get V_{out}:

$$V_{out} = V_X - (I_{feedback} \cdot R_{feedback});$$

$$V_{out} = V_2/21 - \left(\frac{V_1 - V_2/21}{380k} - \frac{V_2/21}{20k}\right) \cdot 380k$$

$$= V_2/21 - (V_1 - V_2/21 - (V_2/21) \cdot 19)$$

$$= V_2/21(1 + 1 + 19) - V_1 \;=\; V_2 - V_1 \;.$$

This is no surprise, after the two special cases that we saw, where we grounded first one, then the other of the two inputs. But the result is reassuring.

6W.3 Problem: odd summing circuit

Suppose you would like to sum the signals provided by two transducers, giving the two transducers *equal* full-scale "weights." The summed output is to drive an analog-to-digital converter (ADC)[4], and the input range of that device defines the permitted output range for your *summer*. We will assume the transducers provide *DC* signals of just one polarity, as stated below.

Here are the relevant specifications:

ADC input range: 0 to +2.5V
Signal source A: **DC voltage range:** 0 to 0.5V
    Output impedance: $\leq$ 10MΩ
Signal source B: **Amplitude range:** 0 to $-$1mA (current sink)
    Output impedance: $\geq$ 10MΩ
    Output voltage compliance: $\leq$ 0.1V

6W.3.1 A solution

Getting our bearings: The list of specifications in §6W.3 may be hard to get a grip on, at first glance – but not once one notices that *A* is a voltage source, while *B* is a current source. That makes the large, important contrast clear; from there we can proceed to the details expressed in the numbers.

The *summing* circuit that we are to build will use an op-amp in its inverting configuration, as usual. Such a circuit sums *currents*, fundamentally; when we feed *voltages* to such a summer we place input

---

[4] You needn't understand what this gadget does in order to answer this question.

resistors between signals source and the summing point, *virtual ground*. Those resistors convert input voltages to input *currents*. We will do that, as usual, for signal A.

But B, the signal that is a *current*, requires no such input resistor. Signal B, a current, goes straight to the point where currents are summed: virtual ground.

A skeleton circuit: A circuit with no component values will look like Fig. 6W.8.

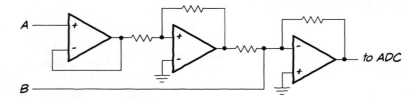

Figure 6W.8 A solution lacking component values.

Drawing the skeleton circuit is what requires some thought in this problem. Here are a couple of points that to us seemed not obvious.

- The tight "output voltage compliance" specification for B requires that B drive an op-amp summing junction. That's no surprise, or burden, since you probably were inclined to use this standard summing configuration, anyway. You must, of course, resist the tempting reflex (based on your experience with summers) to place a resistor between B and virtual ground.

    Why? – by the way: what goes wrong if you put, say, a 1k resistor between B and virtual ground? This R does not interfere with the summing of currents at virtual ground. Rather, the R moves the voltage at B too far from ground – 1V, at maximum current: far outside the 0.1V compliance range.

- A needs a high-impedance buffer – the input *follower* shown in Fig. 6W.8, because the high output impedance of A otherwise interacts with the modest input impedance of the summing circuit.

    "Modest?," you protest? "Why should it be modest? I can make it huge, by using huge R values in the inverting stage that takes signal A." (The inverting stage is needed to cancel the inversion that is inherent in the summing circuit.)

    But you answer yourself: "Yes, modest. Even at 100MΩ those Rs could introduce a 10% error, and we probably ought to limit ourselves to R values of about 10M, in order to avoid running into nasty side effects."

    Those side effects include large errors that result from I_{bias} flowing in unmatched resistive paths and, probably more troublesome, the introduction of an unintended *low-pass* effect within the feedback loop. Such a low-pass causes lagging phase shifts that can make the circuit unstable.[5]

    And you are too clever to fall for the following spurious argument: "I'll take advantage of A's R_{out} to form the inverting stage: I'll use a feedback resistor of 10M." That doesn't work because A's R_{out} is not specified to be 10M. Instead, as usual, the R_{out} value is given only as a *maximum*. You have seen this sort of specification many times before.

6W.3.2   . . . Calculating component values

Two component values matter:

- R_{FB} sets the full-scale V_{out};
- R feeding virtual ground sets the gain for signal A relative to signal B.

[5] We will devote most of Lab 9L to just such stability issues.

6W.3 Problem: odd summing circuit

Figure 6W.9 finishes the design. This part of the design task doesn't require intelligence; just a little arithmetic.

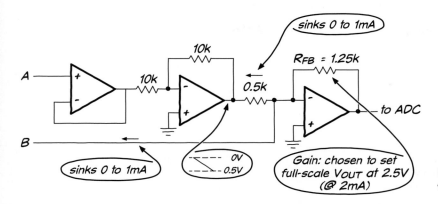

Figure 6W.9 Complete summing circuit, with component values.

Here are the arithmetic arguments:

R_{FB}: When the two signals, A and B, both are at their maximum we want $V_{out} = 2.5\text{V}$. We want to give the two signals equal weight, so B's 1mA should produce $V_{out} = 1.25\text{V}$. This sets the value of R_{FB} at 1.25k.

R feeding virtual ground: In order to give the two signals equal full-scale weights, this R should convert A's full-scale $V_{in} = 0.5\text{V}$ to 1mA. An R of 0.5k will do this.

7N Op-amps II: Departures from Ideal

Contents

7N.1	**Old: subtler cases, for analysis**	**281**
	7N.1.1 Do the Golden Rules apply?	281
	7N.1.2 Improving this active rectifier	282
7N.2	**Op-amp departures from ideal**	**284**
	7N.2.1 Should we care about these small imperfections?	285
	7N.2.2 Offset voltage: V_{OS}	286
	7N.2.3 Bias current: I_{bias}	288
	7N.2.4 Offset current: I_{offset}	290
	7N.2.5 Slew rate and roll-off of gain	290
	7N.2.6 Output current limit	291
	7N.2.7 Noise	292
	7N.2.8 Input and output voltage range	292
	7N.2.9 Some representative op-amp specifications: ordinary and premium	294
7N.3	**Four more applications**	**294**
	7N.3.1 Integrator	294
	7N.3.2 Taming the op-amp integrator: prevent drift to saturation	296
	7N.3.3 A resistor T Network	297
	7N.3.4 An integrator can ask a lot from an op-amp, and can show the part's defects	298
	7N.3.5 An example of a case where periodic integrator reset is adequate	299
7N.4	**Differentiator**	**300**
7N.5	**Op-amp Difference Amplifier**	**301**
	7N.5.1 A curious side virtue: wide input range	301
7N.6	**AC amplifier: an elegant way to minimize effects of op-amp DC errors**	**301**
7N.7	**AoE Reading**	**302**

Why?

We want to solve the problem of optimizing circuit performance by selecting from the great variety of available op-amps. We will try to make sense of the fact – not predictable from our first view of op-amps as essentially ideal – that there are not one or two op-amps available but approximately 37,000 listed (on the day of this writing) on one distributor's website (Digikey).

Admitting that op-amps are not ideal marks the end our honeymoon with them. But we continue to admire them: we look at more applications, and as we do, we continue to rely on our first, simplest view of op-amp circuits, the view summarized in the *Golden Rules*.

After using the Golden Rules to make sense of these circuits, we begin to qualify those rules, recognizing, for example, that op-amp inputs draw *a little* current. Let's start with three important new applications; then we'll move to the gloomier topic of op-amp imperfections.

Big ideas from last time:

- Feedback and high gain allow elegantly simple design and analysis with the "Golden Rules."

7N.1 Old: subtler cases, for analysis

- Some important applications: amplifiers, current source.
- Generalizing: op-amp circuits either exploit or hide the "dog" placed within the feedback loop.

7N.1 Old: subtler cases, for analysis

AoE §4.3.2D

Here is an issue we only glanced at last time: do the Golden Rules apply to a particular op-amp circuit? What about more general ones?

7N.1.1 Do the Golden Rules apply?

Before we begin to exploit the wonderfully useful Golden Rules, we should check that they apply. Do the rules apply to the circuit in Fig. 7N.1? The answer is a wishy-washy "Yes and No." Or, more precisely, they apply for positive inputs only.

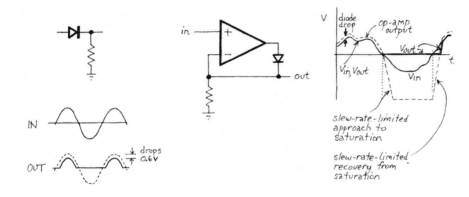

Figure 7N.1 A case that's a bit harder to analyze with the Golden Rules.

...Yes...: Positive inputs take the op-amp output positive (it helps to recall that the op-amp is a differential amplifier; take advantage of your understanding of the op-amp's innards). The positive output is able to feed current back to the inverting terminal, holding that terminal equal to V_{in}, as we would expect in a circuit obeying the Golden Rules. The circuit looks like a follower, with some extraneous elements added – a diode that's conducting, and an R to ground the has no effect except to load the output slightly. $V_{out} = V_{in}$.

...And no: But for negative inputs, feedback fails: the diode blocks the signal from the op-amp output – which now is below ground. Lacking feedback, the op-amp simply snaps down to negative "saturation" (a volt or so above the negative rail, for an ordinary op-amp like the '411), and sits there.

That may sound like a disaster, but in this instance it is not. With the diode blocking current, the op-amp now is out of the circuit that links input to output. V_{out} is simply tied to ground through a resistor. The net effect of these two behaviors – Golden Rule follower in one case, disconnected op-amp in the other case – is to produce an output that is a rectified version of the input.

We included a passive rectifier in Fig. 7N.1 to remind you of the diode drop that the active version hides. The active version, having hidden the diode drop (the usual "dog") is capable of rectifying even signals whose amplitude is much smaller than 0.6V. That is its great virtue. Its weakness is that it works only at moderate frequencies: its output includes a brief glitch.

282 Op-amps II: Departures from Ideal

Trying it out: scope images The glitch is predictable. Fig. 7N.2 shows what it looks like: somewhat worse than one might predict, because it's not just *slewing* time that we must wait for, but also extra time to get out of the "saturated" state, in which the op-amp rests when feedback fails.

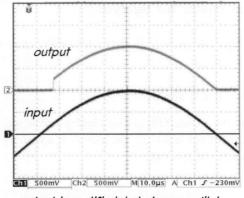
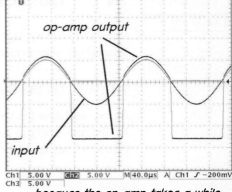

Figure 7N.2 Simple active rectifier produces an output glitch while recovering from saturation (scope gain: 500mV/div.

output is rectified, but shows a glitch...

...because the op-amp takes a while to come up from saturation

The glitch duration is more than five times what a calculation of slewing-time alone would predict if one simply found the op-amp's specified *slew rate* (15V/μs). The slewing that occurs here will be a good deal slower than that specified rate, which assumes extreme overdrive at the op-amp input.

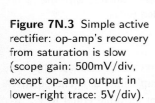
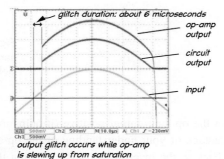
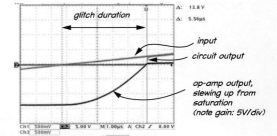

Figure 7N.3 Simple active rectifier: op-amp's recovery from saturation is slow (scope gain: 500mV/div, except op-amp output in lower-right trace: 5V/div).

output glitch occurs while op-amp is slewing up from saturation

You'll notice in §7N.2.5 that the maximum slewing occurs when the entire *tail* current flows in just one side of op-amp's input differential stage. This occurs only when a large *difference* voltage is applied between the two inputs. This does not occur during this brief glitch, where the voltage difference felt by the op-amp at first is very small, gradually growing to about 0.25V.

It would be good if we could redesign the circuit to avoid saturating the op-amp; and we can.

7N.1.2 Improving this active rectifier

AoE fig.4.38

We can improve the circuit's performance – diminishing the glitch duration by at least 10 fold – by arranging things so that feedback *never* fails, and the op-amp therefore never saturates. The rectifier circuit in Fig. 7N.4) redesigns the rectifier for this purpose. One diode or the other always conducts.

For positive inputs, this is simply an inverting amplifier. For the other case – a case like the one that led the simpler rectifier of Fig. 7N.1 to saturate[1] – this circuit gets feedback through the second diode.

---

[1] The polarities are reversed, between the two rectifiers, since the second one inverts.

While you're considering this circuit, use it as a chance to check your faith in op-amps. Specifically, does the diode between op-amp output and circuit output mess things up? (Have faith!)

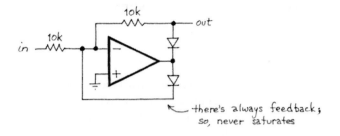

Figure 7N.4 One more diode improves the rectifier by preventing saturation.

This rectifier inverts. The left-hand image in Fig. 7N.5 shows the usual virtue of an active rectifier, a small input rectified without the subtraction of a diode drop, The right-hand image shows the changed op-amp output waveform that reduces the glitch: the op-amp output does not need to travel far (note the low scope gain: 0.5V/div).

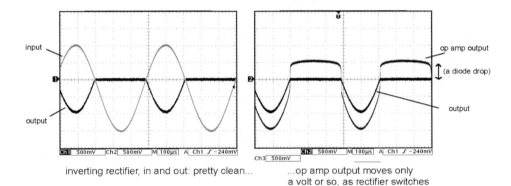

inverting rectifier, in and out: pretty clean...   ...op amp output moves only a volt or so, as rectifier switches

Figure 7N.5 Improved active rectifier.

Figure 7N.6, showing the transition at a higher sweep rate, confirms that the glitch is much reduced. Even at the high sweep rate of 400ns/div in the right-hand image of Fig. 7N.6 – 25 times faster than in Fig. 7N.3 – the glitch is barely apparent.

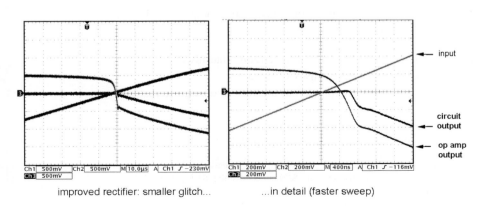

improved rectifier: smaller glitch...   ...in detail (faster sweep)

Figure 7N.6 Details of improved rectifier transition: nearly glitch-free.

R_{out} for the improved rectifier? This circuit shows a curious R_{out}. Although the op-amp always provides feedback (and thus diminishes the glitch of the simpler rectifier), the *circuit output* does not always show the low R_{out} that we expect from an op-amp circuit. When you have spotted the case for which this larger R_{out} applies, add another circuit fragment that solves that problem.[2]

7N.2 Op-amp departures from ideal

AoE §4.4.1

Let's admit it: op-amps aren't quite as good as we have been telling you. That's the sad news. But the happy news is that an amazing variety of op-amps exists – literally thousands of types – competing to optimize this or that characteristic. You can have low input current, low voltage-error (V_{offset}, coming soon in these notes), low noise – you just can't optimize all characteristics at once. You can find op-amps that run on high voltages (hundreds of volts), others that will run on a total supply of 0.9V; some will deliver tens of amps as output current, others will subsist on nano-amps of supply current. Some are very fast.

Op-amp imperfections fall into two categories: DC and dynamic.

Dynamic errors: The dynamic errors affect performance at higher frequencies.

- Gain rolloff: the high gain that is essential to achieve the virtues of feedback dissipates as frequency rises. It is gone at the frequency f_T, where gain is down to unity. This characteristic often is specified as the "gain-bandwidth product," GBW.
- The limited *slew rate* of the op-amp output imposes a limit on frequency and amplitude available *together* (small signals can be reproduced at frequencies higher than those available for larger signals).
- Noise injection (this is a subtlety that never becomes important in our lab circuits).

We will return to these errors later in these notes.

DC errors: The errors that concern us most are two DC errors which reveal that the Golden Rules exaggerate a bit.

- The inputs *do* draw (or source) a little current. This current, called I_{bias}, is the average of the currents flowing at the two inputs.
- The inputs are not held at precisely equal voltages. The principal cause of this inequality is V_{offset}, the op-amp's misconception concerning equality.[3]

Probably you are not surprised to hear this. Fig. 7N.7 gives three circuits that always deliver a saturated output after a short time. None would do this if op-amps were ideal.

No doubt you can figure out why they saturate. But the explanations are various enough so that you won't mind hearing them spelled out.

(a) This circuit would produce zero out only if V_{offset} were zero, as it cannot be.

---

[2] Take a look at AoE if you're stumped, but if you don't have it to hand here's a powerful hint: the case that worries us is the one in which feedback flows through the lower diode. In that case, the upper diode is not conducting. So R_{out} is a disappointing 10k. A follower can remedy this defect.

[3] A second cause, usually smaller in magnitude, is the limited gain of the op-amp. We calculated the difference needed between the two inputs of the op-amp to generate an 11V output, for example, back in §6N.6.2.

7N.2 Op-amp departures from ideal

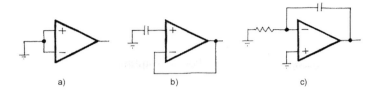

Figure 7N.7 Three circuits sure to saturate. Why?

(b) This circuit would produce zero out only if I_{bias} were zero, as it cannot be. In the case of a bipolar input stage, I_{bias} is the transistor's *base* current, and this current gradually will charge the capacitor. If the input stage of the op-amp is a FET, then I_{bias} is the transistor's *leakage* current. Either way, this current will charge the capacitor, so that the voltage at the non-inverting input must drift toward one supply or the other. (A '411 wired this way would drift toward the negative supply; see the discussion of a sample-and-hold made with a '411, in §12N.7.1).

(c) If V_{offset} and I_{bias} were zero, the voltage at the inverting terminal would be zero, and this case would be the same as Case (a). As we mentioned in discussing the integrator, the capacitor provides *no feedback* at DC (which you might also call "long-term feedback"). So the capacitor cannot save this output from saturation.

7N.2.1 Should we care about these small imperfections?

If one recites typical numbers for an op-amp's deviations from the ideal, those deviations sound pretty small. Here, for example, are some typical values for the '411: $V_{offset}=0.8$mV, $I_{bias}=50$pA, slew rate 15V/μs; gain (open-loop) not infinite but 200,000 at DC. Listen to those units: "pA," "millivolts"? These sound so tiny! Do these departures from the ideal matter? Our answer is a wishy-washy "sometimes."

The number of op-amps available is enough to remind us that these small differences *can* matter. If they did not, one or a few op-amps would survive. Instead, there is room for an amazing variety of op-amps – and new ones are born every few months. For, optimizing one specification entails relaxing another. So it is worth learning to pay attention to the specifications – even while continuing to recognize that for a great many applications one can treat the op-amp as ideal, even if it is a mere "jellybean" device like the '411.[4]

Here are some more indications of the number available: several leading manufacturers websites show what each offers.

Texas Instruments More than 1800; their collection now includes all of National Semiconductor's couple of hundred op-amps.
Linear Technology about 350.
Analog Devices about 320.

And even a single manufacturer can make so many op-amps that it needs to sort them by type in order not to overwhelm its customers with one giant list. TI uses the following categories for its ordinary op-amps:

- Low Offset Voltage;
- Low Power;
- Low Noise;
- Low Input Bias Current;

---
[4] "Jellybean" is a strange piece of jargon meaning cheap and common.

- Wide Bandwidth;
- Wide Supply Voltage;
- Single Supply Voltage.

And this set excludes more specialized op-amp types such as "Power Amplifiers and Buffers." The variety is daunting, despite this breakdown into sub-species.

§7N.2.9 includes a table from AoE giving the range of op-amp specifications available. A glance at this list reminds us – if we need reminding – that we can't have all virtues at once. Bipolars, for example, can beat MOSFET devices in *offset voltage* (V_{offset}) by a factor of perhaps 50 – but at the cost of input current (I_{bias}) several thousand times larger than for the CMOS part. That's to be expected, of course: an instance of electronic justice.

Another AoE table, in its §4.7, shows 34 "representative" op-amps, including the two that we rely on in most of our lab exercises – the LF411 and LM358. The set of specifications listed in that table is a bit wider than in the other – including input and output swing and power supply current. Does anything else in life come in so many tasty flavors?

Two circuits that make small imperfections matter: In two worked examples, we will look at two cases where the small DC errors *can* matter:

- a sensitive DC amplifier (§7W.1);
- an integrator (§7W.2).

Now, let's look more closely at the most important of the op-amp's deviations from ideal performance.

7N.2.2 Offset voltage: V_{OS}

AoE §4.4.1A

The offset voltage is the difference in input voltage necessary to bring the output to zero.

This specification describes the amp's delusion that it is seeing a voltage difference between its inputs when it is not. One can represent this error as a DC voltage added to one of the inputs of an *ideal* zero-offset internal difference amp, as in the leftmost image of Fig. 7N.8.[5]

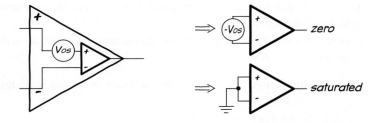

Figure 7N.8 V_{offset} represented as a voltage added to one input of an idealized op-amp; ... and consequences.

This separation of idealized and realistic characteristics recalls the Thevenin model of Chapter 1N. Two consequences are suggested in Fig. 7N.8 – though the thought that one might hold the output at *zero* by applying just the right counter-error exaggerates what is possible.

The amplifier makes this mistake because of imperfect matching between the two sides of its differential input stage. To achieve the impossible $V_{\text{offset}} = 0$, the op-amp would have to match perfectly not only the transistors of its differential amplifier but also the transistors of the current-mirror that serves as that first-stage's *load*. These elements are shown in a simplified op-amp circuit diagram, Fig. 7N.11. The full '411 schematic appears in Fig. 9S.31 on page 396. Fig. 7N.9 shows the V_{OS} specifications for our principal op-amp.

[5] Thanks to David Abrams for suggesting this redrawing.

7N.2 Op-amp departures from ideal

Symbol	Parameter	Conditions	LF411A min	LF411A typ	LF411A max	LF411 min	LF411 typ	LF411 max	Units
V_{os}	Input offset voltage	$R_S = 10\Omega, T_A = 25°C$		0.3	0.5		0.8	2.0	mV

Figure 7N.9 '411 spec: V_{offset}.

You can compensate for this mismatch by deliberately drawing more current out of one side of the input stage than out of the other, in order to balance things again. This may sound like a rash thing to do – after all, the manufacturer has done its best to achieve matching; is there any hope that your heavy hand can reach into the brain of the op-amp and improve things?!

Actually, the answer is, "Yes," because you will make the correction while watching the op-amp output with a sensitive voltmeter. Thus you will know when you have optimized V_{offset}.[6] This correction is called *trimming offset*, and you will do it in Lab 7L. But this trimming is a nuisance, and the balancing does not last: time and temperature-change throw V_{offset} off again (see AoE §4.4.1B). It's a reasonable way to improve the performance of a "one-off"[7] lab instrument; it is not a reasonable approach for a circuit produced in large quantities (imagine the cost of adding a human adjustment step at the end of an assembly line; and even if the assembly line automates this adjustment, the operation adds to cost).

AoE §4.7

Self-trimming op-amps: "auto zero": Some op-amps are able to trim their own offsets by including – integrated with the main amplifier – a second "nulling" op-amp dedicated to "auto zeroing" the main amp.

The nulling amp works in two cycles – running at a few tens or hundreds of hertz – and thus is said to "chop" between the two conditions. This periodic nulling process, "chopper-stabilizing," is not the same as the early "chopper" op-amp design, which periodically disconnected the main amplifier's inputs, and thus provided a *sampling* effect that severely limited bandwidth. Chopper stabilizing, in contrast, leaves the main amplifier connected at all times, and therefore does not diminish high-frequency performance.[8]

The main benefit that results from this nulling is, of course, extremely low V_{offset}: as low as about $1\mu V$ (typical) and $5\mu V$ (max).[9] Secondary benefits include enormous open-loop gain and very good immunity to all DC and low-frequency errors: power-supply rejection, V_{offset} drift with time and temperature.

You will meet such a chopper-stabilized op-amp, the LTC1150, in Lab 7L. And we hope that the exercise of trimming down the '411's V_{offset} in that lab will not obscure the general truth that usually it will be wiser to:

- use a good op-amp, with low V_{offset};
- and design the circuit to work well with the V_{offset} of the amp you have chosen.

---

[6] You probably recognize that you – the relatively clumsy human twiddling a potentiometer – are able to perform this delicate surgical adjustment only because you are, while watching the output voltmeter, a part of a *feedback* loop.

[7] "One-off" is jargon for a circuit you make just *one* of, in contrast to something produced in quantity.

[8] In case you're curious to hear a quick account of how this nulling works, here goes.
The nulling op-amp first nulls its own offset. This it does by tying its two inputs together and driving its output back to its own offset-null terminal (a single input that accomplishes what we do by hand on the '411, with a pot between two terminals). Then the nulling amp connects its inputs to the main amp's inputs and uses its output to drive the *main* amp's offset-null input. Capacitors store the two nulling voltages, which must be maintained during the periodic chopping. Negative feedback thus will drive the difference between the main amplifier's inputs very close to the ideal of equality.

[9] A very good explanation by an ADI engineer appears in an *Analog Dialog* note, "Demystifying Auto-Zero Amplifiers-Part 1", **34**, no.2, March 2000,
http://www.analog.com/en/analog-dialogue/articles/demystifyingauto-zero-amplifiers-part-1.html.

Figure 7N.10 offers a silly reminder that what we're trying to do when we trim offset is to undo an asymmetry built into the op-amp. And Fig. 7N.11 shows a more literal diagram of what trimming does: the trimming network reaches into the input stage of the op-amp, and draws more current out of one side than the other – just enough to offset the built-in imbalance of currents.

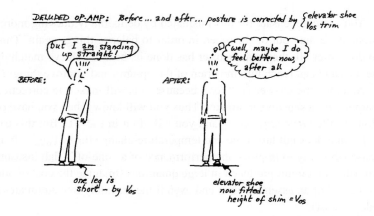

Figure 7N.10 Offset trim as shim in an elevator shoe.

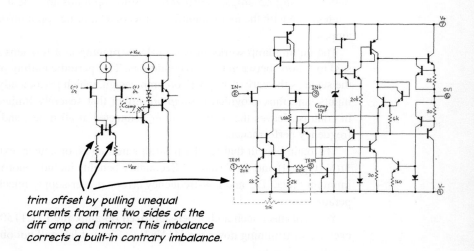

Figure 7N.11 Inside the '411. Schematics: simplified, and detailed views show how offset trim works.

trim offset by pulling unequal currents from the two sides of the diff amp and mirror. This imbalance corrects a built-in contrary imbalance.

7N.2.3 Bias current: I_{bias}

AoE §4.4.1C

The bias current, I_{bias}, is a DC current flowing in or out at the input terminals (it is defined as the average of the currents at the two terminals); see Fig. 7N.12.

For an amplifier with bipolar transistors at the input stage, I_{bias} is base current; for a FET-input op-amp like the '411, I_{bias} is a leakage current – tiny, but growing rapidly with temperature: see Fig. 7N.13.

See AoE Fig. 4.55

Balancing resistive paths to minimize the effects of I_{bias}: The bias current flows through the resistive path feeding each input; it can, therefore, generate an input error voltage, which may be amplified highly to generate an appreciable output error. The high-gain DC amplifier in Fig. 7N.14 shows a

7N.2 Op-amp departures from ideal

Symbol	Parameter	Conditions		LF411A			LF411			Units
				min	typ	max	min	typ	max	
I_{os}	Input offset current	$V_S = \pm 15V$	$T_J = 25°C$		25	100		25	100	pA
			$T_J = 70°C$			2			2	nA
			$T_J = 125°C$			25			25	nA
I_B	Input bias current	$V_S = \pm 15V$	$T_J = 25°C$		50	200		50	200	pA
			$T_J = 70°C$			4			4	nA
			$T_J = 125°C$			50			50	nA

Figure 7N.12 '411 Specs: I_{bias} and I_{offset}.

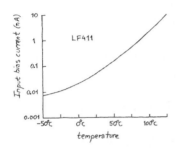

Figure 7N.13 '411 bias current: tiny, but grows fast with temperature.

deliberate mismatch of resistive paths driving the two inputs. We sometimes used this circuit in order to demonstrate the effects of I_{bias} and V_{offset}:

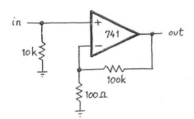

Figure 7N.14 Demonstration circuit: uses high-gain dc amp to make errors measurable.

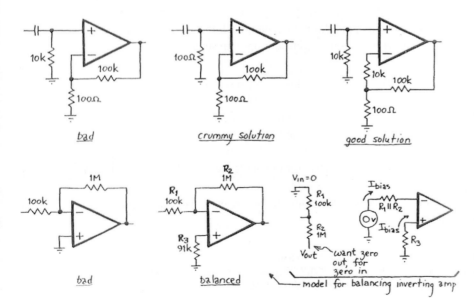

Figure 7N.15 Balanced resistive paths minimize output errors resulting from bias current.

The bias current of 80nA (typical) flows through 10k in one path, through 100Ω in the other,[10] developing a difference signal of about −0.8mV at the non-inverting input.[11]. This pseudo-signal would produce a spurious output of about −0.8V, assuming no V_{offset} (the assumption that $V_{\text{offset}}=0$ is not realistic; but we like to think about one problem at a time).

...But you may not need to balance paths if I_{bias} is very low: But the circuit of Fig. 7N.14 doesn't show you I_{bias} effects unless you take the trouble to use an antique op-amp – a *741* rather than a '411. A '411 gives no measurable effect from I_{bias}. The fact that this demonstration required a '741 suggests that often you will *not* need to worry about the effects of bias current; and for less than a dollar you can find an op-amp with a bias current 10^3 smaller than the '411's. The objection that balancing paths may not be necessary is fair; but you should know how to judge whether or not to worry about the effects of I_{bias}.

If I_{bias} effects are troublesome, then you should take the trouble to minimize this disturbance by matching the resistances of the paths to the two op-amp inputs. Fig. 7N.15 gives some examples of circuits that do or do not balance paths.

Once you have balanced these resistive paths, I_{bias} no longer causes output errors. But a difference between currents at the inputs still does. That difference is called I_{offset}.

7N.2.4 Offset current: I_{offset}

AoE §4.4.1D

Offset current is the difference between the bias currents flowing at the two inputs. For the '411 the I_{OS} specification is about $\frac{1}{2}I_{\text{bias}}$; for the bipolar op-amps I_{OS} is smaller relative to I_{bias}. But recall how tiny I_{bias} is for the '411 and other FET-input devices.

I_{OS} predicts an error even when the resistances seen by the two inputs are balanced. Remedy? Use resistances of moderate value (under a few 10s of MΩ; see the argument for the clever T resistor trick noted later in §7N.3.2.)

7N.2.5 Slew rate and roll-off of gain

AoE §4.4.1K

The limited values of both slew-rate and gain turn out to be caused by a gain-killing capacitor designed into the op-amp. This is a component that we indicated, but did not explain, back in Fig. 7N.11. We will talk about this *compensation* device in Chapter 9N when we consider op-amp stability.

Gain rolloff: For the moment, we will only note that the result of this "compensation" capacitor (so-called probably because it "compensates" for the circuit's tendency to oscillate spontaneously) is to roll off the op-amp's gain at −6dB/octave. The op-amp behaves, in other words, as if its output had been passed through a simple *RC* low-pass. In effect, this *has* happened, though in the internal second stage. So, the chip's very high gain, necessary to make feedback fruitful, evaporates steadily with increasing frequency – and is *gone* at a few MHz (typically 4MHz for the '411; see Fig. 7N.16).

AoE §4.4.1J

The frequency, f_T, at which gain has fallen to unity, defines a frequency limit on the usefulness of *any* op-amp circuit.[12] This limit explains why not every circuit should be built with op-amps, wonderful though the effects of feedback are.

---

[10] This is just R_{Thev} for the divider that drives the inverting input.
[11] The voltage is negative because the '741's I_{bias} is base current drawn by its NPN input transistors.
[12] f_T is closely related, and usually identical, to "gain–bandwidth" product – "GBW." The two differ only in uncompensated or decompensated op-amps, which cannot stably run at unity gain. See Walter G. Jung, *IC Op-Amp Cookbook*, SAMS (3d ed., 1997), pp. 67–70.

7N.2 Op-amp departures from ideal

Symbol	Parameter	Conditions	LF411 min	LF411 typ	LF411 max	Units
A_{VOL}	Large-signal voltage gain	$V_S = \pm 15V$, $V_O = \pm 10V$, $R_L = 2k$, $T_A = 25°C$	25	200		V/mV
		over temperature	15	200		V/mV
GBW	Gain-bandwidth product	$V_S = \pm 15V$, $T_A = 25°C$	2.7	4		MHz

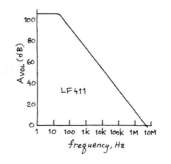

Figure 7N.16 '411 gain roll-off: spec and curves.

Slew rate: The compensation capacitor also explains the op-amp's slew rate. The slew rate, the maximum dV/dt of the op-amp output, is achieved when the maximum input-stage current – the entire *tail* current, issuing from a radically unbalanced differential stage – charges this compensation capacitor. For the '411, we can work backwards from the specified slew rate and the size of the compensation cap to infer the first-stage "tail" current, which is the maximum output from the differential stage:

$$I = CdV/dt = 10 \times 10^{-12} \times (15V/\mu s) = 150\mu A.$$

Fig. 7N.11 shows the compensation capacitor between base and collector of the high-gain second stage of the op-amp: the common-emitter amplifier.[13]

But we don't want to leave you with the impression that the '411's GBW of 4MHz is a general ceiling on what you can expect. No, op-amps are available with almost any characteristic optimized (always at the expense of other traits). If you need an op-amp with GBW = 1GHz, you can have one.[14]

7N.2.6 Output current limit

AoE §4.4.1H

The current limit protects the small transistors of the output stage. The limit prevents overheating that otherwise would result when some clumsy user overloaded the amp. Most op-amps, including the '411, suffer no harm even from the blunder of shorting the output not just to ground but to either of the power supplies. We will postpone looking at how this current-limiting circuitry works until we reach Chapter 11N on *voltage regulators*.

You saw the effect of this current limit in a curve that appeared in Lab 6L. Fig. 7N.17 shows this behavior again.

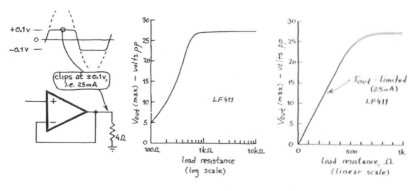

Figure 7N.17 Output current limit: output clips under load, despite very low R_{out}.

---

[13] Well, that's almost a fair description; the circuit is complicated slightly by the presence of a follower driving the base of the common-emitter amp.
[14] For example, the TI/Burr–Brown's OPA640 promises f_T of 1.3GHz. Current-feedback devices can run even faster.

Op-amps II: Departures from Ideal

As you'd expect, high-current and high-voltage op-amps are available. But often you can get the high current you need by letting an op-amp serve as brains for a brawny current booster like the push–pull follower that you built in §6L.8. If however you insist on your right to be lazy, and want tens of amps at hundreds of volts, you can try Apex/Cirrus Logic: for example, MP38 offers 10A, 200V, 125W.

7N.2.7 Noise

AoE Chapter 8

Any amplifier adds some noise, as it works, and this effect – much too subtle to appear in our lab exercises – is specified for each op-amp. Two sorts of noise are specified: *voltage noise* (e_n) and *current noise* (i_n). The effects of the two add, but particular ranges of source resistance can make one or the other important. For high source resistance, current noise is the more important (the current noise generates a noise signal as it flows in $R_{\rm SOURCE}$). The '411 data sheet boasts of its current-noise spec, among the "features" listed at the head of its data sheet: $i_n = 0.01{\rm pA}/\sqrt{\rm Hz}$. Part of that head is reproduced in Fig. 7N.18.

Figure 7N.18 LF411 data sheet boasts of its current–noise spec (courtesy of National Semiconductor).

7N.2.8 Input and output voltage range

AoE §§4.4.1F, 4.4.1G, 4.4.1H

Ordinary op-amps (like the '411)

Most op-amps cannot handle input voltages close to one or both power supplies, and cannot swing their outputs very close to one or both supplies. The '411 illustrates these limitations (power supplies assumed: '411A: ±20V, '411: ±15V). Note $V_{\rm O}$ and $V_{\rm CM}$ in Fig. 7N.19.[15]

The consequence of exceeding "common-mode input range" can be startling. The '411 (like some other similar op-amps[16]) does a very weird thing if you take one or both inputs below the lower limit

---

[15] Complicating the specifications shown in Fig. 7N.19, the LF411 data sheet states, among its application hints, that the '411 will operate with a common-mode input voltage all the way to the positive supply but "... the gain bandwidth and slew rate may be reduced in this condition."

[16] Our informant, the late Jim Williams (see below), lists several that do this same nasty trick: LF147, LF351, LF156. All are JFET-input devices, like the '411.

DC Electrical Characteristics (Note 5)

Symbol	Parameter	Conditions	LF411A Min	LF411A Typ	LF411A Max	LF411 Min	LF411 Typ	LF411 Max	Units
V_O	Output Voltage Swing	$V_S = \pm 15V$, $R_L = 10k$	±12	±13.5		±12	±13.5		V
V_{CM}	Input Common-Mode Voltage Range		±16	+19.5		±11	+14.5		V
				−16.5			−11.5		V

Figure 7N.19 LF411 input and output voltage ranges do not go all the way to the supplies.

Application Hints

The LF411 series of internally trimmed JFET input op amps (BI-FET II™) provide very low input offset voltage and guaranteed input offset voltage drift. ...

Exceeding the negative common-mode limit on either input will force the output to a high state, potentially causing a reversal of phase to the output. ...

Figure 7N.20 LF411 data sheet confesses a weird hazard.

of that input range. This behavior is not advertised on page 1 of the data sheet, but instead is buried on page 8 among "Application Hints," shown in Fig. 7N.20:

We got stung by this misbehavior – which can turn negative feedback into positive – when we were breadboarding the "PID" motor-control circuit of Lab 10L. Eventually, we caught on, and substituted an op-amp (the LM358) that does not show this pathology.

The engineering community is aware of this strange behavior, and one company thought it might make some money by redesigning a popular op-amp to eliminate this phase reversal. The late Jim Williams, a longtime applications engineer and analog wise man at Linear Technology Co., told a striking story about this effort. His company made such an improved op-amp, and sent samples and simulation models to General Motors, which had been using a '411-like op-amp in the transmission control for its diesel locomotives.

GM reported back that when they simulated use of the "improved" op-amp, their locomotives now and then would jump into reverse. Everyone agreed to drop plans to substitute the "improved" op-amp. Apparently, GM had adapted its software to the nasty phase reversal, and when this reversal did not occur, trouble occurred. The case reminds one of the old folk tale (plausible enough) of a person who wakes up because a clock is silent – fails to chime the hour. GM's locomotives awoke (jumping into reverse) when their op-amps failed to flip phase.

Williams' story also reminds us why no one is in a hurry to plug an "improved" IC into a working design. In turn, this disinclination explains the curious fact that today's version of an old IC like the LM741 op-amp is much better than the original – but cannot boast of this on its data sheet. If it made such a boast – changing the data sheet – then it would no longer be a '741, and any designer proposing to buy this *improved* part for an existing design would need to persuade her bosses that it was worthwhile to run all the tests required before the new part could be adopted.

Rail-to-rail op-amps

AoE §4.4.1F

Some op-amps show common-mode input range better than the '411s. "Single supply" types (like the '358 that you will meet today in Lab 7L) have a common-mode input range that includes the "negative supply" – usually just *ground* in this context. Another class of op-amps does better than that – the so-called "rail-to-rail" types.

Some accept *inputs* rail-to-rail; some can drive their *outputs* rail-to-rail; a few do both. Those are nice features – but like all other optimizations they come at the price of some other characteristic, such as offset voltage. So, you'll normally choose a rail-to-rail type only when you need that behavior. But we should admit that the need for this behavior is becoming more frequent as typical power supply voltages fall. With supplies of ±15V, giving up a volt or so close to each rail isn't so bad. Giving up that much is unacceptable when the supply is, say, a single 3V coin cell.

A configuration permitting a *circuit* to show huge common-mode input range

A "difference" or "differential" amplifier made with an op-amp can permit a common-mode input range that extends far beyond its power supplies. That claim may seem to contradict the tight constraints mentioned in §7N.2.8, but it does not: it describes the wide swings permitted at *circuit* inputs rather than *op-amp* inputs. If that seems strange, recall that an ordinary *inverting* op-amp circuit could do the same trick, since V_{in} is not applied to the op-amp itself.

Using the difference-amp configuration, an op-amp IC from Texas Instruments[17] (INA149) promises an astonishing common-mode input range of ±275V. The differential configuration, described in §7N.5 permits this result.

7N.2.9 Some representative op-amp specifications: ordinary and premium

AoE §4.4.1

Here is part of a table from AoE, illustrating the range of values for op-amp specifications. The "premium" values are the best available, and as AoE warns us, you can't combine the best in several specifications simultaneously; you must choose your op-amp according to what characteristics are most important to you.

	bipolar		JFET-input		CMOS		
Parameter	jellybean	premium	jellybean	premium	jellybean	premium	Units
V_{OS} (max)	3	0.025	2	0.1	2	0.1	mV
I_B (typ)	50nA	25pA	50pA	40fA	1pA	2fA	25°C
f_T (typ)	2	2000	5	400	2	10	MHz
SR (typ)	2	4000	15	300	5	10	V/μs

You may recognize the specifications for "JFET-input... jellybean" as almost exactly those of the '411 op-amp that you use in most of our labs' op-amp circuits.

7N.3 Four more applications: integrator, differentiator, rectifier, difference amplifier, AC amplifier

7N.3.1 Integrator

To appreciate how very good an op-amp integrator can be, we should recall the defects of the simple RC "integrator" you met in Chapter 1N.

Passive RC integrator

AoE §1.4.4

To make the RC behave like an integrator, we had to make sure that

$$V_{out} \ll V_{in} .$$

This kept us on the nearly straight section of the curving exponential-charging curve, when we put a square wave in. The circuit failed to the extent that V_{out} moved away from ground. But the output *had* to move away from ground, at least a little, in order to give an output signal: see Fig. 7N.21.

---

[17] The design originated with Burr-Brown, which TI acquired.

7N.3 Four more applications

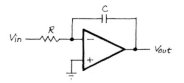

Figure 7N.21 *RC* "integrator:" integrates, sort-of, if you feed it the right frequency.

AoE §§4.2.6, 4.5.5

Op-amp version

The op-amp integrator solves the problem elegantly, by letting us tie the cap's charging point to 0V, while allowing us to get a signal out: "virtual ground" lets us have it both ways, see Fig. 7N.22.

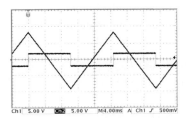

Figure 7N.22 Op-amp integrator: virtual ground is just what we needed.

As the scope image in Fig. 7N.23 shows, the circuit output can be larger than V_{in} without distorting the integration.

Figure 7N.23 Op-amp integrator eliminates requirement $V_{out} \ll V_{in}$.

The op-amp integrator is so good that one needs to prevent its output from sailing off to saturation (that is, to one of the supplies) as it integrates error signals: over time, a tiny lack of symmetry in the input waveform will accumulate; so will tiny op-amp errors. The scope image in Fig. 7N.24 shows the integrator output near negative saturation, then working for a while as the DC level drifts upward, till it bumps against the positive limit of the op-amp's range.

We should admit, therefore, that the integrator of Fig. 7N.22 was a defective circuit. We showed only a capacitor in the feedback path to keep the circuit conceptually simple. We will need to do more to make it usable.

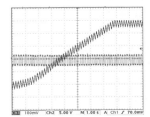

Figure 7N.24 Integrator without DC feedback drifts to saturation.

A simple way to describe what's wrong with the integrators in Figs. 7N.22 and 7N.7 is to point out that the circuit includes *no DC feedback at all* (this is a point we noted back in §7N.2). Such an arrangement guarantees an eventual drift to saturation.

So, practical op-amp integrators include some scheme to prevent the cap's charging to saturation.

7N.3.2 Taming the op-amp integrator: prevent drift to saturation

AoE Fig. 4.66

One remedy: a large resistor: A large resistor in parallel with the cap prevents saturation. This R this leaks off a small current, undoing the effect of a small input error current. The right-hand circuit of Fig. 7N.25 shows a so-called "T network," which can obviate the use of very large R values; this network is explained in §7N.3.3.

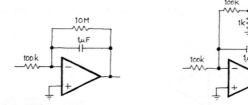

Figure 7N.25 Integrator saved from saturation by resistor parallel C_{feedback}.

When $R_{\text{FEEDBACK}} = 100 \times R_1$, as it is in Fig. 7N.25, then the leakage current through R_{FEEDBACK} is only 1% of the signal current when V_{out} is comparable to V_{in}. Often, that imperfection in the integrator's performance is acceptable.

Effects of the resistor: Evidently, the resistor compromises performance of the integrator. We can figure out by *how much*. There are several alternative ways to describe its effects:

The resistor limits DC gain: In the circuit above, where R_{in} = 100k and R_{feedback}=10M, the DC gain is −100. So a DC input error of ±1mV ⇒ output error of ±100mV. The integrator still works fine, apart from this error.

The resistor allows a predictable DC leakage: Suppose we apply a DC input of 1V for a while; when the output reaches −1V, the error current is 1/100 the input or "signal" current because, as we justed noted, $R_{\text{feedback}} = 100 \times R_{\text{in}}$. This error grows if V_{out} grows relative to V_{in}.

The resistor does no appreciable harm above some low frequency: At some low frequency, X_C becomes less than R, and soon R is utterly insignificant. As you know, $X_C = R$ at $f = 1/2\pi RC$. For the components shown in Fig. 7N.25 (where R_{feedback}=10M, C=1μF) that frequency is about 0.02Hz!

The resistor can become troublesome...: *At first, a bit of curvature shows the distortion...*
The effect of the feedback resistor becomes troublesome when V_{out} is much larger than V_{in}, as in Fig. 7N.26, where the curvature shows that we are not watching true integration, which should show a straight *ramp* waveform.
   ...at an extreme, the limited DC-gain becomes obvious: the circuit is then not *an integrator*
Pushing on in the same direction, if we let V_{out} exceed V_{in} by a great deal – an effect we get if we apply a very small input at very low frequency – then the feedback resistor destroys the integrating effect. The waveform in Fig. 7N.27 shows the compromised integrator revealing itself as no more than a ×100 DC amplifier, after a while.

But don't let these distortions scare you away from using a feedback resistor in an integrator: to show these distortions we had to go out of our way, applying extremes of low input amplitude and low frequency.

7N.3 Four more applications

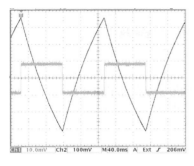

Figure 7N.26 Distortion of integral by feedback R becomes apparent if V_{out} is much larger than V_{in}. (Scope gains: input (small square wave), 10mV/div; output (triangular), 100mV/div).

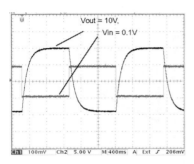

Figure 7N.27 Pushed to an extreme, the integrator is hopelessly compromised by the feedback resistor. (Scope gains: input (small square wave), 100mV/div; output (exponential), 5V/div).

7N.3.3 A resistor T Network

[AoE Fig. 4.66]

When a design calls for very large R values – as may occur for an integrator's feedback resistor, for example – a so-called "T network" permits use of modest resistor values, avoiding some of the harms that enormous Rs can introduce.

Figure 7N.28 offers a diagram intended to persuade you that the clever T resistor arrangement of Fig. 7N.25 can indeed make a 100k resistor look about $100\times$ as large.

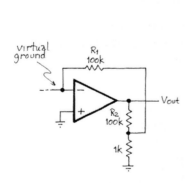
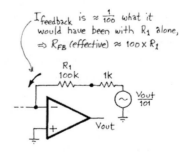

Figure 7N.28 How the T arrangement enlarges apparent R values.

Neat? The scheme is useful because the lower R_{Thevenin} of the T feedback network has two good effects:

- it generates smaller I_{bias} errors than the use of giant resistors would (see §7N.2.3);
- it drives stray capacitance at the op-amp input better (avoiding unintentional low-pass effects in the feedback); this is not important here, but is in a circuit where no capacitor sits parallel to a big feedback resistor.

The T scheme does present one drawback: it amplifies input *voltage* errors – V_{offset}, §7N.2.2, even when the signal source is a current source. If this remark is cryptic, reconsider it after looking at §7W.1 where we contrast the case of two integrator drift rates, for *voltage* sources versus *current* sources.

Another remedy: a switch

A switch in parallel with the cap also can keep the integrator from drifting to saturation. This has to be closed briefly, from time to time, and this requirement can be a nuisance.

AoE Fig. 4.18

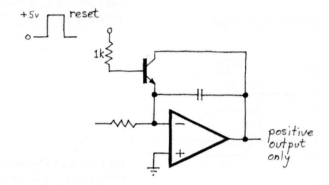

Figure 7N.29 Integrator saved from saturation by discharge switch.

The switch permits a more perfect integrator (no leakage current through a feedback R compromises its integration). Whether it is worthwhile to gain this advantage at the expense of rigging a resetting signal will depend on your application. Note the restriction that the transistor *reset* of Fig. 7N.29 fails if the integrator output goes negative. Only when we meet the *analog switches* made with MOSFETs, in Chapter 12N, will we be able to implement an transistor reset switch that works for both polarities of integrator output.

7N.3.4   An integrator can ask a lot from an op-amp, and can show the part's defects

A high gain amplifier like the one mentioned in §7N.2.3 can demonstrate effects of small input errors: V_{offset} and I_{bias}. So can an integrator, since it accumulates the effects of errors.

A demonstration circuit: integrator shows effects of op-amp imperfections

As a demonstration, we set up a contest between an old bipolar op-amp, the LM741, and an ordinary FET-input op-amp, the LF411. We built the circuit in Fig. 7N.30 twice:

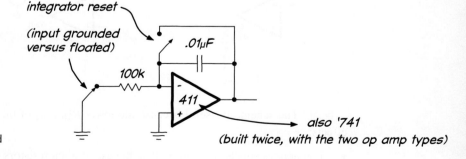

Figure 7N.30 Integrator circuit used to watch effects of V_{offset} and I_{bias}.

We first *grounded* the inputs to both of the twin integrators, and watched the drift. This appears in the leftmost image of Fig. 7N.31. The '411 wins – but not by a great margin: perhaps 15:1. Then we

7N.3 Four more applications

opened both inputs – mimicking what a *current* source would look like if connected to the integrator input. This produced an extreme victory for the FET-input (low I_{bias}) '411. In order to see *any* drift in the '411 integrator, we had to crank the scope sensitivity way up (to 50mV/div) – and we had to be patient.

7N.3.5 An example of a case where periodic integrator reset is adequate

A former student showed us this case, Fig. 7N.32, where manual resetting was quite practical (the student was collecting a measure of total response of a frog muscle to a stimulating pulse).

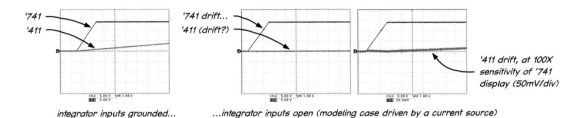

Figure 7N.31 Drift rates for competing integrators: FET-input device wins – but much more dramatically when driven by current source. (Scope gains: all 5V/div except '411 drift in rightmost image, 50mV/div)

Worked Example 7W.1 looks in greater detail at this contrast between the two op-amps and their errors,.

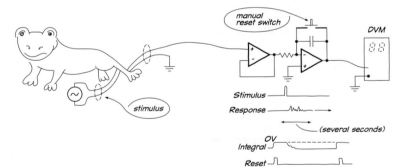

Figure 7N.32 A slow integration can make manual reset practical.

The duration of the integration was long, and the pace of the experiment slow enough, that it made sense for this circuit to start with a manual reset (to clear the integrator), then apply the stimulus, then record the integrated value, all over the course of many seconds. An accurate – low-drift – integrator is required of course (and it was in pursuit of this that the former student returned to our lab).

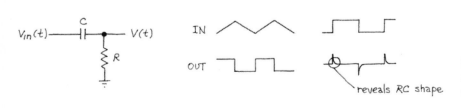

Figure 7N.33 RC "differentiator:" differentiates, sort-of, if RC kept very small.

7N.4 Differentiator

AoE §4.5.7

Again, the contrast with a passive differentiator helps one appreciate the op-amp version. Fig. 7N.33 reminds us of the passive circuit, which you met in Lab 2L. To make the passive version work we had to make sure that

$$dV_{out}/dt \ll dV_{in}/dt \ .$$

Again the op-amp version exploits *virtual ground* to remove that restriction.

But the differentiator is less impressive than the op-amp integrator. The op-amp version of the differentiator must be compromised in order to work at all: its gain must be ruined at high frequencies in order to keep it stable.

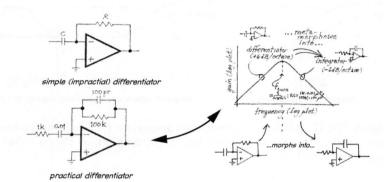

Figure 7N.34 Differentiators, idealized and practical: stabilized version morphs into integrator at high frequencies.

A practical differentiator, shown in Fig. 7N.34, turns into an integrator (of all things!) at some high frequency. This scheme is necessary to prevent unwanted "parasitic" oscillations (we will look more closely at this topic in Labs 9L and 10L). Fig. 7N.35 is a preview of the stability problem that we look into later.[18] The simple, impractical differentiator is marginally stable, and shivers in response to a steep waveform (the input signal is a square wave). That shiver is unacceptable, so we trade away high-frequency performance for decent operation in a restricted range.

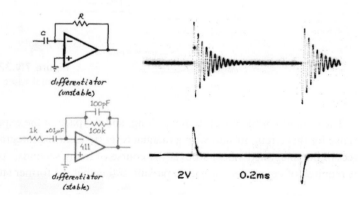

Figure 7N.35 Preview of stability problems: op-amp differentiator must be slowed down to keep it stable.

---

[18] This image is taken from Chapter 10N.

7N.5 Op-amp Difference Amplifier

AoE §4.2.4

Figure 7N.36 shows a standard differential amp circuit made with an op-amp. The circuit's gain is R_2/R_1 (this is *differential* gain, of course; ideally, the common-mode gain is zero). Its common-mode rejection can be very good if the ratios of the two dividers are well matched, as they can be when the resistor sets are integrated. In TI's INA194 diff-amp IC, for example, the resistors are integrated with the op-amp and laser-trimmed toward equality. CMR[19] for the INA194 is specified at 90dB.

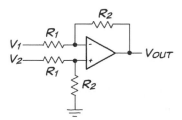

Figure 7N.36 Simple differential amp, made with an op-amp.

In §6W.1 we offer some informal analysis of this standard difference amplifier, hoping to make its behavior comfortable to your intuition. We won't reproduce all that here, of course. But we can make a quick start in the direction of making sense of the circuit by noting that if the two input voltages are *equal*, then the two dividers must show equal voltages at their right-hand ends. One is grounded, so the op-amp output must also sit at zero volts. In other words, equal inputs produce 0V out. That make sense. You may want to try out some other particular cases: try grounding one input; try grounding the other. And so on.

7N.5.1 A curious side virtue: wide input range

A difference amp can tolerate a surprisingly-wide common-mode input range. Recall that common-mode input voltage implies *no* V_{out}; a small *difference* between inputs sharing a large common-mode signal implies only a *small* V_{out}. Example 6W.1, and Fig. 6W.7, discussed an integrated difference amplifier, INA149, with an amazing common-mode input range of ±275V.

7N.6 AC amplifier: an elegant way to minimize effects of op-amp DC errors

AoE §4.2.2A

If you need to amplify AC signals only, you can make the output errors caused by V_{os}, I_{bias} and I_{os} negligible in a clever way: just cut the DC gain to unity, illustrated in Fig. 7N.37.[20]

This is the circuit used in Lab 7L to amplify a microphone's signal while using just a single power supply rather than the usual "split supply" that you are accustomed to in op-amp circuits. The use of one supply obliges us to bias the circuit input positive, presenting a case just like the one in Fig. 7N.37 on the right: a small wiggle rides a DC offset or bias. We want to amplify the wiggle, but not the DC bias.

What is f_{3dB} for the circuit shown in Fig. 7N.37? If one ignores the "1" in the gain expression, output amplitude is down 3dB when the denominator in the gain expression is $\sqrt{2} \times R_1$. But that

---

[19] "CMR" in this case is the same as the more familiar "CMRR," the ratio of good gain to bad gain – since "good gain," differential, is 1.

[20] The subscript *v* in the gain expression of Fig. 7N.37 indicates *voltage* gain in contrast to, say, *current* gain. Usually we omit this notation.

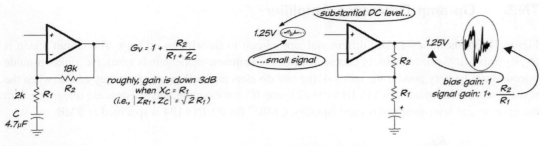

AC amp: general case... ...and as applied to lab's single-supply microphone amplifier

Figure 7N.37 AC amplifier: neatly makes effects of small errors at *input* small at *output*.

happens when $X_C \equiv |Z_C| = R_1$, and this occurs, as you well know, at $f_{3dB} = 1/(2\pi R_1 C)$. (Don't be fooled by the graphical detail that the capacitor is tied to ground, making the RC look something like a low-pass. You're not fooled. You recognize it as a *high-pass* in sheep's clothing.)

The RC analysis we need here is the same one that you used in choosing the emitter-bypassing capacitor for an amplifier back in Chapter 5N. Fig. 7N.38 is a reminder of that circuit, a high-gain common emitter amp. The *single-supply* op-amp design, with its need for a *bias* voltage, is exceptional in our op-amp work, but you know it well from your work with discrete transistors. [AoE §2.3.5A]

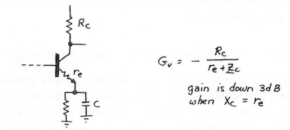

Figure 7N.38 Gain of AC amp is down 3dB when denominator (series impedance of R and C) is up to $R\sqrt{2}$.

7N.7 AoE Reading

- Reading
    - §4.4: a detailed look at op-amp behavior.
    - §4.5: a detailed look at selected op-amp circuits.
    - §4.6: op-amp operation with a single power supply.
    - §4.7: other amplifiers and op-amp types.
- Problems:
    - Additional Exercises 4.31–4.33.

7L Labs: Op-Amps II

This lab introduces you to the sordid truth about op-amps: *they're not as good as we said they were last time!* Sorry. But after making you confront op-amp imperfections in the first exercise, §7L.1, we return to the cheerier task of looking at more op-amp applications. There, once again, we treat the devices as ideal.

On the principle that a person should eat his spinach before the mashed potatoes (or is it the other way round?) let's start by looking at the way that op-amps depart from the ideal model. The integrator exercise does what it can to make the potentially dull topics of *offset-voltage* and *bias current* less dull by letting you undertake a demanding task of integration – one that you could not carry out without first overcoming the effects of these two op-amp imperfections.

We want you to get to know the integrator as a circuit important in its own right. But it also serves well to demonstrate the possibly troublesome effects of op-amp errors that can *sound* negligibly small: *bias* currents of picoamps, *offset* voltages below a millivolt. The integrator is unforgiving: it accumulates the effects of such errors, of course, holding a grudge for picocoulombs of charge delivered in the distant past, many milliseconds ago. We've also rigged a setup in which you get to integrate a signal from a piece of hardware (a disc drive or DC motor), for a refreshing change from feeding always from a function generator. One could almost imagine that this disc-drive motion integrator is a useful circuit.

7L.1 Integrator

Try the integrator: Construct the active integrator shown in Fig. 7L.1 using a '411 op-amp. The pushbutton is redundant at the moment; the 10M feedback resistor will keep the output from wandering far astray. Soon, we will remove the feedback resistor and rely on a manual pushbutton reset. But not yet. Keep this circuit set up after you finish this integrator section. You will use it again with the differentiator in §7L.2.

Try driving your integrator with a signal in the range 50Hz to 1kHz – all three waveforms listed below. This circuit is sensitive to small DC offsets of the input waveform (its gain at DC is about 500);

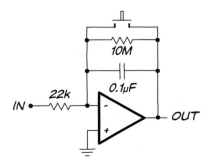

Figure 7L.1 Integrator.

the output is likely to go into saturation near the 15V supplies. To prevent this, you will have to adjust the function generator's *DC-offset* control. That adjustment will be more manageable if you set the function generator's attenuation to 20dB.

Try a square, sine and triangle waves. From the component values, predict the peak-to-peak triangle wave amplitude at the output that should result from a 2V(pp), 50Hz square wave input. Then try it.

Include feedback resistor: integrator tamed...: The 10M resistor tames this integrator – and somewhat corrupts it. The resistor can keep the output from wandering off to saturation, if you adjust the function generator carefully. But it makes the integration not quite honest – and not good enough for longer-term integration like what we mean to attempt soon. The feedback resistor permits a current to leak off the capacitor. For some applications this may not matter (the ratio of R_{feedback} to the input R is almost 500 to 1, so when V_{out} is comparable to V_{in} the error current caused by R_{feedback} is about 0.2% of the signal current). You'll probably be grateful for this feedback resistor while you try the integrator's response to various waveforms.

...Remove feedback resistor: integrator less tame but truer: When you remove the 10M resistor, the circuit will help you get a real gut feeling for the meaning of an integral. In fact, the attempt to adjust the generator's DC offset to keep V_{out} from saturating could drive you crazy. Don't suffer for long. Don't forget that the pushbutton can at least restore V_{out} to zero, for a fresh start.

7L.1.1 Integrator used to let one infer I_{bias} and V_{offset}

Drift; causation ambiguous: Ground the integrator's input, zero the output with the pushbutton, then release the button and watch the drift of V_{out}. You'll need a very slow scope sweep rate. Try 1 sec/div. (This is a good time to switch to using a digital scope, if one is available.) Note the drift rate. But note, also, that both I_{bias} and V_{offset} can contribute to this effect. To make things worse, the two effects may reinforce each other, or one may subtract from the other. Let's peel them apart.

Infer I_{bias} from drift rate with the effect of V_{offset} removed: Removing the effect of V_{offset} happens to be extremely easy: just float the input. Now V_{offset} does not force a current through the input resistor. So, V_{offset} does not contribute to the current that charges the capacitor. Given this simplification, see if your drift rate implies a value for I_{bias} that is in the '411's specified range: 50pA typical, 200pA max. You will need to be patient to measure this drift rate. Note it once you get a reading, and use the rate to calculate the value of I_{bias}.

Infer V_{offset} from drift rate: Now you should be in a position to estimate V_{offset} from the overall drift rate you saw as you started this lab. (You'll need to pay attention to the *sign* of each of these drift rates.) Compare your calculated value to the '411's specification: 0.8mV typical, 2.0mV max. What's your inferred V_{offset}?

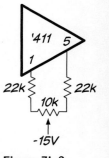

Figure 7L.2 Offset-trim network for '411 op-amp.

7L.1.2 Making a low-drift integrator in two ways

Trim V_{offset} to a minimum ('411): We hope you found that V_{offset} accounts for most of your integrator's drift. Let's now solve that problem. Add the offset trim network of Fig. 7L.2. Ground the integrator's input, zero the output initially, with the reset pushbutton, then try to minimize the drift rate of the output.

The trimming is a fussy task. Start with modest scope sensitivity – say, 1V/div – and gradually increase sensitivity as you adjust V_{offset} down toward zero. Reset the integrator from time to time with the pushbutton. See if you can get the drift rate down to a few millivolts/second. When you've got it about as low as you can manage, see what residual V_{offset} your drift rate implies.

Now, lest you think that trimming is a really-satisfactory remedy for bad V_{OS} specs, heat or cool the '411 (with a heat gun, perhaps, or a can of component cooler), and see how good the drift looks. (In fairness to the '411, we should admit that even the good low V_{offset} chopper, coming soon, doesn't much like big changes of temperature. If you're energetic, you can repeat this test when you've wired up the chopper.)

Use a self-trimming op-amp: "chopper" amp, LTC1150: A "chopper" like the LTC1150 uses a second op-amp to drive a"nulling" input on the main amplifier – a single input that accomplishes what our manual offset-trim did, driving the difference between the two inputs of the main amplifier toward zero. Try this amplifier in place of the '411.

The pinout of this op-amp is the same as for the '411, so you can drop it into the circuit you wired for the '411 – except that *you must remove the trimming potentiometer, so as to leave pins 1 and 5 open.*

Ground the input – at the non-inverting input.[1] Watch the drift rate.

As for the '411, at this point the cause of drift is uncertain. Separate the effects of I_{bias} and V_{offset}, by eliminating the effect of V_{offset}, as you did for the '411. Infer the op-amp's V_{offset} from the drift rate that you saw with input grounded and keep a note of it. The LTC1150 promises V_{OS} of $\pm 0.5\mu V$ (typ), $10\mu V$ (max). We were chagrined to get values slightly *above* the specified maximum. We hope you'll do better![2]

7L.1.3 Apply the integrator: drive-motor position sensor

A person might reasonably ask, after watching an integrator in action, "But why would I want to integrate?" We have contrived one example of a signal that seems to want integrating to begin to answer this legitimate question. We'll admit that we cooked it up because it's kind of fun, rather than because it's useful. But it may set your imagination to conceiving other possible uses for an integrator.

Casting about for a signal whose integral might be interesting, we[3] thought of signals from coils moving in a changing magnetic field. The output voltage should be proportional to the rate at which the flux through the coil changes. We first tried pushing on a big audio speaker – and then thought of using the "voice coil" of a hard disc drive. The voice coil (so-called because the basic design was borrowed from audio speakers) is a simple motor that drives the drive head along the radius of the disc, so as to find or follow a disc track.

If opening up a dead disk drive is too much trouble, you can watch the same effect with the low-budget version shown on the left side of Fig. 7L.3. Those are two cheap little DC motors, to which we glued an arm cut from the plastic rotor at the far left. That rotor came from a hobby "servo motor" (a device that we will meet again in Lab 24L). The smaller motor is a 9V device; the larger is a 3.3–5V device.[4] Almost any DC motor will do, since you can adjust the integrator's sensitivity to your convenience (how?[5]).

We will use the DC motor not as a motor but as a *generator*. Instead of applying a voltage to move the disk-drive's "head" – or to spin the little DC motor – we will move the head by hand to generate

---

[1] We found that the location of our grounding connection on the breadboard caused drift-rate variations of more than a factor of two. The small currents flowing in the ground line seem to account for these microvolt variations at the input.
Connecting directly to the non-inverting input point on the breadboard provides the most honest shorted-input connection: zero in the sense that the op-amp sees zero.
[2] When we figure out what inflated our drift rate, we'll let you know.
[3] Well, "we" means the Royal Paul.
[4] If you want to buy a motor that we have tried, you can get the larger one (a Mabuchi FF-130RH) from many sources. We happened to buy ours from All Electronics, for a little more than $1.
[5] We know you know the answer: by adjusting the value of the integrator's input resistor.

poor man's DC-voltage source...   ...fancy DC-voltage source: disk drive 'voice coil'

Figure 7L.3 Alternative signal sources: DC motors; hard disc, with voice coil that positions drive head.

a voltage proportional to *velocity* of the head, or to the spin rate of the little motor. Integrating this signal, we should get a voltage proportional to the head's *position* (or the motor shaft's).

Query: Moving the head fast, or spinning the shaft fast, produces a large voltage input, moving either one slowly produces a small voltage input. If you move the head quickly from position A to position B – or rotate the motor shaft from angular position C to D – does the *rate* at which you make this movement affect the integrator's output?[6]

In this exercise our goal is to show this position on the scope screen. We don't want the display to wander appreciably, over the span of several seconds. So, we need a low-drift integrator like the two that we have just put together.

Let's apply your integrator – probably the most recent circuit, using the LTC1150, since that should still be wired up – to the signal put out by the disc-drive's voice coil. Watch the input signal (the signal from the two wires brought out from the disc's coil, or from the DC motor) and the integrator output. Try 2V/division on the integrator output. Zero the integrator, and then move the disc head, inward and outward on the radius. We hope you like what you see coming from your integrator. We hope that you find you have built a circuit that displays the head's radial position – or the rotary position of the DC motor's shaft.

If you're not worn out, swap in the '411 for the chopper amplifier, and restore the trimming pot. (If you're lucky, your trim may still be good; if not, adjust it till you get an acceptably slow drift from your integrator, and then apply the disc-drive's signal.) See how the trimmed '411 does at displaying the drive's head position.

7L.2 Differentiator

The circuit in Fig. 7L.4 is an active differentiator. Try driving it with a 1kHz triangle wave.

The differentiator is most impressive when it surprises you. It may surprise you if you apply it to a *sine* from the function generator: you might expect a clean cosine – unless you noticed the contrary result back in Lab 3L. In fact, some generators (notably the Krohn–Hite generators that we prefer in

---

[6] Nope. Quick movement does drive the integrator harder – but for a shorter time than the slower movement (by definition of "quick" and "slower"!). So, the integrator output honestly measures the *position* of head or motor shaft. It does not care about rate of change of that position: it integrates the input voltage, but that voltage is the time-derivative of position – in case you like speaking of simple things in fancy mathematical terms, as we do not.

7L.2 Differentiator

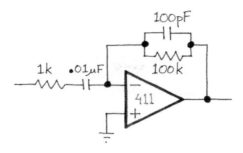

Figure 7L.4 Differentiator, including extra R and C for stability.

our lab) will show you, when their sine is differentiated, a waveform that reveals the purported *sine* to be a splicing of more-or-less straightline segments. This strange shape reflects the curious way the sine is generated: it is a triangle wave with its points whittled off by a ladder of four or five diodes. The diodes cut in at successively higher voltages, rounding it more and more as the triangle approaches its peak, see Fig. 7L.5.

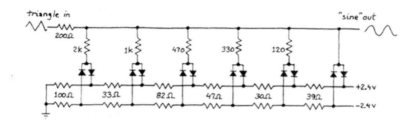

Figure 7L.5 Sketch of standard function-generator technique for generating sine from triangle.

You may even be able to count the diodes revealed by the output of the differentiator – though we've never been able to spot evidence of *all* the diodes.

AoE §4.5.7

A note on stability (preview of work to come): Here we are obliged to mention the difficult topic of stability, a matter treated more fully later in Day 9. Fig. 7L.6 shows a simple differentiator – one with a single R and a single C. It necessarily lives at the edge of instability (sounds like a soap opera, doesn't it), because such a differentiator has a gain that rises at 6dB/octave ($\propto f$). To build the differentiator as simply as in the figure would violate the stability criterion for feedback amplifiers (see AoE §4.9.3.)

In order to circumvent this problem, it is traditional to include both a series input resistor and a capacitor parallel to the feedback resistor, as shown in Fig. 7L.4. This additional R and C convert the differentiator to an *integrator* above some cut-off frequency. You saw a sketch of this frequency-response in Fig. 7N.34.

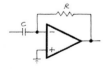

Figure 7L.6 Simple differentiator (unstable).

This compromising of the differentiator's performance is disappointing. You can watch the effect of this network by noting the phase shift between input and output as you gradually raise $f_{\rm in}$ from zero toward the breakover frequency and beyond. At breakover the phase shift should go to zero. At still higher frequencies you should see the phase shift characteristic of an integrator.

Incidentally, a faster op-amp (one with higher $f_{\rm T}$) would perform better: the switchover to integrator must be made, but the faster op-amp allows one to set that switchover point at a higher frequency.

Integrate the derivative? A more intriguing way to see the imperfection of the differentiator is to feed its output to the integrator you built earlier, then compare *original* against *output* waveforms.

Ideally, they would be identical – at least in phase (that is, apart from *gain* artifacts). Are they? Does the answer depend on input frequency?[7]

Sweeping frequencies (optional): You might also enjoy watching the frequency response of the op-amp differentiator, to confirm the pattern predicted in the sketch above: does the circuit show first gain that rises with frequency (differentiator behavior), then gain that falls (integrator behavior)?

7L.3 Slew rate

Begin by measuring slew rate and its effects with the circuit in Fig. 7L.7. We ask you to do this in two stages.

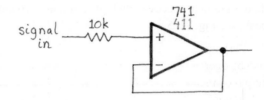

Figure 7L.7 Slew rate measuring circuit. (The series resistor prevents damage if the input is driven beyond the supply voltages).

7L.3.1 Square wave input

Drive the input with a square wave, in the neighborhood of 1kHz, and look at the output with a scope. Measure the slew rate by observing the slope of the transitions. Note that full *slew rate* appears only when the op-amp's input stage is strongly unbalanced. So, if the slew rate seems low, make sure your input amplitude is sufficient.

7L.3.2 Sine input

Switch to a sinewave, amplitude a volt or so, and measure the frequency at which the output waveform begins to distort (this is roughly the frequency at which amplitude begins to drop as well). Is this result consistent with the slew rate that you just measured using a square wave? Watch out for a refinement, here: you will not see the op-amp's full *slew rate* until the op-amp inputs are radically unbalanced. A large square wave achieves this effect easily; a sinewave may not, unless it is both large and quick.

Now go back and make the same pair of measurements (slew rate, and sine at which its effect appears) with an older op-amp: a 741. The 741 claims a "typical" slew rate of 0.5V/μs; the 411 claims 15V/μs. How do these values compare with your measurements?

7L.4 AC amplifier: microphone amplifier

7L.4.1 Single-supply op-amp

In this exercise you will meet a "single-supply" op-amp, used here to allow you to run it from the +5V supply that later will power your computer. This op-amp, the '358 dual (also available as a "quad,"

---

[7] It should. The integrator is honest, but the differentiator differentiates only up to its breakover frequency, which you can calculate from its *R*s and *C*s. Use either pair.

7L.4 AC amplifier: microphone amplifier

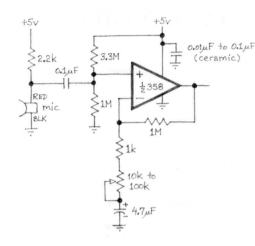

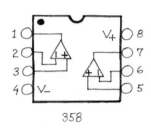

Figure 7L.8 Single-supply microphone amplifier.

the 324), can operate like any other op-amp, with $V_+ = +15V$, $V_- = -15V$. But, unlike the '411, it can also be operated with $V_- = $ GND, since the input operating common-mode range includes V_-, and the output can swing all the way to V_-.

Our application here does not take advantage of that hallmark of a single-supply op-amp, its ability to work right down to its negative supply. Often that is the primary reason to use a single-supply device. The op-amp millivoltmeter described in Worked Example 7W does exploit this capacity of the '358, unlike today's microphone amplifier.

Note Build this circuit on a private single breadboard strip of your own, so that you can save the circuit for later use: it will feed your computer.

In Fig. 7L.8 the '358 is applied to amplify the output of a microphone – a signal of less than 20 mV – to generate output swings of a few volts. The "AC amplifier" configuration, you will notice, is convenient here: it passes the input bias voltage to the output *without amplification* because gain at DC is *unity*.

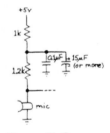

Figure 7L.9 Quieting power supply to microphone.

The microphone is an "electret" type (the sound sensor is capacitive: sound pressure varies the spacing between two plates, thus capacitance; charge is held nearly constant, so V changes with sound pressure according to $Q=CV$). The microphone includes a high-input-impedance field-effect transistor (FET) within the package, to buffer the electret.

The FET's varying output current is converted to an output voltage by the 2.2k pullup resistor. So the output impedance of the microphone is just the value of the pull-up resistor: 2.2k.

Try isolating the power supply of the microphone as in Fig. 7L.9: You may find, after your best efforts, that your amplifier still picks up pulses of a few tens of millivolts at 120Hz. Fig. 7L.10 shows what the pulses look like.

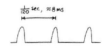

Figure 7L.10 Ground noise on PB503 breadboard: caused by current pulses recharging filter capacitor.

Probably you will have to live with these, unless you want to go get an external power supply (the adjustable supply you used in Lab 1L will do fine here). These pulses show the voltage developed in the ground lines when the power supply filter capacitor is recharged by the peaks of the rectifier output. They *shouldn't* be there, but they are hard to get rid of. They appear because of a poor job of defining ground in powered breadboards like the PB503, and you can't remedy that defect without rewiring the innards of the PB503.

7S Supplementary Notes: Op-Amp Jargon

Bias current (I_{bias}): average of input currents flowing at op-amp's two inputs (inverting, non-inverting).
Frequency compensation Deliberate rolling-off of op-amp gain as frequency rises: used to assure stability of feedback circuits despite dangerously-large phase shifts that occur at high frequencies.
Gain–bandwidth product (f_T): a constant describing gain and frequency-response of an op-amp; it equals the extrapolated frequency where op-amp open-loop gain has fallen to unity.
Hysteresis As applied to Schmitt trigger comparator circuits: the voltage difference between upper and lower thresholds.
Offset current Difference between input currents flowing at op-amp's two inputs (inverting, non-inverting).
Offset voltage (V_{offset} or V_{OS}): op-amp's input stage mismatch voltage: the voltage that one would need to apply, between the inputs, in order to bring the op-amp output to zero.
Open loop Circuit wired without feedback.
Rail-to-rail Said of inputs and outputs of an op-amp. "Rail" is jargon for power supply. Some devices allow rail-to-rail input range, some come close to this output range; some have both capabilities.
Saturation Condition in which op-amp output voltage has reached one of the two ($\pm$) output voltage limits, usually within about 1.5V of the two supplies, but much nearer in rail-to-rail-output devices.
Schmitt trigger Comparator circuit that includes positive feedback.
Single-supply op-amp Device that accepts inputs close to (and sometimes below) its negative supply, which normally is ground. Device also can drive its output close to the negative supply (ground).
Slew rate Maximum rate (dV/dt) at which op-amp output voltage can change (assumes substantial voltage difference between op amp input terminals).
Summing junction Inverting terminal of op-amp when op-amp is wired in inverting configuration: inverting terminal then sums currents. That is, $I_{feedback}$ is algebraic sum of all currents input to the summing junction.
Transresistance amplifier Current-to-voltage converter.
Virtual ground Inverting terminal ($-$) of op-amp when non-inverting terminal is grounded. Feedback tries to hold ($-$) at 0V, and current sent to that terminal does not disappear into "ground," but instead flows through the feedback path (hence the ground is only "*virtual*").

7W Worked Examples: Op-Amps II

7W.1 The problem

Here's a statement of the design task, and some questions:

Design it: Design an integrator, using operational amplifiers, to the following specifications:
- it will ramp at $+1$V/ms when a $+1$V signal is fed to it;
- R_{out} for the signal source is unknown;
- you should include protection against saturation caused by long-term integration of small DC errors.

Questions:

Frequency response: Assume an integrator with a resistor paralleling the feedback capacitor. Approximately what is the *low-frequency* limit of your circuit's response, if we define that limit as the frequency where the response is reduced by approximately 3dB?

Integrator errors: Assume an integrator with reset switch rather than one with a resistor paralleling the feedback capacitor. Given your integrator design, at what rate does its output move (V/second) when the input is in either of two conditions: *grounded* or *open*?

Assume you are using an LF411 op-amp, and use *typical* values.

Case 1: Input grounded (as if fed by a transducer that is a *voltage* source).

Case 2: Input open (as if fed by a transducer that is a *current* source). Note that for this case you should *remove* any input buffer you may have placed before the integrator.

Choose a better op-amp: Now find an op-amp from the small set shown in Fig. 7W.11 that will give around 5× better performance (integrator error $\leq 1/5$ the value you found for the '411 with input *grounded*).

Specify the op-amp, and calculate its resulting dV/dt error (input grounded).

7W.1.1 A solution

A skeleton circuit: We need a standard op-amp integrator, with protection against saturation. Let's use a feedback resistor, $100 \times R_{integrator}$, as usual. We also need two other op-amp circuits:

- an inverter to give the requested sense of output (positive out for positive in);
- an input buffer (a follower) because "R_{out} for the signal source is unknown."

...Component values: For RC our plan to make the feedback resistor $100 \times R_{integration}$ steers us toward an integration resistor (as we're calling the R that feeds the integrator) that is not extremely large. If, as usual, we put a 10M ceiling on our R values, then $R_{feedback} = 10$M and $R_{integration} = 100$k. With an R value chosen, we know what current will flow for the specified 1V input: $I_{full-scale} = 1\text{V}/100\text{k} = 0.01$mA. This full-scale current lets us choose C to be $I/\frac{dV}{dt} = \frac{0.01\text{mA}}{1\text{V/ms}} = 0.01 \times 10^{-6} = 0.01\mu\text{F}$.

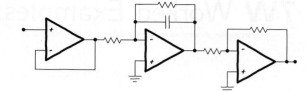

Figure 7W.1 Skeleton circuit for integrator (no component values).

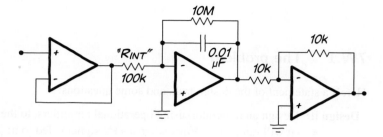

Figure 7W.2 Integrator with component values.

Frequency response: The feedback resistor will reduce V_{out} by 3dB when $|Z_C| = R$.[1] That equality occurs at what we usually think of as "f_{3dB}": $1/(2\pi RC)$. The frequency takes the name "f_{3dB}," as you well know, because there the output of an RC filter (where R and C are in series) is down 3dB.

But we should not think that we are looking at the same case, here. In this circuit, where R and C are in parallel, this frequency would better be called "$f_{\text{crossover}}$": this is the frequency where the magnitudes of currents flowing in R and C are equal. Below this frequency boundary, more of the current flows in the R; above this frequency, more of the current flows in the C. But this frequency is also the one where the integrator "will reduce V_{out} by 3dB" *relative to what it would have been without the feedback resistor*. So, the reference to 3dB remains appropriate.

After that long-winded digression, the calculation is straightforward. The feedback resistor will reduce V_{out} by 3dB at $f = 1/(2\pi RC) \approx 1/(6 \times 10 \times 10^6 \Omega \times 0.01 \times 10^{-6} F) = 1/(6 \times 0.1) = 1.6\text{Hz}$. This is very low, as we would hope. So, at most frequencies, the feedback resistor should not compromise the integration substantially.

Integrator errors (predict output drift rate, with no feedback resistor): Here are typical LF411 values: $V_{\text{offset}} = 0.8\text{mV}$; $I_{\text{bias}} = 50\text{pA}$.

Input grounded: The offset voltage V_{OS} produces an error in both the follower and integrator of Fig. 7W.2. Both errors contribute to the integrator's drift (error in the output inverter does not contribute to drift). Worst case, the two V_{OS} values may be of the same sign, and thus will add.

The *current* that they cause to flow, charging the integrator capacitor – let's call it "I_{error}" – results from this double V_{OS} (worst case) across the integration resistor, R_{int}:

$$I_{\text{error}} = 2 \times V_{\text{OS}}/R_{\text{int}} = 1.6\text{mV}/(100\text{k}\Omega) = 16\text{nA} .$$

This current causes a drift:

$$dV/dt = I/C = 16\text{nA}/(0.01\mu F) = 1.6\text{V/s} . \tag{7W.1}$$

The effect of I_{bias} (at 50pA) is negligible compared with the 16nA current caused by V_{OS}.

---

[1] You may want to take our word for this – or you may want to calculate the value of $Z_C \parallel R$ when $|Z_C| = R$.

7W.1 The problem

Input open: For this case, we are modeling the behavior of an input transducer that is a *current source*. To simulate such a source we remove the follower that buffers the input stage, because a *current source* is happy to see the low input impedance that it finds at the inverting terminal (to put things anthropomorphically).

We also remove R_{int}, which serves no good purpose when the input signal is a current, and in fact can cause mischief. R_{int} simply moves the circuit input voltage away from ground, as currents flow. Such displacement can disturb a current source (one with small "output voltage compliance," a notion you met in Lab 4L).

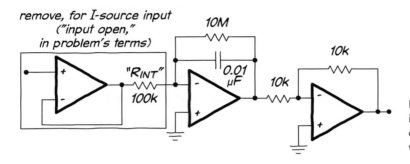

Figure 7W.3 Integrator input would be simplified if one assumed signal source was a *current source*.

Now V_{offset} has no effect on currents (because V_{offset}, which moves the inverting input away from ground, does not alter the magnitude of current arriving from the current-source that drives the integrator). Only I_{bias} causes the integrator output to drift:

$$dV/dt = I/C = 50\text{pA}/(0.01\mu\text{F} = 10\text{nF}) = 5\text{mV/s}\,.$$

This is radically better (about 300 times better) than the result we got for a *voltage* source driving the same integrator. If you have a choice, you will choose a current source as your transducer.

Improve performance with a better op-amp: If your signal comes from a voltage source and you need better performance, you will need to find an op-amp better than the '411. To get a drift rate 5 times better than what the '411 provides – as the problem asks – we need to reduce the drift rate from the 1.6V/s (see equation (7W.1)) to about 0.3V/s.

Since we found that the '411's drift was caused by V_{offset} alone (the effect of Ibias was negligible), we need only shop for an op-amp with V_{offset} no more than 1/5 that for the '411, along with comparably low I_{bias}. We need $V_{\text{offset}} \leq 0.8\text{mV}/5 = 0.16\text{mV} = 160\mu\text{V}$, and low I_{bias}.

The drift rate for this op-amp (assuming, once again, that the integrator sees $2 \times V_{\text{offset}}$ because of the presence of the follower ahead of the integrator) should be

$$\begin{aligned}\frac{dV}{dt} = \frac{I}{C} &= \frac{(2 \times V_{\text{offset}})/R_{\text{integrator}}}{C} \\ &= \frac{0.1\text{mV}/0.1\text{M}\Omega}{0.01\mu\text{F}} \\ &= 1\text{nA}/10\text{nF} = 0.1\text{V/s}\,.\end{aligned}$$

Yes (no surprise): the OPA111B does the job.

Op-amps with much lower V_{offset} are available if you need them. You will meet such a need in the next section. But there's no point choosing an op-amp much better than what we need, so we'll be content with the LF411 for the present task.

7W.2 Op-amp millivoltmeter

Problem Given

- a meter movement: 1mA full-scale, 100Ω,
- op-amp(s) of your choice,
- other parts as needed,

design a voltmeter to the following specs:

- use a single 15V supply, if possible; if you can't manage that, then use split supplies, ±15V
- full-scale (input-) voltage: 10mV
- accuracy 1% of full scale
- input resistance: $\geq 1\text{M}\Omega$
- reading for input grounded or open: 0 (to the usual 1% of full-scale range);

and if you're not exhausted after that – or perhaps along the way – add some desirable features:

- protection for the meter movement (from consequences of overdriving the input)
- really fancy: valid reading for *either* input polarity, along with a polarity-indicating LED

7W.2.1 Solution: millivoltmeter

See AoE Fig. 4.56

A start: amplify...: Comparing input range to the output voltage range that we need to drive the meter movement will let us determine what gain we need. We'll also need to consider what form of amplifier is suitable.

Gain needed: Let's review the problem specifications that bear on gain:

- input voltage range: 0–10mV;
- output voltage needed: 0–100mV, since meter movement drops; 100mV full-scale. It is described as 100Ω and 1mA full-scale

So a gain of 10 will do it.

What amplifier configuration: Should we use inverting or non-inverting? Either could satisfy the input-resistance requirement (1M). Let's see what the circuits would look like.

Non-inverting amp: This is the obvious choice, since this is a single-supply instrument and we want input impedance that is quite high. Fig. 7W.4 shows a sketch of the circuit.

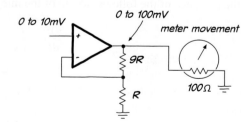

Figure 7W.4 Non-inverting amplifier, to drive meter movement.

The amplifier will need to be a "single-supply" type: its input must understand voltage levels down to zero, and its output must be able to go close to zero. A device like the LM358 that we use in §7L.4's microphone amplifier satisfies these requirements,

7W.2 Op-amp millivoltmeter 315

Current-source version of non-inverting design: It is also possible to apply a *current* rather than *voltage* to the meter movement; see Fig. 7W.5.

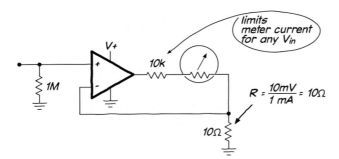

Figure 7W.5 Current, rather than voltage, applied to meter movement.

The circuit of Fig. 7W.5 includes a current-limiting resistor (the 10k), in case the input is overdriven. This protection anticipates that described in §7W.2.2.

Inverting amp: Could the job be done with an inverting amplifier? Yes, but this design is awkward if done with a single supply. It's straightforward if done with split supply; see Fig. 7W.6.

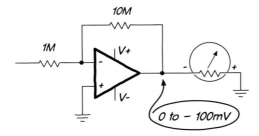

Figure 7W.6 Inverting design: straightforward if we use a split supply.

Use of a single-supply makes the task harder, because the circuit must operate close to its "positive" supply (well, the *more* positive supply – in this case, *ground*); see Fig. 7W.7.

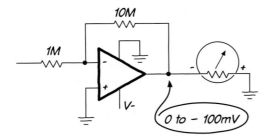

Figure 7W.7 Inverting amplifier, to drive meter movement.

The *inversion* means that the op-amp output will always be negative, so the op-amp will have to be powered in the curious way shown. Input of the op-amp now lives at the *positive* supply rail; the output runs from that rail to 100mV below it. This circumstance rules out an ordinary single-supply op-amp like the '358. We would need a rail-to-rail op-amp, able to understand inputs all the way to the positive supply. This we can find: e.g., ADI's OP282 or AD8531 (the latter's power supply is limited to 6V, a restriction common in CMOS parts; we could use a supply voltage lower than 15V: say, 5V).

The op-amp would also need to be able to take its output right up to the positive supply. Here, the configuration we have set up for ourselves is so odd that the characteristic does not seem to be specified; but probably the devices can sink current close to the positive supply (what is specified is the offset from positive supply while *sourcing* current).

So, it is possible to find an op-amp that fits the configuration – though we'll soon learn that this configuration is ruled out by the specifications of the amps that can handle what we require for this configuration: high-side, rail-to-rail input and output.

The op-amp specs the designs would require: So far, we have shown only how to get a gain of 10. We now need to look at the harder part of the task: getting accurate results.

We are asked to keep errors at $\pm 1\%$. In addition, the meter should read *zero* (to within 1% of full-scale) when the input is *open*. We don't want to annoy the meter's users with a wandering output at times when nothing is connected to the millivoltmeter.

No wandering, with input open: The non-inverting design of Fig. 7W.4, with its astronomical R_{in}, would wander, if we did not add a resistor as shown in Fig. 7W.8. After a time, the input, and therefore output, would drift all the way to an extreme, pinning the meter movement at full-scale. To prevent this result, we must include a resistor to ground, as shown in Fig. 7W.8.

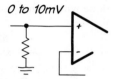

Figure 7W.8 A resistor to ground prevents drift, in the non-inverting design.

We chose the resistor value to satisfy the original problem specification, $R_{in} \geq 1\text{M}\Omega$. The inverting design already includes this taming resistor; no change is needed in order to prevent that configuration from wandering.

What magnitude of input error is tolerable? We are to keep errors under 1%. Then we can tolerate $V_{\text{error,input}} = 0.01 \times V_{\text{full-scale}} = 0.1\text{mV} = 100\mu\text{V}$. This value we might call our "error budget."

What V_{offset} is tolerable? This total *DC* error will result from the sum of two effects: V_{offset} and I_{bias}. Since these two effects may have the same sign, we will allow each to use up only half the total error budget: we will allow $50\mu\text{V}$ apiece.[2]

V_{offset}: We need to shop for an op-amp with V_{offset} around $50\mu\text{V}$. That value is low – about 15 times smaller than what the '411 shows us. It is attainable. But we should not expect to find it in an op-amp that has been designed to optimize some other feature.

Hunting for a satisfactory op-amp, we find that this V_{offset} goal knocks out of contention the single-supply rail-to-rail op-amps required by the inverting design above. The two parts that could handle inputs up to the positive rail showed V_{offset}'s of 3mV and 25mV, respectively. So, let's abandon the inverting design – which always seemed a bit perverse, anyway.

$V_{\text{offset}} \leq 50\mu\text{V}$? Only a very few parts satisfy this specification, in the collection of "candidate op-amps" (see Fig. 7W.11):

- the LT1012A (at $25\mu\text{V}$ max);

---

[2] Usually, one would assign budget fractions after seeing what dominates, rather than assume at the outset that both constraints are equally difficult.

7W.2 Op-amp millivoltmeter

- the MAX420E ($5\mu V$, max) – a "chopper" – the sort that continually readjusts its offset to zero it, many times each second.
- LMP2014MT ($5\mu V$) (another chopper) ; for this one we'd have to drop V_+ from 15V to 5V.

Two other parts comes close: OPA627b at $100\mu V$ max, and OPA336, at $125\mu V$ max – slightly exceeding our error budget. The OPA336 *typical* value of $60\mu V$ would be acceptable. We will not here go into the hard question of when it might be all right to use something other than worst-case specifications.[3] We might keep these two in mind then.

If we are to do the task single-supply, then only the LMP2015 will do.

I_{bias}: The bias current flows in the 1M input resistor, generating an error voltage. Again, we'll assume that we can use up just half the error budget with this error: $50\mu V$. The bias current that we can tolerate, then, is

$$I_{bias} \leq 50\mu V/1M\Omega = 50pA .$$

This is a small value, but certainly available (even the humble LF411 comes close, showing 50pA as its typical I_{bias}). The challenge will be to find a part that delivers this good specification while also giving low V_{offset}. If we mean to do the task single-supply, then we need also the peculiar virtues of an amp that can handle inputs and outputs close to ground.

Only the choppers keep the total error under $100\mu V$. The OPA627b at $100\mu V$ is close enough, with its 5pA I_{bias} (max), if we mean to use a split supply. The LMP2015 is the only part that does it all: keeps the input error under $100\mu V$ and allows use of a single supply.

7W.2.2 Refinements

Protect meter movement against overdrive: The designs of Figs. 7W.4 and 7W.6 could drive the meter movement all the way to the 15V power supply. The op-amp's output current, typically limited to a few tens of milliamps, might not damage the movement. But it is easy to modify the circuit, just slightly, to protect the meter movement. Simply putting a series resistor ahead of the movement and then boosting the amplifier's gain makes sure that the movement could not be driven to more than 150% of full-scale; see Fig. 7W.9.

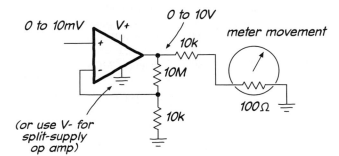

Figure 7W.9 Slight change to limit overdrive of meter movement.

---

[3] A knowledgeable friend of ours who designed digital circuits for a large computer maker told us that his company would have been unable to compete if they had not assumed that parts in their designs would work somewhat better than their worst-case, full-temperature-range specifications.

Allow both positive and negative inputs: This change requires use of a split supply. If the meter movement is of one polarity only,[4] then a four-diode *bridge* (familiar to you from power supplies) will put current through the movement in the forward direction regardless of the sign that reaches the bridge. The design of Fig. 7W.10 uses a *current-source* scheme (like that of Fig. 4N.3.1). Letting the circuit set a *current* rather than a voltage makes the design indifferent to voltage drops across the bridge diodes. The indicator LEDs are necessary to indicate input polarity.

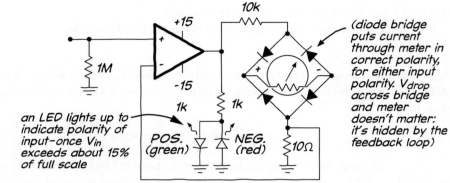

Figure 7W.10 A design permitting both positive and negative inputs.

an LED lights up to indicate polarity of input—once V_{in} exceeds about 15% of full scale

7W.2.3 The point of this exercise...

This exercise was fussy. What we wanted to show was that a very simple task becomes challenging if the design can tolerate only very small errors. We hope this example will help to give you a feel for why such an amazing variety of op-amps is available – with new ones continually arriving.

Candidate op-amps: All but one of those listed in Fig. 7W.11 are unsuitable for single-supply work. If you designed the meter with a split supply, all are eligible (though the single-supply part is limited to +5V total voltage).

	V_{os} (μV)		TCV_{os} (μV/°C)		I_b (pA)		Comments
	typ	max	typ	max	typ	max	
LM741C	2000	6000	–	15*	80,000	500,000	BJT, old, crummy
LF411	800	2000	7	20	50	200	JFET jellybean
OPA111B	50	250	0.5	1	0.5	1	precision FET
OPA627B	40	100	0.4	0.8	1	5	low-noise prec JFET
LT102A	8	25	0.2	0.6	25	100	low-bias BJT, prec.
OPA336	60	125	1.5	–	1	10	CMOS, V_s=5.5V max
MAX420E	1	5	0.02	0.05	10	30	chopper CMOS, V_s to $\pm$15V
Single Supply							
LMP2015	0.8	5	0.015	0.05	3	–	chopper, CMOS, 5V

* for "premium" grade only, otherwise not specified

Figure 7W.11 Candidate op-amps; most are eligible for split-supply only.

---

[4] Some movements are bipolar, starting with the needle at the center of its range of movement.

8N Op-Amps III: Nice Positive Feedback

Contents

8N.1	**Useful positive feedback**	**319**
8N.2	**Comparators**	**320**
	8N.2.1 What's a comparator?	320
	8N.2.2 Trouble with noise	321
	8N.2.3 Stabilizing with positive feedback	324
	8N.2.4 A good comparator circuit *always* uses positive feedback	324
	8N.2.5 An alternative way to provide hysteresis: AC feedback only	326
8N.3	***RC* relaxation oscillator**	**327**
	8N.3.1 Op-amp relaxation oscillator	327
	8N.3.2 The numbers...	328
	8N.3.3 A lazier way to build an *RC* oscillator	328
	8N.3.4 '555 *RC* oscillator/timer	329
	8N.3.5 Newer IC oscillators	330
8N.4	**Sine oscillator: Wien bridge**	**331**
	8N.4.1 Wien bridge's positive feedback network	331
	8N.4.2 Wien bridge's negative feedback network	331
8N.5	**AoE Reading**	**335**

Why?

Some effects we would like to achieve with today's circuits:

- make a decisive comparison circuit;
- make oscillators;

We will apply positive feedback for these purposes.

8N.1 Useful positive feedback

To this point we have treated positive feedback as a bad thing – as a result that you get when your pen slips, interchanging senses of feedback.

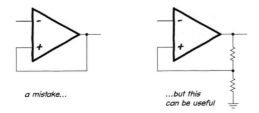

Figure 8N.1 Positive feedback isn't always a mistake.

Op-Amps III: Nice Positive Feedback

It turns out, however, that although negative feedback remains overwhelmingly the more powerful idea, with more applications, positive feedback can be useful. In this chapter we look at cases where it makes circuits decisive, and then at cases where it allows us to form *oscillators*. In Day 9 we will look again at positive feedback, but this time less cheerfully. We will meet some of the many instances in which positive feedback sneaks up on us and makes a well-behaved circuit begin to act crazy. Today, though, all our positive feedback is benign.

8N.2 Comparators

8N.2.1 What's a comparator?

AoE §4.3.2A

A comparator is just a high-gain differential amplifier: such an amplifier "compares" its two inputs, though we have not described its performance that way till now. What's distinctive about the *comparator* is not so much the *device* as the use to which it is put. An op-amp can serve as a comparator, but – as you will confirm in the lab – an op-amp makes a second-rate comparator: it is slow. A chip designed for the narrow purpose of *comparing*, a special-purpose comparator like the '311, works almost 100× faster. And other comparators of more recent design are available with speeds nearly 100 times better than the '311s.

Comparator versus op-amp: Although an op-amp can sometimes do the job of a comparator, the contrary is not true: a comparator cannot serve in an op-amp's *negative-feedback* circuit. A comparator in such a circuit would be unstable. You would likely find its output oscillating at a high frequency. The op-amp has been tamed so that it's stable in a negative feedback loop. A comparator is not so restrained; it is designed for speed, and since it does not expect negative feedback, it is not designed to work with it.

Many comparators, like the LM311 that you will meet today, differ from an op-amp also in offering a versatile output stage.[1] This output stage is not a push–pull, as in an op-amp, but a *switch*.[2] In the '311, both emitter and collector are brought out for the user to connect; see Fig. 8N.2.

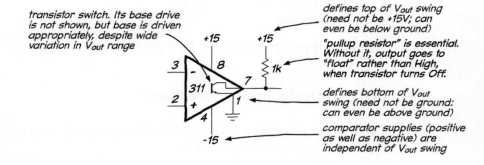

Figure 8N.2 '311 comparator, showing some differences from op-amp features.

A student may find the need to take care of emitter and collector a nuisance, at first (it's one more thing to think about – or two more). But comparators are designed this way for a good reason: to allow the user to determine both the top and bottom of the output swing. So, for example, the '311 makes

---

[1] Some comparators do use a push–pull switch in the output stage, for the greater speed permitted by its low output impedance. See, for example, Linear Technology's 4ns comparator, LT1715.
[2] Strictly, it doesn't have to be used as a switch; it can be wired as a follower, by placing a resistor on the emitter rather than collector. We will not try this variant, whose virtue is its lower output impedance.

it easy to provide a traditional "logic-level" swing of 0V to 5V, rather than an op-amp-like swing of ±15V. Incidentally, the diagram shows the output transistor's *base* just hanging, but of course it doesn't just hang; the IC's designers took care to make it work properly with a wide range of output voltages – and independent of the power supplies for the differential amp that is the guts of the comparator.

Trying comparator versus op-amp: When a comparison task is not demanding, the op-amp and comparator perform (if you'll forgive the word) *comparably*. Fig. 8N.3 shows the responses of a '411 and a '311,[3] when both are fed a large, low-frequency sinusoid. The comparator swing is quicker, but both comparator and op-amp do the job.

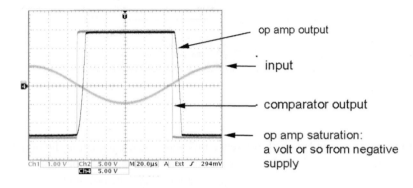

Figure 8N.3 Both op-amp and comparator can handle an easy task: large, low-frequency input. (Scope gains: input, 1V/div; outputs, 5V/div.)

Given higher-frequency input, the comparator keeps doing its job while the op-amp is left behind, see Fig. 8N.4.

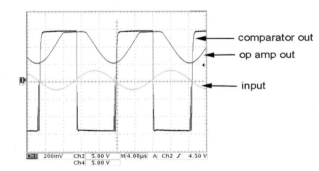

Figure 8N.4 Op-amp falls behind, with input at higher frequency. (Scope gains: input, 100mV/div; outputs, 5V/div.)

The output of the op-amp fed a high frequency sinusoid no longer looks like a square wave; it isn't quick enough even to swing its output as far as ground. The comparator output is square – but it is showing some signs of potential trouble, too. Both edges of the comparator output – rising and falling – show multiple transitions where there ought to be one. We'll see lots more of this misbehavior in a minute.

8N.2.2 Trouble with noise

We have sketched in Fig. 8N.5 what you might see if you fed these two simple comparators – implemented with '411 and '311 – a noisy input like the one shown. Assume that you want the comparator

---

[3] The similarity of part numbers is not significant.

to switch on the big, slow waveform's "zero-crossing." You do not want to switch on the little wiggles. (To make this hypothesis plausible, you might imagine that the slow waveform is 60Hz, and our goal is to let a computer tell time by counting the 60Hz zero-crossings.)

Compare AoE
Figs. 4.31 and 4.33

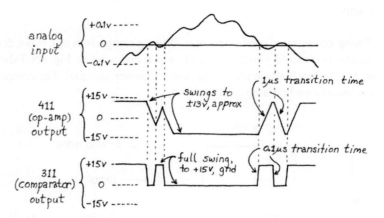

Figure 8N.5 A comparator without feedback will misbehave; the op-amp is *slow* as well.

Try it out: If you were to try making a 60Hz clock reference this way – say, attenuating the output of a "line" transformer – you might get trouble. In Fig. 8N.6, we used a function generator rather than transformer to drive the comparator. Both images shows a comparator output – the bottom trace – that is pretty messy. Evidently, if this '311 output were our time reference, our clock would run fast!

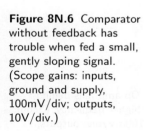

Figure 8N.6 Comparator without feedback has trouble when fed a small, gently sloping signal. (Scope gains: inputs, ground and supply, 100mV/div; outputs, 10V/div.)

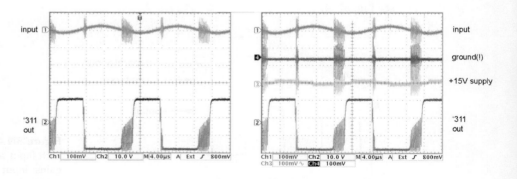

Our hand-drawn figure, Fig. 8N.5, and the scope images in Fig. 8N.6, might lead you to think that the trouble comes from a defective input. After all, you can see wiggles on the input; maybe all we need is a better signal source. Or maybe we should low-pass filter the input to get rid of that fuzz? Plausible, but wrong.

The flaw in this view is that the causation runs backwards: in fact, it is the *output* activity that causes the noise on the *input*. The right-hand image in Fig. 8N.6 adds two significant traces: *ground* and the *positive supply*, +15V. With ground itself shaking, even a glassy-smooth input waveform will seem, to the comparator circuit, to be shaking. Given the noisy ground and power supplies (we haven't shown the negative supply but it looks similar), it's not surprising that the comparator keeps changing its mind.

The case reminds me of my inglorious days as a grade-school outfielder. A fly ball would rise into the blue sky smoothly enough – but as it began to descend, and earnest outfielder began to run to meet it, the baseball would begin to misbehave. It would start to dance up and down, making itself hard to catch: same problem as for the '311:

Figure 8N.7 Baseball dances the way comparator input dances – once the outfielder's reference frame starts jumping.

If we look closely at the *input* to a chattering comparator circuit, we see an effect a lot like the dancing baseball. The function generator's sinewave seems to get fuzzy at the worst possible time, just as the input crosses the comparator's threshold (in this case, ground). The two images in Fig. 8N.8 show that fuzz at two different time scales.

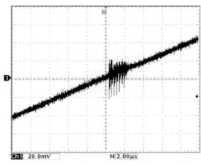

input begins to shiver, just at the wrong time: as it crosses switching threshold. Bad luck?

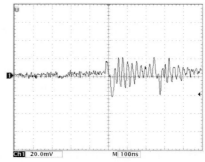

shivering input looks like the baseball that bounced in the air, as the fellow ran

Figure 8N.8 Shiver of input signal, just at threshold-crossing.

But this is not "bad luck," or the result of a perverse and defective function generator. The seeming-shiver is caused by the shuddering of *ground*. The function generator is not at fault (nor was the baseball actually hopping up and down as it descended).

The noise on the ground and supply lines of the '311 results from the intermittent surges of current that occur each time the output begins to switch. The output switches, thus sinking or sourcing a surge of current through supply and ground. This surge (in the inductance of the lines – more important than resistance) causes supply and ground to wiggle. That wiggle, felt by the high-gain differential input makes the comparator change its mind and decide not to switch. But shutting off the current disturbs the lines again, causing the switching to resume; and so on.

If, during these disturbances, the input signal remains close to the switching threshold, the circuit will switch repeatedly, and we will get the resulting craziness. You can solve this problem however. The standard solution is to provide some *positive* feedback.

8N.2.3 Stabilizing with positive feedback

Feedback variations: As we observed back in §8N.1, we need to broaden our view of positive feedback. Positive feedback is not just the feedback you get when you're confused. The three circuits in Fig. 8N.9, may begin to ease you toward a more nuanced view. One of the three case is a silly error – an utterly useless circuit; another is a second-rate, but not useless, circuit; and one is quite healthy.

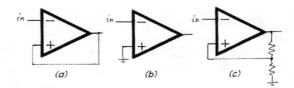

Figure 8N.9 Three ways to wire a comparator: only one is healthy.

Of the circuits in Fig. 8N.9, circuit (a) is the one that is doomed to live at saturation. If decisiveness is the mark of a good comparator circuit, then this *decider* carries the ideal of decisiveness too far; it is downright fanatical – and useless. It is almost impossible to get this circuit to change its mind.

Circuit (b), which uses no feedback, either positive or negative, is not necessarily useless, but is dangerous, as we have just seen. It can handle only the *easy* comparisons where the input moves rapidly across the threshold. The circuit is very likely to oscillate when fed a gently-sloping input.

Circuit (c) shows the usual, healthy comparator circuit. It applies *some* positive feedback to achieve decisiveness (our diagram of course hedges on the question *how much* positive feedback is used here). All our practical comparator circuits will use this tactic.

8N.2.4 A good comparator circuit *always* uses positive feedback

It turns out to be easy to make a comparator circuit ignore small wiggles like those shown causing mischief in Figs. 8N.5 and 8N.6. Just feed back a small fraction of the output swing, and – here is the novelty – feed it back in a "positive" sense, so that the output swing tends to confirm the comparator's tentative decision.

Such *positive feedback* makes the comparator decisive: the comparator uses positive feedback to pat itself on the back, saying (in the manner of many of us humans), "Whatever you've decided to do must be right." The act of switching tends to reinforce the decision to switch.

A comparator circuit that includes positive feedback is called a Schmitt trigger.[4]

Figure 8N.10 sketches the argument that a little hysteresis should let a comparator ignore small disturbances on the input.

> Again compare AoE Fig. 4.33

Fig. 8N.11 shows what happened when we tried applying this remedy to the gently sloping sinusoid that produced so much fuzz back in Fig. 8N.6. We have fed a small fraction of the output back to the non-inverting input (so, the feedback is, indeed, *positive*). Alongside the traces of the stabilized circuit we've included the earlier no-feedback scope image for comparison. A very little hysteresis does the job, in this case: about 20mV is enough.

Hysteresis is familiar in non-electronic life: Hysteresis[5] and applications of positive feedback may be unfamiliar to you in electronics – but the notion is familiar to you in the rest of life. It is a form of dishonesty: changing the question after you decide. You agonize, for example, over whether to go

---

[4] Named after Otto H. Schmitt, an American graduate student who described such a circuit, made with vacuum tubes, in 1934. Does this humble circuit deserve to have one person's name attached to it?

[5] From the Greek for "falling short." Nothing to do with hysteria.

8N.2 Comparators

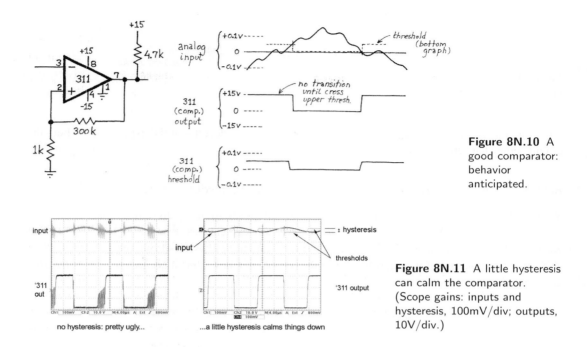

Figure 8N.10 A good comparator: behavior anticipated.

Figure 8N.11 A little hysteresis can calm the comparator. (Scope gains: inputs and hysteresis, 100mV/div; outputs, 10V/div.)

to Harvard or Stanford; after you decide on Harvard, you tell yourself, "Of course. All that sunshine gets to be so tedious."

Strong hysteresis appears in college admissions from the college's point of view, too: when the college considers your application they ask themselves, "Is his oboe playing really exceptional? – or is he just another valedictorian with perfect SATs?" Once you're in, the question whether to expel you might be, "Was your drug bust a felony, or just another misdemeanor?" The standards change, because admissions committees like to tell themselves that whatever they decided must be correct – just as stable comparators do.

Getting elected to congress versus getting thrown out shows the same pattern: once you're in, you stay. Positive feedback makes a person feel good; makes him decisive and self-satisfied; negative feedback – the inclination to think the decision must be wrong because it's *my* decision – makes a person indecisive and, at an extreme, neurotic. Positive feedback can make you a contented dictator; lack of it can make you a Woody Allen.

How much hysteresis? If hysteresis makes the comparator decisive, how much should you design in? As much as possible? What do you lose as you enlarge hysteresis? Hysteresis is effective medicine, but you should apply no more of it than you need because of its side effects; see Fig. 8N.12. It provides immunity to *noise* – blindness to disturbances on the input, up to some magnitude; but this same immunity is also blindness to *signal*. So a zero-crossing detector with symmetric hysteresis of 200mV will not switch at all for an input whose peak-to-peak swing is less than 200mV.

Often you will be obliged to guess at how much hysteresis you need. In the lab circuits we build, 100–200mV usually is enough. If you are building a single, custom circuit, you can make hysteresis adjustable, as we do in lab, and find by experiment what level of hysteresis is required for stability.

A second side-effect of hysteresis sometimes matters: delay. So, even for an input signal large enough to switch the comparator, the switching will come *late*: the circuit will not be a true *zero-*

Figure 8N.12 Relation between noise and hysteresis.

crossing detector. The trick described next can get around that defect – at the expense of making the circuit somewhat sensitive to input frequency.

8N.2.5 An alternative way to provide hysteresis: AC feedback only

We can eliminate the *delay* that hysteresis normally imposes by using hysteresis that works only for a short time just after a transition: AC-coupled feedback achieves this, see Fig 8N.13. The two zero-crossing detectors shown below provide the same amount of hysteresis, but the AC-coupled one shifts the threshold only for a short time after the output switches. Thus it accurately detects the zero crossing, without delay.

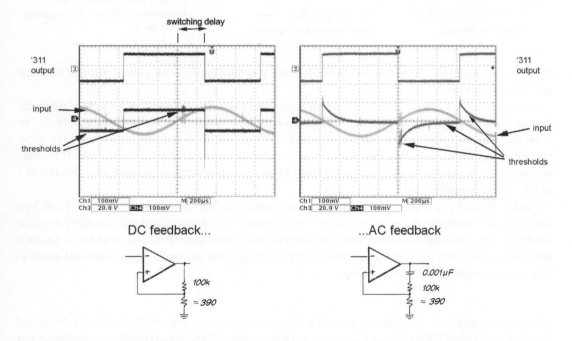

Figure 8N.13 AC-coupled feedback can eliminate the delay that DC hysteresis imposes. (Scope gains: inputs and thresholds, 100mV/div; outputs, 20V/div.)

This circuit would fail, chattering, if the input signal were very slow and gently sloped, so that it was still close to threshold at a time when the *RC* had decayed away, eliminating the threshold shift.

Oscillators: Curiously, we seem to be backing into a discussion of oscillators, having met some nasty, accidental repetitive switching before we ever tried to build an oscillator. The multiple transitions or fuzz produced by a comparator circuit that lacks hysteresis (for example, Fig. 8N.6) are, indeed, oscillations. Now, we want to learn how to produce such waveforms, but in a form that is controlled and predictable.

It's pretty obvious that a circuit that produces *purposeful, controlled oscillations* can be useful. Mankind has been using such things ever since starting to measure time with the help of a pendulum rather than a sundial or water clock.

Conceptually, an oscillator is something that keeps talking to itself – and contradicting itself. But the contradiction must include a delay, as well. The circuit on the left in Fig. 8N.14, contains insufficient delay; it talks to itself, but it is so quick to contradict itself that it barely quivers around the switching threshold (we concede that the quiver is an oscillation; but we'd like something larger). The circuit on the right does oscillate with a full swing, having been given extra delay, a delay provided by a couple of inverters placed in series.

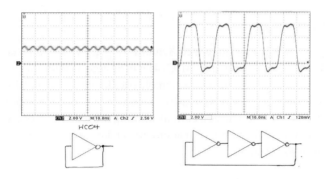

Figure 8N.14 Self-contradiction, with sufficient delay, brings oscillation.

The inverting amplifiers shown are *logic* gates, essentially single-supply comparators, with their switching thresholds placed at one half the supply voltage.

8N.3 RC relaxation oscillator

8N.3.1 Op-amp relaxation oscillator

Figure 8N.15 gives the lab circuit, which you recognize as just a Schmitt trigger like the one in Fig. 8N.10 (here, given unusually large hysteresis) – but this Schmitt trigger is wired in a new way: it is feeding *itself*:

AoE §7.1.2A

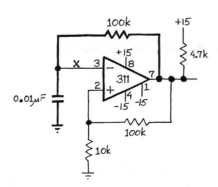

Figure 8N.15 *RC* relaxation oscillator.

8N.3.2 The numbers...

What are the thresholds? How long does V_{CAP} take to move between the thresholds?

If you're inclined to answer this question by using $e^{t/RC}$ to calculate the time for the RC formed by 100k and $0.1\mu F$ to charge between one threshold and the other (about $1/11 \times |V_{PositiveorNegative:15V}| \approx \pm 1.4V$), you're not wrong. But you are working too hard.

The excursion of approximately 2.8V in a span of about 16V is a small enough fraction so that the current is nearly constant.[6] That makes the problem amenable to the easier formula $I = CdV/dt$, with I the midpoint value, 15V/100k. As we say in Lab 8L, the answer obtained with this shortcut is within about 1% of the more exact answer that emerges from the laborious calculation of the exponential charging time. We further simplify the problem by ignoring the effect of the pullup resistor, present on charge-up, but absent from discharge-down. We can afford that 5% error on one swing.

8N.3.3 A lazier way to build an *RC* oscillator

When you want an oscillator in a hurry, you can plug an integrated Schmitt trigger inverter in place of the comparator circuit of Fig. 8N.3. This logic gate (called a 74HC14) has hysteresis built in, and thresholds at approximately half the supply voltage, as for the earlier simple oscillators of Fig. 8N.14. Fig. 8N.16 shows such an oscillator in two forms: one, on the left, uses a resistor as in the comparator relaxation oscillator of Fig. 8N.15. The circuit on the right uses a current-limiting diode in place of the resistor to produce a *sawtooth* waveform. Note that the output of that circuit is taken at the capacitor, *not* at the output of the gate, as for a square-wave oscillator.

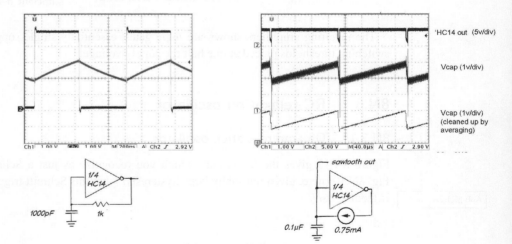

Figure 8N.16 *RC* relaxation oscillator made with integrated Schmitt inverter rather than op-amp.

The weakness of this simple oscillator is that its frequency is not very predictable. Here are the specifications, taken from Lab 8L:

- threshold voltages:
  - negative-going: 1.8V (typical), 0.9V (min), 2.2V (max)
  - positive-going: 2.7V (typical), 2.0V (min), 3.15V (max)
- hysteresis: 0.9V (typical), 0.4V (min), 1.4V (max)

---

[6] The 16V is the maximum voltage difference between the capacitor's starting voltage – about 1.4V positive or negative – and the voltage at the other end of the 100k feedback resistor – about 15V positive or negative. We are not describing the *output* voltage swing of the comparator.

8N.3 RC relaxation oscillator

Which of these matters to determine f_{OSC}?[7] An oscillator built with the HC14 is not highly predictable, but it sure is simple!

The op-amp "relaxation oscillator" of Fig. 8N.15 is a classic, and sometimes provides the best source of a square wave at moderate frequencies.[8] And the relaxation oscillator made with a logic gate provides an easier way to get an approximately equivalent result. But often the easiest method of all is to use an IC dedicated to the task of implementing oscillators.

8N.3.4 '555 RC oscillator/timer

AoE §7.1.3

The '555 is a peculiar, not-particularly-elegant square-wave oscillator that is even older than the '741 (the '555 was born in 1971) – but it may be the most popular IC ever produced.[9] Its dominance is fading, but it is very widely sourced, and one still can find small books of '555 application notes. In the lab you will use an improved version: a part labeled either "7555" or "...C555," or "...555C," made of CMOS MOSFET's[10] for low power, and capable of running faster than the original, bipolar device.

Because the '555[11] contains a *flip-flop*, a device you have not met, it's a little hard to explain. We'll try, though: Fig. 8N.17 shows a diagram of its insides, and two ways to wire the part.

The left-hand figure shows the output driving the timing capacitor, exactly as in the op-amp relaxation oscillator of Fig. 8N.15 (and as in the logic-gate version of Fig. 8N.16). We hope that this scheme is beginning to look familiar.

Compare AoE Fig. 7.8 with Fig. 7.9

The right-hand circuit of Fig. 8N.17 shows the traditional '555 wiring – a bit strange. This scheme does not connect the timing *RC* to the output at all, relying instead on a dedicated *discharge* transistor to discharge the capacitor as needed. This arrangement makes an already-ornate circuit still more complex; but it has the virtue of isolating the timing circuitry from any effect of a *load*. The left-hand circuit of Fig. 8N.17, in contrast, would change frequency if a load attenuated its V_{out}.

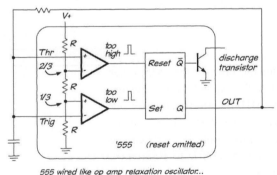
555 wired like op amp relaxation oscillator...

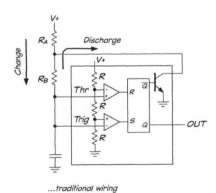
...traditional wiring

Figure 8N.17 '555 oscillator.

[7] The magnitude of hysteresis matters. Just where this span is placed – whether it goes from 0.9V to 1.8V or from 1.8V to 2.7V – does not matter.

[8] One of our heroes, the late Bob Pease, formerly of National Semiconductor, used this design, for example, in order to make a super-low-power 1Hz oscillator. His circuit is a single-supply version of the relaxation oscillator of §§8N.3.1 and 8N.3.3, using a low-power comparator and drawing 1.4µA at 6V.
http://electronicdesign.com/analog/whats-all-oscillator-stuff-anyhow.

[9] According to its designer, it was the top selling IC for around 30 years, and by 2004 had sold more than a billion units. See
http://semiconductormuseum.com/Transistors/LectureHall/Camenzind/Camenzind_Index.htm.

[10] Oof! That's a mouthful of acronyms, is it not? Spelling it out, though, is even worse: "CMOS" is Complementary Metal Oxide Semiconductor; "MOSFET" is Metal Oxide Semiconductor Field Effect Transistor.

[11] We'll call the chip '555 even though the version you will use is CMOS and may like to be called *7555*; for the purpose of this explanation of its operation, the version of the '555 does not matter at all.

330 Op-Amps III: Nice Positive Feedback

As in the op-amp relaxation oscillator, the capacitor voltage here moves between two thresholds, always frustrated. The relaxation oscillators of §§8N.3.1 and 8N.3.3 achieved two thresholds by the use of hysteresis. The '555 uses a simpler scheme: two comparators, each with a fixed threshold (and no hysteresis, incidentally).

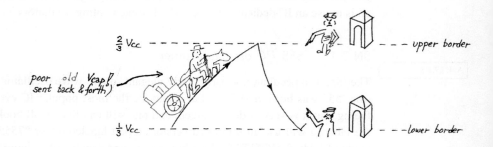

Figure 8N.18 A very informal explanation of the '555's operation.

The two comparators act like unfriendly border guards at the "$\frac{1}{3}V_{CC}$, $\frac{2}{3}V_{CC}$ boundaries: when the capacitor voltage crosses either frontier,[12] it finds itself sent back toward the other – like some sad, stateless refugee. In the traditional wiring shown on the right side of Fig. 8N.17, the border guards send the capacitor voltage up and down by turning *on* or *off* a *discharge* transistor: this transistor sinks charge out of the cap, or – when off – allows the capacitor to charge up toward the positive supply.

You can vary the '555's waveform *at the capacitor* by replacing the resistors R_A and R_B with other elements: a current source or two, or – for R_B, but not R_A – even a piece of wire. Can you picture the waveforms that result – at the capacitor, not at the '555 "output" terminal – for the several possible configurations?[13]

The '555 has other applications, too: it can form a *voltage-controlled* oscillator. We invite you to try this in Chapter 8L, and you will use this technique in the final analog project. The '555 also can be wired to put out a single pulse when triggered (behaving like a so-called "one-shot"). And with some cleverness you can tease it into doing a great many other tricks.

8N.3.5 Newer IC oscillators

AoE §7.1.4A

You will not be surprised to hear that by now, about forty years after the birth of the '555, other IC oscillators have appeared. Some are just easier to wire: the LTC6906, for example, integrates the capacitor, so that a single resistor sets f_{osc} (frequency range 10kHz to 1MHz). Unfortunately for our purposes, this part is not offered in a "through-hole" DIP package. Many newer parts run at voltages lower than the '555's minimum of 4.5V. Some can be programmed from a microcontroller (an extreme case is one from Dallas/Maxim that lives in a 3-pin T0-92 package: DS1065 (30kHz to 100MHz)). Lots of alternatives are available; but the '555 remains a useful workhorse.

---

[12] Would it be too silly to think of the upper border voltage as the frontier of Upper Volta?

[13] Of course you can. So can we: using a *wire* rather than a resistor in the discharge path produces a quick sharp discharge, while the slow RC-curvy charge persists. One disgruntled student complained that he didn't see anything useful about such a "shark-fin" waveform. He had a point.

The same circuit with *current source* replacing the charging resistor, while keeping the wire in the discharge path produces the quite-respectable waveform called *sawtooth*. These are widely useful: analog pulse width modulation (PWM), demonstrated in §8L.4, applies such a sawtooth. The digital equivalent is demonstrated in §22L.2.7.

If you put current sources in both charge and discharge paths, the waveform on the capacitor becomes a *triangle*. See AoE §7.1.3E.

8N.4 Sine oscillator: Wien bridge

AoE §7.1.5B

This oscillator – the only sine generator you will build in this course – is clever, and fun to analyze. How does it put out a sine, rather than the usual square or triangle waves (those are the waveforms that are easy to generate)? Its clever gain control prevents the *clipping* that would produce the usual square output. Its positive feedback network favors just one frequency above all others. Only this "fittest" frequency survives – and if only one frequency survives, this single surviving frequency will, by definition, be a pure sinusoid. This circuit, like many oscillators, uses both positive and negative feedback. For the purpose of analysis, we have separated these two feedback senses below, so that we can consider them separately.

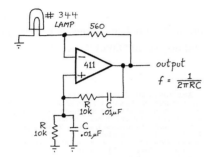

Figure 8N.19 Wien bridge sine oscillator.

8N.4.1 Wien bridge's positive feedback network

The positive feedback network is frequency selective, and at the most favored frequency the network passes back to the + input the maximum, 1/3 of the output swing. It treats all other frequencies – above and below its favored frequency – less kindly. Fig. 8N.20 sketches the behavior.

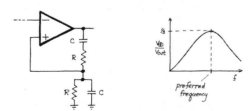

Figure 8N.20 Wien bridge positive feedback: fraction of output fed back.

At that favored frequency, the phase shift also goes to zero. The preferred frequency – the one at which the oscillator will run – turns out to be $1/(2\pi RC)$.[14]

8N.4.2 Wien bridge's negative feedback network

The negative feedback – redrawn in Fig. 8N.21 to look more familiar – adjusts the gain, exploiting the lamp's current-dependent resistance (the lamp is rated at 14mA at 10V; this suggests an approximate

---

[14] Yes, this seems almost too good to be true. We won't prove the point to you. To prove that this frequency gives the maximum feedback signal is a messy exercise in phase-sensitive calculus. But you can pretty readily confirm, at least, that at this frequency the impedance of the series RC is $\sqrt{2}R$, and the impedance of the parallel RC is half that. Thus 1/3 is fed back.

value for its *resistance* – but the key to the circuit's cleverness is that the *R* value is not a constant). The use of the lamp was the key contribution of Hewlett to the design of a low-distortion sine generator. His patent application of 1942 describes the elegantly simple scheme.[15]

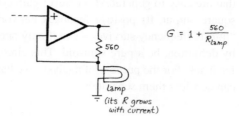

Figure 8N.21 Self-adjusting gain: negative feedback of Wien bridge, redrawn.

He sounds almost embarrassed by the simplicity of his solution ("a small incandescent lamp, or similar device..."). But simple is neat – with this invention he and his friend David Packard launched a company.

Figure 8N.22 Excerpt from Hewlett's patent on use of a lamp to regulate output amplitude.

> Amplitude control to prevent the oscillations from building up to such a large value that distortion occurs, is obtained according to my invention by non-linear action in the amplifier circuit. In order to produce this non-linear variation, I provide for resistance R_3 a small incandescent lamp, or similar device in which the resistance increases rapidly with increased current flow, the lamp being heated by the plate current of the tubes 10 and 11 or by an auxiliary means, so such a temperature that its resistance will vary rapidly with a small change in current. Thus, when the oscillation amplitudes tend to increase, the temperature of the lamp R_3 increases with a resulting increase in resistance thereby causing a greater negative feedback, thus reducing the amplification. Similarly, as the oscillations decrease in amplitude, the current through the lamp is reduced permitting the lamp to cool with an accompanying decrease in resistance and reduction of the negative feedback, thus increasing the amplitude of the generated oscillations. As a result the system operates at substantially a constant amplitude which is preselected to be below the value at which grid current flows. As a result no distortion of the wave form takes place.

The lamp's peculiar *I* versus *V* curve: You may, conceivably, remember doing an experiment back in the first lab of this course, in which we asked you to plot *I* versus *V* for two "mystery boxes." One plot showed a straight line: that was a resistor. The other showed a bend: that was a small incandescent lamp bulb.

The lamp (#344) used in today's lab circuit is not the one used then (#47), but shows a similar bend. This bend reflects the effect of filament heating, since the resistance of a metal grows with temperature. We have inserted two straight lines in Fig. 8N.23, to show the lamp's effective resistance at a two arbitrary *average I*-versus-*V* points. The lamp's resistance in your lab circuit lies between the two lines that we have drawn.[16]

---

[15] US patent 2, 268,872 (1942). His use of a lamp to control gain was anticipated, by about one year, by Meacham. *Bell System Technical Journal* **17**, cited in Wikipedia article on the Wien Bridge.

[16] We saw amplitude out of our Wien bridge at about 9V. This is an *rms* value of about 6.4V, and rms is what matters for the lamp plot, which describes essentially the lamp's response to temperature.
To sustain a 6.4V oscillation at the output, the Wien bridge input – what the lamp "feels" – is 1/3 as large: about 2.1V. At this rms value, the plot shown in Fig. 8N.23 would predict a lamp resistance of about 330Ω. Such an *R* value would then call for a feedback *R* of 660Ω in order to give the required gain of three. We use 560Ω in the lab, not 660Ω, so the lamp for

8N.4 Sine oscillator: Wien bridge 333

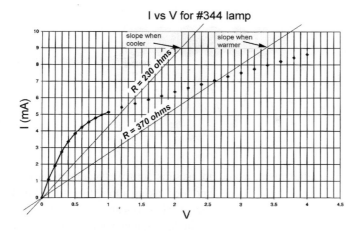

Figure 8N.23 I versus V for #344.

Note that the curve that is plotted shows not the response of the lamp to quick changes of I or V but rather the I,V combinations that the lamp settles into when allowed many seconds at a given *average* – long enough for its temperature to stabilize.[17] The temperature results from the heating effect of that particular I,V product. This notion is rather subtle, and may make your head spin briefly, as you first consider what the lamp's plot shows you.

Convince yourself that the sense of variation in lamp resistance does tend to stabilize gain at the necessary value. Don't forget that Fig. 8N.23 plots I on the vertical axis, as usual; a plot of *resistance* would interchange the axes, and would show slope *rising* with I and V (and temperature).

What value of gain *is* necessary to sustain oscillations without clipping? It may take you several seconds to answer; it takes the op-amp much less time.[18]

Rubbery behavior: The detail of performance that is hardest to explain is the so-called "rubbery behavior" that results if one touches the non-inverting terminal while the circuit is running. Fig. 8N.24 is a sketch, while Fig. 8N.25 is a scope image showing the response to a finger's touch.

 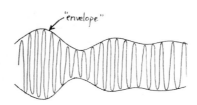

Figure 8N.24 Rubbery behavior" of Wien bridge oscillator when one touches the non-inverting terminal: at two sweep rates.

The finger's touch disturbs the circuit probably because the finger's capacitance upsets the carefully adjusted equilibrium of the *gain*-setting circuit.[19]. To explain why the circuit goes rubbery when poked, recall that although the negative feedback network is stable, it is only marginally stable, because the lamp (with its long time-constant of response) looks like an extreme low-pass filter within the loop. This situation is reminiscent of the jumpiness of the differentiator, which you saw back in

---

the oscillator that ran at 9V amplitude must have followed a curve slightly different from that in Fig. 8N.23 – but only slightly different.

[17] Strictly, the I and V values plotted show the *rms* average, since what concerns us is its heating effect. This is precisely the significance of rms values: the rms value of a sinusoid, for example, is $1/\sqrt{2} \times V_{\text{peak}}$, and this is the *DC* voltage that would deliver the same power as the rms value of the sinusoid.

[18] Now that several seconds have elapsed, let's announce the result: yes, we need a gain of 3 to sustain an oscillation.

[19] In lab, you might try a 10pF capacitor to ground, as a stand-in for your finger. We found its effect very similar to a finger's.

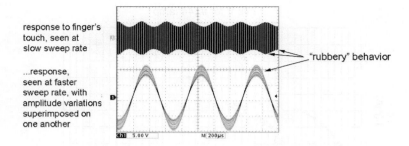

Figure 8N.25 Wien bridge lives close enough to the edge of instability so that a finger's touch makes it shiver.

Day 7. This circuit, with its automatic gain adjustment, would not be satisfactory as a general purpose amplifier; it is too jumpy, overshooting in its gain adjustments, and then taking a long time to stabilize again. It turns out that the jumpier the circuit – that is, the longer it takes to stabilize after a disturbance, because "damping" is slight – the more perfectly it holds a single frequency when not disturbed (electronic justice once again?). The oscillator is satisfactory in our application, because normally we do not disturb it: we do not poke it with a finger.

The circuit is also sensitive to mechanical vibration, and this vulnerability explains why electronic analogues often are used in place of the lamp.[20]

How good is our Wien bridge? The answer is "pretty good." We compared the lab circuit's frequency spectrum against that of the function generator on the powered breadboards, see Fig. 8N.26. The lab sinusoid looked much better than the breadboard's.

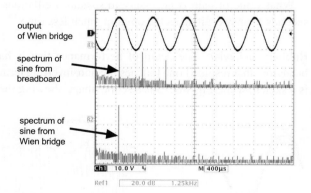

Figure 8N.26 Frequency spectra of lab's Wien bridge versus breadboard's sine generator.

We should admit that we put up our champion – the lab circuit – against a pretty weak competitor: an IC called an ICL8038, an inexpensive part that produces a sinusoid by whittling the points off a triangle. This is the same strategy used by our good lab function generators; the latter do their whittling more discreetly, as is not surprising, given that they cost well over 100 times as much as the '8038.

But the good quality of the Wien's sine has not let it take over the world. The whittled-triangle-wave method remains dominant in analog generators, because this method makes it easy to provide a very wide range of frequencies (0.3Hz to 3MHz on our lab generators, for example). And such methods now compete with a more recent method: digital synthesis, the playing out of a table of sine values to a digital-to-analog converter (a "DAC"). We'll meet DAC's later in this course.

---

[20] See, for example, AoE §7.1.5B, Fig. 7.20B, or Jim Williams' Wien bridge oscillator in his application note "High Speed Amplifier Techniques," *Linear Technology* AN47, August 1991, p. 49, Fig. 114), each using a JFET as voltage-controlled resistance, in place of the lamp.

> AoE §7.1.5A

Sine from square – but not now: Also in the digital part of the course you will meet an analog–digital hybrid method that produces a clean sinusoid: feed a square wave to a steep analog filter, while adjusting the filter's f_{3dB} to preserve only the square wave's fundamental – a pure sine.

For now, the Wien bridge gives us our cleanest sinusoid.

8N.5 AoE Reading

Reading
- Chapter 4 (op-amps):
  §4.3.2A: comparators
  §4.3.2B: Schmitt trigger (the usual comparator circuit)
  §4.3.3: triangle-wave oscillator
  §4.6.4: voltage-controlled oscillator
- Chapter 7 (oscillators):
  From start through §7.1.5B, Wien bridge oscillator
  For lab, the '555 is probably the most important; the Wien bridge is the most interesting
  §7.2.1E: long pulse using '555 again
  §7.1.7A, Table 7.2 listing a wide variety of oscillators
- Chapter 12 (logic interfacing):
  §12.1.7A: comparators (more detail than in Chapter 4)

Problems:
- Additional Exercise 4.34 (comparator, using slow op-amp)

8L Lab. Op-Amps III

Positive feedback: good and bad

Until now, as we have said in Chapter 8N, we have treated *positive* feedback as evil or as a mistake: it's what you get when you get confused about which op-amp terminal you're feeding. Today you will qualify this view: you will find that positive feedback can be useful: it can improve the performance of a comparator; it can be combined with negative feedback to make an oscillator ("relaxation oscillator": positive feedback dominates there); or to make a negative impedance converter (this we will not build, but see AoE §4.107, Fig. 4.104: there, negative feedback dominates). And another clever circuit combines positive with negative feedback to produce a *sinewave* out (this is the "Wien bridge oscillator").

Later, in Day 9, we will see what a pain in the neck positive feedback can be when it sneaks up on you.

8L.1 Two comparators

Comparators work best with positive feedback. But before we show you these good circuits, let's look at two poor comparator circuits: one using an op-amp, the other using a special-purpose comparator chip. These circuits will perform poorly; the faster of the two, the restless circuit, will help you to see what's good about the improved comparator that *does* use positive feedback.

8L.1.1 Op-amp as comparator

You will recognize the "comparator" circuit in Fig. 8L.1 as the very first op-amp circuit you wired, where the point was just to show you the "astounding" high gain of the device. In that first glimpse of the op-amp, that excessive gain probably looked useless. Here, when we view the circuit as a comparator, the very high gain and the "pinned" output are what we want.

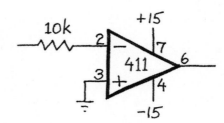

Figure 8L.1 Op-amp as simple comparator.

Drive the circuit with a sinewave at around 100kHz, and notice that the output "square wave" output is not as square as one would hope. Why not?[1]

---

[1] The rise and fall are slow, moving at a rate defined by the op-amp's *slew rate*, a limitation that you recall from Lab 7L:

8L.1.2 Special purpose *comparator* IC

Now substitute a 311 comparator for the 411. (The pinouts are *not* the same.) You will notice from Fig. 8L.2 that the output stage looks funny: it is not like an op-amp's, which is always a push–pull; instead, two pins are brought out, and these are connected to the *collector* and *emitter* of the output transistor, respectively. These pins let the user determine both the top and bottom of the output swing. If one uses +5 (with an external "pullup" resistor) and ground here, for example, the output provides a "logic level" swing, 0 to +5V, compatible with traditional digital devices. In the later circuits, you will keep the top of the swing at +15V, and the bottom of the swing to −15V. So arranged, the '311 output will remind you of op-amp behavior.

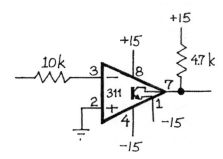

Figure 8L.2 '311 comparator: no feedback.

Does the '311 perform better than the '411?[2]

Parasitic oscillations: A side-effect of the '311's fast response is its readiness to oscillate when given a "close question" – a small voltage difference between its inputs. Try to tease your '311 into oscillating, by feeding a sinewave with a *gentle slope*. With some tinkering you can evoke strange and lovely waveforms that remind some of the Taj Mahal in moonlight – but remind others of the gas storage tanks on the Boston's Southeast Expressway. Judge Figure 8L.3[3] for yourself.

Minimizing oscillations: remedies *other than* positive feedback: You can sometimes stabilize a comparator *without* the remedy of *hysteresis*, which we are about to promote. Let's see how far these efforts can carry us. First, to get some insight into *why* the '311 is confusing itself, keep one scope probe on the oscillating output and put the other on the +15V power supply line, close to the '311. *AC-couple* this second probe, and see if the junk that's on the output appears on the supply as well. Chances are, it does.

Keep those ugly '311 oscillations on the scope screen, and try the following remedies:

- Put ceramic *decoupling* capacitors on positive and negative power supplies. If the oscillations stop, tickle them into action again, by reducing the slope of the input sine.
- Short pins 5 and 6 together: these adjustment pins often pick up noise, making things worse. Again tease the oscillations back on, if this remedy temporarily stops them.

15V/μs for the '411. Thus a rail-to-rail transition takes a couple of microseconds – and a bit more because the op-amp needs a little time to recover from *saturation*, the state in which this no-feedback circuit obliges it to spend most of its time.

[2] This question reminds us of a question posed to us by a student a few years ago: 'I can't find a '411. Is a '311 close enough?' This person no doubt was recalling the many times we had said, "Put away that calculator! We'll settle for 10% answers; 2π is 6," and so on.

[3] Two undergraduates produced this handsome waveform: Elena Krieger and Belle Koven, October 2004.

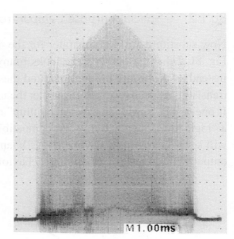

Figure 8L.3 '311 indecision: gas storage tanks? Taj?

8L.1.3 Schmitt trigger: using positive feedback

The *positive feedback* used in the Fig. 8L.4 circuit will eliminate those pretty but harmful oscillations. See how little hysteresis you can get away with. As input, use a small sinewave (<200mV) around 1kHz, and adjust the pot that sets hysteresis until you find the border between stability and instability. Then watch the waveform at the *non-inverting input*. Here, you should see a small square wave (probably fuzzy, too, with meaningless very high frequency fuzz *not* generated by your circuit; perhaps radio, visible now that you have the gain cranked way up). This small square wave indicates the *two* thresholds the '311 is using, and thus (by definition) just how much hysteresis you are using. Finally, crank the hysteresis up to about three times that borderline value (for a safety margin).

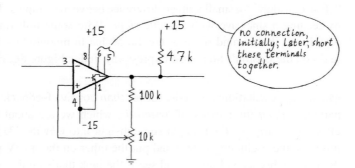

Figure 8L.4 Schmitt trigger: comparator with positive feedback (and *hysteresis*).

If you describe your circuit as a "zero-crossing detector," how late does it detect the crossings? Could you invent a way to diminish that lateness?[4]

Leave this circuit set up for the next experiment.

8L.2 Op-amp RC relaxation oscillator

Fix hysteresis at its maximum value, in the circuit of Fig. 8L.4, by replacing the potentiometer of the preceding circuit with a 10k resistor. Then connect an *RC* network from output to the comparator's

---

[4] You could – except that the world has already invented the method. See §8N.2.5, especially Fig. 8N.13, describing *AC* positive feedback.

inverting input, as shown in Fig. 8L.5. *This feedback signal replaces any external signal source; the circuit has no input.* Here, incidentally, you are for the first time providing *both* negative and positive feedback.

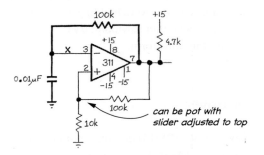

Figure 8L.5 *RC* relaxation oscillator.

Predict the frequency of oscillation, and then compare your prediction with what you observe. You can save yourself time by assuming that the capacitor *ramps*, as if fed a constant current equal to $15\text{V}/R_{\text{feedback}}$ (as pointed out in §8N.3.2).

8L.3 Easiest RC oscillator, using IC Schmitt trigger

Here's another way to get a square wave output, timed by an *RC*. It is very similar to the *RC* relaxation oscillator of the previous section but uses an IC that has hysteresis built in. The output is a "logic-level" swing, 0 to 5V. The circuit's weakness is the imperfect predictability of the part's hysteresis, an uncertainty that makes the frequency of oscillation uncertain; but logic output and simplicity make it a circuit worth meeting.

8L.3.1 RC feedback

Figure 8L.6 shows all there is to it. Use a capacitor of 1000pF,[5] and choose a feedback *R* to get $f_{\text{osc}} \approx 1\text{MHz}$. As you choose *R*, note the following specifications:

- threshold voltages:
    - negative-going: 1.8V (typical), 0.9 (min), 2.2 (max)
    - positive-going: 2.7V (typical), 2.0 (min), 3.15 (max)
- hysteresis: 0.9V (typical), 0.4 (min), 1.4 (max)

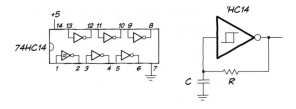

Figure 8L.6 *RC* relaxation oscillator.

The range of possible values makes prediction of f_{osc} look discouraging – but use *typical* values to choose your *R*, and see how close your result comes to the target of 1MHz (or "period of 1μs," to speak in terms more appropriate to your scope use, where you will see period rather than frequency).

---

[5] This value was chosen to be large relative to the scope probe's roughly 10pF load, so that you can probe the capacitor without changing f_{osc} appreciably.

8L.3.2 Sawtooth?

One student complained that the waveform of §8L.3.1 looks silly – like a sharkfin. Like that fellow, you may prefer an oscillator that produces a *sawtooth*. That you can easily achieve by replacing the feedback R with a *current-limiting diode*: the 1N5294 that you met as you were building a differential amplifier, in Lab 5L It is rated at 0.75mA. A larger C will produce a prettier sawtooth: try $0.1\mu\text{F}$. Note the range of the sawtooth's swing, because we are about to use this waveform.

8L.4 Apply the sawtooth: PWM motor drive

You now have met the two elements of a "pulse-width modulation" (PWM) circuit: a ramping oscillator output, and a comparator. Let's put them together to make a PWM driver. We'll use it to spin a DC motor. Fig. 8L.7 has the circuit, with some details omitted.

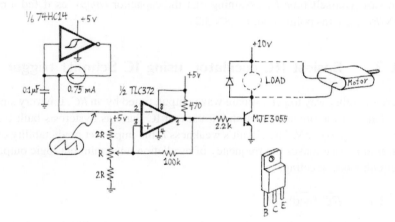

Figure 8L.7 PWM circuit; some details left to you.

Give your comparator circuit a little hysteresis: 50–100mV should suffice. Note that the transistor loads the circuit somewhat, so that output swing rises not to 5V but to about 4V.

While you are testing the circuit, and perhaps adjusting values, you may want to install a *resistor* in place of the motor – 100Ω or more. Look at the *sawtooth* and *comparator output* waveforms as you vary the threshold using the potentiometer.

Are you able to adjust the comparator-output's "duty cycle" over a wide range – from perhaps 10% to 90%? ("Duty cycle" is jargon for "percentage of period for which the output is high.") If the range is not so wide as you'd like it to be, tinker with the values of the resistors above and below the threshold potentiometer. Then check that the transistor switch is doing its job.

When you are satisfied, replace the resistor that has been serving as test *load* with a small DC motor. You can use the 3–12V motor you used to drive the integrator in Lab 7L.[6] We have shown a 10V supply to be sure we don't overheat the 12V motor. But you'll get away with +15V if you don't have a 10V supply handy.

You may want to stick a bright pointer on the motor shaft – or perhaps the vane is still in place from Lab 7L – to make it easier to judge how fast it spins as you adjust the duty cycle. Try loading the motor – applying some braking with your fingers. You should find torque that's not bad, at low duty

---

[6] We use the Mabuchi FF–130SH–11340. But any motor that can stand 10V will do.

cycles, whereas simply dropping the voltage to the motor – as your grandmother's sewing machine *rheostat* did – gives poor torque at low speed.

8L.5 IC RC relaxation oscillator: '555

The '555 and its derivatives have made the design of moderate frequency oscillators easy. There is seldom any reason to design an oscillator from scratch, using an op-amp as we did above. The ICM7555 or LMC7555 or TLC555 is an improved '555, made with CMOS. (We will refer to the CMOS part as '555 for brevity, though you may be using the LMC part.) It runs up to 1MHz (7555) or 3MHz (LMC7555), versus 100kHz for the original, bipolar '555, and its very high input impedances and rail-to-rail output swings can simplify designs.

8L.5.1 Square wave

Connect a 7555 in the '555's classic relaxation oscillator configuration, as shown in Fig. 8L.8. Look at the output. Is the frequency correctly predicted by

$$f_{\rm osc} = \frac{1}{(0.7[R_A + 2R_B]C)}?$$

Now look at the waveform on the capacitor. What voltage levels does it run between? Does this make sense?

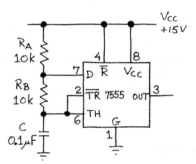

Figure 8L.8 7555 relaxation oscillator: traditional '555 configuration.

Sharkfin (?) oscillator, again: Now replace R_B with a short circuit. What do you expect to see at the capacitor? At the output? (A detail: the cap voltage now will fall a good way *below* the lower threshold value of $V_{CC}/3$; this occurs because the 7555 takes a while to respond to the crossing of that threshold, and the voltage now is slewing down *fast*).

50% duty cycle: The 7555 can produce a true 50% duty-cycle square wave, if you invent a scheme that lets it charge and discharge the capacitor through a *single* resistor.[7] See if you can draw such a design, and then try it. *Hint:* the old '555 could not do this trick; the 7555 can because of its clean rail-to-rail output swing. When you get your design working, consider the following issues:

- In what way does the output waveform of this circuit differ from the output of the traditional '555 "astable" (as an oscillator sometimes is called)?

[7] Well, "invent" may be a bit exaggerated if you have read Chapter 8N.

Lab. Op-Amps III

- Is the oscillator's period sensitive to loading? See what a 10k resistive load does, for example.

If your design is the same as ours, then the frequency of oscillation should be

$$f_{\text{osc}} = \frac{1}{(1.4RC)},$$

a result that is the same as for the "classic" configuration, except that it eliminates the complication of the differing charge and discharge paths. The "classic" design yields the same equation, to a very good approximation, as long as R_A is much smaller than R_B. Does the value of f_{osc} that you *measure* for your design match what you would predict?

Finally, try $V_{\text{CC}}=+5\text{V}$ with either of your circuits to see to what extent the output frequency depends on the supply voltage.

8L.5.2 Triangle oscillator

Two 1N5294s can give you a triangle waveform at the capacitor. A full-wave bridge and one 1N5294 could also do the trick. If you're in the mood, try out your design for the *triangle* generator.

8L.6 '555 for low-frequency frequency modulation ("FM")

Here is a quick preview of a technique we will use in the analog project lab, 13L, to send audio across the room as flashes of light. The rate at which a light-emitting diode (LED) flashes will convey the audio information. A '555 can do the *encoding*, converting time-varying voltages (the music waveform) to variations in frequency. A simple parallel *RLC* circuit like the one you built in Lab 3L can decode the FM, converting it back to a time-varying voltage (the music waveform, recovered).

The '555 can do this modulating because it brings out (to pin 5) a point on the 3-resistor divider that defines the upper comparator threshold. If a signal pushes that pin above its normal resting voltage of $2/3 \times V_{\text{CC}}$, the '555 oscillation slows; if it pushes pin 5 down, the oscillation speeds up.

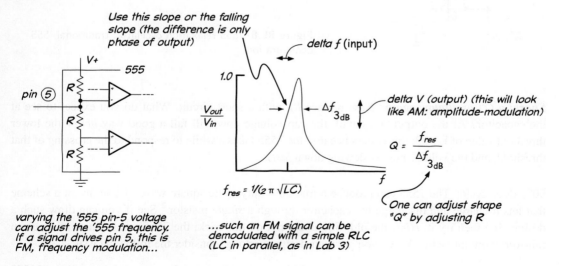

Figure 8L.9 A '555 can implement frequency-modulation; a parallel *RLC* can demodulate such a signal.

If you're short of time, you may want to limit yourself to the easy task of confirming that this FM scheme works. You can drive pin 5 with a low-frequency sinusoid (try 100Hz at amplitude of 0.5V) while watching the '555 output on a scope. Don't forget that you need a *blocking capacitor* between the function generator and pin 5, since pin 5 normally rests at 10V when the '555 is powered by +15V. You will see what looks like jitter on the '555 output. If it is hard to interpret, try reducing the driving frequency well below 100Hz.

8L.7 Sinewave oscillator: Wien bridge

Curiously enough, the sinewave is one of the most difficult waveforms to synthesize. (Your function generators make a sine by chipping the corners off a triangle, as you may recall.) The Wien bridge oscillator does it by cleverly adjusting its own gain to prevent clipping (which would occur if gain were too *high*) while keeping the oscillation from dying away (which would occur if gain were too *low*).

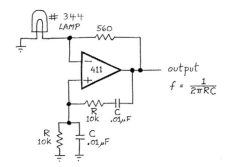

Figure 8L.10 Wien bridge oscillator.

The frequency favored by the positive feedback network should be

$$\frac{1}{2\pi RC}.\qquad(8L.1)$$

See whether your oscillator runs at this predicted frequency.

At this frequency, the signal fed back should be *in phase* with the output, and 1/3 the amplitude of V_{out}.[8] The negative feedback provided by the other path (the one that includes the lamp) adjusts the gain, exploiting the lamp's current-dependent resistance. Convince yourself that the sense in which the lamp's resistance varies tends to stabilize gain at the necessary value. What gain is necessary to sustain oscillations without clipping?[9]

You can reduce the output amplitude by substituting a smaller resistor for the 560Ω in the negative feedback path.[10] Try poking the non-inverting input with your finger and note the funny rubbery behavior at the output. Try sweeping the scope slowly as you poke: now you can watch the slow dying away of this oscillation of the sine's *envelope*.

If you're energetic, you can confirm that this sine is much cleaner than the function generator's, by putting it through an op-amp differentiator. The differentiator fed by the Wien oscillator should show

---

[8] At this frequency, where {the magnitude of X_C} = R, the impedance of R and C in series (the upper branch of the divider) is $\sqrt{2}R$ whereas the impedance of the R and C in parallel (the lower branch of the divider) is $\frac{\sqrt{2}}{2}R$. Hence the result that 1/3 of V_{out} is fed back, at this favored frequency. More surprising perhaps is the fact that this divider delivers a waveform in phase with the input.

[9] Yes, you're right: exactly *three*.

[10] Pretty smart circuit, eh? Adjusts lamp's R value to suit your new feedback resistor!

an output that looks like a very convincing sinusoid, not like the *zigguratoid*[11] that you saw when you differentiated the function generator's "sine" back in Labs 7L and 2L. Or, if you happen to be using a digital scope that includes an FFT, instead of using a differentiator you can simply compare the frequency spectra of the two sinusoids – function generator versus your little Wien bridge, as we did in Chapter 8N.

---

[11] We hope you will not spend your afternoon seeking this waveform in an electronics reference work.

8W Worked Examples: Op-Amp III

8W.1 Schmitt trigger design tips

8W.1.1 First easy case: setting thresholds

Setting particular thresholds exactly can be a pain in the neck. The process may put you through tedious algebra. But there are two easy cases.

Thresholds symmetric about zero: Suppose, for example, that you want thresholds at ± 1V, and output swing is ± 15V. The feedback divider pulls threshold equally far above and below ground: the foot of the feedback divider is tied to ground.

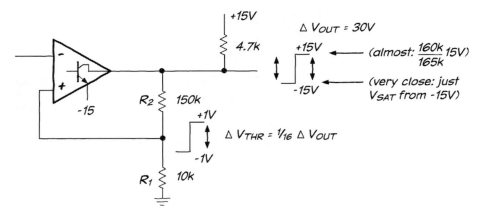

Figure 8W.1 First easy case: thresholds symmetric about zero.

The divider is to deliver 1V out of 15 $\Rightarrow R_2 \approx 15 \times R_1$ (or 14×, if you want to be more precise). The key notion is that the foot of the divider is put at the midpoint of the output swing. You'd get the same tidy result if the thresholds were to be symmetric about 2.5V while the output swung between 0 and +5: say, thresholds at 2.4V and 2.6V.

Second easy case: thresholds very far from symmetric: In this case you may, for example, want thresholds at 0V and +0.1V, and output swings 0 to +5. Here the feedback divider pulls *up* but not *down*. So just design a divider that pulls the threshold up to 0.1V:

$$R_1/(R_1+R_2) \times 5\text{V} = 0.1\text{V},$$

That means V_{Thresh} is about one part in 50, and R_2 is about $50 \times R_1$.

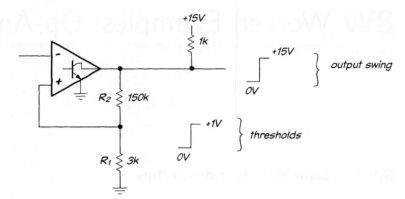

Figure 8W.2 Another easy case: thresholds pulled in only one direction ('very far from symmetric').

8W.1.2 An easier task: determining hysteresis

Suppose the task given is not "*put thresholds at 1V and 1.1V*," but instead "*set hysteresis at 0.1V; put thresholds close to 1V*."

The two formulations look just about equivalent, but the second task turns out to be much easier than the first. Let's try it. Suppose output swing is 0 to +5V, see Fig. 8W.3.

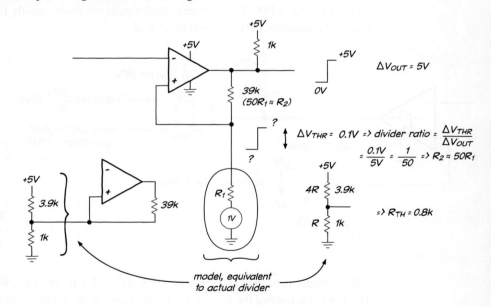

Figure 8W.3 Aiming for particular hysteresis, near a target voltage: threshold and hysteresis considered *separately*.

Here's the process:

Hysteresis: This is determined entirely by the divider ratio. Again we want 0.1V/5V: one part in 50; again $R_2 \approx 50 \times R_1$.

Thresholds "close to 1V:" This is determined by the voltage to which the foot of the feedback divider is tied (at the moment, let's assume we have a handy source of this 1V, an ideal voltage source; in fact, we are going to use a voltage divider, and then we will treat V_{Th} as that 'voltage at the foot...' If this issue isn't yet worrying you, forget this comment until later).

You'll provide this 1V with a voltage divider – but it has some $R_{Thevenin}$. Does that mess things up? No: let $R_{Thevenin}$ be the value you want for R_1.

8W.1 Schmitt trigger design tips

This settling for approximate thresholds may seem like a cheap trick. Often it is not: often what you want is not named threshold voltages, but an approximate threshold, and appropriate hysteresis. Then this shortcut fits the formulated goals nicely.

A wrinkle: adjustable threshold: Can you see a way to make threshold adjustable while holding hysteresis constant? (Possible help: recall that op-amps are cheap.)

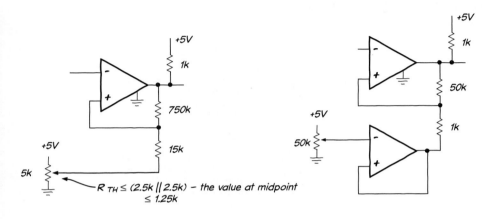

Figure 8W.4 Two ways to hold hysteresis constant despite changes of threshold.

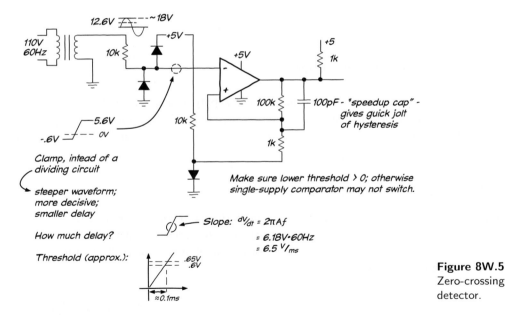

Figure 8W.5 Zero-crossing detector.

A potentiometer whose $R_\text{out} \ll R_1$ works, as in the left-hand figure of Fig. 8W.4. If you are too lazy to think about R_Thev, the op-amp buffer of the right-hand figure also solves the problem.

Problem (Design a zero-crossing detector) We want to know when the AC line voltage crosses zero (within a few 100μs). We are to use a single-supply comparator, powered from +5V, and we have a 12.6V transformer output available; the transformer is powered from the "line" (60Hz).

Compare AoE Fig. 4.85

Solution See Fig. 8W.5.

Worked Examples: Op-Amp III

Problem (Schmitt trigger, thresholds specified) Design a Schmitt trigger, using a '311 powered from ±5V, to the following specifications:

- output swing: 0 to +5V
- thresholds: ±0.1V (approximate)

Solution See Fig. 8W.6. This problem could be painful if we were not willing to approximate; so let's feel free to use approximations. It helps to consider the two cases separately: what happens when the '311 output is low (0V); and what happens when the output is high (5V).

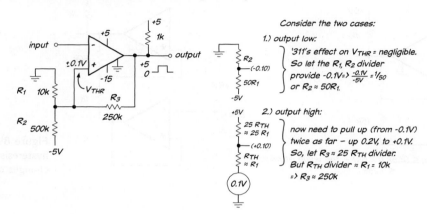

Figure 8W.6 Solution: aiming for particular thresholds, but willing to miss by a bit.

8W.2 Problem: heater controller

The goal is to design a comparator circuit – using a single-supply device (TLC372) rather than the usual '311 – that will keep your coffee at a temperature that you like. We'll put a temperature sensor on the hotplate, and will hold the plate's temperature at a level that is adjustable, and always somewhat below the boiling point.

8W.2.1 Temperature sensor and comparator

The temperature sensor, an LM50, puts out a voltage proportional to temperature (celsius) – with a half-volt constant tacked in. We will use this sensor to monitor the heater's temperature.

Our single-supply comparator has the bottom end of its output switch internally tied to ground, unlike the '311, which lets us define that emitter voltage.

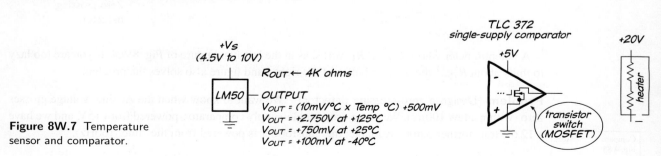

Figure 8W.7 Temperature sensor and comparator.

8W.2.2 Comparator circuit (your design)

Show how to use this sensor, and a potentiometer controlled manually, to turn the heater coil ON or OFF as the temperature moves below or above the pot's setpoint. Here are the details:

- temperature adjustment range: 50–100°C;
- hysteresis: 5°C;
- heater is a resistive and somewhat-inductive immersion heater. It likes to be driven with 20V DC.

8W.2.3 A Darlington transistor pair

In case you need such a device,[1] here is some data that you may use in your design (*typical* specifications, rather than the usual min and max.), if you decide to use the part.

A Digikey search came up with a TI part, TIP110, promising lots of β – the betas of the two transistors multiply, since the first transistor's I_E provides base current for the output transistor. Fig. 8W.8 shows some *typical* curves for this part. Let's use these – rather than the usual worst-case specs, just for the novelty of taking data from curves rather than from a list of numbers.

The resistors integrated with the transistor pair speed *turn-off* (not an issue in this case). The diode that parallels the Darlington pair normally will not conduct.

Though we propose to use *typical* specifications, let's try to give our design a factor-of-two safety margin. In your solution, tell us approximately what portion of the 20V supply you expect to see dropped across the load.

8W.2.4 Solution: heater controller

Temperature range to voltage range: The sensor adds a 500mV offset to its reading of 10mV/°C. So the temperature range, 50–100°C, maps to this voltage range:

$$\text{Lowest} = \text{offset} + \text{minimum reading} = 500\text{mV} + (50°\text{C} \times 10\text{mV}/°\text{C}) = 1\text{V}$$

$$\text{Highest} = \text{offset} + \text{maximum reading} = 500\text{mV} + (100°\text{C} \times 10\text{mV}/°\text{C}) = 1.5\text{V}\ .$$

So the manual control – a potentiometer – needs to be wired to span this range. We have plugged in some arbitrary values in Fig. 8W.9 (chosen to make our arithmetic easy).

Comparator circuit: We'll need some positive feedback, to get the required hysteresis, so the skeleton circuit will look like Fig. 8W.10.

Component values: Now let's choose some values.

- R_pullup: oddly enough, we'll need to postpone this choice until we've designed the switching circuit. For now, it's enough to say that this resistor will be so much smaller than R_2, in Fig. 8W.10, that we will be able to neglect it as we calculate hysteresis.
- R_1, R_2. These will set the amount of hysteresis – and once again we need to postpone their choice until we have calculated R_Thevenin for the manual-adjustment potentiometer.
- So we'll need to do that R_Thevenin calculation:
    - As usual, we need to consider *worst case*, and then make sure that the variation in R_Thevenin as one adjusts threshold doesn't appreciably alter the amount of hysteresis.[2]

---

[1] Does this sound like a hint?!
[2] We will not be so careful as to calculate the fractional *variation* in hysteresis that can result with varying pot settings. We'll settle for the usual, simpler "times-ten" rule: $R_\text{Thevenin} \ll R_1$.

Typical Characteristics

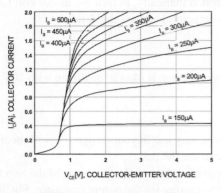

Figure 1. Static Characteristic

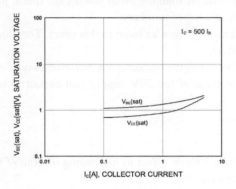

Figure 3. Base-Emitter Saturation Voltage
Collector-Emitter Saturation Voltage

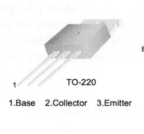

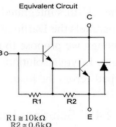

Figure 8W.8 Some typical spec curves for a Darlington transistor pair, TIP110.

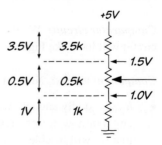

Figure 8W.9 Potentiometer sets voltage, 1–1.5V, temperature, 50–100°C.

- Worst case is the *maximum* $R_{Thevenin}$. That will come when the two "paths" seen from the slider – up to the +5 supply, and down to ground – are most nearly equal. That will occur with the slider at the top.

  At that point the two paths look like $3.5k \parallel 1.5k \approx 1k$

- Choose R_1: knowing $R_{Thevenin} \approx 1k$ lets us choose R_1. As usual, we want it much larger than what's driving it. Make $R_1 = 10k$.

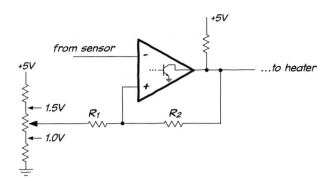

Figure 8W.10 Skeleton comparator circuit (no values).

- Now R_2: we want 5°C hysteresis, which translates to 50mV. So we want to feed back a small part of the 5V output swing: 50mV/5V = 50/5000 = 1/100. So $R_2 \approx 100 \times R_1$. Let R_2=1M.

8W.2.5 Heater driver

Can the '372 drive the load? We want to be able to put 1A through the heater. The comparator itself can't do that, so we'll use a transistor switch. Our rule of thumb for driving a switch calls for $I_B \approx I_C/10$ – *not* I_C/β, because we want to drive the switch strongly into saturation.

That rule would call for 100mA from the comparator – and that is more than the small output transistor of the TLC372 can handle (it can sink 20mA max). So, we'll need to use either a MOSFET switch (a device that it's not fair to invoke till we've met them, in Lab 12L), or a Darlington transistor pair.[3] We've already tipped our hand, in the *problem* statement, so let's assume we'll use the TIP110 Darlington pair.

8W.2.6 Using the Darlington switch

Curve 2 in Fig. 8W.8 shows plenty of gain – near the part's maximum – at 1A: current gain (β) better than 2000. So, we'll assume a β of 1000. By our normal rule-of-thumb, we'd drive the transistor 10× harder, as if β were 100. But that rule does not quite translate to the Darlington, since the main (output) transistor cannot saturate; only the first can. Further, we don't need to rely on a rule of thumb because the "static characteristic" plot in Fig. 8W.8 shows a plot of input current (I_B) versus output current (I_C). From these curves we can see just what base drive is needed (and then we'll double that drive, to be cautious).

It appears that 300μA in gets 1A out (that's β of about 3000). We can double the input drive to 600μA. We probably should be even more generous, since doubling base current does not, in general, produce a doubling of the Darlington's I_C. Let's go to 1mA (a nice, round number – and our favorite current, in this course).

That value permits us, at last, to calculate what value we need for R_{pullup}. The input voltage for the Darlington will be about 1.5V.[4] So, we'll put about 3.5V across the comparator's R_{pullup} when it is driving the switch ON. To get the 1mA that we want, therefore, we can use $R_{\text{pullup}} = 470\Omega$, and the base resistor $\approx$3.5V/1mA=3.5k. We'll use 3.3k.

---

[3] Or we could use the similar Szilaki transistor pair, in principle – but you won't find these for sale on Digikey as ICs. So, we'll stick with Darlington.

[4] A quick view of the schematic would indicate that the input voltage would be "two $V_{\text{BE}} \approx$ 1.2V;" a close look at the V_{BE} saturation curves, however, in Fig. 8W.8 suggests that the input voltage is somewhat higher. This is getting very fussy, indeed; but we might as well use the detailed information that we happen to have.

Finally, we need a *protection diode* across the load, in order to protect the Darlington from voltage spikes each time the switch turns OFF. The inductive load otherwise may damage the Darlington.

8W.2.7 The complete circuit

And here it is, all put together in Fig. 8W.11. Values shown in parentheses are 10% or 1% resistor values – "3.5k" happens to be a strange value to specify.

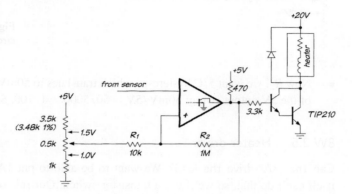

Figure 8W.11 Full heater-control circuit.

...What part of the +20V will be dropped across the load?

This is one way to ask "what is the ON voltage of the switch." We learn, from two sources, that this ON voltage should be under 1V at 1A:

- the third plot in Fig. 8W.8 shows about 0.9V "collector–emitter saturation voltage."
- The first plot shows the same thing for strong base drive and $I_C=1$A.

So about 19V of the 20V supply will be applied to the load: pretty good efficiency.

9N Op-Amps IV: Parasitic Oscillations; Active Filter

Contents

9N.1	Introduction	353
9N.2	Active filters	354
9N.3	Nasty "parasitic" oscillations: the problem, generally	356
9N.4	Parasitic oscillations in op-amp circuits	356
	9N.4.1 Occasionally, an odd object in the loop flips phase, at all frequencies...	357
	9N.4.2 ...but much more often, feedback changes flavor only at high frequencies	358
	9N.4.3 Why the op-amp includes a 90° lag	360
	9N.4.4 ...Given the op-amp's phase lag, what can you do for stability?	360
9N.5	Op-amp remedies for keeping loops stable	361
	9N.5.1 Roll-off high frequency gain	361
	9N.5.2 The photodiode circuit	361
	9N.5.3 Driving a long coaxial cable	362
9N.6	A general criterion for stability	365
9N.7	Parasitic oscillation without op-amps	367
9N.8	Remedies for parasitic oscillation	370
	9N.8.1 First remedy: quiet the supply (shrink the disturbance fed back):	370
	9N.8.2 Another remedy: kill high-frequency gain	370
9N.9	Recapitulation: to keep circuits quiet...	372
9N.10	AoE Reading	372

Why?

What are we trying to do today? We want to try two tasks, of which the second is the larger and more fundamental:

- Problem: make an improved filter, using op-amps;
- Problem: fend off unintended ("parasitic") oscillations.

9N.1 Introduction

Today we glance at *active filters* to remind you that such filters are available when you need a filter better than an ordinary *RC* such filters are available. Then we give most of our attention to *nasty* oscillations.

In the previous chapter we saw that positive feedback can be useful – though it is the underdog in feedback circuitry, less important than the great strategy of negative feedback. It can make a switching circuit decisive, and it allows construction of *oscillators*. Now we turn to the dark side of positive feedback: those cases where it sneaks up on us, causing oscillations that we didn't want (and usually did not expect). A bitter formulation of Murphy's Law in this special context says – describing the perverse ways that "parasitic" oscillations are inclined to pop up – "Oscillators won't, amplifiers will."

We will look at some cases of parasitic oscillations in order to get familiar with the problem. Then we will look at remedies, circuit changes that can prevent such behavior. It is the remedies, of course, that we are after. If nasty fuzz pollutes the output of a circuit in your lab, and you can make a few changes that make it go away, you will be a hero. This is a black art – sometimes involving literal hand-waving,[1] as well as some hand-waving explanations – and it is a skill that is right at home in a course based on *The Art of Electronics*.

9N.2 Active filters

AoE §6.2.4

As we point out in §9S.2, the motive for using these filters rather than a simple *RC* is to approach nearer to the ideal in filter behavior: flat *passband*, abrupt *roll-off*, then much attenuation in the *stop band*. An op-amp-assisted two-stage *RC* can improve both characteristics – flattening the passband and steepening the roll-off – as the frequency sweep in Fig. 9N.1 illustrates.

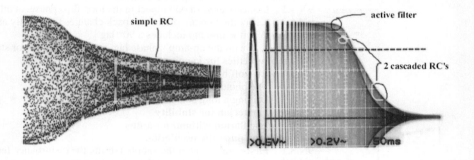

Figure 9N.1 Active versus passive low-pass: frequency responses contrasted.

The circuit in Fig. 9N.2 uses aptly placed positive feedback to achieve both improvements: it flattens the passband by giving the signal a boost just below f_{3dB}, where the passive *RC* is droopy; it provides a $1/f^2$ ultimate slope, as even a simple cascading of two *RC*s would; more magically, it provides a steeper transition between passband and stopband than the passive cascade could deliver.

AoE §6.3.1

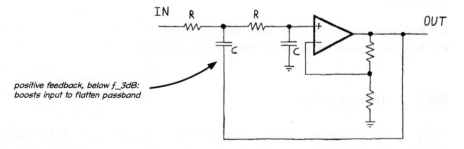

Figure 9N.2 VCVS active filter (2-stage or "two-pole").

positive feedback, below f_3dB: boosts input to flatten passband

AoE §6.2.5

Furthermore, simply by altering the gain of the amplifier one can choose to optimize one or another filter characteristic: flatness of passband (best in the so-called "Butterworth" filter) or steepness of roll-off (best in "Chebyshev" design) or waveform shape preservation (best in "Bessel").[2] The filter

---

[1] "Literal" hand-waving? Yes. Waving your hand near a circuit that shows parasitic oscillations while observing the effect of your hand's proximity (important for its stray capacitance) sometimes can provide a clue to where in the circuit an oscillation originates.

[2] The Bessel filter preserves waveform shape by providing phase shift that varies linearly with frequency, for "constant time delay:" AoE 6.2.6B.

of Fig. 9N.2, which is the design you will try in lab, is a "two-pole" design. When you need an even better filter, you can cascade additional stages.

Unfortunately, the method we have promoted elsewhere in this course, in which the several stages of our circuits are independent modules to be strung together, works poorly for filters. You would not design a four-pole active filter by cascading two identical two-pole filters. Instead, you would be obliged to do the harder work that ordinarily we eschew: analyze the entire four-pole problem. In practice, this may mean consulting a table like those in AoE for the gains of the several stages: see AoE §6.3.2.

Or you can make the process even more mechanical than that. You can download a program from any of several IC vendors: Analog Devices, Microchip, LTC, or Texas Instruments, then let that program do the design work for you.[3] In Fig. 9N.3 we show some screen prints from TI's FilterPro. You describe the filter performance that you'd like, as on the left-hand image.

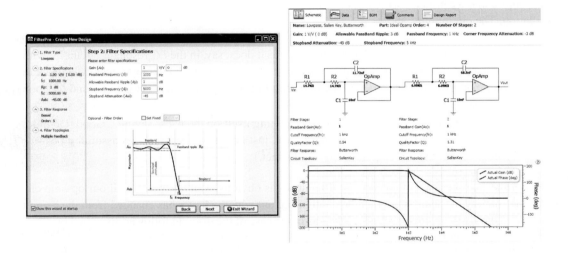

Figure 9N.3 TI "FilterPro" program will do the design work for you. We offer the figure for a glimpse of response curve and circuit diagram; we concede that details of specifications may not readable in this figure.

FilterPro draws you the circuit, with component values, as on the right side of Fig. 9N.3. The specified filter behavior was demanding enough to call for a 4-pole filter. So FilterPro cascaded two stages resembling the VCVS that we showed in Fig. 9N.2, except using unity gain (the filters of Fig. 9N.3 are called "Sallen-and-Key" configurations).[4] FilterPro holds your hand to an embarrassing degree: not only does it draw the circuit, specifying part values that you'd otherwise have had to look up in a table, it also politely asks whether you'd like to see component values shown with *exact*, 1%, or 5% tolerances, and then produces a bill of materials showing all the parts you need to buy.

Why is TI so nice to us? Well, they may hope that we will buy their op-amps, once we're using their tool, but probably their highest hope is that we'll buy their IC dedicated to implementing active filters, the UAF42.[5] The UAF42 is a neat part, with on-chip capacitors trimmed to 0.5% tolerance – but it is expensive ($16 for one, as we write). So, you may want to use the TI software, then shop for your own op-amps.

---

[3] Analog Devices: http://www.analog.com/designtools/en/filterwizard/\#/type; Microchip: www.microchip.com, search for "FilterLab;" TI: http://www.ti.com/tool/filterpro.

[4] For Sallen-and-Key filters see AoE §6.2.4E.

[5] Inherited from Burr–Brown, a company that TI bought.

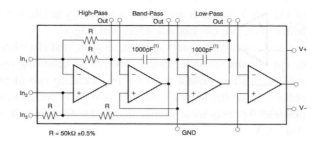

Figure 9N.4 TI's active-filter IC, UAF42.

9N.3 Nasty "parasitic" oscillations: the problem, generally

The conditions that lead to unwanted oscillations of course resemble those that lead to wanted oscillations – those that we produce with purposeful *oscillators*. The requirements are just two, for sustained oscillation. The circuit must show...

Gain:[6] otherwise the disturbance has to die away (as it does, say, in the *LC* resonant circuit hit with a square wave). More specifically, as the disturbance appears at the output and is fed back to the input, then appears again at the output, it must not get smaller: net gain around the loop must be at least 1.

Positive feedback: the circuit must talk to itself, patting itself on the back, saying, "Good. Do more of what you're doing." (In speaking of *gain*, just above, we were indeed assuming positive feedback; but the two requirements are distinct.)

Sometimes the presence of some kind of feedback is obvious, as it has been in all our op-amp circuits; the subtlety of the problem then lies in figuring out why the flavor has changed from the benign *negative* to the nasty *positive*. Occasionally – as in the case of the discrete follower discussed in §9N.7 – it is far from obvious that any feedback at all is present. If the circuit nevertheless oscillates, we will be obliged to do some experimenting, and perhaps even *thinking*, to determine how the feedback occurs. Our ultimate goal will always be to find remedies that prevent the oscillation.

9N.4 Parasitic oscillations in op-amp circuits

Op-amp circuits, which ordinarily rely on negative feedback, are evidently ripe for trouble. If ever something comes along to change the "flavor" of feedback from negative to positive, a placid amplifier can become an oscillator.

Any old op-amp circuit looks like the one sketched in Fig. 9N.5. We're familiar with the notion that we can tease the op-amp into doing what we like by putting this or that blob within the loop – but today we will find that the op-amp can rebel, going crazy when asked to accept blobs of certain types.

---

[6] It is *voltage* gain that is required. Strictly, it would be better to say "power gain," since a transformer can provide voltage gain, but it cannot sustain an oscillation. The inventors of the transistor were keenly aware of the voltage versus power distinction. Walter Brattain, one of the transistor's three inventors, recorded in his lab notebook, Dec. 15, 1947, "got voltage amp about 2 but not power amp." A day later, he noted, "... power gain 1.3 voltage gain 15." An important step forward. From Bell Laboratory archives, cited in Wikipedia article on history of the transistor.
We refer in this discussion to "voltage gain" in the hope that it is a clearer and more familiar concept. Unless output current is reduced proportionately to rise of voltage, as in a transformer, voltage gain will be accompanied by the necessary *power gain*.

9N.4 Parasitic oscillations in op-amp circuits

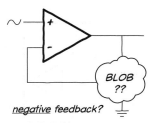

Figure 9N.5 Is it stable? Depends on the blob.

9N.4.1 Occasionally, an odd object in the loop flips phase, at all frequencies...

Does the circuit in Fig. 9N.6 apply negative feedback? "Sure," you may be inclined to answer. "The feedback goes to the inverting terminal." But what if inside the blob lives an *inverter*? "Oh," you answer. "That's silly – it would never happen. But if it did, we'd just swap inputs."

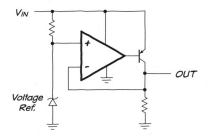

Figure 9N.6 This probably looks right; but it isn't: the feedback, here, is *positive*.

You're pretty much right: such a case is at least extremely rare – but the circuit shown *is* just such a case (this is a "low-dropout" voltage regulator that you will meet in Lab 11L). And you're right that the remedy is to swap inputs. We've done that in Fig. 9N.7 – and the circuit *looks* shocking, but in fact does make sense. This feedback makes sense because within the loop lies an inverting circuit: a PNP common-emitter amplifier.

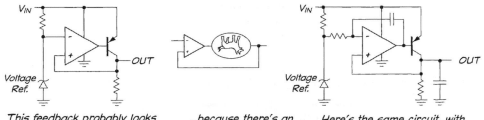

This feedback probably looks wrong, but it's not...

...because there's an inverter in the loop.

Here's the same circuit, with stabilizing feedback added.

Figure 9N.7 This probably looks wrong. It is a very rare circuit – where a full-time inverter sits within the feedback loop.

To try to persuade you that this circuit is useful, and not just a stunt that we cooked up, let's spell out its virtue: in contrast to the usual voltage regulator, in which a discrete follower is included within the feedback loop, this one avoids inserting a V_{BE} drop between V_{IN} and V_{OUT}. As a result, the difference between V_{IN} and V_{OUT} can be as low as 100mV. Hence the circuit's name, low-dropout..

On the right side of Fig. 9N.7 we have shown some extra stabilizing that normally must be added. But the strange-looking feedback of both circuits is correct, in any case; not crazy.

An example of inverted feedback often is recited in courses on controls: two people share an electric blanket with dual controls, one for each side of the bed. Accidentally, the controls for the two sides get swapped. As one person begins to feel chilly, he turns up the heat. You can imagine the uncomfortable night that ensues.

9N.4.2 ...but much more often, feedback changes flavor only at high frequencies

Figure 9N.8 shows a much more common case: the dog is not upside-down, but only *slow*. The signal fed back arrives a bit late. Fig. 9N.8 shows the sleepy dog lying within the loop. Fig. 9N.9 shows the electronic equivalent: a low-pass inside the loop.

Figure 9N.8 Something with slow response, inside loop, can cause trouble.

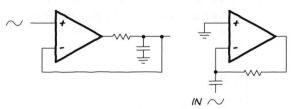

Figure 9N.9 Lowpass in loop can cause trouble.   Lowpass in loop?   This may look implausible.   Is this more plausible? Yes, it's a diffenrentiator (and a jumpy one)

At low frequencies, the lowpass does no harm. Since the input is changing slowly, the signal fed back, though a little out of date, is still quite similar to an up-to-date description. So feedback works as usual.

But if the input and output are changing fast, then an out-of-date description can be entirely misleading: it can be as much as 90° wrong as it goes into the op-amp, and as much as 180° wrong when one includes the additional 90° lag imposed by the op-amp itself. (This lag in the op-amp is surprising, we admit; see §9N.4.3). So, the simple differentiator shown in Fig. 9N.9 shivers when fed a square wave. The square wave is especially disturbing because of the high-frequency components in its edges, see Fig. 9N.10. .

§4.5.6, and Fig. 4.69

Figure 9N.10 Differentiator response to square wave: unstable when extra R and C are omitted: low-pass in feedback brings trouble.

That jumpiness is not acceptable, so the differentiator normally is tamed by the addition of the additional RC pair shown in Fig. 9N.11. These kill the circuit's gain at high frequencies. An alternative way to say this is that they halt the phase lag that grows dangerously with frequency – and then reverse that phase shift.

The differentiator that you built in §7L.2 included the stabilizing second RC pair shown in Fig. 9N.11. So in that lab exercise you did not see the shivering instability.

You will notice that we were not able to solve the differentiator problem the way we handled the odd case of a simple inverter within the feedback loop, the rare instance of Fig. 9N.4.1. In the differentiator circuit we cannot we cannot swap inputs as we could for the low-dropout regulator, because swapping

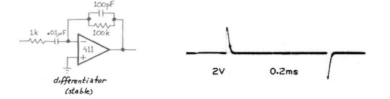

Figure 9N.11 Differentiator response to square wave: stable when extra R and C are added.

inputs on the differentiator would give positive feedback at low frequencies and for a DC input.[7] At DC this positive feedback would lock up the circuit, with the output stuck at one rail[8] or the other.

It's not hard to think of homely analogies to this problem, where information concerning the results of actions is slow getting back to the person in control.

Imagine, for example, that you've decided to play a perverse game of seeing whether you can drive a car using someone else's vision. You wear a blindfold, and your passenger tells you whether to steer left or right to stay in your lane. If you don't drive too fast, and if the passenger gives continual quick advice, you may manage. You won't manage if a car suddenly swerves in front of you. Your response won't be quick enough for that. But for a slow drive on a country lane, the process might work.[9]

But if your passenger is a slow-talking and perversely calm cowboy, who won't be hurried and says something like, "Well, young feller, I think you need to just mosey a shade, just a shade I'm saying, not a lot, over westward, that is to say leftward....," you won't be able to keep up, unless you're driving along at a walking pace. The incoming advice is likely to be out of date. Soon a drastic correction in the opposite direction will be called for – and the car will be weaving right and left all over the road: oscillating.

This is a time-domain way of describing a problem that may be clearer in frequency domain. Looking at the low-pass in the feedback loop in Fig. 9N.9, one can see that at low frequencies the feedback signal will be pretty much in phase with the op-amp output, and the capacitor will be having almost no effect: feedback is negative and healthy. At high frequencies, however (what we mean by "high" depends, of course, on the value of RC), the phase shift between op-amp output and the signal that reaches the op-amp's inverting input can be considerable, approaching $-90°$.

But, you may protest, $90°$ does not flip a signal; we need a $180°$ change before we're in trouble. Yes, you're right. But the catch, the circumstance that makes $90°$ deadly, is the fact – usually hidden from us as we watch op-amps in circuits – that the op-amp includes a $90°$ lag *within itself*. The op-amp, strange to tell, behaves like an *integrator*, beginning at a very low frequency. For the LF411, integrator behavior begins at about 20Hz (not 20kHz, but 20Hz).

AoE §§4.9, 2.5.4B

Frequency compensation to stabilize an op-amp: Figure 9N.12 is a plot showing gain roll-off of a "compensated" op-amp, and the associated phase shift between input and output. We have inserted the numbers that fit the LF411 op-amp that you have been using in the labs. The plot shows how the '411's gain rolls off with frequency, and shows the resulting phase shift between non-inverting input and output (note that this description assumes we are operating the op-amp *open-loop*, as we almost never do; but this open-loop behavior of the amplifier is crucial to its proclivities in the usual,

AoE §4.9.1

closed-loop circuit). We explore in §9S.1 how op-amp *frequency compensation* is achieved.

---

[7] You may fairly object, "But I'm not going to look for the derivative of a DC level!" True, but we cite DC as the extreme of "low frequency," just providing a case that is easy to analyze.

[8] You've probably picked up this piece of jargon: "rail" to mean either "limiting voltage, close to power supply," as here, or "power supply," as in *rail-to-rail-output op-amp*, the type that can drive its outputs close to both supplies.

[9] Please do not try this; this is a *thought* experiment!

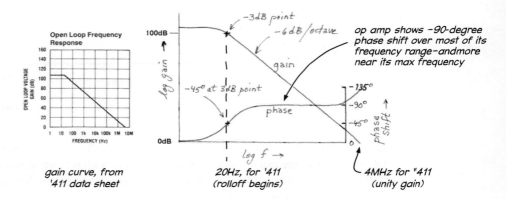

Figure 9N.12 A compensated op-amp like the '411 shows 1/f roll-off and −90° phase shift.

9N.4.3 Why the op-amp includes a 90° lag

Even stranger, at first hearing, is the fact that this 90° lag within the op-amp, and the sad $1/f$ roll-off of gain, is deliberately *imposed* by the op-amp's designers. Why? A short answer is that without *compensation* things would be worse: the amplifier's several stages form a cascade of low-pass filters, and these filters eventually would provide a deadly 180° lag that would make the amplifier unusable. So a 90° lag may be surprising, but it's much better than what we would get without compensation.

The effect of the op-amp's built-in 90° lag (and normally it is even a shade more[10]) is that in order to produce accidental positive feedback we need add only about another 90°, *not* 180°. And any low-pass filter can provide that 90° phase lag.

9N.4.4 ...Given the op-amp's phase lag, what can you do for stability?

If the feedback network is purely resistive, there is no problem: the network causes no phase shift, so the circuit is unconditionally stable. But when a low-pass characteristic sneaks in, we have to take care.

A non-remedy: "I won't apply high frequencies": A naive first thought surely might be to try to exploit the truth that feedback trouble comes only at relatively high frequencies. So, why not just *apply only low frequencies*?

Because you can't do that. It may be true that a particular op-amp circuit goes unstable only above 50kHz, and that you plan to use the amplifier for signals below 10kHz. That plan will not save you.

The problem is that the circuit responds not just to the signals that we choose to apply (perhaps rather low frequencies). It responds also to *noise*, which one must assume is present *at all frequencies*. This is one of the sad truths that underlies the gloomy dictum, "amplifiers will...." The consequence is that we must design all our amplifier circuits so that they will not oscillate even when driven with high-frequencies.

Some valid remedies: The "compensated" op-amp was designed to be stable with resistive feedback, not to be stable with a low-pass in the loop. So, when a low-pass filter is present, we need to forestall oscillations. Here are some ways to do that:

- Impose a roll-off of high-frequency gain, in addition to the roll-off built into every compensated op-amp. This method is illustrated in the photodiode circuit of Fig. 9N.14.

---

[10] As much as about 45° more, at f_T. See Fig. 9S.8.

- Bypass the troublesome blob at the higher frequencies where it would introduce dangerous phase lag. We will see this method, which we call "split feedback," later on page 364.
- Adopt an extreme remedy, altering the op-amp itself. Design a peculiar custom op-amp that can undo harmful phase lags. This is the strategy used in the "PID" loop,[11] a design that you will apply in Lab 10L.

We will illustrate the first two of these remedies below. Altering the op-amp we will postpone until you meet Lab 10L's PID loop.

9N.5 Op-amp remedies for keeping loops stable

9N.5.1 Roll-off high frequency gain

The circuits in Fig. 9N.13 are vulnerable to parasitic oscillations, in various degrees, because of lagging phase shifts (low-pass effects) within the feedback loop.

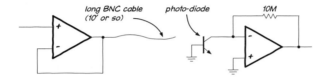

Figure 9N.13 Two circuits vulnerable to parasitic oscillations.

9N.5.2 The photodiode circuit

In the photodiode circuit, the implicit or accidental low-pass is far from obvious. There, it is formed by the large feedback R that drives the stray capacitance at virtual ground (a few pF). The 10M R is not reduced by the paralleled photodiode, because the photodiode is a current source. Its dynamic resistance, $\Delta V/\Delta I$, is very large.

Suppose that stray capacitance at the inverting input is, say, 3pF. In that event, the large feedback resistor forms a low-pass in the feedback loop with f_{3dB} of around 5kHz. At that frequency, phase shift in the feedback loop would be a lagging $45°$ – not enough to bring on oscillation. But at higher frequencies, the phase lag grows, approaching the deadly $90°$.

For the photodiode circuit (one you built in Lab 6L), the best remedy is to parallel the large feedback resistor with a small capacitor – say, 10pF, or even smaller, see Fig. 9N.14. This capacitor will compromise the circuit's response to quick optical signals. Circuit gain is down 3dB at a little above 1kHz (that is where $X_C \approx$ 10M). To put the same point differently, above that frequency the feedback circuit no longer forms a low-pass, with its lagging phase shift. Instead, it forms a C–C divider, with its phase shift moving back toward *zero* as frequency climbs beyond the crossover frequency.

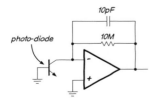

Figure 9N.14 Photodiode circuit, stabilized.

---

[11] PID stands for Proportional, Integral, Derivative, describing the three functions of feedback *error* that are applied in such a control loop. We devote Lab 10L to this technique.

9N.5.3 Driving a long coaxial cable

AoE §4.6.2

The low-pass in the case of the long cable is formed from the considerable stray capacitance of a BNC cable (about 30pF/foot) and the op-amp's small but non-zero output impedance – on the order of 50–100Ω,[12] but crucially altered by feedback. This output *resistance* in fact is transformed by feedback into a seeming output *inductance*, forming a low-pass with a resonant frequency much lower than the f_{3dB} that a simple *RC* would provide. It is this seeming *inductance* that explains the frequency of the oscillation brought on by the capacitive load. A resistive simple R_{out} sounds like a plausible explanation of the oscillation; but the numbers don't work.

See AoE Fig. 4.53

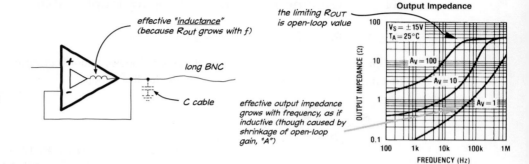

Figure 9N.15 Op-amp's Z_{out} behaves as if inductive, pushing down frequency of dangerous phase lag.

If one assumed that this R_{out} paired with the cable's capacitance formed the low-pass effect that brings on oscillation, one would predict oscillations at a frequency far above the amplifier's f_T At such a frequency, the amplifier's gain is long gone, and oscillation is not possible. Plugging in the numbers for the present case – $R_{out} \approx 40Ω$, C_{load}=450pF (assuming a 15-foot cable) – one would calculate the accidental-filter's $f_{3dB} \approx 8$MHz. At this frequency, the '411 op-amp has no gain at all. No oscillation could occur.

But a view of the amplifier's output impedance as *inductive* does fit the observed oscillation. The op-amp output acts *inductive* not because of actual inductance in the IC but because the impedance of the closed-loop circuit grows linearly with frequency – as the impedance of an inductor does.[13] This inductance driving C_{load} forms a resonant circuit, and resonant at a frequency consonant with the observed oscillation.

The oscillation in Fig. 9N.16 occurs at around 3.4MHz. This is close to the op-amp's maximum frequency – where its "phase margin" – the difference between its internal phase shift and the deadly 180° – is down to perhaps 50°. If resonance is close to this frequency, the resonant circuit can provide the necessary extra phase lag (as we've said, a little less than 90° will suffice, close to the op-amp's f_T). Figure 9N.16 shows an LF411 follower driving a square wave into 15 feet of coaxial cable.

Figure 9N.16 illustrates how spooky and whimsical such oscillations can be: the square wave only *sometimes* sets the circuit into oscillation. The square wave is the most disturbing signal we can provide, since its edges include stimulation over a wide range of frequencies. But the square wave

---

[12] Op-amp data sheets normally do not specify a hardware R_{out} (that is, a value for the IC itself, apart from feedback). But the '411 data sheet suggests a value of about 40Ω. This we infer from the plot of R_{out} (Fig. 9N.15), which levels off at around 40Ω at high frequencies if B, the fraction fed back is small (0.1 or 0.01). At these frequencies, the AB product is small, so the effective R_{out} – as improved by feedback – hits the limit of the R_{out} apart from feedback.

[13] This occurs – as you will see in §9S.5 – because the circuit's effective R_{out} is the barebones "hardware" R_{out} divided by the quantity $1+AB$, and A falls with frequency, at the rate $1/f$. The forthcoming AoE – the x-Chapters explores this point thoroughly, showing scope images to demonstrate that the op-amp's output impedance truly is inductive, even including the expected phase *I*-to-*V* phase difference. In *Troubleshooting Analog Circuits*, Newnes, (1991, 1993, p. 100), Bob Pease warns, similarly, against assuming that an op-amp output is simply resistive.

in Fig. 9N.16 brings on oscillation only for signals that are below ground: apparently, the output impedance of the circuit differs above and below ground (where a different set of transistors drives the output, in the push–pull stage). And the right-hand image in Fig. 9N.16 shows that not even the negative edges always bring on oscillation.

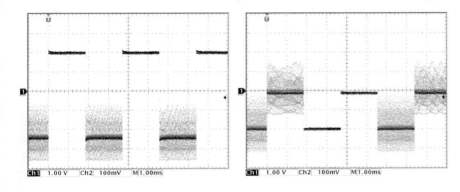

Figure 9N.16 A long coaxial cable can push an op-amp follower into oscillation.

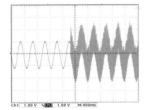

Figure 9N.17 Sinusoid may not bring on oscillation – but then it may.

A sinusoid is less disturbing, and may get through without provoking the oscillation. It seems to be getting safely through, in the left-hand half of the trace in Fig. 9N.17. But halfway through the scope's sweep we managed to bring on oscillation simply by pressing the function-generator's range button: the momentary interruption of the signal was enough to send the circuit off the rails. The lesson seems to be that capacitive loading of op-amp circuits is hazardous. We need remedies.

See AoE Fig. 4.102

Remedies: 1. Feed back less: One solution is oddly counterintuitive: instead of using a follower, use an amplifier to drive the cable. Give it, say, a gain of five and the circuit may calm down.

At first, you're likely to protest, "More gain? But gain is just what an oscillator requires." Yes, but it's *loop gain* – the gain going round the loop – that matters. This is a notion that the Wien bridge illustrates nicely: the Wien circuit sustains an oscillation at the frequency where it is attenuated to 1/3 in the feedback loop – if the circuit has a gain of 3. Attenuation greater than 1/3 in the feedback loop, at that gain, would kill the oscillation.

Giving the cable driver "gain" necessarily *feeds back less*, since the gain is $1/B$, where B is the fraction fed back. Figure 9N.18 shows the effect of gradually feeding back less than one. By the time we've fed back just 1/5 (for a gain of 5), the circuit is utterly placid.[14]

[14] In case you are intrigued by the point that the op-amp output acts inductive, you may enjoy describing differently the effect of "feeding back less" – that is, of changing from follower to amplifier. Feeding back less – reducing B – raises the closed-loop output impedance, increasing the value of the apparent "inductance." This change tends to push $f_{\text{resonance}}$ down, by a factor $\sqrt{2} \times$ ratio-of-apparent-L-values. By itself, this change would be destabilizing. But, at the same time, a contrary, stabilizing effect outweighs this change. Feeding back less pushes the unity-gain frequency down by a factor proportional to the shrinkage of the fraction fed back – outweighing, or "outrunning" – the rate at which the resonant frequency has fallen. All this is well explained, with helpful plots, in the forthcoming *AoE – the x-Chapters*, on advanced topics in operational amplifiers, to be published in 2016.

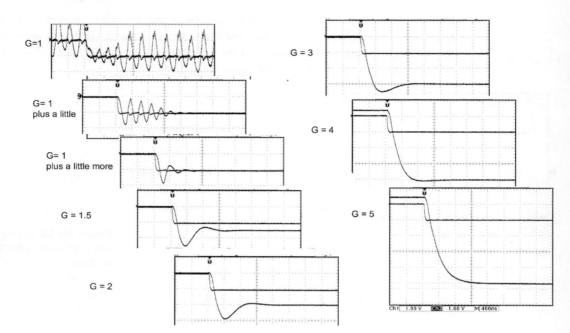

Figure 9N.18 Op-amp driving a capacitive load can be quieted by feeding back less – giving it gain.

These scope images give us our first chance to see that there are *degrees* of stability; it is not just a question of Yes/No: it does/does not oscillate. The second trace of Fig. 9N.18, for example (gain of one "...plus a little"), does not sustain an oscillation. But this ringing probably would be troublesome enough to call for a remedy.

We will meet this notion that there are degrees of stability in two other settings: in op-amp compensation (treated in §9S.1), where *phase margin* describes how closely the op-amp approaches the brink of disaster, the $-180°$ shift; and in the PID loop of Lab 10L, where we will be able to adjust the gain of our home-made "op-amp" to flirt with such instability.

Remedies: 2. Push the phase-shifted point outside the feedback loop: A series resistor of 100Ω or so can provide a very simple remedy for the long cable problem.

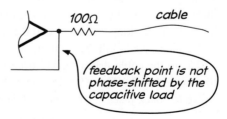

Figure 9N.19 Simple remedy for long-cable problem: series resistor.

The phase shift caused by the cable's capacitance still occurs – and, in fact, is aggravated. But the point in the circuit where most of this dangerous phase-shift occurs now lies *outside* the feedback loop. We have paid a price for stability: the circuit's R_{out} is degraded by the series resistor.

9N.6 A general criterion for stability

Remedies: 3. A refinement of long-cable remedy: "split" feedback: With a few more parts, we can have stability along with the original follower's good low R_{out} at lower frequencies. Fig. 9N.20 shows the modified long-cable circuit, "splitting" the feedback path.

AoE Fig. 4.76

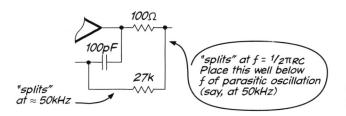

Figure 9N.20 Refined remedy for long-cable problem: "split" feedback.

Place the "crossover" frequency – at which most of the signal passes through the capacitor, bypassing the phase-shifting network – safely below the frequency at which the circuit oscillates.

Such a frequency might sound excessively high for audio, for example. But note that you don't want to push the crossover frequency lower than necessary. The benefits of feedback are vitiated at that crossover frequency. The correction of push–pull distortion, for example, would be spoiled by too low an $f_{crossover}$. This low crossover would slow the op-amp's drive that is needed so as to compensate for the push–pull's effects. You may recall the snap above or below $|0.6V|$ that the op-amp needed to provide in order to cure crossover from §6N.9.

The "split feedback" remedy is a powerful one. We use it in the push–pull driver circuit that concludes Lab 9L, and again in the motor-driver circuit of the PID in Lab 10L.[15]

See AoE §4.6.2 once again

One more long-cable remedy: use a specialized driver IC: Some followers are designed specifically to tolerate heavy capacitive loads (for example, Analog Devices' AD817, or TI/National's LM8271/2). This type simply slows down when connected to a heavy capacitive load: the value of the "compensation" capacitor that reduces high-frequency gain is effectively increased by C_{load}.[16] So, the circuit is self-stabilizing.

This solution may look appealing – but when you want to stabilize a circuit in a hurry, adding a series resistor may beat waiting a day or so for delivery of a specialized IC.

9N.6 A general criterion for stability: loop gain where phase shift approaches 180°

AoE §4.9.1A

"Loop gain" describes what it sounds like: the gain on a trip around the feedback loop: through the amplifier IC (whose gain is called A), then through the feedback network (whose "gain" – often an *attenuation* – is called B). The gain all the way round the loop, then, is AB. In general, we like high loop gain. It is what gives feedback circuits their several virtues – constancy of gain, low R_{out}, and so on. But it is also what makes op-amp circuits vulnerable to oscillation.

In order to keep a circuit stable, one must be sure that the loop gain, AB is less than one at the frequency where the phase shift reaches the deadly 180°.

Assume there is noise around at all frequencies, disturbing the circuit input. That disturbance – call it ΔV_{in} – is amplified to produce a ΔV_{out} that is A times as big; then some or all of that ΔV_{out} is fed back to the input. A fraction B is fed back, so $AB\Delta V_{in}$ is fed back. If, at a frequency where 180° of

---

[15] Other, subtler remedies also are available. Some are described in Bob Pease's Application Note "Linear Brief", National Semiconductor LB-42, and more briefly in his book *Troubleshooting Analog Circuits*.
[16] If this notion is puzzling, take a look at §9S.1.

Op-Amps IV: Parasitic Oscillations; Active Filter

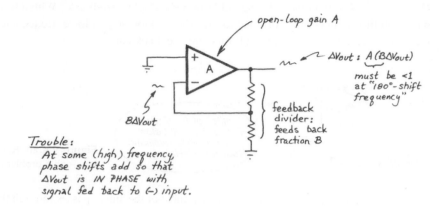

Figure 9N.21 Stability conditions summarized.

phase shift occurs, this fed-back signal is as big as the original, or bigger, the circuit oscillates. In other words, $AB \geq 1$ at $-180°$ brings oscillation.

This view suggests remedies:

- limit the size of A;
- limit the size of B;
- or limit both.

In any case, we need to limit the product, AB. Let's look separately at the ways to limit A, then B.

Shrinking A: "frequency compensation": The gain curve in Fig. 9N.22 of the "uncompensated" amplifier shows too much gain at the frequency where phase-shift reaches $180°$. An uncompensated op-amp wired as a follower ($B=1$) would oscillate.

AoE §4.9

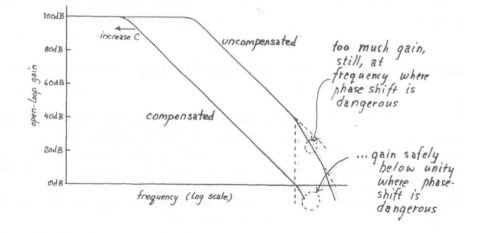

Figure 9N.22 Compensated versus uncompensated op-amps.

The "compensated" amplifier is made stable through the deliberate *early* roll-off of its gain with frequency. Its gain has been reduced, so as to fall to unity before the frequency that would impose that deadly phase shift. Op-amp compensation is discussed more thoroughly in §9S.1. In Lab 10L we will find that one can tinker with more than just the gain–roll-off of A; one can even alter its phase-shifting. Exciting complications to come!

Shrinking B: We thrashed this point pretty thoroughly on page 363: give the circuit gain, and it is less inclined to oscillate.

In summary: limit loop gain, *AB*: For stability, you need to hold the *loop* gain below unity at frequencies where dangerous phase shifts appear. Fig. 9N.23 shows an op-amp's gain plot reminding us of the meaning of *loop* gain in contrast to *circuit* gain.

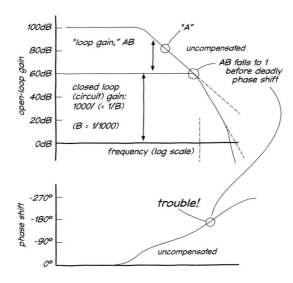

Figure 9N.23 *AB* determines feedback stability, as well as feedback virtues (illustrated with uncompensated op-amp, barely stable at gain of 1000).

Fig. 9N.23 reiterates a point we tried to illustrate with Fig. 9N.21, summarizing conditions for stability. It is the value of *AB* at the dangerous phase-shift frequency that determines whether a feedback circuit is stable or not. High *AB* can be worrisome.

But *AB* also measures the potency of the *good* effects of feedback; it measures the capacity of the op-amp to improve circuit performance. In §9S.5 we look more closely at the way in which loop gain depends on both circuit configuration and chip properties: the way it depends, in other words, on *B*, and on *A*.

You have felt this sort of tension before, where reaching far for one thing you want (here, performance improvements through feedback), you may give up something else that you need (here, circuit stability). In discrete common-emitter amplifiers, for example, you were able to trade off *gain* versus *linearity*. Of the two concerns attracting us here, stability is the more urgent. With that paramount goal in mind we must hold *AB* safely low. We will settle for the circuit performance that results.

9N.7 Parasitic oscillation without op-amps

Surely this circuit cannot oscillate…: Feedback need not be explicit, as it is in op-amp circuits, in order to cause trouble. Fig. 9N.24 is a circuit that oscillates because of very sneaky effects. Wouldn't you nominate this circuit as *least likely to oscillate*? Isn't it a sure-fire dud?[17] No voltage gain, and no feedback: surely, such a limp thing cannot sustain an oscillation.

---

[17] Oxymoron, anyone?

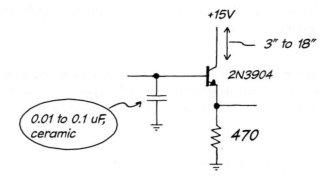

Figure 9N.24 Follower as *oscillator*.

...but it does oscillate: It oscillates, and fast enough so that we like to use it to jam FM broadcasts in the lab[18] – just to convince everyone that these parasitic oscillations can, indeed, be troublesome. This big oscillation completely silences the FM broadcast in the lab (we hope we're not jamming more than our lab room).

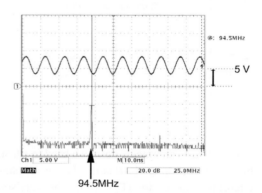

Figure 9N.25 Jamming Jammin' 94.5 with the discrete follower. (Scope settings: 5V/div, 10ns/div.)

Symptoms of a parasitic: One of your circuits might oscillate at a time when you were not watching it with a scope. What clues might you get to the fact that a parasitic buzz was occurring? Two possible clues come to mind.

First clue: if you're very lucky, you may be listening to FM radio, and may notice reception going bad. This is not likely.

Another clue, in the present case, appears if you happen to use a DVM to check the circuit output. You may notice "impossible" readings, as we did, recently, see Fig. 9N.26.

Emitter more positive than base? It looks that way. But this is a fake reading: the oscillation has fooled the DVM, which accidentally has rectified a part of the fast sinusoid.

If you're inclined to protest that you'd not be likely to pick up either of these clues, you may be right. Good evidence lies in a story that Pease tells, in which this very circuit caused a parasitic oscillation that went undetected until the computers that included the circuit failed their FCC radio-frequency emissions tests (see footnote on page 370).

How can it oscillate? To answer that riddle one must answer two subsidiary riddles:

---

[18] There happens to be an aggressive local station that calls itself "Jammin' 94.5," and with that name it just seems ripe for jamming.

9N.7 Parasitic oscillation without op-amps

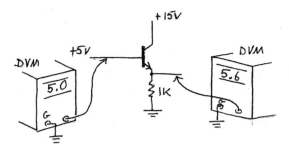

Figure 9N.26 "Impossible" DC readings may reveal a parasitic oscillation.

- How does the circuit achieve voltage gain (to sustain oscillations)?
- How does the circuit provide positive feedback?

The answers emerge if one takes into account imperfections of the circuit. The idealized follower cannot oscillate; a real one can.

- the gain appears if one draws explicitly the inductance always implicit in the power supply,[19] along with stray capacitance, see Fig. 9N.27.

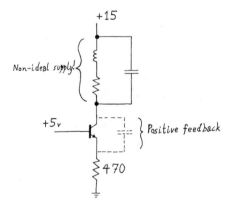

Figure 9N.27 Follower redrawn to include feedback, and inductance that can provide gain.

When you build this circuit in §9L.2, we encourage you to use an outrageously long power-supply lead, to exaggerate the stray inductance present in even well-built circuits. You can confirm that the collector supply line looks like a resonant circuit (a parallel *LC* like the one you built in Lab 3L): put your hand close to the wire, or even grab it. The parasitic frequency will change, because of the changed stray *capacitance*, now enhanced by your hand.

- The feedback appears if one notices that a disturbance on the power supply, which here is the collector, can disturb the emitter (through CE capacitance) in a sense that increases the collector disturbance: in other words, here is positive feedback.[20]

Cf. AoE §7.1.5D esp. Fig. 7.30

This circuit, as redrawn, is nearly identical to the current source whose oscillations are explained in AoE. There the circuit is likened to a purposeful oscillator called a Hartley *LC* oscillator.

Most of us settle for learning some rules of thumb that stop oscillations when they appear; it is not easy to model a circuit in detail, including stray inductance and capacitance. Another of those

---

[19] Ballpark value? Perhaps 500nH/meter, though the value depends on distance to other conductors.
[20] Here we mean that one "notices" this path conceptually. You will not be able to observe these details with your scope. Attaching a scope lead to the collector is almost certain to stop the oscillation, as noted below in §9N.8.1.

electronic Murphy's laws holds, however: if the circuit can find a frequency at which it could sustain an oscillation, it will find that frequency and oscillate – irritating its designer.[21]

9N.8 Remedies for parasitic oscillation

9N.8.1 First remedy: quiet the supply (shrink the disturbance fed back):

You may be able to stop the oscillation by quieting the collector: this "shrinks the disturbance," in the terms we used just above. Two rather obvious ways to do this come to mind:

- keep the power supply lead (+15V) short (this reduces the inductance of the line, putting the resonant frequency so high that the transistor lacks gain at that frequency);
- install a small ceramic *bypass* capacitor on the collector.

In fact, the oscillation is likely to be strangely fragile – but this characteristic can make it the more troublesome. You may suspect that the collector is oscillating, and decide to probe it to find out. But no, you find it's not oscillating. But the act of probing (that is, of attaching the probe's 10pF of capacitance to ground) may have stopped the oscillation – though only while you probe it, of course. You may need to investigate by attaching a short length of wire to the probe tip and then "flying over" the circuit, at a distance of a centimeter or two to sense oscillations without killing them. Spooky, indeed.

9N.8.2 Another remedy: kill high-frequency gain

Even when your circuit is stuck with some disturbance fed back, of a phase that could bring oscillation, the oscillation will not occur if the circuit gain is insufficient. You may see some transient junk, but not a sustained oscillation. The base resistor in a follower circuit has this effect: spoiling the circuit's gain at the high frequencies where oscillation otherwise could begin.

One explanation of the *base resistor* remedy: redraw as common base amplifier: Figure 9N.28 is a sketch of our transistor, viewing the circuit as a *common base* amplifier, an amplifier driven from its emitter. You probably have not seen an amplifier driven this way before (except in passing, when you may have seen in AoE, a Miller-killer made with a common-base amp, base firmly fixed. See AoE Fig. 2.84(C). Here, as suggested in Fig. 9N.28, we are doing the opposite of what that high-frequency configuration achieved. We *want* to spoil the high-frequency gain.

The gain of this *common base* circuit is

$$\text{gain} = \frac{R_C [\text{or } Z_C]}{r_e + [(\text{resistance driving base})/(1+\beta)]}.$$

Here, with the base driven (or "held") by a stiff voltage source, the denominator of the gain equation is just "little r_e;" but we can diminish the gain in a way equivalent to the way we cut the gain of the common-emitter amp: by tacking in a resistor that enlarges the denominator of the gain equation. Here, we insert a resistor in series with the *base*; in the common-emitter we achieved an equivalent reduction of gain by inserting a resistor in series with the *emitter*.

AoE §2.4.4B

---

[21] This circuit is not just a pedagogue's concoction. It seems to have caused real mischief to some engineers who used essentially this circuit to drive a computer's "Reset" pin. The computers then failed FCC interference-radiation tests because each computer included this unintended little radio broadcasting machine! R. Pease, *Troubleshooting Analog Circuits*, p. 109, Newnes (1991).

9N.8 Remedies for parasitic oscillation

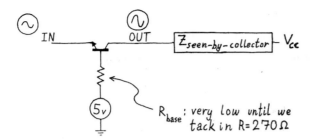

Figure 9N.28 Follower redrawn as common-base amplifier.

Another explanation of the *base resistor* remedy: Miller effect: There is another way to describe what the base resistor achieves; perhaps you'll prefer this other description, though we have barely mentioned *Miller effect* in this book. Fig. 9N.29 shows our follower with two competing *feedback* paths drawn in – C_{CB} and C_{CE}. The latter provides *positive* feedback, as you know; C_{CB} provides *negative* feedback by moving the base voltage. It is this feedback, tending to spoil the gain of a high-frequency inverting amplifier, that is called Miller effect.

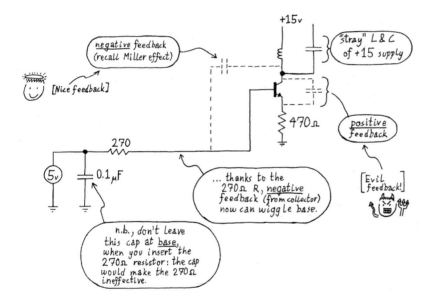

Figure 9N.29 Competing feedback paths: stabilizing versus destabilizing.

How much negative feedback? The fraction of a collector-wiggle fed back is set by the voltage divider formed by C_{CB} and R_{TH} that is driving the base. By increasing the latter from a value close to zero to 270Ω, we much increase the strength of the negative feedback. Thus we tip the feedback competition in favor of the good guys, *negative* – and the oscillation should stop.

Remedies: The remedies, then, are fundamentally the same as those we apply when we need to stop or prevent op-amp oscillations:

- shrink the disturbance that is fed back; or
- diminish the circuit's (high-frequency-) gain.

9N.9 Recapitulation: to keep circuits quiet…

Some of the methods that tend to prevent parasitic oscillations are obvious; some are not. Among the obvious we might classify these:

- keep power supply leads short and wide (minimize inductance);
- decouple power supplies (as usual).

Among the less obvious protections we might classify these:

- include a base resistor on any discrete transistor follower;
- provide a "split" feedback path in an op-amp circuit that is to drive a troublesome load – a load that can impose substantial phase lags (see page 364)

In Lab 10L we will return to these stability questions, and there will try yet another remedy: we will design a sort of custom op-amp that can *undo* some troublesome phase lags. Parasitic oscillations always will lurk, threatening your circuits. You'll stand a better chance of fending them off if, boyscout-like, you are *prepared*.

9N.10 AoE Reading

Reading:

- AoE:
    Chapter 7: §7.1.5E: parasitic oscillations
    Chapter 4:
        §4.9: op-amp frequency compensation
        §4.5.7: differentiators
        §4.6.2: capacitive loads

9L Labs. Op-Amps IV

Today we look first at one more *benign* use of positive feedback, an *active filter*, and then spend most of our time with circuits that oscillate when they should not. In this lab, of course, they "should," in the sense that we want you to see and believe in the problem of *unwanted* oscillations. On an ordinary day, the oscillations that these circuits can produce would be undesirable, and would call for a remedy. Some of you have met these so-called "parasitic oscillations" in earlier labs. Consider yourselves precocious. In the exercises below, you will try first to bring on oscillations, then to stop them.

If any topic in electronics deserves the title "Art of...," taming oscillations must be that topic. With some trepidation, we invite you to try to make these nasty events occur. We quoted, in Chapter 9N, the standard variation on Murphy's Law that says, "Oscillators won't; amplifier circuits will." Since today's circuits purport to be amplifiers rather than oscillators, probably they "will...."

9L.1 VCVS active filter

We ask you to build two versions of a two-pole filter: a simple cascade of two passive RCs, and an active lowpass. The two filters have equal f_{3dB}s.[1]

If you have a *resistor substitution box*, use that to set the value of the active filter's R_{gain}. The substitution box makes changing values easy and, unlike a *variable resistor*, lets you know what value you are applying.

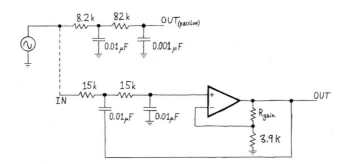

Figure 9L.1 VCVS: a particular implementation, with cascaded RCs added for comparison.

Set R_{gain} to 2.2k, providing the response with the flattest passband ("Butterworth"). Confirm that both circuits behave like a low-pass filters; note f_{3dB}, and note approximate attenuation at $2f_{3dB}$ and

---

[1] They may look as if they will not show the same f_{3dB}: indeed, the RC values of active and passive filters are *not* the same: 2kHz for the passive, 1kHz for the active. Yes, the RCs are unequal. But only the active filter has $f_{3dB} = 1/(2\pi RC)$. The *passive* filter's two elements share an RC value. But the entire passive filter shows an f_{3dB} that is about half the f_{3dB} of each stage, because where each filter stage drops $3dB$, the entire filter drops $6dB$.

Filter Type	R_{Gain}	Gain
best time delay (Bessel)	1k	1.3
flattest (Butterworth)	2.2k (2.3k)	1.6
steep (Chebyshev)	4.7k (4.3k)	2.1
nasty peak (no one claims this one!)	6.8k	almost 3
OSCILLATOR!	10k	3.5

$4f_{\text{3dB}}$. We hope you will find the -12dB/octave slope that is characteristic of a 2-pole filter, though you won't see that full steepness in the first octave above f_{3dB}.

We hope, also, that the simple cascaded RC looks wishy-washy next to the active filter. The simple RC and the active filter should show the same f_{3dB}.

9L.1.1 Sweep both filters

Presumably you have been watching the outputs of both filters on the scope, as you drive the two filters with a common input. For a vivid display of the two filters' frequency responses you will want to *sweep* frequencies automatically. You have done this before, using the function-generator's *sweep* function, but this time you may have to do the task a little differently from the way you did it earlier if you used an analog scope.

This time, you *cannot use* the X–Y display mode. Instead, use a conventional sweep (this allows you to watch two output signals, not just one), while *triggering* the scope on the function generator's RAMP output (use the steep falling edge of the ramp).

9L.1.2 Effects of varying the amount of positive feedback: other filter shapes

While *sweeping* frequencies, try altering the shape of the active filter's response by changing R_{gain}. Since you are changing not only shape but also output amplitude, you will need to adjust scope gain each time you change R_{gain}. It is useful to keep the trace sizes of active and passive filters comparable, because the passive response provides a convenient reference against which to judge the active responses. Your substitution box may not offer exactly the values shown in the table (we chose the shown values using AoE's Table 6.2, §6.3.2). Do the best you can. In parentheses we have shown the more-exact value that is desirable, in case you're a perfectionist.

The last two cases, in which we deliberately overdo the positive feedback, are pointless in a filter – but for today's lab they may be useful. They remind us of the boundary we are moving toward in this lab, where positive feedback becomes harmful, and which we'll soon reach.

9L.1.3 Step response; waveform distortion

Watch the circuit's response to a 200Hz square wave, and note particularly the *overshoot* that grows with circuit gain. If you are feeling energetic, you might test also the claim that the tamest of the filters (with $R=1$k), which shows the best step response, also shows the least waveform distortion. The $R=4.7$k filter should show most distortion. Try a triangle as test waveform. The contrast will not be very striking: we saw only a little distortion, from the worst of the filters.

9L.2 Discrete transistor follower

How can the familiar and mild-mannered circuit in Fig. 9L.2 oscillate? To answer that riddle one must answer two subsidiary riddles (as you have seen in §9N.7).

9L.2 Discrete transistor follower

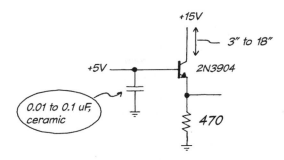

Figure 9L.2 Follower as *oscillator*.

- How does the circuit achieve voltage gain (to sustain oscillations)?
- How does the circuit provide positive feedback?

The answers depend on the truism that power supplies are not perfect, and that stray capacitance can let the transistor talk to itself.

Build the circuit, allowing three to six inches of wire on the +15V lead, and watch the *emitter* (if you try to watch the collector instead, you are likely to kill the oscillation with the scope probe's capacitance). See what "ground" is doing at the foot of the emitter resistor, as well. (To measure this voltage you should ground the scope lead some distance away.)

If the circuit fails to show oscillations, try worsening your supply by making the path for +15V to your circuit more circuitous: pass it through a wire a couple of feet long. When the path is crummy enough (that is, inductive enough), the oscillation should begin.

Remedies

You saw, we hope, a high-frequency oscillation (when we tried the circuit, we saw a sinewave of a couple of volts at about 60MHz; with care, and shortening the supply lead, we were able to coax it up toward 100MHz). Try the remedies that are in your bag of tricks. We propose these below.

Quiet the supply (shrinking the disturbance fed back): You can stop the oscillation by quieting the collector: this "shrinks the disturbance," in the terms we used in Chapter 9N. There are two ways of doing this.

Shorten supply leads: This isn't much fun, and you may believe that it works even before you try it. But a short +15V supply lead probably will stop the oscillation. If you do this, revert to the long, ugly leads, so as to keep the trouble coming.

Decouple the noisy supply: You can quiet the collector with a capacitor to ground. This cap "isolates" the power lines from the transistor circuit. Or, to say the same thing in other terms, provides a local source of charge when the transistor begins to conduct more heavily: the power supply and ground lines need not provide this surge of current, and therefore will not jump in response; instead, the local bucket of charge will provide the needed current.

Try a *ceramic* capacitor, about 0.01–0.1μF. If this fails, take a look at our notes on stability in §9S.3.

But, again, remove that decoupling cap after trying it. We want oscillations, to test the more interesting remedies that follow.

9L.2.1 Diminish high-frequency gain

With the parasitic oscillation screaming away, now let's see if we can stop it by spoiling the high-frequency gain of this follower (which, after all, was not intended as an amplifier!).

Remedy 1: base resistor: See if the remedy works: add a resistor of few hundred ohms between the +5V source and the base. Make sure you *remove* the decoupling cap that was making the base voltage "stiff" at high frequencies. We hope the base resistor stops your oscillation. If not, try larger values of R.

Remedy 2: ferrite bead: Another way to tame this oscillation – arguably more elegant, because it kills the high-frequency oscillation without degrading low-frequency R_{out} – is to place a *ferrite bead* on the base lead (or in series with the lead, if the bead is surface-mount, as most now are). This increases the base lead's inductance, and also makes it "lossy," so that at high frequencies it looks resistive rather than inductive (inductive is risky, because it can resonate with stray capacitance, bringing on oscillations again). A typical bead's frequency response is shown in Fig. 9L.3[2]

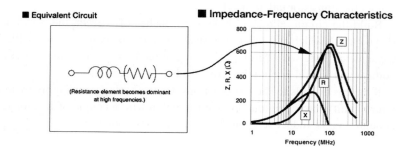

Figure 9L.3 Frequency response of a Murata BLMP21P ferrite bead.

We will not use a surface mount part, of course, but rather a cylindrical bead. Our usual hookup wire, 22 gauge, is too heavy to fit through the bead. Use thinner wire, for example 30-gauge Kynar. If you form two loops, you will quadruple the effective impedance[3] Fig. 9L.4 shows what the bead looks like on a wire.

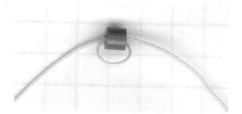

Figure 9L.4 Ferrite bead with fine wire looped through it.

9L.3 Op-amp instability: phase shift can make an op-amp oscillate

9L.3.1 Simple op-amp amplifier, with various loads

Low-pass filter in the feedback loop: The circuit in Fig. 9L.5 is a familiar non-inverting amplifier, with a heavy capacitive load that makes it marginally stable. As load, use about ten feet or so of BNC

---
[2] The curves are typical, presented in an "engineering note" by Vishay: http://www.vishay.com/docs/ilb_ilbb_enote.pdf.
[3] A second loop doubles the magnetic field, for a given current, and doubles the response to the magnetic field.

9L.3 Op-amp instability: phase shift can make an op-amp oscillate

cable (30pF/foot). If you don't find a long BNC, join several shorter sections with links or with a BNC "Tee".[4]

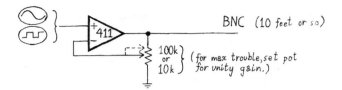

Figure 9L.5 Load capacitance can make a feedback circuit unstable.

Your first task is to coax this circuit into oscillation. To do this, *minimize* the gain: turn the pot to feed back *all* of the output signal.

The circuit may not oscillate, at first. If it does not, drive it with a square wave. That should start it, and once started it is likely to continue even when you turn off the function generator, or change to a sinewave.

Remedy 1: Shrink the disturbance fed back: Now – with the circuit oscillating – gradually turn the pot so as to shrink the fraction of V_{out} that is fed back. (What are you doing to *gain*, incidentally?) When the oscillation stops, see how the circuit responds to a small square wave (make sure the output does not saturate). If you see what looks like ringing (or like the decaying oscillation of a resonant circuit), continue to shrink the fraction of V_{out} that is fed back until that ringing stops.

We hope that you now have confirmed to your satisfaction that stability can depend on the gain of a feedback circuit. This fact explains why so-called "uncompensated" op-amps are available: if you know you will use your op-amp for substantial gain, you are wasting bandwidth when you use an amp compensated for stability at unity gain.

Other remedies: But suppose you want unity gain, and need to drive a long coaxial cable? You could try any of several solutions. Laziest would be to shop for an op-amp or buffer designed to be stable when loaded with a lot of capacitance ("a lot" is something above, say, 100pF).

Try another pair of remedies, here. Try them separately. See if either is sufficient. If neither is, you might try them together.

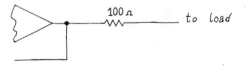

Figure 9L.6 Series R can take shifted point outside feedback loop.

- Series R: take the load outside the feedback loop. Put a 100Ω to 220Ω resistor in *series* with the BNC cable, and *outside* the feedback loop, as in Fig. 9L.6. See if this stabilizes the amp.
- Snubber: connect a small-valued R AC-coupled to ground, as in Fig. 9L.7.

The snubber is a little mysterious. It appears to kill high frequencies in an unusually crude way: by loading the output heavily at high frequencies – a range where Z_{out} is rising, as the impedance of the snubber shrinks toward its minimum, the value of R alone. (The impedance of the snubber shown in Fig. 9L.7 falls to $\sqrt{2}R \approx 14\Omega$ at around 160kHz).

---

[4] At high frequencies, this might be a poor idea, but the T works fine at our op-amp frequencies.

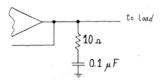

Figure 9L.7 Snubber.

9L.4 Op-amp with buffer in feedback loop

The circuit in Fig. 9L.8 resembles others that you have built before. The difference this time is that we'll be looking for trouble.

Use big power transistors: MJE3055, MJE2955. You used these back in the push–pull exercise of §6L.8: terminals are BCE seen from the front (the side with the part number labeling).

The key change that makes trouble very likely is our adopting an op-amp that has gain higher than '411's. We ask you to use an LT1215, a device whose gain starts higher (10^6 at DC, versus the '411's 200,000) and extends out to 23MHz (versus the '411's 4MHz). So, the op-amp has not only more virtue but also more potential vice, at frequencies where phase-shifts inside the feedback loop become dangerously close to converting negative feedback into positive. We hope it will buzz for you, so that you can try a new remedy.

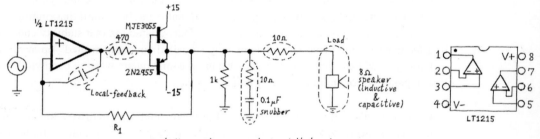

Figure 9L.8
Push–pull buffer in feedback loop.

(Note that the pinout of this dual op-amp matches that of the LM358, standard for duals, but does not match that of the more familiar LF411.) Try the circuit without speaker attached: feed the circuit a sinewave of a volt or so, at around 1kHz, and confirm that the circuit follows, without showing crossover distortion[5].

Now, as load, add an 8Ω speaker. (If the sound is obnoxious, lower the amplitude.) We hope you will see oscillations at the output of your amplifier (they may not affect the *sound*, however).

Persuade your circuit to show parasitic oscillations, brought on by the low-frequency sinewave. Determine the approximate *frequency* of the parasitic fuzz, so as to decide who's to blame, op-amp or transistors. If the frequency is in a range where the op-amp still has substantial open-loop gain, you can safely assume the op-amp is to blame. Oscillations caused by the transistors are likely to be much faster. (We got the handsome oscillations shown in Fig. 9L.9.)

Cheat, if you must, to get an oscillation: If you cannot get your circuit to misbehave with speaker as load, try an installed capacitive load: say, $0.001\mu F$ to ground. Experiment with the C_{load} that brings on oscillation. But be warned that too large a C, curiously enough, will prevent oscillation, simply by overburdening the op-amp, in the manner of the *snubber* circuit.

---

[5] If you're "lucky," you may be blessed with a parasitic oscillation even in this preliminary test.

9L.4 Op-amp with buffer in feedback loop

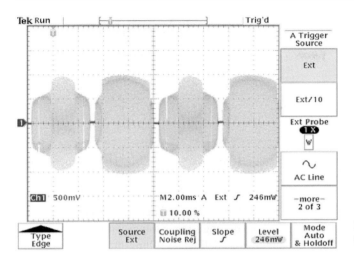

Figure 9L.9 Push–pull fuzz: no split feedback, speaker attached.

Remedies: Then try (separately) the following remedies; see if you can stop the oscillations.

- Decouple the power supplies (if you haven't already done so).
- Add a base resistor of a few hundred ohms, if you suspect the transistors.
- Add local (or "split") feedback: 100pF direct from op-amp output to the inverting input. You will need an R in the *long* feedback path from circuit output, to mate with this new C. The idea is to let the R provide feedback at moderate frequencies, C to provide feedback at the higher frequencies where phase-shifts make the circuit unstable. Choose R to equal X_C at some frequency below the frequency at which you have observed the oscillation.
- Add a snubber RC (10Ω, 0.1μF, again). See if the snubber *alone* can stop the oscillation (remove "split feedback" to do this test).
- Once the op-amp is stabilized – with split feedback or with snubber – see if removing the base resistor disturbs the circuit.
- Try a small resistor (start with a value under 10Ω) *outside* the feedback loop, in series with the speaker or C_{load} (nothing new to you: see Other remedies on page 377).

We hope you can quiet the amplifier. If you can't, ask for help!

9S Supplementary Notes. Op-Amps IV

9S.1   Op-amp frequency compensation

AoE §4.9

AoE §2.5.4B

We have noted elsewhere that all the op-amps we meet in this course use internal "frequency compensation" that makes them stable – at least, if we refrain from putting strange things within their feedback loops. Frequency compensation, surprisingly enough, means deliberate rolling-off of the amplifier's gain. This may seem perverse – especially given Black's fundamental insight that "excess gain" is exactly what is necessary to achieve the benefits of negative feedback.[1] But the need for stability is so fundamental that every other concern must give way in its pursuit.

The '741 op-amp, a hugely successful part, seems to owe its success to the fact that it was the first internally compensated op-amp – and therefore delightfully easy to use.[2] About 45 years after its introduction by Fairchild (the company that first developed the first fully integrated circuit) the '741 still occasionally appears in new designs – at least, in designs from developing countries, where its wide availability and low price may be especially appealing. The '411 op-amp that we use in most of our labs uses a design fundamentally the same as the '741's, and includes internal compensation, now standard. The major difference between the two parts is the '411's use of field-effect transistors at the input stage.

A look inside the '411 circuit shows a capacitor planted between output and input of the high-gain stage; this capacitor performs just as the cap does when placed between output and inverting terminal of an op-amp: it forms an *integrator*.

This rolloff and phase shift (or "integrator behavior") is imposed because it is better than the alternative: excessive phase shifts that would bring instability.

9S.1.1   Waveforms to persuade you the op-amp really *does* integrate

A skeptical reader might well reject what we have just said: haven't you seen, over and over in the lab, that an op-amp does *not* show phase shift? You have watched input and output of followers and amplifiers on the scope. A follower *follows*, faithfully reproducing the input waveform, does it not?

You are entitled to be skeptical. But the way to reconcile what you have observed with we have just claimed is to recall that every op-amp circuit you have seen – with the unimportant exception of a couple of comparator circuits – *included negative feedback*. That feedback hides the op-amp's internal integration, just as it hides any other annoying transfer function that are put within the loop. It hid the push-pull's crossover distortion; in general, it hides the "dog" that we sketched in §6N.9.

---

[1] By now, the language of Harold S. Black's patent probably is becoming familiar to you (patent 2,102,671 [1937]): "...by constructing a circuit with excess gain and reducing the gain by negative feedback, any desired degree of linearity between output and input...can be realized, the limiting factor being in the amount of gain that can be attained...."

[2] Walter Jung, Chapter H, "Op-amp History," in Analog Devices' http://www.analog.com/library/analogDialogue/archives/39-05/op_amp_applications_handbook.html. So, engineers gobbled up the part that did not require adding one small capacitor. I always take comfort in evidence that the rest of the world is as lazy as I am!

9S.1 Op-amp frequency compensation

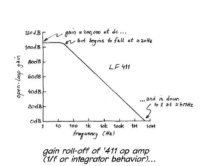
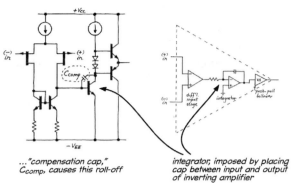

Figure 9S.1 Gain rolloff for compensated LF411 op-amp – and its cause.

Only if one runs the op-amp *open loop* does the op-amp's integration become visible. Doing this experiment is a bit of a chore, so we have not asked you to build this circuit. One must carefully adjust the DC-offset of the input to keep the op-amp from saturating; then apply a small wiggle. When we did this, we got the waveforms of Fig. 9S.2, which expose the op-amp's integration. At the lowest frequencies – at 4Hz, in the figure – only a little phase shift appears; at 100Hz, the full $-90°$ shift of the integration is evident.

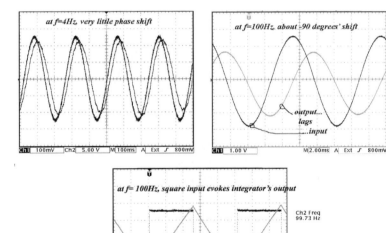

Figure 9S.2 Op-amp's internal integration becomes visible if one runs it *open loop*.

9S.1.2 Why compensation? Considering the hazards to an uncompensated op-amp

AoE §4.9

The uncompensated op-amp would need help: This rolling-off of gain must occur in order to keep the circuit stable – whether the roll-off is done for us ("internal compensation") or by our attaching an external "compensation" capacitor. To see why it is necessary, we need to note that the op-amp, a multi-stage circuit (see Fig. 9S.3), behaves like a series of lowpass filters.

It behaves like this because its several stages form a series of non-zero source resistances – two of

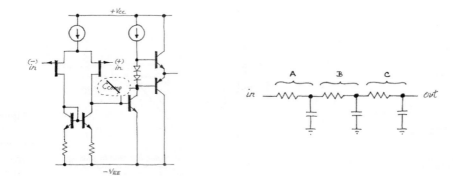

Figure 9S.3 An uncompensated op-amp behaves like two or three lowpass filters, cascaded.

them actually *current sources*[3] – driving stray capacitances. If we didn't intervene, we would find that the cumulative phase shift grows alarmingly, see Fig. 9S.4.

AoE §4.9.1

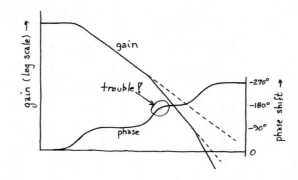

Figure 9S.4 An uncompensated op-amp would show deadly phase shift.

We cannot tolerate a phase shift as large as 180° under conditions where *loop gain* – the signal amplified and then fed back – is as much as unity. Loop gain, you may recall usually is written as the product AB, where A is the op-amp's open-loop gain (the characteristic plotted in Fig. 9S.4) and B is the fraction fed back.

Two possible solutions...: Two solutions are available, in principle (this we have said in Chapter 9N), for stabilizing such an op-amp: reduce B, by feeding back less than all of V_{out}; or reduce A by deliberately degrading the amplifier's gain as frequency climbs.

AoE §4.9.1A

Limit B: even an uncompensated op-amp can be stable provided very little is fed back...: The first of these remedies would look like Fig. 9S.5: reduce the fraction fed back by putting a divider in the feedback loop (this is a gain-of-1000 amplifier).

... but uncompensated op-amp as follower goes crazy: An uncompensated version of the classic 741 op-amp has too much gain where its phase-shift hits $-180°$, and wired as a follower it does, indeed, go berserk, see Fig. 9S.6.

---

[3] Of course, the current sources here do fit the description, "non-zero source resistances."

9S.1 Op-amp frequency compensation

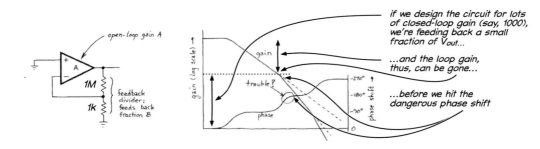

Figure 9S.5 An abnormal solution: even an uncompensated op-amp could work provided little enough is fed back.

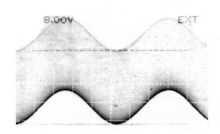

Figure 9S.6 Uncompensated op-amp, wired as follower, oscillates.

...Limit A: but the normal method is to limit high-frequency gain: As you can imagine, designers prefer to work with an op-amp that is stable even when not asked for very high gain. Best of all is an op-amp (like all that we use in lab) that is stable even when the *entire* output voltage is fed back (in other words, when it is wired as a follower). To tolerate 100% feedback ($B=1$), the op-amp must have its gain ruined before phase shift reaches $-180°$. Figure 9S.7 shows a follower and the roll-off of gain that it requires.

AoE §4.9.1

Phase margin: As the compensated op-amp approaches its maximum frequency – and its gain approaches unity – its phase shift flirts with trouble: moves beyond $-90°$, but stopping short of disaster.

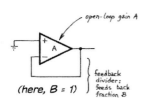

the most challenging circuit: a mere follower: since it feeds back 100%, it requires conservative gain roll-off

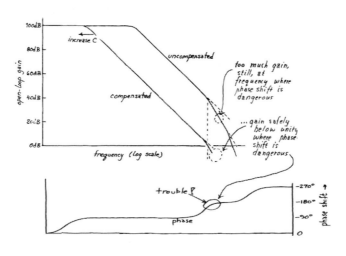

Figure 9S.7 The usual solution: roll-off open-loop gain (A) for stability even at $B=1$.

Supplementary Notes. Op-Amps IV

The distance from "disaster" – from $-180°$ – is called the op-amp's "phase margin." Most op-amps put that at between 45° and 60°. The plot in Fig. 9S.8, from the '411's data sheet, shows about 50° margin (wrongly labeled "phase" rather than "phase margin").

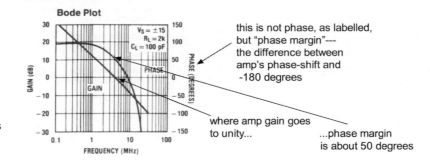

Figure 9S.8 Op-amp phase shift approaches 180°, but stops safely short.

This detail reminds one why even somewhat less than a $-90°$ shift imposed by something put into the feedback loop can send a circuit off the rails.

9S.1.3 Decompensated and uncompensated op-amps

AoE §4.9.2B

Uncompensated op-amps: One can buy an op-amp that leaves it to the user to install that gain-killing capacitor (external to the IC, of course). Early op-amps worked this way; after the '741 they became exceptional. Such an "uncompensated" amplifier is offered because internally-compensated op-amps give up gain in order to make the devices stable in the hardest case, the follower circuit. Fig. 9S.7 contrasted the gain rolloff of a generic compensated versus uncompensated op-amp. At unity gain, the uncompensated op-amp would be unstable; its compensation could be adjusted to fit the closed-loop gain in a particular application.

Decompensated op-amps: An intermediate possibility is a "decompensated" op-amp:[4] one that would be unstable at unity gain (where you feed back 100%) but stable at some closed-loop gain such as 10 (where you feed back only 1/10). If you know that your circuit will not push circuit gain below 10, then you can afford higher open-loop gain at a given frequency. The Linear Technology LT6230, for example, a 215MHz part, is offered in two versions: one compensated the other decompensated – rigged to be stable for gains of 10 or more. The curves in Fig. 9S.9 (pretty hard to read, because they are so dense with information) contrast the two versions.[5]

Since the circuit with gain of 10 runs with reduced B, the circuit can be stable with open-loop gain, A, boosted, as well. You'll notice that the decompensated part shows higher gain at every frequency. It can afford that, because it is the *loop gain* – the product AB – that matters, and a gain-of-ten circuit holds this product to a level that is safe for this op-amp.

9S.2 Active filters: how to improve a simple RC filter

Figure 9S.10 shows a linear amplitude-versus-frequency plot for an ordinary *RC* lowpass filter. This response disappoints us in two ways:

---

[4] AoE points out that a better name would be "under-compensated."
[5] Linear Technology: LT6230/LT6230-10 datasheet.

9S.2 Active filters: how to improve a simple RC filter

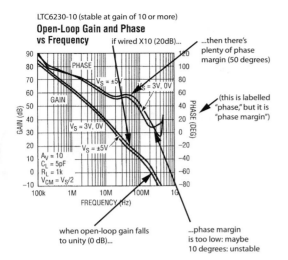

two versions of an op amp: DEcompensated (unstable at unity gain)...

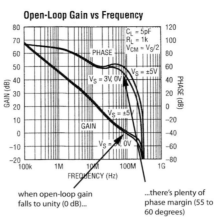

...versus COMPENSATED (stable at unity gain)

Figure 9S.9 This op-amp comes in both compensated and decompensated versions. The latter offers higher gain, but is unstable as a follower.

Figure 9S.10 *RC* lowpass response (linear plot).

- its pass-band is not as flat as we would like. That's evident in the plot shown: not much like the flat pass-band of a textbook drawing!
- its fall-off is not as steep as we would like. We have gotten used to "−6dB/octave," characteristic of our simple *RC* lowpass, but sometimes we'd like more.

Simply cascading *RC*s does not solve both problems: it makes the ultimate fall-off twice as steep (−12dB/octave) – but it does only a very little to boost the droopy passband.

At first glance, the cascading seems to make the *pass-band* even worse: droopier. This is illustrated and confirmed with a scope photo in Fig. 9S.11[6] comparing a single *RC* with $f_{3dB}=1$kHz, against a cascade of two *RC* lowpass filters, each with $f_{3dB}=1$kHz.

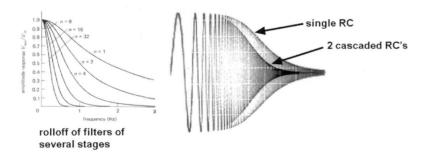

rolloff of filters of several stages

Figure 9S.11 Cascaded *RC*s look pretty bad: the passband looks droopy...

---

[6] The left picture is taken from AoE Fig. 6.2; the two parts of Fig. 9S.11 can't be compared directly, because the scope image is *log–linear* (log frequency, but linear amplitude), whereas AoE plots here are linear on both axes.

Supplementary Notes. Op-Amps IV

The frequency that was f_{3dB} now is f_{6dB}: the output is down 6dB. But to compare passbands fairly, one ought to compare filters with the same f_{3dB}'s. A comparative plot so "normalized," makes the cascade look better; see Fig. 9S.12. In fact, the cascade's passband is *less* droopy than the single RC's – a result that may contradict your intuition.

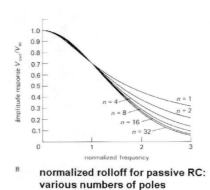
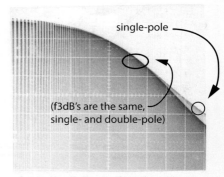

Figure 9S.12 Normalized plot of frequency response of single versus 2-cascaded RC lowpass filters: cascade looks better. The left-hand image reproduces AoE's Fig. 6.2B.

normalized rolloff for passive RC: various numbers of poles

detail of rolloff: single-pole vs 2-pole

If nothing better were available, we would live with these droopy passbands, in order to get the steep ultimate roll-off. However, LC filters can give better performance – and, as you will see when you do Lab 9L, so can op-amp *active filter* circuits, which are able to mimic the performance of LC filters without the use of inductors.

To get a better filter *shape* – that is, frequency response – we need a design trick that lets us flatten the passband while getting a good steep roll-off.

In order to flatten the pass-band of a simple cascaded two-RC lowpass, we need to give the filter output a boost (a kick in the pants) just where it is most wishy-washy: just below f_{3dB} (*below*, in the case of a lowpass, anyway). On the other hand, in order to steepen the ultimate fall-off, we need only duplicate the effect of 2 similar RCs in series. Soon, you may find yourself getting fussier, and caring about finer points, like the steepness of the *initial* roll-off (close to f_{3dB}). But for a start, let's try whether we can see intuitively how a particular active filter manages to flatten the passband of an RC lowpass while achieving the -12dB/octave roll-off of a "2-pole" filter (the ultimate slope you would see from two RC lowpass filters in series).

9S.2.1 An intuitively-intelligible active filter: VCVS

AoE §6.3.1

AoE calls the filter in Fig. 9S.13 "even partly intuitive," implicitly acknowledging that most active filters are depressingly non-intuitive: their designs reflect messy math, and therefore force one toward cookbook design methods that are not instructive. The VCVS[7] "Sallen–Key" filter shown in the figure makes both of the improvements that we called for.

Active filter flattens the passband...: Positive feedback boosts the passband, feeding back what amounts to a *bandpass* kick. Why "bandpass?": the feedback is *high-pass*, but the source of the feedback is the output of a *low-pass* (the basic circuit we are building); the product of the two is bandpass, hence the well-placed and well-shaped kick in the pants.

What we have described as a bandpass kick could also be described as akin to bootstrapping C_1 – and then some: not only does C_1 not burden the signal fed by R_1 at frequencies below f_{3dB}, it actually

---

[7] VCVS: "voltage controlled voltage source:" a name that does nothing to demystify this circuit!

boosts boosts the signal that is fed to the passive lowpass formed by R_2C_2. This is because, in this passband frequency range, the far side or "foot" of C_1 is driven *in-phase* with the input signal and at *larger amplitude*.

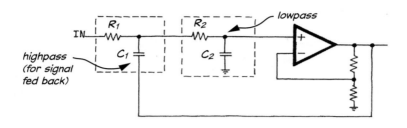

Figure 9S.13 2-pole active filter: *VCVS* (lowpass), one variant of Sallen–Key filter.

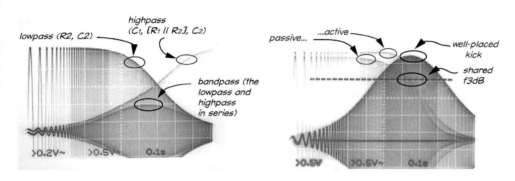

Figure 9S.14 Active filter seen as bandpass, in passband.

In the "stop band", the first capacitor, C_1, *loads* R_1 instead of boosting it, since the circuit output – applied to the foot of C_1 – is small on account of the attenuation imposed by the R_2C_2 lowpass. Deep into the stop band, the filter behaves like the cascade of two simple *RC* lowpass sections. (It's not quite as clean as that, since the two sections do not show the 1:10 impedances that would prevent their interaction.) Hence the steep ultimate roll-off (-12dB/octave).

...Active filter can achieve very-steep rolloff: A mystery not explained by this 'kick in the pants' argument is the very steep rolloff achievable close to the cutoff (or f_{3dB}) frequency. Phase shifts, ganging up fortuitously explain it, we suppose: not a very satisfactory answer, but about all that's available in this non-mathematical view.

Figure 9S.15 shows the filter's response plotted against the response of a simple *RC* and two cascaded *RCs* all sharing a common f_{3dB}. The 2-pole active filter flattens the passband while steepening fall-off.

See AoE Fig. 6.30, comparing responses of three filter types

The active filter can be *tuned* to optimize what you choose: Adjusting the gain lets one optimize one filter characteristic or another, always at the expense of some other filter behaviors. One gain gives flattest passband (called a "Butterworth" filter); another gain (higher) provides steepest rolloff (called a "Chebyshev" filter); yet another gain (lower) preserves waveshape best ("Bessel"). In Fig. 9S.16 we show the effect of three gain settings.

Since it is *positive feedback* that boosts the passband, we need to use this medicine judiciously: too much, and we'll first make a bump in the passband, in place of droop – and then, if we really overdo it, we'll kick the circuit into uncontrolled, continuous oscillation. Fig. 9S.16's oscilloscope images show a range of possibilities. These images result from various amounts of positive feedback (circuit gain varies, too; these plots have not been normalized for constant amplitude in the passband).

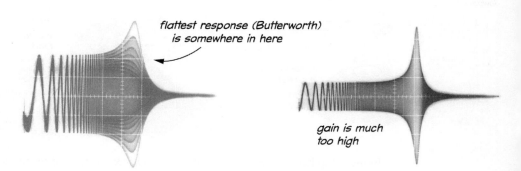

Figure 9S.15 2-pole VCVS filter: compared to simple RC, and to a pair of RCs cascaded.

Figure 9S.16 Filter's frequency response depends on strength of positive feedback: multiple exposure shows several response shapes.

Higher-order filters: If you want a 3- or 4-pole or higher-order filter, you can cascade active 2-pole sections. Here our intuitive grip slips almost entirely: the several sections are *not* alike, and the explanations for the strange coefficients (gains, in the VCVS form, and f_{3dB} values) become mercilessly mathematical.

We find some comfort in the thought that the big, ugly equation that describes the desired response of, say, a 4-pole filter (which shows powers up to four in frequency) can be factored into the somewhat cozier quadratic form that you have already built – a modest 2-pole filter. But apart from that, these fancy filters are distressingly mysterious. If the mystery doesn't offend you, you'll find you can build good active filters by using the tables provided in AoE (table in §6.3.2: lowpass, four alternative shapes).[8]

For the ultimate in cookbookery, as pointed out in Chapter 9N, you can invoke TI/Burr–Brown's filter design program, FilterPro. You may feel like a dope, but the program will deliver a good design. However you proceed, we hope that today's glimpse of an active filter in lab will embolden you to use these circuits.

AoE §6.3.3

Compare AoE Fig. 6.27

Transient response; shape preservation: Fig. 9S.17 shows responses of the VCVS 2-pole lowpass, at three gain settings: tamest (best time delay and shape preservation); flattest; and steepest. (These are the gains you will try for yourself, in the lab.) You can see that overshoot grows with gain, and

---

[8] And here's a friendly and clear book, if you feel like learning more about active filters: Don Lancaster's *Active Filter Cookbook* (Sams, 1975).

that the *steepness* of initial rolloff also grows with gain. As usual we pay a price in once characteristic when we improve some other.

The bottom image in Fig. 9S.17 shows a less striking contrast among the three filters' treatment of a triangle. The bottom image shows the Bessel doing a somewhat better job of preserving the triangle's shape, as advertised.

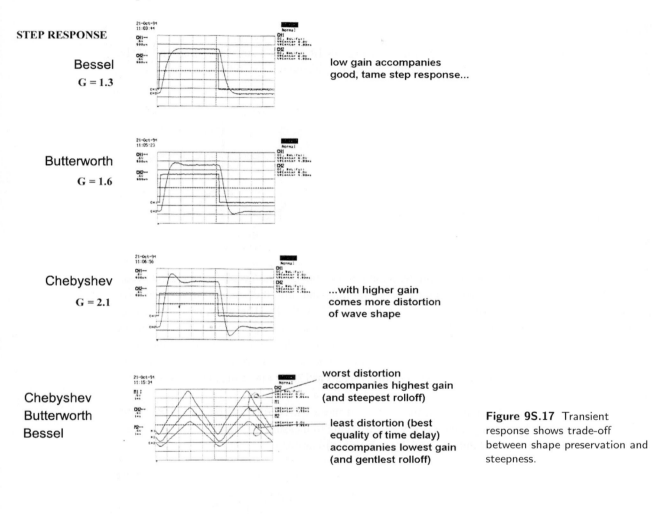

Figure 9S.17 Transient response shows trade-off between shape preservation and steepness.

9S.3 Noise: diagnosing fuzz

Who's to blame if your circuit output (and sometimes *input*, as well!) looks fuzzy? 'An oscillation,' probably – though not certainly. But what's oscillating? Usually, you can find out pretty easily. Sometimes, it's *not* easy!

9S.3.1 Preliminary question: is it a parasitic oscillation?

Is the oscillation "parasitic?" Not necessarily. *Something* is oscillating, but the fuzz may not be your circuit's fault. Consider a few other likely causes. It will help to know the frequency of the fuzz but, oddly enough, it may not be easy to judge at once whether the offending noise is at a frequency much *lower* or much *higher* than your signal's. You may, at a glance, know only that there's fuzz. Adjusting the scope's sweep rate should settle the question.

Then, if the frequency is much lower, usually the problem is 60Hz pickup; if the frequency is very much higher, probably you're seeing pollution by radio or TV broadcast. In-between frequencies may indicate parasitic oscillations.

9S.3.2 One pollution: line noise

An example: The lower trace in the scope image of Fig. 9S.18, shows a sinewave at around 1kHz, with a displeasing fuzziness or thickening that the focus knob would not fix. What's wrong?

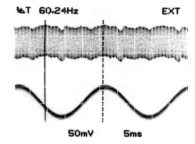

Figure 9S.18 Stable low-frequency bumps on your higher-frequency signal waveform indicate *line* noise. (Sweep rates: top trace, 5ms/div, bottom trace, 0.2ms/div.)

It could be either high-frequency or low-frequency noise. 60Hz "line noise"[9] is a troublemaker whose work is easy to spot: try triggering the scope on *line*, and drop the scope's sweep rate down to a few ms/div. That's what we did to get the top trace of Fig. 9S.18. The wobbles stabilized for us. When this occurs, the effect will be evident at a glance. In addition, the waveform in the image repeats at 60Hz, as the cursors indicate (along with the $1/\Delta T$ calculation displayed above the upper trace). That's not surprising, but does tell us a little more than we can learn from the fact that *line* triggering stabilizes the display. Sometimes *line* noise repeats at 120Hz rather than 60Hz, as we'll see in a moment.

Incidentally in Fig. 9S.19 is the circuit that produced these waveforms: we put in a 10M series resistor, then watched the op-amp input (as if we were trying to measure R_{in}, say). The measurement point is driven very flimsily, and thus is specially vulnerable to noise.

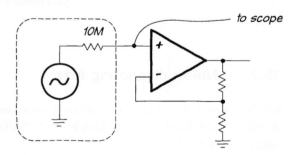

Figure 9S.19 High-impedance drive at op-amp input makes that point vulnerable to noise pickup.

The troubleshooting point, reiterated: If you see stable bumps when triggering on *Line*, then your bumps are *synchronized* with the 60Hz power supply, and almost surely they are *caused* by that supply.

---

[9] We expect that by now you have picked up this piece of jargon meaning "noise resulting from the 60Hz oscillation of the power line voltage" (50Hz in Europe).

- If the bumps repeat at 60Hz (or 120Hz, the repetition rate of the full-wave-rectified waveform put out by the power supply), still there is more than more one likely cause:
    - your circuit may be picking up the 60Hz, at some *high-impedance* input (we just saw this in the circuit of Fig. 9S.19, and it's the more common cause);
    - alternatively, you may be looking at the effect of ground noise – a voltage difference between the grounds on your several instruments, caused by currents flowing in long ground "returns." Make sure that all your instruments are plugged into a single power strip or, anyway, a single power outlet. This "ground loop" problem is explained a bit further in the next subsection.
    - If the bumps are of *one polarity and repeating at 120Hz*, you're probably looking at the effect of ground- and supply-current surges that occur when your power supply filter capacitors recharge, each half-cycle of the *line* input. This you cannot fix – except with the sidestep of finding a quieter supply, as we suggest in §9S.3.3.

9S.3.3 Another pollutant: ground noise from ground loop

Figure 9S.20 illustrates the second cause, not pickup but ground noise resulting from plugging instruments into differing outlets. The top trace in the figure shows a sinewave as it comes into a lab breadboard from a function generator. The trace looks sinusoidal, but fuzzy. Slowing the sweep and triggering on *line*, as usual, we see what the second trace shows. The bumps stabilize, so that this seems to resemble the *pickup* case.

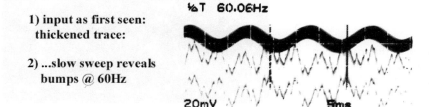

Figure 9S.20 Ground noise pollutes every trace.

But look at the circuit: Fig. 9S.21. The fuzz is not the fault of the generator, and it is not pickup. This time the impedance at the input of the circuit is *low*: not the 10MΩ of the earlier case, but the 50Ω of the function generator. So "pickup" is not likely.

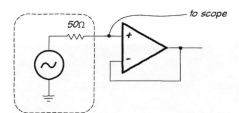

Figure 9S.21 Circuit showing bad 60Hz noise at input: this one is not vulnerable to pickup.

Moving the oscilloscope power plug from a power strip a couple of feet down the bench into a power strip shared with function-generator and power supply solved the problem. The input now looks like trace 3 in Fig. 9S.22. (The sweep rates in Figs. 9S.20 and 9S.22 happen to differ). The line noise is barely apparent in the latter.

The ugliness of Fig. 9S.20 resulted from the fact that two instruments, scope and function generator, by chance were plugged into separate room outlets. These separate outlets may not merge their third

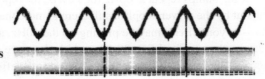

Figure 9S.22 Ground loop fixed: traces lose their 60Hz noise.

3) improved ground makes sine cleaner...

4) ...and slow sweep shows no obvious 60Hz noise

"world-ground" lines for a considerable distance. So, this is a ground loop at work, one formed by the long ground lines. When this loop is "cut" by a 60Hz magnetic field (caused by currents flowing in other power lines in the building), currents are induced in this loop. The noise results from a voltage drop as these currents flow in the non-zero impedance of the ground lines.

Another ground loop instance: The scope image in Fig. 9S.23 shows the same ground loop behavior, but to more spectacular effect. "What is this mess?", you may find yourself exclaiming, if your carefully built circuit evokes such a display.

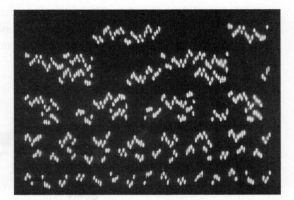

Figure 9S.23 Ground noise that can occur when instruments do not share a common ground definition: it can make a display unintelligible!

The image in Fig. 9S.23 is trying to show us a 4-trace "scope multiplexer" display (something you may build in a digital lab later in this course). The image in Fig. 9S.24 shows what it ought to look like. The traces are *chopped*, and the display in the figure is correct: this is what a 4-bit counter should look like. The hideous image that we showed in Fig. 9S.23 comes, once again, from a scope that is plugged into an outlet strip different from the one used for breadboard and function generator: pretty awful!

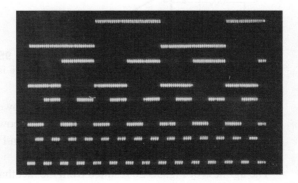

Figure 9S.24 4-channel scope display as it should look.

9S.3 Noise: diagnosing fuzz

More commonplace ground noise: poorly implemented ground: Here's the *third* case – where the bumps are of *one polarity and repeating at 120Hz*. The PB503 breadboard has a flimsy ground that jumps during the supply-current surges that occur when its filter capacitors recharge, each half-cycle of the *line* input. For this problem, there's not much you can do, short of substituting a quieter power supply. Our PB503 powered breadboards show a lot of this ground noise; a good lab power supply, like our Agilent units, shows no such defect. Lab 7L suggests such a change of supply in the discussion of the microphone amplifier circuit.

Figure 9S.25 illustrates this effect. The top trace shows what you're likely to see first: again a thickening or fuzziness of the output trace. (This time, the thickening is slight.) Trigger on line, slow the sweep rate, and the wiggles stabilize; then the 120Hz repetition rate reveals the cause is not 60Hz pickup, but the power-supply problem. The most revealing signal is the bottom trace: the ground line itself is bumpy! Time to reach for a better supply.

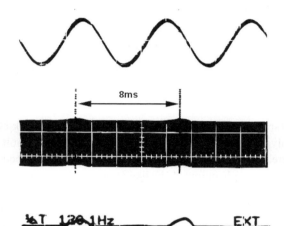

Here's the initial observation: a slightly thickened signal trace.

Now if you suspect the noise is line or power supply, trigger on "line" and slow the sweep rule. If the bumps stabilize, the supply is the cause.

A close look at the ground line itself shows the supply noise (caused by the current surges as the filter cap is charged on each half cycle of 60Hz).

Figure 9S.25 Ground noise pollutes every waveform observed in the polluted circuit (sweep rate much higher in top trace than in lower two).

9S.3.4 Another pollutant: RF pickup

At the other frequency extreme, you may notice fuzz – usually low-amplitude – that you can resolve by sweeping the scope very fast. The lower trace in Fig. 9S.26 shows such indeterminate fuzz. If you can trigger on this fuzz and sweep fast, you can confirm that it's in the multi-MHz range. This we have done in the upper trace of the figure. You then can complete your confirmation of *radio* as the culprit by *turning off* your circuit and watching its input. If your circuit is not to blame, the input noise will persist.

There's no easy remedy for this problem. In principle, what you want to do is shield your circuit – or at least minimize wire lengths at high-impedance points. Low impedance, too, makes your circuit less vulnerable to such pickup. It is useful, in any case, to know that the fuzz is not your circuit's fault: you won't now spend time chasing through your circuit to find causes for its misbehavior.

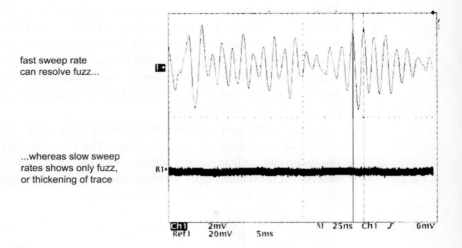

Figure 9S.26 Vague fuzz sometimes will resolve into high-frequency sine, if you sweep fast (top trace shows fast sweep, bottom trace slow sweep). Note differing scope gains on the two traces.

9S.3.5 True parasitic oscillations

If it *is* an oscillation, what's oscillating? Often the *frequency* of oscillation can settle this question.[10] Let's look at two examples.

Example 1: Discrete-transistor follower: Look back at Fig. 9L.2, which shows a case where there's only one suspect.[11] If the output trace is fuzzy, the fault is the transistor's; no detective work required, here, but this easier case may help us to solve the second case, in Fig. 9S.28. Fig. 9S.27 shows what the follower's output looks like at two sweep rates.

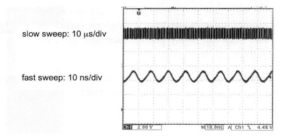

Figure 9S.27 Follower's fuzz resolves into RF frequency — but this time it is *not* pickup!

Example 2: Op-amp and power push–pull: Here's a harder case: who's to blame for the fuzz at the output of the circuit in Fig. 9S.28? At a glance, looks as if either op-amp or transistors could be to blame. Let's look at the the output of this circuit.

First, in Figure 9S.29, there's the rather spectacular craziness that appears when we sweep fairly slowly – as we would if we were looking only for the audio output waveform.

A faster sweep reveals the frequency of oscillation: see Fig. 9S.30. This oscillation frequency – a few hundred kilohertz (500kHz) – is implausibly low for a transistor's parasitic oscillation, as in the follower that we saw a little earlier (in §9L.2). So, we blame the op-amp, and apply op-amp remedies

---

[10] For this observation and for a good discussion of stability problems, we are indebted to T. Frederiksen, *Intuitive IC Op Amps*, p. 154ff. National Semiconductor Corporation (1984).

[11] Could the fuzz be radio pickup? Not likely, because the point of observation, the transistor's emitter, should present a very low impedance.

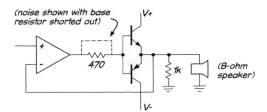

Figure 9S.28 A circuit whose oscillations could be blamed on either op-amp or transistors.

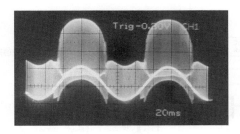

Figure 9S.29 Wild parasitic fuzz on push–pull output, driving speaker.

(a shorter path that bypasses the push-pull at high frequencies: see Fig. 9L.8). Apparently it's the op-amp that's buzzing. We're in a frequency range where accidental lowpass elements in the feedback loop can cause phase shifts that approach their maximum values, and the op-amp still has gain. So, negative feedback has turned positive.

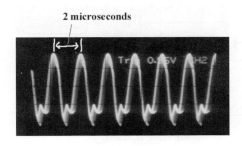

Figure 9S.30 Fast sweep reveals frequency of parasitic oscillation once more.

Why the standard remedy, decoupling, sometimes fails: Sometimes an oscillation gets *worse* when you add the usual small ceramic decoupling capacitor to the power supply. Probably, by bad luck, the capacitance that you just added formed a *resonant circuit* with the supply-line's stray inductance. Try a ceramic of $10\times$ different value; try a tantalum cap, instead of the ceramic. If all else fails, try a small series resistor between supply and cap – though here you of course give up some of the nice low impedance you were after when you put the cap in. If the cap does form part of a resonant LC, this R – usually considered a vice in a capacitor – lowers the Q of the unwanted resonant circuit, quieting it.

9S.4 Annotated LF411 op-amp schematic

You very seldom *need* to know in detail what's going on within an op-amp, but it's fun and satisfying to look at a scary schematic, for example Fig. 9S.31, and realize that you can recognize familiar circuit elements.

If the sheer quantity of circuitry here didn't scare you off, you might recognize at least the following elements without our help: a differential amp at the input stage; a common emitter amp, next; a push–

pull output – and a hall of mirrors (if you've dared to look at §5S.2). The current limit at the output mimics a trick you will see in voltage regulators (Chapter 11N).

Some of the details are subtle, though, so we've done what we can to explain through our annotations. We offer this schematic both as a reviewing device and as a reward for the hard work you put into discrete transistors. We hope that this exercise makes you feel knowledgeable – lets you feel that you are beginning to be able to read schematics that would have meant nothing to you just a few weeks ago.

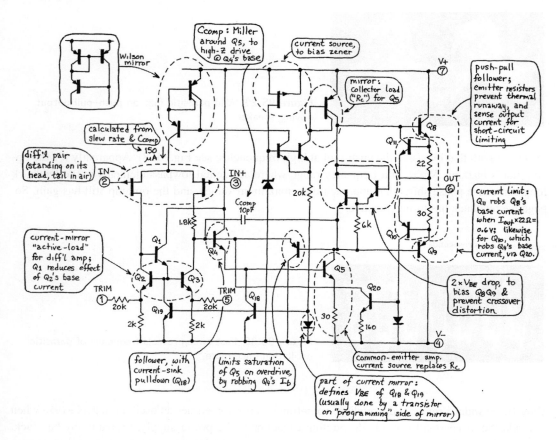

Figure 9S.31 Annotated schematic of LF411 op-amp.

9S.5 Quantitative effects of feedback

AoE §2.5.3

We will leave to Example 9W.1 detailed calculations of the way that *circuit* gain varies with *open-loop* gain. Here, we will settle for an intuitively appealing sketch of the way feedback does its magic. Figure 9S.32 is a generalized model of feedback.

AoE §2.5.2

It is easier to understand if we redraw the B block to show a fraction of V_{out} fed back, as in the usual case. Fig. 9S.33 probably looks pretty familiar.

- B = *fraction fed back* (a characteristic of the circuit, not the chip).
- A = *open-loop gain* (a characteristic of the chip).

9S.5 Quantitative effects of feedback

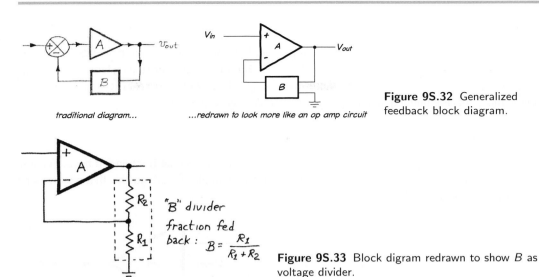

Figure 9S.32 Generalized feedback block diagram.

Figure 9S.33 Block digram redrawn to show B as voltage divider.

And let's take note of the product AB, called *loop gain*. Since high values of both A and B enlarge the benefits of feedback, you'll not be surprised to find (in § 9S.5.1 and later) that when we measure feedback's effects we will use that AB product.

These notions are enough to let us speak quantitatively of circuit characteristics that until now we could only describe with adjectives: "big," "small," "very good." These adjectives might describe $R_{\rm in}$, $R_{\rm out}$, and constancy of gain.

W will derive the expression for circuit *gain*, and try to show only qualitatively how feedback improves input and output impedances. Try $R_{\rm in}$ or $R_{\rm out}$ for the non-inverting amp: see Fig. 9S.34. (You saw such an argument for $R_{\rm OUT}$ in §6N.5.2.)

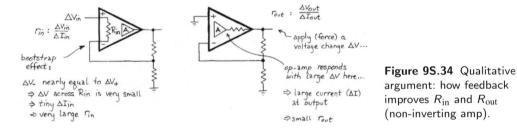

Figure 9S.34 Qualitative argument: how feedback improves $R_{\rm in}$ and $R_{\rm out}$ (non-inverting amp).

9S.5.1 $R_{\rm in}$

AoE §2.5.3B

AoE §2.4.3

$R_{\rm in}$ is *bootstrapped*: wiggle $V_{\rm in}$ and the inverting terminal wiggles about the same way; so, ΔV across the input is very small, and so, therefore, is ΔI. Differential $R_{\rm in}$ is improved by the factor $(1+AB)$, it turns out. This is bootstrapping, an effect we mentioned back in §9S.2.1 when discussing the VCVS filter. The argument is that if the far side of some component moves as the near side does (being pulled up "by its bootstraps"), then the component disappears electrically. But this argument becomes silly when $R_{\rm in}$ is 10^{12}, as it is for the '411: then $C_{\rm in}$ dominates the input impedance. Unfortunately, most of $C_{\rm in}$ is *not* bootstrapped.

9S.5.2 R_{out}

The argument for R_{out} is similar: wiggle the output (apply a little ΔV); the amp responds by moving point X a lot, putting a large voltage across its $R_{out-chip}$. But this sources or sinks a large current, I_{out}. No doubt you can see the nice result that follows – magically-low effective R_{out}. Using A and B we can now make this argument quantitative, and we will do that in §9S.5.4.

AoE §2.5.3C

9S.5.3 Gain

We have said that op-amp circuit gain turns out to be $1/B$, where B is the fraction fed back. Feeding back 1/10 in a non-inverting amp evokes a gain of 10, for example.[12] Let's justify that result, using just our definitions of A and B.

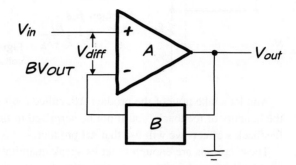

Figure 9S.35 Getting a general expression for circuit gain.

By "gain" we mean v_o/v_i.[13] We know that the op-amp is a differential amplifier that amplifies v_{diff} by its open-loop gain, A. Let's just look at how it amplifies that difference, and what feedback does to its gain:

$$v_o = A \times (v_i - Bv_o) = Av_i - ABv_o$$
$$v_o(1+AB) = Av_i$$
$$\text{Gain} = \frac{v_o}{v_i} = \frac{A}{(1+AB)} = \frac{1}{(1/A+B)}.$$

When A is large – and $AB \gg 1$ – then Gain $\approx 1/B$. This is the usual case to which we are accustomed – and it was this constancy of gain that interested Black most of all as he began his thinking about using feedback to stabilize telephone "repeater" amplifiers. Of course, A shrinks with frequency, and as it does the wonderful effects of feedback shrink too.

9S.5.4 A quantitative example

Try R_{in} or R_{out} for the non-inverting amp, assuming the following characteristics.

Assumed op-amp specifications:

- Op-amp's open-loop R_{in} (differential, between its two inputs) = 1MΩ;

---

[12] You may or may not recall, from a time when you were younger – back in Day 6 – that this evoked behavior is what we called the op-amp's ability to generate "inverse dog." The "dog" was the transfer characteristic of whatever was put inside the feedback loop.

[13] The lower-case v, describing the amplifier's "small-signal" response, is equivalent to what we more usually write: $G = \Delta v_o/\Delta v_i$. We need not make much of the difference between V and v or ΔV here, because this is a DC amplifier in any case. Thus it is unlike the discrete-transistor amplifiers that have seen, which were AC-coupled. Only ΔV (properly also called v) interested us when we worked with those.

9S.5 Quantitative effects of feedback

- Op-amp's open-loop $R_{out} = 100\Omega$;
- the amplifier circuit has nominal closed-loop gain of 10;
- the chip's open-loop gain is 1000 at 4kHz[14].

AoE §4.4.2A

R_{in} and R_{out} are each *improved* by the factor $(1+AB)$. (This argument is strongly reminiscent of the argument that a bipolar transistor improves what's on the far side of it by the factor $(1+\beta)$ this is what we called the "rose-colored lens" effect.)

How large is this factor $(1+AB)$? The term A is given as 1000; and B is the fraction of the output swing that is fed back: here, it is divided by 10 (that is, B=0.1; this is, of course, how the amp achieves its gain of 10). So, the product AB is 100. Therefore the hardware characteristics of the chip are improved by that factor:

R_{in} (differential) is boosted from 1MΩ to 100MΩ.

R_{out} is *reduced* from 100Ω to 1Ω.

Notice that these effects are not quite the same as the effect of the transistor's β, which operated on whatever impedance was on the far side of the transistor; the op-amp's $(1+AB)$ operates on what's inside the feedback loop: one does not look *through* the op-amp.[15]

Important points: The virtues of a feedback circuit depend on both A and B: a circuit works best where AB, the loop gain, is large. The quantity A falls with frequency; B is greatest in a follower, least when you look for very high gain (electronic justice, once again: if you want lots of circuit gain, you'll give up some benefits of feedback).

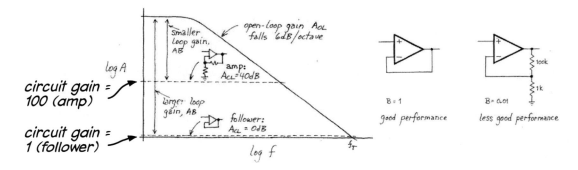

Figure 9S.36 A falls with frequency; B is highest for follower.

A refinement that carries good news: The gain rolloff imposed on "compensated" op-amps makes them behave – open-loop – like integrators, above a few tens of hertz. This notion is familiar to you. That phase shift, $-90°$, looks at first as if it would do nasty things to our gain calculations. We recited a rule saying that gain for a finite-gain op-amp is $A/(1+AB)$, and we used this formula to calculate effective circuit gain in § 9S.5.4 on the preceding page. What happens to this calculation if we complicate it by admitting the truth that for almost all frequencies the open-loop gain, A, lags the input by 90°? In complex or phasor notation that means describing the open-loop gain as $-j|A|$.

At first glance, this certainly seems like a nasty complication – and a disappointment in this course,

---

[14] This is true for the LF411.
[15] A partial exception, perhaps, to this rule occurs for the inverting configuration, which makes R_{source} relevant to the calculation of B, the fraction fed back.

Supplementary Notes. Op-Amps IV

where in general we have been assuring you that phase shifts usually don't concern us. Already, we have carved a big exception to that reassurance: phase shifts must concern us when we consider the stability of feedback circuits. So far, that has given us challenges. Now, for a change, the phase shift is going to pay us a surprising dividend, *improving* the gain of a finite-gain op-amp.

AoE makes this point by proposing a hypothetical case, which we will restate here, spelling out the steps in AoE's calculation.

> AoE §2.5.4B

A hypothetical case: gain-of-10 circuit, limited A: A and B are similar to those stated in the example of § 9S.5.4 on page 398. But this time, A is less: $A = 100$, $B = 0.1$.

Circuit gain, ignoring phase shift: Circuit gain: $G = \frac{A}{1+AB} = \frac{100}{1+10} = 9.1$

This is a 9% error: disappointing; given the Golden Rules, we hoped for something better. It turns out that we are entitled to such hopes.

Circuit gain, acknowledging phase shift: Now let's accept the ugliness of complex notation, in hopes of finding better gain – something closer to the idealized view of op-amp behavior expressed by the Golden Rules. Let's write the open-loop gain not as "100" but as "$-100j$," and then let's see what results.

Circuit gain: $G = \frac{A}{1+AB} = \frac{-100j}{(1+(-100j)\times(0.1))} = \frac{-100j}{1-10j}$.

So far, this looks a lot like the no-shift calculation. But now let's allow the $-j$ to do its work.

To find the magnitude of this gain, G, we need two steps:

1. clean up the denominator, getting rid of the j, by multiplying numerator and denominator by the complex conjugate of the denominator, $1+10j$:

$$G = \frac{-100j}{1-10j} \times \frac{1+10j}{1+10j} = \frac{-100j+1000}{1+100} = \frac{-100j+1000}{101} = -0.99j+9.9;$$

2. given the complex result, $a+bj$, find its magnitude: $|G| = \sqrt{a^2+b^2}$ Here, $G = \sqrt{9.9^2+0.99^2} = \sqrt{98.01+0.98} = 98.99 = 9.95$

This error is just 0.5% – considerably better than the way things looked when we ignored *phase*.

9W Worked Examples: Op-Amps IV

9W.1 What all that op-amp gain does for us

We use the Golden Rules to calculate *gain* if, say, we feed back one part in 100. The Golden Rules rely on an assumption that op-amp gain is very high (because, in Black's words, "...improvements are attained in proportion to the sacrifice that is made in amplifier gain..."). Assuming enormous gain, as we do when we rely on the Golden Rules, we say the gain of the circuit in Fig. 9W.1 is 100.

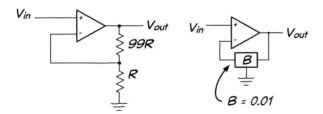

Figure 9W.1 Feed back 1%.

Let's see what happens as we assume several less-than-ideal values for op-amp *open-loop* gain, A, in the circuit from Fig. 9W.1.

Problems: circuit gain at several frequencies

9W.1.1 Assume no phase shift in A

First, let's assume that the op-amp's open-loop gain A shows *no phase shift*.

Problem What is the circuit's gain for:

- for $A = 100$?
- for $A = 1000$?
- for $A = 100,000$?

Solution In general, as you know, $G = A/(1+AB)$. Here, $B=0.01$, so, for the several values of A the values of the product AB are 1, 10 and 1000.

At the highest gain of 100,000, we get $G = 100\mathrm{k}/(1+1000) = 100$ (to within 0.1%).

At $A=1000$, we get $G = 1000/(1+10) = 91$.

At $A=100$, we get $G = 100/(1+1) = 50$. This sounds very sad. In a minute, we'll see that this overstates the degradation of gain.

9W.1.2 Two cases contrasted: no phase shift versus $-90°$ phase shift in A

Now we will ask similar questions, but this time trying two contrasting assumptions about the op-amp's open-loop gain. First we will assume no phase shift, as in the preceding question, 9W.1.1. Then we will assume, more accurately, that A includes a 90° lag, as it does above a few tens of hertz. At first glance this phase lag sounds like bad news. But it actually improves results for circuit gain. It turns out that the results of §9W.1.1 are gloomier than real life.

Problem Design a non-inverting amp with gain of 20 (using the Golden Rules). Assume an ideal op-amp as usual. Calculate gains for finite A.

Solution The circuit feeds back one part in 20:

$B = 0.05$.

We assume this time that $A = 100$.

This hypothetical gain is lower than what we usually have seen in our lab circuits. For the '411 op-amp this would be its open-loop gain at 40kHz. We set A low in this example in order to emphasize the helpful effect of the op-amp's phase shift.

The product $AB = 5$.

Problem Calculate gains assuming no phase shift.
What is circuit gain if op-amp $A = 100$?

Solution *Assuming no phase shift.*

$$G = \frac{A}{(1+AB)} = \frac{100}{(1+5)} = 16.$$

This is 20% below the hoped-for "Golden Rule" gain.

Problem Calculate gains assuming $-90°$ phase shift.
What is circuit gain if we acknowledge the phase lag, and put the open-loop gain $A = -100j$?

Solution *Assuming $-90°$ phase shift.* Now we write the open-loop gain in the complex notation that indicates the lagging phase shift: $A = -100j$.

$$G = \frac{A}{(1+AB)} = \frac{-100j}{(1-5j)}.$$

We did a similar calculation in §9S.3. Here, we can find the magnitude of this complex quantity using a shortcut: the gain is the ratio of *magnitudes* $V_{\text{out}}/V_{\text{in}}$. Then $G = 100/\sqrt{(1+25)} = 19.6$. Much better: just 2% below the Golden Rule gain. Taking account of the phase shift tends to restore a person's faith in the Golden Rules, does it not?

9W.2 Stability questions

Problem (Three marginally-stable circuits) Figure 9W.2 shows three circuits that may buzz, either on transitions or continuously, because of parasitic oscillations. For each, we'd like to hear a brief explanation of what circuit elements are likely to cause trouble, and we'd like you to show the most important remedies.

For the third of the circuits – the push–pull follower within the feedback loop – assume an LF411 op-amp, and consider two hypothetical frequencies for the parasitic oscillations: these frequencies may suggest alternative explanations for the trouble. Let us suppose two cases showing two widely differing frequencies: $f_{\text{parasitic}}$ at 100kHz as one case, 20MHz as the other.

9W.2 Stability questions

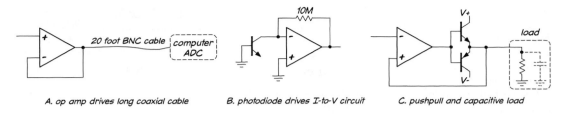

Figure 9W.2 Three marginally stable circuits.

Solution

Why it's likely to buzz: Incidentally, as we acknowledge in Chapter 9N, such a long cable would need to be treated as a "transmission line" in order to work properly at high frequencies (above a few MHz) or with high edge-rates. Here, we will assume that our frequencies are low enough so that we can neglect such effects – as the frequencies in our lab circuits nearly always are.

Figure 9W.3 Op-amp drives long coaxial cable.

The capacitance of the coax cable, about 30pF/foot for 50Ω cable, forms a low-pass filter when this stray capacitance is driven by the non-zero R_{out} of the op-amp – an impedance that in fact is even more troublesome than it seems when we call it "R_{out}." The low-pass can add a lagging phase shift approaching 90° (see Chapter 9N). Added to the op-amp's built-in 90° lag, that shift can turn negative feedback positive.

The op-amp output impedance actually looks *inductive* because the value of R_{out} rises with frequency, as the op-amp's open-loop gain falls, diminishing the benign effects of feedback.[1] This inductive behavior brings dangerous 90°-lagging phase shifts into the frequency range where the op-amp still has substantial gain.

Remedies: long cable:

Simplest Remedy: A small series resistor (say, 100Ω) calms this circuit by putting the phase-shifted point *outside* the feedback loop: see Fig. 9W.4. The price of this improvement, however, is a much-enlarged output impedance for the circuit (100Ω or so, in this case), whereas one of the great virtues of a follower should be its low Z_{out}.

Figure 9W.4 Simple stabilizing remedy for op-amp driving a long coaxial cable: series resistor.

Split-feedback Remedy: A "split" feedback network can give stability without degrading R_{out}: see Fig. 9W.5. The frequency at which the signal "crosses over" from the resistive to the capacitive path is determined by the *RC* value of R_{split} and C_{split}.

If for example, we saw a parasitic oscillation at 1MHz, we might put $f_{\text{crossover}}$ around 100kHz.

---

[1] See §9S.1.

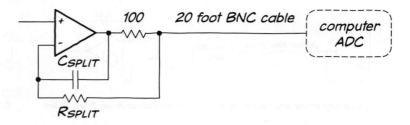

Figure 9W.5 Improved stabilizing remedy for op-amp driving a long coaxial cable: two feedback paths, for high versus low frequencies.

$$\Rightarrow RC_{\text{crossover}} = 1/(2\pi \cdot f_{\text{crossover}});$$
$$\approx 1/(6 \cdot 0.1 \times 10^6) = 1.6 \times 10^{-6}.$$

Many RC combinations would do, such as $C=100\text{pF}, R=15\text{k}$.

At high frequencies, the feedback network reduces to that in Fig. 9W.4. At low-frequencies, feedback hides the 100Ω resistor, giving the low output impedance that one looks for in a follower.

Higher gain (a cheap trick remedy): giving the circuit some gain entails reducing the fraction of the output signal that is fed back. That may be enough to stabilize the circuit. Fig. 9N.18 illustrates the increasing *degree* of stability that results from increasing gain in this circuit.

Special op-amp (an even cheaper trick – though it may cost more): A heavy capacitive load can be made manageable also by simply buying the unusual op-amp that is designed for such a load. These devices permit the output loading simply to slow down their output changes. See §9N.5 and AoE §4.6.2.

Op-amp *I-to-V* converter driven by photodiode: Sketching in the stray capacitance at the "virtual ground" (see Fig. 9W.6) reminds us that the large feedback resistor meeting a current source (the photodiode) forms a lowpass within the feedback loop. The effective R is 10M, driving this small stray capacitance (perhaps 10pF: the sum of the effects of the op-amp input, the photodiode, and the circuit layout). There is no voltage source at the input – as there would be in, say a $\times 100$ inverting amplifier. Such an amplifier, in contrast to this circuit, would drop the value of the effective R that drives the stray C. The amplifier would use an input resistor of 10k. This low R then would drive C_{stray}. In the photodiode circuit, the 10M driving even just 5pF puts f_{3dB} for this accidental lowpass at 300kHz: low enough so that oscillations are likely.

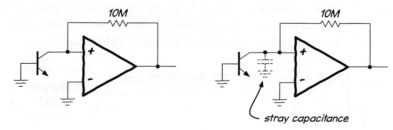

Figure 9W.6 Stray C makes photodiode circuit vulnerable to oscillation.

Remedy, photodiode circuit: feedback capacitor: A small C to parallel the large feedback R tames the circuit: see Fig. 9W.7. Even a tiny C will be effective. $C_{\text{feedback}}=1\text{pF}$ would dominate the feedback network (with effects noted just below) above about 16kHz.

The effect of the feedback capacitor could be described in a couple of ways.

- For the first, how this additional C alters the signal that is fed back. C_{stab} forms a C–C divider, at

9W.2 Stability questions

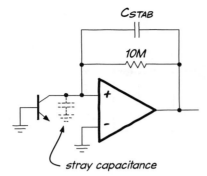

Figure 9W.7 A feedback C can quiet the photodiode amplifier.

frequencies where $X_{C-\text{stabil}}$ is much less than the 10M of R_{feedback}. This C–C divider imposes *no* phase shift, so leads to no stability problem.

- A second way is to consider circuit gain: without the feedback C, the circuit shows gain rising with frequency until this rise collides with the falling-gain curve of the op-amp itself. This condition brings instability, just as it did in the simplest op-amp differentiator (see Chapter 9N, and AoE §4.5.7).

The feedback capacitor limits the circuit gain to a constant value, above frequencies where X_C dominates the feedback network. This condition is stable (unlike the differentiator, this circuit does not go so far as to roll off circuit gain, integrator-like).

Op-amp drives push–pull and capacitive load: Capacitive loading is the principal problem, once again, but the presence of the push–pull makes the *follower* another suspect – as we saw with the single-transistor follower of §9L.2.

The problem statement proposes two hypothetical cases, with parasitic oscillations at differing frequencies: $f_{\text{parasitic}}$ is 100kHz or 20MHz.

The significance of the frequency of parasitic oscillation: The frequency of the parasitic oscillation can give a strong – and even conclusive – clue, if we're wondering whether to blame op-amp or discrete transistors.

A 20MHz parasitic cannot be blamed on a '411 op-amp. The '411's open-loop gain falls to unity at 4MHz.

A 100kHz parasitic is not at all likely to arise from the discrete transistor circuit, since we rely on the impedance of stray inductance to account for voltage gain in this configuration (see Chapter 9N). At this low frequency, the impedance of such inductance is about 1000 times lower than at the frequencies we observed in the discrete-follower exercise from §9L.2. The op-amp is much more likely the cause of the 100kHz oscillation.

We will tailor the remedies to the likely cause of the trouble.

Remedy, push–pull: at 100kHz apply op-amp remedies: The most effective op-amp remedy – once we have decoupled the power supplies, because we'd be embarrassed to be caught forgetting that standard precaution – probably is the "split" feedback strategy that works also on the problem of the long coaxial cable.

The R and C values in the split-feedback network define the crossover frequency, as you know (we applied this in the case of a long cable on page 403). We could put the crossover at about 50kHz with the values shown in Fig. 9N.20: $R=27$k, $C=100$pF.

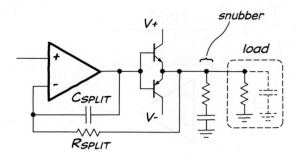

Figure 9W.8 Push–pull stabilizers, if op-amp is to blame.

The "snubber" values are quite standard and not critical, reflecting a conventional guess at the values that will so load the circuit as to calm it. We'll use 10Ω and $0.1\mu F$ as usual. Not very interesting, but sometimes helpful.

Remedy, push–pull: at 20MHz apply discrete transistor remedies: Here, the strategy is to spoil the high-frequency gain, and for this purpose an R_{base} of a few hundred ohms is effective. We hope you demonstrated this remedy to yourself in the Lab, see §9L.4.

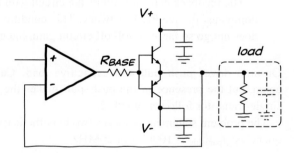

Figure 9W.9 Push–pull stabilizers if discrete-transistors are to blame.

There is no harm, of course, in applying both sets of remedies, in advance. Certainly we ought always to decouple power supplies and keep supply and ground paths short and of low-impedance. Including a base resistor on a follower also should now be habitual, too. We separated the two sorts of remedies in this exercise: those appropriate to op-amp oscillations and those appropriate to discrete-transistor parasitics. There is no reason to be so rigid in applying protections to forestall parasitics. We might as well be generous with our protections.

10N Op-Amps V: PID Motor Control Loop

Contents

10N.1	**Examples of real problems that call for this remedy**	**408**
	10N.1.1 PID applications	408
10N.2	**The PID motor control loop**	**408**
	10N.2.1 Potentiometer-to-motor loop: it looks very simple	408
	10N.2.2 The integration within the loop	409
10N.3	**Designing the controller (custom op-amp)**	**410**
10N.4	**Proportional-only circuit: predicting how much gain the loop can tolerate**	**412**
	10N.4.1 "Plant" frequency response (system gain: A_S)	412
	10N.4.2 Controller gain: A_C	413
	10N.4.3 Degrees of stability: phase margin	413
10N.5	**Derivative, D**	**415**
	10N.5.1 An intuitive explanation for what the derivative does for the loop	416
	10N.5.2 How to calculate the needed derivative gain	417
	10N.5.3 But you don't want too much D	419
	10N.5.4 Integral	420
10N.6	**AoE Reading**	**420**

Why? The problem PID sets out to solve

The problem is a familiar one: how to keep a feedback loop stable, despite lagging phase shifts within the loop. We saw a collection of troublesome circuits in Lab 9L. We found that even the lag introduced by a simple lowpass filter could upset a feedback loop. To stabilize such circuits, we learned several techniques:

- cut the amplifier's high-frequency gain – say, by paralleling the feedback path with a small capacitor;
- "split" the feedback paths, so that at the troublesome high frequencies feedback bypasses the thing that causes a lag in the loop;
- enhance negative feedback relative to positive, as in the discrete follower.

The novelties in today's challenge are two-fold:

- this time, it is not just a lowpass filter but an *integrator* that is placed within the feedback loop;
- and, given this difficulty – for which none of our earlier remedies was sufficient – we need to do more than tinker with what's in the loop: *we need to modify the control amplifier itself.*

This sounds pretty radical, and it is. We don't plan to crack open an integrated IC, of course. Instead, we will put together a circuit that mimics an op-amp, but with characteristics that we can control.

10N.1 Examples of real problems that call for this remedy

The circuit that you will put together in Lab 10L is called a "PID" loop, where PID stands for Proportional, Integral, Derivative. These names describe the three functions of loop *error* – difference between input voltage and the signal fed back – that are provided by the PID's homebrew "op-amp." If this is puzzling, hang on; we'll be spelling out all three elements in the pages that follow.

10N.1.1 PID applications

PID loops are used in a variety of settings where something rather sluggish is to be controlled. An automatic elevator, for example, is designed to stop with elevator floor lined up with floor level – and without much overshoot; a car's "cruise control" keeps speed nearly constant, without much oscillation about the target speed; industrial processes such as chemical mixing or heating use PIDs;[1] fly-by-wire airplanes use electronics rather than cables to link cockpit controls with tail and wing surfaces, and responses must be fast, but stable.

In Lab 10L we mimic what might be a fly-by-wire control: we try make a DC motor's shaft position match the shaft position of a potentiometer turned by hand.

The Wien bridge recalls criteria for oscillation, and for stability: As you know well, having grappled in Lab 9L with circuits that wanted to buzz, lagging phase shifts within the loop cause trouble. Rather than drum again at the points made in that lab, let's get at the criteria for stability through the back door: by recalling how we had to design the Wien Bridge oscillator in order to get a sustained sinusoid from the circuit.

The Wien bridge sustained an oscillation if net gain on a trip around the loop was just *unity*. If gain was less than one, the oscillation died away. If gain was more than one, the circuit oscillated – but also clipped its output, spoiling the sinusoid.

Similarly, in any feedback circuit, an oscillation will die away if loop gain is less than one at the frequency where phase shift hits 180 degrees.[2] In our encounters with op-amp *compensation*, we met one way to ensure stability: roll off the open-loop gain as frequency rises. Let's start with a primitive version of this remedy, since gain-cutting is the most familiar, among our standard remedies. This we will describe as limiting the "P," or Proportional, gain. Later, we will move on to fancier solutions.[3]

10N.2 The PID motor control loop

10N.2.1 Potentiometer-to-motor loop: it looks very simple

The task we undertake here looks easier than it is. Certainly the goal is modest. Fig. 10N.1 shows the scheme.

What could be simpler? Not much, on paper. But the challenge turns out to lie in keeping the circuit *stable*. The issue is fundamentally similar to the one you met in Lab 9L when you noticed that a humble lowpass in an op-amp's feedback loop could turn negative feedback into *positive*. The

---

[1] The history of PID use in industry accounts for some of the strange language in this field. As you will see, for example, the thing controlled often is called "plant."
[2] Always −180 is the hazard, *not* +180; but inversion of the sense of feedback is the main point, so let's stress that by speaking of the deadly 180, rather than emphasize the sign, here.
[3] In this discussion, as in the lab, we will add these elements not quite in the order mentioned in the name PID: we will use P first, then D, finally I.

10N.2 The PID motor control loop

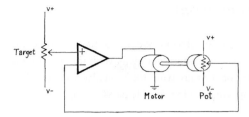

Figure 10N.1 Basic motor-position control loop: very simple!

problem there arose from the −90° phase shift imposed by the op-amp itself (and even greater phase shift, close to its unity-gain frequency). That −90° meant that we couldn't afford to insert much lag into the loop.

Today's circuit is harder. We are stuck with an *extra* −90° shift, or integration – quite apart from the one ordinarily imposed by the op-amp itself. This additional integration forces us to alter our methods.

This time, the integration comes from the nature of the stuff we are putting inside the loop: a motor whose shaft position we are sensing. Since we cannot afford the phase shift of an ordinary op-amp, we must build ourselves a sort of *custom* op-amp: one that provides modest gain and no phase shift.

10N.2.2 The integration within the loop

The "integration" or *lag* inside the loop results from a phase-shift that is inherent in the design of this particular *load*. Specifically, the lag results from the fact that we aim to control the motor's *position*, whereas we drive the motor with an error signal that is a *voltage*.[4] An implicit integration results.

That follows because a DC motor spins at a rate proportional to the applied voltage;[5] this spin rate, persisting for a while, causes a change of position. Hence the unavoidable integration within the loop.

To make this last point graphically vivid, Fig. 10N.2 shows how the position pot responds to a *square-wave* input to the motor. The triangular output looks a lot like what you saw in the integrator of Lab 7L, doesn't it? – apart from the inversion inserted by the lab integrator.

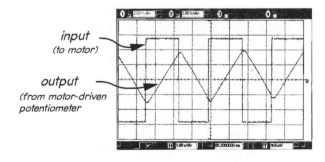

Figure 10N.2 Motor-drive to position-sensing potentiometer forms an *integrator*.

---

[4] This voltage is an amplified "error signal:" a measure of the deviation of circuit output from target. In the motor control circuit, both target and measured outputs are DC voltages.

[5] An easy way to convince yourself of this, if it's not already plausible enough, is to recall that if you spin the shaft of a DC *motor* with your fingers, the motor acts like a DC *generator*. It generates a voltage proportional to its spin rate. So, when a voltage drives a DC motor, the motor responds with a voltage – so-called "back EMF" ("electromotive force," or voltage). That back EMF almost equals the applied voltage; the spin rate levels off where the difference between applied voltage and back-EMF permits a current just sufficient to replace the energy the motor loses to friction, its mechanical loading, and to the motor's heating.

10N.3 Designing the controller (custom op-amp)

We need to go back toward first principles, as we design this control loop. We know that we need gain to get the benefits of negative feedback – but we suspect that we cannot get away with the usual large gain of a standard op-amp (we may need to "roll off" gain; or we may want the gain always low). And we know we cannot afford the phase lag of an ordinary op-amp – a lag of at least 90°. Can we design such an amplifier?

Yes. As we learned when we built an op-amp out of transistors, the device is essentially a differential amplifier. Everything else is, relatively, a detail. So let's start by building a differential amplifier that does *not* show an op-amp's −90° lag.

Figure 10N.3 shows a standard differential amplifier – the one you will use in the PID lab. It provides just *unity* gain, so we'll need to follow it with a gain stage.

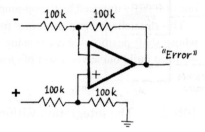

Figure 10N.3 Differential amp, unity gain.

No phase shift...: Perhaps you don't believe our claim that this differential amplifier can show no phase shift. How can it? – given that it was built with an ordinary op-amp, and we have been insisting that an op-amp includes a 90° lag. The lack of phase shift in the diff amp is rather amazing – though perhaps not to you who have built followers and other op-amp circuits that did not betray the integration hidden within the op-amp. But the claim is correct: the phase shift of the op-amp is just one more regrettable "dog" that feedback dutifully hides from us. Our PID loop will not see the op-amp's internal phase-shift; feedback has made that shift harmless.

...and modest gain: Even after we have eliminated our home-made op-amp's phase shift, the feedback loop will be vulnerable to instability, because of delays or lags that may arise in addition to delay caused by the motor-pot's *integration*. So we will give our amplifier much less gain than is provided by the usual device (like the '411, with its gain of 200,000 at DC). In addition, it will be convenient to make our amplifier's gain adjustable. See Fig. 10N.4 for a sketch of how the circuit would then look.

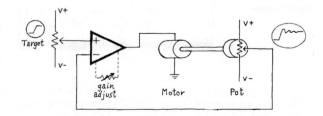

Figure 10N.4 Proportional-only drive: now we're ready to see how much gain our loop can tolerate.

Now we have a set up that should allows us to make a stable loop. But how are we to determine how much gain to apply? In the lab, our answer is entirely experimental: we increase the gain gradually until we see instability – and then we back off. In this section and the next we will try to describe what is happening as we approach that borderline of instability.

10N.3 Designing the controller (custom op-amp)

Aside for a conceptual diagram of the loop

Here is a way to describe the loop that is more formal than what we usually like to impose on you. Still, you may be slightly interested to see the terms that control-theory people like to use. You'll notice, here, the curious use of the word "plant" to describe the thing that is controlled. (In our case, 'plant' is motor and potentiometer.) The word reflects the industrial beginnings of control theory, as we have said earlier.

The "controller" is our homemade control amplifier. If Fig. 10N.5 puts you off, don't lose heart; in a moment we will draw it to look more like the op-amp loops we are accustomed to. Figure 10N.6 is the same diagram redrawn in a more familiar form: the motor-pot "plant" is the usual "dog" within the loop.

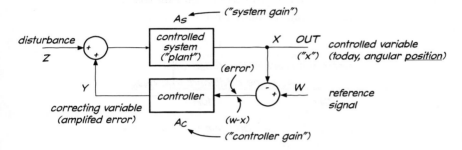

Figure 10N.5 Loop to control motor shaft position: block diagram in form used in class discussion.

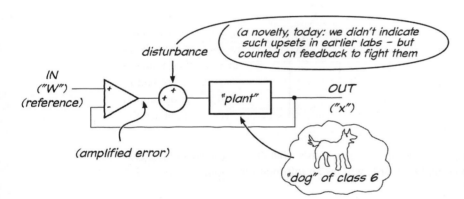

Figure 10N.6 Loop to control motor shaft position: block diagram in form familiar from op-amp discussions.

These two diagrams do not tell you anything you didn't know from a statement of the original problem. But the first diagram, particularly (Fig. 10N.5), may help you to see that the signal that travels around the loop passes through the "plant" and the "controller," and that stability will depend upon what a trip around the loop does to a signal – or to a disturbance.

Giving explicit attention to a "disturbance" is an incidental novelty, here. So is the naming of the transfer function or *gain* for each of the two blocks: *plant* or "system" gain is labeled A_S while *controller* gain is named A_C. This assignment of labels allows us to write a simple expression for the gain applied to anything that makes it around the loop ("signal" or "noise" or "disturbance"). This product, defined in equation 10N.1, is what we called "loop gain" in the usual op-amp setting. The loop gain here is

$$A_{\text{loop}} = A_C \cdot A_S : \quad \text{controller gain times system gain .} \tag{10N.1}$$

In this loop, we are stuck with A_S – the behavior of the "plant." The challenge, as we set up and trim the PID loop, will be to adjust judiciously the other gain, A_C – the *controller* gain.

We will first adjust the *magnitude* of A_C to maintain stability. Then, when we add the derivative D, we will be adjusting also the *phase* of A_C. It will turn out that phase and magnitude interact: proper phase behavior permits higher gain.

10N.4 P: proportional-only circuit: predicting how much gain the loop can tolerate

10N.4.1 "Plant" frequency response (system gain: A_S)

Figure 10N.7 plots the frequency response of a hypothetical motor-to-position element (this is A_S only; we're not yet looking at the "loop"). The plot is hypothetical, but quite close to what you're likely to see in the behavior of your PID lab.

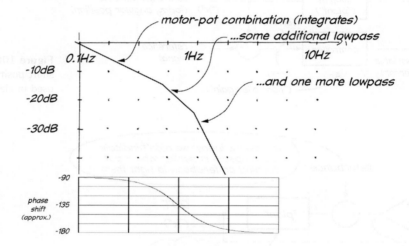

Figure 10N.7 Gain and phase response of motor-pot combination: $-90°$ lag, from the lowest frequencies.

If this plot seems rather abstract, here's an attempt to say in words what the downward-sloping *gain* plot says about the motor-pot combination. Our words will try an informal time-domain explanation for the frequency-domain rolloff.

The downward-sloping plot indicates that for a given input amplitude (a sinusoid), at low frequencies the motor shaft will have time to turn quite a long way before the input signal reverses sign. At higher frequencies, it will turn only a little way in the available time (roughly a half period). If this sounds like what we say about the voltage on a capacitor in an *RC* circuit, when we consider it in time domain, that's appropriate. The processes are very similar.

You'll notice that the plot differs from the usual op-amp frequency response plot in that its rolloff begins, at the lowest frequencies, with a 90° degree lag. This is simply a consequence of the *voltage-drive/position-sense* mechanism we noted above; there is no phase-shift free region, as there is for an op-amp. But more important than this difference is the similarity to the op-amp plot: both plots allow us to anticipate where trouble could begin.

We can expect that we *may* get trouble where *plant* shift approaches 180°. This happens at the bend in the gain plot of Fig. 10N.7, where a *second* 90° lag is added to the integrator's constant $-90°$. When this second 90° kicks in, *negative* feedback may have been transmogrified into *positive*.

But the fact that the *plant* can provide this scary phase shift does not settle whether the circuit will oscillate. Whether this phase shift causes oscillation depends on whether we provide enough

10N.4 Proportional-only circuit: predicting how much gain the loop can tolerate

controller gain. Recall the Wien bridge design: in order to sustain an oscillation we needed gain sufficient to make up for attenuation in the positive feedback path.

In designing our PID loop, we will not go right to that edge of instability; but we need to know where that edge is. We want to keep our eye on that "corner" frequency, the boundary of danger.

10N.4.2 Controller gain: A_C

In Fig. 10N.8 we have added to the A_S plot that you just saw, two hypothetical values of A_C. The term A_C gain, which multiplies A_S, slides the A_S curve up, forming the loop gain $A_C \cdot A_S$ in the plot of Fig. 10N.8. In both cases, A_C is flat with frequency.

In one case, A_C is low enough so that loop gain falls to less than unity before the dangerous *corner* frequency. In the other case, where A_C is set higher (labeled $A_{C'}$), the loop gain exceeds unity at the corner frequency. This higher A_C would make the circuit oscillate.

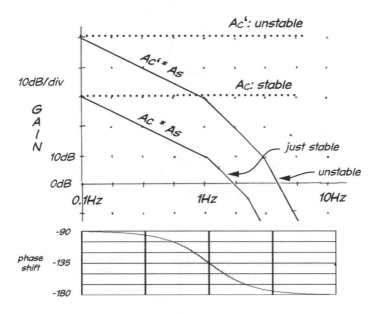

Figure 10N.8 Proportional loop: stable at modest P gain, unstable at a little more P gain.

In the lab circuit, you will discover the corner frequency by finding the frequency of "natural oscillation." This is PID jargon for the frequency where oscillation sets in, when the gain is just a little too high. You discover this frequency by gradually increasing A_C (here, just P gain), until oscillation begins. In our lab circuit it is just a few Hz: perhaps 3Hz. In lab you will be able to restore stability by reducing the controller gain, A_C.

10N.4.3 Degrees of stability: phase margin

You will also find that stability is a matter of *degree*: as you adjust the gain upward, approaching the oscillation condition, the loop will become more and more jumpy. This jumpiness is easiest to see in response to a disturbance deliberately imposed.

In the lab, you can disturb the circuit in either of two ways. You can change the "target" input voltage abruptly (applying a *step* change either by giving a sharp twist to the pot or by applying a slow square wave from a function generator). An alternative (perhaps more fun) is to apply a step "disturbance:" use your fingers to turn the motor shaft away from its resting position and watch the circuit try to recover. When the loop is close to oscillation, the output will overshoot repeatedly, before settling to its final position.

Given a plot of A_S ("system" or "plant" frequency response, including phase shift) one can see just how far the loop phase shift lies from the deadly $-180°$, at a given A_C (controller gain). That distance, in degrees, is called the "phase margin." This is a notion you have seen before, in discussion of op-amp *frequency compensation*. Op-amps are "compensated" to provide between 45 and 60 degrees of margin.

Figure 10N.9 illustrates the response of a loop to a square wave, at several phase margins. In life, it may be more likely that you would run the process in reverse: observe the response and infer the phase margin.

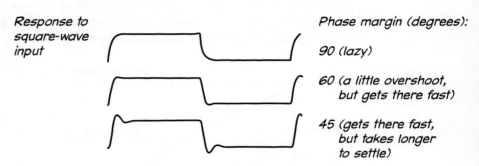

Figure 10N.9 Step response shows effect of various phase margins on stability. After Tietze, U. & Schenk, C., *Electronic Circuits: Handbook for Design*, Second edition, Springer (2008), p. 1105.

In the lab PID circuit, you will find that at lower A_C the circuit is, indeed, tamer, less disposed to overshoot. But low A_C also has some harmful effects: the loop response is quite slow, and the circuit tolerates a good deal of residual error.

Figure 10N.10 shows response for low P gain. The response looks like an *RC*: makes sense, since the motor drive diminishes as the *error* diminishes (while the position draws nearer to the target).

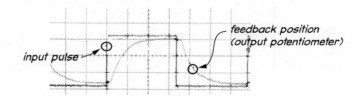

Figure 10N.10 Low P gain (four) keeps stability, but at a price.

We can increase P gain somewhat, and get better performance in some respects (smaller residual error, and quicker response). But at this higher P gain the circuit shows some overshoot: see Fig. 10N.11.

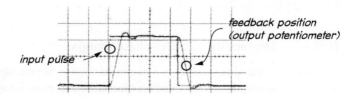

Figure 10N.11 Middling P gain (sixty) produces some overshoot.

And if we push P gain still higher, we go close to the edge of instability. In Fig. 10N.12, the downward step produces several cycles of ringing, and the upward step produces continuous ringing. With P gain that high, the circuit is not usable. But it can be made stable, even at this higher gain, if we add in some D, the derivative of the error signal.

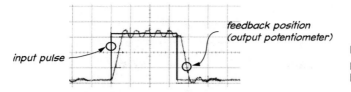

Figure 10N.12 High P gain (eighty) produces some overshoot and ringing: barely stable.

10N.5 Derivative, D

Since it is the *lagging* phase shifts that cause the mischief, can we improve the loop's response by injecting some phase shift of the other flavor, *leading*? Yes, we can, if we do it carefully. Let's start by looking at what we would *like* to see as an effect of injecting such *derivative*, and then we'll look at how we might determine how much D to add.

Figure 10N.13 shows the original plot of loop gain, $A_C \cdot A_S$. This time we have added a dotted line to show how we'd like the corner to get straightened out by the A_C response.

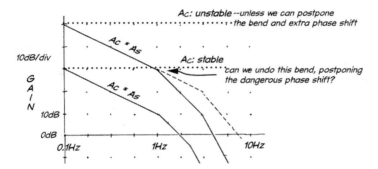

Figure 10N.13 What we'd like: to undo another phase lag.

In order to get that result – in the $A_C \cdot A_S$ product – we need to put an *upward* tilt into the controller gain, A_C. That is, we need to make the controller response look like the *derivative* beginning at the frequency where trouble otherwise would begin.

How do we get that result? We generate the derivative, D, of the error term, and add that to the P error, amplifying both to get a bent A_C curve rather than the flat ones of Fig. 10N.8. We will use the sum of these two error functions[6] to provide the correction signal, in our negative feedback loop.

To put the upward bend at the right frequency – the frequency where the system's corner occurs – we arrange things so that at this frequency the magnitude of D equals P. In Fig. 10N.14, D grows steadily with frequency, and when summed with P, it takes over at the corner frequency. Thus it jumps in just in time to save the day: in time to undo that second 90° lag. Adding D to this circuit has permitted us to use higher gain than we could manage with P alone, and the higher gain strengthens feedback, enhancing its usual benefits.

---

[6] The curve in Fig. 10N.14 will look strange to you if you have not been looking at log plots recently. The derivative, D, seems to *take over* from the P rather than to *sum* with it. At lower frequencies, A_C looks like P alone, ignoring D; at higher frequencies, it looks like D alone.
 Yes. But that makes sense if we recall that, on the log plot, as soon as D climbs noticeably above P its value is so much the larger of the two that their sum is virtually the same as D alone. So the lower frequencies are totally dominated by the P term, the higher frequencies totally dominated by D.

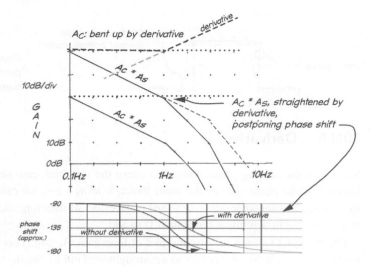

Figure 10N.14 D term added to P term makes stable a loop that otherwise would oscillate.

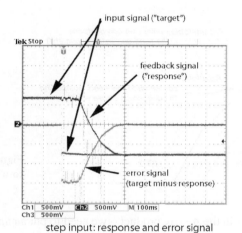

step input: response and error signal

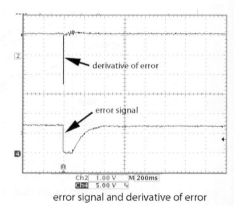

error signal and derivative of error

Figure 10N.15 Loop response to a step change of input: error and derivative of error. (Scope gains: 500mV/div in left-hand image; 1V/div for derivative trace, 5V/div for error signal in right-hand image.)

10N.5.1 An intuitive explanation for what the derivative does for the loop

If you find the frequency-domain account we have just given a bit abstract, consider the rough arguments made in this subsection, trying to make sense of the benefits that the D term provides.

The problem is easiest to get a grip on if one considers a loop that is quiet and settled – and suddenly driven by a step change at the input. That abrupt change of input makes the *error* function, P suddenly step from zero to gain × input-step. What does the *derivative* add? It adds a strong additional signal in the same direction, because not only has the error appeared, but it has appeared rapidly, climbing abruptly from zero to its new value. Fig. 10N.15 shows such waveforms.

The loop does not seem, at first, to respond at all; then it gradually drives the response to match the input, and the error is driven down toward zero once more. The right-hand image in Fig. 10N.15 shows the strong kick that the derivative gives, as the error steps. This kick tends to speed the response of the loop.

If we sweep the scope faster to see the step response in greater detail, as in Fig. 10N.16, we see the derivative in this case giving a kick several times stronger than the correction signal provided by the

P error alone (note the different scope gains for the *error* versus *derivative* displays). The left-hand image in Fig. 10N.16 squeezes the large derivative signal on-screen by reducing the scope gain for that channel by a factor of five. The right-hand image uses a 1V/division scale for all channels, showing how strong the D signal is relative to P.

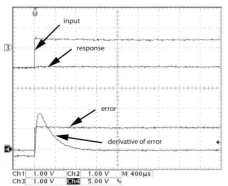
input, response, error and deriv (deriv compressed)..

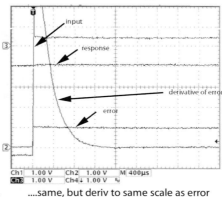
....same, but deriv to same scale as error

Figure 10N.16 Faster scope sweep, and D correction signal stronger than the bare P. (Scope gains: 1V/div except derivative in left-hand image.)

And a homely interpretation of what D does for the loop: The D term – the derivative of the error – is a little hard to grasp intuitively. Here's a try: suppose the problem that the loop means to solve is to move a car into a parking place that has a concrete wall at the front. Our controllers appear, in this drama, as two valet-parking attendants. The first is a P-only controller – a cautious little old man, well past retirement age. His strategy for getting the car parked and not hitting the wall, is simply to go very gently and slowly. He gets there, eventually – but as he sees the wall growing nearer (the *error* term shrinking), he drives more and more slowly, finally just inching along.

The other valet-parker is a cooler P-plus-D controller. He's a teenage punk, wearing a baseball hat backwards. He hops into the driver's seat, guns the car, squirts into the parking place and hits the brakes. He gets the car parked fast.

That may seem metaphoric nonsense. Fig. 10N.17 shows waveforms from Fig. 10N.15 that we claim describe the same behavior, perhaps making the valet-parkers more plausible. The main effect is a strong boost at the start – hitting the gas hard when a new error appears. The lesser effect is a gentle braking near the end of the process. Perhaps the high D gain puzzles you, given that we have designed the loop to make D and P gains equal at the corner frequency. The D gain is high where the error suddenly steps because the steep edge of this error waveform contains high frequencies – as Fourier taught.

10N.5.2 How to calculate the needed derivative gain

How do we calculate the necessary differentiator gain? To avoid complications, let's assume that the gain of the P path is unity. Then our goal is to arrange things so that the derivative contribution, D, is equal to the P contribution at the frequency where trouble otherwise would occur. The D should keep the loop stable, until yet another lowpass cuts in; at that point, we should have arranged to make the loop gain safely low: less than unity, so that a disturbance must die away.

Let's start by reminding ourselves what we mean by differentiator "gain;" then we'll calculate what RC we need for stability. Gain for a differentiator, by definition is

$$\frac{V_{\text{out}}}{dV_{\text{in}}/dt}$$

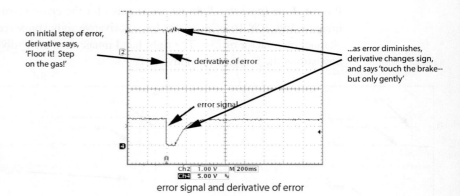

Figure 10N.17 Derivative effects likened to valet-parker's stepping on gas, then on brake. (Scope gains: error 5V/div; derivative 1V/div.)

error signal and derivative of error

We know that, for the op-amp differentiator,

$$V_{\text{out}} = I \times R_{\text{feedback}}$$

and this I is just

$$C \times \frac{dV_{\text{in}}}{dt}.$$

So (neglecting the sign of the gain; the op-amp version inverts)

$$\text{Gain}_{(\text{Deriv})} = \frac{V_{\text{out}}}{dV_{\text{in}}/dt} = \frac{R \cdot C \cdot dV_{\text{in}}/dt}{dV_{\text{in}}/dt} = RC.$$

So RC defines the differentiator's *gain*.[7]

If $V_{\text{in}} = A\sin(\omega t)$, then $V_{\text{out}_{\text{Deriv}}} = \omega RCA \cos(\omega t)$. We want to find the value of RC, the differentiator's gain, that would set $V_{\text{out}_{\text{Deriv}}}$ equal to $V_{\text{out}_{\text{Proportional}}}$. Let's treat P gain as unity; then we want both P and D to equal V_{in}.

If we set the D gain equal to unity, then

$$V_{\text{out}_{\text{Deriv}}} = V_{\text{out}_{\text{Proportional}}} = V_{\text{in}},$$

and, equivalently,

$$\omega RCA \cos(\omega t) = A\sin(\omega t).$$

Consider just the maximum amplitudes of V_{in} and dV_{in}/dt (where sin and cos terms equal 1). We want to set these amplitudes equal to each other, and both are equal to the input amplitude, A. Then

$$\omega RCA = A \quad \text{hence} \quad \omega RC = 1, \quad \text{or} \quad RC = \frac{1}{\omega}.$$

In other words,

$$RC = \frac{1}{2\pi f}.$$

To paraphrase this equation in words: RC should be about 1/6 of the period of natural oscillation.

Figure 10N.18 is a scope image to show the beneficial effect of adding some D. Here the P gain is high enough so that if P were the only feedback the circuit would oscillate continuously. Now, with D added, the circuit gets where it's going fast, and does not overshoot.

---

[7] Perhaps you find this a rehearsal of the obvious! Just a look at units make it very plausible that RC should be the differentiator's gain: input is volts/second; output is volts; the conversion factor needs units of seconds.

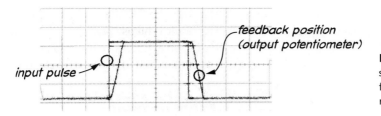

Figure 10N.18 Derivative can stabilize loop, and get you there faster (overall gain: 120; D resistor: 220k).

10N.5.3 But you don't want too much D

Too much D makes the circuit timid...: As you might expect, and can see in Fig. 10N.19, too much D brings trouble again: first, the circuit becomes timid. As it approaches its destination it loses its nerve, puts on the brakes too early.

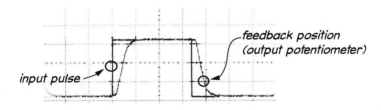

Figure 10N.19 Too much D makes the circuit timid, slow to get where it's headed (overall gain: 120; D resistor: 680k).

...still more D makes the circuit oscillate: More surprising, if you inject still more D, the circuit becomes unstable once again. Fig. 10N.20 tries to make an argument to show why this is so (phase shift plot shows exaggerated abrupt change of phase, misleading if you take this image too literally).

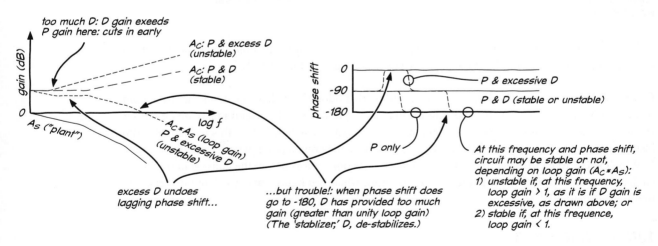

Figure 10N.20 Excessive D can bring on oscillations, again, by pushing total $A_C \times A_S$ gain too high.

We introduced the *derivative* of error as a stabilizing device. In Fig. 10N.14, D permitted overall gain that otherwise would have caused instability by undoing a lagging phase shift. That worked because controller gain, A_C, was low enough so that when multiplied by the "plant" gain, A_S, the loop gain had fallen below unity before we reached the deadly phase shift.

Where D gain is excessive, in contrast, as in Fig. 10N.20, D cuts in too early (phase shifts in this figure, incidentally, are shown as unrealistically abrupt).[8] The overall controller gain, A_C, thus is excessive, and as a result the loop gain is excessive as well. The result is an oscillation, as it was in the earlier case when we used P alone and high P gain made A_C excessive. Excessive loop gain brings trouble at the frequency where phase shift hits $-180°$, regardless of the cause of the excessive gain.

10N.5.4 Integral

Adding the last of the three error functions, an *integral*, I, of the error, confers benefits usually less important than those provided by adding D. It does not help stabilize the circuit. Instead, it diminishes long-term errors, driving these toward zero.

The effect of I is evident if we look at what it can do for a very-low P gain circuit – one that we saw before in Fig. 10N.10. On the left of Fig. 10N.21 is that low P-gain response. On the right is the same circuit with an integral of error added in.

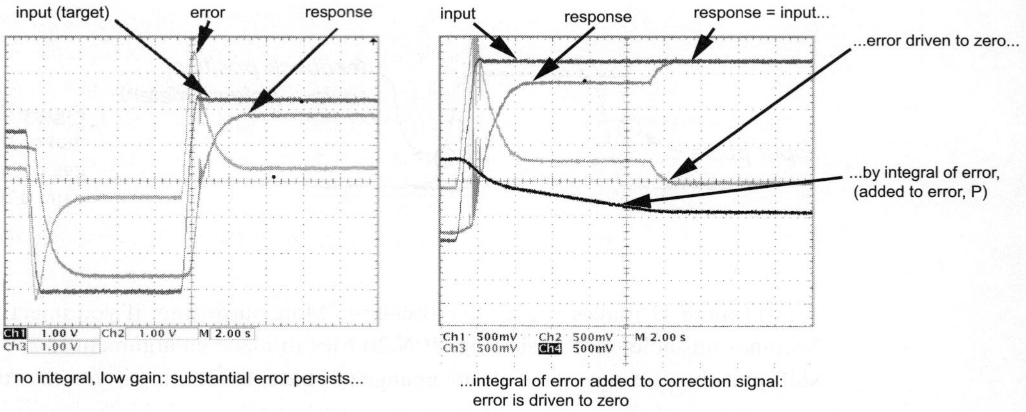

Figure 10N.21 Integral can drive long-term errors down, even at low gain. (Scope gains: 1V/div in left-hand figure; 500mV/div in right-hand figure.)

The integral of error drives the loop tight in against the target – but only after a delay of almost five seconds in the present example, shown in the right-hand image of Fig. 10N.21.[9]

Good question: how can the PID loop be stable, with I added? We have made much of the point that this loop includes an inherent 90° lag, so that adding another $-90°$ would make the circuit unstable. Yet, the *integral* term does insert another $-90°$ (lagging) signal. It does this without bringing instability because the integral's effect is long gone before the crucial point, where gain goes to unity.

Why it is OK to go to dangerous phase shift, as long as one returns before the unity-gain frequency, we leave "as an exercise for the reader."[10] Let us know, when you figure out a good intuitive explanation for this marvel![11]

10N.6 AoE Reading

- §15.6.2 (analog PID)

---

[8] "Cuts in early" is just another way to say "D gain is excessive," since D "cuts in" on the graph where D gain exceeds P gain.
[9] This delay is caused by "static friction" or "stiction," as noted in Lab 10L. The motor does not begin to move until its drive voltage reaches some minimum level.
[10] We do this, of course, because we're not sure of the right answer!
[11] Wise people tell us that it *is*, indeed, OK to go to 180° shift and beyond, as long as one returns to less than 180 at the unity-gain point. You will see this truth demonstrated again when you meet phase-locked loops in the digital part of this course. So far, intuition fails us here: we can't explain it because we can't really understand it.

10L Lab. Op-Amps V

10L.1 Introduction: why bother with the PID loop?

Today's feedback loop...: Today's circuit looks straightforward: a potentiometer sets a *target position*; a DC motor tries to achieve that position, which is measured by a second potentiometer. *Lags* cause the difficulty: the correction signal is likely to arrive too late to solve a problem that the circuit senses. If that happens, the *remedy* can make things worse.

This motor control circuit is a classic feedback network called a PID circuit. Its response ultimately will include three functions of the circuit *error* signal: P (proportional) I (integral); and D (derivative). Stability is the central issue.

...And why it interests us: This feedback problem holds two sorts of interest for us.

- Pedagogical appeal:
  - It gives us a chance to apply – and to apply in concert – a collection of circuits that you have seen either only as fragments, or only on paper:
    o differential amplifier, made of op-amps (this we have met only on paper);
    o differentiator;
    o integrator;
    o summing circuit;
    o high-current driver (with motor as load).
- PID confronts a classic control problem; it provides a scheme with many practical applications

Putting the pieces together: Students often tell us that they like to build circuits that *do* something – in contrast to circuits that produce just images on a scope screen. Today's circuit qualifies: it guarantees to make a little DC motor squirm ("squirm" when it's unstable; tamely spin, then stop, when it's stable).

This PID circuit is by far the most complex in this course to date. That makes it a good setup for improving your debugging skills (this is a "glass-half-full" way to say that you're very likely to make some wiring errors today). We often boast that in this course, *bugs are our most important product*. This boast becomes most convincing near the end of the course, when you may choose to put together a computer from ICs; but even today the circuit is complex enough to make it a challenge. (The debugging will be especially challenging if your sloppy lab partner fails to keep leads short and color-coded, and forgets to bypass power supplies. We know *you* wouldn't make such errors.)

Finally, we're happy to let this lab reinforce a concern first discussed in Lab 9L: stability. Although today's circuit problem is singular – integrator within the loop – and the remedy more subtle than usual, the general stability problem is one that confronts us in almost every circuit that has gain.

10L.2 PID motor control

The task we undertake here is one we have described in Chapter 10N. We will repeat some of what appears there to save you the trouble of refering back when you do the lab. But 10N is more thorough than what we write here.

Our goal, most simply stated, is just to use a feedback loop to get one DC voltage to match another; and since both voltages come from potentiometers, the goal can also be described (maybe sounding more exciting) as making the position of one potentiometer shaft mimic that of another. We want to be able to use our fingers to turn a potentiometer by hand and see a motor-driven pot mimic our action. Such controls are sometimes offered on fancy audio equipment, so that the equipment can be controlled either by twisting a knob, or by using a remote that controls the knob from across the room. A more impressive PID application is remote surgery.

10L.2.1 The motor-pot assembly

The motor-pot gadget comes in two forms. The scope images and details of frequency response in most of Lab 10L and Chapter 10N use the rotary-potentiometer version (on left side in Fig. 10L.1. The other version, shown to the right in that figure uses a *slider* pot rather than rotary, and responds much faster. Use whichever is available.[1]

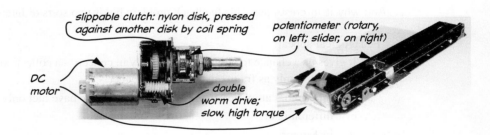

Figure 10L.1 Two motor-pot assemblies: rotary and slider potentiometers.

The clutch in the rotary version serves two purposes: it permits a human hand to control the potentiometer directly when the motor is stopped. It also protects the motor against stalling and overheating if the motor drives the pot to one of its limits. The slider version lacks this feature; a stalled motor will draw around 800mA – but apparently without damaging the motor. Because of the lack of a slip-clutch in the slider version, we recommend the rotary.

10L.2.2 The motor control loop

Figure 10N.1 on page 409 was a minimal sketch of today's circuit. The special difficulty we're facing now comes from the fact that the motor-to-potentiometer block, our circuit's load, *integrates* voltages. So, we can't use an ordinary op-amp as the *triangle* in the feedback loop shown the figure. The extra $-90°$ shift, or integration, in this circuit forces us to alter our methods. Since we cannot afford the phase shift of an ordinary op-amp, we build ourselves a *custom* amplifier: one that provides modest gain and no phase shift. We will apply the radical remedy: altering not the load in the loop but the op-amp itself. Fig. 10L.2 is a reminder of Fig. 10N.4 – redrawn to suggest that we now will be able

---

[1] The rotary type can be, for example, ALPS RK16812MG099; the slider version shown is COM-10734 from Sparkfun Electronics. The potentiometer "taper" is not critical. Linear is good; audio taper is OK (the ALPS part uses a B3 audio taper).

to tinker with the amplifier's gain. Fig. 10L.2 only shows the simplest configuration – "P". Later, we will alter also the phase behavior.

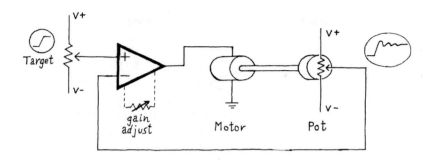

Figure 10L.2
Proportional-only drive will cause some overshoot; gain will affect this.

We will first try this circuit with its gain adjusted *low*, and we expect to find the circuit fairly stable. Then, as we increase gain, we should begin to see overshoot and ringing; if we push on to still higher gains, we should see the circuit oscillate continuously. At the end of this chapter we append some scope images illustrating just such responses to variations in simple "proportional" gain.

10L.2.3 Motor driver

Let's start with a subcircuit that is familiar: a high-current driver, capable of driving a substantial current (up to a couple of hundred milliamps). We'll use the power transistors you've met before: MJE3055 (*npn*) and MJE2955 (*pnp*). The motor presents the kind of troublesome load likely to induce parasitic oscillations, as in the final exercise of Lab 9L. We need therefore the protections that we invoked there: not only decoupling of supplies, but also both a *snubber* and *split feedback* that bypasses the troublesome phase-shifting elements.

We are trying hard here to *decouple* one part of the circuit from the others: the 15μF caps should prevent supply disturbances from upsetting the *target* signal. Similar caps at the ends of the motor-driven potentiometer aim to stabilize the feedback signal.

You may also want to use an *external power supply* to provide the *motor's* $\pm$15V supplies if you have such an extra supply handy. We suggest this not for decoupling but because the motor's maximum current exceeds the breadboard's 100mA rated output and might disturb those supplies even if one inserted plenty of decoupling caps. The external supply, unlike the breadboard supply can provide the necessary current. (But we have also built this lab happily *without* this separate power supply.)

Note: Your motor pot's resistance value may differ from the 10k shown in Fig. 10L.3. If so, scale the resistors appropriately. If your motor's potentiometer has value 100k, for example, just scale the 4.7k Rs up by a factor of ten.

Note also that you *must not use '411 op-amps*. The '411 has the nasty property – common to "bi-fet" devices – that it can flip its output phase if the input voltage goes below its specified common-mode input range.[2] The result is that if an input to any of the amplifiers in this loop momentarily swings to within about a volt of the negative supply, the loop is very likely to get hung up by this nasty positive feedback.[3]

Wire up the two potentiometers as well as the motor-driver itself. The resistors at the ends of the

---

[2] The op-amp output is forced *high* in this event. If the input going too far negative is the non-inverting, this changes the flavor of feedback from negative to positive.

[3] This hazard is not hypothetical; we first breadboarded this circuit with '411s – and were forcibly reminded of the part's nasty phase-inversion by the occasional lock-up failure of the loop.

Lab. Op-Amps V

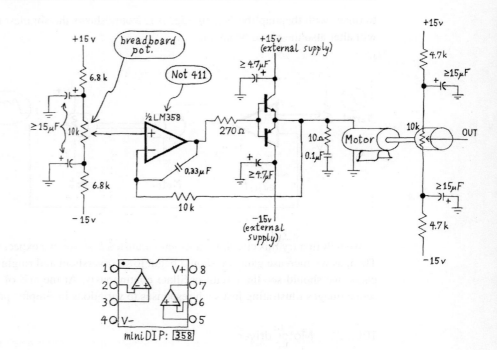

Figure 10L.3 Motor-driver.

two potentiometers – 6.8k resistors on input, 4.7k resistors on the motor pot – restrict input and output range to a range of about ±7V to keep all signals well within a range that keeps the op-amps happy. The difference in R values makes sure that the input range cannot exceed the achievable output range.

You can test this motor driver by varying the input voltage and watching the voltage out of the motor-driven pot. Don't be dismayed if you see a good deal of hash on the scope screen. This hash may look very much like a parasitic oscillation, familiar to you from Lab 9L. Fig. 10L.4 shows what we saw when watching the motor drive with the motor moving.

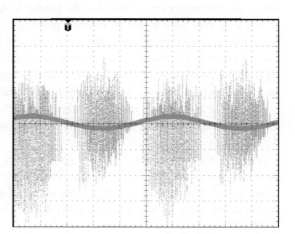

Figure 10L.4 Motor drive hash looks like a parasitic oscillation.

But if we look at this hash more closely, as in Fig. 10L.5 we find some clues that it is not the usual parasitic oscillation at work. The spikes seem to be the effects of the DC motor's brushes breaking contact periodically with the motor's commutator. One clue is the fact that noise is not continuous, but seems to be a set of narrow spikes at a low repetition rate. The other clue – pretty conclusive – is the fact that the spike voltages exceed the power supply: this effect looks a lot like the behavior of an inductor (the motor winding), angry each time the commutator switches the current *off*. So, don't let this hash worry you. It's ugly, but we'll live with it.

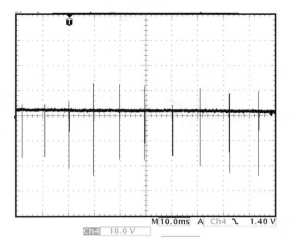

Figure 10L.5 Motor drive hash seen in greater detail: not parasitic oscillation after all.

Any V_{IN} more than a few tenths of a volt should evoke a change of output voltage. You will hear the motor whirring, and will see the shaft slowly turning (the motor drive is geared down through a two-stage worm- and conventional-gearing scheme).[4] After perhaps 20 seconds, the pot will reach its limit and will cease turning. But that's all right: a clever clutch scheme, mentioned back in §10L.2.1, allows the motor to slip harmlessly when the pot reaches either end of its range. If the *signs* of V_{IN} and the *change* in V_{OUT} *do not match*, then be sure to interchange leads of one of the pots, to make them match. We don't want a hidden inversion, here. It would upset our scheme when we later close the loop.

10L.2.4 Pseudo op-amp

Now we do a strange thing: we use three op-amps to make a rather crummy op-amp-like circuit, as in Fig. 10L.6. This is the circuit we have sometimes referred to as our "control amplifier."

The first stage you recognize as a standard differential amp: it shows unity gain. The second stage simply inverts;[5] the third stage seems to be doing no more than *undoing* the inversion of the preceding circuit, along with permitting adjustment of gain. That is true at this stage; but we include this circuit because soon we will use it, fed by two more inputs, as a *summing* circuit. So used, it will put together the three elements of the *PID* controller: Proportional, Integral, and Derivative. In the present P-only circuit[6] we also use the third amplifier to vary the overall gain of our home-made op-amp.

The entire circuit then is simply a differential amplifier with adjustable gain. And this gain is always low relative to the very high values we are accustomed to in op-amps. We need the modesty of this gain, and we need its lack of phase shift between input and output. Both characteristics contrast with those of an ordinary op-amp as you know. The fixed, high gain of an ordinary op-amp, along with its *integrator* behavior beginning at 10 or 20Hz, would get us into trouble today, turning negative feedback into positive.

---

[4] See photo of motor-pot innards in Fig. 10L.1.
[5] This inversion is included to let this signal share a polarity with the "Derivative" and "Integral" signals to be generated shortly; these signals will come from circuits that necessarily invert.
[6] We call this "P," as we call the other signals, soon to be added, "D" and "I," although all are inverted. Strictly, then, this is "−P." We omit the "minus" in these labels, thinking it easier to refer to P than to minus P.

Lab. Op-Amps V

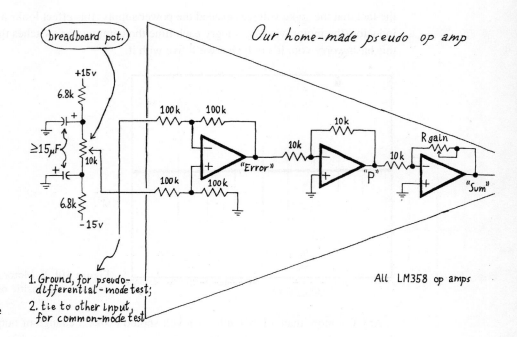

Figure 10L.6 Differential amp followed by gain stage and an inversion.

Check common-mode and differential gains

Common-mode gain...: We suggest that you use a resistor substitution box to set the *summing* circuit's gain. Set the gain at ten, and see whether a *common-mode* signal – a volt or so applied from the input pot applied to both inputs – evokes the output you would expect. (Do you expect *zero* output?)

...(Pseudo-)differential gain: Then ground one input (the 100k that feeds the first op-amp's *inverting* input) using the level from the potentiometer as input. Watch that input, and the circuit output, with the *R* substitution box value set to 100k: see if you get the expected gain of +10.

A couple of features of this test may bear explanation.

- Yes, the gain is *positive* when the input pot drives the *non-inverting* input to this home-made op-amp, since *two* inverting stages follow the diff-amp.
- We are applying a "pseudo-differential" signal by grounding one input of the diff-amp and driving the other. (You did this also in Lab 5L, as you drove the home-made "op-amp.") Since the *differential* gain is so much higher than the *common-mode*, this *pseudo-differential* signal works almost as a true differential signal would. An applied signal of *v* appears as a *differential* signal of magnitude *v*, combined with a *common-mode* signal of magnitude $v/2$. Given even a mediocre CMRR, this modest *common-mode* signal mixed with the *differential* is harmless.

A DVM may be handier than a scope, at this point, to confirm that the output of this chain of three op-amp circuits shows a pseudo-differential gain of +10, while you drive the input with the *input* potentiometer voltage. When you finish this test, leave the output voltage close to zero volts.

10L.2.5 Drive the motor

You have already tested the motor driver. Let's now check the three new stages – those that form the pseudo op-amp – by letting their output feed the motor-driver. *Ground* the inverting input for this test, as shown by a short dotted line in Fig. 10L.7. You do *not* yet need to make the connection to that terminal from the output potentiometer – a connection shown as an alternative, longer dotted line in that figure.

Confirm that you can make the motor spin one way, then the other, as you adjust the input pot slightly above and then below zero volts. (The motor-driven pot, as we have said, fortunately can take the pot to its limit without damaging pot or motor.)

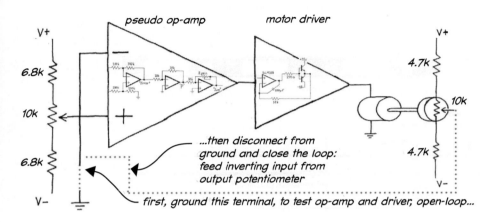

Figure 10L.7 Op-amp plus driver: first try open loop to test diff-amp, gain stage, sum and motor drive; then close the loop.

10L.2.6 Close the loop

Now let's close the loop. Reduce the *gain*, using the *R substitution box*: set it to about 1.5 (R_{gain}=15k). Replace the *ground* connection to the inverting input of our "pseudo op-amp" with *the voltage from the output potentiometer*. This connection is shown as the longer dotted path in Fig. 10L.7. Make sure to disconnect the inverting input from the *ground* that you tied it to in §10L.2.5.

Watch V_{in} on one channel of the scope, $V_{\text{output-pot}}$ on the other. *If a digital scope is available, this is a good time to use it* because a very slow sweep rate is desirable: as low as 0.5 second – or even 1 second – per division.

Several ways for testing the loop: Two or three methods are available to you for testing the new setup.

- **Two ways to drive the input**
  - **Square wave from function generator**. A function generator can provide a small square wave (±0.5V, say) at the lowest available frequency (about 0.2Hz on our generators). This input can temporarily replace the *manual* input potentiometer. This is probably the best choice since it provides a consistency you cannot achieve by hand.
  - **Manual step input**. You may, however, prefer the simplicity of *manually* applying a "step input" from the *input* pot: a step of perhaps a volt.

  The *output* pot should follow – showing a few cycles of overshoot and damped oscillation.
- **An alternative test: disturb the output, and watch recovery**. A second way of testing the circuit's response is available if you prefer (and you may want to try this in any case, after looking at the response to a step input): leave the input voltage constant, then manually force the pot away from its resting position simply by turning the knob of the *output* pot. Let go, and watch the knob return to its initial position – showing some overshoot and oscillation as when the change was applied at the input pot.

You start with a very low gain (1.5), which should make the circuit stable, even in this P-only form. Now use the substitution box to dial up increasing gain. At $R_{\text{gain}} = 220\text{k}$ (| gain | =22) we saw some overshoot and a cycle or two of oscillation, evident in the motion of the motor and pot shaft. If this shaft were controlling, say, the rudder of an airplane, this effect would be pretty unsettling. The circuit works – but it would be nice if we could get it to settle faster and to overshoot less.

Increasing the gain, at R_{gain}=680k (| gain | =68), we were able to make out several cycles of oscillation (the bigger, uglier trace in Fig. 10L.8 shows the motor drive voltage; there the oscillation is more obvious).

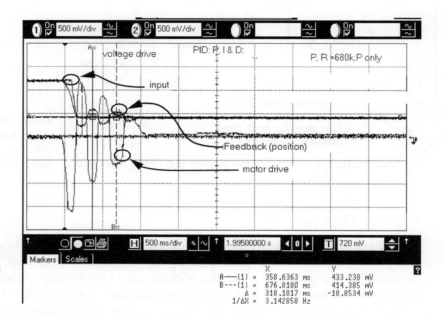

Figure 10L.8 P only: gain is high enough to take us to the edge of oscillation.

With a little more gain (R_{gain}=1M, in our case; | gain | =100) and the application of either a step change at the input, or a displacement of the output pot by hand, we see a continuous oscillation. Find the gain that sets your circuit oscillating, and then *note the period of oscillation* at the lowest gain that will give sustained oscillation. We will call this the period of "natural oscillation," and soon we will use it to scale the remedies that we'll apply against oscillation.

10L.3 Add derivative of the error

Well, of course we *can* get it to settle faster; we *can* improve performance. (If we couldn't, would the name of this sort of controller include the I and D in PID?) We can speed up the settling markedly, and even crank up the P gain a good deal once we have added this derivative. Thinking of the stability difficulty as a problem of taming the phase shifts of sinusoids – as we did for op-amps generally – we can see that inserting a derivative into the feedback loop will tend to *undo an integration*.

The integrations are the hazard, here: one is built in – the translation from motor speed to motor position. Additional integrations resulting from lagging phase shifts can carry us to the deadly $-180°$ shift that converts *nice* feedback into *nasty* – the sort that brings on the oscillation you have just seen.

10L.3.1 Derivative circuit

The standard op-amp differentiator in Fig. 10L.9 can contribute its output to the summing circuit. Here we show the entire prior circuit (Fig. 10L.7), with the differentiator added. The differentiator's gain is rolled off at about 1.5kHz. Again we recommend that you use a *resistor substitution box* to set the D gain if such a box is available.

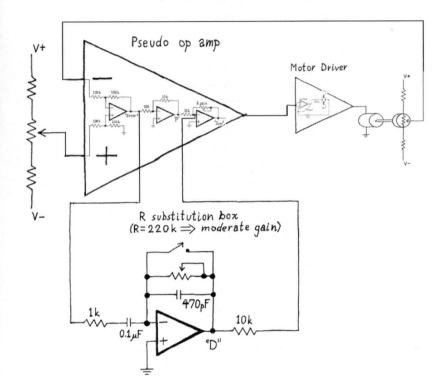

Figure 10L.9 Derivative added to loop.

Figure 10L.9 also shows a switch in the feedback path that permits you to kill the derivative when you choose to.

How much derivative? Our goal in adding derivative is to cancel the extra phase shift otherwise caused by a low-pass effect that brings on instability. How do we know at what frequency this trouble occurs, and therefore how to set the frequency-response or (equivalently) *gain* of the differentiator?

It turns out that you already have this information: you got it by measuring the frequency (or period) of "natural" oscillation, back in §10L.2.6. There, as you know, you gradually increased the P-only gain till you saw that an input disturbance would evoke either an output that took a long time to settle or else a continuous oscillation. (When *we* ran that experiment, for example, we got a "natural oscillation" period of roughly 0.6 second using the rotary motor-pot. With the slider pot, that period was about 40ms. If you are using a slider version, adjust your D gain accordingly – do not use the values we note below, which apply to the rotary version.)

We aim to make the derivative contribution, D equal to the P contribution, at the "corner" frequency where sustained oscillation would occur. RC defines the differentiator's *gain*. You'll find an argument for this proposition in Chapter 10N in case you need to be persuaded. In order to make the D gain equal to the P gain at the frequency of "natural oscillation," we want $RC = 1/(2\pi f)$, where f is that oscillation frequency.

A scaling rule of thumb: frequency of natural oscillation dictates D gain: If, as this formula suggests, RC should be about 1/6 of the period of natural oscillation, then for our $T_{oscillation}=0.6s$ we'd set RC to about 0.1s, or a bit less.[7] If we use a convenient C value of $0.1\mu F$, the R we need is about 1M.

Let's make this value adjustable, though, because we want to be able to try the effect of more or less than the usual derivative weight: if you have a second *resistor substitution box*, use it to set the differentiator's gain (RC). Otherwise, use a 1M variable resistor. Watching the position of the rotator will let you estimate R to perhaps 20%; the midpoint value certainly is 500k, and 750k is close to the 3/4-rotation position. The differentiator's output goes into the summing circuit installed earlier, through a resistor chosen to give this D term weight equal to the P's.

We hope you will find this D to be strong and effective medicine. Once it has tamed your circuit's response – eliminating the overshoot and ringing – crank up the P gain, to about twenty ($R_{SUM\_GAIN}=220k$) or more. Is the circuit still stable? Try more D. Does an excess of D cause trouble? The scope image of the circuit's response will let you judge whether you have too much or too little D: too little, and you'll see remnants of the overshoot you saw with P-only; too much D, and you'll see an RC-ish curve in the output voltage as it approaches the target: it chickens out as it gets close. And if you keep increasing the D gain still further, as we said in the Chapter 10N, especially §10N.5.2, the circuit goes unstable once again: it oscillates.

Switch: The toggle switch across the feedback resistor will let us cut D in and out; the switch seems preferable to relying, say, on a very-large variable R to feed the summing circuit. We find it can be hard to keep track of multiple pot settings to know whether we're contributing D or not. A switch makes the ON/OFF condition easier to note.

10L.4 Add integral

Adding the third term – the I of PID – can drive residual error (a difference between the input pot voltage and the output pot voltage) to zero. Fig. 10L.10 is a diagram of the full PID circuit with the integrator added.

Two details of the integrator may be worth noting:

Two polarized caps placed end-to-end: This odd trick works to permit use of *polarized* capacitors in a setting that can put either polarity across the capacitance. The effective capacitance is, of course, only one half the value of each capacitor. We use polarized caps only because large-value caps like these $15\mu F$ parts are hard to find in non-polarized form.

Seeming absence of DC feedback: at first glance, this integrator seems doomed to drift to saturation, since the integrator includes neither of our usual protections against such drift – feedback resistor or momentary discharge switch. But neither is necessary here because overall feedback – all the way around the large loop, from input pot to output pot – makes such unwanted drift impossible. In short, there *is* DC feedback, despite appearances to the contrary.

---

[7] See, e.g., Tietze and Schenk *Electronic Circuits: Handbook for Design*, Second edition, Springer (2008). A less formal approach appears in David St. Clair's *Controller Tuning and Control Loop Performance*, Straight-Line Controls, Inc.; from the author's(members.aol.com/pidcontrol/) one can download a simulator that allows one to try his rules. The easiest simulator, along with a good tutorial, appears in http://newton.ex.ac.uk/teaching/CDHW/Feedback/. The *simulation* lets you try (as you would expect!) the effect of varying P gain and of adding in D and I – just as we do today.

10L.4 Add integral

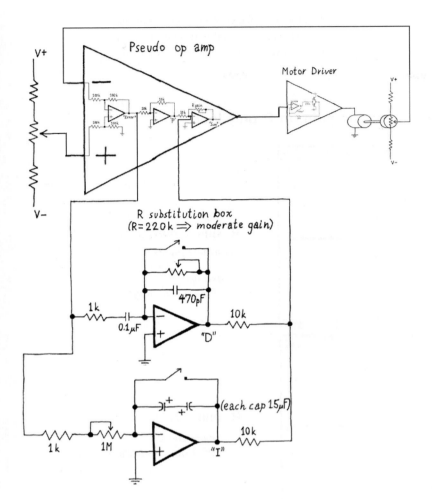

Figure 10L.10 Integral added to complete the PID loop.

Watching the effect of I: In today's circuit, the residual error is hard to see on the scope, so adding I will not reward you as adding D did. Your best hope will come if you cut the P gain very low: try $R_{\text{gain}}=100\text{k}$, so that the circuit feedback ought to tolerate a noticeable residual error, when not fed an I of the error. If you have been using a function generator to provide step inputs to your circuit, now replace that signal source with the manually-adjusted pot input. Slow the scope sweep rate, to a rate that permits you to see the multi-second effect of the integration.

If you are using a digital scope, you will be able to watch input ("Target"), output ("Motor pot"), and *Integrator* signals responding to a step input applied from your input potentiometer. If you are patient, you can even make out the effects of the motor and pot's "sticktion" (a cute term for "static friction"): the motor and pot do not move smoothly in response to a slowly-changing input (here, the I term). Instead, the motor fails to move till the I voltage reaches some minimal level; then output voltage jumps to a new level, and waits for another shove. You can see these effects in some of the scope images in §10L.5. at the end of this chapter.

Too much I again brings instability: It sounds dangerous, doesn't it? – tacking in an integral term when integration, plus other lagging phase shifts, are just what threatens the circuit's stability. It *is*

432   Lab. Op-Amps V

dangerous, as you can confirm by overdoing the I. You should be able to evoke continuous oscillation, as in the dark days before you knew about the stabilizing effect of D. Yet, remarkable though this fact is, some I does improve loop performance – driving long-term error toward zero – and need not bring on instability.

10L.5 Scope images: effect of increasing gain, in P-only loop

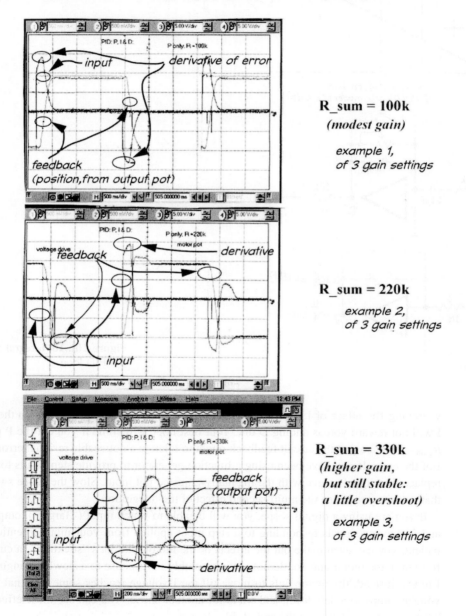

R_sum = 100k
(modest gain)

example 1,
of 3 gain settings

R_sum = 220k

example 2,
of 3 gain settings

R_sum = 330k
*(higher gain,
but still stable:
a little overshoot)*

example 3,
of 3 gain settings

Figure 10L.11
Increasing P-only gain brings increasing overshoot.

11N Voltage Regulators

Contents

11N.1	**Evolving a regulated power supply**	**434**
	11N.1.1 Unregulated supply	434
	11N.1.2 Zener sets V_{out}	434
	11N.1.3 Zener plus discrete-transistor follower	435
	11N.1.4 Zener or reference plus op-amp follower	435
	11N.1.5 Reference plus op-amp follower plus "pass" transistor	436
	11N.1.6 Stabilized circuit	436
	11N.1.7 Current limit: a complete regulator circuit	436
	11N.1.8 Dropout voltage	437
11N.2	**Easier: 3-terminal *IC* regulators**	**439**
	11N.2.1 Fixed output: 78xx	439
	11N.2.2 Variable output: 317	440
11N.3	**Thermal design**	**441**
	11N.3.1 Thermal transfer, in general	441
11N.4	**Current sources**	**443**
	11N.4.1 An IC for intermediate currents	443
	11N.4.2 A bipolar IC for low currents	443
	11N.4.3 JFET as current source	443
11N.5	**Crowbar overvoltage protection**	**444**
11N.6	**A different scheme: switching regulators**	**445**
	11N.6.1 Preliminary: getting used to inductors' peculiar proclivities	445
	11N.6.2 Three switching configurations: "Boost," "Buck" and Invert	446
	11N.6.3 Efficiency	447
	11N.6.4 Implementing the feedback	448
	11N.6.5 Switchers aren't always what you want	449
11N.7	**AoE Readings**	**450**

Why?

What problem do we meet today? We try to design a circuit that provides an output *supply* voltage that is constant despite fluctuations that may arise in both input voltage and output current loading.

Voltage regulators: One could argue quite plausibly that this is not a topic in its own right; it is only one more application of negative feedback, and could fit quite well into the op-amp chapter. Isn't a regulator just a follower driven by a reference? Yes, it is – though we'll refine the *follower* soon. But this function is needed so often that specialized ICs have evolved to do just this job, so that one almost never does use an op-amp. And power supplies and their regulators are so universal in instruments of all kinds that AoE assigns them a chapter of their own. We follow this scheme, giving them a day in the lab.

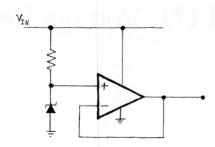

Figure 11N.1 Voltage "regulator:" just another use of feedback.

11N.1 Evolving a regulated power supply

AoE §9.5

11N.1.1 Unregulated supply

In the beginning – well, in Chapter 3N and Lab 3L, anyway – there was the unregulated power supply. Look back to Fig. 3N.28: it provided a DC level from the AC "line" voltage. Why isn't this good enough? How does it fall short? Let us count the ways:

- You can see from Fig. 11N.2 that output shows some ripple – and it is not a good idea to try to solve that problem by boosting capacitor size. Doing that would reduce ripple, but at the cost of increasing transformer heating, calling for a larger and heavier transformer.

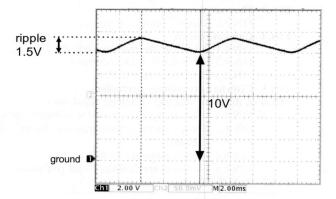

Figure 11N.2 Unregulated power supply shows substantial ripple. (Scope gain: 2V/div.)

- V_{out} varies somewhat, with loading ...
  - The transformer's output impedance is mediocre: in Lab 3L, for example, we saw about 20% droop between light and full loading from our nominal 6.3V transformers;
  - Even apart from the reduction of transformer voltage under load, loading reduces average V_{out} by increasing ripple.
- V_{out} can vary widely because of large variation in the *line* voltage (nominally 120V, but varying by about $\pm 10\%$, and occasionally more at times of heavy loading.[1])

11N.1.2 Zener sets V_{out}

A zener diode can stabilize V_{out}: see Fig. 11N.3. What's wrong with this scheme?

---

[1] One state's regulations (Illinois), for example, allow 10% variation in voltage supplied for "power" purpose, about 6% variation for lighting: from 113–127V on nominal "120V" lines. http://answers.google.com/answers/thread view/id/525096.htm

11N.1 Evolving a regulated power supply

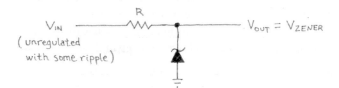

Figure 11N.3 Zener voltage source.

- It's inefficient: to provide 100mA of output current at full load, for example, it must idle with 110mA flowing through the zener (110 because we want at least 10mA through the zener even under full load).
- The zener's voltage varies somewhat with variation in its current, a variation that necessarily occurs as loading varies.
- Tolerances for even the *nominal* zener voltage are only mediocre.

11N.1.3 Zener plus discrete-transistor follower

AoE §2.2.3

A follower helps: it permits large I_{load} and large variation in this load current without calling for large currents in the zener: see Fig. 11N.4. This is better, but still disappointing.

- V_{BE}, somewhat variable, causes variation in V_{out} even for constant V_{zener}.

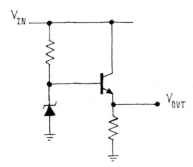

Figure 11N.4 Zener plus discrete follower: regulator?

11N.1.4 Zener or reference plus op-amp follower

An op-amp can hide the V_{BE} drop and its variations, and can give us much lower R_{out}: see Fig. 11N.5.

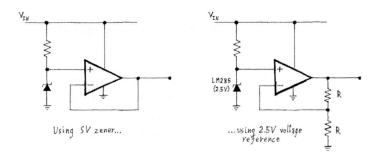

Figure 11N.5 Zener plus op-amp follower: regulator?

On the right of Fig. 11N.5 we have made a small improvement: we replaced the zener with an IC voltage reference (2.5V: same as Chapter 11L) providing a more precise initial voltage ($\pm 3\%$ in the grade you'll meet in lab), and good constancy over wide variations in current (20μA to 20mA).

This circuit lacks some refinements we will want to add – and it has one great flaw:

- its output current is small: limited to the op-amp's 25mA (typical for the '411; current limits for all ordinary op-amps are similar)

11N.1.5 Reference plus op-amp follower plus "pass" transistor

This is almost what we need. It can provide a large current. One could boost that further by using a *Darlington* pass transistor configuration (with its beta-squaring), or a power MOSFET (a field effect transistor that needs essentially zero input current). The op-amp – with its feedback that encompasses the pass transistor – gives the circuit very low R_{out}.

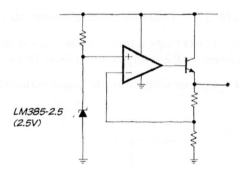

Figure 11N.6 Zener plus op-amp plus pass transistor: regulator?

However, it has a flaw that makes it unpredictable, and perhaps unusable. The output drives a power supply line which will be studded with decoupling capacitors. These are very likely to cause phase lags that push the op-amp into oscillation – as you know from your experience in Lab 9L.[2]

11N.1.6 Stabilized circuit

To keep the circuit stable despite capacitive loading, we use the methods of Lab 9L – primarily the *splitting* of the feedback path, see Fig. 11N.7.

The effective R that pairs with the feedback C to set $f_{\text{crossover}}$ is R_{Thevenin} for the feedback divider. This is a useful circuit. We'll add one more feature and call it done: we'll add a *current limit*.

11N.1.7 Current limit: a complete regulator circuit

To make the circuit foolproof, you should design it to survive abuse by fools: people who short the output to ground, from time to time.[3] All reasonably designed power supplies include this protection, implemented through the addition of one more transistor: see Fig. 11N.8.

The *current-limiting* scheme: Q_{limit} begins to direct current away from the pass transistor if the output current grows too large. Note that this scheme depends on the truth that the op-amp's output

AoE Fig. 9.2

---

[2] You might think we could say oscillations are certain, with these capacitive loads. Oddly though, extreme capacitive loading can stop oscillations. So we probably ought to say only that this circuit is vulnerable to parasitic oscillations, and must be stabilized.

[3] This collection of "fools" encompasses all of us of course.

11N.1 Evolving a regulated power supply

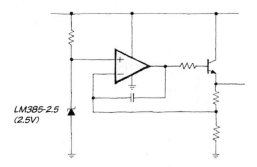

Figure 11N.7 Stabilized circuit.

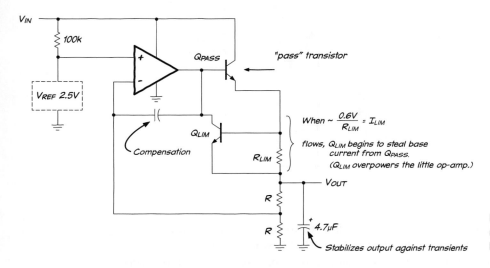

Figure 11N.8 Homebrew regulator, including current limit.

current is limited; if not for that limit, the new transistor would actually *add* to the output current, making things worse.

Op-amps use the same current-limit scheme: The output stage of op-amps uses the same sort of current limit. You will recognize in Fig. 11N.9 the limit circuit in the output stage of the LF411, for example.

> Full '411 circuit appears in §9S.4.

11N.1.8 Dropout voltage

Any regulator of this type ("linear," rather than the "switching" type that you will see in §11N.6) needs some minimum difference between input and output voltage. This is called the "dropout voltage," because the output drops out of regulation if you don't fulfill this requirement.

Most regulators need 2–3V. You will find this minimum plausible when you recall the V_{BE} drop in the follower, the need of most op-amps for a volt or so between V_{out} and the positive supply, and then the current-limit circuitry.

Specialized low-dropout regulators can get by on a few tenths of a volt drop between V_{in} and V_{out}. This feature can be important in battery-powered designs, where the battery voltage may gradually droop. In such a circuit a low-dropout regulator could function long after a conventional linear regulator would have dropped out. But note that *switching* regulators may be better still for battery-powered devices. See §11N.6.

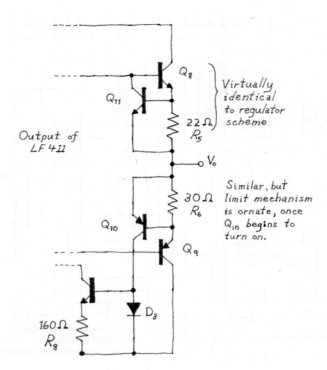

Figure 11N.9 Current limit in op-amp ('411) looks like regulator's limit circuit.

Low-dropout regulator: The strategy used to reduce the dropout voltage is to reconfigure the *pass* transistor. In a conventional regulator design like that of Fig. 11N.1.7, where the pass transistors is wired as a *follower*, a V_{BE} drop lies between input and output and the op-amp itself must lose some voltage between its positive supply and its output.

AoE §9.3.7

The low-dropout regulator wires the pass transistor not as follower but as what might be described as an inverting amplifier: see Fig. 11N.10.

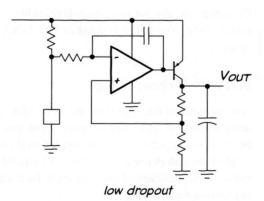

Figure 11N.10 Low-dropout regulator: pass transistor is reconfigured. Note funny feedback!

This is a circuit we mentioned back in Chapter 9N, as a rare case with an inversion within the feedback loop. The feedback capacitor is required – providing *split* feedback. The resistor between voltage reference and inverting input permits feedback to that terminal despite the "stiffness" of the good voltage reference.

A large bypass capacitor on the output also tends to tame the circuit's tendency to be jumpy.[4] When you don't need the low-dropout feature you ought to use the ordinary configuration, which is more stable.

11N.2 Easier: 3-terminal IC regulators

Now that you have paid your dues by re-inventing the regulator, we'll let you consider some regulators that are much easier to use.

11N.2.1 Fixed output: 78xx

AoE §9.3.2

The whole circuit of Fig. 11N.8 – reference, op-amp and pass transistor – plus somewhat more, is available on one chip. The simplest of these regulators is embarrassingly easy to use: this is the *three-terminal* fixed-output type in Fig. 11N.11.

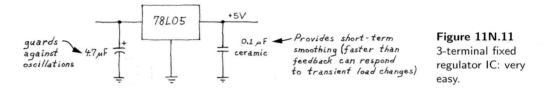

Figure 11N.11 3-terminal fixed regulator IC: very easy.

You may find it gratifying to notice – in the schematic of Fig. 11N.12 – that the designers of the 78xx regulators seem to have been looking over our shoulder as we evolved the linear regulator.

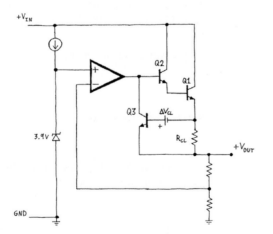

Figure 11N.12 Simplified circuit of the classic 3-terminal regulator, LM7805, looks a lot like our recent design.

The circuit is just what we designed, apart from use of a Darlington pass transistor and the application of a peculiar boost (labeled ΔV_{CL}) to the base voltage of the current-limiting transistor. The diagram also omits the necessary stabilizing elements.

This device, like other IC regulators, limits not only its output current but also its own *temperature*. Such a thermal shutdown protects the regulator when what is excessive is not *current* alone but *power*. (You will demonstrate this protection in the lab, by putting a large voltage drop across the regulator:

---

[4] Compare AoE §9.3.2 noting role of large capacitor to tame the negative regulator with similar pass-transistor configuration, LM7905.

Vin $\gg V$out). The thermal limit also protects you, the designer, against the possibly destructive effects of inadequate heat-sinking (see §11N.3).

11N.2.2 Variable output: 317

AoE §9.3.3

The *317* is almost as easy to use as the simpler 78L05, and is more versatile: it allows you to adjust V_{out}. So you need not stock an IC for each voltage that you may need. It can also be used to rig up an easy current source for values above about 5mA.

The left-hand sketch in Fig. 11N.13 shows a simplified version of the '317's circuitry.[5] The right-hand sketch shows the device wired to source 10mA into a load returned to ground (or to a negative supply).

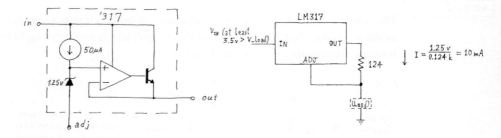

Figure 11N.13 LM317 variable regulator: simplified circuit, and use as current source.

The '317 likes to hold a *constant* voltage (1.25V) between its two output terminals, as the diagram above indicates; it uses no "ground" terminal.

We have shown the '317 as current source not because this is the most common use for the part; it isn't. Instead, we find this the easiest way to understand the '317's *voltage* regulation: just let it source this fixed current through a resistor to ground. The fixed current times this resistance defines the voltage at ADJ. Then $V_{out} = V_{ADJ} + 1.25$. So, the circuit in Fig. 11N.14, for example, permits adjustment of V_{out} through the range 5–10V.

The notion that the device feeds a fixed current to R_1 is only one way to see its operation. You may prefer the formula

$$V_{out} = 1.25V(1 + R_2/R_1),$$

where R_1 is the current-setting resistor, R_2 the resistor to ground. But tidy formulas always threaten to obscure how the circuit achieves what it does.

A circuit example: adjust V_{out}: Figure 11N.14 shows a sample circuit to bring these abstractions down to earth. The two resistors below ADJ let the 10mA current drop 3.75V to 8.75V, to give the advertised V_{out} range.

But don't get carried away: V_{out} range is limited: One might be tempted to think the '317 all-powerful: make a 1000V supply by installing a 100k R to ground? No. V_{out} is limited to about 37V because of a 40V limit on difference between V_{in} and V_{out}: if the output is shorted to ground with V_{in} very large (substantially greater than 40V) the part will be destroyed.

'317 peculiarities: maintain minimum I_{out}: Part of the '317's cleverness lies in the fact that it lacks a ground terminal.[6] But the side effect of the lack of a ground terminal is the fact that the '317 is

---

[5] A more detailed version showing current limit and Darlington pass transistor appears in AoE Fig. 9.9.
[6] "Why is this clever?," you may protest. Well, this peculiarity allows it to act as a floating current source and thus to allow adjustment of V_{out} simply through adjustment of a single resistor.

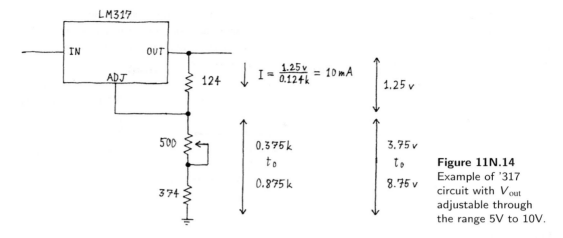

Figure 11N.14 Example of '317 circuit with V_{out} adjustable through the range 5V to 10V.

powered by the current that passes through the device – from IN terminal to OUT. So, the regulator fails if the user sets that current too low. The *minimum* output current is specified as sometimes 5mA, sometimes 10mA.[7] Use 10mA, to be safe; it follows that R_2 should be about 120Ω: or 124Ω if you're using 1% values.

11N.3 Thermal design

Overheating is an issue we have not considered until now (except perhaps in specifying large transistors for the push–pull drivers of several op-amp labs). We're bound to consider it when designing power supplies, because the currents are much larger than those we have been using to this point. To keep a part from overheating, we need to let it dissipate heat at the same rate that its electrical power generates heat.

So much is obvious. But you might not anticipate that the rate at which heat can be dissipated is calculable with methods exactly analogous to the use of Ohm's law. All we need to do is accustom ourselves to the changed units.

11N.3.1 Thermal transfer, in general

AoE §9.4

A look at the *units* will reveal the analogy between thermal and electrical flows and resistances. In place of voltage (by which we always mean voltage difference) we use temperature difference. Temperature difference drives the flow of heat, just as voltage difference drives the flow of current. The rate of thermal flow, a rate of energy or heat transfer, we measure as power: in watts, the W in the equation below.

The resistance to heat transfer is analogous to electrical resistance. Here, its units are °C/Watt: temperature difference per rate-of-heat-transfer. In the familiar electrical case, resistance in ohms is the equivalent: voltage difference per current (which is rate-of-charge-transfer).

AoE §9.4.1A

Electrical $I = V/R$

Thermal   Rate of heat transfer = (Temp Difference/Thermal Resistance)
            $W = (\text{Tempdifference}/\Omega_{\text{thermal}})$

---

[7] The data sheet from National Semiconductor (now absorbed into Texas Instruments) is ambivalent on this question: as it specifies device properties, in its tables of "...Electrical Characteristics," the data sheet recites $I\text{out} \geq 10\text{mA}$. But it uses half that value in all the suggested circuits that appear later within the same data sheet.

Figure 11N.15 represents the analogy graphically, while, to make things less abstract, Fig. 11N.16 is an exploded sketch of an IC and its heat sink, showing where these thermal resistances appear. The subscripts in the figure bear explaining. For example, $R_{\theta JC}$ is the thermal resistance between *junction* (the guts of the IC) and the *case*.

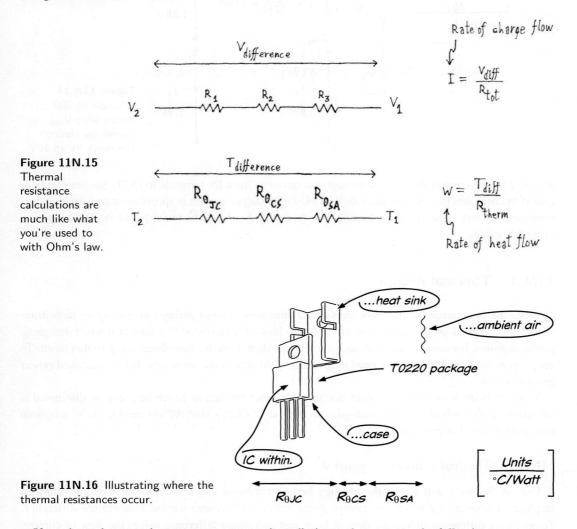

Figure 11N.15 Thermal resistance calculations are much like what you're used to with Ohm's law.

Figure 11N.16 Illustrating where the thermal resistances occur.

If you know how much power your part needs to dissipate, then you use the following:

$$R_{\text{Thermal}} = \frac{T_{\text{Junction}} - T_{\text{Ambient}}}{\text{Power}}$$

to determine what total thermal resistance your circuit can tolerate. You may be stuck with some of the resistances – such as *junction-to-case*, a value determined by IC design. But you have a choice of heat-sinks, and you can choose one effective enough to keep the total R_{Thermal} acceptable. We go through such an exercise in 11W.

11N.4 Current sources

AoE §9.3.14A

We have noted that the '317 can serve as a current source – but only for currents of about 5mA and up. Let's look at some current-source ICs that are more versatile.

AoE §9.3.14B

11N.4.1 An IC for intermediate currents

The LT3092 (see Fig. 11N.17) can *sink or source* 0.5–200mA. It can do both because it is a *two terminal* device. Placement of the *load* determines which job the part performs. This versatility contrasts with the behavior of bipolar transistor current sources that we have seen, where *npn* can only *sink*, and *pnp* can only *source*. (On the other hand, you *have* met a two-terminal current-limiting device: the JFET described in §11N.4.3, the part that you used to define the "tail" current of Lab 5L's differential amplifier.)

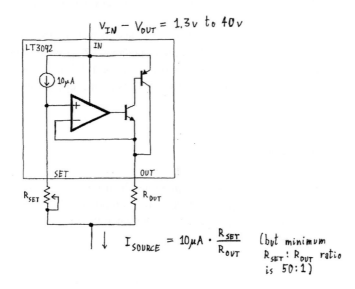

Figure 11N.17 LT3092 two-terminal current limiter.

The dynamic impedance (R_{out}) of the LT3092 is spectacular at DC (100MΩ) and good at low frequencies: about 3MΩ @ 1kHz.

11N.4.2 A bipolar IC for low currents

For lower currents (50–400μA) the REF200 (see Fig. 11N.18) is very neat, and provides enormous R_{out} at DC, as does the LT3092. In §11W.2 we have posed some puzzles asking how to apply the three elements of this part to sink or source a variety of currents: 50, 100, 200, 300, 400μA. The mirror, as indicated in the figure, holds its right-hand current equal to the current fed into its left-hand side.[8]

11N.4.3 JFET as current source

You met this two-terminal current source in Lab 5L, now shown in Fig. 11N.19. It was pleasantly easy to use: it comes in a glass package that looks like a diode. Within the diode-like package is a JFET – a junction field-effect transistor, a transistor type that we will not otherwise meet in this course (except

---

[8] The mirror is drawn as a simple mirror, but in fact is a Wilson or cascode mirror, as noted in Chapter 5S.

444 Voltage Regulators

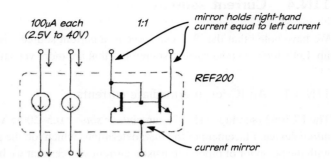

Figure 11N.18 REF200 low-value current-source IC.

to say, in passing, that JFETs give the '411 op-amp its high input impedance). The JFET's two control terminals are shorted together within the package, and in this configuration the JFET runs at a fixed current.

You have experienced what is appealing about this device: it is very easy to use. It does have weaknesses. Compared with the LT3092 and the REF200 its dynamic impedance (R_{out}) is good but less spectacular: about $1M\Omega$ @1mA. Its serious drawbacks are its price, (several dollars, varying with the value of *current*[9]) and its poor tolerance ($\pm$ about 20% for the SST-505 and for the Central Semiconductor parts like 1N5294, as well).

Figure 11N.19 JFET (Junction Field Effect Transistor) two-terminal current source.

AoE §9.1.1C

11N.5 Crowbar overvoltage protection

The failure of a regulator could feed an excessive voltage to a lot of precious electronics. The most familiar hazard is the possibility of overvoltage (or "surge") applied to an entire computer.[10] Fig. 11N.20 shows a circuit that shuts down the supply (clamps it to approximately 1V) when the voltage climbs too high. (ICs are available to do the voltage sensing, too.)

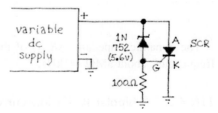

Figure 11N.20 Crowbar overvoltage protection (the circuit Fig. 11L.8).

The word "crowbar" apparently refers to the image of someone (courageous? foolhardy?) shutting down a huge power supply by shorting it to ground with a massive piece of steel.

---

[9] We were charmed to discover, in the strange pricing structure of these parts, the first objective evidence ever adduced for the proposition that most practicing engineers are as lazy as we. When 1mA versions of this part were offered, they carried a dramatic premium over parts like the 0.75mA value that we use. Why? Because everyone likes to do Ohm's law arithmetic with the value *1*, and we lazy people drive the price of that part way up.

[10] In the concluding labs for this course, in which they had constructed a computer from many ICs, one pair one pair of students accidentally demonstrated this vulnerability in horrible form: they used a variable supply for their home-made computer, adjusting it always to 5V: well, *almost* always. One day when they failed to check this 5V value, they cooked many ICs in their computer. Finding and replacing the destroyed parts was a painful lesson in the hazards of overvoltage. The present version of the microcomputer lab breadboard includes a *crowbar* clamp like the one described in this section. So such disasters are now very much less likely.

Since the SCR – the diode-like element on the right in Fig. 11N.20 – turns ON when its control terminal ("gate," G) reaches about 0.6V, much as a transistor does, the circuit of Fig. 11N.20 fires and clamps its output if the input reaches about 6.2V. If this circuit is to succeed in protecting downstream circuitry, the clamp itself must be stronger than the failing upstream regulator: a fuse should be placed upstream, to blow when the clamp fires.

A peculiarity of the SCR distinguishes it from the somewhat similar power transistor: once turned on, it remains *on* after the gate drive has been removed. Thus the crowbar circuit operates like a circuit breaker, locking the supply in shutdown state. Only turning off the SCR current, by shutting down the faulty power supply (or by blowing an upstream fuse), can release the SCR.

11N.6 A different scheme: switching regulators

AoE §9.6.1

A linear regulator is doomed to inefficiency, since its strategy for holding V_{out} constant as V_{in} varies is simply to soak up the difference. A linear regulator fed a ripply 10V can put out a very constant 5V – but the power wasted in the regulator, in this case, would equal the power delivered to the load. Efficiency in such a case is 50%. One cannot do much better, because uncertainty about the *line* voltage, and the need to keep ripple from dipping below the regulator's dropout voltage obliges one to put V_{in} 4 or 5V above V_{out}. What is to be done?

The answer is to adopt a method that is quite different. Don't soak up the difference between V_{in} and V_{out} in order to maintain V_{out} at a given level. Instead, use a *switch* to send intermittent gushes of current into a storage element as needed. When a gush is not needed, leave the switch turned *off*.

11N.6.1 Preliminary: getting used to inductors' peculiar proclivities

Because we have not used inductors much in this course, we may owe you a reminder of the inclination of inductors to keep a current going, once it is flowing. All the switching power supply configurations (shown below in §11N.6.2) exploit this behavior.

We hope you demonstrated this property for yourself in a switch exercise back Lab 4L. In class we repeat this exercise as a demonstration, provoking an inductor to show off its flywheel-like behavior. The circuit we use is just a switch to ground (it happens to be a MOSFET – the field-effect transistor that you will meet next time), with an inductor linking the switch to a positive supply. The positive supply is a single AA cell, at about 1.6V: see Fig. 11N.21.

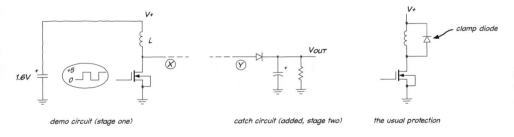

Figure 11N.21 Inductor switching demonstration: voltage spikes can be caught to generate large V_{out}.

We wire it first as in "stage one." The square wave input turns the FET switch ON; current ramps up in the inductor; the switch turns OFF – and the inductor wants to keep current flowing. Fig. 11N.22 shows the inductor's response.

As the switch opens, the inductor (indignant!) insists on keeping the current flowing, by driving the switch voltage (at **X** in Fig. 11N.21) higher and higher – until, at about 76V, the switch "breaks down" and conducts.

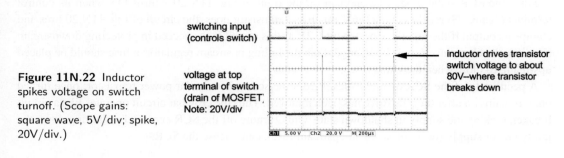

Figure 11N.22 Inductor spikes voltage on switch turnoff. (Scope gains: square wave, 5V/div; spike, 20V/div.)

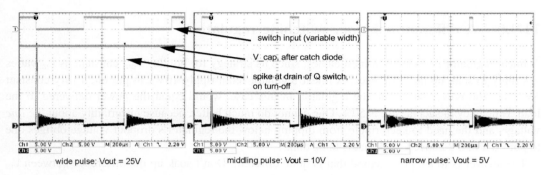

Figure 11N.23 A diode and capacitor can be used to catch the inductor's spikes (three pulse widths shown). (Scope gain: 5V/div.)

This breakdown is not good for a transistor switch; a large inductor could destroy the transistor in a single cycle, though power MOSFETs (but not bipolar transistors) include an internal protection diode whose tolerance for reverse current is specified on a data sheet. Normally, one would protect the switch with a diode as shown on the far right of Fig. 11N.21. For purposes of the demo, however, we want to show these large spikes – and in a moment we will put them to use. Note that the *supply* voltage is a modest 1.6V.

This perverse behavior of the inductor can be put to use. The voltage spikes can be used to source current into a capacitor in repeated surges. The capacitor thus can charge to a voltage much higher than the original supply voltage – as in the demonstration.[11]

Fig. 11N.23 shows such an application of the seemingly mischievous voltage spikes. The capacitor voltage rests close to the voltage of the repeated spikes, which are shown for three input pulse widths.

By varying the width of the input pulse, we varied V_{out}, as shown in Fig. 11N.23, from 5V to 25V. The ON time of the switch, determined by the pulse width, determines how close the inductor comes to its maximum current. At the pulse repetition rate shown, we found that further widening of the pulse did not raise V_{out}.

11N.6.2 Three switching configurations: "Boost," "Buck" and Invert

AoE §9.6.4

The demonstration – which took 1.6V up to 20V or so – used one of three basic circuit configurations available to switching regulators. Appropriately, that configuration is called "boost," labeled "B" in

---

[11] Your camera's flash circuit probably works this way: it starts with a battery voltage of perhaps 3V, and when you turn on the flash you may hear the circuit singing. If so, you are hearing the switching of an inductor, used to charge a storage capacitor up to 300 to 400V to fire the flash when you press the shutter release.

11N.6 A different scheme: switching regulators

Fig. 11N.24. (All three circuits are drawn showing Schottky diodes for their low forward voltage drops; for even greater efficiency, a MOSFET switch can be used in place of the diode).

The first configuration is a step-down, called "buck." As help in getting a grip on inductor behavior we offer a sketch of voltage and current waveforms in the buck circuit. Note that the transistor switch is turned ON by taking its control input to ground (the transistor is a *p*-channel MOSFET, analogous to a *pnp* bipolar transistor).

Figure 11N.24 Three switcher configurations.

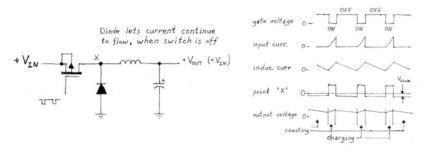

Figure 11N.25 Current and voltage waveforms in a buck (step-down) switcher.

The *ripple* on the output voltage of Fig. 11N.25 may look familiar – it may look like the 120Hz ripple of an unregulated supply. Yes and No. Note the radical difference in frequencies: the switching frequency usually is in the range 100kHz to 1MHz, to allow use of small inductors and capacitors. The "droop" time here is on the order of a microsecond rather than the 8ms of a powerline-driven full-wave rectifier.

11N.6.3 Efficiency

AoE §9.6.1A

You can convince yourself with a few seconds' thought that a *switch* that is either fully ON or fully OFF ideally dissipates no power at all. Consider the two cases in Fig. 11N.26.

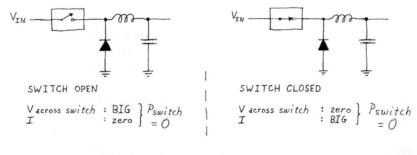

Figure 11N.26 Idealized switching regulator: zero power lost in switch.

In life, things aren't quite so good. Some power does get dissipated: in the inductor; in the switch, with its non-zero R_{on} and capacitance; in the diode, which conducts while the switch is OFF (up from *ground* – startling those of us unaccustomed to inductors' strange habits). Nevertheless, the results can come close to the ideal; much closer than a linear regulator can. 80% to 95% efficiency is feasible.

Figure 11N.27 contains part of a table describing National Semiconductor's[12] *boost* switching reg-

---
[12] National was swallowed up by Texas Instruments.

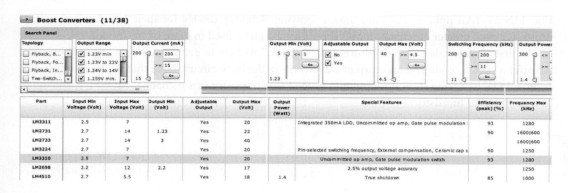

Figure 11N.27 Excerpt from National Semiconductor table of switching regulators.

ulators (those that generate $V_{out} > V_{in}$). Note the impressive efficiencies – perhaps difficult to read, here, but all at 85% or more.

Efficiency is the great strength of switchers, but they have other virtues: they can generate $V_{out} > V_{in}$, as in the examples we just looked at. They can change the *sign* of a voltage input – or can do both transformations at once. Those tricks allow powering circuits at a variety of voltages (such as 3.3V, 2.5V, 1.8V) from a single source.

For battery-powered devices, like a cell phone, the importance of efficiency is obvious: it permits long battery life. For line-powered circuits the importance is less obvious. The efficiency is valuable there rather because it permits supplies that are lighter and smaller and less hot than a linear supply.

11N.6.4 Implementing the feedback

We have not mentioned, to this point, how the switch is controlled in a switching regulator. Negative feedback compares V_{out} against a reference voltage. In this, the switcher behaves like the linear regulator. But the feedback circuit does not simply hold the switch steadily ON when V_{out} is low, OFF when V_{out} is higher than the reference. Such a scheme could produce very large ripple and could require huge inductors and capacitors.

Instead, the switching occurs continuously – or nearly so – and what is varied is the *duty cycle* of the switch: the percentage of time it spends ON in each cycle. (It is also possible to vary frequency but varying duty cycle is the more usual scheme.)

The switching regulator that you meet in the power-supply lab provides a hybrid scheme, varying not duty cycle or frequency but the duration of a burst of high-frequency switching cycles (it describes itself as a "gated oscillator switcher;" others call this design "Pulse Burst Modulation" or "hysteric conversion"[13]).

When V_{out} falls a little below the reference voltage (by an amount set by a built-in hysteresis circuit), the regulator fires a burst of 20kHz cycles, driving V_{out} up. Once V_{out} rises high enough, all switching terminates and the regulator's power consumption falls close to zero.

A result of this scheme is ripple at frequencies well below the 20kHz switch rate: as low as 2kHz in the cases shown in Fig. 11N.28. This is in a frequency range that can be very annoying to human ears,[14] so it seems a strange choice by the designers of the part.

The motives for this scheme are two. One is to minimize the regulator's power dissipation: when the

---

[13] See a good note by Analog Devices: http://www.analog.com/static/imported-files/tutorials/ptmsect3.pdf. This note points out the good feedback stability, i.e. freedom from loop oscillation, of PBM.

[14] "Ears", you may protest? Who is listening to this switching frequency? Well, ideally, no one. We were unable to hear our

11N.6 A different scheme: switching regulators

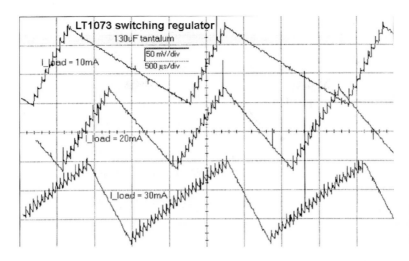

Figure 11N.28 Ripple on output of 1073 switch-burst regulator.

switching shuts off, current draw falls from hundreds of milliamps to about $100\mu A$, and under very light loading the circuit can be idle for "seconds at a time," the data sheet promises. A second virtue of this sort of design – in which the feedback loops relies on hysteresis with bang–bang behavior – is that it is not vulnerable to the instabilities that can trouble continuous-feedback designs. You know, from your experience in the *nasty oscillators* and PID labs, that ordinary feedback loops can be upset, especially by capacitive loading.

11N.6.5 Switchers aren't always what you want

They are noisy. . . : For digital devices, which shrug at low-level noise, switchers are the right choice. But the switching noise voltage that always appears on their outputs can rule them out for sensitive analog circuits.

AoE §9.5.5, especially Fig. 9.51

. . . and can be difficult. . . : And we should mention another reason why many people reach for a good old linear regulator: a switching supply can be subtle and difficult to design. It can be hard to keep them stable. And it is not only students in an introductory course like this one who might shy away from switchers. IC manufacturers know that switching power supplies have a reputation that tends to scare engineers.

AoE §9.6.5

. . . So manufacturers offer help: National Semiconductor/TI offer a series that they call "Simple Switchers," and since the name may not be enough to persuade its customers it also provides a web design program that steers you to particular ICs offered by them[15] given your design goals, and finishes the design for you (much as TI's FilterPro does for active filters): see Fig. 11N.29. This web application goes so far as to make up a "Bill of Materials" (BOM) for you: see Fig. 11N.30.

You do feel a bit dimwitted when you use this service. But you're not required to admit that this is how you "designed" your switching supply!

---

lab circuit. But inductors can sing at the frequency which they are driven; so can capacitors, and both effects can cause the singing of CRT video monitors. So it's good to put switching noise above the audible range.

[15] http://www.ti.com/ww/en/analog/webench/. For some reason their web service doesn't seem aware of their competitors' parts.

450 Voltage Regulators

Figure 11N.29 National/TI's Webench leads the timid by the hand, in switcher design.

Figure 11N.30 Webench makes it really easy.

And, yes, switchers can be hard even for smart engineers: Linear Technology, also feeling the designer's pain, offers a long application note by their late wizard, Jim Williams, entitled *Switching Regulators for Poets: A Gentle Guide for the Trepidatious*[16] Williams, who was not a beginner at electronics, wrote:

Before this effort, my enthusiasm level for switchers resided somewhere between trepidation and terror. This position has changed to one of cautiously respectful optimism.

If Jim Williams could feel optimistic, so can you.

11N.7 AoE Readings

Chapter 9:

> Unregulated supply: §9.5;
> Evolving a linear regulator: §9.1;
> IC regulators: §9.3;
> Switching regulators: §9.6.1ff.

---

[16] This note was written thirty-odd years ago, but the fundamental difficulties of switching regulators have not changed much since then – except for the arrival of delightful crutches like National's Webench. The note is AN25fa (1987).

11L Lab: Voltage Regulators

This lab begins with a trial of your design for a home-made voltage regulator. There is nothing very new here: the only elements you have not seen in a previous lab are

(1) the voltage reference – a super-duper zener, in effect;
(2) the current-limit included in the bipolar version described below.

The MOSFET version – an alternative regulator design – allows you to try the new transistor type as you try the regulator. Both home-made regulators raise stability issues that you will recognize from your experience with Lab 9L's "nasty oscillators." This exercise is not realistic: you are not at all likely to design a regulator from parts; we hope, though, that designing one once will give you insight into how a linear regulator works.

In the remainder of the lab, you will try first IC *linear regulators* – very straightforward; then an IC *switching regulator*. The switcher uses feedback to stabilize V_{out}, as the linear regulator does, but it regulates the output voltage in a way that is unfamiliar to us in this course: by switching an inductor, and exploiting the inductor's efforts to keep current flowing when the switch opens. The switching regulator can achieve effects that at first glance seem magical; V_{out} greater than V_{in}; V_{out} negative, for V_{in} positive. We hope you'll be impressed by this little IC.

11L.1 Linear voltage regulators

11L.1.1 Voltage reference

The homemade voltage regulator that we ask you to build in §11L.1.2 is required to put out a voltage that remains constant despite variations in loading. This behavior is familiar to us from all the low R_{out} devices we have used and built, including all our voltage *followers*. But today we ask our regulator to do something we have not seen before: hold its output constant despite variation in its *input* voltage.

A zener can provide this constancy, as you know. We will use a zener later in §11L.1.6. But for better constancy of V_{out}, which we want from our homemade regulators, we will rely on a *voltage reference* IC, the LM385–2.5.

Its behavior resembles that of a zener – but is much better. It shows very good constancy of V_{out} as its current varies over a very wide range. This you can see in the left-hand plot of Fig. 11L.1.

As you read the plot, notice that the scales marked on left and right vertical axes, indicating voltage variation for the two devices, differ by a factor of almost 200.

11L.1.2 Voltage regulator: your design

Please design, then try out, a voltage regulator built to satisfy either of the descriptions set out below. In addition to performance specifications, each design problem includes a specification of available

Lab: Voltage Regulators

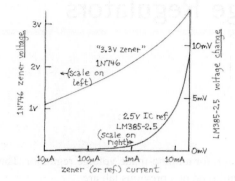
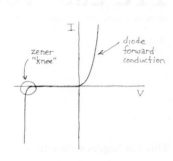

Figure 11L.1
I–V curves for zener and IC voltage reference compared.

parts. After specifying the task, just below, we will state more specifically what we mean when we ask you to *try out* the circuit.

Design task

Option A: Bipolar 5V regulator, with current limit
- V_{out}: 5V (±5%);
- V_{in}: 8–25V;
- Current limit: 100mA, more or less;
- Kit of parts:
    - op-amp: 1/2 LM358 (dual op-amp, single-supply; see Lab 9L);
    - "pass" transistor: 2N3055 (pinout: BCE, seen from front);
    - current-limit transistor: 2N3904;
    - voltage reference: LM385–2.5, a 2.5V super-zener ;
    - frequency compensation: start with *none*; add it when you've seen the problem.

Hints
- Look at the '723 IC's schematic, AoE §§9.2 and 9.2.1: in effect, you are building a homebrew '723;
- when you calculate the resistor you need to get a 100mA limit, note two complicating facts:
    - the '358 provides a rather high maximum output current: 40mA, typical; most of this will be *added* to your circuit's output current, under current-limit conditions;[1]
    - when this is occurring, the current-limit transistor will be passing enough current so that the usual rule $V_{BE}=0.6V$ will not hold. Note that $V_{BE}\approx 0.6V$ at 1mA, for a '3904; V_{BE} rises about 60mV per decade of current, so at, say, 35mA you may see V_{BE} of about 0.7V, and in your particular circuit you may see as much as 0.75V.

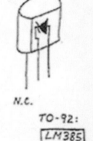

Figure 11L.2
Pinout of LM385 voltage reference.

Or, an alternative design (a little more ticklish, because of its tendency to *oscillate*).

Option B: Low-dropout 5V regulator
This design differs in omitting the current-limit (normally very desirable!), so that you can minimize the *dropout voltage* – the minimum difference between regulator input and output. That difference defines the minimum power wasted in a

---
[1] You could limit the magnitude of the current that the current-limit transistor could add to I_{out}: insert a resistor on the base lead of the *pass transistor*. Put this resistor upstream of the I_{limit} transistor tap. 300Ω, for example, would limit current to about 25mA for a V_{in} of 8 to 10V.

linear regulator, and – just as important, in a battery-powered instrument – determines just how low V_{in} can fall before the output fails.
- V_{out}: 5V (±5%);
- V_{in}: from slightly more than 5V up to 25V;
- Kit of parts:
    - op-amp: 1/2 LMC6482 (dual op-amp, single-supply, rail-to-rail; for pinout, see LM358 in Lab 7L);
    - "pass" transistor: *Bipolar:* 2N3906 (same pinout as 2N3904: EBC) or *MOSFET*: BS250P (R_{ON}=14Ω; pinout: SGD, seen from front) or VP01 (R_{ON}=8Ω) or IR-LIB9343PBF (R_{ON}=0.12Ω; pinout: GDS);
    - voltage reference: LM385–2.5, a 2.5V super-zener;
    - frequency compensation: start with *none*; add it when you've seen the problem.

Hint • Beware the inversion by the transistor circuit, which is *not a follower*. Instead, this is an inverting amplifier. We will need to take feedback from the transistor's *drain* (FET drain is equivalent to bipolar collector): I_D grows as gate voltage *falls*. That fact needn't cause you any difficulty, as long as you are aware of it; if you know about this inversion, you know how to keep the sense of feedback negative *in fact* (though the feedback may look shocking *on paper!*).

11L.1.3 Trying your circuit

Oscillations: Your circuit probably does *not* work well if you have followed our instructions: we have asked you to omit "frequency compensation," and your circuit output is very likely to oscillate, at least when presented with a capacitive load. Add a ceramic capacitor in the range 0.01–0.1μF, between V_{out} and ground. Such a cap normally stabilizes a power supply, but it is likely to de-stabilize the bipolar regulator, at least, looking like a capacitive load. Incidentally, make sure that you watch the output with a *scope*: if you make the mistake of thinking of this as a DC circuit which one can therefore understand with DC voltmeter alone, you will find that oscillations can cause DC errors that are puzzling indeed.

Once you have seen the oscillation problem, try to solve it. In both regulators, a feedback capacitor between op-amp output and the inverting input[2] should do the job (killing the high-frequency gain[3]). This is the remedy we have called "split feedback." What value is necessary? Note the relevant R: the R of the lower-frequency feedback path (here, R_{Thev} for the feedback divider). Choose RC to bypass the nasty stuff well below the frequency of oscillation that you have observed.

In addition, you should place a couple of capacitors to ground, at the output: a small ceramic (for high-frequencies; say, 0.01–0.1μF) and also a big tantalum (a little less good at high frequencies: try 4.7μF or bigger).

Vary V_{in}: A regulator should provide constant output despite variations in input – including the 120Hz "ripple" that rides an unregulated input. To save time, instead of applying *ripple*, just vary V_{in} (from a variable lab power supply) by hand, and watch V_{out}. You're not likely to see any variation in V_{out}, if your circuit is working right – until you take V_{in} so low that the circuit "drops out" of regulation.

---

[2] Note that even for the low-dropout design – where the feedback from circuit output goes to the *non-inverting* input because of an inversion effected by the PMOS circuit – the *high-frequency* feedback provided by the feedback capacitor must go to the *inverting* input in order to give feedback the proper sense.

[3] You may prefer to say, as we did in Lab 9L, that the cap provides a separate feedback path, bypassing the network that produces dangerous phase shifts at high frequencies.

Measure the dropout voltage: A regulator's "dropout voltage" is just the minimum difference between V_{in} and V_{out} required in order to let the regulator hold V_{out} where it ought to be. Measure $V_{dropout}$ for your homemade regulator. What characteristic of what device(s) in your circuit imposes this minimum on V_{in}?[4]

The answers differ for the two design options, as you would expect. $V_{dropout}$, the minimum difference between V_{in} and V_{out}, is the sum of...

Option A, using *npn* pass transistor
- minimum difference between the op-amp's positive supply and its output: about 1.5V...
- ...plus V_{BE} for the pass transistor (roughly 0.6V)...
- ...plus the drop across R_{lim} in the current-limiting circuit (up to 0.6V).

Option B, low-dropout version
- Assuming that the you use a PMOS FET, which needs several volts between its input terminal ("Source") and control terminal driven by the op-amp ("Gate"), we don't need to worry about the specification we called "minimum difference between the op-amp's positive supply and its output" above; op-amp output will be well below the positive supply voltage.
- Instead of V_{BE} for the pass transistor, we need consider only its minimum voltage between in and out terminals (its "Source" and "Drain"). For low values of V_{DS} the MOSFET behaves like a small-valued *resistance*, so this minimum depends on output current and the transistor's R_{ON}.
  In this circuit you can see that this R_{ON} value is important. The small MOSFET would drop almost 1.5V at 100mA; the big one would drop a mere 4mV.
- Since the low-dropout version lacks circuitry to limit its output current, there is nothing further to consider in accounting for the dropout voltage of this version

Loading: Now see to what extent V_{out} remains constant independent of *loading*. As "load," use 100Ω resistors, placed in parallel (why not smaller ones?). (A current meter placed in series between V_{out} and this load will let you measure the load current.) Don't go overboard in loading the low-dropout version, by the way, since it has no built-in current limit. Feel free to go overboard with the current-limited version – once you are convinced that the current limit works: you can even *short* the output to ground.

You should find – if you care to look – that the low-dropout regulator's dropout voltage varies with loading. Once you have decided what's imposing the minimum dropout voltage, you'll recognize why this relation between dropout and loading is necessary.

11L.1.4 Three-terminal fixed regulator

This device, shown in Fig. 11L.3, is embarrassingly easy to use. It is so handy, though, that it's worth your while to meet it here. It protects itself not only with current limiting, but also with a thermal sensor that prevents damage from excessive power dissipation ($I_{out} \times [V_{in} - V_{out}]$), an overload that could occur even though the current alone remained below the limiting value. You will watch this thermal protection at work, and incidentally will see the effect of *heat sinking* upon the regulator.

Watch this thermal protection by providing a load that draws less than the chip's maximum current of 100 mA. Use two 120Ω resistors in parallel to ground; see Fig. 11L.4. (Check, with a quick calculation, that you are not overloading these 1/4-watt resistors.)

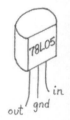

Figure 11L.3
78L05 3-terminal 5V regulator.

---

[4] This is a rhetorical question! See below.

11L.1 Linear voltage regulators

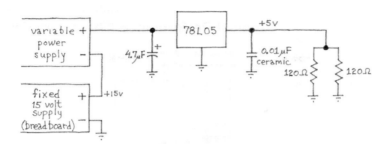

Figure 11L.4 78L05: demonstration of thermal protection.

Note: We suggest you *stack* the variable supply on top of the breadboard's fixed +15V supply. This scheme *requires* that you *float* the negative terminal of the variable supply; that terminal must *not* be tied to world ground.

A calculation, for the zealous: This demonstration is fun even if you don't predict what V_{in} will bring on current limiting, but you can make the experiment especially satisfying if you try to predict what voltage across the regulator will bring on thermal limiting. That is not hard, though it might take you a while to dig it out of the datasheet. The thermal specifications state a *maximum permissible junction temperature* of 125°C and give the *thermal resistance* between junction and ambient, which in this case means both the ambient air and the circuit board to which the regulator's leads are attached. You will be using the TO92 package – plastic – which dissipates substantial heat through its leads (even the length of these leads matters, you'll notice, from the curves shown in Fig. 11L.6).

To calculate the maximum power the package can dissipate before overheating the junction, plug in the values here.

Figure 11L.5 Push-on heat sink for T092 package.

78L05: Thermal Specifications

Package	Thermal Resistance (typical, max)	
	$R_{\theta JC}$	$R_{\theta JA}$
TO-39	20°C/W, 40°C/W	140°C/W, 190°C/W
TO-92	180°C/W	190°C/W

Definitions:

$R_{\theta JC}$ = thermal resistance, junction to case

$R_{\theta JA}$ = ... junction to ambient: includes JC and CA (case to ambient) in series

Assume that the device begins to limit when its junction temperature reaches about 150°C – but don't be shocked if your calculation indicates the device may be tolerating a higher temperature, perhaps as high as 200°C. Note units of *thermal resistance*, R_θ: °C/watt.

Experiment: dropout voltage and thermal self-protection: Gradually increase V_{in} from close to 0V. As you proceed, note:

- the dropout voltage;
- the V_{in} that actually evokes the chip's thermal self-protection.

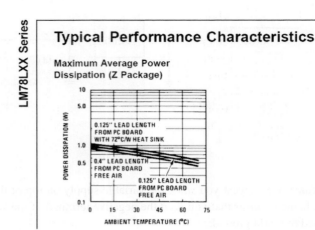

Figure 11L.6 78L05 max power dissipation depends on lead length (T092 package).

By the way, *what* does the chip do to limit its power dissipation? How will you know when the chip has begun to take this protective action?[5]

When you notice the chip limiting its own power dissipation, you will be able to call off this self-protection by cooling the chip. Try putting a bit of wet tissue paper to the 78L05's fevered brow. It should recover at once. Then try a push-on heat sink instead. You may have to crank up V_{in} to see the self-protection begin again. Now fan the regulator, or blow on it. Does the output recover, once more?

11L.1.5 Adjustable three-terminal regulator: 317

This regulator allows you to select an output voltage by use of two resistors. You can make a variable output supply by replacing one of them with a variable resistor. The 317 lets you stock one chip to get all the positive supplies you need (at least, up to 1A output current); it also lets you trim to exactly 5V, for example, if the 78L05's 5% tolerance is too loose. In addition, the 317 is easy to wire as a current source. In other respects this regulator is much like the 78L05: it includes both current and temperature sensing to protect itself from overloads.

Wire up the circuit in Fig. 11L.7. Try $R=750\Omega$; what should V_{out} then be? Measure it.

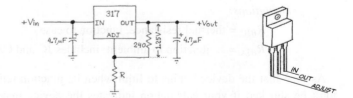

Figure 11L.7 317 voltage regulator circuit.

You should be aware of a *quirk* in the '317's specifications (a quirk that stung Tom, recently): the device does not regulate properly unless the current from its output is at least 10mA. So you must avoid large resistor values in the feedback path, unless you can be sure that a substantial load current will be drawn at all times.

Replace R with a 1k pot, and check out the 317's performance as an adjustable regulator. What is the minimum output voltage ($R=0$)?

---

[5] Yes: the IC limits its output current to limit its $I \times V$ power dissipation. The evidence will be the fall of V_{in} from its normal level of 5V.

11L.1.6 Crowbar overvoltage protection

Here's a little circuit that can protect against the potentially horrible effects of a power supply failure, by clamping the supply voltage close to ground in case that voltage exceeds some threshold. Here, we have set the threshold around 6V: about right, as the start of danger in a 5V supply (we chose 5V because it is a standard computer supply voltage). The "softness" of the zener's *knee*[6] gives this circuit only approximate control of the threshold. You might prefer, in practice, to use an integrated overvoltage sensor that includes a precise voltage reference[7]; some of these ICs include the SCR, as well.

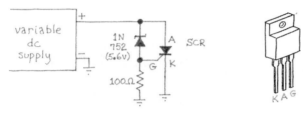

Figure 11L.8 Crowbar circuit, and pinout of SCR: G = gate; K = cathode; A = anode.

The *silicon controlled rectifier (SCR)* shown in Fig. 11N.20, is a device you have not seen before in these labs. It behaves more or less like a power transistor: to turn it on you need to provide some current at its gate which will accept current if you bring its voltage up to around 0.6V above the cathode.

The SCR differs from a transistor in *latching itself ON* once it begins to conduct. To turn the device off, you must stop the flow of current by some external means: in this circuit by shutting off the power supply. Evidently the SCR is well-suited to this application: trouble turns it *ON*; only someone's intervention then can revive the power supply. Note that the crowbar does *not* shut off if the supply voltage simply attempts to revert to a safe level, such as 5V in the present case.

In a practical circuit, you would add a capacitor to ground at the SCR's gate, to protect against triggering on brief transients. Today, your power supply should be quiet – since it powers nothing other than this protection circuit! So, we have omitted that protection capacitor.

Try the circuit by gradually cranking up the supply voltage, using either an external variable supply or the breadboard's 15V supply, which (on the PB503) is adjustable with a built-in potentiometer. When you finish this experiment, incidentally, you'll probably want to set the breadboard's adjustable supply back to +15V, since you're likely to expect that voltage next time you use this supply.

11L.2 A switching voltage regulator

The switching regulator's strategy is wonderfully simple: instead of soaking up the difference in voltage between V_{in} (unregulated) and V_{out} (regulated), it alternately connects and disconnects V_{in} and V_{out} (with a suitable filter to smooth V_{out}). In principle, the regulator – now reduced to a clever *switch* – need not dissipate power. At all times one or the other of the switch's V or I is zero. (That's the *ideal*; *real* switching regulators can give 80% efficiency routinely, 95% in some designs.)

---

[6] Some lonely Dilbert must have invented this piece of jargon.
[7] For example, ON Semiconductor MC3423. See AoE §9.13D.

11L.2.1 Switching regulator IC: LT1073

Figure 11L.9 shows a block diagram for the LT1073 switching regulator, a low-current device that is remarkably easy to use (the data sheet boasts "no design required" as its first selling point: a bit embarrassing for us aspiring designers – but also evidence that switching regulator design has scared away lots of engineers!).

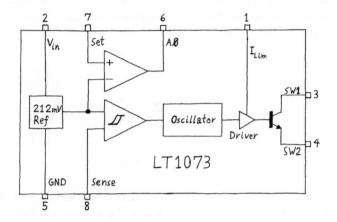

Figure 11L.9 LT1073 switching regulator: block diagram.

11L.2.2 Step-up switching regulator

Figure 11L.10 shows a stunt you can't do with a linear regulator: get +5V from a single AA cell. It's feedback as usual – but what's not usual is that this time feedback controls an oscillator that gives a kick to the inductor (at about 20kHz) when V_{out} falls below the target.

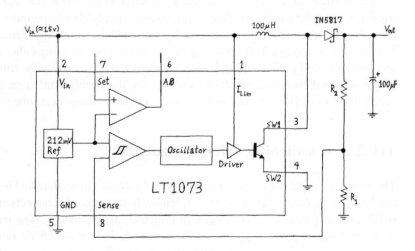

Figure 11L.10 Step-up ("boost") regulator, using LT1073.

A design choice: feedback divider: Choose R_1 and R_2 to give an output voltage of about 5V, given the value of the internal voltage reference (212mV). You'll get clean output waveforms if you keep R values under 100k, though much larger values are permitted.

11L.2 A switching voltage regulator

Output capacitor: Use a tantalum 100μF capacitor on the output, or a Sanyo OS-CON type if you can find one: this capacitor, with its lower series-resistance (and somewhat lower inductance), should give your circuit less of the spiky noise that appears when a jolt of current is injected into the capacitor, on switch turn-off. To check your circuit's response to variable loading, use the circuit in Fig. 11L.11, including a meter that will let you measure I_{load}. *Note* that the variable resistor must have a higher power rating than our usual pots (which are 1/2W parts). This circuit calls for a 1W pot.[8]

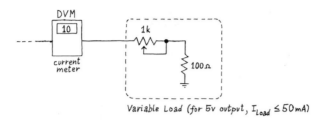

Figure 11L.11 Variable LOAD circuit. (Potentiometer should be rated at 1W or more.)

Use a scope to watch V_{out} and the switch terminal, SW1 (pin 3). The ripple of about 150mV on V_{out} (embroidered with somewhat spiky steps) reflects the *hysteresis* of the comparator on the '1073: the regulator lets V_{out} droop a bit, then kicks it upstairs.

Query: if the comparator hysteresis is 5mV (typical), why is the apparent hysteresis here so much larger?[9] Figure 11L.12 shows what we saw for three load currents.

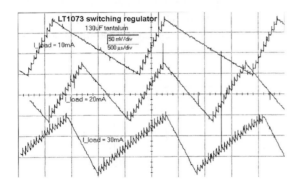

Figure 11L.12 Ripple of switching regulator, and the stairstep rise in response to regulator's 40kHz switching. (Scope gain: 50mV/div, sweep rate: 500μs/div.)

Reducing ripple (*optional*): The 1073 includes a ×1000 amplifier, and this can be used to reduce the output hysteresis that the circuit tolerates. All you need do is insert the amplifier into the feedback path (the 470k resistor in Fig. 11L.13 is simply a *pullup*: the amplifier's output is not the usual push-pull stage).

---

[8] The calculation of the power rating for a variable resistor is simple, but also subtle. One might think that if we adjust the variable resistor in Fig. 11L.11 to, say, 10Ω, this 10Ω value will dissipate little power (about 20mW, at 50mA). But the flaw in this reasoning is that the fraction of the pot where this power is to be dissipated is so small that it cannot unload the heat. So, one should calculate power that would need to be dissipated with a full 50mA load current through the *entire* pot, even though this will not occur in this circuit (the variable resistor at 500Ω would pass much less than 50mA). It is the the *heat per unit length* of the resistive element that must not exceed specification.
Hence our result of about 1W: we calculate I^2R power when 50mA flows through the 0.5k pot (about 1.25W) Power is lower at all but the minimum R value.

[9] That's right: because we're feeding back only a fraction of V_{out}, where we are watching the ripple that "reflects" the effects of the comparator's hysteresis.

Lab: Voltage Regulators

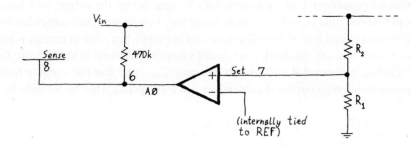

Figure 11L.13 Op-amp in feedback path reduces output ripple.

Query: How does this amendment reduce ripple?[10]

11L.2.3 Step-down switching regulator

> AoE Fig. 9.60: typical waveforms for such a step-down switcher

The circuit in Fig. 11L.14 is similar to the step-up, but this time the transistor switch lies between V_{in} and the inductor, rather than between inductor and ground. When the switch in this step-*down* configuration turns off, the inductor draws current through the diode, from ground. To do this, it takes the voltage at the diode slightly *negative*: pretty weird, if you're not accustomed to seeing inductors at work.

Feed this circuit from a variable external power supply. It should work with an input that varies from a few volts positive to 30V (the max permitted). The capacitor on the input helps stabilize the supply when the switch connects to the inductor. Again, let R_1 and R_2 provide a +5V output.

Can you infer from the behavior of the output waveform what the regulator is doing to hold V_{out} constant, as you vary V_{in}?[11]

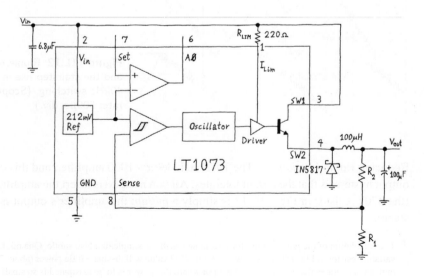

Figure 11L.14 Step-down ("buck") switching regulator.

---

[10] Nothing fancy, here: the internal amplifier's gain of 1000 makes the circuit responsive to changes at the SET input 1000 times smaller than those at the SENSE input. This would seem to reduce our observed output ripple of about 120mV to well under a mV. The datasheet's promise, however, is more modest: "a few millivolts." This modesty is puzzling. See Switching Regulator datasheet: see especially the first 10 pages: http://cds.linear.com/docs/en/datasheet/1073fa.pdf .

[11] Probably you'll see a variation in the switch's *duty cycle*.

11L.2.4 Negative from positive

The step-down can be tricked into providing a *negative* output from a positive input. All you need do is redefine "ground" in the circuit of §11L.2.3 and Fig. 11L.14 as *OUT*, while redefining the the signal labeled V_{out} in Fig. 11L.14 as *ground*. Strangely simple, eh?

Try it – remembering that *all* your instruments that define ground (power supply, scope) must agree on this redefinition of ground.

11W Worked Examples: Voltage Regulators

11W.1 Choosing a heat sink

Problem Choose a heatsink adequate to unload the heat developed by a '317 regulator that must handle a 5V drop at 1A (for example, we may assume $V_{in}=10V$, $V_{out}=5V$). Along the way, choose the best '317 package.

This problem is very similar to one done with slightly different numbers in AoE §9.4.1A.

Solution

Find thermal resistance specifications for our parts: Our goal will be to unload the 5W (5V at 1A). We need to establish some values:

- maximum permitted junction temperature;
- temperature difference between junction and ambient;
- total permitted thermal resistance in that path given those two values.

When we have done that, we can calculate how big a heat sink will be required to balance the books: that is, to take out heat at the rate it is being generated.

- Junction temperature: the '317 can stand a junction temperature of 125°C. We'll not design right to the limit. Let's treat the design maximum as 100°C.
- Ambient: you might expect 25°C, but recall that this regulator is likely to live in a box with other electronics. So, let's be conservative and expect ambient of 50°C.

What thermal resistance can we tolerate? Given the temperature difference (50°C) and the power to be dissipated (5W), we can calculate what total thermal resistance is acceptable.

$$R_{Thermal} = \frac{T_{Junction} - T_{Ambient}}{Power}$$

$$R_{Thermal} = \frac{50°C}{5W} = \frac{10°C}{W}.$$

This is the total thermal resistance that we can tolerate between the hot junction and the outside world.

Without a heat sink? The datasheet in Fig. 11W.1 tells us thermal resistance of the part itself, junction to ambient, assuming no heat sink.[1]

The TO-220 case gives a thermal resistance of 80°C/W, junction to ambient. That won't do. To make the no-heatsink case still worse, the specification assumes ambient at 25°C, unrealistically low. We need a heat sink.

---

[1] Data is from the Fairchild Semiconductor datasheet for the LM317 in a TO-220 package.

Thermal Characteristics

Values are at $T_A = 25°C$ unless otherwise noted.

Symbol	Parameter	Value	Units
P_D	Power Dissipation	Internally Limited	W
$R_{\theta JA}$	Thermal Resistance, Junction to Ambient	80	°C/W
$R_{\theta JC}$	Thermal Resistance, Junction to Case	5	°C/W

Figure 11W.1 Thermal resistances from 317 data sheet.

Heat sink is required: Using a heatsink places several thermal resistances in the path from junction to ambient. The steps will be *junction-to-case*, *case-to-sink* and, finally, *sink-to-ambient*. We will use a thermally-conductive gasket (with its R_{CS}) for insulation, because the case is at the V_{out} voltage, then a heat sink (with its R_{SA}):

$$R_{\text{Thermal\_total}} = R_{JC} + R_{CS} + R_{SA}.$$

The total must be kept under 10 (let's drop the units for the moment). We know that R_{JC} uses 5, leaving us 5. Now it is time to turn to a catalog of available parts. We can use the Digikey website searching their tables that describe available gaskets and heat sinks.

There we find a gasket at 0.07. The heat sink must be a shade under 5 to keep our total under 10. We used the "Natural" column – meaning no fan assumed. We find a couple at 4.4. Fig. 11W.2 shows fragments of the part search screen. We could use either of these two heatsinks. Now our total thermal resistance is about 9.5, so safely under 10.

Compare Parts	Image	Digi-Key Part Number	Manufacturer Part Number	Manufacturer	Description	Thermal Resistance @ Forced Air Flow	Thermal Resistance @ Natural	Material	Material Finish
☐		HS410-ND	7023B-MTG	Aavid Thermalloy	BOARD LEVEL HEATSINK 1.95" TO220	2.5°C/W @ 400 LFM	4.4°C/W	Aluminum	Black Anodized
☐		6398BG-ND	6398BG	Aavid Thermalloy	HEATSINK TO-220 PIN BLACK	2°C/W @ 400 LFM	4.4°C/W	Aluminum	Black Anodized

Figure 11W.2 Searching for a heat sink.

This was a somewhat fussy design task – but in your real design life, doing these calculations is worth the trouble. It's nice to avoid the embarrassment of seeing your circuits melt, or go up in flames.[2]

11W.2 Applying a current-source IC

Problem Use the REF200 current-source IC to generate several output currents.

---

[2] A former student sent us a note asking that we add the topic of thermal management to the course. He went through our class without ever hearing about the issue – and he reported that the first circuit he built at his new job embarrassed him (and *us*, we should say) by overheating.

Worked Examples: Voltage Regulators

We introduced the REF200 IC in Chapter 11N: see Fig. 11N.18. The manufacturer promises that the its two 100μA sources and the *current mirror* can be used to sink or source a variety of currents: 50, 100, 200, 300, 400μA.

Some details:

Current Mirror: We have not discussed mirrors in the mainstream of this book, but the short description of a mirror's behavior is as simple as what is stated in Fig. 11N.18: it holds equal the currents sunk by its top two terminals. The left-hand current serves as the "programming" current; the right-hand current matches that current. For an explanation of mirrors see §5S.2.

Sink versus Source: The 100μA elements are *two-terminal*, equally adept at sourcing or sinking. In the proposed problem, the 50μA circuit-solutions will not be so versatile, by the way: different designs are needed for *sink* versus *source*. The configurations for the other current values can be true 2-terminal designs.

Some of these designs are difficult.

Solution *A variety of output currents from the REF200.* Aren't some of these solutions quite subtle? — the 50μA source, and the 300μA and 400μA, especially.[3]

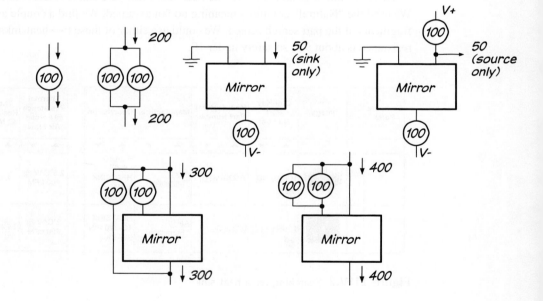

Figure 11W.3
Applications for the REF200 dual current source: 50μA through 400μA.

Was this exercise fun, or frustrating? I don't think I came up with the 50μA source solution without peeking at the TI datasheet.

---

[3] These applications appear on the datasheet for the REF200 on the Texas Instruments site.

12N MOSFET Switches

Contents

12N.1	**Why we treat FETs as we do**	**465**
	12N.1.1 How we'll get away with treating FETs in a day: some ideas carry over from bipolars	466
	12N.1.2 …Just another three-terminal "valve"…	466
	12N.1.3 Some competing symbols for the MOSFET	467
	12N.1.4 Why bother with FETs?	468
12N.2	**Power switching: turning something ON or OFF**	**469**
	12N.2.1 How on is ON?	469
	12N.2.2 How hard is the thing to drive?	470
	12N.2.3 Effects of FET capacitances on switching	470
12N.3	**A power switch application: audio amplifier**	**471**
	12N.3.1 Waveforms	472
12N.4	**Logic gates**	**473**
	12N.4.1 A primitive logic inverter…	473
	12N.4.2 …An elegant logic inverter: CMOS	473
12N.5	**Analog switches**	**474**
	12N.5.1 A first-try: single-MOSFET analog switch	474
	12N.5.2 An improved analog switch: CMOS	474
	12N.5.3 Imperfections	475
12N.6	**Applications**	**475**
	12N.6.1 Integrator reset	475
	12N.6.2 Lots of other applications…	477
	12N.6.3 A particularly intriguing application: switched-capacitor filter	477
	12N.6.4 Sample-and-hold	478
	12N.6.5 An overview of considerations in choosing components	478
	12N.6.6 Standard design issues	479
	12N.6.7 Summary of sample-and-hold errors	480
12N.7	**Testing a sample-and-hold circuit**	**480**
	12N.7.1 Its behavior (showing defects)	480
	12N.7.2 Charge injection	482
	12N.7.3 Bigger C can make charge injection harmless	483
	12N.7.4 Speed limits during sampling	483
12N.8	**AoE Reading**	**485**

12N.1 Why we treat FETs as we do

AoE §3.1.5

Are we pushing the breathless pace of this course too far in proposing to dispose of Field Effect Transistors (FETs) in a day? We gave more time to bipolar transistors, and much more to operational amplifiers.

  We think we can get away with this hasty treatment, because the FETs that are most important operate as switches, and we think you can pick up those *switching* applications quickly. Linear transistor applications, like those we struggled with a few weeks back, are much harder – as we suspect you noticed.

The FETs that you will try in Chapter 12L are the insulated-gate type or MOSFET.[1] This is the type that wholly dominates digital electronics, and it is within digital devices that the overwhelming preponderance of transistors live. From Chapter 14N on, we will begin seeing MOSFETs in *digital gates*. But for today we will see MOSFETs used in two other settings: first, as *power switches* (where they compete with the bipolar switch, an application you met back in Lab 4L) then as *analog switches* (also called "transmission gates").

12N.1.1 How we'll get away with treating FETs in a day: some ideas carry over from bipolars

Some fundamental similarities between bipolar transistors and FETs give you a big head-start, as you meet these new devices. We will say almost nothing in this chapter about one large class of FETs: the *junction* type, JFETs.[2] Chapter 12S gives a cursory introduction to those devices. We introduce today the other class of FETs, the sort with an insulated gate. The symbols for the two types indicate the differences: in one case, a diode junction at the input, in the other case an insulator (the "oxide" in the MOS sandwich).

12N.1.2 ... Just another three-terminal "valve"...

AoE §3.1.1A

A FET, like a bipolar transistor, is a three-terminal *valve* controlled by a voltage between two of the terminals. And if we concentrate on the type most often used – the so-called "*n*-channel" type, analogous to *npn* bipolars – we find the polarities of the control voltages and current flow are familiar. Fig. 12N.1 shows symbols for the new part, and your old friend the *npn*.

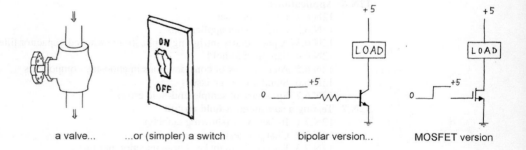

Figure 12N.1 MOSFET switches resemble bipolar switches.

a valve... ...or (simpler) a switch   bipolar version...   MOSFET version

The symbol for the MOSFET, which includes a gap, makes visible the FET's great virtue: astronomically high input resistance. When you met the bipolar *follower* we trumpeted its ability to boost input impedances by the factor β. The MOSFET does dramatically better. The gap shown in the symbol represents an insulating SiO_2 layer. In class we sometimes do a silly demonstration, illustrated in Fig. 12N.2, of the MOSFET's high input impedance by using a student as a "wire" to convey a little charge from teacher to MOSFET.

The teacher touches one hand to +5V or ground to control the lamp that the MOSFET drives. Not

---

[1] This type is named strangely, as if to provide a recipe for the geek stranded on the proverbial desert island and wanting to building a transistor: "Metal Oxide Semiconductor" FET: MOSFET. The British acronym, IGFET ("Insulated Gate..."), makes more sense.
[2] You met these in passing as current-limiting diodes in 5L, and 11N.

12N.1 Why we treat FETs as we do

charged teacher can light lamp, using student as conductor

Figure 12N.2 Class demonstration meant to persuade people of MOSFET's high R_{in}.

a very useful circuit, but a fun demo. If the student lets go of the MOSFET input while the lamp is lit, the lamp stays lit.(Why?) [3]

The FET works more like a vacuum tube than like a bipolar transistor (not that "vacuum tube" is likely to mean much to you[4]): the presence of an insulator at the MOSFET's input means that only the "field effect" can reach and control the *channel*, the conducting region of the transistor.

12N.1.3 Some competing symbols for the MOSFET

Unfortunately, a variety of MOSFET symbols coexist, some more helpful than others. Usually we will use the leftmost of the three symbols in Fig. 12N.3.

simplified symbol – but don't take the diode-like symbol seriously!

showing body diode, as fourth terminal

showing body diode, internally connected, as usual

Figure 12N.3 MOSFET schematic symbols.

All of these symbols – unlike some which we'll refrain from showing – help the reader by indicating which terminal is source (the one nearer the gate input), and which the drain. The leftmost symbol is appealingly simple, but we worry that the diode-like symbol on the source terminal might mislead you: no such diode exists in the device (see Fig. 12N.5 for a literal picture of what's in the device). The right-hand two symbols show the "bulk" or "body diode," the implicit junction between substrate and channel.

AoE §3.5.4C

The middle symbol, showing the body as a fourth terminal, is appropriate to some integrated-circuit schematics. But power MOSFETs always have body tied to source, as in the rightmost symbol. This internal connection implements an implicit *diode* between source and drain, as shown in Fig. 12N.4.

AoE §3.1.3

Because of this diode connection, a power MOSFET cannot be used as a bidirectional switch: drain voltage must not be allowed to go a diode drop below source.

---

[3] This effect only underlines the fact of the MOSFET's enormous input resistance. The lamp stays lit because stray capacitance at its input (gate) holds whatever charge is placed on it, since the MOSFET's input current is almost zero. Many semiconductor memories exploit this behavior: "dynamic" and "flash" memories, EPROM's and EEPROM's, for example. Some of these devices remember by holding a small charge on the gate of a MOSFET for many *years*. We will meet these FET-memories again in Chapter 17N.

[4] No; you're too young. But the man who taught me in my first electronics course had learned with tubes and despised bipolar transistors because of their addiction to input (base) current. When FETs came along he breathed a great sigh of relief and recognition: a proper electronic valve had returned! It's probably not by chance that FETs were the devices the researchers at Bell Labs were trying to implement when they stumbled onto the bipolar transistor. The FET's behavior was more familiar – though the mechanism, one should admit, is not much like the tube's.

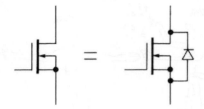

Figure 12N.4 "Body" or substrate forms a diode linking source to drain.

A glimpse of the MOSFET's innards: A hasty sketch of the MOSFET's structure may help to make less mysterious the odd "body diode" sometimes included in the MOSFET schematic symbol. This diode is always present and is an implicit result of the MOSFET's construction, not a diode added to the device. Fig. 12N.5 is a sketch of the structure of an *n*-channel MOSFET.

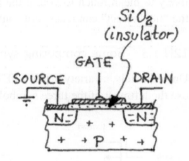

Figure 12N.5 MOSFET construction. "Body" diode is implicit substrate-to-channel junction (no channel shown, here).

The *p*-type body forms a diode with the channel – and must never be forward biased. So the body of an *n*-channel FET must be at least as negative as the source, which is more negative than the drain.[5]

Having braced you with a sense that you already have a partial familiarity with FETs, let's push on into some of their peculiarities.

12N.1.4 Why bother with FETs?

Exhausted by your struggles with bipolar transistors, but triumphant, perhaps you wonder why we should bother you with another sort of device. We consider them because they can do some jobs better than bipolar transistors. Consider MOSFETs when you want:

- very high input impedance;
- a bidirectional "analog switch";
- a power switch.

AoE §3.1

The first of these FET virtues is the most important: enormous input impedance. Well, strictly, it is the input *resistance* that is enormous. The considerable input *capacitance* of a large MOSFET, like the power switches you will meet in today's lab, can make input *reactance* troublesome at high frequencies.

---

[5] If the "drain" terminal happens to be carried more negative than the "source," this more negative terminal *becomes* the effective *source*. See AoE §3.1.3. The near-symmetry of the FET permits this result, drain and source differing only in capacitance. But note that an implicit diode prevents taking drain more than a diode drop below source (*n*-channel), in the case of a power MOSFET. See §12N.1.3 and AoE §3.5.4C.

12N.2 Power switching: turning something ON or OFF

AoE §3.5

You built a bipolar switch long ago in Lab 4L; today you'll build the equivalent circuit with a "power MOSFET" (= just "big, brawny" MOSFET). Fig. 12N.6 shows both; what's to choose between them?

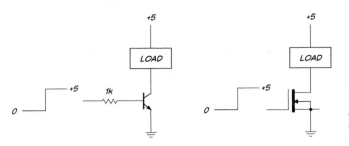

Figure 12N.6 Two power switching circuits: bipolar and MOSFET, doing the same job.

The most important issues are:

- how much power the switch wastes when ON; and
- how easy or hard the thing is to drive.

12N.2.1 How on is ON?

An ideal power switch puts all power into the load, dissipating none itself. So the two transistors compete in how low they can hold their outputs when ON. The contrasting specifications are the bipolar's *saturation voltage* V_{CE-sat} versus the FET's R_{on}.(watch out for the nasty fact that "saturation" means something quite different when applied to FETs! See AoE 3.1.1A).

The bipolar wastes some power also in its V_{BE} junction, whereas the FET wastes none in its gate where no current flows.

AoE §3.5.6A

MOSFET switches are easily paralleled: The FET also shows a nice behavior that makes it easy to parallel MOSFETs for high currents (and low R_{on}): at high currents and high V_{GS}, MOSFETs have the good sense to *reduce* their currents as they heat, whereas bipolars do the opposite. This difference leads to two contrasts: first, several MOSFET *switches* can be paralleled without "ballast resistors" – emitter resistors that provide negative feedback to a bipolar switch, see Fig. 12N.7. The bipolars, if paralleled without these resistors, would give more and more of the shared current to the hottest transistor. Soon it would be getting little or no help from its partners, and would burn out.

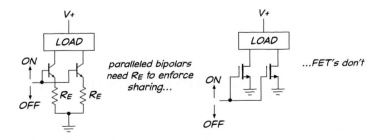

Figure 12N.7 MOSFET *switches*, unlike bipolars, can be paralleled without "ballast" resistors

The bipolar's perverse response to temperature – the hotter it gets, the more current it wants to pass – leads to "current hogging" not only among several paralleled parts but also to hogging on the microscopic scale: if a particular location in a bipolar junction gets hotter than its neighboring regions, it passes more than its share of current, and gets hotter still.

AoE §3.5.7C

Local current hogging on the bipolar transistor makes the device behave like a transistor with a lower power rating once the voltage across the device is appreciable (AoE's Fig. 3.94 shows the effect beginning at just 10V).

Bipolars are, in effect, workaholics – whereas MOSFETs behave more like the rest of us: they are inclined to slough off work instead of hogging it. This doesn't mean a bipolar cannot do a particular job; it just means that you will need a bigger transistor than you might expect, because of this effect. Lest you get too excited about the MOSFET's good sense, however, be warned that the rule permitting easy paralleling of MOSFET *switches* does not apply to MOSFET *linear* circuits. These usually run at lower currents where they show the nasty positive temperature response of bipolars: they pass more current as they get hotter.

12N.2.2 How hard is the thing to drive?

We have made much of the MOSFETs giant input impedance *at DC*. You'll see evidence of this in the lab, when you find that your finger can turn the switch on or off (if you touch your other hand to +5, then ground). But away from DC, the FET's higher *capacitances* begin to tip things against it.

Strange to tell, the large capacitances of big MOSFETs have given rise to specialized MOSFET-driver ICs (the sort of strange and highly specialized thing that Darwin used to come upon in the Galapagos). These can deliver a very large current for a short time, in order to charge the FET's input capacitance.

AoE §3.5.2

AoE re MOSFET driver ICs: §3.5.3

12N.2.3 Effects of FET capacitances on switching

The values of input capacitance – C_{gs} and C_{gd} – are big (100s of pF for a MOSFET that can handle a few amps at low R_{on}). But feedback makes things even worse. The C_{gd} gets exaggerated by the quick swing of the drain as the device switches: that big dV/dt causes a large flow of current in C_{gd}. You'll see that effect in the lab. It appears as a strange kink or hesitation in the movement of gate voltage, as the MOSFET switches. You see it only when you provide weak gate drive, as in Fig. 12N.8.

AoE §3.5.4A

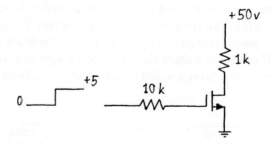

Figure 12N.8 Weak gate drive can expose slowing effect of MOSFET capacitances.

Now the 10k resistor slows the charging of stray capacitances at the gate, and the FET's switching is slow; see Fig. 12N.9.

The exaggeration of C_{gd} by the slewing of drain voltage in a direction opposite to the movement of V_g is an instance of *Miller effect*. That is a problem that bedevils anyone trying to make a high-frequency inverting amplifier. (And it is a topic that we have not pursued in this book, though you have glimpsed Miller effect as a way to understand how a base resistor stabilizes a discrete emitter follower back in §9N.8.2.)

AoE §3.5.4B and Fig. 3.102

AoE §3.5.8A

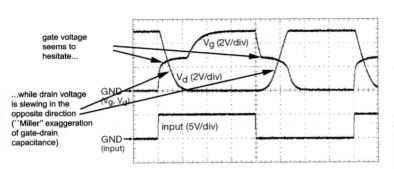
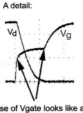

Figure 12N.9 MOSFET capacitances can slow switching.

IGBTs: We will note only in passing an integrated hybrid of MOSFET and bipolar transistors called "Insulate Gate Bipolar Transistors" (IGBT). These show the high input impedance of a MOSFET, but better ON voltage than a MOSFET, for high-voltage loads. Since we'll not meet such loads in this course, we'll not be trying these devices, and refer you to AoE if you need to know more.

12N.3 A power switch application: audio amplifier

You have seen the efficiency that a *switching* voltage regulator can achieve in the LT1073 of Lab 11L. Ideally, the switch dissipates no power whereas a linear regulator is obliged to soak up the difference between V_{in} and V_{out}. A linear regulator is not likely to exceed about 50% efficiency; a switching regulator can approach 90%.

An audio amplifier can achieve a similar gain in efficiency by replacing the valve-like behavior of a push–pull amplifier with ON/OFF switching. (Such an amplifier is labeled "Class D," in contrast to the "Class AB" biased push–pull.) The difficulty that results is the generation of switching noise, as you will see in Lab 12L when you try such an amp.

This electrical noise need not produce *audible* noise, however. The switching frequency is placed far above audio range (into at least hundreds of kilohertz). Nevertheless, a look at the waveforms in Figs. 12N.10 and 12N.12 may make you uneasy. The noise that appears on the sinusoid at the top of each scope image is noise on the circuit *input*. The abrupt drive pulses disturb the ground line, and radiate high-frequency electrical noise. This disturbance occurs despite the usual efforts to quiet the supplies: decoupling capacitors of $1\mu F$ and several of $0.1\mu F$. The application notes for the part confess the hazards posed by the narrow and steep output pulses:

> From an EMI standpoint, this is an aggressive waveform that can radiate or conduct to other components in the system and cause interference.

If your application cannot stand such noise, you will prefer a less efficient (and less "aggressive") push–pull output stage.

The switching amplifier puts out very short *pulses* at the full supply voltage (5V in today's application, §12L.3).

These pulses are effectively *filtered* by the speaker's inductance – applying brief ramps of current to the speaker. This exploitation of the speaker's inductance allows a "filterless" design (thus small and cheap), with high efficiency: 70% to almost 90% best case. Small battery-powered devices like cellphones or portable music sources need that sort of efficiency.

12N.3.1 Waveforms

Sinusoid: Figure 12N.10 shows an input sinusoid at about 1kHz and the output pulse streams that the LM4667 applies to the two ends of the speaker. The switching noise is alarming to look at – but not to listen to. On a properly designed printed circuit board, in contrast to the breadboarded circuit here, the switching noise would be much less extreme. See, for example, the clean waveforms in AoE Fig. 2.73 The two output waveforms (the traces labeled 2 and 3 in the figure) show first one then the other dominating, to drive the speaker coil first in one direction and then in the other, as the voltage of the input waveform crosses its midpoint. The waveforms somewhat resemble the *pulse-width modulated* waveforms that you saw in Lab 7L, though "pulse-density modulation" would be a more accurate description of these pulse streams.

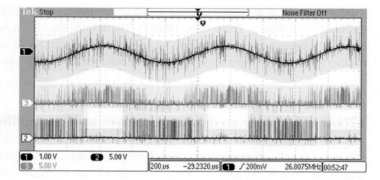

Figure 12N.10 Switching audio amplifier showing input (polluted by switching noise) and two output pulse streams.

Detail of the pulse stream, and noise: Figure 12N.11 shows the pulses in more detail, along with the disturbance that they cause to the entire circuit, including the displayed *input* sinusoid. The somewhat irregular rate at which the pulses occur reveals the quite subtle circuitry within the amplifier, which uses a technique called "delta–sigma" modulation to produce the output pulse streams, using these pulses to achieve a correct *average* speaker current, over a history of many pulses. Note that the pulse rate lies far above the frequency of the audio signal that is to be reproduced.

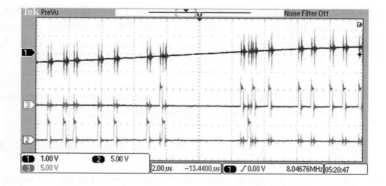

Figure 12N.11 Detail showing output pulses and noise coupled to input sinusoid.

A music waveform: Finally, Fig. 12N.12 shows a music waveform passed by this amplifier. The switching hash is very obvious – but we hope that when you try this in lab you can confirm that the audible effect of this hash is slight.

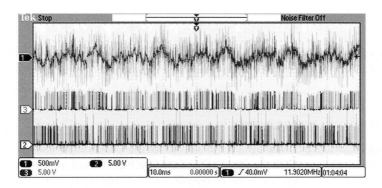

Figure 12N.12 Music waveform as input; pulse streams that result.

12N.4 Logic gates

A power switch like the one shown in Fig. 12N.6 is essentially the same circuit as the simple *logic inverter* on the left-hand side of Fig. 12N.13. The inverter differs from the power switch, however, in that the purpose of the *logic* device is not to put power into a load; quite the contrary.

Instead, the purpose of the circuit is only to generate an output voltage level (high or low) that is determined by a high or low input, while minimizing power dissipated everywhere in the circuit. The input of a MOSFET clearly is well suited to saving power, since the gate draws almost no current. But some cleverness is required in order to make sure that the remainder of the circuit also operates without dissipating power. Such cleverness appears in the better logic gate shown on the right in Fig. 12N.13.

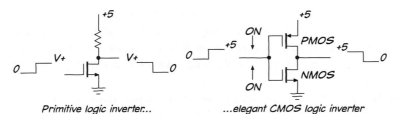

Figure 12N.13 Two logic inverters.

12N.4.1 A primitive logic inverter...

The very simple "logic gate" on the left in Fig. 12N.13 works – but not well. As you will confirm in Lab 14L, its output cannot rise fast because stray capacitance at its output must be charged through the pullup resistor. This makes the switching slow. In addition, the circuit is inefficient because when V_{out} is *low*, the resistor dissipates power. If that's not bad enough, the gate also has such high output impedance in the high state that its noise immunity is poor. So, such an inverter rarely is used.

12N.4.2 ...An elegant logic inverter: CMOS

The right-hand inverter in Fig. 12N.13 is an elegant circuit – and the simplest example of the sort of MOSFET logic that has taken over the world of digital electronics. The name for this pairing of *n*-channel and *p*-channel MOSFETs is CMOS, with "C" standing for Complementary. You will demonstrate the superiority of this inverter in Lab 14L. But even without trying the circuit one can see from its symmetry that it shows none of the three weaknesses of the single-transistor inverter. We see in the next section another good application of CMOS within *analog switches*.

12N.5 Analog switches

MOSFETs compete with bipolars as power switches; as "analog switches" – devices that are to pass or block a *signal* rather than just the flow of current to a load – the MOSFETs have the application to themselves.

12N.5.1 A first-try: single-MOSFET analog switch

AoE §3.5.4C

One can appreciate the CMOS analog switch by comparing it – as we did for the CMOS logic inverter – to a less clever design. Fig. 12N.14 shows such a simple gate: a single MOSFET used to pass or block a signal. In order make sense of a detail shown in the rightmost circuit of the figure, we must acknowledge the MOSFETs "body" diode. This is an element of the FET that we have not used before.

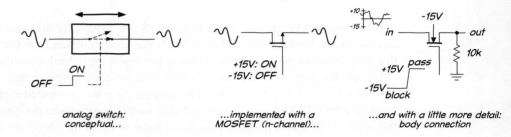

Figure 12N.14 An analog switch done with one MOSFET.

analog switch: conceptual...

...implemented with a MOSFET (n-channel)...

...and with a little more detail: body connection

The "body" diode is shown connected to -15V. Since that is the most negative point in the circuit, this connection makes sure that this diode never conducts.

In the case of the analog switch, source and drain swing as widely as the input and output signals, and typically both can swing positive and negative. So one must take care to tie the body to a voltage that can never take the wrong sign relative to the in and out signals. If body were instead tied to *source* as in a power MOSFET, then a negative input signal, while the switch was OFF, could cause the body diode to conduct.

The analog switch made with one MOSFET works only as long as the input signal remains safely below the positive control voltage that is applied to the *gate*. Otherwise, when the input signal swings close to the positive power supply, the MOSFET will feel its V_{GS} shrinking and eventually will turn itself off.

12N.5.2 An improved analog switch: CMOS

AoE §3.4.1A

The neat remedy for the limitation of the single-MOSFET design is to add a second MOSFET, of the complementary *p* type, in parallel. This second transistor is driven with the complement of the control signal that drives the gate of the *n*-MOSFET so the two transistors are either both OFF or both ON.

It is a nifty Jack Sprat circuit, as you can confirm if you imagine applying an input signal that swings all the way from negative to positive supply. At one extreme of input swing – near the positive supply, say – *n*-MOS fades, but *p*-MOS is fully on; at the other extreme of input swing, Jack and Mrs. Sprat exchange roles. And in the middle, the two transistors both work fairly well. Good CMOS switches – like the one you will meet in the lab – handle such wide signal swings happily, with nearly constant R_{ON}.

12N.6 Applications

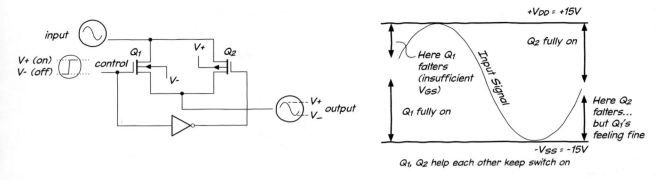

Figure 12N.15 CMOS analog switch: paralleled *n*-MOS and *p*-MOS help each other out.

12N.5.3 Imperfections

AoE §3.4.2

A perfect transistor analog switch would be fully ON or fully OFF: it would behave like the mechanical switch. A real analog switch, like a real power switch, falls short of course.

When ON, it looks like a small resistor (the DG403, which you'll use in the lab looks like 30Ω or less); when OFF, it looks like a small capacitance linking input to output (DG403: about 0.5 pF); it also leaks a little: a steady current may flow (DG403: ≤0.5nA). Whether or not these imperfections matter to you depends on the circuit.

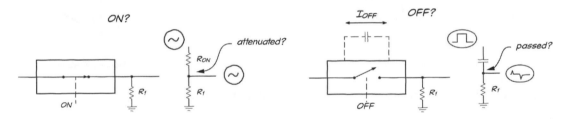

Figure 12N.16 Analog switch departures from simple ideal: ON or OFF.

To get a sense of when such imperfections may be troublesome, we will look at a particular case – one that we do as an in-class demonstration. We will see the effects of some of the DG403's characteristics on a particularly-demanding circuit, a *sample-and-hold*, in §12N.6.4. But before we do that hard work, let's have some fun, just noting some applications for these devices.

12N.6 Applications

Analog switches will solve many problems that otherwise would be difficult. Here is small set of applications, to start your imagination stirring.

12N.6.1 Integrator reset

Here's a task that seems utterly straightforward: reset an integrator.

Manual reset: It is easily done with a mechanical pushbutton. The "No" on the right-hand sketch in Fig. 12N.17 warns against a common error. It is a *plausible* error: just force the integrator output to zero. What's wrong with that?[6]

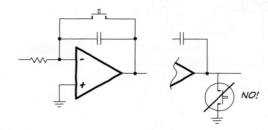

Figure 12N.17 Manual integrator reset is straightforward – if you do it right.

Transistor reset: Letting a control signal effect a reset turns out not to be so straightforward. Would those in Fig. 12N.18 work? Yes, *half* the time: when the integrator output was positive. When, instead, it went negative, reset would lose control: the more negative terminal would become the effective reference voltage, and both sorts of transistor would begin to turn ON even though the control signal were held at 0V – a voltage intended to turn the switch off.

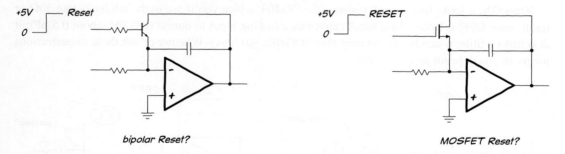

Figure 12N.18 Two attempts at transistor reset for integrator.

An easy reset that works: Replacing the single-transistor switch with an analog switch solves the problem. The device in Fig. 12N.19 can handle an integrator output that goes to either supply extreme, assuming the switch is powered from the same supply voltages as the op-amp.

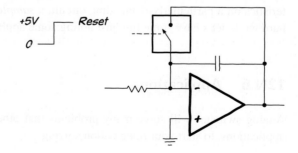

Figure 12N.19 Analog switch easily solves the reset problem, handling both output polarities.

---

[6] Well, doing that does not discharge the capacitor. Instead, it forces the far end of the capacitor away from virtual ground. The capacitor may or may not eventually discharge depending on what is driving the integrator. A bad reset scheme, in any case.

12N.6.2 Lots of other applications...

Multiplexers and their uses: An analog switch can steer a signal here or there. Doing that for the purpose of sharing a resource is said to "multiplex" that resource. Examples abound – such as a telephone line shared among several subscribers who take turns in a scheme mediated by a switching office. In today's lab you will build a humbler "mux" circuit: a "scope multiplexer" that lets two input signals share a single scope channel. In a computer a device like an analog-to-digital converter (ADC) can be shared among many sources, since ordinarily it is used for just one process at a time. Simply route all those signal sources through an analog multiplexer that feeds the ADC; convert one source, then move on to the next. The resource is time-shared.

Once you have met the analog switch you will think up lots of other applications. It is a good addition to your bag of tricks, solving problems that, like the integrator reset, would be difficult or impossible without this device.

12N.6.3 A particularly intriguing application: switched-capacitor filter

Late in Lab 12L you will have a chance to try a novel sort of low-pass filter – one that foreshadows the replacement of analog functions with digital, a change coming soon in the schedule of this course. This filter uses analog switches to meter out charge in discrete packets to simulate the behavior of a resistor. This scheme allows a binary digital signal (ON versus OFF) to control the effective resistance by varying the *frequency* with which the "doses" are delivered.[7] This probably sounds very abstract. A diagram will show how simple the notion is. Fig. 12N.20 shows the lab circuit.

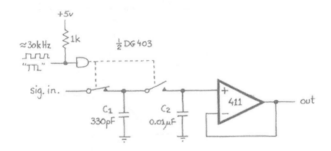

Figure 12N.20 Switched-capacitor filter (lab circuit).

The square wave labeled "TTL" transfers a small dose of charge between the smaller cap and the larger, on each cycle. These transfers gradually carry the output voltage toward the level of the input voltage.

Figure 12N.21 illustrates a homely analogy in which voltages are represented (as usual) by water levels. An energetic person with a soup spoon tries to keep the bucket's water level equal to the level of water in the tidal basin (well, given the speed of tides, he doesn't need to be very energetic; but maybe you see what we're trying to show).

If you're thinking that this seems a cumbersome way to mimic the behavior of a resistor, consider how easy it is for a computer to vary the frequency of a square wave. That is all that's required to control f_{3dB} of this filter.

If you have time at the end of today's lab, you can play with an integrated version of this sort of

---

[7] The delivery of charge in discrete packets has affinities with digital methods, but we don't mean to suggest that this switched-cap filter is a digital device. It is not, since the controlling input signal is a *continuously*-varying frequency. And the doses of charge that are delivered, though discrete, can vary continuously in their magnitude. The circuit output, more obviously, is an analog quantity, a voltage.

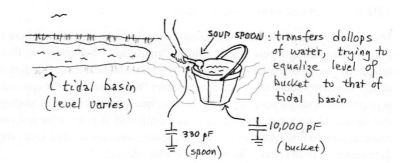

Figure 12N.21
Switched-cap circuit likened to spoon and bucket.

filter. It is an extremely steep active filter,[8] but it uses the switched-cap scheme rather than resistors, so that you can control f_{3dB} simply by twiddling the frequency knob on a function generator. This easy control of f_{3dB} will be useful in later labs, as well, where you will use the filter to remove artifacts introduced by analog-to-digital conversion.

12N.6.4 Sample-and-hold

This is an important circuit and provides a good test bed for consideration of the effects of switch imperfections.

§3.4.3C, Fig. 3.84, §4.5.2

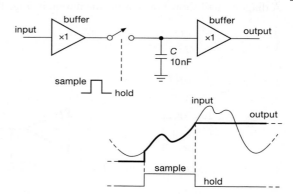

Figure 12N.22 Sample-and-hold: simple scheme.

The sample-and-hold is useful whenever an analog value must be held constant for a while. The most frequent need for this function is in feeding an analog-to-digital converter, a device that (in most of its forms) takes a substantial time to carry out the conversion. While converting, the ADC likes to look at an unchanging input voltage.

Normally a buffer is included on the input side, as shown in Fig. 12N.22, in order to drive the capacitor promptly to the *sampled* voltage. In the lab circuit we omit the left-hand buffer for simplicity, relying instead upon the low source resistance of the function generator. The output buffer is always included to prevent leakage of the charge that is *held* on the storage capacitor.

12N.6.5 An overview of considerations in choosing components

Before we look at a particular case, with part values, let's first scan the concerns that will guide the choice of the storage capacitor.

---

[8] This is an 8-pole *active* "elliptic" filter (MAX294). Its radical fall-off with frequency is shown in Fig. 12L.14.

Choosing C: What is at stake in the choice of C? What good and bad effects result from use of a large C? – from use of a small C?

A large C makes it easy to *hold* a voltage, but hard to *sample* it (hard to *acquire* the voltage quickly).

Sampling: During *sampling* we want $V_{\rm cap}$ to come close to $V_{\rm in}$, and to do this fast. The storage cap is charged through an R value that is the sum of all resistances driving C, usually dominated by $R_{\rm ON}$ of the analog switch (much larger than the $R_{\rm out}$ for the op-amp circuit that buffers the signal source).

Simply evaluating RC allows a first approximation of *acquisition* time: count RCs to see how long the *sample* pulse must persist in order to drive error below a particular level – say 5 RCs to get under 1% for example. Subtler effects may also apply: a large storage cap may call for a substantial *current*, perhaps beyond what an ordinary op-amp can provide. And you may need to specify an op-amp capable of driving a capacitive load – the sort of load that, as you know, can bring instability; see AoE §4.5.2.

Hold: During the *hold* interval, *droop* is the concern, and this droop will result from the sum of all error currents flowing into or out of the capacitor. Those currents are the switch's leakage current, $I_{\rm OFF}$, the op-amp's $I_{\rm BIAS}$, and the capacitor's self-discharge (probably negligible compared to those two other currents).

AoE §3.4.2E

Charge injection: This is a nasty complication – one that may push the necessary capacitor size far above what would be necessary considering only *droop*. Charge injection describes the jolt of charge the switch delivers as it is switched OFF. This occurs because of imperfect symmetry in the switch: internal control voltages, driving the gates of the two paralleled MOSFETs, swing in opposite directions in order to turn off both FETs. Those signals are capacitively coupled to the output. If their couplings were just equal, the two effects would cancel. To the extent the couplings do not match, a net charge is delivered to the storage capacitor – and at the worst possible time: just when it is disconnected from the signal source. So the injected charge adds a voltage step to the saved value. The smaller the storage capacitor, the larger the voltage step caused by charge injection.

Later, in §12N.7.2, we will look at some errors observed in a demonstration sample-and-hold.

12N.6.6 Standard design issues

Here is a recapitulation of the design questions likely to arise in design of a sample-and-hold.

- What size C? (We know now that we lose something at either extreme.) Big enough to keep droop tolerable: use $I = CdV/dt$.
- I is the sum of all leakage currents (analog switch, capacitor's self-discharge, op-amp buffer's bias current);
- dt or Δt is the hold time – usually just long enough to allow an ADC to complete its conversion;
- dV or ΔV is the tolerable change of voltage during conversion: the judgment of what is tolerable must be somewhat arbitrary; let's suppose we will let the sample-and-hold contribute a total error of 1/2 the resolution of the converter. We have already used up about 1/2 this total "error budget" in the sampling stage: $V_{\rm cap}$ did not quite reach $V_{\rm in}$, remember? So we can allow droop to soak up the remainder of the budget: 1/4 resolution of the converter.

This result puts a lower limit on C. The *sampling* concerns put an upper limit, for a given sampling time. As usual, we are caught between two competing concerns.

Effects of "charge injection" in the lab analog switch (DG403): For the DG403, the charge injected is 60pC. That doesn't sound big until you look at the ΔV it causes when dumped into a small cap. If we use $C = 100 pF$, 60pC is troublesome:

$$Q = CV; \quad \Delta V = \Delta Q / C = 60\text{pC}/100\text{pF} = 0.6\text{V}.$$

AoE §3.4.2E

The hazard of charge injection, therefore, pushes us toward larger cap size; so does the problem of *feedthrough*: the transmission of high-frequency signal when the switch is turned off.

In short, we need a cap a good deal bigger than what we might choose if we considered only the problem of droop.

12N.6.7 Summary of sample-and-hold errors

Figure 12N.23 summarizes sample-and-hold errors with switching delay added. The effect of charge injection, called "hold-step," here is a negative-going error; in the example later in this chapter (see Fig. 12N.26), the sign of the effect is the opposite. The switching *delays* provide a further reason why fairly long sample or acquisition times are necessary. The DG403 takes around 100ns to switch; pretty clearly we would be kidding ourselves if we did a droop calculation and an *RC* acquisition calculation and concluded that we could acquire the data in 10ns.

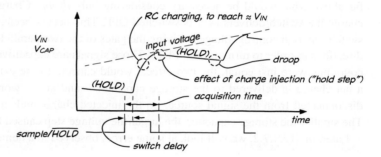

Figure 12N.23
Sample-and-hold errors.

Finally, to give you some sense of what values to expect, here are some specifications for a better analog switch and a good integrated sample-and-hold. The ADG1211 analog switch specifies charge injection at 1pC – 60 times better than the DG403 that we use in our lab circuit (see §12N.7.2). The LF198 integrated sample-and-hold promises a maximum *hold-step* of 1mV – about 100 times better than what we achieve with our first try at a homemade sample-and-hold in §12N.7.2.

12N.7 Testing a sample-and-hold circuit

Figure 12N.24 shows a sample-and-hold, made with '411s and the analog-switch that you'll meet in today's lab, the DG403. We'll look at some waveforms that describe how well it performs.

12N.7.1 Its behavior (showing defects)

Droop: A small C charges up quickly, but also loses its charge quickly. Fig. 12N.25 shows the circuit's *droop*. From the *droop rate* one can infer the discharge current.

12N.7 Testing a sample-and-hold circuit

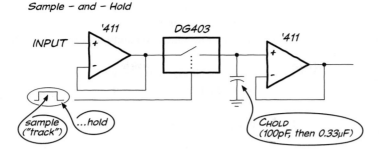

Figure 12N.24 Sample and hold as built for demo.

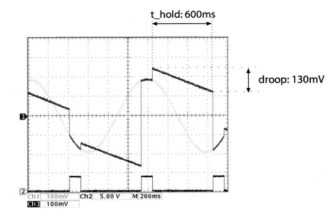

Figure 12N.25 Droop rate reveals leakage current ($C = 100$pF). (Scope gains: 100mV/div except Sample pulse on Channel 2: 5V/div.)

$$I_\text{discharge} = CdV/dt = 100\text{pF} \times 130\text{mV}/600\text{ms} \approx 100\text{pf} \times 215\text{mV/s}$$
$$= 100 \times 10^{-12}\text{F} \times 0.215\text{V/s} \approx 22\text{pA}.$$

Is this current caused by switch leakage, op-amp bias current, or both? The consistent downward slope of V_CAP reveals the answer: apparently op-amp bias current dominates.

Switch leakage could not account for the droop because V_CAP *falls* in the region low on the image in Fig. 12N.25 between first and second *sample* pulses; it falls here even though V_IN lies *above* V_CAP for most of this *hold* period. So the cap is not discharging toward V_IN, across the switch.[9]

The consistent downward droop also fits the *sign* of the '411s bias current.[10]

This argument is convincing – but let's check by trying an op-amp with extremely low bias current, a CMOS part, then look at the droop again. We replaced the '411 (JFET input: I_BIAS=50pA, typical

---

[9] It is possible, but unlikely, that the switch is discharging the cap toward the switch's negative supply voltage. The DG403 data sheet is not so specific as to indicate the sign or path of the OFF-state leakage current, but I'm told that leakage through the channel is likely to be much larger than leakage across oxide to gates, or to substrates. The datasheet supports this view too by implication, as it specifies "Switch OFF Leakage Current." It lists two input conditions, with switch input and output at opposite extremes, ±15V, and in each of two orientations; the leakage currents are shown to go to extremes of *both signs*. Apparently, the signs of the leakage currents correspond to the orientation of the voltage across the switch; if not, specifying these extremes of V-across-switch would not seem relevant.

[10] The '411s p-channel JFETs offer a junction that is always back-biased, and that junction, if back-biased, would leak toward the two other JFET terminals, both more negative than gate: *source* and *drain*. Leakages to substrate or package could contribute, but the fact that I_BIAS grows exponentially with temperature along with a note in the '411 datasheet support the view that *junction* current dominates. See the '411 I_BIAS versus temperature datasheet curve under DC Electrical Characteristics, and Note 7 in that datasheet. National's datasheet can be seen at http://www.ti.com/general/docs/lit/getliterature.tsp?genericPartNumber=lf411-n&fileType=pdf.

(200pA, max)) with an LMC662CN (CMOS: I_{BIAS}= 2fA, typical (2pA, max)). Fig. 12N.26 gives the result contrasted with the '411s. The droop occurs over about 0.6 seconds.

The droop rate for the circuit using the CMOS buffer is very low. One can barely make it out, even if one waits 20 seconds, as in Fig. 12N.27.

The current this droop-rate implies is extremely low. Using two horizontal cursors we estimated the droop at about 12mV in 18 seconds. We can calculate the implied discharge current:

$$I_{\text{discharge}} = C \times dV/dt = 100 \times 10^{-12}\text{F} \cdot 0.66 \times 10^{-3}\text{V/s} = 66 \times 10^{-15}\text{A} = 66\text{fA}.$$

This is in a plausible range for I_{BIAS} of the CMOS op-amp.[11] It seems to follow that the analog switch is leaking hardly at all.

12N.7.2 Charge injection

A good op-amp buffer can minimize droop, even with a small capacitor. But, as we noted back in §12N.6.5, a small C makes the circuit vulnerable to another error that is harder to handle: the effects of *charge injection*. When the switch opens, it delivers a jolt of charge (60pC, for the DG403, typical).

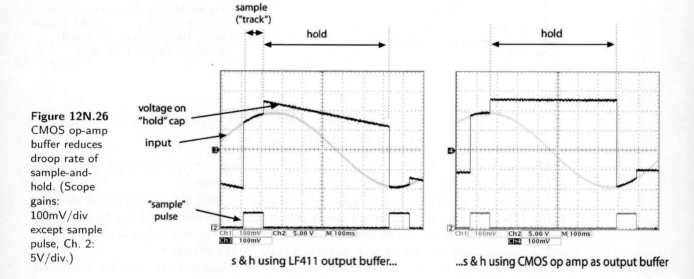

Figure 12N.26 CMOS op-amp buffer reduces droop rate of sample-and-hold. (Scope gains: 100mV/div except sample pulse, Ch. 2: 5V/div.)

s & h using LF411 output buffer... ...s & h using CMOS op amp as output buffer

Delivered to a small cap – 100pF in this case – this produces a big, spurious step (0.6V we'd predict; the step visible in our scope images is smaller, but still very troublesome). The "hold-step" is apparent in Figs. 12N.26 and 12N.28.

The voltage hop just at the moment when the circuit ought to be recording the input level is bad. To make things worse, the magnitude of the *charge-step* is not constant with input voltage. The step is large at the top of the input waveform's swing, much smaller at the bottom of the sinusoid. So it cannot be subtracted away (say, to correct the digital value read by an ADC fed by a sample-and-hold). This variation is apparent in the scope image, Fig. 12N.28.

---

[11] 66fA is surprisingly far above *typical* I_{BIAS}, but it is just a few percent of the *maximum* specified value. Probably the cause is leakage in our solderless breadboard.

12N.7 Testing a sample-and-hold circuit

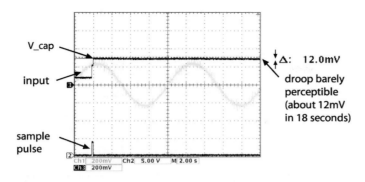

Figure 12N.27 Sample-and-hold droop with CMOS buffer: barely measurable, at almost 20s hold. (Scope gains: 200mV/div except sample pulse, Ch. 2: 5V/div; note time scale change, relative to Fig. 12N.26: here, 2s/div.)

12N.7.3 Bigger C can make charge injection harmless

To improve the sample-and-hold, it looks as if we need a much larger C. That's true: it will diminish the effect of charge injection, and will slow the droop. Fig. 12N.29 shows that the *hold-step* effect of charge injection disappears when we replace the 100pF storage C with one much larger, $0.33\mu F$: this works but it exacts a price as we'll see next.

12N.7.4 Speed limits during sampling

The much larger *hold* capacitor solves the charge injection problem; but it is bound to cost us something. Yes: it slows acquisition of the sample. In both images of Fig. 12N.30 the storage C is now $0.33\mu F$ rather than the original 100pF. V_{CAP} never quite catches up with V_{IN} during the brief acquisition time provided by the sampling pulse.

We can make out two causes for delay in *acquisition* of V_{IN}.

The ramp: The straight ramp looks, at first glance, like an op-amp's *slew-rate* limit. But the value of its slope ($0.04V/\mu s$ for the left-hand image of Fig. 12N.30) is much too low for this op-amp (LMC662: slew rate about $1V/\mu s$). The cause must be something else: yes, it's the op-amp's limited output current, which must charge or discharge the cap. One can infer $I_{OUT-MAX}$ from the slope of this curve:

$$I_{OUT-MAX} = C dV/dt = 0.33\mu F \times 0.04V/\mu s \approx 0.013A = 13mA .$$

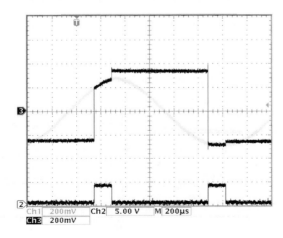

Figure 12N.28 Charge injection produces large error with small C (100pF). (Scope gains: 200mV/div except sample pulse, Ch. 2: 5V/div.)

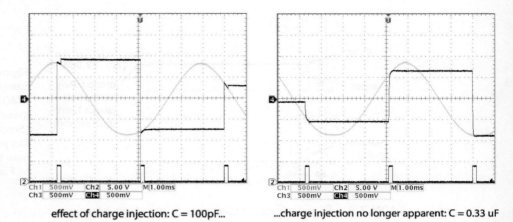

Figure 12N.29 Big *hold* capacitor slows droop and shrinks the effect of charge injection. (Scope gains: 500mV/div except sample pulse, Ch. 2: 5V/div.)

effect of charge injection: C = 100pF...   ...charge injection no longer apparent: C = 0.33 uF

That's about right for the LMC662.[12]

The curvy final section: The curvy section at the end of the ramp you recognize as an *RC*: here the FET switch is in its resistive region and its R_{ON} drives the big cap. If *RC* is about 10μs (a very rough estimate – for this piece of an *RC* curve is hard to judge), then

$$R = RC/0.33\mu F = 10\mu s/0.33\mu F \approx 30\Omega\ .$$

This value fits quite well too.[13]

A sampling pulse needs to be brief, since everyone's always in a hurry to get a digital conversion done. So the slowing effect of the large *C* is troublesome. We are reminded once again of a general truth: design usually is constrained so that optimizing one characteristic (here, suppressing the charge-step effect) exacts a price in some other characteristic (here, acquisition time). Using a large *hold* capacitor is not a good solution.

You will have gathered that it is difficult to build a good sample-and-hold. We really want a switch with low charge injection, rather than a mediocre switch and then a big *C* to suppress the effect of the injection.

In fact, nearly always we will do laziest, easiest thing to get a good sample-and-hold: we'll buy one from people who make these for a living. The next sample-and-hold that you meet will be hidden

---

[12] LMC662C I_{out} limit spec'd at 13mA for 5V supply. Our circuit uses a split supply, ±5V.
[13] R_{ON} for the DG403 is spec'd at 20Ω, typical, 45Ω, max.

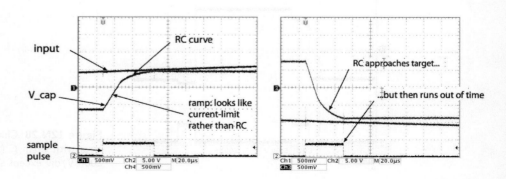

Figure 12N.30 A big *C* slows acquisition: a brief sampling pulse doesn't let V_{CAP} reach V_{IN}. (Scope gains: 500mV/div except sample pulse, Ch. 2: 5V/div.)

within the ADC that you will apply in a microcomputer lab (Analog Devices' AD7569 or the ADC integrated with the SiLabs controller). The designers of the ADC take care to keep the hold-step effect below the level that matters, and to keep the storage cap small enough so that acquisition is pretty fast: adequate, anyway, to meet the requirements of the specified conversion time.[14]

12N.8 AoE Reading

Chapter 3: An introductory overview: §3.1 (introduction, FET characteristics) through 3.14 (linear and saturated regions).
Gate current, especially "dynamic gate current": §3.2.8C.
Most important is a very long section on FET switches, §3.4. Ignore JFET switches, here.
A short discussion of the lab's switched-capacitor filter design appears in §6.3.6.

---

[14] The AD7569 part does an 8-bit conversion in 2μs, using something like 200ns – one clock cycle – to acquire the analog level in its on-chip sample-and-hold; the SiLabs ADC needs 5μs but resolves the signal 16 times better, so the sample-and-hold's performance is somewhat more impressive.

12L Lab: MOSFET Switches

This lab opens with a quick look at a "power MOSFET" ($\approx$ "big MOSFET") as power switch. The MOSFET offers an alternative to the bipolar power transistor that you met back in Lab 4L. The power MOSFET exercise should go quickly: we want you to see the MOSFET's enormous DC input resistance, but also the large input capacitance that makes it hard to switch *fast*. Then the remainder of the lab is given to trying applications for the so-called *analog switch* or *transmission gate*: a switch that can pass a signal in either direction, doing a good job of approximating a mechanical switch – or, more precisely, the electromechanical switch called a *relay*.

The most important of the circuits that uses this analog switch uses an op-amp as well, and that fact is one reason why we chose to postpone this FET lab till now (MOSFETs might seem to belong just after the two bipolar transistor labs). This combined application is the *sample-and-hold*, often used in a circuit that converts a signal from analog into digital form. The sample-and-hold holds an input value constant during the conversion process. The sample-and-hold also serves us, here, as a good test bed for the analog switch: it reveals the imperfections of the device.

12L.1 Power MOSFET

This exercise repeats a task you carried out back in §4L.5 using a bipolar transistor. Here you will use a MOSFET to do the job. Both of the MOSFETs listed below are designed to turn ON with only a "Logic" level *high* input voltage applied to V_{GS}. In some respects the MOSFET is much better than the bipolar equivalent. Its great strength, as you are about to confirm, is its huge input impedance, at least at DC.

Part	R_{DSon} (millohms, max)	$V_{GS(thresh)}$ max	Mfr
BUK9509–40B	10	4.5V	NXP
IRLZ34	70	4.0V†	IR
†. $V_{GS(thresh)}$ max is specified at 2.0V but that threshold voltage specification is not very useful, because it's specified at a low current (0.25mA for IRLZ34). You're not likely to be running this power MOSFET at such a tiny current. A better indication of necessary V_{GS} drive is the fact that R_{ON} is specified at V_{GS} of 4V and 5V We have here adopted 4V as $V_{GS(thresh)}$.			

12L.1.1 Power

Both transistors are high-current devices (75A for the NXP part, 30A for the IR) – but this big number may be misleading. These specifications assume that the *case* temperature somehow has been held at 25°C: more or less impossible at high current. The *power* limits are more significant.

Again, these can be confusing. The power limits for the devices are high: 157W for the NXP, 68W

12L.1 Power MOSFET

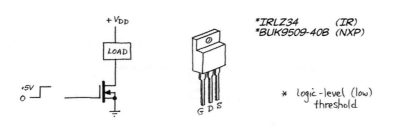

Figure 12L.1 Power MOSFET transistor switch (pinout echoes that of power BJT, BCE).

for the IR. But, again, these specifications assume the case is held at 25°C. With the case at 75°C, for example – more likely, at high current – the NXP part can dissipate about 100W, the IR 45W; with the case at 100°C NXP can dissipate about 80W, the IR 44W.

So we are obliged to worry about the temperature rise caused by electrical power in the transistor. The data sheets specify the *thermal resistances* for the transistor, allowing us to calculate what power it can dissipate with and without heat-sinking.

Let's look at the NXP part in detail: without a heat sink, junction-to-ambient ($R_{\Theta JA}$), when standing vertical, is $R_{\Theta JA} = 60\text{K/W} = 60°\text{C/W}$ (typical).[1]

Adding a heat sink helps a lot, of course. A biggish heat sink for the 220 package offers 6.5°C/W (Aavid 7022BG). The thermal gasket's R_Θ adds about 0.6°C/W, giving a total thermal resistance, from junction to ambient, of

$$R_{\Theta JA} = R_{\Theta JC} + R_{\Theta CA} = (0.95 + 7.1) \approx 8.1°\text{C/W}.$$

If we run the transistor junction temperature to 150°C (25°C short of the limit) and assume (conservatively) that the transistor may live in a box at 50°C, the temperature difference is 100°C. Let's see what power the transistor can unload, with and without heat sink.

- Without heat sink (or fan), transistor standing vertical:

$$P = \text{rate-of-heat-transfer} \cdot \text{temp-difference} = (1/R_\Theta) \cdot \Delta T$$
$$= (1/R_{\Theta JA}) \cdot \Delta T = (1/60) \cdot 100 = 0.016 \cdot 100 \approx 1.6\text{W}$$

- With heat sink: $P = (1/8.1) \cdot 100 \approx 12\text{W}$.

These values are a long way from the impressive number, 157W, specified with the case at 25°C.

12L.1.2 Input impedance

Build the circuit shown in Fig. 12L.2. Use a #47 lamp as load and confirm that the FET will switch when driven through the 10k resistor at low frequencies (switch the input between 0 and +5V, moving the input wire *by hand*). High input impedance is the MOSFET's great strength, as you know.

To get a more vivid sense of what "high input impedance" means, let the input side of the 10k resistor float, then touch it with one hand, and touch your other hand alternately to ground and the +5 supply. Impressive? Now try letting go of the 10k resistor after switching the FET on or off. Why does the FET seem to remember what you last told it to do?[2] This exercise, frivolous though it seems, foreshadows some of the strange results you will get if you ever forget to tie the input of a MOS logic device either high or low: misbehavior that is spookily intermittent!

---

[1] Our note on thermal resistivity uses the units °C/W rather than K/W. The two are equivalent. Note the convention that omits "degrees" (°) from the Kelvin value.

[2] Your particular MOSFET may or may not "remember." The maximum leakage rate is rather high. See datasheet.

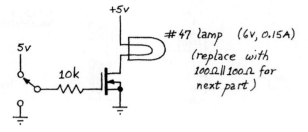

Figure 12L.2 Setup for "measuring" input impedance. (Rotate each leg of the TO-220 package 90° to ease insertion into breadboard.)

12L.1.3 Switching at higher frequencies

Effect of input capacitance. Now use a signal generator, with a 0 to +5V square wave, to drive the MOSFET switch. Keep the 10k resistor in place, for the present. Watch V_{GATE} and V_{OUT} as you increase the driving frequency from about 10kHz. What goes wrong?[3] The waveform at the gate should look pretty strange.

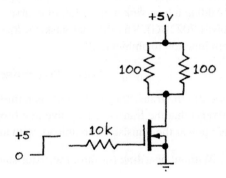

Figure 12L.3 FET switch, again: speed limit.

Solve the problem by replacing the 10k resistor with a value that works better at high frequencies. Now see how the switch looks at a few hundred kHz.

12L.1.4 Power dissipation: bipolar (BJT) versus MOSFET

Quien Es Mas Macho? Here, to make an otherwise humdrum experiment a little more exciting, we propose that you treat this exercise as a contest between brawny champions of two competing types, MOSFET and bipolar. In some respects, the transistors are similar: they use the same big package (called TO–220), and carry some ratings that are at least in the same ballpark. The bipolar is specified to carry as much as 10A continuous (again, apart from limits on heating, as we noted of the MOSFETs in §12L.1.1).

You can't tell from the specifications we have noticed so far – maximum current and power – which transistor will win. The answer will depend mostly on the voltage drop across the transistor, a measure that is specified in different ways for FETs and bipolars. But the MOSFET should look pretty good in a contest where the goal is to put available power into the load rather than into the transistor switch.

Set up the contest by putting the same heavy load current through both transistors (about 0.5A),

---

[3] Yes, the 10k has trouble driving the gate capacitance – especially C_{GD}, which is exaggerated (doubled) by the circumstance that the output (drain) swings in a direction opposing the input (gate) swing. §12N.2.3 makes this point.

and measure the voltage across each. *Use a resistor rated at 5W or more*: if you forget this advice and use a .25W resistor, you may be reminded by smoke or even by burned fingers.

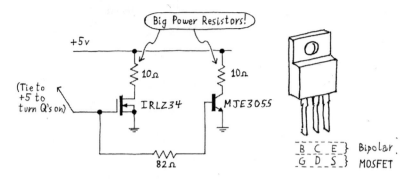

Figure 12L.4 MOSFET versus bipolar switch contest. (Rotate each leg of the transistor 90°, to ease insertion into breadboard.)

How ON is the ON switch? MOSFET versus bipolar: From this number – V_{DS} or V_{CE} – you can calculate R_{ON} for the FET – and an equivalent value for the bipolar if you like, though the term "R_{ON}" is not used for bipolars. Then (making the FET look still better) you can calculate power dissipation in each transistor, remembering to include the considerable power dissipated as a result of *base* current.

For a vivid confirmation of the fact that most of the power in these switch circuits is going into the load and not the transistor switch, you can put a finger on load resistor, then transistor. You should find the resistor much warmer than the transistor

12L.2 Analog switches

The CMOS analog switch is likely to suggest solutions to problems that would be difficult without it. This lab aims to introduce you to this useful device. You see from Fig. 12L.5 that schematically it is extremely simple: it passes a signal or does not.

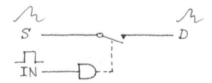

Figure 12L.5 Analog switch: *generic*.

The switch we are using has especially nice properties: it is switched by a standard logic signal, 0 to +5 (High, +5 = ON). But it can handle an analog signal anywhere in the range between its supplies, which we will put at ±15 volts. It also happens to be a *double-throw* type, nicely suited to selecting between two sources or destinations.

Figure 12L.6 shows the switch and its pinout. As signal source use an external function generator; as source of the "digital" signal that turns the switch on or off use one of the slide switches on the breadboard. The easiest to use is the 8-position switch: put one of its slides in the ON position; now that point will be high or low, following the position of the slide switch just to the right of the "DIP" ("dual in-line package") switch.

Lab: MOSFET Switches

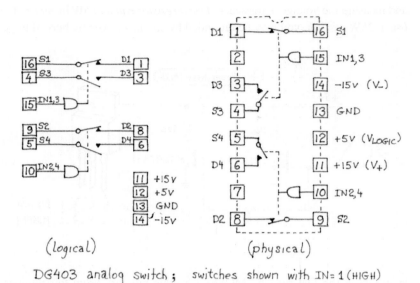

Figure 12L.6 DG403(Maxim) analog switch: block diagram and pinout.

Caution: each DG403 package contains two switches. Tie the unused "IN" terminal to ground or to +5 (this makes sure the logic input to the switch does not hang up halfway, a condition that can cause excessive heating and damage.)

And a *reminder*: this IC uses *three power supplies*: connect all of them!

12L.2.1 Switch imperfections

We'd like you to see that the switch isn't perfect. Specifically, we ask you to measure its *resistance* when ON and its *feedthrough capacitance* when OFF. But the fun and the main interest of the remainder of today's lab lie in the *applications* of analog switches. So, hurry through these preliminary measurements.

R_{on} (lab): Ideally, the switch should be a short when it is ON. In fact, it shows a small resistance, called R_{ON}. Measure R_{ON}, using the setup in Fig. 12L.7. Use a 1kHz sine of several volts amplitude. Confirm that the switch does turn ON and OFF, and measure R_{ON}.

Figure 12L.7 R_{ON} measurement.

Feedthrough: The circuit of Fig. 12L.8 makes the switch look better: its R_{on} is negligible relative to the 100k resistor. Confirm this.

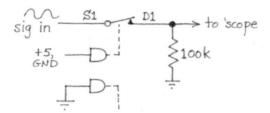

Figure 12L.8 More typical application circuit (R_{ON} made negligible).

When the switch is OFF, does the signal pass through the switch? Try a high-frequency sine ($\geq$100kHz). Try a square wave. If any signal passes through the OFF switch, why does it pass?[4]

Note: As you do this calculation, don't forget that you are looking at the output with a scope probe whose capacitance (to ground) may be more important than its large R_{in}: you're really watching a *capacitive divider* at work. As long as you don't forget this probe capacitance, you should have no trouble calculating the switch's C_{DS}.

12L.2.2 Applications for the analog switch

These applications offer the fun in today's lab. You may not have time for all of these circuits. We hope you'll try the sample-and-hold application anyway.

Chopper circuit: Figure 12L.9 gives a cheap way to turn a one-channel scope into two channel (and on up to more channels if you like). (Query: what are the limitations on this trick?)[5]

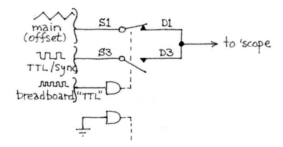

Figure 12L.9 Chopper circuit: displays two signals on one scope channel.

For a stable display, trigger on one of the input signals (preferably the square TTL[6] signal) not on the chopper's output, where the transients will confuse your scope. If you are using a digital scope that offers a "dots only" display option, use that: it draws no line linking the two waveform traces. The images in Fig. 12L.10 show the contrasting cases.

---

[4] Capacitance between input and output couples the two terminals though the switch is OFF.
[5] The main limitation is speed. The scheme will entirely miss a brief transient that occurs while the circuit has chopped to the other input. In addition, R_{on} combined with the scope's capacitance imposes a minimum rise and fall time.
[6] TTL stands for Transistor–Transistor Logic, a now-obsolete logic family, fabricated from bipolar transistors. You will encounter TTL in Lab 14L.

Lab: MOSFET Switches

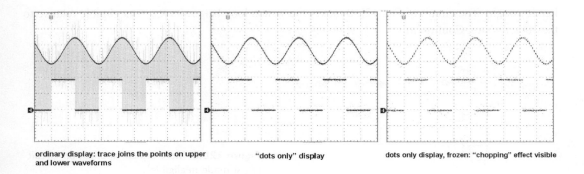

ordinary display: trace joins the points on upper and lower waveforms

"dots only" display

dots only display, frozen: "chopping" effect visible

Figure 12L.10 Digital scope can hide traces between the two displayed waveforms.

The rightmost image in the figure shows the chopper with scope *stopped*. It makes the distinct points of the chopped display visible: a dot on the sinusoid, then on the square wave, then back to the sinusoid....

Sample-and-hold: This application is much more important. It is used to *sample* a changing waveform, *holding* the sampled value while some process occurs (typically, a conversion from analog into digital form).

Try the circuit of Fig. 12L.11. Can you infer from the droop of the signal when the switch is in *hold* position, what leakage paths dominate? (This will be hard, even after some minutes of squinting at the scope screen; don't give your afternoon to this task.)

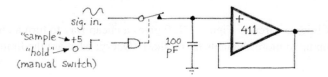

Figure 12L.11 Sample-and-hold.

Query: How does one choose a value for C? What good effects, and what bad, would arise from choice of a cap that was: (a) very large; (b) very small?[7]

Can you spot the effect of *charge injection* immediately after a transition on the control input? Compare the specified injection effect and the voltage effect you would predict, given the specification ($\leq$60pC, typical, for the DG403) and the value of your storage capacitor.

Optional: a dynamic view of charge injection: If you inject charge periodically, by turning the switch on and off with a square wave, you can see the voltage error caused by charge injection in vivid form.[8] The *held* voltages, when we did the exercise, rode above the input by a considerable margin; you will notice the margin varies with the waveform voltage. Why?[9]

Good sample-and-hold circuits evidently must do better, and they do. See, e.g., the AD582 with charge injection of $\leq$2pC; and see AoE§3.4.2E.

---

[7] A very large cap droops only slowly and doesn't show large jumps in response to charge injection; but it is slow to charge to the input voltage, requiring a long *sample* pulse. That long pulse slows the sample-and-hold process.
[8] Thanks to two undergraduates for showing us this technique: Wolf Baum and Tom Killian (1988).
[9] Apparently the capacitances between the IC's two internal switching signals and its output – capacitances that ideally would be equal so that their effects would cancel – vary somewhat with voltage across the switch.

12L.2 Analog switches

Negative supply from positive (flying capacitor): You know that a switching regulator can generate a negative output from a positive input, with the help of an inductor (you may have built such a circuit in Lab 11L). For low-current applications the circuit in Fig. 12L.12, which requires no inductors, is sometimes preferable. The trick is to do a little levitation by shoving "ground" about in a sly way. A similar use of "flying capacitors" can generate a voltage larger than the input voltage. This circuit provides only a small output current, as you can confirm by loading it.

Such circuits often are put onto an integrated circuit that would otherwise require either a negative supply or a second positive supply, higher than the main supply [10].

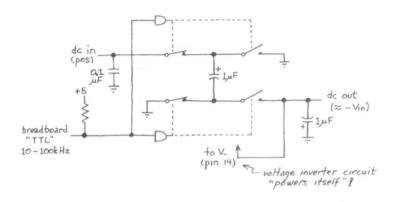

Figure 12L.12 Flying capacitor voltage inverter.

AoE §6.3.6A

Exercise. Switched-capacitor filter I: built up from parts: Use the function generator to feed a sinusoid to the filter of Fig. 12L.13 while clocking the circuit with a *TTL* square wave from the powered breadboard. This filter's f_{3dB} is set by the clock rate. This makes it a type convenient for control by computer.

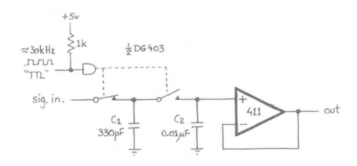

Figure 12L.13 Switched-capacitor lowpass filter.

The effective *RC* simulated by this spoonful-at-a-time charge circuit depends on the relative sizes of the two capacitors (how big the spoon is, relative to the bucket) and on how fast the transfers are made ($f_\text{switching}$):

$$f_{3dB} = \frac{C_1/C_2}{2\pi} \times f_\text{clock} . \tag{12L.1}$$

---

[10] Semiconductor memories, for example, sometimes require such a supply. Once upon a time, some memories demanded that the user supply +12V as well as +5V. The flying-capacitor trick ended this demand placed on the user; the chip solved the problem internally, and the ICs that lacked the internal step-up were driven out of the market

Lab: MOSFET Switches

Given the values used here, this formula predicts

$$f_{3dB} = (0.03/2\pi) \times f_{clock} . \tag{12L.2}$$

Try the circuit and compare its f_{3dB} with the predicted value. Does the filter behave generally like an *RC* filter: does it show the same phase shift at f_{3dB} that you would see in an *RC* lowpass? Does f_{3dB} vary as you would expect with clock frequency? Because this is a "sampling" filter, you can expect it to get radically confused when the input signal changes quickly relative to the rate at which the square wave transfers samples (the "sampling rate"). This radical confusion is called "aliasing."

We'll look in depth at sampling-rate issues, including aliasing, when we reach analog-to-digital conversion in Lab 18L. For now, be warned that you should expect trouble when the frequency of a sinusoidal input approaches one-half the "sampling rate" applied to the analog switch control inputs.

Do you see feedthrough of the clock signal? Sometimes a passive lowpass filter is placed at the output of the switched-cap filter to attenuate that noise.

Exercise. Switched-capacitor filter II: integrated version: This integrated filter is more complicated than the one you just built up from parts, as you might expect. Integrated switched-capacitor filters are available in several forms. Essentially, they are op-amp *active filters* like the one you glimpsed in Lab 9L, except that they use switched capacitors to simulate the performance of a resistor. This design permits easy control of the effective *RC*, as you saw in the preceding exercise. Some integrated filters allow the user to determine the filter type.

The one you will meet here is committed as a *lowpass*; that makes it easy to wire. It is an *eight-pole elliptic* filter – one that sacrifices other virtues in order to achieve spectacularly steep rolloff. You will use this filter in Lab 23L on analog–digital interfacing. Here we just want you to discover how easy to use it is and how sharp its rolloff.

Note: If you build this circuit now, please build it on a private breadboard; you will use the circuit in Lab 23L. If you don't build it now, you can build it at the time when you need it.

Like your simple switched-cap filter, this one shows an f_{3dB} proportional to the filter's *clock* rate. That is the feature that later will be especially useful to us when we will want to vary f_{3dB} very widely without rebuilding circuits.

Build the circuit shown in Fig. 12L.14, clocking the device with the breadboard TTL signal. Let the external function generator provide an analog input to the filter. Make sure to add a *DC offset* to the input signal, so that the signal stays between about 1V and 4V. This filter, you'll notice, uses a single supply, not the split supply that we used with your home-made switched-cap filter on page 493. Hence the need for a DC offset.

To test the filter, try setting the breadboard TTL-oscillator frequency to about 100kHz, placing f_{3dB} around 1kHz ($f_{3dB} = f_{clock-filter}/100$, as you will have gathered). Feed the filter a sinewave centered on 2.5V; keep the amplitude below about 1V; vary f_{in} from far below f_{3dB} to far above. Again, the filter should fail when the input frequency climbs above about half the TTL-oscillator frequency (which sets the filter's "sampling rate"): aliasing, once again.

You should be able to see the filter's abrupt roll-off. You will also notice the discrete character of the filter's output: the filter output shows "steps" rather than a smooth curve.

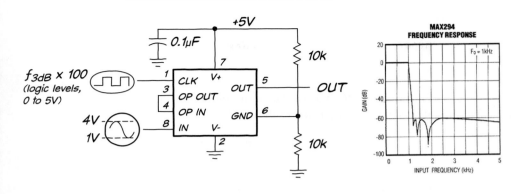

Figure 12L.14 MAX294 switched-capacitor lowpass filter (MAX294 frequency response is shown relative to f_{3dB}, which is adjustable).

Two optional filter tests, for fun:

Peel away frequency components of a square wave: As you know from your experience in *compensating* scope probes, a square wave can quickly reveal a circuit's frequency response. Try this: feed your filter a square wave at around 100Hz, and take f_{clock} gradually down from its maximum. Can you make out the frequency components $5 \times f_{\text{square}}$, and then $3 \times f_{\text{square}}$, shortly before you strip the square wave down to its fundamental sine?

Fancier demonstration of controllable f_{3dB} (for the energetic): The prettiest picture of the filter's performance appears if you *sweep* the input frequency, and watch the effect of varying f_{clock} to the filter. f_{3dB} should be variable over a wide range: up to about 1kHz if you use the *breadboard* oscillator as its clock; up to around 25kHz if you use a higher clock rate from an external function generator's TTL output.

Is *feedthrough* of the chip's clock noticeable at the output? Can you confirm the steep rolloff that is claimed: 48dB/octave, close above f_{3dB}? Don't be shocked if you find some bumps in the "stopband:" Fig. 12L.14 shows that the MAX294 rolls off like a rock – but then bounces a bit (like a rock hitting a marble floor?).

12L.3 Switching audio amplifier

A traditional audio amplifier, like the push–pull followers that we are accustomed to, is obliged to soak up power as it delivers power to the load. For example, if a push–pull is powered from ±2.5V and delivers a sinusoid of 2.5V peak-to-peak, then at best half the supply voltage is dropped across the load (±1.25V), the rest across the push–pull's transistors. Efficiency, therefore, is below 50%.[11] A higher supply voltage would impose still lower efficiency at this low amplitude.

A switching amplifier gets around that problem, using the strategy that you met in the switching regulator of Lab 11L. The switch, which is ON or OFF and never in-between, ideally dissipates no power. All power goes to the load.

We invite you to try an integrated amplifier that uses this method, a National/TI LM4667. It provides *two* output pins to drive both ends of the speaker. This technique doubles the maximum voltage available to drive the speaker, relative to what would be available with a single-ended output, which would tie one end of the speaker to ground.

---

[11] Efficiency works out to be $(\pi/4) \cdot (V_{\text{pk}})/(V_{\text{supply}})$. In this case that comes to about 40%. You can work this out for yourself if you enjoy solving an integral.

You have seen scope images in Chapter 12N that provide foretaste of the electrical noise you can expect to see from this amplifier.

12L.3.1 The circuit

The circuit of Fig. 12L.15 is one of the first *surface mount* ICs you have used (in contrast to "through-hole" parts),[12] mounted on an adapter that allows plugging it into a breadboard.[13]

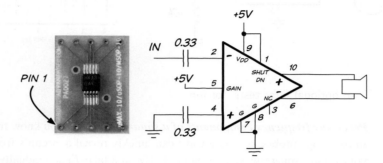

Figure 12L.15 LM4667 switching amplifier, wired for singled-ended input and 6dB gain.

Wire the circuit of Fig. 12L.15 on a single breadboard strip that you can put aside and keep. This audio amplifier will be useful later in the course when you will want to hear audio signals coming from your computer.

Add *decoupling* capacitors as close to the part as you can place them (to minimize the stray inductance of the supply lines): use $1\mu F$ and several $0.1\mu F$ ceramics.

Drive the circuit with a small sinusoid: perhaps half a volt amplitude at a few hundred hertz,[14] and listen as you install the decoupling caps. Experiment with their number and placement. You are likely to be persuaded that decoupling is worthwhile.

Figure 12N.10 showed scope images of this circuit's waveforms, and as we said in Chapter 12N, the switching noise is extreme. But we hope that with the help of judicious supply decoupling you can make the switching noise barely *audible*.

---

[12] You may have used such a version of the CA3096 transistor array in Lab 5L.
[13] The input capacitors may bother an astute observer: isn't this just like the *bad* circuits we have warned you about? The amplifier input seems to lack the required *DC path to ground*. It looks that way, but internal to the amplifier there is such a path to ground. This does not mean that the input is tied to zero volts; instead, in this single-supply circuit, it is *biased*, as usual. The inputs are biased to about 1.2V.
[14] Values chosen so as not to torture people who may be working near you.

12S Supplementary Notes: MOSFET Switches

In Chapter 12N we treat MOSFET switches and some of their applications. This note says a little more about both MOSFETs and the junction type, JFETs. You can safely ignore this note if you're only concerned with MOSFET applications, as we are in our lab exercises.

12S.1 A physical picture

The operation of a FET is easier to describe than the operation of a bipolar transistor. You will recall that we did not even try to describe *why* a bipolar transistor behaves as it does.

AoE Fig. 3.5

A FET, in contrast, just cries out for diagramming: a glance at Fig. 12S.1 will remind you of the FET's greatest virtue, its very high input impedance. The input terminal looks like either an *insulator* (so-called MOSFET type) or a *p–n* junction (in the so-called JFET) – one that is *never forward-biased*, unlike the bipolar transistor's base-emitter junction.

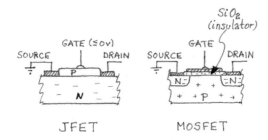

Figure 12S.1 A first view of JFET and MOSFET: a slab of semiconductor.

So, no *current* flows at the control terminal: you just apply a *voltage*; the electric *field* modifies the channel. Hence the name, of course.

The MOSFET diagram in Fig. 12S.1 makes the transistor looks as if it would not conduct, drain to source – since a positive voltage at drain would reverse-bias the right-hand *n:p* junction. This is correct – until we give it some help.

The MOSFET begins to conduct if one applies a positive voltage to the gate to induce a layer of *n*-type (electron-rich) region that can link the two *n* regions at drain and source. The "enhanced" MOSFET then resembles the JFET that has not been depleted: a conducting "channel" links source to drain. The MOSFET plotted in Fig. 12S.2 (like most, but not all) is OFF until you apply a gate voltage to turn it ON. Such a device is said to be of *enhancement mode*, conducting only when you give it some help (when you "enhance" its conduction).

The JFET or MOSFET – with a slab of doped semiconductor between its end terminals, source and drain[1] can be expected to conduct if we apply a voltage difference between its drain and source

---

[1] Perhaps the plumbing analogy suggested by the names bothers you: current goes from drain to source? Isn't that what

Supplementary Notes: MOSFET Switches

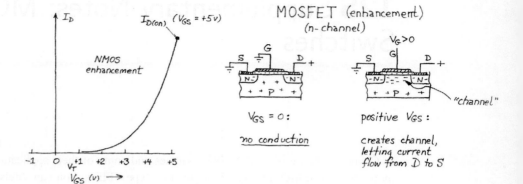

Figure 12S.2 MOSFET conducts if positive V_{gate} induces region to link drain and source *n*-type regions: it's OFF till you turn it ON.

terminals. At low voltages[2] across the device (V_{DS}), JFET or MOSFET behavior is *Ohmic*: current proportional to V_{DS}.

So far the behavior makes some intuitive sense. We'll soon (§12S.1.1) tackle the harder question of why at larger V_{DS}, *current source* behavior replaces Ohmic.

For comparison with the MOSFETs curve, Fig. 12S.3 shows the I_{D}-versus-V_{GS} curve for a JFET: ON until you do something to turn it OFF (perhaps only in degree). A negative voltage applied to the gate narrows the conducting channel.

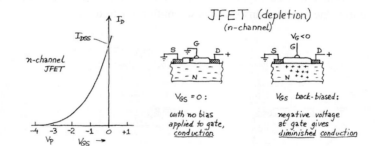

Figure 12S.3 JFET *I–V* curve looks like MOSFETs – but slid to quadrant left of $V_{\text{GS}}=0$: it's ON till you turn it OFF.

Such devices are said to operate in "depletion mode." All JFETs and a few MOSFETs operate this way.

The need of a *depletion-mode* device for a negative control voltage dooms it as a general-purpose switch. The *enhancement-mode* MOSFET, in contrast, can work with a single supply to generate an output capable of switching another switch of the same type; see Fig. 12S.4.

Thus one can build a large system of single-supply MOSFET switches, using enhancement-mode devices (we're hoping this suggestion brings computers to mind). The switches we have drawn are not of the type used in large arrays of logic devices. We saw in Chapter 12N that a *CMOS* configuration is much more efficient (AoE §3.4.4A).

> happens when the sewer gets blocked? The explanation for these strange names must be that the fellows developing the FET – the same people who named the bipolar's "collector" and "emitter," though here the case is muddied by the *pnp* configuration where a flow of "holes" dominates – were thinking like physicists, not like engineers: they seem to have considered electron flow rather than conventional current. This seems to be the case even though this transistor type is named "bipolar" because current flow depends in part on flow of both "holes" and electrons.
>
> [2] Low voltages here means less than $V_{\text{GS}}-V_{\text{T}}$. That *difference* voltage measures the degree to which the gate is driven ON. V_{T} is the threshold voltage at which I_{D} rises from zero to a small but measurable value – usually specified as a few nano-amps. See AoE §3.2.7.

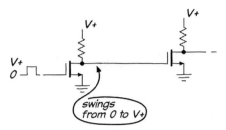

Figure 12S.4 A MOSFET switch (enhancement) can drive another with just one supply.

One cannot do this with depletion-mode transistors.

12S.1.1 A go at explaining the FET's constant-current behavior

The style of this course surely is to ignore the details of physics that might explain what's going on within a transistor, but FETs do tempt one to hope for intuition concerning their behavior. We'll give a try to this effort, while not expecting serious success.

Linear region versus current-source regions: *resistive* behavior at low V_{DS}... We claimed in §12S.1 that it is plausible that a continuous slab of doped semiconductor should conduct. It is also plausible that the current it passes should be proportional to the voltage applied (between drain and source). This is the behavior one sees for low V_{DS}, as one can see in the leftmost, the "linear," region of Fig. 12S.5.

AoE §3.2.7

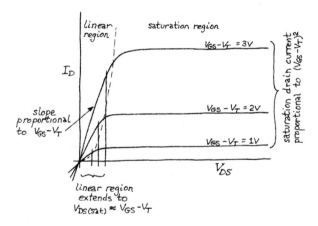

Figure 12S.5 FET is resistive at low V_{DS}, but behaves like a current-source at higher V_{DS}.

This family of curves looks much like the set of curves that describe I_C versus V_{CE} for a bipolar transistor, where each curve would be determined not by V_{GS} but by either V_{BE} or I_B. In Fig. 12S.6 are bipolar and FET curves, placed side by side for comparison (the contrast in current-region slopes is a bit unfair to the bipolars: note the much more sensitive current scale for the bipolar). The horizontal axis shows the voltage across the transistor. Each curve is defined by a particular input control level: V_{GS} or V_{BE}.

You may wonder whether a bipolar transistor might not be capable of the same *resistive* behavior. The answer is that its I–V is less linear than the FETs, at low voltages: see Fig. 12S.7.

A simple circuit trick can further straighten the FETs curves, it turns out – making the FET a useful voltage-controlled resistor.[3]

---

[3] The "trick" is to use an R:R voltage divider to feed half of V_{DS} to the gate. See AoE §3.2.7A, if you're curious.

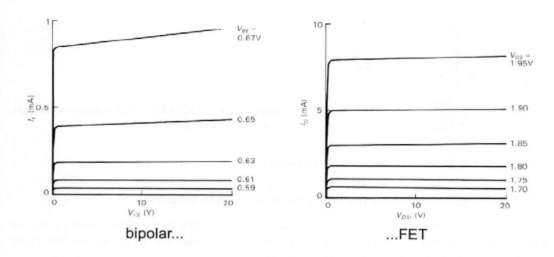

Figure 12S.6 Families of *I* and *V* curves, bipolar compared with FET (AoE Fig. 3.2).

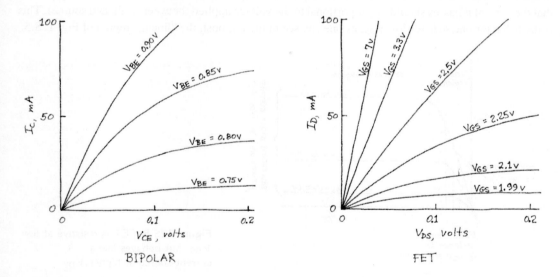

Figure 12S.7 At low voltage across transistor, FET is more linear than the bipolar transistor (AoE Fig. 3.2).

So far, so simple: the resistive behavior at low voltage sounds reasonable. A harder challenge is to get some understanding of the constant-current behavior that is much more typical of the FET.

...Saturation region: *current-source* behavior at higher V_{DS} Resistive behavior, intuitively comfortable for us resistor-philes, is exceptional for an FET. Ordinarily the FET, like a bipolar transistor, passes a *current* determined by its control signal – V_{GS} for the FET. This current-source behavior appears in Fig. 12S.5 over all except the lowest range of V_{DS} values. The nearly horizontal current plot resembles what one would see for a bipolar transistor. Why? We'll attempt an informal explanation.

As one gradually increases V_{DS}, the rising voltage near the drain diminishes the width of the conducting region there, and eventually that conducting channel "pinches off." The leftmost image of Fig. 12S.8 sketches the non-conducting case with $V_{GS}=0$. The two to the right show conduction, first in the linear region, then in the pinched off region. It is the rightmost image that attempts to picture the *constant current* or "saturation" region of FET operation.[4]

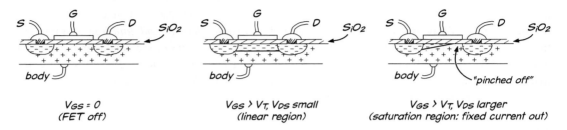

Figure 12S.8 MOSFET conduction: three regimes.

A first (and quite reasonable) guess would be that current I_D would fall to zero when V_{DS} rose enough to pinch off the channel. But I_D does not do that. Instead, if V_{DS} continues to rise, I_D levels off. The region between pinch-off point and drain is not as rich in carriers as is the induced "channel". But the field there is strong enough to carry electrons that have drifted through the channel on to the drain.

Here's our attempt at an intuitive understanding of this effect. If this is too silly for your tastes, take a look at the texts cited below.

The pinched-off region forms a bottleneck, a narrow region where a kind of traffic jam occurs, see Fig. 12S.9. As V_{DS} rises, much of this drop occurs over the short length of that pinched section, pushing up the current density in this bottleneck; then, if V_{DS} grows further, increasing the field (drivers at the drain end lean on their car horns) the bottleneck region grows longer, and traffic flow or I_D levels off at its *saturation* value. In short, further increases in V_{DS} cause two opposing effects that nearly cancel: stronger field, but reduced carrier mobility.

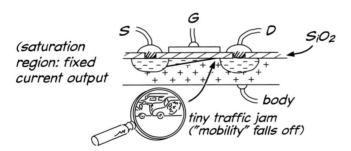

Figure 12S.9 Tiny traffic jam to explain FET "saturation:" mobility falls off as field strengthens.

---

[4] You will recognize what a bad choice of language this is: to use the word "saturation" to describe normal FET operation, while the word means something radically different when applied to bipolar transistors or operational amplifiers. But we are stuck with this usage.

So current stays roughly constant.[5]

12S.1.2 Saturation not quite constant: channel-length modulation resembles Early effect

AoE §3.1.1A

You may have noticed that the I–V curves across FETs resemble those for bipolar transistors not only in rough outline, as we suggested on page 499 and Fig. 12S.6, but also in their upward tilt. That tilt, described by Early effect for bipolars, results from *channel-length modulation* in an FET. The mechanism is similar to that for Early effect. As V_{DS} grows, the length of the conducting channel is diminished by the growth of the pinched-off region. So current slightly increases (*very* slightly: the dynamic impedance of the 1N5294, a JFET current-limiting diode that we use in several labs, for example, is high: above 1MΩ [specified at $V_{DS} = 25V$]).

12S.1.3 An application for constant-current behavior

AoE §3.2.2 and Fig. 3.20

A JFET, with its ability to hold current constant even at $V_{GS} = 0$, makes a handy two-terminal current-limiting device – as you may recall from Lab 5L, when you used such a thing in the "tail" of the differential amplifier. The curves you saw back in Fig. 12S.5 remind us that we cannot expect *constant-current* behavior however, unless we take care to keep V_{DS} above the level where *resistive* behavior ceases. For the 1N5294 that minimum voltage is 1.2V.

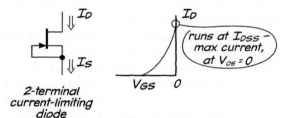

Figure 12S.10 JFET forms simple current-limiting diode.

---

[5] For a fuller explanation and handsome diagrams see Burns, S. and Bond, P., *Principles of Electronic Circuits*. West (1987), §5.2; see also Jaeger, R. and Blalock, T., *Microelectronic Circuit Design*. McGraw-Hill (4th ed., 2011), §4.24.

13N Group Audio Project

Contents

13N.1 Overview: a day of group effort	**503**
13N.1.1 Circuits that do something, again...	503
13N.1.2 ...and circuits of your own design	503
13N.1.3 A description of the technical task	504
13N.1.4 The elements of the audio-transmission chain	504
13N.2 One concern for everyone: stability	**506**
13N.3 Sketchy datasheets for LED and phototransistor	**507**

Why?

Part of what we aim for today is a review, since the circuit includes an unusually broad variety of elements. What you will achieve is the wireless transmission of an audio signal, using optical encoding.

13N.1 Overview: a day of group effort

Today's lab and class differ from all the others in the course. In class, we want to hear you do the teaching: tell your classmates about the piece of the project that you designed. We want your audience to tell you (politely) how you might improve your design. The lab differs from all the others in that it is a collaboration. Your fragment won't do anything very entertaining until it is joined with the work of your classmates.

13N.1.1 Circuits that do something, again...

Our motive for trying this class and lab format was to respond to students' pleas that they be given a chance to build circuits that "do something." Apparently, people get tired of seeing only scope traces. The earlier build-an-op-amp and then PID labs also tried to respond to this yearning. Today's lab offers a similar chance to put many circuit fragments together. This lab differs from PID and the others, however, in asking you to do all the *design* work.

13N.1.2 ...and circuits of your own design

Generally, we have not dared to write labs this way. We feared that the process would be too slow: you would draw a design, find its defects, correct them, then try the amended design. It doesn't sound like work that could fit within an afternoon. Trying that process every day would have made this quite a different sort of course – a good one, but not one that could gallop through the large amount of material that we attempt. We do, however, dare to try this method on this one day, because we are confident that you will come up with working designs: each of the several circuit elements is well within the range of your skills.

We hope that the exercise will serve several purposes. It will be fun, and should provide satisfying confirmation that you have learned something in the past weeks.

13N.1.3 A description of the technical task

Figure 13N.1 is a sketch of what we want to build. The notes below will spell out what the several boxes ought to do.

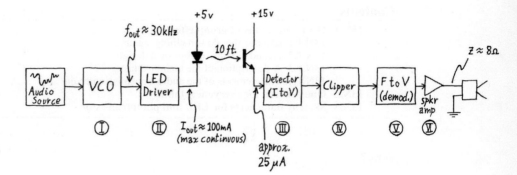

Figure 13N.1 A block-diagram of the group audio project.

You will send music across the room – or, at very least, across a gap of a few feet – "wirelessly." In Marconi's day, and again more recently, wireless meant "using radio." Here it does not. We will transmit using infrared "light."[1] More particularly, we will encode the audio *information* onto a *carrier* that is a frequency above human hearing – but far short of radio. We put our carrier at about 30kHz. We will encode the information – the time-varying voltages – as *frequencies*. This scheme is called, as you know, Frequency Modulation, FM.

FM is appealing here because of its good noise immunity and also because it allows us to drive the LED at full brightness (when ON), permitting maximum range.[2]

The variation in frequency need not be large. In order to convey an audio signal up to about 3kHz, a frequency variation of well under 3kHz can be sufficient.

13N.1.4 The elements of the audio-transmission chain

Here is a short catalog of the elements of today's circuit. In some cases, our descriptions will be purposely vague because we don't want to short-circuit the design process that we hope you will go through.

On the transmission side:

Modulator: This is a voltage-controlled oscillator (VCO). We use a '555 oscillator to convert an audio signal of time-varying voltage into a square wave of varying frequency. This circuit's voltage-to-frequency function won't be linear, but works quite well enough for today's purposes. You could use a more perfect VCO (such as, say, the VCO included in the 74HC4046 that you will meet in Lab 18L). But for this extra effort you would get no audible improvement over today's rougher design.

---

[1] Is it "light" if humans can't see it? A nice question, which we'll not stop to worry about.
[2] Perhaps you will protest that "noise immunity" and "range" are two ways of saying the same thing. You have a point.

Details: (Questions of interest to designers of this stage; perhaps too detailed for the rest of you).

The needed frequency variation is slight and we can tolerate a somewhat nonlinear response – because our standards for this task are not especially high, and because we anticipate nonlinearities in the demodulator, in any case.

There are easy and hard ways to vary the frequency of a '555 oscillator. The hard ways probably would use a voltage-controlled current source to vary a sawtooth frequency. The easy way takes advantage of the *control* terminal (pin 5 for the DIP package), letting the audio signal vary the threshold voltages.

Consider duty cycle: does your design achieve close to 50% duty cycle (that is, percentage of period during which the waveform is *high*)? Does duty cycle matter?

Make the center frequency adjustable by manipulation of a potentiometer wired as variable resistor. Don't forget to provide a constant R along with the adjustable so that the circuit doesn't go crazy if you happen to twiddle the pot to an extreme.

The carrier frequency: We chose a frequency of around 30kHz, intermediate between extremes that could have caused difficulty:
- we wanted it not too low: not in a range audible to humans, because some of the carrier is likely to persist after demodulation – noise mixed with the audio signal;
- we wanted it not too high: not in a range where our op-amps and comparators begin to falter.

LED driver: This takes the relatively delicate square wave that is output by the VCO (a '555 oscillator) and makes it strong enough to drive a high-current infrared LED. The LED is capable of running 100mA continuous current, and drops about 2V when conducting.

Details, LED driver:

LED current is specified (maximum continuous current; for current specification, see sketchy datasheets in §13N.3.) Does it follow that your driver should be a *current source* (or "sink")?

Consider what the signal from the '555 output looks like electrically, and what the load looks like (LED dropping about 2V).

Try to minimize power dissipation. The current is large; you don't want a voltage higher than necessary. If you use a resistor, make sure it won't get hot enough to burn your fingers.[3]

On the receiving side:

Photodetector: This converts the current passed by an infrared-sensitive phototransistor to a voltage. The voltage swing should be about as large as you can manage, taking care always to avoid clipping the waveform.

Details:

Note that ambient light will include some infrared, so that the phototransistor current never will fall to zero (despite the transistor's dark case designed to block visible light). Note also that the phototransistor current flows in one direction only.

You're likely to take the current from the phototransistor's *emitter* (compare the photo circuits of Lab 6L). But don't infer that the circuit behavior you see will resemble that of an emitter follower (Lab 4L) where you last took an output from emitter.

Clipper: This circuit takes the detector output, perhaps small, and transforms it into a square wave swinging ±15V. (The large amplitude is helpful to the later stages.) The uniform amplitude

---

[3] Do this not by wearing thick gloves but by considering power dissipated versus the resistor's power rating.

imposed by this circuit is important: variations in amplitude that appear in the *detector's* output would corrupt the ultimate demodulated audio output if allowed to persist.

Details:

Note that the output of the detector will include a DC component (unless the detector designers have removed that for you). That DC level is not information; only the frequency of the varying voltage carries information.

To keep the output clean, you will need hysteresis. Make this adjustable right down to zero since you'd like your circuit to be as sensitive as possible – permitting maximum range for the entire circuit.

Demodulator: This takes the square wave of varying frequency and converts it, in two stages, into a time-varying amplitude: an audio signal like the one that drove the VCO on the input side.

- Stage one: uses a filter with steep skirt to convert changes in frequency to changes in amplitude.
- Stage two: uses a conventional AM demodulator to produce the audio waveform.

Details:

Any filter with a steep response will do, but the easiest circuit probably is an *RLC* resonant circuit. You will recall from Lab 3L, especially §3L.1.2, that this circuit's $\Delta_{\text{amplitude}}/\Delta_{\text{frequency}}$ response can be made very steep.

For best conversion of Δ_frequency to Δ_amplitude you'd like a response curve that is close to a straight-line ramp. We suggest that you not strain to calculate the best shape for this response curve. Instead, in the rough-and-ready style of this course, make the *shape* of the circuit's frequency response adjustable.

Recall what circuit elements in Chapter 3L's resonant circuit affected the shape of the response: large R provided large Q values (steep curve); but large R also provided a lower output amplitude (entire curve became small, though peaky). In short, large R produces a small pointy Matterhorn; smaller R produces a large gently-sloping Blue Ridge Mountain.

Consider phase: does the demodulator's output voltage rise or fall as frequency rises? Does the answer to this question matter?

Amplifier: Boost demodulator output to drive an 8Ω speaker.

Details:

This is a circuit you have built before. Make its gain adjustable (since a loud output can be maddening during the time when the group is trying out and adjusting the whole circuit). You probably need not set minimum gain below one (and this choice will permit use of a high-impedance form of the amplifier).

Note that you do not want to put any *DC* current through the speaker. Only wiggles produce sound; a DC level produces only heat.

The major challenge will be to keep this circuit *stable*.

13N.2 One concern for everyone: stability

The several elements of the circuit will be wired each on its own breadboard strip. Then these will be brought together, linked and powered. This is a setup peculiarly vulnerable to parasitic oscillations. Against this hazard:

- *decouple* all power supplies, on every individual strip, with ceramic caps of 0.01–$0.1\mu F$;

- do all you can to keep power supply leads *short*. (In our course, we provide power through bus connectors pulled off old breadboard strips; the low inductance and substantial capacitance of power buses helps: see Fig. 13N.2.)

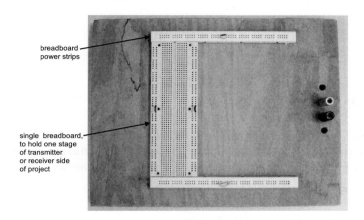

Figure 13N.2 Power bus strips help to minimize power supply impedance, against parasitic oscillations.

13N.3 Sketchy datasheets for LED and phototransistor

The LED and phototransistor work in the near-infrared, nearly matching peak sensitivity to peak emission wavelength. Both are narrow-angle devices, chosen to maximize range – but having the drawback that the emitter must be well aimed. Here are some of their specifications.

Device	central wavelength (nm)	emission/reception Angle	current	rise & fall time (μs)
LED: TSTS7100	950	$\pm 5°$	250mA (max cont.)*	0.8
PhotoQ: QSD124	880	$\pm 12°$ (half power angle)	6mA (ON, min.)	7
* But 100mA is more usual for this part				

And here are pinouts.

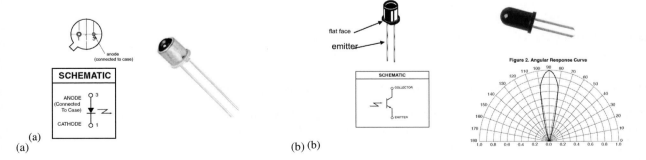

Figure 13N.3 (a): Infrared LED pinout: TSTS7100. (b): Infrared phototransistor pinout: QSD124.

13L Lab: Group Audio Project

13L.1 Typical waveforms

Figure 13L.1 gives a preview of what's coming in the form of waveforms at several points. Before you have built the circuits, these waveforms may be a bit cryptic; but trying to understand these plots may help you to get a grip on the whole project. Then we'll finish these lab notes with some suggestions on how to test your circuits.

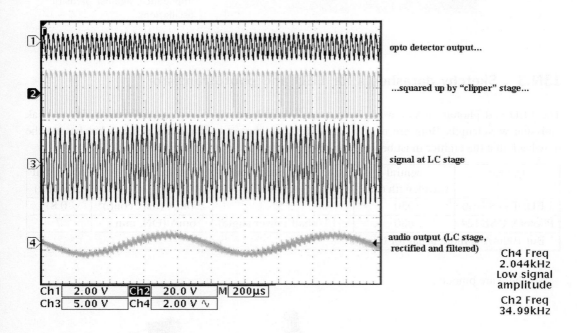

Figure 13L.1 Waveforms at several stages of the IR transmission chain.

13L.1.1 How much do you need to know about stages that abut yours?

Not much, we expect. If your abutters surprise you, we expect you'll be nimble enough to adjust on project day. It's certainly fair to ask them, beforehand – but probably not necessary.

13L.1.2 How much advice do we want to give you beforehand?

Not much because it's helpful to the whole group to have design issues aired before the group rather than settled backstage. We hope that the design you present will *not* be perfect. Perfect designs are not very instructive. And please don't test your design beforehand by building it; a real discussion stopper is a line like "I don't know why the value is 10k – but I built it and it worked." For reviewing purposes, it's much better to discuss rationales.

13L.2 Debugging strategies

Here are some debugging tips that can simplify our task: we'll pair modules that talk to each other; when these pairs work, we'll string together all the elements of the chain. You'll find a block diagram of that entire chain in Fig. 13N.1.

13L.2.1 Debug VCO (V-to-f) and demod (f-to-V)

These adjustments are particularly ticklish. VCO people can adjust the center frequency. Use a DVM as frequency-meter, to watch this frequency. Note what frequency works best (that is, produces the cleanest sine from the demod block) in case you mess up an adjustment and need to return to that center frequency. We find a 1k resistor in series with the pin 5 input helps to protect the CMOS '555 from damage. It does not affect circuit performance (the internal divider's resistors present an equivalent resistance of 33k).

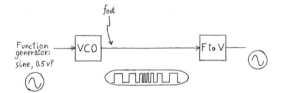

Figure 13L.2 Testing VCO and demodulator.

Before testing with the VCO, demods will have done the best they could on their own: demods can do their preliminary test by feeding their circuit a square wave whose frequency is swept. The demod circuit should output a fairly good replica of the *ramp* shape that evokes the *swept* square wave.

In testing the linked VCO-to-demod pair, VCO people will try to find their comfortable place on the slope of the "mountain" that is the shape of demod's frequency response. Demod people meanwhile, can adjust shape of that mountain: they can adjust the Q of their circuit, looking for a region that is pretty linear.

Don't drive yourselves crazy by letting both VCO and demod people twiddle at the same time: take turns! When demod puts out a pretty good reproduction of the source sine, leave the adjustments as they are. It will be harder to optimize response at the later stage when you will be working with the entire chain of circuit elements.

13L.2.2 Optical link only

Optical detector people probably have at least a gain-control to adjust; perhaps also a DC-offset. If you're not sure whether the LED is alive, use an infrared detector card (if it has just come from storage in a dark drawer, it will need a few minutes in room light for activation).[1]

---

[1] Laboratory quality IR detector cards are expensive, but you're likely to find inexpensive versions on eBay. Your cellphone camera may or may not show IR, depending on its camera's filtering.

510 Lab: Group Audio Project

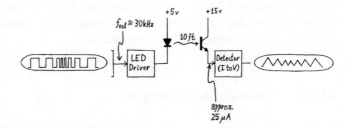

Figure 13L.3 Testing optical link only.

13L.2.3 Optical recovery, squared-up

Occasionally, attaching the comparator as load will destabilize the detector. If this happens (indicated by fuzz on the detector output), make sure you have decoupled all your power supplies, and that you have minimized power-supply lengths.

Clipper people will have a hysteresis-adjust to play with. They want minimum hysteresis (for max sensitivity) that is consistent with stability.

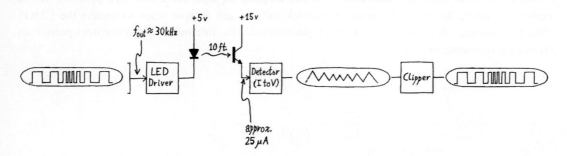

Figure 13L.4 Adding clipper to optical-link test.

13L.2.4 Sine-to-sine, with audio amp

Again, stability is the hardest issue here: the amp is straightforward (make sure the transistors aren't tiny 2N3904s and 2N3906s). Testing the amp linked to the demod block gives a better test than just taking a function generator as source for the amp.

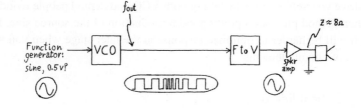

Figure 13L.5 Trying to recover audio, testing speaker driver.

Part IV

Digital: Gates, Flip-Flops, Counters, PLD, Memory

14N Logic Gates

Contents

14N.1	**Analog versus digital**	**513**
	14N.1.1 What does this distinction mean?	513
	14N.1.2 But why bother with digital?	514
	14N.1.3 Alternatives to binary	516
	14N.1.4 Special cases for which digital processing obviously makes sense	517
14N.2	**Number codes: Two's-complement**	**518**
	14N.2.1 Negative numbers	518
	14N.2.2 Hexadecimal notation	519
14N.3	**Combinational logic**	**520**
	14N.3.1 Digression: a little history	520
	14N.3.2 deMorgan's theorem	522
	14N.3.3 Active-high versus active-low	522
	14N.3.4 Missile-launch logic	524
14N.4	**The usual way to do digital logic: programmable arrays**	**526**
	14N.4.1 Active-low with PLDs: a logic compiler can help	526
14N.5	**Gate types: TTL and CMOS**	**528**
	14N.5.1 Gate innards: TTL versus CMOS	528
	14N.5.2 Thresholds and noise margin	529
14N.6	**Noise immunity**	**530**
	14N.6.1 DC noise immunity: CMOS versus TTL	530
	14N.6.2 Differential transmission: another way to get good noise immunity	532
14N.7	**More on gate types**	**533**
	14N.7.1 Output configurations	533
	14N.7.2 Logic with TTL and CMOS	535
	14N.7.3 Speed versus power consumption	535
14N.8	**AoE Reading**	**535**

Why?

We aim to apply MOSFETs to form gates capable of Boolean logic and to look at some such binary operations.

14N.1 Analog versus digital

14N.1.1 What does this distinction mean?

The major and familiar distinction is between *digital* and *analog*. Along the way, let's also distinguish "binary" from the more general and more interesting notion, "digital."

- First, the analog" versus digital distinction.

Logic Gates

- An *analog* system represents information as a continuous function of the information (as a voltage may be proportional to temperature, or to sound pressure).[1]
- A *digital* system, by contrast, represents information with *discrete* codes. The codes to represent increasing temperature readings could be increasing digital numbers (0001, 0010, 0011, for example) – but they could use any other code that you found convenient. The digital representation also need not be *binary* – a narrower notion mentioned just below.

• Now the less important binary versus digital distinction: binary is a special case of digital representation in which only two values are possible, often called True or False. When more than two values are to be encoded using binary representations, multiple binary digits are required. Binary digits, as you know are called *bits* (*bi*nary dig*it*)

A *binary* logic gate classifies inputs into one of two categories: the gate is a comparator – one that is simple and fast, see Fig. 14N.1.

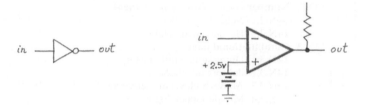

Figure 14N.1 Two comparator circuits: digital inverter and explicit comparator, roughly equivalent.

The digital gate resembles an ordinary comparator (hereafter we won't worry about the digital/binary distinction; we will assume our digital devices are binary).

How does the digital gate differ from a comparator like, say, the LM311?

Input and output circuitry: the '311 is more flexible (you choose its threshold and hysteresis; you choose its output range). Most digital gates include *no* hysteresis. The exceptional gates that include hysteresis usually are those intended to listen to a computer bus (assumed to be extra noisy), and the inputs to some larger ICs such as the PAL that you will meet in our labs, XC9572XL, which includes 50mV of hysteresis on all of its inputs.

Speed: the logic gate makes its decision (and makes its output show that 'decision') at least 20 times as fast as the '311 does.

Simplicity: the logic gate requires no external parts, and needs only power, ground and In and Out pins. It can work without hysteresis because the logic signals presented to its inputs transition decisively and fast.

14N.1.2 But why bother with digital?

Is it not *perverse* to force an analog signal – which can carry a rich store of information on a single wire – into a crude binary high/low form? Let's consider the cons and pros. A caricaturing of "digital audio" appears in Fig. 14N.2, where the sound of the Stradivarius is converted to arcade-game quality. We doubt this represents your understanding of *digital audio*.

---

[1] We don't quite want to say that the representation – say, the voltage – is *proportional* to the quantity represented since a log converter, for example, can make the representation stranger than that.

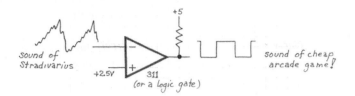

Figure 14N.2 Naive version of "digital audio": looks foolish!

Disadvantages of digital:

    Complexity: more lines required to carry same information[2]
    Speed: processing the numbers that encode the information sometimes is slower than handling the analog signal

AoE §10.1.1

Advantages of digital:

    Noise immunity: the signal is born again at each gate; from this virtue flow the important applications of digital:
- allows stored-program computers (may look like a wrinkle at this stage, but turns out, of course, to be hugely important);
- allows transmission and also unlimited processing without error – except for round-off/quantization; that is, deciding in which binary bin to put the continuously variable quantity;
- can be processed out of "real time:" at one's leisure. (This, too, is just a consequence of the noise immunity already noted.)

Can we get the advantages of digital without loss? Not without *any* loss, but one can carry sufficient detail of the Stradivarius' sound using the digital form – as we try to suggest in Fig. 14N.3 – by using *many* lines.

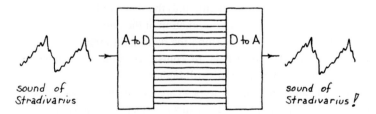

Figure 14N.3 Digital audio done reasonably.

A single bit allowed only two categories: it could say, of the music signal, only "in high half of range" or "in low half." Two bits allow finer discrimination: four categories (top quadrant, third quadrant, second quadrant, bottom quadrant). Each additional line or bit doubles the number of slices we apply to the full-scale range.

Our lab computer uses 8 bits – a "byte"[3] – allowing $2^8 = 256$ slices.[4] Commercial CDs and most contemporary digital audio formats use 16 bits, permitting $2^{16} \approx 65,000$ slices. To put this another

---

[2] "More lines" often does not apply to *transmission* of signals because serial transmission is widely used. But at least in its inner workings, digital processing does encode values as multiple parallel bits.

[3] Some purists insist on calling an 8-bit quantity an "octet," recalling a long-ago time when byte had not come to mean 8 bits, but depended on word length in a particular computer. Byte now is the standard way to refer to an 8-bit quantity.

[4] The SiLabs controller that some people will choose in the micro labs can cut 16 times finer, with 12 bits. But we cannot make practical use of that resolution on our breadboarded circuits.

way, 16-bit audio takes each of our finest slices (about as small as we can handle in lab, with our noisy breadboarded circuits) and slices that slice into 2^8 sub-slices. This is slicing the voltage very fine indeed.

In Fig. 14N.4 we demonstrate the effect of 1, 2, 3, 4, 5, and finally 8 bits in the analog–digital conversions (both into digital form and back out of it, to form the recovered analog signal that is shown).

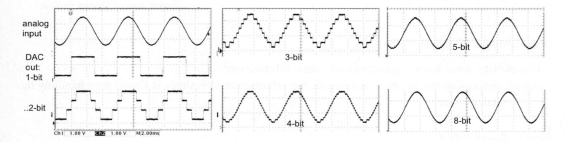

Figure 14N.4 Increasing the number of bits improves detail in the digital representation of an analog quantity.

14N.1.3 Alternatives to binary

AoE §14.5.5B

Binary is almost all of digital electronics, because binary coding is so simple and robust. But a handful of integrated circuits do use more than two levels. A few use three or four voltage levels, internally, in order to increase data density: NAND flash for example, uses four levels or even eight. Data storage density is increased proportionately.[5] This density comes at the cost of diminished simplicity and noise immunity. The better noise immunity of the older 2-level parts sometimes is promoted as one of their selling points, but multi-level now dominates NAND flash memory.

More important instances of non-binary digital encoding occur in data *transmission* protocols. Telephone modems ("modulator–demodulators") confronted the tight limitation of the system's 3.4kHz frequency limit, and managed to squeeze more digital information into that bandwidth by using both multiple-levels (more than binary) and *phase* encoding. One combination of the two called "16QAM" squeezes 4 bits worth of information into each "symbol," pushing the data rate to 9600 bits/second. Does that not sound quite *impossible* on a 3.4kHz line? It is, of course, possible. The process of increasing coding complexity has continued, pushing data rates still higher.

A plot like a phasor diagram – with "real" and "imaginary" axes – can display the several phases as well as amplitudes of single-bit encodings. The plot in Fig. 14N.5 shows "4QAM:" single-amplitude, four-phase-angle encoding that permits four unique "symbols" (or, "two bits per symbol") by summing two waveforms that are 90° out of phase (the "quadrature" of the QAM acronym).[6]

Denser encodings are possible using the same scheme. The constellation diagrams in Fig. 14N.6 show 16QAM which carries 4 bits of information in each "symbol," one symbol being represented by one dot on the diagram.[7]

---

[5] Fun fact (according to Wikipedia): a quaternary digit sometimes is called a "crumb," joining its foody companions, bit, byte and nybble (a 4-bit binary value).

[6] Source: National Instruments tutorial, "Quadrature Amplitude Modulation:" http://zone.ni.com/devzone/cda/tut/p/id/3896. This tutorial lets one watch, in slow motion, the relation between the time-domain waveform and this phase-and-amplitude constellation plot. It helps.

[7] Source: Acterna tutorial: chapters.scte.org/cascade/QAM_Overview_and_Testing.ppt. Acterna was acquired by JDSU, Milpitas, CA, in 2005.

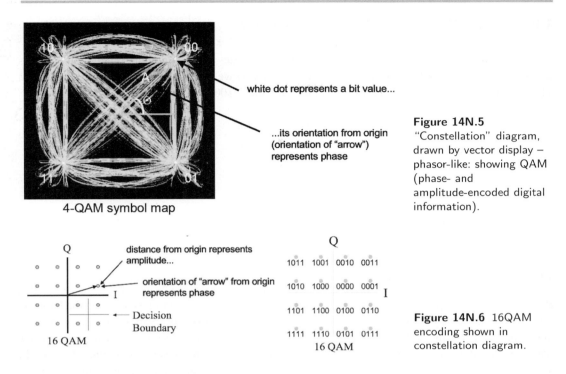

Figure 14N.5 "Constellation" diagram, drawn by vector display — phasor-like: showing QAM (phase- and amplitude-encoded digital information).

Figure 14N.6 16QAM encoding shown in constellation diagram.

The QAM waveforms (seen in time-domain) are exceedingly weird: see Fig. 14N.7.[8]

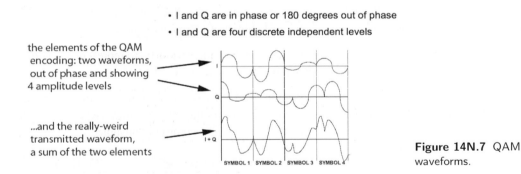

Figure 14N.7 QAM waveforms.

Complex encoding schemes like QAM are appropriate to data transmission, but — luckily for us — the storage and manipulation of digital data remains overwhelmingly *binary*, and therefore much more easily understood.

14N.1.4 Special cases for which digital processing obviously makes sense

In some cases the information is *born* digital so no conversion is needed. The information may be numbers (as in a pocket calculator) or words (as in a word processor). Since the information never exists in analog form (except perhaps in the mind of the human), it makes good sense to manipulate the information digitally: as sets of codes. To do otherwise would be perverse (store your term paper as a collection of voltages on capacitors?).

---

[8] Source: Acterna, tutorial, again.

14N.2 Number codes: Two's-complement

Binary numbers may be familiar to you already: each bit carries a weight double its neighbor's. That's just analogous to decimal numbers as you know: it's the way we would count if we had just one finger. The number represented is just the sum of all the bit values: $1001 = 2^3 + 2^0 = 9_{10}$.

```
        DECIMAL                BINARY
      10²  10¹  10⁰          2²  2¹  2⁰

       2   1   3              1   0   1

    200 + 10 + 3 = 213      4 + 0 + 1 = 5₁₀
```

Figure 14N.8 Decimal and binary number codes compared.

14N.2.1 Negative numbers

But 1001 *need not* represent 9_{10}. Whether it does or not in a particular setting is up to us, the humans. *We* decide what a sequence of binary digits is to mean. Now, that observation may strike you as the emptiest of truisms, but it is not quite Humpty Dumpty's point. We are not saying 1001 means whatever I want it to mean; only that sometimes it is useful to let it mean something other than 9_{10}. In a different context it may make more sense to let the bit pattern mean something like "turn on stove, turn off fridge and hot plate, turn on lamp." And, more immediately to the point, it often turns out to be useful to let 1001 represent not 9_{10}, but a *negative* number.

The scheme most widely used to represent negative numbers is called "two's-complement." The relation of a 2's-comp number to positive or "unsigned" binary is extremely simple:

> the 2's-comp number uses the most-significant-bit (MSB) – the leftmost – to represent a *negative* number of the same weight as for the unsigned number.

So 1000 is +8 in unsigned binary; it is -8 in 2's-comp. And 1001 is -7. Two more examples appear in Fig. 14N.9: the 4-bit 1011 can represent a negative five, or an unsigned decimal eleven; 0101, in contrast, is (positive-)5 whether interpreted as "signed" (that is, 2's-comp) or "unsigned."

```
                  Unsigned           2's Comp

         1011    8 + 2 + 1 = 11₁₀    -8 + 2 + 1 = -5₁₀

         0101    4 + 1 = 5₁₀          ... = 5₁₀
```

Figure 14N.9 Examples of 4-bit numbers interpreted as *unsigned* versus *signed*.

This formulation is not the standard one. More often you are told a rule for forming the 2's-comp representation of a negative number, and for converting back. This AoE does for you:

> To form a negative number, first complement each of the bits of the positive number (i.e., write 1 for 0, and vice versa; this is called the "1's complement"), then add 1 (that's the "2's complement").

It may be easier to form and read 2's-comp if, as we've suggested, you simply read the MSB as a large negative number, and add it to the rest of the number, which represents a (smaller) positive number interpreted exactly as in ordinary unsigned binary.

2's-comp may seem rather odd and abstract to you just now. When you get a chance to *use* 2's-comp it should come down to earth. When you program your microcomputer, for example, you will sometimes need to tell the machine exactly how far to "branch" or "jump" forward or back as it executes a program. In the example below we will say this by providing a 2's-comp value that is either positive or negative. But first a few words on the *hexadecimal* number format.

14N.2.2 Hexadecimal notation

Let's digress to mention a notation that conveniently can represent binary values of many bits. This is *hexadecimal* notation – a scheme in which 16 rather than 10 values are permitted. Beyond 9 we tack in 5 more values, A, B, C, D, E and F, representing the decimal equivalents 10 through 15.

The partial table in Fig. 14N.10 illustrates how handily hexadecimal notation – familiarly called "hex" – makes multi-bit values readable, and "discussable."

Binary	Hexadecimal	Decimal
0111	7	7
1001	9	9
1100	C	12
1111	F	15
10011100	9C	156

Figure 14N.10 Hexadecimal notation puts binary values into a compact representation.

You don't want to baffle another human by saying something like, "My counter is putting out the value One, Zero, Zero, One, Zero, One, One, Zero." "What?," asks your puzzled listener. Better to say "My counter is putting out the value 96h." One can indicate the *hex* format, as in this example, by appending "h." Or one can show it, instead by writing "0x96."[9]

Now, back to the use of 2's-comp to tell a computer to jump backward or forward. The machine doesn't need to use different commands for "jump forward" versus "jump back." Instead, it it simply adds the number you feed it to a present value that tells it where to pick up an instruction. If the number you provide is a negative number (in 2's-comp), program execution hops back. If the number you provide is a positive number, it jumps forward.

Perhaps the example in Fig. 14N.11 will begin to persuade you that there can be an interesting, substantial difference between "subtracting A from B" and "adding negative A to B." Subtraction – if taken to mean what "subtraction" says – requires special hardware; addition of a negative number, in contrast, uses the same *hardware* as an ordinary addition. That's just what makes the use of 2's-comp a tidy scheme for the microcomputer's *branch* or *jump* operations.

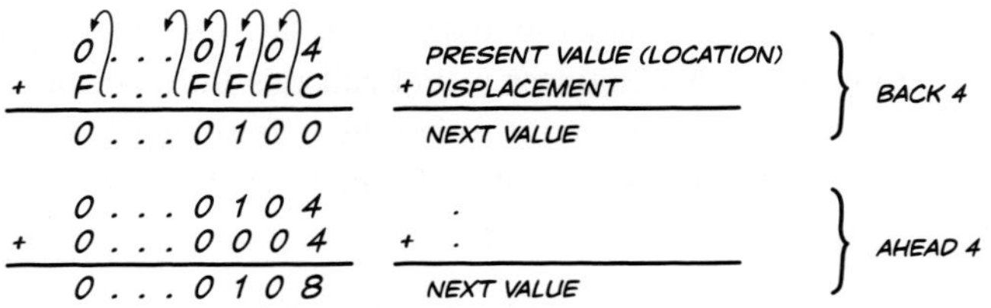

```
    0...0104       PRESENT VALUE (LOCATION)  ⎫
  + F...FFFC     + DISPLACEMENT              ⎬ BACK 4
    0...0100       NEXT VALUE                ⎭

    0...0104                                 ⎫
  + 0...0004     + ·                         ⎬ AHEAD 4
    0...0108       NEXT VALUE                ⎭
```

Figure 14N.11 Example of 2's-comp use: plain *adder* can add or subtract, depending on sign of addend: computer jump displacement.

---

[9] The RIDE assembler/compiler that you soon will meet recognizes both conventions.

14N.3 Combinational logic

Explaining why one might want to put information into digital form is harder than explaining how to manipulate digital signals. In fact, digital logic is pleasantly easy, after the strains of analog design.

14N.3.1 Digression: a little history

Skip this subsection if you take Henry Ford's view that "History is more or less bunk."

Boole and deMorgan: It's strange – almost weird – that the rules for computer logic were worked out pretty thoroughly by English mathematicians in the middle of the 19th century, about 100 years before the hardware appeared that could put these rules to hard work. George Boole worked out most of the rules in an effort to apply the rigor of mathematics to the sort of reasoning expressed in ordinary language. Others had tried this project – notably, Aristotle and Leibniz – but had not gotten far. Boole could not afford a university education and instead taught high school while writing papers as an amateur, years before an outstanding submission won him a university position.

Boole saw assertions in ordinary language as propositions concerning memberships in *classes*. For example, Boole wrote, of a year when much of Europe was swept by revolutions,

...during this present year 1848 ...aspects of political change [are events that] ...timid men view with dread, and the ardent with hope.

And he explained that this remark could be described as a statement about *classes*:

...by the words timid, ardent we mark out of the class... those who possess the particular attributes...[10]

Finally, he offered a notation to describe compactly claims about membership in classes:

...if x represent "Men", y "Rational beings" and z "Animals", the equation

$$x = yz$$

will express the proposition

"Men and rational animals are identical"

So claimed Mr. Boole, the most rational of animals! And he then stated a claim that suggests his high hopes for this system of logic:

It is possible to express by the above notation any categorical proposition in the form of an equation. For to do this it is only necessary to represent the terms of the proposition ...by symbols ...and then to connect the expressions ...by the sign of equality.[11]

Given Boole's ambitions, it is not surprising that the sort of table that now is used to show the relation between inputs and outputs – a table that you and I may use to describe, say, a humble *AND* gate – is called by the highfalutin name, "truth table." No engineer would call this little operation list by such a name, and it was not Boole, apparently, who named it so but the philosopher Ludwig Wittgenstein, 70-odd years later.[12] Since Boole's goal was to systematize the analysis of *thinking*, the name seems quite appropriate, odd though it may sound to an engineer.

---

[10] "The Nature of Logic" (1848), in I. Grattan-Guinness and G. Bornet, eds., *George Boole: Selected Manuscripts on Logic and its Philosophy*, Birkhäuser, (1997), p. 5.
[11] "On the Foundations of the Mathematical Theory of Logic ..." (1856), ibid., p. 89.
[12] Wittgenstein's sole book published in his lifetime is the apparent source of the name. If "truth table" sounds a bit intimidating, how about the title to Wittgenstein's book : *Tractatus Logico-Philosophicus* (1921)!.

Before we return to our modest little electronics networks, which we will describe in Boole's notation, let's take a last look at the sort of sentence that Boole liked to take on. This example should make you grateful that we're only turning LEDs ON or OFF, in response to a toggle switch or two. He proposes to render in mathematical symbols a passage from Cicero, treating conditional propositions.

Boole presents an equation, $y(1-x) + x(1-y) + (1-x)(1-y) = 1$, and explains,

> The interpretation of which is: – *Either Fabius was born at the rising of the dogstar, and will not perish in the sea; or he was not born at the rising of the dogstar, and will perish in the sea; or he was not born at the rising of the dogstar, and will not perish in the sea.*

One begins to appreciate the compactness of the notation, on reading the "interpretation" stated in words. (You'll recognize, if you care to parse the equation, that Boole writes $(1-x)$ where we would write $\bar{x}$ or x^* (that is, x false).[13].

So, if we call one of Boole's propositions B (re Fabius' birth), the other P (re his perishing), in our contemporary notation we would write the function, f as

$$f = (B \cdot \bar{P} + \bar{B} \cdot P) + (\bar{B} \cdot \bar{P}) \ .$$

This rather indigestible form can be expressed more compactly as a function that is false only under the condition that B and P are true: $\bar{f} = B \cdot P$.

The truth table would look like:

B	P	f
0	0	1
0	1	1
1	0	1
1	1	0

And probably you recognize this function as NAND.

AoE §10.1.4E

Comforting truth #1: To build *any* digital device (including the most complex computer) we need only the three logic functions in Fig. 14N.12. All logic circuits are combinations of these three functions, and only these.

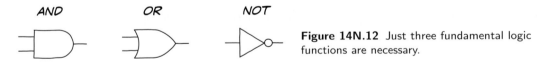

Figure 14N.12 Just three fundamental logic functions are necessary.

Comforting and remarkable truth #2: Perhaps more surprising, it turns out that just *one* gate type (not *three*) will suffice to let us build any digital device. The gate type must be NAND or NOR; these two are called "universal gates."

Figure 14N.13 Universal gates: NAND and NOR.

DeMorgan – a penpal of Boole – showed that what looks like an AND function can be transformed into OR (and vice versa) with the help of some inverters. This is the powerful trick that allows one gate type to build the world.

---
[13] S. Hawking, ed., *God Created the Integers*. Running Press (2005), p. 808.

14N.3.2 deMorgan's theorem

This is the only important rule of Boolean algebra you are likely to have to memorize (the others that we use are pretty obvious: propositions like $A + A^* = 1$).

Theorem (deMorgan's theorem) You can swap *shapes* if at the same time you invert all inputs and outputs. For the graphical form see Fig. 14N.14.

Figure 14N.14 deMorgan's theorem in graphical form.

When you do this you are changing only the *symbol* used to draw the gate; you are not changing the *logic* – the hardware. (That last little observation is easy to say and *hard* to get used to. Don't be embarrassed if it takes you some time to get this idea straight.)

So any gate that can *invert* can carry out this transformation for you. Therefore some people actually used to design and build with NANDs alone back when design was done with discrete gates. These gates came a few to a package, and that condition made the game of minimizing package-count worthwhile; it was nice to find that any leftover gates were "universal." Nowadays, when designs usually are done on large arrays of gates, this concern has faded away. But deMorgan's insight remains important as we think about functions and draw them in the presence of the predominant *active-low* signals, signals of a type described in §14N.3.3 below.

This notion of DeMorgan's, and the "active-low" and "assertion-level" notions will drive you crazy for a while if you have not seen them before. You will be rewarded in the end (toward the end of this course, in fact), when you meet a lot of signals that are "active-low." Such signals spend most of their lives close to 5V (or whatever voltage defines a logic high), and go low (close to 0V) only when they want to say "listen to me!"

14N.3.3 Active-high versus active-low

Signals come in both flavors. Figure 14N.15 shows two forms of a signal that says "Ready".

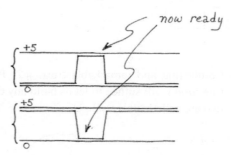

Figure 14N.15 Active-high versus active-low signals.

A signal like the one shown in the top trace of Fig. 14N.15 would be called "Ready;" one like the lower trace would be called "$\overline{\text{Ready}}$," the "bar" indicating that it is active low: "true-when-low." Signals are made active low not in order to annoy you, but for good hardware reasons. We will look at those reasons when we have seen how gates are made. But let's now try manipulating gates, watching the effect of our *assumption about which level is true*.

AoE §10.1.7

Effect on logic of active level: active-high versus active-low: When in §14N.2.1 we considered what the bit pattern 1001 *means* when treated as a *number*, we met the curious fact – perhaps pleasing – that we can establish any convenient convention to define what the bit pattern means. Sometimes we want to call it 9_{10} ([positive] nine); sometimes we will want to call it -7_{10} (*negative* 7). It's up to us.

The same truth appears as we ask ourselves what a particular *gate* is doing in a circuit. The gate's operation is fixed in the hardware (the little thing doesn't know or care how we're using it); but what its operation means is for us to interpret.

That sounds vague, and perhaps confusing; let's look at an example (just deMorgan revisited, you will recognize). Most of the time – at least when we first meet gates – we assume that high is true. So, for example, when we describe what an AND gate does, we usually say something like

"The output is true if both inputs are true."

The usual AND truth table of course says the same thing, symbolically.

A	B	A·B		A	B	A·B
0	0	0		F	F	F
0	1	0	or:	F	T	F
1	0	0		T	F	F
1	1	1		T	T	T

AND gate using *active-high* signals – and truth
table in more abstract form, levels not indicated

The right-hand table – showing Truth and Falsehood – is the more general. The left table is written with 1s and 0s which tend to suggest *high* voltages and *low*. So far, so familiar.[14]

But – as deMorgan promised – if we declare that 0s interest us rather than 1s, at both input and output, the tables look different, and the gate whose behavior the table describes evidently does something different.

A	B	A·B		A	B	A·B
0	0	0		T	T	T
0	1	0	or:	T	F	T
1	0	0		F	T	T
1	1	1		F	F	F

Example of the effect of *active-low* signals: AND
gate (so-called!) doing the job of OR'ing lows

We get a 0 out if *A* **or** *B* is zero. In other words we have an *OR* function if we are willing to stand signals on their heads. We have then a gate that *OR's lows*; Fig. 14N.16 shows its behavior. This is the correct way to draw an AND gate when it is doing this job, handling active-low signals.

Figure 14N.16 AND gate drawn to show that it is OR'ing lows.

---

[14] AoE declares its intention to distinguish between 1 and high in logic representations (10.1.2A). We are not so pure. We will usually write 1 to mean high, because the 1 is compact and matches its meaning in the representation of binary numbers.

It turns out that often we need to work with signals "stood on their heads:" active-low. Note, however, that we call this piece of hardware an *AND gate* regardless of what logic it is performing in a particular circuit. We call it *AND* even if we draw it as in Fig. 14N.16. To call one piece of hardware by two names would be too hard on everyone.

We will try to keep things straight by calling this an AND *gate*, but saying that it performs an OR *function* (*OR*'ing lows, here). Sometimes it's clearest just to refer to the gate by its part number: "It's an '08." We all should agree on that point, at least!

14N.3.4 Missile-launch logic

Here's an example – a trifle melodramatic – of signals that are active-low: you are to finish the design of the circuit in Fig. 14N.17 that requires two people to go crazy at once in order to bring on the third world war:

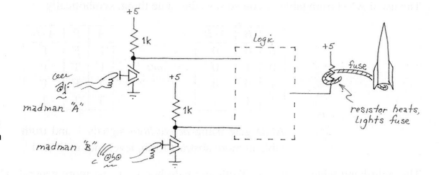

Figure 14N.17 Logic that lights fuse if both operators push "Fire" at the same time.

How should you *draw* the gate that does the job?[15] What is its conventional *name*?[16]

Just now, these ideas probably seem an unnecessary complication. By the end of the course – to reiterate a point we made earlier – when you meet the microcomputer circuit in which just about *every* control signal is active-low, you will be grateful for the notions "active-low" and "assertion-level symbol."

Why are we doing this to you?! Why active-lows?

In Fig. 14N.18 is a network of gates that you will implement if you build the microcomputer from many ICs in Lab 20L (this is logic that you will "wire" in a programmable array of gates, a *PAL*).[17]

Did we rig these signals as active-low just to give you a challenge, to exercise your brain (as, admittedly, we did in the missile-launch problem of Fig. 14N.17)?

No. Our gates look funny (assuming that you share most people's sense that all those bubbles look like a strange complication) because we were *forced* to deal with active-low signals. It was not our choice.

You can see this in the names of the signals going in and out of the nest of gates shown in

---

[15] You need a gate that responds to a coincidence (AND) of two *lows*, putting out, for that combination, a *low* output. This is a gate that ANDs *lows*, and you should draw it that way: AND shape, with bubbles on both inputs and on the output.
[16] The name of this gate is OR even though in this setting it is performing an AND function upon active-low signals.
[17] PAL is an acronym for "Programmable Array Logic," a rearrangement of the earlier name, "Programmable Logic Array." Whereas the acronym "PLA" sounds like the action of spitting out something distasteful, "PAL" sounds rather obviously like a gadget that is user-friendly. We should admit, however, that a PAL is not just a renamed PLA. The name PAL describes a subset of PLAs, a particular configuration: an OR-of-AND terms. You'll find more on this topic in Appendix A.

14N.3 Combinational logic

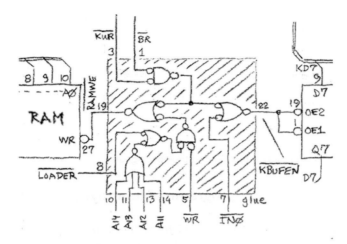

Figure 14N.18 Strange looking nest of active-low signals – imposed on us.

Fig. 14N.18: *every* signal is active-low, except the "A" and "D" lines, which carry address and data (and therefore have no "inactive" state: low and high, in their case, are of equal dignity, equally "true"). The RAM *write* signal is "WE*;" the enables of the chip on the right both show bubbles that indicate that they, too, are active-low.

Figure 14N.19 is a corner of the processor that is the brains of the computer: its output control signals, WR*, RD* and PSEN*[18] also are active-low. So we need to get used to handling active-low signals.

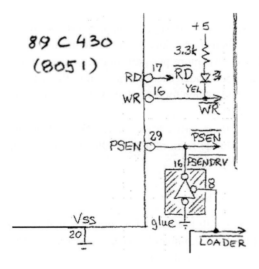

Figure 14N.19 Control signals from the processor also are active-low.

We will be able to explain in a few minutes *why* control signals typically are active *low*. First, however, let's give a glance at "programmable logic," the method normally used to do digital logic. This method has displaced the use of "standard logic" in small packages of gates like the ones described in §14N.5.1, below. We will discover the happy fact that a logic compiler can make the pesky problem of "active low" signals very easy to handle.

---

[18] In case you're impatient to know what these signal names mean, PSEN is "Program Store Enable," a special form of *read* especially for reading code rather than ordinary data. You'll get to know these signals in the microprocessor (micro) labs.

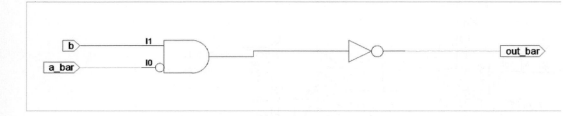

Figure 14N.20 Schematic of simple logic mixing active-high and active-low signals.

14N.4 The usual way to do digital logic: programmable arrays

Small packages of gates – like the four NANDs in a 74HC00 gate, which you'll meet in the Lab 14L – do not provide an efficient way to do digital logic. It is better to build digital designs by programming a circuit that integrates a large number of gates on one chip. In this course we will use small versions of such an IC, integrating about 1600 gates on an IC (that's small by present standards). After this course you may meet some of the larger parts.

The advantage of these ICs are two-fold: first, they are more flexible than hard-wired designs. You can change the logic after wiring the circuit if you like. (You can reprogram the part even while it is soldered onto a circuit board.) Second, they are much cheaper than hard-wired designs simply because of their scale (a few dollars for these 1600 gates).

First, a hasty sketch of the sorts of parts that are available:

- Gate arrays. These come in two categories...
  - Application-specific ICs (ASICs). These are expensive to design (perhaps $100,000 to implement a design), and their use pays only if one plans to produce a very large number.
  - Field-programmable gate arrays (FPGAs). These cost more per part, but present no initial design cost beyond the time needed to generate good code. These devices store the patterns that link their gates either in volatile memory (RAM, which forgets when power is removed), or in non-volatile memory (ROM, usually "flash" type).
- Programmable logic devices (PLDs – often called by their earlier trade-name, PAL). These are simpler than FPGAs, and each of its many outputs is in the form of a wide OR of several wide AND gates. We'll be using these, soon.

14N.4.1 Active-low with PLDs: a logic compiler can help

One programs a PLD (or FPGA) – as you soon will do – with the help of a computer that runs a "logic compiler," a program that converts your human-readable commands into connection patterns. The logic compiler can diminish the nastiness of "active-low" signals. In this course you will use a logic compiler – or "Hardware Description Language" (HDL) – named Verilog. Verilog competes with another HDL named VHDL.[19]

The technique that can make active-lows not so annoying is to create an active-high equivalent for each active-low signal: we define a signal that is the complement of each active-low signal.[20] Then

---

[19] VHDL, a bit hard to pronounce, is better than what it stands for: "Very [High Speed Integrated Circuit] Hardware Description Language." It was developed at the request of the US Department of Defense, whereas Verilog arose in private industry.

[20] This newly defined signal exists only "on paper" for the convenience of the human programmer; no extra hardware implementation is implied.

14N.4 The usual way to do digital logic: programmable arrays

one can write logic equations in a pure active-high world. One can forget about the active levels of particular signals. Doing this not only eases the writing of equations; it also produces code that is more intelligible to a reader.

Figure 14N.20 is a simple example: a gate that is for ANDing two signals, one active-high, the other active-low. The gate's output is active-low. It's easy enough to draw this logic.[21]

In Verilog, the equation for this logic can be written in either of two ways.

An ugly version of this logic: It is simpler, though uglier, to take each signal as it comes – active-high or active-low. But the result is a funny looking equation. Here is such a Verilog file:

```
module actlow_ugly_oct11(
    input a_bar,
    input b,
    output out_bar
    );

// see how ugly things look when we use the mixture of active-high and
// active-low signals:
    assign out_bar = !(!a_bar & b);
endmodule
```

The file begins with a listing of all signals, input and output. We have appended "_bar" to the names of the signals that are active-low simply to help us humans remember which ones are active-low. The symbol "!" (named "bang") indicates logic inversion; "&" means "AND."

The equation to implement the AND function is studded with "bangs," and these make it hard to interpret. None of these bangs indicates that a signal is false or disasserted; all reflect no more than the active level of signals.

A prettier version of this logic: If we are willing to go to the trouble of setting up an active-high equivalent for each active-low signal (admittedly, a chore), our reward will be an equation that is easy to interpret.

Here, we define a pair of active-high equivalents after listing the actual signals as in the earlier file; we do this for the one active-low *input* and for the active-low *output*. (The odd line "wire..." tells the compiler what sort of signal our homemade "out" is):

```
module actlow_pretty_oct11(
    input a_bar,
    input b,
    output out_bar
    );
 wire out;

// Now flip all active-low signals, so we can treat ALL signals as active-high as
// we write equations
  assign a = !a_bar;   // makes a temporary internal signal (a) never
                       // assigned to a pin
```

---

[21] We allowed the logic compiler to draw this for us; a human would not have drawn that final inverter; a human would have placed a bubble at the output of the AND shape.

```
assign out_bar = !out; // this output logic looks reversed. But it's correct
                      // because the active-high  "out" will be an input to
                      // this logic, used to generate the OUTPUT

// then see how pretty the equation looks using the all-active-high
// signal names

assign out = a & b;
endmodule
```

The equation says what we understand it to mean – active levels apart: "make something happen if both of two input conditions are fulfilled."

In short, the logic compiler allows us to keep separate two issues that are well kept separate:

(1) what are the active levels – high or low? This is disposed of near the start of the file (as in the "pretty" version, above);
(2) what logical operations do we want to perform on the inputs? (This is the interesting part of the task).

If this Verilog code is hard to digest on first reading, don't worry. We will look more closely at Verilog and its handling of active-low signals next time. This all takes getting used to.

14N.5   Gate types: TTL and CMOS

14N.5.1   Gate innards: TTL versus CMOS

A glance at Fig. 14N.21 should reveal some characteristics of the gates.

AoE §12.1.1

AoE §10.2.2

AoE §10.2.3

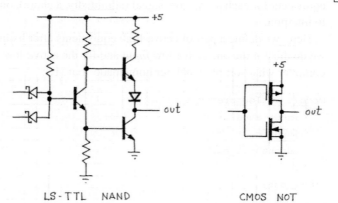

Figure 14N.21 TTL and CMOS gates: NAND, NOT.

Inputs: You can see why TTL inputs *float high*, and CMOS do not.
Threshold: You might guess that TTL's threshold is off-center – low, whereas CMOS is approximately centered.
Output: You can see why TTL's high is *not* a clean 5V, but CMOS' is.

Power consumption: You can see that CMOS passes *no* current from +5 to ground, when the output sits either high or low; you can see that TTL, in contrast, cannot sit in either output state without passing current in (a) its input base pullup (if an input is pulled low) or (b) in its first transistor (which is ON if the inputs are high).

14N.5.2 Thresholds and noise margin

All digital devices show some noise immunity. The guaranteed levels for TTL and for 5V CMOS show that CMOS has the better noise immunity: see Fig. 14N.22.

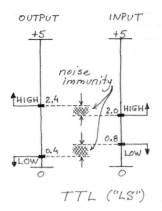
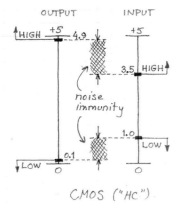

Figure 14N.22 Thresholds and noise margin: TTL versus CMOS.

Curious footnote: TTL and NMOS devices are so widely used that some families of CMOS, labeled *74xCTxx*, have been taught TTL's bad habits on purpose: their thresholds are put at TTL's nasty levels ("CT" means "**C**MOS with **T**TL thresholds"). We will use a lot of such gates (74HCTxx) in our lab microcomputer, where we are obliged to accommodate the microprocessor whose output *high* is as wishy-washy as TTL's (even though the processor is fabricated in CMOS). When we have a choice, however, we will stick to straight CMOS, though the world has voted rather for HCT. More functions are available in HCT than in straight HC.

Answer to "Why is the typical control signal *active low*?" We promised that a look inside the gate package would settle this question, and it does. TTL's asymmetry explains this preference for active low. If you have several control lines, each of which is inactive most of the time, it's better to let the inactive signals *rest high*, let only the active signal be *asserted low*. This explanation applies only to TTL, not to CMOS. But the conventions were established while TTL was supreme, and they persist long after their rationale ceases to apply.

Here's the argument: two characteristics of TTL push in favor of making signals active-low.

A TTL *high* input is less vulnerable to noise than a TTL *low* input Although the *guaranteed* noise margins differ by a few tenths of a volt, the *typical* margins differ by more. So, it's safer to leave your control lines high most of the time; now and then let them dive into danger.

A TTL input is easy to drive high In fact, since a TTL input *floats high*, you can drive it essentially for free: at a cost of no current at all. So, if you're designing a microprocessor to drive TTL devices, make it easy on your chip by letting most the lines rest at the lazy, low-current level, most of the time.

Both these arguments push in the same direction; hence the result – which you will be able to confirm when you put together your microcomputer (big-board version), where *every* control line is active-low.[22]

14N.6 Noise immunity

All logic gates are able to ignore noise on their inputs, to some degree. The test setup below shows that some logic families do better than others in this respect. We'll look first at the simplest sort of noise rejection, *DC noise immunity*; we'll find that CMOS, with its nearly-symmetric high and low definitions does better than TTL (the older, bipolar family), with its off-center thresholds. Then we'll see another strategy for resisting noise, *differential* transmission.

14N.6.1 DC noise immunity: CMOS versus TTL

Figure 14N.23 shows the test setup we used to mix some noise into a logic-level input. We fed this noisy signal to four sorts of logic gate, one TTL, three CMOS.

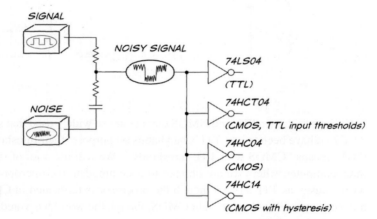

Figure 14N.23 DC noise immunity test setup: noise added to signal, fed to 4 gate types.

Progressively increase noise level ... : First, moderate noise: all gate types succeed, in the left-hand image of Fig. 14N.24, all gates ignore the triangular noise. These gates, in other words, are showing off the essential strength of digital devices.

In the right-hand image of Fig. 14N.24, the TTL and *HCT* parts fail. They fail when the noise is added to a *low* input. HCT fails when TTL fails and this makes sense since its thresholds have been adjusted to match those of TTL.[23] CMOS did better than TTL because of its larger noise margin.

---

[22] The phrase "control line" may puzzle you. Yes, we are saying a little less than that every *signal* is active low. Data and address lines are not. But every line that *has* an active state, to be distinguished from inactive, makes that active state *low*. Some lines have no *active* versus *inactive* states: a data line, for example, is as active when low as when high; same for an address line. So, for those lines designers leave the active-high convention alone. That's lucky for us: we are allowed to read a value like 1001 on the data lines as 9; we don't need to flip every bit to interpret it. So, instead of letting the *active low* convention get you down, count your blessings: it could be worse.

[23] You may be wondering why anyone would offer HCT whose noise-immunity is inferior to garden-variety HC CMOS. HCT exists to allow upgrading existing TTL designs by simply popping in HCT, and also to form a logic-family bridge between TTL-level circuitry and CMOS. We use HCT that way in one version of the micro labs that conclude this course, §§20L.1 and 21L.1: to accept signals from the handful of parts that deliver TTL output levels rather than CMOS. See, for example, the HCT139, a 2:4 decoder that receives TTL-level signals from the microcontroller (Lab 21L and Fig. 20L.1 showing the full micro circuit).

14N.6 Noise immunity

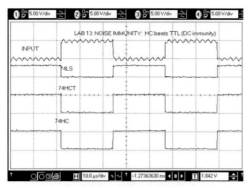
moderate noise: all digital gates ignore it

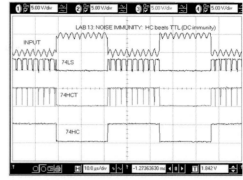
more severe noise: TTL and TTL-imitator fail

Figure 14N.24 Moderate noise: all gate types ignore the noise; more noise fools TTL gates. (Scope settings: 5V/div.)

Much noise: all but Schmitt trigger fail: When we increase the noise level further, even CMOS fails. But we included one more gate type, in the test shown in Fig. 14N.25, a gate that does even better than the CMOS inverter. That one gate not fooled by the large noise amplitude, shown on the bottom trace, is one with built-in hysteresis (a 74HC14).

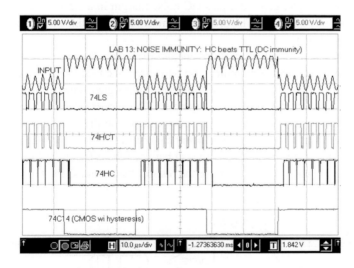

Figure 14N.25 Much noise: all gate types fail except a gate with hysteresis (HC14). (Scope settings: 5V/div.)

Seeing this gate succeed where the others fail might lead you to expect that all logic gates would evolve to include hysteresis. They don't though, because hysteresis slightly slows the switching and the concern for speed trumps the need for best noise immunity. This rule holds except in gates designed for especially noisy settings. Gates designed to receive ("buffer") inputs from long bus lines, for example, often do include hysteresis.

Below, in §14N.6.2, we'll meet a gate type that takes quite a different approach – and, in fact, is proud of its small swing. Fig. 14N.26 shows a chart in which this gate type, called LVDS (Low Voltage Differential Signaling) boasts of its tiny signal swing:[24]

The small swings of LVDS gates afford two side benefits: low EMI (emissions: "Electro Magnetic

---

[24] National Semiconductor: http://www.national.com/appinfo/lvds/files/ownersmanual.pdf, §5=-4.

532 Logic Gates

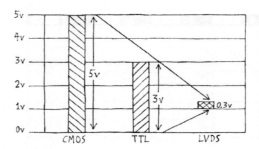

Figure 14N.26 Small signal swing can be a virtue, but requires differential signaling.

Interference") and reduced disturbance of the power supply. We call these "side" benefits because the fundamental great strength of the low swings is that the logic can give good noise immunity at low supply voltages, as we'll argue in §14N.6.2.[25]

14N.6.2 Differential transmission: another way to get good noise immunity

Recent logic devices have been designed for ever-lower power supplies – 3.3V, 2.5V and 1.8V – rather than for the traditional +5V. The trend is likely to continue. Such supplies make it difficult to protect gates from errors caused by noise riding a logic level. The 0.4V DC-noise margin available in TTL and HCMOS is possible only because the supply voltage is relatively large. Compressing all specifications proportionately would, for example, give 50% less DC noise margin in a 2.5V system – and things would get worse at lower supply voltages. The problem is most severe on lines that are long, like those running on a computer backplane.

A solution to the problem has been provided by *differential* drivers and receivers. These send and receive not a single signal, but a signal and its logical complement, on two wires; see Fig. 14N.27 (we first met this method in §5N.7, Fig. 5N.27). Typically, noise will affect the two signals similarly, so subtracting one from the other will strip away most of the impinging noise. This is a process you saw in §5L.1, the analog differential amplifier lab; here we find nothing new, except that the technique is applied to digital signals. The differential drivers called LVDS transmit *currents* rather than voltages; these currents are converted to a voltage difference by a resistor placed between the differential lines at the far end of the signal lines, the receiving end. (This is nice, for the resistor "terminates" the lines, forestalling "reflections." See Appendix C.)

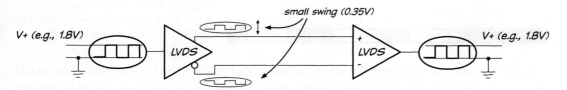

Figure 14N.27 Differential signals can give good noise immunity despite low power supply voltages.

In Fig. 14N.28 we put a standard 0/5V logic (TTL) signal into a differential driver and injected about a volt of noise (a triangular waveform).[26]

---

[25] Fig. 14N.26 is derived from a figure in the National Semiconductor LVDS guide: http://www.national.com/appinfo/lvds/files/ownersmanual.pdf.

[26] We did this in a rather odd way: by driving the *ground* terminal on the transmitter IC with this 1V (peak-to-peak) triangular waveform.

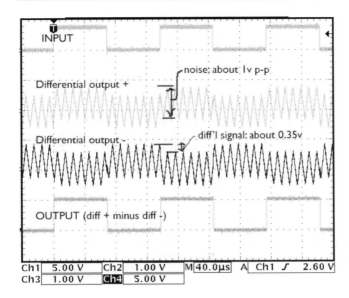

Figure 14N.28 LVDS signals: differential signals can survive noise greater than the signal swing.

The two middle traces in Fig. 14N.28 show the output of the driver IC: a differential pair of signals, one in-phase with the TTL input, one 180° out of phase. The driver converts the *voltage* TTL input to *currents* flowing as diff$_+$ and diff$_-$; at the receiver, a terminating resistor (100Ω) converts the currents back into voltages.[27] The differential swing is small: about 0.35V – dwarfed by the triangular noise in Fig. 14N.28. But since the differential receiver looks at the *difference* between diff$_+$ and diff$_-$, it reconstructs the original TTL cleanly, rejecting the noise.

This success, with the small differential swing, illustrates how differential signaling can provide good noise immunity to logic that use extremely low supply voltages. The LVDS specification requires a voltage swing of only 100mV at the receiver.

These differential gates show two other strengths: (1) they are fast (propagation delays of transmitter and receiver each under 3ns); and (2) they emit less radiated noise than a traditional voltage gate (as noted above). They do this first because of the small voltage swings; and further, because of the symmetric *current* signals that travel in the signal lines (currents of opposite signs, flowing side-by-side), signals whose magnetic fields tend to cancel.[28]

14N.7 More on gate types

14N.7.1 Output configurations

Active pullup: All respectable gates use *active pullup* on their outputs, to provide firm highs as well as lows. You will confirm in the lab that the *passive-pullup* version (labeled NMOS in Fig. 14N.29) not only wastes power but also is *slow*. Why slow?[29]

AoE §10.2.4C

Open-collector/open-drain: Once in a great while "open drain" or "open collector" is useful.[30] You have seen this on the '311 comparator. It permits an easy way to OR multiple signals: if signals A, B

---

[27] The terminating resistor serves another good purpose: matching the "characteristic impedance" of the transmitting lines, it prevents ugly spikes caused by "reflection" of a waveform that might otherwise occur as it hit the high-impedance input of the receiving gate. Again we refer you to Appendix C for more on this topic.
[28] See National Semiconductor LVDS guide: http://www.national.com/appinfo/lvds/files/ownersmanual.pdf.
[29] Slow because the inevitable stray capacitance must be driven by a mere pullup resistor, rather than by a transistor switch.
[30] "Open-collector" for bipolar gates; "open-drain" for MOSFET gates.

534 Logic Gates

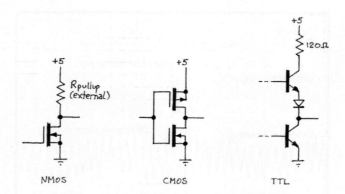

Figure 14N.29 Passive versus active pullup output stages.

and C use active-low open-drain outputs, for example, they can share a single pullup resistor. Then any one of A, B or C can pull that shared line low. This arrangement often is called "wired OR." We will meet this possible application when we treat *interrupts* in Chapter 22N.

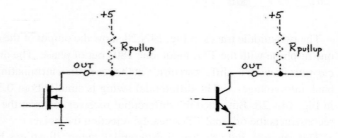

Figure 14N.30 Open-drain or open-collector: rarely useful.

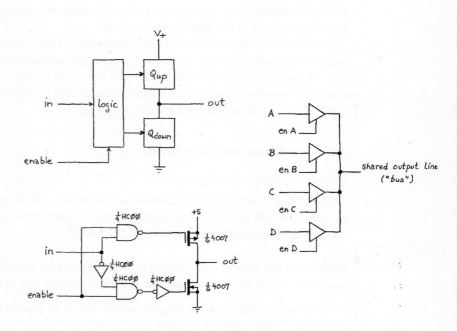

Figure 14N.31 Three-state output: conceptual; the way we build it in the lab; driving shared bus.

AoE §10.2.4A

Three-state: *Very* often *three-state* outputs are useful:[31] these allow multiple drivers to share a common output line (then called a "bus"). These are widely used in computers.

Beware the misconception that the "third state" is a third output voltage level. It's not that; it is the *off* or disconnected condition. Fig. 14N.31 shows circuitry to implement this output stage, first conceptually, then the way we'll build it in Lab 14L.

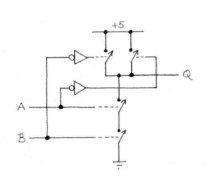 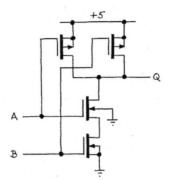

Figure 14N.32 NAND gate built with CMOS.

14N.7.2 Logic with TTL and CMOS

The basic TTL gate that we looked at a few pages back in §14N.5.1 was a NAND; it did its logic with diodes. CMOS gates do their logic differently: by putting two or more transistors in series or parallel as needed. Fig. 14N.32 shows the CMOS NAND gate you'll build in the lab, along with a simplified sketch, showing it to be just such a set of series and parallel transistor switches. The logic is simple enough for you not to need a truth table. You can see that the output will be pulled low only if both inputs are high – turning on both of the series transistors to ground. Thus it implements the NAND function.

14N.7.3 Speed versus power consumption

AoE Fig. 10.26

The plot in Fig. 14N.33 shows tradeoffs available between speed and power-saving for some "standard logic" families. As you can see from this figure, everyone is trying to snuggle down into the lower left corner, where you get fast results for almost nothing. Low voltage differential signalling and faster CMOS (TI's "LVC," and the still faster AUC, for example) seem the most promising, at the present date.

And arrays of gates – PALs and FPGAs – can show speeds better than those indicated in Fig. 14N.33 because stray capacitances can be kept smaller on-chip than off.

14N.8 AoE Reading

Chapter 10 (Digital Logic I):
- §10.1: Basic Logic Concepts)
- …§10.2: Digital ICs: CMOS and bipolar (TTL)

---

[31] You will often hear these called "Tri-State." That is a trademark belonging to National Semiconductor (now absorbed by Texas Instruments), so "three-state" is the correct generic term.

Logic Gates

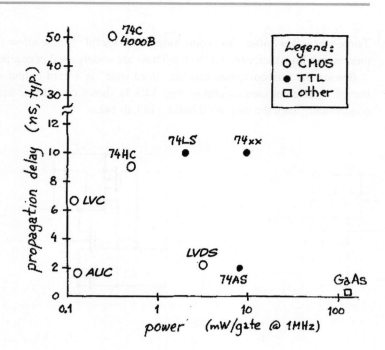

Figure 14N.33 Speed versus power consumption: some present and obsolete logic families.

- for big picture of logic family competition see Fig. 10.22
- §10.3
  - §10.3.1 logic identities are useful; deMorgan's theorem is important
  - don't worry about Karnaugh mapping: §10.3.2
  - ...and don't study the long catalog of available combinational functions: §10.3.3
  - take a quick look at §10.6: some typical digital circuits

Chapter 11 (Digital Logic II):
- §11.1: history of logic families
- §11.2.1: PALs
- ...in §11.2, omit §11.3, the complex example, byte generator
- ...except take a quick look at the contrasted *schematic* versus *HDL* design entry methods (§§11.3.3A and 11.3.4)
- postpone §11.3.5: microcontroller
- take a look at summary advice in §§11.4.1 and 11.4.2

14L Lab: Logic Gates

The first part of this lab invites you to try integrated gates, black boxes that work quite well, to carry out some Boolean logic operations.

The later sections ask you to look within the black box, in effect, by putting together a logic gate from transistors. The point here is to appreciate why the IC gates are designed as they are, and to notice some of the properties of the input and output stages of CMOS gates. We will concentrate in this lab, as we will throughout the course, on CMOS. To overstate the point slightly, we might say that we will treat ordinary TTL as a venerable antique.

But all of the work we do in this lab is rather antique – because logic now is seldom done with little packages of a few gates. Normally, logic networks are built from large arrays of gates that are programmed to carry out a particular function. Soon you will be taking advantage of a modest version of such devices (1600 gates in a package). But today we'll use just one or a few gates at a time because that's a good way to start getting a grip on what a gate does.

14L.1 Preliminary

Some *ground rules* in using logic:

1. Never apply a signal beyond the power supplies of any chip. For the logic gates that we use, that means...
2. ...keep signals between 0 and +5V. (This rule, in its general form – "stay between the supplies" – applies to analog circuits as well; what may be new to you is the nearly-universal use of single supply in digital circuits.)
3. Power all your circuits from +5V. and ground only – until we reach the CPLDs[1] or PALs,[2] some of which are powered with 3.3V instead. This applies equally to CMOS (in its traditional 5V form[3]) and to TTL. Power and ground pins on ordinary digital parts (not complex ICs like PALs and microcontrollers) are the diagonally-opposite corner pins, as in Fig. 14L.1.

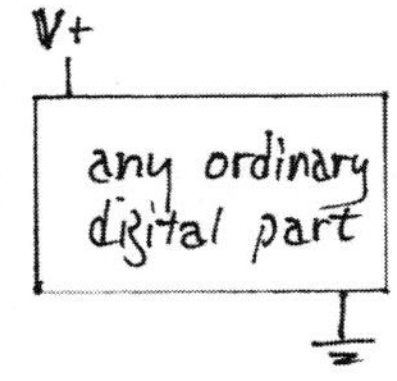

14L.1.1 Logic probe

Figure 14L.1 Most ordinary digital parts take power and ground at their corner pins.

The logic probe is a gizmo about the size of a thin hot dog, with a cord on one end and a sharp point on the other. It tells you what logic level it sees at its point; in return, it wants to be given power (+5V and ground) at the end of its cord (n.b., the logic probe does *not* feed a signal to the oscilloscope!).

If you find a BNC connector on the probe cord, connect +5V to the center conductor, using one of the breadboard jacks. If, instead, you find that the cord ends with alligator clips or a pair of strange-looking grabbers, use these to take hold of ground and +5V.

---

[1] Complex Programmable Logic Devices.
[2] Programmable Array Logic, as you may recall.
[3] Lower supply voltages are widely displacing 5V. At the time of writing, 3.3V, 2.5V and 1.8V are common supplies. For the most part we will use 5V logic.

How to use the probe: Most logic probes use different colored LEDs to distinguish high from low – and to distinguish both from "float" (simply "not driven at all; not connected"). This ability of the probe is extremely useful. (Could a voltmeter or oscilloscope make this distinction for you?[4])

Use the probe to look at the output of the breadboard function generator when it is set to *TTL*. Crank the frequency up to a few kilohertz. Does the probe blink at the frequency of the signal it is watching? Why not?[5]

14L.1.2 LED indicators

The eight LEDs on the breadboard are buffered by logic gates. You can turn on an LED with a logic high, and the gate presents a conveniently high input impedance (100k to ground).

To appreciate what the logic probe did for you earlier, try looking at a fast square wave, using one of these LEDs rather than the logic probe: use the breadboard oscillator (TTL) at a kilohertz or so. Does what you see make sense? You may now recognize that the logic probe stretches short pulses to make them visible to our sluggish eyes: it turns even a 30ns pulse into a flash of about one-tenth of a second (the faster probes can do this trick with even narrower pulses).

14L.1.3 Switches

Switches available on the powered breadboards: The PB503 includes three sorts of switch on its front panel:

- two debounced pushbuttons (at the lower left corner of the breadboard, marked PB1, PB2) These deliver an *open collector* output, and that means that they are capable of pulling to ground *only*, never to the positive supply. To let such an output go to a logic *high*, you will need to add a *pullup* resistor, to 5V: see Fig. 14L.2.

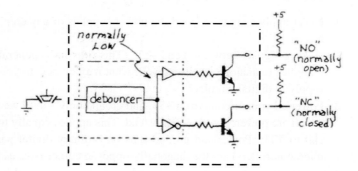

Figure 14L.2 Pullup required on open-collector output (debounced pushbuttons on PB503).

  Note that what looks like a discrete transistor in the figure is included on the breadboard; you need add *only the pullup resistor*.
- An 8-position DIP switch (not debounced), fed from a +5V/0V slide switch (marked DIP switch in Fig. 14L.3) This looks as if it could give you 8 independent outputs, but it cannot; at least, not conveniently. The DIP switch on most breadboards is simply a set of eight in-line switches that deliver the level set by the slide switch, if closed – and *nothing* (a *float*: neither high nor low) if the DIP switch is open.

---

[4] No! The special virtue of the logic probe is the ability to distinguish a logic low from the plain old "zero volts" that a DVM might show you. The DVM cannot distinguish a *float* – no connection to anything – from a good solid connection to ground, which the logic probe understands is what we mean by low .

[5] Why not? Because your eyes are not quick enough to see a 1kHz blink rate (you probably can't notice a blink rate beyond about 25Hz). The logic probe slows fast switching to the lazy rate of a few Hz, so that we humans can see it.

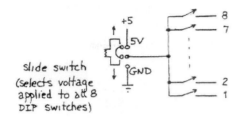

Figure 14L.3 DIP switch in-line with output of *one* common slide switch.

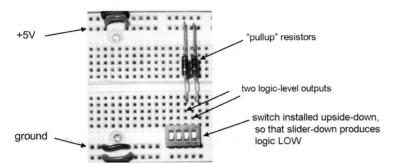

Figure 14L.4 DIP switch installed on breadboard can provide logic levels.

To get eight independent levels you need eight *pullup* resistors, while setting the slide switch always *low*. That's a nuisance, so most of the time you'll probably want to use this 8-position switch to provide just *one* logic level. To get that, *close* the DIP switch; to avoid fooling yourself later it's probably a good idea to leave all the DIP's switches *closed*.

Newer PB503s provide an active *high* from these DIP switches, rather than *open* versus *ground*, and the eight switches are independent. These changes make the switches more useful. But the high voltage that defines the HIGH for each DIP switch is selectable; this provides a nasty danger to digital circuits. A slide switch determines whether the high available to all eight switches is +5V or the innocuous sounding "+V".

If you have such a PB503, *beware*: "+V" is whatever positive voltage happens to be present on the positive adjustable rail (1.3–15V). Any voltage well above 5V will *destroy* an ordinary digital part. We recently discovered this by accident, when smoke alerted us to this new feature provided by Global Specialties.

- two uncommitted slide switches (SPDT) These are on the lower right, and are *bouncy* (not debounced, anyway). To make them useful, tie one end to ground, the other to +5, and use the *common* terminal as output. You might as well wire these now, use them today, and then leave them so wired for use in later labs.

A good way of generating digital inputs: An easy way to provide levels where you want them is to wire a DIP switch so that one side ties directly to the ground bus. The DIP switch should be inserted *upside-down*, so that the slider *down* position (which looks like a low) generates a logic low. The switch shown in Fig. 14L.4 is providing a 00 combination.

If the switch is wired directly to an ICs inputs, the pullups may have to run sideways. Again, the switch in Fig. 14L.5 is shown providing a 00 combination, this time going to the two inputs of a NAND gate, in a 74HC00.

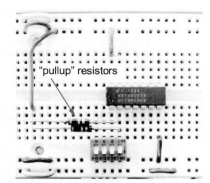

Figure 14L.5 DIP switch can provide logic levels just where you want them.

14L.2 Input and output characteristics of integrated gates: TTL and CMOS

14L.2.1 Output voltage levels

Power and ground: As we suggested back in §14L.1, a 14-pin DIP like the ones you meet in this lab takes power and ground at its corner pins: Northwest and Southeast when part is oriented horizontally: pins 14 (V_{CC}: +5V) and 7 (ground).

Input signals: Use a DIP switch or other source to provide 0 or +5V to the two inputs of a NAND gate, driving both TTL and CMOS, simultaneously. If you want to be really fancy and lazy, show the two inputs on two of the powered-breadboards LEDs, and the output on a third LED. Doing that helps you to see inputs and outputs simultaneously, as if you were looking at one line of a truth table.

74XX00 pinout: In Fig. 14L.6 the TTL part is 74LS00.[6]
The CMOS part is 74HC00.[7]

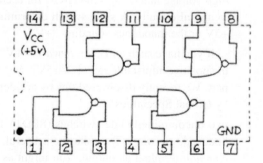

Figure 14L.6 NAND gates: TTL and CMOS.

Note: for the CMOS part (but not TTL), tie all the six unused input lines to a common line, and temporarily ground that line. Take the trouble to tie the unused inputs together, rather than ground each separately, because soon we will want to drive them with a common signal (in §14L.3.1).

Now note both *logic* and *voltage* levels out as you apply the four input combinations. (Only one

---

[6] "LS" stands for low power Schottky, a process that speeds up switching. At the time when this *LS* prefix was chosen (1976), TTL was thought to go without saying; thus there's no *T* in the designation in contrast to CMOS, the late-bloomer, which always announces itself with a *C* somewhere in its prefix: HC, HCT, AC, ACT, etc. See the 74HC00, just below.

[7] The "74" shows that the part follows the part-numbering and pinout scheme established by the dominant logic family, Texas Instruments' 74xx TTL series; "C" indicates CMOS; "H" stands for "high speed": speed equal to that of the then-dominant TTL family, 74LS.

logic-out column is provided below because here TTL and CMOS should agree.) As *load*, use a 10k resistor to ground.

INPUT		OUTPUT Logic Levels	Volts: TTL	Volts: CMOS
0	0			
0	1			
1	0			
1	1			

14L.3 Pathologies

14L.3.1 Floating inputs

TTL: Disconnect both inputs to the NAND, and note the output *logic* level (henceforth we will not worry about output voltages; just logic levels will do). What input does the TTL "think" it sees, therefore, when its input floats?[8]

AoE §10.8.3B

CMOS: Here the story is more complicated, so we will run the experiment in two stages.

1. Floating input: *effective logic level in*: Tie HIGH one input to the NAND, tie the other to 6 inches or so of wire; leave the end of that wire floating, and watch the gate's output with a logic probe or scope as you hold your hand near the floating-input wire, or take hold of it at an insulated section. (Here you are repeating an experiment you did with the power MOSFET in §12L.1.2.) Try touching your other hand to +5V, to ground, to the TTL oscillator output, or hold your free hand near the transformer of the breadboard's internal power supply. We hope that what you see will convince you that floating CMOS inputs are less predictable than floating TTL inputs, although we urge you to leave *no* logic inputs floating.

2. Floating input: *Effect on CMOS power consumption*:
   You may have read that one should not leave unused CMOS inputs floating. Now we would like you to *see* why this rule is sound (though, like most rules, it deserves to be broken now and then[9]).

Tie the two NAND inputs to the other six, earlier grounded; disconnect the whole set from ground, and instead connect it to a potentiometer that can deliver a voltage between 0 and +5V. Rotate the pot to one of its limits, applying a good logic level input to all four of the NAND gates in the package.

Now (with power off) insert a current meter (VOM or DVM) between the +5V supply and the V+ pin (14) on the CMOS chip. Restore power and watch the chip's supply current on the meter's most sensitive scale. The chip should show you that it is using very little current: low power consumption is, of course, one of CMOS' great virtues.

Now switch the current meter to its 150mA scale (or similar range) and gradually turn the pot so that the inputs to all four NAND gates move toward the threshold region where the gate output is not firmly switched high or low. Here, you are frustrating CMOS' neat scheme that assures that one and

---

[8] It thinks it sees a high – though a high that is vulnerable to noise. A look at the innards of the gate shown in Fig. 14N.21, reveals the internal resistive pullup on the base of the first transistor. That transistor therefore is held on when inputs are open, as if driven by a genuine high input.

[9] In particular, we don't want you to feel obliged, when breadboarding circuits in this lab, to drive all unused inputs. That's required in a circuit that you build in permanent form, but your lab *time* is more valuable than the extra *power* that your circuit may consume when inputs are left floating – and your time is more valuable even than the IC that conceivably might be damaged by your failure to drive all its unused inputs.

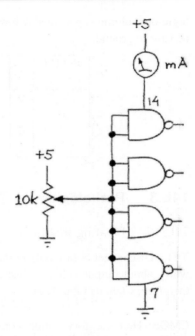

Figure 14L.7 Test setup: applying intermediate *input* level raises power consumption.

only one of the transistors in the output stage is on. When input voltage is close to $V_{DD}/2$, the typical switching threshold, both transistors can be partially on. You can see on the current meter dial the price for this inelegance: power consumption thousands of times higher than normal.

Floating inputs thus are likely to cause a CMOS device to waste considerable power. Manufacturers warn that this power use can also overheat and damage the device. In this course we will sometimes allow CMOS inputs to float while breadboarding, as we have said. But you now know that you should never do this in any circuit that you build to keep.

14L.3.2 Effect of failing to connect power or ground to CMOS logic gate

Now we ask you to do, purposely, what students often do inadvertently,[10] as they breadboard logic circuits: omit power and ground connections. The effect is not what one might expect. Perhaps you would expect an un-powered IC to deliver a constant low output, or a floating output.

But the protection clamps on input and output of each gate complicate the behavior. Fig. 14L.8 shows what the clamps look like for 74HC parts[11].

The clamps allow inputs (and sometimes outputs) to provide current that can power the IC. So, a *high* input can feed the +5V supply line through the upper clamp diodes (though these don't *look* connected); a low can feed the ground line through the lower clamp diode – though both +5V and ground so driven would reach compromised levels. So if you apply both a high and a low to inputs somewhere on the chip, you have powered the chip! As you can imagine, the result is a very strange pattern of misbehavior: everything seems to be working – until an unlucky combination of input levels occurs.

If you let the output of 74HC00 NAND gate drive a 10k load resistor (other end tied to ground) and put that output on one of the breadboard's buffered LEDs (to make it easy for you to see the logic

---

[10] You may be sure that *you* wouldn't make this mistake – but many others do. One term, the error became so common in our class that the teaching assistants began to distribute candy rewards to the students who did not make this error: who applied power and ground first, rather than postpone the chore till after they had wired the more interesting connections.

[11] Fairchild/National App. Note AN248.

14L.4 Applying IC gates to generate particular logic functions

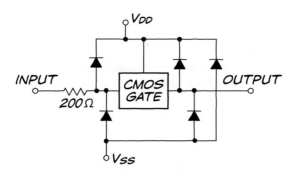

Figure 14L.8 74HC logic gate protection circuitry.

level out) you may see the HC00 NAND generating an XOR function. Why?[12] We should confess we can't promise the XOR result: variations in protection circuitry among manufacturers make the outcome unpredictable.

14L.4 Applying IC gates to generate particular logic functions

Before we look into the gritty details of what lies within a logic IC, let's have some fun with these gates. First we'll do three tasks with NANDs to get used to the remarkable fact that with NANDs you can build *any* logic function. Then we'll invite you to apply any of the standard logic functions, including XOR, to make a digital *comparator* in two forms.

Most people prefer these brain teasers to a session of wiring. Don't let these problems bring your lab work to a dead *stop*! Please don't give any of these problems more than ten minutes of your precious lab time. You can always finish these brain teasers at home.

14L.4.1 NAND applications

BOTH: Use NANDs (CMOS or TTL) to light one of the breadboard's buffered LEDs when both inputs are high.

EITHER: Use NANDs (CMOS or TTL) to light one of the LEDs when either of the inputs is low (here we mean a plain *OR* operation, not exclusive-or, by the way). (Trick question! Don't work too hard.)

14L.4.2 Digital comparators using any gates you like

2-bit equality detector: Use any gates to make a comparator that detects equality between two 2-bit numbers. (This circuit, widened, is used a lot in computers, where a device often needs to watch the public "address bus," to respond upon seeing its own distinctive "address.")

Hint: the XOR function is a big help here. (XOR is 74HC86; pinout is same as for '00: in fact, *all*

---

[12] The argument for XOR is that only this input combination provides both +5 and ground and therefore makes the gate put out the HIGH that a NAND ought to put out in response to a 01 or 10 input. 00 input provides no +5, so the gate cannot put out a high (as the NAND should). More problematic is 11: the gate is not powered, but some 74HC00's (Motorola's, for example) put out a HIGH for this input, while others (National's) put out a low. In neither case is the gate functioning. The HIGH from the Motorola chip seems to be simply passed through, with the entire IC presumably hanging close to +5V.

2-input 74XX gates use the same pinout – except the oddball '02, which is laid out backwards. See Appendix J.) You can think of XOR as a 1-bit *equality*/inequality* detector.[13]

2-bit *A>B* detector: Use any gates to make a comparator that detects when one of a pair of two bit number (call it *A*) is larger than the other (call it *B*). You need not make the circuit symmetrical: it need not detect *B>A*, only *A>B* versus $A \leq B$.

Warning: this circuit can get pretty complicated. We'd like to urge you, once again, not to use a lot of lab time on this problem.

14L.5 Gate innards; looking within the black box of CMOS logic

In the following experiments, we will use two CD4007 (or CA3600) packages; this part is an array of complementary MOS transistors, as in Fig. 14L.9.

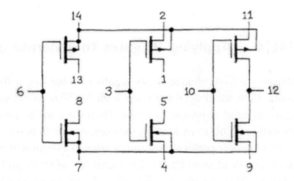

Figure 14L.9 '4007 (or "CA3600") MOS transistor array.

14L.5.1 Two inverters

Passive pullup: Build the circuit in Fig. 14L.10 using one of the MOSFETs in a '4007 package. Be sure to tie the two "body" connections appropriately: pin 14 to +5V, pin 7 to ground. This will look familiar to you: it is one more instance of the convention we mentioned earlier, in §14L.1: corner pins carry power and ground. (In fact, you will find that this is automatic for the particular FETs in the package that we show below; but you should be alert to this issue as you use MOSFETs.)

Confirm that this familiar circuit does invert, as you drive it with a TTL level from an *external function generator* (the breadboard oscillator is too slow to make this circuit look really bad, as we soon will want it to). The function generator's TTL output provides a V_{OH} that is not high enough to satisfy CMOS's preference for a high of close to 5V (though TTL usually *will* make the CMOS gate switch).[14] The usual remedy is to "pull up" the TTL output, with a resistor of a few kΩ to +5V.[15] Watch the output on a *scope* (voltmeter or logic probe will not do, from this point on).

Now crank up the frequency as high as you can. Do you see what goes wrong, and why?[16] Draw what the waveform looks like.

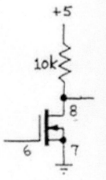

Figure 14L.10 Simplest inverter: passive pullup.

---

[13] This may be your first encounter with the convention that uses an asterisk ("equality*") to indicate a signal that is active-low. The same low/high meanings could be written with an overbar: "equality/inequality."

[14] If you don't recall TTL's V_{OH} value, you might look again at the V_{OH} numbers you got in §14L.2.

[15] This pullup may work on your function generator; it did not work on ours (Krohn–Hite) because their outputs labeled "TTL" are not true TTL: their TTL output cannot be pulled above 4V. A real TTL gate rises easily to +5V when pulled up. But 4V is sufficient to drive the CMOS gates.

[16] The rise is slow, because the usual stray capacitance is driven by the 10k pullup resistor, forming a lazy *RC*.

14L.5 Gate innards; looking within the black box of CMOS logic

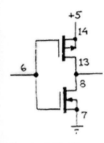

Figure 14L.11
Active pullup:
CMOS inverter.

Active pullup: CMOS Now replace the 10k resistor with a *p*-channel MOSFET, to build the inverter in Fig. 14L.11. Look at the output as you try high and low input frequencies.

The high-frequency waveform should reveal to you why all respectable logic gates use *active pullup* circuits in their output stages. (The passive-pullup type, called *open collector* for TTL or *open drain* for MOS, appears now and then in special applications such as driving a load returned to a voltage other than the +5V supply, or letting several devices drive a single line. Usually that last case is better handled by a *3-state*. See below.)

14L.5.2 Logic functions from CMOS

CMOS NAND: Build the circuit in Fig. 14L.12 and confirm that it performs the NAND function.

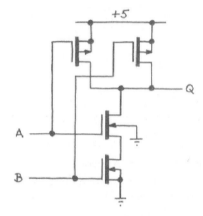

Figure 14L.12 CMOS NAND.

CMOS three-state: The three-state output stage can go into a *third* condition besides high and low: *off*. This ability is extremely useful in computers: it allows multiple *drivers* to share a single driven wire, or *bus* line. Here, you will build a buffer (a gate that does nothing except give a fresh start to a signal), and you will be able to switch its output to the OFF state. It is a "three-state buffer."

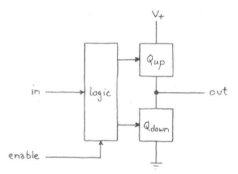

Figure 14L.13 Three-state buffer: block diagram.

The trick, you will recognize, is just to add some logic that can turn off *both* the *pull-up* and *pull-down* transistors. When that happens, the output is disconnected from both +5 and ground; the output then is off, or "floating." One usually says such a gate has been "three-stated" or put into its "high-impedance" state.

If you're in the mood to design some logic, try to design the gating that will do the job using NANDs along with the '4007 MOSFETs. Here's the way we want it to behave:

Lab: Logic Gates

- If a line called Enable is low, turn *off* both the *pull-up* and *pull-down* transistors. That means:
  - drive the gate of the upper transistor *high*;
  - drive the gate of the lower transistor *low*.
- If Enable is high, let the input signal drive one *or* the other of the upper and lower transistors on: that means:
  - drive the gates of the upper and lower transistors the same way – *high* or *low* – turning one on, the other off (because one is *p*-channel, the other *n*-channel, of course).

That probably sounds complicated, but the circuit is straightforward. If you're eager to get on with building, borrow our solution, shown in Fig. 14L.14.

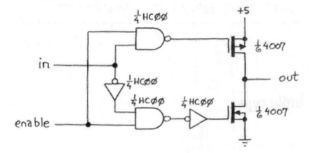

Figure 14L.14 Three-state circuit: qualifies gate signals to Q_{up} and Q_{down} using 74HC00 NAND gates.

Use a slide switch to control the 3-state's Enable, as shown in Fig. 14L.15. A second slide switch drives a 100k resistor tied to the 3-state's output. The slide switch can be set to ground or +5V: this slide switch, feebly driving the common output line through the 100k, is included so that we can *see* whether the 3-state is enabled or not. A scope or voltmeter by itself is not able to detect the "off" condition, as you know.

Test your circuit by driving the input with the breadboard oscillator, and watching the effect of Enable. When the 3-state is *disabled*, even the feeble 100k should be able to determine the output level. When *enabled*, the 3-state should drive its square wave firmly onto the common output line. This "common output line" provides our first glimpse of a *data bus*.

At the moment, before you have seen applications, this trick – the 3-state's ability to disappear, electrically – may not seem exciting. Later, when you build your computer, you will find 3-states invaluable if not exciting.

Figure 14L.15 Circuit to demonstrate operation of 3-state buffer.

14L.5 Gate innards; looking within the black box of CMOS logic

We include Fig. 14L.16 only to suggest why 3-states might be useful. It is meant to illustrate a single line of a "bus" or shared line, whereas a real parallel computer bus is at least 8 lines wide, and often much wider.[17]

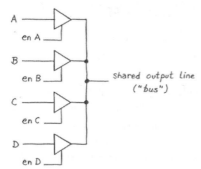

Figure 14L.16 Three-states driving a shared "bus" line (*do not bother to build this*).

---

[17] Recent designs have begun to exploit fast *single* line buses sending data *serially*.

14S Supplementary Notes: Digital Jargon

Active-high /-low Defines the level (high or low) in which a signal is "True," or – better – "Asserted" (see next term). We avoid the former because many people associate "True" with "High," and that is an association we must break.

Assert Said of a signal. This is a strange word, chosen for its strangeness, to give it a neutrality that the phrase "Make true" would lack (because many people seem predisposed to think of True as High).

We can say, with equal propriety, "Assert WR* (or $\overline{WR}$)," and "assert RESET51." In the first case it means "take it low;" in the second case it means "take it high" (because the signal names reveal that the first signal is *active low*, the second *active high*)

Assertion-level symbol This is a very strange phrase, meant to express a notion rather hard to express: it is a logic symbol, of the *two* that deMorgan teaches us always are equivalent, *chosen to show active levels*. So, a gate that ANDs lows and drives a pin called EN*, should be drawn as an AND shape with bubbles at its inputs and output, even though its conventional name may be OR. (The conventional name, note, assumes that signals are *active-high*.)

Asynchronous Contrasted, of course, with synchronous: asynchronous devices do not share a common clock, and may have *no* clock at all. Most flip-flops use an *asynchronous* clear function, also called "jam-type." Usage is complicated by the possibility that an *asynchronous function*, such as RESET*, can be an element of a *synchronous* circuit such as a counter. Such a combination occurs, for example, in the 74HC161 counter.

Clear Force output(s) to zero. Applied only to sequential devices. Same as *Reset*.

Combinational Same as "combinatorial" said of logic. Output is a function of present inputs, not past (except for some brief propagation delay); contrasted with *sequential* (see below).

Decoder A combinational circuit that takes in a binary number and asserts one of its outputs defined by that input number. (Other decoders are possible; this is the one you will see in the Labs.) For example, a '139 is a dual 2-to-4 decoder that takes in a 2-bit binary number on its two select lines. It responds by asserting one of its four outputs..

Demultiplexer ("demux") . Combinational circuit that steers a single time-shared input to one or another destination (this input may be a bus rather than a single line, as on a classic 8051 like the Dallas DS89C430: a multiplexed AD bus [the eight Address/Data lines]).

Edge-triggered Describes flip-flop or counter *clock* behavior: the clock responds to a transition, not to a level. By far the most widely used clocking scheme.

Enable The meaning varies with context, but generally it means "Bring a chip to life": let a counter count; let a decoder assert one of its outputs; turn on a 3-state buffer; and so on.

Float Said of input, or of a 3-stated output: driven neither high nor low.

Flop Lazy baby talk for "flip-flop," which is baby talk for "bistable multivibrator." But babies in this case express themselves better than old-fashioned physicists or engineers, who were inclined to say all those ugly syllables. (Flop is not to be confused, incidentally, with the acronym FLOP, a piece of computer jargon that stands for "Floating Point Operation".)

Hold time Time *after* a clock edge (or other timing signal) during which data inputs must be held stable. On new designs hold time normally is *zero*.

Jam clear, jam load Asynchronous clear, load: the sort that does not wait for a clock.

Latch Strictly, a *transparent latch* (see below), but often loosely used to mean *register* (edge-triggered) as well.

Latency Delay between stimulation and response. Often arises in response to *interrupt*, where prompt response can be important.

Load A counter function by which the counter flops are loaded as if they were elements of a simple *D-flop* register.

Multiplexer ("mux") Combinational circuit that passes one out of n inputs to the single output; uses a binary number code input on *select* lines to determine which input is so routed. For example, a 4:1 mux uses 2 select lines to route one its four inputs to the output. A "demultiplexer" does the complementary operation (see above).

One-shot Circuit that delivers one pulse in response to an input signal (usually called *trigger*), which may be a level or an edge. Most *one-shots* are timed by an *RC* circuit; some are timed by a clock signal, instead.

Preset Force output(s) of a sequential circuit high. Same as *Set*.

Propagation delay Time for signal to pass through a device. Timed from crossing of logic threshold at input to crossing of logic threshold at output.

Register Set of D flip-flops, always edge-triggered

Reset Same as *Clear*: means force output(s) to zero

Ripple counter Simple but annoying counter type; asynchronous – the Q of one flop drives the clock of the next. Slow to settle, and obliged to show many false transient states.

Sequential Of logic: circuit whose output depends on past as well as (in some cases) on present inputs. Example: a T (toggle) flip-flop: its next state depends on both its history and the level applied to its T input before clocking. Contrasted with *combinational* (above).

Set Same as *Preset*: means set output high (said only of flops and other sequential devices)

Setup time Time *before* clock during which data inputs must be held stable. Always *non-zero*. Worst-case number for a 74HC: about 20ns.

Shift register A set of D flops connected Q-of-one to D-of-the-next. Often used for conversion between parallel and serial data forms.

State Condition of a sequential circuit, defined by the levels on its flip-flop outputs (Qs). A counter's state, for example, is simply the combination of values on its Qs.

State machine Short for finite state machine or FSM. A generalized description for any sequential circuit, since any such circuit steps through a determined sequence of states. Usually reserved for the machines that run through non-standard sequences. A toggle flip-flop and a decade counter are FSMs, but never are so-called. A microprocessor executing its microcode is a state machine. So is an entire computer.

Synchronous Sharing a common clock signal (syn-chron means same time). Synchronous functions "wait for the clock." (In the counter of Lab 16L, for example, SYNC_LD* behaves this way. A second LD* ("load") function on the same counter is designed to be *asynchronous* and takes effect as soon as it is asserted.)

Three-state Describes a logic *output* capable of turning *off* instead of driving a high or low logic level. Same as *tri-state*.

Tri-state A trade name for *three-state*: Registered trademark of National Semiconductor, now part of Texas Instruments.

14W Worked Examples: Logic Gates

14W.1 Multiplexing: generic

AoE §10.3.3C

In this section we look at three ways of designing a small *multiplexer*:

- using ordinary gates;
- using 3-states;
- using analog switches

The notion of multiplexing, or *time-sharing*, is more general and more important than the piece of hardware called a multiplexer (or "mux"). You won't often use a mux, but you use multiplexing continually in any computer, and in many data-acquisition schemes.

Figure 14W.1 illustrates multiplexing in its simplest, mechanical form: a mechanical switch might select among several sources. The sources might be, for example, four microphones feeding one remote listening station of the secret police. Once a month (or once each hour, if he's ambitious) an agent flips the switch to listen to another of the four mikes. The motive for multiplexing, here, as always, is to limit the number of lines needed to carry information.

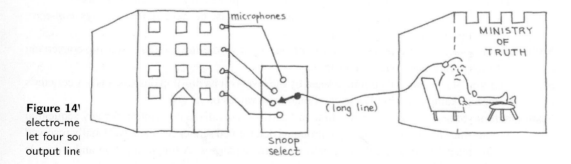

Figure 14... electro-me... let four so... output line...

This case is a little far-fetched. More typically, the "scarce resource" would be an analog-to-digital converter (ADC). And the most familiar example of all must be the *telephone* system. Without the time-sharing of phone lines, phone systems would look like the most monstrous rats' nests: picture a city strung with a pair of wires (or even one wire) dedicated to joining each telephone to every other telephone! Fig.14W.2 illustrates what even a network of 8 phones would look like.

A similar argument makes sharing lines efficient in a computer, where the shared lines are called "buses." Here the claim is complicated by the recent emergence of fast *serial* links, in preference to parallel lines, to link a computer's processor to memory and peripherals. These serial links now predominate. Such point-to-point communication is much less vulnerable to transmission line reflections than is the "multi-drop" scheme of a parallel bus. Thus the serial links permit higher data rates.

AoE §14.6.1

14W.1 Multiplexing: generic

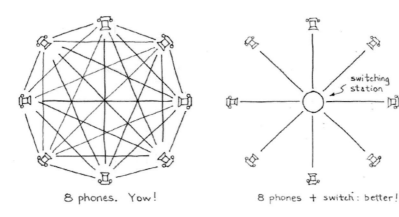

Figure 14W.2 Argument for multiplexing: limit the number of wires running here to there in a telephone system.

But even where the external links use just a few wires (a differential pair in each direction, for example, in PCI express) still parallel buses persist *within* the IC, out of sight, as the right-hand image of Fig. 14W.3 tries to suggest.

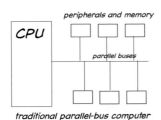
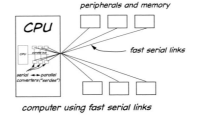

Figure 14W.3 Computers traditionally used parallel buses; now more often serial, point-to-point.

AoE §14.6.2C

14W.1.1 Multiplexing: hardware

A mechanical multi-position switch can do the job; so can the transistor equivalent, a set of analog switches. Logic gates can also do the job – though these allow flow in only one direction unlike the other two devices. The digital implementation requires a little more thought if you haven't seen it before. And even the analog switch implementation requires a *decoder* as soon as more than two signals are to be multiplexed.

We need two elements:

1. *Pass/Block* circuitry, analogous to the closed/open switch;
2. *Decoder* circuitry, that will close just one of the pass/block elements at a time.

Let's work up each of these elements, first for the implementation that uses ordinary gates.

Pass/Block An AND gate will do this job more or less: see Fig. 14W.4. To *pass* a signal, hold one input high; the output then follows or equals the other input; to *block* a signal, hold an input low. This case is a little strange because this forces a *low* at the output – and this forced *low* is not the same as the result of opening a mechanical switch.

Joining outputs The outputs of the AND gates may *not* be tied together. Instead, we need a gate that "ignores" lows, passes any highs (since the "blocking" ANDs will be putting out lows). An OR behaves that way and is just what we need. Fig. 14W.6 shows a 2:1 mux made with gates:

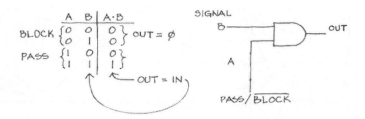

Figure 14W.4 AND gate can do the Pass/Block* operation for us.

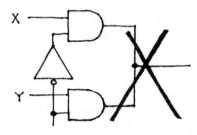

Figure 14W.5 Don't do this! The outputs of ordinary gates may *not* be tied together.

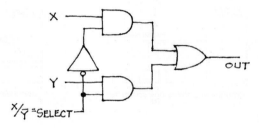

Figure 14W.6 2:1 mux.

Decoder If the mux has more than two inputs we need a fancier scheme to tell *one and only one* gate at a time to *pass* its signal. The circuit that does this job, pointing at one of a set of objects, is called a *decoder*. It takes a binary number (in its *encoded* form, it is compact: n lines encode 2^n combinations, as you know) and translates it into 1-of-n form (*decoded*: it's not compact, so not a good way to transmit information, but often the form needed in order to make something happen in hardware).

The decoder's job is to detect each of the possible input combinations. Fig. 14W.7, for example, is the beginning of a 2-to-4-line decoder.

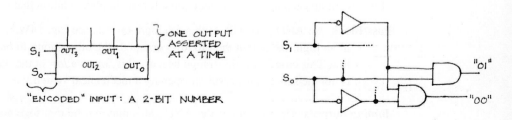

Figure 14W.7 2-to-4-line decoder: block diagram; partial implementation.

Decoders are useful in their own right (you will use such a chip in your microcomputer, for example, if you build up the computer from parts: a 74HCT139). The decoder is *included* on every multiplexer and in every multiplexing scheme that goes beyond two signals.

14W.1.2 Worked example: Mux

Problem (4:1 Mux, designed 3 ways) Show how to make a 4-input multiplexer using (a) ordinary gates, (b) transmission gates, and (c) gates with 3-state outputs.

Solution

Decoder: All three implementations require a decoder, so let's start with this.

We need to detect all four possibilities and might as well start, as in the earlier example, by generating the complement of each *select* input. (Does it go without saying that we need *two select* lines to define four possibilities? If not, let's say it. If this isn't yet obvious to you, it will be soon.)

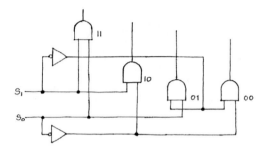

Figure 14W.8 1-of-4 decoder.

Pass/Block gating: ordinary gates: You know how to do this job with AND gates. The circuit in Fig. 14W.9 generates all the four input combinations, and applies them to ANDs as Pass/Block* as in Fig. 14W.4.

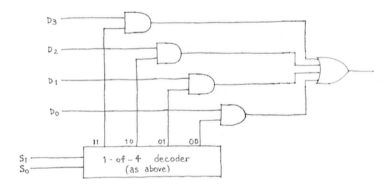

Figure 14W.9 4-1 mux, gate implementation.

Pass/Block gating: transmission gates: This is easier. (You *do*, of course, need to remember what these "transmission gates" or "analog switches" are. If you don't recall them, from Lab 12L, go back and take a second look.)

Again we use the decoder. This time we *can* simply join the gate outputs, since any transmission gate that is *blocking* its signal source *floats* its output, unlike an ordinary logic gate, which drives a *low* at its output when blocking. This scheme is sketched in Fig. 14W.10.

Three-states: This is a snap after you have done the preceding case. Again use the same decoder, and again you *can* join the outputs. A 3-state that is off, or blocking, does not fight any other gate. That's the beauty of a 3-state, of course.

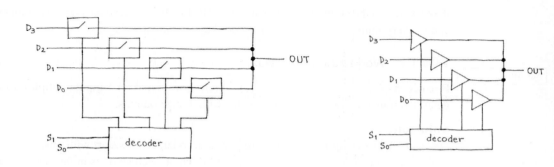

Figure 14W.10
4-1 mux: transmission-gate and 3-state implementations.

Problem Under what circumstances would (b) [the transmission-gate implementation] be preferable?"

Solution In general it is preferable *only* for handling *analog* signals. (For this purpose it is not just *preferable*; it is *required*.) For digital signals it is usually inferior, since it lacks the all-important virtue of digital devices: their noise immunity, which can also be described as their ability to clean up a signal: they ignore noise (up to some tolerated amplitude), and they put out a signal stripped of that noise and at a good low impedance. The analog mux, like any analog circuit, cannot do that trick; to the contrary, the analog mux slightly degrades the signal. At very least, it makes the output impedance of the signal source worse, by adding in series the mux's R_{on} (around 100Ω).

We should not overdo this point: it is not wrong to use an analog mux, or a single analog switch, to route digital signals. In fact, analog switches with low R_{ON} can provide the fastest way to connect or disconnect to a bus – outrunning a conventional three-state at the expense of losing signal regeneration. These devices are available, marketed as "Zero Delay Logic" (the QuickSwitch[1]). And transmission gates are used within fully-custom ASICs (application specific ICs). But analog switches normally are used – as the name surely suggests – for analog signals.

The analog version, incidentally, can pass a signal in either direction, so it works as a *demultiplexer* as well as mux. But that is not a good reason to use it for digital signals. If you want to demux digital signals, use a *digital* demux.

14W.2 Binary arithmetic

This worked example looks at five topics in binary arithmetic:

Two's-complement: versus *unsigned*, and the problem of *overflow*
Addition: a hardware design task, meant to make you think about how the adder uses *carries*
Magnitude comparators : done laboriously by hand, then easily with Verilog
Multiplication: an orderly way to do it contrasted with a foolish way
ALU and flags: foreshadowing the microprocessor, this exercise means to give you the sense that the processing guts of the CPU are simple, made of familiar elements

---

[1] This is a product of the company, Integrated Device Technology. It promises very low propagation delay (exaggerated as "Zero Delay"), because of its extreme simplicity. In fact, delay will occur, but caused mostly by the *RC* product of the switch's R_{on} and stray capacitance of the line that is driven.

14W.2.1 Two's-complement

Here's a chance to get used to 2's-comp notation. We want to underline two points in these examples.

- A given set of bits has no inherent meaning; it means what we choose to let it mean, under our conventions (this is a point made in Chapter 14N as well);
- A sum (or product or other result), properly arrived at, may nevertheless be wrong if we overflow the available range.

Both of these points are rather obvious; nevertheless, most people need to see a number of examples in order to get a feel for either proposition. "Overflow," particularly, can surprise one: it is not the same as "a result that generates a carry off the end." Such a result may be valid. Conversely, a result can be wrong (in 2's-complement) even though no carry out of the MSB occurs. This is pretty baffling when simply *described*. So let's hurry on to examples.

Let's suppose we feed an *adder* with a pair of four-bit numbers. We'll note the result, and then decide whether this result is correct, under two contrasting assumptions: that we are thinking of the 4-bit values as *unsigned* versus *two's-complement* numbers.

Problem (Two's-complement)   Suppose that you feed a 4-bit adder the 4-bit values A and B, listed below. Please write–

1. The maximum value one can represent in this four-bit result –
   - unsigned
   - in 2's-complement.
2. The sum.
3. Then what the inputs and outputs mean in decimal, under two contrasting assumptions:
   - first, the numbers are (4-bit-) *unsigned*;
   - second, the numbers are (4-bit-) *two's-complement*.
4. Finally, note whether the result is valid under each assumption.

We have done one case for you as a model.

IN:	A	B	A plus B	Valid?
Binary:	0111	1000	1111	
Decimal:				
Unsigned:	7	8	15	yes
2's-comp.:	7	–8	–1	yes
Binary:	0111	0111		
Decimal:				
Unsigned:				
2's-comp:				
Binary:	0111	1010		
Decimal:				
Unsigned:				
2's-comp:				
Binary:	0111	0100		
Decimal:				
Unsigned:				
2's-comp.:				
Binary:	1001	1000		
Decimal:				
Unsigned:				
2's-c:				

Try these. Then see if you agree with our conclusions, set out below.

Solution (Two's-complement)

The maximum values that one can represent with four bits

- Unsigned: 15_{10}.
- In 2's-complement: $+7_{10}, -8_{10}$.

IN:	A	B	A plus B	Valid?
Binary:	0111	0111	1110	
Decimal:				
Unsigned:	7	7	14	yes
2's-c:	7	7	−2	no
Binary:	0111	1010	0001	
Decimal:				
unsigned:	7	10	1	no
2's-c:	7	−6	+1	yes
Binary:	0111	0100	1011	
Decimal:				
unsigned:	7	4	11	yes
2's-c:	7	4	-5	no
Binary:	1001	1000	001	
Decimal:				
unsigned:	9	8	1	no
2's-c:	-7	-8	+1	no

Problem What is the rule that determines whether the result is valid?

Solution For unsigned, isn't it simply whether a carry-out is generated? It is.

For *two's-complement* the rule is odder: *if the sign is altered by carries*, the result is bad. A carry into the sign bit *can* indicate trouble – but only if there is no carry out of that bit; and vice versa.

In other words, an XOR between carries in and out of MSB indicates a 2's-complement overflow. The microprocessor you will meet in Lab 20L uses such logic to detect just that overflow (indicated by its "OV" flag, in case you care at this point). You can test this *XOR* rule on the cases above if you like.

14W.2.2 Addition

You may find it entertaining to reinvent the *adder*. As a reminder, the table below shows the way a single-bit *full-adder* should behave (the name "full" indicates that the adder can take in a *carry-in* signal so that such full-adder modules can be cascaded).

14W.2 Binary arithmetic

INPUT			OUTPUT	
Carry-in:	A	B	Carry-out	Sum
0	0	0	0	0
0	0	1	0	1
0	1	0	0	1
0	1	1	1	0
1	0	0	0	1
1	0	1	1	0
1	1	0	1	0
1	1	1	1	1

You will notice that you can, if you like, look at the *output* as a two-bit quantity rather than look at it as distinct *sum* and *carry*. The advantage of the latter view is just that it invites extension to larger word widths: when a 4-bit sum overflows, it too generates a *carry*.

Problem (Design a one-bit full adder.) Use the truth table above, if you need it, and design a full adder. Fig. 14W.11 illustrates its block diagram.

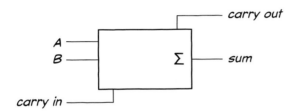

Figure 14W.11 Full adder.

Solution The *sum* is A XOR B when carry-in is low; it's the complement of XOR when carry-in is high. Does that suggest what the sum function is, as a function of the *three* input variables that include carry-in? Yes, it's XOR of all three (output 1 if an *odd* number of inputs are high): see Fig. 14W.12.

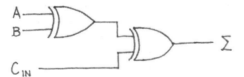

Figure 14W.12 Sum is C_{in} XOR A XOR B.

Carry-out? You may be able to see the pattern after staring a while at the truth table. Or you might reason your way to the function: carry-out should be asserted when two or more of the input bits are asserted. You might conceivably want to learn the antique method of Karnaugh mapping (see Fig. 14W.13 and let the map reveal patterns: patterns that show chances to ignore a variable or two).

One way or another, you can arrive at a sketch of the one-bit full adder (Fig. 14W.14).

Solution (Alternative: let Verilog design it.) We don't want you to work too hard getting good at Boolean manipulations because logic compilers are available that are willing to let you describe your design at a very high level of abstraction. The compiler takes care of the gating.

Here is a Verilog file defining a one-bit adder:

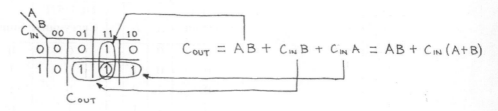

Figure 14W.13 Carry-out Karnaugh map.

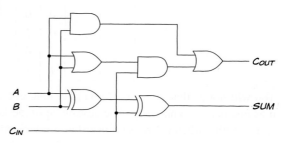

Figure 14W.14 Full adder.

```
module adder_assign(
    input a,
    input b,
    input c_in,
    output sum,
    output c_out
    );

assign {c_out,sum} = a + b + c_in;
endmodule
```

Can the design task be that easy? Apparently so; here's the result of simulating this design:

```
Finished circuit initialization process.
{a,b,c_in} = 000, {c_out, sum} = 00
{a,b,c_in} = 100, {c_out, sum} = 01
{a,b,c_in} = 010, {c_out, sum} = 01
{a,b,c_in} = 001, {c_out, sum} = 01
{a,b,c_in} = 101, {c_out, sum} = 10
{a,b,c_in} = 111, {c_out, sum} = 11
{a,b,c_in} = 011, {c_out, sum} = 10
{a,b,c_in} = 001, {c_out, sum} = 01
{a,b,c_in} = 000, {c_out, sum} = 00
```

Looks right. Fig. 14W.15 shows the bizarre logic that Verilog used to get this good outcome. No human would design the circuit this way; sum is normal, but carry-out is bizarre:

$$c\_out = ((a\, xor\, b)c\_in)\, xor\, (ab)$$

Weird, but correct. We don't really care if it's weird. Verilog does the hard work.

Once we have defined this single-bit function with a carry-in and carry-out line, we can string these little beads forever making an adder as big as we like (though very long strings might become *slow*). We will encourage you to look for a similarly-repeatable pattern in designing a multiplier later.

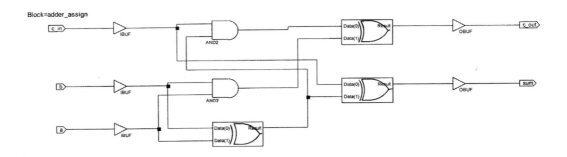

Figure 14W.15 Verilog's logic to implement a 1-bit full adder.

But once again, if you do the design in Verilog, expanding the adder from one bit to more is utterly straightforward. All you need do is define the input a, which before was a single bit, as a set of bits. Do the same for b and sum, and the job is done. Here is the Verilog for a 4-bit adder:

```
module adder_assign_4bit(
    input [3:0] a,
    input [3:0] b,
    input c_in,
    output [3:0] sum,
    output c_out
    );

assign {c_out, sum} = a + b + c_in;
endmodule
```

14W.2.3 Digital comparators

AoE §10.3.3E

Detecting equality: Many computer peripherals need to know when a value arriving on the address bus is equal to the peripheral's reference address. A set of XOR functions can detect such equality, bit by bit (see Fig. 14W.16). We invited you to design an equality detector in Lab 14L, and no doubt you figured out that all you needed was to detect equality, bit-by-bit, with XOR gates fed by corresponding bits of a multi-bit word: A_1 XOR B_1, A_0 XOR B_0. Then AND lows from the several XORs (the name of the gate that AND lows is "NOR," though the name, of course, does not describe the gate's function in this case).

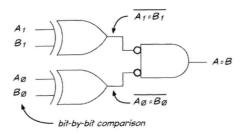

Figure 14W.16 XORs can detect equality, bit-by bit.

Worked Examples: Logic Gates

Detecting inequality: $A > B$ ("magnitude comparator"):

Problem (2-bit greater-than circuit.) Use any gates you like (XORs are handy but not necessary) to make a circuit that sends its output high when the 2-bit number A is greater than the 2-bit number, B. Fig. 14W.17 shows a black box diagram of the circuit you are to design.

Figure 14W.17 2-bit greater-than circuit.

Note that this problem can be solved either systematically, using Karnaugh maps, or with some cleverness: consider how you could determine equality of two one-bit numbers, then how to determine that one of the one-bit numbers is greater than the other; then how the more- and less-significant bits relate; and so on. We'll do it both ways.

Solution (1. Systematic.) Figure 14W.18 works, though we don't suppose anyone wants to learn Karnaugh mapping – a technique briefly described in AoE §10.3.2 – for the pleasure of solving such problems.

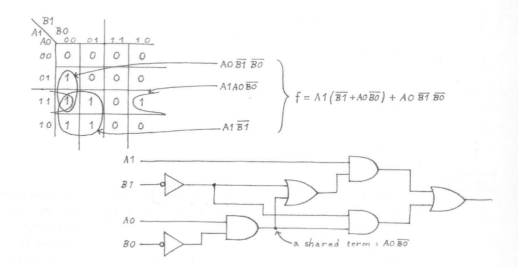

Figure 14W.18 2-bit greater-than circuit: systematic solution.

$$f = A1\,(\overline{B1} + A0\,\overline{B0}) + A0\,\overline{B1}\,\overline{B0}$$

Solution (2. Brainstorm the problem.) Figure 14W.19 shows a more fun way to do it. The brainstormed circuit is no better than the other. In gate count it's roughly a tie. Use whichever of the two approaches appeals to you. Some people are allergic to Karnaugh maps; a few like them.

Solution (3: let Verilog do it.) As usual, Verilog rather spoils the fun by doing the brain work for you. (You have to go through the chore of learning Verilog, of course.) Here's a Verilog file that does the job.

```
//////////////////////////////////////////
module comparator_2bit(
    input [1:0] a,
    input [1:0] b,
    output reg a_beats_b
```

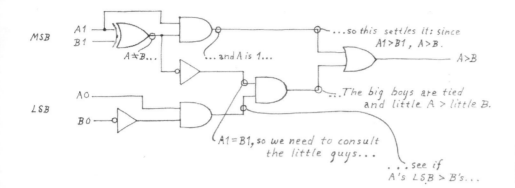

Figure 14W.19 2-bit greater-than circuit: brainstormed solution.

```
  );

  always @(a, b)
    begin
  if (a > b)
     a_beats_b = 1'b1;
  else
     a_beats_b = 1'b0;
    end

endmodule
```

Verilog implements this, once again, with a very strange circuit:[2] see Fig. 14W.20. But it works. Fig. 14W.21 shows the result of the simulation.

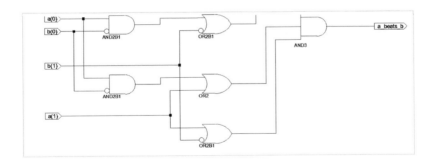

Figure 14W.20 Verilog's implementation of 2-bit $A{>}B$ circuit.

Figure 14W.21 Verilog simulation shows its strange implementation works.

---

[2] This compilation and simulation was done with Xilinx ISE 10, whereas the others in this chapter are done with ISE 11. Schematic and simulations differ in the two versions though the simulation results are the same.

Verilog easily expands the design: Designing a magnitude comparator for more than two bits would be a chore, done by hand. It's embarrassingly easy with Verilog. In the source file, just change the definitions of a and b to make them 8-bit rather than 2-bit quantities:

```
//////////////////////////////////////////
module comparator_8bit(
    input [7:0] a,
    input [7:0] b,
    output reg a_beats_b
    );

  always @(a, b)
   begin
 if (a > b)
    a_beats_b = 1'b1;
 else
    a_beats_b = 1'b0;
   end

endmodule
```

As you see in Fig. 14W.22, the schematic for the circuit that implements this design is pretty nightmarish: but again, we don't mind if Xilinx is willing to do all the work.

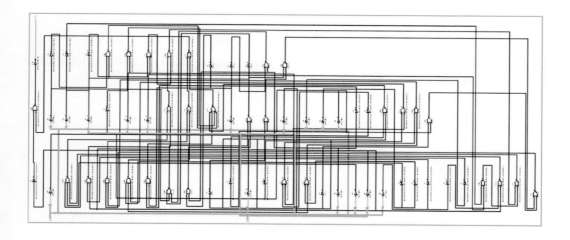

Figure 14W.22 Alarming schematic of 8-bit $A{>}B$ circuit.

Expanding in other senses would be straightforward, too: adding other outputs such as $A{=}B$ and $A{<}B$ and $A{\geq}B$. We won't do that. But you may be getting a sense that you quite easily could get Verilog to do this for you.

14W.2.4 Multiplication

Multipliers are much less important than adders; you should not feel obliged to think about how multiplication works. (Multipliers are important in digital signal processors, but not in the work we will do in this book.) Think about it if the question intrigues you.

One way to discover the logic needed to *multiply* would be to use Karnaugh maps or some other method to find the function needed for each bit of the product.

14W.2 Binary arithmetic

But even this small example may be enough to convince you that there must be a better way. A 3×3 multiplier (with its *six* input variables) would bog you down in painful Karnaugh mapping; but larger multipliers are commonplace. There is a better way.

This turns out to be one of those design tasks one ought *not* to do with Karnaugh maps, nor with any plodding simplification method. Instead, one ought to take advantage of an orderliness in the product function that allows us to use essentially a single scheme, iterated, in order to multiply a large number of bits. The pattern is almost as simple as for the adder.

Binary multiplication can be laid out just like decimal if we want to do it by hand – and the binary version is much the easier of the two: see Fig. 14W.23.

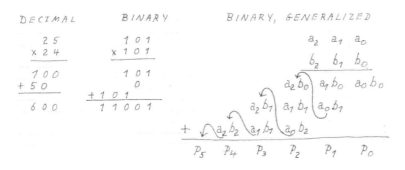

Figure 14W.23 Decimal and binary multiplication examples – and a generalization of the pattern; binary is easier!

Once you have seen the tidy pattern, you will recognize that you need nothing more than 2-input AND gates, and a few *adders*. With this method, you don't have to work very hard to multiply two 4-bit numbers – a task that would be daunting indeed if done with Karnaugh maps. (You need only *half* adders, not *full* adders; but you might choose to use an IC adder like the '83, rather than build everything up from the gate level.) Let's try that problem.

Problem (3×3 multiplier.) Design a 3 × 3 multiplier along the same lines, this time using two 4-bit full adders (74HC83) and as many 2-input gates as you need.

Solution Figure 14W.24 is a circuit you'll want to buy, not build! But you can see that you could easily extend this to make a 4 × 4 – or 16 × 16 multiplier.

14W.2.5 Arithmetic logic unit (ALU)

As a wrap-up for this discussion of arithmetic here is a device central to any computer: a circuit that does any of several logical or arithmetic operations. To design this does not require devising anything new; you need only assemble some familiar components.

Problem (Problem: design a 1-bit ALU.) Figure 14W.25 is a block diagram of a simplified ALU. Let's give it a carry-in and a carry-out, and let that output bit be low in any case where a carry is not generated (for example, *A* OR *B*).

Two select lines of the set in Fig. 14W.25 should determine which operation the ALU performs (use any select code you like).

Solution (Design it with gates and a mux.) All we need is to combine standard gates and an adder with a multiplexer: see Fig. 14W.26.

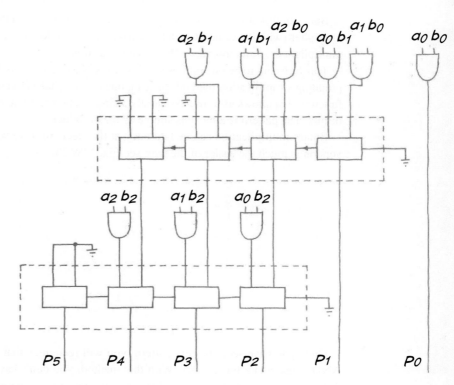

Figure 14W.24 3×3 multiplier using IC adders.

C_{IN}	A	B	AND	OR	XOR	ADD C_{OUT}	SUM
0	0	0	0	0	0	0	0
0	0	1	0	1	1	0	1
0	1	0	0	1	1	0	1
0	1	1	1	1	0	1	0
1	0	0				0	1
1	0	1				1	0
1	1	0				1	0
1	1	1				1	1

Figure 14W.25 1-bit ALU: block diagram.

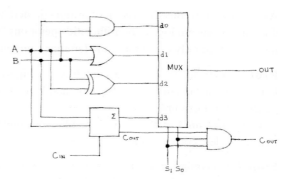

Figure 14W.26 A simple ALU.

Solution (Design it with Verilog.) Again, this is also quite easily done in Verilog. We can implement the multiplexer using CASE statements (as we do in a counter design described in Appendix A). Here is Verilog code to form the ALU.

```
module ALU(
    input [1:0] S,    // this is the two-bit select code, choosing an ALU operation
    input A,
    input B,
    input CIN,
    output reg OUT,
    output reg COUT
    );

        parameter AND = 'b00, OR = 'b01, XOR = 'b10, SUM= 'b11;
// definitions that allow us to use descriptive names for the function selects

        always@(S, A, B, CIN)
          begin
            case (S)
              AND:
                begin
                  OUT <= A & B;
                  COUT <= 0;
                end
              OR:
                begin
                  OUT <= A | B;
                  COUT <= 0;
                end
              XOR:
                begin
                  OUT <= A ^ B;
                  COUT <= 0;
                end
              SUM:
                begin
                  OUT <= A ^ B ^ CIN;
                  COUT <= CIN & (A | B);
                end
            endcase
         end
endmodule
```

14W.2.6 A refinement: "flags"

A computer's ALU always includes flip-flops that keep a record of important facts about the result of the most recent operation – not the result itself, but summary descriptions of that result. These records may include:

- was a *carry* generated?
- was the result *zero?*
- is the result positive or negative, in 2's-comp notation?
- was a 2's-comp overflow generated?
- and sometimes a few other items as well (for example, the 8051's "half-carry," used to help convert binary values to binary-coded-decimal

Note: You should understand at least the *D* flip-flop before trying this problem.

Problem (Add flags to ALU.) Add to the ALU designed earlier *flags* to record the following pieces of information concerning the result of the ALU operation.

Carry Was a *carry* generated?

Zero Was the result *zero*?

Sign Is the result positive or negative, in 2's-comp notation? (Assume that the single-bit output of this stage makes up the MSB of a longer word.)

Overflow Was a 2's-comp overflow generated? (Make the assumption noted: *this* output is the MSB of a longer word, and thus constitutes the *sign* bit when this word is treated as a 2's-complement value.)

Assume that a timing signal is available to clock the flag flops a short time after the ALU output has settled.

Solution See Fig. 14W.27.

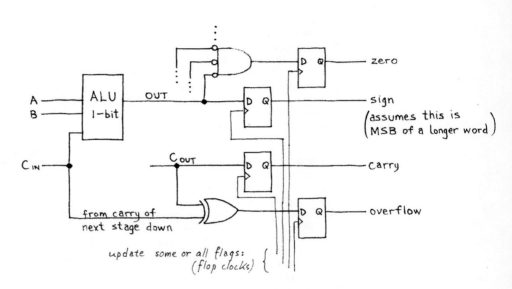

Figure 14W.27 Flags added to ALU.

15N Flip-Flops

Contents

15N.1 Implementing a combinational function	**568**
15N.1.1 Simplifying or minimizing logic	569
15N.2 Active-low, again	**569**
15N.2.1 Diagram it with gates	570
15N.2.2 Do it with a logic compiler: Verilog	571
15N.3 Considering gates as "Do this/do that" functions	**573**
15N.3.1 AND as "IF"	574
15N.4 XOR as Invert/Pass* function	**574**
15N.5 OR as Set/Pass* function	**575**
15N.6 Sequential circuits generally, and flip-flops	**575**
15N.6.1 The simplest flip-flop: the "latch," made From a few transistors	575
15N.6.2 The simplest flip-flop again: the latch made with two gates	576
15N.6.3 Switch debouncers	577
15N.6.4 But SR usually is not good enough	577
15N.6.5 Edge-triggering	579
15N.7 Applications: more debouncers	**582**
15N.7.1 D-flop and slow clock	582
15N.7.2 Positive-feedback debouncer	582
15N.8 Counters	**583**
15N.8.1 Ripple counters	583
15N.9 Synchronous counters	**584**
15N.9.1 What synchronous means	584
15N.9.2 Synchronous counting requires a smarter flip-flop	584
15N.9.3 The counter behavior we want	585
15N.10 Another flop application: shift-register	**586**
15N.11 AoE Reading	**587**

Why?

In this chapter we meet circuits that remember states of binary signals. Such circuits can be more versatile than the merely *combinational* circuits that we met earlier. First, a recap of Day 14.

- Digital gives good noise immunity.
- Noise immunity allows most of what's wonderful about digital:
  - *binary* coding of information permits electronics of extreme simplicity, thus small size and low cost;
  - noise immunity is fundamental to data storage, and transmission through adverse conditions;
  - the fact that information is encoded permits processing to detect and correct errors (e.g., in digital music playback).
- Since noise immunity is fundamental, better noise immunity is important: 5V CMOS beats old TTL. For signal transmission, differential signaling is best, and works at the low voltages now common.

568   Flip-Flops

- Arrays of gates – PALs, PLDs and FPGAs – are taking over from small packages of just a few gates.
- And a logic compiler, used to program such parts, normally takes care of minimization.
- Most digital control signals are *active-low* (for historical reasons).
- And a logic compiler can hide the confusion that can result when active-high and active-low signals are mingled.
- Binary numbers: 2's-complement represents negative numbers. Simple scheme: MSB is negative, in 2's-comp. But signed/unsigned is a matter for humans to interpret; the binary representation does not reveal which is intended.

15N.1 Implementing a combinational function

Most of the time the combinational logic you need to design is so simple that you require nothing more than a little skill in *drawing* to work out the best implementation. For such simple gating, the main challenge lies in getting used to the widespread use of *active-low* signals.

Figure 15N.1 shows an example: "Clock flop if RD* or WR* is asserted and Now* is asserted."

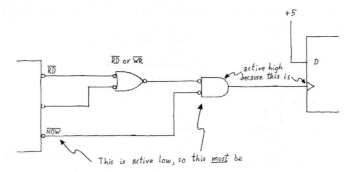

Figure 15N.1 Easy combinational logic; some signals active low: don't think too hard; just *draw* it.

You can make this problem difficult for yourself if you think too hard: if you say something like, "Let's see, I have a low in and if the other input is low I want the output high – that's a NOR gate..." – then you're in trouble at the outset. Do it the easy way, instead.

Here is the easy process for drawing clear, correct circuits that include active-low signals:

1. draw the *shapes* that fit the description: AND shape for the word AND, regardless of active levels;
2. add inversion bubbles to gate inputs and outputs wherever needed to match active levels;
3. figure out what gate types you need (notice that this step comes *last* – and comes *never* if you use a logic compiler and PLD).

In general, bubbles meet bubbles in a circuit diagram properly drawn: a gate output with a bubble feeds a gate input with a bubble, and your eye sees the cancellation that this double-negative implies: bubbles pop bubbles. This usual rule does *not* apply in the exceptional case when your logic is looking for *dis*-assertion of a signal. We'll see such cases in §15N.2.1. It also often does not apply to edge-triggering, where it is *timing* that determines one's selection of a clock edge ("leading" versus "trailing").

15N.1.1 Simplifying or minimizing logic

When a logic network is not utterly simple, you may wonder if your implementation is minimal – as simple as possible. Once upon a time, achieving that simplest design was a serious part of a designer's work. One might use Boolean algebra to simplify an equation – factoring, for example. If, in those dark old days, you were given a truth table like the one in Fig. 15N.2,[1] how would you implement it?

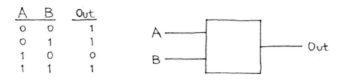

Figure 15N.2 A truth table to implement.

1. You might simply stare at the table for a while, and discover the pattern. Some people like to work that way.
2. You might write the Boolean expression for each input combination that gives a **1** out:

$$f = A*B* + A*B + AB .$$

Then you could use Boolean algebra to simplify this expression. Try factoring:

$$f = A*(B*+B) + AB = A* + AB .$$

Thus you discover that you can toss out one variable, A, from two of the terms. After that, if you are on your toes, you recognize a chance to apply another Boolean postulate, and you end up with an agreeably simple equation:[2]

$$f = A* + B .$$

But such skills in using Boolean algebra are obsolete. So are the graphic aids to simplification provided by *Karnaugh maps*. Both are obsolete because anyone with access to a computer can take advantage of an automated logic compiler. The compiler implements a hardware description language (HDL), normally VHDL or Verilog. In this course we use Verilog.

The compiler cheerfully simplifies what you type (or draw – some compilers will accept graphical input). So let's not give any more attention to a problem that mercifully has been removed from a designer's life. We can save our intellectual energy for more interesting tasks.

15N.2 Active-low, again

Last time we met an example using signals that were *active low* in the somewhat silly missile-launch circuit. Here, let's look at an example that is a little more realistic. This case includes some signals that are active-high but a larger number that are active-low, as most control signals are.[3]

---

[1] A fair and relevant question is whether life is likely to deliver a problem in this form. The truth table arises only if you find this a convenient way to think about the design task. If you start instead by drawing gates to describe the design, this design-from-table problem does not arise.
[2] You might find it simpler still to write an equation for the single case that produces a *low* output: $f* = AB*$.
[3] The reason for the predominance of *active-low* signal is simply historical. Such signals were advantageous in the era of bipolar logic, as we argued in Chapter 14N.

15N.2.1 Diagram it with gates

Draw gates that will:

Assert $\overline{\text{DOIT}}$ if:
- $\overline{\text{WRITE}}$, VALID and $\overline{\text{TIMED}}$ all are asserted, as long as neither $\overline{\text{PROTECTED}}$ nor $\overline{\text{BLOCK}}$ is asserted;
- or if either $\overline{\text{WILLY}}$ or NILLY is asserted.

The only hard part: agreeing on what we mean: Curiously enough, this exercise presents the only point in the course when students sometimes get *angry*, disgusted with us teachers for talking nonsense, or worse. It's as if we're suddenly reciting blasphemous and inconsistent remarks about the students' religions.

Why? Because the phrase, "$\overline{\text{WRITE}}$ is asserted" is ambiguous. We don't want to insist that our understanding of this phrase is the only reasonable one; we only want to ask you to join us in what we admit is only a *convention* – a way of understanding this phrase. We want you to understand it as we do; that way, we'll not misunderstand each other.

We take "$\overline{\text{WRITE}}$ is asserted" to mean that the signal named "$\overline{\text{WRITE}}$" is *low*. The *bar* over the signal's name here means that the signal is *active-low*.

AoE §10.1.2A

Indignant students sometimes protest that "$\overline{\text{WRITE}}$ is asserted" is to say that WRITE is false. If WRITE is false, it is low, and its complement, "$\overline{\text{WRITE}}$" is *high*.

Such a protest is not wrong; it just does not follow the convention that we are imposing – and which is widely recognized: many signals (in fact most, by a wide margin) are *active-low*, and one indicates this active-low quality with a "bar" or asterisk. Such a signal normally is *born* active-low, rather than being the complement of an active-high signal of the same name.

We did write the word "WRITE" (without a bar over its name) to the left of the boundary and circle (often called a "bubble"), in Figure 15N.3. But we did not thereby mean to imply that any such signal *exists*. Probably it does not exist, and in those drawings there is no point indicated where one could apply a probe to test an active-high WRITE signal. Only the active-low "$\overline{\text{WRITE}}$" is available for probing.

In the case of "$\overline{\text{DOIT}}$," we used this active-low labeling because the label is placed to the left of the bubble, out in the world where the signals run and are available for probing. On the leftmost part of the diagram, in contrast, we used labels without bars (showing "WRITE," for instance), because a bubble intervenes between this fictitious point and the outside world where "$\overline{\text{WRITE}}$" travels.

Well, we have expended a lot of words to make a small point – but it is one on which we must agree, if we are to able to talk to one another about signals that are active-low.

Figure 15N.3 shows an implementation of the logic that we described in words.

It's not *quite* as simple as "Let bubbles meet bubbles." For most of the signals shown in Fig. 15N.3, "bubbles meet bubbles," so that one can see inversions cancelling each other. But you shouldn't apply such a rule mechanically.

The simple-minded rule does work when you are using gates to detect *assertion* of signals. It does *not* work – as we noted back in §15N.1 – when a gate is to detect *dis*-assertion. The leftmost two-input AND gate in the figure is used to detect such a *disassertion* of the signals $\overline{\text{PROTECTED}}$ and $\overline{\text{BLOCK}}$. So, in this case, bubbles do *not* meet bubbles. The mismatch, obvious to the eye, reveals at a glance that the job of this AND gate is to sense falsehood or disassertion.

DeMorgan sometimes allows two equally good symbols: DeMorgan teaches that two alternative symbols always are available for drawing a logical function, and we have shown these alternatives in

15N.2 Active-low, again

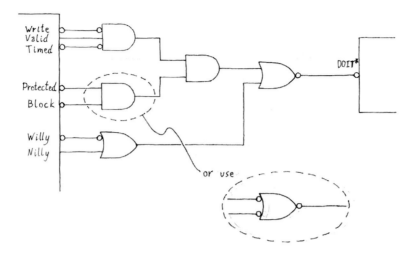

Figure 15N.3 Gates drawn to implement the logic described earlier.

Fig. 15N.3 for the logic used for sensing disassertion or falsehood. Usually, one of the two deMorgan-equivalent symbols shows more clearly than the other what is going on in the circuit. You should make the choice that seems to you to represent best what your circuit is doing.

Once in a while, however – as in the case of the logic applied to $\overline{PROTECTED}$ and $\overline{BLOCK}$ – the two alternative symbols are equally good. The one we drew first senses that *both signals are false*, i.e., disasserted. The alternative form, shown lower, senses that *either signal is asserted* and uses that event to *disassert* the input to the next gate. The results are logically identical of course. Use whichever you prefer. The OR shape does the better job of reflecting the language of the design specification, "neither …nor" …. But the AND shape, sensing "both signals false," shows logic that may be easier to think about.

15N.2.2 Do it with a logic compiler: Verilog

AoE §11.3.4B

One can force Verilog to suppress active-lows: One can do it by introducing extra signal names that are purely active-high – the complements of any nasty active-low signals that the world imposes on us. We have done this in the Verilog listing below. The happy consequence is that when we write the equation that implements the logic, we can forget about whether signals are active-high or active-low. In the sheltered world we have created for ourselves, there is no such thing as active-low. In this sheltered place, a bang means what one would hope it could mean: that a signal is disasserted or false. That's the case for prot and block in the equation for *doit*, below:

```
module act_lo_verilog(write_bar, valid, timed_bar, prot_bar, block_bar,
                    willy_bar, nilly, doit_bar);
    input write_bar;
    input valid;
    input timed_bar;
    input prot_bar;
    input block_bar;
    input willy_bar;
    input nilly;
    output doit_bar;

    wire doit;
//  all inputs and outputs are ordinary signals ("wire"),
```

```
// versus the sort that holds a value ("reg") by default;
// "doit" -- one never brought out to a pin but used as input to a function --
//   is explicitly called "wire"

// Now flip all active-low signals, so we can treat ALL signals as active-high
// as we write equations
 assign write = !write_bar;   // makes a temporary internal signal (write) never assigned
                              // to a pin
 assign timed = !timed_bar;
 assign prot = !prot_bar;
 assign block = !block_bar;
 assign willy = !willy_bar;
 assign doit_bar = !doit; // this output logic is reversed because the active-high "doit"
                          // will be an input to this logic, used to generate the OUTPUT

assign doit = write & valid & timed & (!prot & ! block) |
                  (willy | nilly);    // ( & means AND, | means OR)
endmodule
```

The ugly active-low signals, with names like `willy_bar`, do not appear in our equations. Our equation looks simple and tidy.

Some Verilog users scorn such a crutch. But we think it's nice not to have to worry about active levels when writing equations. In an active-high world, whenever you see a "bang" ("!" or "~"), you can take that to mean *dis*assertion of a signal.

If you don't force signals into active-high form, then the best you can do is to give the variable a name that reveals whether it is active low. Following such a scheme, the equations are not so neat: they include *bangs* to show assertion of signals that are active-low:

```
module act_lo_verilog(write_bar, valid, timed_bar, prot_bar, block_bar,
                      willy_bar, nilly, doit_bar);
    input write_bar;
    input valid;
    input timed_bar;
    input prot_bar;
    input block_bar;
    input willy_bar;
    input nilly;
    output doit_bar;

 wire write_bar,valid,timed_bar,prot_bar,block_bar,willy_bar,nilly,doit_bar;
// specify all inputs and outputs and intermediate "doit" as ordinary signal
// ("wire") versus the sort that holds a value ("reg")

assign doit_bar = !( (!write_bar & valid & !timed_bar) & !(!prot_bar | !block_bar) |
(!willy_bar | nilly) );
// NOR of assertions of prot_bar* and block_bar*, this time

endmodule
```

You can judge for yourself which of the two schemes is the less confusing.

The two versions evoke exactly the same designs from the logic compiler, and Verilog will draw a *schematic* gate diagram to show its implementation.[4] For a simple design, it may be worthwhile to

---

[4] Verilog offers two schematic versions of its implementation, under "Synthesize." RTL (Register Transfer Logic – referring apparently to the not-very-interesting point that the typical circuit includes "registers" (flip-flops) linked to other registers – describes a generic design and is usually the simpler and clearer. "Technology schematic" shows the way the design will be implemented in the particular piece of hardware chosen for the circuit. Sometimes RTL is quite useless: Xilinx Verilog sometimes will use an *adder* to design a counter in RTL, when told "count <=count+1." The technology schematic, in

15N.3 Considering gates as "Do this/do that" functions

look at the schematic: it's reassuring, when it looks right, and the graphic representation can render obvious some errors that are not evident in the text of the source file.

Figure 15N.4 shows Verilog's drawing of the `doit_bar` logic: no surprises, but reassuring. Verilog uses a wide AND function where the hand-drawn figure used two levels; Verilog's design is better, imposing less delay than the hand-drawn version would. Xilinx's ISE11 and later allow you to decide which signals you want to see in the schematic. That's a useful feature in a design with lots of signals.

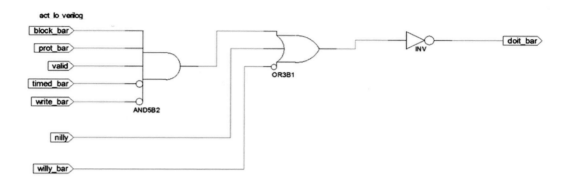

Figure 15N.4 Schematic gate diagram drawn by Verilog compiler (RTL version).

15N.3 Considering gates as "Do this/do that" functions

A truth table describes a logic function fully, so you may be inclined to think that you can leave your understanding of gates at that level: just write out (or memorize) the truth tables for the fundamental functions: perhaps AND, OR, and XOR. But sometimes it helps to think of a two-input gate as using one of its inputs to operate upon the other. That probably sounds cryptic. Some examples will make the point.

AND as Pass/Block* function: Treat input A as *control*. Then the AND gate will *pass* B if A is high, *block* B if A is low: see Fig. 15N.5.

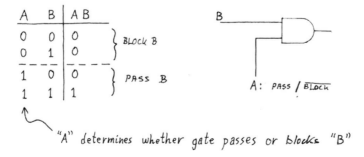

Figure 15N.5 AND gate viewed as Pass/Block*.

contrast, shows Verilog coming to its senses, drawing the simpler traditional counter design. So, to play it safe, look at the technology schematic if RTL is far from what you expected.

You will see AND used this way to implement a *synchronous clear** in a counter described in Chapter 16N. You will see it again when we treat software *masking* in Chapter 22N.

Note that when the gate does what we describe as "block," it is driving its output *low*, not simply passing "nothing" as a 3-state buffer would. So such a blocked signal must be fed to an OR gate, if combined with a "passed" signal, as in a 2:1 multiplexer: see Fig. 15N.6.

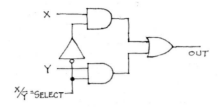

Figure 15N.6 AND gates, when "blocking" must be fed to an OR, which ignores lows.

The OR gate has the good sense to ignore a *low*. Simply joining the two AND outputs would fail of course, setting up a fight between the two gates.[5]

15N.3.1 AND as "IF"

Here's another way of describing how the AND function occasionally can help: think of the AND function as passing or permitting a signal *IF* a condition is fulfilled (pass *B* if *A* is asserted).

Ordinary language occasionally uses the word "and" to mean "if" – well, it did in Shakespeare's time, anyway. No doubt you recall these lines from Romeo and Juliet[6]

> And you be mine, I'll give you to my friend;
> And you be not, hang, beg, starve, die in the streets

For "an" and "and" in this passage we would, of course, write "if."

15N.4 XOR as Invert/Pass* function

An XOR gate can do the useful work of acting as a *controllable inverter*. The XOR will *invert B* if *A* is high, *pass B* if *A* is low: see Fig. 15N.7. You will see the XOR put to this use on page 585.

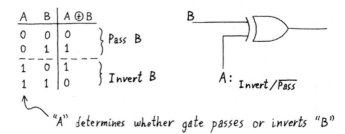

Figure 15N.7 XOR gate viewed as controllable inverter.

---

[5] This point is noted in §14W.1.
[6] Just kidding. If you remember these lines (from Romeo and Juliet, Act III, Scene 5), you've probably stumbled into the wrong classroom or have picked up the wrong book.

15N.5 OR as Set/Pass* function

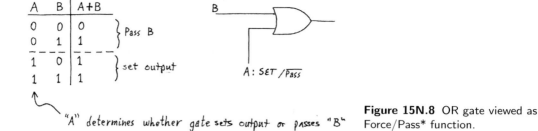

"A" determines whether gate sets output or passes "B"

Figure 15N.8 OR gate viewed as Force/Pass* function.

The OR gate will *set* the output (force output high) if *A* is high, *pass B* if *A* is low: see Fig. 15N.8. You will meet this use of OR in the discussion of software *masking* in Chapter 23N.

15N.6 Sequential circuits generally, and flip-flops

The *combinational* circuits that we looked at in §14N.3 "live in the present," like the resistive circuits we met on (*analog*) Day 1. *Sequential circuits* offer a contrasting behavior that neatly parallels the contrast we saw between Days 2 and 1: *RC* circuits care about the *past*, and thus offer behaviors that are more complex and more interesting than those of purely resistive circuits (§2S.2, Fig. 2S.11). The same is true of sequential digital circuits in contrast to combinational.

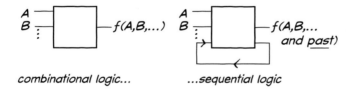

combinational logic...   *...sequential logic*

Figure 15N.9 Combinational versus sequential logic circuits: block diagrams contrasted.

    If combinational logic were the whole of digital electronics, the digital part of this course would now be just about finished. You could close the book and congratulate yourself.

    But, no: the interesting part of digital lies in sequential circuits. These permit building counters, memories, processors – and computers.

    Flip-flops are easy to understand. A harder question arises, however, when we look at flops in detail: why are *clocked* circuits useful? We will work our way from primitive circuits toward a good *clocked* flip-flop, and along the way we will try to see why such a device is preferable to the simpler design.

15N.6.1 The simplest flip-flop: the "latch," made From a few transistors

The feedback shown in the block diagram in Fig. 15N.9) appears in Fig. 15N.10 on the right as the link between the output of each transistor switch and the gate controlling the other MOSFET switch. This is a circuit that will lock itself into one "state" or the other: one or the other switch will be held ON while the other is held OFF.

    At first glance, this circuit seems to be quite useless: a circuit that is stuck forever – and, even worse, unpredictable. But the circuit is not really stuck, because it is easy to force it to a desired state. If you turn OFF the transistor that is ON for a moment (by connecting the gate of the ON transistor to ground), then the ON switch turns off and, as a result, the switch formerly OFF turns ON. Thus the

<small>AoE §14.5.3; compare Fig. 14.20</small>

Flip-Flops

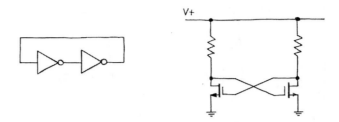

Figure 15N.10 Simplest flip-flop: cross-coupled inverters.

momentary grounding of an FET's gate can *flip* the circuit; doing the same to the other transistor can *flop* it back to its other state.

The name flip-flop provides a charmingly simple description of the circuit's behavior. (Engineers know how to speak simply; more hifalutin' people like to obscure the circuit's behavior by calling it a "bistable multivibrator" (!)).

This transistor latch can store one bit of information; in an integrated version, the pullup resistors would be implemented with transistors, as shown in Fig. 15N.11. Two more transistors would permit both writing and reading (that is, *writing* by forcing the cross-coupled transistors to the desired state, *reading* by simply connecting the Qs to the outside world).

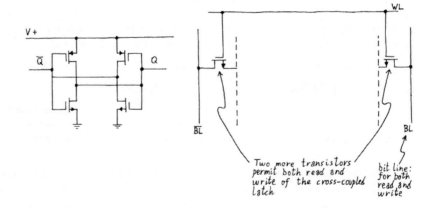

Figure 15N.11 Simplest flip-flop as applied to store one bit in a static memory (RAM).

15N.6.2 The simplest flip-flop again: the latch made with two gates

Figure 15N.12 shows the cross-coupled latch done with gates rather than plain transistors. This is the version you will build in lab.

AoE §10.4.1

The bubbles on the inputs of the NAND gates remind us that these gates will respond to *lows*: it is a *low* at the S input[7] that will "set" the flop (that is, will send *Q high*); a low at the R input will "reset" the flop (that is, will send *Q low*).

As was true for the equivalent latches built with transistors, the useful property of the circuit is its ability to *hold* the state into which it was forced once the forcing signal is removed. What input levels,

---

[7] Strictly speaking, the S input should be labeled $\overline{S}$, since it is active low. We didn't dare label it that way, fearing people would start thinking that this was the complement of some signal. Perhaps we should have been more courageous. But this flop is conventionally is called SR rather than S-bar R-bar," so maybe this labeling is tolerable.

15N.6 Sequential circuits generally, and flip-flops

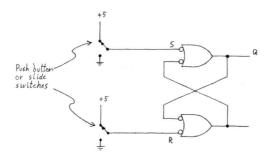

Figure 15N.12 Cross-coupled NANDs form a simple SR latch.

then, let it perform this *holding* task?[8] (Those input levels might be said to define the circuit's resting state.)

This device works, but is hard to design with. It is useful to *debounce* a switch; but nearly useless for other purposes, in this simplest, barebones form.

15N.6.3 Switch debouncers

AoE §10.4.1A

There are several ways to debounce a switch. The SR provides one way – for *double-throw* switches only (not for the simpler sort of switch that simply makes or breaks contact between two points). Fig. 15N.13 shows an SR latch wired as debouncer, and waveforms at S∗ and R∗, and at the debounced output. It shows the SR flop's ability to *remember* the last state into which it was forced. When S∗ rises to +5, and bounces low a few times, Q does not change in response to this bounce; the bounce simply drives the flop alternately into SET and REMEMBER modes. When R∗ first goes low, it forces Q low; when R∗ bounces high, Q does not change in response, because the flop simply remembers the low state into which it was forced. Again, it alternates between RESET and REMEMBER modes.

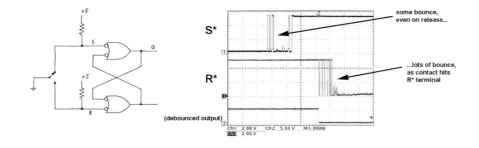

Figure 15N.13 Switch bounce: first hit sets or resets flop: bounce does not appear at output.

15N.6.4 But SR usually is not good enough

But in other settings this flop would be a pain in the designer's neck. To appreciate the difficulty, imagine the circuit in Fig. 15N.14 made of such primitive latches and including some feedback. Imagine trying to predict the circuit's behavior for all possible input combinations and histories. Painful.

Designers wanted a device that would let them worry less about what went on at all times in their circuits. They wanted to be able to get away with letting several signals change in uncertain sequence,

---

[8] The input combination for "holding" or "remembering" the last state into which it was pushed, is 11: high on both inputs, disasserting both $\overline{S}$ and $\overline{R}$.

578 Flip-Flops

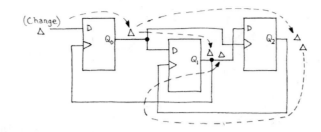

Figure 15N.14 Example meant to suggest that asynchronous circuits are hard to analyze or design.

for part of the time. Fig. 15N.15 shows the basic scheme they had in mind. This permits less perfect circuit behavior without alarming results.[9]

Figure 15N.15 Relaxing circuit requirements somewhat: a designer's goal.

The *gated SR latch*, or *transparent latch* takes a step in the right direction. This circuit is just the NAND latch plus an input stage that can make the latch indifferent to signals at S and R: the input stage works like a camera's *shutter*.

The two-input SR version shown on the left-hand side of Fig. 15N.16 normally is simplified to its D form shown on the right. This single input is more convenient and eliminates the problematic input case for the SR, in which *both* inputs are asserted.

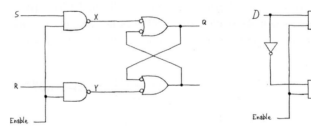

Figure 15N.16 Clocked or gated SR latch (left), and the more useful transparent D latch (right).

The circuit in Fig. 15N.17 achieves more or less what we wanted; but not quite. The trouble is that it's still hard to design with this circuit. Consider what sort of clock signal you would want, to avoid problems with feedback. You would need an ENABLE pulse of just the correct width: wide enough to update all the flops, but not so wide that an output change would have time to wrap around and call for a second state change. Ticklish.

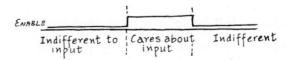

Figure 15N.17 Transparent latch moves in the correct direction: now we can stand screwy input levels part of the time.

Because the simple NAND latch is so hard to work with, the practical flip-flops that people actually use nearly always are more complex devices that are *edge-triggered*.

---

[9] Does this waveform remind you of the average undergraduate's week, except that a student's "do right" duty cycle may be lower: perhaps a narrow pulse on Sunday evening? If so, the student's week is more like the timing of the *edge-triggered* flop of §15N.6.5.

15N.6.5 Edge-triggering

Edge-triggered flip flops care about the level of their inputs only during a short time just *before* (and in some rare cases after) the transition ("edge") of the timing signal called "clock:" see Fig. 15N.18.

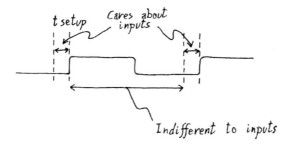

Figure 15N.18 Edge-triggering relaxes design requirements.

An older design, called by the nasty name, *master–slave*, sometimes behaved nastily[10] and was generally replaced by the edge-trigger circuit. The master–slave offers the single virtue that it is easy to understand, but we will not examine it here. (If you are curious to meet this circuit, see AoE §10.4.2).

The behavior called *edge-triggering* may sound simple, but it usually takes people a longish time to take it seriously. Apparently the idea violates intuition: the flop acts on what happened *before* it was clocked, not after. How is this possible? (*Hint:* you have seen something a lot like it on your scopes, which can show you the waveform as it was a short time *before* the *trigger* event. How is *that* magic done?[11])

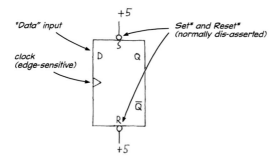

Figure 15N.19 74HC74 edge-triggered D flip-flop.

D (for data) flip-flop: Figure 15N.20 shows its crucial timing characteristics. A D-flop only *saves* information: it does not transform it. But that simple service is enormously useful.

Digression on the signal name "clock": The name "clock" for the updating signal introduced in §15N.6.5 is potentially misleading. It suggests, to a thoughtful reader, several notions that do not apply:

- the term suggests something that keeps track of, measures, or accumulates time;
- or, even short of that, it suggests at least something that *ticks* regularly: a sort of strobe or heartbeat.

---

[10] Some master–slave *JK* flops showed a pathology called "ones-catching": they were sensitive to events that occurred while the clock was at a single level, high, rather than being responsive only to a *change* in clock level.

[11] An answer to this riddle appears in §16N.1.1, especially Figs. 16N.7 and 16N.8.

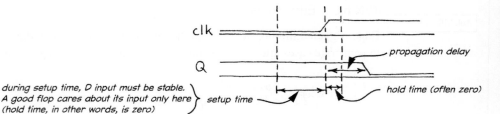

Figure 15N.20 D-flop timing: setup time and propagation delay; hold time should be *zero*.

Neither notion applies to the *clock* of a flip-flop. (A clock may be – and fairly often *is* – a steady square wave; but it need not be.)[12] A clock is no more than a trigger signal that determines *when* the flip-flop updates its memory.

A better word for the signal might be "update." But we are stuck with the term "clock." Just get used to disbelieving its metaphor. The engineers who named it were not competent poets.

Triggering on rising- versus falling-edge: The D-flop shown on the left in Fig. 15N.21 responds to a *rising* edge. Some flops respond to a *falling* edge instead. A flop never responds to *both* edges.[13] As with gates, the default assumption is that the clock is *active high*. An inversion bubble indicates active low: a *falling edge* clock.

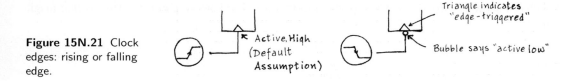

Figure 15N.21 Clock edges: rising or falling edge.

Edge- versus Level-sensitive flops and inputs: Some flip-flops or flop functions respond to a clock-like signal, but not to the *edge* itself; that is, the device does not lock out further changes that occur after the level begins. The **transparent** latch is such a device. We met this circuit back in Fig. 15N.16.

Examples of level-sensitive inputs:

- A "transparent latch," like those used in the LCD display board described in Lab 16L).
  The character displays we sometimes use in these labs work as shown in Fig. 15N.22.
- The latches included on the AD558 Digital-to-Analog Converter, which you will meet in §18L.1.1, work as shown in Fig. 15N.23. The transparent behavior allows one to ignore the input register, getting a "live" image of the digital input, simply by tying En* low.
- Reset on a flip-flop. The Reset* and Set* signals reach into the D-flop's *output latch*, and prevail over the fancy edge-trigger circuit that precedes the out latch; see Fig. 15N.24. This so-called "jam type" takes effect at once.

    This set-or-reset scheme is universal on flip-flops, not quite universal on counters. But a reset is never treated as a clock itself – even in the cases (called "synchronous") where the action of

AoE §10.4.2A

---

[12] It is not a steady wave, for example, when the clock signal comes from a pushbutton – as in the single-step logic of the lab microcomputer. In that circuit, a user presses a button; a debounced version of this button-press applies a rising edge to a D-flop whose D input is tied high. Such schemes are quite common.

[13] Well, *hardly* ever. Flip-flops using both edges seem to be confined to programmable logic such as the Xilinx Coolrunner II.

15N.6 Sequential circuits generally, and flip-flops

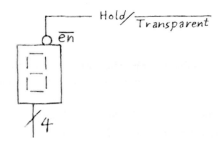

Figure 15N.22 Level-sensitive: transparent latch: Output follows or displays input while En* is low.

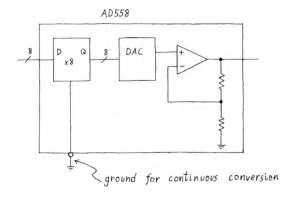

Figure 15N.23 Transparent latch on DAC allows one to use the DAC as continuous converter, if necessary.

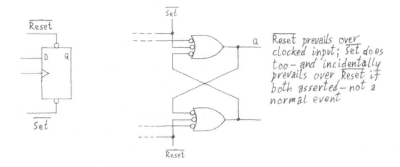

Figure 15N.24 Reset* and Set* on 74xx74 D-flop.

resetting "waits for the clock." The edge-triggered behavior is reserved for *clocks*, with only the rarest exceptions (like the 8051's interrupt-request lines).[14]

Edge-triggering usually works better than level-action (that is, makes a designer's tasks easier); edge-triggering therefore is much the more common configuration.

---

[14] These INT* lines are not quite true edge-triggered inputs, but "pseudo-edge sensitive;" they respond to a transition, detected as a difference between levels before and after a processor clock. Their behavior is, however, very close to edge-sensitive.

15N.7 Applications: more debouncers

15N.7.1 D-flop and slow clock

A D-flop can debounce a switch if you have a slow clock available to you. The strategy is to make sure that the flop cannot be clocked twice during bounce. If it cannot, then the output – Q – cannot double back.

Unlike the SR debouncer, this method introduces a delay between switch-closing and output. Occasionally that matters (for example, in the reaction timer described in §19L.1.2); usually it doesn't. This debouncing method has the virtue of working with the simpler, cheaper switch: the single-pole type (SPST). Fig. 15N.25 shows the dead-simple circuit:

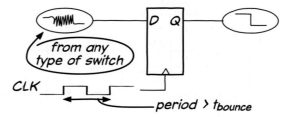

Figure 15N.25 Another debouncer: D-flop and slow clock.

In Figs. 15N.26 and 15N.27 are some waveforms making the point that the clock period must exceed bounce time. When the clock runs too fast, as in Fig. 15N.26, the flop *passes* the bounce to the output: not good.

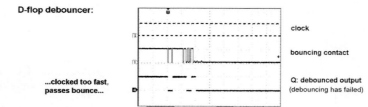

Figure 15N.26 D-flop debouncer must use a clock period greater than bounce time: here, clock is too fast.

A slower clock, as in Fig. 15N.27 does the job: the main drawback of this debounce method – in addition to the uncertain *delay* – is the need to generate the slow-clock signal. If you need to debounce a single switch, it is not an appealing method. If you need to debounce several lines – as in our powered breadboards, where two switches are debounced, or in the microcomputer keypads that you soon will meet, where five lines need debouncing – the slow-clocked flop method makes sense.

15N.7.2 Positive-feedback debouncer

The simplest debouncer, for a double-throw switch ("SPDT") is just a non-inverting CMOS gate with feedback: see Fig. 15N.28. At first glance, this may look too simple to work – but work it does. The key notion is that the gate input *never floats*; the gate input is always driven by the gate output, even when the switch is in transition, not touching either +5 or ground.

15N.8 Counters

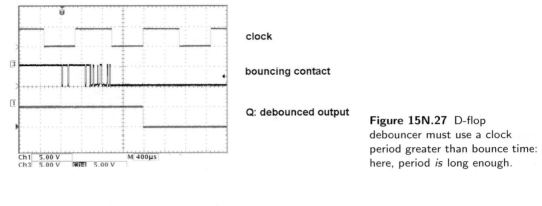

Figure 15N.27 D-flop debouncer must use a clock period greater than bounce time: here, period *is* long enough.

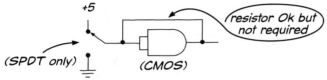

Figure 15N.28 Another debouncer: positive-feedback with CMOS gate.

15N.8 Counters

15N.8.1 Ripple counters

Ripple counters are simple, but their weakness is that their Qs cannot change all at the same time. The progressing waveform visible in Fig. 15N.30 gives the counter its name.

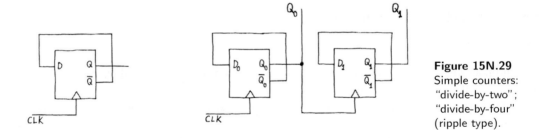

Figure 15N.29 Simple counters: "divide-by-two"; "divide-by-four" (ripple type).

Because the several flops change state in succession rather than all at once, not only are they slow to reach a stable state but also they necessarily produce *transient* states that ought not to be present.

The two-bit counter in Fig. 15N.29 is a *down* counter, because the second flop changes only when the first is *low* before the clock edge. It is easily transformed into an *up* counter.[15]

An integrated ripple counter, like the 74HC393, forms an *up* counter by using a falling-edge clock. Its ripple delays are less than those shown by the counter built from 74HC74 D-flops (Fig. 15N.30). The '393 flop-to-flop delays are about 4ns (Fig. 15N.31).

---

[15] How? Just treat the $\overline{Q}$ as outputs; or, equivalently, treat Q as an output but use $\overline{Q}$ to do the ripple *clocking*.

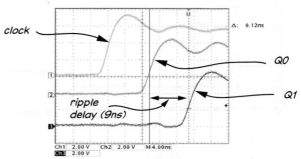

Figure 15N.30 2-bit ripple counter shows delay between Qs – thus false transient states. (Scope settings: 2V/div, 4ns/div.)

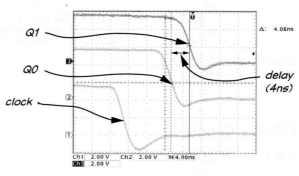

Figure 15N.31 Integrated ripple counter shows less delay between Q's – but still ripples. (Scope settings: 2V/div, 4ns/div.)

15N.9 Synchronous counters

15N.9.1 What synchronous means

A synchronous circuit sends a common clock to all of its flip-flops; so they all change at the same time.[16] This is the sort of design one usually strives for.

The signature of the synchronous counter, on a circuit diagram, is just the connection of a common clock line to all flops (hence syn-chron: same-time). The synchronous counter – like synchronous circuits generally – is preferable to the ripple or *asynchronous* type. The latter exist only because their internal simplicity lets one string a lot of flops on one chip (e.g., the "divide by 16k" counter we have in the lab – the '4020, or the more spectacular '4536: divide-by-16M).

Figure 15N.32 shows a generic synchronous circuit; there's no clue to what it may be doing, but its synchronous quality is very obvious.

15N.9.2 Synchronous counting requires a smarter flip-flop

Making a counter that works this way requires a flop smarter than the one used in the ripple counter. Whereas the ripple counter's flops change state whenever clocked, the sync counter's flops are clocked continually, and should change only sometimes. More specifically, the second flop in the timing diagram Fig. 15N.34 and circuit Fig. 15N.35 – Q_1 – should change on every second clock; if there were

---

[16] Except to the small extent that propagation delays of the several flops may differ.

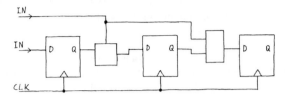

Figure 15N.32 Generic synchronous circuit: common clock is its signature feature.

a third flop, Q_2, it should change on every fourth clock, and so on. In order to get this behavior from flops that are clocked continually we need a flop (to implement Q_1 here) that will, under external control, either *hold* its present state or *toggle* (flip to the other state). That is called "Toggle Flop" or "T-flop" behavior.

And XOR can make a D-flop into a T: It turns out that an XOR gate is just what we need in order to convert a humble D-flop into the smarter T-flop. We noted, back in §15N.4, that the XOR function can be described as one that lets one input determine whether the other input is *passed* or *inverted*. With the help of the XOR, then, we can make a D-flop into a T – one that changes on the next clock only if its T input is driven high: see Fig. 15N.33.

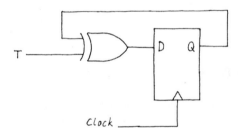

Figure 15N.33 An XOR as pass*/invert can convert a D-flop into a T.

15N.9.3 The counter behavior we want

Figure 15N.34 shows a timing diagram for a synchronous divide-by-four, noting the cases where we want Q_1 to change and those where it should *hold*. We will connect D- and T-flops as required, to get this desired behavior.

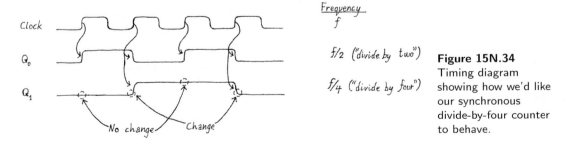

Figure 15N.34 Timing diagram showing how we'd like our synchronous divide-by-four counter to behave.

Since the low-order flop, Q_0, changes on every clock, it doesn't need to be smart: just tie Q_0* back to D_0, as in the ripple counter. It's Q_1 that needs to be a little smarter.

In two cases shown in Fig. 15N.34 Q_1 holds its present state after clocking (these cases are labeled

"No change," in the figure); in two other cases Q_1 changes state. What we need to arrange then, is to have the second flop's T input driven high each time we want change on the next clock, and driven low when we want Q_1 to hold on the next clock.

Setup-time: an important subtlety: In making a synchronous design, it is essential to drive all the inputs properly *before* the clock edge (in our case, the rising edge). It won't do therefore, to look at Fig. 15N.34 and say, "Oh, I see: Q_1 should change whenever Q_0 falls." That information – "Q_0 has fallen" – comes too late. Our circuit needs to drive T_1 appropriately *before* the next clock edge. More precisely, it needs to drive T_1 early enough so that D_1 is in the correct state by *setup-time* before the clock edge. (For the 74HC74, t_{setup} is 16ns, max; the *typical* value – versus *max* – is just a few nanoseconds.)

So Q_1 will behave as we want if we tell it, "Change on the next clock if Q_0 is high." The simple wiring in Fig. 15N.35 implements that design.

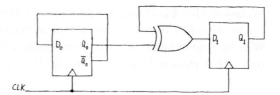

Figure 15N.35 Synchronous counter.

15N.10 Another flop application: shift-register

AoE §10.5.3

The circuit in Fig. 15N.36 is about as simple a flop circuit as one could imagine: just a sort of daisy chain.

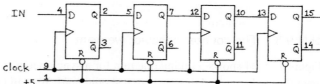

Figure 15N.36 Lab shift register.

A shift-register generates predictable, orderly delay; it shifts a signal *in time*; it can convert a *serial* stream of data into *parallel* form, or vice versa. Serial to parallel is the easy direction: see Fig. 15N.37. Such a circuit is useful because it permits using a small number of lines to transmit many bits of information (the minimum number of lines for a simple shift register would be two: data and clock; some schemes – more complex – can omit the clock).[17] Converting from parallel to serial is not quite so simple. It requires *multiplexer* logic feeding each D (except the first): one must be able first to steer parallel data into each D "loading" the shift register, then to revert to the normal daisy-chained shift-register wiring in order to shift out the serial stream.

In the lab you will use a shift register for a different purpose: to generate a pulse; that is, to act like a so-called "one-shot." A NAND gate added to the shift-register does the job. (We use two because

---

[17] We will glimpse these clock-less serial protocols later in the course: we will see it in the traditional RS232 serial scheme (§24N.4.1), and in the more recent USB (which, however, uses two data lines because it uses differential signalling for good noise immunity at modest voltages).

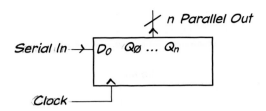

Figure 15N.37 Shift-register can convert data from serial form into parallel.

in a later lab we need a double-barreled one-shot). See if you can sketch its timing diagram, and then what happens if you wire a NAND gate, as in the lab, to detect $Q_0\overline{Q_1}$.

15N.11 AoE Reading

Reading:
- Chapter 10 (Digital Logic I):
  - §10.4: sequential logic
  - in §10.5, Sequential Functions available as ICs, see especially
    - §10.5.1: latches and registers
    - §10.5.4: PLDs
  - §10.8: logic pathology, especially
    - §10.8.2: switching problems
    - §10.8.3: weaknesses of TTL and CMOS
- Chapter 11 (Digital Logic II):
  - §11.1: PLD history
  - §11.2.1: PAL structure
  - give just a hasty glance at the FPGA (§11.2.3)

Specific advice:
You need not worry about Karnaugh mapping. It was a useful device in the era before computer logic compilers. These have now made Karnaugh mapping obsolete.
Concentrate on flip-flops and counters in §10.4.

15L Lab: Flip-Flops

15L.1 A primitive flip-flop: SR latch

This circuit, Fig. 15L.1, the most fundamental of flip-flop or memory circuits, can be built with either NANDs or NORs. We will build the NAND form. It is called an *SR* latch because it can be "Set" or "Reset." In the NAND form it also is called a "cross-coupled NAND latch." Build this latch, and record its operation. Note, particularly, which input combination defines the "memory state;" and make sure you understand why the state is so called.

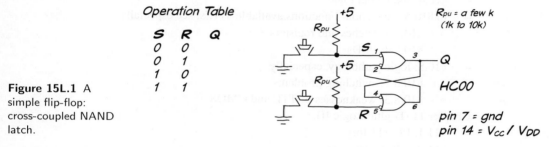

Figure 15L.1 A simple flip-flop: cross-coupled NAND latch.

Leave this circuit set up. We will use it shortly.

15L.2 D type

Practical flip-flops It turns out that the simple *latch* is very rarely used in circuit design. A more complicated version, the *clocked* flip-flop is much easier to work with. The simplest of the clocked flip-flop types, the D, simply saves at its output (Q) what it saw at its input (D) just before the last clocking edge. The particular D flop used below, the 74HC74, responds to a *rising* edge.

The D flop is the workhorse of the flop stable. You will use it 100 times for each time you use the fancier JK (a device you may have read about, but which we are keeping out of the labs because it is almost obsolete). Perhaps you will never use a JK.

15L.2.1 Basic operations: saving a level; reset

The D's performance is not flashy, and at first will be hard to admire. But try.

Feed the D input from a breadboard slide switch. Clock the flop with a "debounced" pushbutton. The pushbutton switches on the left side of the breadboard will do. Note that these switch terminals need *pull-up* resistors, since they have *open-collector* outputs. (This you saw last time; but perhaps you've forgotten.)

Warning: a surprising hazard. The value of the resistor used as *pull-up* on CLOCK turns out to matter. A large pull-up value (say, ≥ 10k) is likely to cause mischief: that is explored on page 590.

15L.2 D type

Dis-assert $\overline{\text{Reset}}$ and $\overline{\text{Set}}$ (sometimes called $\overline{\text{Clear}}$ and $\overline{\text{Preset}}$), by tying them high.

Note that the '74 package includes *two* D-flops. Don't bother to tie the inputs of the unused flop high or low: that is good practice when you build a permanent circuit (averts possible intermediate logic state that can waste power, as you saw last time) – but would slow you down unnecessarily, as you breadboard circuits.

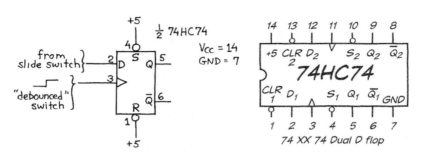

Figure 15L.2 D-flop checkout.

- Confirm that the D-flop ignores information presented to its input until the flop is clocked.
- Try asserting $\overline{\text{Reset}}$. You can do this with a wire; bounce is harmless here. (Why?[1]) What happens if you try to clock in a high at D while asserting $\overline{\text{Reset}}$?
- Try asserting $\overline{\text{Set}}$ and $\overline{\text{Reset}}$ at the same time (something you would never purposely do in a useful circuit). What happens? (Look at *both* outputs.) What determines what state the flop rests in after you release both?[2] (Does the answer to that question provide a clue to why you would not want to assert both $\overline{\text{Set}}$ and $\overline{\text{Reset}}$ in a circuit?)

15L.2.2 Toggle connection: version I: always change or "divide-by-two"

The feedback[3] in the circuit in Fig. 15L.3 may trouble you at first glance. (Will the circuit oscillate?) The *clock*, however, makes this circuit easy to analyze. In effect, the clock breaks the feedback path. Build this circuit and try it.

Figure 15L.3 D-flop biting its own tail.

- First, clock the circuit manually (but see page 590 concerning the value of the pullup resistor).
- Then clock it with a square wave from the function generator (the breadboard generator is less good than an external generator, with its higher f_{max}). Watch Clock and Q on the scope. What is the relation between f_{clock} and f_Q? (Now you know why this humble circuit is sometimes called by the fancy name "divide-by-two.")
- Crank up the clock rate to the function generator's maximum, and measure the flop's *propagation delay*. In order to do this, you will have to consider what In and Out voltages to use as you measure the time elapsed. You can settle that by asking yourself just what it is that is "propagating." If the answer to this question is that it is "a change of logic level" that propagates, then what is the appropriate voltage at which to measure propagation delay?[4]

---

[1] The first low achieves the reset. Bounce – causing release and then reassertion of Reset* causes no further change in Q.
[2] The result is unpredictable, because it depends on which of the two inputs – S* or R* – has the last word: that is, which is the last to cross its threshold, rising from low to high.
[3] See Chapter 15N for difficulties that feedback introduces into non-clocked sequential circuits.
[4] Since it is logic-level *change* that is propagating, we should measure time the voltage that typically defines a level change. For 5V CMOS that is 2.5V.

Looking for Trouble?

Here's a side excursion for the adventurous.

We said "clock the circuit manually," and assume that you would carry out this step by using one of the PB503's *debounced pushbuttons* – probably the NC, so that you get a clock upon pressing the pushbutton:

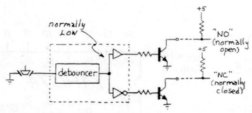

Breadboard's debounced switches: these require a pullup resistor.

You need to choose a value for the pullup resistor. At first glance, the value doesn't seem to matter, as long as it's not so tiny that it overloads the transistor switch. But in fact the pullup value *does* matter. To help you see how the flop is behaving, display its Q on one of the breadboard's LED's. Now, try $R_{pullup}=1k$. This should work well: the flop should toggle each time you press the button.

Now try $R_{pullup}=1M$. This should work badly: you should see the Q *sometimes* toggle – and at other times appear not to respond at all.

This occurs because of the slow rising edge provided by the over-large R_{pullup}. This big R, driving stray capacitance, produces a slowly-rising edge that an edge-triggered device finds troublesome. The problem is exactly the one that you saw causing weird instability in the LM311 comparator, back in §8L.1.2 (remember the "Taj Mahal by moonlight"?).

You will find a scope image detailing the effect of such a slow edge in §16N.2.1. Use a scope to watch clock and Q, triggering on Q, and see if you can make out what goes wrong when the value of R_{pullup} is too large.

Then restore the small R_{pullup} of 1k to clock your later circuits properly.

15L.2.3 Toggle connection. Version II: change when told to or T-flop

A more useful *toggle* circuit uses an input to determine whether the flop should change state on the next clock. This behavior is properly called "T" or "Toggle." (The preceding circuit – which toggles always – is not called a T-type; the best name for it is probably "divide-by-two.")

Figure 15L.4 shows how the circuit should behave. Show how to use an XOR gate and a D-flop to make such a T-flop. Note that an XOR can serve as a controllable inverter if you use one input to the XOR as Signal, the other as Control as you saw in Chapter 15N.

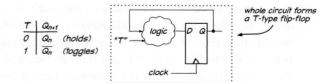

Figure 15L.4 T-flop behavior.

To exercise your T-flop, clock it from the function generator, control it with a manual switch, and watch clock and Q on a scope.

Keep this T-flop set up; we will use it again.

15L.3 Counters: ripple and synchronous

15L.3.1 Ripple counter

If you wire a flop to toggle on *every* clock (the easiest way to do this is with a D-flop[5]) and cascade two such flops, so that the Q of one drives the clock of the next, you form a "divide by four" circuit. One flop changes on *every* clock; the other changes on *every other* clock. Evidently, you could extend this scheme to form a divide-by-a-lot. Today, we won't go further than divide-by-four.

Wire a second D-biting-its-tail flop; cascade the two flops to form just such a divide-by-four *ripple* counter. Show what the circuit looks like.

- Watch the counter's outputs on two LEDs while clocking the circuit at a few hertz. Does it divide by four? If not, either your circuit or your understanding of this phrase is faulty. Fix whichever one needs fixing.

  This counter behaves oddly, in one respect: it counts *down*. An easy way to make it count *up* (or are we only making it *appear* to count up?) is just to watch the $\overline{Q}$ outputs rather than the Q's. (Ripple counters usually are built with *falling-edge* clocks, instead; then the Qs do count up.)

- Now clock the counter as fast as you can, and watch Clock and first Q_0 then Q_1 on the scope. Trigger on Q_1.

- Watch the two Qs together and see if you can spot the rippling effect that gives the circuit its name: a lag between changes at Q_0 and Q_1. (If you are using an analog scope, you will have to sweep the scope about as fast as it will go, while clocking the counter fast to make the display acceptably bright.)

15L.3.2 Synchronous counter

The timing diagram in Fig. 15L.5 shows how a *synchronous* divide-by-four counter ought to behave.

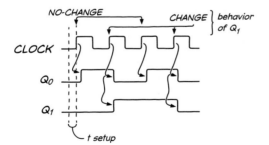

Figure 15L.5 Timing diagram showing behavior wanted from synchronous divide-by-four counter.

As you know, the diagram is the same as for a ripple counter – except for one important point: in the synchronous design, *all flops change at the same time* (at least, to a tolerance of a few nanoseconds). Achieving this small change in behavior requires a thorough redesign of the counter.

You will need a T-flop to make a synchronous counter. Then the key is to find, on the timing diagram, the pattern of pre-clock conditions that determine whether Q_1 ought to change (toggle). To permit synchronous behavior, we must look at conditions *before* the clock edge (during set-up time); no fair to say something like "let Q_1 change if Q_0 falls;" that scheme would carry us back to the ugly ripple behavior.

---

[5] You don't need the fancier "T" behavior to build a ripple counter. You *will* need it to make a synchronous counter.

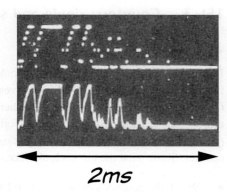

Figure 15L.6 Microswitch bouncing from high to low (pulled up through 100k. Note that this pullup value is too large to provide a healthy fast-rising clock edge).

Once you have discovered the pattern, drive T_0 appropriately, and try your design. Again, *keep this counter set up*.

See if you can use the scope to confirm that the *ripple* delay now is gone: watch Q_0 and Q_1 again. Synchronous counters are standard, ripple counters rare. In many applications, the ripple counter's slowness in getting to a valid state and its production of transient false states are unacceptable.

15L.4 Switch bounce, and three debouncers

Figure 15L.6 is a photograph of a storage scope[6] showing a pushbutton bouncing its way from a high level to low.

To see the harmful effect of switch bounce, clock your divide-by-four counter with a (bouncy-) ordinary switch such as a microswitch pushbutton as in Fig. 15L.7. Watch the counter's outputs on two LEDs. The bouncing of the switch is hard to see on an analog scope (easy on a digital), but its effects should be obvious in the erratic behavior of the counter.

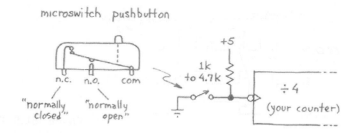

Figure 15L.7 Switch bounce demonstration.

15L.4.1 Watching switch bounce (Optional: for scope enthusiasts)

Switch bounce is hard to see on an analog scope, because it does not happen periodically and because the bounces in any event do not occur at exactly repeatable points after the switch is pushed. (On a *digital* scope, on the other hand, freezing an image of recorded switch bounce is dead easy.)

You can see the bounce, at least dimly, however, if you trigger the analog scope in normal mode with

---

[6] Incidentally, this "storage scope" was an ordinary analog scope fed by a microcomputer of the kind you will build later in this course. The computer took samples during the bouncing process, stored them in memory, then played them back repeatedly to give a stable display. You will have a chance to try this, if you like, during a microcontroller lab, §24L.1.3 or §25L.2.3.

a sweep rate of about 0.1ms/cm. You will need some patience, and some fine adjustments of trigger level. Some switches bounce only feebly. We suggest a nice snap-action switch like the microswitch type.

15L.4.2 Eliminating switch bounce: three methods

AoE §10.4.1A

Cross-coupled NANDs as debouncer Return to the first and simplest flip-flop (which we hope you saved), the cross-coupled NAND latch, also called an SR flop. As Input use the bouncy pushbutton. Ground the switch's *common* terminal, and make sure to include pullup resistors on both flop inputs (they should be there, still); see Fig. 15L.8.

Why does the *latch* – a circuit designed to "remember" – work as a debouncer?[7]

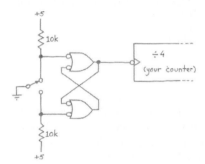

Figure 15L.8 NAND latch as switch debouncer.

CMOS buffer as debouncer (positive feedback) Figure 15L.9 shows a much simpler way to debounce when you use a double-throw switch (SPDT) like the one used in the preceding exercise, debouncing with NANDs, as in Fig. 15L.8: just use a non-inverting CMOS gate[8] with its output talking to its input.

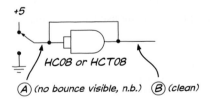

Figure 15L.9 CMOS AND gate as switch debouncer.

This one is no fun to *watch* – because *no bounce* will be visible, even at the input. Once you see why this is so, you will have understood why this circuit works as a debouncer. In particular, what happens during the time when the switch input to the AND is connected to neither +5 nor ground?

D-flop as debouncer; discovering duration of switch bounce A plain D-flop can debounce, if clocked appropriately. Wire up this circuit to test the notion. Fig. 15L.10 suggests using one flop of a 4-flop '175 because you are about to use this part in another circuit.

Start with a high-frequency square wave driving the D-flop clock (say, 100kHz or more). You should see evidence of bounce in the counter's misbehavior. Now lower the clock rate to 100Hz. Bounce should cease to trouble the counter. Raise the clock frequency till evidence of bounce reappears. Note

---

[7] It does because *bounce* makes the inputs revert to their *memory* state, in which they simply hold the state into which they were pushed when the particular input earlier went *low*.

[8] This circuit does not work reliably with TTL: it's a CMOS special.

Lab: Flip-Flops

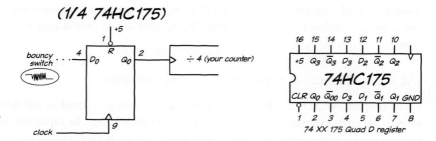

Figure 15L.10 D-flop as debouncer.

the clock period (you'll get the best information by watching the clock on the scope of course; you'll get a ball-park *frequency* from the dial on the function generator). That clock period reveals how long your switch is bouncing.

15L.5 Shift register

This is an important device – but *some people will have run out of time around this point*. Don't worry if this happens to you. The circuit is not hard to understand even if you don't get a chance to build it. The digital *one-shot* of §15L.5.3, based on the shift-register, is used in a later circuit, a "capacitance-meter." This circuit is described in Lab 16L and also appears among suggested digital projects. But you can build all of this later if you find you need the one-shot and don't have time for it right now. Some people will not build the capacitance-meter and thus will not need the one-shot.

With that in mind, build the circuit in Fig. 15L.11 (a *digitally-timed one-shot* that evolves from the shift-register) on a *private breadboard*; you may use this circuit next time. We use a convenient structure in this circuit that includes four flops with a common clock. Such a configuration is called a "register," and here it is applied to the particular use as *shift-register*. The circuit delays the signal called "IN," and *synchronizes* it to the clock. Both effects can be useful. You will use this circuit in a few minutes as a *one-shot* – a circuit that generates a single pulse in response (maintaining the metaphor[9]) to a "trigger" input (here, the signal called IN).

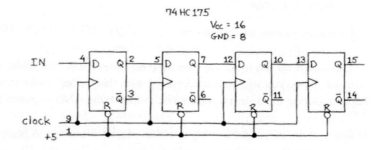

Figure 15L.11 Shift-register.

Clock the circuit with a logic signal from an *external* function generator; use the breadboard's oscillator to provide IN. Let f_{clock} be at least $10 \times f_{\text{IN}}$.

---

[9] As you may have gathered already, engineers have at least some of the characteristics of good poets: they keep the images simple and vivid. They call the pedant's "monostable multivibrator" a one-shot; they call the pedant's "bistable multivibrator" a "flip-flop," as you know. They call the active-pullup output stage of a TTL gate "totem pole," because that's what it looks like. The synchronizing color-burst in a TV signal sits on the waveform's "back porch." There's lots of this vivid imagery in engineering. Can the language of the social sciences offer any comparable pleasures?

15L.5.1 One flop: synchronizer

- Use the scope to watch IN, and Q_0 (Q of the leftmost flop); trigger the scope on IN.
- What accounts for the *jitter* in signal Q_0?
- Now trigger on clock, instead. Who's jittery now?
- Which signal is it more reasonable to call jittery or unstable? (Assume that the flops are clocked with a system clock: a signal that times *many* devices, not just these 4 flip-flops.)[10]

15L.5.2 Several flops: delay

- Now watch a later output – Q_1, Q_2, or Q_3, along with IN. (We'll leave the triggering to you, this time.)
- Note the effect of altering f_{clock}.

15L.5.3 Several flops plus some gates: double-barreled one-shot

A little logic – used to detect particular *states* of the shift-register – can produce pulses of fixed duration in response to an input pulse of arbitrary length (except that the pulse must last long enough to be sure of getting "seen" on a clock edge: in other words, the input pulse must last longer than a clock period). The one-shot's input signal is called the 'trigger.

In Fig. 15L.12 we show a pair of output pulses we would like you to produce. If you fill in the shift-register waveforms, you will discover the logic you need in order to produce those pulses.

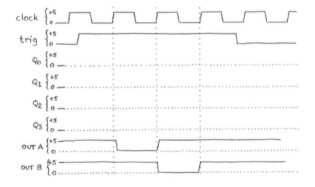

Figure 15L.12 Timing diagram for digitally-timed one-shot.

Incidentally, this all-digital circuit – whose output pulse is timed by the clock – is not what people usually mean when they say "one-shot:" the most common form uses an *RC* to time the pulse width. Such a one-shot is sometimes more convenient, but lacks the great virtue of *synchrony* with the rest of the digital circuit.

Draw your own design and checkout the following.

- Slow-motion: first use a manual switch to drive *Trigger*, and set the clock rate to a few hertz. Watch the one-shot outputs on two of the breadboard's buffered LEDs. Take *trigger* low for a second or so, then high. You should see first one LED then the other flash low, in response to this low-to-high transition.

---

[10] Since we treat the clock as the reference timebase, when we trigger on clock we notice the uncertainty of the input waveforms timing. It makes sense to say that the input is "jittery."

- Full-speed: when you are satisfied that the circuit works, drive *trigger* with a square wave from one function generator (the breadboard's) while clocking the device with an external function generator (at a higher rate).

How would you summarize the strengths and weaknesses of this one-shot relative to the more usual *RC* one-shot?[11]

---

[11] Well, in case you're interested, here are our views: Strength: the digitally-timed one-shot's great virtue is that its output is *synchronous* with the system clock, so its pulse begins and ends at a predictable time, shortly after the clock edge, and safely away from t_{setup}. The only weaknesses are the greater complexity of the circuit relative to a traditional *RC*-timed one-shot, and the circuit's *latency*: the output can be delayed as much as one full clock cycle from the rise of trigger. The digital version also will not respond at all to a trigger signal that is too brief.

15S Supplementary Note: Flip-Flops

15S.1 Programmable logic devices

Hearing "programmable," you may think of memories, which are indeed "programmable logic devices." But PLD is a term reserved for devices organized differently from memories, and designed, broadly speaking, to replace discrete gates rather than to store data and code as memory ordinarily does.[1] Programmable devices have superseded the small packages of gates like those you met in Lab 14L: packages that provide four 2-input NANDs, for example 74HC00. A PLD is much more versatile than a cabinet full of 74xx gates, and very much denser, and therefore makes a design cheaper to implement and much easier to amend. The appeal of PLDs is irresistible.

15S.1.1 Varieties of programmable devices

AoE §11.1

AoE §11.2.1

PALs and GALs PLDs began to appear in the mid-1970s. After some experimentation with the form of internal logic, one dominant scheme appeared: the so-called PAL (a trade name for Programmable Array Logic – a clever rearrangement of letters to form a user-friendly acronym). This is a bunch of wide AND gates feeding an OR ("sum of products" form): see Fig. 15S.1.

The simplicity of this structure made the parts fast – somewhat faster than little packages of gates ("Small Scale Integration," SSI), because internal capacitances were lower than printed-circuit capacitances. Reduced capacitances not only let logic levels change faster but also permitted the gates to omit driver stages. The result was greater speed.

A variety of PALs appeared: some with active-low outputs, others active-high; some with flip-flops included; some with a few wide ANDs, others with narrower but more numerous ANDs feeding the OR. This variety was good in a way, and also a nuisance, requiring users to stock lots of parts. Early PALs, like early Read Only Memories (ROMs) were programmed by selectively melting or not melting a tiny fuse on each signal line; hence use of the term "burn" for *program*. Later versions of PALs were reprogrammable.

Figure 15S.1 PAL cell.

To solve the problem caused by the excessive variety of PALs that a user had to stock, another company developed a more versatile standard circuit: a "generic" design that could be configured as convenient: active-high or -low, flop included or not. These, see Fig. 15S.2, got the tradename GAL.[2]

Early PALs held a few hundred gates. Such PALs (or PLDs) nowadays are labeled PLD, in contrast to more recent PLDs which can house thousands of equivalent gates. These proudly call themselves CPLDs: "Complex PLDs."[3] The elements that make up the CPLD are called "macrocells."

---

[1] It is possible to use memories to replace logic: address can serve as inputs, stored data as outputs. But this is not often done because of the unnecessary complexity of a memory's innards. One class of programmable logic device, however, forms a large exception to this remark: Field Programmable Gate Arrays (FPGAs) do indeed use small memory blocks – so-called "lookup tables" – to form logic functions. A bit more on FPGAs appears below.

[2] The name surely was chosen to appeal to a lonely Dilbert in his cubicle: even more user-friendly than a PAL.

[3] The PLD that we use in this course holds 1600 gates. In a few years, CPLD will stand, no doubt, for "Cute (little) PLD." The growth of complexity is relentless.

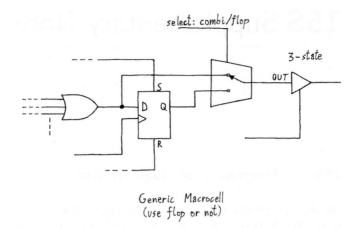

Figure 15S.2 GAL cell.

15S.1.2 FPGAs

Meanwhile, fancy designs were implemented on big arrays of transistors – programmable in a sense, but not reconfigurable: these were designed to do one particular job, and then were manufactured in large quantities in order to recover the high cost of their design.[4] Because they were custom parts, they were called "Application Specific Integrated Circuits," ASICs (this is acronym land, all right!). The gulf between the costs of PLDs (even CPLDs) and ASICs was enormous (the latter might cost $100,000 to design; the former were available for a couple of dollars, to be designed and programmed in hours). Only in large quantities did the ASICs cost less, per unit.

Then in the late 1980s arrived an intermediate part, programmable, but with an underlying structure more flexible than the PALs: a gate array rather like an ASIC, but *programmable* by the user: in the field, rather than the factory, so to speak. These were called "Field Programmable Gate Arrays": FPGAs. The largest of these now contain so many gates (*millions*) that they can incorporate microprocessors and their peripherals. They are somewhat more difficult to work with than PLDs, whose delays are quite predictable because of the rigidity of their structure. Timing in the FPGA depends on how the elements are laid out – and even delays in signal paths matter, now that gate delays have been cut so low. So the fine points of FPGA design are difficult.[5]

But for the less demanding tasks, FPGAs or CPLDs keep their details unobtrusive: one can write a design in a high-level hardware description language like Verilog or VHDL, and implement the design in either CPLD or FPGA. For the less complex designs one normally would prefer the less complex and cheaper part – and the one with the more predictable timing: the CPLD. In this course we will use these, since our designs don't rise to the scale that requires an FPGA.

AoE §11.2.3

15S.1.3 The innards of a PLD

Figure 15S.2 shows the design of a macrocell, one of the logic sub-units included in a smallish PLD, the Xilinx XL/XC9572, which includes 72 such blocks.[6] As you'll recognize, it is a versatile GAL, since it can be programmed to include or bypass the flip-flop, and to select among several other characteristics (such as active level).

---

[4] So-called "Non-recurring Engineering Costs," NRE.
[5] For an introduction to FPGAs that is unusual for being entertaining as well as authoritative, see Clive Maxfield's *The Design Warrior's Guide to FPGAs: Devices, Tools and Flows*, Newnes (2004).
[6] Material based on or adapted from figures and text owned by Xilinx, Inc., courtesy of Xilinx, Inc.: "XC9500XL High-Performance CPLD Family Data Sheet" © Xilinx (2009).

15S.2 Flip-flop tricks

You'll recognize that this is a versatile piece of hardware. To use it, we need to become acquainted with the logic compiler language that allows one to take advantage of this capable device.[7] We have chosen Verilog. Appendix A offers an introduction to Verilog that we hope you will consult as you try out PLDs.

15S.2 Flip-flop tricks

In case it's useful to see in one place most of the flip-flop configurations that you have met in the course, here are important ones. Despite this section's title, we don't mean to disparage these circuits. They are extremely useful. We've included Verilog code to implement each of these circuit fragments.

Simplest: set/reset (SR) flop: asynchronous set and reset: Figure 15S.3 shows a SET*/RESET* circuit done with a standard D-flop (usually that's easier than building the thing with gates). We've also shown the SR as you might build it with gates.

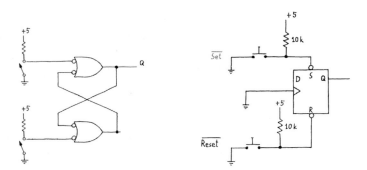

Figure 15S.3 SR flop, made with a D flop IC.

This simple circuit can be designed in Verilog in either of two ways, more or less mimicking the two circuits sketched in Fig. 15S.3:

- one design literally implements the cross-coupled NAND logic, as a combinational circuit;
- the other uses a transparent latch with asynchronous Preset.

Cross-coupled NAND implementation: Figure 15S.4 shows what the first design looks like. This design is literally faithful to the original and simplest SR latch design. It makes Verilog nervous, because it involves a "loop," or a self-contradiction. Verilog warns "the following signal(s) form a combinatorial loop: qbar, q." But it works.

Transparent-latch implementation: We can start with a flip-flop instead of gates. Verilog uses a transparent latch. The circuit in Fig. 15S.5 is not quite the one sketched on the right of Fig. 15S.3. Rather than use an asynchronous Reset* as we did, Verilog uses a transparent latch fed an input LOW. Verilog is smart enough to see that this is exactly equivalent to driving Reset*. Take a look at the innards of a transparent latch if you doubt this.

Figure 15S.6 shows what a transparent latch with a LOW D input looks like. The LOW on D drives the Reset* of the output SR latch, as required.

---

[7] And we do mean "acquainted," not "intimate with" or "expert at using." We do not have time, in this hurried course, to introduce more than a smattering of the language we have chosen.

Supplementary Note: Flip-Flops

The flop version is preferable: Though both designs work, the flop version is better, because the cross-coupled NAND circuit – despite its extreme simplicity – uses two of the PAL's *macrocells*. Each NAND uses an entire cell. That won't hurt if you've lots of these to spare, but it seems foolish to use two cells where one is sufficient.

What's good and bad about this simple SR latch?

Virtues: Simplicity; useful as a debouncer for a double-throw switch (SPDT)
Vices: Reset fails if Set persists during attempted Reset. So edge-triggered variation (Fig. 15S.7 below) is usually preferable.

Edge-sensitive "flag": async edge set, async reset The Q of this flop often serves as a "flag" – a signal that persists until deliberately knocked down.

Virtue: Flop can be Reset despite persistence of the signal that set Q high.

```
module s_r_latch(
    input s,
    input r,
    output q,
    output qbar
);

    assign q = ~s | ~qbar;
    assign qbar = ~r | ~q;

endmodule
```

Figure 15S.4 SR latch implemented with cross-coupled NANDs.

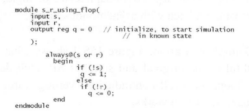

```
module s_r_using_flop(
    input s,
    input r,
    output reg q = 0   // initialize, to start simulation
                       // in known state
);
    always@(s or r)
        begin
            if (!s)
                q <= 1;
            else
            if (!r)
                q <= 0;
        end
endmodule
```

Figure 15S.5 SR flop implemented with Verilog transparent latch.

Sticky flag: synchronous set, async reset If the asynchronous clock of Fig. 15S.7 isn't satisfactory – for example, because the SET line in that figure may carry glitches – then a *synchronous* flag is required, as in Fig. 15S.8.

Virtue: Synchronous clocking protects against vulnerability to glitches on SET line of Fig. 15S.7 (assuming that glitches occur *after* clocking of the flop in Fig. 15S.8, as is usual).
Virtue: As for async clocked circuit of Fig. 15S.7, flop can be reset despite persistence of the signal that set Q high.

15S.2 Flip-flop tricks

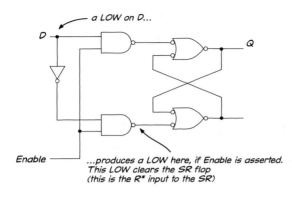

Figure 15S.6 Transparent latch with D low drives RESET* of SR when EN is asserted (as Verilog knows!).

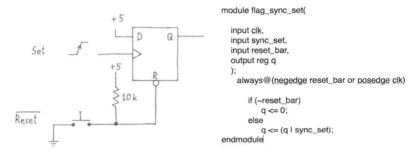

Figure 15S.7 Edge-triggered flop can be Reset while the signal that set Q high persists.

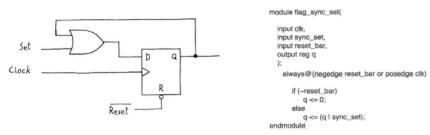

Figure 15S.8 Sticky flag, synchronous set.

Another sticky flag: synchronous set, synchronous reset One more gate (see Fig. 15S.9) makes the sticky flag of Fig. 15S.8 fully synchronous: the Reset also waits for the clock.

Virtue: Fully synchronous.

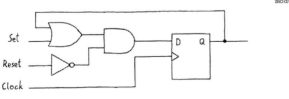

```
module flag_sync_set_reset(
    input sync_set,
    input sync_clear,
    output reg q,
    input clk
);

always@(posedge clk)
    if (sync_clear)
        q <= 1'b0;
    else
        q <= (sync_set | q);

endmodule
```

Figure 15S.9 Sticky flag II, synchronous set, synchronous reset.

A generalized circuit: synchronous Do_This, Do_That flop circuit Redefining the logic that feeds the flop's D input as a 2:1 multiplexer, one can draw – and think of – the circuit in Fig. 15S.10 as

Supplementary Note: Flip-Flops

something more general than the preceding ones, whose only aim was to set or reset the flop. This circuit could be used to distinguish, say, parallel-load versus serial operation for a shift register.

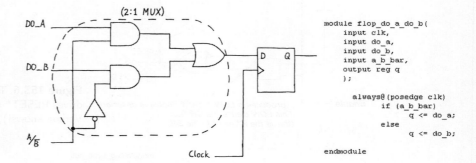

Figure 15S.10 Generalized flop circuit: mux provides synchronous Do_This, Do_That circuit.

```
module flop_do_a_do_b(
    input clk,
    input do_a,
    input do_b,
    input a_b_bar,
    output reg q
    );

    always@(posedge clk)
        if (a_b_bar)
            q <= do_a;
        else
            q <= do_b;
endmodule
```

Virtue: Fully synchronous again.

16N Counters

Contents

16N.1	**Old topics**	**603**
	16N.1.1 Flip-flop characteristics: recapitulation	603
	16N.1.2 A recapitulation: counters from flip-flops: ripple and synchronous:	606
16N.2	**Circuit dangers and anomalies**	**607**
	16N.2.1 A flip-flop vulnerability: slow clock edge	607
	16N.2.2 A glitchy circuit: timing diagram can help	608
16N.3	**Designing a larger, more versatile counter**	**610**
	16N.3.1 We need four T-flops...	610
	16N.3.2 ...We tell the several Ts when to change, and we add $Carry_{in}$ and $Carry_{out}$...	610
	16N.3.3 A fancier, cleaner $Carry_{out}$: synchronous	612
16N.4	**A recapitulation of useful counter functions**	**614**
16N.5	**Lab 16L's divide-by-N counter**	**615**
16N.6	**Counting as a digital design strategy**	**616**

Why?

The problem we'd like to solve today is how to measure and record a number of digital events (that is, we would like our circuit to "count").

16N.1 Old topics

16N.1.1 Flip-flop characteristics: recapitulation

Why clocks? Because "breaking the feedback path" eases design and analysis of sequential circuits. See Fig. 16N.1 for example. The circuit we built using the transparent latch oscillated at a bit more than 60MHz: see Fig. 16N.2. The edge-triggered self-contradictor, on the right in Fig. 16N.1, is well-behaved, as you know from Lab 15L.

AoE §10.4.2C

The beauty of edge-triggered, synchronous circuits: The *clock* edge is a knife cleanly slicing between *causes* of changes – all to the left of the clock edge, in the timing diagram – and the *results* of changes, all to the right of the clock edge. The idea is illustrated in Fig. 16N.3.

In saying this, we assume zero *hold* time, though this specification is not universal. Different manufacturers of one part (74HC74), for example, made different choices: some provide zero hold time (STMicrosystemns, and Texas Instruments); others specify non-zero hold time (3ns) (On Semiconductor and NXP/Philips) – see Fig. 16N.4. The logic array (CPLD) that we use in this course, Xilinx's XC9572XL, specifies zero hold time.

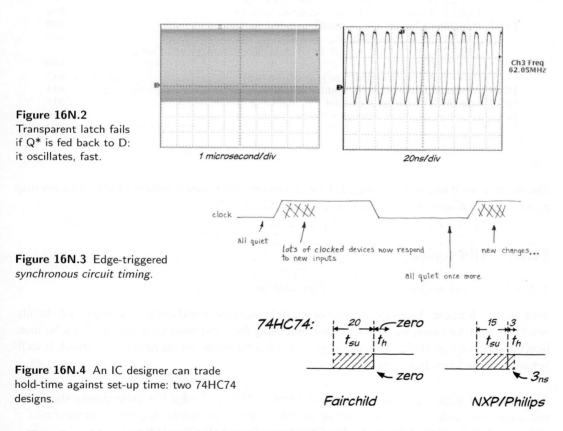

Figure 16N.1 Transparent latch can be unstable. Clocked device (edge-triggered) makes feedback harmless: instability is impossible.

Figure 16N.2 Transparent latch fails if Q* is fed back to D: it oscillates, fast.

Figure 16N.3 Edge-triggered synchronous circuit timing.

Figure 16N.4 An IC designer can trade hold-time against set-up time: two 74HC74 designs.

There is nothing magical about achieving $t_h=0$: the IC designers simply adjust the relative internal delay paths on the *data* and *clock* lines to shift the "window" of time during which data levels matter (called "aperture" in some other settings, such as analog–digital conversion). Other things equal, sliding the window to the right to introduce hold time, shortens set-up time. But zero hold-time is desirable; keeps timing simple and worry-free, whereas a non-zero hold-time obliges one to specify and worry about such exotica as *minimum* propagation delay, a characteristic not normally shown on data sheets.

A particular example: The circuit in Fig. 16N.5 works fine, as we noted back in §16N.1.1 and as you demonstrated in Lab 15L – unless you try to push the clock speed very high. In that case, trouble reappears: see Fig. 16N.6.

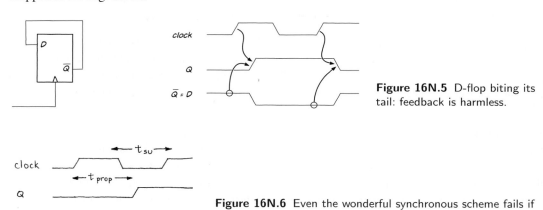

Figure 16N.5 D-flop biting its tail: feedback is harmless.

Figure 16N.6 Even the wonderful synchronous scheme fails if you try to run it too fast: t_{setup} violation.

When it's clocked too fast this fails, because D changes during t_{setup}. This produces an unpredictable output. The circuit may even hang up, refusing to make up its mind for a strangely long time (a device so hung up is called "metastable;" see AoE waveforms at 10.4.2C).

Set-up time: "Set-up time" sounds like a technical detail, but it's a concept you need in order to get timing problems right.

AoE §10.4.2A

Detailed view (in the special case of the 74HC74 D-flop): **Set-up time** is the time required for a change of D level to work its way into the flop – to the point labeled with a star symbol in Fig. 16N.7. Incidentally, while you're looking at the guts of the '74 flop you can gather from this figure how the "edge-trigger" effect works.

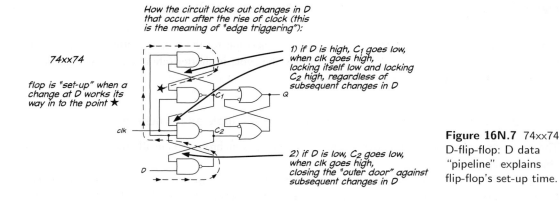

Figure 16N.7 74xx74 D-flip-flop: D data "pipeline" explains flip-flop's set-up time.

- Edge-triggered (positive or *rising* edge, in the present case – and in most flip-flops) means that information *enters the flop when the clock rises*. The diagram shows that only when CLOCK rises to a logic High can the output SR latch be changed – Set or Reset, depending on which of its two inputs, C_1 or C_2, is forced low.

- Edge-triggered also means that information *does not enter the flop after the clock has risen*. This lockout occurs because the C_1 and C_2 signals are fed back; one or the other of those two inputs to the SR latch will be low, and that signal wrapped back locks this low in place (by preventing a low at the other C), regardless of later changes in the level of D.

The new information needs to work its way through a "pipeline" that is two gates deep in order to set up the flop to act properly on a clock. You might guess that this process could take as long as two gate delays, around 20–25ns, and that is in fact about the value specified for the flop's setup time. (In fact, internal gate delays are less than the delay of a packaged gate; but the scale remains about right.)

A close analogue of set-up time: analog scope's data pipeline: You may have noticed the strange (and amazing?) fact that your old analog oscilloscope can show you information from *before* the trigger event. How does it manage to look *back* in time? The answer is analogous to what provides the flip-flop's setup time: a data pipeline carries the signal that is to be displayed, delaying its arrival at the scope screen – whereas the *trigger* signal responds fast. Just for fun, we opened up an old analog scope to look for its pipeline. There it was, see Fig. 16N.8 – the pipelining method is almost too simple to be believed.

Figure 16N.8 Scope delay line: simple, and effective (an old Tek 2213 scope).

AoE §H.4.3, Fig. H.19

We expected the cable to be about 100 feet long, to account for the scope's ability to display data that came in 100ns *before* the trigger event. In fact, the cable is shorter (about eight feet), thanks to clever construction that reduces the signal speed.

16N.1.2 A recapitulation: counters from flip-flops: ripple and synchronous:

The ripple counter is the easier to design and build, but the synchronous is in all other respects preferable: see Fig. 16N.9.

Figure 16N.10 is a scope photograph showing two bits of two integrated counters in action. The left-hand traces show a ripple counter's two low-order Qs, (you saw this image in §15N.8.1); the right-hand traces show a synchronous counter's Qs. You can gather which signal serves as clock for each flop – and you can see why synchronous counters are preferred.

AoE §10.4.3F

Virtues of synchronous counter
- Settles to valid state faster, after clock (wait for just *one* flop delay).

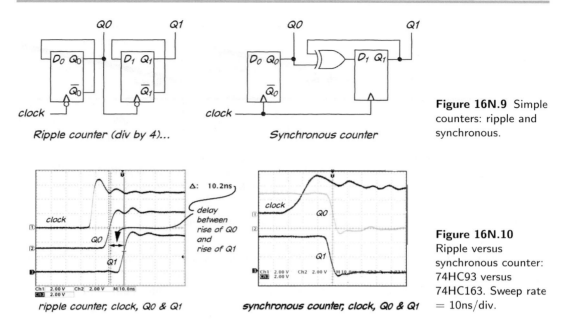

Figure 16N.9 Simple counters: ripple and synchronous.

Figure 16N.10 Ripple versus synchronous counter: 74HC93 versus 74HC163. Sweep rate = 10ns/div.

- Shows no false intermediate states[1].

Virtues of ripple counter
- Simple. Therefore, an IC counter can hold more ripple stages than synchronous (but note that this difference disappears for a counter implemented with a PLD with its rich stock of gates).
- In some applications, its weaknesses are harmless. For example,
    - frequency dividers (where we don't care about relative timing of in and out, and don't want to look at the several Q outputs in parallel, but care only about the relative frequencies, out versus in),
    - slow counters (driving a display for human eyes, for example: we humans can't see the false states).

16N.2 Circuit dangers and anomalies

16N.2.1 A flip-flop vulnerability: slow clock edge

The fact that an edge-triggered *clock* input responds to each edge (either rising or falling – not both, for a particular part) makes such a terminal vulnerable to *switch bounce*; one rarely debounces a signal unless it goes to a clock. But edge-sensitive inputs are also vulnerable to a subtler hazard: a clock edge that is not steep enough – an edge is *slow*.

Figure 16N.11 shows what a divide-by-two counter did when the clock came from a switch pulled up by a resistor that was too large (100k). The large pull-up R conspired with stray capacitance to produce an edge whose slope was too gentle.

This is very similar to the trouble we saw back in the comparator circuit of Lab 8L when we fed a gently-sloping waveform into a circuit that lacked hysteresis. As we noted back in Chapter14N, a

---

[1] Even a synchronous circuit can show false transients states if *skew* is substantial among the propagation delays of its several flip-flops. Unequal capacitive loading can have this effect. See AoE 10.4.3F.

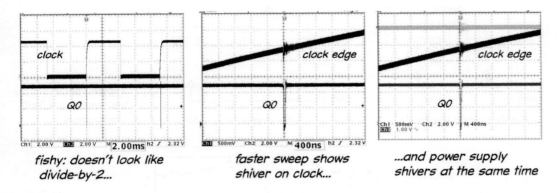

fishy: doesn't look like divide-by-2...   *faster sweep shows shiver on clock...*   *...and power supply shivers at the same time*

Figure 16N.11 Slow edge can cause trouble for an edge-sensitive input.

few logic devices *do* include hysteresis. The PALs that you meet today include 50mV of hysteresis on every input. We found this hysteresis not sufficient, however, to prevent indecisiveness just like that shown in Fig. 16N.11 when we used a slow edge to clock a counter built with such a PAL (XC9572XL).

16N.2.2 A glitchy circuit: timing diagram can help

Figure 16N.12 is a commonplace and useful combinational circuit, a 2:1 multiplexer, designed with conventional gates, and built using a PAL.

AoE §10.4.4B

See §14W.1

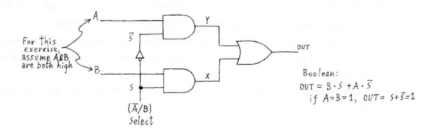

Figure 16N.12
Two-to-one mux; glitchy case discussed here assumes both data inputs high.

In order to illustrate a potential problem, we have wired the circuit for the dullest possible operation. Common sense – or a Boolean analysis, if you want to be so academic about it – says that the output must be a high always (the mux's job is to select *A* or *B*: high or high). Since a piece of wire to V_+ would give the same output, we should think of the wiring of Fig. 16N.12 as only illustrating a particular input combination that might be present at a particular time.

A timing diagram, drawn carefully enough as in Fig. 16N.13 to show gate delays, predicts that the circuit will not be so well-behaved.

A scope looking at S (select) and OUT might hide the glitch from you – if you weren't looking for it. The slow sweep rate in the leftmost image of Fig. 16N.14 makes the brief low pulse very hard to see. It appears in the middle image, swept faster. And it is measurable – though evidently very brief (this mux was made with a PAL, and the internal gate delays that produce the glitch are small, smaller than if it were built with discrete gates). Brief though it is, it probably is sufficient to be detected by another device made with the same technology.

Remedies? Integrated muxes do not produce this glitch – even though their published circuit schematic looks just like ours. Why? We're not sure. It may be because the designers of ICs can match delays

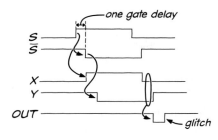

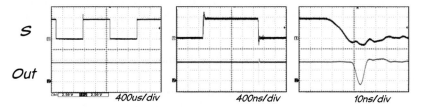

Figure 16N.13 A timing diagram for the mux predicts a glitch.

Figure 16N.14 Muxglitch: whether it's noticeable or not depends on the scope's sweep rate.

where they need to. The IC mux may include a delay in the non-inverted-S path, a delay that matches the delay imposed by the inverter in the $\overline{S}$ path.

But the designers have another option available, to prevent the glitch – and if you are building the mux yourself this is your only option. One can add what sometimes is called a "redundant" gate to prevent this particular glitch (see Fig. 16N.15).[2] This gate – not truly redundant, as preventing a glitch is a necessary operation, hardly "redundant" – forces OUT high when both inputs are high, regardless of the level of S.

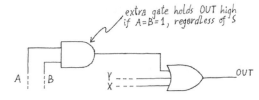

Figure 16N.15 Extra gate can eliminate an anticipated glitch.

The lesson? We offer this example not to make you a nervous glitchophobe. Glitches won't jump out of your circuits at random. We hope this example will illustrate several points.

1. If transients on the outputs of a circuit could cause mischief, one needs to consider the circuit dynamically, not just in Boolean, static terms (there are plenty of circuits, of course, where you know output transients would be harmless).
2. A timing diagram can help to disclose the possibility of such glitches.
3. An output glitch like the one illustrated in Fig. 16N.14 results from what is called a "hazard."[3] The cause of the problem is the need for two internal signals to change at the same time. The logic compiler that produces code for the PAL will eliminate any gate that is *logically* "redundant," like the extra gate installed to prevent this glitch – unless you tell the compiler's minimizer to back off and permit this non-minimal implementation.

---

[2] Old-timers who recall Karnaugh maps, which use shapes or "covers" to group or cover logic *Ones*, may call this remedy a "redundant cover."

[3] And this hazard, producing a transient false state, is called a "static hazard," in case anyone cares. The difference between the condition technically called a "hazard" and one called a "race" doesn't interest us much. We *are* interested in the resulting glitch.

16N.3 Designing a larger, more versatile counter

In §15N.9 you designed a 2-bit synchronous counter, and we hope you built it in lab. Let's carry on, moving toward a counter that is more versatile and thus closer to the sort of integrated counter one might buy (if you chose to buy such a smallish IC rather than build the part from a PLD). Here are some features we'd like in our improved counter, which we propose to build at the modest size of four bits (thus it will be a "divide-by-16" if we don't mess with its "natural" count length).

- It should be easily "cascadable" – that is, easily linked with other similar counters to form a counter of larger capacity.
- It should include a *clear* function, and this function should be *synchronous*.
- It should include a *load* function – that is, one that allows the user to load the four flops in parallel to permit counting from an arbitrary initial value.

We'll work our way to this result in stages.

16N.3.1 We need four T-flops...

Figure 16N.16 are four T-flops, made from Ds. The shared clock line makes the flops' timing *synchronous*. We have made even the LSB flop a T type, as we did not in the simpler two-bit sync counter of §16N.1.2. In that simpler counter we made the LSB flop toggle on every clock, willy-nilly. The design we are working toward this time is not so simple: our Carry$_{in}$ function will require a controllable T on the LSB.

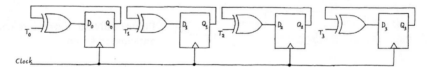

Figure 16N.16 Four flops awaiting some logic to drive the several Ts.

And Fig. 16N.17 shows a timing diagram to illustrate how we want the counter to behave. To save ourselves effort in drawing, we have started the diagram not at zero, but at count 12_{10}.

16N.3.2 ...We tell the several Ts when to change, and we add Carry$_{in}$ and Carry$_{out}$...

As in the case of the little 2-bit counter you designed last time, the rule is like the rule that governs a car's odometer: let the higher digit (in this case, a *binary* digit: a *bit*) change only when all the lesser bits roll over to zero. So, in order to get synchronous behavior, we need to tell a flop, "change on next clock" whenever a rollover is imminent. This rollover is imminent when all the lesser flops are high.

So, we just AND lesser Qs to drive any particular T. While we're at it, we'll include in our circuit sketch in Fig. 16N.18 the Carry$_{in}$ and Carry$_{out}$ that will allow cascading this counter with other similar parts.

One detail of the Carry$_{out}$ logic may strike you as odd: the Carry$_{out}$* AND (more or less a NAND gate) includes not only the four Qs but also the Carry$_{in}$ for this counter stage. The logic includes Carry$_{in}$* because that will be driven by a Carry$_{out}$* from a lesser stage if the counter is part of a cascade.

16N.3 Designing a larger, more versatile counter

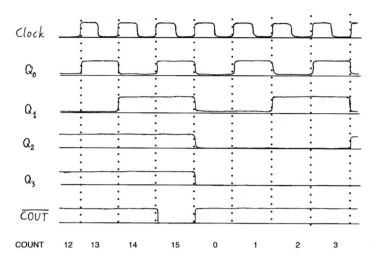

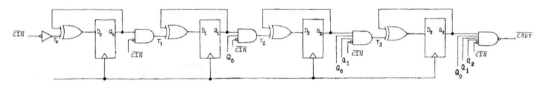

Figure 16N.17 Timing diagram describes the way we want our 4-bit UP counter to behave.

Figure 16N.18 Four T-flops driven to achieve synchronous UP count; Carry$_{in}$ and Carry$_{out}$ added to allow cascading.

Thus the Carry$_{out}$* that includes Carry$_{in}$* says not only "I'm full" (speaking for this single stage) but also "We're all full" (speaking for itself and for all lesser stages similarly designed). This simple and neat scheme permits cascading multiple counter stages simply by tying Carry$_{out}$* of each stage to Carry$_{in}$ of the next stage.

Cascading (Carry$_{in}$ and Carry$_{out}$): Any respectable counter allows "cascading" several of the devices to form a larger counter. The carries (in and out) implemented in §16N.3.2 would permit us to form a 12-bit counter, for example, by stringing together three 4-bit blocks of counters like the one in Fig. 16N.18.

- **Carry$_{in}$** This is an enable: when asserted, it tells the counter to pay attention to its clock. Notice that all chips are tied to a common clock line in Fig. 16N.19. Do *not* drive one counter's *clock* with a carry out. If you do, you're reverting to a ripple scheme, not a fully-synchronous counter.
- **Carry$_{out}$** This signal warns that the counter is *about to roll over* or overflow. In the case of a natural-binary up counter, it detects the condition *all ones* on the flop outputs. In a *down* counter, Carry$_{out}$ would detect the condition *all zeros*.

  Notice that this signal must come *before* the roll-over, not after, because it has to tell the next (more-significant) counter what to do on its next clock.

Figure 16N.19 illustrates how easy it is to cascade three counters with properly designed Carry$_{out}$* and Carry$_{in}$*. Such cascading is easy – but if carried on for many stages would limit the counter's speed because of "ripple delay" of the carries. External ANDing of the several Carry$_{out}$'s of the counters would allow faster clocking.

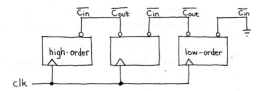

Figure 16N.19 Cascading three integrated counters: *easy!*.

A refinement in C_{OUT}*, worth noting: The need to include C_{IN}* in the gating for C_{OUT}* appears only in a case where more than two counters are cascaded as in Fig. 16N.19. We want the *third* counter – the high-order – to increment only if *both* the lesser counters are full. We don't want it to increment just because its immediate neighbor is full. This notion is entirely familiar to all of us who have watched car odometers roll over. Fig. 16N.20 shows such an odometer – really a mechanical counter from a tape deck. The figure is meant to remind us that the third digit advances only after the two lesser digits fill up, at 99.

Figure 16N.20 Odometer or mechanical counter: third digit should advance only when two lesser digits both are at max (099).

090 rolls to 091: third digit does not change   *099 rolls to 100: third digit does change*

16N.3.3 A fancier, cleaner Carry$_{out}$: synchronous

The carry we have described usually works well – because it is applied to a *synchronous* input, Carry$_{in}$*. It may not work well if it is put to an *asynchronous* use. The problem is that the combinational Carry$_{out}$* can show glitches. This can occur if the several Qs on which it depends show sufficient *skew* – that is, change at different times, despite their common clock. A combinational Carry$_{out}$* for a PAL counter like the one used in Lab 16L showed such glitches: The detailed image, on the right in Fig. 16N.21, makes clear that the timing of the glitch – *after* the clock edge – makes it harmless to a *synchronous* circuit, which samples its inputs only during *setup time* just before the active clock edge.[4] This immunity is the great virtue of synchronous circuitry.

But when one wants to use Carry$_{out}$* to drive a circuit that responds to an edge of Carry$_{out}$* itself – like the toggling flop shown as the bottom trace of Fig. 16N.21, or like an overflow-detecting flip-flop, clocked by the Carry$_{out}$* – then the glitches can cause trouble, and we need a remedy.[5]

To clean up the Carry, one must make it *synchronous* before using it. Instead of detecting the state *maximum-count* (which warns of a rollover on the next clock), a *synchronous* Carry uses a combinational circuit to detect the state one *before* the maximum state (for an up counter; minimum state, for a down counter), and feeds that to the D of a flop clocked by the counter clock.

---

[4] Some devices show non-zero hold time, as you may recall. But even these parts rarely show hold times that would reach so far as the glitch in Fig. 16N.21.

[5] We were surprised to find these glitches severe in *down* counting by Lab 16L's Up/Down counter, and barely apparent in *up* counting. Apparently, some mismatch between rise-times and fall-times accounts for this difference – since the Carry$_{out}$ depends on *lows* for the Down direction, *highs* for Up.

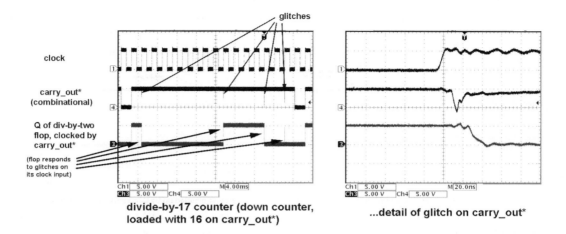

Figure 16N.21 Combinational, asynchronous Carry$_{out}$* can show glitches – substantial enough to clock a flop.

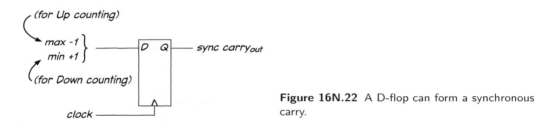

Figure 16N.22 A D-flop can form a synchronous carry.

This synchronizing technique is simple enough; it is not so easy to make a satisfactory synchronous Carry$_{out}$ for an up/down counter – and in our Lab 16L counter we did not attempt to solve the subtle problems that it raises.[6]

Such a Carry$_{out}$*, being synchronous, is clean: it does not pass glitches that occur in the disorderly time soon after the system clock edge. Here, in Fig. 16N.23, is such a synchronous Carry (this is the design we adopted, for the PAL counter installed in Lab 16L).

Noise still appears on the Carry$_{out}$* line – but the noise pulses are not substantial enough to matter, as Fig. 16N.23 attests: the flip-flop ignores those noise pulses, responding only to the legitimate Carry$_{out}$* signal (the flop is a divide-by-two, clocked by Carry$_{out}$*). Contrast the tidy behavior of the divide-by-two here with the nasty effects of carry glitches in Fig. 16N.21.

The right-hand image in Fig. 16N.23 shows the prettifying effect of adding a decoupling capacitor on the supplies of the PLD counter.[7] This decoupling does not affect the behavior of the circuit, since the noise was harmless – but it does make the waveform prettier. (And we hope you are getting into the habit of always bypassing your power supplies in any case.)

...Add a synchronous clear....: A counter usually includes a Clear function, and often a *synchronous* version is preferable to the *jam* type. Such a clear is easy to implement, see Fig. 16N.24.

---

[6] Nearly always, the scheme sketched in Fig. 16N.22 works. It can fail if one changes the count direction on either Max or Min count, since the circuit detects only the state preceding Max or Min.

[7] The decoupling capacitor is $0.1\mu F$ ceramic, applied at the PLD socket.

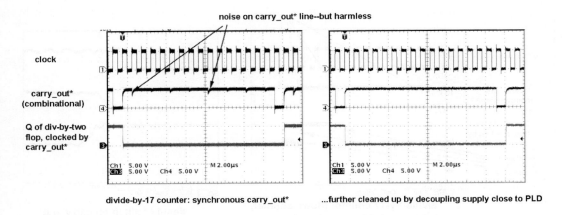

divide-by-17 counter: synchronous carry_out*   ...further cleaned up by decoupling supply close to PLD

Figure 16N.23 Synchronous Carry$_{out}$* is clean, unlike the asynchronous combinational carry.

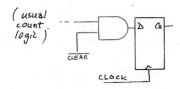

Figure 16N.24 Synchronous clear: requires only an AND gate.

16N.4 A recapitulation of useful counter functions

Integrated counters are available with a variety of features; and you can implement any of these features yourself, using a PAL.

Cascading counters to form a larger device: We noted, back in §16N.3.1, that we like to be able to string counters together, much as we string flops together, to form a device with larger capacity. Synchronous counters always permit this: one can simply feed Carry$_{out}$ from one stage to Carry$_{in}$ of the next. And the properly designed Carry$_{out}$ logic will be asserted only when all prior stages are full, as required (and as illustrated in the odometer of Fig. 16N.20).

Loading: Many counters allow you to load a value "broadside" into the flops:

Load When you assert LD*, the counter is transformed into a simple register of D-flops: on the next clock edge those flops simply take in the values presented on the data inputs. (This description fits so-called "synchronous" load; "asynchronous" or "jam" load also is available; it works like the jam clear described below.) The PAL counter of Lab 16L includes both types of load. The async version is convenient for loading initiated by a manual pushbutton; the sync version is better when the load is initiated by a circuit signal (as in the music maker of Lab 16L).

When you release LD*, the counter becomes a counter once more. The behavior of Load may be hard to grasp when you simply hear it stated. We'll look at an example of the use of load in Lab 16L's "Divide-by-N" counter.

16N.5 Lab 16L's divide-by-N counter

> A selection of counters showing sync/async functions appears in AoE §10.5.2

Clearing: Clearing could be called a special case of *loading*, but it is so often useful that nearly all counters offer this function (more than offer *load*), and they offer it in two styles:

Asynchronous or "jam" clear. The clearing happens a short time after Clear* or Reset* is asserted (say, 5–10ns); the clearing does *not* wait for the clock. This type of clear is familiar to you from your experience with flip-flops, whose clears always are *jam* type.

Synchronous clear. The clearing is *timed* by the clock: on assertion of Clear* or Reset*, *nothing happens* in response until the next clock edge. We saw how to implement such a Clear on page 613.

Query: which sort of clear would you like in a programmable divide-by-*N* counter, like the one you will build in the lab? Figure 16N.25 shows some standard oldish MSI counters with their clear functions.

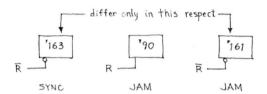

Figure 16N.25 Three integrated counters: some offer jam clear, others synchronous.

In Chapter 16W, you will find some counter-application problems detailed. Here, we will concentrate on the counter applications that you will meet in the lab.

16N.5 Lab 16L's divide-by-N counter

The counter that you install on the micro breadboard is loadable, as you know. Eight of its 16 lines are controlled by the keypad, six are fixed *low* within the IC, and two are controlled by a two-position DIP switch. Fig. 16N.26 is a sketch from the "big picture" computer schematic in Appendixschematic.

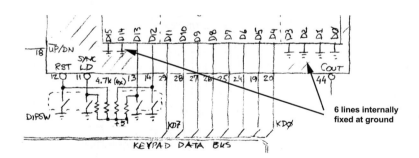

Figure 16N.26 Data inputs of counter: 8 controlled by keypad.

If we let the counter reload itself each time it hits a limit – zero, when counting down – then the keypad can determine how many states the machine will count through, on each cycle: see Fig. 16N.27.

Counting *down* lets us take advantage of the lines internally tied to ground, to let the keypad set cycle-counts that are not very long (17 states for the lowest keypad value, the key value *one*:[8] 17 states because the trailing zeroes, internally defined, make a "1" from the keypad load a 16_{10}, and the zero state provides the 17th step). The keypad value 2 provides 33 states; and so on.

---

[8] Loading zero doesn't work: it puts the counter at its destination, at the outset.

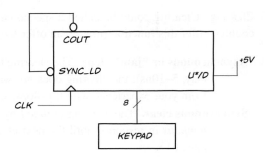

Figure 16N.27 Counter wired to reload itself each time it hits zero.

16N.6 Counting as a digital design strategy

Because it is easy to make a digital circuit that counts, it often turns out that a good way to make a digital device designed to measure some quantity is to build what in effect is a *stopwatch* measuring the duration of a cleverly generated pulse. More specifically, here's the idea:

1. build a counter that counts clock edges (this is a sort of "watch");
2. add gating that lets you start and stop the watch (making it a "stopwatch");
3. build some circuitry that provides a waveform whose period is proportional to a quantity that interests you (call this quantity "input");
4. use the stopwatch to measure period, and thus to measure "input."

Here is an example to illustrate the technique.

Digital voltmeter (or ADC; voltage input): The circuit of Fig. 16N.28 measures the time a ramping voltage takes to reach the voltage labeled "Input." The higher the input, the longer the time and the larger the resulting readout. This is a "single-slope" ADC. A slightly more complex design, called "dual slope," is preferable (see §18N.4.3); but its strategy is essentially the same as for the single-slope design, shown in the figure.

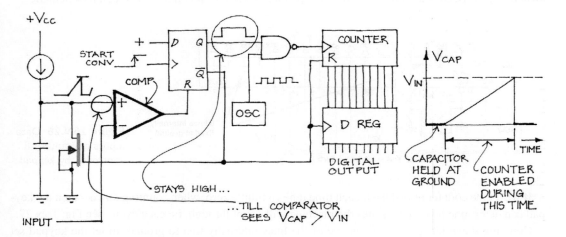

Figure 16N.28 Example: measure period to measure *voltage*.

16L Lab: Counters

16L.1 A fork in the road: two paths into microcontrollers

From here on, this course offers alternate ways of exploring microcontrollers. We expect that you will use one or the other (though a zealot with extra time might do both). We want to describe these two paths briefly, and suggest advantages for each. If you are doing these labs on your own, you can of course choose the one that suits you. If you are doing these labs as part of a course, your teachers may want to steer you down one path or the other.

16L.1.1 Path One: build up a microcomputer from parts

This is what we have done in our course for the past twenty-five years or so: in a series of labs we add the elements of a small computer, building the circuit up from nine integrated circuits. This path uses the counter and RAM that you will put together in Labs 16L and 17L. The RAM will store the computer's code and data for most of our applications.

- Lab 20L: we add the processor/controller (an 8051 derivative), and its attendant "glue" PAL: a part that links the 8051 to its peripherals and memory, and that allows us to *single-step* programs.
- In a set of labs we add peripherals and exercise them with small programs:
    - Lab 21L: byte input and output (keypad and display); subroutines;
    - Lab 22L: bit input and output – polled and using interrupt.
    - Lab 23L: ADC and DAC.
    - Lab 24L: Store and playback of analog data; serial bus peripheral; timers
    - Lab 25L: Standalone controller trial (here, big board mimics SiLabs scheme).

16L.1.2 Path Two: start with a self-contained microcontroller

This path relies on a different version of the 8051 microcontroller, one that can be controlled and interrogated by an ordinary personal computer (a "PC"). This arrangement obviates most of the building that is required on Path One – but does require a PC for each setup. Using this alternative, the lab exercises look different – and the series of labs is shorter by two days (or more, if people following Path One don't keep up to the one-day-per-lab pace; most people slip by at least half a day as they wire the first micro lab, Lab 20L).

- SiLabs 20L.3: debugger setup; bit output, delay.
- SiLabs 21L.2: byte operations, (keypad and display), subroutines.
- SiLabs 22L.2: PWM; comparator.
- SiLabs 23L.2: interrupt, ADC and DAC.
- SiLabs 24L.2: serial buses: UART & SPI.
- SiLabs 25L.2: data table; SPI RAM.

16L.1.3 Then the two paths converge

The two paths converge in Chapter 26N that asks students to apply the 8051 as a *standalone* controller to implement some scheme of your own invention.

16L.1.4 Advantages of each path

What we like about Path One: This method puts minimum magic between you and the computer. The 8051 knows nothing you don't teach it. It includes no mysterious monitor program. You control the machine directly – even the *single-step* is done in simple hardware. This hardware makes the computer pause, to allow you time to scrutinize address and data buses and any control signals that particularly interest you. It makes the computer pause by the simple expedient of choking off its clock source. To a large extent, the working of the computer is made transparent – though its "brain," in the form of the 8051 that executes instructions one at a time, remains opaque.

Figure 16L.1 shows what the circuit looks like as you single-step through the In/Out program of Lab 21L, for example.

The values of address and data buses are displayed. A logic probe touching the 8051's RD* pin in Fig. 16L.1 shows this signal asserted, indicating that the processor is taking in data from the data bus. The keypad is the source of this data, and it is reassuring to see that the data display does, indeed, match the value applied by the keypad. This sort of low-level detailed information about signals works to demystify the operation of the computer.

We like this simplicity. And we like providing you with the opportunity to make wiring errors – and then to chase them down. In Chapter 20W we spell out some of the intriguing detective work that can

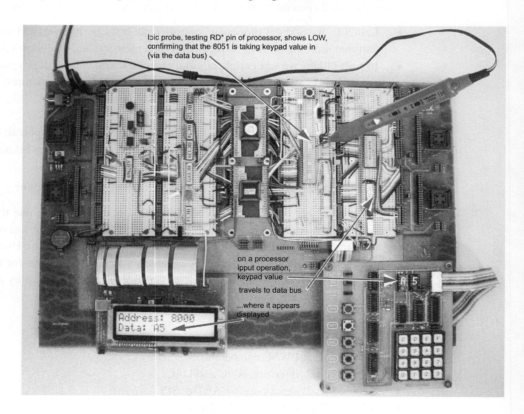

Figure 16L.1
Breadboarded micro makes computer transparent.

come up as one tries to make sense of a particular malfunction. Putting together this little computer will give your debugging skills a workout.

What we like about Path Two, in which you tie a controller to a PC: This – let's admit it – is the way everyone else in the world works with microcontrollers. This method gets you results fast. Every development kit (or "evaluation board," as these setups often are called) seems designed to blink an LED for you within five minutes after you open the box. Everyone likes to see evidence that the thing works – and that you (little old you, on your first afternoon with a controller) can make things happen. We're as susceptible to the charm of this figurative pat on the back as the next person. So, we understand very well why people like to work this way.

And since this is surely the way you will work with controllers *after* this course ends, there's an obvious appeal to starting right in at working as you will work in the afterlife.

Figure 16L.2 shows what the little controller we use on this path looks like. It is a small 32-pin surface-mount part that is soldered to an adapter so that one can plug it into a breadboard. This little device squeezes in almost all of what is wired on the large circuit board of Fig. 16L.1 (it lacks only the 32K RAM of the big circuit – and includes some elements not included there).

Figure 16L.2 C8051F410 microcontroller mounted on DIP adapter for breadboarding.

We don't know which path you will prefer. We like them both. If you had time the best approach would be to wire the big "transparent" computer first, then play with the compact '410.

16L.1.5 Then convergence: do something of your own devising

As we said above, we hope that when you've had a chance to get used to the controller you will use it "standalone" to implement some design of your own:

- Chapter 26N: apply the standalone controller.

This you can do with either the Dallas version or the Silabs '410. Big-board people can program their 8051's on-chip ROM from a PC; or they may want to use the SiLabs '410 when it's time to put the controller all alone on a breadboard. The interactive debugging that the '410 allows makes it easier to work with than the loadable-but-not-debuggable Dallas part.

16L.2 Counter lab

In this lab you move up from the modest "divide-by-four" of Day 15 to a 16-bit "fully synchronous" counter. We will let it show off some of its agreeable features, notably its *synchronous load*, and then we will put it to use in two circuits: a programmable *divide-by-n* machine, then a *period-measurer*, which can operate as a capacitance meter, with just a little help.

At first you will use only scope and logic probe to watch the counter's performance; then you will add a hexadecimal display that should make the counter's behavior easier to follow. A keypad will let you control the counter and load it.

Note: The keypad is not a standard commercial part. See Appendix E for ordering information and Appendix 14N.20 for the schematic, in case you want to build your own. The display board that we use also is a custom-made circuit, using an LCD display to show 16 lines of address, and 8 or 16 of data.

On Day 17, people who have chosen to build the "big-board" computer will use this counter and display to provide an *address* to a memory. The keypad will let you write 8-bit values into any memory location. In a later lab, counter and memory will serve as foundation of the microcomputer. So today *big board* people are beginning to build the little computer.

If you will not be building the big-board version, then you can build today's counter in a single strip. That arrangement is described in §16L.3.2.

Note: If you choose the big-board computer option, you will build today's circuits on your own private breadboard. Big-board lab notes will assume that you are using a printed circuit holding several breadboard strips mounted together. This breadboard will become the foundation of the microcomputer that you soon will be putting together.

If you don't have this printed-circuit board you can, of course, build the big-board computer on four or five breadboard strips. The printed circuit only eases the wiring somewhat.

In an effort to make the big-board task less daunting than it might otherwise be, we have laid out a printed-circuit board that defines *buses* for the computer, though leaving to you the connections between these buses and your circuitry. The benefits of these ready-made buses are two-fold.

1. First, we hope they will make the concept of a bus less abstract than if you were to see it only as "whatever lines go to many places." The buses are labeled, and defined by the traces that run the length of the board.
2. Second, the buses allow (as in any computer) tidy wiring among circuit elements. You should seldom find the need to run long wires about your circuit. The bus lines run close to a ground plane, and are less vulnerable to cross-talk than long, wandering wires (and they're less ugly).[1]

Figure 16L.3 shows what the buses look like on the board's top and bottom layers (it is a two-layer printed circuit).

The right-hand image is included to remind you that some of the external connections are already wired to the dedicated buses:

- a display board is wired to the 16-line *address bus* and to the 8-line *data bus*;
- a keypad is wired to the 8-line "keypad data bus" (which is *not* the same as the general data bus; this difference, obscure now, will make sense soon, once the data bus has begun to function in Lab 17L).

A preview of the set-up: Figure 16L.4 is a glimpse of the way the breadboard will connect to display and keypad.

---

[1] You may be inclined to protest that traces that run close to other traces for eighteen inches are *more* likely to cause crosstalk than spaghetti wires wandering above the board. You have a point – but we anticipated this problem and placed a *ground* trace between adjacent signal lines on the buses of the pc board. These ground traces should greatly reduce crosstalk.

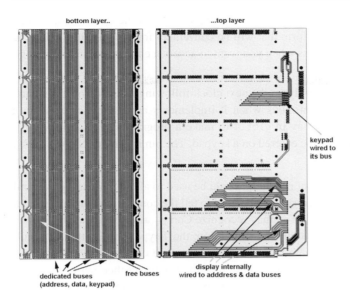

Figure 16L.3 Buses defined by printed circuit that underlies the usual breadboard strips.

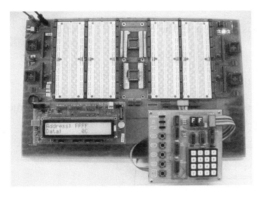

Figure 16L.4 Display and keypad connected to the breadboard.

16L.3 16-bit counter

The counter you install today is a versatile 16-bit device designed to define the address in our lab microcomputer whenever we want to determine this address manually. It is made from a PAL much like the PAL you will design and perhaps program to implement the "glue" logic that links the processor to the remainder of the computer circuitry.[2] Here are its features.

- It counts up or down – and this behavior is handy for our purposes: soon (in Lab 17L) you will use the counter to take you to a particular memory location or address.
- It includes three-state outputs: this feature will let us tie the counter outputs directly to the computer bus (specifically, to the *address* bus).
- It can be "loaded" with an initial value. Today, we will use LOAD* to let us make a *divide-by-strange-number* counter; later, in the micro labs, we will use LOAD* to let us hop to a particular starting point in the memory's address space.

  The *load* function is implemented in two alternate ways, available through use of two control pins:

[2] More specifically, the counter is made from a Xilinx XC9572XL (3.3V supply).

- Asynchronous load: this takes effect as soon as the signal is asserted (here, that occurs when you press a pushbutton). This is the form that will be convenient for your use in forcing the counter to a particular starting address when you want to access a chosen region of memory. The pin is named "LOAD*".

- Synchronous load: this takes effect only on the next clock following assertion of this signal. This form works better than asynchronous when it implements an automatic reload in a counting loop. We will use it that way in §16L.4 to make a programmable "divide-by-N" counter, with N determined by a value entered on a keypad. The pin is named "SYNC_LD*".

- It includes Carry$_{in}$* and Carry$_{out}$* pins that normally would be used to allow "cascading" several of these units, to form a bigger counter. Carry$_{in}$ can be used to tell the counter to count or, instead, to hold its present value despite clocking. (You may recognize Carry$_{in}$ as just equivalent – on the counter-wide scale – to the T input that controls a single flop: the T-type flip-flop that you built in Lab 15L from a D-flop and an XOR gate.) We will take advantage of this Carry$_{in}$ function – which can also be described as an *enable* – in forming the stopwatch and then period-meter of §§16L.5.1 and 16L.5.2. Carry$_{out}$* will become useful to those who choose to use this counter to implement a reaction timer, in the digital project lab, 19L.

- The counter includes a RESET* that is synchronous, like one of the two LOAD functions.

16L.3.1 A preview of the counter and its signals

Figure 16L.5 is a sketch of the counter as a fragment of the big-board schematic. Exactly the same signals are used whether the counter is wired in the big-board or on a single breadboard.

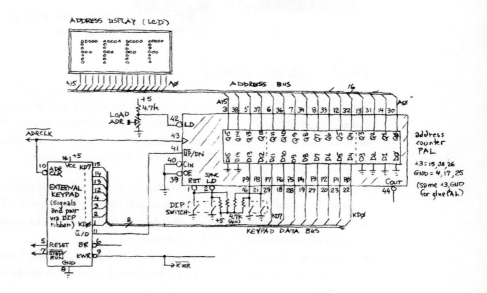

Figure 16L.5 16-bit PAL counter and its control.

This PAL counter is another *surface mount* part that has been mounted on an adapter board to permit breadboarding with it. (You may have seen such a carrier in Labs 5L and 12L). Fig. 16L.6 shows the PAL on a carrier. We hope that you will add labels to the carrier, identifying each pin.

Figure 16L.6 PAL (TQFP package) mounted on a carrier, to permit breadboarding.

16L.3.2 Single-breadboard option

Figure 16L.7 shows what our breadboard looked like, when we eschewed the big-board, using just a socket and single breadboard strip (the photo shows the PAL in a package different from the one you will meet today; this one is *PLCC* whereas today's PAL is in a QFP surface mount package):[3] The figure shows only the input *control* lines wired along with parallel-input lines P9, P8, shown grounded by the DIP switch. Later, *data* input lines P7...P0 will be fed by the keypad's 8 output lines (see §16L.4). The 16 output lines (the counter's $Q_{15}...Q_0$, labeled "$A_{15}...A_0$" in Fig. 16L.7) will be fed to the 16 *address* lines on the LCD card.

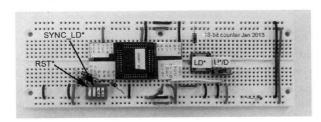

Figure 16L.7 Counter on a single breadboard.

(The PAL shown in Fig. 16L.7 used pin assignments slightly different from those used in today's PAL, assignments that are shown in Fig. 16L.8.)

16L.3.3 Big-board option

Figure 16L.8 shows the PAL counter mounted on a carrier.

The signal names shown closest to center, in the rightmost image within Fig. 16L.8, are the power, ground and programming pins. You may ignore the programming pins, TMS, TDI, TDO, and TCK.

16L.3.4 Wiring the PAL

Wire power and ground first: The PAL uses six lines for power and ground, three for each. We suggest that you wire all of these initially. Everyone tends to find wiring power and ground tiresome – labor worthy of a lab partner but not of oneself. But try to get the habit of doing these first.[4] If your counter is a 3.3V part – XC9572XL – make sure to take power from the 3.3V supply rather than 5V.

Since your PLCC is likely to take *ground* and perhaps also V_+ from the white breadboard strips, don't omit or postpone making the necessary connections to the strips' vertical power and ground buses, from the "Vcc" and "G" points available at both ends of each white breadboard. The positive supply is 5V, even if you are using a 3.3V counter, since these strips will power 5V logic.

We suggest that you invest five minutes now, to make all those connections to the breadboard strips – and to *bridge* the midpoint of each of the ten power buses. If you postpone that little chore, you will save a few minutes now, and probably will suffer a longer stretch of frustration later on, when some part of your circuit misbehaves for lack of power or ground.

---

[3] The package is called "Thin Quad Flat Pack," TQFP.
[4] Omitting power and ground sounds unlikely, but is a common error, as we have noted before.

Lab: Counters

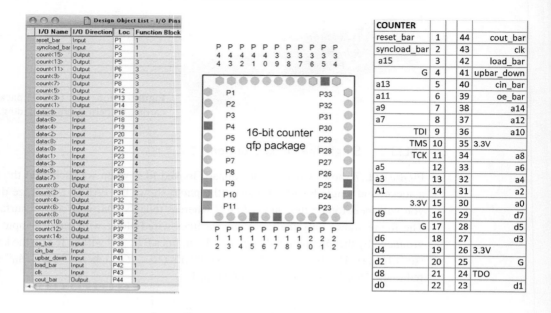

Figure 16L.8 Pinout of PAL counter.

Control lines: Next, we suggest that big-board users should wire the counter's control lines from the PAL's PLCC connector to some free bus lines on the breadboard. You will find these uncommitted bus lines – labeled "free buses" in Fig. 16L.3 – near the top and bottom of the pc board. There are enough of these lines so that you may find it worthwhile to write or print labels, attaching labels either to the wires or to the breadboard itself.

The following are these control signals:

- CLK (rising edge, as usual);
- CIN* (carry in, or enable);
- COUT* (carry out, asserted on FFFFh for up count, on 0000h for down count);
- OE* (3-state control);
- LD* (asynchronous load: on assertion of this signal, counter is loaded with 00 in top two bits, two bits from DIP switch that you must wire, eight bits from keypad, and 0000 into lowest-order bits);
- SYNC_LD* (synchronous load: on next clock, counter is loaded as for async load);
- RESET* (synchronous reset: forces counter to all zeroes);
- U*/D (up/down control: low causes counter to increment on next clock, high to decrement).

If you prefer a diagram, Fig. 16L.9 is a reminder of these signals along with the data inputs and counter Qs.

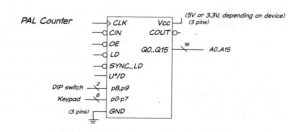

Figure 16L.9 Signals in and out of PAL counter.

16L.3 16-bit counter

DIP switch controls four inputs...: We used a 4-position DIP switch to control 2 of the lines, SYNC_LOAD* and RESET*, and a momentary pushbutton to drive the asynchronous LOAD* pin. (Two other positions of this DIP switch drive the top 2 bits of data into the counter: P9, P8.) Fig. 16L.10 is the DIP switch, controlling its four signals.

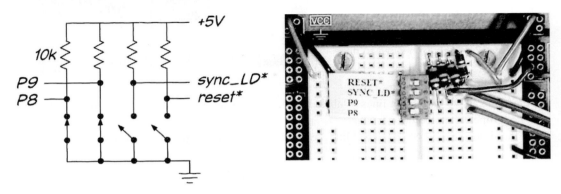

Figure 16L.10 DIP switch used to control four counter signals: big-board and single-strip versions.

The same sort of DIP switch controls these four lines also in the single-breadboard version of this circuit, illustrated on the right in Fig. 16L.10, and also in Fig. 16L.7.

...pushbutton controls load function: A pushbutton drives the counter's asynchronous *load* function – which transfers a value into the counter in parallel (doing this, the counter is behaving like a register of flip-flops, not like a usual counter). The value comes from the keypad (8 bits), plus another two bits controlled by the DIP switch described in Fig. 16L.10. (We seldom change those DIP switch settings on the "P9, P8" bits, incidentally. Most of the time we leave them *low*.)

The pushbutton, with its four legs connected to just two terminals, can be puzzling, so in Fig. 16L.11 we've sketched the way it works. We have also tried here to remind you to flatten and rotate the switch's legs. If not rotated, they are too wide for the breadboard.

Figure 16L.12 shows what it looks like wired to drive the LD* terminal of the counter.

Counter output lines and inputs: The counter's 16 output lines – A0...A15 – run to the 16 lines of the address bus. If you are using the counter on a single strip, then the 16 lines go directly to the LCD card (see Fig. 16L.15).

...Keypad feeds counter's parallel inputs: Another eight lines feed the counter from the keypad. Note that these lines from the keypad DIP connector are *internally* wired, on the micro breadboard, to the bus labeled "KD0...KD7." *Note also* that these *keypad* data lines – labeled "KD0...KD7" – are

Figure 16L.11 Pushbutton details.

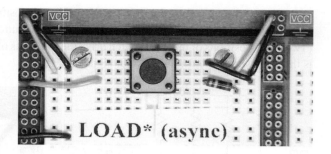

Figure 16L.12 Pushbutton wired to drive LD*.

not the same as the *main* data lines – labeled "D0...D7." That principal data bus is not to be wired until next time, when memory arrives.

On the single-strip, those 8 keypad lines feed P0...P7, as we noted back in §16L.3.2.

...Control inputs: Take care of the counter's several control inputs. Specifically:

- drive CLK with the keypad's ADR_CLK line;
- drive the PAL's U*/D line with the keypad's signal of the same name;
- set the DIP switch so as to *disassert* RESET* and SYNC_LD*, while tying P9 and P8 low;
- wire CIN* (an enable) and OE* (the three-state control) so as to assert these signals continuously;
- leave COUT* open, for the present.

That is a lot list of signals to worry about; it may help to look again at Fig. 16L.9.

Counter as wired on big-board: Figure 16L.13 shows what the wiring of the PLCC may look like once address, keypad and the several control lines have been connected. We tried to keep the wiring *low*, so that no wire is likely to get knocked out as we progress. The boards will be stored and brought out many times. We like wiring that can stand this travel. You will, too.

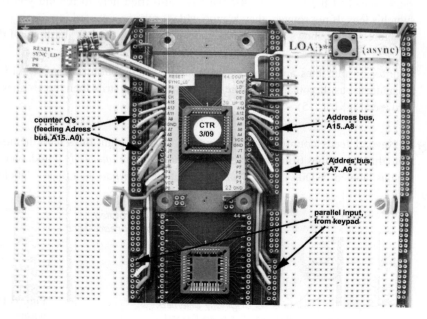

Figure 16L.13 All necessary lines connected to PAL counter.

16L.3.5 Add the display board

We'll soon be watching the counter on an oscilloscope, but let's set up the other way of watching it before proceeding further.

Counter display as wired on big-board: The LCD card plugs into a DIN connector[5] on the big green breadboard PCB. This connector provides:

Power: slide the LCD power select switch on top right to "external" to take 5V power from the big green board.
Address: the DIN connector links these 16 display inputs to the 16 lines of the PCB *address* bus. Those lines are driven by the Qs of the counter's 16 flip-flops.
Data: either 8 or 16 lines can be displayed. Today, we have no data to show, but the LCD will display these floating inputs nevertheless. Next time, they will carry data stored in RAM.

The counter output can be displayed in either hexadecimal or binary format. Fig. 16L.14 shows hexadecimal; Fig. 16L.15 demonstrates both formats. Hex is much better to let one recognize a number's value at a glance. But as you begin to play with the counter, we'd suggest that it's better to start with a binary display: especially when the counter is clocked slowly, the *binary* display lets you recognize the counting pattern.

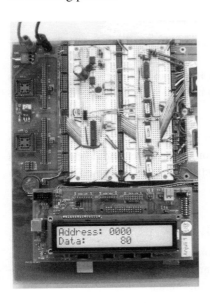

Figure 16L.14 LCD can show counter output.

Counter display as wired on single strip: Attach the counter's 16 Qs to the two 8-bit *address* connectors on the LCD card. Fig. 16L.15 shows the counter board inverted. This is not necessary, but allows us to avoid putting a half-twist into the flexible cables that carry the 16 signals.

The low-order bits look garbled on both displays in Fig. 16L.15. This effect just reflects the quick counting and the slow shutter speed of the camera.

---

[5] DIN is Deutsches Institut für Normung, German Institute for Standardization. It has defined thousands of technological standards for the Europe. This 48-pin connector is just one such standard.

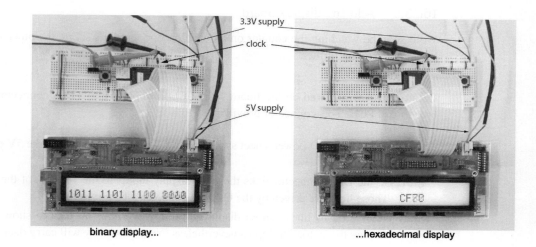

Figure 16L.15 Display of counter output using LCD card tied to single-breadboard.

binary display... ...hexadecimal display

16L.3.6 Attach the keypad

Big-board: Attach the keypad using its 16-line DIP cable. Once plugged into the DIP socket, the keypad's KD0...KD7 lines drive the counter's parallel input lines P0...P7, lines that you earlier connected to the micro breadboard KD0...KD7 bus.

Single strip: Plug the keypad into the breadboard strip and feed its eight data lines to inputs P0...P7 of the counter.

Those eight lines drive the *middle* eight lines of the counter's 16 data input lines.[6] These eight lines become significant only when the counter does a Load operation.

Does the counter count? We hope so. You'll find out if you push the keypad's INC button. The counter should increment its value. If you press the DN button, the counter should decrement its value. The RPT key clocks the counter at about 20Hz, with direction UP by default, but DOWN if you press DN and RPT together.

Try loading and Reset*: Try the LOAD* pushbutton. It should make the keypad value appear on the middle two digits of the hex display.[7] Now try the *synchronous* load: flip the DIP switch to drive SYNC_LOAD* low (doing this shows no immediate effect) and then clock the counter from the keypad's ADCLK* line. After testing SYNC_LOAD*, disassert that signal.

Try RESET*, using the DIP switch: note whether its behavior is synchronous or asynchronous.[8]

---

[6] This is a bit strange. But the keypad provides only eight lines, and placing these in the counter's middle – to feed Q4...Q11 – strikes a compromise. The counter can load a value that is on a 16-count boundary (soon to mean a "16-address" boundary). From that point you have to walk up or down to get to any particular address. You can never be farther than an 8-step walk from your destination.
At the same time, the keypad cannot carry out very large hops in address space since it does not control the counter's high four lines. (The DIP switch's P9 and P8 lines can help.) But we don't think you'll find the limitations of the counter's LOAD function troublesome.

[7] The display can show you binary or hex as we said back in §16L.3.5. A slide switch on the display board, lower left, selects the display format.

[8] Like all the counter's functions except LOAD*, it is synchronous: it waits for the clock.

16L.3.7 Try the counter at higher speed

Now replace the keypad's ADR_CLK* signal, which was clocking the counter, with a logic-level signal from the function generator. Try f_{clock} of a few kHz.

Watch clock and Q_0, then clock and Q_5 (triggering in both cases on the Q – and the *slowest* Q when watching more than one).

In the unlikely event that your counter's behavior is erratic you may want to clean up the clock, by passing the function generator's TTL signal through a local logic gate that has hysteresis: see Fig. 16L.16.

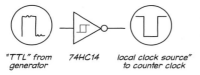

Figure 16L.16 HC14 can clean up a messy clock.

Then watch Q_0 and Q_{15}: do you see any delay of the higher-order Q relative to the lower-order, as you did in even the small ripple counter you built in Lab 15L?[9] Now take a look at C_{out}*.[10] You will need to trigger on the carry – and note that it is only one clock wide, so may look very narrow indeed if your scope screen's sweep covers many clocks.

If you use a couple of spare scope channels to watch Q_0 and Q_1 you can see the difference between carry on an UP count and carry on a DOWN count. If you feel like playing with the LCD display again, drop the clock rate way down so that you can see individual states. Check that UP and DOWN behave as advertised.

16L.4 Make horrible music

If you let the counter's C_{out}* line drive its SYNC_LOAD* pin, then it will reload itself each time it overflows. It will load itself with a 16-bit value formed from two lines of the DIP switch (P9 and P8: let's set these both low), 8 lines from the keypad, and four LSBs set to constant zeroes within the counter's innards. So, the keypad value can determine where the counting starts.

If you count DOWN the pattern is pretty straightforward so we suggest you temporarily tie UP*/D to +5, as you do this subsection.

SYNC_LOAD* is driven by COUT* so these signals are asserted only when the counter is about to "roll over;" since the counter is set to count *down*, it reloads each time it hits *zero*. So this circuit lets the keypad set the number of counts that occur between *loads*; thus the keypad sets period and frequency of the COUT* pulse waveform.

Watch COUT* and Q3 of the low-order counter as you vary the keypad input value. Does the response fit what you expect? The keypad value is 8 bits, and a single keystroke updates only the lowest 4 bits. You may want to keep the high nybble at zero, as you experiment. You could temporarily tie those four lines to ground so that the keypad feeds only P0...P3 to the counter; that would make your instrument behave more like a musical keyboard – as we're about to ask it to do.

Hearing the effect of LOAD*: strange modulus counter: To make more vivid the power of this loadable counter to vary its *modulus* – the number of states it steps through – let's listen to the counter's

---

[9] We hope not.
[10] In an UP/DOWN counter like this one, this carry sometimes is called "borrow" on the down count, by analogy to subtraction. We'll call it just "carry."

output frequency. Let C_{out} drive a transistor switch (a power MOSFET is easiest), which in turn drives the breadboard's speaker, as shown in Fig. 16L.17. If you want to annoy your neighbors with a louder tone, let COUT* drive a toggling flip-flop: the 50%-duty-cycle signal that comes out of the flop makes more noise than the narrow COUT* pulse does. Without this flop, the low duty-cycle of the COUT* pulse can make it barely audible for some count-lengths.

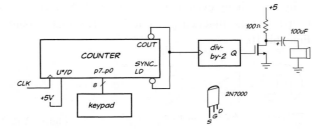

Figure 16L.17 Music maker: divide-by-two makes output volume greater, and independent of frequency.

Fixing an interesting flaw in this circuit:

The problem: glitches on COUT:* When we tried this, we were surprised to find that the COUT* signal – which detects the counter's state *zero* during *down* counting – carried nasty glitches, glitches substantial enough to clock the *divide-by-two* flip-flop that we were using to provide a 50:50 duty cycle. The glitches made nonsense of our music machine: see Fig. 16L.18.

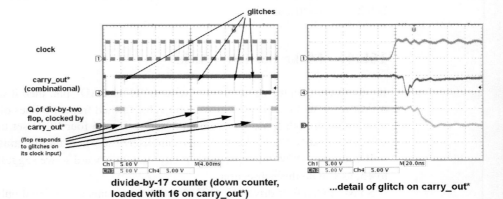

Figure 16L.18 Glitches on carry can cause disaster when COUT* drives a clock.

The remedy: synchronize the COUT signal with a flop:* You'll notice that the glitches in Fig. 16L.18 occur *after* the clock edge. This is the usual case, since it is the changing signals, brought about by clocking, that trigger the glitches. Such glitches are entirely harmless to a flip-flop clocked by the signal that evoked the glitches – in this case, the clock of the main counter.

Figure 16L.19 illustrates a remedy for these bad carries: a synchronizing circuit that blocks the glitches. The glitches will not pass through the flop, because they occur *after* the edge that updates this synchronizing flip-flop.

If your music machine was troubled by carry glitches, include such a synchronizer upstream of the divide-by-two that is to drive the speaker, in your music machine. Sketch your clean divide-by-two (so that if you look back in a month or so, you'll rediscover how clever you were).

Once things are working right, the keypad should act like the keyboard of a crude musical instrument. Do you hear the pitch fall by an *octave* when you change from key X to key 2X?

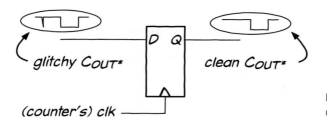

Figure 16L.19 Synchronizer flop can clean up COUT*.

Out of tune because of the zero state: For very low keypad values, your instrument goes out of tune because of the extra state *zero* included in each down-count cycle.

For example, if you load a 1 from the keypad, the counter receives a 16_{10} because of the constant zeroes in the LSB positions. It counts down from 16 to zero, then reloads 16. So, it goes through 17 states, not 16. When you load a 2 from the keypad you might expect a doubling of period – but a 2 evokes 33 states – not quite an octave different from 17. In fact, that small difference amounts to almost exactly a half-tone, as your ear may confirm for you.[11]

16L.5 Counter applications: stopwatch

We know this is a long lab. We offer these exercises because we think you'll find them fun, and instructive. Do what you can. Don't feel guilty if you don't build the stopwatch. You may get a chance to build it (and apply it as a capacitance meter) in the project lab, 19L.

A very slight alteration of the 16-bit counter will allow you start and stop the counter: you need only add a manual switch to control the level of the Carry$_{in}$*. So altered, your counter would be a primitive stopwatch. The addition of a few flip-flops and two NAND gates can make this stopwatch more convenient: first by letting you latch the counter output into the displays (that way you need not watch the counting-up process), then by clearing the counter automatically after the result has been latched.

The circuit sketched in Fig. 16L.20 can measure the length of time a signal spends low (as we have wired it). This period-measuring circuit then could be put to any of a number of uses. In Lab 19L for example, we propose that you might use it to measure the period (or half-period, to be a little more accurate) of the waveform coming from a 555 RC oscillator. If you then hold the *R* constant and plug in various *C*s, you will find that you have built a *capacitance meter*. You might find this a satisfying payback for about a half-hour's wiring.

When you have had some fun with the circuit, we ask you to restore your counter and RAM to their earlier form: so, flag the changes you make to the circuit as you go along – you may want to plug in odd-looking wires at the points where you remove a wire. (There are only a few of these points.)

16L.5.1 Stage One: simple stopwatch

Remove the lines that now drive Clock and SYNC_LD*; remove the line that grounds C_{in}*. Drive Clock with a TTL signal from the function generator (not the *breadboard* generator; you will soon need a frequency higher than its 100 kHz); temporarily tie SYNC_LD* high.

First confirm that you can start and stop the counter by taking C_{in}* low, then high, using a manual switch. You can clear the counter, as you confirmed earlier in this lab, by pressing the RESET* pushbutton. You now have a clumsy stopwatch.

---

[11] You may recognize the interval "2" to "1" as a major seventh.

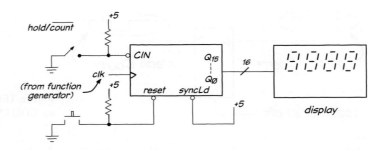

Figure 16L.20 Stopwatch block diagram.

16L.5.2 Stage Two: automatic period-meter

The stopwatch becomes a period-meter if we add the automatic output-latching and counter-clearing mentioned above. Fig. 16L.21 sketches the scheme.

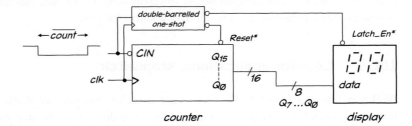

Figure 16L.21 Period-meter block diagram.

To save the result and then clear the counter for next pass we need a circuit that will generate pulses timed thus as in Fig. 16L.22. The save pulse comes earlier than the clear to prevent *clear* from clobbering the data we're trying to catch. You may recognize this pair of pulses, evoked by the rise of Count* (here shown as "trig," so named for the terminal on the one-shot) as precisely the output of the digitally-timed one-shot that you built in Lab 15L. We hope you saved that circuit. (Fig. 16L.23 offers a reminder.) If you didn't, you can rebuild it in about five minutes.

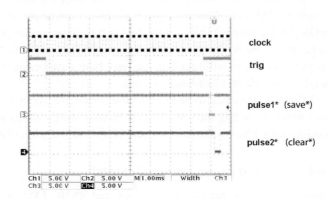

Figure 16L.22 Save then Clear signals needed.

The first pulse – labeled Latch_En* in Fig. 16L.21, "save" in Fig. 16L.22 – will update the display. It drives the terminals labeled Latch_En* on the LCD board (two Enables*, each controlling an 8-bit latch). You should join the two Latch_En*'s for a 16-bit update. You should also make the appropriate display choices:

- no labels;

- 16-bit display;
- two-line display;
- the 8-position DIP switch on top right of the LCD board should be set to 16 (not 'Mux').

The transparent latch will take in new information when En* is low, then will hold this information after En* goes high.

Notice that because the display uses transparent latches rather than edge-triggered registers we need to generate a *pulse*, not the usual *edge*. This example illustrates the clumsiness of such pseudo-clocking. Edge-triggering is much neater. ("Why didn't we design it in?," you may ask.[12])

The second pulse clears the counter when the count has been safely stored in the latches.

Generating the required pulses: double-barreled one-shot: Figure 16L.23 is the one-shot that you built in Lab 15L.

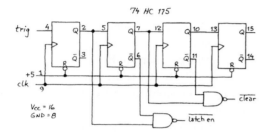

Figure 16L.23 Double-barreled one-shot: built in Lab 15L we hope.

Use the signal called Latch_En* to drive the Latch_En* of the displays. Set the keypad value to zero to load zeroes, then let Clear* drive LD*, which you will recall, is the *asynchronous* Load. Here, we use it as an async *clear*: this is slightly preferable to using the synchronous RESET*, since LD* doesn't require a clock to implement this version of *clear*.

Use a square wave from the function generator to clock both one-shot and counter. (The frequency now may be as high as you like. Try 1MHz).

Drive the one-shot input (*Trig*, above) *and* the counter's C_{in}* with the manual switch. You should find that the circuit measures the duration of the time you hold C_{in}* low. (If you clock at 1MHz, then the duration is measured in microseconds, of course.) Note that the count you see is in *hexadecimal* – a little unfamiliar to most of us ten-fingered creatures.

When you have had enough of your period-meter, restore the counter and display connections to their former state, so that we can use the counter next time for its usual purpose: to provide an address for the memory that you will install in Lab 17L.

---

[12] And you probably recall the answer: we use a transparent latch because when we want a "live" display of the input levels we need only ground En*, and this is our usual preference. In order to produce an almost-live display using edge-triggering, we would need to drive Clock with a fast square wave. That would be a nuisance since transparency is the usual case.

16W Worked Examples: Applications of Counters

16W.1 Modifying count length: strange-modulus counters

Once upon a time designing with counters was a chore that sometimes entailed designing the counter itself. Nice integrated counters then made the work pretty easy. Now programmable parts make the design *really* easy.

In this chapter we'll use the somewhat dated method of applying an IC counter, because these let us confront useful timing issues. Particularly, we'd like to see the difference between *synchronous* and *asynchronous* functions. Integrated counters are available with both sorts of Reset*: the 74HC161 is a 4-bit natural binary counter with asynchronous Reset*; the 74HC163 is the same counter with synchronous Reset*. Then we'll look at the more interesting set of problems that come up when we use the counter to make an instrument that measures something.

We'll build a divide-by-13 counter. The *fully-synchronous loadable* counter makes it now almost as easy to rig up one of these as to pull a divide-by-16 from the drawer. Not quite so easy; but almost.

First, let's consider synchronous versus asynchronous Load or Clear. It's not hard to *state* the difference: a *synchronous* input "waits for the clock," before it is recognized; *asynchronous* or *jam* inputs take effect at once (after a propagation delay of course); asynchronous functions do not wait for the clock. Either synchronous or asynchronous load or reset overrides the normal counting action of the counter.

But it is hard to see why the difference matters without looking at examples. Here are some.

Problem (Divide-by-13 counter)   Given a divide-by-16 counter (one that counts in natural binary, from 0 through 15), make a divide-by-13 counter (one that counts from 0 through 12, or through another set of 13 states). Decide whether you want to use Clear or Load, and whether you want these functions to be synchronous or asynchronous.

Solution (A poor design: use an asynchronous clear)   Figure 16W.1 gives a plausible but bad solution: detect the unwanted state **13**; clear the counter on that event.

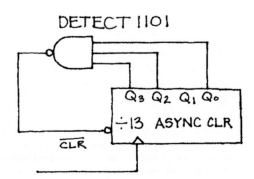

Figure 16W.1 A poor way to convert natural binary counter to divide-by-13.

16W.1 Modifying count length: strange-modulus counters

Why is this poor? The short answer is simply that the design obliges the counter to go into an *unwanted* or false state. There is a glitch: a brief invalid output. You don't need a timing diagram to tell you there is such a glitch; but Fig. 16W.2 shows how long the false state lasts.

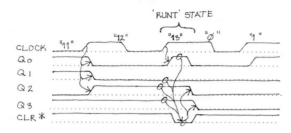

Figure 16W.2 Poor divide-by-13 design: false 14th state between 12 and 0.

In some applications you might get away with such a glitch (after all, ripple counters go through similar false transient states, and ripple counters still are on the market). You could get into still worse trouble though: the CLR* signal goes away as soon as state "13" is gone; the quickest flop to clear will terminate the CLR* signal; this may occur before the slower flops have had time to respond to the CLR* signal; the counter may then go not to the zero state but instead to some unwanted state (12, 9, 8, 5, 4, or 1). That error would be serious; not just a transient. (A very similar example is treated in AoE §10.5.2D.)

Solution (A proper design: synchronous) A counter with a synchronous Clear or Load function – like the 74HC163 – allows one to modify count length cleanly without putting the counter into false transient states and with no risk of landing in a wrong state. It is also extremely easy to use.

To cut short the natural binary count, restricting the machine to 13 states, we again need to detect a final state and drive RESET*. This time we need to detect the **12** state, not the 13, as before. On detecting 12, the logic tells the counter *before* the clock to clear on the next clock.

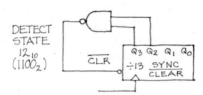

Figure 16W.3 Synchronous divide-by-13 from divide-by-16 counter.

In case you need convincing, Fig. 16W.4 is a timing diagram showing the clean behavior of this circuit. The RESET* signal sits asserted for a full clock cycle before it is acted upon. The synchronous-clear divide-by-13 circuit works nicely. It's too bad though that it requires a NAND.[1] Can we do any better?

Use of the Load function instead of the Clear can achieve nearly the same result with a single inverter instead of the NAND. That difference is not important, but let's show off our versatility by making the divide-by-13 this way.

This use of Load rather than Clear has some funny side effects.

- Using Load can oblige one to use a strange set of states (starting from three, say, and counting up

---

[1] We admit this "one extra gate" becomes ridiculous when one uses a PLD. But let's assume we really want to do the job with this IC counter.

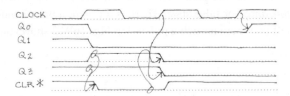

Figure 16W.4 Proper count modification: using synchronous Clear*.

to 15, then loading three again in order to define 13 states); this would be all right if the frequency alone interested you, but it would not be all right if you wanted to see the counts 0 through 12.
- Or, if we want to use states 0...12, then use of Load requires a *down* counter. This is the technique we use in Lab 16L to make a counter of variable modulus: load the initial value; count to zero, then use the Carry/Borrow signal to load once more. ("Borrow," by the way, is just a *down*-counter's "Carry" signal).

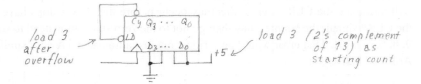

Figure 16W.5 Divide-by-13 counter using synchronous Load function.

We should not make too much of these odd effects: the number of states a counter steps through always bears a slightly funny relation to the value loaded or detected: if you load and count down, *states* = (*count*+1); if you load and count up, *states* = (2's-complement of count loaded); if you detect and clear, *states* = (*count-detected*+1). So, things are tough all over, and it doesn't matter much which scheme you choose.

Synchronous load and clear functions are nice: they support the ideal of fully-synchronous design. But synchronous functions have not simply *replaced* asynchronous because sometimes the synchronous type is a decided nuisance. On the counter of Lab 16L, for example, we provide an *asynchronous* Load so that a user need only press a button in order to Load the counter.[2] But most of the time, synchronous functions remain preferable.

16W.2 Using a counter to measure period, thus many possible input quantities

Counting as a digital design strategy: We have noted already in Chapter 16N that we can build a variety of instruments using the following generic two-stage form:

- an application-specific "front end" generates a pulse whose duration is proportional to some quantity that interests us;
- a "stopwatch" measures the duration of that front-end pulse.

---

[2] We had learned from sad experience that our students were not clever enough to carry out the two-step process required by a *synchronous* Load function. Asking them to press a Load button with one hand, then Clock the counter with the other hand turned out to be too much for some. Sometimes we would observe a student repeatedly pressing Load and muttering "It doesn't work!"

16W.2 Using a counter to measure period, thus many possible input quantities

Here we will look examples of circuits that fail in an attempt to use this arrangement, and then we will go through a longer design exercise where we try to do the job right.

16W.2.1 Three failed attempts

First failure: digital piano-key speed-sensor: Figure 16W.6 is a flawed scheme: part of a gadget intended to let the force with which one strikes the key of an electronic instrument determine the loudness of the sound that is put out. The relation between digital count and loudness shows a nasty *inverse* relation.

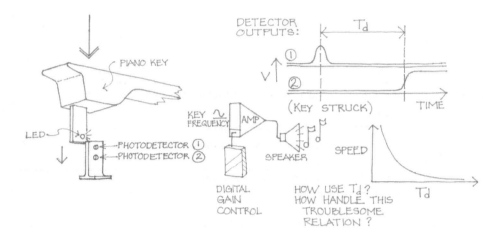

Figure 16W.6 First failure: measuring period to measure how hard a key was hit.

We offer this as a cautionary example: probably this piano design is a scheme worth abandoning. The design could be made to work with some clever post-processing of the signal; but to call for that suggests there must be a better arrangement.

Second failure: latching a final count: In Fig. 16W.7 a small error makes this big circuit useless. The clearing and latching are driven by a shared pulse. So the cleared value is saved: zeros forever.

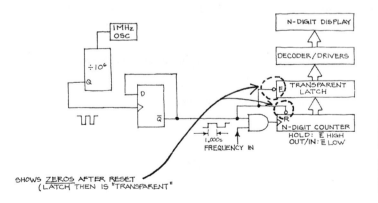

Figure 16W.7 Second failure: a transparent latch catches zeros.

Third failure: again missing a final count: As the note in Fig. 16W.8 says, this circuit is not perfect. It is much better than the transparent-latch design of Fig. 16W.7, but once in a while it may show a spurious count. The output register is clocked while the counter is clocking, so the count value may change during the register's setup time. Not much is at stake: an occasional odd display lasting one second is the danger.

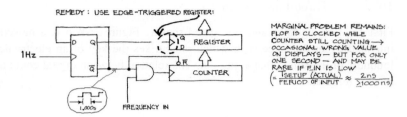

Figure 16W.8 Third failure: an edge-sensitive register violates setup time.

16W.2.2 Trying to get it right: sonar ranger

Now that you're getting good at designing these circuits (or at least good at finding fault with the designs of others), let's do an example more thoroughly.

An ultrasonic sonar sensor[3] generates a high pulse between the time when the sensor transmits a burst of high frequency "ultrasound" and the time when the echo or reflection of this waveform hits the sensor. The timing diagram in Fig. 16W.9 says this graphically.

So the duration of the high pulse is proportional to the distance between the sensor and the surface from which the sound bounces. Only a couple of bursts are sent per second.

Problem Design hardware that will generate a count that measures the pulse duration, and thus distance. Let your hardware cycle continually, taking a new reading as often as it conveniently can. Assume that the duration of the pulse can vary between $100\mu s$ and 100ms. You are given a 1MHz logic-level oscillator.

In particular:

- Make sure the counter output is saved, then counter is cleared to allow a new cycle.
- Choose an appropriate clock rate, so that the counter will not overflow, and will not waste resolution.

Solution We do this in several steps. First, just draw the general scheme; the block diagram in Fig. 16W.10. It looks almost exactly like all the earlier examples. The difference lies only in the device that generates the *period* to be measured by our "stopwatch."

The rest of our work is only to take care of the timing details. As usual, it is these that present the only real challenge.

---

[3] It was designed originally for use in Polaroid cameras; now used often in robotics.

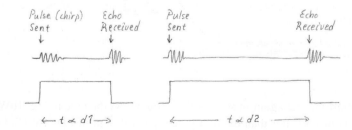

Figure 16W.9 Timing of sonar ranging device.

16W.2 Using a counter to measure period, thus many possible input quantities

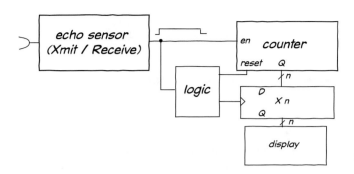

Figure 16W.10 Sonar device: digital readout. A block diagram.

We need to save the final count, then clear the counter, at the end of each cycle. You do exactly this in Lab 16L, but there your job is complicated by the circumstance that the register you use is the nasty *transparent latch* on the display board. We handled that difficulty by adding the complexity of a double-barreled one-shot. Here we will choose *edge-triggered* flip-flops instead. Nearly always, edge-triggered devices ease a design task.

Let's suppose we mean to use the PAL counter of Lab 16L. This device offers a synchronous clear and an asynchronous load; we will have to decide which to use for the clearing operation. The counter also offers a Carry$_{in}$* that will allow a logic signal to start and stop the counter. How's Fig. 16W.11 look, using the counter's synchronous RESET*? Is the timing scary? Will the count get saved, or could the *cleared* value (just zeros) get saved? The timing diagram above says it's OK: the D-flops certainly get their data before the clearing occurs on the next clock edge.

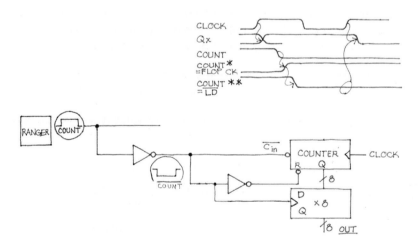

Figure 16W.11 Trying to make sure the count is saved before it is cleared.

To get some practice in thinking through this problem, look at another case, Fig. 16W.12, that's a little different: this hypothetical counter uses an *asynchronous* ("jam") clear. The problem is that we clock the D register while the counter still is counting. This is the defect that worried us in the circuit of Fig. 16W.8.

Probably this will work like an elaborate equivalent to the cheaper circuit of Fig. 16W.13. We should stick with the earlier circuit then: the one that clocks the D-flops slightly before clearing the counter and that uses a *synchronous* clear. A ghost of a problem remains however. Whether we should worry about it depends on whether we can stand an occasional error.

Worked Examples: Applications of Counters

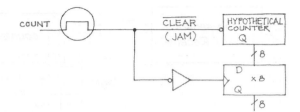

Figure 16W.12 Another example: failing at the same task.

Sometimes the counter Qs will be changing during the *setup time* of the D register. That can lead to trouble. The most bothersome trouble would be to have some of the register's D-flops get *old* data (from count N) while others get *new* data (from count $N+1$). That's worse than it may at first sound: it implies not an error of *one* count, but a possibly huge error: imagine that it happens between a count of 7Fh("h" indicating "hexadecimal") and 80h: we could (if we were very unlucky) catch a count of 8Fh. That's off by nearly a factor of two.

This will happen *very* rarely. (How rarely will the Qs change during setup time? *Typical* time during which flop actually cares about the level at its data input ("aperture," by likeness to a camera shutter, apparently) is 1–2ns (versus 20ns for *worst case* t_{setup}); if we clock at 1MHz, that dangerous time makes up a very small part of the clock period: 1 or 2 parts in a thousand. So, we may get a false count every thousand samplings. If we are simply looking at displays that does not matter at all. If, on the other hand, we have made a machine that cannot tolerate a single oddball sample, we need to eliminate these errors.

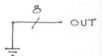

Figure 16W.13 Sad equivalent to the bad save and clear circuit.

A very careful solution: Here is one way to solve the problem: *synchronize* the signal that stops the counter (with one flop); delay the *register* clock by one full clock period, to make sure the final count has settled. Perhaps you can invent a simpler scheme than that shown in Fig. 16W.14.

Problem (A nice addition: overflow flag.) Can you invent a circuit that will record the fact that the sonar ranger has *overflowed* – i.e. warn us that its latest reading is not to be trusted?

Hint: flip-flops remember. A good circuit would clear its warning as soon as it ceased to apply: when a valid reading had come in.

Solution Here's one way.

- Let the *end* of the Carry∗ pulse clock a flop that is fed a constant high at its D input. Call the Q of the flop *overflow*.
- When the period finally does end, let the *overflow* Q get recorded in a *Warning* flop that holds the warning until the end of the next measurement.
- Meanwhile, to set things up properly for the next try, let the end of the period clear the *overflow* flop so that it will keep an open mind as it looks at the duration of the *next* period.

This scheme (in Fig. 16W.15) looks pretty ornate. See if you can design something tidier.

A really careful designer might notice that in this design we are tolerating the possibility of a glitch on Carry. Since the counter's Carry is simply an ANDing of several Qs, it could glitch on a transient state of the counter. We saw this possibility in §16N.3.3, where a scope image (Fig. 16N.21) shows some of these events. Such a glitch would trigger a spurious overflow indication.

The very cleanest design would use a *synchronous* carry – detecting a state one-before overflow (max-count minus one) and feeding that to a synchronizing D-flop. But we don't want to wear you

16W.2 Using a counter to measure period, thus many possible input quantities

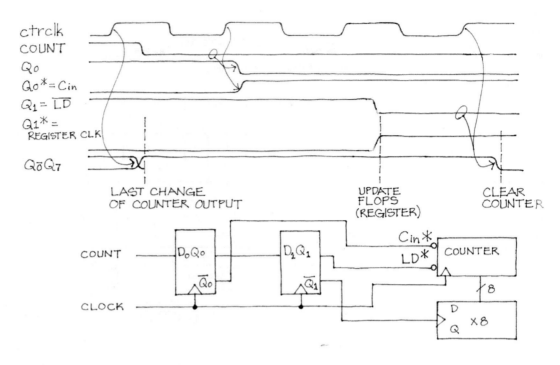

Figure 16W.14 Delaying flop clock to make sure we don't violate setup time.

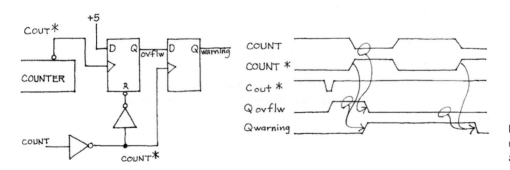

Figure 16W.15 One flop records fact of overflow; another tells the world.

out; we only want to alert you to the sort of timing issues that make digital design not so obvious as it appeared in Chapter 14N, when all was combinational, and timing problems did not arise.

The details of this design are fussy, but we will see one of the methods used here several more times: feeding a constant to a D input to exploit the nice behavior of an *edge-triggered* input. You will see this again in a "Ready" key, Lab 22L, and then in interrupt hardware, Lab 23L.

16W.2.3 Recapitulation: full circuit of sonar ranger

Figure 16W.16 is, with few words, a diagram of a no-frills solution to the sonar ranger problem. The diagram omits the overflow warning logic drawn above, and the "very careful..." logic. It uses small 4-bit decoders and displays: not a likely choice now that LCD and even pretty OLED ("organic LED") displays are available.

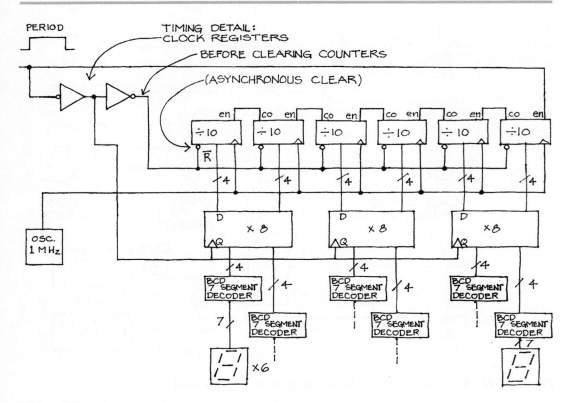

Figure 16W.16 Sonar period-measuring circuit: full circuit.

16W.3 Bullet timer

Here's an application of flip-flops and counters: a circuit that will measure muzzle velocity of a bullet. The bullet is fired at two wires that are 10cm apart. Breaking the first wire starts a counter; breaking the second stops the counter. A competition rifle has a muzzle velocity of about 3000 feet per second. Your instrument could compare rifle performances to that speed. (This problem resembles the *reaction timer*, a problem posed in the Digital Project Lab, Chapter 19L, but is a bit simpler.)

Here are some specifications for the circuit.

- To sense the bullet's passage, use a thin wire with one end grounded, the other pulled up to +5V through a resistor. When the bullet breaks the wire, the voltage rises to +5V. Assuming that stray capacitance is 50pF, choose a pullup value that will not introduce an appreciable error.
- Decide whether you need to worry about the equivalent of bounce – intermittent connection to ground as the bullet passes, perhaps scraping along the severed wire.
- you want $0.1\mu s$ resolution. Decide how many bits your counter needs. (You'll need to estimate the time for the bullet of the "competition rifle" to move 10cm).
- Cascade 4-bit counters (74HC162s: synchronous BCD counters – 0 to 9 count – with synchronous Clear* and Load*) to do the counting for you. Use enough '162s to measure perhaps 30% variation in bullet speed. Make your counter synchronous.
- Your time-keeping clock is a 30MHz crystal oscillator. Divide it down to get the needed resolution.

- Include a manual pushbutton that generates a signal named $\overline{\text{ARM}}$. Pushing this button clears a flop; this flop will go high when the bullet hits the first wire.
- Let this flop start the counter when the bullet hits the first wire.
- Let the bullet's breaking the second wire clear the flop and stop the counter.
- Display the result using displays like the one you use in your computer.
- Include a circuit to detect *overflow* of your multi-digit counter. Light a red LED on overflow; let assertion of the $\overline{\text{ARM}}$ signal clear the overflow signal. Assume you are allowed to use an *asynchronous* Carry$_{\text{out}}$ signal.

16W.3.1 Bullet timer solution

We'll present possible designs for several fragments of a circuit that would perform as demanded – and then we'll sketch the whole circuit. You may find the explanations offered with each fragment excessive. If you do, skip to the one-page solution at the end, which includes short explanations of most design choices.

Starting and stopping the counter: The bullet will break one wire to start the counter, another to stop it. To keep the counter going between these two events we need an element that will hold this "counting" state: a flip-flop.

The bullet needs to switch the level of a signal, twice. Let it break a wire that ties a signal to one level, while a resistor pulls to the other level.

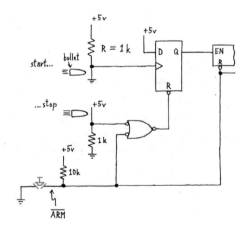

Figure 16W.17 A flop will start counter, then stop it.

Generate the clock signal: Given our 30MHz oscillator, and our need to resolve to 0.1μs, which implies need for a 10MHz clock rate, we need a divide-by-three counter. Fig. 16W.18 is a circuit that uses a counter with *synchronous* clear in order not to generate nasty glitches[4].

Details: timing questions: The time we are trying to measure is quite brief, and we are trying to resolve it to 0.1μs, so we should make sure the delays in our circuitry don't compromise its performance.

---
[4] A 74HC162 counter is shown. This is BCD or "divide-by-ten." A '163 natural binary would work as well. Both offer *synchronous* Clear*.

Worked Examples: Applications of Counters

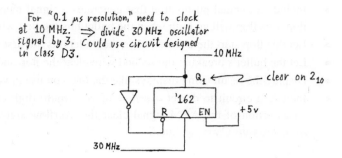

Figure 16W.18 10MHz clock from 30MHz oscillator.

RC **delay when wire breaks** The pullup/pulldown resistor forms an *RC* with stray capacitance which we have estimated at 50pF. How small should we make the *R*? Let's keep *RC* well under the 0.1μs resolution. Keep it at, say, a half or quarter of that – 50ns or 25ns: see Fig. 16W.19. A pullup of 1k probably is OK; 470Ω is a little better.

Figure 16W.19 Keep delay small from R_{pullup} and C_{stray}.

Chatter or bounce, as bullet passes? Is this harmless? It is. The bullet's length is not negligible relative to the 10cm distance between Start and Stop wires, so if one imagines the bullet scraping contact during that time, perhaps we have a problem to solve: one that looks like switch bounce. But in fact there is no problem: the flip-flop, as drawn, responds only to the first transition (on clock) then first low (on the reset); the flop acts like a debouncer. This is a sufficient argument; but in addition it seems very probable that the wire breaks contact cleanly on the initial break so that the electrical signal does *not* show chatter (bounce-like noise) in any event.

How many 4-bit counters needed? How long may the bullet take to travel the 10cm gap between Start and Stop wires? Average speed is given as 3000 feet/s; average time then (see below) is about 110μs. If we want to resolve to 0.1μs, we'll need four decimal digits (though one will not show more than the value "1"; on a DVM this would be called "3 1/2 digits"), so we'll use four 4-bit counters, each a "divide-by-ten."

Once we know the range of times we need to measure (said to be 30% more or less than the average), and resolution wanted, the counter use is pretty pedestrian. (We have sketched in Fig. 16W.20 4-bit hex displays; probably you will want to use the LCD's 16-bit display instead.)

Detect overflow: Detecting overflow is more interesting. We need to use the highest-order counter's *carry-out* signal to sense overflow – because an overflow warns us that we cannot trust the recorded

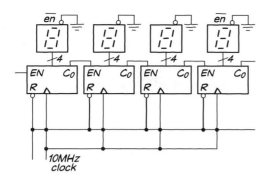

Figure 16W.20 Four divide-by-ten counters will measure time for us.

count. (We admit that overflow is wildly unlikely in this design. But it's a technique worth learning; so, here goes, though you may object that you have seen this at least twice before.)

We need to get straight whether the carry indicates an overflow or *impending* overflow. In fact, for a properly-designed synchronous counter, the carry indicates the latter: it means that overflow will occur on the *next* clock. So, it is the termination of *carry-out* that will indicate overflow. And, since we want the lighting of the overflow indicator to persist, we'll need a flip-flop to keep the LED lit till the user turns it off.

Simplest overflow detector: async use of Carry$_{Out}$: Figure 16W.21 shows a straightforward way to watch for overflow. The termination of a Carry$_{Out}$ signal indicates the counter has overflowed (Carry$_{Out}$ is asserted on maximum count – one short of overflow).

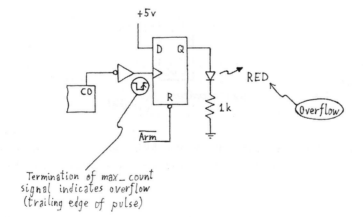

Figure 16W.21 Overflow sensor keeps LED lit after overflow.

This plan *may* work but we have noted in Chapter 16N that glitches on the Carry$_{Out}$ line could throw off the simple overflow detector that we have sketched here. The detector can malfunction if the counter's Qs, though synchronous, show enough *skew* so as to change not-quite-simultaneously. Such a timing mismatch could produce a spurious Carry$_{Out}$ signal.

One can solve this problem by using a *synchronous* carry to indicate overflow. If one designs the counter from scratch, one can design such a carry into the counter. This we could have done in the PAL counter that you use in Lab 16L. Here is Verilog code to implement such a carry (which we chose *not* to include in the lab counter):

```
if (upbar_down) // a high on this line takes us down
  begin
    count <= count - 1;
    if (count == 16'b 0000000000000001) // minimum count plus one
      cout_sync_bar <= 0; // this will assert cy on next clock, state zero
    else
      cout_sync_bar <= 1;
  end
else
  begin
    count <= count + 1;
    if (count == 16'b 1111111111111110) // maximum count minus one
      cout_sync_bar <= 0; // this will assert cy on next clock, state max
    else
      cout_sync_bar <= 1;
  end
```

The code is not very elegant – and does not behave properly if the direction-of-count (determined by the upbar_down signal) is changed while the counter is at an extreme. For this reason, we did not include this fancier sync_carry.

Using commercial IC counters, we don't have that option, but can make a near-equivalent by adding a *synchronizing* flip-flop that will not pass glitches. Such a circuit is sketched in Fig. 16W.22.

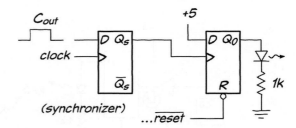

Figure 16W.22 Synchronizer added to counter's carry-out can protect overflow flop from glitches.

Glitches on the asynchronous Carry$_{out}$ line will occur soon *after* the clock edge; they will not pass through the synchronizer. Fig. 16W.23 makes this point graphically: the signal *sync'd_Carry_out* does not pass the glitches that appear on Carry$_{out}$.

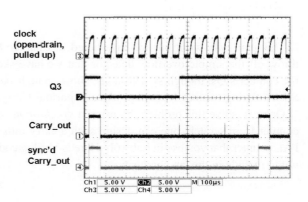

Figure 16W.23 Glitches on Carry$_{out}$ can be cleaned up by a synchronizing flop.

16W.3 Bullet timer

This remedy may well not be necessary. To produce the glitches shown in Fig. 16W.23 we had to try hard.[5] But the synchronizing strategy is one worth remembering. Asynchronous circuits – ready to respond at any time – cannot tolerate noise on their clock lines; synchronous circuits can – so long as the noise does not occur during setup time, as normally it does not.

A full solution (circuit diagram): Figure 16W.24 pulls together the several pieces of the design – with most of the explanations included.

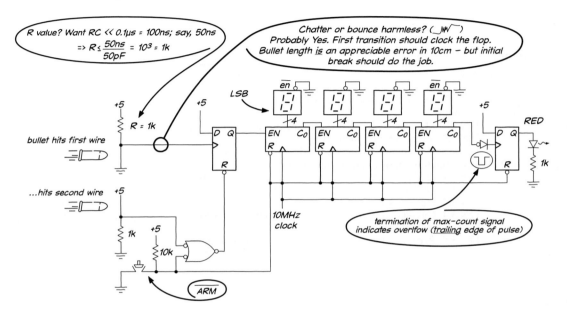

Figure 16W.24 Full circuit for a solution.

---

[5] We had to tease the circuit by slowing the clock edge. This slow edge apparently clocked the several flops at slightly different times. Capacitively loading some of the outputs (Q0...Q2) was ineffective, probably because the flop outputs are buffered before being brought out to the pins. And inducing the glitches by slowing the clock edge is a bit meretricious: the slow edge becomes unreliable as a synchronizing signal. Just as it can clock the several flops within a synchronous counter at different times, it can also clock the synchronizer at a time different from the time when it clocks the counter. Such a mismatch can undo the cleanup – passing the glitch to the output of the synchronizer, if the synchronizer's clock is a little late.

17N Memory

Contents

17N.1	**Buses**	**648**
	17N.1.1 When are three-states needed?	648
	17N.1.2 Example of 3-state use: data bus	649
17N.2	**Memory**	**651**
	17N.2.1 Memory describes a way of organizing storage	651
	17N.2.2 Memory types; memory jargon	652
	17N.2.3 How memories are specified	654
17N.3	**State machine: new name for old notion**	**655**
	17N.3.1 Designing any sequential circuit: beyond counters	655
	17N.3.2 Designing a divide-by-three counter	656
	17N.3.3 Same task done with Verilog	658
	17N.3.4 Circuits requiring complicated combinational logic: a *memory* can replace the combinational gating	659
	17N.3.5 Recapitulation: several possible sequential machines	660

17N.1 Buses

These are just lines that make lots of stops, picking up and letting off anyone who needs a ride. The origin of the word is the same as the origin of the word for the thing that rolls along city streets[1].

Power buses carry V+ and ground to each chip; data and address buses go to many ICs in a computer. Some control signals may run to many parts: a R/W* (read/write) signal might, for example.

17N.1.1 When are three-states needed?

Answer: whenever more than one *driver* (or "talker," to put it informally) is tied to one wire.

Where in Fig. 17N.1 are three-states needed? (In case you need reminding, a three-state is an output stage capable of disconnecting entirely, rather than switching the output to ground or positive supply, as in a conventional gate.) The presence of more than one *receiver* (or "listener") does *not* call for 3-states. (Think of a telephone "party line" – one shared by several telephones – if you find yourself confused on this score: the little old eavesdropper on your conversations may be annoying, but is harmless so long as he is silent.) Try the rule on the circuit of Fig. 17N.1.

Let's start with the data bus. RAM and ROM and CPU (2, 4, 7) – all talkers – need 3-states where they meet the data bus. Notice that ROM needs it as much as RAM and CPU; bidirectional or single-direction doesn't matter. Keyboard (9) needs it; DAC (digital-to-analog converter) does not (only listens).

---

[1] If Latin isn't too Harvardy for you: "bus" is just short for "omnibus," "to or for all." The city bus, an innovation in early 19th century England, was given this name to contrast it against a carriage that one would hire for a single point-to-point ride.

17N.1 Buses

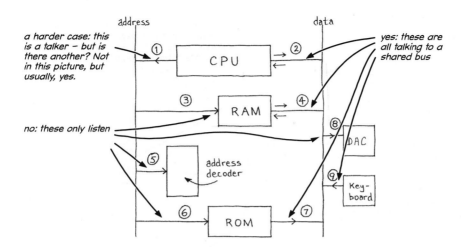

Figure 17N.1
Where are 3-states needed?

On the address bus: RAM, ROM and decoder (3, 6, 5) only listen; so no 3-states there. How about the CPU (1)? It's a talker. But it looks, in this diagram, as if it's the only talker. If so, no need for a 3-state. This was the view taken by the designers of early processors – including the 8051 that you will soon meet. But after those first parts were launched designers began to notice that it sometimes was useful to be able to share the address bus.

A fancy case would be one using more than one processor; but the garden-variety case was one in which the processor relinquishes control from time to time to allow a fast transfer that bypasses the processor. This is so-called "Direct Memory Access" (DMA). A disk drive, for example, can pump data into the RAM directly using a dedicated address counter; it can do this much faster than if each data word had to pass into the processor and then out again as in a normal *copy* or transfer.

Does our little lab computer need DMA capability? Surely not – it's such a simple machine. Not so: it is simple, but it does use DMA. When we load data into memory by hand we are doing DMA, though in glacially slow motion. How do we do this if the 8051's designers did not anticipate DMA? We fake 3-state behavior by putting all the processor's output pins into their highest impedance state. They are not *off*, as in a true 3-state; instead, they are weakly pulled up so that any external device can control the lines. (We force this weak pullup by asserting the 8051's RESET pin.)

17N.1.2 Example of 3-state use: data bus

Figure 17N.2 shows a part of the circuit you will build in Lab 17L. One 3-state is shown (a parallel set of eight, the '541); another is implicit (within the RAM).

The '541 keeps the keypad from monopolizing the data bus. What keeps the RAM from hogging the bus all the time? (Note that the RAM's OE*, its explicit 3-state control, is tied low – asserted – all the time).

Answer: the RAM's WE* pin turns off the RAM's 3-states when asserted (low).

RAM includes 3-states too: Figure 17N.3[2] illustrates the innards of the static RAM that you will meet in today's lab. This is a traditional *asynchronous* RAM. *Synchronous* static RAMs can achieve higher speeds, but are less easy to use and we will not look into them in this book. You will find them discussed in AoE §14.5.3C. The figure shows the roles of the RAM's CE* ("chip enable"), OE*

---
[2] Cypress Semiconductor CY7C1399BN.

650 Memory

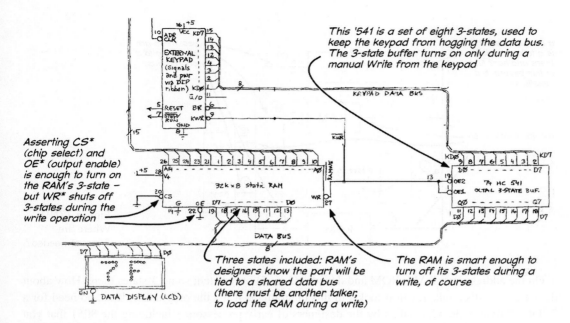

Figure 17N.2 Example of 3-state use: keypad meets memory.

(output enable), and WE* (write enable). The fact that assertion of WE* turns off the 3-states permits the continuous assertion of CE* and OE* that we use in Lab 17L (as shown in Fig. 17N.2). In later labs, once the circuit has been complicated by the arrival of the CPU (Central Processing Unit, the brains of the machine), we use these signals in less rigid and more usual ways.

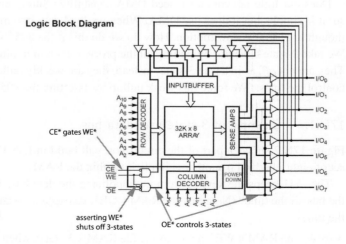

Figure 17N.3 RAM detail (32K × 8), showing roles of its three enables, CE*, OE* and WE*.

Figure 17N.3 shows the relations one would expect among the three enables:

WE*: Asserting WE* (on a write) turns off the 3-states as required: the RAM stops talking in order to *listen*.

CE*: CE* gates write. So WE* is ineffective when the chip is disabled. The big-board computer exploits this behavior. To keep its gating simple, it permits WE* to reach the RAM on I/O writes (non-memory actions). These are harmless because CS* is disasserted in I/O address space.

Figure 17N.3 wrongly suggests that CS* does not condition data *reads*. The data sheet from which this figure comes makes clear that CS* must be asserted to permit any memory action: read or write.

OE*. This directly controls the 3-states. The response to OE* is quicker than response to CS*, which controls more complex functions in the RAM.

17N.2 Memory

AoE §14.5

We have spoken rather glibly of a "memory" without explaining what it is. Let's remedy that omission. A *memory* is an array of flip-flops (or other devices each of which can hold a single bit of information), organized in a particular way to limit the number of lines needed to access this stored information. A memory is used most often, as you know, to hold data and program for a computer; it can also be used to generate a combinational logic function – as you will see in Lab 17L.

17N.2.1 Memory describes a way of organizing storage

8 bits can be stored in a *register*. The register is just a collection of flip-flops, with a common clock input, and all Ds and Qs brought out in parallel. If those 8 bits are stored, alternatively, in *memory* form, then inputs and outputs are multiplexed each onto one line; and a further improvement allows abolishing even the IN versus OUT distinction, so that IN and OUT become the bidirectional DATA line.

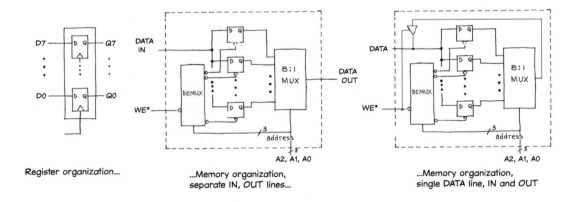

Figure 17N.4 Memory as a way of organizing data storage, minimizing external connections.

At first glance the memory looks silly: a lot of extra internal hardware just to save a few pins? Let's count pins:

Function	**Number of Lines if using Regiser**	**Memory**
data IN:	8	1
data OUT:	8	1 or 1 shared between IN and OUT
clock/write	1	1
power, ground	2	2
address	0	3
total	19	7 or 8

Pin count: register versus memory: The memory advantage doesn't look exciting – until you crank up the number of stored bits to a reasonable level. Then the memory reveals that it is the *only* feasible way to store the information (try counting pins on a 1Mbit memory reorganized as a *register*, for example).

Memories normally squeeze away one more pin, as shown on the right in Fig. 17N.4, combining input and output into one pin. WE* turns off the 3-state that normally would be showing the memory's content, as we saw when we looked into the RAM in Fig. 17N.1.2.

Memory organization allows access to only one stored value at a time, whereas the register delivers all its bits in parallel. But the memory scheme matches nicely with the sequential nature of computer processing. Note however – speaking of parallel versus sequential – that the RAM we have sketched in Fig. 17N.4 is just *one bit wide*. The RAM you meet in lab is eight such structures in *parallel*, storing an 8-bit "word" at each location. This is the normal arrangement. The one-bit-wide memory is a scheme we offered only for its simplicity.

That's all there is to the *concept* of a memory. Now on to the *jargon*.

17N.2.2 Memory types; memory jargon

One can slice the universe of memory types in several ways:

AoE §14.4

- **RAM versus ROM: can one write to the memory?**
- If **No** it's called a **ROM** ("read-only memory"). Only the manufacturer can write to a ROM. The name obviously overstates the part's character: a memory that no-one could write to would be thoroughly useless.
- If **No, not unless you work pretty hard**, it's called a **PROM** ("programmable read-only memory"). A user can write to a PROM, though not so easily as one can write to a RAM (see below).
  - **PROMs come in several flavors** (none of them nowadays called PROM)
    - **EPROM** (E for Erasable): the oldest. To erase these you'd have to remove them from the circuit and put them in a UV box for 15 minutes or so.[3] That was a nuisance, so they have been supplanted by EEPROMs (electrically-erasable PROMs) of various sorts. These put a small charge on the floating gate of a MOSFET. In this they resemble dynamic RAMs – but the charge persists for 10 or 15 years, rather than for a handful of milliseconds.
    - **EEPROM**: these can be erased a byte-at-a time, and can be erased in-circuit. That is obviously more convenient than the EPROM scheme. Our PALs use this technology.
    - **Flash memory** (so-called because it is erased quickly): good for mass storage, but not byte-erasable – must be erased in larger blocks. The memory on the 8051 is of this type. Everybody's "memory stick" uses flash.

AoE §§14.4.5B, C

- If **yes, easily** the memory is called "RAM," a misnomer, since *all* these semiconductor memories are "random access memories."[4] But the name has stuck.

---

[3] If you think the acronyms are excessive, in this field of RAMs and ROMs, consider this one, perhaps the mother of all excessive acronyms: the sort of EPROM that was offered in a cheaper package omitting the quartz-erasing window. This packaging reduced the part price more than tenfold. Lacking a window, it could not be erased. The idea behind this packaging was that a user could work out a design using an EPROM, then when satisfied with the programmed content would specify the cheaper package. These parts were called OTP EPROMS: "one-time-programmable erasable programmable read-only memories." Luckily for the world, these are obsolete.

[4] The name may seem puzzling. It does not mean that what the memory delivers is a randomly-chosen entry of course. It means that entries may be recalled or stored, in any order, and access times for all stored values are substantially equal. This equality contrasts most radically against a medium like tape and also, in lesser degree, against all types of disk storage.

17N.2 Memory

- **RAMs come in two main flavors.**

  Storage technology: flip-flops, called "static" RAM; versus capacitors, called "dynamic" RAM. One can tell that these names must have been invented by the manufacturers of the dynamic memories: the ones that use capacitors. Cap memories aptly could be named "forgettories" (to steal a word from AoE), since they forget after a couple of *milliseconds*! These memories need continual reading or pseudo-reading, to "refresh" their capacitors. Despite their seeming clumsiness, *dynamic* RAMs enjoy the single advantage that their remembering unit is very small and uses little power; so most large memories are dynamic. In our lab computer we prefer static RAM because it requires no refreshing. In fact, this memory is so simple and power-thrifty that we simply leave it powered continuously from week to week using a 3V coin cell. This arrangement holds data for a year or so.

  Blurring the static/dynamic distinction: "Pseudo-static" RAM blurs this contrast, providing the ease of use of a static RAM with almost the full density of dynamic: the refresh process is handled internally and is not apparent to the user. These have largely displaced traditional static RAM except where its extreme low power consumption is important (as in our big-board computer's battery-backed memory) or where extreme speed is needed (as in fast *cache* memory).

  - **Synchronous RAMs.** Both static and dynamic RAMs come in a more complex form with a clocked state machine governing the asynchronous innards. We will not look into these but instead refer you to AoE §§14.5.3C and 14.5.4B.

- **Volatility:** Does it remember, without power? In general ROM and PROMs of all sorts do remember (that is, they are "non-volatile"); RAMs do not (they are "volatile"). Again, some memories straddle this boundary:

  - **Chubby RAM** (more official name: Battery Backup SRAM, BBSRAM). Fig. 17N.5 shows DIP and surface-mount versions of this device. You can guess why the package is fat.[5]

Figure 17N.5 Chubby RAM acts like RAM, but remembers for 10 years. Figure used with permission of Dallas/Maxim.

These are as easy to use as ordinary RAM, but are expensive ($10 to above $30: for example, STM M48T35, Dallas DS1244Y).

  - A variation on this scheme is a battery *socket* with battery and power-down/up circuitry.

  - Other backup schemes are used, too, such as "shadow-RAM": it's a RAM with a non-volatile memory attached that can be updated from the RAM before power-down (for example, Cypress CY14B101KA.) These memories offer packages smaller than those for the battery-backed devices, unlimited number of RAM store cycles, and automatic non-volatile storage on power-down. A capacitor sustains the supply voltage long enough to accomplish the back-up storage.

---

[5] Well, in case you don't like guessing, we'll reveal that the fat package houses a lithium coin cell.

17N.2.3 How memories are specified

Organization: "#-of-words × word-length." For example, the RAM you will use in the lab is 32K × 8 or 32KiB × 8: 32,768 words (or locations), each 8 bits wide.[6]

Units: "K" and "KiB" versus "k": The use of the term "kilo-" in "kilobyte" offers potential for ambiguity. How many bytes in a kilobyte? This may sound like a silly question – but is a kilobyte 2^{10} bytes (1024) or is it 10^3 bytes?

One answer – not very satisfactory – might be, "Oh, who cares; the two values differ by only 2.4%." One response to that answer might be, "Well some lawyers care." At least two lawsuits complained about the use of 10^3 to define kilobyte, and recovered some damages.[7] Even within the electronics industry, manufacturers don't agree. Some makers of hard disk drives say a "kilobyte" means 10^3 bytes. Doing this made their drives seem a little bigger than if they had adopted the 2^{10} convention (but not much bigger) – hence one of the lawsuits. Memory manufacturers generally use "kilobyte" to mean 2^{10}. Flash drive sizes normally are stated in Gigabytes: 2^{30} bytes, not a power of ten.[8]

But it would be nice to be rid of that ambiguity. An upper-case K is one convention that can distinguish the computer "Kilobyte," 2^{10}, from the scientific meaning of "kilo" as 10^3. Another convention, established in 1998 but not yet widely used, uses the unit "KiBi," abbreviated "KiB." "KiBi" is easier to distinguish from an ordinary kilo. With it arrived the corresponding "MiBi" and "GiBi," and so on. In this book you will find us using upper-case K to denote 2^{10} bytes.

Memory part numbers tend to mention the number of *bits*, perhaps because this is a more impressive number. But it is a fair description of total memory size. The Cypress 32K × 8 of the sort that we use in Lab 17L, for example, calls itself CY62256N (256K bits).

Memory timing: The most important memory characteristic, once you have settled on a type and size, is *access time*: the delay between presentation of a valid address ("your" job) and delivery of valid data (the memory's job).

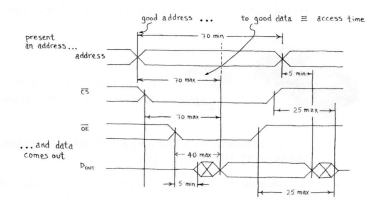

Figure 17N.6 Read timing for 70ns static RAM.

AoE §14.5.3A

Static RAM timing: Figure 17N.6 shows a timing diagram for the 32K×8 RAM used in the micro labs, assuming it is 70ns speed grade (middling, among the speeds available for this part). Don't let all the details overwhelm you. The number that matters most is the time between presenting an *address*

---

[6] If this is your first encounter with the unit "KiB," see an explanation just below.
[7] See http://en.wikipedia.org/wiki/Binary_prefix#Orin_Safier_v._Western_Digital_Corporation.
[8] The exceptional flash drive company that chose 10^3 paid some damages for overstating drive size by 7%.

and getting stable *data*: this is *access time*, in its usual sense. Another time, a little less important, is the time between asserting *Chip Select** (or *Enable**) and getting stable *data*. For this RAM, those times are the same. Since CS* usually is generated as a function of address, it's the CS*-to-data time that will tell you how soon data can be ready.

When you have wired your microcomputer you will be able to *see* the access time. (This applies only to the big-board computer; the single-chip SiLabs normally uses no external RAM.) In our design the relevant access time is the somewhat quicker OE* response. If you're feeling energetic when you reach that point in the lab, take a look at the time between assertion of OE* and the presentation of good data.

17N.3 State machine: new name for old notion

By now you have seen a variety of counters: ripple and synchronous, MSI ICs, and a bigger counter made from a PAL. A counter is a special case – approximately the simplest – of the more general device, *state machine*. This term is short for the more formal "Finite State Machine (FSM) ."[9]

An FSM is a sequential device that walks through predictable sequences of "states." (A state, you will recall, is defined as the set of outputs on the device's flip-flops.) That sounds oddly abstract; but if you apply the notion to a simple counter you will see that the idea is simple and, indeed, familiar. As a reminder, Fig. 17N.7 is a 2-bit up-counter along with a "state diagram" or "flow diagram" and a "state table" that show – rather abstractly, we admit – how the counter behaves.

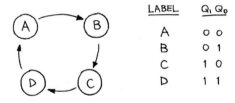

Figure 17N.7 2-bit counter: flow diagram and state table.

17N.3.1 Designing any sequential circuit: beyond counters

AoE §10.4.3

Figure 17N.8 is a highly-general diagram showing any clocked sequential circuit.[10]

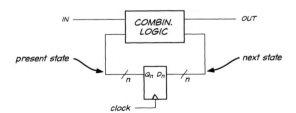

Figure 17N.8 Sequential circuit: general model.

As you can see, the D flops will take the levels at the D inputs and transfer them to their Qs after the

---

[9] Such a machine is distinguished from a "Turing machine" – a theoretical model of a general-purpose computer. A Turing machine is assumed to have infinite number of states (and memory) as an FSM does not.

[10] An FSM need not be clocked. The more general model would define the block we have shown as a set of D flops as just a "delay" block. Design of non-clocked FSMs is more difficult than what we undertake in this book, and only rarely necessary.

next clock. Thus the Q of circuit will go from its *present state* to the values on the Ds on the next clock; the values on the Ds thus show the circuit's *next state*. AoE shows how to use this notion to design a divide-by-three counter. We will do nearly the same task – but using different state assignments.

Figure 17N.8 shows *outputs*; in counters – the only state machines we have worked with, so far – the outputs have been simply the flip-flop Q's, apart from sometimes a Carry$_{out}$. The outputs may depend on the machine's *state* alone, or may depend upon *state and inputs*. The two types are given the names "Moore" (state alone) versus "Mealy" (state and inputs).[11]

A counter with a Carry$_{out}$ that sensed only the levels of the counter's Qs would be a Moore machine. A counter whose Carry$_{out}$ depended also on Carry$_{in}$ (as it usually does) would be a "Mealy" machine. We concede that learning such names is not edifying. But you may want to be able to recognize the terms.

17N.3.2 Designing a divide-by-three counter

Figure 17N.9 is what looks like the description of an ordinary counter, but we have deliberately made the design somewhat odd. Not only does it use just three states (not a power of 2, as in most counters we have seen) but also the sequence is not what you would get from an ordinary 2-bit counter, if you truncated the count sequence.

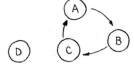

Figure 17N.9 Divide-by-three counter, using slightly odd sequence.

PS	NS
0 0	0 1
0 1	1 1
1 0	X X
1 1	0 0

The count sequence goes 0, 1, 3, 0,.... The flow diagram shows a fourth state, "D" floating outside the sequence. The present-state, next-state table also shows this unused state, 10 – and declares that we don't care what comes next ("XX" is entered as next state). This is a little dishonest, since we do care to a limited extent: we'll admit shortly that we do care that the counter should not get stuck in this unused state, 10. But it was traditional, in the days when people still designed such circuits by hand, to use don't-cares: to specify *don't care* as the next state for any unused state. The motive was to constrain the circuit as little as possible, to keep the logic minimal.

The circuit will, of course, *care* about that next state: when the design is finished that next state will be as fully determined as any other. In fact in a few minutes we will see what that next state is in our design.

These days, with large numbers of gates usually available to one on a logic array (PLD or FPGA) such minimizing is not worthwhile, and the present-state/next-state table might as well specify all next states. We won't do that, just now, because we would like to show the danger that can result from such use of don't-cares.

The PS–NS table shows the flip-flop states before and after any particular clock. Since we plan to use D-flops to design this circuit, *Next-State* also defines the pair of values we must feed to the D inputs of the two flip-flops, and we have so labeled the Next-State columns.

---

[11] Mealy and Moore wrote important papers that defined these ideas in the mid-1950s. Moore was a professor of computer science; Mealy worked at Bell Labs.

17N.3 State machine: new name for old notion

Present State	Next State
Q1 Q0	D1 D0
00	01
01	11
10	XX
11	00

This circuit is simple enough so that you can probably see the function by inspection. Recall that the XX means we can assign High or Low, as we please, to make the circuit easy to build.

If, in a particular case. you cannot see at a glance what functions the Ds need, you needn't be clever. The simple-minded way to find the function is just to list all the input combinations that can generate a 1, then OR those input combinations. In this case, for D1, for example, one might write D1=($\overline{Q1}$ Q0)+(Q1 $\overline{Q0}$). This happens to be XOR, but one could build it by OR'ing the two AND functions.

Here the two functions probably do jump out at you. But either by inspection or by some plodding process you will see the equations:

- D1 could be the XOR function of Q1Q0;
- D0 is simpler still: it is high while Q1 is low, so it's just $\overline{Q1}$.

We got these functions by assuming convenient levels for the Xs. Now let's see what Next State we have provided for the unused state, "D". D1 is high, D0 is low. Oops! That means that state D goes to D.

Why should we suddenly care? We said at the outset that we *don't care*. But we didn't quite mean it, as we admitted earlier. In fact, there is one outcome that we do care about avoiding: the machine must not get stuck in one of its unused states, like state D.

And if you're inclined to protest, "But it will never go to state D," then you're being too optimistic. On power-up the machine could go to D. And even if you build a power-on reset circuit to force the counter to a legitimate state (say, state A) when you turn on power, a glitch can carry the circuit into D. So, it's never safe to design a machine that would get stuck in an unused state.

The remedy is simple: enter some other next state in place of XX, and redesign the circuit's logic. If we make state A (coded as 00) the next state from D (coded as 10), then the function for D1 may look less tidy than XOR. This next state means that in the PS–NS table only the 01 combination – $\overline{Q1}$ Q0 – produces a high for D1. This function in fact is easier to implement than XOR. It is just an AND of $\overline{Q1}$ with Q0. Flip-flops ordinarily provide their complemented outputs, so the D1 logic requires only a single AND gate. Fig. 17N.10 shows the divide-by-three circuit designed as a state machine.

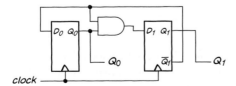

Figure 17N.10 Divide-by-three counter, from manual design.

This – or the similar design problem embedded in Lab 17L – presents probably your last occasion to do such a task by hand. Logic compilers like Verilog will handle the detail work for you.

17N.3.3 Same task done with Verilog

Verilog can do the fussy work for you in the design of a state machine. Here's the Verilog code for the counter we asked you to design (it differs from your design in including one extra signal, named "startup," used to provide known initial values for simulation). After some dull preliminaries, the Verilog file is pretty much just a state table, showing next state for each of the four states.

```
module div3_state(
  input clk,
  output reg [1:0] Q = 0
  );
parameter zero = 'b00, one = 'b01, two = 'b10, three = 'b11;

  always@(posedge clk)
    begin
    case (Q)
      zero:
        Q <= one;
      one:
        Q <= three;
      two:// the unused state, D
        Q <= zero;
      three:
        Q <= zero;
    endcase
  end
endmodule
```

Henceforth, you can work at the lofty level that Verilog permits. Fig. 17N.11 is a schematic of the logic that Verilog produced. This schematic is the same as the one we drew with our hand-made design; no surprise that – but reassuring.

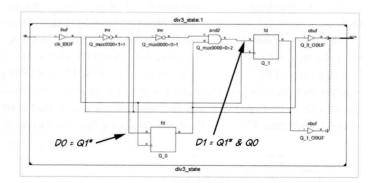

Figure 17N.11 Schematic of divide-by-three counter produced by Verilog.

Simulation of the Verilog design produced the waveforms in Fig. 17N.12, which look like what one would hope to see. The next time you design a state machine – once this lab is over – you're likely to do the task with the help of a logic compiler like Verilog which lets you work at a high level of abstraction. This divide-by-three was an atypical state machine, chosen only for its simplicity. We worked through the design of a more plausible state machine, a "bus arbiter," in Chapter 15S.

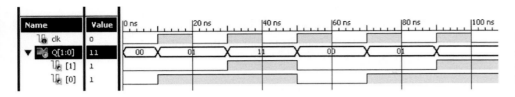

Figure 17N.12 Simulation showing behavior of counter produced by Verilog.

17N.3.4 Circuits requiring complicated combinational logic: a *memory* can replace the combinational gating

A logic compiler and PLD or FPGA makes the designer's job pretty comfortable in producing a state machine. But a *memory* offers another solution. This solution is exceptional. It was offered on a few dedicated ICs, like Cypress' CY7C258/9 2K×16 registered PROM. These parts now seem to be obsolete. We pursue it in Lab 17L, however, because the hardware that you install there is almost perfectly adapted to the scheme.

The *address* lines of the memory can carry the *input* variables and present-state bits; the *data* lines of the memory can deliver the *output functions* while also generating the *next-state* information. Memory makes it easy to provide many outputs: just use a memory with a wider data word. To provide for longer sequences of operations, just increase the number of flip-flops. To build a machine that does many alternative tricks, increase the number of inputs – which means only adding address lines: that is, use a memory that stores more data words.

A circuit like the one shown in Fig. 17N.13, if given eight inputs, would behave like the microcontroller that you soon will meet. Its eight input lines permit a repertoire of roughly 256 different tricks.[12] This state machine is the processor's *instruction decoder*. The "instructions" are eight-bit values fed to the decoder's inputs. Each evokes a particular sequence of operations.

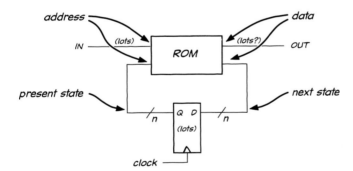

Figure 17N.13 FSM using memory in place of combinational logic.

The outputs of the instruction decoder implement the responses to the instructions. The response is a sequence of fine-grained operations necessary, say, for copying the contents of one on-chip register to another. If the instruction means "copy the contents of register 0 to register 1," then the FSM outputs might carry out three steps, something like this:

- turn on 3-state for Reg 0, putting Reg 0 contents onto an internal data bus;
- clock Reg 1, capturing the value from the internal data bus;
- turn off 3-state for Reg 0.

---

[12] "Roughly?" you protest. "Why not *exactly*?" It would be exactly 256 if all codes were implemented, and no code were used to allow adding a second byte that can expand the instruction set.

This hypothetical operation is a very simple one. Complex operations, such as the processor's arithmetic *divide*, involve a great many successive actions. Later, in the processor that we use in the micro labs, the *divide* instruction, "DIV" requires 40 clock cycles in contrast to four for the simplest operations like the register-copy process described just above.[13]

17N.3.5 Recapitulation: several possible sequential machines

If there are *no* external inputs, the circuit knows only one trick; it might be a counter, for example (it could be binary or decade; it could even count in some strange sequence; but it can perform just one operation). Its next state will depend upon its present state alone.

If there is *one* external input: then the FSM can do either of two operations. This could be an up/down counter for example.

If there are *eight* external inputs, as for the 8051 processor, then this sequential machine would be capable of running many alternative sequences: 256 of them. The eight input lines select one out of its large repertoire of tricks; these eight input lines are said to carry "instructions" to the processor. A sequence of instructions, as every toddler nowadays knows, constitutes a program.

---

[13] The comparison is valid but counting clock cycles for an 8051 is complicated by the fact that different variants use different numbers of clock cycles to perform a given action. The DS89C430, for example – the Dallas variant that we use in the big-board labs – itself varies the number of clocks that it uses, depending on the source of its code. On-chip code can run at one clock per bus-access cycle; off-chip code takes four clocks. The original 8051 required 12 clock cycles for each cycle.

17L Lab: Memory

This lab is in two parts and, like Lab 16L, proposes somewhat different paths for people planning to build the big-board computer versus those expecting to do the single-chip controller labs.

- The first part, necessary for those building the big-board but optional for others, asks you to add memory – a 32K×8 RAM – to the counter circuit that you wired last time. This memory and its manual addressing and loading hardware will form the foundation of the computer you soon will build, if you choose the big-board series of labs rather than the single-chip controller labs.
- The second part invites you to play with *state machines*.
    - One of the state machines is to be formed from the circuit you have constructed, exploiting RAM and the loadable counter that provides the RAM's address.
    - A second state machine, a sequential lock, is to be made from a PLD programmed in Verilog. Those not busy installing the RAM (these are the single-chip people) will have time free for this task. People building this sequential lock will need Verilog code to do the task, and may want to consult the Worked Example 17W which describes such a lock. You will then need to program a PAL with your design to do that part of today's lab. So this lab is unusual in requiring some preliminary work.

These state machines provide an introduction to the microprocessor that you will meet in Lab 20L.

The first part is essential, if you intend to build up the big-board computer rather than use a one-chip implementation. You will make that choice when you reach Lab 20L. The later stages of big-board construction cannot proceed until you have installed the RAM.

The second part of this lab is not essential to any later work: it demonstrates a concept fundamental to computers but you can proceed without building these circuits. If you are very short of time and plan a big-board, then evidently you should do the first part and then squeeze in what you can of the second part.

Those not planning to build the big-board computer should do the state machine design of §17L.2.1. Then you should have time to return to a task of Lab 16L that most people did not reach: adding to the counter the elements that can turn it into a *period meter*.

17L.1   RAM

The memory that your 16-bit counter is to address is a big array of CMOS flip-flops: 32K bytes, also known as 32KiB, 256K flops.[1] These flops are arranged as 32K "words" or address locations, each word holding eight bits. The memory is "static:" it uses flip-flops, not capacitors, to store its values. It therefore requires no refreshing. It is "volatile," in the sense that it forgets if it loses power. But it requires so little power that it can easily be kept alive with a 3V lithium coin cell, as you will show

---

[1] In case anyone cares – and normal people do not – 32KiB equals 32,768 bytes, 262,144 flops.

later when you begin to use the RAM to hold programs. Thus we will make it pseudo-non-volatile simply by attaching a lithium coin cell to keep it ever powered.[2] Today we will omit this battery backup, because you are not likely to write anything valuable during this session.

Because its outputs are equipped with three states, the memory does not need separate In and Out lines. Its common data lines serve as outputs until WE* is asserted; at that time the internal 3-state drivers are turned off and the eight data lines serve as inputs: the RAM begins to listen. This arrangement you have seen detailed in Chapter 17N.

The data bus circuit shown takes advantage of this tidy scheme: we turn on *our* three-state drivers (an 8-bit buffer: the '541) with the same line that asserts WE*. Thus the RAM shuts off its 3-states just as we turn ours on, and vise versa.[3]

17L.1.1 Data buffer and RAM linked to counter

Now install the RAM and the 3-state buffer that sits between keypad and data bus.

The 3-state buffer prevents clashes between the keypad and other devices that sometimes drive the data bus: the RAM is the only "other device" today, but soon the CPU and peripherals will also want their turns driving the bus. The RAM and buffer connect to the micro breadboard's several buses: address, data and keypad.

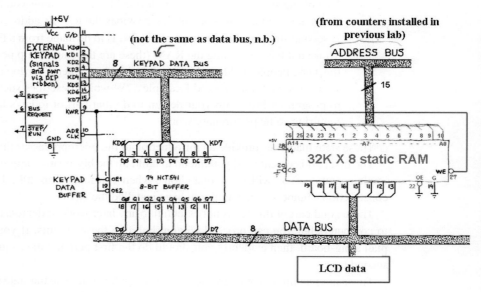

Figure 17L.1 RAM and 3-state buffer: added to counters installed last time.

Note We suggest that you photocopy or cut out the pinout label – a copy of which appears on page 670 – and paste it onto the RAM. The label will save you a good deal of pin-counting.[4]

The RAM *address* lines are to be driven, as you know, by the PAL counter that you wired last time.

---

[2] You will see in Lab 21L that "attaching" the battery slightly oversimplifies the arrangement. We use a battery-backup IC that handles the switchover from main to battery supply.

[3] This is roughly true, but not true to the nanosecond: the RAM, made of relatively slow CMOS, turns off more slowly than our driver turns on; so, the two drivers may clash for ten to twenty nanoseconds. But this "bus contention" is brief enough to be harmless.

[4] A curious fact: we have redefined some of the address and data pins relative to the industry-standard pinout. We did this in part to put them in sequential order (though interrupted by some control and power pins), but also to comport with the layout of the address buses on the micro breadboard: we have placed the more-significant signals higher (closer to the end

17L.1.2 Checkout

Confirming that the memory remembers is as easy as it sounds.

- Set up a value on the keypad by hitting two keys, *A*, *B*, say. *AB* appears on the keypad display.
- Push the WR* button: *AB* should appear on your LCD *data display*. If it does, you have wired the 3-state buffer properly (the '541): the *AB* is driven from keypad to data bus.
- When you release WR*, *AB* should remain on the display: on release of WR*, the RAM resumes driving the data bus, showing us its stored byte.
- Change the keypad value, without hitting WR*. The breadboard data value should not change.
- Increment the address. The data display should change – since you now are looking at a different RAM location. Now decrement. Confirm that the *AB* stored earlier remains.

Works? Congratulations.

17L.2 State machines

17L.2.1 Warmup: A Simple State Machine: Divide-by-3 Counter

The next exercise, incidentally, does not use the RAM just installed. We will return to that memory later in this lab.

Here, we'd like to try generalizing our notion of the sort of sequential circuits we can design. To this point, nearly all our sequential circuits have been counters.[5] A counter is a special case of the more general device, *state machine*.

A generic state machine: Figure 17N.8 (repeated in Fig. 17L.2) is a block diagram of any old state machine. If this represented, say, an up/down counter, then the input would be a single line, UP*/DOWN, and the outputs would be simply the flip-flops *Q*s.

A particular state machine – a humble counter, from flops and gates: The first state machine we ask you to consider is just a counter: a familiar case, and relatively easy to design. We ask you to design and build a divide-by-three counter, using D-flops and whatever gates you need. In order to avoid putting you to sleep, we have asked for an eccentric count sequence, one you can't easily get with the counter-design techniques you have used to date.

To make the problem fun, let's make this divide-by-three different from a couple of others you have seen. (We don't want you simply to open the book to find AoE's solution, or to recall an in-class demo where we made a div-by-three using a counter with synchronous clear.) This circuit is to count *down* from 3 to 1.

Figure 17L.2 along with a reminder of the generic state machine hardware is a present-state/next-state table showing the desired count sequence.

Recall that the "XX (don't care)" values for *next state* mean that *we* don't care what the circuit does next after the unused state; we don't care as we set out to design this circuit. The *circuit*, however, will "care" once designed, in the sense that it will go from the 00 present state to just one fully-determined

---

with pins 1 and 28). This should help your wiring a little, assuming that you mount the RAM with pin 1 away from you (away from display and keypad).

A few minutes' thought will convince you that you can shuffle address lines with address lines, and data lines with data lines, as you please – when working with a RAM. One cannot do this with a ROM programmed outside the circuit and using a non-shuffled set of pin assignments.

[5] Exceptions: single flip-flops, shift register, and digital one-shot.

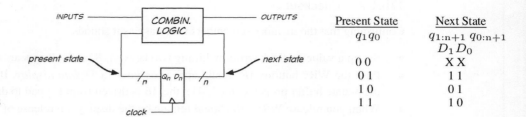

Figure 17L.2 Generic FSM block diagram and state table for divide-by-3 down counter.

next state. It will not proceed at random from state 00. (In Chapter 17N you saw a counter that could have got stuck in its unused state.)

We hope that staring at this table you'll see how you might generate the two functions D_1 and D_0. You needn't be clever; the simple minded way is just to list all the input combinations that can generate a 1, then OR those input combinations.

Incidentally, we slightly spoiled the game for you by decreeing that you'll need two flip-flops, and that one state is wasted (the "don't care"). But you knew that anyway. When you finish your design you should check – on paper – to see that the unused state did not by chance get assigned as XX = 00. Why? Here you discover that despite our bravado in writing "don't care" a couple of paragraphs back, we didn't quite mean it: we *do care* that the circuit should not get stuck in an unused state.

This example was simple (though perhaps not *easy*: no task is easy the first time you have to try it). The RAM-based state machine described in §17L.2.3 is less simple – but much easier to design: such a machine requires no gating at all, leaving all that work to the RAM instead.

17L.2.2 Aside: same task done with Verilog

Incidentally (no work required of you, in this subsection), we thought you might want to see that Verilog can do the fussy work for you in the design of a state machine. Here's what the Verilog code for the counter we asked you to design would look like (it differs from your design in including one extra signal, named "startup," used to provide known initial values for simulation):

```
module ctr_down_3(clk, startup, Q);
    input clk,startup;
    output [1:0] Q;

    wire clk, startup;
    reg Q;
    parameter zero = 'b00, one = 'b01, two = 'b10, three = 'b11;
    always@(posedge clk, posedge startup)
begin
    if (startup)
        Q <= three;
    else
        case (Q)
        three:
            Q <= two;
        two:
            Q <= one;
        one:
            Q <= three;
        zero:
            Q <= three;   // unused state steered to an (arbitrary) initialization
        endcase
    end
endmodule
```

17L.2 State machines

As you can see, Verilog relieves you of almost all the work – but we hope you'll find it gratifying to do the gritty work of figuring out what functions are to be fed to each flop *once* in your life anyway. Fig. 17L.3 is a schematic of the logic that Verilog produced.

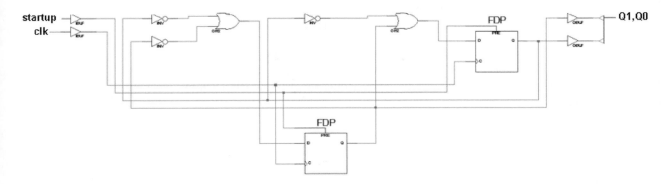

Figure 17L.3 Schematic of down-counter div 3 produced by Verilog.

We wonder how close this design is to your own. Yours is likely to be simpler – since your circuit lacks a signal to force it to a known startup state (here implemented by a simple *jam set* of both flops), and can take advantage of the design's "don't care's."[6] Simulation of the Verilog design produced the waveforms in Fig. 17L.4 which look like what one would hope to see. Once this lab is over you'll probably do the task the easy way with the help of a logic compiler like Verilog.

17L.2.3 RAM-based state machine: register added, to let RAM data define next state

We offer this exercise – which is feasible only for people who have installed the RAM – for two reasons: to illustrate one way to make a complex state machine (though a method much less common than the use of an HDL like Verilog); and to foreshadow the *microprocessor*. We hope to let you feel the truth that the processor is only a specially fancy version of devices you have seen before. It is a fancy state machine, cousin to plain counters. It differs in being spectacularly more versatile than any counter.

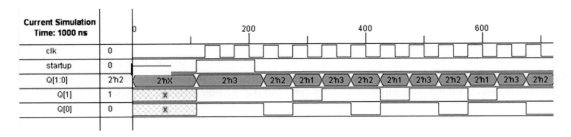

Figure 17L.4 Simulation showing behavior of counter produced by Verilog.

We assume that when you begin this section of the lab you have wired a circuit that allows you to address any location in RAM, and to write a value from the keypad into that location. As you have begun to prove to yourself (with the little counter of §17L.2.1), one can build an arbitrary "state

---

[6] Xilinx calls *set* "Preset."

machine" by feeding *next state* values to the Ds of some flops, then clocking the flops. Here you will get a chance to try out the very easiest way to generate those next state values: with a memory.

The RAM is well adapted to this job: address evokes data; that's in the nature of a memory. Applied in a state machine:

$$\text{ADDRESS (= present state)} \longrightarrow \text{(evokes)} \longrightarrow \text{DATA (= next state)}.$$

Today we use a RAM because it is available. Normally, a memory-based state machine would use a ROM (a non-volatile memory). Here's the idea. We've sketched it in faulty form on the left of Fig. 17L.5; then we've corrected this unstable design by adding a register to stabilize it by "breaking the feedback path."

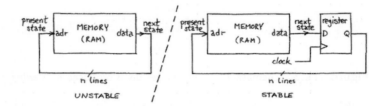

Figure 17L.5 RAM is well adapted to the present – next task; it needs a register of flops to make it stable.

The flops are solving a problem you have seen before. The toggling edge-triggered D-flop (on the right in Fig. 17L.6) is stable, since edge-triggering breaks the feedback path. This is a familiar circuit, one that you built in Lab 15L. It changes state on every clock.

The transparent version, shown on the left side of Fig. 17L.6 is unstable, as you saw demonstrated in Chapter 16N. It oscillates spontaneously while enabled.

Counter used as register: Ordinarily, one would use a D register to hold the "present state." (Some chips, now obsolete, integrated memory and register onto a chip, as we noted in Chapter 17N.) To save ourselves extra wiring we use our pseudo-D register, the XC9572 16-bit counter already installed.

Recall that as long as the counter's SYNCLD* pin is asserted, the counter does not count; instead it uses the clock to *load* the data presented at its 10 data inputs. (The six other data lines are set internally to all zeros.) Therefore, when we want the counter to serve as a register, we will simply hold SYNCLD* low.

A DIP switch installed in Lab 16L controls SYNCLD*. It now can serve as a "Run*/Program" switch. We will use it to assert SYNCLD* only we when we want to *Run* the state machine.

RAM data feeds counter "Register": Figure 17L.8 is a reminder of the overall scheme we now implement: we form a classic state machine, a loop, in which RAM data drives the counter (locked in its *register* mode by the grounding of SYNCLD*), and the counter drives RAM address lines.

Figure 17L.6 Transparent latch can be unstable. Clocked device (edge-triggered) makes feedback harmless: instability is impossible.

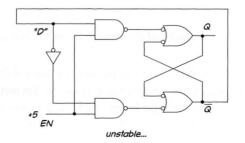
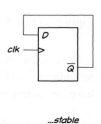

unstable...   *...stable*

17L.2 State machines

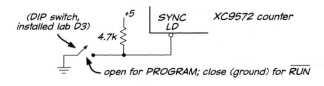

Figure 17L.7 DIP switch installed in Lab 16L now can serve as a Run*/Program switch.

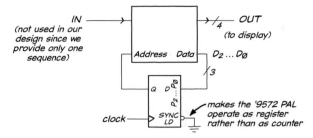

Figure 17L.8 RAM and loadable counter can form a classic state machine.

Load the "Program" sequence into RAM:

Load while keypad still drives all eight counter "Pn" inputs: Load the program before making any of the wiring changes that soon will apply (shown in Fig. 17L.9): leave the keypad's KD... lines driving the counter's P... inputs. As usual, the SYNCLD* pin mentioned in Fig. 17L.8 should be high.

Sample "program": Load the following data into RAM, at the eight listed locations (these are not the lowest eight because the keypad controls only lines D4..D11; the lowest four bits of loaded address are a constant, zero). You should set the high-order addresses to zeros, using the keypad and the DIP switch that controls the top two available lines. Table 17L.1 below is hard to digest, so let's start by looking closely at two steps in its sequence of states.

Suppose the machine starts, by chance[7], at address zero. The first line of the table shows an address – A15..A0 – of zero, and 8-bit data of 03h. Because these three low-order data lines are wired, as shown in Fig. 17L.8, to address line A6..A4,[8] the stored data value "three" takes the machine to a *next state* of not *three* but *30h*. Similarly, the "2" stored at address 30h takes the machine next to address *20h*.

Note – a point that is rather hard to see in the table – that the eight entries of your table do not go into addresses 0..7. Instead, they go into addresses 10h apart: 0, 10h, 20h,..., 60h, 70h.

- at address 0, store 03h
- at address 10h, store D5
- at address 20h, store A1
- at address 30h, store B2
- at address 40h, store 07
- at address 50h, store F4
- at address 60h, store D3
- at address 70h, store 06

In order to store this scattered data table, use the asynchronous LOAD* pushbutton to get to the specified address. For example:

---

[7] ...or by your use of RESET*.
[8] A6..A4 are fed by pins labeled "P2..P0" because the low-order four addresses, not accessible from outside the PAL, are grounded internally. The first accessible address input thus is A4.

Lab: Memory

Table 17L.1 Data and Next-Address Loaded into RAM.

Address (hexadecimal)			from keypad KD7 (P7) ... KD0 ... P0			Data (arbitrary)	next address
A15–A14	A13–A12	A11–A7	A6–A4	A3–A0		D7–D4	D3–D0
(fixed)	(DIP sw)	(kpd)	(kpd)	(fixed)		(RAM)	(RAM)
0	0	0	0	0		0	3
			1	0		D	5
			2	0		A	1
			3	0		B	2
			4	0		0	7
			5	0		F	4
			6	0		D	3
			7	0		0	6

- to get to address 10h, set up 01 on the keypad; press the *Load* pushbutton; then write the data that is listed (D5h);
- to get to address 20h, set up 02 on the keypad; press the *Load* pushbutton; then write the data that is listed (A1h);
- ... and so on.

The process is a bit laborious, and you may protest that the results are modest. But we hope this little exercise helps to demystify state machines.

Make the hardware change, and run the "program": Once you have loaded the "program" values into RAM, make the change shown in Fig. 17L.9: disconnect the bottom three KD...lines from the counter's P2..P0 inputs, letting the main data bus three low-order lines drive those inputs instead.

Figure 17L.9 shows the three RAM *data* lines driving the counter's three data-loading lines, P2..P0 (the RAM is not shown). In this circuit, the "counter" is acting not as a counter, but as a 16-bit *register*, whenever SYNC_LD* is grounded. The three lines driven from the data bus are driven, in this case, by the RAM data lines D2..D0. You can pick up these three lines from the micro breadboard's data-bus connector.

...Try it out: Try the circuit and program above, using the following procedure:

Assuming that (1) you earlier loaded the data into RAM, and that (2) you have moved the counter's three low-order data lines, D0..D2, from keypad (their usual source) to the low three bits of the DATA bus,

1. ground the SYNC_LD* line using the DIP switch: this is the Run* position;
2. clock the counter, using the "INC" button on the keypad.

The data bus should show you the next address in the low nybble, and a sequence of letters in the high nybble. See if you can make out a message in the MSD's "arbitrary" data. (The message in this lab is pretty silly; but then we have only the letters A through F to work with.) You can eliminate the distracting "next address" information from the display by covering the right-hand display, perhaps with a scrap of paper or an idle IC.

We will stop here, hoping that this is enough to let you imagine the way the scheme could be

17L.3 State machine using a PAL programmed in Verilog

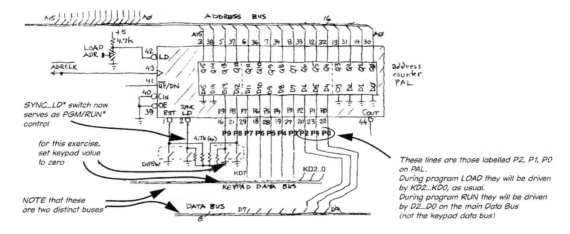

Figure 17L.9 Octal D register (counter with LD* grounded) fed by three bits of RAM data: now feedback loop is closed.

extended: adding one *input* as an address line would allow you to select two alternative sequences; two inputs would allow four sequences or 'tricks;' and using eight input lines (as in the '8051, due to arrive in Lab 21L) would allow 256 tricks. So the '8051 is a big state machine; in principle, given enough time, you now could design such a microprocessor. But don't worry: that is *not* the next lab exercise.

17L.3 State machine using a PAL programmed in Verilog

As we remarked back in §17L.2.2, Verilog makes the design of a state machine quite straightforward, allowing one to describe the machine pretty much the way one thinks of it: one lists all the states, showing for each of these both outputs and *next state*s. Here we invite you to demonstrate such a state machine – one that is *not* simply a counter, as the example of §17L.2.2 was.

We assume that you have designed the *sequential lock* described in Example 17W and have programmed a PAL to implement this design. If not, we can hand you such a part (that's less fun for you than if you did it all yourself – but better than not having a chance to a state machine in action).

If you have used the pin assignments that we used – which apply inputs to the lower left corner of the carrier and take outputs from the top left corner – then your breadboard may look something like ours, shown in Fig. 17L.10.

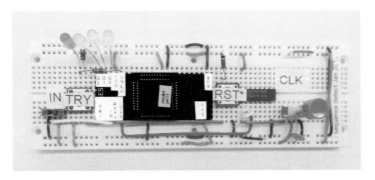

Figure 17L.10 Sequential lock, breadboarded.

Lab: Memory

Again we use a carrier for the surface mount package, as in Lab 16L.[9] We debounced the CLK, using the easy trick of a non-inverting gate fed back (we used an HC08). This method requires a double-throw switch (SPDT) to generate the clock. You might prefer to sidestep the debouncing by clocking your circuit with a very-low-frequency TTL signal from a function generator.

We included three LEDs that show the machine's *state*. A working version of a lock would hide these but they are very helpful when one is debugging – or simply trying to recognize what the machine is doing. Watching only the UNLOCK pin is much less illuminating.

No doubt you will want to try making some errors as you enter values to open your *lock*, to see how the machine responds to incorrect inputs. Once it has asserted UNLOCK, this signal should persist, despite clocking, as long as TRY is asserted. A release of TRY and then a clock edge should terminate the UNLOCK signal. Does your circuit follow this script? We hope so.

Figure 17L.11 provides RAM pinouts for you to copy and attach to your device; resize if necessary so that the pin labels are $0.1''=2.54$mm apart.

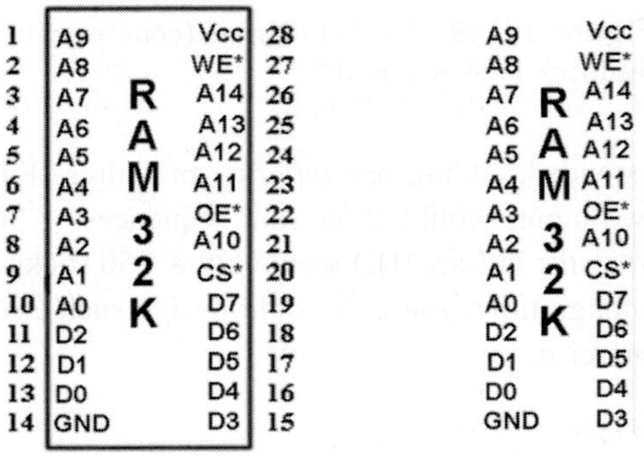

Figure 17L.11 RAM pinout.

	RAM 32K				RAM 32K	
1	A9	Vcc	28	A9		Vcc
2	A8	WE*	27	A8		WE*
3	A7	A14	26	A7		A14
4	A6	A13	25	A6		A13
5	A5	A12	24	A5		A12
6	A4	A11	23	A4		A11
7	A3	OE*	22	A3		OE*
8	A2	A10	21	A2		A10
9	A1	CS*	20	A1		CS*
10	A0	D7	19	A0		D7
11	D2	D6	18	D2		D6
12	D1	D5	17	D1		D5
13	D0	D4	16	D0		D4
14	GND	D3	15	GND		D3

---

[9] The figure shows a different package mounted on a carrier. You will be using the 44 TQFP package, not the PLCC that is shown.

17S Supplementary Notes: Digital Debugging and Address Decoding

17S.1 Digital debugging tips

Here are a few thoughts on debugging digital circuits.

Use a logic probe, not DVM or – worse – your *eyes* This should go without saying, but we're not sure it yet does. We find it *boring* to stare at a wire, trying to see if it *goes* where it should. It's more fun, faster, and more instructive to probe that same wire seeing whether it *behaves* as it should.

Probe chip *pins,* not the breadboard: Usually, but not always, breadboard and chip match: an IC's pin may be folded under, or it may be failing to make contact with the breadboard because the breadboard socket contact has spread with rough use. So, be skeptical: poke the pins.

Figure 17S.1 A case where probing the breadboard would not be good enough to reveal the IC's behavior.

Disconnect loads from a misbehaving signal: Suppose one bit of the '541 3-state buffer won't follow its data input; it shows neither High nor Low when the buffer is enabled. You can detect this most easily with a logic probe but a scope would show a voltage intermediate between legal logic levels. Whose fault is this fishy output level? The '541's? Or is it something "downstream" on that data line? The downstream device may be driving the line, producing "bus contention" – a fight between a High and a Low that produces the intermediate level.

It's easy to settle that question if you pull out the line that the '541 pin drives. Now enable the '541 and watch its output pin: OK? look at the breadboard at that position (see Fig. 17L.1): OK? Look at each of wires that earlier was plugged into that slot. Often you will find '541 OK, but one of the wires asserting a high or low when it should be only listening. If you do, you know to continue your search *downstream*.

Check power supplies: Here's a dull rule – dull, but useful: before looking for subtler causes, make sure your misbehaving chip has *power*.

That may sound silly to you. You may be inclined to reject this advice, thinking, "If the chip didn't have power it would act dead, and my circuit is misbehaving in stranger ways. It doesn't just act dead." Such a thought rests on an error: it is hard to remember – but necessary – that a CMOS circuit can work *without power applied to its V_{DD} and ground pins*. This is a point you may recall from the Lab 14L.

CMOS can do this perverse trick because protection diodes on the input lines allow *inputs* (or outputs – but these are much less likely to drive the chip) to provide power and ground (at slightly degraded levels). Fig. 17S.2 shows the typical input- and output-protection scheme.

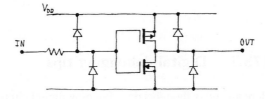

Figure 17S.2 CMOS input-protection diodes can power a chip through *inputs*.

Missing power and ground connections can produce strange intermittent failures: If your circuit *sometimes* fails, perhaps it lacks power, ground, or both. It may be taking the missing connection through an input. So, to take a simple example, imagine that you are using one gate of an AND package, as in Fig. 17S.3.

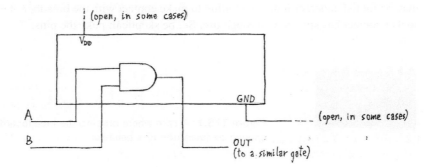

Figure 17S.3 Hypothetical use of a gate in an AND package.

If you omit one or more power and ground connections, you may find the following puzzling behavior patterns, as you may have seen in Lab 14L:[1]

Input	Will this gate work?		
A B	Both V_{DD} & GND open	Only V_{DD} open	Only GND open
0 0	NO	NO	YES
0 1	YES	YES	YES
1 0	YES	YES	YES
1 1	NO	YES	NO

If you fail to provide power through the power and ground pins then the gate will work *sometimes*. It will work by stealing power from a data line – as long as it finds both highs and lows available at either input or output. The cases that *don't work*, above, are the ones where *ins and outs are unanimous*.

So when a gate is showing weird misbehavior you may save a good deal of time if you can humble yourself sufficiently to poke power and ground pins early on to see if they're connected to the power buses.

---

[1] As we said there, we found that National Semiconductor 74HC00's behave this way (now TI/National Semiconductor); Motorola's give a high for the High–High input combination.

17S.1 Digital debugging tips

Strange *floats*: three common causes: A logic probe sometimes will show you a *float* that you don't expect. (We are using the word "float" as shorthand for *an intermediate logic level*, as in §17S.1: neither High nor Low. Often such a level is not the result of a truly *floated* output, which can be achieved only by an OFF 3-state.) Such a float is a sign of a specific kind of trouble. Sometimes it looks even stranger: you see not just a float, but a fast *oscillation*. To detect the float you should use a logic probe, and if it has a switch you need to set it to the appropriate technology (CMOS versus TTL) (I recently got fooled by ignoring that switch). The probe will detect the fast oscillation too if you set its switch (if any) to Pulse rather than Memory.

If you do detect an *oscillation*, then it's time to get a scope: see §17S.1.1.

A strange float usually results from one of the following three causes.

1. The source has been disconnected by mistake (a bad connection; perhaps the stripped length of a wire stuffed into a hole is too short).
2. The source logic chip may lack power or ground. Don't forget (as noted above) that a CMOS part lacking power and ground can appear to work fine, part of the time, by stealing power through its inputs.
3. Two outputs may be fighting. To check this, just disconnect all but one source, as suggested above.

Probe the fishy bit once you have discovered which one is fishy. Often the stubborn "1" will turn out to be not a true "1" but a *float* (the data displays may treat floating inputs as logic Highs, in the manner of TTL, or they may simply show a value earlier driven and now held on stray capacitance).

The most common cause for the surprising float is the dullest: #1 above: simply a bad (open) connection in the signal path.

Strange values on data or address displays: treat these as valuable clues:
One likely pattern: 0 goes to 2; 4 goes to 6; 8 goes to A... Switch the LCD from its usual *hexadecimal display* to *binary* and step through the sequence slowly. Write the sequence down. Usually the misbehavior jumps out at you once you see the sequence in binary. If the pattern does not declare itself then try writing an additional column of values: the *expected bit pattern*. Table 17S.1 below gives some sample misbehaviors made intelligible by the binary listing:

Hypothetical hexadecimal-character sequences, where plain increment was expected:
Another likely pattern: 0 goes to 0; 1 goes to 8; 8 goes to 1; 3 goes to C...; or the count sequence is strange... Again, write out the pattern or sequence. Usually you will find bits interchanged. In the example we have just described, data lines D3 and D0 are interchanged; so are D2 and D1. In other words, the ribbon cable feeding the four data lines needs to be rotated 180°.

17S.1.1 A harder puzzle: oscillations

A mysterious *oscillation* is probably the second-hardest problem to debug. (The hardest problem is a malfunction that is *intermittent*.) In digital circuits, surprise oscillations do not result from the usual "parasitics" that you learned to know and love in the analog section: since digital circuits normally are immune to small noise signals, they won't amplify and propagate these as an analog circuit can.

But it's not too hard to make an unintended oscillator by wiring an unintended feedback loop – one that contradicts itself. That sounds like negative feedback, which in our analog experience would be self-stabilizing. But in a digital circuit, with its fast transitions, negative feedback with the inevitable gate-delay often produces an oscillation. You can call it "output contradicting input." That name makes the oscillation seem plausible. The *period* (or frequency) of oscillation often will provide a clue to what's the likely feedback path. Below are two cases to illustrate the point.

Table 17S.1 Misbehaviours

Sequence:		
Hex	**Binary**	⇒ Diagnosis
3	0011	
2	0010	
3	0011	
6	0110	D1 always high;
7	0111	(probably floating)
6	0110	
7	0111	
A	1010	
Hex	**Binary**	⇒ Diagnosis
0	0000	
1	0001	Look at how fast each bit toggles:
8	1000	D0 changes every time: Good!
9	1001	
4	0100	There is a bit that changes every second clock –
5	0101	it behaves, in other words, like D1; but the bit that behaves
C	1100	like D1 is in the wrong position: it's the *leftmost* bit.
D	1101	Meanwhile, the slowest bit is in the D1 position – whereas it ought to be D3. Apparently, D1 and D3 are swapped; to put it another way, the top 3 bits have been rotated (This mistake was easy to make on an HP hex display with its screwy pinout)

Case 1: A fast oscillation indicates a short feedback path: If we see an oscillation as fast as the one shown in Fig. 17S.4 – period just a couple of gate delays, we can guess that the feedback path runs through *a single gate*. Recall that the period shows *two* passes through the gate, so *two* gate delays. The HC14 waveform reproduces a case we saw recently in the lab: someone tied output to input, on this gate, an inverting Schmitt trigger. The HC14's typical gate delay (from its spec sheet) is about 11ns at 5V. This oscillation runs faster than that; but we're in the right range.

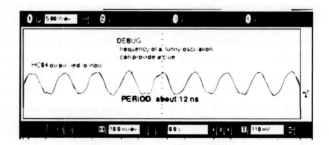

Figure 17S.4 Case 1. A nasty oscillation: path is through a single gate.

Case 2: A slower oscillation indicates a feedback path through a slower device: About the only point that's easy about Case 2 is the conclusion that this loop passes through far more than a single gate. The period again should reflect *two* contradictions, so we'd expect the path delay to be half the period, or around 60–65ns. This is too slow for a gate, and too fast for a processor cycle (the CPU clock period is about 90ns; the quickest processor action takes several clocks). This delay is about right for *memory*.

In fact, we generated the waveform in Fig. 17S.5 by passing a contradiction through *memory*: we tied A0 to D0, then wrote in data that made D0 "contradict" A0: at address 0 we wrote *1*; at address 1 we wrote *0*. The contradiction should take t_{access} to propagate. A simple short, we're sad to admit,

17S.2 Address decoding

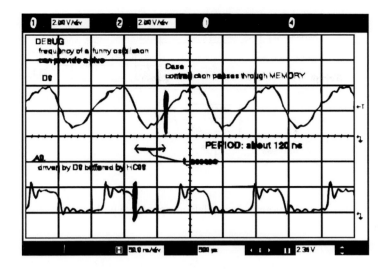

Figure 17S.5 Case 2. A second nasty oscillation, this one too slow to blame on one or two gate delays.

did not produce an oscillation for this case: the address drive (CPU) is stronger than the data drive (CMOS RAM). So we cheated a little, buffering D0 with a 74HCT08. That makes it a fair fight and produces the oscillation (the 'T08 also contributes perhaps an additional 6–9ns of delay, as well).[2]

In general, don't suffer too long (>5 minutes) with a pesky problem. Ask for help!

17S.2 Address decoding

17S.2.1 What is address decoding?

Address decoding is logic that makes this or that device respond when a computer puts out a particular *address*. Usually, the full signal decoding job includes *control* signals as well as address: signals that mean "do this" or "do that."

It is called "decoding" because the computer puts out this information in information-dense ("encoded") form, as a binary number – but a particular device often needs to be given a simple command on a single line: "Turn On," or "Clock" or "Write". The job of the *decoding* logic is to make this transformation.

Figure 17S.6 gives the idea schematically. Sometimes the decoder is to detect a *single* address, or "location." That's the case in the figure for the keyboard's 3-state (an input device) and for the DAC's register (an output device). In other instances, the decoder is to enable a device for a range of addresses. That's the case in the figure for the 32K RAM.

17S.2.2 Incomplete or "lazy" decoding

Complete decoding: In Fig. 17S.6 we've indicated 16 address lines going into the decoder (to keep things simple the figure omits control signals). Assuming that these are all of a computer's address lines (true for the '8051 processor that you will use in your lab computer[3]), including all 16 lines as

---

[2] The oscillation is much faster than the specified t_{access} would let one predict for this 100ns RAM; but such devices usually run faster than this *worst-case* spec. So again we're close enough to the expected value so that the period of oscillation helps our diagnosis.

[3] As you know, more recent processors – and your PC – use many more lines. Some recent versions of the '8051 use 24 address lines; the venerable '68000 used 32 address lines. The motive for adding address lines, perhaps obviously, is to

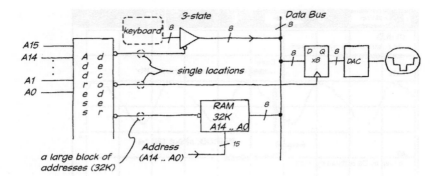

Figure 17S.6
Address decode: generic example.

inputs permits specifying a single location unambiguously. In our lab computer we will save ourselves some wiring by being much lazier.

Lazy decoding: In our lab computer we do the ultimate lazy decode by using *one* address line – A_{15} – to distinguish *Memory* from *I/O* (input/output), which is everything other than memory: keyboards, displays, ADCs, DACs, motors, pushbuttons... and anything else you might hook up to a computer. That means that we allow half of address space[4] for memory, half for I/O.

Figure 17S.7 Simplest decode: split address space in half.

For the memory, this seems to make sense: 32K locations for a 32K RAM. But for I/O? Do we plan to install 32K input devices and 32K output devices in our lab computer? No. We plan to install *four* of each. Our "lazy" incomplete address decoding pays attention to A_{15} alone to distinguish I/O, then just *two* additional lines (A_0 and A_1) to specify which of our four I/O devices we intend.[5] As a result, each of our *four* devices reappears 8K times. This effect sometimes is called *"aliasing"* of an address – not to be confused with the notion from *sampling* theory.

17S.2.3 An example: RAM and ROM to share address space

Suppose that we decide to install RAM and ROM in our computer, and also some I/O. We don't want to be as lazy as in our P123 machine: we'll confine I/O to the top 1K of address space (that assignment

---

permit easy access to large memories. With 16 lines it is possible to access more than 2^{16} locations, but only by use of the clumsy device of "paging," as in the early IBM PC's – and the DS89C430 processor that you may meet in our micro labs. We will not use this feature.

[4] "Address space" is just jargon for "the locations that are distinguishable using the available address lines."
[5] In principle, one could use *any two* address lines to distinguish our four devices, and could have used *any one* address line to distinguish I/O from memory. But any choices other than the ones we made would produce a complicated patchwork of address uses (I/O interleaved with memory, and so on). There's no good reason to do anything so perverse.

17S.2 Address decoding

still allows for a large cellarful of gadgets). Let's put ROM in the bottom 32K, and RAM into the top 32K, but ducking out of the way in I/O space.

- ROM? That's easy; it's the same as for the simplest I/O versus Memory split: ROM lives wherever the MSB, A_{15}, is low.
- RAM: it occupies all of the top half, where A_{15} is high – unless the address falls into I/O space, defined below.
- I/O: this is a little harder. We usually do such tasks in reverse: we ask "How many lines are needed to distinguish the number of locations we want to distinguish?" Here, the answer for "1K" (which is 2^{10}) is that we need *ten* lines. Those 10 – the ten *low order* lines, $A_0 \ldots A_9$[6] must be left free to take every possible value. The other six lines – $A_{10} \ldots A_{15}$ – must be high.

An address map: Figure 17S.8 shows what the resulting address map looks like. The address decode for the ROM is a single address line (A_{15} low); the I/O decode requires a 6-input AND function.[7] The RAM decode requires just a two-input AND function, piggybacking on the work done to decode *I/O*.

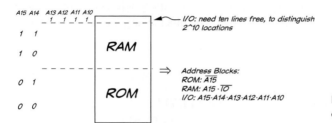

Figure 17S.8 Address map for example placing ROM, RAM and I/O.

Gates to effect the decoding: Figure 17S.9 shows gating that would carry out the address decoding. These address decode signals presumably would be further conditioned – by control signals such as RD*, PSEN* and perhaps WR*. For this reason we have shown outputs that are *active-high* as we would not if these signals were to go directly to a chip's Enable*.

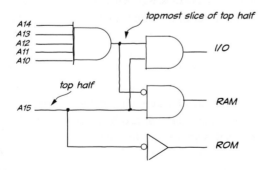

Figure 17S.9 Gating to allocate address space among ROM, RAM and I/O.

---

[6] Don't forget number "zero"! Engineers believe in zero.
[7] We're avoiding the word "gate," because the active-lows that are likely – especially on ROM and RAM enables – mean that "AND function" will be performed by a NAND rather than AND. We don't want to clutter this discussion with such detail.

17W Worked Examples: Memory

17W.1 A sequential digital lock

Here, we'd like you to do one task three ways. This is a favorite device of exam-writers; this kind of question lets the teachers feel that there's some coherence in the digital material. Students may not feel the same way about these questions.

Because this design will be needed by some people doing Lab 17L, we present this problem early – at a time when the previous method, using a microcontroller, is premature. Ignore that section of this chapter until you feel comfortable with controllers, a time that we expect will come late in the course.

We'll first try to describe how the circuit should behave. The block diagram for the Problem I may help you see what we're getting at.

Here's the scheme.

- The lock faces the user with three switches:
  – a toggle switch lets the user choose a HI or LO input LEVEL;
  – a pushbutton switch, CLOCKIT, lets the user clock in that LEVEL;
  – a second pushbutton switch, TRY, lets the user say that she thinks she has put in a valid code.
- The machine is to respond as follows:
  – each time the user pushes CLOCKIT the machine takes in the current LEVEL;
  – if the user presses TRY and then CLOCKIT, the machine judges whether the sequence of bits put in matches the lock's code:
    o if the sequence *does* match, the circuit's output bit, UNLOCK, opens the lock;
    o if the sequence *does not* match, the machine is reset to its initial state (and, of course, the lock is not opened).

17W.1.1 Problem I. draw the circuit

Figure 17W.1 is a block diagram/sketch of the circuit. Please detail all the circuits, including how you would generate the signals CLOCKIT and TRY. The secret code is determined by a 3-position DIP switch, whose output is fed to the COMPARE circuit (we do this so we can change the key code).

17W.1.2 Problem II: Verilog

Below we have started a Verilog file that could do the same job (minus the compare circuit): it is to implement a "state machine" (rather than a *shift register*) that takes in a 3-bit sequence, and behaves like the hand-drawn circuit.

The "case" design seems well-suited to this task. An example of a state machine designed with Verilog *case* statements appears as §A.11

17W.1 A sequential digital lock

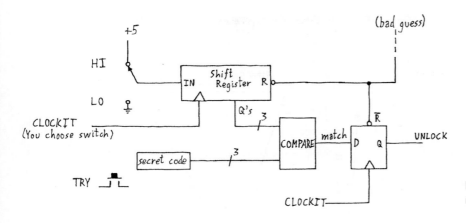

Figure 17W.1 Sequence detect lock: rough sketch.

```
`timescale 1ns / 1ps
//////////////////////////////////////////////////////////////////
// Company:
// Engineer:
//
// Create Date:    11:36:07 04/23/2008
// Design Name:
// Module Name:    sequence_detect
// Project Name:
// Target Devices:
// Tool versions:
// Description:
//
// Dependencies:
//
// Revision:
// Revision 0.01 - File Created
// Additional Comments:
//
//////////////////////////////////////////////////////////////////
module sequence_detect(clk,reset_bar,in,try, unlock,state);
// winning sequence of inputs is 011 (continuing try keeps it unlocked)
    input clk,reset_bar,in, try;
    output unlock;
 output [2:0] state;

    wire clk,reset_bar,in,unlock;
 reg [2:0] state;

 parameter A = 3'b000, B = 3'b001, C = 3'b010, JACKPOT = 3'b011, OPEN = 3'b100;
                   // 4 correct button-presses unlocks it

 assign unlock = (state == OPEN);  // a Moore output, since
       //  it depends on state only

end

endmodule
```

The solution below does this with CASE statements. Probably that's the easiest way.

In Fig. 17W.2 we show the results of a simulation using the posted testbench. These waveforms will help to describe the behavior we hope to see.

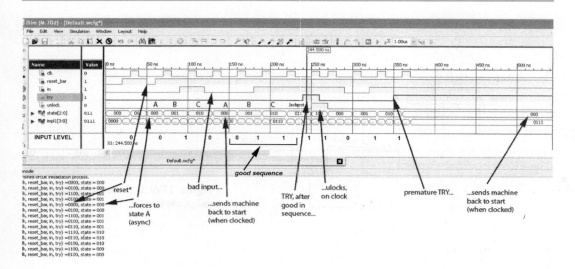

Figure 17W.2 Result of simulating a Verilog version of the 3-bit lock using the posted testbench.

17W.1.3 Problem III: 8051 hardware and code

Assembly language? The task is also well-suited to a microcontroller (no need for speed, here). Again, we have started a file, below, assigning variables to particular pins. We did this task with shift-register strategy, though you can do it any way you like.

Some reminders: it is possible to move a single bit into the CARRY bit ("C"), and from there one can shift or rotate it into the accumulator.

```
; sequence_detect_408.a51  detect particular bit input sequence, to form digital lock

    $NOSYMBOLS
$INCLUDE (Z:\MICRO\8051\RAISON\INC\REG320.INC)    ; Raison's DS310 register defs. file
$INCLUDE (Z:\MICRO\8051\RAISON\INC\VECTORS320.INC) ; Tom's vectors definition file

STACKBOT EQU 07Fh ; put stack at start of scratch
;   indirectly-addressable block (80h and up)
SOFTFLAG EQU 0 ; a flag that ISR will set
INBIT EQU P1.0 ; user's input level (code bit)
TRY EQU P1.1 ; user's try-to-open bit
UNLOCK EQU P1.2 ; a bit that controls the lock

KEY EQU 011b ; set up key code

ORG 0 ; tells assembler the address at which to place this code

LJMP START ;  here code begins -- with just a jump to start of
;    real program.  ALL our programs will start thus

ORG 280h ;  ...and here the program starts

START: MOV SP, #STACKBOT

END
```

17W.2 Solutions

17W.2.1 Solution I: draw the circuit

The design in Fig. 17W.3 includes no Reset pushbutton. To reset the machine requires the rather awkward procedure of clocking with Try once more (deliberately causing a failure that evokes a Reset*). It might have been nicer to include a separate Reset* pushbutton.

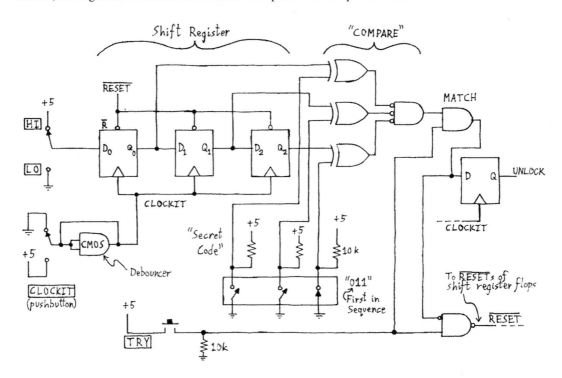

Figure 17W.3 Sequence detector done with shift register and gates.

17W.2.2 Solution II: Verilog

Below is the code that produced the simulation shown in Fig. 17W.2. This is not a shift-register design, but rather a classic state machine.

```
// Verilog Code for Sequential Lock
`timescale 1ns / 1ps
//////////////////////////////////////////////////////////////////////////////////
// Company:
// Engineer:
//
// Create Date:    11:36:07 04/23/2008
// Design Name:
// Module Name:    sequence_detect
// Project Name:
// Target Devices:
// Tool versions:
// Description:
//
```

```verilog
// Dependencies:
//
// Revision:
// Revision 0.01 - File Created
// Additional Comments:
//
//////////////////////////////////////////////////////////////////////////////////
module sequence_detect(clk,reset_bar,in,try, unlock,state);
// winning sequence of inputs is 011 (continuing 1 keeps it unlocked)
    input clk,reset_bar,in, try;
    output unlock;
      output [2:0] state;

      wire clk,reset_bar,in,unlock;
      reg [2:0] state;

      parameter A = 3'b000, B = 3'b001, C = 3'b010, JACKPOT = 3'b011, OPEN = 3'b100;
                                // 4 correct button-presses unlocks it

      assign unlock = (state == OPEN);   // a Moore output, since
                                         //  it depends on state only

      always @(posedge clk, negedge reset_bar)
      begin
          if (~reset_bar)
              state <= A ;

          else
              case (state)
                  A:
                  if (try)
                     state <= A;     // premature try takes you back to start
                  else
                      if (~in)
                          state <= B;
                      else
                          state <= A; // start over if you err

                  B:
                  if (try)
                     state <= A;   // premature try takes you back to start
                  else
                      if (in)
                          state <= C;
                      else
                          state <= A; // start over if you err

                  C:
                  if (try)
                     state <= A;    // premature try takes you back to start
                  else
                      if (in)
                          state <= JACKPOT;
                      else
                          state <= A; // start over if you err

                  JACKPOT:
                  if (try)
                     state <= OPEN;  // unlock if push TRY and clock it
                  else
                     state <= A;  // start over if you haven't the nerve to try
```

```
                OPEN:
            if (try)
                state <= OPEN;  // if you're still pushing try and clocking, it stays unlocked
            else
                state <= A;     // start over (and lock up)
        endcase
end

endmodule
```

The schematic that this code produces is formidable, as Fig. 17W.4 shows. But since Verilog is doing the hard work, we don't mind.

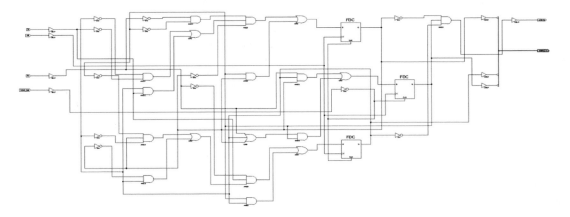

Figure 17W.4 Schematic of circuit produced by Verilog code.

17W.2.3 Solution III: 8051 hardware and code

Hardware The hardware is pretty obvious, but we've drawn it in Fig. 17W.5. The hardware should help to make the code intelligible.

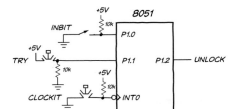

Figure 17W.5 8051 hardware connections for sequence detector.

Code: assembly language This code looks less complicated than the Verilog code. Like the first, hand-drawn circuit of Fig. 17W.3, this design uses a shift register. After bringing in three bits, the program just does a *compare* (CJNE) between the current state of the shift register and the reference *unlock* code. The solution shows, commented out, a masking of key input that would permit *the most recent* three inputs at any time to serve as unlock code. Without this commented masking, one would be permitted no mistakes after the machine was started up (that is, after a hardware controller reset).

Worked Examples: Memory

```
; sequence_detect_511.a51   detect particular bit input sequence, to form digital lock

     $NOSYMBOLS
          $INCLUDE (C:\MICRO\8051\RAISON\INC\REG320.INC)   ; Raison's DS320 register defs. file
          $INCLUDE (C:\MICRO\8051\RAISON\INC\VECTORS320.INC)  ; Tom's vectors definition file

          STACKBOT EQU 07Fh   ; put stack at start of scratch
                              ; indirectly-addressable block (80h and up)
          SOFTFLAG EQU 0      ; a flag that ISR will set
          UNLOCK EQU P1.2     ; a bit that controls the lock
          INBIT EQU P1.0      ; user's input level (code bit)
          TRY EQU P1.1        ; user's try-to-open bit

          ; CLOCKIT pushbutton drives INT0*

          KEY EQU 011b        ; set up key code

          ORG 0               ; tells assembler the address at which to place this code

          LJMP START   ; here code begins -- with just a jump to start of
                       ; real program.  ALL our programs will start thus

          ORG 280h     ; ...and here the program starts

START:    MOV SP, #STACKBOT
          ACALL INITS
          ACALL INTSETUP

REINIT: ACALL INITS           ; fresh start
TEST: JNB SOFTFLAG, TEST      ; hang here till user hits the CLK button
      CLR SOFTFLAG            ; set up next pass
NOW: JNB TRY, TAKEIN          ; see if user thinks his code is ready
; next line would allow a person to keep trying inputs,
; since only the most recent 3 bits would matter
;       ANL A, #07h           ; knock out all but most recent three bits
        CJNE A, #KEY, FLUNK   ; see if input matches template
        SETB UNLOCK           ; that's it!
HERE: JB TRY, HERE            ; keep it unlocked so long as TRY is pressed
        SJMP REINIT

TAKEIN: MOV C, INBIT          ; pick up current input bit
        RLC A                 ; take in current bit at LSB
        SJMP TEST             ; ...and go await next input

FLUNK:  CLR A                 ; land here after bad input.  Start over
        SJMP REINIT

; ISR: JUST SET A SOFT FLAG
        ORG INT0VECTOR   ; this is defined in VECTORS3210.INC, included above.
                         ; It is address 03h, the address to which micro hops
                         ; in response to interrupt zero
ISR: SETB SOFTFLAG
        RETI

INITS: CLR A                ; clean start for input register
       CLR UNLOCK           ; lock it (active high bit)
       CLR TRY
       RET

; ----NOW ENABLE INTERRUPTS----
INTSETUP: SETB IT0          ; make INT0 Edge-sensitive (p. 22)
```

```
            SETB EX0         ; ...and enable INT0
            SETB EA          ; Global int enable       (pp.31-32)
        RET
            END
```

Code: C language The code in C looks perhaps more complicated than the assembly code. The oddest feature is a kluge letting C do a register *rotate*, an operation not built into C. The code is

```
SEQUENCE = (SEQUENCE << 1) | (SEQUENCE >> (8-1));   // rotate left, takes in MSB at LSB
```

This is odd enough to bear explaining. The left-shift is straightforward:

```
SEQUENCE = (SEQUENCE << 1).
```

The next step is the clever part (not ours; we picked this up from Wikipedia): ...

```
| (SEQUENCE >> (8-1)).
```

This does a right-shift by seven places, putting MSB in LSB position, then ORs this shifted value with the result of the earlier left-shift. So MSB comes in as LSB after the shift. In other words, the 8-bit register appears to have been "rotated" left.

You may find tidier ways to implement this sequence detector.

```
; sequence_detect_511.a51   detect particular bit input sequence, to form digital lock
    $NOSYMBOLS
        STACKBOT EQU 07Fh       ; put stack at start of scratch
                                ; indirectly-addressable block (80h and up)
        SOFTFLAG EQU 0          ; a flag that ISR will set
        UNLOCK EQU P1.2         ; a bit that controls the lock
        INBIT EQU P1.0          ; user's input level (code bit)
        TRY EQU P1.1            ; user's try-to-open bit
        ; CLOCKIT pushbutton drives INT0*
        KEY EQU 011b            ; set up key code

        ORG 0                   ; tells assembler the address at which to place this code
            LJMP START  ;  here code begins -- with just a jump to start of
                                ; real program.  ALL our programs will start thus
        ORG 280h                ; ...and here the program starts
START:      MOV SP, #STACKBOT
            ACALL INITS
            ACALL INTSETUP

REINIT: ACALL INITS             ; fresh start
TEST:   JNB SOFTFLAG, TEST      ; hang here till user hits the CLK button
        CLR SOFTFLAG            ; set up next pass
NOW:    JNB TRY, TAKEIN ; see if user thinks his code is ready
; next line would allow a person to keep trying inputs, since only the
;   most recent 3 bits would matter
;       ANL A, #07h             ; knock out all but most recent three bits
        CJNE A, #KEY, FLUNK ; see if input matches template
        SETB UNLOCK             ; that's it!
HERE:   JB TRY, HERE            ; keep it unlocked so long as TRY is pressed
        SJMP REINIT
    TAKEIN:     MOV C, INBIT  ; pick up current input bit
        RLC A                   ; take in current bit at LSB
        SJMP TEST               ; ...and go await next input
FLUNK:      CLR A               ; land here after bad input.  Start over
        SJMP REINIT

; ISR: JUST SET A SOFT FLAG
            ORG INT0VECTOR  ; this is defined in VECTORS3210.INC, included above.
                            ; It is address 03h, the address to which micro hops
                            ; in response to interrupt zero
```

Worked Examples: Memory

```
ISR: SETB SOFTFLAG
       RETI
INITS: CLR A             ; clean start for input register
       CLR UNLOCK        ; lock it (active high bit)
       CLR TRY
       RET
; ----NOW ENABLE INTERRUPTS----
INTSETUP: SETB IT0       ; make INT0 Edge-sensitive (p. 22)
          SETB EX0       ; ...and enable INT0
          SETB EA        ; Global int enable (pp.31-32)
          RET
               END
```

Part V

Digital: Analog–Digital, PLL, Digital Project Lab

18N Analog ↔ Digital; PLL

Contents

- **18N.1 Interfacing among logic families** — **689**
  - 18N.1.1 Five-volt parts: trouble driving CMOS — 690
  - 18N.1.2 Parts powered by lower voltages — 692
- **18N.2 Digital ⇔ analog conversion, generally** — **693**
  - 18N.2.1 Slicing across the vertical axis: *amplitude or voltage resolution* — 694
  - 18N.2.2 Slices across the horizontal axis: sampling rate — 694
  - 18N.2.3 Effect of an inadequate sampling rate (aliasing) — 696
- **18N.3 Digital to analog (DAC) methods** — **697**
  - 18N.3.1 Thermometer DAC — 697
  - 18N.3.2 An op-amp current-summing circuit — 697
  - 18N.3.3 *R–2R* ladder — 697
  - 18N.3.4 Switched-capacitor DAC — 699
  - 18N.3.5 1-bit DAC; pulse-width modulation (PWM) — 700
- **18N.4 Analog-to-digital conversion** — **701**
  - 18N.4.1 Open-loop — 701
  - 18N.4.2 Dual-slope — 703
  - 18N.4.3 Closed-loop — 704
  - 18N.4.4 A binary search: example — 706
  - 18N.4.5 Delta–sigma ADC — 706
- **18N.5 Sampling artifacts** — **712**
  - 18N.5.1 Predicting sampling artifacts in the frequency domain — 712
  - 18N.5.2 Aliasing is predictable — 714
- **18N.6 Dither** — **714**
- **18N.7 Phase-locked loop** — **716**
  - 18N.7.1 Phase detectors: simplest, XOR — 716
  - 18N.7.2 Phase detectors: edge-sensitive — 717
  - 18N.7.3 Applications — 720
  - 18N.7.4 Stability issues much like PID's — 720
- **18N.8 AoE Reading** — **723**

Why?

Problem: convert a range of voltages to a range of digital codes. The complementary process – digital to analog – is intellectually less challenging and less various, but useful.

18N.1 Interfacing among logic families

This topic is dull, but sometimes necessary. There is no question that a 74HCxx gate can drive a 74HCxx gate – and if you find yourself using antique TTL gates (say 74LSxx to 74LSxx), they will understand each other. But sometimes you will be obliged to mix logic families, usually in order to get a function not available in the main logic family of your design. Then you need to know whether the two devices can drive each other – and if they cannot, how to solve the problem.

18N.1.1 Five-volt parts: trouble driving CMOS

AoE §12.1.3

Most of the time you are likely to use CMOS logic. You'll recall from Lab 14L the nice properties of CMOS inputs and outputs: output Highs and Lows are clean – within a few tenths of the positive supply or ground; input thresholds are comfortably far from those good output levels. The inputs draw no DC current to speak of (a microamp max over full temperature range; typically very much less than that); they present a load that is a modest capacitance (10pF, max[1]) They can drive a substantial load, sinking or sourcing 4mA (74HCxx). They're a pleasure to work with.

But if you were to ask an old TTL part to drive a 74HCxx part it might not work. The problem is that the TTL output *high* voltage (V_{OH}) is inadequate: less than the required CMOS input *high* voltage (V_{IH}). TTL's V_{OH} is only 2.4V; HC's V_{IH} is a little more than 2/3 of the supply voltage: 3.15V when powered at 4.5V. Fig. 18N.1 illustrates the problem.

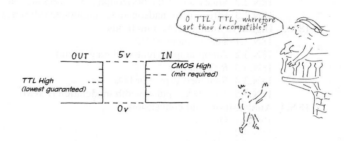

Figure 18N.1 TTL V_{OH} inadequate to drive CMOS V_{IH}.

What is to be done? TTL clearly needs a boost, or CMOS needs to reach lower. Both remedies are possible. CMOS can "reach lower" if we substitute 74HCTxx for 74HCxx. In order to solve exactly the problem we are discussing, HCT uses TTL input thresholds at the expense of giving up CMOS' symmetry and its superior noise rejection. Alternatively, TTL can "get a boost" if we use a resistor to *pull up* its output to 5V. To understand the effectiveness of the simple pullup one must notice that the inadequate TTL *high* voltage (2.5V worst case, about 3–3.5V typical) is not locked at that level; it is not a low-impedance source like the TTL *low*. Rather, it is a source that runs out of gas at this TTL *high* voltage – but does not object if some kindly pullup resistor wants to come along and finish the job. In the case of TTL parts, the "running out of gas" results from the use of an NPN pullup transistor (in contrast to the PMOS pullup of a CMOS output).

As you choose that pullup resistor, just avoid extremes: a very large R would make the rise slow, because the R must charge stray capacitance; using a very small R would overload the driving gate, asking the TTL output to sink more current than it can handle (this maximum current is specified as I_{OL}). A few kΩ is fine for LS TTL; with the weaker signals available from some microprocessors you need to beware overloading: you must pay close attention to that I_{OL} limit (see Lab 21L for examples). A standard 8051 output can sink 1.6mA (this applies for the Dallas part; the SiLabs controller can sink 8.5mA).

AoE §12.1.2A

Much that isn't TTL behaves like TTL: You may be inclined to protest that you're not so old-fashioned to use TTL (which is indeed nearly obsolete, even in its late "LS" form – low-power Schottky). Maybe so – or at least the occasions when you meet TTL will be rare. But a strange fact makes

---
[1] These are specifications for ON Semiconductor's 74HC00.

18N.1 Interfacing among logic families

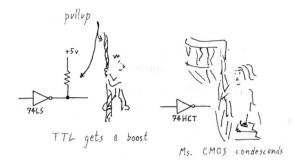

Figure 18N.2 TTL can meet CMOS – with a little help.

this "TTL" problem more pervasive than it seems: many devices that are not made of the bipolar transistors used in true TTL parts use TTL input and output levels nevertheless. NMOS devices – those made with n-channel transistors only, omitting the p-channel that lets CMOS do such a good job of pulling its output high – understandably cannot provide good V_{OH} levels. But the startling bad news is that many devices made of CMOS use the lower TTL levels, presumably because that choice slightly eases the designer's and fabricator's job, letting them use NMOS pullup transistors rather than PMOS.

5V parts using TTL Levels: Important instances of this slumming by parts that ought to know better than to behave like TTL are:

- 5V CMOS PALs/CPLDs, like Xilinx's XC9572XC, a part you may meet in lab (if instead you use the 3.3V part, XC9572XL, the same issue arises when it is to meet 5V logic; see §18N.1.2 below);
- some 5V CMOS processors like the DS89C430, an 8051-style microcontroller that you will meet if you do the big-board labs.

In the big-board microcomputer circuit you will notice 74HCT parts as well as 74HC. Those HCT parts appear wherever TTL-level signals go into CMOS logic: signals coming from the 8051-style processor and from the CPLD – even though both of those ICs using TTL levels are fabricated in CMOS.

Figure 18N.3 is an instance: the '139 I/O decoder is spec'd as HCT (not straight HC) in order to accommodate the wishy-washy HIGHs from three CMOS parts: two CMOS PALs and the 8051 processor. We could have used pullup resistors to make these interfaces work. But I'm sure you'll agree that HCT provides an easier solution. That's why there are more HCT parts in the world than straight HC. This accommodating quality of HCT can be very handy.

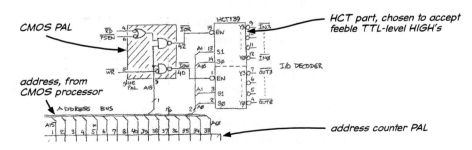

Figure 18N.3 Example of HCT use to accommodate TTL levels out of 5V CMOS parts (big-board computer detail).

18N.1.2 Parts powered by lower voltages

The old TTL levels live on in some low-voltage parts: As supply voltages come down from 5V, the lukewarm logic Highs are disappearing – the V_{OH} that runs out of gas far below V_{supply}. In a 3.3V design, no one dares to waste 1.5V as TTL logic did. But TTL logic levels remain important quite far from their origin in bipolar logic. Not only do some 5V CMOS parts use TTL output or input levels, as we just saw, but also 3.3V CMOS can handle TTL levels at both input and output.

As you design a circuit powered from a voltage lower than 5V – say, 3.3V – you may run into cases where you need a function that is available only in a 5V version. We meet this case in the big-board computer where the DIP-packaged controller is a 5V part. Sometimes such interfacing is easy; in other cases, where voltage disparities are larger, a special translator gate will be required.

Some devices make it easy to mix 5V and 3.3V parts: The PALs that we use in lab make such mixing of 5V and 3.3V parts painless:

3.3V PAL can accept and drive 5V logic. Some 3V devices, such as the XL PALs, make it easy to mix 5V and 3.3V. They can be driven by 5V signals (such inputs are called "5V tolerant"), and can drive 5V logic at TTL levels. The outputs of these parts powered at 3.3V are high enough to drive 5-volt 74HCT inputs, though not 74HC.

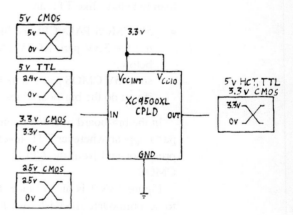

Figure 18N.4 Interfacing 5V to 3V logic is easy, with "5V-tolerant" low-voltage logic (figure is derived from Xilinx note DS054, "XC9500XL High-Performance CPLD Family Data Sheet.").

AoE §12.1.3

We use this adaptability in our labs:

Inputs. The 3V PAL takes input signals from 5V parts of two sorts:
- 74HCTxx CMOS ($V_{OL} \leq 0.4V$, $V_{OH} \geq 3.7V$ – specified at 4.5V supply. An example is the keypad KRESET signal, inverted by an 74HCT14 on its way to the GLUE PAL.
- TTL levels ($V_{OL} \leq 0.4V$, $V_{OH} \geq 2.4V$). Examples are signals, such as RD*, coming from the 5V controller DS89C430.

Outputs. The 3V PAL drives 5V parts that use TTL input levels. Examples are the several signals to DS89C430 controller (clock, reset and PSENDRV*) and to the 5V RAM.

5V PAL can use 3.3V I/O. A 5V version of the Xilinx PAL that we have used in the labs – called XC95xxXC – can do this trick: though the PAL's innards must be powered by 5V, it provides a separate V_{CCIO} pin that can be powered by 3.3V. That done, the part can understand and drive either 3.3V parts or 5V parts at TTL levels.

When the supply mismatch is more extreme, a *translator* IC will be required to let the low-voltage part drive the higher-voltage part. Maxim, for example, makes a "universal" bidirectional translator for linking parts supplied from voltages as disparate as 1.2V and 5.5V (MAX3370 is the single-line part). NXP (formerly Philips Electronics) developed a simpler bidirectional translator using a single MOSFET. This ingenious circuit is described in §18W.2.

18N.2 Digital ⇔ analog conversion, generally

AoE §13.2

Analog versus digital: The task of converting from either form to the other can include challenging *analog* tasks. In order to convert from analog to digital (ADC, A/D) a circuit must classify an input (usually a voltage), putting it into the correct digital category; often the circuit must do that fast.

Running in the other direction (DAC, D/A) is conceptually easier. But it is hard to do this precisely (making the conversion good to many bits), and the digital processing will have added extraneous *non*-information, like "steppiness" in the sampled and recovered waveform; this junk must be cleaned away.

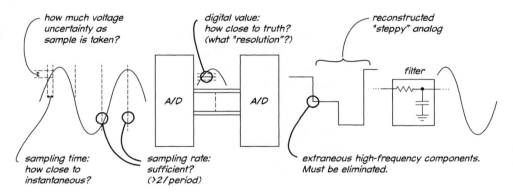

Figure 18N.5 Analog in, to ADC; then digital out, to DAC: analog reconstructed.

That cleaning job can present a substantial analog task; but it is one that is accomplished by the low-pass filter that cleans up the output of every digital-audio player. The filtering can be accomplished in part within the digital domain. Some of this processing, called "oversampling," involves interpolating pseudo-samples between actual samples.[2] Digital filtering, including the "oversampling" that boosts the apparent sampling rate, eases the task of the final analog stage, a smoothing low-pass filter.

What *sampling* does not mean: We will return to the question of how one must sample a waveform, in order to convert it to digital form. But let's dispose of one possible misconception at this early point: if, as in Fig. 18N.5, the sinusoid is sampled three times during one period, the *sampled* value is not an *average* of the waveform value during that time interval. Instead, it is the value present during the very short sampling pulse. The ideal sampling would be instantaneous; an actual sampling pulse may last a few tens of nanoseconds.

ADC: how fine to slice, in voltage (vertical) and time (horizontal): If you mean to convert a sinewave to digital form then back again, you run at once into the problem of *how much information* you need to carry into the digital domain. You must be content with a limited number of slices both in the vertical direction – amplitude resolution – and in the horizontal direction, where the fineness of "slicing" is determined by the sampling rate.

---

[2] Interpolation is one way of describing the process of digital filtering.

18N.2.1 Slicing across the vertical axis: *amplitude or voltage resolution*

This is a pretty straightforward issue. You decide, as usual, how big an error is tolerable. In this course, nearly all the converters you build or use in the lab will be *8-bit* devices; thus we'll settle for 256 voltage slices, each worth between 10 and 16mV (how large the slice is depends on the full-scale voltage, since each slice is $V_{\text{full-scale}}/2^{(\text{number-of-bits})}$).

Commercial digital audio, at 16 bits, would cut each of our 8-bit slices into another 256 sub-slices (*16*-bit resolution $\Rightarrow$ 1 part in 64K).[3] Our breadboarded circuits would bury any such pretended resolution in noise, as you know if you recall how big the usual fuzz is on your scope screen when you turn the gain all the way up. In one exercise, §23L.2.2, we do invite people using the SiLabs controller to try out the full 12-bit resolution of the controller's DAC. But the point we intend to show even in that exercise is mostly that such resolution is wasted in our breadboarded circuits. And in general, the level of noise present in the analog signal sets a limit on what resolution is useful in a converter. In audio conversion there is an additional boundary: one gets no advantage by resolving an audio waveform to a level below what a human can hear.

Quantization error: Since going from analog to digital entails forcing a continuous input into bins – a sort of procrustean fitting of a smoothly-varying original into a limited set of alternatives – the process necessarily introduces errors. Fig. 18N.6 sketches an analog input, and the levels (the nearest available) to which the input was assigned, in an imagined conversion into digital form.[4]

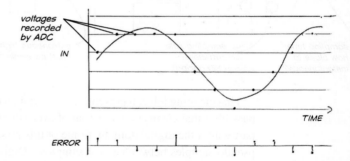

Figure 18N.6 Quantization error: a maximum of 1/2 the smallest voltage slice (after Pohlmann *Principles of Digital Audio*, McGraw-Hill, 2011).

This figure is offered in order to persuade you that the maximum error for the conversion process will be 1/2 the size of the smallest "slice." Such errors are an inevitable result of the classification of continuously varying *analog* voltages into discrete *digital* bins. Thus, quantization necessarily adds this noise at the level of 1/2 LSB.

18N.2.2 Slices across the horizontal axis: *sampling rate*

Here the result is surprising to anyone who has not considered the question before: Nyquist pointed out the curious fact that a shade more than *two* samples per period of a sinewave will carry a *full*

---
[3] The size of the smallest "slice" (least significant bit, LSB) is fantastically tiny on, say, a portable player, whose full-scale voltage may be under 2V. For example, 16-bit resolution and 2V full-scale implies an LSB value of 30μV.
[4] This figure is modeled on one offered by Pohlmann, K., *Principles of Digital Audio*, 6th ed. McGraw Hill (2011), Fig. 2.7, p. 31. The darkened points represent samples taken – and the representation may confuse you since each dot shows in *analog* form the supposed *digital* representation of the original input. Nevertheless, we think the information this figure offers does serve to make less abstract the notion of "error <1/2 LSB".

description of that sinewave.[5] Thus one can reconstruct the original with just that pair of samples (this works only if one has many cycles of those two samples). Shannon recited the rule thus:[6]

Theorem *If a function $f(t)$ contains no frequencies higher than W cps, it is completely determined by giving its ordinates at a series of points spaced $1/2W$ seconds apart.*
  This is a fact which is common knowledge in the communication art.

What Nyquist does *not* say: We would like to inoculate the reader against a possible misunderstanding of Nyquist's rule: we would like to underline the truth that exactly two cycles per period is not sufficient. Aldrich makes one sufficient argument for this truth: given just two points, it is possible to draw a sinusoid at that frequency and *of any amplitude*. So the information is not sufficient to satisfy the essential requirement: that just one waveform can be drawn through the sampled points.[7]

If reproducing a sinusoid of a single frequency strikes you as a rather sterile academic exercise, recall Fourier's wisdom, noting that any waveform can be expressed as a sum of sinusoids – and note that if we can sample adequately the *highest* frequency component of a signal we necessarily get plenty of information about lower frequencies.[8]

Here's the idea, illustrated in Fig. 18N.7 with the eight-bit ADC and DAC used in §23L.1(AD7569). We are not pushing for extreme stinginess in sampling here; instead, we pick up about *four* samples per period. The scope image shows a round trip: analog input converted to digital and back again.

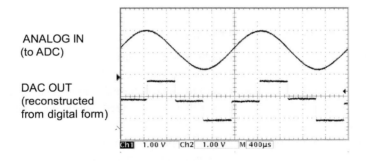

Figure 18N.7 Analog in, to ADC; analog out reconstructed from DAC.

It is not hard to believe that those steppy edges could be smoothed away. Indeed they can, by a good low-pass filter – a filter much better than a simple *RC* low-pass. Fig. 18N.8 shows the steppy output smoothed by such a low-pass. The filtered output does, indeed, look like the original analog input.

We will return to the question of what sampling rate is required, and how good a low-pass filter is needed later in this chapter and again but more thoroughly in Chapter 18S.

You will get a chance to reproduce this effect in Lab 23L when you control sampling rate with your microcomputer. Evidently, the sampling rate required depends upon the frequency of the analog

---

[5] The rule, suggested but not clearly stated by Nyquist in 1928, was proven by Claude Shannon in 1949, and sometimes is called the Nyquist–Shannon Sampling Theorem.
[6] Claude Shannon, *Proceedings of the Institution of Radio Engineers*, **37**,(1), 10–21, (1949), quoted by Nika Aldrich in his superlative book, *Digital Audio Explained for the Audio Engineer*, 2nd ed. Sweetwater (2005), p. 111.
[7] Aldrich, op.cit. p. 123. Don't forget the important assumption, which Aldrich emphasizes, that the only permitted waveform is a *sinusoid*. No fair improvising other odd shapes that could pass through the sampled points. Such odd shapes would contain frequencies higher than the sinusoid that is assumed. Consult M. Fourier if this point is unfamiliar (see, for example, Chapters 3N and 3L).
[8] We usually state Nyquist's rule as applying to the highest frequency in the input signal (see, e.g., AoE §13.5.10C). It is more accurate – though only rarely different – to state the rule as requiring a sampling rate a little more than twice the signal's *bandwidth*. When signal frequencies begin close to zero, as they do for audio, "highest frequency" and "bandwidth" are the same. But in unusual cases one could, for example, sample a narrow bandwidth at a high frequency using a modest f_{sample}. This might come up, say, in digitizing a radio signal.

waveform: if the waveform includes many frequency components, then it is the *highest* that concerns us; the lower frequencies are easier to catch. Commercial digital audio again provides a useful case in point. That system aims to store and recover signals up to 20kHz. To do this, it samples at 44.1kHz. This can be described as "10% oversampling:" a sampling rate just 10% above Nyquist's theoretical minimum.

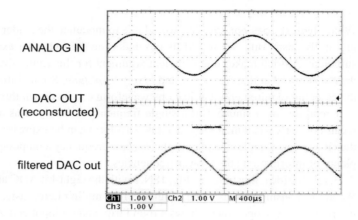

Figure 18N.8 Analog in, to ADC; analog out reconstructed from DAC, steppy waveform smoothed by low-pass filter.

18N.2.3 Effect of an inadequate sampling rate (aliasing)

If you don't do what Nyquist told you to do, you get into trouble. You don't just fail to get what you meant to get; you get nonsense. Sampling at too low a rate, you get an artifact, a fake signal that will be irreversibly commingled with the true signal. The fake is said to be an "aliased" signal because it is translated from its true frequency. Strange to tell, the true signal also is captured, but the presence of the alias usually corrupts the signal hopelessly.[9]

Figure 18N.9 shows a 1kHz analog input sampled at 1.2kHz: sampled well below the theoretical minimum "Nyquist" rate of >2kHz. We have marked the sampling points with small circles, and show the reconstructed waveform (the DAC output), before and after lowpass filtering. The spurious sinusoid in the fourth trace is convincing – but it is a false artifact. (The phases of sampled points, reconstruction and filtered output do not match; this results from delays in the processing that produced these signals.)

The reconstructed waveform may look funny (and not even sinusoidal). But if one lowpass-filters this funny waveform, one gets a sinusoid, as you can see in Fig.18N.9. This sinusoid was not in the original signal but is indistinguishable from a genuine input at 200Hz (the alias appears at $f_{sample} - f_{in} = 1.2\text{kHz} - 1\text{kHz} = 200\text{Hz}$). We will look more closely at such sampling artifacts in Chapter 18S.

To protect against aliasing, you need a good low-pass filter *ahead* of the ADC input – in addition to the one used to smooth the reconstructed signal. This input filter makes sure that no indigestible frequencies ever are fed to the converter. This is a standard feature for ADC systems – but it is a feature that we deliberately omit from Lab 23L, in which you play with sampling rates. We omit it because aliasing is a strange and interesting effect that you ought to meet at least once under laboratory conditions where it does you no harm. Having seen it, you will know that you must forestall *aliasing*.

---

[9] In some exceptional cases, aliasing can be harmless or even useful. If the signal that would be aliased is a single frequency, then its alias might be removed with a narrow bandstop filter. And the predictable behavior of aliasing also allows purposeful "folding" of high frequencies to a lower range. This technique can be used to convert a high-frequency radio *carrier* to an "intermediate frequency" (IF) for demodulation. See Aldrich, p. 35.

18N.3 Digital to analog (DAC) methods

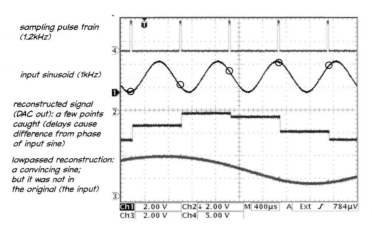

Figure 18N.9 A sinusoid inadequately sampled: Nyquist defied. (Scope settings: 2V/div except top trace, 5V/div; 400μs/div.)

Enough generalities. Let's get back to hardware: let's look at some ways to carry out the conversion, in both directions.

18N.3 Digital to analog (DAC) methods

A DAC is conceptually simple: All we need is a way to *sum* a binary-scaled set of voltages or currents: if an input *1* at the MSB should generate output 1V, the next bit by itself should generate 0.5V; both together should generate 1.5V; and so on. Some methods are pretty self-explanatory; others are not.

18N.3.1 Thermometer DAC

AoE §13.2.1

A voltage divider tapped by switches that are controlled by the digital inputs can do the job.[10] This design has a wonderfully dignified pedigree: it was designed by Lord Kelvin,[11] though it's not as musty as that fact might suggest. It is a design now used for digital potentiometers and some small DACs. It requires only a resistive divider, some switches, and some decoding logic.

With good analog switches, a 16-bit divider can do the job – requiring about 65,000 precise resistors (e.g., TI's DAC8564). As the DAC8564's data sheet advertises, the design "minimizes undesired code-to-code transient voltages (glitch)," in contrast to DAC's whose monotonicity depends on precise matching of resistor ratios.

18N.3.2 An op-amp current-summing circuit

Compare AoE fig.13.3

An op-amp current-summing circuit, and resistors of values $R, 2R, 4R\ldots$, can form the DAC's binary-weighted sum. Cozily familiar though it is to those of us comfortable with op-amps, this method rarely is used. It requires the fabrication of many values of resistor to high precision.

18N.3.3 R–2R ladder

AoE §13.2.2

Instead, one can get the same result – a binary scaling of currents – with an ingenious circuit called an *R–2R* ladder, a circuit that requires just *two* resistor values. This pair of values is easier to fabricate

---
[10] See Aldrich at pp. 148–152; Analog Devices Tutorial:
http://www.analog.com/media/en/training-seminars/tutorials/MT-014.pdf.
[11] This surprising provenance we take from Analog Device's online note.
http://www.analog.com/library/analogDialogue/archives/39-06/Chapter%203%20Data%20Converter%20Architectures%20F.pdf

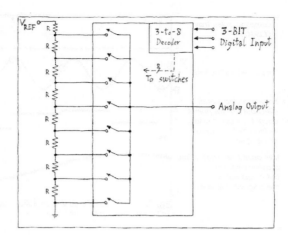

Figure 18N.10 "Thermometer" DAC (after Analog Devices online tutorial).

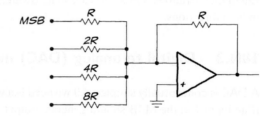

Figure 18N.11 A possible DAC – but hard to fabricate.

than the n different values required for an n-bit converter like that in Fig. 18N.11. Those n values would need to span a wide range and would require great precision in order to give many bits of resolution.

How an R–$2R$ ladder divides down a voltage: The ladder is made up of units that look like the one on the left of Fig. 18N.12. Since the unit looks like $2R$, seen from its left side (its input), we can plug in another exactly similar unit in place of its right-hand $2R$ – and so on, extending the chain as far as we like. At the right-hand end of each of the resistors labeled R the voltage is down to half what it was at the left end of the resistor.

Figure 18N.13 contrasts the clever R–$2R$ against a dopey divider that makes a binary-scaled divider by following our usual loading rules. R_{in} for each succeeding stage is $100 \times R_{OUT}$ for the stage before to make the division voltages good to 1%. This is pretty mediocre performance (equivalent to fewer than 7 bits of precision); but you'll notice that the resistor values quickly blow up, and with them so does the output impedance at the successive division points. Meanwhile, the smarter R–$2R$ gets the job done with just two R values, and holds R_{OUT} constant. Whereas *loading* causes errors in the dumb divider, loading achieves the desired division in the clever R–$2R$ divider.

AoE §13.2.3

Applying the R–$2R$ ladder: Figure 18N.14 shows two DAC circuits exploiting the R–$2R$ scheme. The left-hand circuit uses such a ladder to source current into the summing junction of an op-amp; the right-hand circuit (the schematic of an IC DAC) omits the op-amp, so the output is a *current*. Both DACs use just *two* resistor values.

A detail of Fig. 18N.14 may alarm you at first: the feedback to the op-amp on the right side. But on reflection you'll realize you have seen this oddity before, in the low-dropout regulator of Chapter 11N. In both cases an *inversion* within the feedback loop transforms what looks like positive feedback.

18N.3 Digital to analog (DAC) methods

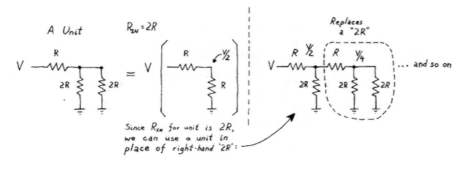

Figure 18N.12 R–2R ladder.

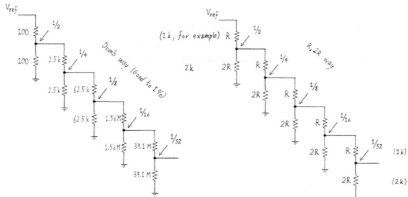

Figure 18N.13 Dumb divider contrasted with the clever R–2R binary divider.

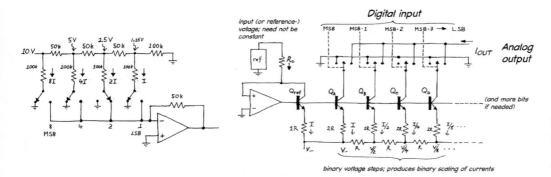

Figure 18N.14 Two DAC designs using R–2R networks: V out, I out.

18N.3.4 Switched-capacitor DAC

Capacitors, rather than resistors, can be used to form a DAC's binary-weighted output. IC fabricators favor using capacitors over resistors, especially for CMOS designs. The circuit in Fig. 18N.15 – for a 3-bit DAC – resembles the op-amp summing circuit but is simpler.[12] The capacitive design sums *charges* directly whereas the op-amp summing circuit, which sums currents, requires either scaled current sources or a reference voltage and scaled resistors. (Compare AoE Fig. 13.4.) The capacitive design can leak, but works well in a DAC whose output is used only briefly, as in an SAR ADC (see §18N.4.3).[13]

Each of the binary-scaled capacitors, after being reset to zero, is connected either to ground or to

---

[12] R. Jacob Baker, *CMOS Circuit Design, Layout and Simulation*, IEEE Press, Wiley (1998), p. 806.
[13] See http://www.analog.com/media/en/training-seminars/tutorials/MT-015.pdf.

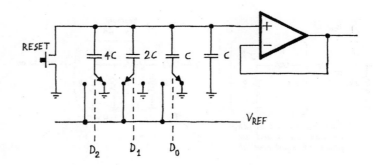

Figure 18N.15 3-bit switched-cap or "charge-scaling" DAC.

V_{REF} according to the level of the digital input bit. When all bits are high, V_{out} equals the value of $V_{REF} \times \frac{7}{8}$. In other words "111" would produce a value of 7V if V_{REF} were 8V. In Fig. 18N.16 we have sketched the capacitive dividers formed by this and a couple of other input values.

Figure 18N.16 3-bit switched-cap: capacitive dividers formed for three input combinations.

18N.3.5 1-bit DAC; pulse-width modulation (PWM)

AoE §13.2.8

A good DAC is hard to build: Though it is easy to draw a diagram for a DAC, it is difficult to make a DAC that works with good resolution – that is, a DAC that responds properly to a large number of bits. In this course we will use modest *8-bit* DACs, which imply resolution down to 10mV given a 2.5V range. This we can manage, even in our breadboarded circuits.

A low-resolution digital signal – even just a one-bit ON/OFF signal – can achieve high analog resolution when averaged. If this sounds a bit cryptic, recall that you have seen this effect in action, back when you built a PWM circuit in §8L.4. An ON/OFF signal whose duty cycle is varied can produce very high resolution when the pulsing waveform is smoothed by an averaging circuit.

Early personal computers used this method to provide audio. It is also the method used in so-called "Class D" amplifiers, whose outputs are simply switches, usually to a positive or negative voltage. You met one of these amplifiers, the LM4667, in Chapter 12L. Class D amplifiers can achieve very high efficiency and thus are favored in battery powered audio circuits. They also lend themselves to motor control (as you saw in Chapter 8L), where PWM provides smoother power modulation and better torque than can be achieved by varying the voltage to a motor.

Figure 18N.17 is a set of scope images showing PWM done with the circuit of §8L.4. Alongside each duty-cycle waveform is an image of an LED driven with that waveform. The LED is driven ON by a logic *low*, so our listing of *duty cycle* in the figure inverts the usual sense of the term.

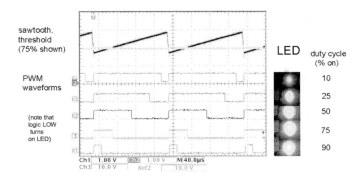

Figure 18N.17 Static PWM (duty-cycle) waveforms, and consequent brightness of driven LED. (Scope settings: sawtooth trace, 1V/div; five PWM traces, 10V/div.)

This is a static case, showing a technique well-suited to letting a microcontroller regulate not only the brightness of an LED but also the *color* of a bi- or tri-color LED, by varying the relative strength of two or three contributing colors. You will find a chance to try this technique in §22L.2.9.

PWM that varies with time can produce a waveform. Fig. 18N.18 is a hypothetical PWM waveform and the sinusoid that would result if the PWM were averaged by a low-pass filter.

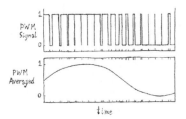

Figure 18N.18 PWM, time-varied, can produce a time-varying waveform.

The fact that a *one-bit* data stream can describe a waveform establishes the foundation for subtler forms of one- and low-bit conversion. In §18N.4.5 below, we will meet the counter-intuitive *delta–sigma* conversion scheme which can use a single bit output stream to provide high resolution.

18N.4 Analog-to-digital conversion

Going from the continuous world of analog into the discrete categories of digital is harder than going in the other direction and the task has evoked a wide and various set of solutions. Table 18N.1 shows the leading methods for comparison at a glance.

Another representation of alternative ADC methods appears in a good, concise graph in Tietze and Schenk, *Electronic Circuits: Handbook for Design and Applications*, 2nd ed. Springer (2008) §18.12, Fig. 18.68.

18N.4.1 Open-loop

AoE §13.6

Flash: We will start with the method that is conceptually simplest, though most difficult to fabricate: *flash* (or parallel) conversion. (The design may remind you of the "thermometer" DAC design of §18N.3.1, whose process it runs in reverse.) The comparator outputs provide a thermometer-like result

Table 18N.1 leading analog to digital conversion methods

Method	Speed	Resolution	Applications
"flash" (parallel)	fast $\geq$5ns at 10 bits $\geq$1ns at 8 bits	low $\leq$10 bits	storage scope, RF signals
binary search	intermed. $\geq 1\mu s$ at 20 bits*	quite high $\leq$20 bits	general purpose very widely used
delta–sigma	low intermed. $\geq$4k samples/sec at 31 bits**	high to very high $\leq$32 bits	general purpose becoming widely used
dual slope	slow a few samples/sec at 24 bits	high $\leq$24 bits	DVMs (digital voltmeter) now must compete vs. $\Delta\Sigma$

* LTC378-20
** LTC344x: 24 bits, no latency; with one-cycle latency, 8k samples/sec. A spectacular 31 bits is offered at 4k samples/sec by TI's ADS1282.

(1s up to the level of the input voltage). That inconvenient code must be compressed by an *encoder* circuit.

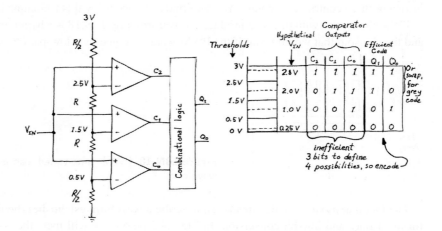

Figure 18N.19
Parallel or "flash" ADC.

This method calls for a lot of analog circuitry to achieve a modest number of bits. A 9-bit converter needs 511 comparators. That may not sound like a lot of parts, if you are accustomed to reading about million-transistor digital parts, and multi-gigabyte memories. But the fact that flash converters now are limited to about 9 or 10 bits underlines the point that *precise, fast analog* circuitry is hard to build, whereas fast digital circuitry is relatively easy.

Each of the outputs of a 10ns memory, say, needs to decide only whether the stored data is closer to High or to Low. Each comparator in a 9-bit, 2.5V-range converter, in contrast, needs to make a decision good to about 5mV; and it needs to do it fast – in somewhat less than 10-odd ns, if it is to run at the same speed as the memory. The comparators' job evidently is much the harder of the two.

The *encoder* is essential. The need for it is not obvious in a tiny example like the one in Fig. 18N.19. That encoder compresses seven lines to three: useful, but not a spectacular improvement in efficiency. At 8 bits, the need for the encoder becomes very obvious: rather than take 255 lines from the ADC – the outputs of all comparators – we take the *encoded* version of this information, on 8 lines.

The device looks as if it could operate continuously. Normally, however, it is not used that way. Catching a value when the comparators were in transition could be disastrous: say, partway through a transition from 01 to 10 we could get an output of 00 or 11. Instead, it is operated with a sampling clock, like other ADCs. A sample-and-hold keeps the comparator inputs constant while the comparators settle, and the (digital-) output is caught in a register.[14]

Flash ADCs are rather specialized devices. They are power hungry, low-resolution converters. As standalone parts, they are reserved for the very impatient. But they are used more frequently within "subranging" ADCs that form a higher-resolution answer at somewhat lower speed, by repeatedly applying a flash converter to fractions of the full range.

Pipelining: A subranging flash device requires multiple clock cycles, in order to carry out a conversion – but it can run faster than the need for multile cycles would suggest if it is fed a stream of new data while earlier samples are being processed.

The input is first subjected to a "coarse" 3-bit conversion using a 3-bit flash ADC. Then that ADC value is fed to a DAC and converted to a value representing the top three bits, and this coarse level is subtracted from the input. The result of the subtraction, a "residue," is passed to a second 3-bit flash ADC. The outputs of the two flash ADCs provide a 6-bit answer, and the fact that the two conversions occur one *after* the other mean that if the MSBs from sample N are saved, to be combined with LSBs when the latter are ready, a new MSB conversion, of sample $N+1$, can be started while the N bits are being completed.

A 16-bit ADC from Linear Technology, for example (LTC2207) achieves the astonishing sampling rate of 105Msps (megasamples per second): 16 bits converted in about 10ns. But it pipelines its samples, using five flash ADCs, each time converting a fraction of the total range. Adding some housekeeping time, it needs seven cycles to carry out a single conversion. But this is still an extremely fast conversion – 16 bits in about 65ns, even if one did not take advantage of pipelining, and simply used the device to do a single conversion. The seven-clock delay of the pipeline is called its "latency."

18N.4.2 Dual-slope

AoE §13.5.7

We looked at *single-slope* converter in §16N.6, Fig. 16N.28, on counter applications[15]. The single-slope measured the time a ramp took to reach V_{in}. The dual slope converter is similar: it lets V_{in} determine the size of a current that feeds a capacitor for a fixed time; when that fixed time is up, the capacitor is switched to a fixed discharging current. A counter measures the time for the discharge to take V_{cap} back down to zero.

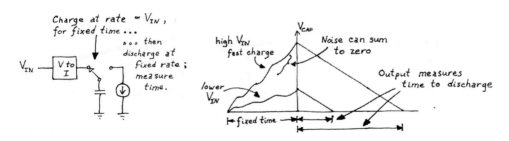

Figure 18N.20 Dual slope converter: charge for fixed time at rate determined by V_{in}; measure time to discharge.

The dual slope differs however in one important respect from the other converters described up till now: it is an *integrating* converter, in contrast with the sampling types. Because of this behavior, variations above and below the average input voltage level tend to get swallowed up, not recorded as the true level. The swallowing-up is perfect if the converter takes in an *integral* number of periods of the interfering noise (the integral of that noise then is zero).

---

[14] See, e.g., Linear Technology's App. Note: http://www.maxim-ic.com/appnotes.cfm/appnote_number/810/CMP/WP-17.
[15] It also appears in AoE §13.5.5.

As a result you can get the right answer in the presence of periodic interference. In particular, if you expect periodic interference at 60Hz and its harmonics (as usually you can expect from United States power lines), you can let a conversion include an integral number of cycles of this signal; then the bumps above and below the average value will cancel. This is the method used in the DVMs we use in our teaching lab.[16]

The second virtue of a dual-slope ADC is its cancellation of DC errors in its components: in order to know when the circuit has ramped V_{cap} back down to ground, the circuit needs an accurate comparator. To resolve the input voltage to, say, 0.1mV it might seem that the comparator would need an offset voltage smaller than that. But the dual-slope design eliminates that requirement. The comparator need only return V_{cap} to its starting point, not to *zero* volts. If the starting point was, say, 1.5mV (supposing this to be the comparator's offset voltage, V_{OS}), a return to that starting point of 1.5mV introduces no error. The effect of V_{OS} is cancelled.

Dual-slope converters now are getting competition from so-called "delta–sigma" ADCs (see §18N.4.5 below). Like the dual-slope, these are integrating types capable of very high resolution and they can be configured to reject 60Hz. They would serve well in a DVM.

18N.4.3 Closed-loop

A closed-loop[17] converter, and in particular one that uses a binary search, beats either flash or dual-slope as a general purpose converter. The flash costs too much, and provides only mediocre resolution; the dual-slope is slow.

The closed-loop converter works like a discrete-step op-amp follower: it makes a digital "estimate;" converts that to its *analog* equivalent and feeds that voltage back; a single comparator decides whether that estimate is too high or too low (relative to the analog input of course); the comparator output tells the digital estimator which direction it ought to go as it forms its next, improved estimate. Negative feedback drives the digital estimate close to the input value.

Fig. 18N.21 shows contrasting waveforms at the *feedback* DACs of two sorts of feedback ADC. The waveforms show the analog estimates made by the two converters as they strive to drive their estimates close to the input value. (The waveforms show analog voltages reflecting the digital estimates, but let's not forget that it is the *digital* values that we are after, not these analog equivalents. These are, after all, Analog-to-Digital converters.)

Figure 18N.21 Tracker versus binary-search or successive-approximation converter: trying to follow a step input.

---

[16] Such converters designed for the European market, where the power line frequency is 50Hz, use a longer integrating time. In principle, one could set the integration time to be an integral multiple of both 50Hz and 60Hz periods. But such a choice could impose unacceptable delay. 0.1s is an integration time often used. This is 6 periods of US line noise, 5 periods of European line noise.

[17] This term is not standard, but only our way of contrasting the converters that follow versus the flash and dual-slope types.

18N.4 Analog-to-digital conversion

Successive-approximation and tracking: The block diagram in Fig. 18N.22 illustrates the implementation of feedback ADCs using the two forms of *digital estimator*, estimators whose efforts to home in on the analog input appear on the right-hand side of Fig. 18N.21. One form is smart; the other is not.

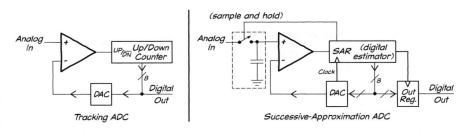

Figure 18N.22 Two closed-loop ADC converters: tracking and binary-search.

Note the presence of a *sample-and-hold* circuit in the block diagram for the binary search type (right-hand circuit of Fig. 18N.22). You met (and built) an S&H in the MOSFET lab, 12L. The tracker, which operates continuously, updating on every clock, does not need a sample-and-hold.

Tracking ADC: The tracking converter – the ADC sketched on the left in Fig. 18N.22 – is dopey: it forms and improves its estimate simply by counting up or down. Although it is not widely used, the tracking ADC persists because of two virtues: first, it always provides the best current answer, unlike the binary-search type which needs multiple clocks to arrive at its answer. Second, it is relatively insensitive to glitches that could cause large errors in the smarter SAR: a spurious clock edge causes an error of only one LSB. So the humble tracker can be the best for really plodding assignments like converting the output of a slowly rotating shaft (a so-called "resolver" application.[18]). Its LSB must continually flip back and forth once the tracker has found its best estimate of a constant input.

AoE §13.7

Binary-search or "SAR" ADC: By contrast, the successive-approximation estimator (SAR) is clever: it does a binary search starting always at the *midpoint* of the range and asking the comparator which way to go next. As it proceeds, it goes always to the midpoint of the remaining range. That sounds complicated; in fact it is easy for a machine that is by its nature already binary. The comparator tells the binary-search device *bit-by-bit* whether the most recent estimate was too high or too low, relative to the analog input.

Because of this gradual way of composing its answer, such a converter needs an output register to catch its last, best estimate: its digital estimates look funny until the process is complete so they should not be shown to the outside world. In addition, the analog input must be passed through a *sample-and-hold*, which freezes the input to the converter during the conversion process. We omit this S&H in our lab circuit, in order to keep the circuit simple. We lose certainty in sampling time, and that uncertainty translates to considerable voltage *errors* in our conversions.[19]

Sampling should occur at regular intervals: Any processing of the signal – including the simplest: playback through a DAC and low-pass filter – assumes *periodic* sampling. When the S&H is omitted, the converted value equals the input at *some* point in the sampling interval, but one cannot know

---

[18] See Analog Devices note,
http://www.analog.com/en/other/multi-chip/ad2s44/products/tutorials/CU_tutorials_MT-030/resources/fca.html.
[19] The converted value will correspond to the input voltage at *some* point during the conversion process. But during that time the input waveform may have changed level by a significant amount, often by far more than the desired ADC resolution of 1/2 LSB.

exactly when. Sampling time thus becomes vague, wandering, in successive samples. We'll see in §18N.5 that sampling at a controlled and regular rate is normally required.[20]

We present the *tracker* only to make the binary-search estimator look good. There are very few applications for which anyone would consider a tracker.

As we suggested earlier, it is hard to make precise analog parts: hard to make, say, a DAC good to many bits. It's easy to make digital parts, and easy to string them together to handle many bits. For example, it would be easy to build an up/down counter or SAR of 100 bits. So it is the *analog* parts – the DAC and comparator – that limit the resolution of the closed-loop converter. They also limit the converter's *speed*, as illustrated in Fig. 18N.23 (cf. Fig. 18L.7).

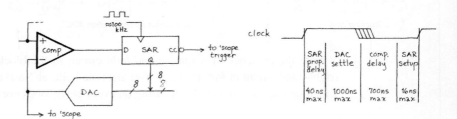

Figure 18N.23 Closed-loop ADC: clock period must allow time for all delays in loop.

18N.4.4 A binary search: example

Here, for anyone not already convinced, is a demonstration of the strength of the *binary search* strategy, used by the successive-approximation register (SAR) that you use in Lab 18L. If someone chooses a number in the range 0 through 255 and tells you whether each of your successive SAR-like estimates is *too low*, you can get to the answer in eight guesses. If you write out the binary equivalents of those guesses (as the "binary value" table does in Fig. 18N.24), you can see how the SAR forms its answer, bit by bit.

It proceeds from MSB to LSB, always setting the next bit *low*. After each guess the comparator tells the SAR whether the current guess is *too low*. If it is, the SAR sets the bit just put *low* back to a *high*, while forcing the next bit low. When the "travelling zero" arrives at the low-order end of the estimator, and has been forced high or left low, the search is over. In the case sketched in Fig. 18N.24 we have arbitrarily assumed that the answer is 156.

Figure 18N.25 is a scope image showing such a search: the SAR drives the DAC output to home in on the analog input. The conversion value differs from one sketched in Fig. 18N.24 in its d3 and d1 values, but the patterns in the two figures are obviously quite similar.

18N.4.5 Delta–sigma ADC

AoE §13.9, especially §13.9.4A

The binary-search or successive-approximation converter, long the most versatile and widely-used design, now is challenged by another good conversion strategy, usually called "delta–sigma."[21]

---

[20] We've hedged by using the word "normally," having been chastened by Shannon's observation that "samples can be unevenly spaced," though in that case "the samples must be known very accurately" and "the reconstruction process is also more involved...." (Claude Shannon, op. cit., p. 12.) Not only input levels but also exact timing of the sampling points would need to be known "very accurately." We will steer wide of such subtleties.

[21] Confusingly enough, it sometimes is called "sigma–delta," even when the sequence of operations is not truly interchanged. The labels vary from one manufacturer to another. Historically, sigma–delta may be the more accurate, matching the name applied by the method's inventors. See http://www.beis.de/Elektronik/DeltaSigma/DeltaSigma.html.

18N.4 Analog-to-digital conversion

binary search tree

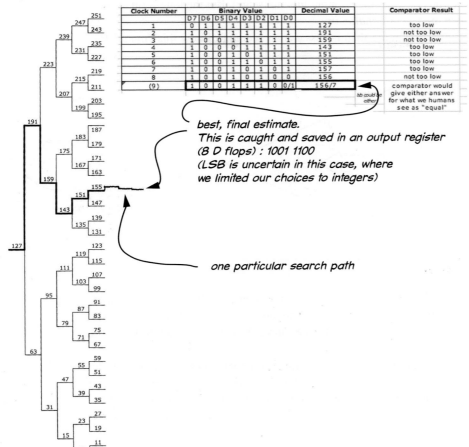

Figure 18N.24 One particular binary search: eight guesses to get 8-bit answer.

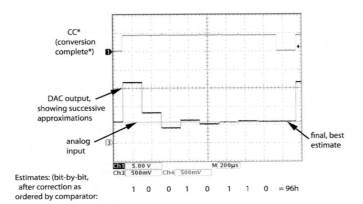

Figure 18N.25 SAR homes in on analog input. (Scope settings: 500mV/div, except CC (pulse, 5V/div; 200 μs/div.)

Figure 18N.26 is a sketch of the "modulator" that implements this arrangement. We have drawn a triangle around the whole thing to make the point that this is a feedback loop familiar from your experience with op-amps. The feedback signal fed to the "difference" block is a strangely crude signal, banging to the full-scale limits of the input range – either +1V or –1V in this hypothetical case. It cannot match *analog in* at any particular time (unless input hits full scale). But the long-term average of the output bitstream *can* match the level of *analog in*.

Figure 18N.26 gives in detail (perhaps more than necessary) an account of how this "modulator" works.

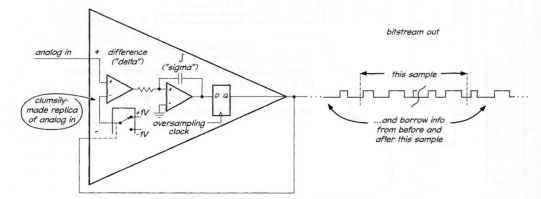

Figure 18N.26
Delta–sigma modulator.

1. The circuit notes the sign of the difference between the analog input and the present "estimate" – in Fig. 18N.26, an estimate that is just a High or Low extreme, +1V or –1V, determined by the one-bit digital output. This difference, appearing at the output of the differential amplifier, is the *delta*. Then...
2. In response, the modulator injects a dollop of charge of appropriate sign and magnitude into an analog summing circuit (that's the *sigma*), trying to drive the fed-back estimate close to the level of the analog original. The feedback loop attempts to hold the integrator output close to zero. Doing this, the converter produces a bitstream output whose average value matches that of the analog input, over many cycles.

So far this scheme is not surprising. The bitstream output roughly resembles PWM (§18N.3.5). And the loop is just another negative feedback scheme, somewhat resembling the crude *tracker* ADC on page 705, which injects a positive or negative increment into the *digital* summer, the counter, an accumulator that holds the current best estimate. The output of the tracker was formed in a very simple way: it was simply the current multi-bit value to which the Up/Down* counter had been driven. The tracker was very slow, requiring 2^n clocks to find an n-bit digital result, after a full-scale change of the analog input.

The delta–sigma is much smarter than that – and much more surprising. Fig. 18N.27 is a block diagram of a delta–sigma ADC, its elements so generic that we don't expect you to get excited by this description – not yet.

Note that the "low-pass filter" in this diagram is a *digital* circuit; its output is a series of n-bit

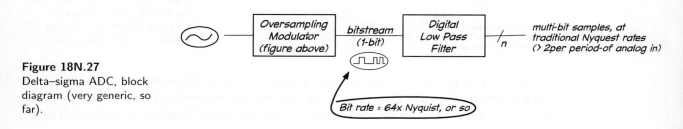

Figure 18N.27
Delta–sigma ADC, block diagram (very generic, so far).

samples, assembled from the fast-flowing stream of single-bit data, produced by the "oversampling modulator," the circuit of Fig. 18N.26. As AoE says, that is where the magic resides.

If a delta–sigma converter used 2^n clocks to get an n-bit result, as a tracker does (worst case) and as a simple PWM bitstream would do, there would be not much to admire or explain in delta–sigma's operation. But practical delta–sigma designs do not do this. Instead, they manage to get high resolution with quite modest oversampling rates, such as 64 (implying a clock rate that is quite manageable). With this limited oversampling they nevertheless can achieve 16 or 24 bits of resolution.

The magic of delta–sigma is extremely counter-intuitive. AoE's great contribution to the literature on this topic is to *admit* that this conversion method resists intuition. The conventional "explanations" of the method abound in glib evasions of explanation – sudden retreats into frequency-domain abstractions[22] – as AoE points out.

Analog Devices' (ADI's) explanation begins, as these delta–sigma clarifiers often do, by acknowledging that everyone else's explanation is hard to follow: "...most commence with a maze of integrals and deteriorate from there. Some engineers...are convinced, from study of a typical published article, that it is too difficult to comprehend easily."[23] To the credit of ADI, the company also provides an interactive time-domain animation of the behavior of a delta–sigma modulator, which one clocks by hand in order to get a sense of the way that the circuit encodes a DC input.[24] This animation does help.

Maxim's application note explaining delta–sigma (which they choose to call "Sigma–Delta") remarks that "Designers often choose a classic SAR ADC instead, because they don't understand the sigma–delta types." "...and no wonder!" was our feeling, when we hit the point in this mostly clear note that says, without explanation, "...the integrator acts as a low-pass filter to the input signal and a highpass filter to the quantization noise."[25] This crucial observation recurs in all discussions of the conversion method – and is explained, at last, in AoE, and below on page 710.[26]

Before we reach this high-pass/low-pass issue, let us note a preliminary point in favor of the one-bit converter:[27] the quantization noise is spread over a wide frequency range, and can be reduced in a particular, narrower range of interest. This initial thought is almost a commonplace.

Trading word length for sampling rate: oversampling: Sampling at more than the Nyquist rate (which, as you know, is slightly more than twice the top input frequency) is called *oversampling*. It is possible to convey a given amount of information in shorter "words" (fewer bits per sample) by sending these words at a higher rate.[28] This simple claim certainly is plausible; no magic here (the plot in Fig. 18N.28 describes, for example, how a sub-ranging converter can convert a wide word in multiple cycles at narrower conversions). But if one tries to apply this analysis to an ADC that uses *one-bit* to provides a high-resolution output – say, resolved to 16 bits – one gets a preposterous result. Sixteen bits would require sampling at about 65,000 times the Nyquist rate. That cannot be done, with present technology.

---

[22] ...the last refuge of a scoundrel – in the view of some of us abstraction-o-phobes.
[23] ADI application note: http://www.analog.com/en/analog-to-digital-converters/ad-converters/.
[24] Kester, Walter, op. cit., and ADI application note: http://designtools.analog.com/dt/sdtutorial/sdtutorial.html.
[25] Maxim Application Note 187: http://www.maxim-ic.com/an1870.
[26] We also want to acknowledge the wonderfully thorough explanations offered by Nika Aldrich in his book which is rich with waveform images that help to make this difficult topic intelligible.
[27] Some use more than a single bit, and are properly described as "low-bit" converters. Their workings are fundamentally similar to those of the one-bit designs.
[28] See John Watkinson, *The Art of Digital Audio*, 3d ed. Focal Press (2001), p. 253.

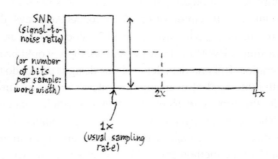

Figure 18N.28 One can trade word-width against sampling rate, holding information rate constant.

...but long words are best for efficient data storage: Although the low-bit strategy can pay in conversion processes, it does *not* pay in data storage. As Watkinson points out,[29] a bit can convey far more information in a longer word: in a 4-bit word, with its 16 possible codes, each bit carries "four pieces of information," as he puts it. In a 16-bit word, with its 65,000-odd codes, each bit carries 4096 "pieces of information." So, for example, no-one advocates storing digital audio as 4-bit samples.[30]

Delta–sigma design gets high resolution with quite modest oversampling rates, such as 64 (implying a clock rate that is quite manageable, under 3MHz, for a 16-bit audio signal). It achieves this by spreading quantization noise over the wider oversampled frequency range, pushing much of that noise into higher frequencies where it can be filtered out.[31]

As the lowest image of Fig. 18N.29 suggests, still greater reduction of noise is possible if the quantization noise can be "shaped" by high-pass filtering in the conversion process, then eliminating most of that noise with a low-pass at the output. Some such "shaping" is inherent in the modulator's operation, as we argue below.

The technique is very successful, and now provides the cheapest way to get high-resolution conversion at low frequencies (for example, Analog Devices' AD7748 provides 24-bit results in 20ms). The method also dominates audio conversion.

How delta–sigma puts quantization noise where it is less harmful: The point that delta–sigma depends on "noise shaping," some of which is inherent in the modulator's operation, is often reiterated in discussions of delta–sigma – as we have complained. The process is sketched in Fig. 18N.29, and is explained in AoEsee also the fine expostion in a web posting by Uwe Beis.[32]

AoE §13.5.10E

Here, we try to diagram the effects described in AoE. The delta-modulator sketched in Fig. 18N.30 is an all-analog version that shows an injection of quantization noise (v_{qn}) where the comparator and flip-flop of a digital modulator would appear. The contrast between the loop's treatment of noise (high-passed) versus its treatment of signal (low-passed) becomes more evident if we redraw the loop for the two cases.

How the noise is high-passed: The quantization noise is injected after the differencing and integration. We have redrawn those two elements as a single triangle – an *op-amp*-like object. That redrawing

---

[29] Watkinson, op. cit., p. 254
[30] But we should concede that Sony's Super CD, which stores a single-bit delta–sigma bitstream, stands as a striking exception to this generalization favoring long words for storage. The Super CD never was widely adopted, but persists among audiophiles.
[31] Gosh! Haven't we just uttered the formula that we complained about, above? But we will not stop with this remark. We will try to justify the claim.
[32] http://www.beis.de/Elektronik/Electronics.html

18N.4 Analog-to-digital conversion

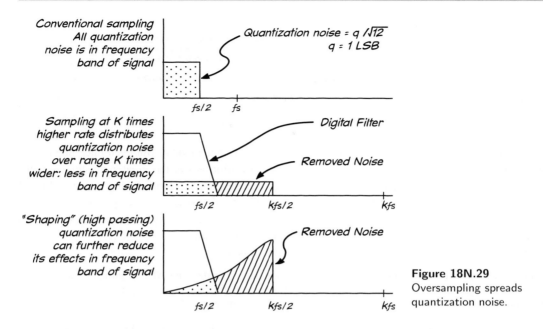

Figure 18N.29 Oversampling spreads quantization noise.

should not alarm us, because a normal, compensated op-amp is exactly that: a difference amplifier whose gain rolls off as an integrator does.

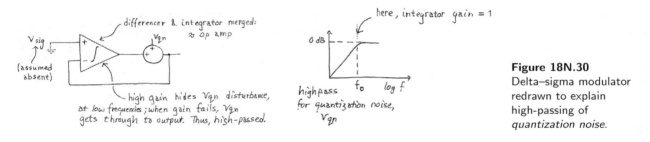

Figure 18N.30 Delta–sigma modulator redrawn to explain high-passing of quantization noise.

If you think of the injected quantization noise as just one more "dog" within the feedback loop (mentioned in §6N.9.1), then the "op-amp" hides this dog (is quantization noise its barking?) – but only as long as the amp has sufficient gain to do its feedback magic.

Then, as the gain begins to fall, more and more of the quantization noise survives. Eventually it passes right through to the output, undiminished. In other words it is *high-passed*. Much of the noise normally falls above the range of frequencies that we consider *signal* for the ADC.

How the signal is low-passed: The modulator, at the same time, is said to *low-pass* the signal. Thus it keeps what we want, while it pushes what we don't want outside the frequency range that interests us. Fig. 18N.31 gives the argument that says the modulator can low-pass the signal. Again, we have redrawn differencer and integrator as a single op-amp-like triangle. Notice that we are not claiming that the entire *follower-like* circuit in the figure is itself a filter. Rather, the claim is that the triangle – the modulator circuit fragment that we have likened to an op-amp – implements the low-pass function. We don't normally describe an op-amp that way; but we do know that any ordinary ("compensated") op-amp does, indeed, attenuate higher frequencies. It is a low-pass filter, or an integrator, though overall feedback often hides this fact from our view.

 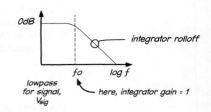

Figure 18N.31
Delta–sigma modulator redrawn to explain low-passing of *signal*.

The remarkable result: high resolution from a modest number of bits: But even once one understands (or accepts) this point about the shifting of quantization noise, the greater mystery remains (as AoE points out). Delta–sigma offers high resolution (for example, 16 bits) using oversampling of only, say, 64× (that is, 64 times the minimum "Nyquist" rate of two samples per period of the input signal). But that would imply that just 128 sequential bits could encode a value to one part in 64K – a process that ought to take 64K sequential bits. How is this possible?

A note from Analog Devices concedes how puzzling this result is:

> For any given input value in a single sampling interval, the data from the 1-bit ADC is virtually meaningless. Only when a large number of samples are averaged, will a meaningful value result. The Σ–Δ modulator is very difficult to analyze in the time domain because of this apparent randomness of the single-bit data output.[33]

Any single 16-bit sample constructed by the delta–sigma ADC, then, uses the digital low-pass of Fig. 18N.27 to tap information from a great many more bits than the nearest 128. The digital filter looks before and even *after* the present digital sample.[34] In other words the bitstream is, in fact, much longer than the 128 bits we announced just above. In fact, the digital filter may take information from *several hundred* or even a thousand successive bits in the stream or "taps".[35] AoE argues, in addition, that bit position matters, so that a 128-bit stream carries much more information than simply its average (which would provide only 7-bit resolution).[36]

It follows that a delta–sigma ADC requires a stream of samples; it cannot do its magic on an isolated sample.

18N.5 Sampling artifacts

18N.5.1 Predicting sampling artifacts in the frequency domain

We have recited Nyquist's rule that one must acquire a little more than two samples per period of an input sinusoid in order to get a full representation of the waveform (assuming we sample multiple cycles): see §18N.2.2 above. We now can present Nyquist's rule in a way that makes it easier to design a sampling system, a way that allows us to predict the artifacts (often called "images") that will be caused by sampling.

In general, if one samples an analog signal of frequency f_{in} (let's assume a single input sinusoid for simplicity), sampled at frequency f_{sample}, the sampling process will produce artifacts at predictable frequencies:

$$f_{\text{ARTIFACTS}} = (n \times f_{\text{sample}}) \pm f_{\text{in}}; \tag{18N.1}$$

---

[33] Kester, Walter, ADC Architectures III: Sigma–Delta ADC Basics, Tutorial MT-022, p. 6.
[34] This is a peculiarity of digital filters: they can "look ahead in time," having samples available in both directions relative to the "present" sample at the cost of delaying the output.
[35] ADI tutorial: http://www.analog.com/static/imported-files/tutorials/MT-022.pdf.
[36] §13.9.4C of AoE says, speaking of a 64-bit stream, "Because the individual bits are weighted differently, there are many more than 64 possible results.".

18N.5 Sampling artifacts

here n is any integer. We devote Chapter 18S to the subject of sampling artifacts, and we try to explain there why the artifacts appear at the specified frequencies; we will not try to justify equation (18N.1) now.

Figure 18N.32 is a sketch of the first two sets of images for a hypothetical input signal range. The "images" sit above and below f_{sample} and above and below $2 \times f_{\text{sample}}$, as if reflected about the sampling rate and its integer-multiples.

In Fig. 18N.33, scope frequency-spectrum images show a particular case, where we applied a single input frequency (rather than the more realistic *range* of frequencies sketched in Fig. 18N.32). This f_{in} is constant at 500Hz. The sampling rate is reduced progressively from top trace down, and the sampling artifacts appear in these frequency-spectra.

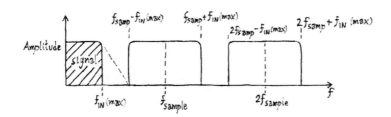

Figure 18N.32 Sampling artifacts sketched: above and below f_{sample} and its multiples.

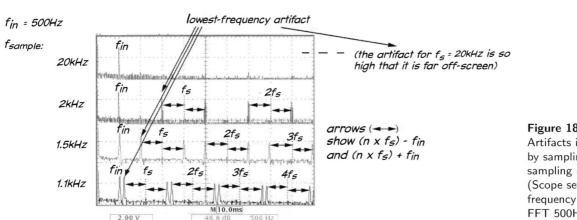

Figure 18N.33 Artifacts introduced by sampling: four sampling rates. (Scope settings: frequency spectrum FFT 500Hz/div.)

In Fig. 18N.33 we have marked the *lowest-frequency artifact*, because it turns out this is the only one that need concern us. We do need to eliminate all the other artifacts as well, as we reconstruct the signal, but a filter that knocks out the lowest image necessarily takes out the higher artifacts along with the lowest.

So equation (18N.1), predicting endless repetition of false images, is correct, but carries more information than we need. The lowest image – the one at $f_{\text{sample}} - f_{\text{in}}$ is the target. Let's say this boldly, because it is a truth more useful than the more general formula of equation (18N.1). Here is the artifact that we need to worry about – and choose a filter to eliminate:

$$f_{\text{lowest-artifact}} = f_{\text{sample}} - f_{\text{in}}. \tag{18N.2}$$

As the sampling rate gets stingier, the filtering task becomes more difficult, since the lowest artifact moves closer to the true signal, f_{in}.

The bottom case of Fig. 18N.33, for example, where the lowest image is at 600Hz and the true signal at 500Hz, presents a challenge for a filter. The other cases, at higher sampling rates, make it easy to eliminate the lowest image. The original CD audio standard presented a similar challenge (and even a little more difficult): with the sampling rate at 44.1kHz and audio up to 20kHz, the lowest image occurred at 24.1kHz – and the industry found that it could not build so steep a filter without introducing audible distortion.[37] This failure led to the introduction of the "oversampling" strategy that took over digital audio after that initial disappointment. It is also true that such digital filtering soon became cheaper than fashioning excellent analog filters.

The difficulty of filtering out the lowest artifact is sketched in Fig. 18N.34 where the lowest figure mimics the stingy sampling used in the early CD implementation.

The second generation of CD players adopted $2\times$ *oversampling* (§18N.2), achieving an apparent sampling rate of 88kHz. Later CD players pushed this oversampling higher still. These changes made the task of the analog reconstruction filter manageable.[38]

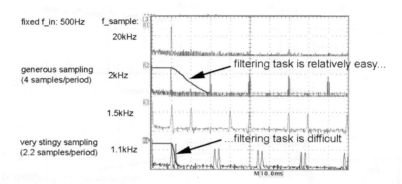

Figure 18N.34 Stingy sampling can make the reconstruction filtering difficult or even impossible. (Scope setting: frequency spectrum FFT at 500Hz/div.)

18N.5.2 Aliasing is predictable

The frequency-domain view described in §18N.5.1 allows one to predict just where *aliased* artifacts will arise as well. It turns out that frequencies too high for proper conversion will get "folded" to a lower frequency; see Fig. 18N.35.

One can see how this happens by recalling that an image forms at $f_{\text{sample}} - f_{\text{in}}$. If f_{sample} is not more than double f_{in}, then this lowest image must lie below f_{in}. Once this has occurred there is no remedy. A low-pass cannot eliminate the falsehood without also eliminating the true input. This is the effect we saw earlier in a time-domain scope image of aliasing, Fig. 18N.9.

18N.6 Dither

Here is another counter-intuitive truth: adding some random noise (analog or digital) to a signal can improve the performance of an ADC or DAC. Random noise so used is called "dither."[39] The need

---

[37] Well, *audible* to certain golden-eared audio critics, at least. See, for example, this comment on an audiophile site: "...ALL of these '14 bit' players tended to all but remove subtle ambient clues .... The sound was rather 'flat' spatially, the decay of reverb all in the speaker plane, rather than front-to-back. ...subsequent Sony players ...showed, for me anyway, a substantial improvement in reproduction of venue acoustics."
http://www.harbeth.co.uk/usergroup/showthread.php?2137-First-generation-CD-players-how-good. Certainly we are entitled to wonder whether such complaints were well founded.
[38] See Pohlmann, p. 94.
[39] The word is an archaic one meaning "...to tremble, quake, quiver, thrill. Now also in gen. colloq. use: to vacillate, to act indecisively, to waver between different opinions or courses of action" (Oxford English Dictionary). Either sense describes its function here, pretty well: it makes the input hop back and forth across a threshold.

18N.6 Dither

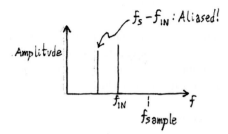

Figure 18N.35 Aliased image where f_{sample} is too low (set at $1.5 \times f_{\text{in}}$).

for this addition appears only for low-level inputs to an ADC (or outputs of a DAC). The good signal-to-noise ratio of an ADC at full-scale – almost 100dB, for a 16-bit part – shrinks toward nothing as the signal becomes small. A sinusoid whose amplitude is close to one LSB will evoke a small square wave at the reconstruction DAC, if no dither is added. With this square wave come higher-frequency harmonics, artifacts introduced by sampling.

Dither allows – amazingly enough – resolution well below the level of one LSB.[40] We look more closely at this effect in Chapter 18S, but Fig. 18N.36 gives the scope image indicating the benefits that result from adding noise to a low-level input.

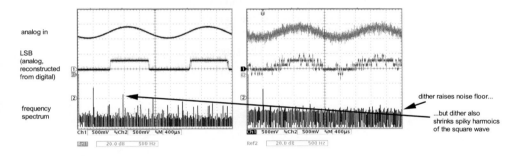

Figure 18N.36 Dither can increase converter resolution. (Scope settings: 500mV/div; frequency spectrum FFT, 500Hz/div.)

The square wave on the middle trace, left side of Fig. 18N.36, shows the converted value that results from a low-level input that moves between two levels.[41] The bottom trace shows the many (spurious) harmonics in the frequency spectrum of this square wave: strong and audible harmonics added to the original signal by the sampling process.

If one did not notice the frequency of the square wave, one might make the mistake of assuming that the reconstruction low-pass filter at the DAC output could eliminate the harmonics. But that is

---

Nika Aldrich tells a wonderful story, for the origin of dither – a story that we hope is true. British mechanical navigation computers for World War II aircraft were observed to work more accurately when airborne than on the ground. Someone hypothesized that engine vibration was freeing up elements of these machines, pushing cogs, gears, and levers back and forth across mechanical boundaries. Having confirmed this behavior, the British installed vibrator motors in their mechanical navigators: they added random noise, which came to be known as "dither." Aldrich, op. cit. p. 250.
Others tell many similar stories: devices that worked in propeller-driven B29s failed when adopted in jet bombers; gun controllers worked on battleships in Vietnam, jostled by the vibration of firing, but failed on trials in Boston harbor. All very plausible – and hard to confirm.

[40] Pohlmann, op. cit. p. 42, reproducing work of Vanderkooy and Lipshitz. The technique assumes a converter that can handle frequencies well above the frequency of the signal to which the dither is added. The dither frequency can be placed between the max signal frequency and the Nyquist frequency, $f_{\text{sample}}/2$.

[41] To get these images we used a rather large low-level step to simulate LSB. You will notice that this step is not tiny: it is about 250mV. But the magnitude does not matter for the purpose of this demonstration of dither.
The right-hand image of Fig. 18N.36, after the addition of dither, shows small sub-steps smaller than the large step that we have labeled LSB. This could not occur using normal dither and is a side-effect of our showing not the true LSB but a square wave larger than our demo DAC's LSB.

not so: the harmonics lie well within the audible range, will survive filtering, and will, indeed, sound unpleasant.

The right-hand image in Fig. 18N.36 shows the effect of adding dither: the analog input shows the thickening caused by this added noise. The frequency spectrum of the reconstructed waveform shows this "thickening" as a raised "noise floor." The LSB (a DAC output) in the middle trace now chatters above and below the midpoint level; averaged, this jaggedy waveform would produce a sinusoid close to the original.

Perhaps the raised noise floor distresses you. Probably it need not. It could be low enough to be insignificant in, say, an audio signal. But even if it were audible, this *white noise* would be much less disturbing than the strong patterned artifacts that are the harmonics of the undithered signal. Dither also can be used to randomize low-level noise other than sampling artifacts, masking effects that otherwise would be audible.[42] An excellent discussion of dither appears in Nika Aldrich's book.[43]

18N.7 Phase-locked loop

A PLL uses feedback to produce a replica or a multiple of an input *frequency*. It is a lot like an op-amp circuit; the difference is that it amplifies not the *voltage* difference between the inputs, but the frequency or *phase* difference. Once the frequencies match, as they do when the circuit is "locked," the remaining error is only a *phase* difference. The phase *error* signal is applied to a VCO[44] in a sense that tends to diminish the phase error, driving it toward zero.[45]

This sounds familiar, doesn't it: like a discussion of an op-amp circuit. The scheme is simple. The only difficulty in making a PLL work lies in designing the *loop filter*. We'll postpone that issue for a little while.

AoE §13.8.1

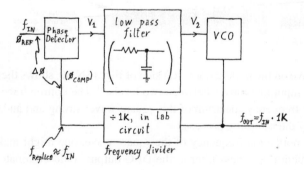
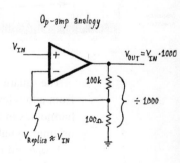

Figure 18N.37
Phase-locked loop.

18N.7.1 Phase detectors: simplest, XOR

AoE §13.8.2A

The simplest phase detector is just an XOR gate – a gate that detects *inequality* between its inputs. (We assume digital inputs; sinewave inputs require phase detectors that are different, though conceptually equivalent.)

The phase difference evokes pulses from the XOR output; averaged, these produce a DC level that

---

[42] Pohlmann, op. cit. p. 39.
[43] Aldrich, op. cit., Chapter 15. Normal dither adds 3dB to the noise floor, whereas our homebrew example in Fig. 18N.36 adds much more. But we hope our simplified version serves to illustrate the principle.
[44] You met a VCO – Voltage Controlled Oscillator – back in Chapter 13L. There you used a '555 as a rough-and-ready VCO.
[45] Some phase detectors, like the simple XOR described in §18N.7.1, cannot take the phase difference to zero; others, like the edge-sensitive detector of §18N.7.2, can take the error to zero.

18N.7 Phase-locked loop

Figure 18N.38 The simplest phase detector: an XOR gate.

drives the VCO. When the loop locks, there must be some phase difference in order to provide a non-zero feedback signal.

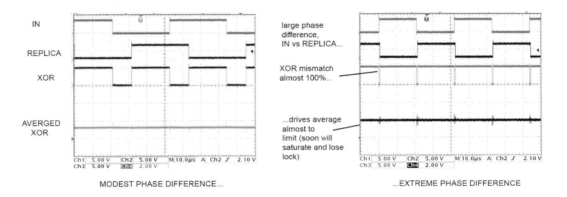

Figure 18N.39 XOR phase detector: phase difference is required to generate VCO feedback signal; phase-difference extremes can lose lock. (Scope settings: 5V/div on all except bottom trace, showing average: 2V/div.)

If the phase difference approaches either zero or 180° (as in the right-hand image of Fig. 18N.39), the VCO input signal approaches the limiting value; pushed a little further the PLL loses lock – just as an op-amp loses feedback once its output saturates. A second weakness will appear shortly in §18N.7.2: the XOR detector can let the loop lock when the replica frequency is a multiple of the input.

18N.7.2 Phase detectors: edge-sensitive

The CMOS '4046 that you will meet in the lab includes a fancier phase detector (called "Type II"). It shows two principal virtues: (1) it cannot lock onto multiples of the input frequency; (2) it locks with zero phase difference between the two waveforms. It also uses less power, by shutting off its drive to the loop filter when it is happily locked. It is inferior to the XOR detector in just one respect: a glitch – a spurious edge – disturbs it much more seriously than a glitch would disturb the XOR detector, which would simply swallow the pulse into its averaging process. It is because of this superior *noise immunity*, important in some applications, that the '4046 PLL offers, as an alternative to its clever "Type II" edge-sensitive phase detector, a simple XOR ("Type I").

The better behavior of the edge-sensitive detector: The scope images in Fig. 18N.40 contrast the responses of the two phase-detectors. The XOR response is already familiar to you.

  The left-hand image shows a case, using the XOR as phase detector, where the loop is correctly locked. A phase-difference persists, as we expect, but otherwise the loop is working properly.

  This phase difference leads the *edge-sensing* phase detector to generate pulses that would try to drive the phase difference toward zero. (They would succeed if used; they are not applied, however, in this demonstration.) These pulses arise as logic *highs* during the brief time when IN leads REPLICA. These signals, if applied to the VCO, would tell it "Speed up!," driving the phase-difference toward zero. IN and REPLICA would stabilize in-phase.[46]

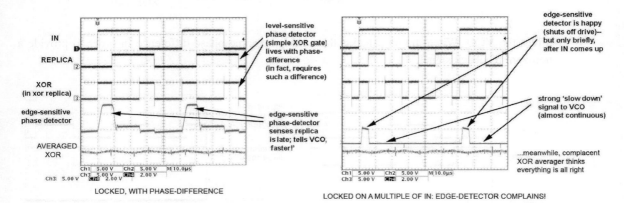

Figure 18N.40 Contrasting responses of 2 phase detectors: XOR needs phase difference, and locks on a multiple of f_{IN}. (Scope settings: 5V/div on all except bottom trace, showing average: 2V/div.)

The right-hand image shows the edge-detector's response to the case where $f_{REPLICA}$ is a multiple of f_{IN}. The XOR loop is satisfied (and locked). The edge-detector is far from satisfied; it delivers a nearly-continuous "Slow down!" signal (a logic low), low except for a short time just after the rising edge of IN (when the detector thinks everything may be OK).

How the edge-sensitive phase detector works: The edge detector in Fig. 18N.41 is sensitive to *edges* rather than levels. It implements a state machine that generates correction signals during the time when one square wave's rising edge has led the other.

When phase difference goes to zero, the edge-sensitive phase detector shuts off its 3-state. At that time, the filter capacitor just holds its voltage, acting like a sample-and-hold, really; not like a filter at all. Figure 18N.42 shows a *flow diagram* for this state machine.

The detector becomes unhappy when it finds one waveform rising before the other. The state machine notices which rising edge comes first. If A – the input signal – comes up before B, for example, the detector says to itself, "Oh, B is slow. Need to crank up the VCO a bit." So it steps into the right-hand block, where it turns on the upper transistor, squirting some charge into the filter capacitor.

Then B comes up; A and B both are up, so this detector (like the XOR) sees equality, and goes back to the middle block, saying, "Everything looks OK for the moment; A and B are behaving the same way." In the middle block, the output stage is *off* (*3-stated*); the capacitor just holds its last voltage.

This pattern tends to force the VCO up, bringing B high a little earlier. But that reduces the time the detector spends in the right-hand block. The machine spends more and more of its time in the middle block – the "Everything's all right" block. This is where it lives when the loop is locked. You will see this on the '4046, when you watch an LED driven by a '4046 pin that means "The output is 3-stated." That is equivalent to "I don't see much phase difference," or simply, "The loop is locked."

---

[46] When the edge-detector is "happy," not trying to change the VCO frequency, its output is neither High nor Low, but Off (3-stated). AoE points out that this absence of correction signal can cause "hunting," an error corrected in improved versions of the edge-sensitive detector. See AoE §13.8.2A, especially Fig. 13.89.
In the scope image of Fig. 18N.40 the edge-detector output looks funny: stepping downward rather than sitting at a constant level. This is an artifact caused by loading imposed by the test circuit, and is not significant.

18N.7 Phase-locked loop

Figure 18N.43 is the simple circuit that behaves this way.

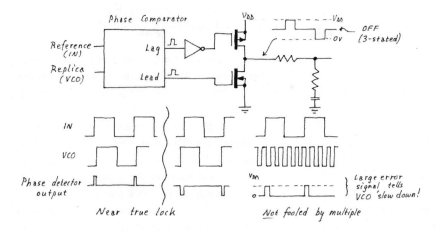

Figure 18N.41 '4046 edge-sensitive phase detector; it's smarter than XOR, and cannot be fooled.

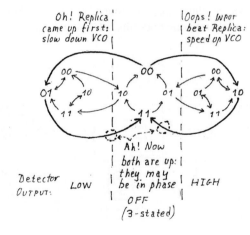

Figure 18N.42 '4046 state diagram (after National Semiconductor data).

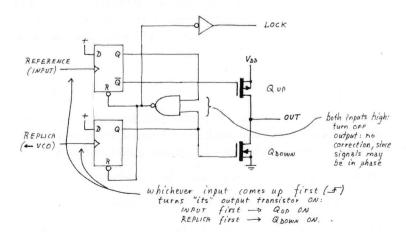

Figure 18N.43 '4046 phase detector: just two flip-flops; the winner of the race turns on "its" output transistor till both inputs go high.

18N.7.3 Applications

The PLL lets one generate a signal that is a precise multiple of an input frequency. Fig. 18N.37 showed a loop used to generate $f_{out} = 2^{10} \times f_{in}$. All that's required is a divide-by-1K circuit placed within the feedback loop.[47]

This trick is often used within integrated circuits to drive internal circuitry at a clock rate higher than the off-chip clock frequency. For example, Analog Devices' ADuC848 multiplies the modest oscillator frequency – a standard "watch" frequency of 32.768kHz[48] – by 384 to get a fast on-chip clock rate of more than 12MHz (cited in AoE §15.10).

You can use this tactic in a late big-board lab (23L), where we need a clock that is a $64\times$ multiple of the *sampling rate* in order to drive a switched-capacitor output filter. The goal is let the PLL automatically adjust filter rolloff to fit a sampling rate that we can vary. The filter's f_{3dB} is proportional to its clock rate ($f_{3dB} = f_{clock}/100$). So using the PLL to make the filter clock track the sampling rate allows us to vary the sampling rate freely while the filter automatically adjusts its clock rate as needed.

Figure 18N.44 sketches the scheme. Here are some details of the circuit (drawn more exactly in Lab 18L): the signal labeled "sample" is a narrow pulse, so it is put through a divide-by-two flop to give a 50%-duty-cycle square wave to the PLL (the PLL performs poorly when given a narrow spike as input). The PLL multiplies this by 64, driving the switched-capacitor filter at $32 \times f_{sample}$. The filter's f_{3dB} is $1/100 \times f_{clock}$, so this arrangement puts the filter's f_{3dB} at about 1/3 of f_{sample}. This permits inputs up to about 2/3 of the theoretical maximum that Nyquist permits – 1/2 of f_{sample}.[49]

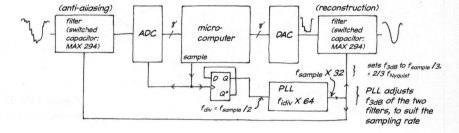

Figure 18N.44 PLL in Lab 23L adjusts filter automatically as sampling rate varies.

The PLL also can demodulate an FM signal: see Fig. 18N.45 – just watch the input to the VCO (which is the filtered *error* signal, describing variations in input frequency). This exercise is proposed in §18L.2. We could have used a PLL to demodulate the low-frequency FM we used in the group audio project in Chapter 13L – if we had known about PLLs back then.

18N.7.4 Stability issues much like PID's

AoE §13.8.3A

The PLL is vulnerable to oscillations because of an implicit 90° shift within the loop, an *integration* imposed by the phase-detector. An additional simple low-pass filter could impose a total shift of 180° at some frequency, and such a shift could produce oscillations – an endless, restless hunting for *lock*.

---

[47] In the silly terms of Chapter 6N, this div-by-1K is a dog that induces the feedback circuit to generate "inverse dog," a 1K× multiplication.

[48] If you're wondering why such a strange number, the answer is that this is 2^{15}Hz, so a watch can derive a 1Hz "tick" just by putting the crystal frequency through a 15-bit binary ripple counter.

[49] Perhaps we should admit that most students do *not* get around to trying the PLL in Lab 23L, because they are busy with other issues. Still, it's a rather neat application of the PLL so we hope some of you will try it out.

18N.7 Phase-locked loop

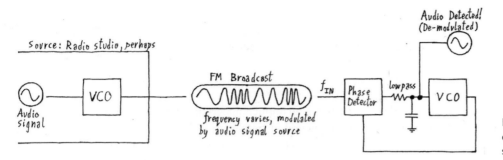

Figure 18N.45 PLL demodulating an FM signal.

This is essentially the same as the hazard you recall from our earlier discussions of op-amp stability and compensation. It is also much like the problem of the motor-control PID lab (10L), which placed an integrator within the feedback loop.

The second resistor in the low-pass is designed to eliminate this hazard by pushing the low-pass phase shift toward zero as frequency rises: see Fig. 18N.46.

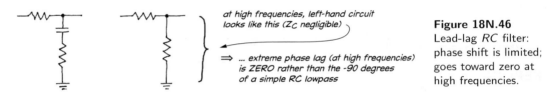

Figure 18N.46 Lead-lag *RC* filter: phase shift is limited; goes toward zero at high frequencies.

Figure 18N.47 shows, on the left, the rollofff of the lead-lag filter: it falls like $1/f$, as an ordinary low-pass would – but then levels off. The effect of this levelling-off upon the PLL is sketched on the right. Rolloff proceeds for a while at a scary -12dB/octave (like $1/f^2$) – the effect of the implicit integration of the PLL combined with the $-90°$ of the filter. But then the whole loop's gain returns to a safe -6dB/octave (like $1/f$).

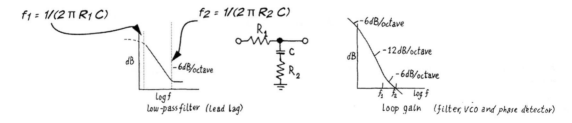

Figure 18N.47 Lead-lag low-pass filter keeps PLL stable by limiting phase lag.

Figure 18N.48 describes the same filter but showing only phase shifts: first (on the left) in the filter alone, and also (on the right) in the PLL loop that includes this filter.

You'll notice that the PLL loop gain goes to the $-180°$ shift that we fear – but makes it back to $-90°$ before closing time, and that's good enough (surprisingly).[50]

---

[50] You may recall we remarked on the marvel that such a phase lag is permissible back in Lab 10L. In fancy language, this sort of filter makes a "pole-zero network."

Figure 18N.48 Phase shift in lead-lag and entire PLL heads toward trouble – and then backs off.

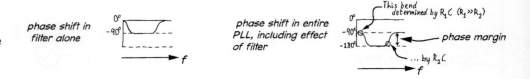

The phase-detector's integration: The phase-detector performs an implicit integration, because phase (the quantity it measures and applies within the loop) is an integral of frequency. This is an idea that is easy to recite but perhaps harder to get a grip on. A look at units may help to cement the notion that phase is the integral of frequency: if frequency is measured in radians/second, integrating this with respect to time gives radians, a measure of phase.[51]

And in Fig. 18N.49 is a scope image that may help to make this idea less abstract.[52] The images shows the output of the phase-detector[53] as IN and REPLICA run at slightly different frequencies. One signal gradually slides past the other – and the *phase* difference signal shows the ramping we expect an integrator to show when fed a constant difference. This image may convince you that the *integration* occurs – and thus that we face a stability hazard reminiscent of what we met in the PID lab, 10L.

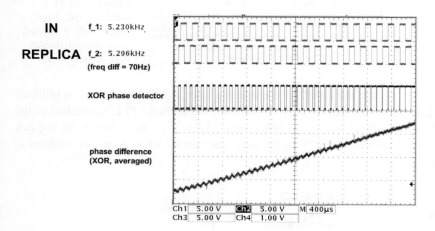

Figure 18N.49 Evidence that phase detector integrates frequency-difference: ramp for constant difference.

---

[51] It may also help to think of a similar claim that phase-difference is integral of frequency-difference. Imagine that the PLL's f_{in} and f_{replica} differ by some small amount, and this difference remains constant for a while. A scope image of the two signals – a time-domain image – would show one slowly creeping past the other. A phase-difference detector would measure this as a ramping difference (until the difference hit a maximum and rolled over to zero, then started its upward ramp again...). This is the effect illustrated in Fig. 18N.49.

[52] An alert student objected to this argument, saying that the integration appeared to be achieved by the *RC*, not by the phase detector. This objection, though plausible, is not correct. It is true that the *ramping* in Fig. 18N.49 is visible as an analog waveform only because of the *RC* that converts the *duty cycle* output of the XOR phase detector to a voltage. But the *integration* carried out by the phase detector is complete before this conversion occurs. The XOR output expresses as *duty cycle* the phase difference between its inputs. This phase difference is an integral of the frequency difference fed to the detector.

[53] This is the *RC*-averaged output of the simple XOR detector.

18N.8 AoE Reading

- Chapter 12: Logic Interfacing
  - §12.1.1: Logic family chronology – a brief history
  - §12.1.2: Input and output characteristics
  - §12.1.3: Interfacing between logic families
- Chapter 13: Digital Meets Analog
  - DAC and ADC
    - §13.2: Digital-to-analog converters (DACs)
    - §13.5: Analog-to-digital converters (ADCs)
    - §13.5.2: successive approximation (important to us: this is Lab 18L's method)
    - §13.5.10: Delta–sigma converters (a difficult topic, bravely explained here)
    - Review of Chapter 13: a good, compact summary of the many ADC and DAC methods.
  - Phase-locked Loops
    - §13.13: Phase-locked Loops
    - §13.13.2A: The phase detector
    - §13.13.4A: Stability and phase shifts
    - §13.13.6C: FM detection
    - §13.14: take just a quick look at §13.14.2: Feedback shift register sequences. (We used this method to generate the dither demonstrated in this chapter).

18L Lab: Analog ↔ Digital; PLL

This lab presents two devices, both partially digital, that have in common the use of feedback to generate an output related in a useful way to an input signal. The first circuit, an analog-to-digital converter, uses feedback to generate the digital equivalent to an analog input *voltage*. The second circuit, a phase-locked loop (PLL), uses feedback to generate a signal matched in *frequency* to the input signal – or to some multiple of that frequency.

18L.1 Analog-to-digital converter

The A/D conversion method used in this lab, "successive approximation," or "binary search," is probably the method still most widely used, though now competing with delta–sigma designs. It provides a good compromise between speed and low cost. It substitutes some cleverness for the brute force (that is, large amounts of analog circuitry) used in the fastest method, *flash* or parallel conversion.[1]

Note that the converter you will build today with four chips normally would be fabricated on a single chip. We build it up using an essentially obsolete *successive-approximation register* (SAR), so that you will be able to watch the conversion process. In an integrated converter the approximation process is harder to observe because the successive analog *estimates* are not brought out to any pin, though the stream of bits that forms that estimate may be. We also omit, for simplicity, the *sample-and-hold* that normally is included.

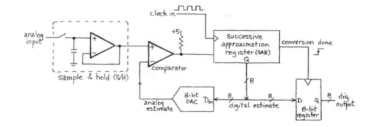

Figure 18L.1
Successive-approximation A/D converter: block diagram.

18L.1.1 D/A converter

The process of converting digital to analog is easier and less interesting than converting in the other direction. In the successive-approximation ADC, as in any closed-loop ADC method, a DAC is necessary to complete the feedback loop: it provides the analog translation of the digital *estimate*, allowing correction and improvement of that digital estimate. As a first step in construction of the ADC we will wire up a DAC, the Analog Devices AD558 of Fig. 18L.2.

---

[1] We admit that it is an unusual brute who could put together a flash ADC.

18L.1 Analog-to-digital converter

This DAC integrates on one chip not only the DAC itself but also an output amplifier and an input latch. The latch is of the *transparent* rather than edge-triggered type, and we will ignore it in this lab, holding the latch in its transparent mode throughout (grounding pins 9 and 10, CS* and CE*, keeps the latch continuously transparent).

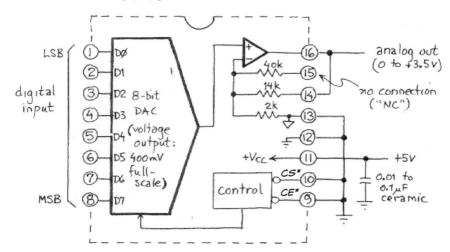

Figure 18L.2 AD558 DAC.

Checkout: (Hurry through this checkout. The fun comes later!)

Confirm that the 558 is working by controlling its two MSBs and its LSB with a DIP switch or simply with wires plugged into ground or +5. Hold the other five input lines low. (You may find a ribbon cable convenient, though less tidy than a wire-at-a-time; in any case, you will need to feed all eight lines in a few minutes.)

Note the relation between switch settings and V_{out}. Full-scale V_{out} range should be about 0–3.8V. What "weight" (in output voltage swing) should D7 carry? D6?[2]

Note the weight of the LSB. Here, if you look at the output with a scope you are likely to find noise of an amplitude comparable to that of the signal. If the noise is at a very high frequency, however, it may be quite harmless. How does your DVM appear to treat this noise? How do you expect the comparator to treat it when you insert the DAC into the ADC loop?[3]

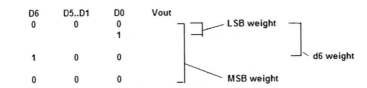

Figure 18L.3 DAC checkout: voltage weights for particular input bits.

18L.1.2 ADC: watching the conversion process

When you are satisfied that the DAC is working, add the rest of the successive-approximation converter circuit: comparator and SAR along with DAC. If you are using a 74LS503 rather than 74LS502,

---

[2] D7 should carry the weight of one half full scale; D6, one quarter.
[3] We mean to suggest that very high-frequency noise – radio frequency that one often sees when scope gain is very high – may be too quick to be sensed by this ADC circuit, just as it is too quick to make the DVM respond. Large high-frequency signals, however, can be accidentally rectified by a DVM, and can produce spurious results, as you may possibly have seen back in Lab 9L when the discrete follower sometimes can make a DVM claim to see a voltage beyond the positive supply.

you must *ground pin 1*, an ENABLE* pin; we do not use pin 1 in the '502 circuit,[4] and can leave it open if using a '502.

If you have neither '503 nor '502 but are using a custom-programmed PAL that emulates the '502 please note that its pinout will not match that of the other two chips. Consult its own datasheet. Connect the eight DAC inputs to the eight breadboard LEDs so that you can watch the estimating process.

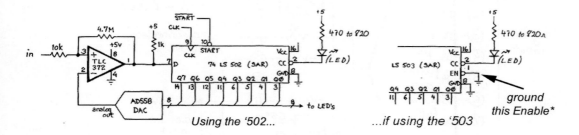

Figure 18L.4 Successive-approximation A/D converter: slow-motion checkout: layout.

Slow motion checkout: Use a debounced switch for Clock and another switch (which may bounce) for Start* (S*). Watch Conversion Complete* (CC*) on an LED.

- You will have to limit the current to this ninth LED using a resistor. We chose the resistor value considering the following points: an LS-TTL output can sink 8mA, plenty for a high-efficiency LED. The LED drops about 2V when lit.

Note: the behavior of S* may strike you as odd:

- First, the SAR is "fully synchronous:" asserting S* by itself has no effect. The SAR ignores S* unless you clock the SAR while asserting S*.
- Second, "Start*" is poorly named. It should be called something like "initialize*," because conversion will not proceed beyond the initial guess until you *release* S*.

Ground the analog input[5] and walk the device through a conversion cycle in slow motion. Watch the *digital* estimates on the eight LEDs and the analog equivalents, the *analog* estimates, on DVM or scope.

As you clock slowly through a conversion cycle, the DAC output (showing the analog equivalent of the SAR's digital estimates) should home in on the correct answer, 0V. Do you recognize the *binary search* pattern in these successive estimates? If you get a digital *1* rather than *0* as your converter's final, best estimate – that is, if the LSB ends up HIGH rather than LOW – make sure that you have grounded the input close to the converter. Then use the scope to look closely at ground and +5V lines, watching for noise. A digital ...*01* outcome need not shock you, given that an LSB is only about 15mV, and the comparator's V_{OFFSET} is spec'd at 5mV (max); a small drop in the ground line added to this V_{OS} could tip the LSB.

---

[4] On the '502 it serves as a synchronized version of the D input; not useful for our purposes.
[5] Note that the "analog input" is not the non-inverting terminal of the comparator, but one end of the 10k resistor; that resistor is necessary to maintain hysteresis.

18L.1 Analog-to-digital converter

Operation at normal speed: Now make three changes.

1. Connect "Conversion Complete*" (CC*) to "Start*" (S*). This lets the converter tell itself to start a new conversion cycle as soon as it finishes carrying out a conversion. (Disconnect the pushbutton that was driving S*, of course.)

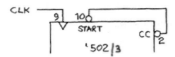

Figure 18L.5 CC* wired to drive start*: SAR starts self.

2. Feed the converter from a pot (2.5k or less: (Why?[6]) rather than from ground: see Fig 18L.6.

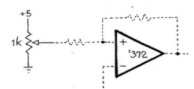

Figure 18L.6 Variable DC analog input.

3. Clock the converter with a TTL-output oscillator, rather than with the pushbutton. (The oscillator built into your breadboard is convenient.) Let $f_{\rm clock} \approx 100$kHz.

Watch the DAC output and ADC input on the scope. (If you want a rock-steady picture, trigger the scope on CC*.) Vary the pot setting (analog input voltage), and confirm that the converter homes in on the input value.

Displaying full search "tree": You have watched the converter put together its best estimate, homing

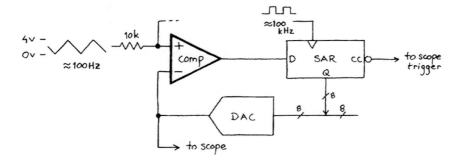

Figure 18L.7 Displaying binary search tree.

in on the input value. If you feed the converter *all possible* input values, you can get a pleasing display as in Fig. 18L.7 that shows the converter trying out every branch of its estimate "tree." Feed the converter an analog input signal from an external function generator: a triangle wave spanning the converter's full input range (0 to about 4V). Set the frequency of the triangle wave at about 100Hz. Trigger the scope on CC*.

If necessary, tinker with the frequency and amplitude of the input waveform, until you achieve a display of the entire binary search tree. This can be a lovely display (you may even notice the "quaking aspen" effect). You are privileged to see the binary search in such vivid form – while it remains to most other people only an abstraction of computer science. (An exceptionally noise-free image of the search tree produced by this lab circuit appears as AoE's Fig. 13.31.)

---

[6] You will recognize here our usual concern that $R_{\rm source}$ does not mess things up. Here, an over-large pot would mess things up by causing a significant increase in hysteresis.

18L.1.3 Speed limit

The ADC completes an 8-bit conversion in nine clock cycles. Evidently, the faster you can clock it the faster it can convert. The faster it can convert, the higher the frequency of the input waveform that you can capture.

How fast can you clock your converter? Fig. 18L.8 shows what must happen between clock edges. These numbers suggest a maximum clock frequency of a little less than 600kHz.

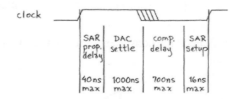

Figure 18L.8 ADC speed limit: what must be accomplished within one clock period.

Feed the converter a DC level and gradually crank up the clock rate as you watch the analog estimates on the scope. This time use the *external* function generator as clock source (the breadboard's top frequency, 100kHz, is too low). At some clock frequency you will recognize breakdown: the final estimate will change because the clock period will no longer be allowing time for all levels to settle. Chances are, this will happen at a frequency well above the worst-case value of 570kHz (above 1MHz, in most cases we have seen.)

18L.1.4 Completing the ADC: latching the digital output

Up to this point we have been looking at the converter's feedback DAC output. Do not let our attention to this *analog* signal distract you from the perhaps obvious fact that the feedback–DAC output is not the *converter's* output. We use an ADC in order to get a *digital* output, of course, and on a practical IC ADC, as we observed at the start of this chapter, the analog estimate is not even brought out to any pin.

We now return our attention to the normal subject of interest: the ADC's *digital* output.

Output register: An integrated ADC normally includes a register to save its output (and incidentally, in the age of microcomputers, such a register routinely includes 3-state outputs, for easy connection to a computer's data bus).

Now we will add an 8-bit register of D flip-flops to complete the ADC. We need to provide a clock pulse, properly timed, that will catch the converter's best estimate and hold it till the next one is ready. Timing concerns make this task more delicate than it looks at first glance.

Conversion Complete* certainly *sounds* like the right signal. It turns out that it is not – not quite. The trailing edge (rise) of CC* comes too late; the other edge (which, inverted, could provide a rising edge) comes too early.

At the beginning (fall) of CC*, the SAR is putting out its initial estimate for the LSB; it has not yet corrected it (set it high) if such correction is necessary. Thus you would lose the LSB data, getting a constant Low, if you somehow used the start of CC* to latch the output.

At the end (rise) of CC*, the SAR is already presenting the first guess of its next cycle (0111 1111).[7]

What we need is a pulse that ends well away from both edges of CC*: see Fig. 18L.10. A single gate can do the job.

---

[7] Why? Because the rise of CC* and the initial guess both come in response to the SAR clock, and CC* happens to come up a little *later* (see 74LS502 data sheet: t_{PLH} is slower than t_{PHL}. How's that for fine print?)

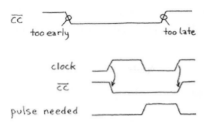

Figure 18L.9 CC* timing.

Figure 18L.10 Output register clock needed.

Add this gate and feed its output to the clock of the output register (74HC574). Let the register's outputs drive the breadboard's eight LEDs. (Don't forget power and ground, not shown in Fig. 18L.11. They are the corner pins, as usual for digital devices.)

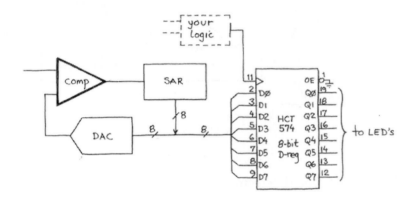

Figure 18L.11 Output register added to SAR ADC (output-register clocking left open for your addition).

Watch the converter's digital output and confirm that it follows the analog input applied from the potentiometer. It may be indecisive, by 1 bit. Is this indecision avoidable?[8]

18L.2 Phase-locked loop: frequency multiplier

AoE §13.8.4

You should build the PLL on your private breadboard – either the micro board or on a single strip. You may want to use the one that holds a mike amplifier built back in Lab 7L.

You will have a chance to apply this PLL, if you like, in Lab 23L. In that lab we will need to generate a multiple of the frequency at which the computer samples a waveform (we want a square wave at 32× the sampling rate); we will use that multiple to regulate $f_{3\mathrm{dB}}$ of a low-pass filter. This adjustable filter is of the switched-capacitor type, like the one we built in Lab 12L.

Thus we will make the filter follow our sampling rate (to clear away the spurious high-frequency elements in the DAC's *steppy* output waveform). This future application for the PLL explains some elements of its design: particularly its frequency range, set to cover most of the range of sampling rates you're likely to adopt.

We will first apply the loop, however, as if we were using it to generate a multiple of the line frequency, 60Hz. This example is discussed in detail in AoE §13.8.4, and the circuit in Fig. 18L.12 is the one designed in that discussion, except that we have altered the VCO component values to permit a wider range of operation.

---

[8] We have nothing subtle in mind; we're just referring, as in §18L.1.1, to the likelihood that noise may be comparable to the value of an LSB, in your breadboarded circuit.

Construct the circuit shown in the figure. The phase detector and VCO are drawn as separate blocks; but note that they are *within one 4046 chip*: you do *not* need two 4046s. The 4040 is a 12-stage ripple counter.[9]

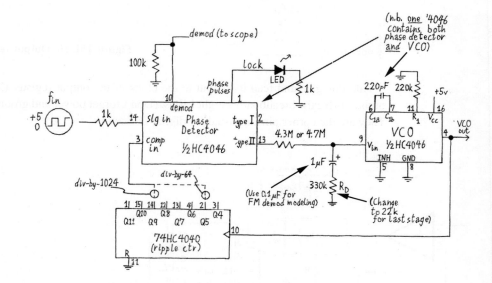

Figure 18L.12
Phase-locked loop frequency multiplier circuit (using 74HC4046 PLL).

18L.2.1 Generating a multiple of line frequency: (Type II detector)

Take the *replica* signal from Q9 of the 4040 (pin 14) (Q9 divides f_{clock} by 2^{10}. In case this seems odd, recall that Q0 divides by 2, not 1.) Set the function generator, which drives the input, to around 60Hz. Use the scope to compare this *input* signal with the synthesized *replica* at 4046 pin 3. Confirm that the PLL *locks* onto the input frequency within a few seconds. Are the two waveforms in phase? The *lock* LED should light when the loop is locked (a logic *high* at pin 1 indicates that the phase detector output is *3-stated*: that is, the detector is satisfied, rarely seeing reason to correct the *replica* frequency).

See if the replica follows the input as you change frequency slowly; then try teasing the PLL by changing frequency abruptly. You should be able to see a brief *hunting* process before the loop locks again.

The limit of input frequency *capture* and *lock* ranges (identical, for this PLL) is determined by the VCO range: from about 40kHz to 250kHz for the specified values of R_1, C_1: 220k, 220pF. Since the counter in the feedback path divides by 2^{10}, the input frequency range runs from approximately 40Hz to 250Hz.

The loop filter: The hard part of PLL design comes in designing the low-pass filter to maintain stability. The PLL uses a feedback loop strikingly reminiscent of the loop in the PID motor-control lab, 10L: in the PID exercise, the motor-to-position transducer placed an integration inside the loop. This 90° lag obliged us to use care to avoid injecting another 90° lag at frequencies where loop gain was above unity.

---

[9] The pinout labeling shown follows the usual convention, in which the low-order Q is labeled Q0 on the part, but not shown in the figure. NXP/Philips and others use this convention. National Semiconductor and Fairchild oddly started their Q numbering at *1* (CD74HC4040, on TI site). If you consult a National data sheet you may run into this discrepancy. TI dodges the labeling question by using *letters* rather than numbers: Q*A*, Q*B*…(SN74HC4040).

In the PLL, again an integration lies within the loop, and again it threatens us with instability. This time the integration results from the fact that we are sensing a characteristic (phase) that is an integral of the characteristic that our loop means to control (frequency). Here, the integration is harder to localize than in the PID loop where it clearly occurred in the motor-potentiometer unit. In the PLL the integration – the equivalent of the motor-pot combination – is performed by the VCO–phase-detector pair. The detector's output is a signal that measures phase error.[10] For a constant frequency difference, the *phase error* (difference between input signal and the replica generated by the VCO) *ramps* (see Fig. 18N.49); in other words, it integrates the frequency error. You will find a scope image that demonstrates this ramping effect in Chapter 18N. This behavior closely resembles what we saw in the motor-pot combination: a constant error signal fed to the motor-pot produced a ramping output voltage: integration, once more.[11]

We need a lowpass filter in the loop to smooth the signal into the VCO – but we cannot afford another 90° lag. What is to be done? The answer turns out to be a simple circuit amendment: use a low-pass whose phase shift at high frequencies goes toward *zero* rather than toward $-90°$. A low-pass with an extra resistor between cap and ground does the trick, as we argue in §18N.7.4.

If you replace the filter's 330k resistor with 22k, you should find that the loop hunts much longer – overshooting, backtracking.... Watching the *demod* signal at pin 10, which shows the VCO input voltage, may make the hunting and overshooting vivid for you.

By reducing that R value, you have made a dangerous reduction in the *phase margin* – the safety margin between the phase shift around the loop (at the frequency where the loop gain goes to unity) and the edge of the cliff: the deadly $-180°$ shift. This is a notion you will recall from Lab 9L and from the PID exercise; see, especially Fig. 10N.9.

If you want to live really dangerously, short out the lower resistor; now you have reduced the phase margin to zero, and the circuit may hunt forever. When you have seen this effect, restore the 330k, to restore stability.

What frequency should be present at the VCO output when the loop is locked? Check your prediction by looking with the scope at that point (pin 4 of the 4046). Why is this waveform jittery? (In a world without noise it would *not* be.)[12]

Now look at the output of the phase detector (pin 13 of the 4046). This is the Type II detector, edge-sensitive. You'll notice a string of brief positive going pulses, each decaying away exponentially when the detector reverts to its 3-stated condition (you may also see smaller negative-going pulses to the extent that the circuit is troubled by noise).

Theory predicts that these correction pulses should vanish in the steady state. But the 10MΩ load of the scope probe you are using is discharging the filter capacitor fast enough to cause the pulses; the current the scope draws also causes the VCO to be slightly under-driven, causing the VCO to run a little late; a lagging phase difference persists: the loop now *needs* a phase difference to provide the

---

[10] You may be inclined to ask, "Why do this? Why not be more straightforward and measure frequency error directly?" The answer is that we have no way to do that quickly. To measure frequency even indirectly, inferring it from *period*, requires waiting for a full period of the two waveforms. Phase error information is available more promptly: after each of the respective rising edges in the Type II detector.

[11] If you're not exhausted by this discussion, you may want to consider a complication: a *second* integration performed by the RC, at least as used in the edge-sensitive "Type II" detector: a constant phase difference produces a ramping output from the phase detector. This second integration is analogous to the I used in the PID circuit. The I term drives the PID error to zero; the Type II detector, similarly, uses this integration to drive the phase difference to zero.

[12] The jitter arises because the corrections to VCO frequency come only once in every 1024 cycles of the VCO output. Slight wanderings of frequency between these checks go uncorrected. As usual, being college teachers, we are reminded of the fact – probably beneficial to humans, though unsettling in the PLL analogy – that the college checks whether students can do the course work only now and then: at best, weekly; the big test comes only once per term. In the meantime, who knows what goes on in the student's life? No doubt that life would be full of jitter, if one were so rude as to look at it between checkpoints.

positive pulses that feed the scope's R to ground. If you're feeling energetic, interpose a '358 op-amp follower between the capacitor and the probe. The positive pulses should become narrower – and now the phase difference is reversed: the VCO output comes a little *early*.[13]

18L.2.2 Try FM demodulation (in slow motion)

If you watch the input to the VCO (which is available in buffered form at pin 10), you can see a measure of VCO frequency. That's not very interesting when the input frequency is constant – but becomes interesting when a variation in input frequency carries information, as it does in an FM radio broadcast (or in our analog project, where we sent music across the room with an LED's flashes). We aren't going to pause long enough to do such an exercise, here; but we suggest that you get a glimpse of the way it might work trying a two-minute demonstration.

Replace the 1μF capacitor, temporarily, with a 0.1μF part (to make the loop adjust faster), and watch the *demod* signal (pin 10) as you try to vary f_{IN} sinusoidally by slowly varying the function generator frequency. You should sweep the scope very slowly – using "roll" mode (perhaps 1s/div) if you have a digital scope. We hope the demod signal will look like the sinusoid that varies or "modulates" the PLL's input frequency. Too bad we didn't know about PLLs when we did the group audio-LED project. When you've had enough fun with this, restore the 1μF capacitor.

Type I detector (exclusive-OR): The 74HC4046 includes three phase detectors. AoE describes two of them (Types I and II) in §13.13.2A.[14] The third detector on the HC4046, Type III, is simply a variation on the SR latch, with S and R driven by positive *levels* on the input and replica lines, respectively. We will *ignore* the Type III detector:[15] *two* detectors seem plenty to consider on a first encounter with a PLL. But feel free to check out the Type III if you like: its output appears at pin 15.

The Type I detector output is at pin 2, and the inputs are the same as for the Type II detector; to use the Type I, simply move the wire from pin 13 to pin 2 (and then to 15 for the Type III, if you *must!*). For both Type I and Type III you should be able to see the fluctuation of the VCO frequency over the period of the input, which you can exaggerate by reducing the size of the 1μF loop filter capacitor.

If you make a sudden, large change in the input frequency, you should be able to fool the Type I circuit into locking onto a *harmonic* of the input frequency (a multiple of the input frequency). For our purpose in §23N.5.2 such an error would make the circuit useless, so we will use the Type II detector. You are likely to make the same choice in most applications.

Note also the phase difference that persists between input and feedback signal in the locked state. This simple phase detector (like the Type III) *requires* such a phase difference; this difference generates the signal that drives the VCO. If the phase difference ever goes to zero (or to π), the loop loses feedback: it can no longer correct frequency in both senses as required. When it hits that limit it's like an op-amp feedback loop that fails because the op-amp output hits saturation and thus cannot make a further correction. The Type II detector is altogether classier: it requires no errors to keep the loop locked; it is able to use the capacitor as a *sample-and-hold*, once the loop is locked, rather than as a conventional filter.

When you have finished looking at the behavior (and *mis*behavior) of the Type I detector, revert to the earlier circuit, using the Type II (output at pin 13).

---

[13] Why? It's a fussy detail, but now, instead of discharging as the scope probe did, the '358's I_{BIAS} *injects* some current from its PNP input transistor's base.

[14] AoE's discussion treats the similar CD4046 rather than the 74HC version.

[15] The Type III detector seems to have been added as a sort of afterthought; early versions of the 4046 did not offer it. Like the simple XOR (Type I) it requires a phase difference between input and replica signal.

18L.2.3 Expanded lock range: ×64 rather than ×1k

Now let's set up the PLL as you will want to find it the next time you use it in Lab 23L: change the *tap* on the 4040 from Q9 to Q5 (pin 2). Now you are generating, at the VCO output, a modest $64 \times f_{in}$. Because we are feeding back a larger part of the output frequency, we need to attenuate more in the loop filter. (We are worrying about the *loop gain* as we did for op-amps.)

Feeding back 1 part in 64, rather than 1 in 1024 – about $16\times$ as much – calls for reducing the fraction preserved by the loop filter proportionately. So we reduc the 330k that was below the $1\mu F$ cap by a factor of 16, to about 22k. The two sets of dividers – analog filter and digital counter – are sketched in Fig. 18L.13 for the two stages: stage one, where the loop multiplies by 1024; and stage two, where the loop multiplies by 64. The fraction fed back is held roughly constant (to about 12%) in the two cases.

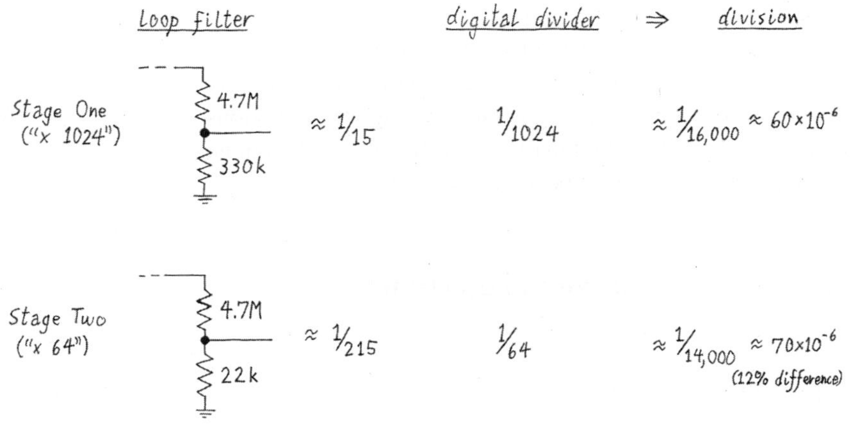

	Loop filter		digital divider	⇒	division
Stage One ("× 1024")	4.7M / 330k	≈ 1/15	1/1024		≈ 1/16,000 ≈ 60×10^{-6}
Stage Two ("× 64")	4.7M / 22k	≈ 1/215	1/64		≈ 1/14,000 ≈ 70×10^{-6} (12% difference)

Figure 18L.13 PLL filter divider is adjusted to hold fraction fed back roughly constant in the two stages.

You may notice that we have drawn the "loop filter" as if it were simply a resistive divider. We have done that because at the high frequencies where we anticipate challenges to stability X_C is insignificant relative to the R values.

Over what range of input frequencies does the PLL now remain locked? The range should be wide; we need this range in order to make the sampling scheme of Lab 23L flexible. The '4046 is capable of *capture* and *lock* over a frequency range of about 6:1.[16] We will be content with a range that accepts an input between about 600Hz and 4kHz. Capture and lock range are the *same*, for the Type II detector; for the less clever detectors, capture range (the range over which the loop will be able to *achieve lock*) is narrower than lock range (the range over which the loop will *hang on* once it has locked). This ability of the Type II to capture any frequency it can hold seems to follow from its immunity to harmonics: you can't fool the Type II.

---

[16] This is the approximate range as one drives the VCO over its permitted range from 1.1–3.4V (the data sheet specifies this range with a 4.5V supply; at 5V each voltage is presumably about 10% higher).

18S Supplementary Notes: Sampling Rules; Sampling Artifacts

18S.1 What's in this chapter?

Here are reminders of Nyquist's sampling rule and some consequences, with scope images of artifacts that sampling can introduce.

The images in §18S.3, particularly Figs. 18S.3, adequate, and filtered, show four sampling rates applied to a 1kHz sinusoid – sampling rates that are *lavish, adequate, stingy*, and *inadequate*.[1] But before we look at those images, let's recall the general notions that govern sampling. These notions that will help explain why the frequency spectra look as they do. Finally, in §18S.4, we'll attempt an intuitive explanation of the effects of sampling, particularly, the appearance of images.

18S.2 General notion: sampling produces predictable artifacts in the sampled data

Here, for a start, are some truths about sampling in a tiny nutshell:

- Sampling produces spurious images of the input at $f_{\text{sample}} \pm f_{in}$ and these images repeat, appearing in pairs above and below each integral multiple of f_{sample}: in general, $f_{\text{image}} = (n \cdot f_{\text{sample}}) \pm f_{in}$.
- One result of the appearance of these images is expressed as Nyquist's sampling rule: one must sample at a rate slightly more than double the maximum input frequency.[2]

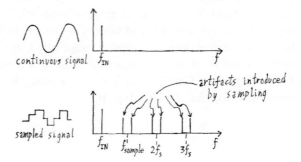

Figure 18S.1 Sampling creates multiple images of the true spectrum at higher frequencies.

It is not obvious why these spurious images appear. Toward the end of this chapter we give an argument that we hope will appeal to your intuition; but you may prefer to take it as a given truth.

---

[1] To conserve space here, we omit Sleepy and Grumpy.
[2] This follows from the preceding comment about images because the first (false) image appears at $f_{\text{sample}} - f_{in}$; to make sure that this image does not overlap with f, one must put f_{sample} above $2 \times f_{in}$. More on this later.

Of all these spurious images, only one is troublesome: the image that is lowest in frequency at $f_{sample} - f_{in}$. In principle, a low-pass filter can remove this and all other false images – as long as these images lie above the true signal (that is, as long as we have sampled adequately: more on that below). Sometimes, however, this filtering is hard to achieve: it is hard when the spurious image lies close to the signal frequency, as we'll see in one example below.

18S.2.1 Aliasing

Violation of Nyquist's sampling rule – sampling too infrequently – brings on a disastrous error named "aliasing:" this is the generation of artifacts that cannot be disentangled from the true signals.

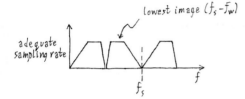 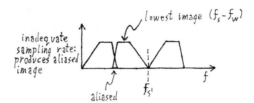

Figure 18S.2 Schematic representation of the problem of aliasing.

We're in trouble if the image lowest in frequency, at $f_{sample} - f_{in}$, falls into the frequency range of our signal, the analog input. Once this happens, an output filter cannot save us. The left-hand image in Fig. 18S.2 shows a sampling rate that is just adequate: the lowest part of the lowest image lies slightly above the input signal. So, a good low-pass filter should be able to eliminate the image.

The right-hand image in Fig. 18S.2 shows the mishap called aliasing. It occurs when one does not sample fast enough to satisfy Nyquist: that is, when f_{sample} is not more than double the maximum input frequency. Sample at a rate $>2 \times f_{in(max)}$ *or else*! Or else you'll get nonsense: false images mixed in with the truth. Each image is folded from its true frequency (f_{in}) down to an aliased image (at $f_{sample} - f_{in}$). When an aliased image appears we can't bomb it with a filter, because it has invaded friendly territory.

18S.2.2 Filtering

CD players reproduce 20kHz music that was sampled at only about 44kHz: just 10% above the theoretical minimum rate. They face a challenging problem in filtering because the lowest sampling artifact lies very close to the highest genuine signal. The CD industry learned to solve this problem by faking a higher sampling rate – by interpolating digitally-calculated pseudo-samples between the genuine samples. This trick is called "oversampling" as you know. The first time you hear the claim, the assertion that a CD player can "oversample" the music sounds like a fraud. Wasn't the sampling rate fixed at the time the CD was made? Yes, it was. But once you understand what's intended, you discover that "oversampling" is a harmless figure of speech. Some people call the operation "digital filtering"[3] or even "digital interpolating." These phrases sound less magical.

18S.3 Examples: sampling artifacts in time- and frequency-domains

The images in this section come from the test setup used in our class demonstrations in which an *analog* sinusoid was fed to an 8-bit ADC, and that *digital* signal then was fed to a DAC. The reconstructed *analog* signal was then low-pass-filtered.

[3] The digital filtering operation can be more complex than this mere interpolation.

These are the traces in the scope images:

1. the top scope trace is the input;
2. the second trace is the unfiltered output of the (8-bit-) DAC that has reconstructed the waveform from the converted digital values;

The four sampling rates illustrate four cases:

- *lavish* sampling: 100kHz (extravagantly excessive sampling, in fact);
- *adequate* sampling: 4kHz (four samples/period). This is good enough if a good output filter is applied.
- *stingy* sampling: 2.2kHz (providing just 10% more than the theoretical minimum of two samples per period). A steep filter (8-pole elliptic–MAX294) is able to knock out the lowest image.
- *inadequate* sampling: 1.5kHz, then 1.2kHz (barely more than one sample per period). This brings on "aliasing."

Lavish sampling rate: Figure 18S.3 illustrates lavish sampling; the sampling rate is so high that the lowest artifact, at 99kHz, is far out of the picture (far off the right side of the frequency spectrum shown).

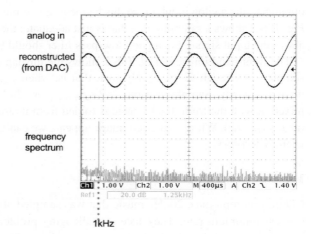

Figure 18S.3 Lavish sampling: 100 samples/period. (Scope settings: 1V/div; frequency spectrum FFT, 1.25kHz/div.)

As the reconstructed waveform shows, this way of sampling *works* in one sense: it allows us to reconstruct the analog waveform accurately. But it is a ridiculous way to achieve this result, producing almost fifty times as much digital data as we need to do the job. If this were the way we stored data on a disc, a CD would be at least the size of a large pizza.

Adequate sampling rate: Figure 18S.4 shows more reasonable sampling The artifacts are far enough above the true signal so that filtering them out does not look difficult – and it isn't.

The image in Fig. 18S.5 shows the effect of the filter applied to the reconstructed waveform: all artifacts are removed from the frequency spectrum – and the steppy quality of the reconstruction disappears from the filtered reconstruction. This smoothing is evident in the third trace in the right-hand image of the figure. The original sinusoid is recovered.

18S.3 Examples: sampling artifacts in time- and frequency-domains

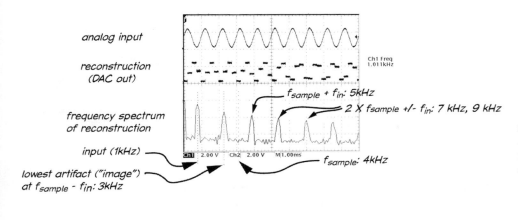

Figure 18S.4 Adequate sampling: artifacts now apparent. (Scope settings: 2V/div; frequency spectrum FFT, 1.25kHz/div.)

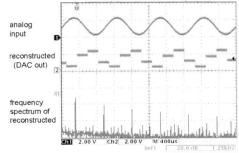 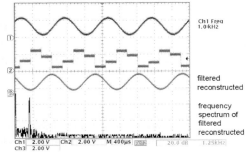

Figure 18S.5 A low-pass filter can remove artifacts introduced by sampling. (Scope settings: 2V/div; frequency spectrum FFT, 1.25kHz/div.)

Stingy sampling rate: In general, one aims to sample as stingily as feasible to acquire no extraneous digital samples. The samples must be stored and perhaps transmitted; we want to minimize complexity and cost. Nyquist says we need slightly more than two samples per period. But how much more? The answer depends on the quality of the low-pass filter used to remove sampling artifacts. A good filter, like the MAX294, an 8-pole elliptical active filter that you will use in labs, lets one come quite close to the minimal Nyquist sampling minimum of two samples per period of the analog input.[4]

In Fig. 18S.6, the sampling rate used to acquire the 1kHz input is just 10% higher than the theoretical minimum of 2kHz: 2.2kHz. (Here, we are approximating the stinginess of CD sampling: 44.4kHz to acquire signals up to 20kHz: again, about 10% above the theoretical minimum Nyquist rate.) The filter is barely able to knock out the lowest image, and in fact attenuates the signal slightly as it eliminates that image.

The left-hand scope screen shown in Fig. 18S.6 shows how very close the lowest image lies to the true signal: at 1.2kHz, just 200Hz above the signal. This filtering task, though challenging, is not as difficult as the imposed CD audio standard, because CD resolution is 256 times finer than what we are demonstrating with this 8-bit conversion, and therefore calls for greater filter attenuation.[5] Soon after establishing the CD standard, the CD's developers[6] in effect conceded that they had given themselves

---

[4] To keep things simple, in this discussion we usually treat the analog input as a single pure sinusoid. Real inputs almost never are so simple. But if the analog input includes many frequencies, one need consider only the highest included frequency – perhaps 20kHz for music, for example. If the sampling rate is adequate for that frequency, it is adequate for the entire audio input.

[5] One might fairly protest, however, that a full-amplitude signal at 20kHz is impossible in music, so the filter attenuation that is required is actually not as great as the contrast between our 8-bit conversion and CD-audio's 16-bit conversion.

[6] The CD standard was developed by Philips, who had introduced the large "laser disc" as a medium for movies, and Sony, who had introduced the portable audio device, the "Walkman" (it was a tape player with headphones).

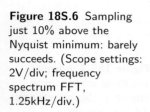

Figure 18S.6 Sampling just 10% above the Nyquist minimum: barely succeeds. (Scope settings: 2V/div; frequency spectrum FFT, 1.25kHz/div.)

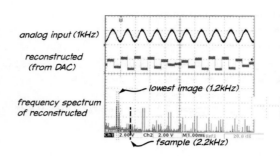
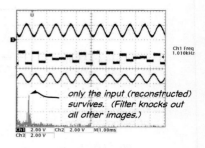

a task too difficult for analog filters. The industry began, in so-called "second generation" CD players, to use digital "oversampling," interpolating the digitally-calculated samples that we referred to in §18S.2.2.

Inadequate sampling rate: Inadequate sampling – violating Nyquist's requirement that one pick up a little more than two samples per period of the highest frequency of interest[7] – leads to a disaster called "aliasing," as we said in §18S.2.1. When this occurs, an input frequency gets "folded" from its true high-frequency to a spurious low-frequency image.

Aliasing illustrated: Nyquist warns us that we had better sample a 1kHz signal at better than 2kHz. If we defy him, we get into trouble. In the case shown in Fig. 18S.7, $f_{in}=1$kHz, $f_{sample}=1.2$kHz. This sampling rate is far below Nyquist's minimum required rate and produces an *alias*: an artifact at 200Hz. This is an image at at $f_{sample} - f_{in}$. This image lies below the truth, and this fact makes it damaging.

Figure 18S.7 Inadequate sampling rate produces aliasing, an artifact too low to filter out. (Scope settings: 2V/div, 2ms/div; frequency spectrum FFT, 500Hz/div.)

The third trace in Fig. 18S.7 shows a convincing sinusoid at 200Hz – but it is obviously a phoney. We can see that the only true signal is the 1kHz sinusoid shown on the top trace, and in the output

---

[7] The phrase, "highest frequency of interest" may bear some explaining.
First, an elementary point: while we illustrate sampling cases in this chapter by applying a single sinusoid as input, real applications normally apply a range of frequencies as input to the ADC. If the highest frequency is sampled adequately, then so necessarily are all lower frequencies, frequencies, as we said in §18N.5.1.
Second – a much subtler point – it is not invariably true that one need sample at better than twice the highest input frequency. Strictly, instead, one must sample at better than twice the frequency *bandwidth* of the signal applied to the ADC. When the input frequency range runs from essentially zero to f_{max}, as in audio applications, the two formulations lead to the same result. But cases do arise where the difference between the two formulations matters. Sometimes one's goal is to sample *a narrow bandwidth at high frequencies*, and in this case the distinction between the two formulations becomes important. If, for example, one needs to convert signals between 1MHz and 1.1MHz, one needs a sampling rate of a bit better than 200kHz, rather than 2.2MHz. See AoE §13.6.3.

frequency spectrum. The fact that under-sampling captures the input is quite counterintuitive; it might seem that we are not getting enough information to describe the input fully. But the sampled points, which define the alias, must lie on the 1kHz original as well; that is where they came from, so 1kHz is another component of the reconstructed output waveform.[8] Sampling has created the 200Hz fraud, which of course shows up also in the frequency spectrum. The alias is deadly: a low-pass filter cannot remove it without also eliminating the genuine *signal*.[9]

18S.3.1 Aliasing: details

The filtered signal includes alias... and you can hear it: If this reconstruction is low-pass-filtered to include the genuine signal, as normally it would be after it emerged from a DAC, the filtered result also includes the alias. Fig. 18S.8 shows that, in this case, the 200Hz alias is larger than the true signal at 1kHz. Together, they sound like a low tone with a harsh, metallic quality (contributed by the 1kHz): not what a person wants to listen to.

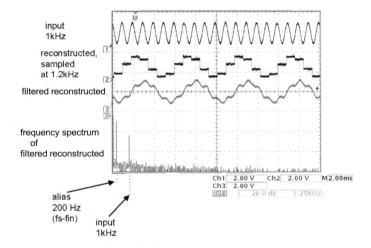

Figure 18S.8 Effect of alias: in this case, second tone forming a chord with the true signal; in the usual case, a range of aliased frequencies would have its frequency spectrum *inverted*. (Scope settings: 2V/div; frequency spectrum FFT, 1.25kHz/div.)

Normally, aliasing is precluded by insertion of a lowpass filter ahead of the ADC. This "anti-aliasing" filter blocks any frequency that cannot be sampled adequately. In these illustrations we omit such a filter because, today, aliasing interests us. Usually it only threatens us.

18S.4 Explanation? The images, intuitively

The rest of this chapter is an attempt to give an intuitive explanation for the fact that sampling introduces artifacts ("images") that appear at $n \cdot f_{\text{sample}} \pm f_{\text{in}}$.

---

[8] Thanks to David Abrams for spelling out this point to us – a point that is surprising, and rarely mentioned in the literature on sampling.

[9] This bad news ought perhaps to be qualified for certain special cases: if an alias is entirely predictable – as it is if the high-frequency input is in a known and narrow frequency range, then perhaps a narrow band-stop filter could be applied. And there are rare cases where the aliasing is useful: a radio carrier frequency can be deliberately aliased, accomplishing what an RF "mixer" ordinarily is used to achieve. See AoE §13.6.3.

18S.4.1 Sampling resembles amplitude modulation

Radio amplitude modulation – "AM" – provides an analogy that helps.[10] In AM radio, a signal "modulates" the carrier. We will argue in the coming pages for the similarity between such modulation and the sampling of an analog signal. Amplitude modulation – the oldest and simplest way to impose a signal upon a radio "carrier" – multiplies the carrier by the "modulating" signal; see Fig. 18S.9.

Figure 18S.9 AM modulation multiplies modulating frequency times carrier.

$$\text{AM signal} = \underbrace{(1 + \underbrace{m\cos\omega_m t}_{\text{modulation}})}_{\text{this preserves the carrier itself}} \underbrace{\cos\omega_c t}_{\text{carrier}}$$

A little trigonometry indicates that the multiplication applied in amplitude-modulation will produce (along with the carrier itself) a sum of the two frequencies, and their difference:

$$\cos\omega_1 \cdot \cos\omega_2 = \frac{1}{2}\cos(\omega_1 + \omega_2) + \frac{1}{2}\cos(\omega_1 - \omega_2),$$

or, more compactly,

$$f_{\text{sidebands}} = f_{\text{carrier}} \pm f_{\text{modulation}}.$$

The "$1 + \cdots$" in the equation of Fig. 18S.9 appears because (as the figure says) ordinary AM radio deliberately *preserves the carrier* in order to ease the receiver's job. This is convenient, but not essential.

A straight multiplication, without the $1+\cdots$, would produce just the *sidebands*.[11] No *carrier* would appear in the output.

That result is very similar to the effect of our digital sampling, which produces "images" that are sum and difference of input and sampling-rate. Input takes the role of the modulating signal in AM; sampling pulses take the role of carrier. In sum, we argue that our sampling "images" at $f_{\text{sample}} \pm f_{\text{in}}$ are equivalent to the AM sidebands at $f_{\text{carrier}} \pm f_{\text{modulation}}$.

We'll also discover soon that sampling includes an offset term, a sort of "1" like the $1+\cdots$ of AM, in order to preserve the original input that is to be recovered from the digital form; but let's work up to that point gradually.

Figure 18S.10 shows images of AM radio in which a *carrier* is amplitude-modulated by the signal. The sum and difference frequencies appear, in the frequency spectrum of the product, as "sidebands:"[12]

Figure 18S.10 Amplitude modulation produces "sidebands" that straddle the carrier frequency. *The AARL Handbook for Radio Amateurs*, 17th ed., (1993), pp. 9–4, 9–5. Images, copyright AARL, used with permission.

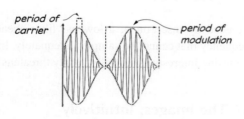
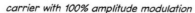
carrier with 100% amplitude modulation

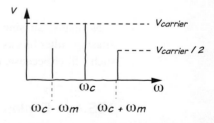
fourier components of 100% modulated carrier

Scope images are more persuasive than sketches, so here are some AM waveforms. Fig. 18S.11 first shows an unmodulated carrier (not very interesting), then, to the right, what modulation looks

---
[10] Simpson, R., *Introductory Electronics for Scientists and Engineers*, 2nd ed. Allyn & Bacon, (1987), pp. 720–721.
[11] Some radio modulation schemes do dispense with transmitting the carrier; others suppress one of the two sidebands. Both variations are more power-efficient, but less easy to demodulate.
[12] Simpson, op. cit. p. 120.

18S.4 Explanation? The images, intuitively

like in the time domain (familiar to you from Lab 3L). Below those scope traces are frequency spectra showing first the carrier alone, then the carrier and sidebands: sum and difference. The sidebands are at $f_{\text{carrier}} \pm f_{\text{signal}}$, where the "signal" in this case is the modulating audio.

The modulation shown in Fig. 18S.11 differs from that sketched in Fig. 18S.10 in that the modulation of the scope image is not a single frequency, but the (more typical) case: a collection of frequencies constituting the signal. In this case, the signal is speech.

In the coming pages we will show some images of AM-like multiplication, done in the lab with an analog multiplier chip. The goal of this examination of the effects of *multiplication* of two waveforms is to try to persuade you that sampling is a very similar process, and thus not much more mysterious than radio's amplitude modulation – a technique that, by now, is an old friend of yours.

unmodulated carrier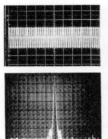

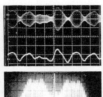 modulated carrier

recovered audio ("envelope" of carrier)

frequency spectrum of carrier

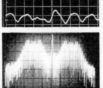 frequency spectrum of modulated carrier, showing sidebands (sum and difference: carrier plus and minus audio signal)

Figure 18S.11 AM radio: "sidebands" straddling carrier resemble conversion "images" that straddle f_{sample}.

18S.4.2 A reminder: spectrum of a sine versus pulse train

More specifically, sampling resembles multiplying an input by a sequence of narrow *pulses* at the sampling rate.[13] So it is worth recalling what the frequency spectrum of such a pulse train looks like.

Sinusoid versus pulse train: The frequency spectrum of a sinusoid (a single frequency) is familiar: one spike. In the left-hand scope image in Fig. 18S.12, such a solitary spike appears at 1kHz:

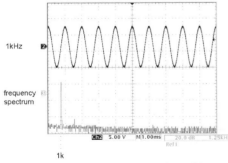

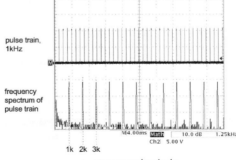

frequency spectra: sinusoid... ...versus pulse train

Figure 18S.12 Frequency spectra contrasted: sine at 1kHz versus narrow pulse repeating at 1kHz. (Scope settings: 5V/div; frequency spectrum FFT, 1.25kHz/div; sweep rates: left-hand image 1ms/div, right-hand image 4ms/div.)

In the image to the right in Fig. 18S.12 appears the spectrum of a train of pulses, about $20\mu s$ wide, repeating at 1kHz (the sweep rate is 4× slower in the pulse image). Again we see a spike in the

---

[13] For the closest analogy to AM radio, we should perhaps state this point backwards: "sampling resembles multiplying a sequence of narrow pulses – whose role resembles the role of the radio *carrier* – by the modulating analog input." We have stated it in the other form, because we usually think of sampling as a process applied to an input, not to the sampler. But the two notions are, of course, equivalent.

frequency spectrum at 1kHz, as for the sine – but in addition the spectrum shows spikes at 2kHz, 3kHz,..., repeating forever. That is, they repeat at integral multiples of f_{in}, where f_{in} is the repetition-rate of the input pulses.

Effect of multiplying a signal by another waveform:

Sine × sine (like AM): You recall that if we multiply two sinusoids we get sum and difference frequencies. Fig. 18S.13 gives two images saying that. One shows a simple case: $f_1=1$kHz, $f_2=4$kHz. The difference and sum products appear: 1kHz and 5kHz. (Neither contributing signal survives).

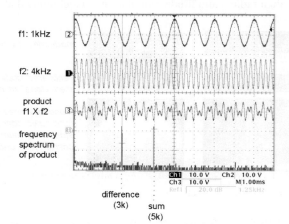

Figure 18S.13 Product of two sinusoids: sum and difference. (Scope settings: 10V/div, 1ms/div; frequency spectrum FFT, 1.25kHz/div.)

Sine × sine with offset: If we multiply a sine at 1kHz by another sine at 4kHz – the case we just looked at – but we add a constant to the latter, then we get the sum and difference frequencies, as before, and in addition we get the *original*, the 1kHz sine. That sounds as if it might be useful, does it not? – given that our goal is to recover an *original* sampled waveform!

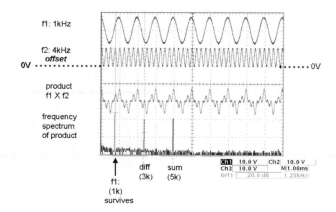

Figure 18S.14 Sine multiplied by sine-with-offset: the result is images – and also the *original*. (Scope settings: 10V/div, 1ms/div; frequency spectrum FFT, 1.25kHz/div.)

This result, in Fig. 18S.14 does, indeed, resemble what we get – and require – in sampling. The original is what we're after; the sum and difference frequencies are artifacts, and these we plan to strip away.[14]

---

[14] We hope you're not rattled by the difference, here, from AM radio. There, it is the *carrier* that is preserved by the offset of the modulating signal. Here it is the modulating signal that is preserved by the offset in the carrier-like pulse train. We are arguing that sampling *resembles* AM, not that the processes are identical.

18S.4 Explanation? The images, intuitively

Sine × pulse train (more like sampling): We've already shown that in the frequency domain the pulse train looks like sinusoids at f_{in}, $2f_{in}$, $3f_{in}$, ... So you'll not be surprised to see that the spectrum of the product of sine × pulse-train looks like sine × sine endlessly repeated – resulting in repeating differences and sums. The offset applied to the pulse-train makes sure that the original 1kHz sine also appears.

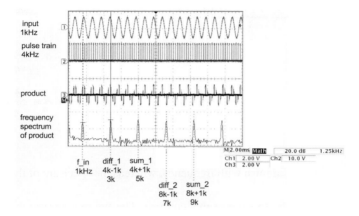

Figure 18S.15 Product: sine × DC-biased pulse train → lots of sum-and-difference pairs – *plus the original sine.* (Scope settings: 2V/div, 1ms/div; frequency spectrum FFT, 1.25kHz/div.)

The survival of the original is surely good news! When we *sample*, this is our *entire* goal: to recover that original signal from among the artifacts. To clear away the artifacts, we need only a good lowpass filter.

To recapitulate: in A/D conversion, the *sampling pulses* play the role of the AM *carrier*; these pulses (whose amplitude is constant) are modulated by the analog input, which plays the role of the *modulating signal* in AM.

The repetition of images around integral multiples of f_{sample} – repetition that does not occur in AM radio – occurs because the sampling pulses are not sinusoidal, but instead are narrow pulses with their many Fourier components: components that keep reappearing at f_{sample}, $2f_{sample}$, $3f_{sample}$, ...

Readers adept at mathematics may recognize this as simply the convolution of spectra (in the frequency domain) of the product of the sampled waveform with a periodic delta function (in the time domain), a consequence of the convolution theorem.

18S.4.3 The punchline: the sampled data looks a lot like the multiplied data

Figure 18S.16 shows two scope images that support the claim that sampling resembles multiplication by a pulse train. (The sampled waveform has been reconstructed with a DAC, as you can see.)

The frequency spectra match: the spectrum of the waveform that was *sampled and then reconstructed as analog* matches the spectrum of the *product of sine and pulse train*. This is the result we were pursuing in this chapter.

18S.4.4 A refinement or correction to the claim

The sampled and reconstructed output (from an output DAC) differs from the multiplied waveforms we have just been discussing in one respect: whereas the sampled output is sustained between *sampled* points, the product (as in Fig. 18S.16) is not sustained, but reverts to zero between multiplying pulses.

The sustaining of the sampled level, as in normal reconstruction of analog from digital, has the effect of low-pass filtering the reconstructed result. The frequency spectrum of the DAC reconstruction

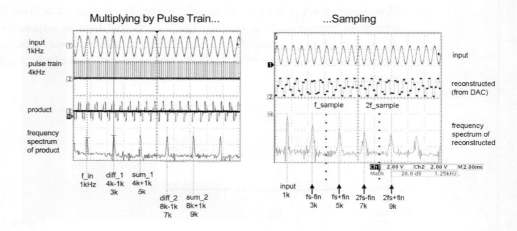

Figure 18S.16 Sampled sine (1kHz, $f_{sample}=4\text{kHz}$) versus multiplied sine (1kHz × pulses at 4kHz). (Scope settings: 2V/div, 1ms/div; frequency spectrum FFT, 1.25kHz/div.)

shows this lowpass effect: the image amplitudes diminish with frequency, whereas the spectrum of the *multiplied* waveform shows no such fall-off.

Figure 18S.17 attempts an argument by analogy to support this contrast. The figure contrasts the frequency spectra of *pulse train* versus *square wave*. Granted, the square wave is not the same as the sustained "steppy" waveform produced by a DAC; but it resembles it more closely than it resembles the pulsed output. This is only an argument by analogy, trying to make plausible the observed effect; the rolloff of the frequency components of the DAC output.[15]

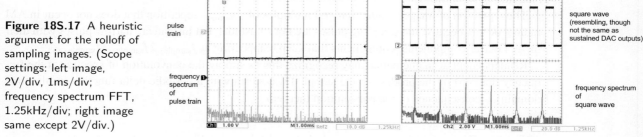

Figure 18S.17 A heuristic argument for the rolloff of sampling images. (Scope settings: left image, 2V/div, 1ms/div; frequency spectrum FFT, 1.25kHz/div; right image same except 2V/div.)

To compensate for this rolloff – which applies to signals in the frequency band of interest as well as to the artifacts that are to be eliminated, a filter applied to the DAC output should include a slight high-frequency boost in the passband to compensate. Output filters for digital audio devices do include such a boost.

---

[15] Thanks to our colleague Jason Gallicchio for suggesting this heuristic argument – though we can't blame him if you find the argument a stretch.

18W Worked Examples: Analog ↔ Digital

18W.1 ADC

18W.1.1 Size of *error* versus size of *slice*

If a voltage range is sliced into n little slices, each labeled with a digital value, the value will be correct to $1/2$ the size of the slice: 1/2 LSB, in the jargon. Perhaps this is obvious to you. If not, try this example.

Example How many bits are required for 0.01% resolution?

Solution 0.01% is one part in 10,000. If a converter spans a 5V input range for example, we mean that when we give a digital answer, like "4.411V," we expect to be wrong by no more than 0.5mV (1 part in 10,000 of the *full-scale* range: 0.5mV/5V).

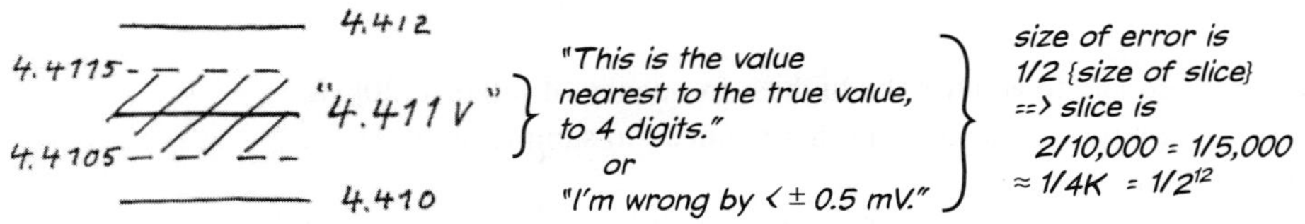

Figure 18W.1 What I mean when I claim to be correct to 0.01%: I say the answer is 4.411V; I could be wrong by ±0.5mV.

How many bits does the converter need? We can tolerate slices that are *two* parts in 10,000 wide, or 1/5k. 12 bits give 4K slices (4096), and give an *error* of 1/8K or 0.012%: this does not quite satisfy the specification. One more bit – 13 – would cut that error in half to 0.006%, going well beyond the minimum requirement. So use 13 bits.

18W.1.2 ADC application: digital audio

CDs established a standard that still serves for pretty-good audio: 16-bit samples at 44kHz (about 10% above the theoretical minimum to get 20kHz music). Let's calculate about how much data had to be stored on a standard CD.[1]

Example How many bytes had to be stored on a 74-minute audio CD?

Solution Two bytes per channel; four bytes per stereo sample. So,

number of bytes = 2bytes/channel × 2 channel/sample × 44k sample/sec × 74 min × 60sec/min

$\approx$ 780 million bytes $\approx$ 745M or 745MB.

That's a lot of data. A CD actually holds about three times that much, devoting the remainder to error correction and other overhead.[2]

---

[1] According to legend, the 74 minutes was chosen by Sony to permit recording Beethoven's 9th Symphony: a curious concern, given the likely audience for most CDs. See Ken C. Pohlmann, *Principles of Digital Audio*, 3rd ed., (1995), p. 265.

[2] Pohlmann, op.cit. p. 265.

18W.1.3 A more interesting application; noise and filter considered

Converting a noisy signal: Suppose someone asks you to specify an ADC to transmit some music digitally. The music amplitude variations are to be distinguished to about 50dB (−50dB relative to full-scale, "−50dBFS:" much lower resolution than the CD audio standard), but you are to reproduce the full conventional 10Hz to 20kHz frequency range. Here are the details.

Example If the resolution we look for is about 50dB, between loudest and quietest passages, how many bits do we need to use?

Solution We want the quietest music – and the magnitude of the least significant bit (LSB) – to be 50dB below full-scale. We can plug that number into the definition of dB:

LSB/full-scale = −50dB;   −50dB = $20\log_{10} A2/A1$; $\log A2/A1 = -50/20$; $A2/A1 = 10^{-(5/2)} = 1/316$.

So, we need enough bits to provide about 315 slices. Eight bits allow 256 slices: not enough. So we'll use nine bits, enough for 512 slices.

We could get the same result using a shortcut from the starting *dB* specification. We could use the fact that each additional bit halves the size of a slice, shrinking it by about 6dB. So, 50dB requires a little more than eight bits, by this calculation: nine to be safe.

Straightforward case: no noise: Let's suppose the signal is pretty clean: nothing substantial above 20kHz.

Problem What sampling rate?

Solution The sampling rate must be adequate to acquire the highest frequency of interest, 20kHz in this case. A naive view of Nyquist's sampling rule would say that we need to sample at 40kHz to get two samples/period. That answer is wrong.

The sampling rate must take account of the limited slope of any filter that is assigned the task of keeping the highest signal while attenuating the lowest image. Such filtering is sketched in Fig. 18W.2. The filter we use in this problem is one that you will meet in lab (if you didn't get a chance to try it in Lab 12L). It is, indeed, steep, as Fig. 18W.3 shows.

The filter will need to attenuate the lowest image to under 1 LSB. The filter's response curve in Fig. 18W.3 shows attenuation of −60dB at about 20% above the frequency where the filter begins to roll off (the "corner frequency," roughly f_{3dB}). That is sufficient (and we probably can't squeeze a more precise answer from the plot of rolloff). So let's make sure that the lowest image is 20% above the highest frequency of interest. That would put the lowest image at 24kHz.

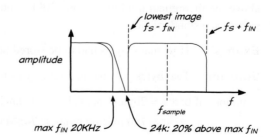

Figure 18W.2 Filter rolloff dictates how close image may come to top input frequency.

What sampling rate puts the lowest image there? Recall the rule that the lowest image appears at $f_{sample} - f_{in(max)}$. Plugging in the numbers in our case, we find $f_{sample} - f_{in(max)} = 24$kHz; and $f_{sample} = 24$kHz$+f_{in(max)} = 44$kHz. It's not be surprising that we should land close to the sampling rate of the audio CD – but it was by chance that we arrived as close as we did.

Note that the MAX294 plot in Fig. 18W.3 shows $f_{3dB}=1$kHz, but f_{3dB} is adjustable from under 1Hz to 25kHz; the plot can be scaled to whatever f_{3dB} you choose.

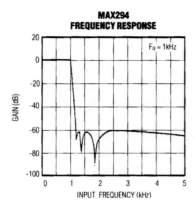

Figure 18W.3 MAX294 frequency response.

A nastier case, noise is added: Now suppose that your converter is fed the strikingly noisy signal shown in Fig. 18W.4, rather than the clean music signal you were led to expect. You may recognize the scope image as a portrait, in time and frequency domains, of music transmitted and received via the infrared audio project on Lab 13L. The signal is polluted by vestiges of the 30kHz *carrier*.[3]

Let's suppose that our goal is, as before, to convert this and similar signals to digital form, for transmission (by wire) over a longer distance.

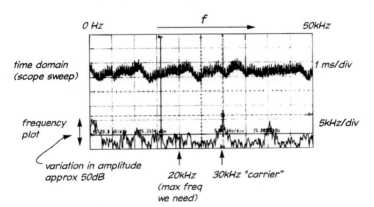

Figure 18W.4 Frequency spectrum of some music.

Problem What happens if you omit an anti-aliasing filter?

Given the sampling rate you prescribed for the "clean" case, qualitatively speaking how would the 30kHz *carrier* noise appear or sound when your digital transmission was re-converted to analog? What happens to the other large *noise* peak that appears above the carrier frequency, in the frequency spectrum?

Solution *You* wouldn't forget to put an anti-aliasing filter ahead of the ADC; but we did. Assuming a clean signal (or the presence of an anti-aliasing filter), we prescribed a sampling rate of 44kHz. The

---

[3] The alert student will also recognize the underlying music as a phrase from a solo by Miles Davis in *So What?*, on the classic album, *Kind of Blue*.

30kHz carrier frequency shows up as a large peak just right of the center in Fig. 18W.4. This would produce its lowest *image* at $f_{\text{sample}} - f_{\text{in(max)}} = (44\text{kHz} - 30\text{kHz}) = 14\text{kHz}$. It would be a high but audible whine. The smaller peak at about 40kHz would produce an image at 4kHz – even more troublesome. We should not have omitted the anti-aliasing filter!

18W.2 Level translator

This is not really a worked example – because the task is so difficult that none of us could expect to come up with such an elegant solution. This is more a chance to marvel at an ingeniously simple design.

Logic devices powered from differing voltages may need to communicate. Letting one drive the other is a problem we have discussed in Chapter 18N. We found, for example, that 5V 74HCT parts could receive 3.3V 74HC outputs successfully. 3.3V parts could accept 5V inputs if they were designed to be "5-volt tolerant," as many are.

But the present problem is slightly different.

Problem Design a *bidirectional* translator that will allow a single line linking a lower-voltage part to a higher be driven from *either end*. The bidirectional quality is what makes this task not standard.

Such bidirectional lines are not common, but they do come up at least in the I-squared-C ("Inter-Integrated Circuit") interface – which you will find briefly described in Chapter 24N. Even before you feel a need for such an interface, you may enjoy seeing the ingenuity of this one.

Solution This solution, described by Philips in an application note AN97055, uses a single MOSFET and a pair of resistors to link digital parts powered by differing supplies.[4] Fig. 18W.5 is a sketch of how the interface works.

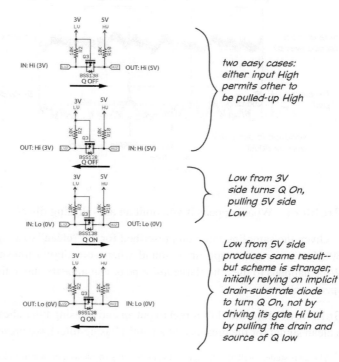

Figure 18W.5 Philips bidirectional logic-level translator. (Source: Philips/NXP application note AN97055.)

---

[4] The note by Philips is posted by Adafruit: http://www.adafruit.com/datasheets/an97055.pdf, and Adafruit and Sparkfun among others have wired up implementations of this interface idea.

19L Digital Project Lab

Contents

19L.1	**A digital project**	**749**
	19L.1.1 Capacitance meter ("C-meter")	749
	19L.1.2 Reaction timer	750
	19L.1.3 Sinewave generator	751
	19L.1.4 4-Channel scope multiplexer	752

19L.1 A digital project

Today is different from most others in that, like Day 13, we'd like you to have the fun of building something of your own design. Unlike Day 13 though, we have no new information to deliver, so we jump right into the lab itself, drawing on circuits we've covered in Part IV. Below we've sketched several tasks that should be fun and manageable. We'd like you to write up a solution and build the circuit that you have designed.

Two of the designs take advantage of the Lab 17L displays and we hope that feature will save you time.

19L.1.1 Capacitance meter ("C-meter")

This circuit is a small extension of the *period meter* proposed as an exercise at the end of Lab 16L. As you know, the period of a '555 oscillator is proportional to RC. If you hold R fixed, then you can infer the value of C from the oscillator's period (or half-period if that happens to be easier to work with).

Your circuit should run continuously, displaying its result on the LCD that has been a part of your counter circuit, installed in Lab 16L. If you use the *data* lines rather than the *address* lines, you can take advantage of a transparent latch built into the data display. You can show the capacitance in some convenient units. These could be, say, hundredths of a microfarad; it's your choice, but don't make it so weird that your meter baffles its user. By making the '555's frequency adjustable over a wide range you can "calibrate" your meter so that, for example, a $0.1\mu F$ cap might read 0100.

You will find in §15L.5.3 a digital *one-shot*, "double-barrelled" (that is, it has two outputs), and that circuit is just what you might use to latch a result into the data displays. If you use the address displays (which would allow four digits rather than two), you would need to add registers to save the meter's count for display. So, using the data latches will save you some effort.

We suggest that you build your circuit on a single breadboard strip, not to clutter the board that will house your growing computer. We suggest also that you do *not* use the computer's counter because this counts in *hexadecimal*. In its place, we suggest either of two BCD counters ("Binary Coded Decimal:" 0 to 9). The 74HC160 is a synchronous 4-bit counter much like the 74HC163 that you met in §16W.1, except in terminating its count on 9 rather than Fh (all ones). The 74HC390 is a dual ripple

BCD counter. Since we will be content with only a two-digit result, you can manage with the '390 alone. (Pinouts for these counters appear in Appendix J.)

Tasks and timing: You will recognize that, if the meter is to run continuously as we would like it to, your circuit will need to time things right:

- Start and stop the counter using the '555 output (for this purpose, you can use the counter's Carry$_\text{In}$ if you use a 74HC160, or you'll need to gate the clock if you use a ripple counter (the 74HC390).
- After stopping the counter, latch its output into the eight *data* lines of the display (this port, unlike the 16 address lines, includes a *transparent latch* – and because we have been holding them in *transparent* mode until now, you may not have been aware that a latch was included. The LATCH_EN* pin is pulled low by default (through a 100k resistor). Your logic easily will override this pulldown.
- Finally, clear the counter to get ready for the next cycle. Both '160 and '390 counters offer an asynchronous *clear* function.
- As clock for the counters we suggest you use a crystal oscillator (available in your parts kit at 11MHz[1]). But, if you are short of time, you can resort to the slightly cheap trick of adopting a function generator's TTL output as your clock source.

19L.1.2 Reaction timer

This you may want to do in either of two versions, counters-gates-and-flops or counters-and-PAL.

The problem: Here are some specifications for the circuit.

- One pushbutton, *A*, starts the timing; another pushbutton, *B*, stops the timing; *A* also turns on a green LED (it wants 3mA at 2V), to which the other person is supposed to respond by pressing *B*.
  - You may use an SPDT pushbutton if you need to debounce; SPST will suffice if you decide bounce is harmless.
- You are given an 11MHz TTL oscillator: consider this for the paper design; but when you *build* the circuit you can save time by skipping this part: just take a TTL clock from a function generator, using it to generate the 1KHz that you need.
- 1ms resolution is sufficient, for your timer.
- Debounce only where necessary.
- Let a third pushbutton, *C*, reinitialize the machine. When the *C* button is pushed, the machine clears all displays and awaits another measurement cycle.
- Measure the "reaction time" in either of two ways:
  - use the 16-bit counter you wired up in Lab 16L (this will be the quicker method);
  - or, if you want a *decimal* display rather than hexadecimal, use a 12-bit counter made up of cascaded 74HC160s – *decimal* or "*BCD*" ("binary-coded decimal") counters with *synchronous* load, Carry$_\text{In}$ and Carry$_\text{Out}$, and *asynchronous* Clear.
- The "reaction time" can be displayed on your LCD card or (if you like a standalone circuit) you can use TIL-311 hexadecimal LED displays (these displays will show only decimal in this application).

---

[1] Strictly, the frequency is 11.059 MHz.

19L.1 A digital project

- If person B cheats by hitting his button before A, turn on a RED "cheat" LED.
- If the counter overflows, light a yellow "overflow" LED.

One solution: build the reaction timer with flops and gates: We hope you will put your circuit on a private breadboard that can survive if you're pleased with your work.

An alternative solution: build the control logic with a PAL: You may prefer to take on the challenge of putting *the controlling flops and gates* (not the counters) on a PAL. This would give you a little experience doing something you have not yet done with Verilog: using flip-flops one by one.

Here are the ingredients we suggest.

- Combinational/sequential PAL: use an XC9572 (or the 3V part XC9572XL), the part that you used in the Glue PAL. It provides 72 flip-flops, each with its own programmable clock, set and reset.
- Use the PAL to control the counter. See the book's website for a Verilog Digital Project Lab file with pin assignments and test vectors. You need to write only the equations.
- This design is in one way easier than the C-meter: it runs just once, and the counter stops, allowing one to see the final count without any effort to catch and hold that final count in another register

19L.1.3 Sinewave generator

One can use the strategy suggested late in Lab 12L to make a sine from a square wave: apply a very steep low-pass filter to the square wave to keep only the *fundamental*. To make the design not excessively easy we'd like you to include a phase-locked loop to generate the clock for the MAX294 filter, and that clock will be a multiple of the frequency of the input square wave.

Here are the ingredients we suggest.

- Switched-capacitor filter: MAX294 (8-pole elliptic); its f_{3dB} is $1/100\times$ the frequency of the clock fed to this filter IC. You can find a data sheet for the '294 online, and there you'll see typical circuits.
- Counter: the straightforward way to arrange things is to start with a square wave at the frequency of the *clock* input to the filter, and divide that down to provide a *signal* input to the filter.

    You want the filter's f_{3dB} to lie above the frequency of the input square wave but below the frequency of its first harmonic (a harmonic that you may recall occurs at $3\times f_{squareinput}$). You might divide down the filter clock by perhaps 128. Your PAL counter – installed in Lab 16L – can do the job.
- PLL: Once you have demonstrated that method – the more straightforward – you could try the harder scheme: instead of starting with the high-frequency filter clock and dividing down, you can start with a square wave at the frequency of the sinusoid that you want to generate, and use a PLL to generate the MAX294 clock. You might set the filter clock at $128\times f_{sinusoid}$. That should put f_{3dB} under one half the frequency of the first harmonic of the square wave. You'll find the PLL circuits shown in Lab 18L are close to what you'll need for this task.
- The frequency of the sinusoid should be adjustable over as wide a range as you can manage: at least 50Hz to 5kHz.
- Single-supply: probably you'll want to run this on a single 5V supply.

19L.1.4 4-Channel scope multiplexer

This scope mux is fun, and will get you used to several useful notions: (1) normal binary-count sequences, and their timing diagrams (perhaps already familiar); (2) use of a 3-state; (3) multiplexing (time-sharing of a single signal line).

This gadget should display the output of a counter (four digital lines) on one channel of a conventional scope as in Fig. 19L.1.

Along the way, we will ask you to do several tasks:

- wire up a dual 4-bit ripple counter ('393);
- put together a '139 2:4 decoder (or '138 3:8 decoder) and a quad 3-state ('125) to form a 4:1 multiplexer;
- build a primitive 2-bit D/A that will add an offset to your digital signals, to let you look at all four of them on the scope screen;
- arrange things so that the multiplexer advances to the next display at appropriate times.

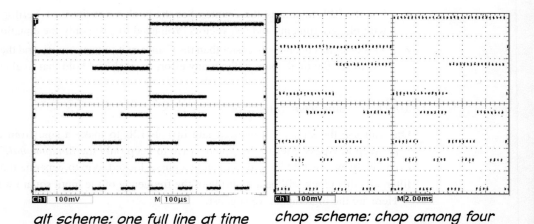

Figure 19L.1 Mux display: the two possible versions.   *alt scheme: one full line at time*   *chop scheme: chop among four traces, on each scan*

We have tried to make this job manageable by describing the elements of this scope mux in stages. But we don't mean to ask you to draw your design more than once: please just show us a full schematic. (You need not show a construction plan, with pin numbers.) Here are the elements of the circuit that you'll need to design.

- Two 4-bit counters.
  Consult the '393 pinout in Appendix J, then wire up both halves. One counter forms a part of the scope multiplexer – let's call that the MUX COUNTER in this discussion; the other counter simply provides four lines of "*data*" – something for the mux to display. Let's call that the DATA COUNTER. (You can show the counter wiring on your complete schematic; never mind drawing the piece here.)
- A 4:1 multiplexer built from parts.
  Show how to combine part of a '139 2:4 decoder (or '138 3:8 decoder) and one quad 3-state ('125) to form a 4:1 multiplexer: show this either here or on the complete drawing. Let two bits of the MUX COUNTER drive the two SELECT lines of the '139. (You'll find a description of the '139 in Lab 21L and in Appendix F. Pinout of the '125 is in Appendix J.)

The signal that you choose to clock the MUX COUNTER will depend on which of the two strategies you choose – CHOP or ALT (more on that below). For one strategy, you would use the fastest available clock – here, the raw input clock; for the other strategy, you would use one of the Qs of DATA COUNTER. (The first time you read this paragraph, you're bound to be baffled; don't let these cryptic remarks baffle you. Just draw your design, and then reread this paragraph to see whether it makes sense.)

- Use our primitive DAC circuit.

Figure 19L.2 is our design for a primitive 2-bit *digital-to-analog converter*. Q_n and Q_{n+1} come from a counter, and drive the output through a 4-step voltage ramp, endlessly repeating. What's shown here provides only the endlessly-cycling *offset* or baseline steps. A *signal* from the mux will soon be added, as shown shown in Fig. 19L.3, feeding current through the 7.5k resistor to the *offset* baseline provided by the cycling counter.

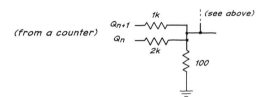

Figure 19L.2 Primitive digital-to-analog converter.

- Put the pieces together.

Show how to wire the whole mess so as to display the 4-bit output of one of the two counters on the scope. Fig. 19L.3 is our sketch of the scheme that we would like you design in detail.

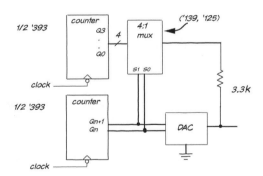

Figure 19L.3 Block diagram of scope multiplexer.

Parts:

- 74HC393 dual 4-bit counter
- 74HC139 dual 2-to-4 decoder
- 74HC125 quad 3-state.

Part VI

Microcontrollers

20N Microprocessors 1

Contents

20N.1	**Microcomputer basics**	**757**
	20N.1.1 A little history	757
20N.2	**Elements of a minimal machine**	**760**
20N.3	**Which controller to use?**	**762**
	20N.3.1 Microprocessor versus microcontroller, again	763
20N.4	**Some possible justifications for the hard work of the big-board path**	**764**
	20N.4.1 Why use the 8051?	765
20N.5	**Rediscover the micro's control signals...**	**765**
	20N.5.1 ...and apply the control signals	767
20N.6	**Some specifics of our lab computer: big-board branch**	**771**
	20N.6.1 Von Neumann versus Harvard Class	771
	20N.6.2 Address multiplexing	772
	20N.6.3 The single-step logic	772
20N.7	**The first day on the SiLab branch**	**773**
	20N.7.1 Little test program for the standalone branch: blink an LED	774
	20N.7.2 A full version of this simple program	775
	20N.7.3 The first tiny program: blink an LED	775
	20N.7.4 Enable I/O	776
	20N.7.5 Watchdog timer	777
20N.8	**AoE Reading**	**778**

Why?

Here's today's problem: make a state machine that is fully general and programmable: transform memory and glue logic into a computer. Computers are so familiar to you that you may not think of this as much of a challenge. Once upon a time it was.

20N.1 Microcomputer basics

20N.1.1 A little history

Prehistory: before the microprocessor: Yes, there was a time when computers roamed the earth, but were not based on microprocessors. In the 1930s electromechanical computers were built, using relays; some were true "Turing machines," fully programmable. At the same time, the last giant analog computers were growing, unaware that their era was ending (notably, Vannevar Bush's differential analyzer, built in the late 1920s at MIT). The first fully-electronic computer, using vacuum tubes, was the Colossus, put into operation in 1943 as part of the British cryptanalytic work at Bletchley park.[1] Because of its application to classified work it remained unknown to the public until the 1970s. In 1946 the first American fully-electronic machine – not secret, and therefore often thought to be the first fully-electronic computer – was launched. This was the ENIAC. It was *big* (see Fig. 20N.1), and could run at first a few hours between disabling tube failures, later – with better tubes and the

---

[1] See the good summary history at http://plato.stanford.edu/entries/computing-history/#Col.

expedient of never turning the machine off – for about two days between failures. But it was fast – about 1000 times faster than its electromechanical predecessors.

Neither Colossus nor ENIAC was quite the general-purpose machine that we think of as a computer. Both were powerful calculators that had to be set up by arranging physical wiring, using switches and jumper wires. They had not incorporated Turing's conception of a stored program machine.[2]

Figure 20N.1 ENIAC: the first US fully electronic computer (US army photo).

A processor on a circuit board: Transistors relieved the painful difficulty caused by vacuum-tubes' short lifetime. Integrated circuits made digital circuitry much denser and cheaper than it had been. Fig. 20N.2 shows a Data General Nova computer's processor board, ca. 1973. The board is about sixteen inches square, and is implemented with small- and medium-scale TTL ICs. Note the afterthoughts implemented with hand-wired jumpers. (This was a production board, not just a prototype.)

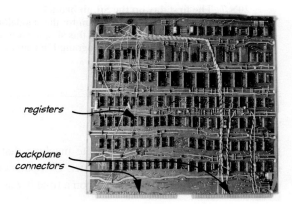

Figure 20N.2 Processor on a PC board: Nova computer, ca. 1973.

The first microprocessors: Intel produced a 4-bit processor called the 4004 in 1970. It was not conceived as a general purpose device, having been designed to implement a calculator for a single customer. Intel gave it more general powers, but no one took much interest. Two years later, Intel produced the first 8-bit microprocessor, the 8008. It had 18 pins, dissipated a watt, and cost $400, and was not widely adopted.

In 1974 the 8080 arrived, and this one enjoyed some success – though it was not quite the one-chip processor that one might expect: the 8080 required *twenty* other chips to implement its interface to the rest of the computer. These early processors found their way into games, notably Pong[3] and Space Invaders. The 8080 also was the brain of the early hobbyist's home computers.

---

[2] Ibid.

[3] Pong was a huge hit – even though its designer scorned it. He said that after the failure of a more complex game, he decided to "build a game so mindless and self-evident that a monkey or its equivalent (a drunk in a bar) could instantly understand it." (see M. Malone, *The Microprocessor: A Biography*. Springer-Verlag (1995), p. 137).

20N.1 Microcomputer basics

An issue of *Popular Electronics* from 1975 showing the Altair is said to have convinced Bill Gates and his friend Paul Allen that something was afoot that they ought not to miss. They rushed to write a BASIC interpreter for the 8080, and Gates quit college to start a company to develop this product. Meanwhile, competition was hot among the fancier parts, the *microprocessors* – the brains of full-scale personal computers: in 1976, defectors from Intel moved across the street and formed Zilog, which put out the Z80 (so-called because it was to be *the last word* in 8080s). As often happens, it did not displace the inferior design.

The following year, Apple launched the Apple II computer, its first mass-produced model, using a 6502 processor made by a small company named MOS Technology. In 1979, Intel's 8086 became the industry leader by getting itself designed into the IBM PC, even though Motorola's new 68000 was a much classier processor (adopted in the underdog Macintosh computers).

Figure 20N.3 shows a '386 *motherboard*.[4]

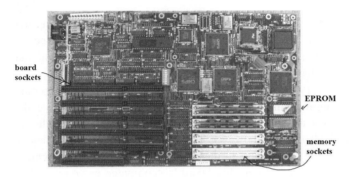

Figure 20N.3 IBM PC motherboard ('386/87 processor), ca. 1990).

Microcontrollers: Well before anyone imagined a mass market for a "personal" computer, it was not hard to envision uses for a versatile one-chip device that might implement a fancy calculator, or perhaps a game. We have noticed Intel's decision to provide, in the 4004, a part that was more capable than what the single customer's calculator required. And a game like Pong could be made cheaper if it used fewer ICs.

So it is not surprising that the first *microcontroller* – a self-contained one-chip computer – was born in 1976, just a few years after the first microprocessor. This was again an Intel design: the 8048. Indeed, the surprise for Intel and the world was the rapid spread of the "personal computer." Very few people had envisioned any need for such a gadget. At first it was only a few hardcore hobbyists who thought they needed to own a home computer.

The 8051, Intel's second try at a microcontroller, was born in 1980 and sold in huge volume: 22 million in the first year.[5] Two years later, Motorola brought out the 6805, the best-selling microcontroller of its era (about 5 billion sold by the year 2001). Motorola remained for many years the market leader in *controllers*, though not in *processors*, primarily because it dominated the growing automotive market. Later companies now outsell both these original giants in the *microcontroller* market (for example Renesas – a Hitachi company – and Microchip outsell Motorola at the time of writing).

Of course, microcontrollers keep getting more and more "*micro*." The largest of the packages in Fig. 20N.4 is the one you will meet in Lab 20L on the big-board (Dallas IC) branch of the labs: a wide-DIP 8051. The PLCC next to it is the same part in the somewhat smaller PLCC package.

---

[4] The motherboard is the printed circuit into which peripheral cards and memory could be plugged. A bare motherboard was a computer, but not a useful one, as it might lack essentials: for example, video display driver or a mass storage device such as a disk drive.

[5] Malone, op. cit.

All of the devices shown in Fig. 20N.4 are controllers of one design – the venerable 8051 that you soon will meet in lab. The tiniest one, on the right, measures 2mm square.[6]

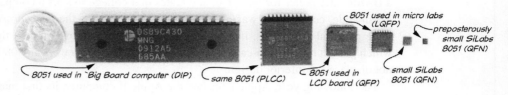

Figure 20N.4 Shrinking microcontrollers: 8051 in four packages.

As packages shrink, so do prices. In quantity 1000 some small controllers now cost 25–30 cents. Some *untested* controllers cost a good deal less than that (we heard a recent estimate of 7 cents) – and by the time you are reading this, no doubt prices will have fallen further.

Recalling what we mean by "a computer": The computers we are accustomed to are Turing machines, state machines that advance, one step at a time, from state to state following a sequence and taking actions at each state. The inputs that call for taking a particular action, and that steer the machine to a particular set of next states, are what we consider to be the machine's *instructions*.

The abstract model proposed by Turing in 1936 was intended to define the limitations of computing, rather than be a design for for a computing machine.[7] Most contemporary computers are also called *von Neumann* designs, meaning that they use a single store for both data and instructions. An alternative arrangement, called "Harvard class," also is used however. This even happens to be the scheme envisioned by the designers of the processor that we use in these labs, the 8051. Harvard architecture places instructions and data in separate stores. In our big-board lab computer, as you soon will see, we impose the more conventional von Neumann architecture on the 8051, despite its designers' intentions. More on this later.

Stored program machine: The processor is a *state machine* (short for "finite state machine" or FSM) within a *state machine*: the computer itself is an FSM. It reads, from memory, one instruction at a time, executes it, then goes back to memory to find out what to do next.[8] The sequence of instructions form a *program* (or "code"), stored in memory.

The smaller-scale state machine that is the *microprocessor* is fed "instructions" – 8-bit codes in our case – that tell it which of its several tricks to do (2^8 tricks in principle). As it executes a single instruction, the processor steps through the multiple states or steps that are required for this execution. Fig. 20N.5 shows an "instruction decoder" drawn in the form usual for state machines. As it executes each trick – stepping through a multi-clock sequence of states – it can turn on 3-states to drive signals here or there, can clock registers to catch signals, and so on.

20N.2 Elements of a minimal machine

An ordinary *microprocessor* requires a good deal of support in order to do anything useful – in order to form a "microcomputer." It needs memory at very least. And it needs some sort of input/output, to act on the outside world. Fig. 20N.6 sketches a block diagram of a generic computer.

---

[6] The parts are, from left: Dallas DS89C430, in DIP and PLCC packages; SiLabs 8051F700 in QFP; SiLabs 8051F410 in LQFP (this is the part we use in the SiLabs microcontroller labs), SiLabs C8051F990 in QFN and the tiny SiLabs C8051T606-ZM in the 2mm QFN package.

[7] Isn't this oddly reminiscent of Boole's development of his algebra with the goal of making philosophical analysis more rigorous?

[8] Most processors use a *pre-fetch* or *pipeline* to get one or more further words from memory, while busy *decoding* the current instruction. The 8051 does this, ancient though its design may be!

20N.2 Elements of a minimal machine

Which of these elements are required, which optional? Could one omit RAM? ROM? A disc drive? Keyboard? Display? (Surely we could omit ADC and DAC.) Yes, we could omit any of those elements, though not all.

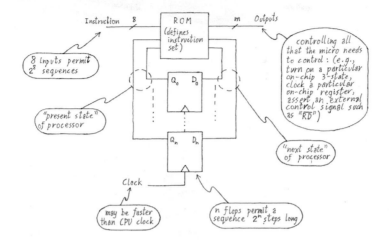

Figure 20N.5 Processor as *state machine*; it is the brains of the larger state machine that is the computer.

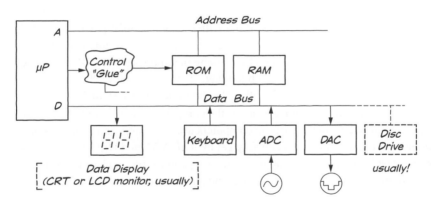

Figure 20N.6 Any computer: which elements are essential?

- RAM: RAM is important for general-purpose computers, but is not required for a dedicated controller.

  In a full-scale computer, RAM can hold program code as well as data, and both code and data normally are loaded from an external disc drive (at present disc drives remain usual, though they are giving way to solid-state storage. But a controller used, for example, to run a vending machine will store its code not in RAM but in ROM. (The machine needs to be able to remember short-term variables; but the amount of memory needed for this purpose may be so small that it can better be called "registers.)

- ROM: ROM allows a computer to run at startup – before any code has been loaded into RAM from an external store. Once upon a time, before ROM, a computer lost all its code on power-down, and had to be nursed back to life when turned on. Fig. 20N.7 shows the front panel of the old Nova computer model that we once used daily.[9] Why all those toggle switches? They were used to enter instruction codes, one word at a time. Put enough in, and you could then run the little program that read in a larger program from the paper tape.

[9] Source: Wikipedia: http://upload.wikimedia.org/wikipedia/commons/5/58/Nova1200.agr.jpg Used under Gnu GPL license.

762 Microprocessors 1

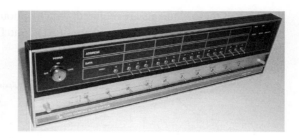

Figure 20N.7 Front panel of a Data General Nova computer (born 1969).

Figure 20N.8 Miracle of the 1960s: cowboy, like computer, lifts himself by bootstraps.

"Bootstrap" loading: It's a bit hard to imagine working with such a machine (having paid about $8000 for it): startup was laborious. It's for that reason that a computer that simply *ran* when you turned it on seemed almost miraculous. It appeared to be doing the impossible feat of lifting itself up by its own bootstraps. Hence the term "bootstrap loading," now shortened to "booting" of the computer, to mean "starting up," or (more specifically) "loading operating system."

- I/O: there must be some sort of input/output hardware if the computer is to accomplish anything; but no particular form is required. General-purpose computers use keyboards and visual displays, but a controller regulating a process might conceivably get by with ADC and DAC, and perhaps a few knobs (read by the ADC). A still simpler application – say, control for an electric toothbrush – might need no more than an LED, a transistor switch, and a pushbutton.

20N.3 Which controller to use?

The variety of available microcontrollers is indeed daunting. It might seem that a person would have to put in a month of study before beginning. Usually, though, things are not that bad – largely because most people are not so rational (or is it *compulsive*?) as to canvass and evaluate all alternative devices. Instead, most of us ask around, taking the advice of those working near us.

That makes sense, because it's a big help to be able to ask advice from someone who has used the particular variety that you undertake to use. This sort of conservatism – even among people who take pride in working close to the cutting edge of new technology – accounts for the remarkable fact that the *8051* design that we adopt in these labs remains the design most widely sourced (that is, offered by the largest number of manufacturers) more than thirty-five years after its introduction by Intel in 1980.

In earlier years, the primary reason for this conservatism may have been the desire to take advantage of a large volume of *legacy code* written for the particular processor or controller. This was important when coding was done in assembly language (because a change of processor required a code rewrite).

AoE §15.3

AoE §15.10.4

20N.3 Which controller to use?

It is not nearly so important now that coding is done in higher-level languages (for controllers, most often in *C*). Apart from some C-extension required for each processor, the higher-level code can be ported to a new processor without the pain of a full rewrite. But it remains convenient to stick with one controller in successive projects. Small differences among controllers do call for learning details that it is pleasant to learn once, not repeatedly.

20N.3.1 Microprocessor versus microcontroller, again

If you take the *Dallas* big-board path, your lab computer will look like what's shown in the block diagram on the left in Fig. 20N.9 – and like the big, pretty messy circuit board on the left of Fig. 20N.10:[10] processor plus memory plus I/O devices, and the glue logic to make this all work. But the brains of your computer, the processor – even if you do build the big-board version – is in fact a *microcontroller*. On the big-board, as in the single-chip SiLabs controller, the brain is an 8051.[11] The elements sketched within the dotted line in the right-hand diagram of Fig. 20N.9 form a full computer – and are included in the 8051 that we use, at first, only as a *processor* rather than as self-sufficient controller. The diagram also describes the single-chip controller that people taking the SiLabs path will use.

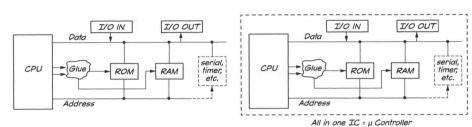

Figure 20N.9 Microprocessor plus stuff = computer; one-chip ⇒ microcontroller.

Even better than a block diagram, photographs of the two alternatives – the breadboarded computer you are in the process of building versus a single-chip microcontroller – show how very tidy and dense the controller circuit is.

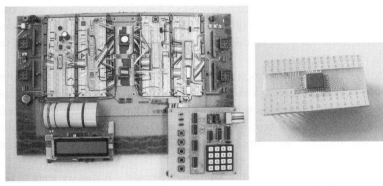

breadboarded micro...   ...single-chip micro

Figure 20N.10 Breadboarded big-board versus one-chip controller.

The SiLabs controller is not offered in a DIP package, so we soldered its small surface-mount 32-pin package ("LQFP") to a DIP adapter. Fig. 20N.11 shows a more specific X-ray view of the 8051.

---
[10] Well, these actually are wired unusually tidily. But it's very hard to breadboard a truly tidy circuit on this scale.
[11] We use "8051" as the controller's generic name, though we use two of many variants of Intel's basic design. The big-board uses a Maxim/Dallas part; the single-chip design uses a Silicon Laboratories device.

(Half of the 16-bit address bus is brought out on time-shared (*multiplexed*) pins, as you will see in Lab 20L if you are taking the big-board branch.)

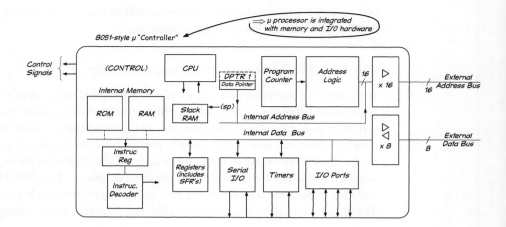

Figure 20N.11 What's within the 8051 controller.

20N.4 Some possible justifications for the hard work of the big-board path

Before asking you to burrow into the detailed work of wiring a computer from a handful of ICs, we did confess that there are easier ways to use a microcontroller. We invited you at the start of Lab 16L to take a quicker route. The normal way to play with a microcontroller (hereinafter, we'll often call it a "micro"), as we have said, is to buy a "development" or "evaluation" board for perhaps $50; plug it into the USB port of an ordinary computer; install the development software – and 5 minutes later watch an LED blink on the evaluation board. (The makers of evaluation boards know how impatient we all are for a reward – and for evidence that our machine works.) AoE describes several such kits in §15.9, the kits' appealing packaging illustrated in Fig. 15.24. We are as impatient as the next person to see evidence that our circuits work – and yet, on this Dallas path, we deny you the pleasure of such quick results. Why?

We can offer several reasons. One is that we think it's satisfying to build a machine with a minimum of mysterious magic humming behind the scenes. Your little computer will include no clever operating system, nor even a "monitor" program; it will know no tricks that you don't teach it. (The processor itself, we should admit, will remain a black box; we didn't have the nerve to ask you to build a processor from gates and flops. Building up a processor could, in itself, occupy a good part of a course.)

A second reason – somewhat distinct from the goal of minimizing magic – is that we want to make the normally opaque computer more nearly transparent, as it runs: the circuit you will build shows its activity on the two *buses – address* and *data*. At full speed, such activity would be unintelligible, but as you single-step through code[12] you can scrutinize all the machine's signals at your leisure. Hardware freezes execution until you are ready to proceed. We think this close look at the machine's operation will make notions like memory addressing and instruction pre-fetch no longer just abstractions.

---

[12] "Single-step" in your computer means "execute one bus-access at a time." This is a slightly finer-grained single-step than the more usual meaning of single-step: "execute one instruction at a time."

A third reason, related but again distinct, is that we are giving you a wealth of opportunities to *make mistakes*. That may not sound like a gift that you want, but when we say, facetiously, in the Worked Example 20W that bugs are our most important product,[13] we say it with some pride. You have been debugging the circuits that you build all through this course. But this micro circuit, with its 400 to 500 connection points, will present harder puzzles than the earlier, smaller circuits.

20N.4.1 Why use the 8051?

AoE §15.3
AoE §15.10.4

The original 8051 design is older than almost any of our students.[14] There are faster devices with more features and more peripherals – though the Silicon Labs version of the 8051 includes a pretty rich collection of peripherals, including ADCs and DACs (see AoE Fig. 15.7). We chose the 8051 because it is the most widely sourced design (that is, it is made by a great many manufacturers); you will find it offered, for example, as a processor module that can be loaded into an FPGA (8051 "cores" from Cast or Hitech Global can be loaded into Xilinx and other FPGAs) and you will see it included within a programmable "analog front end" IC from Analog Devices (ADuC842); it is one of two alternative processors offered on Cypress Semiconductor's recent PSoC designs, which include programmable analog parts along with a controller (see Chapter 26N).

The 8051 is a classic among microcontrollers, illustrating much that you will find in any other controller that you meet. Some of its quirks reflect its age: its restricted off-chip addressing modes, for example, and the fact that all off-chip data must pass through the accumulator register.

When operating upon on-chip registers, the 8051 is much less quirky – and this is the way a controller ordinarily is used. The SiLabs branch of our microcontroller labs work that way.[15] The 8051 remains a lively design base – including some innovations that have not yet reached other processors, like SiLabs's controller that runs from the startlingly low supply voltage of 0.9V. And we hope that even if your next controller is a more recent one, its architecture will look familiar, since it derived from the design of this, the original microcontroller.

20N.5 Rediscover the micro's control signals...

All happy processors are alike, more or less: their control signals resemble one another's. Each unhappy computer is unhappy in its own way – this you will discover as you debug your breadboarded lab computer, and debugging is a large part of the fun and challenge of putting together this machine. §22W.2 notes some standard debugging techniques, and a few common problems we have seen in this lab computer. Here, we invite you to rediscover the control signals that any processor would need, in order to communicate with the devices attached to it.

The processor is the boss, or the conductor of the orchestra (or, maybe it's only a chamber group; not so many elements, in our examples). It determines when each event should occur, and sends out signals to make these events occur. The processor must say what sort of operation it means to do, with whom it wants to do it, and exactly *when*. In the table below, we list the elements of information that seem required – and alongside these generic descriptions, we've listed the signals used by three processors or bus standards.

---

[13] If you think you hear an echo of General Electric's old advertising slogan, you're right.
[14] It is the *original* 8051 design that is old. The versions of the 8051 that we use, from Dallas/Maxim and from SiLabs, are much younger.
[15] The sole exception, the single case where the SiLabs programs run "off chip" is the program that stores data in MOVX RAM. Even this is, in fact, *on chip*, but the addressing mode is the same as for true off-chip accesses, and partakes of the oddnesses of that mode. Address must be provided by a dedicated *data pointer* register, most notably.

Here are the three buses we'll discuss:

- the old IBM PC 16-bit ISA/EISA bus – now gone from desktop machines but persisting in industrial computers in the PC104 standard;
- the 8051 controller/processor's signals – when used, as in our lab computer – to control an external bus[16];
- Motorola's classic 68000 processor,[17] to illustrate another way of handling timing signals.

AoE §§14.3, 14.4.1, Table 14.2

Kind of Information	ISA bus	8051	68000
Direction of transfer	MEMR*,MEMW*, IOR*, IOW*	RD*,WR*	R/W*
Timing	included	included	Data Strobe*, Addr Strobe*
Category: Mem vs I/O	included	doesn't care ...but we care, so we can assign ...an address line to distinguish ...mem from I/O (we use A15)	doesn't care
Which, within category	Address: 20 lines (10 lines for I/O)	Addr: 16 lines (15 if we've used 1 for I/O vs memory)	Addr: 32 lines

We don't expect you to digest all this information from a dense table. Let's note, less formally, some of what this table has to say. All these processors and buses agree concerning what needs saying.

- "Direction of transfer" describes which way information is flowing: toward the CPU or away. By rigid convention, the description is *CPU-centric*:
  - RD* ("read," and active-low, as usual) means *toward the CPU*;
  - WR* means *away from the CPU*.

Figure 20N.12 CPU-centrism defines the otherwise ambiguous "Read" and "Write".

ISA and 8051 use very similar signals – not so surprising when one recalls that Intel provided the processors behind both designs – with names like RD* or WR*. Motorola did it differently: they provided a *direction* signal, R/W*.[18] That R/W* signal includes *no timing information at all*.

- Timing: the ISA and 8051 signals include the timing with other information: the beginning or end of the signal (sometimes both, but not always) indicates *when* the action occurs. For example, when RD* goes low that means the processor wants a peripheral or memory to turn on its 3-state driver; when RD* goes high again, that's the time when the processor will pick up the data that was provided.

The Motorola scheme is different. Motorola provides a separate timing pulse – called "strobe," as pure timing signals often are. Two lines are used to say the equivalent of the 8051's RD*:

---

[16] Bringing out the buses is decidedly not standard behavior for a microcontroller. No buses can be brought out of the SiLabs part.

[17] This processor, once the brains of Macintosh computers among others, is a 16-bit processor. Only one 68000 variant is still in production (by Freescale, not Motorola).

[18] As you've gathered, by now, such a signal indicates in its naming, which level signals which activity. A low on R/W* indicates a *write*, for example.

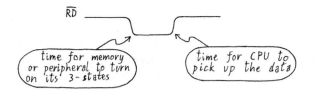

Figure 20N.13 A "RD*" pulse, for example, says not only *what* the processor is doing, but also *when*.

Motorola drives R/W* high and DS* low. Their coincidence is equivalent to Intel's RD*. The Motorola scheme has its partisans (it doesn't waste lines, because strobes can be shared); both schemes work.

- Memory versus I/O: this distinction is useful to us humans. Memory is what it sounds like; I/O is everything else: everything that is not memory. Even devices that you may think of as essential to a computer, like keyboards and displays, are I/O.

  But whether the processor or bus needs to distinguish I/O from memory depends on whether it wants to be able to operate on them differently. The ISA bus does distinguish the two, and this allows it to require fewer address lines for I/O devices (ten lines) than for memory, somewhat easing the hardware designer's work. The 8051 has no special I/O operations or restrictions; anything it can do to something stored in memory it can do to a peripheral as well. This sort of treatment of input and output is commonly referred to as "memory-mapped I/O."

  We make the distinction in our lab computer because doing this allows us to do *extremely* simplified I/O decoding. That works because we have very few I/O devices. We do in extreme form what the ISA bus does for I/O. ISA I/O pays attention to 10 address lines; we pay attention to *two*.

  By contrast, in the category we assign to memory (the lower half of all address space) we preserve normal addressing: we pay attention to 15 address lines in order to address the RAM's 32K locations.

  If these mentions of *decoding* are not yet clear to you, don't worry. We will look closely at our I/O decoding next time. (And you may want to look at §17S.2 on address decoding.)

- Which, within a category: Address lines are used for this job by all processors. The old ISA bus handicapped itself long ago by providing too few address lines for a full-scale computer – 20 lines on the bus; just 16 from the 8086 processor itself). As a result, Intel and IBM then had to devise clumsy ways to enlarge the addressed space ("paging" among 64K blocks). The 8051 is similarly constricted, and new variants offer paging to expand the available space. Motorola's 68000 started big, having modeled itself on larger minicomputers: 32 lines define an address space that is nearly adequate by present standards.

20N.5.1 ...and apply the control signals

Let's try applying these control signals for the ISA and 8051 cases in order to attach a simple peripheral. The hardware we mean to attach is an 8-bit pair of hexadecimal displays that include built-in latches. Those latches will serve to catch what is sent down the data bus to the displays.

AoE §14.2.6

ISA bus case: Let's first choose appropriate ISA bus signals. The operation is an *OUT* – the I/O version of a WRITE – since the flow of information is away from the CPU. So we want one of the two WRITEs. This is not a memory operation, so we need IOW*. That takes care of most of the task: this signal defines direction, timing and I/O-versus-memory. All that remains is to place the display at a single location (so it doesn't tread on the toes of anything else in the computer), and then let the processor send it values.

When using a full-scale computer – like the old ISA-bus computer on which we tried this as a demonstration – we need to take account of what already is installed on the computer, and find an unused I/O address where we can put the new peripheral. The IBM PC we were using has some clear space beginning at hexadecimal address 300h, so let's put it there.

How do we "put it there?" All we mean is "let the peripheral detect when that address arrives from the computer" (carried on the address bus). If the arrival of this address coincides with IOW*, our decoding logic should generate an OUT300h* signal that will catch the data.

300h is the Hexadecimal expression for a *twelve-bit* binary number but the ISA bus specifies that only 10 address lines are used to define an I/O address. So we need decode only 10: A9...A0. Then, in binary, the address 300h will look like

$$A9..A0: \qquad 11\ 0000\ 0000\ .$$

That is the pattern our logic needs to detect. So we need a wide AND gate that will combine this bit pattern with an assertion of IOW* (and a pesky additional signal, AEN, which we want to make sure is *not* asserted, because it indicates an exceptional sort of operation, "Direct Memory Addressing," DMA). Fig. 20N.14 is a timing diagram describing the bus behavior we can expect.

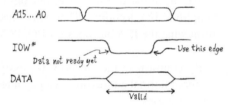

Figure 20N.14 Timing diagram for ISA bus write.

The diagram tells us we should use the later, *trailing* edge of the IOW* pulse, not the leading edge. Put the signals together, into a wide AND gate (probably using a PAL), and it's done: see Fig. 20N.15.

Address decoding; a bus example: We implemented this simple address decoding in a PAL, bringing out one extra line – Address_Match – to indicate less than a complete decode (i.e., ignoring AEN and IOW*). We ran a little Basic program on the PC, putting out incrementing bytes to port 300h, and then watched on the oscilloscope, trying three alternative trigger signals. The results reveal some truths about not just this old PC but about computers generally. Let's take the three cases one at a time:

- Case a), on the left of Fig. 20N.16, (triggering on IOW*).

  We see an overlay of wider and narrower IOW* pulses, and faint ghosts of Address Match (300h) and of OUT300h. These ghosts imply that only a few of the cases where IOW* is asserted coincide with the latter two signals – though the faintness of the traces makes it impossible to be sure of that. Why might there be an IOW* that doesn't coincide with the other signals? Because on a full-scale computer, the little program we wrote to exercise this display is not the only program running. Some other background program is doing I/O writes, as well. So only one of the IOW* signals is *ours*.
- Case b) (triggering on Address Match)

  We took care to see that nothing else is installed at I/O address 300h. But we see 300h matches that do not coincide with IOW* – and in this case there is not even a faint IOW* trace, suggesting that *most* 300h instances are not ours. How can this be?

There is a simple answer: 300h can occur in memory space as well as I/O, and for those cases there will be no assertion of IOW*.

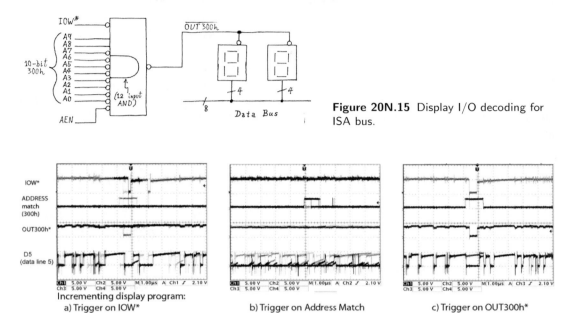

Figure 20N.15 Display I/O decoding for ISA bus.

Figure 20N.16 ISA bus signals as increment-display program loops. (Scope settings: 5V/div; 1μs/div.)

- Case c) (triggering on OUT300h*)

At last things are simple again. The three signals line up nicely: IOW* coincides with address 300h and generates OUT300h*, as planned. The bottom trace, showing line 5 of the data bus, appears both low and high during the OUT300h pulse. This makes sense too, if we assume that we are seeing several scope traces overlaid once more: the incrementing-value program causes any single bit of the incremented value to toggle. So D5 is recorded sometimes high, sometimes low.

What is the rest of the activity of this very busy data line? The waveform is complicated – but appears to be constant at least in this rightmost case. The other two cases look pretty chaotic, and show sometimes ramps rather than good logic levels.

The details of those waveforms need not concern us, but we can recognize that the funny levels show the line at times when *no device* is driving the line. The odd ramps show the cases where the line is not driven high or low but is 3-stated. So it drifts toward a level close to the midpoint of the logic range, pulled probably by the display input.[19]

The major point we want to underline here is that *most of what travels on the data bus is not data* – at least, not data in the strict sense. Instead, most traffic is instructions, in a conventional von Neumann machine like this ISA bus computer. In such a computer, instructions and data (in the narrow sense) live alongside each other in a common memory and travel on the common bus. The constant levels seen on data line D5, in Fig. 20N.16 are bits of *instructions*. They are constant

---

[19] This display input is TTL-like if you use HP LED displays; the display input is weakly pulled up if you use our custom LCD board.

because the program is running a tight loop, and thus repeats as we watch. We'll see in a moment that our little lab computer shares this conventional von Neumann design.

Doing the same thing with an 8051: In some ways, the task we just asked the ISA bus to do is more easily done with an 8051. It is easier because in our small 8051 world there is no need to consult a table that indicates what uses other designers of hardware and code have made of available addresses. Things can be very simple if we like; we are in charge.

For the present example, let's put the data display at a port we call "Port Zero" – using the external buses; and let's decree that we need to distinguish Port Zero from only one other port, Port One. Let us, further, adopt the convention we use in our lab computer, assigning memory to the lower half of all address space, I/O to the higher half; see Fig. 20N.17. A15, then, distinguishes I/O from memory, and "Port Zero" is just the first I/O location.

Figure 20N.17 To keep things simple we split address space equally between memory and I/O.

To decode Port Zero, we need to find the appropriate 8051 control signals, and a few address lines. Fig. 20N.18 shows the candidate 8051 control signals once again.

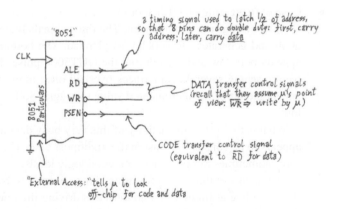

Figure 20N.18 8051 control signals.

We have decided that the msb of the address lines, A15, will mean "I/O" when high. Since this is an *output* operation in which information flows from CPU to the world, it is a write and WR* is the appropriate control signal. Now all that remains is to distinguish port number "Zero" from port number "One." The obvious way to do this is to use the least-significant address line, A0, as shown in Fig 20N.19 (and it would be perverse to use any way *other* than the obvious way; we would only confuse people who used our machine).

How many other ports does this decoding permit? Well, we can have a Port One by detecting A0 *high*. We can also make use of *input* ports at the same two addresses – IN0* and IN1*. So our decoding permits four ports, two IN, two OUT (or, if you prefer, "two bidirectional ports"). This arrangement would not let our computer do much; even our small lab computer will want twice as many ports.

But this simple example begins to let one feel how simple our decoding can be ("lazy" is what we call this sort of decoding – where we deliberately omit many address lines in order to keep our hardware simple). We could define twice as many ports by adding a second address line (A1); and so on. Next time, we will do exactly that.

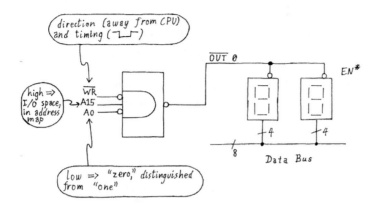

Figure 20N.19 Decoding an OUT0* port for a tiny 8051 controller.

20N.6 Some specifics of our lab computer: big-board branch

We're interested in the little lab computer as a device to introduce some general patterns in the behavior of computers and microcontrollers. But along with these general truths – which reflect the similarities among all contemporary computers – we're obliged to get used to some very processor-specific details as well. These won't carry over to your next design, and do not apply to the SiLabs single-chip labs. The details are necessary only to let big-board builders understand their lab computer. SiLabs people can skip ahead to §20N.7.

20N.6.1 Von Neumann versus Harvard Class

One might expect a computer born at Harvard to be a Harvard Class computer, but our Dallas design is not. In fact, we went slightly out of our way to make sure it was not. We should define our jargon, in case it's not familiar to you. Some computers place *instructions* in a memory separate from *data*; most do not do this. The ones that separate the two are called "Harvard Class" machines.[20] This separation has been the exception until recently when microcontrollers (notably the PIC processors) have made this design choice more common. The 8051's designers envisioned such a separation – but we have defeated that expectation because we didn't want to ask you to install two memories in your lab computer. (We also like to offer you a computer that is conventional in its design.)

The 8051 distinguishes a *code fetch* from a *data read*. PSEN* ("Program Store Enable*") is intended to enable the *code* memory; RD* should enable the *data* memory. Your *glue* PAL merges the two sorts of memory by simply *OR*ing PSEN* with RD* in forming the signal that drives the RAM's 3-state control, its OE* pin: see Fig. 20N.20.[21]

---

[20] They got this name because this design was used in Aiken's Harvard Mark I, one of the world's first automatic digital computer/calculators (1944).
[21] The third signal, BR*, indicates that we humans want to take over the buses; we require dis-assertion of another signal, ALE, to cover an exceptional case, a serial program-load.

Figure 20N.20 OR'ing PSEN* and RD* eliminates the "Harvard Class" distinction between code memory and data memory.

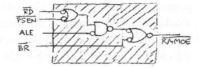

20N.6.2 Address multiplexing

The 8051, like most controllers, makes double use of some of its pins to keep the package small (40 pins in DIP,[22] 44 pins in the more compact PLCC[23] and, consequently, inexpensive. The 8051 saves seven pins by making double use of eight pins for both address and data – one pin is sacrificed to indicate what the eight lines mean.

More specifically, early in every bus cycle[24] these shared lines carry the eight bits of low-order address. A short time later (about 80ns, given the 11MHz clock we use), those lines switch over to carrying the eight bits of *data*. The ALE signal ("Address Latch Enable") serves to catch the *address* information in a latch before it evaporates to be replaced by *data*. Fig. 20N.21 shows the timing (at 11MHz)[25].

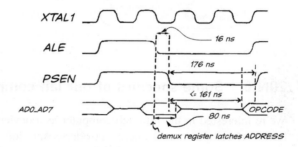

Figure 20N.21 ALE latches multiplexed address information; these shared pins later carry *data* (timing assumes 11MHz clock).

These shared or multiplexed lines are named "AD7…AD0." This sounds like "Address…", but here "AD" means "Address/Data." You may at first be puzzled by the circuit diagram in Fig. 20N.22 showing these lines going to a latch, but also to what is labeled "Data bus."

But this complication is the price we pay for limiting the number of pins on the package. A few other pins, too, serve multiple functions; RST (normally an input, but an output during trouble that forces a system reset), PSEN* (normally an output, but input for the purpose of turning on the on-chip "loader" circuitry). But it is the shared "AD" lines whose multiplexing you will work with every day.

20N.6.3 The single-step logic

As Lab 20L explains, our little computer implements a single-step function entirely in hardware.[26]
Our hardware single-step for the Dallas circuit is simple: it works by sending out just a few clocks

---

[22] DIP, as you know, is Dual Inline Package – our usual breadboarding package.
[23] PLCC is "Plastic Leaded Chip Carrier," a relatively large package by today's standards. It can be soldered as surface mounted or housed in a socket.
[24] Note that this multiplexing applies only when one uses the 8051 with external buses.
[25] The latching shown is standard for the 8051. Some 8051 variants, like Dallas' DS89C4x0 controllers, offer the option of latching the *high* half of address rather than low; in most cases (within any 256-byte "page"), this allows faster bus accesses.
[26] There are other ways to achieve single stepping: one can enable interrupts, then use a level-sensitive interrupt to break into program execution after each instruction (see Intel's "MCS51 Microcontroller's User's Manual," p. 326). And more recent micros permit control of execution from an external computer. This is the way we work with SiLabs 8051: a serial link from a PC allows single-stepping and allows interrogation of on-chip registers – a trick our little Dallas computer cannot do.

at a time, then choking off the clock source. It uses ALE to keep itself synchronized – so that it always stops about halfway through any bus access, at a time when all control, data and address signals are valid.

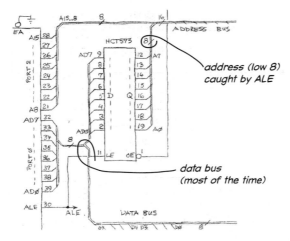

Figure 20N.22 Common lines, AD7..AD0, briefly carry low-half address, then all data.

That behavior makes this single-step very useful for debugging hardware. It also serves to let us debug small programs, by letting us execute their code one line at a time.[27] If, on the keypad, you select STEP rather than RUN then, each time you hit the INC button, the processor is fed enough clocks so that it proceeds about halfway through an operation (which may be an instruction fetch or, for example, an input or output operation.) The STEP PAL does this by shutting off the processor's clock soon after the ALE pulse that marks the beginning of each operation.

Figure 20N.23 shows one such burst of four clocks, terminating soon after ALE; at that time the address has been caught in the '573, and the multiplexed data lines are serving as *data bus*.

Waveforms: Most instruction cycles use four XTAL1 clocks – the 11MHz clock signal fed to the controller.[28] But an instruction that takes substantial execution time will use many more clock cycles. The single-step hardware, by always allowing two clock cycles after the fall of ALE, permits single-stepping regardless of the number of clocks used in a particular instruction.

Schematic: The general idea is simple; the implementation is fussy. To keep clocks intact, not shaved or (worse!) glitchy, the synchronizing flip-flops use the falling edge of the clock. Fig. 20N.24 is an attempt to explain what's going on.[29] Feel free to ignore this gritty detail if you like.

20N.7 The first day on the SiLab branch

Using the SiLabs one-chip design (the '410), you'll have almost nothing to build before you can try out some programs. You need only make the connections from PC to controller, using a pin-sharing

---

[27] "One line..." is a little vague: strictly, it is one bus access at a time.
[28] This is true for bus cycles using the external buses. When running from internal ROM, the processor executes many of its instructions in a single XTAL1 cycle. A further complication (not one you need to worry about): XTAL1, the clock input, is not necessarily the same as the internal clock used by the controller. That internal clock may be a 2× or 4× multiple of XTAL1, stepped up by an on-chip "crystal multiplier" (see *Maxim/Dallas Ultra-High-Speed Flash Microcontroller User's Guide*, pp. 78–79: http://pdfserv.maximintegrated.com/en/an/AN4827.pdf).
[29] We know: it's always hard to make sense of someone else's peculiar design choices, or of someone else's ingenious program code.

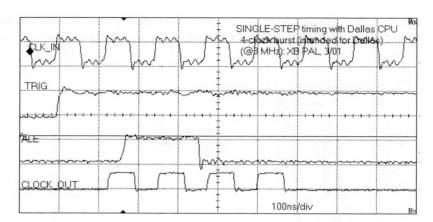

Figure 20N.23 STEP PAL allows single-step by cutting off clock soon after ALE.

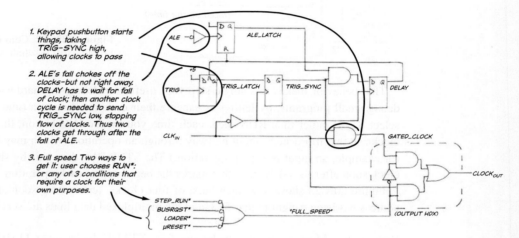

Figure 20N.24 The single-step logic: the details.

scheme that allows debugging control without making a permanent commitment of any controller pins. With just 32 pins, this pin sharing makes good sense – and is preferable to committing four pins full-time as in the more standard JTAG protocol.[30] This proprietary serial link (which SiLabs calls "C2") is described in §20L.3.2.

Since the lab includes next to no hardware work, you will be free to spend your energies getting used to the assembler and to the debugging interface. You will find that you can watch the 8051's registers as you single step. The first day's program (§20L.3.5) does no more than blink an LED – providing the pleasure that we have admitted everyone seems to want from a controller.

20N.7.1 Little test program for the standalone branch: blink an LED

The core of the program: two lines: This program is as simple as the first big-board program (just an endless loop, listed at the end of Lab 20L; we examine this program closely in Chapter 21N) – except that it necessarily includes some background commands used for the SiLabs variant used on this branch.

As we say in Lab 20L, the central part of the program is as simple as this:

---

[30] JTAG is "Joint Test Action Group," the name of an industry group that established a standard way to test circuits by bringing their internal nodes to pins where levels can be read serially.

```
FLIPIT: CPL P0.0 ;   flip LED On, then Off,..., endlessly
        SJMP FLIPIT
```

"CPL" is a single-bit operation that means "complement bit." P0.0 is bit zero of Port Zero, the pin to which the LED is wired.

20N.7.2 A full version of this simple program

AoE §15.2.2

Register initializations: The versatility of controllers is "both a blessing and a curse." To do the simplest task, one may need to set this and that control register appropriately.

You will see in the present program some of these necessary preliminaries. For use of a complex peripheral like a timer, pulse-with-modulator (PWM), or serial port, the initialization chores become daunting indeed. A program using the serial port to send a short message to a PC host computer – a program (serial_message_silabs.a51) that you will meet in §20L.3.7 – requires a preliminary loading of *ten* control registers, for example.

This LED-blink program is not so bad, but does require two initializations in order to run at all:

- enable the output port (turn on the "crossbar" switch, in SiLabs jargon);
- turn off the watchdog timer.

It seems odd that these necessary steps are not the default option; but they are not.

20N.7.3 The first tiny program: blink an LED

Here we will look at that first tiny program. Tiny though it is, it includes some details that are bound to be puzzling at first viewing. Fig. 20N.25 has the assembly language code (on the right) with the machine code that the *assembler* program on your PC generates. This machine code is what gets loaded into the 8051 and what the 8051 executes, picking up these instructions one byte at a time. It is reassuring to see that the code for P0.0 – a single bit – is the same in the lines "SETB P0.0" and "CPL P0.0." That bit address evidently is 80h.

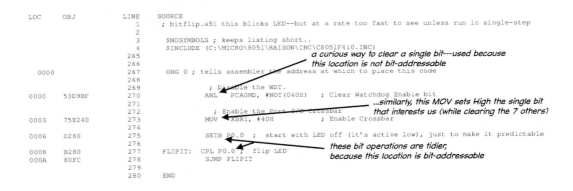

Figure 20N.25 First SiLabs program: blink an LED – but also initialize two essential control bits.

The assembler found this address by referring to the "C8051F410.INC" file, referenced in the $INCLUDE... at the head of the program. That .INC file shows, among many other equivalences, the address of PORT0, 80h. Here is an excerpt from that .INC file:

```
;------------------------------------------------------------
; C8051F410.INC
;------------------------------------------------------------
;
; Copyright 2005 Silicon Laboratories, Inc.
; http://www.silabs.com
;
; Program Description:
;
; Register/bit definitions for the C8051F41x family.
;
; .....


;------------------------------------------------------------
; Byte Registers
;------------------------------------------------------------

P0          DATA        080H            ; Port 0 latch
SP          DATA        081H            ; Stack pointer
; .....
```

The address 80H, listed just above for "P0," is a *byte* address. The distinction between addresses of *bytes* and *bits* can be confusing. Since P0.0 is the zeroth bit in byte 80h, its *bit* address also is 80h. (The bit address of P0.1 would be 81h; and so on.)

Context determines whether 80h is treated as the address of a *byte* or a *bit*. In a *bit* operation like SETB, 80h can only refer to a bit; in a *byte* operation (such as "MOV P1, P0," which would be coded as *858090*), the reference in the machine code to "80" (the MOV destination) can only refer to a byte.[31] In contrast, "P3" would refer to a *byte* rather than to a *bit*. This distinction gets less confusing as you see it in action a few times.

The last line in the program loop, "SJMP FLIPIT," jumps to the *label* FLIPIT that we have placed in the margin. The assembler figures out how far back that is – in this case, four addresses (from 0Ch back to 08h)[32] – so, the assembler plugs in the value FCh, which is *minus four* in 2's-complement. Thus the loop is perpetual (unless you get bored before the end of time, and reset or shut down your machine).

20N.7.4 Enable I/O

The need to turn on the crossbar in order to do any I/O is rather strange that it recalls a Dilbert cartoon (Fig. 20N.26). The '410 output is dead until you "set *thip crinkle and spoit* to 'no' " by setting bit6 high in register XBR1 (see §20N.7.3).

Enabling the output port is necessary in order to take advantage of improvements SiLabs made to the standard 8051 I/O scheme. SiLabs provides a "crosspoint switch" that permits one to steer particular signals to particular pins. This option resembles the freedom available in PAL designs: there, except for some dedicated pins rigidly assigned to JTAG, power and ground, and some preferentially assigned to *global* clocks, each signal may be assigned to any pin.

---

[31] MOV is always a byte transfer, except for a single case, a move into or out of the Carry ("CY") flag (for example, "MOV C, P0.0"). This is the only available *bit* transfer.

[32] You are entitled to be puzzled by our claim that it is *four* steps back. Please take a look at Chapter 21N for a detailed examination of jumps, forward and backward.

20N.7 The first day on the SiLab branch

This crossbar enabling requirement is fussy, but one that is so standard that we'll quickly get used to doing it every time. We will put these very standard settings into a subroutine that we call "USUAL_SETUP."

20N.7.5 Watchdog timer

AoE §15.2.2

The "watchdog timer," which we here disable, does perform a useful service in a final version of any controller code, but we here dodge this complication. The "watchdog" resembles a nervous mother (a "Watch Mom?") much more than it resembles any dog we have met. It wants regular reassurance in the form of continual messages.

Figure 20N.26 Dilbert's photocopier resembles a controller that gives no output till you enable the port. Reproduced with permission from United Features Syndicate, Inc.

If ever the writing to watchdog ceases (no letters from summer camp?), Watch Mom assumes the worst. A failure to write to the timer implies that the program somehow has gone off the rails (the errant child has missed his train?). In that event the controller resets itself (Mom summons the child home?). We do not want to be obliged to include this complication in our test programs so we simply shut off the watchdog (pour Mom a big martini?).

The code that shuts off the watchdog uses a "mask," a programming device we will talk about in Chapter 22N.[33]

```
ADDRESS      CODE       line #       ASSEMBLY CODE              COMMENTS
                        268                                     ; Disable the WDT.
0000         53D9BF     269          ANL   PCA0MD, #NOT(040H)   ; Clear Watchdog Enable bit
```

Detail: assembler directives; $INCLUDE, $NOSYMBOLS, and ORG: Not all the content of the program is directed to the 8051; some of it is addressed instead to the *assembler* program that runs on your PC. Such commands are called "assembler directives."

$INCLUDE The line

    INCLUDE (C:\MICRO\8051\RAISON\INC\C8051F410.INC)

tells the assembler where to find a long table of equivalents, a table that permits us as programmers to use more-or-less intelligible descriptive names for registers and ports, and even bits, on the 8051. Note that on the computer you are using to run the SiLabs IDE ("Integrated

---

[33] In case you can't stand to wait that long, *ANL* AND's the register *PCA0MD* with the value BFh, the complement of the value shown, "040H". The result is to clear a single bit in the register, bit 6.

Design Environment") the path to this essential *.INC* file will differ from what we have shown above. Make sure to learn where your .INC files live, and adjust the "$INCLUDE..." appropriately.

Thanks to the this .INC file, we can write "P0" when we refer to Port Zero, rather than its address (80h). We can write "PCA0MD" (itself a clumsy mouthful) rather than look up its address, which happens to be D9h. The .INC file is useful, but we don't want to get a printout of its long table of equivalents and therefore include one more assembler directive:

$NOSYMBOLS

```
$NOSYMBOLS      ; keeps listing short
```

The comment that accompanies "$NOSYMBOLS" reminds us why we write this line at the head of each of our programs. This line suppresses the assembler's inclination to provide a listing of all equivalences provided by the .INC file.

ORG "ORG 0," means "Origin zero," and tells the assembler to assign address zero to the first executable line of code. This placement was not really a choice of ours, because the 8051 *always* looks to address zero to find what to do after a hardware RESET (implemented by assertion of the RESET* pin). This would be the assembler's default placement, but here is forced by our use of ORG.

To keep things simple, in this first program we let the code fall where it happens to, after the start at address zero. In later programs, at least in big-board code, we will not do this. Instead, we will make our first instruction a *jump* up to a higher address where our programs will begin. We will do that because some particular locations low in memory are dedicated to particular purposes.

Address 03h, for example, is a so-called "interrupt vector," a location to which the 8051 always jumps when interrupted by one particular interrupt source (external interrupt zero). So in general we must not overwrite this and other dedicated locations with our own code. Today, we'll not worry about that, since today we have no use for an interrupt.

We can enjoy the privileges that go with being the dictator in control of our little computer: we need not defer to other possible programs that we may not know about (a program that might, say, need to use interrupt zero). We are in charge; there can be no program that we don't know about.

20N.8   AoE Reading

Chapter 14 (Computers, Controllers and Data Links) :
  §14.1: terminology: microprocessors, microcontrollers; architecture: CPU and data bus; particularly
    §14.1.6: data bus
  §14.2.2B: a detour: addressing
  §14.3.1: fundamental bus signals: data, address, strobe
Chapter 15 (Microcontrollers) :
  §15.1: introduction
  §15.3: overview of controller families
  §15.9.1: software
  §15.10.3: how to select a microcontroller
Web resources re: the 8051:
  Some general sources on the 8051/2, both from website 8052.com:

20N.8 AoE Reading

- Tutorial on the generic 8051: http://www.8052.com/tut8051, secs. 1,2, 4, 6
- Tutorial on the Dallas 8051 variant, 80C320: http://www.8052.com/320tut
- Dallas/Maxim Datasheets, etc.
  "Ultra-High-Speed Flash Microcontroller User's Guide" (henceforth "User's Guide"):
  http://pdfserv.maximintegrated.com/en/an/AN4833.pdf
  features, p. 3
  typical circuit wiring: RAM only, ROM only, pp. 63, 57
  "programming model," pp. 4–6 (memory map)
- DS89C430/50 "Ultra-High-Speed Microcontroller":
  https://www.maximintegrated.com/en/products/digital/microcontrollers/DS89C430.html#popuppd
  signal description, pp. 12–13
  controller block ("functional") diagram, p. 14
  memory map, pp. 19–21

SiLabs Datasheet
- C8051F410 datasheet:
  http://www.silabs.com/Support Documents/TechnicalDocs/C8051F41x.pdf
- C2 programming interface:
  https://www.silabs.com/Support

20L Lab: Microprocessors 1

20L.1 Big-board Dallas microcomputer

This is the first of the microcomputer/controller labs where you can choose to build up the computer from about ten ICs rather tahn start with a single-chip controller like the SiLabs part. We will sometimes refer to this path as the "Dallas" path, acknowledging the company that designed this particular variant on the 8051 design; and sometimes we will refer to this path as "big-board", recognizing the size of the breadboard that this design requires). Alternatively you can start with a single-chip controller. This we do in the SiLabs version of the labs. We will call this the "SiLabs" path. The Dallas route is taken in this, the first section of the lab; the SiLabs path is followed in the final section, §20L.3.

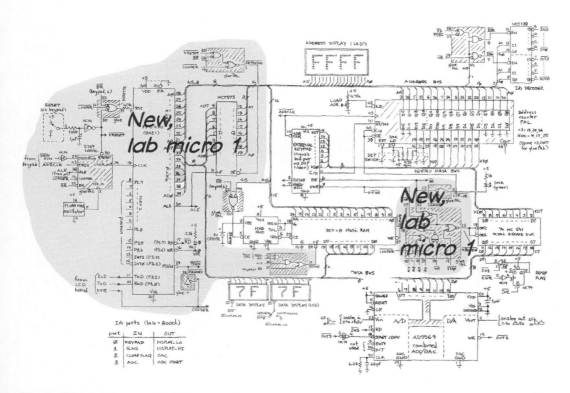

Figure 20L.1 What you'll add to the computer circuit in this lab.

20L.2 Install the GLUEPAL; wire it partially

20L.1.1 What you'll build today

We assume you have completed Lab 17L, and have a working counter and memory, and displays for both. Today, the microprocessor – the brains of the machine – enters, along with a "glue" PAL that should let it talk to the parts you have already built. Within the PAL (which we will call "GLUEPAL") is the purely combinational logic that you designed in Worked Example 17W and also a shift-register-like circuit that implements the micro's single-step function. The single-step circuit feeds the processor a short burst of clocks each time you hit a button ("INC") on the keypad. The termination of this burst of clocks then leaves the processor frozen in mid-cycle, so that you can see a "live" display of address and data values, and so that you can debug hardware that is held in this static condition.

What you'll add in today's lab: Figure 20L.1 is a schematic of the full computer, showing that you've already done more than half the wiring. After today, only odds and ends of hardware will remain.

20L.2 Install the GLUEPAL; wire it partially

20L.2.1 Partial wiring of the PAL

We will postpone wiring some of the PAL's signals: signals that will connect to the 8051 controller, or to parts that are installed in a later lab. We list below the signals that you now *should* connect, and those you need not connect.

At this time, you need to connect some of the GLUEPAL inputs and outputs, but not all.

Inputs Some inputs must be driven by their appropriate sources (BR* is driven by a signal from the keypad, for example). Other inputs must be *disasserted* (LOADER*, for example). Still others may be left open (CLKIN, for example).

Outputs Some will go to their appropriate loads (for example, RAMWE* drives the RAM's *write* pin, as one would expect). Some will go nowhere, for now; these await the processor or other hardware (RESET51, for example). A few will never be connected (TRIG_LATCH, TRIG_SYNC, ALE_LATCH, DELAY, PAL signals that were brought out only for testing).

Fig. 20L.2 summarizes this GLUEPAL wiring.

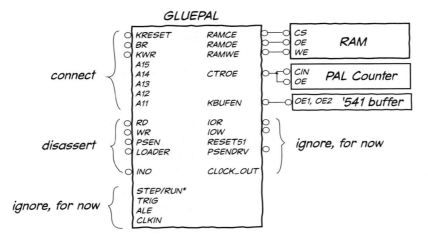

Figure 20L.2 GLUEPAL preliminary wiring: before processor and step logic are added.

Figure 20L.3 shows two details of the PAL wiring: we debounce the reset line from the keyboard (flaky behavior from the processor showed us this was needed). And we use an LED to indicate to us that have control of the buses.

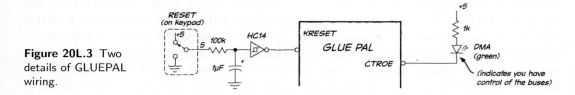

Figure 20L.3 Two details of GLUEPAL wiring.

Figure 20L.4 shows signals in and out of the GLUEPAL. The programming pins labeled "JTAG" you may ignore. The same is true for four signals that you will not use, but which were brought out to facilitate testing during development of the single-step code on this PAL. These four signals are shown close to the center, in rightmost image of Fig. 20L.4: TRIG_LATCH, TRIG_SYNC, ALE_LATCH AND DELAY.

You may want to attach labels to some of these control signals as you connect them to breadboard bus lines (16 uncommitted lines appear at top and bottom of the breadboard) or to ICs on the breadboard.

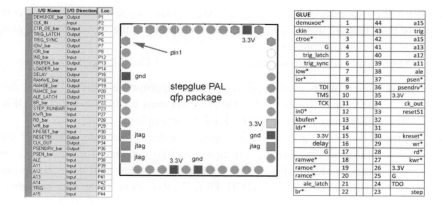

Figure 20L.4 GLUEPAL signal assignments to pins (qfp package).

20L.2.2 Replacing earlier connections

One special difficulty in wiring this computer arises out of the fact that the circuit evolves over several labs, so that later wiring, more complex, sometimes replaces earlier connections. It is easy to overlook these necessary changes. To help you see them, we highlight them here, with illustrations.

GLUEPAL's output CTR_OE* replaces continuous ground that drove counter's CIN* and OE*: BUSRQST* (a signal whose name we often will shorten to "BR" in this discussion) goes into the GLUEPAL, and is slightly conditioned,[1] emerging as CTR_OE*, a signal that now drives the OE* and CIN* pins of the 16-bit PAL counter that you installed in Lab 16L. This signal *replaces* the continuous ground that drove those pins in that lab. See Fig. 20L.5. Generations of students have gotten a kick out of failing to make this change: we recommend that you decline to join them in this error. (You'll hear us kind of thrashing this point later; we hope you're not offended.)

---

[1] As Fig. 20L.5 shows, it is conditioned by the disassertion of the active-low LOADER* signal. Don't worry just now about how LOADER* works. You will use it only at §24L.1, when you begin to pump code directly into memory from a personal computer.

20L.2 Install the GLUEPAL; wire it partially

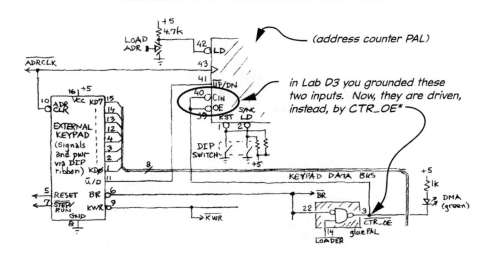

Figure 20L.5 New drive for counter's OE* and CIN* inputs: counter is turned on only when humans have the bus.

GLUEPAL's outputs RAMCE* and RAMOE* replace the constant ground connections driving two RAM enables: In Lab 17L you grounded the RAM's CS* or CE* pin and its OE* (the 3-state control), to enable it continuously (RAM's WE* took care of averting conflicts between RAM and keypad buffer). The GLUEPAL output *RAM CE** now drives RAM CE* or CS* (pin 20); the GLUEPAL's RAMOE* now drives RAM OE* (pin 22). This change is necessary because now the RAM can't be allowed to talk all the time: it competes with the newly-installed *processor*. Through the GLUEPAL, the processor (the 8051) will decide when the RAM should be allowed to drive the data bus. In Fig. 20L.6, a fragment of Fig. 20L.1, we have faded-out the image of the MXD1210 battery-backup IC. Ignore that part; you will install it next time.

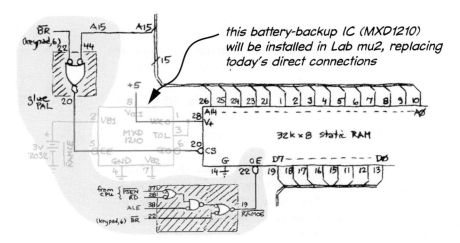

Figure 20L.6 New drive for RAM's CS* and OE*.

GLUEPAL's output "RAMWE*" now drives the RAM's WE* (pin 27): This rewiring replaces the direct connection, made in Lab 17L, from the keypad's KWR* line. KWR* is now just one of several inputs that can generate a RAMWE* signal.

GLUEPAL's output KBUFEN* now drives the '541's OE* pins (1 and 19): This rewiring is shown in Fig. 20L.7.

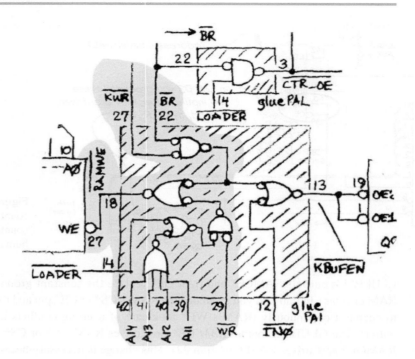

Figure 20L.7 New drive RAM's WE*.

The GLUEPAL thus mediates between the manual write signal, KWR*, from the keypad, which in Lab 17L drove the RAM's WE* pin directly. Again, KWR* now provides only one of two conditions for turning on the keypad buffer. The GLUEPAL permits an alternative: the processor, as well, will be able to control the buffer.

Faking busrequest/busgrant with RESET: Here's an explanation of the way we're using the 8051's reset to fake a bus request. The 8051 does not offer a bus-sharing option like a conventional microprocessor's: we cannot truly "request the bus" for direct-memory-access (DMA), because Intel did not envision a need for this operation back when they designed the 8051.

However, we can fake this function for ourselves by applying RESET (a signal labeled "RESET51" on this lab's circuit diagrams; a rare active-high signal) to the processor when we want to control the buses. This RESET makes the processor shut off all its port drivers (applying a weak high pullup to the pins).

We then jump in to take advantage of this fact: we use BR* to turn on our address counter's three-states even as BR* makes the processor turn off its own drivers. Our drivers easily overpower the weak pullup. Our 16-bit counter then controls the address bus; our '541 3-state buffer now is permitted to drive the data bus with a keypad value (though only when we press the WR* key). In short, the assertion of BR* makes your machine revert to its condition of Lab 17L when you were able to write data into RAM by hand.

Here's the effect of BR*, recapitulated:

- BR* leads to a turn-on of the counter's 3-states;
- BR* is a precondition for a *manual* write to memory;
- BR* permits counting when asserted, freezes the counter's state when disasserted (by controlling the counter's CIN*).

The third of these effects permits us to use the keypad's INC button for miscellaneous purposes while a program is running without upsetting the address counter's state. Lacking assertion of BR* while the computer is running, the counter will ignore the counter-clock pulse that is generated each time you hit INC.

20L.2.3 Testing

When the GLUEPAL is partially wired, as shown in Fig. 20L.2, some testing is possible even though the micro's signals are not yet available.

Can you still control memory by hand? You should find that when you request the bus (assert BR* on the keypad) the green LED lights to indicate that you now control the buses. The RAM should be enabled, and in general you should find that the address counter and memory work as they did in Lab 17L.

By contrast, when you disassert BR*, address and display lines should float: nothing is driving them, and they are likely to show the values FFh because they are weakly pulled up by the LCD board. Later in this lab, when the 8051 enters, it will take over driving the buses as soon as we release BR*.

Does the GLUEPAL's write-protect work? You can test write-protect by going to the border between protected and unprotected RAM and faking a CPU write. Here's the procedure.

- Take the bus, as usual (assert BR*).
- Apply the highest *protected* address to the RAM: 7FFh (7FF "hexadecimal"). Here is a quick way to get there: *load* address 800h then decrement the counter.
- Manually *ground* the GLUEPAL input pin, WR* (this pin has been tied high; you'll need to break that connection, of course, in order to do this temporary grounding). This grounding mimics or fakes a CPU write, (distinguished, please note, from the *manual* write that is signaled by KWR*).
- Watch the RAM's WE* pin with a logic probe. It should *not* go low.
- Now *increment* the address so that you now are applying the first *unprotected* address, 800h.
- Confirm that the RAM's WE* pin now does what it should.[2]

When you are satisfied that the gating works as it should, be sure to *disconnect the temporary ground on the GLUEPAL's WR* line*. In the next subsection you will replace this temporary connection with drive by the processor's WR* signal.

20L.2.4 Install the processor

Now you can install the 8051 – a variant from Dallas Semiconductors, DS89C430. Make sure that you place this chip conveniently close to the bus connectors because it has a lot of lines to tie to that bus. You may want to install it with pin 1 *down*, to make the orientation of its buses match that of the big green breadboard (MSB up).

Note: you now must disconnect three temporary connections to +5 on the GLUEPAL: back in §20L.2 you *disasserted* RD*, WR* and PSEN*. These lines *must now be driven by the 8051*, in place of those three temporary Highs used for the initial tests of the GLUEPAL.

---

[2] The PAL input WR* now should evoke an assertion of the PAL's output RAMWE*, driving the RAM's WE* pin low.

Lab: Microprocessors 1

PSENDRV* detail: Note also the curious fact that the PSEN* on the 8051 serves as both *input* and *output*; thus it *drives* the GLUEPAL's PSEN*, but is *driven by* another output of the PAL, the signal called PSENDRV*.[3]

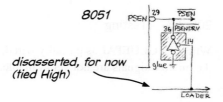

Figure 20L.8 PSENDRV* detail: PSEN* is both an *output* (ordinarily) and an *input* (once in a while).

You will not connect all of the 8051's 44 pins now (and some pins we will never use in these labs). Fig. 20L.9 is a pinout label that you may want to photocopy and glue to the 8051.

Figure 20L.10 is the processor as it appears in a fragment of Fig. 20L.1. We have included the 74HCT573 register, described below. Wire the connections that are shown (some you wired earlier: KRESET*, BR* and LOADER*).

The 8051 RESET wiring: Ground the EA* pin. EA means "External Access," and asserting this signal tells the processor to look to external memory for its code. In Lab 25L we will, for the first time, take EA* high, leading the controller to execute code from its internal ROM. But until then we'll leave EA* grounded.

The processor's RESET line (labeled "RST" or "RESET51" in our circuit diagrams, including Fig. 20L.10) is driven by the GLUEPAL through a 4.7k resistor. Including this series resistor may look odd but is useful. We include the resistor because the RST pin operates occasionally as an *output* from the 8051, and the resistor permits us to probe independently the 8051 RST signal, on the one hand, and on other hand the GLUEPAL signal RESET51 that ordinarily drives RST.

Occasionally the processor itself drives RST (asserting it High) to indicate that the processor has encountered a problem that it cannot make sense of. In desperation, it does what generations of computer users have learned to do when frustrated: it hits Reset. It does this to initialize all the things attached to it; this is the 8051's effort to *reboot*. In today's circuit, the processor's assertion of RST would accomplish nothing useful. But the resistor does help us in our debugging: it allows us to detect whether an assertion of RST comes from us (the normal case) or from the processor (the abnormal case, signalling trouble). If you find the latter case, 8051 asserting RST, you will need to cycle the computer's power, to get the machine out of its hung-up state.

Figure 20L.9 8051 label.

Wire the address/data buses: Wire the multiplexed "ADx" lines (pins 39 to 32) to the 74HCT573 latch. Note that the "**AD…**" label does not mean *address*; the address lines are labeled simply "A…." The "AD…" label was chosen to indicate the dual role of these lines: sometimes they carry **A**ddress; at other times they carry **D**ata.

Don't be rattled by the fact that the eight '573 *input* lines, AD0..AD7, are also to be wired to the 8-bit *DATA BUS*; that's how the de-multiplexing works: one source, two destinations. And note, in Fig. 20L.10, the slightly annoying fact that, on the *output* side of the '573, the *lower* pin numbers are assigned to the *higher* address lines, in this set of eight. Incidentally, you can reverse this arrangement if you like – putting high address lines on the high pin numbers – if this helps in your wiring. If

---

[3] This arrangement may worry you. Are we setting up a logic fight by joining two outputs? No, because a 3-state on the PAL averts any possible clash. PSENDRV* is asserted only when PSEN* itself is sure not to be driving. We're confident that you recognize that these two GLUEPAL signals, PSEN* and PSENDRV*, are *not one signal, but two*.

you do this you must of course take care to apply your reversing to both sides of the '573: input as well as output. Once again, it may help to install the '573 *upside down* if you want to use the signal assignments shown in this lab's circuit diagrams, such as Fig. 20L.10.

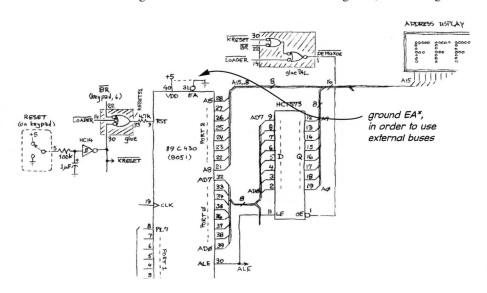

Figure 20L.10 Controller wiring, excluding the STEP PAL signals (which are coming soon!).

The '573 latch is clocked by ALE, and serves to "demultiplex" the lines: while ALE is high, early in the bus-access cycle, this "transparent latch" passes these eight lines through to address lines A0..A7, the lower eight bits of address. On the fall of ALE, the latch catches and holds these eight values. Once those eight address lines are safely latched into the '573, the 8051 can apply those Address/Data lines to their alternate role: defining the computer's 8-bit *DATA* levels.

Fig. 20L.11 details the way the '573 demultiplexes the eight controller lines AD0...AD7. This timing diagram shows a particular case: a processor *write*. This case lets one see that the ADx lines carry first address (low byte), then data. ALE (active *high*) catches the address value in the '573 *transparent* latch, saving it after ALE falls.

20L.2.5 Single-step function in the GLUEPAL

A second section of the PAL is given the task of providing bursts of clocks, usually four, synchronized to the micro's ALE signal, which comes at the start of each bus access. We will use this single-step function whenever we want to watch the processor execute instructions one at a time. This single-step PLD, which is incorporated into the combined GLUEPAL, uses the inputs and outputs shown in Fig. 20L.12. You will recognize that some of these input signals are also used for other functions.

20L.2.6 Signals into and out of the single-step PLD

For now, simply disassert LOADER*. Drive TRIG with ADCLK from the keypad. ALE comes from the micro.

The keypad signals, STEP*/RUN and the debounced RESET, both need squaring up by a 74HC14 Schmitt trigger, as shown in Fig. 20L.12. The HC14 also inverts each signal, and the active levels of the PAL's two inputs take account of this inversion. The PAL inputs thus are named STEP/RUN* and KRESET*. The oscillator's peculiar pin numbering is explained just below in §20L.2.7. CLOCK_OUT goes to the micro's XTAL1 input at pin 19.

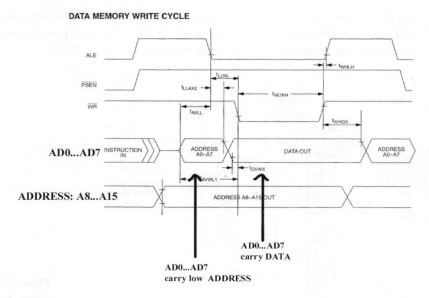

Figure 20L.11 8051 Address/Data multiplexing: time-sharing on eight lines, AD7..AD0.

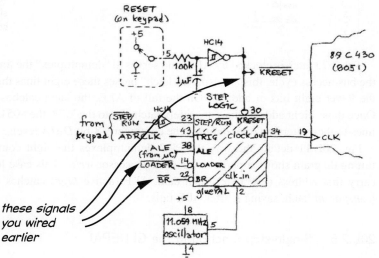

Figure 20L.12 Inputs and outputs of STEP portion of GLUEPAL.

20L.2.7 Clock: 11MHz crystal oscillator

Figure 20L.13 shows the oscillator. It is housed in a square package shaped as if it were an 8-pin "DIP," and the pins are shown numbered as if the package had 8 pins, though in fact it has only four.

The frequency value looks strange (why not give us a nice round number?) but is chosen to facilitate serial communication by the 8051 with a PC. Later, we expect you will use such a link.[4] Take a look at the oscillator's output on a scope – and notice that any ugliness of the edges of this waveform probably is caused mostly by the long ground connection to your scope probe. If you make a special effort to shorten that lead, using the trick shown in Fig. 20L.14, you'll get a fairer view of what your clock looks like. The spring-like coil on the upper probe in this figure shows an adapter from Tektronix;

AoE §12.2 and Fig. 12.32

---

[4] The frequency provides a convenient timebase as the 8051 runs a built-in program that tries to find a transmission and reception rate that matches the likely standard rates offered by the PC. For a partial explanation of use of 11.059MHz and its use to generate a particular baud rate, see http://www.8052.com/tutser.phtml

20L.2 Install the GLUEPAL; wire it partially

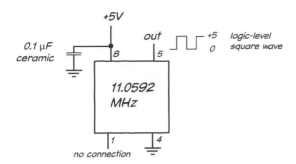

Figure 20L.13 11.0592MHz crystal oscillator.

the lower probe shows the more useful way to minimize ground length: just a short length of wire wrapped around the probe's tip. You won't always have access to the Tektronix accessory; you'll always have a bit of wire. On the other hand, you needn't be such a perfectionist. After all, this is a *digital* signal and, as long as the edges are clean, a clock with ugly wiggles close to ground and +5V shouldn't trouble your computer.

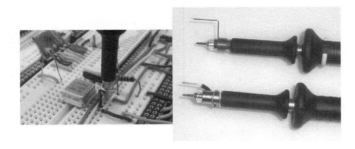

Figure 20L.14 A minimal ground lead minimizes ringing.

20L.2.8 The STEPGLUE PAL, fully wired

Putting together the newly-added STEP signals with the preliminary glue signals shown in Fig. 20L.2, the PAL input and output signals look like Fig. 20L.15.

STEP PAL output waveform: As you know, the STEP PAL is included to allow us to *single-step* the computer. If you select STEP rather than RUN (using the ST*/RUN slide switch on the keypad), then each time you hit the INC button on the keypad the processor is fed enough clocks so that it proceeds about halfway through an operation, as we said back in §20L.1.1. This operation could be an instruction fetch, or could be, for example, an input or output operation. The STEP PAL achieves this clock-metering by shutting off the processor's clock soon after the ALE pulse that marks each operation.

Figure 20L.16 shows a scope image of one such burst of four clocks. At that time, the address has been caught in the '573, and the multiplexed data lines are carrying *data*. You'll be able to watch this process soon, once your computer has demonstrated that it can single-step. A digital scope is useful to get such a display.

The annotations at the bottom of Fig. 20L.16 show the process of "stepping" from one instruction to the next, in the test program of §20L.2.9. As the figure's *address* (labeled "ADDR_lo8") and *data* lines indicate, the instruction fetched in this line is the *80* at address 11(hex).

Here's a *summary* explanation of the STEP logic:

- the signal **ALE** provides a reliable timing marker because it occurs:

- once for each bus cycle;
- close to the start of each cycle.

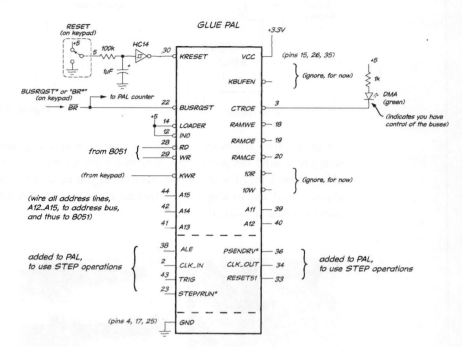

Figure 20L.15 Full wiring of STEPGLUE PAL.

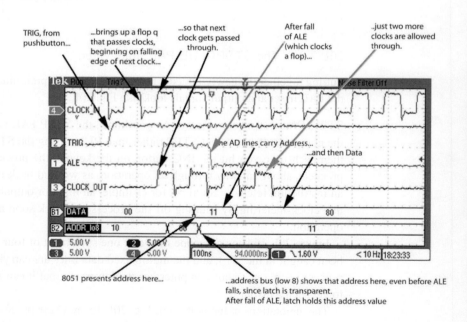

Figure 20L.16 STEP PAL allows single-step by cutting off clock soon after ALE.

- the fall (trailing edge) of ALE marks the time when the full 16-bit address is present (the low-half has been caught in a register);
- soon after that event, the *data* lines become valid.

At that time, our STEP logic chokes off the clocks to the processor since all control and bus lines now are valid, and available for us to examine at our leisure.

20L.2.9 Test program

Enter the program...: Enter this tiny test program, paying attention to the addresses, as you key these values in. The column labeled "LOC" shows the *address*; the column labeled "OBJ" shows the instruction (called "object-") *code* to be executed. When four hex characters, such as 800E, are listed at one address (zero, in this case), that doesn't mean "try to squeeze 16 bits into one memory location." It means "put this pair of bytes into address zero and the location that follows (that is, address one)."

```
MACRO ASSEMBLER FIRST_TEST

   LOC     OBJ         LINE    SOURCE
                       1               ; FIRST_TEST.A51 Lab 18L: confirm that the circuit is a computer!
                       2
   0000                3               ORG 0       ; tells assembler the address at which to place this code
                       4
   0000    800E        5               SJMP DO_ZIP ; here code begins -- with just a jump to start of
                       6                           ; real program.  ALL our programs will start thus
                       7
   0010                8               ORG 10H     ; ...and here the program starts, at hex 10 ("...h")
                       9
   0010    00          10              DO_ZIP: NOP ; the least exciting of operations: do nothing!
   0011    80FD        11              SJMP DO_ZIP ; ...and do it again!
                       12
                       13              END
```

In case you're curious to make sense of the code itself, "80" means "Jump." The "0E" that follows tells the machine *how far* to jump relative to where it is now (at address 2, just after the jump instruction itself). "0E" is, in other words, a displacement or offset value expressed in 2's-complement: $0E_{16} = 14_{10}$. From the reference address of 2 then, this takes the machine to $10_{16} = 16_{10}$. The last instruction in the loop is very similar: "80FD." Here the FD is again a displacement in 2's-complement; FD is -3 and takes the program from 13_{16} back to 10_{16}.

We will look closely at this program again in Chapter 21N, and there we will repeat some of what we've just said. So don't work too hard, yet, to understand the details of this program.

... and run the program: The procedure is simple, once the program is in RAM.

1. Slide the ST*/RUN switch on the keypad to ST*.
2. Assert RESET, by sliding this keypad switch to the right.
3. Disassert BUSREQUEST* by sliding that keypad switch ("BR") to the right. You should see the data and address displays go to all Fs: all bus lines now float.[5] The floats are likely to look like all Fs because the LCD board's inputs include weak pullups.[6]
4. Disassert RESET by sliding that keypad switch left.
5. Hit INC twice.

You should see 0000 on the address display, 80 on the data display.[7]

If everything is working, you should be able to watch the processor walk through the tiny program that you have entered. The processor will look to address 0 for its first instruction, then it will jump to 10h (hexadecimal address 10), where it should execute this tiny loop, marching from 10 through 12.

---

[5] They don't perfectly "float," in the usual sense, but are weakly pulled high. This weak high is designed to permit an external device to drive the line. It's a sort of poor man's 3-state, as we said back in our discussion of "faking" busrequest on page 784.

[6] The LCD board's inputs go, in fact, to another 8051 whose job it is to take the parallel input bytes and send them to the liquid crystal display.

[7] If you don't, try briefly switching ST*/RUN to RUN, after disasserting RESET as suggested above. Then switch back to ST*, and try RESET once more: RESET, release and INC twice. That should do it.

Incidentally, you will see the processor fetch one instruction that is *outside* your loop: it fetches the byte just beyond each of the two *JUMPs* in this little program: the byte at address 2, and the byte at address 13h.

The processor will march from 10 to 13, then back to 10 *forever* – or, till you find the thrill is wearing off. When that happens, try switching from single-step to RUN (slide the keypad's ST*/RUN switch to the right).

When the machine runs this loop, you're entitled to congratulate yourself – and to take a rest from all this wiring! You have done the hard wiring now. Hardware additions in later labs will be minor.

20L.3 SiLabs 1: startup

In this lab through to 25L.2 we invite you to take the quick path to using a standalone microcontroller, as we suggested in the note that described the two alternative paths. After 25L.2, as we have said before, the two paths converge, and at that point we hope you will dream up an application of your own for the microcontroller – whichever route you took to reach this junction.

20L.3.1 Our controller: the SiLabs C8051F410

We chose an 8051-type controller, first (perhaps obviously) to make the two "branches" of this course consistent: one discussion of internal architecture and assembly language covers this 8051 and the Dallas part used in the other branch. A great many 8051 variations are available, of course, and many would have been satisfactory. Here is a summary of considerations that led us to the 'F410:

- modest package size (32 pins), quite easy to solder to a DIP carrier (though not so convenient as a part that is issued in DIP, like the AVR parts).[8];
- good set of included peripherals;
- both ADC and DAC[9];
- analog comparators;
- PWM output
- hardware serial protocols:
  - UART: RS232 standard serial port, useful to let the controller communicate with a full-size computer (just about every controller offers this)
  - SPI (the simplest of serial protocols for communication with peripherals); this scheme is simple enough so that on the other branch of labs we were able to implement SPI in code for that Dallas part that lacks the built-in SPI hardware (see Lab 24L). But SPI is even easier when one finds it implemented in hardware, as it is for the SiLabs '410
  - I²C (a fancier serial protocol).
- on-chip oscillator (accurate enough to permit UART communication and other work, without addition of a crystal oscillator);
- versatile signal and power interfacing:
  - the part includes an on-chip voltage regulator, so it can be powered by +5V while generating the 2.5V that it uses for its "core" logic;

---

[8] DIP parts are becoming so scarce that there may be some value in introducing you to a way to work with these surface-mount parts. As SMT replaces DIP, a few resourceful manufacturers continue to produce adapters or carriers that permit breadboarding a prototype with a DIP. Some soldering skill – or at least patience – is required. But the relatively generous 0.8mm lead spacing of the '410 – tight, but not so tight as the 0.5mm spacing of some other parts – makes the task not difficult.

[9] ADCs are common; DACs less so.

- its V_{io} pin allows one to set the swing of input and output logic levels to one's convenience: ordinarily, we will use +5V; in one exercise – when interfacing to a 3.3V serial RAM – we will apply 3.3V to this V_{io} pin[10]
- reasonable speed: 24.5MHz clock rate, and many instructions execute (as on the Dallas part) in one clock cycle (in contrast to the 12 cycles used by the original 8051)

These are the main features that show the part's competence. But more important than any one of these is the *debugging* facility that SiLabs offers. Like any controller, the '410 can be loaded with code from a full-scale computer (the Dallas part does this too). But the '410 also allows crucial debugging options – without which a controller can become a maddeningly mysterious *black box*. More specifically, here's what the debugging interface permits in addition to the obvious program *load*:

- single-step (this we achieved for the Dallas part with an external piece of hardware: the single-step PAL logic);
- display of the contents of most registers and ports of the 8051, and of internal RAM. This facility is not quite as refined as for the RIDE simulator that you can use to test code before loading it into the controller. But the display capabilities are good. They provide strong clues to what the controller is doing as it executes your code.

20L.3.2 C2: programming pins

SiLabs uses a proprietary 2-wire interface to program and interrogate its small controllers. This interface, called "C2," is good for controllers with few pins because it does not make *full-time* use of any pin, unlike JTAG links, as noted below. The LCD board that you have been using as display in recent labs includes C2 signals – C2D (data) and C2CK – these signals link the controller to a laptop though a USB connection. C2D and C2CK tie to the '410's respective pins 32 and 2. The C2 signals are provided on a 5×2 *header* at the top left corner of the LCD card.

Route USB signals to C2 path: *Important*: In order to use the LCD board for C2 programming, you must make sure that the slide switch labeled "C2" versus "Serial" at lower left of the board is in the *C2* position: see Fig. 20L.17

Figure 20L.17 Switch selects C2 link to controller.

Details of C2 wiring: our breadboarded C2 link: It is possible simply to connect the two C2 pins (clock and data) from the programming pod directly to the '410, as shown in Fig. 20L.18. But we'd like to avoid committing two pins to the debugging function, given the small number of pins available on this controller. For contrast, note that on some of its larger controllers SiLabs uses the pin-greedy but standard JTAG interface. JTAG occupies *four* pins, full-time. But the waste of even two pins is not tolerable for a little 32-pin part like the '410. And we need a RESET* function in any case.

---

[10] The low-voltage of the "core" is characteristic of recent ICs. SiLabs, which offers several parts aimed at low-power designs, includes some controllers that can be powered at 0.9V. This sounds a little more magical than it is: the supply voltage is stepped up, on-chip, by a flying-capacitor charge pump. But 0.9V is pretty spectacular, anyway.

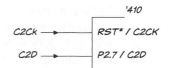

Figure 20L.18 Simplest possible C2 wiring — a scheme that disables two pins. We don't use this arrangement.

So instead of using the simplest wiring that is shown in Fig. 20L.18 we adopt the slightly more complicated arrangement of Fig. 20L.20. This wiring preserves the utility of both debug lines: pin 2 can serve as a manual-reset* input when not in use for C2; pin 32 can serve as a line for general-purpose I/O (GPIO).

The wiring that permits such pin-sharing is quite fussy, and we use a homebrew scheme to make the C2 connection. We do this because we want the '410 controller to feel to you like just another piece of hardware, like the many others that you have breadboarded in this course. We are trying to minimize magic.

Perhaps we ought to admit that there is an easier way to try out a controller, though it is one we will not use today. The easier way is to buy from SiLabs a "daughter card," a small printed circuit with a '410 and a few other parts including points for soldering to its I/O pins. This daughter card can be pressed onto a SiLabs "base adapter" card, which ties to a personal computer through a USB cable. There is nothing at all to build.

After this course we expect you will use such a scheme the next time you use a controller. We have enjoyed the versatility of SiLabs' daughter cards. When we needed many I/O pins, for example, as we did for the LCD card, we just chose a SiLabs part with 60-odd lines and shoved it onto the same base adapter that we had used with the '410, and had used also with a still smaller controller. The development system, like the base adapter, works happily with a wide range of parts from this manufacturer. But today we ask you to work a little harder.

Connector: To link the SiLabs programming connector to a breadboard using its 10-line cable, we bend leads on a *header*, as shown in Fig. 20L.19.[11]

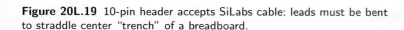

Figure 20L.19 10-pin header accepts SiLabs cable: leads must be bent to straddle center "trench" of a breadboard.

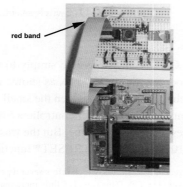

Figure 20L.20 Cable and header pinout: two "C2" signal lines, plus one "readback" line, along with power and ground.

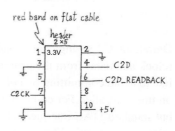

---

[11] The header is a TE Connectivity part 5103310: low-profile, right-angle. Digikey part number is A33179.

20L.3 SiLabs 1: startup

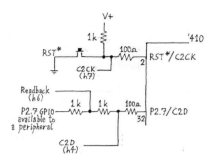

Figure 20L.21 C2 wiring details (rather baroque).

Another view of this header appears in Fig. 20L.24. Fig. 20L.20 shows the cable and connector pinout. Two of the header's ten pins are not used. Do not connect the 3.3V output today.

The cable will deliver power to the '410 (+5V) drawn from the USB connection to the PC. So your circuits will run without any additional power supply. You will not use the 3.3V output until §25L.2, when we meet a serial RAM that needs the lower voltage.

Figure 20L.21 shows the pin-sharing that permits use of the link to a PC without sacrificing two '410 pins. The wiring is odd and fussy, as promised – and some details of this wiring are not at all obvious.

- The two 1k resistors placed between the C2D (data) connection and a peripheral labeled "P2.7 GPIO" are not hard to understand. These resistors provide a simple sort of *isolation*: C2D can drive the '410 without fighting the peripheral.
- The 100Ω resistors between C2 and '410: these are just for protection. They are included to limit current in the '410 pin-protection diodes if C2 signals happen to be applied when the '410 is not powered.
- The connection at the junction between the two 1k resistors in the case of the C2D line: this is a terminal that the programming pod uses to hold the peripheral voltage constant despite activity on the C2 line. This works even when the peripheral is a '410 *output* (a case illustrated in Fig. 20L.22). This feature is not always necessary, but becomes useful if one single-steps a program that uses P2.7. In that case, the C2 lines are intermittently active even as the '410 is driving this P2.7 line used as an output.

One can see in Fig. 20L.22 the effect of this extra connection, which we have labeled "readback." The scope image shows a program that blinks an LED (you'll see this very soon in §20L.3.4). The program is running in *single-step*, controlled by the C2 link.

As the program runs, C2 is communicating with the '410, and these signals appear as the quick pulses that mix with the slow square wave on both top traces. The C2 signals appear at the peripheral in line "b)" of the left-hand scope image of Fig. 20L.22, but disappear when corrected by the additional "readback" connection. The clean LED drive appears in the lower right-hand image of Fig. 20L.22.

This signal cleanup is not of major importance, and in full-speed operation it matters not at all. But it is easily achieved, so let's take advantage of this option.

C2, power and ground to the '410: Power and ground connections for the '410 are few: see Fig. 20L.23. The power connections call for some explanation:

- V_{reg}: +5V: this is the input to an on-chip regulator that generates the core 2.1V or 2.5V used by the '410;

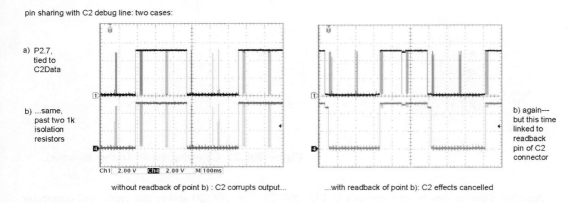

pin sharing with C2 debug line: two cases:

a) P2.7, tied to C2Data

b) ...same, past two 1k isolation resistors

b) again---but this time linked to readback pin of C2 connector

without readback of point b) : C2 corrupts output...   ...with readback of point b): C2 effects cancelled

Figure 20L.22 C2 drive detail: programming pod can hide effect of C2 talk, on "shared" I/O pins.

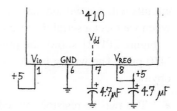

Figure 20L.23 Power and ground connections for the '410.

- V_{io}: +5V: this determines the top of I/O output swing. Setting this to +5V provides input and output properties like those of 74HC: $V_{IH\_min} = 3.5V$; $V_{OH\_min} = 4.5V$ when sourcing 3mA in "push-pull" mode (an enhanced mode, much stronger in *sourcing* current than the generic 8051 output, which, in the style of TTL, is highly asymmetric);
- ground;
- leave V_{dd} open – but decouple it with a tantalum cap to ground (at least $1\mu F$). This is the output of the on-chip regulator. At the moment we are not using it.

Fig. 20L.24 shows the way we wired the header-connector and pushbutton, on a breadboard. The left-hand image shows the board ready to accept the '410 controller. Power and ground connections, along with two decoupling capacitors, also are shown. The LCD board provides the +5 power supply (borrowed from the USB source). The right-hand image shows controller and pod cable in place.

Make sure to keep access to at least one row of connection points at each of the 3 ports – P0, P1 and P2. Even better, give yourself access to *two* rows on top (at P2, P0) because we use some P0 pins for double duty, and it's convenient to be able to leave wires in place rather than remove and later re-install. We hope you will include a label on your '410 carrier: it helps a lot.

The network of resistors is a little convoluted, arranged that way in order to keep it compact. You may find a neater way to do this wiring.

20L.3.3   Pinout of '410 as assigned in these lab exercises

SiLabs gives a user some choice in the assignment of signals to pins, so a '410 pinout does not look like that of an ordinary IC. (In this respect, the '410 resembles the PALs that you have met; there, the freedom to assign signals was even greater than for this microcontroller.) This freedom is provided by a port "crossbar." This crossbar – which is not a native 8051 feature – must be enabled in any program where it is used.

20L.3 SiLabs 1: startup

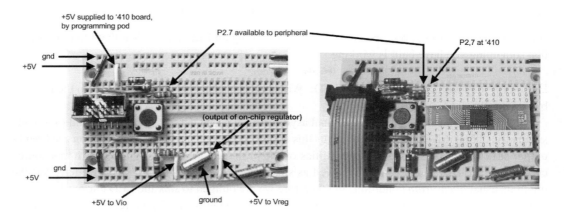

Figure 20L.24 Power, ground, and C2 signal-sharing wiring, shown without and then with controller and pod cable.

Figure 20L.25 shows the way we use the pins of the '410 in Silabs controller labs in Days 20–25. We understand that Fig. 20L.25 shows much more than you need to know today, and refers to some signal names that may be puzzling. We include all these signals so that you will have a reference to go to if later you need it. The numbers shown in parentheses indicate the particular labs for which the pins are so used.

Some of the pins show a second use: "…/ADC0," for instance, at P0.0, the pin that today drives an LED. That second use comes later in §23L.2, and at that time the earlier use will have to be disconnected (in this example, the LED). For some functions, we had complete freedom to assign a pin (LED for example). For others we had no freedom (RX0, TX0 for instance).

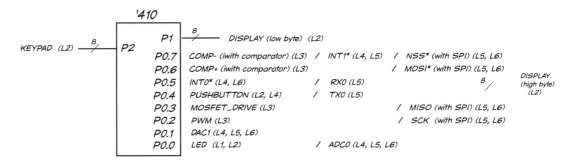

Figure 20L.25 '410 pinout as we have assigned the pins in SiLabs controller labs (Days 20–25).

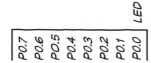

Figure 20L.26 Port pin use in this lab.

Because the '410 pinout is not fixed, we will show pin use near the beginning of each '410 lab. Today's use, see Fig. 20L.26, is the simplest: just an LED at P0.0.

20L.3.4 A very simple 8051 program: code to blink an LED

As we have said before, everyone gets a kick out of seeing an LED blink when that blinking means that something you have built is working.[12]

Hardware: '410 sinks current from an LED: Wire an LED to P0.0 (top right pin) to turn the LED on by sinking current from the +5V supply.[13]

The SiLabs '410 behaves like a traditional asymmetric 8051 by default – but can be told to provide symmetric output drive instead (SiLabs calls this option "push–pull" to describe the symmetric CMOS output structure). The sink/source capabilities of the push–pull are nearly symmetric: it can source current into an LED (3mA @4.5V), as well as sink current (8.5mA @0.4V).[14] But we have left the pin in its usual *open-drain* condition, and therefore we are obliged to *sink* the LED current.[15]

We did this because we think it's useful for you to get a habit of *sinking* current from a load since you will often meet devices that show the asymmetry of TTL-like outputs: good at sinking current, poor at sourcing it. The CMOS PALs are such devices; so is the 8051 in its original output mode. P0.0 sinks current through the LED whose anode is fed, through a 1k resistor, from the +5V supply.

Figure 20L.27
LED wiring: sink current through LED rather than source it.

20L.3.5 Code: blink an LED

The main loop: The core of the program is as simple as this:

```
FLIPIT: CPL P0.0 ;   flip LED, ON, then OFF...
        SJMP FLIPIT
```

CPL is assembly-language shorthand for *complement*; SJMP means *short jump*, where "short" means "short enough so that a single byte will specify how far to jump." Since the "how far" byte is a 2's-complement value, the available range is −128 to +127 (1000 0000b to 0111 1111b).

The assembly language source: Here is that little loop, along with some preliminaries that the SiLabs controller requires; these initializations are explained in Chapter 20N.

```
; bitflip.a51   this blinks LED -- but at a rate too fast to see unless run in single-step

$NOSYMBOLS ; keeps listing short..
$INCLUDE (C:\MICRO\8051\RAISON\INC\c8051f410.inc)


ORG 0 ; tells assembler the address at which to place this code

            ; Disable the WDT.
            ANL   PCA0MD, #NOT(040h)   ; Clear Watchdog Enable bit
```

---

[12] In fact, LED-love goes deeper than this. An LED may not even need to blink to keep us happy. A glowing LED reassures anyone who plugs in a well-designed battery charger. The chargers that save 50 cents by omitting the LED seem likely to fade from the market. Users like reassurance.

[13] This is the standard way to drive a load from an output that shows asymmetric "TTL-like" drive. Old "TTL" logic can sink much more current than it can source. If this applied only to TTL, we could dismiss this as a historical quirk. But we cannot be so dismissive because many CMOS parts choose to mimic TTL. The Xilinx PAL that you met recently does this. So does the Dallas 8051 that appears in the other branch of these micro labs. Such devices can *sink* substantial currents (1.6mA for the Dallas 8051 for example), but can *source* only the feeblest trickle (50µA for the Dallas 8051). Ordinary CMOS parts, by contrast, provide nearly symmetric drive, as you know.

[14] '410 datasheet Table 18.1.

[15] PORT0 of the 8051 is peculiar in that its default configuration is *open-drain*, in contrast to the other ports which provide a weak pullup resistor, and a brief stronger High drive that speeds charging of a load's stray capacitance.

```
            ; Enable the Port I/O Crossbar
        MOV   XBR1, #40h              ; Enable Crossbar

        SETB P0.0 ; start with LED off (it's active low), just to make it predictable

FLIPIT: CPL P0.0 ;  flip LED
        SJMP FLIPIT
END
```

Machine code produced from the assembly-language source: Just above, we listed the *assembly-language* version of this program. The SiLabs (or RIDE) *assembler* program on your PC converts this to code the 8051 can execute. This executable code appears below alongside the original assembly language:

```
ADDRESS     CODE        line #    ASSEMBLY CODE

0000                    266       ORG 0 ; tells assembler the address at which to place this code
                        267
                        268           ; Disable the WDT.
0000        53D9BF      269       ANL  PCA0MD, #NOT(040H)    ; Clear Watchdog Enable bit
                        270
                        271           ; Enable the Port I/O Crossbar
0003        75E240      272       MOV  XBR1, #40H            ; Enable Crossbar
                        273
                        274
0006        D280        275       SETB P0.0 ;  start with LED OFF (it's active low)
                        276
0008        B280        277       FLIPIT: CPL P0.0 ; toggle LED, On, Off, ...
000A        80FC        278          SJMP FLIPIT
```

We noted in §20N.7 that all '410 programs require initializations. That requirement applies even to this tiny two-line LED-blink program.

Here are the register initializations – simple enough, this time, so that there may be no need to list them as we do here. But we want to make this our standard practice. We think it underlines a fact about controllers: the initializations can be more complex than the program itself. Often that will be true in the simple programs that we offer in these labs. When you do something more substantial, it will cease to be true. But we hope that it helps to see the register initializations separated out.

Register	bit/byte-value	function
PCA0MD	d6 (= WDTE)	$0 \Rightarrow$ watchdog timer disabled
XBARE	d6 (= XBARE)	$1 \Rightarrow$ crossbar enabled (permits pins to serve as outputs)
XBR1	40h	this is the *byte* value of the register that includes the XBARE bit

Don't spend any intellectual energy trying to digest these initialization details. You will see many of them, over and over – including the two shown just above – and getting used to them will serve you well enough. Save your brain for more interesting challenges. Next time, we will begin to use the SiLabs "Configuration Wizard," which makes setting up the '410 quite manageable. You will not need to go into the dismaying bit-by-bit detail shown in the table just above. Instead, you will be able to check boxes in order to select particular behaviors: a much more manageable task.

20L.3.6 Try code on a '410

Let's now get on with the fun part: loading and running this code. Follow the procedure described in §20N.7 to "assemble," "make," and download bitflip.a51 to the '410. You may want to use View/Debug Windows/SFRs/Ports to let you watch P0 as the program single-steps. The window in Fig. 20L.28 shows the nested menus that let you choose to watch that port on-screen.

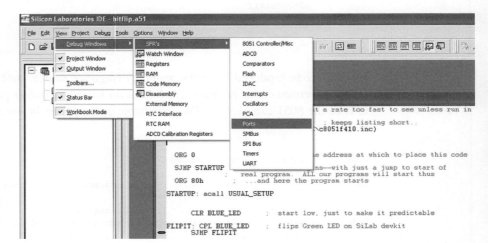

Figure 20L.28 SiLabs debugger lets you display PORT values, among other internal states.

You should see (cf. Fig. 20L.29) P0's value toggle between FEh and FFh (seven lines float high; the LSB toggles as the program runs).

Figure 20L.29 Debug window showing PORT0 levels as program steps.

But the main way to judge success for your test program surely will be to see whether it does, indeed, toggle the blue LED at a few Hz when run in *multiple-step mode*. When it does, you will know that in principle you now can control the world with your little '410.

20L.3.7 Add delay to permit full-speed run

If you run the program of §20L.3.5 at full speed, you will not see any interesting result (as the head of the program warns). The LED will glow at half brightness (since it is on for half the time); the debugger will show you nothing because it cannot monitor signal levels at full speed.

But if we patch in some code that wastes time between toggles, we can run the program at full speed and see the result. Here is the bitflip program, with such a delay included – except that we have not quite finished the program: we have left undecided the number of times the two-byte delay loop should be run. This is a value that will be determined by the value that you load into register R4 in the line that begins MOV R4, #_____. The "#" indicates that the value that will fill the blank is to be treated as a constant, not as the address where some other value is stored. This addressing mode is called "immediate."

Fill in that value, aiming for a delay of about one second each pass through the loop. Assume that the two-byte delay (the time that R4 will multiply) is about 10ms. Please enter your R4 value in *hexadecimal* form, indicated by an *h* following the value as in the line `MOV XBR1, #40h ; Enable Crossbar`.

```
; bitflip_delay_inline.a51 bitflip program, with delay for full-speed operation

$NOSYMBOLS ; keeps listing short
$INCLUDE (C:\MICRO\8051\RAISON\INC\c8051f410.inc)

ORG 0 ; tells assembler the address at which to place this code

        ; Disable the WDT.
        ANL PCA0MD, #NOT(040h)   ; Clear Watchdog Enable bit; Enable the Port I/O Crossbar
        MOV   XBR1, #40h         ; Enable Crossbar
                        ; ...and this time we'll let the processor run full-speed
                        ;   (we omit the usual divide-by-eight line)
        MOV OSCICN, #87h

        MOV A,#0 ; maximize two delay values (0 is max because dec before test)
        MOV B,#0 ; ...and second loop delay value

        SETB P0.0 ;  start with LED OFF (it's active low)

DELAY:   MOV R4, #_____       ; 1 second delay: this multiplies the 64K other loops
INNERLOOP:DJNZ B, INNERLOOP    ; count down innermost loop, till inner hits zero
         DJNZ ACC,INNERLOOP ; ...then dec second loop, and start inner again.
         DJNZ R4, INNERLOOP ; now, with second at zero, decrement the outermost loop

         CPL P0.0             ; toggle LED
         SJMP DELAY
END
```

With these additions, this small program is starting to look quite ugly: the important, central loop is hard to make out because the flow is interrupted by the patch of code that implements DELAY. We will improve this situation next time when we meet the *subroutine* form. That will allow breaking this ungainly program into smaller modules, and will let us give prominence to the part that interests us – here, the bitflip loop.

20L.3.8 Some peculiarities of the DELAY loop

Some details of the DELAY loop are odd, and should be explained. This loop uses three 8-bit registers to generate a long delay (about one second, with the values shown, if you initialize R4 appropriately; a maximum delay would be under three seconds, running the clock at 24.5MHz, as we do here).

Initializations here are strange:

- First, it is odd at first glance to find us initializing two registers to *zero* in a count-down loop. That looks like a value that would produce minimal delay. But the zero value turns out to deliver maximum delay because the count-down operation, DJNZ ("Decrement Jump if Not Zero"), does a decrement *before* testing for zero.
- Second, it may seem strange that we do not re-initialize these two registers within the loop. We do re-initialize R4 (`DELAY: MOV R4, #10h`), but not **A** or **B**. We get away with this simplification because at the end of each DJNZ loop the register that has been decremented is left with value *zero*.
- Finally, a detail so fussy that we're embarrassed to have to mention it: register **A** sometimes

demands to be called by the name "ACC".[16] It likes this name in the DJNZ operation, and we will see, next time, that it also insists on this in the stack PUSH operation. We don't understand why assemblers are not designed to be kind enough to treat "PUSH A" as equivalent to "PUSH ACC."

Try it at full-speed; try a breakpoint: After you download this program, run it at full speed: click on the large green icon ("Go"), at center-left top of the screen in the IDE.

This program also gives us a chance to demonstrate the value of a *breakpoint*. Place a breakpoint somewhere in the bitflip loop: highlight the line where you want to place the break, then right-click, and select "Insert/Remove Breakpoint."

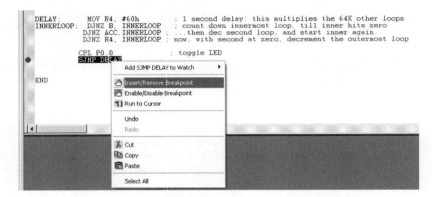

Figure 20L.30 Insert a breakpoint by highlighting a label or instruction and right-clicking.

Once the breakpoint is in place, clicking Go will run the program at full-speed, but only up to the breakpoint, where execution pauses. If you click "Go" repeatedly, you should find the LED toggling each time you do this – with about a 1-second delay as the program runs the Delay code and then again hits the breakpoint.

A better way to delay? But does it not seem perverse to take a fast processor (in this case, clocked at its maximum normal rate, 24.5MHz[17]), and then to slow its operation to a crawl? Yes, it is rather perverse – and there is, indeed, a better way to slow execution than by trapping the processor in a loop for hundreds of thousands of cycles. This better way we will explore in Lab 22L using the controller's hardware *timers*. These can be loaded once with a delay value, and then told to notify the main program when the delay time has expired. This scheme leaves the processor free to do something useful during the timekeeping process: a much better plan than the one we demonstrated today in §20L.3.7.

---

[16] The register referred to by both names is the same. The difference is a difference between addressing modes: direct ("ACC") versus implied ("A," as in "mov A, P0). Thanks to G. Cole for this explanation, http://www.keil.com/forum/3761/definition-of-terminologies/.

[17] One more factor of two in speed would be available if we chose to double the effective clock rate using the CLKMUL register.

20S Supplementary Notes: Microprocessors 1

20S.1 PAL for microcomputers

This PAL implements both "GLUE" logic (purely combinational, to link the processor with RAM and peripherals), and "STEP" logic (sequential, to permit single-stepping the computer). We will present these two elements separately, though they are combined in the single PAL, an XC9572XL (3V version) or XC9572 (5V version).

20S.1.1 GLUEPAL

For the GLUEPAL pinout see Fig. 20L.4.

Logic diagrams:

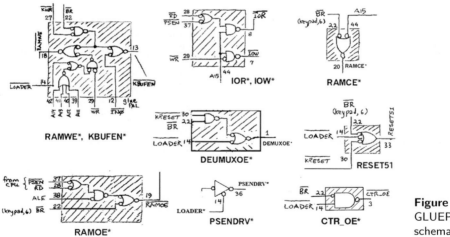

Figure 20S.1
GLUEPAL logic: schematic.

Verilog file: In `stepglue_3v.v` (available on the book's website) we show the Verilog code that implements the combinational GLUE and the STEP logic.

20S.1.2 STEP PAL

A short explanation of the STEP logic:

- The signal **ALE** provides a reliable timing marker, since it occurs:
  - once for each bus cycle;

- close to the start of each cycle.
- The fall of ALE (trailing edge) marks the time when the full 16-bit address is present (the low-half has been caught in a register).
- Soon after that event, the *data* lines become valid.

Figure 20S.2 is a schematic of STEP logic.

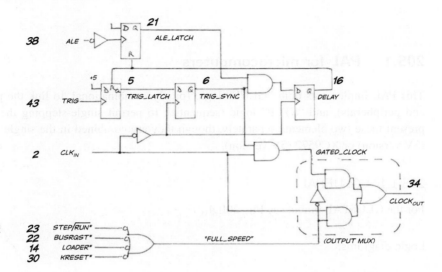

Figure 20S.2 Pinout and schematic of STEP PAL.

STEP PAL: Verilog file:

```
// here's the single-step logic

 initial
DELAY = 1'b0; // this initialization makes the sim work

 always @(negedge ALE, posedge DELAY)
   if (DELAY)
   ALE_LATCH <= 1'b0;
else
   ALE_LATCH <= 1'b1;

 always @(posedge TRIG, posedge DELAY)
   if (DELAY)
   TRIG_LATCH <= 1'b0;
else
   TRIG_LATCH <= 1'b1;

 always @(negedge CLK_IN)
  begin
   TRIG_SYNC <= TRIG_LATCH;
  end

 always @(negedge CLK_IN)
   DELAY <= (TRIG_SYNC & ALE_LATCH);

assign full_speed = (!STEP_RUNBAR | ! BR_bar | !LOADER_bar | !KRESET_bar);
assign CLK_OUT = (CLK_IN & full_speed) | ((TRIG_SYNC &  CLK_IN) & !full_speed) ;
```

20S.2 Note on SiLabs IDE

20S.2.1 Connecting to the '410: PC debugger

Once the programming pod has been connected to the '410 through the C2 lines and to the PC through the USB connector, you can turn on power to the '410 and open the SiLabs IDE.[1]

Connect PC to '410: Open the SiLabs IDE and make a new project or take one of ours, such as *bitflip.wsp*, which encompasses the first test program, one that blinks an LED. The project (whose extension is .wsp) will include an assembly-language file, with a name like bitflip.a51 or bitflip.asm. If you are making up a new project you will need to use PROJECT/ADD FILES TO PROJECT specifying, for example, bitflip.a51. If you double-click on that filename, you will see the .a51 file displayed.

Connection options: Before you try to connect, make sure that the IDE sees and means to use the USB connection. It will fail if it comes up expecting to use an RS232 serial connection instead. A correct link under OPTIONS/CONNECTION OPTIONS will show the USB connection, as in Fig. 20S.3.

Figure 20S.3 Connection options must show the USB link.

If the IDE refuses to allow you to change from serial to USB, you will need to close the IDE and remove and then restore power to the '410.[2] Once you have confirmed that the connection options are correct, click the CONNECT icon (it looks like a USB pod).

Assemble the file: Once connected you can *assemble* the program: invoke PROJECT/ASSEMBLE, or click on the leftmost of the three icons to the left of the *connection* icon. You could now look at the assembled listing – the .LST file – if you chose. This we did in Chapter 20N. But now let's not tarry to look at such details. Let's instead just pump the code into the '410.

---

[1] "IDE" is a generic acronym for controller and processor development software: Integrated Design Environment. The RIDE assembler/compiler/simulator, for example, uses the same term.
[2] When running the IDE using Parallels on a Macintosh, we found it necessary also to execute a Parallels Reset in order to regain control of the connection option.

Download the program to the '410: Now click the download icon – a downward-facing arrow to the right of the *connection* icon. In a few seconds, the code should transfer to the '410. When this has occurred, a blue cursor will appear next to the first executable line of code in the program, and the execution icons to the right of the download icon will appear not grayed-out but bright – as in Fig. 20S.5.

Choose what to monitor through the debugger: Now we get to take advantage of the great value of SiLabs' interactive link to the '410 (the C2 connection). C2 not only delivers code to the controller, it also lets the host PC control the running of the '410's code, and lets the PC monitor the registers and memory internal to the '410. This saves us the pain of the old "burn and crash" mode of debugging: "burn" in some code,[3] then run it to see if it works; if not, call it a "crash," and try again. C2 gives us insight into what the '410 is doing as it runs code.

Single-step, breakpoints and register watch: The IDE allows three important debugging aids:

Single-step: this is what it sounds like – the ability to execute one instruction at a time. This ability is necessary, in order to make the monitoring of internal registers useful.
Set breakpoints: if we place a "breakpoint" marker at a particular line of code, we then can run a program at full speed – and it will stop at the breakpoint. There we can inspect registers, and can proceed in single-step if we choose to. Breakpoints are useful in larger programs, where simply single-stepping becomes tedious.
Monitor registers: as we have said, IDE lets us watch '410 registers and also internal memory locations (both ROM and RAM)
Modify internal registers : we can modify some internal registers (such as P0 in the bitflip program) through the IDE.

Multiple-step rather than single-step: IDE offers a speed intermediate between single-step and full-speed: a step-repeat option that can be useful. We take advantage of it, for example, in the very first test program of this set of labs: *bitflip.a51*. This program toggles an LED attached to P0.0, as probably you have gathered by now. At full-speed, the display is not intelligible: the LED simply glows at half brightness. In multiple-step mode, the LED toggles at a few hertz.

Choosing what to monitor: The selection of what to monitor is done using VIEW/DEBUG WINDOWS..., as in Fig. 20S.4. The most common of the *view* selections – such as the main 8051 registers – are offered also as icons near the top middle of the screen.

Figure 20S.4 IDE lets one select what to view as a program steps.

---

[3] "Burn" is a misnomer for "program," a vestige of the days when ROMs really were programmed by the melting or "burning" of fusible metal links.

20S.2 Note on SiLabs IDE

In the screenshot of Fig. 20S.5, we have selected some of the obvious resources to monitor: a set of registers (which the IDE calls "8051 Controller/Misc"), and the "Ports," including the only one used in the bitflip program P0.

Figure 20S.5 IDE shows value of internal registers.

The bitflip program is so simple that it doesn't offer anything much that is worth watching. Only the LSB of Port 0 changes. But we can see this at least as the alternation of the value of P0 toggling between FE to FF.

"Watch" window: The controller includes such a huge number of registers that the IDE sensibly does not try to list them all, instead inviting us to select a particular register from the program listing, then adding this to a special *Watch* window. In Fig. 20S.6 we have selected a register that includes the watchdog-timer enable bit (not because it is specially worth watching, but because it is a rather obscure register not available among any of the standard sets).

A complication – and stumbling block – appears in Watch selection: we are obliged to choose the correct "Detailed Type," as in Fig. 20S.6. Unfortunately, IDE punishes us cruelly if we make the wrong selection – selecting, say "character" rather than "SFR" for the type of PCA0MD (not an unreasonable choice, since PCA0MD and "character" each occupy one byte). IDE provides no error message. It politely displays, in the watch window, nonsense: not the contents of PCA0MD, but something else entirely (we're not yet sure just what).

A helpful detail that we took advantage of in Fig. 20S.6 is the ability to choose the *number format* of the displayed value. We chose *hexadecimal* as usual. The default format is decimal.

Supplementary Notes: Microprocessors 1

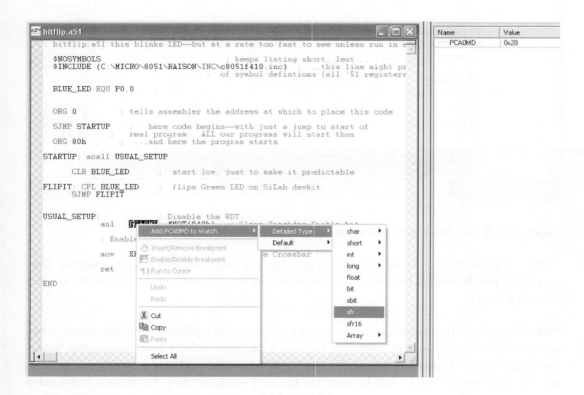

Figure 20S.6 Select a register to add it to a Watch window.

20W Worked Examples: A Garden of Bugs

It struck us recently that it's we teachers who learn most from your difficulties in the lab: you present us some pretty challenging riddles with misbehaving microcontrollers. It seemed sensible to share some of these nice erraticisms with you.

The game is to devise a theory, a *diagnosis*, that could explain the observed misbehavior and then to prescribe a test procedure (in case that isn't obvious from the statement of diagnosis).

In Fig. 20W.1 is a circuit – a demonstration board – that ordinarily would be pretty hard to debug. But on this particular day, as the white arrow indicates, the task was not hard. How the moth happened to die just there is a mystery we have not solved. Possibly a mischievous student had a role in this; but we hope not. We prefer to believe in odd chance events.

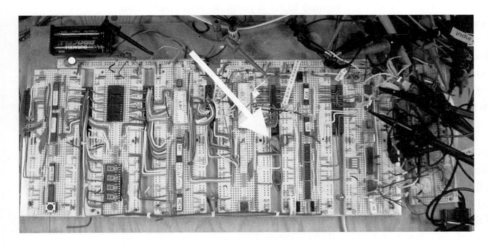

Figure 20W.1 Once in a while, debugging is easy; usually, it's not.

We should admit, while looking at bugs, that we admire them. We sometimes say, only partly facetiously, as we said when we were promoting the Big Board computer, bugs are our most important product. We don't quite mean that, but almost. We ask students to wire this complex circuit while we could have handed them a printed circuit to do the job, giving them a working computer on the first day. We do that because we think it's satisfying to start from scratch. But we do it also because a circuit of this size offers so many chances to be wrong that it is bound to present even a careful person with some challenging bug chases. We hope you enjoy your own, peculiar bugs. Below are some that we have enjoyed.

Bug (Micro seems to refuse to follow its instructions.) We load a program – say, the test program at the end of the Lab 20L – including its startup jump at addresses zero and one: code that says "Jump to 10h"[1]. When the processor reaches 1, then prefetches the code at 2, it jumps – but instead of

---

[1] More literally, the code says, "Hop ahead from where you would otherwise go – 02h – by the displacement 0Eh"

hopping to 10h, it goes somewhere else – perhaps to 11h. There it shows the correct code, once again – but then sails off to some strange high address on the next step.

Explanation   What *we* see on the DATA BUS does not match what *CPU* sees. In other words, the data bus has gotten muddled on its way to the CPU. Lines may be interchanged, or a line may be poorly connected between bus and CPU so that the line floats (this is the case detailed in the next paragraph).

It's easy to check this hypothesis: single-step to the point where the processor should be picking up the vector-constant or instruction that you have observed it to misinterpret. Suppose the processor has been hopping to 11h instead of 10h: in that case, pause at address 1, where the processor appears to pick up the byte *0Eh* from the RAM. Probe the eight lines of the data bus *at the CPU*. Chances are, they will show you 0000 1111 – a hexadecimal *0F* rather than the *0E* that we intended. We got this result, as an example, by pulling out the line that should have joined the CPU's LSB of data (AD0) to the bus; the CPU read that floating AD0 as a high, and thus hopped ahead one more than it should. By itself, that error would not have knocked the program off the rails, since the CPU happened to miss only the NOP at address 10h; but at 11h where the CPU should have picked up 80h (Jump relative), it picked up 81h. This happens to be the code for AJMP which jumps to a strangely-formed destination, one far from what we intended.[2]

Bug (Micro and we disagree over what's in RAM.)   Human loads a program into RAM, and confirms that the code is there (walks through, looking at the code after entering it, all the while keeping the bus).

Program will not run. At full-speed, it crashes. Single-stepped, computer shows values in RAM different from those loaded and confirmed a few minutes earlier. Human takes bus again – and finds the code is correct, and matches what he saw before trying to run the program.

Variation: Sometimes program runs and code matches what human sees; sometimes no match and no run.

Explanation   What *we* see on the address bus does not match what *CPU* drives. In other words, the *address* bus has gotten muddled on its way from CPU to memory.

Test by probing each address line *at the RAM* (not even at the RAM breadboard: poke the RAM pins – one pin may even be folded under the chip!). Use a logic probe, and look for a *FLOAT*. It's best to do this with CPU in control of the buses since this will let you detect not only a failure between bus and RAM, but also elsewhere in the chain – between CPU and '573 demux latch or between '573 and bus.

For the *Variation*, the explanation is similar – but harder to debug, because intermittent: here, *some* address line (usually just one) is floating. Usually the line is a high address line; it floats into one state while you, the human, are loading the program; it floats into another state when the CPU takes over – perhaps because a line next to it switches during that changeover. So we access different blocks of RAM, even though our address-bus *display* remains constant (the displays do not float: they are firmly driven, first by the manual address counters (the 16-bit PAL counter), then by the CPU).

Bug (Computer crashes when one tries to use the "Ready" key.)   This program runs OK – but only until we try to take advantage of the new feature, the Ready key. So apparently the computer can

---

[2] The details of the jump address formulation are boring, but we'll try, in case you're curious: it's an 11-bit displacement from the address at which the machine otherwise would have got its instruction: here, address 13h. The displacement is formed from the opcode's high 3 bits (here, "4'") and the byte that follows (here, "FDh"). This effects a jump to FD10h (if I haven't slipped up in my hex arithmetic).

feel the pressing of the keypad's Write button but responds by crashing, rather than by taking in a key value.

Explanation The keyboard WRITE button, and the KWR* signal that it evokes, remains wired directly either to the RAM's WE* pin or to the '541's ENABLE* pins.

Thus, pushing the WRITE button either corrupts the data on the data bus or alters RAM content at random times during the running of the program. (This hardware error occurs because at an earlier stage – Lab 17L – KWR* *was* wired directly to both WE* and the '541 ENABLEs*. It's hard to remember to get rid of old wiring when installing new.)

In a properly wired machine, it is safe to push WRITE while a program is running, because KWR* is locked out of both WE* and '541-enable paths whenever the CPU has the bus. A gate ANDs the assertion of KWR* with the assertion of BR*; as long as the CPU is running a program, BR* is *not* asserted. Therefore, while the CPU is running a program, KWR* has no effect – except on the READY flop, as intended.

Bug (ADC finds a fight when it tries to drive data bus.) Computer works fine, running all programs up to Lab 23L's check of ADC. At that point, we find the ADC delivering a constant input (happens to be FFh, but many highs, not all floats) to the computer on a read from port 3 (the ADC data port).

Starting to test the ADC, we gave it an analog input of *ground* and ran a cycle. We found some of the data lines from ADC reading *float* during the read cycle, and when we disconnected the ADC from the bus, discovered that the ADC was trying to drive a *low*, but was contending with some other device that was driving the bus *high*. (The fight resolved into a mid-scale *float*-like intermediate level while ADC was tied to the bus.)

Explanation The RAM has been enabled *continuously* all along – and now fights the ADC on an input from I/O (address space in which the RAM should be disabled).

This one is subtle, because the problem appears so late in the life of the computer. Your computer has already succeeded in doing *another* I/O input – from the keypad. How could that work and the ADC input fail? The answer is simply that keypad's '541 is stronger than the AD7569: note the contrast in their I_{OL} and I_{OH} numbers:

Specifications:	I_{OL}	I_{OH}
74HC541	6mA	6mA
AD7569	1.6mA	200μA
RAM	2.1mA	1mA

So the keypad signals passed by the brawny '541 easily overrode the RAM signals despite the bus contention. The '7569 signals don't have such an easy time: in fact, the '7569 drive is slightly weaker than the RAM's and produces a result that's not a legal low or high, at least part of the time (the pairing *'7569 low* versus *RAM high* is a pretty well-matched fight).

The *remedy* of course is to fix the RAM enabling: RAM CS* should be asserted only in *memory* space – where A_{15} is low (as Lab 20L and Fig. 20L.1 show).

Bug (ADC works single-stepped, but not at full-speed.) The ADC test program starts ADC, branches back, reads ADC and shows result (on DAC or on displays). The ADC value is constant or changes erratically. Single-stepped, the ADC works perfectly (tested by feeding it from a potentiometer).

Explanation The ADC's oscillator is running slowly. The important clue here, of course, is the difference in conversion speed that the CPU demands of the ADC in the two cases: when single-stepping, the CPU gives the converter an eternity to complete a conversion; at full-speed, the CPU gives only a little time to spare.

If the conversion is not done when the CPU tries to read the value (and our program does not politely wait to be *told* that a sample is ready; it takes a sample when we *guess* that one is ready) – the ADC gets its revenge by giving bad data. The likely cause for slow conversion is simply *wrong RC* at the ADC's "clk" pin.

You would get a too-slow *RC* if you were a little lazy and plugged in values more standard than those specified – say, 6.8k instead of 6.2k, and 100pF instead of 68pF (*RC* 60% higher than target); you would get radically slow *RC* if you misread a capacitor. The traditional error is to think that "0.068μF" means "68pF." That misconception might lead someone to use 0.1, saying "close enough."

The *remedy* is pretty dull: stare at the *R* and *C* values; correct them if they're wrong. If you wanted an experiment to test whether the ADC were converting too slowly, you could use an external function generator to drive the ADC's *clk* pin; at 4MHz it should work with the program running full-speed. As you crank the ADC clock rate down, the display should go bad, when the ADC conversion exceeds time between START and READ, in your program loop.

21N Microprocessors 2. I/O, First Assembly Language

Contents

21N.1	**What is assembly language? Why bother with it?**	**813**
	21N.1.1 Little test program for the big-board branch	815
	21N.1.2 Details of program listings that sometimes puzzle people: assembler directives, etc.	817
	21N.1.3 Watching the big-board test loop program in slow motion	818
21N.2	**Decoding, again**	**818**
21N.3	**Code to use the I/O hardware (big-board branch)**	**821**
	21N.3.1 What address is "Port Zero"? we must speak in the processor's terms	822
	21N.3.2 Assembly language format	823
	21N.3.3 Complete the tiny program	823
	21N.3.4 The program in action	824
	21N.3.5 The SiLabs standalone controller will not show such details	824
21N.4	**Comparing assembly language with C code: keypad-to-display**	**824**
21N.5	**Subroutines: CALL**	**826**
	21N.5.1 Stack as general-purpose place for short-term storage	827
	21N.5.2 Subroutine CALL and "functions" in C	829
21N.6	**Stretching operations to 16 bits**	**830**
21N.7	**AoE Reading**	**831**

Why?

Our task today is add to a computer the ability to take in and put out byte size information. Before we proceed, here's a reminder of some of the main ideas from Day 20.

- When using *buses*, one must *decode* control and address signals from the processor in order to make memory and peripherals behave properly.
- Control signals put out by a microprocessor, though peculiar to each processor in their details, convey pretty standard information: direction, timing, "which one," and often I/O versus memory.

21N.1 What is assembly language? Why bother with it?

It's the quasi-English that humans use to speak to one another and to "assembler" programs in defining what operations we want a processor to carry out. "MOV R0, #38h," for example, is the assembly language way of expressing what in "machine code" is expressed as the two bytes 78, 38h. (The instruction means "copy into scratch-register zero the value 38hexadecimal," in case you're curious.) Assembly language differs from higher-level languages such as C in that one line of assembly language corresponds to one machine language instruction. In C, a single line of code often expresses a more complex operation, one that requires multiple lines of machine code (and of assembly language code).

Assembly language programming is going out of style because it costs too much in human programming time. In the programming industry you sometimes hear the statement that a person can produce about ten *lines* of fully debugged code in a day. If this is about right, then surely it pays to produce lines that get more done. We use assembly language programming in this course because we like to see the one-to-one correspondence between the code we write and an action by the computer (though sometimes the "action" is invisible to us – as in the example given above where a constant is loaded into a register that is internal to the 8051). Assembly language programs also require no overhead: no "libraries," no special support routines.

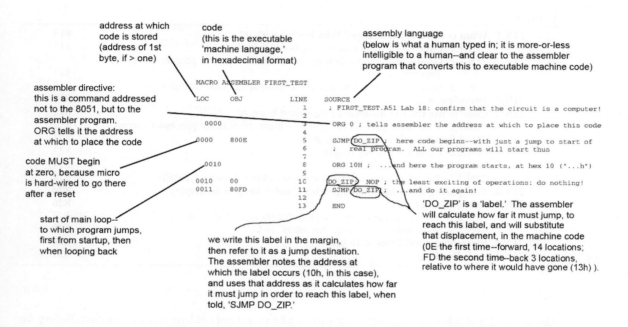

Figure 21N.1 Annotated tiny loop program.

The penalties for using assembly language will become obvious to you as you work: the details are fussy and processor-dependent. The quirks and deficiencies of the 8051's instruction set, in particular, are shocking – reflecting the old age of the 8051's design (it was born around 1976). A higher-level language could hide these uglinesses from you; however assembly language forces you to confront, for example, the amazing fact that the 8051 can *increment* its only 16-bit pointer, but cannot *decrement* that pointer. Shocking, but true. Near the end of these micro labs, when you let a laptop download code to your little lab computer, you will be able to write in C if you like. But for a while we will work entirely in assembly language.

Now we're going to look closely at two small test programs, each used as the first program on one of the two *micro* branches. The descriptions of both are relevant to any use of the 8051, so we ask you to consider both. The description of the particular program that you are using to test your controller will interest you more than the other no doubt. But we will not restate, for example, the discussion of jump *offsets* in explaining the operation of the second test program. We will assume that you have read about the first test program, even though you do not mean to use it in lab.

21N.1 What is assembly language? Why bother with it?

21N.1.1 Little test program for the big-board branch

To confirm that the mass of wires and parts on your big breadboard has morphed into a computer, Lab 20L concludes by asking you to run a test program – about as tiny and simple as we could devise. Here we will dissect this little patch of code, and then we will envision what the processor must do as it runs this program, given in Fig 21N.1 along with some annotations.

This little program does nothing, and does it over and over, forever (or until you get bored). It simply loops.[1].

The SJMP offsets: The assembler does the chore of calculating how far to jump to get to a particular destination. As the example above illustrates, the human programmer need only plant a *label*, and then branch to that label, as needed. But it's probably worthwhile looking at what the processor does as it fills in the *offsets* for those *relative* jump instructions. The relative jump needs to be told how far to deviate from its usual operation. That "usual operation" would call for fetching the code at the very next location.

So the jump from the RESET "vector" (as the forced start at address 000h is called) is a jump *from* the address where the processor would have found its next instruction if this had not been a jump, namely address 2. Why 2? The SJMP instruction itself takes up locations 0000h and 0001h. If this had not been a jump, the next piece of code would have come from address 0002h.

The *offset* measures that difference: where to go next minus where it would have gone. That difference is

```
want to go to:       10h = 0001 0000
would have gone to:  02h = 0000 0010
-----------
difference:          0Eh = 0000 1110
```

If your binary subtraction is rusty, recall that when you "borrow" from the bit to the left, it comes in as a $2_{decimal}$. That's why we get, for example, a "1" as the difference in bit 1. 0Eh is a *positive* number ($14_{decimal}$). You don't need to get good at this work: calculating such offsets is as job normally done by the *assembler* program.

In executing the SJMP, the processor adds the offset, 0Eh, to the present value, 02h, of its *program counter* ("PC"). PC is the 16-bit register that holds the value of the current byte that is to be fetched.

```
program counter now: 02h = 0000 0010
offset:              0Eh = 0000 1110
-----------
sum: goes to:        10h = 0001 0000
```

Since the sum is 10h, the program jumps to address 10h, there to find its next instruction.

The jump from the end of the loop is a backward jump. The SJMP instruction is the same as for the forward jump; execution hops backward, to an earlier address, because the offset this time is a *negative* value (using the 2's-complement convention). The hardware that executes the jump is simply a 16-bit adder operating on the so-called "program counter," the register that tells the processor where to pick up its next instruction.

---

[1] The program "loop" now is metaphorical. On the Harvard Mark I, it was literal. The list of instructions that the Mark I was to execute were listed on a punched paper tape. In order to make the machine repeat a sequence of operations, the programmer taped the end of the tape to its beginning. See http://en.wikipedia.org/wiki/Harvard_Mark_I.

Microprocessors 2. I/O, First Assembly Language

The jump goes, much as in the forward hop,

- *from* the address where the processor would have found its next instruction if this had not been a jump, address 13h;
- *to* address 10h

That difference is

```
want to go to:        10h = 0001 0000
would have gone to:   13h = 0001 0011
-----------
difference:           FDh = 1111 1101
```

FDh is a *negative* number, −3. The 1 in the msb position makes this *2's-complement* value negative. Again, adding *offset* to *program counter* − FDh plus 13h − delivers the sum 10h, the correct jump destination:

```
program counter now:  13h = 0001 0011
offset:               FDh = 1111 1101
-----------
sum: goes to:         10h = 0001 0000
```

A puzzle resolved: 8-bit displacement, 16-bit address? How can an 8-bit value added to a 16-bit value give the needed 16-bit result? Good question. The answer is that the 8051 extends the 8-bit value behind the scenes. We provide the 8-bit displacement "FDh," or "0Eh." The processor, which forms the resulting jump destination by using its 16-bit adder, has the good sense to *extend* the 8-bit values to 16 bits. All it need do is repeat what it finds in the msb (the so-called "sign bit"). So it extends both the displacements that we provided:

```
-----------
0Eh = 0000 1110 is extended to 16 bits by repeating the 0 at the msb:
    0000 0000 0000 1110 = 000Eh
-----------
FDh = 1111 1101 is extended to 16 bits by repeating the 1 at the msb:
    1111 1111 1111 1101 = FFFDh
```

A cheap way to calculate a small 2's-complement value, if you ever want to: Probably this little excursion into 2's-complement arithmetic has cured you of wanting to do this kind of math ever again − but in case it hasn't, here's a cheap trick to help you on the proverbial desert island. Suddenly you need to know the 8-bit hexadecimal expression, in 2's-complement for a small negative value, smaller than 16. Let's try it for the value "three." The trick is to subtract your "three" from the freakish hybrid value "F16" − a value that does not exist, since its left digit is hex, its right digit is decimal. But if you can stand that freakishness, try it.

What is 2's-comp for −3?

```
  ``F16"
-   03h
  -----------
  ``F13" = FDh
```

Yes, it worked.

21N.1.2 Details of program listings that sometimes puzzle people: assembler directives, etc.

We noted these points back in Chapter 20N. We'll just flag these points here:

Multiple bytes on one line: As you know from Chapter 20N, when it looks as if two bytes are listed at a single address…

```
0000      800E         SJMP DO_ZIP  ; here code begins -- with just a jump
```

…they aren't: the 16-bit value is stored at two successive addresses, 0 and 1. Some instructions, which you will meet later, are even longer than two bytes.

Directives addressed to the assembler: ORG and $INCLUDE are examples of such directives
ORG

```
0000      ORG 0  ; tells assembler the address at which to place this code
```

"ORG 0" is addressed to the assembler, not to the 8051.
$INCLUDE

```
4         $INCLUDE (C:\MICRO\8051\RAISON\INC\REG320.INC)
```

INCLUDE tells the assembler to include another file – as if it were listed within the present program.

INCLUDE's benefits, an example: Here is a particular case detailing the way the assembler can use a port names such as "P1" or "P2." Suppose we used built-in ports rather than external buses, to implement the keypad-to-display operation of §21N.4 below.[2]

The single line below does this and assumes that the keypad and display are wired to the two listed ports, P2 and P1 (respectively).

```
COPY: MOV P1, P2    ; in one operation, copy byte from keypad to display
```

For this code, the REG320.INC file that was *included* is essential. It allows the assembler to make sense of "P1" and "P2."

As you know, the assembler looks up the addresses of these symbolic names and plugs those *addresses* into the executable code that it produces. Below is the "listing," showing the assembler's substitutions for P1 and P2, and the corresponding machine code. For "P1," the move destination, the assembler substitutes "144" (decimal); for "P2" it substitutes "160."

```
85A090        ...    COPY: MOV 144 , 160
```

And in the machine code listing in the left-hand column the assembler places – after the *op code*[3] 85 – the hexadecimal equivalent of those two decimal values: $A0_{16}$ for 144_{10}, 90_{16} for 160_{10}.

---

[2] A minor point that we should perhaps acknowledge, in case you are worried that P2 is not available in the big-board computer: we could use P2 and P1 this way with the SiLabs controller or with the Dallas part when it is operated "standalone" rather than in the big-board. The big-board happens to apply P2 to another use, ordinarily; the port is used to define 8 bits of the address bus.

[3] "Operation code."

Descriptive symbols using EQU (another layer of assembler substitution): Even clearer than the port names "P1" and "P2" might be descriptive names. EQU, another *assembler directive*, enables this convenient feature. Once defined, these descriptive names would allow us to make the TRANSFER code almost self-explanatory. We could write it as

```
TRANSFER: MOV DISPLAY, KEYPAD
```

The two instances of EQU in the listing below achieve the first stage of the two-step substitution. After these EQUs, when we write "DISPLAY," the assembler does a text substitution putting in "P1."

```
; byte_in_out_ports.a51

    $NOSYMBOLS    ; keeps listing short
    $INCLUDE (C:\MICRO\8051\RAISON\INC\REG320.inc)

    KEYPAD EQU P2
    DISPLAY EQU P1

    ORG 0h
    SJMP TRANSFER


    ORG 020h
TRANSFER: MOV DISPLAY, KEYPAD
    SJMP TRANSFER
```

...then – this is the second stage of the substitution – the translator consults the *included* .INC file and translates "P1" to address A0 (hex) as you have just seen. The assembler does the same two-step substitution for "KEYPAD," producing the line of code that you saw earlier:

```
85A090          178      TRANSFER:  MOV 144 , 160
```

If you are accustomed to high-level programming, you may not be impressed by this achievement. But you'll probably concede that such substitutions may help to make assembly code more readable.

21N.1.3 Watching the big-board test loop program in slow motion

Figure 21N.2 depicts the steps you would see if you single-stepped the big-board 8051 walking through its little test program. The main points that we hope may emerge from this sketch are

1. the cyclical form of the processor's behavior – it resembles a piston engine, continually pushing and pulling: pushing out an address, pulling back data;
2. because of the pre-fetch behavior, the time-lag that you will observe as you step your lab computer.

21N.2 Decoding, again

Decoding of address and control signals is necessary in a full-scale computer, or even a small one like our big-board lab machine where the computer uses *buses* to carry address and data. Decoding is not necessary when one uses a controller in standalone mode, as we do with the SiLabs controller. This difference is one of several that set apart the two micro branches. The standalone device does its own

21N.2 Decoding, again

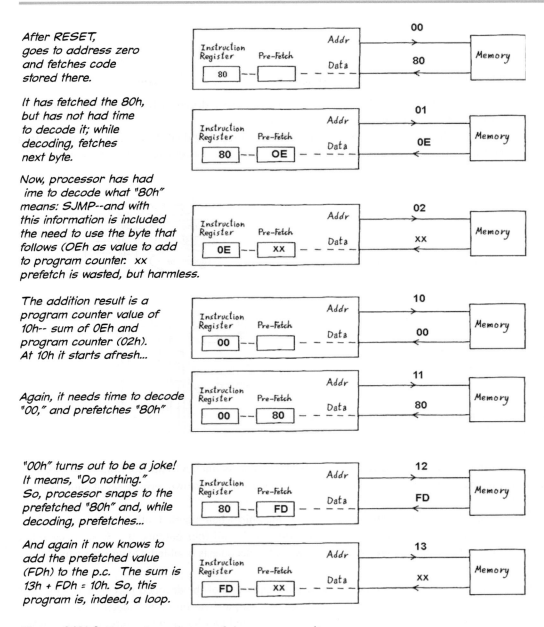

After RESET, goes to address zero and fetches code stored there.

It has fetched the 80h, but has not had time to decode it; while decoding, fetches next byte.

Now, processor has had ime to decode what "80h" means: SJMP--and with this information is included the need to use the byte that follows (0Eh as value to add to program counter: xx prefetch is wasted, but harmless.

The addition result is a program counter value of 10h-- sum of 0Eh and program counter (02h). At 10h it starts afresh...

Again, it needs time to decode "00," and prefetches "80h"

"00h" turns out to be a joke! It means, "Do nothing." So, processor snaps to the prefetched "80h" and, while decoding, prefetches...

And again it now knows to add the prefetched value (FDh) to the p.c. The sum is 13h + FDh = 10h. So, this program is, indeed, a loop.

Figure 21N.2 Freeze-frame images of tiny program as it runs.

decoding, on-chip, so that a user need not be aware of it. Pedagogically, this may make the built-up computer the more instructive; but it certainly makes the standalone the easier to use. We will treat here the decoding necessary for a computer that uses buses. If you are not building such a machine you may safely skip this discussion. But you might be wise to linger long enough to understand decoding, for it is necessary in computers that use a traditional shared ("multidrop") bus.

Last time, we did a simple decoding task with an AND gate. This time we will take advantage of an integrated decoder IC (though in practice you would now be more likely to use a PAL for the task). Recall the available 8051 signals:

- PSEN*,
- RD*,
- WR*,
- 16 address lines

Returning to the task we did last time, let's draw in Fig. 21N.3 decoding for one of four *input* ports, assuming we want to permit four *output* ports as well.

Then let's do a similar decoding job, using instead a 74HC139 dual 2-to-decoder to define four input ports, four outputs, as in Fig. 21N.4. This part is simply a convenient packaging of eight AND functions much like the one we just drew.[4]

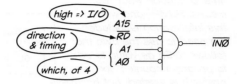

Figure 21N.3 Decoding of one of four input ports.

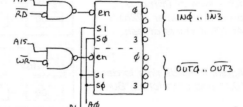

Figure 21N.4 IC decoder used to define four in ports four out ports.

Note that the two identical halves of the '139 share address lines A1 and A0. Some students have made the mistake of thinking that each of the two decoders needs its own, private set of address lines, to avoid conflicting with the other. Not so.

The EN* pin does what it sounds like: enables the chip when asserted, allowing one of the four outputs be asserted (low). When EN* is disasserted (high), all four outputs, 0...3, are disasserted (high). Note that these outputs are *not 3-stated* when the decoder is disabled. Do you see why?[5]

...Apply the decoder: I/O peripherals: Now let's apply two of the decoded port signals. One will bring a byte in from the keypad; another will latch an output byte into a hexadecimal display. For contrast, we have also shown in Fig. 21N.6 how easy In and Out hardware become when one takes advantage of the controller's built-in ports rather than using the data bus.

The 3-state in the bused version is controlled by its OE* (Output Enable*), driven by IN0*. The display latch enable, "latchen*," is driven here by OUT0* and controls a *transparent latch* that can be used to catch a value from the data bus when pulsed low. To this point, we have used the displays as if they included no latch: they have shown whatever is input to them, continuously. A weak pull-down resistor on latchen* (100kΩ) permits this behavior, making the latch *transparent* by default.

---

[4] In order to keep things simple, we omit in Fig. 21N.4 the PSEN* signal that is OR'd with RD* in the big-board's full circuit diagram, Fig. 20L.1. That inclusion of PSEN* is not essential. It permits some addressing modes that are useful only occasionally. We will not use those modes in the code that we provide for these micro labs.

[5] Floating the inactive lines from a decoder would be a very bad plan: they would pick up noise, turning on peripherals at random.

In the right-hand image of Fig. 21N.5, showing the SiLabs controller, we have made the continuous grounding of latchen* explicit and visible.

...I/O on standalone bus-less controller (a contrast): Let's notice what's good and less good about the very simple I/O on a single-IC controller. Keypad to display hardware is as easy as Fig. 21N.6. What is less good is that this simplicity comes at a cost: byte-in, byte-out here gobbles 16 pins for this single task. In the bused scheme, those pins can be shared with an almost unlimited number of other peripherals.

21N.3 Code to use the I/O hardware (big-board branch)

Now that we have the necessary hardware – decoding, and an IN and an OUT peripheral – it's time to write some code to try the hardware. Let's make it as simple as possible: a loop that reads the keypad and sends that value out to the display, over and over. If we were to write just comments, the program would look like this:

```
; pick up a byte from the keypad
; ...write that byte out to display
; do it again, forever
```

Before we start consulting the list of 8051 instructions, let's take a look in Fig. 21N.7 at the insides of the 8051.

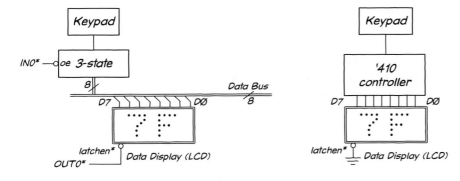

Figure 21N.5 In, Out hardware: keypad in, display out: using data bus (left), versus using controller's built-in ports (right).

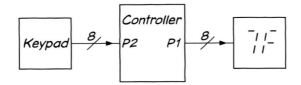

Figure 21N.6 Controller I/O, without use of buses, is radically simple.

This sketch tries to suggest the way the 8051 usually does I/O when using its external buses:

1. The *address* usually must come from the dedicated 16-bit register, Data Pointer (DPTR)[6] whose entire purpose is to provide such addresses.

---

[6] We will soon see that another method is available – sometimes preferable, but not so generally useful: "MOVX @Rn.". See §21S.1.4.

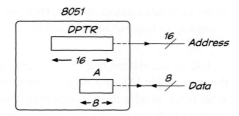

Figure 21N.7 Registers the 8051 uses for I/O when using buses.

2. The *data*, in transfers using external buses, always passes to or from the **A** register – the "Accumulator" – a privileged register. The **A** register is also the privileged source or destination for many other processes, including nearly all arithmetic and Boolean operations.

Given these restrictions, we find that we will need to add one preliminary step at the start of our program: we must initialize the pointer register, DPTR, loading it with the address of the peripherals.

```
; initialize pointer: point to peripherals
```

Only after taking care of that pointer-initialization can we do the I/O operation:

```
; pick up a byte from the keypad...
```

21N.3.1 What address is "Port Zero"? we must speak in the processor's terms

"Port Zero," using the external buses,[7] is a term that makes sense to us humans – but not to the 8051: first, the 8051 does not distinguish memory from I/O in its *bus* operations, so to it "Port" is a foreign concept; second, it has no way of knowing our arbitrary choice of the midpoint of address space as the start of I/O space.

So we must translate what we think of as "Port Zero" into an address. The processor will understand that. Which address? The I/O decoder's wiring determines this: the decoder pays attention to only three address lines: A15 (which must be high) and A1 and A0 (which are used in all their combinations, but which must both be low to define number "zero"). So, Port Zero looks like

```
A15... A11. A7.. A3 A2 A1 A0
1    xxx xxxx xxxx  x  x  0  0
```

Lots of "don't cares" (x's), because we have been so extremely *lazy* in our I/O decoding. Rather than try to recite a lot of x's, we'll treat each don't-care as a zero. That gives us a readable number:

```
A15... A11. A7.. A3 A2 A1 A0
1    000 0000 0000  0  0  0  0
-------
in Hexadecimal:
    8    0    0    0
```

And "8000h" is the way we will refer to *Port 0*, when we write code (the final "h" will tell the assembler that the value is in hex). Hexadecimal is convenient for us, since our displays and keypad also speak Hex.

---

[7] Beware a point on which it is easy to get confused: this "Port Zero" using external buses is *not* the same as PORT 0 (P0), the built-in controller port. In fact that port is not available to us when we are using buses because it has been committed to carrying some of the bus information (AD0..AD7, as it happens). Knowing the likelihood of confusion, we try to specify "...using external buses," to let you know which sort of port we intend.

21N.3.2 Assembly language format

Now that we know how to tell the assembler what address to use for Port 0, we can proceed to write the little program that exercises the I/O hardware. First, let's load the address of keypad and display (Port Zero) into the pointer register, DPTR: see Fig. 21N.8.

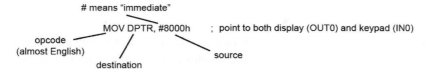

Figure 21N.8 Loading DPTR with a constant: some standard format conventions appear.

This example shows the usual form of assembly language commands.

- The *opcode* (operation code) is more-or-less English (though misspelled); it is also mis-named, really, since it is a *copy* (the original stays where it was).
- Next comes *destination*; then, finally,
- *source*

The order seems odd but we can get used to that. The pound sign, #, invokes a very simple *addressing mode* called "immediate." It means, "take this value itself, not something stored at this address." So the command loads the value 8000h into the DPTR. (To make sure you understand the concept of "immediate," you might ask yourself, "What would *MOV DPTR, 8000h* mean if this operation were permitted (as it is not)?"[8] This move into DPTR is a 16-bit operation, one of the very few permitted on the 8051.

Now, let's finish the code for this task. The DPTR has been loaded, so we can now get data in and out of the 8051 using that DPTR value (it's convenient that keypad and display happen to live at the same address: no need to alter the DPTR value between pickup and delivery of data).

Using the pointer: "@DPTR," indirect

First let's pick up the data from the keypad: see Fig. 21N.9. The MOVX operation, like nearly all 8051 operations, is a *byte* operation, not a 16-bit like the DPTR initialization in §21N.3.2.

21N.3.3 Complete the tiny program

Second, let's send to the display the value we picked up. This time, we'll show the machine code, a matter that normally does not interest us.

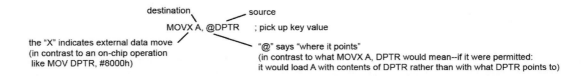

Figure 21N.9 MOVX now makes use of the DPTR initialized earlier.

---

[8] The answer is that it would mean "Take whatever is stored at address 8000h and put that value into DPTR" – not what we intended.

```
[Address    Code        Assembly Language]
0028        908000      STARTUP: MOV DPTR, #8000H ; "point" to both display (OUT0) and keypad (IN0)

002B        E0          GETIT: MOVX A, @DPTR ; pick up key value
002C        F0                 MOVX @DPTR, A ; ...show it on display
002D        80FC               SJMP GETIT
```

You can see the symmetry of the two uses of DPTR: first, pick up what it points to, store the value in **A**; then take the value from **A** to where DPTR points. And we chose to show the machine language so that you could see the payback for the seeming clumsiness of indirect addressing.

It was tedious to have to load DPTR before we could do anything off-chip. But once that is done the code is extremely compact: just one byte for each input and output operation (no need to specify the address for each of the in and out operations). So the code is compact and runs fast (and it may be pleasing to see that a single *bit* in the opcode distinguishes the symmetric operations: IN distinguished from OUT: E0h versus F0h).

21N.3.4 The program in action

Figure 21N.10 is a scope photo showing the IN and OUT operations as the program runs. It is reassuring to see A15 and PSEN* behaving as they should. And we hope that seeing the RD* pulse in the one case (IN from keypad) and the WR* pulse in the other (OUT to display) will convince any skeptics that it is not crazy to pick up and deliver at the single location, 8000h: the machine is not doing *nothing* in making these two visits to one address – as it would be if 8000h were a RAM location. The two peripherals share an address (we guess that makes them roommates), but they are by no means the *same*: a keypad is not a display (and you are not your roommate).[9]

21N.3.5 The SiLabs standalone controller will not show such details

We can offer no equivalent scope image to show what is occurring within the 8051/'410 as it runs its standalone loop. The virtue of a single-chip controller is its ability to run with all its machinery neatly out of sight within its integrated circuit. But that virtue also tends to make it enigmatic and hard to learn from: it works, or it doesn't – and only the SiLabs debugger gives us a way of divining what is happening within.

21N.4 Comparing assembly language with C code: keypad-to-display

Assembly language:

```
        ORG 0
        LJMP STARTUP

        ORG 90h
STARTUP: MOV DPTR, #8000h   ; point to keypad and display
COPY:    MOVX A, @DPTR      ; pick up keypad value...
         MOVX @DPTR, A      ; ...and send it to display
         SJMP COPY          ; keep doing this forever
```

[9] OK, this argument is somewhat fallacious – if only one person could live at any address, then you *would* have to be the same person as your roommate. But we hope you won't notice the flaw in the analogy.

21N.4 Comparing assembly language with C code: keypad-to-display

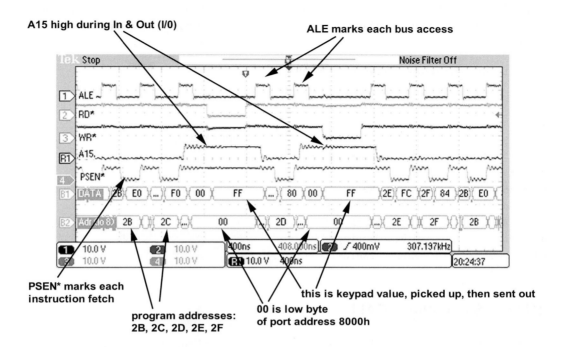

Figure 21N.10 Scope image of program that reads keypad, then sends value to display.

C code:

```
// keypad_to_diplay_408.c
    #include<C:\MICRO\8051\RAISON\INC\REG320.H>

void main()
{
    xdata char volatile *KEY_DISP;

        KEY_DISP = 0x8000;   // ...and show particular address to be used in pointer

    while (1)     // forever...
    *KEY_DISP = *KEY_DISP ;
}
```

You will recognize in the C program the elements of the assembly language code:

- pointer initialization:
  "xdata char volatile *KEY_DISP;" defines a 16-bit ("char") pointer to external memory; "volatile" tells compiler not to optimize away this pointer definition;
  "KEY_DISP = 0x8000;" gives it a value
- the loop done with two instructions of assembly (the two MOVX operations) is done with one line of C:
  " *KEY_DISP = *KEY_DISP ; "
- "while (1)" makes it run forever

The assembly code that the C compiler generates: The compiler's assembly code looks like what a slightly-confused human might write. The essential elements are the same as in our assembly code above. The compiler adds an unnecessary re-initializing of DPTR and an unnecessary save-and-recall of the keypad input value. But unless you're in a big hurry you probably don't mind these extra lines of code.

```
ASSEMBLY LISTING OF GENERATED OBJECT CODE

             ; FUNCTION main (BEGIN)
0000         ?WHILE1:
                                                ; SOURCE LINE # 13
0000 908000         MOV     DPTR,#08000H
0003 E0             MOVX    A,@DPTR
0004 FA             MOV     R2,A
0005 908000         MOV     DPTR,#08000H
0008 EA             MOV     A,R2
0009 F0             MOVX    @DPTR,A
000A 80F4           SJMP    ?WHILE1
```

21N.5 Subroutines: CALL

Suppose you write a block of code, and would like to use it more than once. For example, in Lab 21L we need a software *delay*, in order to slow execution. The program that is to be slowed continually increments the display value, and that incrementing must occur at far below the processor's full-speed rate in order to be intelligible to human eyes.

You *can*, of course, simply write the block of code anew each time you want to use it. The left-hand sketch in Fig. 21N.11 shows such clumsy in-line coding: the delay code is written twice. The right-hand sketch begins to improve on the in-line arrangement – but raises the question "how is execution to resume at the proper place, after delay is invoked?"

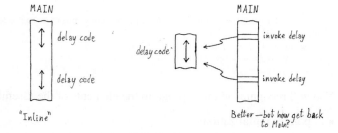

Figure 21N.11 Inline coding can be clumsy; it would be nice to write once, use multiple times.

The computer needs a way to know where it should "return," in computer lingo. Each time the computer goes off to the Delay code – or "routine" – it must make a temporary record of the *return* address. Computers do this by automatically saving the return address on the *stack* – a region in RAM defined as "where the Stack Pointer Points."[10] On the 8051 the Stack Pointer is an 8-bit register, and the Stack RAM must be on-chip 8051 RAM (not the external RAM that we have attached to the processor).

In order to take advantage of the processor's ability to use the stack in this way, all we need to do

---

[10] Does this sound like a circular definition? We don't think it is, but we agree it sounds that way.

21N.5 Subroutines: CALL

is use the *Call* instruction rather than a *Jump* operation. The operation *Call* comes in two varieties on the 8051; *Jump* comes in three.[11]

Figure 21N.12 illustrates the way the Stack can save a *return* address, when one uses Call. The process of saving the return address is entirely automatic. You finish the called "subroutine" with a RET (return). You'll notice the *return address* is the address of the instruction just following the instruction that initiated the CALL – address 402h in this case, which illustrates the *first* of the two CALL operations.

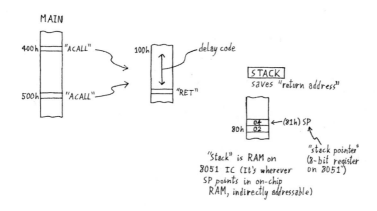

Figure 21N.12 Subroutine Call knows how to get back after running the subroutine.

21N.5.1 Stack as general-purpose place for short-term storage

Call uses the Stack automatically and implicitly. But one can also use the Stack explicitly to store the contents of registers. Often in fact, a subroutine needs to do just that because the routine needs to make use of registers and thus could mess up values needed in the main program. It is especially likely to need **A**, the *accumulator*, the privileged register that always is used in input and output on the external buses. **A** also is involved in most arithmetic and logical operations.

So the Delay routine below saves **A**, which gets overwritten within the DELAY routine.[12]

The code is

```
DELAY: PUSH ACC ; A likes to be called ACC, for some operations.  Annoying!
       MOV A, #80h ; initialize with delay value
KILLTIME: DJNZ A,KILLTIME ; count down till A hits zero
       POP ACC ; restore the register value, "popping" the value pushed onto the stack
       RET
```

The DELAY code above is simpler than the ones used in the lab exercises, by the way. The routine above counts down just an 8-bit register for a very modest delay of the order of a few hundred microseconds. The Dallas lab routine *nests* two 8-bit delay registers to get a longer delay – not *twice* as long but 2^8 times longer. Even so stretched, the delay is not long at 11MHz: about 120ms, maximum. The SiLabs Delay routine uses three nested loops to get its one-second delay.

If it bothers you to see your smart little processor tied up in the degrading work of just killing time

---

[11] ACALL does a relative branch, using an 11-bit offset value, so ACALL can branch within a 2K range, roughly ±1K; LCALL does a branch to a location specified by its 16-bit address and thus can go anywhere in the 8051's address space. AJMP and LJMP present the same contrast; in addition, Jump comes in the short flavor, called SJMP (8-bit offset: range +127, −128).

[12] Register **A** sometimes insists on being called by its longer name, "ACC." It insists on this for PUSH and POP operations. This insistence – really a property of the *assembler* program – is annoying and as far as we know, inexcusable. But we are stuck with it.

– good! It should bother you: your discomfort will allow you to appreciate the alternative way to inject long delays into programs using the controller's built-in *hardware timers*. Later, in §24L.1 and §24L.2, we will take advantage of these.

The CALL operation in action: simulation by RIDE: Figure 21N.13 is an excerpt from a simulation of a program using this CALL. The simulation uses RIDE,[13] a program whose use is described in §23S.1.

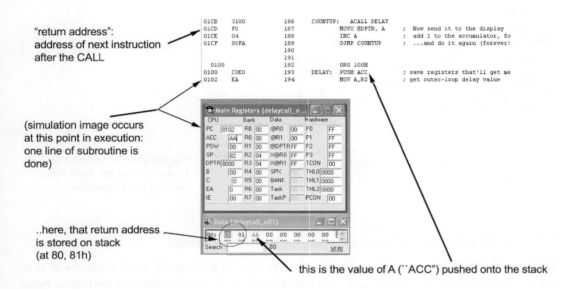

Figure 21N.13 Subroutine CALL invoked: stack shows stored values: return address, and accumulator (**A** register).

We *urge* you to begin using RIDE as you try writing code. The assembler will tell you when you make a silly mistake (writing MOV A, @DPTR, for example, when you meant MOVX A, @DPTR); that way, you will learn assembly language quickly. Then the simulation gives you the gratification of watching your code in action. When your program gets a little complicated, the simulation will also show whether your algorithm seems to work.

CALL in SiLabs program: In the Silabs LED-blink program, modified in §21L.1 to use a DELAY routine, the code is almost the same as in the big-board version:

```
; bitflip_delay_subroutine.a51 bitflip program, with delay for full-speed operation

$NOSYMBOLS ; keeps listing short
$INCLUDE (C:\MICRO\8051\RAISON\INC\c8051f410.inc)

ORG 0 ; tells assembler the address at which to place this code

        ACALL USUAL_SETUP
        SETB P0.0 ;   start with LED OFF (it's active low)
```

---

[13] RIDE is "Raisonance Integrated Design Environment," a good, free assembler/compiler/simulator that you should download and try. We use this assembler/compiler also in the SiLabs programming "design environment" ("IDE").

21N.5 Subroutines: CALL

```
FLIPIT: CPL P0.0            ; toggle LED
        ACALL DELAY ; waste some time
        SJMP FLIPIT ; do it forever

; -- SUBROUTINES ---
DELAY:    PUSH ACC    ; save the registers that this routine will mess up
.........
```

The "main" program is even simpler than the big-board equivalent. We have not listed the full delay code this time. The SiLabs delay code uses three 8-bit registers "nested" in order to generate a substantial delay (1 second) despite the high clock rate of 24.5Mhz.

Interrupt: coming soon – just another kind of CALL: Next time, we will look at a hardware-initiated CALL: *interrupt*. Having understood the operation of the subroutine CALL, you will recognize *interrupts* as only a variation on this theme.

21N.5.2 Subroutine CALL and "functions" in C

You may have recognized from the description of the way *subroutines* work that these are what the C programming language calls *functions*. The C equivalent of a *Main* program that invokes a subroutine called *Delay* could look like the listing just below – after we have stripped out some C-specific preliminaries.[14]

Bit-flip and delay in C: Here, in C, is the main program and the function named Delay:

```
// ---- now, here's the main routine, appropriately named "MAIN"
        void main (void)    // 'void" means it doesn't accept input or deliver output
        {
    while(1)                // this is an odd way to say 'do this forever'
        {
            delay();        // here's the function call
            OUT_BIT = ~OUT_BIT;  // and here's the only action, a bit flip
        }
        }
// ---- and here is the subroutine
        void delay (void)
        {
            for(n= 0; n< 0x8600; n++){}  // 2Hz delay value (250ms delay)
        }
```

Delay in C: The *delay* function just counts up until the value of n reaches the target value. Then execution *returns* to Main.

The assembly language code that the C compiler generates to implement this delay is pretty odd. It does not look like what we would write by hand:

```
            ; FUNCTION delay (BEGIN)
                                    ; SOURCE LINE # 26
0000 750000   R     MOV    n,#000H
0003                ?FOR1:
0003 7480           MOV    A,#080H
0005 B48105         CJNE   A,#081H,?LAB5
0008 E500     R     MOV    A,n
```

---

[14] The "preliminaries" include stating what type of quantity is each of the several variables and constants (so that the C compiler knows how much space to allot to each).

```
000A B42800            CJNE    A,#028H,?LAB5
000D            ?LAB5:
000D 5006              JNC     ?NXT4
000F AA00      R       MOV     R2,n
0011 0500      R       INC     n
0013 80EE              SJMP    ?FOR1
0015            ?NXT4:
                                        ; SOURCE LINE # 28
0015 22                RET
              ; FUNCTION delay (END)
```

But we don't really care that it's odd or awkward. The *for* loop in the *C* program made *our* work easy.

Main in C: The Main program invokes the Delay function with assembly language exactly like what we saw above in §21N.5. Here is the assembly language code that the C compiler generated to implement the C code of the Main loop:

```
ASSEMBLY LISTING OF GENERATED OBJECT CODE

            ; FUNCTION main (BEGIN)
            ?WHILE1:

            LCALL   delay

            CPL     OUT_BIT
            SJMP    ?WHILE1


            ; FUNCTION main (END)
```

Incidentally, a detail of C underlines the equivalence of C functions and assembly language subroutines: in C one can force an exit from a function with the word RETURN. We did not illustrate that here.

21N.6   Stretching operations to 16 bits

The running-sum programs of Lab 21L in both SiLabs and big-board form demonstrate that the 8051 can indeed handle arithmetic beyond the *byte* size of its registers. The 16-bit summing is accomplished with a curious programming device that appears in the code below (this is the SiLabs version).

```
Transfer: mov A, RUN_SUM_LO          ; recall running sum (lo byte)
          add A, KEYPAD              ; form new sum (low byte)
          mov RUN_SUM_LO, A          ; ...save it
          mov DISPLAY_LO, A          ; ...and show it
          mov A, RUN_SUM_HI          ; Get hi byte
          addc A, #0                 ; If a carry from low sum, incorporate it
              ...
```

The details that are far from obvious here involve the use of the CARRY bit – the single-bit "flag" (an internal flip-flop) that records whether a recent operation, such as ADD generated an overflow or "CARRY".

- First, ADD A, KEYPAD ignores any existing carry, because this flavor of ADD is distinguished by this indifference to CARRY. This indifference contrasts with the other flavor of addition named "ADC." ADD ignores any input carry but does update CARRY with the result of this addition operation: an overflow sets the CARRY flag.
- The ADDC A, #0 operation appears at first to be doing nothing: "add zero?!" But adding zero ain't nothing if it also adds in the CARRY flag, as it does.

So the ADDC A #0 command does update the high byte, incrementing it each time a carry occurs from the low byte: just what we need.

21N.7 AoE Reading

Chapter 14 (Computers...)

- §14.2.1: assembly language and machine language
- §14.3.2: programmed I/O: data out

21L Lab: Microprocessors 2

Again in this chapter there are two paths to follow; the big-board to start, and SiLabs in §21L.2.

21L.1 Big-board: I/O. Introduction

Last time, you breathed life into the mess of counters and memory that you had wired in Labs 16L and 17L, by adding a *brain*: the 8051. This mess became a computer, and proved it by running a tiny loop program.

The changes today will not rise to this Frankensteinian level. Today, your little machine learns to talk (through data displays) and to listen (reading the keypad).

And before we do any coding we'll add a battery-backup for your CMOS RAM so that your programs will stay in place even when you turn power off. Of course, the one program you entered so far was tiny. When your programs get bigger, this saving feature will become more obviously worthwhile.

21L.1.1 Battery backup for CMOS RAM

Your CMOS RAM uses extremely little power to hold data – around $0.8\mu A$ – so a lithium coin cell allows it to hold data for years (approximately three). To make this scheme work, however, we need circuitry to take care of two details:

1. we must switch between battery and main supply, feeding the RAM's supply pin – and do this at an appropriate time in the power-down and power-up processes;
2. we must disable the RAM while the supply voltage is low to prevent accidental Writes to the memory during power-down and power-up – overwriting good data.

The MXD1210 "nonvolatile RAM controller" takes care of both tasks. It includes a voltage reference, two comparators, and switches. Note that as you add this IC, you must *remove* the direct connection now present, between the glue PAL's output pin named "**RAMCE***" and the RAM's **CS*** pin. The MXD1210 mediates this connection. Similarly, you must *remove* the direct connection that now joins the main +5V supply to the RAM's power pin, V_{CC}; again, the '1210 comes between these two points.

Two quirks to beware:

1. You need to cycle the main power supply (+5V) ON, then OFF, in order to make the backup work, the first time you use it.
2. Make sure that your "5V" supply really provides 5V. If you are using the old PB503 5V supply, you're probably overloading it – and it may be putting out something less than the MXD1210's *minimum* voltage of 4.5V. In that event, an attempted "cycling" – ON, then OFF – is not really a cycling at all: the MXD1210 would see the +5 as always OFF.

21L.1 Big-board: I/O. Introduction

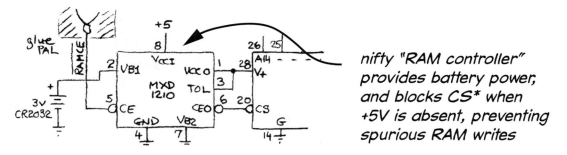

nifty "RAM controller" provides battery power, and blocks CS* when +5V is absent, preventing spurious RAM writes

Figure 21L.1 Battery backup for CMOS RAM.

21L.1.2 Crude output: latched display as the only out port

The data bus has been displayed continuously, till now: we grounded the EN* that controls the *transparent latch* on those displays. If instead we let the processor's WR* signal drive EN*, then the displays will catch and hold what is on the data bus during an OUT operation ("out" is human jargon for "write to I/O"). The program that we use to try this display-latching (listed just below) includes only one CPU *write* operation: this operation will assert WR*, and we'll use that signal to update the displays. This scheme is very crude. You'll notice that we don't even bother to try to distinguish I/O from memory. Instead we rely on the fact that, in the program below, the *only* write is an I/O write.

21L.1.3 Simplest latching hardware: just one out port permitted

Figure 21L.2 shows the hardware. The toggle switch looks too fancy for the present case, and it is; but soon it will be useful. Even in this little exercise it serves a purpose: in one position it preserves the *continuous enable* of the displays that we have used so far. Such enabling is necessary if we are to watch Data Bus activity as we troubleshoot programs in single-step mode. In the other position ("latched"), it lets us use the data displays as an output device, as we're about to do.

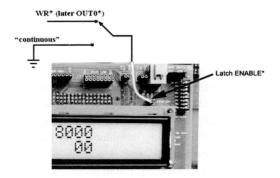

Figure 21L.2 Simplest output display: OK if there's only one I/O device, and never a write to RAM!

And Fig. 21L.3 is a timing diagram to indicate the way WR* can serve to latch the value the micro puts onto the data bus during an OUT:[1]

WR* drives the active-low LATCHEN*, catching and saving the data sent out while WR* is low. The "Port 0" line at bottom details some of the multiplexing of address and data that occur on the eight shared lines AD0..AD7. The assertion of PSEN* signals an *instruction* fetch, here labeled "Next Instruction Read."

---

[1] Figure modifies a fragment of Dallas/Maxim "High Speed Microcontroller User's Guide," Fig. 6-4.

Lab: Microprocessors 2

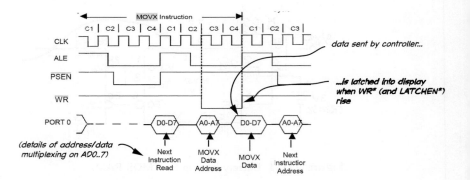

Figure 21L.3 Timing of OUT operation, applied to display latch.

Program to try the latching hardware: In order to try this simple latching hardware, we need a program that will do an "OUT" operation. Here is such a program. It writes to the address defined by the "data pointer" – a 16-bit register on the processor dedicated to this job of "pointing."

```
LOC     OBJ       LINE   SOURCE
                   1     ; IO_LOOP.A51 Lab 20L: to try display latching
                   2
0000               3     ORG 0    ; tells assembler the address at which to place this code
                   4
0000    8016       5     SJMP STARTUP  ; here code begins--with just a jump to start of
                   6                   ; real program.  ALL our programs will start thus
                   7
0018               8     ORG 18H          ; ...and here the program
                   9         ; starts, at hex 18 ("...h")
0018    908000    10     STARTUP:  MOV DPTR, #8000H   ; "point" to an address: loads pointer
                  11     ;  register with address of "port 0," displays and keypad
001B    E4        12          CLR A      ;  This just gives a predictable, tidy start value,
                  13     ;  to display (zero)
001C    F0        14     COUNTUP:  MOVX @DPTR, A  ;  Now send it to the display
001D    04        15          INC A  ;  add 1 to the accumulator, for a little excitement
001E    80FC      16          SJMP COUNTUP ;  ...and do it again (forever!)
                  17     END
```

Add an I/O decoder: The hardware you just tried would be sufficient if we imagined that we would need only one output device. The 8051 in fact can do its own "i/o decoding" when operated in single-chip mode. Later, in Lab 25L you will use it that way. Operating without buses, it can write a byte or even a single bit to one of its several "ports," and can hold that byte there in included flip-flops. But for the present we are doing I/O with the buses, so we need to provide our own decoding.

We now ask you to wire up an explicit "I/O decoder," external to the 8051. It will permit us to attach four output devices, and four input devices as well. (You are not likely add such a decoder to an 8051 again because you are not likely to use *buses* with a controller next time. The controller's great virtue is its ability to get along without help.)

The decoder in Fig. 21L.4 takes an "encoded" 2-bit value, and "decodes" it, to assert just one of its four output lines. It asserts the line corresponding to the number represented by this 2-bit code. "10," for example, evokes an assertion of output line 2. You could of course do such a decoding job with a PAL and these days that's a more likely way to do it. We have used an IC decoder here because its

21L.1 Big-board: I/O. Introduction

operation is easier to anticipate than if we had used a PAL. Using a PAL you might have to consult the PAL's code in order to be sure what to expect.

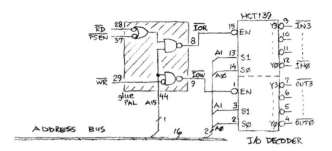

Figure 21L.4 '139 decoder driven by micro's signals.

On the WR* line, add a *blue or red or yellow* LED (some distinctive color: different from the *green* LED that indicates BR*) in order to help distinguish processor *writes* from *reads*.

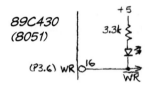

Figure 21L.5 An LED on WR* line helps us distinguish processor writes from reads.

The sharing of port addresses otherwise can be confusing as you single-step a program. The relatively large 3.3k resistor is chosen to load the WR* line lightly (WR* is rated to sink only 1.6mA).

A program to test the decoder: Here's a little program that should tickle all four of the decoder's IN and OUT ports. It uses a handy loop-counter, *DJNZ*, which compresses two operations into one instruction: decrement a specified register, then jump if the register contents have not yet reached zero. When the register *does* reach zero, execution proceeds to the next instruction: "we fall out of the loop," to use the usual jargon.

DJNZ operates only on 8-bit registers, limiting it to small values; but one can "nest" DJNZ operations to produce larger counts or delays (see §21L.1.6). The form of the instruction is to name the register or "directly-addressable byte" that is to be decremented, and to specify the jump destination, as you'll see in the example just below.

```
MACRO ASSEMBLER TRYDCDR_01                           08/01/101 17:27:00

LOC     OBJ       LINE     SOURCE
                   1       ; TRYDCDR_01.A51    Lab 20L: to try I/O decoder: count loops 4/13/01
                   2       ;   removed INC A, 8/01/01
                   3
0000               4       ORG 0   ; tells assembler the address at which to place this code
                   5
0000    801E       6       SJMP STARTUP   ; here code begins--with just a jump to start of
                   7                      ;  real program.  ALL our programs will start thus
                   8
0020               9       ORG 20H    ; ...and here the program starts, at hex 20 ("...h")
                  10
0020    E4        11       STARTUP:   CLR A ; This just gives a predictable, tidy start value,
                  12                        ;   to display (zero)
                  13
0021    908000    14       STARTOVER: MOV DPTR, #8000H ; "point" to a location: loads pointer
```

```
                    15      ; register with address of "port 0," displays and keypad
0024   7804         16          MOV R0, #4H      ; init loop counter
0026   F0           17  COUNTUP: MOVX @DPTR, A   ; Now send it to the decoder
0027   E0           18          MOVX A, @DPTR    ; ...and tickle the IN ports, too.
0028   A3           19          INC DPTR         ; advance the I/O address
0029   D8FB         20          DJNZ R0,COUNTUP  ; ...and do it till all 4 locations are tickled
002B   80F4         21          SJMP STARTOVER   ; Then start over
                    22
                    23          END
```

As you single-step this program you should see first an *OUT* operation, then *IN*, at each of the locations 0..3. These you will recognize on the address display, which here will show the I/O addresses, 8000–8003h. A logic probe on the '139's OUT lines (pins 4...7) and IN lines (pins 12..9 – in that order: IN0 through IN3) should confirm that the decoder, too, sees these operations.

The INx operations will look a little strange when single-stepped: since nothing is yet *enabled* by any of the decoder's INx* lines, when the computer does the IN operation the data bus floats. That probably will look like FFh (since the display inputs are likely to float *high* in the manner of TTL). The output operations will drive out the contents of the accumulator – zero, the very first output, then rather-strange values: whatever the processor has interpreted its floated inputs to be on the dummy input operation. The yellow or blue *WR** LED should help you to distinguish the INs from the OUTs.

21L.1.4 Add keypad input buffer

Let's give the decoder another task so it doesn't get bored: we'll let it turn on the keypad 3-state buffer at appropriate times, to pick up keypad values under program control. Then we'll pair this IN operation with an OUT to the displays.

Hardware, let IN0* turn on the keypad buffer: The hardware change that permits this input is very simple – because you have already built the needed logic into your GLUEPAL. Your KBUFEN* function OR's two sets of inputs:

- KWR* AND'd with BUSRQST* or "BR*" (this AND'ing defines the only legitimate, safe manual write from keypad into memory – safe because BUSRQST* indicates that we humans have control of the buses);
- IN0* from the '139 decoder. This signal will be asserted when the micro is reading from port 0 (hex address 8000h).

Back in Lab 16L, RAM and keypad buffer simply took turns: see Fig. 21L.6.

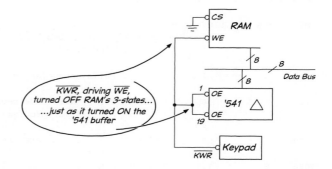

Figure 21L.6 Lab 16L's simple either–or, between RAM and keypad buffer.

In order to allow for the arrival of the processor, you have put the necessary logic onto your PAL. This logic implements the KBUFEN* function: see Fig. 21L.7.

21L.1 Big-board: I/O. Introduction

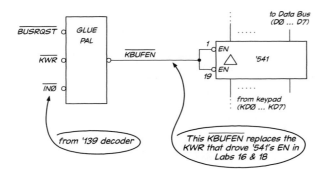

Figure 21L.7 Input buffer driven by GLUEPAL's KBUFEN*.

Now we'll use your logic. The line **IN0*** on the GLUEPAL has been disasserted till now (back in Lab 20L you pulled that line up). Now remove the pullup wire, and drive the GLUEPAL's *IN0** pin with the '139's signal of that name.

Hardware, let OUT0* drive the display latch: In place of WR*, used back in §21L.1.3, let the I/O decoder's OUT0* line drive the LATCHEN* *LOW* byte enable.

Program, in and out, read the keypad: The program below allows you to test the new hardware. The program simply takes in a value from the keypad and spits it out again to the displays. Keypad and display sit at the same address – but they are distinguished, of course, by the difference between IN and OUT: that is, by the levels on the 8051's RD* and WR* lines. To remind you of the scheme, Fig. 21L.8 is a scope photo showing the program below as it runs: the assertion of RD* marks the INPUT operation, the assertion of WR* marks the OUTPUT. We've included ALE and PSEN* (the signal that enables memory during an instruction fetch), in hopes that they may help us get our bearings in reading this loop image.

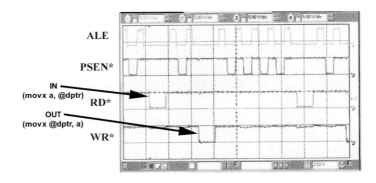

Figure 21L.8 In–out loop: scope image, at full-speed.

Here's the program:

```
MACRO ASSEMBLER INOUT                        03/08/101 18:24:05 PAGE    1


LOC     OBJ        LINE     SOURCE
                     1      ; INOUT.a51   Lab 20L:    read keypad, show on display
                     2
                     3      $NOSYMBOLS   ; keeps listing short, lest...
                     4      $INCLUDE (C:\MICRO\8051\RAISON\INC\REG320.INC)  ; ...this line should
                   170      ; produce huge list of symbol definitions (all '51 registers)
```

838 Lab: Microprocessors 2

```
                    171
0000                172     ORG 0   ; tells assembler the address at which to place this code
                    173
0000    8026        174     SJMP STARTUP ; here code begins--with just a jump to start of
                    175     ; real program.  ALL our programs will start thus
                    176
0028                177     ORG 28H ;  ...and here the program starts
                    178
0028    908000      179     STARTUP: MOV DPTR, #8000H ; "point" to both display (OUT0)
                    180     ; and keypad (IN0)
                    181
002B    E0          182     GETIT:   MOVX A, @DPTR ; read the keypad
002C    F0          183              MOVX @DPTR, A ; ...show it on display
002D    80FC        184              SJMP GETIT
                    185
                    186     END
```

You can see whether the INPUT operation is working, by watching the data-bus value during the IN operation as you single-step the program. Make sure the data display is set to show you everything: CONTINUOUS, not LATCHED. With the LCD showing address 8000h on an IN0* operation, try changing the keypad value. You should find that, while program execution is frozen in the midst of the IN operation, the data bus value shown on the LCD is a *live* display of whatever you type on the keypad.

If you then change to LATCHED and press the REPEAT key, you should see the keypad value echoed on the displays, but updated only once each time around the loop. You will see none of the intermediate instruction codes. This should work at full-speed, as well.

21L.1.5 A more interesting program: add keypad value to running sum

8-bit sum: We will ask you to do this task first showing an 8-bit result. You might perhaps think that an 8-bit computer cannot do better. But no; it's quite easy to piece together 8-bit results to form larger values as we noted in Chapter 21N and as you will demonstrate soon.

8-bit sum, single-stepped version: The program below reads the keypad continually and adds that value to a running sum, displaying the result on the displays. Since the program does not yet include any delay, it can't give you an intelligible display at full-speed; you will need to single-step this program. We'll soon fix that defect.

```
MACRO ASSEMBLER KEYSUM_402                              07/31/102 18:03:55 PAGE     1

LOC    OBJ         LINE    SOURCE
                    1      ; KEYSUM_402.a51  Lab 20L: read keypad, add keyval to running sum:
                    2      ; First version: NO DELAY  4/0
                    3      $NOSYMBOLS ; keeps listing short, lest...
                    4      $INCLUDE (C:\MICRO\8051\RAISON\INC\REG320.INC)  ; ...this line should
                    170    ; produce huge list of symbol definitions (all '51 registers)
                    171
0000                172    ORG 0 ; tells assembler the address at which to place this code
                    173
0000    802E        174    SJMP STARTUP ;  here code begins--with just a jump to start of
                    175    ;   real program.  ALL our programs will start thus
                    176
0030                177    ORG 30H ;  ...and here the program starts
                    178
0030    908000      179    STARTUP: MOV DPTR, #8000H ; "point" to display (OUT0) and to keypad (IN0)
```

```
0033  75817F    180         MOV  SP, #07FH ; init stack ptr to just below bottom of scratch RAM
                181              ; (internal) empty this time, but we'll use it soon
0036  7A02      182         MOV  R2,#02    ; init small delay constant (not useful till next program,
0038  7B04      183         MOV  R3,#04    ; but we include it to spare you re-entering code
                184
                185
003A  7800      186         MOV  R0, #0 ; clear running sum, for predictable start
003C  F0        187         MOVX @DPTR, A ; show sum on display: tidy first-pass
                188
003D  E0        189 COUNTUP:  MOVX A, @DPTR ; read the keypad
003E  28        190         ADD  A, R0    ; form new running sum: new + old, result to A
003F  00        191         NOP    ; just a place-holder, for now
0040  F8        192         MOV  R0, A    ; save result (update running sum)
0041  F0        193         MOVX @DPTR, A ; show sum on display.
                194
0042  80F9      195         SJMP COUNTUP  ; now do it again
                196
                197         END
```

The keypad value FFh (FF, hexadecimal) has a particularly curious effect. Why does the sum change as it does when the keypad provides FFh?[2]

Decimal addition: You can very easily get this program to speak to you in *decimal*, if you like. Simply replace the NOP in the listing above with the instruction "DAA," "Decimal Adjust Accumulator". (The "opcode" for DAA is D4h.) DAA transforms a sum like "0A" into "10". But don't overestimate the cleverness of DAA: it works only immediately after the addition operation, when a half-carry "flag" still is available to describe the value in the accumulator. And it fails if you feed it a non-decimal value from the keypad. Otherwise, it's pretty neat. Try it.

What keypad value now evokes the same curious effect that the pre-*DAA* version showed when you fed it FFh from the keypad?[3]

16-bit sum: single-stepped: Your big-board *Data* bus is 8-bits wide, but the larger LCD board includes *two* 8-bit input latches. These can catch two successive 8-bit values sent by the micro to display a 16-bit result.

Hardware, two latches: The LCD board includes two 8-bit *transparent* latches that can be used to catch the data bus value in two successive OUT operations. Fig. 21L.9 shows the hardware on the LCD board that lets you write to its 16-bit data display in two passes.

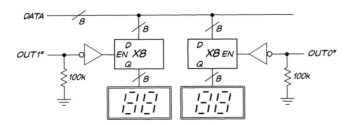

Figure 21L.9 Two latches can demultiplex 16 bits of data from 8-bit bus.

To take advantage of these two latches, you'll need to drive both of the two latch-enable* points shown in Fig. 21L.10. OUT0* continues to drive the *LOW* LATCH EN*; OUT1* from the '139 drives the *HIGH* LATCH EN*.

[2] Recall that FFh is the value *minus 1* in 8-bit 2's-complement notation.
[3] Is it perhaps 99?

Lab: Microprocessors 2

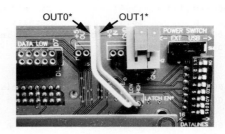

Figure 21L.10 For 16-bit latched data display, OUT0*
and OUT1* drive two latch-enables*.

...and set DIP switch to MUX: The DIP switch shown on the right in Fig. 21L.10 needs to be set to MUX: this setting allows the two latches two time-share the 8-bit data bus as shown in Fig. 21L.9. (The other DIP switch position permits 16-bit parallel input to the two 8-bit latches. That mode is useful when the LCD board is used stand-alone rather than with the micro breadboard.)

Code, 16-bit sum: Here is code to extend the sum from 8 bits to 16:

```
LOC    OBJ      LINE    SOURCE
                1       ; 16_SUM_02.a51  Lab 21: 16-bit running sum & display
                2
                3       $NOSYMBOLS ; keeps listing short, lest...
                4       $INCLUDE (C:\MICRO\8051\RAISON\INC\REG320.INC) ; ...this line should
                170     ;    produce huge list of symbol definitions (all '51 registers)
                171
0000            172     ORG 0 ; tells assembler the address at which to place this code
                173
0000   805E     174     SJMP STARTUP ; here code begins--with just a jump to start of
                175     ; real program.  ALL our programs will start thus
                176
0060            177     ORG 60H ;  ...and here the program starts
                178
0060   908000   179     STARTUP:  MOV DPTR, #8000H ; "point" to low byte of display
                180     ; and to keypad
0063   E4       181     CLR A
0064   F8       182     MOV R0, A ; clear storage registers
0065   F9       183     MOV R1, A
                184
                185
0066   E0       186     GETIT:  MOVX A, @DPTR ; read the keypad
0067   28       187     ADD A,R0 ; form low-half of running sum
0068   F0       188     MOVX @DPTR, A ; ...send it to display (port 0)
0069   F8       189     MOV R0,A ; and save a copy in register
                190
006A   E4       191     CLR A
006B   39       192     ADDC A,R1 ; increment high-byte of running sum
                193     ;    (takes in CY if we've overflowed low byte)
006C   F9       194     MOV R1,A ; ...save result
                195
006D   A3       196     INC DPTR ; point to high-half of display (port 1)
006E   F0       197     MOVX @DPTR, A ; ...send high byte to display (port 0)
006F   1582     198     DEC DPL ; restore pointer to low-byte (there is no 16-bit ptr dec,
                199     ; but this modest dec is safe; done once from 8001,
                200     ;   it can't generate a borrow
0071   80F3     201     SJMP GETIT
                202
                203     END
```

A detail of the program may be worth pointing out: the successive ADD operations use two different versions of ADD. Here we will make a point close to one we made in Chapter 21N:

- the first use is ADD, which does not take in the CY (carry) flag but does update CY as an output;
- the second use is ADDC, which does take in CY.

The ADDC code looks, at first glance, like nonsense, does it not?

```
CLR A
ADDC A,R1
```

This ADDC operation may look like a complicated way of saying "do nothing" since the code means "Add zero (the value of A) to R1 (the high-byte of 16-bit sum)." But the critical difference from "add zero" (meaning "do nothing to the R1 value") is that ADDC *includes any CY* from the operation on the low-order byte. Thus the second ADD – "ADDC" – does extend the 8-bit result properly. You may want to think of the low-order CY as an *overflow* signal; the high-order byte might be said to count overflows from the low-order summing.

Watching the 16-bit result: In order to see the 16-bit result you need to select "16" rather than "8" with the "DATA" select switch: see Fig. 21L.11. To avert boredom, try setting up a fairly large *keypad* value to be summed, such as 21h. As you single-step the program you should soon see the high byte come to life when the low byte overflows.

Figure 21L.11 LCD card allows option of displaying 16 bits of data rather than eight.

2's-comp demystified? The keypad value FFh also is worth trying. When you tried that with an 8-bit display the computer may have seemed clever, as if it somehow recognized FFh as "2's-complement for negative one."

The 16-bit display shows what was going on backstage, so to speak: nothing fancier than addition. No cleverness or understanding of 2's-complement appears does it? The FFh, which we may choose to think of as "minus one," does decrement the low byte. But this result is revealed as only a side-effect of the addition when one can see the 16-bit result. Each time the low byte is decremented, the addition also increments the high byte.

21L.1.6 Keysum with delay: two versions: a first subroutine

Keysum, with delay written in-line: The program listed below appends a few lines of time-wasting code to the 8-bit keysum program on page 838. Note that the new code is to be *appended* to the code you've already entered. Don't overwork yourself by typing in everything anew: instead, start at address 42h. The time-wasting or "delay" code will allow us to run the program at full speed, while still getting an intelligible display.

The delay code again takes advantage of the compact *DJNZ* instruction. Because the 8051 can do this operation only on its 8-bit registers, and counting down 256 times doesn't take long, we have *nested* two of these DJNZ loops for a longer delay. The nesting lets us multiply that inner-loop delay by an 8-bit outer constant. We have initialized these delay registers to *tiny* values so that you can watch the program in single-step without expiring from boredom. Once you've seen it in slow motion, change the delay constants to generate the *maximum* delay. What constants are those? *Hint*: DJNZ decrements the register *before* testing its value.[4]

---

[4] An initial value of zero evokes the longest delay.

```
MACRO ASSEMBLER KEYSUM_DELAY_INLINE_402                 07/21/04  11:43:27 PAGE       1
LOC      OBJ       LINE     SOURCE
                   1        ; KEYSUM_DELAY_INLINE_402.a51   Lab 20L: read keypad,
                   2        ; add keyval to running sum: DELAY IN-LINE  4/02
                   3        $NOSYMBOLS ; keeps listing short, lest...
                   4        $INCLUDE (C:\MICRO\8051\RAISON\INC\REG320.INC)  ; ...this line should
                   170      ; produce huge list of symbol definitions (all '51 registers)
                   171
0000               172      ORG 0 ; tells assembler the address at which to place this code
                   173
0000     802E      174      SJMP STARTUP ; here code begins--with just a jump to start of
                   175      ; real program.  ALL our programs will start thus
                   176
0030               177      ORG 30H ;  ...and here the program starts
                   178
0030     908000    179      STARTUP: MOV DPTR, #8000H ; "point" to display (OUT0) and keypad (IN0)
0033     75817F    180               MOV SP, #07FH ; init stack ptr to just below bottom of scratch
                   181      ; RAM (internal)  empty this time, but we'll use it soon
0036     7A02      182               MOV R2,#02 ; init small delay constant
0038     7B04      183               MOV R3,#04
                   184
                   185
003A     7800      186               MOV R0, #0 ; clear running sum, for predictable start
003C     F0        187               MOVX @DPTR, A   ; show sum on display: tidy first-pass
                   188
003D     E0        189      COUNTUP:  MOVX A, @DPTR ; read the keypad
003E     28        190      ADD A, R0 ; form new running sum: new + old, result to A
003F     00        191      NOP
0040     F8        192      MOV R0, A ; save result (update running sum)
0041     F0        193      MOVX @DPTR, A  ; show sum on display.  Now, go kill some time
                   194
0042     EA        195      MOV A,R2   ; get outer-loop delay value
0043     8BF0      196      INITINNER:  MOV B,R3   ; initialize inner loop counter
0045     D5F0FD    197      INLOOP: DJNZ B, INLOOP ; count down inner loop, till inner hits zero
0048     D5E0F8    198      DJNZ ACC,INITINNER ; ...then dec outer, and start inner again.
                   199
004B     80F0      200      SJMP COUNTUP    ; now do it again
                            END
----------------------------------------
```

It is not easy to calculate the delay offered by this program; it *is* easy to use your scope to *measure* it if you watch a signal that comes once per delay; OUT0* will serve. (We measured the repetition period at about 120ms, with the delay value set to *maximum* – not to the tiny values shown in the listing above.)

The very *tiny* delay values we showed in the listing, for both inner and outer loops, are proposed to let you watch the nested loops as you single-step. (You'll get very bored if you put in max delay values and then single-step.) The listed delay constants are *too small to be useful at full-speed*, however. For longest delay, change both constants to *zero*.

Putting the delay program in-line isn't very tidy. And the DELAY patch is not polite: it messes up the **A** register as it runs, though this messing-up happens not to matter for this particular program. In the next program we will do better in both respects: we'll use a distinct *subroutine* to give us the delay, and we'll take care to restore any registers that we alter.

21L.1.7 Keysum, with delay written as subroutine

Now we'll implement the DELAY more tidily. We'll write it as a self-sufficient little segment of code – a subroutine. This segment can be invoked from the Keysum program, but later can also be invoked

21L.1 Big-board: I/O. Introduction

by any other program – whenever you decide that you'd like to slow execution. We will write it so that, unlike the *in-line* version just above, the routine does *not* permanently mess up registers. That's wise, because those registers may be used by some later program that invokes DELAY.

Again, we have simply altered the code you just entered, so don't do any substantial re-typing. Just change the two lines beginning at address 42h. Then enter the *DELAY* subroutine up at address 100h (an address chosen as a nice round number that we hope to be able to remember when next we need a delay).

Once again, we start with a tiny delay value so that you can watch the program while single-stepping. When you reach the ACALL instruction, the computer will appear to hesitate – or to ignore several of your single-step key-pressings. Then it will hop to the start of the DELAY routine. That hesitation occurs because the processor is busy saving on the *stack* the *return* address (that is, the address of the instruction that it must execute after finishing the subroutine). The stack operations are invisible to us because they take place entirely on the 8051, rather than on the external buses that are visible to us.

The *stack* is just a region in the micro's *internal, on-chip RAM* used for temporary storage of the return address – as well as for other temporary storage. Below, in DELAY, we use the stack not only for the return address (this use is automatic when we use ACALL), but also for temporary storage of registers that will be altered by the DELAY routine: the PSW (flags), and the **A** and **B** registers (for obscure reasons, the assembler requires us to call **A** "ACC" when PUSH'ing and POP'ing). This is a typical use of the stack. It is easy, and the stack takes care of itself – as long as we take care to let every PUSH be paired with an appropriate POP. Notice the sequence of PUSHes and POPs in the listing below.

```
MACRO ASSEMBLER DELAYROUTINE_407
LOC     OBJ     LINE    SOURCE
                1       ; KEYSUM_DELAYROUTINE_407.a51  Lab 20L:   read keypad,
                        ; add keyval to running sum: DELAY SUBROUTINE  4/02,9/02
                2       ; start address changed to allow use as replacement for earlier
                        ; delay-inline 4/07
                3
                4       $NOSYMBOLS ; keeps listing short..
                5       $INCLUDE (C:\TOM\MICRO\8051\RAISON\INC\REG320.INC)
                172
0000            173     ORG 0   ; tells assembler the address at which to place this code
                174
0000    802E    175     SJMP STARTUP  ; here code begins--with just a jump to start of
                176     ; real program.  ALL our programs will start thus
                177
0030            178     ORG 30H ; ...and here the program starts
                179     ;  Note that this listing overlaps with the previous program.
                180     ;  This is intended, so you'll not need to retype any code
                181     ;  before the subroutine call, at 42h
                182
0030    908000  183     STARTUP: MOV DPTR, #8000H ; "point" to both display (OUT0) and keypad (IN0)
0033    75817F  184     MOV SP, #07FH ; init stack ptr, to just below start of indirect RAM
0036    7A02    185         MOV R2,#02 ; init small delay constant
0038    7B04    186     MOV R3,#04
                187
                188
003A    7800    189         MOV R0, #0 ; clear running sum, for predictable start
003C    F0      190         MOVX @DPTR, A ; show sum on display: tidy first-pass
                191
003D    E0      192     COUNTUP: MOVX A, @DPTR ; read the keypad
003E    28      193         ADD A, R0 ; form new running sum: new + old, result to A
003F    00      194         NOP
```

```
0040    F8          195         MOV R0, A ; save result (update running sum)
0041    F0          196         MOVX @DPTR, A ; show sum on display.  Now, go kill some time
                                ; HERE BEGIN THE CHANGES FROM THE IN-LINE-DELAY VERSION:
0042    3100        197         ACALL DELAY
0044    80F7        198         SJMP COUNTUP
                    199
                    200         ; -- SUBROUTINE 'DELAY' ------ -
                    201
0100                202         ORG 100H
0100    C0D0        203         DELAY:  PUSH PSW   ; save registers that'll get messed up.
0102    C0E0        204         PUSH ACC    ; Not necessary, this time; but next time this
0104    C0F0        205         PUSH B      ;  routine is called, those registers might be
                    206         ;  in use in the main routine, so worth saving
                    207
0106    EA          208             MOV A,R2 ; get outer-loop delay value
0107    8BF0        209         INITINNER: MOV B,R3 ; initialize inner loop counter
0109    D5F0FD      210         INLOOP:    DJNZ B, INLOOP ; count down inner loop, till inner hits zero
010C    D5E0F8      211             DJNZ ACC,INITINNER ; ...then dec outer, and start inner again.
010F    D0F0        212             POP B    ; restore saved registers
0111    D0E0        213             POP ACC
0113    D0D0        214             POP PSW
0115    22          215             RET ; Now back to main program.
                    216
                    217         END
```

21L.2 SiLabs 2: input; byte operations

21L.2.1 Port pin use in this lab

Today, in this path, we make use of just one more PORT0 pin in addition to the LED drive of §20L.3: an input pin driven by a pushbutton.

Later, in §21L.2.8, we also add *byte* input and output devices at two other ports, PORT1 (display) and PORT2 (keypad), and finally we also ask PORT0 to drive another 8-bit LCD display. That last addition does not conflict with the two PORT0 uses shown in Fig. 21L.12.

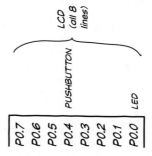

Figure 21L.12 PORT0 pin use in today's lab: a pushbutton input added.

21L.2.2 Bitflip LED blink program, recast with subroutine

Last time, we slowed the LED-blinking program by putting time-wasting DELAY code inline with the bitflip code. This works, but makes the code relatively hard to read and clumsy to write. In a tiny program like Bitflip this hardly matters. But in a more complex program it is useful to break the code into modules, called subroutines, all of which are stitched together by a main *calling* program. Here,

21L.2 SiLabs 2: input; byte operations

we will apply that change in form to the delayed bitflip program of Lab §20L.3. The behavior of the LED will be the same. The code will be easier to read.

The code below is just the old code, rearranged – with a few details added that permit the main program to invoke the two subsidiary subroutines: one subroutine takes care of the SiLabs special initializations (we call this patch of code "USUAL_SETUP" and it reappears in almost all of the programs in this series of labs). The other subroutine, DELAY, simply uses up about one second to set the LED blink rate, as in Lab §20L.3.

We have made one small change in USUAL_SETUP relative to that used in the delay code of §20L.3: we have slowed the system clock by a factor 8, as the '410 permits, in order to save a little power. In this application this slowdown certainly has no appreciable effect. It is not necessary but we thought we'd show this option, which is the IDE's default but here imposed by the explicit command,

```
    ORL OSCICN, \#04h ; sysclk = 24.5 Mhz /8.
```

We'll list the program, and then look at some of its details.

```
; bitflip_delay_subroutine.a51 bitflip program, with delay for full-speed operation

$NOSYMBOLS ; keeps listing short
$INCLUDE (C:\MICRO\8051\RAISON\INC\c8051f410.inc)

ORG 0 ; tells assembler the address at which to place this code

        ACALL USUAL_SETUP
        SETB P0.0 ;  start with LED OFF (it's active low)

FLIPIT: CPL P0.0           ; toggle LED
        ACALL DELAY ; waste some time
        SJMP FLIPIT ; do it forever

; --SUBROUTINES  -- -
DELAY:     PUSH ACC     ; save the registers that this routine will mess up
           PUSH B
           PUSH 4     ; this saves register R4--in the zeroth set of registers

           MOV A,#0 ; maximize two delay values (0 is max because dec before test)
           MOV B,#0   ; ...and second loop delay value
           MOV R4, #10h    ; 1 second delay: this multiplies the 64K other loops

INNERLOOP: DJNZ B, INNERLOOP ; count down innermost loop, till inner hits zero
           DJNZ ACC,INNERLOOP ; ...then dec second loop, and start inner again.
           DJNZ R4, INNERLOOP ; now, with second at zero, decrement the outermost loop

           POP 4
           POP B
           POP ACC
           RET               ; back to main program

USUAL_SETUP: ANL    PCA0MD, #NOT(040h)       ; Disable the WDT.
                                              ; Clear Watchdog Enable bit
; Configure the Oscillator
            ORL   OSCICN, #04h        ; sysclk = 24.5 Mhz / 8
               ; Enable the Port I/O Crossbar
            MOV   XBR1, #40h          ; Enable Crossbar
            RET
END
```

The subroutine is invoked by a CALL operation, here in the shorter ACALL version. (Subroutines and the stack are discussed in Chapter 21N). Subroutine etiquette requires that the subroutine save any values it will mess up, so that the main program can operate properly on a *return* from the subroutine. The saving here is done with PUSH operations; the same registers are restored, before exit from the subroutine, with POP operations. RET pops the "return" address off the stack, so that execution can resume at the main program's instruction that follows the ACALL.

21L.2.3 Some oddities of PUSH AND POP

Potential ambiguity in the name "Rn": One of the PUSHes (PUSH 4) is odd. Why "PUSH 4" rather than "PUSH R4"? Because "R4," surprisingly, is ambiguous: the 8051 has *four* sets of scratch registers named R0..R7. The 8051 resolves the ambiguity by referring to two bits in its PSW (Program Status Word) register. After a reset, these two bits are at zero, so the controller uses the zeroth register set by default.

But the broad-minded assembler program does not dare to presume which of the four register sets is intended. So, it requires that we describe the zeroth R4 by its on-chip *address*. "4" is the (unambiguous) address of the zeroth version of R4.[5]

As we noted in Chapter 21N, PUSH and POP require that we call register **A** "ACC."

The result, we hope, code that's easier to follow: The use of subroutines makes even this simple program easier for a reader to make sense of. The guts of the program sit near the start:

```
FLIPIT: CPL P0.0           ; toggle LED
        ACALL DELAY ; waste some time
        SJMP FLIPIT ; do it forever
```

Appearing early, and all in one place, this code is easier to make sense of than the code of the delayed-bitflip program of Lab §20L.3, where the delay and initialization code was placed inline.

From this point, we will use subroutines regularly in an effort to make our programs readable and also to permit building them up in stages. This effort recalls our methods in the analog part of this course: there, we tried to design and test subcircuits that we then could link as *modules*. The method eased our work in both design and analysis – and the motive for use of subroutines is the same.

21L.2.4 Bit input: "blink LED if..."

In the Lab 20L we made an LED blink. As an application for an intelligent controller, this is not very impressive. A '555 oscillator can make an LED blink. Now we will modify the program just slightly so that the LED blinks *unless* we press a pushbutton. That's still not impressive, you may protest: a pushbutton on the '555 RESET* line could achieve the same result.

Well, OK: fair enough. But we'd like to make a claim that the ability of the processor to do *one thing* under one condition, *another thing* under another condition amounts to a glimmering of *intelligence*. This ability is fundamental to what can make a computer seem smart. So, let's add this capability to the blinker program.

21L.2.5 Bit input

Hardware: When using buses – as you have seen in Chapter 21N – responding to a bit input requires some hardware (a 3-state to get the information onto the data bus) and then a bit-testing operation on the *accumulator* register.

---

[5] We wish the assembler were bold enough to assume that the particular R4 intended is the one in use at the time of the reference. But this is not the way 8051 assemblers work.

Using a controller's built-in port, as in the '410 (which provides no buses), data inputs are much simpler. An input signal (either a *bit* as in the upcoming *bitflip_if.a51*, or a *byte*, as in §21L.2.8) is tied directly to the controller's port pin or pins: no need for a 3-state, because the pin accesses not a public *bus* but a private road into the controller.

The hardware that let's us talk to the '410 with a pushbutton thus is extremely simple: a pushbutton can ground the input signal, while a pullup resistor otherwise pulls it to +5V.[6] If switch *bounce* matters, then at least a capacitor must be added to slow the edge.[7] But let us omit that here, for maximum simplicity, since bounce does not matter in this program.

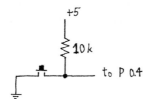

Figure 21L.13 Pushbutton to drive P0.4.

21L.2.6 Code

Subroutines tidy the code: Applying our new wisdom (new in §21L.2.2) we again push the details of initializations and Delay out of the way of the main program loop, making them subroutines.

A formal novelty, introducing* symbols *to make code more intelligible: We have made some small changes to the *bitflip* code, intended to make the code more readable.[8] We have used *symbols* – descriptive names – to stand in for the literal port pins used in this program.

Whereas in the original *bitflip.a51* program we wrote "P0.0," we here will replace the port-pin specification with a descriptive name, "BLUE_LED." Similarly, we will replace the pin designation of the input pin, P0.4, with the description "PB" ("pushbutton).

So, whereas bitflip's loop looked like

```
FLIPIT: CPL P0.0 ;  flip LED, ON, then OFF...
        SJMP FLIPIT
```

the loop in the *flip_if.a51* program gives a reader a little help in divining what the loop does.

At the head of the program we define the two symbols – "BLUE_LED" and "PB" so that we can write these descriptive names rather than list the port pins. We use an *assembler directive*, EQU, to define symbolic names for the two pins:

```
BLUE_LED EQU P0.0
PB EQU P0.4
```

An assembler directive, like EQU, is a command addressed not to the 8051 but to the assembler.[9] Each time the assembler encounters the symbol "BLUE_LED," it will substitute "P0.0" as it translates our assembly code into executable code. It will also do the equivalent substitution for "PB."

Having defined these symbols we can make the flip-if code loop fairly intelligible. We now ask you to finish the program for us.

---

[6] On some 8051/'410 ports even the pullup can be omitted. But at PORT 0, which we use here, no such pullup is present.

[7] As you know, a true *debouncer* for an SPST switch requires more than an *RC* circuit. It also requires a Schmitt trigger to square up the transition of the slow *RC*. But a processor input, unlike a true edge-triggered input such as a flip-flop clock, does not need that squaring up, as we will see in Lab §22L.2.

[8] The use of symbols and labels is sometimes oversold with the description "self-documenting code." We don't claim that labels are quite *that* wonderful.

[9] This topic is treated in Chapter 21N.

Lab: Microprocessors 2

Your task, implement the "If..." in assembly language: Now we have set up the *symbol* "PB" to stand for the bit that is to be tested. We assign a *label*, "FLIPIT," to the location to which we want the program to return when the pushbutton is pressed.

```
FLIPIT:    ____ PB, FLIPIT  ; hang up here, so long as pushbutton is pressed
           CPL BLUE_LED
```

We would like you to fill in the 8051 instruction that will evoke this behavior. You may want to consult the one-page list of 8051 instructions that we have put together in Supplement 20S. The same information, in wordier form, appears at pp. 15–16 of the Philips/NXP programmer's guide for the 8051[10].

When you have filled in the code, the label and symbol offer strong clues to the logic of this loop. Perhaps you find `FLIPIT: ____ PB, FLIPIT` and `CPL BLUE_LED` cryptic still. But the symbols help, do they not?

Program flow conditioned by pushbutton: The novelty in the *behavior* of bitflip_if.a51 – distinct from the novelty in the program's *listing* – is the code's testing of an input. The *conditional* jump command that we asked you to fill in implements the "if" in the "Blink...if...." This conditional ("if"...) keeps the program stuck on this line as long as the pushbutton holds P0.1 at a logic Low. Only when the pushbutton is released, permitting P0.1 to go High does the program flow reach the next line where the LED-flipping occurs.

Here is the full program listing:

```
; bitflip_if.a51  toggle LED, slowly--if pushbutton NOT pressed

        $NOSYMBOLS ; keeps listing short
        $INCLUDE (C:\MICRO\8051\RAISON\INC\c8051f410.inc)

        BLUE_LED EQU P0.0
        PB EQU P0.4      ; pushbutton will determine whether LED is to toggle or not

        ORG 0            ; tells assembler the address at which to place this code

        SJMP STARTUP     ; here code begins--with just a jump to start of
                         ; real program.  ALL our programs will start thus
        ORG 80h          ; ...and here the program starts

STARTUP: ACALL USUAL_SETUP
         CLR BLUE_LED   ; start low, just to make it predictable


FLIPIT: ____ PB, FLIPIT ; hang up here, so long as pushbutton is pressed
        CPL BLUE_LED   ; ...but flip the LED when button not pushed
        ACALL DELAY
        SJMP FLIPIT

; -- -SUBROUTINES  --  --
USUAL_SETUP:                 ; Disable the WDT.
        ANL  PCA0MD, #NOT(040h)   ; Clear Watchdog Enable bit
                                  ; Configure the Oscillator
        ORL  OSCICN, #04h         ; sysclk = 24.5 Mhz / 8, for lower power

        ; Enable the Port I/O Crossbar
```

[10] http://www.keil.com/dd/docs/datashts/philips/p51_pg.pdf.

21L.2 SiLabs 2: input; byte operations 849

```
            MOV    XBR1, #40h           ; Enable Crossbar
                                 RET

DELAY:      PUSH ACC      ; save the registers that this routine will mess up
            PUSH B
            PUSH 4        ; this saves register R4--in the zeroth set of registers

            MOV A,#0  ; maximize two delay values (0 is max because dec before test)
            MOV B,#0  ; ...and second loop delay value
            MOV R4, #10h    ; 1 second delay: this multiplies the 64K other loops

INNERLOOP:  DJNZ B, INNERLOOP   ; count down innermost loop, till inner hits zero
            DJNZ ACC,INNERLOOP  ; ...then dec second loop, and start inner again.
            DJNZ R4, INNERLOOP  ; now, with second at zero, decrement the outermost loop

            POP 4
            POP B
            POP ACC
            RET                 ; back to main program
END
```

We found the blink rate a little low for our tastes. Try speeding it up by, say, a factor of four.

21L.2.7 C language equivalent: "If..."

C code that would achieve the same result – flipping an LED unless a pushbutton grounds an input named "inbit" – looks a lot like the assembly language loop of § 21L.2.6 on page 847. In C the "if" is explicit; in assembly language the "if" is expressed in the conditional jump operation that you filled in above.

```
if(!inbit) outbit = outbit;
else outbit = !outbit;
```

For code as simple as this, C offers no advantage over assembly.

21L.2.8 Byte operations, In and Out

We have used *bit* operations so far because they give such a quick reward for minimal wiring. *Byte* operations – with which we started on the big-board branch of the micro labs – are just as simple to code, but require just a little more wiring effort. Let's make that small effort now, and do some byte operations.

Hardware for Byte In, Byte Out:

Byte input from keypad: Let's use the keypad, with its 16-pin DIP connector, as input device. On a breadboard strip you can convert the DIP's 8 data lines – which unfortunately lie on two sides of the connector – into the 8-in-a-row form that will be convenient for wiring to the '410: see Fig. 21L.14.

From this in-line set of eight data lines, run a flat ribbon cable to the '410's PORT 2, as in Fig. 21L.15.

The D7 line on the cable, you'll notice, comes not from the '410's P2.7 but from the far end of the C2-neutralizing resistive network described in §20L.3 (see especially Fig. 20L.21).

Keypad in, byte display out: We'll first do a byte-in, byte-out transfer, keypad to display. As a way of seeing the byte-wide output from the '410, let's use the LCD display board. It provides 8 bits for data, 16 bits for what, on the Dallas branch, is *address*. For the one-byte display of the current program we

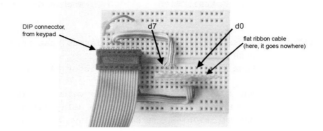

Figure 21L.14 DIP connector from keypad, adapted to in-line flat cable.

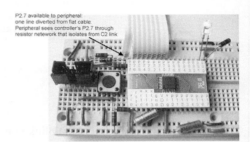

Figure 21L.15 Keypad cable wired to '410: P2.7 split from cable.

Figure 21L.16 Hardware connections for byte-in, byte-out keypad to display.

will use the low byte (D7..D0) of the *data* display – one of the two rightmost byte-wide connectors on the LCD board.

Though PORT0 here drives the display, you need not disconnect the LED at P0.0 or the pushbutton at P0.4 unless you choose to. Neither should interfere with the '410's drive of the display: each applies only a pullup resistor to its pin (unless you are so rash as to press the pushbutton while using PORT0 to drive the LCD display).

21L.2.9 Code for byte-in, byte-out

The code to transfer a byte from keypad to display is as simple as you would expect it to be:

```
TRANSFER: MOV DISPLAY, KEYPAD
         SJMP TRANSFER
```

21L.2 SiLabs 2: input; byte operations

A task, define the symbols: Again we have (mischievously?) not quite finished the code below. You will need to use the EQU *assembler directive* to define the two symbols used in the program: *DISPLAY* and *KEYPAD*. The form will be – as the incomplete listing below indicates –

```
; two EQUATES for you to complete:?
_____ EQU _____
_____ EQU _____
```

The (almost-)complete assembly file:

```
; byte_in_out.a51
      $NOSYMBOLS ; keeps listing short
      $INCLUDE (C:\MICRO\8051\RAISON\INC\c8051f410.inc)
      $INCLUDE (C:\MICRO\8051\RAISON\INC\VECTORS320.INC) ; Tom's vectors definition file
      STACKBOT EQU 80h ; put stack at start of scratch indirectly-addressable block (80h and up)

      ; two EQUATES for you to complete:?
      _____ EQU _____
      _____ EQU _____

      ORG 0h
      LJMP STARTUP

      ORG 080h

STARTUP: MOV SP, #STACKBOT-1   ; "-1" because SP increments before first store
         ACALL USUAL_SETUP

TRANSFER: MOV DISPLAY, KEYPAD
          SJMP TRANSFER

; -- SUBROUTINES -- -
USUAL_SETUP: ANL    PCA0MD, #NOT(040h)      ; Disable the WDT.
                                            ; Clear Watchdog Enable bit
                                            ; Configure the Oscillator
             ORL    OSCICN, #04h     ; sysclk = 24.5 Mhz / 8

             ; Enable the Port I/O Crossbar
             MOV    XBR1, #40h       ; Enable Crossbar
             RET
END
```

When you've satisfied yourself that the program can, indeed, transfer a byte from keypad to display, we'll ask the '410 to do something a shade more exciting.

21L.2.10 Add keypad value to running sum

Perhaps it pains you to see your smart little '410 simply transferring data. OK. Let's allow it to show that it can also *add*.

Code for 8-bit running sum: This program adds the keypad value to a running sum, and outputs that sum to the display. We will do this first with an 8-bit output, then 16-bit (just to show that the 8-bit controller is not necessarily restricted to 8-bit operations). When the running sum overflows, the 8-bit output remains valid "modulo 8," in the jargon. The overflow information is lost.

Here is the core loop of the program:

```
Transfer:    MOV A, RUN_SUM ; recall running sum
             ADD A, KEYPAD ; form new sum.  Note that result lands in A
             MOV RUN_SUM, A ; ...save it
```

```
            MOV DISPLAY, A ; ...and show it
            ACALL DELAY
            SJMP Transfer ; ...forever
```

To make sense of this code, you need to recall that the result of ADD goes to the accumulator, the **A** register. So each pass through the loop updates both the *running sum* (RUN_SUM) and the *display*.

As you try this program, you might start by setting up **01h** on the keypad. The summing then amounts just to continually *incrementing* of course. When you're feeling more adventurous, try keypad value FFh.

```
; keysum_8bit.a51    shows sum of keypad and running sum;
; decimal adjust after initial binary display
;
    $NOSYMBOLS ; keeps listing short
    $INCLUDE  (C:\MICRO\8051\RAISON\INC\c8051f410.inc)

    STACKBOT EQU 80h; put stack at start of scratch indirectly-addressable block (80h and up)

    DISPLAY   EQU   P0      ; so-called Data byte on LCD board
    KEYPAD    EQU   P2
    RUN_SUM   EQU   R7      ; this choice is arbitrary

    DELAY_MULTIPLIER EQU 08h   ; about half-second delay, at div-by-8 clock rate
    ORG 0h
    LJMP STARTUP
    ORG 080h

STARTUP: MOV SP, #STACKBOT-1
    ACALL USUAL_SETUP
        ; Initialize running sum
            CLR    A
                MOV RUN_SUM, A ; clear running sum

Transfer:    MOV A, RUN_SUM         ; recall running sum
             ADD A, KEYPAD          ; form new sum
;            DA A       ; For decimal sum--try after watching binary addition
          MOV RUN_SUM, A ; ...save it
          MOV DISPLAY, A ; ...and show it
          ACALL DELAY
          SJMP Transfer ; ...forever

; -- -SUBROUTINES  -- -
DELAY: PUSH ACC    ; save the registers that this routine will mess up
       PUSH B
       PUSH 4      ; this saves register R4--in the zeroth set of registers

    MOV A,#0 ; maximize two delay values (0 is max because dec before test)
    MOV B,#0 ; ...and second loop delay value
    MOV R4, #DELAY_MULTIPLIER ; a more general way to specify delay:
                        ; this multiplies the 64K other loops

INNERLOOP: DJNZ B, INNERLOOP ; count down innermost loop, till inner hits zero
           DJNZ ACC,INNERLOOP ; ...then dec second loop, and start inner again.
           DJNZ R4, INNERLOOP ; now, with second at zero, decrement the outermost loop

       POP 4    ; restore saved registers
       POP B
       POP ACC
       RET                  ; back to main program
```

```
USUAL_SETUP:                    ; Disable the WDT.
            ANL    PCA0MD, #NOT(040h)      ; Clear Watchdog Enable bit
                                           ; Configure the Oscillator;
            ORL    OSCICN, #04h            ; sysclk = 24.5 Mhz / 8, for lower power

         ; Enable the Port I/O Crossbar
            MOV    XBR1, #40h              ; Enable Crossbar
            RET
END
```

Delay routine marginally amended: We made one small change to the Delay routine above: rather than fix the duration of Delay within that subroutine, as we did earlier (for example, in § 21L.2.2 on page 844), we set the value of the delay *multiplier* at the head of the program. The *multiplier* in this routine determines how many times the 16-bit countdown loop will be repeated (each loop taking about 80ms).

This change makes Delay more versatile, and also illustrates a point concerning the design of programs: it's a good idea to initialize constants up at the start of a program where they are easy to see and to change. If you decide to change the Delay duration from one second to two, you don't want to have to search the program listing to locate the place where that duration is set.

8-bit output amended to *decimal* form: Binary addition makes efficient use of the running-sum and keypad bytes permitting use of all 256 combinations. But when a display is intended for humans – with their many fingers, and their peculiar decimal counting system – a *decimal* output often is preferable. The 8051 knows how to restrict its output to the decimal values as long, as we give it keypad values that also are limited to decimal values.

In order to see this behavior, all we need do is *un-comment* the instruction "DA" in the 8-bit running-sum code above. DA ("Decimal Adjust") operates on the accumulator (**A** register), noting when a "half-carry" from the low nybble occurs. So, for example, if **A** holds the value 09h and we add 1, the next value in *binary* would be 0Ah. But DA notes the "half-carry" and amends the accumulator value appropriately: the "half-carry" tells the processor to roll over the low nybble to 0 while incrementing the high nybble. The result thus becomes 10h, the *decimal* count that indeed must succeed 09h.

We should note what DA cannot do: it cannot transform a hexadecimal value to decimal; it works properly only after an operation, like ADD, that affects flags appropriately – particularly, the *half-carry* flag. The details of flag effects appear in the descriptions of each processor operation in the 8051 instruction set reference.

The first time you do an operation that depends on flag behaviors, you would be wise to check this behavior in the instruction set reference. INC (increment) and DEC affect no flags, for example. That is a behavior one could not anticipate from any principle that we can perceive.

It *is*, of course, possible to do a generalized conversion from a binary value to decimal. But this process calls for a multi-step algorithm, and is neither simple nor quick. DA neatly covers the present task.

21L.2.11 16-bit running sum

Now let's allow the 8051 show that it can handle values larger than one byte. We don't often ask it to do so – and, in fact, are inclined to keep its computation tasks minimal. But at a price of speed, an 8-bit processor can concatenate operations to operate on values larger than what fits in a single byte.

Here, we will keep the running sum to 16 bits rather than 8, and we will display it on the LCD board using four hex digits rather than two, as for the *byte* sum. We'll exercise this hardware with an amended running-sum program.

Lab: Microprocessors 2

Hardware for 16-bit display: On the LCD board you will find two 8-bit 'data' connectors, labeled "D15...D8" and "D7...D0." These will accept your micro's 16-bit output. Set the right-hand DIP switch to "16" rather than "8mux." This makes the two 8-bit ports independent. ("8mux," by contrast, would merge the inputs, so that they could be separated only by the separate assertions of the two latch *ENABLE**s. That is an option you will not use here.)

We will use three byte-wide ports all at once. On the left in Fig. 21L.17 is a reminder of how they can be wired to the '410, and on the right keypad goes in as before, while two bytes go from controller to LCD.

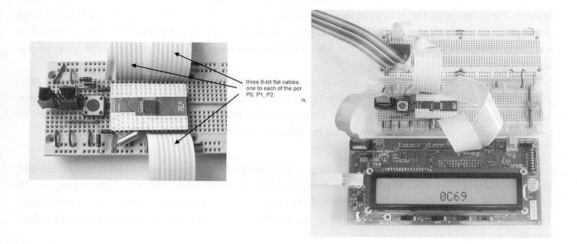

Figure 21L.17 All three '410 ports can be wired with flat cables.

The keypad already is wired to P2, and we will leave it there. Low-byte of data is done, too, wired as just above. You need only add a flat cable between micro's P1 and the LCD's "data high" connector (D0..D15).

LCD switch settings:

- probably you will want to suppress *labels* using the *LABELS* slide switch at the bottom right of the LCD board;
- you'll want to suppress the upper display line, which shows the 16 inputs labeled "address" on the LCD board. Suppress this line by setting the *LINES* switch to "1;"
- in order to show 16 bits of data you need to set the *DATA* switch to "16."

Code for 16-bit display: This version of the keypad-summing program forms a 16-bit *running_sum* by taking advantage of the *Carry* flag (CY) which warns us when an 8-bit sum has overflowed.

The oddity of the way the 16-bit sum is formed is explained in §21N.6, but we'll reiterate the point here, in summary form:

```
MOV  A, RUN_SUM_HI    ; Get hi byte
ADDC A, #0            ; If a carry from low sum, incorporate it
```

Adding *zero* certainly looks pointless at first glance. But it is not so when used with ADDC, the form of addition that incorporates the CY flag as an input. If the prior ADD (forming the low byte of the sum) has generated a Carry, ADDC will increment the high byte. Otherwise, the high byte remains as it was. This is just the behavior we require.

```
; keysum_16bit.a51    shows 16-bit sum of keypad and running sum;
  $NOSYMBOLS         ; keeps listing short, lest...
$INCLUDE (C:\MICRO\8051\RAISON\INC\c8051f410.inc)   ; ...this line might produce huge list
            ; of symbol defintions (all '51 registers)
$INCLUDE (C:\MICRO\8051\RAISON\INC\VECTORS320.INC) ; Tom's vectors definition file
STACKBOT EQU 80h ; put stack at start of scratch indirectly-addressable block (80h and up)

        DISPLAY_HI      EQU    P1           ; high byte of LCD address
        DISPLAY_LO      EQU    P0           ; low byte of LCD address
        KEYPAD          EQU    P2
        RUN_SUM_HI      EQU R7
        RUN_SUM_LO      EQU R6

        DELAY_MULTIPLIER     EQU 06h        ; half-second delay
                                            ; multiplier value stored in R4

ORG 0h
        LJMP STARTUP


ORG 080h

STARTUP: MOV SP, #STACKBOT-1

        ACALL USUAL_SETUP
        ; Initialize running sum (zero it)
        MOV RUN_SUM_HI, #0        ; clear running sum
        MOV RUN_SUM_LO, #0

Transfer: MOV A, RUN_SUM_LO       ; recall running sum (lo byte)
          ADD A, KEYPAD           ; form new sum (low byte)
;         DA A
          MOV RUN_SUM_LO, A       ; ...save it
          MOV DISPLAY_LO, A       ; ...and show it
          MOV A, RUN_SUM_HI       ; Get hi byte
          ADDC A, #0              ; If a carry from low sum, incorporate it
;         DA A
          MOV DISPLAY_HI,A        ; ...and show high byte of 16-bit sum
          MOV RUN_SUM_HI, A       ; ...and save it
          ACALL DELAY
          SJMP Transfer           ; ...forever

; -- -SUBROUTINES  -- -
DELAY:   PUSH ACC      ; save the registers that this routine will mess up
         PUSH B
         PUSH 4        ; this saves register R4--in the zeroth set of registers

         MOV A,#0 ; maximize two delay values (0 is max because dec before test)
         MOV B,#0 ; ...and second loop delay value
         MOV R4, #DELAY_MULTIPLIER     ; a more general way to specify delay:
                                       ; this multiplies the 64K other loops

INNERLOOP: DJNZ B, INNERLOOP      ; count down innermost loop, till inner hits zero
           DJNZ ACC,INNERLOOP     ; ...then dec second loop, and start inner again.
           DJNZ R4, INNERLOOP     ; now, with second at zero, decrement the outermost loop
           POP 4     ; restore saved registers
           POP B
           POP ACC
           RET                    ; back to main program
```

```
; ----------------------------------------------
USUAL_SETUP:              ; Disable the WDT.
        ANL   PCA0MD, #NOT(040h)   ; Clear Watchdog Enable bit
                                   ; Configure the Oscillator;
        ORL   OSCICN, #04h         ; sysclk = 24.5 Mhz / 8, for lower power

                                   ; Enable the Port I/O Crossbar
        MOV   XBR1, #40h           ; Enable Crossbar
        RET
    END
```

    Try the keypad value FFh again, as you did in the 8-bit running sum on page 838. With the earlier 8-bit display, FFh produced a *decrementing* value. Now the 16-bit version makes the magic of "FFh as *minus one*" somewhat less magical: the carries into the high byte now are visible. The lower byte does, indeed, decrement; but at the same time the high byte increments. We stand backstage now, and we can see what the byte-addition magician was doing all along.

    Again, we have planted decimal-adjust (DA) instructions, which you may use if you like, by removing the *comment* semicolons. Once again, DA works properly only so long as the keypad input presents values in the *decimal* range (not fair using A, B..., etc.).

You can disconnect cables, etc. You will not need today's displays and keypad and associated cables in the next lab. You may disconnect these.

    You can leave in place the pushbutton used to disable the LED's blinking in §21L.2.5. You will not need it again for that purpose, and you ought not to press it randomly (some students were startled to find that doing so disabled some later programs; we were startled to find that they were startled). The pushbutton will see use again when you try interrupts in Lab §23L.2.

21S Supplementary Notes: 8051 Addressing Modes

21S.1 Getting familiar with the 8051's addressing modes

You don't need to know a lot about the 8051's addressing; we use most modes, of its modest set, but not all. When you are learning assembly language it's good news that the 8051 isn't very versatile (in contrast, say, to Motorola's 68000 which we used some years ago; it offered *fourteen* addressing modes). There's less to learn than for a more complex machine. On the other hand, when you're trying to write code to get something done, the 8051's restrictions are less pleasing. Many addressing modes that make perfect sense – such as "MOVX @DPTR, #012h" or "CLR R5" or "MOV R3, R4" – just aren't available. (Such peculiarities may provide one more motive to escape into *C*.)

21S.1.1 Internal versus external memory

A microcontroller includes some on-chip memory; most controllers rely on this memory exclusively. The SiLabs controller in the other micro path operates this way. Our big-board computer does not use the controller this way until late in this course; we provide an *external* memory to hold both code and data. Our Dallas 8051 stores all of its program *code* and most of its *data* in this external memory – the 32K RAM. The Dallas controller does include on-chip ROM which you will use later; but for the moment we are hiding it from you. The SiLabs part runs from its 16K on-chip ROM. The 8051 in either version includes only a little on-chip RAM: 2K for the SiLabs part, 256 bytes for the Dallas part, plus additional 8-bit registers.

21S.1.2 The easy way: on-chip transfers

The controller is easy to program when the goal is to operate upon on-chip registers or "scratch" RAM: you just write "MOV destination, source." If P0 is tied to the keypad, P1 to the display, you transfer keypad to display with code that's charmingly simple: `mov p1, p0`. It's the off-chip data transfers that look complicated. We'll look closely at these in the pages that follow.

External memory and I/O access for the 8051: The 8051's way of doing business with the outside world through an external bus, as in our Dallas computer, is extremely rigid. This rigidity is a weakness – but for a student meeting the 8051 for the first time (as we argued more generally right at the start of this chapter), the rigidity may be good news: *one* external mode is available – well, about one and one half.[1]

21S.1.3 The usual method: use DPTR

Specifically, when the 8051 accesses *data* (in contrast to *instructions*) stored in external memory or in I/O devices tied to the external buses, the controller nearly *always* uses the address stored in the "Data

---

[1] The "half" is the MOVX @Rn, a form discussed in §21S.1.4.

Pointer" (DPTR), an on-chip 16-bit register. And the on-chip source or destination of that data transfer *always* is the **A** register – the "accumulator." Fig. 21S.1 shows that same proposition graphically.

So, you know that nearly any input from an external device will be coded as

    MOVX A, @DPTR;

And any *output* to an external device will be coded as

    MOVX @DPTR, A;

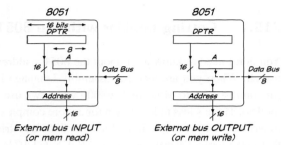

Figure 21S.1 8051 uses DPTR and **A** to do data accesses on external buses.

Note the strange word-order: opcode, then destination, then source: "bites Tom dog." The @ indicates that we want to copy the contents of **A** (an 8-bit register on the 8051) not *into* DPTR[2] (or *from* DPTR, when the transfer is going in the opposite direction) but *into* (or *from*) the location to which DPTR *points*. DPTR is called a "pointer," and this mode of addressing is called "indirect."

Given the 8051's reliance on DPTR, the program's getting data to and from the right place depends not simply on the line of code that does the transfer (we've seen that source and destination forms are rigidly fixed), but also on setting up the DPTR properly beforehand.

Setting up the DPTR: Nearly every program that does a data transfer with the outside world, then, needs to initialize the DPTR. The code to do that always looks like

    MOV DPTR,#PORT_ADDRESS;

where PORT_ADDRESS is a 16-bit address. The # mark indicates an addressing mode known as *immediate*. This mode takes the *value* of PORT_ADDRESS (let's suppose it is 8000h, to make things more concrete), and places that value into the 16-bit register, DPTR.

Notice also, that this time there is no @, because this time we *do* intend to put the value *into* the DPTR (not where it points). "MOV DPTR..." is *not* a transaction with *external* memory or I/O device: it is an operation on a register within the 8051. That explains why the variety of *move* is "MOV" rather than "MOVX," the external form described next.

External moves (MOVX) contrasted with internal operations: That last point reminds us that even when the 8051 is set up to use external buses, as in our lab computer, it still can (and must) do operations on its internal registers. As we have said, MOV DPTR, #PORT_ADDRESS; is one such operation. In addition to such non-external MOVs – the ones that are not MOVX – all other processor operations work on the processor's internal registers: ADDs, logical operations, bit tests, and so on.

External operations are limited to the simple MOVX: just a transfer. In order to do something to a value picked up from an INPUT device – from the keypad for example – one must always bring the value into the 8051 first, storing it in a register; only then can the processor do anything to the value.

---

[2] Incidentally, the operation wouldn't quite make sense, anyway, since **A** holds eight bits and DPTR holds 16.

21S.1.4 A nifty alternative: set base register and use R_n as index

For operations accessing a few locations not too far apart (within 256 of each other) there is another way to proceed: one can hold the *high* half of the address fixed, and then use an 8-bit register to play with the *lower* half of address. This mode is well-suited to I/O addressing, where the several devices are likely to live close to one another, as they do in our little machine.

Using this mode, the 16-bit address, though *off-chip*, is defined *not* by DPTR but by two other registers. The high half is the value held in P2 – the 8051's internally-defined "PORT 2," just an on-chip register, the one wired to the high half of the external address bus. The low half of the address is the value held in one of *two* of the eight internal "Working Registers," $R0$ or $R1$. Fig. 21S.2 illustrates the idea, showing use of Port 2 and register $R1$ to define a 16-bit I/O address.

Here's an example, reading ADC and writing to DAC, at ports 2 and 3 (using the external bus). First, we'll show a listing using DPTR, then a listing using this second method (fixed high half, $R0$ and $R1$ providing lower half of address).

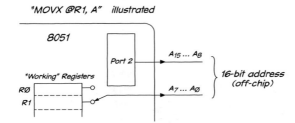

Figure 21S.2 Another way of defining an external address: P2 and Rn.

DPTR:

```
STARTUP: MOV DPTR #8003h   ; point to ADC

PICKUP: MOVX A, @DPTR    ; Read ADC (first pass, this gets bad data)
        MOVX @DPTR, A    ; start ADC for next time (ADC will convert while
                         ; program is busy doing the 5 lines below)
        DEC DPL          ; point to DAC (8-bit decrement of low byte;
                         ; 16-bit decrement of DPTR not permitted)
        MOVX @DPTR, A    ; ...send sample to DAC
        INC DPTR         ; restore DPTR
        SJMP PICKUP
```

Fixed high (P2) and Rn for the low half:

```
STARTUP: MOV P2,#80h ; set up top half of address as I/O base
         MOV R0, #3h ; set up 8-bit index register to ADC
         MOV R1,#2h  ; ...and another register to point to DAC

TFR:     MOVX A, @R0 ; read ADC (bad, first pass, OK ever after)
         MOVX @R0,A  ; start ADC for next pass
         MOVX @R1, A ; ...and send it to DAC

         SJMP TFR    ; and do it all again
```

The *DPTR* code looks a little more compact – but it doesn't run quite as fast as the second method (since it includes an extra INC and DEC operation). That small difference may not matter. A more important point in favor of the *Rn* method is simply that it leaves the DPTR free for other operations.

Supplementary Notes: 8051 Addressing Modes

The hardware that allows this double use of Port 2 to define high 8 address lines: At a glance it seems impossible for us to control Port 2, defining the high address, without interfering with the operation of DPTR, on data transfers, but the scheme does work because of a *switch* (really a 2:1 mux) that selects the source of the high-8 address lines. Most of the time the switch is set to use DPTR. But when one uses the MOVX *Rn* form, the mux is set to use the output of the Port 2 latch.

Figure 21S.3 is the circuit diagram showing the switch. (One could say that the switch *had to be there* to permit this operation; but to see that switch in the diagram is reassuring, is it not?)

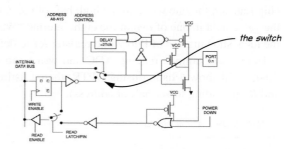

Figure 21S.3 How Port 2 can use *either* DPTR, or Port 2 flops to define high-8 address lines. Adapted from Dallas High Speed Microcontroller's User's Guide, Fig. 10-2, with permission.

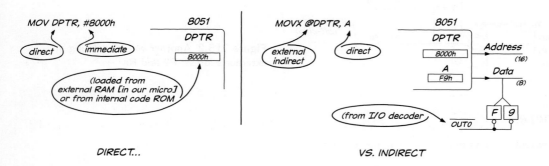

Figure 21S.4 Indirect addressing: familiar example, MOVX @DPTR... to *external* memory or device.

21S.1.5 Another big distinction: direct versus indirect addressing

Chances are, you're already comfortable with this distinction: it's the difference between

- putting a value into a register (MOV DPTR, #8000h) and
- putting a value into a location pointed to by a register (MOVX @DPTR, A).

Familiar case: indirect addressing using "@DPTR": Figure 21S.4 is a reminder of what goes on in the two contrasted cases.

A less familiar case: indirect addressing using internal RAM: You know that $R0$ and $R1$ can be used as pointers to external addresses, where P2 is used to hold the high byte of the 16-bit address. But $R0$ and $R1$ can also point to *internal* RAM (where an 8-bit pointer is sufficient, because of the 8051's modest allotment of RAM). This less familiar case is useful to clarify the contrast between direct and indirect addressing.

We are making the same distinction as in the "@DPTR..." example above – namely

- putting a value into a register (MOV R0, #034h),[3] on one hand, versus...
- putting a value into a location pointed to by a register (MOV @R0, #0F9h)

The first operation puts the value 34h into register *R0*; the second operation puts the value F9h into *the location pointed to by R0*, namely, address 34h. Fig. 21S.5 illustrates the idea.

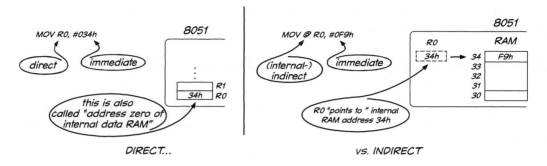

Figure 21S.5 Indirect addressing: 8-bit pointer: to *internal* RAM.

Downright obscure case of indirect addressing: "indexed" table-read from code memory: Instead of moving a pointer by incrementing or decrementing, sometimes it is convenient to let a register value determine which of several entries in a table is to be read. This is called "indexed" addressing: an *index* value is added to a *base* address to form the effective address from which the read is to occur. Some processors offer a rich set of such operations (for example, Motorola's old 8-bit workhorse, the HC11, once a competitor of the 8051). The 8051 knows just two ways to do this trick.

Two addressing modes use the instructions "MOVC A, @A+PC" or "MOVC A, @A+DPTR." These instructions...

- use the **A** register (accumulator) to hold the "index" value;
- use as the "base" address either
  - DPTR or
  - the "program counter" (just the program address, after execution of this indexed operation).

Here's an example using the *program counter* as base address (this scheme offers the advantage that it does not tie up the DPTR). The following program fragment uses a value taken from the keypad (a 4-bit value[4]) as index into a short table that determines what mask to apply to an ADC value.[5]

```
        MOVX A, @DPTR ; pick up keypad value
        ANL  A, #0Fh  ; keep only the low nybble
        ...
        MOVC A, @A+PC ; ...and use that as index into table of mask values
        RET

TABLE:  DB 80H   ; Table of mask values: 1 from keypad picks up this entry (DB means "define byte")
        DB 0C0h  ; 2 from keypad picks up this entry (and so on)
                 ; (0 from keypad is illegitimate choice)
```

---

[3] The leading zero in this case – 034h – is not needed. It is required when first character is non-numeric, as in the next instance, 0F(h. The zero is needed to prevent the assembler's assuming that a value beginning with a character rather than number is a text string.

[4] The high four bits are masked out so that only the most recent key-press matters.

[5] This fragment is a piece of a demonstration program we used in §14N.1.2 to try out the effect of varying the number of bits in a digital sample.

```
DB 0E0h
DB 0F0h
DB 0F8h
DB 0FCh
DB 0FEh
DB 0FFh
```

Note that this table must follow immediately after the command (except for the RET); other intermediate instructions would upset the scheme.

21S.1.6 Examples: I/O operations using external bus versus 8051's internally-defined ports ("portpins")

Using the external buses: Most of the I/O that we do in the big-board labs makes use of the external buses. Here, for example, is a program that takes in a byte from Port 0 and writes the value back to the same port:

```
GETIT: MOVX A, @DPTR ; read the keypad
       MOVX @DPTR, A ; ...show it on display
       SJMP GETIT ; ...again, forever
```

This would work properly of course, only if you had first loaded DPTR to point to keypad and display. You know how to do that.

Using the internally-defined ports: If you chose to use the internally-defined 8051 ports you could do the same task. You could tie the keypad to Port 1 (using it as an input), and the display to Port 3 (using it as output). Then your code would be more compact than the code using the external buses: no need to pass the value through **A**, and no use of the DPTR. (Here, we reiterate an example mentioned in §21S.1.2, except that we use ports available on the Dallas part rather than the SiLabs.)

```
GETIT: MOV P3, P1 ; read the keypad, send to display
       SJMP GETIT ; ...again, forever
```

This might not be a smart way to do the task though: you would tie up half of the 8051's 32 I/O lines. The internally defined ports look better when you use them for operations on *bits*, rather than on *bytes*.

21S.1.7 Examples: BIT operations

BIT operations using the external bus: The external bus provides the *hard* way to do the job but it makes clear what hardware is at work, whereas using the 8051 port pins is easy but hides the flip-flops and drivers that are doing the magic. So let's start with the more explicit *external bus* method. Let's drive an LED at one bit of Port 1 (external bus) and sense an input bit at the same port. We'll toggle the LED if that input bit is high, hold the present LED state if the input bit is low.

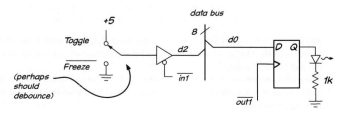

Figure 21S.6 Hardware needed for bit drive and sense using external bus.

21S.1 Getting familiar with the 8051's addressing modes

In order to drive the LED, we need a flip-flop on the *output* bit. Without that flop the computer could only send a level for the very short duration of the OUT1* pulse. The flop sustains the LED drive level. To take in and test an *input* bit the computer needs a *3-state* buffer between the source of that input and the data bus. Fig. 21S.6 sketches the hardware needed. You may be muttering "If you want a switch to control an LED why not wire it to the LED?" Fair enough. Our only defense is that we propose this example only as the simplest possible I/O. In a real case, the computer would respond to the switch in some way more interesting than simply passing the level on to the LED.

External bus requires transferring an entire byte: Using the external bus, the 8051 cannot pick up or send less than a byte. This seems sad. We seem to be obliged to waste the other seven bits.

In fact, the need to move bytes does not mean that we must waste the seven bits that are not involved. We can arrange things so that only a single bit matters. Though we take in a byte, we can make our response conditional on the value of a single input bit. And though we send a byte out, we can take care to leave all the seven unused bits unchanged in case they are needed for other purposes. The code below applies the hardware of Fig. 21S.6 to do this.

```
DOBITS: MOV DPTR, #8001h   ; Point to LED and input bit
        CLR ACC.0          ; set up LED OFF state, for startup

SEND:   MOVX @DPTR, A      ; send bit level to LED drive

TEST:   MOVX A,@DPTR       ; pick up the input bit (plus junk)
        JNB ACC.2,TEST     ; Test the single interesting bit. Leave LED untouched if input is low

        CPL ACC.0          ; ...but toggle that bit if the input is high:
                           ; set up the bit level, to be sent soon...
        SJMP SEND
```

Not bad; but we'll see in a moment that the same is easier to do with the 8051's port pins.

21S.1.8   BIT operations using 8051 port pins

Hardware: practically nothing: If we use the 8051's portpins instead, the hardware is nil: just connect the input switch and LED to two port pins, as in Fig. 21S.7.

One detail worth noticing is the changed LED drive: the 8051 can source almost *no* current (50μA), but can sink a lot more: 1.6mA at Port 3 (still not a lot; Ports 0 or 2 can sink double that – but they're not available to us since we use them for the external bus). So we had to drive the LED by *sinking* current – and we had to boost the *R* value a little because the 8051 I_{OL} is much less than the current available from an HC gate (either High-sourcing, or Low-sinking). (This asymmetric drive – strong sink, feeble source – is characteristic of older devices, TTL and NMOS; the CMOS 8051 has chosen to follow this tradition.) The SiLabs 8051 offers this asymmetric drive, in order to emulate the original 8051, but can provide nearly symmetric drive if you prefer (the option is called "push–pull" and can be selected for any particular output bit).

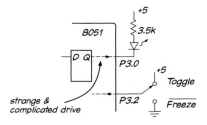

Figure 21S.7 Hardware needed for bit drive and sense using 8051 port pins: practically nothing!

Code: simpler when using portpins: The code to test a bit and toggle a bit, as in §21S.1.7, is simpler than when we used the external bus:

```
        SETB P3.0 ;   start high: LED off

TEST:   JNB P3.1,TEST ; hang here while input bit is low
                      ; ...fall through when goes high:
FLIPIT: CPL P3.0 ;    flips only LSB (bit 0) of port

        SJMP TEST
```

This code also runs faster than the one written for the external bus, requiring far fewer bus accesses (both in execution, and in instruction fetches) – not that this difference matters at all in the present application.

21S.1.9 Seemingly ambiguous cases disambiguated by context

Bit versus byte: If we write

```
SOFTFLAG EQU 0 ; a pseudo op addressed to assembler, not to the 8051, by the way
.
.
.
CLR SOFTFLAG
.
```

the assembler inserts *zero* in place of the human-friendly name "SOFTFLAG." But how does the 8051 know whether to treat this as *byte* zero of internal RAM (which happens to be the other name for register *R*0) or *bit* zero (which happens to be the first bit-addressable location, and lives at byte 20h)?

The answer is boringly simple: since no general CLR byte operation is available (only CLR A is offered), the 8051 sees no ambiguity: this is a *bit* clear, so it clears *bit* zero.

21S.1.10 Multiple locations that go by one name (or address)

Internally defined Port 0 lives at address 80h. But this is also the first byte of our usual *stack* – the "scratch-pad" RAM area on the 8051/2.[6] Is this a conflict? Will we confuse the 8051 if we place the *stack* at 80h?

No. These two locations, both called "80h," are *two* distinct locations, distinguished by the way they are accessed. If the addressing mode is *indirect* (as it is in *stack* use), then it's the scratch RAM that is accessed; if the addressing mode is *direct*, then it is P0 that is addressed.

[6] It isn't quite right to call the controller "8051," here, since it's only the 8052 – along with our Dallas versions imitating it – that does offer RAM at this address. The 8051 offered only half as much on-chip RAM, and thus ran out of RAM exactly at that point, 80h.

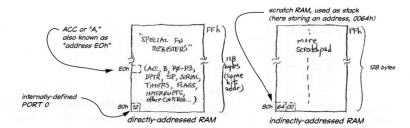

Figure 21S.8 P0 and bottom of stack share an address.

21S.1 Getting familiar with the 8051's addressing modes

We couldn't resist adding to Fig. 21S.8 another detail that we have mentioned in passing: the register that we call **A** (and sometimes ACC) lives at a particular address – and could be referred to that way (if we wanted to be perverse).

Here are a couple of lines of code to remind us of the distinction between *direct* and *indirect* addressing modes. Each line operates on memory at *address 80h*. But the two memory locations sharing that address are *two distinct* locations.

```
MOV P0, #12h ; (DIRECT): write a byte to internally-defined Port Zero (at address 80h)
ACALL DELAY ; (INDIRECT): store return address automatically on stack--which, in our Lab 21L
            ; program, began at address 80h (return address was 064h, as indicated in sketch)
```

80h in internal RAM lives in two places but each is accessible only in its peculiar way – directly (P2) or indirectly (scratch RAM, used as stack in our applications).

And a third and fourth possible meaning for the humble "80h": And yet another – a *third* – 80h exists in our lab computer's version of the 8051: 80h in *external*[7] RAM. If that's not enough potential confusion for you, then consider the alarming fact that a *fourth* multiple-personality is available for 80h in an 8051 that holds to its Harvard-class design: in the external memory of such a machine, 80h defines two distinct locations: one in *code* memory, another in *data* memory. But *you* won't be confused by all of this, because you've been warned!

21S.1.11 The opposite cases: single locations that go by two names

Sometimes, rather than establishing two or more distinct locations that go by what seems the same name (like 80h in §21S.1.10), the 8051 permits one to speak of one location in several ways – or to access it in more than one way.

Access in either of two ways: The RAM region labeled "scratchpad" appears in *both* direct- and indirect-addressing. So the value ABh stored, in this hypothetical case, at 34h can be accessed either way. (This truth probably is not useful, except to protect you from the notion that you can get double use out of this block of memory.)[8]

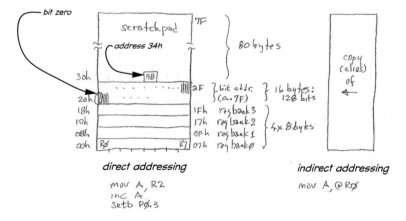

Figure 21S.9 More double-lives of RAM locations.

---

[7] This memory that we are calling "external" sometimes is called "MOVX RAM," instead, and perhaps better so because some versions of the 8051 include on-chip "MOVX RAM" accessed with the MOVX operations. The SiLabs '410 for example does this. See §24L.2.2.

[8] You may recall that this distinction partially explains the two names for the accumulator register: "A" versus "ACC." See §20L.3.8.

A *bit* location can be described in two ways: Bit zero (the first in a block of 128 individually-accessible bits – see Fig. 21S.9) can be referred to in a straightforward way as bit zero: "JNB 0," for example, tests bit zero. Bit zero happens to live in a *byte* at address 20h. So one can refer to that bit as the zeroth bit of byte 20h: "JNB 20h.0." There is no advantage to describing it this way, but it may be reassuring to see that the assembler understands the fact shown by the memory map – that the zeroth bit of byte twenty is, indeed, also bit number zero.

Some *byte* locations can be described in two ways: In Fig. 21S.8 we note that the register that we normally refer to as "P0" (internally-defined Port zero) lives at address 80h. It would be possible to refer to it by its address. It's usually much more convenient to use its functional name, "Port zero." But we'll see next that sometimes it essential to use the address rather than the name, to avoid ambiguity.

21S.1.12 One more oddness: you may be able to do it if you say it nicely

Specifying an internal address rather than register name...: One cannot do the simple operation MOV R1, R2. If you need to do it, you may find yourself writing first, MOV A, R2; then MOV R1, A. That works, but seems very clumsy. You *are* permitted to do what you tried to do, if you will refer to the register by its *address* rather than by its *name*:

```
     MOV R1, 2     ; this is permitted--though address 2 IS R2.  Very strange.
```

This is strange, but not quite crazy (and an assembler can't quite fix it[9]): the two operations, though they give the same results, are coded differently in machine language; in other words, they are *different* operations. This point may become clearer if we compare two *permitted* operations that give the same result but are encoded differently: MOV R1, A is *a different instruction* from MOV R1, 0E0h, even though the two operations do exactly the same action: copy the contents of the accumulator to register R1. Here is the output of the assembler, illustrating this point:

```
F9            MOV R1, A      ; same results, in these two cases...
A9E0          MOV R1, 0E0H   ; ...but not the same machine code
```

The more compact machine code for *MOV R1, A* seems to reflect the privileged character of register **A**. Operations on **A** resemble idioms used very often in a language; they soon acquire a shortened form. "Do not" becomes "don't."[10]

...Another similar case: PUSH R0 fails, but PUSH 0 succeeds: PUSH R0 fails for a good reason: the 8051 provides four sets of scratch registers, R0 through R7. Nothing in the name of the register identifies which of the four "R0s" one intends; it's just "the current R0." (Which? The one specified by two bits in the "program status word," PSW). The PUSH and POP operations aren't allowed to make the assumption usually made that we intend the *current* register set. So,

```
PUSH R0 ; fails: assembler considers this ambiguous
PUSH 0  ; succeeds: 0 is the address at which the lowest of the 4 ''R0's" lives:
        ; unambiguous
```

This is hard to remember – particularly since R0 often is considered *not* ambiguous – as in MOV A, R0. Context determined which rule the assembler chooses to apply. The PUSH and POP operations

---

[9] An assembler *could* decide that when you write the forbidden "MOV R1,R2" it would implement that as the permitted "MOV R1,2;" but that might be too high-handed for most people's tastes: when you write assembly code you expect to get what you specify, not an assembler's clever correction of your code.

[10] Well maybe that simile is a bit strained. The machine code did not *evolve*. It was chosen by Intel's engineers.

apparently were considered cases where programmers might well be switching among register sets, as is likely in an interrupt response. Ordinary MOVs were not so classified.[11]

On the next two pages we have set out some of the addressing modes you will use most often, with sketches to remind you of their operation. If these diagrams all are intelligible, then you're on your way to getting comfortable with the 8051's somewhat baroque addressing.

21S.2 Some 8051 addressing modes illustrated

Direct:

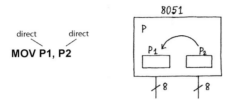

Figure 21S.10 One register to another (direct–direct).

Immediate:

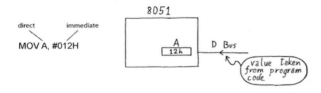

Figure 21S.11 Immediate addressing: a constant loaded from the program listing itself.

Immediate, 16-bit:

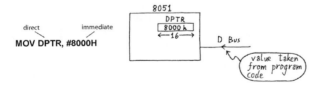

Figure 21S.12 Immediate load of 16-bit register.

External-indirect, direct: Here, the pointer is a 16-bit register, necessary to access all of address space.

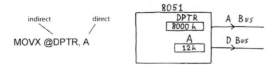

Figure 21S.13 Indirect, off-chip.

External–indirect using Rn, direct: Here, the 16-bit pointer combines P2 (probably fixed) for high byte with R0 or R1 for low-byte. Does *not* use DPTR.

---

[11] Are you suddenly dreaming again of an escape from assembly language? Take heart. Mostly it's not quite so fussy as this.

Supplementary Notes: 8051 Addressing Modes

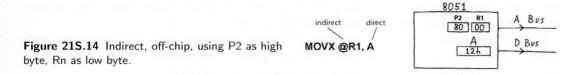

Figure 21S.14 Indirect, off-chip, using P2 as high byte, Rn as low byte.

On-chip indirect: Here, the pointer is an 8-bit register – sufficient, because the address space is small (256 bytes).

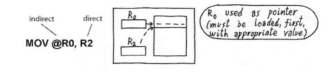

Figure 21S.15 On-chip indirect.

Indexed:

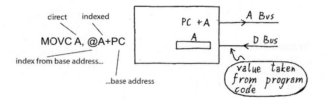

Figure 21S.16 Indexed relative to a base address provided by accumulator (another form uses DPTR to define base).

Summary diagram showing 8051's weird address space:

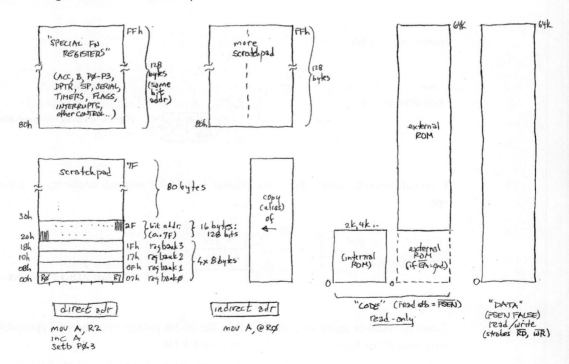

Figure 21S.17 Weird address space.

22N Micro 3: Bit Operations

Contents

22N.1	**BIT operations**	**869**
	22N.1.1 BIT operations using the external bus	869
	22N.1.2 BIT operations using 8051 port pins	870
	22N.1.3 Bit code: simpler when using portpins	872
	22N.1.4 Equivalent bit operations in C	873
	22N.1.5 A hardware "flag:" input hardware to do something *once*	873
22N.2	**Digression on conditional branching**	**874**
	22N.2.1 Branching instructions	874
	22N.2.2 Masking, to operate on more than one bit	875
	22N.2.3 Byte compare: CJNE, and work-arounds for larger *compares*	878
22N.2.3.1	Masking to *Set* More than One Bit	879
	22N.2.4 Flag hardware	880

Why?

Today's goal is to break code into modules – subroutines or "functions." We will also let the computer do single-bit I/O.

22N.1 BIT operations

The 8051 seems feeble when asked to do any operation requiring high precision (meaning values defined to many bits); it is hamstrung by its byte-wide memory and registers. The little controller is much better adapted to dealing with *bits*. Here, it looks nimble. So let's admire it in the domain where it thrives. We'll start by using the external buses, then we'll use the built-in ports, which make both hardware and code a good deal simpler.

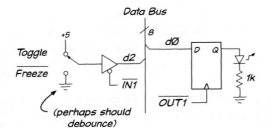

Figure 22N.1 Hardware needed for bit drive and sense using external bus.

22N.1.1 BIT operations using the external bus

The external bus provides the *hard* way to do the job, but it makes clear what hardware is at work (whereas using the 8051 port pins hides the flip-flops and drivers that are doing the magic). So let's start by using the external buses. Let's drive an LED at one bit of Port 1 (external bus) and sense an input bit at the same port. We'll toggle the LED if that input bit is high but hold the present LED state if the input bit is low.

In order to drive the LED, we need a flip-flop on the *output* bit. Without that flop the computer could send a level for only the very short duration of the OUT1* pulse (a little less than 200ns clocking at 11MHz). The flop sustains the LED drive level. To take in and test an *input* bit, the computer needs a *3-state* buffer between the source of that input and the data bus. Fig. 22N.1 sketches the hardware needed.

And here's some code to use this hardware: it's not bad, but we'll see in a moment that the same is easier to do with the 8051's port pins.

```
DOBITS: MOV DPTR, #8001h ; Point to LED and input bit
        CLR ACC.0 ; set up LED OFF state, for startup

SEND:   MOVX @DPTR, A ; send bit level to LED drive
        PUSH ACC      ; save bit level, so we can flip it later

TEST:   MOVX A,@DPTR ; pick up the input bit (plus junk)
        JNB ACC.2,TEST ; Test the single interesting bit. Leave LED untouched if input is low
                   ; (note we're obliged to call the A register ACC for this bit operation)
        POP ACC    ; recall accumulator, including bit level that interests us
        CPL ACC.0 ; ...toggle that bit, since the input is high: set up the bit level,
                   ; to be sent soon...
        SJMP SEND
```

Do these bit operations use the port's full *byte*? The answer is slightly subtle. It would be sad if the answer were a straight, "Yes," meaning that we had to use eight bits to handle one (or, in the present case, 16 bits on the IN and OUT ports to handle just two bits one in, one out). The answer is not that bad.

Yes, we must pick up and send a full *byte*; the processor cannot do smaller transfers.[1] But that does not mean that the unused seven bits at each port (IN1 and OUT1) are wasted. Those other seven bits are free for other uses. The JNB operation tests only the single bit that is specified; the same is true for the CPL (complement) operation.

22N.1.2 BIT operations using 8051 port pins

External bus versus built-in ports: Often, in this course's big-board labs, we ask you to do I/O operations using the 8051's external buses. We do this because this approach obliges us to think about issues that are obscured when we use the controller's built-in ports: on inputs, we must *always* place a 3-state buffer between any signal source and the bus; on outputs, usually we must place a flip-flop or register between the bus and the peripheral that is to be controlled. The same rules apply to the built-in ports, but the designers of the IC have done the work for us. In a sense then, we ask you to use the external buses because they are harder to use.

The external buses offer only one potential advantage over the built-in ports: they allow attaching an almost unlimited number of devices to the controller. But in all other respects the built-in ports are preferable, and provide the normal way to link a controller to the outside world.

Hardware: practically nothing:

---

[1] For the sole, partial exception to this statement, and one that applies only to *on-chip* transfers, see the operations that copy a level into or out of the *Carry* flag in §22N.1.3.

Traditional 8051: sink, don't source: If, instead of using the external buses, we use the 8051's port-pins, the hardware is nil: just connect the input switch and output LED to two port pins: see Fig. 22N.2. One detail here is worth noticing: the changed LED drive. The 8051 can source almost *no* current

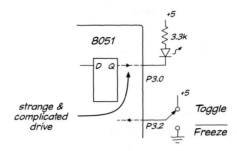

Figure 22N.2 Hardware needed for bit drive and sense using 8051 port pins: practically nothing!

(50μA), but can sink a lot more: 1.6 mA at Port 3 (still a modest current). So we had to drive the LED by *sinking* current – and we had to boost the R value a little because the 8051 I_{OL} is much less than the current available from an HC gate (either high – source; or low – sink). (This asymmetric drive – strong sink, feeble source – is characteristic of older devices, TTL and NMOS; the CMOS 8051 has chosen to follow this tradition. See, for example, Chapter 16N.)

Silabs option: "push–pull" output: Designers of updated 8051s can't resist trying to fix what's annoying about the original. SiLabs thought the asymmetric "TTL-like" drive was annoying, and SiLabs permits us to choose (pin-by-pin) to provide instead a more-or-less symmetric CMOS drive.[2] They call this – appropriately enough – "push–pull." (We have used this term mostly for *emitter followers*, but the term applies well to the CMOS push–pull *switch* also.) Figure 22N.3 shows the traditional 8051 drive, and the SiLabs push–pull drive, along with the analog input/output link (an analog switch).

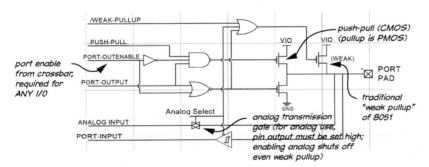

Figure 22N.3 SiLabs port detail: traditional 8051 I/O, "push–pull," and analog in/out.

This diagram may help you to understand some details of SiLabs port setups:

The "weak pullup." This is a MOSFET used as a resistor, value approximately 120kΩ

Precondition for using port as input or for analog. In order to use a port pin as digital input or as analog input or output, we must shut off the strong pulldown transistor.

For this reason, the default state of all I/O pins – the condition after a controller *Reset* – is High (with the usual weak pullup).

And before using a pin for input or for an *analog* application, we play it safe by writing a

---

[2] The *push–pull* remains somewhat asymmetric: better at *sinking* than at *sourcing*, but not so radically as traditional TTL or the standard 8051. It can source 3mA but can sink 8.5mA at similar voltage drops (about 0.5V between output and source – either ground or V_{IOt}). See data sheet table 18.1, p. 163 (C8051F410 datasheet 1.01).

High to the pin. (The weak pullup applied with a High is weak enough not to interfere with a digital input, and is disabled entirely for analog operations.[3])

The code that permits analog use looks like this (here, used to permit use of a ADC at P0.0, the DAC at P0.1)[4]:

```
PORT_SETUP: setb P0.0      ; make sure latch is high (this for ADC)
            setb P0.1      ; ...and this for DAC
            mov pomdin, #0FCh   ; set DAC0 and ADC control bit (d0 and d1) low,
                                ; for analog (this is not pin level)
            mov poskip, #03h ; tell crossbar to skip ADC bit (lsb) and DAC0 bit (bit1)
            ret
```

The loading of POSKIP refers to the scheme the '410 uses for assigning functions to pins. This "crossbar," which SiLabs users have seen in all their labs, is useful because the '410 provides more signals than can be carried by the available pins. The scarce pins must be assigned as needed, to whatever functions are invoked by a particular program. Some of this assigning is done by a default rule that here we need to override.[5]

Crossbar Enable. The "crossbar," as you have seen in all the SiLabs programs, must be turned on: it steers signals to pins, and strangely defaults to OFF, as noted in Chapter 21N.

```
; Enable the Port I/O Crossbar
    MOV XBR1, #40h            ; Enable Crossbar
```

This 1 at bit 6 of XBR1 – a signal named XBARE – enables the crossbar.

22N.1.3 Bit code: simpler when using portpins

Code for testing an input bit, drive an output; built-in ports: The code to test a bit and toggle a bit is simpler than what is required for the external bus. The ports used here are appropriate for the Dallas part (the SiLabs '410 has no Port 3). For the SiLabs part the code would be the same but would be applied to a different port.

```
        SETB P3.0 ;   start high: LED off
TEST:   JNB P3.2,TEST ; hang here while input bit is low
                      ; ...fall through when goes high:
FLIPIT: CPL P3.0 ;    flips only LSB (bit 0) of port
        SJMP TEST
```

This code also runs faster than the one written for the external bus, requiring far fewer bus accesses (both in execution and in instruction fetches).

More importantly, you'll notice that the single 8-bit port can be used in part for *input*, in part for *output*, bit by bit. By contrast, the hardware used for bus I/O (§22N.1.1) obliged us to commit an 8-bit port to input, another to output (though we were free to apply the 16 bits to a variety of uses).

---

[3] On a write of High, the weak pullup briefly is made strong in order to speed the charging of stray capacitance. This is a standard 8051 feature. But after this brief transition drive, the pullup reverts to its weak value. When a pin is assigned to analog use, even the weak pullup is turned off. C8051F410 datasheet, p. 151.

[4] We omit another initialization, one that you will meet in the lab: for the ADC voltage reference, at P1.2.

[5] The default assignments are defined in the data sheet for the C8051F410 (Fig. 18.4, p. 150). The user is obliged, sometimes, to overrule the default assignment of functions to pins using the "skip" command. Writing a **1** to a pin in the *skip* register, causes the '410 to "skip" that pin when doing its usual pin assignments.

Code to transfer a bit; built-in ports:

A privileged bit. "Carry" allows **bit** ***moves and some other operations:*** In general, the 8051 moves only *bytes* at a time. But the *Carry* flag provides the exception to that rule: quite a rich set of operations can *copy* the level of a bit into or out of the *Carry* flag, and some instructions can do Boolean operations on the flag – with the result saved in the flag. Here are some examples:

- ORL C, bit: OR the specified bit with the C bit, result to C, e.g., ORL C, A.0
- ANL C, bit: AND the specified bit with the C bit, result to C, e.g., ANL C, P2.2
- ANL C, /bit: AND the complement of the specified bit with C, result to C, e.g., ANL C, /P1.2
- MOV C, bit: copy the bit level to the C flag, e.g., MOV C, P3.7
- MOV bit, C: copy the level of the C flag to the named bit, e.g., MOV P1.2, C

There is no equivalent operation when using the external buses; only a byte transfer is permitted on bused input and output.

22N.1.4 Equivalent bit operations in C

Bit transfer in C: A bit transfer in C, using the built-in ports, looks a good deal like the assembly language – and, in fact, is somewhat wordier:

```
// bit_transfer_port.c

    #include<Z:\MICRO\8051\RAISON\INC\REG320.H>

    sbit  outbit = P1^0; // declare these to be bit variables ("s" is "special,"
          // meaning that the bit is located in a "special function register," SFR)
    sbit  inbit = P3^2;
void main()
    {
      while (1) // forever...
      {
        outbit = inbit; // ...copy KWR* level to LED, using two built-in ports
      }
}
```

And it is reassuring to find that the compiler generates (for once!) exactly the assembly code that we wrote by hand:

```
ASSEMBLY LISTING OF GENERATED OBJECT CODE

          ; FUNCTION main (BEGIN)
0000              ?WHILE1:

0000 A2B2         MOV     C,inbit
0002 9290         MOV     outbit,C
0004 80FA         SJMP    ?WHILE1
```

22N.1.5 A hardware "flag:" input hardware to do something *once*

Input hardware, from a pushbutton, is pretty simple when using the external buses, and very simple when using the built-in ports.

But using a pushbutton to get the processor to do something *once* is more involved, and we have devoted Worked Example 22W to this problem – mostly to give you practice in seeing through nonsense, but also to introduce the notion of an edge-triggered "flag" for the processor to test. Fig. 22N.4 is the relatively complex hardware that can serve this purpose when using the external buses. To see how this design evolved, take a look at that note. The general notion that this particular hardware illustrates is the use of a hardware *flag*.

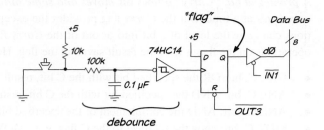

Figure 22N.4 Hardware to let a pushbutton make the processor do something once.

Incidentally, the tasks done by debouncer and flip-flop can be done in software; but since we aim to keep the coding simple, we prefer to burden the hardware to do this job. (This circuit appears in §22L.1.)

22N.2 Digression on conditional branching

JNB, in §22N.1.1, was the first *conditional branch* that we met in this course. Such operations – giving the ability to do one thing IF... and another operation OTHERWISE – give computers their seeming intelligence. (Just summing numbers isn't impressive: any mechanical desk calculator used to do that trick.) The *conditional* operations permit this cleverness. The 8051 has a limited set – but is admirably nimble in one subset of these operations: those that test *single bits*.

22N.2.1 Branching instructions

- JNB, JB ...: jump if a specified *bit* is low, or high, respectively (and JBC offers an exotic variation: jump conditional, and clear the flag if it was high).
- JNC, JC ...: jump if the *carry* flag is low, or high, respectively.
- CJNE A ...: jump if the accumulator is not equal to some specified value; also can be applied to a register (for example: CJNE R2, D0_SOMETHING).
  - CJNE affects the Carry/Borrow flag, so CJNE with a subsequent test of CY can also test not just for equality/inequality, but also for *less than*.
- DJNZ Rn ...: Decrement register Rn and jump if the result is not zero.
- JNZ, JZ jump if accumulator zero, ... if accumulator not zero. (JZ is not a general-purpose flag like Carry.)

Bit test using built-in port: hardware

Lab 22L uses a PORT3 input bit to determine whether a PORT1 output bit should toggle. The hardware is almost nothing.

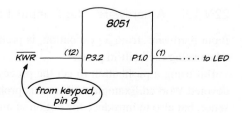

Figure 22N.5 Single-bit input using controller's Port pin to control bit output.

22N.2 Digression on conditional branching

Program: branch on input bit:

Assembly-language bit test: The program below, from Lab 22L, toggles the LED – unless one presses the keypad's "WR" key. That key asserts KWR*, and the low on that line hangs up the program in a tight loop that misses the LED-toggling:

```
                             1   ; PORTBRANCH_402.a51 Branch conditional of controller's PORT pin,
0082    30B2FD             181   CHECKIT:    JNB P3.2,CHECKIT   ; hang here till bit 2 comes up
                                                                ;(hang if WR button pressed)
0085    63B001             182               XRL P3,#01H        ; another way to flip LSB (bit 0) of
                                                                ; Port 3, while WR button NOT pressed
0088    B2B0               183               CPL P3.0           ; a more straightforward way to flip
                                                                ; the output LED
008A    80F6               184               SJMP CHECKIT
```

To the lab listing, which used XRL (XOR), we have added a straightforward way to flip a bit, CPL, which is the code the C compiler uses, below.

C equivalent, bit test: This C code does the same conditional bit-flip:

```
sbit inbit = P3^2;
sbit outbit = P1^0;

void test_bit_io (void)

{
while(1)
 if(!inbit) outbit = outbit;
 else outbit = !outbit;
}
```

The compiler generates the assembly implementation listed below. The compiler's code is not quite equivalent to ours: instead of hanging in a loop while the button is pressed the compiler always reads in the LED level and writes it back to the LED, but flipping it or not between read and write operations.

```
0000               ?WHILE1:
                                                        ; SOURCE LINE # 14
0000 20B206        JB      inbit,?ELSE1
0003 A2B3          MOV     C,outbit
0005 92B3          MOV     outbit,C
0007 80F7          SJMP    ?WHILE1
0009               ?ELSE1:
                                                        ; SOURCE LINE # 15
0009 B2B3          CPL     outbit
000B 80F3          SJMP    ?WHILE1
```

22N.2.2 Masking, to operate on more than one bit

Assembly language: The ready-made bit-test instructions, JB and JNB, offer the tidiest way to branch on the condition of one bit. But sometimes it is useful to look at more than one bit, though something less than a full byte. In that case, a technique called *masking* is necessary, to separate the interesting bits from the uninteresting (the grain from the chaff).

For example, the lab keypads store a byte formed from the *two* most recent key-pressings. If you want to branch on just the *most recent* key, then you need to "mask out" the effect of the preceding key: you need to make the high four bits not matter. You can, in effect, erase those bits if you AND each of them with a 0. You can preserve the low four bits by ANDing each of those with a 1.

Figure 22N.6 is a program fragment that clears the top four bits, preserving the low nybble. The program first loads a byte, *AB*, from register R0 ("MOV A, INPUT"). Then it knocks out the left-hand nybble, keeping only the *B*. The right-hand image shows a simulation of the effect: the *A* that came in on the high four bits has been cleared in the **A** register.

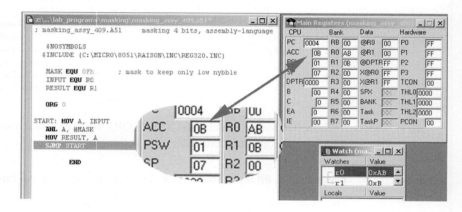

Figure 22N.6 Program fragment to clear top nybble of a byte.

The code below makes use of such a mask to implement the most-recent-key test mentioned above. The program branches to a label named "DO_A" if the *A* key was the most recent key pressed, to "DO_B" if it was *B* pressed. Otherwise, it exits this patch of code proceeding to actions not specified.

```
        MASK EQU 0Fh  ; shorthand that allows us to alter mask in head of program
        TARGET1 EQU 0Bh ; ditto for a value we're looking for
        TARGET2 EQU 0Ah  ; ...and another target
; ------------------
        MOV DPTR, #8000h ; point to keypad
        MOVX A, @DPTR    ; pick up key value
        ANL A, #MASK     ; mask out top 4 bits (high nybble)
                         ; (all 1s in the low nybble pass the key value)
        CJNE A, #TARGET1, TRY_A   ; ...if it was not B, see what did come in
                         ; ...otherwise, fall through to go DO_B
        SJMP DO_B        ; go and do what you should do in response to key "B"
TRY_A : CJNE A, #TARGET2, EXIT    ; check whether character is "A" ; if not, then give up
        SJMP DO_A        ; go and do what you should do in response to key "A"
EXIT: ...
```

Figure 22N.7 sketches the masking operation: the mask byte is AND'ed bitwise with the byte picked up from the keypad, forcing zeros in the high nybble while passing the low nybble. The sketch assumes that what came in from the keypad was a hexadecimal *C* in the low nybble, and anything at all in the high nybble.

C-language equivalent: multi-bit mask: The masking operation in C is very close to what it is in assembly language: a bit-wise AND (using "&" – not "&&," in case you know C). The C compiler's implementation in assembly is virtually identical to what we wrote by hand above. Fig. 22N.8 is a program fragment that applies a mask to a variable named INPUT, generating RESULT;

We next show equivalents in C code for the assembly-language operations just above: applying a mask and then checking for characters *A* or *B*. The C program fragment could be designed in either of two ways.

22N.2 Digression on conditional branching

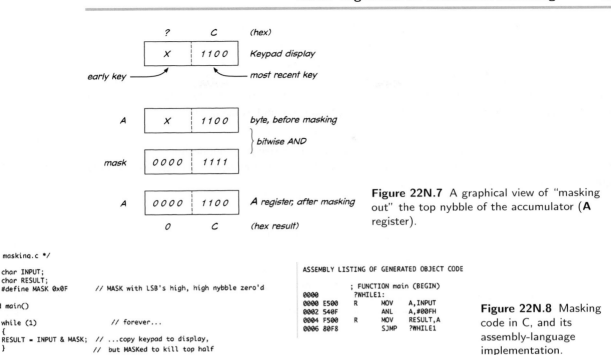

Figure 22N.7 A graphical view of "masking out" the top nybble of the accumulator (**A** register).

Figure 22N.8 Masking code in C, and its assembly-language implementation.

```
/* masking.c */
    char INPUT;
    char RESULT;
    #define MASK 0x0F     // MASK with LSB's high, high nybble zero'd
void main()
{
    while (1)             // forever...
    {
    RESULT = INPUT & MASK; // ...copy keypad to display,
    }                      // but MASKed to kill top half
}
```

```
ASSEMBLY LISTING OF GENERATED OBJECT CODE
              ; FUNCTION main (BEGIN)
0000          ?WHILE1:
0000 E500  R     MOV   A,INPUT
0002 540F        ANL   A,#00FH
0004 F500  R     MOV   RESULT,A
0006 80F8        SJMP  ?WHILE1
```

One C coding method: mimic assembly code with "IF...ELSE" compares: The code could look a lot like the assembly code we just saw earlier. (You may notice also how close it is to some Verilog code, which also uses the If...Else form. See, for example, all the counters that include a reset in Appendix A.) Here is some C that works that way. This program loops forever, setting one output bit or another if particular keys are pressed.

```
/* masking_if.c */
// apply constant mask to port 0 input
// indicate result by setting one or another bit of P1

        #include<C:\MICRO\8051\RAISON\INC\REG320.H> /* defines registers by name */

        sbit  ABIT = P1^0; // output bits, declared as single bits
        sbit  BBIT = P1^1;
        xdata char *KEYPAD; // byte-sized variable, pointed to by DPTR

void masking_if ()
{
        int MASK = 0x0F; // mask with LSB's high, high nybble zeroed
        KEYPAD = 0x8000; // init the pointer

    while (1) // forever...
    {
    if((*KEYPAD & MASK) == 0x0A)   // test for "A" key; "&" is bitwise AND
        {ABIT = 1;   // if it's A, then set that output pin...
          BBIT = 0;}

    else if ((*KEYPAD & MASK) == 0x0B) // test for "B" key
        {ABIT = 0;
         BBIT = 1;
        }
        else
```

```
                {ABIT = 0;
                 BBIT = 0;}
        }
}
```

The compact form "*KEYPAD" refers to the value pointed to by the variable "KEYPAD", so that the code *KEYPAD & MASK both picks up the 8-bit key value and applies the mask.[6]

Another C coding method: use CASE statements: An alternative program design – similar to one that you have seen in some Verilog state machine examples, if not elsewhere – can test for particular inputs by listing them as successive possible "CASEs."

```
    sbit   ABIT = P1^0; // output bits, declared as single bits
    sbit   BBIT = P1^1;
    xdata char *KEYPAD; // byte-sized variable, pointed to by DPTR
    char MASK;
    char MASKED_INPUT;   // ...the input value after masking
 #define KEYAD 0x8000 // address of keypad
 #define MASKVAL 0x0F // mask with LSB's high, high nybble zeroed

void masking_case ()
{
    MASK = MASKVAL;   // init mask
    KEYPAD = KEYAD;   // init pointer value

  while (1) // forever...
    {
  MASKED_INPUT = (*KEYPAD & MASK);  // pick up key value and mask it

     switch (MASKED_INPUT)   // ...now test the masked key value
        {
         case 0x0A:    // test for "A" key
            ABIT = 1;
            BBIT = 0;
            break; // ...and if you find a match, don't execute the alternative branches

         case 0x0B:   // test for "B" key
            ABIT = 0;
            BBIT = 1;
            break;
         default:
            ABIT = 0;
            BBIT = 0;
       }
    }
}
```

22N.2.3 Byte compare: CJNE, and work-arounds for larger *compares*

Test for equality: CJNE: The code in the assembly language masking examples used the only 8-bit *compare* that the 8051 provides:

---

[6] This C code differs subtly, and perhaps unimportantly, from the earlier assembly language code. The assembly code takes in one byte and applies successive tests; the C code takes in a value anew with each "*KEYPAD & MASK" operation.

```
            CJNE A, #TARGET1, TRY_A      ; ...if it was not B, see what did come in.
```

CJNE is one of the 8051 instructions that gets a lot done with one instruction (another was DJNZ, which you have seen as a loop counter). It does a *compare* then a branch if the two operands are *not equal*.

Test for difference: greater-than, less-than: assembly: Sometimes CJNE alone is not versatile enough. It detects only *equality/inequality*. If you want to discover whether one quantity is *greater than* another, you can follow CJNE with a test of the Carry flag (CY), a flag that is better described as a *borrow* flag, when used after a compare. To call it a "flag," you recognize, is just to say that it is a level held in a dedicated flip-flop. The value persists until overwritten.

Here is some code to determine whether the value in the **A** register is EQUAL to a value that we'll call REF, or whether, instead, it is LESS_THAN that value, or GREATER_THAN. The code begins with the usual use of CJNE then looks at the Carry flag.

```
            CJNE A, #REF, NOT_EQUAL     ; the comparisons begin
EQUAL:      SJMP DO_EQUAL               ; land here if A equals REF

NOT_EQUAL:  JC LESS_THAN                ; A not equal REF; if CY/Borrow flag is asserted, then A < REF
GREATER_THAN: SJMP DO_GT                ; land here if no Cy/Borrow: ==>A > REF

LESS_THAN:  SJMP DO_LT                  ; land here if A < REF
```

Test for difference: greater-than, less-than: C code: Doing the same task in C is a little more readable, though not simpler. This program displays its three possible results with three output patterns, which could be lighting single LEDs, for example.

```c
    while (1) // forever...
      {
      if (*KEYPAD == REF)
        RESULT_OUT = EQUAL;
      else
        if (*KEYPAD > REF)
          RESULT_OUT = GT;
        else
          RESULT_OUT = LT;
      }
```

As was the case above, `*KEYPAD` is the byte that comes in from the keypad. The asterisk denotes "pointer," and "KEYPAD" is an address used by that pointer. As you know, in the 8051 that pointer is DPTR.

22N.2.3.1 Masking to *Set* More than One Bit

A ready-made instruction will *set* a single bit: "`SETB P1.1`," for example, sets bit 1 of the built-in port. But if you want to set more than one bit, again a mask is helpful. Here it is the OR operation, rather than AND, that does the job. To set bits 1 through 3 of P1 (while leaving the remaining bits unchanged), one would first have to get P1's value into the accumulator, then force the selected bits:

```
    MOV A, P1       ; get port byte, in order to apply the masking
    ORL A, #00Eh    ; force bits 1..3 high; leave the others untouched
    MOV P1, A       ; ...and send the result back to the port
```

The XOR operation can be used to similar effect with a mask: one can use XOR and a mask to flip only specified bits, setting up a 1 at the position of each bit that is to be flipped: `XRL A, #88h`, for example, flips the msb and bit 3, leaving the other bits unchanged.

22N.2.4 Flag hardware

A "flag" is an indicator bit that stays asserted (as a rural mailbox flag stays up after the farmer has put it up to tell the mailman, "I've mail for you to pick up"). A flag stays asserted until it deliberately is cleared, by the device to which the *flag* information was addressed. In our rural RFD analogy, the mailman knocks the flag down after picking up the mail.

In hardware, a flag is the output of a flip-flop. The flag is forced High by a Set or by an edge on the flop clock while D is held high; it is cleared by a Reset (we are assuming that the flag is active high). You'll find a collection of wrong ways to set up a flag, and finally a right way in Lab 22L.

Whether done with an external bus or with built-in ports, implementing a hardware *flag* requires adding an external flip-flop so the flag will stay up till cleared: see Fig. 22N.9.

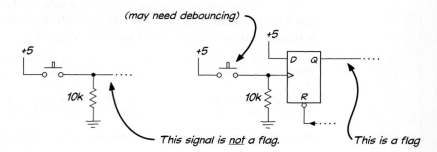

Figure 22N.9 Hardware flag requires use of a flip-flop; if set with clock (edge-sensitive), the flop needs a reset.

Flag-testing code: Flag-testing code looks just like the bit tests mentioned in §22N.2.1 except that it must include in addition an operation that can clear the flag (probably using an OUT operation to drive the flag-flop's RESET*).

In Day 23, when we meet *interrupts*, we will return to this topic – and will find that the 8051 can implement an edge-triggered flag without the use of either external flip-flop or the three-state required for a *bused* flag.

Interrupt, advance warning: The big-board lab in 22L concludes by asking you to try an "interrupt." In order to run that test program, all you need to appreciate is that in responding to *interrupt* the 8051 jumps to a fixed, predetermined location, where it executes whatever it finds. That code conventionally is called an "interrupt service routine." It is a slightly-special form of subroutine. We will look more closely at *interrupt*, and will find that it is a hardware-initiated CALL. Having understood the operation of the subroutine CALL, you will recognize *interrupts* as no more than a variation on that theme.

22L Lab Micro 3. Bit Operations; Timers

22L.1 Big-board lab. Bit operations; interrupt

Introduction: This lab lets the micro do what it does well: asks it to operate on single *bits*, rather than on bytes. This ability sometimes lets the micro look quite capable compared with a full-scale computer. An ordinary computer is designed to move large chunks of data on a relatively wide data bus. A little microcontroller, which does that kind of task poorly, is nimble in *bit* operations – more or less like the mouse that was able to remove the splinter from the paw of the legendary lion. No doubt some day the little micro will be rewarded.

22L.1.1 Easy I/O: use a micro port pin for bit operations

Normally, a microcontroller makes things happen with its pins, not through a bus brought off the chip; the usual controller shows *no bus* to the outside world. Though our machine does have an external bus, and we'll take advantage of it when we want to do byte operations, let's watch the micro do what it's especially good at: driving a single bit, and then checking a single bit.

A single-bit output:

Hardware: port bit drives LED: To watch this action, tie an LED to one of the available *port* pins: we'll use the LSB of Port 1. Because the ports provide a very feeble *high* (a mere 50µA source at 2.4V), but a pretty good *low* (1.6mA sink at 0.45V), we will let the pin *sink* current from the LED, not source current into it.[1] The 3.3k resistor limits the current to about one mA (the LED drops about 2V).

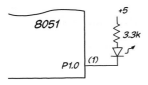

Figure 22L.1 Port bit used as output.

Program: toggle one port output bit: The program below tests this hardware by toggling that port bit, bit 0 of Port 1. We have used the processor's *complement* operation, *CPL*. Because the port is bit addressable, the code is very compact.[2]

---

[1] Until CMOS took over the digital world, this was the standard way of driving an LED; TTL, like this micro and unlike ordinary CMOS, was good at sinking and poor at sourcing current. A CMOS gate can sink or source the few mA that will light an efficient LED.

[2] A more laborious, but more general, method would have been to take the byte value into the accumulator, then XOR with a high in the position of the bit we want to change. The zeros elsewhere in that 8-bit "mask" value (01h, in this case, where

Lab Micro 3. Bit Operations; Timers

```
MACRO ASSEMBLER PORTPIN_408

LOC     OBJ         LINE    SOURCE
                      1     ; PORTPIN_408.a51  Lab 21: use controller's PORT pin  BIT operation
                      2
                      3     $NOSYMBOLS ; keeps listing short
                      4         $INCLUDE (C:\MICRO\8051\RAISON\INC\REG320.INC)    170
                    171
0000                172     ORG 0 ; tells assembler the address at which to place this code
                    173
0000    8076        174     SJMP STARTUP ; here code begins--with just a jump to start of
                    175     ; real program.  ALL our programs will start thus
0078                176     ORG 78H ;  ...and here the program starts
                    177
0078    C290        178     STARTUP: CLR P1.0 ; start low (LED lit), just to make it predictable
                    179
007A    B290        180     FLIPIT:  CPL P1.0 ;  flips only LSB (bit 0) of port 3
007C    80FC        181     SJMP FLIPIT
                    182     END
```

A single-bit input:

Hardware: one bit from keypad: The 8051 is just as good at responding to a single bit as input. To demonstrate this, we can tie one of our debounced signals from the keypad, KWR*, to a pin on another port: Port 3, data line 2, this time: P3.2. This choice looks strangely arbitrary; it turns out to be a convenient choice, as we'll learn when we meet our first use of *interrupt* in §22L.1.3. See also Fig. 22N.5.

Program: branch on input bit: The program below toggles the LED – unless one presses the keypad's "WR" key.

```
MACRO ASSEMBLER PRTBRANCH_402                         09/04/102 15:54:33 PAGE    1

LOC     OBJ         LINE    SOURCE
                      1     ; PORTBRANCH_402.a51 Branch conditional of controller's PORT pin
                                                  ; driven by KWR*
                      2
                      3     $NOSYMBOLS        ; keeps listing short, lest...
                      4         $INCLUDE (C:\MICRO\8051\RAISON\INC\REG320.INC)
                    170
                    171
0000                172     ORG 0 ; tells assembler the address at which to place this code
                    173
0000    807E        174     SJMP STARTUP ; here code begins--with just a jump to start of
                    175     ; real program.  ALL our programs will start thus
                    176
0080                177     ORG 080H ;  ...and here the program starts
                    178
0080    C290        179     STARTUP:   CLR P1.0 ; start low, just to make it predictable
                    180
0082    30B2FD      181     CHECKIT:   JNB P3.2,CHECKIT ; hang here till bit 2 comes up
                                                ; (hang if WR button pressed)
0085    639001      182     XRL P1,#01H ; another way to flip LSB (bit 0) of port 1
                                                ; while WR button NOT pressed
0088    80F8        183     SJMP CHECKIT
                            END
```

the bit we want to toggle is in the LSB position) would leave the other seven bits unchanged. XOR permits operations on more than a single bit, at one stroke.

22L.1.2 An *enter* key: a *hardware Flag*

The mismatch between computer speed and human slowness can make for trouble when one tries to tell a computer to do something once. In the preceding program, for example, it's easy enough to use the KWR* signal to determine whether or not to blink the LED. But how would you use KWR* to toggle the LED *once*? With the computer running at full speed, you could not do so. You would need to add a flip-flop (to record the fact that you had requested the action – say, by pressing the WR* button), and then would need to let the computer clear that flop when it had received the request and set about acting upon it. That is the purpose of the "Enter" key shown below.

Ready flop hardware: Figure 22L.2 is a fragment of Fig. 20L.1 showing a flop used for this purpose. The flip-flop is doing two tasks:

1. it records the request (the pressing of the button): this can be useful in case the computer is busy in some time-consuming process, so that the button might be released before the computer gets around to checking the level on the button;
2. it permits the computer to clear the request 'flag' (the flop's Q), even though the human may still be leaning on the button.

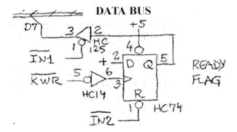

Figure 22L.2 Ready flag flip-flop.

The second of these tasks is the more important of the two, since the computer – even the little 8051 running at 11MHz – can do a lot while a human finger presses a button. Even if you try to press a button for just a moment, a "moment" is probably at least 0.1s, in which time the processor can execute perhaps a hundred thousand instructions. So the computer could do the requested task on finding the button pressed, come back and find it still pressed, do it again. It would respond many times for each button-press, if we did not provide the *edge-triggered* function with the external flip-flop. The computer must, of course, clear the flag once it has found the flop Q high. The circuit above uses IN2* to carry out this clearing.

It is important that the KWR* signal is *debounced*: if it were not then the clearing would occur while bounce continued and a new "Ready" signal would arise, eliciting another response. Such issues are noted in Worked Example 22W.

Code to try the ready flop: The program GETREADY_1208.a51, below, amends Lab 21L's keypad-to-display program: instead of always reading the keypad, the program reads it just once after a press of the keypad's WR button, taking the keypad value into the accumulator. The processor continually sends the value of the accumulator out to the displays.

```
LOC     OBJ             LINE    SOURCE
                          1     ; GETREADY_1208.a51  Lab 22:  branch conditional on READY flag
                          2     ;   puts out a display value that is constant
                          3     ;     till changed by WR key that causes new read of keypad
                          4
                          5             $NOSYMBOLS ; keeps listing short
```

```
                      6         $INCLUDE (C:\MICRO\8051\RAISON\INC\REG320.INC)
0080                174         STACKSTART EQU 080H ; starts stack just above SFR's
0000                175         SOFTFLAG EQU 0H ; this is bit-addressable location
                    176
0000                177         ORG 0       ; address at which to place this code
                    178
0000   020090       179         LJMP STARTUP     ; here code begins -- with jump to start of
                    180         ;    real program.  ALL our programs will start thus
0090                181         ORG 090H    ; ...and here the program starts
0090   758180       184         STARTUP: MOV SP , # 127  ; source code is "STARTUP: MOV SP,
                                        ; #STACKBOT-1." Assembler inserts "127"
0093   11A3         185         ACALL PTRINITS
                    186
0095   E4           187         CLR A       ; clear display value, for predictable startup
0096   C200         188         CLR 0       ; clear softflag, for first pass
                    189
0098   11AF         190         GETKEY: ACALL READYCHECK ; take new character?
                                ; updated by subroutine, tells this program
                                ; whether or not to read the keypad value anew
009A   300003       191         JNB 0 , USEOLD       ; if flag is low, don't update key value.
009D   C200         192         CLR 0       ; ...but if you land here, flag was high. Clear it...
009F   E0           193         MOVX A, @DPTR   ; ...read the keypad (flag was set)
00A0   F0           194         USEOLD: MOVX @DPTR,A  ; and output to display
00A1   80F5         195         SJMP GETKEY      ; look for another key...endlessly
                                ;----SUBROUTINES------------------------------------
00A3   908000       200         PTRINITS: MOV DPTR, #8000H  ; "point" to display and keypad
00A6   75A080       201         MOV P2, #80H     ; for MOVX @Rn operation, this sets the
                    202                          ; high half of address as base of I/O
00A9   7801         203         MOV R0, #01H     ; low half of address of READY port
00AB   7902         204         MOV R1, #02H     ; low half of address of READY-CLEAR port
00AD   E3           205         MOVX A, @R1      ; send a pulse to clear flop-
00AE   22           206         RET
                    208         ;----------------------------
                    209         ; READYCHECK subroutine: if Ready, SETS BIT ZERO as FLAG
00AF   C0E0         211         READYCHECK: PUSH ACC ; save scratch register
00B1   E2           212         MOVX A, @R0 ; get Ready flag
00B2   30E703       213         JNB ACC.7,NOCHANGE ;  if flag is low, don't clear the flag
00B5   D200         215         CHANGE:  SETB 0
00B7   E3           216         MOVX A,@R1 ; ...and send a pulse to clear flop
00B8   D0E0         218         NOCHANGE: POP ACC ; restore saved scratch register
00BA   22           219         RET
                    221         END
```

22L.1.3   Interrupt

The preceding program obliged the processor to keep checking a pin in order to determine whether or not to do a new keypad read. That may not be a good arrangement if you'd like the processor to be doing something more productive. An alternative scheme would let the processor devote itself to some process without continually asking a dull question like, "Do you want me to increment the display?"

Instead, the peripheral that wants attention would drive a pin on the processor – more or less the way the someone might tug on a bell-pull that summons the butler. Only when that happened would the processor go over to the requested task, in this case incrementing an internal register. (Meanwhile, the butler could be getting something done; perhaps polishing the silver.) In the next program we'll try this program design.

Here are two novel elements in the program's behavior, using an *interrupt*:

22L.1 Big-board lab. Bit operations; interrupt

- we must tell the processor that we want to pay attention to the interrupt. We must do this both –
  - in general: we enable the "global" recognition of interrupts. This we do by setting one bit (one flip-flop output level) that is dedicated to this purpose. This is EA below.[3]
  - in particular: we enable the particular interrupt that we plan to use: "external" interrupt zero ("external" means "applied to a pin on the processor, not generated as response to some internal event such as a timer's rolling over").
- we must put the program that we want executed (incrementing the display, in this case) at the particular location dedicated to "interrupt 0 response." In this case that's address 03h – also labeled "INT0VECTOR" in the listing below.

Hardware: none! You don't need to add anything; the KWR* line already is wired to the 8051's pin 12 (P3.2), where it was connected in §22L.1.1. That pin happens[4] to be the pin that is assigned to INTERRUPT0* requests. Fig. 22L.3 shows how the hardware looks.

Program: interrupt causes increment of display: Interrupt uses *hardware* to initiate a CALL. Because pulling a pin does not, in itself, specify a destination for the CALL, the interrupted computer needs some strategy to find out where to go when interrupted. The 8051 uses the simplest possible scheme: it always goes to a particular address in response to a particular type of interrupt. As in an ordinary software CALL, the processor saves on the stack *the address of the next instruction* before going off to the patch of code to which it is sent by INTERRUPT.

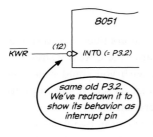

Figure 22L.3 KWR* wired again to P3.2, this time serving as INT0* request pin.

```
MACRO ASSEMBLER INTDSPLY_1204                    12/08/04 16:36:00 PAGE    1

LOC    OBJ              LINE     SOURCE
                          1      ; INTDSPLY_1204.a51 let interrupt inc display
                          2
                          3      ; should increment display each time interrupted
                          4
                          5          $NOSYMBOLS
                          6      $INCLUDE (C:\MICRO\8051\RAISON\INC\REG320.INC)
                        172      $INCLUDE (C:\MICRO\8051\RAISON\INC\VECTORS320.INC)
                                 ; Tom's vectors definition file
007F                    196      STACKBOT EQU 07FH ; put stack at start of scratch
                                 ; indirectly-addressable block (80h and up)
                        197
                        198
0000                    199      ORG 0H
0000   0200D0           200      LJMP STARTUP
                        201
```

---
[3] Unfortunately "EA" is also the name of the pin on the Dallas 8051 that tells the controller to look to external buses to find its code. That EA is not related to this one.
[4] Well, OK: we rigged it that way!

886 Lab Micro 3. Bit Operations; Timers

```
00D0                202         ORG 0D0H
                    203
00D0    75817F      204         STARTUP: MOV SP , # 127
00D3    908000      205         MOV DPTR, #8000H
                    206
                    207         ; ---NOW ENABLE INTERRUPTS---
00D6    D288        208         SETB IT0 ; make INT0 Edge-sensitive (p. 22)
00D8    D2A8        209         SETB EX0 ; ...and enable INT0
00DA    D2AF        210         SETB EA           ; Global int enable       (pp. 31-32)
                    211
00DC    E4          212         CLR A ; (for clean startup, as usual)
                    213
00DD    F0          214         STUCK: MOVX @DPTR, A   ; show display--constant,
                                                       ; till interrupt inc's it
00DE    80FD        215         SJMP STUCK        ; (responds to falling edge--
                                                  ;pseudo-edge sensitive,
                    216                           ;   so you must clock several times
                                                  ;   while high, then low)
                    217
                    218         ; --------------------
                    219         ; ISR: JUST INCREMENT A
0003                220         ORG INT0VECTOR   ; this is defined in VECTORS3210.INC,
                                                 ; included above.
                    221                          ; It is address 03h, the address to which micro hops
                    222                          ;    in response to interrupt zero
0003    04          223         ISR: INC A
0004    32          224         RETI
                    225
                    226         END
```

Most of this program will be easy to follow. As in all programs that use subroutines, if you single-step you will see the micro appear to hesitate after interrupt, as it does after the CALL instruction. Then it hesitates again at the RET (return from CALL) and RETI. We wish you could watch the stacking – but here we've hit a limit in what the controller is willing to let us see. It's been pretty generous in letting us watch its address and data buses, for most of its operations.

This Interrupt Service Routine (ISR) is unusual in being so tiny that it can sit in two bytes beginning at the *vector* address, 03h. Only eight bytes are reserved here, so more often the ISR code will immediately jump out to a location where there's room to write a longer routine. We'll see that case in later labs. An ISR also usually changes to a new set of scratch registers, R0 through R7 (there are four sets). We didn't bother to do that, here, since the ISR uses none of those registers.

22L.2 SiLabs 3: Timers; PWM; Comparator

22L.2.1 PORT0 pin use

Again we use one PORT0 pin to drive an LED, and we will add other peripherals at four other pins: see Fig. 22L.4.

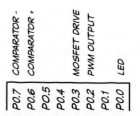

Figure 22L.4 PORT0 pin use in this lab.

22L.2.2 Timer: blink an LED, again

We confessed, back in Lab 20L, that it was a bit dopey to tie up a controller in a loop of a few hundred thousand cycles in order to slow the blinking of an LED to a rate that a human could notice. Now we invite you to try the controller's better way to do such a task using its built-in hardware timer.

The 8051, like almost any controller, includes a collection of such timers. We will first try one whose design is pretty standard (though it is an enhancement to the very first version of the 8051). It is called Timer2. The '410 offers three or four other timers (depending on whether you consider a "Real time clock" to be a timer). As usual, we will have to initialize a set of registers. But once that is done the program will occupy almost none of the processor's attention. The SiLabs '410 also provides a 16-bit counter, the core of its "Programmable Counter Array" (PCA), that can implement particular operations including pulse-width modulation. This we will try in §22L.2.7.

The usual way to use a timer is to treat it like an oven timer: set a delay time and ask the timer to let us know when time is up. Usually, "time's up" will trigger an interrupt since the point of using the timer is to free the processor for other operations more interesting than keeping time or measuring delay time. The program in §22L.2.4 behaves this way. It invokes an interrupt service routine to toggle an LED each time its timer overflows. It then automatically reloads the delay value and the cycle begins again.

It is also possible to let a timer drive an output pin directly: the PCA Counter/Timer can do that in "High-speed output mode." We have not done that here. Our ISR is so simple that it is virtually equivalent to the PCA direct-drive. Our ISE simply toggles the LED, but it could, of course, do something (even-) more exciting.

As usual, our program is extremely simple. So most of our programming labor therefore goes into reading the datasheet to find what register initializations are needed. This time we will ask you to determine these initializations.

22L.2.3 Initialize TMR2CN...

Configuring a microcontroller: bad news, good news: The bad news is that a controller offers so many options for the use of its built-in peripherals, that the user is obliged to make a great many choices before even beginning to write any code. We illustrate this discouraging wealth of choices with a rather modest example: the setting up of a 16-bit timer, done bit-by-bit, using a detailed description of a configuration register (§22L.2.4). The good news is that you don't need to proceed this way – bit-by-bit. Instead, you can take advantage of an easy programming tool called Configuration Wizard.

For this particular program, apart from the USUAL_SETUP, almost the entire task consists in initializing just one register, named Timer2 Control (TMR2CN). As usual, the details of even this single register can look quite overwhelming (so much information!)

We would like you to set up this register as follows:

- 16-bit timer;
- enable the timer/counter;
- clock source: system clock / 12;
- timer initial value (used only on first pass, which you may want to watch in single-step): FF F0h;
- timer reload value (used on all subsequent passes; this is the value re-loaded after overflow): 00 00h;
- interrupt on overflow (called Timer2 Interrupt).

We invite you to do this task using either of the two methods:

Laborious way: proceed bit-by-bit. The name makes this sound bad. But in fact this process probably will take you only two or three minutes. By scrutinizing each bit of the TMR2CN register described below, you can determine what value that register needs.

It may seem perverse to look at this register bit-by-bit, and we doubt that you will ever do this again once you have met the Configuration Wizard. But we think that seeing the functions of the several bits can help to demystify the controller. After all, the wizard calls itself by that name just because the translation of choices into bit patterns seems pretty magical.

Less laborious way: invoke the Configuration Wizard. Start up the Configuration Wizard on your laptop (the wizard, named Wizard2 in its present incarnation, is a separate utility – not accessible from within the SiLabs IDE). Then follow the process in §22L.2.5, describing the Wizard's use.

22L.2.4 Configuring registers bit-by-bit

Our goal here is to set up Timer2 (one of four general-purpose timers) to operate as a 16-bit timer that will interrupt each time it rolls over. Most of the timer's functions are controlled by one register, TMR2CN. Fig. 22L.5 is a page from the data sheet for the '410 showing the function of each bit in this register.

Timer register initializations: We reprint this in full detail not because you need this much information, but to give you a sense of the initial scariness of a controller's data sheet (a "sheet" that in this case is 270 pages long!). If you find this detail concerning a single register daunting, then you're like us. We hope though that a few minutes' inspection will reveal that – at least this time – we need know only a little of this overwhelming detail.

Then we examine each bit in TMR2CN, to see what we ought to do with it. Proceeding bit by bit:

d7: TF2H Yes, we need to pay attention to this bit, which will indicate when the timer has "timed out." But this is not a bit we need to initialize, except in the sense that once it has been set by an overflow it is up to us to clear it (as the last sentence in the TF2H description says explicitly. This *flag* bit can serve to generate an *interrupt*. We use it so in this program.

Do we need to clear this on startup? No. The "Reset Value" shown at top right of Fig. 22L.5 shows that this bit, like all other bits of this register, starts out low after a Reset.

d6: TF2L We an ignore this bit, which indicates an 8-bit overflow – overflow of the low byte of the 16-bit timer2.

d5: TF2LEN We'll leave this bit, which could interrupt on an 8-bit overflow, disasserted. The Reset condition, zero, gets us the result that we want.

d4: TF2CEN No, we aren't using the "capture" mode (and won't here go into what it might mean; if you're interested, you can see it used in one of the two period-measuring programs shown in §25L.1.5 that asks you to "do something with the standalone controller"). So we leave the bit Low, at its inactive Reset level.

d3: T2SPLIT We need this bit low to configure the timer as a single 16-bit timer, rather than as two 8-bit timers. We leave it in its Reset condition, Low.

Continuing with the remaining bits:

d2: TR2 Yes, this is important: this bit we must set high, to enable Timer2.

d1: T2RCLK This bit does not matter to us since it has no effect unless we have selected "capture" mode (by setting TF2CEN High), which we have not done.

SFR Definition 24.8. TMR2CN: Timer 2 Control

R/W	R/W	R/W	R/W	R/W	R/W	R/W	R/W	Reset Value
TF2H	TF2L	TF2LEN	TF2CEN	T2SPLIT	TR2	T2RCLK	T2XCLK	00000000
Bit7	Bit6	Bit5	Bit4	Bit3	Bit2	Bit1	Bit0	Bit Addressable

SFR Address: 0xC8

Bit7: **TF2H**: Timer 2 High Byte Overflow Flag.
Set by hardware when the Timer 2 high byte overflows from 0xFF to 0x00. In 16 bit mode, this will occur when Timer 2 overflows from 0xFFFF to 0x0000. When the Timer 2 interrupt is enabled, setting this bit causes the CPU to vector to the Timer 2 interrupt service routine. TF2H is not automatically cleared by hardware and must be cleared by software.

Bit6: **TF2L**: Timer 2 Low Byte Overflow Flag.
Set by hardware when the Timer 2 low byte overflows from 0xFF to 0x00. When this bit is set, an interrupt will be generated if TF2LEN is set and Timer 2 interrupts are enabled. TF2L will set when the low byte overflows regardless of the Timer 2 mode. This bit is not automatically cleared by hardware.

Bit5: **TF2LEN**: Timer 2 Low Byte Interrupt Enable.
This bit enables/disables Timer 2 Low Byte interrupts. If TF2LEN is set and Timer 2 interrupts are enabled, an interrupt will be generated when the low byte of Timer 2 overflows. This bit should be cleared when operating Timer 2 in 16-bit mode.
0: Timer 2 Low Byte interrupts disabled.
1: Timer 2 Low Byte interrupts enabled.

Bit4: **TF2CEN**. Timer 2 Capture Enable.
0: Timer 2 capture mode disabled.
1: Timer 2 capture mode enabled.

Bit3: **T2SPLIT**: Timer 2 Split Mode Enable.
When this bit is set, Timer 2 operates as two 8-bit timers with auto-reload.
0: Timer 2 operates in 16-bit auto-reload mode.
1: Timer 2 operates as two 8-bit auto-reload timers.

Bit2: **TR2**: Timer 2 Run Control.
This bit enables/disables Timer 2. In 8-bit mode, this bit enables/disables TMR2H only; TMR2L is always enabled in this mode.
0: Timer 2 disabled.
1: Timer 2 enabled.

Bit1: **T2RCLK**: Timer 2 Capture Mode.
This bit controls the Timer 2 capture source when TF2CEN=1. If T2XCLK = 1 and T2ML (CKCON.4) = 0, this bit also controls the clock source for Timer 2.
0: Capture every smaRTClock clock/8. If T2XCLK = 1 and T2ML (CKCON.4) = 0, count at external oscillator/8.
1: Capture every external oscillator/8. If T2XCLK = 1 and T2ML (CKCON.4) = 0, count at smaRTClock clock/8.

Bit0: **T2XCLK**: Timer 2 External Clock Select.
This bit selects the external clock source for Timer 2. If Timer 2 is in 8-bit mode, this bit selects the external oscillator clock source for both timer bytes. However, the Timer 2 Clock Select bits (T2MH and T2ML in register CKCON) may still be used to select between the external clock and the system clock for either timer.
0: Timer 2 external clock selection is the system clock divided by 12.
1: Timer 2 external clock uses the clock defined by the T2RCLK bit.

Figure 22L.5 Bit-by-bit details of the timer control register that we need to initialize.

d1: T2XCLK Yes, this matters. It chooses a clock source for the timer – but since we are not using the capture mode we will leave the bit in its Reset state, giving a clock rate that is System_Clock / 12.

You may need a pencil and paper to record the bits that need your attention (I need such a crutch). When you have done that you need to decide how to get the bits to the levels that you want.

Specifically, should you use a MOV into the register (such an operation determines the levels on every bit)? Or should you instead use *bit* operations? A preliminary question is whether bit operations are permitted for this register. The answer to that question is *yes*, as the line "Bit Addressable," near the top right corner of the register description in Fig. 22L.5, indicates.

Often *bit* addressing is useful, permitting the change of only bits that interest us while leaving the others untouched. But here a MOV seems a better idea for two reasons: first, we know the initial condition of all bits (the Reset value) so we need not fear overwriting some useful information with a MOV; second, some bits need to be cleared (or kept Low), others need to be set (High). A MOV can take care of both Setting and Clearing at a stroke.

So you should figure the byte value that you want for TMR2CN (probably expressed in hexadecimal just to be consistent with the other values that are listed). Then MOV that value into the register.

One more register involved, albeit marginally: CKCON: It's pretty much true that the single register TMR2CN is enough to initialize the timer; that's a nice change from the more usual case. One other register could be said to be involved, as Fig. 22L.5 suggests: CKCON. The T2MH bit (d5) in CKCON could, if set, determine the Timer2 clock rate. We leave this bit in its Reset state, Low, and doing that allows TMR2CN to set the clock rate. A line in the present program reiterates the default setting, unnecessarily but harmlessly: `ANL CKCON, #DIV12 ; allow TMR2CN to set timer clock rate`.

We mention this not to irritate you but to remind you that register initializations usually involve more than a single register, and that (more perplexing) the significance of a bit on one register may depend on a setting in a different register (as T2MH here depends on the level of T2XCLK in the TMR2CN register). All this register-initializing is, as we have said before, the price we pay for the controller's versatility.

22L.2.5 The easy method: take advantage of the Configuration Wizard

Wouldn't it be nice if you could just tell your assistant, "Please do the right things to lots of fussy bits in several registers to let the 16-bit counter interrupt when it rolls over?" It turns out you can be almost as chatty as that if you set up the timer using the Configuration Wizard. We'll use it below to set up Timer2.

Start the configuration process: First, specify the part and say that you mean to write in assembly language: see Fig. 22L.6.

Specify the peripheral: Then we'll choose *timers*, and among them, TIMER2 in Fig. 22L.7.

Figure 22L.6 First choose the device and assembly-language format.

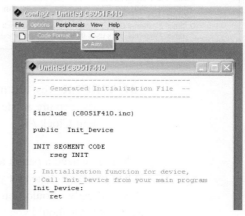

Set up the peripheral: The wizard shows us, on a single page, most of the options that we need to consider. Many correspond to the single bits specified in TMR2CN (§22L.2.3). But these descriptive items are a lot easier to grasp. And on this page we are not restricted to a single register.

To finish setting up the timer, we need to leave the timer page to turn on interrupts for Timer2. To get us to that interrupt page, Fig. 22L.9, the Wizard lets us click on "Configure Timer Interrupts."

And the fruits of this setup process are a few lines of assembly code:

22L.2 SiLabs 3: Timers; PWM; Comparator 891

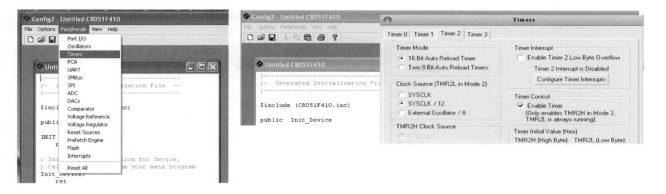

Figure 22L.7 Choose which peripheral and then which one within the category.

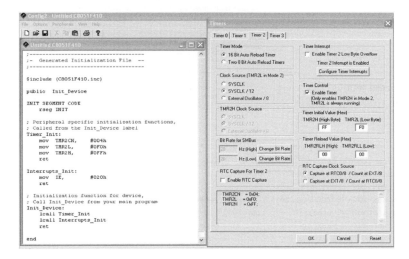

Figure 22L.8 We can set up the timer by clicking boxes, nicely described. We're even shown assembly code to do the job.

```
; Peripheral specific initialization functions,
; Called from the Init_Device label
Timer_Init:
    mov    TMR2CN,    #004h
    ret

Interrupts_Init:
    mov    IE,        #020h
    ret
```

These lines take care of the several registers found to be involved: a lot better than the bit-by-bit process that we began in §22L.2.4. And not only is the wizard easier to use; it is also more likely to get each bit right.

A debugging opportunity, watch the overflow generate an interrupt...: Below, in the listing of timer2_bitflip.a51, we set the first two values, HITIME and LOTIME, at FF FAh, extremely close to the top count, FF FFh. We used this peculiar one-time start count to facilitate debugging. We wanted you to be able to look at the timer in action: you can, after a reset, single-step the program and watch the program as it approaches its overflow.

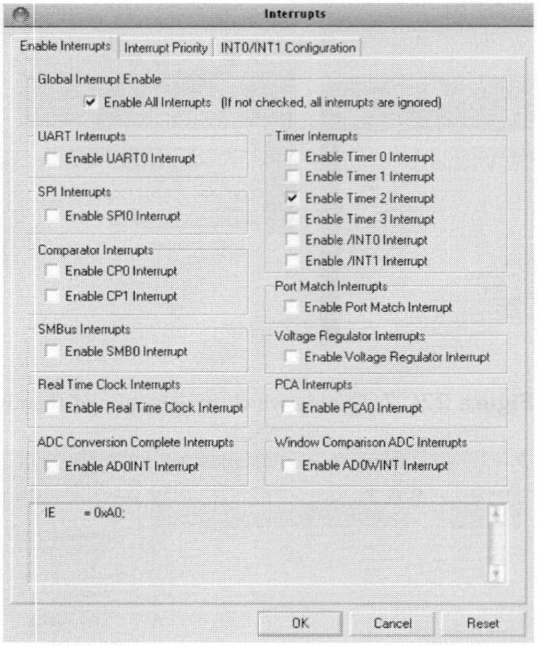

Lab Micro 3. Bit Operations; Timers

Figure 22L.9 Configuration wizard lets us enable timer interrupts.

The bit-flip timer program

```
; timer2_bitflip.a51 use Timer2 to toggle LED, on interrupt by timer overflow
    $NOSYMBOLS      ; keeps listing short
    $INCLUDE (C:\MICRO\8051\RAISON\INC\c8051f410.inc)
    $INCLUDE (C:\MICRO\8051\RAISON\INC\VECTORS320.INC)  ; Tom's vectors definition file
    STACKBOT EQU 80h       ; put stack at start of scratch
    DIV12 EQU 0DFh         ; timer clock mask to div by 12, in CKCON
    TIM2INTEN EQU IE.5
    TIM2_ENABLE EQU TMR2CN.2
    SOFTFLAG EQU 0         ; software flag that ISR uses to talk to Main
    BLUE_LED EQU P0.0      ; LED to toggle
    GLOBAL_INTEN EQU EA    ; an easier mnemonic for the overall interrupt enable
    HITIME SET 0FFh        ; set start count close to the overflow val
    LOTIME SET 0FAh        ; ...and this is the low byte
    RELOAD SET 0h  ; these are reload values, for maximum delay
; port use:
;       LED at P0.0
        ORG 0
    SJMP STARTUP

    ORG 80h
STARTUP: MOV SP, STACKBOT-1  ; initialize stack pointer
  ACALL USUAL_SETUP
    ACALL TIMER_INITS
    SETB TIM2_ENABLE

STUCK:    SJMP STUCK ; await interrupts.  All the action is there
```

Adding watch: As you may have read in the IDE note, you can select a register to *watch* in the debugger by highlighting the register in the window, then right-clicking to get a menu that offers *Add...to Watch*. If you do this for TMR2RLH and TMR2RLL you can *view* the two timer registers, as in Fig. 22L.10.

Then a few more single-steps will cause the timer to overflow, from FF FFh to 00 00h, generating an interrupt request that causes the program hop to its ISR: see Fig. 22L.11.

Figure 22L.10 Single-stepping, we can watch timer increment approaching overflow.

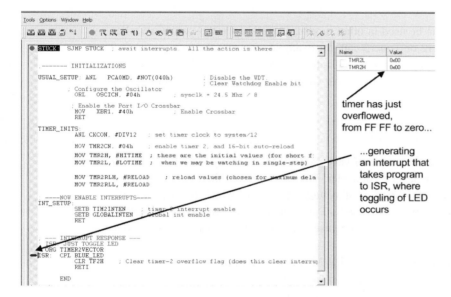

Figure 22L.11 A timer overflow takes program to the ISR.

This first overflow – the quick one that we rigged by initializing the timer to nearly its full count – occurred after just a few increments of the timer. This process generates no useful delay when run at full-speed. It was useful only to let us watch the timer at work.

The usual case: timer counts many cycles between overflows: But on all later passes – once the counter has overflowed and been reloaded with the reload values (TMR2RLH and TMR2RLL, which we initialize to zero) – the counter then provides maximum delay, with a PWM start value of 00 00h.

Even this delay is under one second, but long enough to make the blink rate of the LED comfortably visible.

22L.2.6 The bit-flip timer program

Download the program, `timer2_bitflip.a51`, from the book's website. Here is the program, without initializations:

```
        ORG 80h
STARTUP: MOV SP, STACKBOT-1   ; initialize stack pointer
   ACALL USUAL_SETUP
```

```
        ACALL TIMER_INITS
        SETB TIM2_ENABLE

STUCK:  SJMP STUCK       ; await interrupts.  All the action is there

;------------------------------
; ISR: JUST TOGGLE LED
        ORG TIMER2VECTOR
ISR:    CPL BLUE_LED
        CLR TF2H   ; Clear timer-2 overflow flag (does this clear interrupt flag?)
        RETI
```

The delay available from this timer as configured is only about 0.5 seconds. One could stretch this delay by letting the ISR include a software count. A single register there could extend the delay to a couple of minutes. For really long delays, the so-called "Real Time Clock" is made to order. We will not try that peripheral in these labs, but you may want to on your own.

Timers do various tricks: The timer can be used in other modes, as well as for the "wake me when you're done" scheme of §22L.2.2. Timer2 or Timer3 can, for example, measure the frequency of an external signal relative to system clock using "External...Capture Mode."[5] Timers 0, 1, and 3 can also count edges on an input pin rather than measure time.[6] We will not take time now to explore these many options.

22L.2.7 Pulse width modulation: *dim* an LED

PWM, analog and digital versions: The technique of varying drive to a load by varying *duty cycle* on an output pin is quite easy to implement on a controller. The hardware required is almost nil, in contrast to what is required for continuous variation of voltage or current, which requires a DAC. PWM can be done entirely in software. But the '410 offers PWM in an easier form. The '410 offers a Programmable Counter Array (PCA), a timer that can be used to count an 8- or 16-bit value, and to set a bit when the count exceeds a reference value.

Thus in purely digital form it can mimic the analog PWM method in which a waveform ramps (sawtooth or triangle) and triggers a comparator when ramp value exceeds an analog reference value. You may want to look at Fig. 8L.7 on page 340 to refresh your recollection of the scheme.

Figure 22L.12 is the digital equivalent, which the '410 uses to implement an 8-bit PWM. When *Count = Reference*, the output bit goes High; when Count rolls over, output bit goes Low.

For constant duty cycle hold the reference register constant (this REFERENCE is PCA0CPHn in Fig. 22L.12 and PCA0CPH0 in the program `pwm_by_wizard_nov10.a51`, on the book's website). The COUNT register increments at a rate set by the 'PCA timebase.' When COUNT exceeds REFERENCE, PWM output goes high; when COUNT rolls over to zero again, PWM goes low. REFERENCE thus determines duty cycle (from about 0.4% to 100%).[7]

22L.2.8 PWM code: slow ramp of LED brightness

The program `pwm_by_wizard_nov10.a51` implements the gradual brightening of an LED. It does this by slowly increasing the *reference* value (over a span of about three seconds), then rolling it over, so that the *Reference* value forms a slow sawtooth, sweeping brightness from minimum to maximum. Here is the core of the tiny program omitting intializations :

---

[5] See SiLabs C8051F410 datasheet, §§24.2.3, 24.3.3.
[6] PCA takes as input a signal named ECL, p. 249, which can be assigned to a pin using the crossbar, as shown at p. 149.
[7] Code can take duty cycle to 0%, by writing to a separate control register.

22L.2 SiLabs 3: Timers; PWM; Comparator

```
STARTUP: MOV SP, #STACKBOT
    acall USUAL_SETUP
    acall Init_Device
    mov PCA0CPH0, #0    ; clear output, for orderly startup

UP:     acall DELAY
    inc PCA0CPH0
    sjmp UP
```

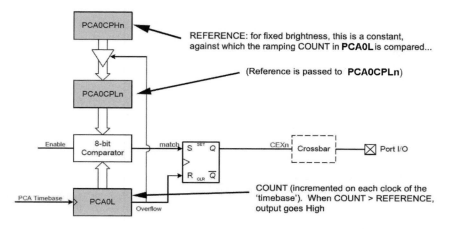

Figure 22L.12 '410 implements PWM in digital form. Taken from Fig. 25.8 of the datasheet of C8051F410, with permission.

PWM configuration using wizard: Using the wizard, first we turn on the PCA and choose PCA0 (the only counter array available on the '410), setting it up as 8-bit PWM. In Fig. 22L.13 we have also (arbitrarily) set the duty cycle at 50%. Do you see how?[8] Then we place the PWM output at a particular pin.

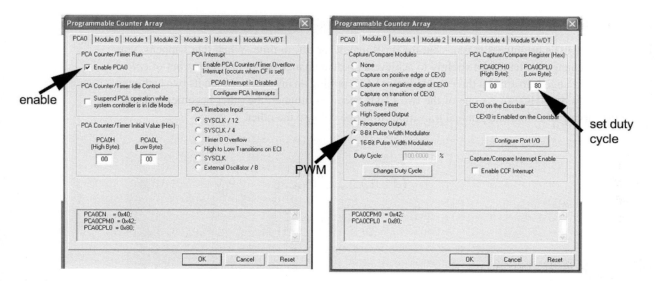

Figure 22L.13 First turn on the PCA (the counter array) and 8-bit PWM...

[8] For this 8-bit PWM we have set the REFERENCE value at the midpoint: 80h.

We set up the PWM output as push–pull, and place this function on a free PORT0 pin, P0.2, skipping two pins that will be used in the next lab.[9]

The PWM program: This program, `pwm_by_wizard_nov10.a51`, can be obtained from the book's website.

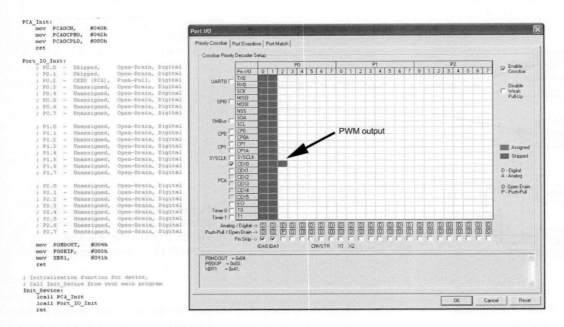

Figure 22L.14 ...then we place the PWM output.

22L.2.9 Improvements 1: change color

Ramping the LED's brightness is fun – but the thrill may wear off, in time. When that happens, you may want to change the hardware slightly to achieve a gradual change not of *brightness* but of *color*. To make that change, substitute a three-lead bicolor LED, putting an inverter between its two cathodes (use a 74HC14 or 74HC04). As one of the two LEDs becomes dimmer, the other becomes brighter; hence the mixing of colors and gradual transition from one color to the other.

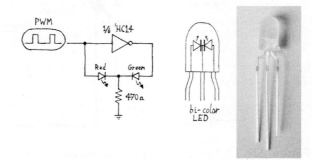

Figure 22L.15 Bicolor LED changes color as PWM varies brightness of green versus red LEDs within package.

---

[9] P0.0, used last time for an LED, will be used in Lab 24L as input to the ADC; P0.1 will carrry the DAC output.

22L.2.10 Improvements 2: ramp up, ramp down

Instead of applying a *sawtooth* waveform to modulate the brightness of the LEDs, it may be prettier to apply a *triangle*. That is, let the ramping *down* in brightness be as gradual as the ramping *up*. Try that, modifying the code of the program in §22L.2.7. A CJNE instruction could watch the ramping ACC value, reversing the ramp sense when ACC hits either extreme – zero or 0FFh.

22L.2.11 Comparator: an oscillator as a start on something more interesting

Among the analog peripherals included in the '410 are two comparators. They can be configured as simple analog parts: two analog inputs, one logic-level output at a port pin. Hysteresis is programmable.

More interestingly, the comparator output can be polled or can be used to generate an interrupt. In the program of §22L.2.12 below, the comparator does no more than zero the timing capacitor. It could do something more interesting, as §22L.2.13 suggests.

We will start modestly, using just a resistor in place of the photodiode that we propose in §22L.2.13, so the circuit will replicate the op amp *RC* oscillator of Lab 8L – except that it runs on a single supply.

Oscillator hardware: Figure 22L.16 shows the hardware.

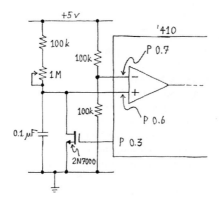

Figure 22L.16 Oscillator formed with '410 comparator.

Comparator configuration:

Configuration using wizard: Figure 22L.17 shows comparator configuration choices. In order to place the comparator inputs where we want them – at P0.6, P0.7 – we must tell the crossbar to skip all the prior bits (the crossbar places the inputs as low as it is permitted to). The MOSFET drive bit, P0.3, is shown set to *push-pull* so that we need not add a pullup to the MOSFET gate drive.

Configuration choices, noted bit-by-bit: In case you like to see the choices bit-by-bit[10], they are in Table 22L.1.

22L.2.12 Code: *RC* oscillator

The code, comparator_oscillator_jan11.a51, that implements an oscillator with the hardware of Fig. 22L.16 can be found on the book's website. Here is the core of the program. We have omitted initializations so that you can concentrate on what is more interesting. You'll notice that the subroutine CHECK_END needs to be completed by you.

---

[10] Not likely, we admit. Reading these register descriptions is about as much fun as reading the US tax code.

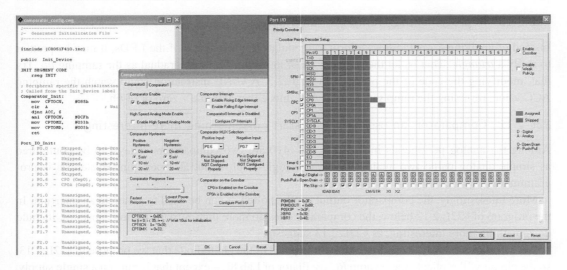

Figure 22L.17 Comparator configuration wizard: simple *RC* oscillator.

Table 22L.1 Register descriptions.

Register	bit/byte-value	function
CPT0CN	D7 (= CP0EN)	1 $\Rightarrow$ enable comparator 0
...	D6 (= CP0OUT)	1 $\Rightarrow$ output high if {input +} > {input -}
...	D3,D2 (= CP0HYPx)	01 $\Rightarrow$ pos hysteresis 5 mV
...	D1,D0 (= CP0HYNx)	01 $\Rightarrow$ neg hysteresis 5 mV
CPT0MX	D7..D4 (CMX0Nx)	0011 $\Rightarrow$ neg input assigned to P0.7
...	D3..D0 (CMX0Px)	0011 $\Rightarrow$ pos input assigned to P0.6
CPT0MD	D1,D0 (=CPT0MDx)	11 $\Rightarrow$ comparator to slowest and lowest power setting
P0MDOUT	D3 (byte 08h)	1 $\Rightarrow$ set P0.3 to push-pull, to drive discharge MOSFET
P0MDIN	D7,D6 (byte 3Fh)	0 $\Rightarrow$ set up pin (P0.7, P0.6) as analog input
P0	D7,D6 (byte 0C0h)	1 $\Rightarrow$ set latch high, to permit pin use as input
P0SKIP	D0..D5	1 $\Rightarrow$ skip these to place comparator inputs at D6, D7 in crossbar I/O assignments

```
STARTUP: MOV SP, #STACKBOT-1

         ACALL USUAL_SETUP
         ACALL Init_Device
         ACALL MISC_INITS

OSC_LOOP: MOV A, CPT0CN        ; get this where we can test bit
         JNB ACC.6, OSC_LOOP   ; branch on comparator output (not flag): till Vcap > Vref
         SETB P0.3             ; discharge cap (turn on MOSFET)
         ACALL delay           ; ...allow enough time for full discharge of cap
         CLR P0.3              ; ...turn off MOSFET
         MOV P1, R7            ; display hi byte of cycle counter
         SJMP OSC_LOOP         ; carry on (we ask you to replace this with code that
                               ;   signals DONE when COUNT_high equals KEYPAD input

;----SUBROUTINE THAT WE'D LIKE YOU TO WRITE:
CHECK_END:         ; get current count for comparison
                   ; if not yet done, keep counting
DONE:       ; light LED or turn on buzzer, to indicate exposure complete
HALT:       SJMP $  ; once COUNT_high equals KEYPAD, hang here till Reset
```

22L.2 SiLabs 3: Timers; PWM; Comparator

```
;-------------------------------
; ISR for comparator interrupt
      ORG COMP0_INT

            ANL CPT0CN, #0DFh    ; clear comparator interrupt flag
            LJMP COMPARATOR_INT_RESPONSE

      ORG 150h
COMPARATOR_INT_RESPONSE: PUSH ACC ; save scratch register
16_BIT_INC: MOV A, #1            ; set up an increment that will affect CY flag
            ADD A, R6            ; increment low byte of count
            MOV R6, A
            MOV A, R7            ; get hi byte
            ADDC A, #0           ; increment hi byte if there was a carry from low byte
            MOV R7, A
            POP ACC
            RETI
```

22L.2.13 Apply the oscillator: "Suntan Alarm"

This circuit and code does nothing exciting as it stands. It does nothing you couldn't achieve with a single-supply comparator or op-amp – or with a '555 oscillator. Except that it uses one supply, the circuit is essentially the same as that of the first *RC* oscillator in Lab 8L. But this program could be the foundation for something more interesting – like AoE's suntan monitor (§15.2).[11]

To apply this oscillator to that task we would need just two changes:

- replace the fixed and variable *resistors* to V+, shown in Fig. 22L.16, with a phototransistor, see Fig. 22L.18, so that the frequency of oscillation is proportional to light-intensity;
- let the comparator output do more than simply zero the capacitor: each time it discharges the cap, the program could increment a many-bit counter. At 16 bits and a 1kHz oscillation, such a counter would allow measuring about a minute's duration; at 24-bits, the counter would allow much longer exposures.

phototransistor, BPV11

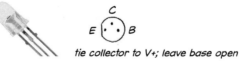
tie collector to V+; leave base open

Figure 22L.18 Phototransistor to replace resistor in oscillator circuit.

When the count reached a target value (set perhaps by keypad, perhaps by a pot feeding an ADC), the program could do whatever you think appropriate: sound an alarm, or perhaps deliver a mild electric shock to the user.[12] Better than a shock, probably, is an audible alarm, which we later propose that you might install.

Figure 22L.19 is one of the small 5V buzzers that your "Done" signal might turn on:[13]

---

[11] That AoE suntan monitor was born as an exam question that we wrote as a joke. You may imagine our surprised satisfaction then, when we discovered that someone has *patented* just such a suntan monitor. No kidding. It is US Patent # 4,428,050, "Tanning Aid" (Pellegrino, 1984): the invention's description includes these remarks: "...Finally, the device includes an alarm for giving an appropriate warning when the preset dose for each session and the total dosage are achieved, and a preset "turnover" feature which can be used to divide the session ... for the purposes of tanning the front of the body and the back of the body to the same extent." We are sad to report that the invention does not seem to have made Mr. Pellegrino rich.

[12] Just kidding. We don't recommend that scheme.

[13] CUI CEM-1205C, shown, draws 35mA. Available from Digikey.

CUI CEM-1205C

Figure 22L.19 Small 5V buzzer: could be used to signal that tanning exposure is complete.

22L.2.14 Code for suntan alarm

The program `comparator_osc_display_jan11.a51`, available on the book's website, adds the features suggested in §22L.2.13:

- It relies on a phototransistor (current proportional to illumination; this is a part you met back in Lab 6L).
- It increments a 16-bit counter on each cycle of the oscillator.
- We'll ask you to finish the code by writing a subroutine that compares the high byte of the 16-bit *count* against the keypad value. When these are equal, the program turns on a DONE signal. This signal, as we have suggested, could be discreet – turning an LED – or brash – turning on a buzzer.
- ... The program then halts, awaiting a reset from its user.

Try the suntan integrator: You should see the LCD data display (driven by PORT1) slowly counting up – at a rate proportional to light intensity sensed by the photoresistor/transistor. Try shading the sensor with your hand, to simulate a cloud passing over the sunbathing geek.

22L.2.15 Improvements 3: improve the monitor by adding the DONE signal

Now try writing the subroutine labeled CHECK_END in `comparator_osc_display_jan11.a51`. This is the routine that compares the high byte of the 16-bit COUNTER (let's call it "COUNT_HI") against the value on the KEYPAD. When the two match, the program should indicate that sunbathing is complete (by turning on LED or a buzzer) and then should get trapped in an endless loop (from which only a RESET* can break out). As long as the two values, COUNT_HI and KEYPAD, do not match, the program should let the oscillations continue. You will find the command CJNE useful.

22W Worked Examples. Bit Operations: An Orgy of Error

"Let a hundred weeds bloom!" declared Chairman Mao – and here we do just that. The point is partly to show off how many ways one can be wrong. But the more wholesome purpose is to sharpen your skills in detecting and correcting nonsense.

Here is a set of bad designs, each with its (spurious!) rationale. The rationales usually will indicate what was wrong with the preceding design. Finally, we will reach a good design.

22W.1 The problem

We would like to add to the microcomputer a pushbutton that will make the machine do an action *once* when the button is pushed. It might be called an "Enter" or "Ready" key.

22W.2 Lots of poor, and one good, solutions

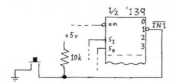

Rationale:
I want to send information to the processor, so I need an IN port. A "port" is a door (or a window? a sort of "port hole"), and I'll just feed this information through the pin labelled appropriately on the I/O decoder, the '139.
I'll use IN1*.

Figure 22W.1 Send it into "IN0*" of a decoder.

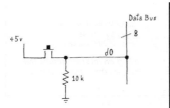

Rationale:
Oh--so the pin labelled "IN1*" on the decoder is an output from the '139? (Are these people trying to confuse me?)
Yes, I see that this pin is used to enable another device, so this pin is not itself an input.
So, instead, I'll use the data bus. I'll use one line of the 8 available "data" lines (surely, data lines are for data, and here's some data). This way, I'll have a simple, direct way to let the CPU (or computer) know when I want something done.

Figure 22W.2 Direct connection to data bus.

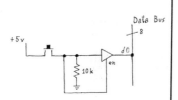

Rationale:
Oh, yes--I forgot the 3-state. Need a 3-state whenever driving a bus. (Nearly all the traffic on the "data" bus is not data at all, but instructions).
So, forcing a "data" line to a particular level for a while, as #2 does, is guaranteed to corrupt instructions, crashing the computer.
OK: this 3-state lets me leave the bus alone until the moment when I need the computer's attention.

Figure 22W.3 3-state to data bus.

Worked Examples. Bit Operations: An Orgy of Error

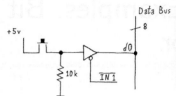

Figure 22W.4 3-state, controlled by computer.

Rationale:
OK, so #3 was exactly as bad as #2. Well, now I see that it's the CPU and not I that must decide when data should be driven onto the data bus.
So, I'll let the CPU turn on the 3-state when it's ready. I'll use IN1* from the I/O decoder.

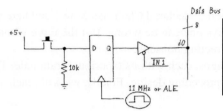

Figure 22W.5 ...add a flip-flop.

Rationale:
I'd better add a flop, so that
1) the CPU won't miss my button-push, if it's busy when I hit it, and
2) the CPU won't do what I ask 1000 times, in case my finger stays down for a while.

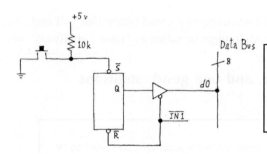

Figure 22W.6 ...use an SR flip-flop.

Rationale:
Yes, you're right: that flop in #5 was clocked much too often. It did nothing at all to solve my problem.
This design is better, because it lets me set the flop (so CPU can't miss my press), and then lets the CPU clear the flop (so what I ask will not be done thousands of times).

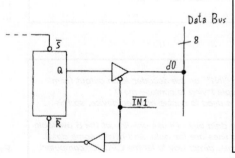

Figure 22W.7 ...use an SR flip-flop, timing RESET* differently.

Rationale:
Nope: the IN1* that clears the flop in #6 comes too early—before the CPU reads the data bus, a read that occurs close to the end of RD* and thus close to the end of IN1*:

CPU takes data from Data bus close to end of IN1* pulse

I can solve that problem by inserting an inverter before the Reset*. This way, the Clear will happen after the CPU has read the data.

22W.2 Lots of poor, and one good, solutions

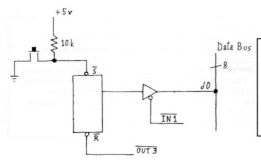

Rationale:
  Not quite! The inverter causes a continuous Reset*, at all times except during the IN1* assertion. OK, there's no neat way to use IN1* to clear the flop.
  So, I'll use a separate signal to clear the flop. My pushbutton will set the flop; the CPU then can clear the flop when it's ready to.

Figure 22W.8
...use an SR flip-flop with independent RESET*.

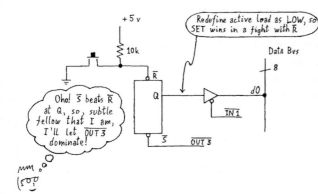

Rationale:
  No, #8 won't work. The CPU's Reset*, in #8, would fail if my finger were still down.
  So, I'll rearrange things so that the CPU sends a Set*—since I recall that Set* beats Reset* (in determining the level of Q) iin case they're both asserted.

Figure 22W.9
...use an SR flip-flop, relying on SET* to beat RESET*.

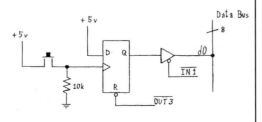

Rationale:
  Now I see that nothing I do with Set* or Reset* can work. Since my finger may still be down after the CPU has tried to clear the flag, the clear will fail. This is true of both #8 and #9. The only difference between the two is that in #9 the IN1* does clear the flag; but that's hardly useful, since the clearing does not persist.
  Edge-triggering can solve the problem: the Reset* will succeed even if the pushbutton remains down.
  By the way, in case anyone worries about bounce, note that the first rising edge takes Q up, so no need to debounce the clock signal.

Figure 22W.10
...use an edge-triggered D flop.

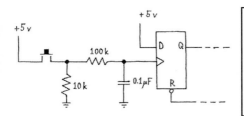

Rationale:
  OK, you have a point: I'd better debounce, since bounce time is long (> 1ms) relative to CPU response time (< 10 microseconds). So, the clear could happen, and then many more bounce edges could occur, simulating new 'do it' requests.
  In addition, the switch will bounce on release, generating new edges.

Figure 22W.11
...use an edge-triggered D flop, and debounce with an RC.

Figure 22W.12
... use an edge-triggered D flop, and debounce properly — squaring up the edge.

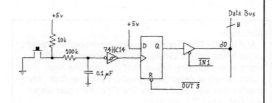

Rationale:
 The "debouncer" shown in #11 was not complete. It was only a slow-edge generator, and slow edges can produce multiple edges at a clocked input. So, slow edges at a clock are about as bad as bounce.
 The Schmitt trigger--HC14 with about 0.4V hysteresis--solves the problem. It will accept a slow edge as input and put out a clean, fast transition on its output, driving the clock properly.
 This design, at last, should work!

22W.3 Another way to implement this "Ready" key

Exhausted? You'll soon see that you can sidestep most of this hard work (both the thinking and the doing) by taking advantage of the 8051's edge-triggered *interrupt*. You still would need to debounce the switch, but otherwise there would be nothing to build. You could even skip the debouncing hardware if you were willing to work harder at writing code: you could debounce in software. We won't do that in these labs because we like to keep the code minimal; but you may do that in some other setting. Code is cheap.

23N Micro 4: Interrupts; ADC and DAC

Contents

23N.1	**Big ideas from last time**	**905**
23N.2	**Interrupts**	**906**
	23N.2.1 Why interrupts?	906
	23N.2.2 *How* to interrupt?	907
	23N.2.3 Enabling interrupts; edge- or level-response	909
	23N.2.4 Priority among interrupts	909
	23N.2.5 The difference between RET and RETI	910
	23N.2.6 A demonstration program	910
23N.3	**Interrupt handling in C**	**911**
23N.4	**Interfacing ADC and DAC to the micro**	**912**
	23N.4.1 The IC Version of ADC, DAC Hardware: AD7569	913
	23N.4.2 ADC and DAC on SiLabs '410	915
	23N.4.3 Code	916
23N.5	**Some details of the ADC/DAC labs**	**917**
	23N.5.1 PLL to Adjust f_{clock} of Reconstruction Filter	920
	23N.5.2 An audio amp, so you can listen	920
	23N.5.3 Or use the switching amplifier of Lab 12L	920
	23N.5.4 What you might listen to	921
23N.6	**Some suggested lab exercises, playing with ADC and DAC**	**921**

Why?

We want to know how to handle analog values as inputs and outputs for a computer; and how to use *interrupts* to break into program flow.

23N.1 Big ideas from last time

Subroutine CALL:

- Allows sharing a block of code among multiple programs;
- allows writing modular code: main program invokes multiple fragments, each to do a task

Bit operations:

- Set or clear single bits (allows controlling 8 devices at one 8-bit port);
- test single bits;
- especially easy when using the built-in ports rather than external bus;

23N.2 Interrupts

An interrupt, as the name suggests, breaks into the flow of one program in order to allow another program to run. This is a special sort of CALL: one initiated by a *hardware* event rather than by a line of code.[1] Let's first recall *why* one might invoke an interrupt, and then we'll look at *how*.

23N.2.1 Why interrupts?

AoE §14.4.8

The name certainly suggests the answer: interrupt when you're in a hurry. There's a sense in which this is true, and another in which it is not.

The alternative to using interrupts is to *poll* – to ask the "Ready" key of Lab 23L, for example, "Are you ready?" Would it be better to let the "Ready" key just break in, as if ringing a doorbell, leaving the machine free to do what it needs to do until the doorbell rings? Sometimes.

When you need a prompt response: A "Ready" *interrupt* might be a good idea, and one can think of other cases where a prompt response seems still more urgent: when the computer's power-supply voltage begins to fall for example (it won't do for the computer to say, in effect, "The sky is falling? Well, don't bother me with that until I've finished calculating pi to 20 places"). Sensing the power-failure (like HAL in 2001), the computer could save some information to a non-volatile storage medium. The 8051 version that we use, DS89C4xx, does include such an interrupt on power failure. So does the SiLabs version. There, the interrupt is named "Voltage regulator dropout." But each must be enabled in software. The 8051 includes no "non-maskable" interrupt.

Or, you might want to interrupt in another case that we've all met: the case when the computer gets hung up running some program, and you'd like to regain control.[2] A "watchdog" interrupt can be used to reset the processor if it is set up to do so. The "watchdog" is a counter that must be cleared from time to time, under program control. If the program fails to clear it, then something has gone wrong. In that event, the counter overflows and generates a reset that forces the machine to start up afresh.

In Lab 23L, where your computer takes data from an ADC, we will use interrupt to permit regular sampling. Interrupt lets us set the sampling rate without concern for the rate at which particular blocks of code may run. In the case of the lab program, because the sampling rate is far below the time required for any operation by the program, we could about equally well have used *polling*. We used the interrupt there mostly because of the *edge-triggering* available on the interrupt. This feature makes it easy to implement the desired result: one sample taken for each cycle of the *take-a-sample* input, a square wave.

A typical application for interrupt: FIFO for incoming data: An input buffer can satisfy two needs that might call for use of interrupt:

- the need for prompt response to the external device;
- the need to leave the main program free to do something useful while awaiting the interrupting event.

---

[1] Many processors add a "software interrupt" which behaves like an ordinary interrupt, but is initiated by a line of code rather than by a hardware event. The 8051 does not include this feature.

[2] Breaking into a program, willy nilly, would require a so-called "non-maskable" interrupt – a sort that the 8051 lacks. Lacking this, we can break into a runaway 8051 program with the hardware RESET. This resembles an interrupt in one respect: it forces the machine to begin executing code at a dedicated address; but it differs from interrupt in not attempting to save any information about the program that was running when RESET was invoked.

An ADC normally needs to take data at a regular rate. An "Interrupt Service Routine" (ISR), a response to an interrupt request, can do this. The ISR can take in a sample each time an interrupt occurs, and can store the data in a "circular buffer" – a RAM storage block that can be written into and read from at different rates.[3] Such a buffer also can be described as a FIFO buffer – First In, First Out – since the reading should always be from the oldest item.

The interrupting ADC routine can keep up with the rigid sampling rate without imposing that strict timing on the main program. The main program need not respond right away when a sample comes in; it need only keep up with the *average* rate at which the data arrives. In short, the main program can act like a normal student: fall behind, doing things more interesting than studying; then make an extra push to catch up; go off and do something entertaining; then return and catch up in another flurry of activity.[4]

Data coming into the computer through a serial port can present a similar challenge and can call for the same remedy. We don't want the computer to *miss* a character coming in on the RS232 serial port (both the Dallas and SiLabs 8051s, like almost all controllers, include these venerable ports); but we don't want to ask the main program to drop everything just because one more character has arrived. A circular FIFO buffer can accommodate both needs.

Latency: Although an interrupt often can get a response faster than if one had to wait till the program got around to polling, there is a lag between the event calling for action and the response. This "interrupt latency" can be troublesome particularly because it may be variable, depending upon what the processor is doing when the interrupt request occurs.

Another typical application for interrupt: real-time clock: This is a less obvious case for interrupt. It would be sad to make the computer continually ask a real-time clock[5] – in the manner of a child asking the driver on a long trip, "Are we there yet?" – "Has a tenth of a second elapsed? Has it? Has it?" Better to let the timer interrupt the computer when a time unit *has* elapsed; the computer then updates its time record and resumes what it was doing. Such an interrupt illustrates also the other rationale for interrupting: we don't want the update to be delayed long or the computer's timekeeping will be slack.

23N.2.2 *How* to interrupt?

We have said that the interrupt resembles a CALL. That means that the response to the interrupt, like the response to the CALL instruction, leads the processor to save on the STACK the address of the next instruction, before going off to execute the program that was invoked (either the CALLed subroutine or the so-called "interrupt service routine," ISR). So resumption of the program that was interrupted can be as automatic as the return from a subroutine.

However, the two cases differ in two important respects: the CALL includes information about where to go in order to execute the CALLed subroutine. The interrupt, requested by pulling one pin, does not explicitly say where the computer should go to find the ISR. As you might expect, there are simple and complicated ways to steer program flow when an interrupt occurs, among different

---

[3] Other details of such a circular buffer, in case you're interested: the ISR would store the current sample and update the buffer's pointers. It should also provide a flag, testable by the Main program, indicating whether the buffer is empty or full.

[4] A FIFO, incidentally – like most students – can never get *ahead*.

[5] A "real time clock" (RTC) usually describes a hardware counter designed to keep track of time. It is a standard peripheral, one that can be read by the computer – and which can be set to interrupt the processor. The one that we propose here is implemented in software instead. The SiLabs controller includes a hardware RTC on-chip.

processors and controllers. The 8051 uses the simplest scheme – and we're happy to report that this simple method seems to be standard among controllers.

CALL and interrupt differ in one other respect, as well: the program designer knows what the computer is doing at the time when the CALL occurs. Not so for an interrupt. Therefore an interrupt response normally must save at least some information that the main program may require: often the 8051 would save the so-called *Program Status Word* (PSW) which stores flags that record the results of recent operations. In addition, the interrupt routine can invoke a dedicated set of scratch registers by switching from the set used by *Main* to another of the four sets available.

> Compare PC104 vectoring, AoE §14.4.7

The simplest interrupt response: autovectoring: When it responds to an interrupt (more on *whether* the processor will respond below), the 8051 saves its return address on the stack and then begins executing code at a dedicated location, one associated with the particular interrupt. These addresses are called *interrupt vectors*. INT0*, for example, causes execution to begin at address 03h; INT1* goes to address 13h. Eight bytes are allocated there – but if the ISR is not tiny enough to fit into eight bytes then the normal action at the dedicated address is just to jump out to a location where there is space for the needed code. That jump, of course, slows the process.[6]

The 8051 was born with two external interrupts: the Dallas version used in the big-board labs has added three more. SiLabs, instead, has done its expansion in the '410 by adding many *internal* interrupt sources.

When each interrupting device can be assigned its own private interrupt pin, things are pleasantly simple. At address 03h one writes the response to INT0*; at address 13h one writes the response to INT1*, and so on. If several sources must share one interrupt-request line, then additional hardware, and some code, is needed to determine "who rang the doorbell."

In Fig. 23N.1 the two cases are contrasted.

> AoE §14.4.8A

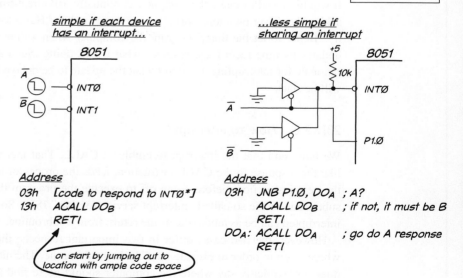

Figure 23N.1 Interrupts are easy if each device can be assigned its own interrupt; less easy when sharing is required.

The code for the shared interrupt includes necessary polling. Therefore its response cannot be quite as quick as the implementation using separate interrupts. Incidentally, if we were aiming for quickest

---

[6] The ARM controller offers a clever variation on this jump-out scheme. Recognizing that this jump takes time, ARM allows a single *jumpless* ISR for use in cases that need the fastest response. This "Fast Interrupt Response" (FIQ) vector is the *last* in the set of vectors; all other vectors require a jump out to the ISR. The FIQ routine simply starts executing, using the free space at the end of the table of jumps.

response, we would use a JUMP rather than a CALL from the interrupt vector (then the RETI – Return from Interrup – would finish the code to which the program had JUMPED).

23N.2.3 Enabling interrupts; edge- or level-response

Enable, when you're ready: After a RESET, by default, all interrupts are disabled. That makes sense because normally one needs to initialize the machine before interrupts can be handled correctly. When you're ready to let the processor accept interrupts, you *enable* them. You do it in two steps:

1. enable the particular interrupt (say, external interrupt zero, INT0*);
2. then, when you've taken care of all the particular interrupt sources you want to permit, enable interrupts *globally*.[7]

Decide whether to make it edge-sensitive: *Edge*-sensitive interrupts offer several advantages over *level*-sensitive.

- Edge-sensing makes it easy to get *one* response for one input event (the pressing of a button, or edge of a square wave).
- The machine keeps a record of an interrupt request that it has not been able to respond to, when the interrupt is edge-sensitive. For example, the machine may not respond to a request while it is busy processing another interrupt; when it finishes that process, the edge-sensed request will have been saved on a flip-flop.
- Response to an edge-sensitive interrupt automatically clears the internal "interrupt flag" that was set by the request. It clears the flag upon servicing the request, reasonably enough. The level-sensitive version requires a line of code to clear the flag: not difficult, but pesky.

23N.2.4 Priority among interrupts

If you use more than one interrupt, the processor sometimes will be faced with two or more *simultaneous* interrupt requests. Which should it honor first? And if the processor is responding to one interrupt request and gets a second, should it honor the latter? At once? After finishing the first response? Every processor provides an orderly way to resolve such problems of contention.

The processor checks for interrupts late in every instruction cycle. If it finds more than one request, it responds to the one with the highest priority; so much you knew already (isn't this, in fact, no more than a tautology? The one with priority gets priority?). But we can give content to this remark by saying more specifically what we mean by *priority*. It comes in two flavors, "natural" and programmed.

Natural priority among interrupts: Natural priority (which has precious little to do with nature!) is an arbitrary assignment of priority among the processor's several interrupt sources. The Dallas controller (DS89C420/30) implements thirteen interrupt sources, six of them external; the SiLabs '410 provides eighteen sources, two of them external.

Dallas gives the *power-fail* interrupt highest priority for obvious reasons. An on-chip supply-voltage monitor circuit generates this interrupt when V_{CC} falls below about 4.4V. Curiously, SiLabs gives the low-voltage interrupt next-to-lowest priority. Each interrupt is assigned a priority level. For example, interrupt INT0* has priority over higher-numbered external interrupt sources.

How is "priority" applied? If, when it checks for interrupts, the processor finds requests from both

---

[7] The Dallas power-fail interrupt, though requiring specific enabling, does not depend on the global interrupt enable, EA.

INT0* and INT1*, it serves INT0*. So much is straightforward. What it does when it has finished serving INT0* depends on whether the INT1* was set up as *edge-sensitive*.

If it was edge-sensitive, then – as we noted in §23N.2.3 – an internal flip-flop has saved the request, and this request will get a response as soon as INT0*'s service routine terminates.[8] If INT1* had not been edge-triggered, the request would not have been honored after the INT0* service unless the INT1* request still were being asserted (likely, but not certain).

Using natural priority (that is, the default setup), no interrupt is allowed to break into the service of another interrupt, despite having higher priority.

Programmed priority among interrupts: One can assign priority among interrupts, departing from the natural order (INT1* could be given priority over INT0*). And such higher priority also allows an interrupting device to break into the service of the lower-ranked interrupt. In such a case, service of the lower priority resumes when the higher is completed. In short, the lower-priority service routine is treated just as if it were a main program with interrupts enabled.

23N.2.5 The difference between RET and RETI

An interrupt service routine is terminated with RETI, not with the garden-variety RET. The difference comes from the need to keep track of the current level of interrupt being serviced – information essential in order to implement priority among interrupts.

RETI does what RET does, and in addition it updates the record of current-interrupt service. For example, RETI at the end of an INT0 response makes it possible for INT1 to get serviced, while natural priority would have blocked service during the INT0 response.

This record might be called the "interrupt mask," since it determines whether another interrupt request will get a response. This restoration is important in a case like the one described in the program of §23N.2.6 where more than one interrupt request can occur.

23N.2.6 A demonstration program

To make these notions less abstract here is a program – useless except to show the fussy details of priority – giving INT1* priority over INT0*. The two interrupts contend: one incrementing the display value, the other decrementing it.

```
; INTPRIOR_D03.a51 show programmed priority for one int over another

; increment display each time interrupted by INT0, decrement for INT1;
            ; INT1 can break into INT0 service routine
       STACKBOT EQU 07Fh   ; put stack at start of scratch indirectly addressable block (80h and up)

       ORG 0h
       LJMP STARTUP
       ORG 0D0h

STARTUP: MOV SP, #STACKBOT
       MOV DPTR, #8000h

; ---NOW ENABLE INTERRUPTS---
       SETB IT0       ; make INT0 Edge-sensitive (p. 22)
       SETB IT1       ; ditto for INT1
       SETB PX1       ; give higher priority to INT1 (SFR "IP0", see user's guide);
```

---

[8] In case you care about the gritty details, program flow returns to the main program at the end of the first ISR; the processor executes a single instruction and then goes off to service INT1*.

```
                        ; name is from RIDE list, "IP"...
        SETB EX0        ; ...and enable INT0
        SETB EX1        ; ...and INT1
        SETB EA         ; Global int enable      (pp.31-32)

        CLR A ;  (for clean startup, as usual)

STUCK: MOVX @DPTR, A ; show display--constant, till interrupt inc's it
       SJMP STUCK ; (responds to falling edge--pseudo-edge sensitive,
             ;   so you must clock several times while high, then low)

; --------------------
; ISR0: This is response to INT ZERO: INCREMENT A
ORG INT0VECTOR ; this is defined in VECTORS3210.INC, included above.
             ; It is address 03h, the address to which micro hops
             ;   in response to interrupt ZERO
ISR0:   NOP        ; pointless--but gives time to interrupt with higher priority
        INC A
        RETI

; ISR1: This is response to INT ONE: DECREMENT A
        ORG INT1VECTOR    ; this is defined in VECTORS3210.INC, included above.
                          ; It is address 13h, the address to which micro hops
                          ;   in response to interrupt ONE
ISR1: NOP
      DEC A
      RETI
END
```

Because this program modifies the "natural" priority between the two competing interrupts, not only would INT1 come ahead of INT0 in case both were requested at the same time, but also INT1 would be allowed to break into the INT0 ISR. This is a departure from the natural scheme that always permits the current response to finish. This two-interrupt demonstration program appears also in the SiLabs interrupt lab §22L.2, differing only in the way the display is handled on the bus-less '410.

23N.3 Interrupt handling in C

Here is C code for a program a little simpler than the one in §23N.2.6. This one uses a single interrupt, INT0*, to cause incrementing of the display.

```
// interrupt_test.c   increments display when interrupted
// even simpler than assembly Inc/Dec demo
#include<C:\MICRO\8051\RAISON\INC\REG320.H>

data volatile char counter;   // this 8-bit variable will be kept on-chip
xdata char *display;          // another byte, pointed to by pointer "display"
void setupEx0 (void);         // initializations: this line warns compiler that this function
                              // is coming-- since it's described only after Main:

void main (void)
{
        setupEx0();
        while (1)

          {
            *display = counter; // let display show current value, inc'd only on interrupt
          }
        }
```

Micro 4: Interrupts; ADC and DAC

```c
/* External interrupt 0 */
    void it_ext0 (void) interrupt 0 using 1    // using register bank 1, for 'context switch'?
                                               // no need, this time, but a nice idea
    {
      counter++; // ISR simply increments COUNTER--the value to be displayed. (ISR does not show it)
    }

    void setupEx0 (void)
      {
        display = 0x8000; // point to display
        *display = 0; // initial clear of display
        counter = 0; // ...and clear count register for startup
        EX0 = 1; // enable ext int 0
        IT0 = 1; // ...make it edge sensitive
        EA = 1;  // global int enable
      }
```

The C program looks pretty complicated compared with assembly language. It includes initializations that look very much like those in the assembly code. The interrupt initializations, such as `EA = 1` are especially close to assembly language, where we wrote `SETB EA`.

The line at the head of the ISR is rather cryptic: `void it_ext0 (void) interrupt 0 using 1`. This says that this function is located at "interrupt 0" vector location, and we added "using 1" in order to say "switch to another set of scratch registers: use Set #1 rather than #0." We do this here not because it is necessary but only to remind ourselves that this change is permitted. We did not do this in the assembly-language examples, but we could have.

The motive for such a partial "context switch" is to save time. Instead of PUSHing multiple registers that the ISR may mess up, we just invoke a whole new set. There was no need to save any registers in this case. All the ISR does is increment one variable. But note that swapping register sets provides only a *partial* context switch. Many registers cannot be swapped: **A** and DPTR, for example, have no alternative versions like those of the eight scratch registers. These would have to be PUSHed in order to get saved, if the routine used them.

Curiously, the C compiler dutifully does save the **A** register even though the ISR does not use it. Here is the ISR produced by the compiler:

```
                ; FUNCTION it_ext0 (BEGIN)
                                            ; SOURCE LINE # 23
0000 C0E0              PUSH    ACC
0002 C0D0              PUSH    PSW
0004 75D008            MOV     PSW,#008H
                                            ; SOURCE LINE # 27
0007 0500    R         INC     counter
0009 D0D0              POP     PSW
000B D0E0              POP     ACC
                                            ; SOURCE LINE # 28
000D 32                RETI
```

The PUSH, loading and POP of PSW is what changes the register bank to set #1. Clearly in this instance, the "time-saving" swap of register sets only *wastes* some time.

23N.4 Interfacing ADC and DAC to the micro

You know that for a computer that uses buses, as the Dallas version does, bringing data into the micro will require an intermediate three-state, and driving a DAC will require an intermediate register. Even

23N.4 Interfacing ADC and DAC to the micro

if you are using the SiLabs '410, with its integrated ADC and DAC, consider the case that uses the buses; it will raise issues that the highly-integrated '410 tends to obscure.

In addition to input and output buffers (one a 3-state, the other a register), let's add to the ADC interface two features that are quite standard:

- a START function, letting the micro determine when an ADC conversion is to begin: the ADC requires just a brief pulse;
- a DONE* signal from the ADC that we are to convert into a *FLAG*. Let's call this flag "NEW_DATA."

The START pulse we can generate easily using the external buses: any line from the I/O decoder will suffice. If we mimic the choice in the lab computer, we can use OUT3*.[9] Because we need only a pulse, not a sustained level, this is the rare case in which we do *not* need the flip-flop or register that ordinarily is required in order to catch an output. There is no output to catch this time; on an OUT operation data does indeed get driven onto the data bus – but in this case that data goes nowhere. It just falls on barren ground.

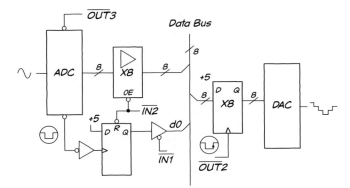

Figure 23N.2 ADC and DAC wiring using external buses.

The code to send a pulse using the external buses is no more than `MOVX @DPTR, A`; to get the same result using a built-in port (say, bit 1 of PORT3) one would need two lines of code: `CLR P3.1, SETB P3.1`.

The flag hardware is familiar to you – the same as that used in the "Ready" key of Lab 22L. In this case, DONE* could as well have driven the flop's SET*, since the DONE* pulse is brief and will be gone by the time the processor tries to reset the flop (whereas a finger on a pushbutton is likely to remain long after the attempted RESET*).

23N.4.1 The IC Version of ADC, DAC Hardware: AD7569

The IC that big-board people will add to the computer in Lab 23L implements all that we just designed – as if Analog Devices had been looking over our shoulder as we drew Fig. 23N.2. Fig. 23N.3 shows a block diagram of the part. And Fig. 23N.4 shows the ADC7569 as it is wired in the big-board lab computer.

We don't use the AD7569 *flag*, INT*, because the converter is fast enough (1.6μs) that there is not much point in asking "Are you ready?" The process of asking takes enough time so that the answer would be, "Yes."[10]

---

[9] In the lab version, we invert OUT3* in order to satisfy a peculiar requirement of the AD7569 ADC: START* must rest in the low state though it responds to a falling edge.

[10] Reassigning the DPTR to the INT* flag port, bringing in the value and checking it takes about 3μs; restoring the pointer pushes the time to about 4μs – a losing exercise for the 1.6μs ADC. (These values assume a clock rate of 11.059MHz, as in our lab computer.)

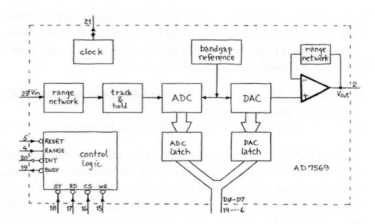

Figure 23N.3 AD7569 block diagram: looks a lot like our homebrew ADC and DAC.

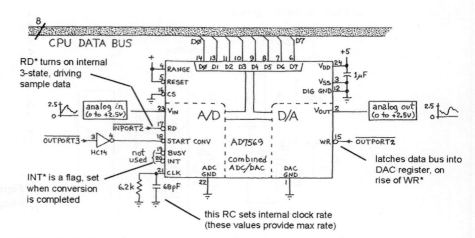

Figure 23N.4 AD7569 as it is wired in the big-board lab circuit.

The name of the signal INT* certainly suggests its use to interrupt the processor. We did not use it that way, for the same reason that we did not *poll* INT*. The delay in interrupt response ("latency," mentioned earlier in §23N.2.1) makes interrupting much slower than simply writing the code so that inherent program delays allow enough time for conversion.[11] At 11MHz, the processor *could* outrun the converter if the program did nothing much between a START and a READ. But it is not hard to arrange things so that instead, the program can be doing something useful during the conversion process.

Code must allow time for conversion: In Lab 23L for example, we allow plenty of time for conversion by reading the ADC *before* starting it on each interrupt *then* starting the converter for next time. So between START and READ lie not only the interrupt overhead (stacking and un-stacking the return address), but also anything else that the main program must do. Below we show the ISR that reads the ADC, writes to DAC and starts the converter for next time. This ISR is used first for the simple IN, OUT program that lets one explore sampling effects.

This program uses the compact "MOVX @Rn..." instructions rather than the usual DPTR I/O. If

---

[11] Interrupt response is one to two μs (@11MHz) depending on what operation the processor is performing when the interrupt occurs. If, as normally is required, one saves registers that the ISR will alter, the ISR response time becomes still longer.

23N.4 Interfacing ADC and DAC to the micro

this is unfamiliar, take a look at Chapter 21S. This use of "MOVX @Rn..." assumes that the pointers have been initialized, as shown in the listing that follows this fragment.

```
        ; VECTOR, FROM INTERRUPT:
        ORG INT0VECTOR

ISR:    MOVX A, @R1     ; read ADC (bad, first pass, OK ever after)
        MOVX @R1,A      ; start ADC for next pass
        MOVX @R0, A     ; ...and send sample to DAC
        RETI
```

The ADC control signals in action: The demo program below is essentially the same, though it does not use *interrupt*. It starts and reads the converter in a tight loop. This loop, whose scope traces are shown in Fig. 23N.5, allows just enough time for a conversion.

```
        MOV P2, #80h    ; init top half of port address (I/O base)
        MOV R0, #02h    ; ...low half: DAC
        MOV R1, #03h    ; ...low half: ADC (both Start and Data: out and in)

TRANSFER: MOVX A,@R1    ; read ADC (bad, first pass)
        MOVX @R1, A     ; start ADC, for next pass
        MOVX @R0, A     ; send sample to DAC
        SJMP TRANSFER
```

BUSY* is a pretty useless signal that means what it sounds like. INT* you recognize as a *flag* almost identical to the flag we designed in our home-brew ADC circuit of Fig. 23N.2. It differs only in that is cleared not by the *assertion* of the data-read signal (named RD* – but not the same as the processor's RD*), but instead by *termination* of that signal, the rising edge.

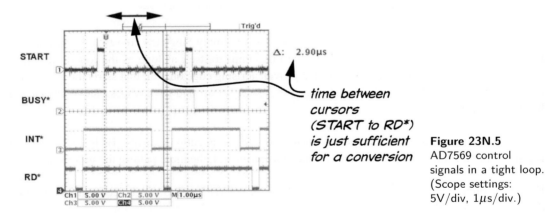

Figure 23N.5 AD7569 control signals in a tight loop. (Scope settings: 5V/div, 1μs/div.)

23N.4.2 ADC and DAC on SiLabs '410

ADC and DAC on the '410 are integrated with the 8051, like all the other peripherals. Their details appear in §23L.2. The option to *left-justify* the converter values allows us to use them as either 8-bit or 12-bit parts, as noted below.

Start and done in the SiLabs version: The SiLabs ADC, like the AD7569, needs to be *started* (by a rising edge on one line of an internal register: bit 4 of AD0CN), and our lab programs test a flag bit to know when a conversion has been completed (the equivalent to the AD7569's INT* flag is another bit of the SiLabs internal register, AD0CN: bit 5). To update the DAC we must provide a rising edge also, much as both the homebrew and integrated DACs required in §§23N.4 and 23N.4.1.

Voltage ranges, ADC and DAC, SiLabs version: ADC and DAC ranges differ slightly from those for the AD7569:

ADC range: 0–2.2V for the '410 versus 0–2.55V for the AD7569;
DAC range: 0–1.3V for the '410 (in lab, we limit max to 1.2V) versus 0–2.55V for the AD7569.

Resolution, SiLabs converters: The SiLabs converters are capable of 12-bit resolution – but in the labs we use them nearly always as 8-bit parts,[12] because noise in our breadboarded setup normally buries the effect of the low four bits.

Getting 8-bit resolution from a 12-bit part: You will see in §23L.2 how easy it is to ignore the low four bits when you choose to. It is easy because we are permitted to choose which way to "justify" the 12-bit quantity: see Fig. 23L.9. If we choose *left* justification then we can use the high eight bits, ignoring the low four bits. Thus we can treat the 12-bit converters as if they were 8-bit parts.[13]

23N.4.3 Code

The code is simple...: The code that gets a sample from the SiLabs ADC is simple: start a conversion, wait till told a sample is ready, take in the sample:

```
GET_SAMPLE:   CLR CNVRT_START ;  low on AD0BUSY (to permit rising edge)
              SETB CNVRT_START ; rising edge on AD0BUSY starts conversion
              JNB CNVRSN_DONE, $; hang here till conversion-done flag
              MOV SAMPLE_HI, ADC0H ; now pick up high byte from ADC and save it
              RET
```

The line of code, JNB CNVRSN_DONE, $, includes an assembly-language convention you may not have seen before: $ means "this line." Thus the line jumps to itself as long as the tested bit is low. GET_SAMPLE includes a pair of commands that provide a rising edge on a bit that starts the ADC, plus a test loop that waits till a sample is ready. The bits, CNVRT_START and CNVRSN_DONE are defined at the head of the program as usual.

The initializations are not so simple: As usual, lots of initialization choices. Table 23N.1 has the register-initialization details. The more complex the peripheral, the more painful the initializations!

But the Configuration Wizard helps a lot: Those fussy ADC and DAC choices can be specified descriptively using the Configuration Wizard. Figs. 23N.6, 23N.7, and 23N.8 show how we have used the wizard to do the simpler task of configuring just the ADC – but even that is not simple.

Port I/O...: First, see Fig. 23N.6, we set the ADC input (P0.1) as *analog*, and do the same for the V_{ref} output at P1.2. As usual, we also enable the *crossbar*; this is essential for any I/O.

---

[12] We do invite you to show off the good 12-bit resolution by reading the DAC output on a DVM, while using a debugger *breakpoint* to stop program execution just after the DAC write. This rather silly way of using the DAC gives the DVM time to average out the noise on the breadboarded DAC output. Doing this, we found we were able to make out the LSB steps of about 0.5mV. Without such heavy averaging, neither DAC nor ADC can achieve 12-bit resolution, given the controller's own electrical noise, even apart from the noisiness of our breadboarded circuitry.

[13] If you are a perfectionist, you may object that we ought to round the high eight bits according to whether the low four bits are above or below their midpoint value. But we are not perfectionists, and do not do this in §23L.2.

23N.5 Some details of the ADC/DAC labs

Table 23N.1 Register initializations.

Register	bit/byte-value	function
ADC0CN	d7	ADC0 enable: 1 ⇒ enable
...	d5	AD0INT: 1 ⇒ sample ready (this is a flag we test)
...	d4	AD0BUSY: rising edge starts conversion
...	d2	AD0LJST: 1 ⇒ Left Justify
...	d1, d0	selects event that starts conversion
...	00b	selects rising edge on AD0BUSY as event that starts conversion
ADC0MX	02h	ADC0 mux assignment ⇒ to P0.2
ADC0TK	d3...d0	tracking mode
...	FBh	dual-mode[a]
ADC0CF	d2, d1	number of conversions per start-convert
...	00b	⇒ one sample/start-convert
P0	d2, d1	both bits high, to set latch to weak pullup
P0MDIN	0F9h	ADC & DAC pins set to analog (P0.2, P0.1)
P0SKIP	46h	skip ADC and DAC, d2, d1
REF0CN	13h	enable Vref, set full-scale to 2.2V

[a] Value borrowed from sample program SiLabs ADC.

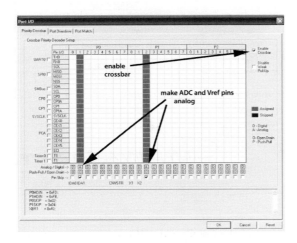

Figure 23N.6 ADC and Vref set to analog mode in wizard's port I/O initialization.

And further ADC configuration: Then we take care of left-justifying, of how the ADC is to be started, and of its V_{ref}: see Fig. 23N.7. Still fussy – but much better than working bit-by-bit from the data sheet.

23N.5 Some details of the ADC/DAC labs

Why here? We set out here details that normally would appear in the lab notes. We put them here so that they can be applied to the ADC and DAC experiments in either set of micro labs: Dallas or SiLabs.

Finally, we set up the voltage reference, defining the converter's full-scale range.

Reconstruction filter for DAC: MAX294 switched-capacitor lowpass: A good lowpass filter on the output can clean up a very-steppy reconstructed waveform, as you have seen in the Chapter 18N demo scope images. The MAX294 is a spectacularly-steep lowpass filter that you have met before, in Lab 12L.

This filter is well-suited to the present task because we can easily adjust its $f_{3\text{dB}}$. The filter's $f_{3\text{dB}}$

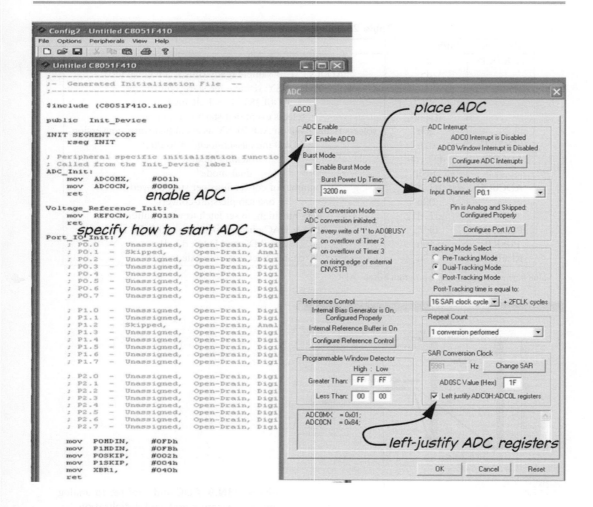

Figure 23N.7 Several ADC choices made with the help of the Configuration Wizard.

is determined by the frequency of the *clock* applied to the filter ("clock"? you ask? Yes, this is a "switched capacitor" filter; you saw the idea illustrated back in Lab 12L.)

The wiring for the MAX294 is shown in Fig. 23N.9. The voltage divider provides a pseudo-ground at the midpoint of the +5V supply, allowing this device to take a unipolar input like the one we are going to feed it (0 to 2.2V [SiLabs] or 0 to 2.55V [Dallas]).

The *input* of the '294 is AC-coupled and biased because the filter's input range does not include ground: input must be about a volt away from the supplies, positive and negative (and here the "negative" supply is ground).[14] The biased input here, centered on 2.5V, does satisfy that requirement.

As you may recall from Lab 12L, the plot of the filter's response, shown in Fig. 23N.9 shows an f_{3dB} of 1kHz, but you can generalize this response to any cutoff frequency in the MAX294's wide range (100Hz to 25kHz).

This good, steep filter allows stingy sampling. Fig. 23N.10 shows the recovery of the original sinusoid from a DAC output that looks pretty sadly sparse.

---

[14] The MAX294 data sheet states this requirement that signals not go too close to the supplies in a rather elliptical way. The data sheet "applications information" on *Input-signal Amplitude Range* sends one to a plot of typical distortion versus signal amplitude.

23N.5 Some details of the ADC/DAC labs

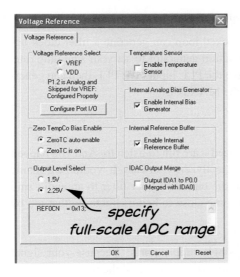

Figure 23N.8 Wizard defines ADC range using voltage reference menus.

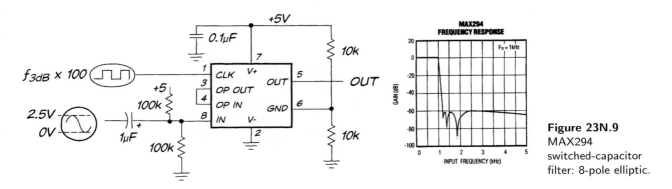

Figure 23N.9 MAX294 switched-capacitor filter: 8-pole elliptic.

The MAX294's additional virtue as filter for this application is the fact that you can easily control its f_{3dB} (more usually called "f_{cutoff}"). If you use a function generator to drive the filter's clock, then you easily can adjust the filter's response.

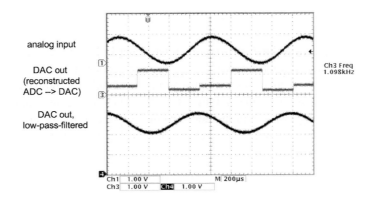

Figure 23N.10 Good reconstruction lowpass filter knocks out sampling artifacts even at stingy sampling rate.

23N.5.1 PLL to Adjust f_{clock} of Reconstruction Filter

As you may recall from §§18L.2.3 and 19L.1.3, a phase-locked loop can generate a filter clock appropriate to your computer's sampling rate. If you use a function generator to set your sampling rate, driving the computer's interrupt, the generator can also drive the PLL input. The PLL then can generate an appropriate multiple of f_{sample} with which to clock the MAX294.

23N.5.2 An audio amp, so you can listen

LM386 amp: The effects of sampling – the artifacts introduced – are fun to listen to. And you happen to be carrying with you a very effective frequency-spectrum analyzer: your ears. It would be a shame not to take advantage of this ability of yours. So let's feed the output of your filter to an audio amp capable of driving an 8Ω speaker. (The MAX294 doesn't come close to being able to do that job by itself.) The LM386 audio amp is designed for just this task.

General truth: we need to block the DC-bias of these single-supply parts: Because your signal source (from the MAX294) is unipolar (centered on 2.5V) and so is the output of the internally-biased LM386, you need *twice* to insert blocking capacitors to knock out this DC bias. Fig. 23N.11 is pinout and typical wiring for the part.

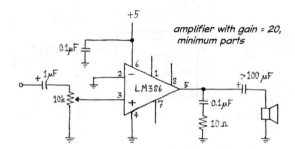

Figure 23N.11 LM386 audio amplifier.

The LM386 likes a small input signal symmetric about ground – strange to tell – even though it uses only a positive supply.[15] So, use a blocking capacitor to feed the 10k "volume" control pot ahead of the '386.

The blocking cap between the output of the '386 and the speaker is very large, because this capacitor forms a high-pass filter with the very-low 8Ω resistance of the speaker. Don't forget to *decouple* the '386 power supplies, using both a ceramic (0.1μF) and a larger tantalum (at least 1μF), close to the +5V supply pin of the IC. The '386 is predisposed to parasitic oscillations.

23N.5.3 Or use the switching amplifier of Lab 12L

If you prefer use the switching amplifier, the LM4667, that you may have met in Lab 12L. If you use that in place of the LM386 you can make louder noises – so it may be prudent to include a potentiometer like that shown in Fig. 23N.11 to attenuate the signal from your DAC. You probably will need therefore *two* successive blocking capacitors: the first (as in Fig. 23N.11) bringing the signal down and centering it on ground; the second passing it to the non-zero bias voltage of the IC at its pin 2. The IC's input resistance is on the order of 90kΩ.[16] Fig. 12L.15 on page 496 shows the wiring diagram.

---

[15] It understands a signal as much as 0.4V below its negative supply, which ordinarily is *ground*.
[16] This is specified as its *differential* input resistance.

You should expect alarming-looking switching noise from this circuit. But this noise should be barely audible.[17]

Lots of oscillators at work: We now need *three* oscillators. The easy way, if your lab is replete with these devices, is to haul three function generators to your bench. If you don't have that luxury, then you can wire up a '555 oscillator to generate the *filter clock* square wave. See Lab 8L for typical '555 circuits. Make this '555 frequency *adjustable* in a range from about 10–500kHz. Use a 5V supply.

To recapitulate, here are the signals we'll need to generate:

Analog input Use the main generator to provide the input sinusoid. Make sure – once again – that the analog levels lie within the range of the ADC: 0–2.55V (Dallas path) or 0–2.2V (SiLabs path). (You can harm the ADC only by going beyond the *supplies* – 0 to +5V.) You'll need to turn on the function generator's *DC-offset*

Sampling rate You can use the function generator on your powered breadboard, TTL output, to interrupt the 8051 (INT0*). If you like, you can watch this on a DVM that includes a frequency meter.

Filter clock Use a third logic-level oscillator to drive the clock of the MAX294 filter. This clock could come from your homebrew '555 oscillator. The oscillator on the powered breadboard is too slow for this task (that oscillator's max frequency is 100kHz, implying max filter f_{cutoff} of only 1kHz).

23N.5.4 What you might listen to

A sinusoid from the function generator is the obvious first candidate to put through your ADC–DAC machine. Such a single frequency signal makes the effects of sampling easiest to understand. (Take a look at Chapter 18S if you want a reminder concerning the expected effects of sampling.)

After you have played with a sine, you may want to hear what sampling does to a more complex waveform such as music. For example, aliasing for such a source produces weird effects, the result of an *inversion* of the frequency spectrum of all the aliased signals.

And some people get a big kick out of hearing sampling applied to their own voices. It can be surprising to discover how low a sampling rate you can apply to speech while still producing an intelligible output. The good MAX294 filter should allow sampling male speech at well under 1kHz, for example. (Women need a somewhat higher sampling rate.) To sample your own speech (or singing?) pull out the microphone amplifier that you built back in Lab 7L. Its output, centered on 1.25V, can go directly into the ADC on either the Dallas or the SiLabs machine.

23N.6 Some suggested lab exercises, playing with ADC and DAC

Is Nyquist right? Your ears should answer this question. If you want to go close to the minimum number of samples/period, you may find it helpful to put a DVM frequency-readout on both f_{in} (the sinusoid) and f_{sample} (the square wave that drives the 8051's interrupt line). If you're bold, *alias* the input signal. Have some fun with this setup. It's not every day you get to confirm these sampling notions, which often seem rather abstract and arcane.

---

[17] On a cleanly laid out printed circuit it would not be audible. On your breadboard it may be.

Modifying waveforms: Here's a chance to write some more of your own code to alter waveforms caught by the ADC that you add in today's lab; the results will go out to your DAC. Technically, you'll be doing "signal processing" – but you'll also be learning why an 8051 is not called a "signal processor"! It's slow at this work, and painful to program. Still, it'll give you some practice in writing small programs – with a gratifying payoff. Altered waveforms on a scope screen may be even more exciting than the blinking LED of §20L.3.

Full-wave rectify: Let a sine input (which must lie within the ADC's range, 0–2.55V [Dallas] or 0–2.2V [SiLabs]) generate a full-wave-rectified output. "Rectify" about the midpoint of the voltage range of ADC and DAC, as in Figure 23N.12.

Figure 23N.12 Full-wave rectifier.

Full-wave rectify code: Dallas version: The sample, brought into the accumulator (here called ACC), is tested by checking its MSB (that bit is low in the bottom half of the range):

```
MACRO ASSEMBLER FULLWAVE_401B
FULLWAVE:   JB ACC.7,AS_IS  ; check msb, to see if we're above midpt
            CPL A           ; ...if not, flip every bit
            NOP     ; Placeholder, to ease writing of half-wave program, coming next
AS_IS:      DEC DPL         ; point to DAC
            MOVX @DPTR,A    ; ...and send sample out to DAC, either way
            ...
```

Note that the midpoint of the ADC & DAC ranges is the hexadecimal value 80h. The MSB flips as the input signal crosses this midpoint.

Full-wave rectify code: SiLabs '410: Here the strategy is the same.

```
; full_wave_silabs.a51  : 8-bit full-wave rectify
; now full-wave rectify
        MOV A, SAMPLE_HI ; get the (byte-) sample
        JB ACC.7, AS_IS ; if above midpoint, leave it alone
        CPL A ; ...but if below midpoint, flip every bit
AS_IS: MOV IDA0H, SAMPLE_HI ; ...and send it to DAC
        ...
```

The SiLabs version bears the usual relation to the Dallas: program loop is simpler, because the DPTR is not involved. But the initializations are much more cumbersome.

AoE §10.1.3C

ADC data formats: offset-binary versus 2's-complement: The code above handles the "rectify" operation in offset-binary format, the format used for SiLabs '410 and for the AD7569 when single-supplied, as in the big-board circuit. You may feel like trying the alternative 2's-complement format – whose difference from offset-binary is startlingly simple. To go between the two formats, one need only flip the MSB.

The AD7569 cleverly changes between the two formats as its *power supplies* indicate to it whether it lives in a bipolar or uni-polar world: if you apply a negative supply (–5V) to the V_{SS} pin (pin 3), ADC and DAC will speak in 2's-complement. Pretty smart, no?

But, if you like, you can work in 2's-comp without bringing in a second supply by tacking in a line of code that changes the data format. In the rectifier, the benefit from using 2's-comp is subtle indeed – perhaps invisible on the scope screen – but possibly interesting, nevertheless.

The offset binary program transforms 07Fh to 80h, and leaves 80h alone; thus two original values map to one output value (two "zeroes"). 2's-comp does not do this: zero would be untouched (as 80h is in the program above); minus one (0FFh) would be transformed not to zero but to plus one (01h). This seems cleaner. If you do change to 2's-comp in software, don't forget to change back to offset-binary before sending the data out to the DAC.

And, speaking of designs that are cleaner or neater, how about pure analog, using four diodes?[18] But we never claimed that signal processing with an 8051 was more than an exercise.

Half-wave rectify: Here's a challenge: write code that substitutes for a resistor and *one* diode. Again, in Fig. 23N.13 we mean to rectify about the midpoint voltage – corresponding to 80h. On the AD7569, these voltages are 1.27 or 1.28V at 10mV per bit; on the SiLabs ADC, midpoint voltage is 1.10V. (If you prefer, you can convert the ADC readings into 2's-comp, and then treat signals below the midpoint as if they were negative values.) Certainly this code can be very similar to the full-wave program.

Figure 23N.13 Half-wave rectification centered about waveform midpoint.

Lowpass filter: It turns out to that a program to simulate a lowpass filter is straightforward,[19] though, once again, the performance of this primitive version compares poorly against that of 50-cents' worth of analog parts. Proper digital signal processing, done usually with dedicated processors or with large arrays of logic, like super-PALs, can produce extremely good filters. Such devices do what we're doing, but evidently in a fancier and smarter way. Take a look at Chapter 23S.

Let the processor average the current sample with the previous average, and output the result. (Give the most recent sample a weight equal to the previous average). Test your routine by feeding it a square wave of low frequency (100Hz or lower). Does the shape look roughly like

$$V_{\text{out}} = V_{\text{in}} \left(1 - e^{-t/RC}\right) ?$$

If you're not sure, does it climb fast at first then slowly? Does the waveform travel in each step just halfway to its destination (about like Zeno's hare)? The answer to all these questions should be yes.

Test the circuit's treatment of a sinewave. Do you find the usual lowpass filter phase shift effects? (You will see a constant delay; this is an artifact of the digital processing caused mostly by the ADC conversion time, and this is different from the phase shift that characterizes the analog lowpass filter.)

If you know the sampling rate, you know the rate at which the averaging of old and new occurs, and the time between "steps." Since we know that at every step the output should go halfway to the destination (from where it was to V_{in}), and we learned back in Lab 8L that the '555 waveform goes halfway in $0.7RC$, can you infer the effective "RC" of this filter, at a given sampling rate? To try the idea, suppose that the averaging steps come at 100 microsecond intervals, and see if you can calculate the "virtual RC."[20]

---

[18] A good equivalent analog rectifier would need an op-amp or two to hide the diode drops.
[19] Well – the idea is straightforward. You may not like the coding labor, with the 8051's limited instruction set. But you enjoy challenges. That's why you're here, right?
[20] Not hard, we think you'll agree. Time for each step is $100\mu s$, so RC is about $140\mu s$.

Micro 4: Interrupts; ADC and DAC

Some thoughts on how you might do this averaging: The average is just the sum divided by two, of course. Forming the sum is not difficult. You can use any of the eight R_n registers, for example, to hold the running sum. The result of the summing of that and the new sample (which will be in the accumulator **A**) will go to the accumulator. So far so straightforward. But what if the result overflows the 8-bit capacity of the accumulator? 80h plus 80h, for example, sum to **0h**, in eight bits. And half of that – which you can find by shifting or rotating right – is still zero! What is to be done?

You must take account of the overflow bit – the *carry*. Find an instruction that uses that when you do your "divide by two" – which should be a shift or rotate, not literally a divide. *Hint*: there really is such an operation available; we're not just teasing you.

Tune the filter? You're not obliged to give *old-average* and *new-sample* equal weights. Shifting can rapidly divide by powers of two; but there's also a *multiply* operation. This can multiply a byte by a byte. The new input sample could be placed in B, and the attenuator value in A; the more significant half of the result goes into B. The attenuator works as follows: FFh means "barely attenuate[21], 0h means "kill it," 080h means "divide by two," and so on. The "old average" should be attenuated by the complementary fraction: if you cut new to 1/4, keep 3/4 of the old. The assembler could figure this out for you (you might write MOV A, #(0FFh - NEW_WEIGHT)); or, as in the program fragment below, you could let the program itself compute the complementary value (here, using "CPL A"). You could write, for example,

```
            NEW_WEIGHT EQU 040h ; weight of new: 080h ==> equal to old average, e.g.; 040h
                               ; gives NEW one third the weight of OLD



            MOV R3, #NEW_WEIGHT

FILTER:     ACALL GETONE ; go pick up a sample (in accumulator)
            MOV B,A ; now set it up to be multiplied by weighting value
            MOV A, R3 ; set Multiplier value for new sample (fraction)
            MUL AB ; shrink new, to adjust its weight rel to old: B holds result
            MOV R5,B ; save diminished new

            MOV B, R7 ; get old average
            MOV A,R3 ; set up weight of OLD as complement of weight of NEW
            CPL A
            MUL AB ; re-weight OLD
            MOV A,R5 ; recall weighted NEW
            ADD A,B ; form sum of weighted new plus weighted old
```

Use of multiply makes the filter more versatile – but also introduces round-off errors. Assuming that you're obliged to attenuate new, then old before adding the two, you'll find (if your program resembles the one we wrote) that heavy attenuation of *new* sample causes the final value to fall short of the large input values, on both positive and negative excursions (error noticeable with attenuation value set to 020h – about one part in eight). Rounding up the MSD of multiplication result (by checking MSB of lower-byte of result) helps. You may find a more ingenious solution.

---

[21] It would be cleaner if FFh meant "Don't attenuate at all." Unfortunately, FFh means multiply by 255/256 – close to unity, but about 0.4% short. (Thanks to Myunghee Kim for pointing out this difference.)

23N.6 Some suggested lab exercises, playing with ADC and DAC

An illustration: tunable lowpass: Figure 23N.14 is a scope image – actually several, superimposed – showing response to a step,[22] with attenuation set to four alternative values. This program uses the multiply instruction and applies to the new sample attenuation values of 1/2, 1/4, 1/8 and 1/16. The exponential shape is most evident for the easiest case, 1/2 (multiplier value of 080h).

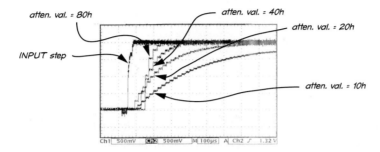

Figure 23N.14 Response of lowpass to step input with multiply used to vary attenuation of new sample relative to running average.

---

[22] The *step* input looks a little odd: that's an effect of *averaging* of the scope image. In fact, the input was a clean edge.

23L Lab Micro 4. Interrupts; ADC and DAC

23L.1 ADC → DAC

Interrupt: Here we will use an interrupt for a second time. Last time you used it to tell the processor when you wanted to increment a value, which was then sent out to the display. This time we'll put it to better use: we will do one conversion – analog to digital, then back out to analog – on each interrupt. Rather than interrupt with a pushbutton (the keypad's WR button) as in Lab 22L, we'll use the breadboard's TTL oscillator to interrupt. Thus we will control the sampling rate.

This program, of course, is not a practical one (if we wanted analog, why bother to go in and out of digital form?). But it does provide a setup that will let you play with the effect of *sampling rate* on the re-converted signal. You will be able to see – and hear – the artifacts introduced by sampling; and then, using a good adjustable low-pass filter, you will be able to remove most of those artifacts.

23L.1.1 Install ADC and DAC

Now we'll add ADC and DAC, so that the micro can operate on waveforms. First, we'll use an extremely simple program that only takes in a sample and then spits it out at once to the DAC. The program is not interesting – but its behavior is. Because the sampling is timed by an INTERRUPT signal applied from a function generator, you can very easily control the sampling rate. The steep output filter, also adjustable, completes the setup for this experiment.

ADC–DAC hardware:

AD7569 converter: The AD7569 is an integrated ADC/DAC pair that is wonderfully easy to use. It is an 8-bit binary-search (or "SAR") device, like the one you built in Lab 18L. It differs in requiring a START signal: instead of *free-running* – starting a new conversion as soon as it finishes the last – it converts only when told to do so. This feature requires an extra line or two of code, but gives the computer control over sampling, allowing it to be sure of picking up data at a time when a new, valid sample is available. Fig. 23L.1 shows what the device includes; Fig. 23L.2 shows the way we wire it into the lab computer. The START signal requires an inverter. You have spare inverter sections available in a 74HC14 you installed earlier.

Some details of the way we apply the AD7569: The wiring we have chosen might puzzle a thoughtful observer.

1. We ignore the two signals that might tell the computer whether a conversion has been completed: INT*, and BUSY*. One might expect that the computer should test one of these levels, taking a sample only when the signal had reached the proper level (so-called "programmed" or "polled" I/O), or that the processor ought to take in a sample only when interrupted by the appropriately

23L.1 ADC → DAC

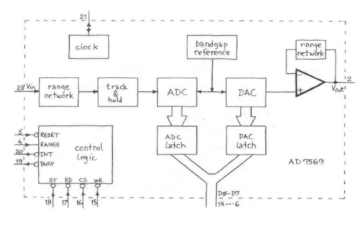

Figure 23L.1 AD7569 block diagram.

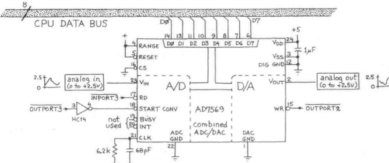

Figure 23L.2 AD7569 8-bit ADC/DAC, as wired to lab microcomputer.

named INT* signal. Such schemes are tidy but inefficient when the converter is as fast as this one: the processor would take about as long to ask whether INT* was low as the ADC takes to complete a conversion. The CPU's interrupt response is slower still. So we take a simpler approach: let the processor take in a sample any time – recalling, as we program, that we must not try to pick up a sample within 1.6μs of the time when we told the converter to start converting. This restriction turns out to be no hardship at all; in fact, you would have to try quite hard to violate this requirement.

2. We hold CS* constantly low. CS* is only a redundant gating signal, AND'd with both RD* and WR*; CS* does nothing by itself.
3. We use OUT3** (the I/O decoder's OUT3*, inverted) to tell the converter to start: a falling edge does the job.

23L.1.2 Program to confirm that the DAC works

Though DAC and ADC are integrated on one chip, it's a good idea to test them separately: the ADC is a bit more complicated than the DAC. Let's start with the DAC. If it works, you'll know that you've done much of the wiring correctly.

Write a program that clears a register, puts it out to the DAC, increments the register and puts it out again; and so on, endlessly. If your DAC is working properly, your program will generate a repeating waveform at the DAC output, ramping from zero to about 2.55V. If the ramp is not smooth, you can make some quick inferences about what's gone wrong. Suppose you see one of the kinky ramps shown in Fig. 23L.3.

You can at least tell what is the highest-order DAC data line that's not driven properly or that is

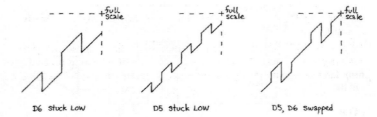

Figure 23L.3 Defective DAC ramps: count the kinks to infer which data line is screwy.

misbehaving on the DAC: one kink (at midpoint of range) implies MSB is bad; two kinks (at 1/4 and 3/4 of range) implies d6 is bad; and so on. This sort of information gives you a clue to what lines you should probe as you troubleshoot. It may not reveal subtler points, such as whether the line is stuck high or low or floating (floats cause the strangest waveforms).

23L.1.3 Program to confirm that the ADC works

Write a program that takes a value from the ADC, shows it on the data displays, then does this all again, endlessly. Feed the ADC with a DC level, from the 1k or 10k potentiometer on the breadboard, as shown in Fig. 23L.4.

Figure 23L.4 DC input for testing ADC.

When you clock the processor at full-speed – 11MHz – you need to think about the ADC's timing requirements: make sure you leave at least 1.6μs between the falling edge of ST* (= OUT3**) and the assertion of RD* (the ADC doesn't like to be asked for data before it is ready). Arrange things so that you tell the converter to Start at a time when some necessary operations lie between START and READ. If you can do that, you can avoid the ugliness of killing time with do-nothing instructions (NOPs).[1]

ADC–DAC: quick convert and re-convert: Version 1, using @R_n to address ADC and DAC: The program below takes in and puts out a sample, each time it is interrupted. The "Main" program does exactly nothing. You may find this program offensive – it makes such degrading use of your clever computer. It is, indeed, equivalent to an ADC and DAC, linked.

Note that you must *interrupt* at the sampling rate: use a second function generator – perhaps the logic output from your powered breadboard's oscillator – to set this sampling rate. And *note* that you must remove the line that in Lab 22L was used to drive INT0* from the keypad's KWR* line; otherwise, you'd be asking the new connection – function generator – to fight the old.

Until now we have relied almost entirely on DPTR to define addresses for off-chip memory and I/O accesses. Another method is available however and it is illustrated below. This version is slightly more compact than the DPTR version which is shown after this listing.

This @R_n addressing *mode* uses the combination of Port 2 (for high byte of address: constant at 80h) and $R0$ and $R1$ (for the low byte of address). You'll notice that this code requires initializations different from those you are accustomed to seeing in programs that use DPTR.

---

[1] The easy trick here is to put the START operation just *after* the READ so that ADC can be doing its conversion during the time the program spends looping back to the next READ.

23L.1 ADC → DAC

Incidentally, the 8051 provides *four sets of scratch registers* (four "banks"): R0..R7. It sometimes would be convenient to switch to a new set within an interrupt response, as in this program. If we did that we would leave registers $R0$ and $R1$ free for other use in the main program.

We have not used this technique here because there is no need for such double use of the $R0$ and $R1$ registers. But you may run into other cases where you will want to swap register banks. One does this by writing to a 2-bit field in the PSW (program status word).

```
MACRO ASSEMBLER ADCDACINT_RN_504                       05/05/04 16:06:07 PAGE     1


   LOC     OBJ      LINE    SOURCE
                       1    ; ADCDACINT_Rn_504.a51 XFER from ADC to DAC on interrupt,
                       2    ; using P2 and Rn   5/04
                       3    ; read ADC, write to DAC, start ADC each time interrupted
                       4
                       5         $NOSYMBOLS
                       6         $INCLUDE (C:\MICRO\8051\RAISON\INC\REG320.INC)
                     172         $INCLUDE (C:\MICRO\8051\RAISON\INC\VECTORS320.INC)
   0080              196         STACKBOT EQU 80H  ; not necessary, but a good place for it
                     197
   0000              198         ORG 0H
   0000   01E0       199         AJMP STARTUP
                     200
   00E0              201         ORG 0E0H
                     202
   00E0   75817F     203    STARTUP:  MOV SP , # 128 - 1
   00E3   11ED       204         ACALL    INITDACPTRS    ; GO INITIALIZE ptrs for ADC & DAC
                     205
                     206    ;   ---NOW ENABLE INTERRUPTS ---
   00E5   D288       207         SETB TCON.0 ; make INT0 Edge-sensitive
   00E7   D2A8       208         SETB IE.0 ; enable INT0
   00E9   D2AF       209         SETB IE.7 ; Global int enable
                     210
   00EB   80FE       211    STUCK:   SJMP STUCK ; AWAIT INTERRUPTS
                     212
                     213
                     214    ;** INITS ***************
                     215
   00ED   75A080     216    INITDACPTRS:  MOV P2,#80H ; set up top half of address as I/O base
   00F0   7803       217         MOV R0,#3H ; set up 8-bit index register to ADC
   00F2   7902       218         MOV R1,#2H  ; set up another register to point to DAC
   00F4   F2         219         MOVX @R0,A ; For first pass: send anything -- just a pulse to START ADC
   00F5   22         220         RET
                     221
                     222    ;** ISR: ADC-->DAC- *****
                     223
                     224    ; VECTOR, FROM INTERRUPT:
   0003              225         ORG INT0VECTOR
                     226
   0003   E2         227    ISR: MOVX A, @R0 ; read ADC (not quite current timing, first pass;
                                                 ; OK ever after)
   0004   F2         228         MOVX @R0,A ; start ADC for next pass
   0005   F3         229         MOVX @R1,A ; ...and send sample to DAC
   0006   32         230         RETI
                                 END
```

ADC–DAC: quick convert and re-convert: Version 2, using DPTR to address ADC and DAC: Here is the same program written to use DPTR. We have placed it in the location used by the alternative form above – so if you enter one version, then the other, the second will over-write the first. We doubt however that you're likely to want to type the code in twice! Ignore this version if you like.

```
LOC    OBJ       LINE   SOURCE
                 1      ; ADCDACINT_504.a51 XFER from ADC to DAC on interrupt
                 2      ;      4/01: uses amended port assignments, and just one
                 3      ;         data pointer: 5/04 place all ISR at vector location
                 4
                 5      ; should read ADC, write to DAC, start ADC each time interrupted
                 6
                 7          $NOSYMBOLS
                 8          $INCLUDE (C:\MICRO\8051\RAISON\INC\REG320.INC)
                 174        $INCLUDE (C:\MICRO\8051\RAISON\INC\VECTORS320.INC)
0080             198        STACKBOT EQU 80H ; not necessary, but a good place for it
                 199
0000             200        ORG 0H
0000   01E0      201        AJMP STARTUP
                 202
00E0             203        ORG 0E0H
                 204
00E0   75817F    205    STARTUP:   MOV SP , # 128 - 1
00E3   11ED      206        ACALL    INITDACPTRS        ;GO INITIALIZE ptrs for ADC & DAC
                 207
                 208    ; ---NOW ENABLE INTERRUPTS ---
                 209
00E5   D288      210        SETB TCON.0 ; Here's my first attempt to use a bit
                 211        ; operation: make INT0 Edge-sensitive
00E7   D2A8      212        SETB IE.0 ; ...and enable INT0
00E9   D2AF      213        SETB IE.7 ; Global int enable
                 214
00EB   80FE      215    STUCK:  SJMP STUCK ; AWAIT INTERRUPTS
                 216
                 217
                 218    ;** INITS ***************
                 219
00ED   908003    220    INITDACPTRS:  MOV DPTR, #8003H  ; Here init ADC data pointer,
00F0   F0        221        MOVX @DPTR, A ; For first pass: send anything -- just a pulse to START ADC
00F1   22        222        RET
                 223
                 224    ;** ISR: ADC-->DAC *****
                 225    ; VECTOR, FROM INTERRUPT:
0003             226        ORG INT0VECTOR
                 227
0003   E0        228    MOVX A, @DPTR ; Read ADC (port 3)
0004   F0        229    MOVX @DPTR, A; send anything -- just a pulse to START of ADC, for next pass
0005   1582      230        DEC DPL ; now point to DAC (port 2) (this DEC operates on low-byte only)
                 231
0007   F0        232        MOVX @DPTR, A ; ...send the sample to DAC
0008   0582      233        INC DPL ; Revert to ADC data pointer, for next pass
000A   32        234        RETI
                 235
                 236    END
```

Confirm that either program works: feed a sinusoid to the ADC and watch the sinusoid the DAC puts out. Make sure the input sine lies within the voltage range of the ADC: 0–2.55V. The *reproduction* of a sinusoid will be *steppy*, unless you are taking a great many samples per period. We'll now add a filter to minimize that steppiness.

23L.1.4 MAX294 switched-capacitor fFilter

A good lowpass filter on the output can clean up a very steppy reconstructed waveform, as you know.

We set forth the wiring of the MAX294 lowpass filter in §23N.5 (Fig. 23N.9) because the circuit applies equally to the SiLabs and Dallas versions of today's lab.

An audio amp so you can listen: §§23N.5.2 and 23N.5.3 also show wiring for an audio amplifier that will let you hear the effects of sampling. They also suggest ways that you might modify waveforms taken from the ADC, feeding the results to the DAC: we propose rectifiers, half- and full-wave, and a lowpass filter.

23L.1.5 Modifying waveforms

Here's a chance to write some more of your own code. We have listed in `FULLWAVE_401b.a51` a full-wave rectifier program; get it from the book's website. We leave to you the half-wave and the filter. Hints concerning these little "signal processing" programs appear in §23N.6.

Full-wave rectify: Let a sine input (which must lie within the ADC's range, 0–2.55V) generate a full-wave-rectified output. "Rectify" about the midpoint of the voltage range of ADC and DAC, as in Fig. 23N.12. Note that the midpoint of the ADC and DAC ranges is the hexadecimal value 80h.

Half-wave rectify: Certainly this code can be very similar to the full-wave program. We leave this to you.

Low-pass filter: Chapter 23N, again, describes how you might program such a filter. Test your routine by feeding it a square wave of low frequency (100Hz or lower) and watching the response to an input step. Does the output shape look roughly like

$$V_{\text{out}} = V_{\text{in}} \left(1 - e^{t/RC} \right) ?$$

Tunable low-pass for the ambitious: As we said in Chapter 23N, if you have some extra time you can use the *multiply* operation to tune the response of your filter. You may enjoy trying this – or you may shrug and say, "There's a job for a DSP."

23L.2 SiLabs 4: Interrupt; DAC and ADC

Port pin use in this lab: Two pins serve as *analog* lines in and out of the converters: an input to an ADC, an output from a DAC. One – the ADC at P0.0 – replaces the LED that has occupied this pin in §§20L.3 and 21L.2. Disconnect that LED. Of the two interrupts shown in Fig. 23L.5, one will reappear (INT0*, used again in §25L.2); one will not (INT1*).

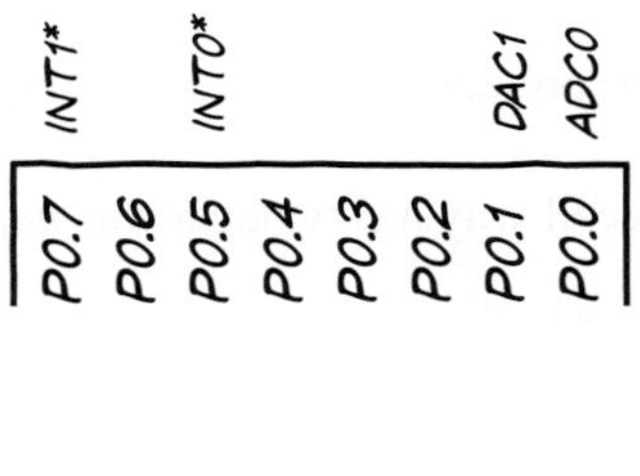

Figure 23L.5 Port 0 pin use in this lab.

23L.2.1 Interrupts

As suggested in Chapter 23N, it takes some effort to devise a program that surely benefits by use of interrupts. Many of our lab programs that use interrupts will do the strange trick of letting the interrupt set a software *flag* that the main program *polls*. This technique seems to vitiate the value of interrupt, but does not entirely do so: apart from the value of being able to break into a running program, *interrupt* also can offer the utility of an on-chip flip-flop that can record the event of an interrupt request. The presence of this flop then permits two useful results.

- The flop amounts to a *flag* that will stay pending until the interrupt request is honored. This will seem unnecessary if you assume that an interrupt always gets a quick response; but some circumstances can prevent such an immediate response. Two are quite common:
  - interrupts may have been temporarily disabled, within the main program. This is the case in the *storage scope* exercise of Lab 24L.2;
  - or (a more common case) the processor may be busy responding to another interrupt, and may be set up to finish that routine before responding to another request. This is always the case when interrupts are running with their so-called "natural" priority. It is also the case when a lower-priority interrupt request occurs during response to a higher-priority request (these cases may or may not overlap with the "natural priority" cases).
- The edge-triggering of the internal flop makes it easy to implement a do-it-once response to the transition of a request line, whereas ordinary *polling* would require extra code to achieve something like edge-sensing. We will take advantage of interrupt's *edge* response in all of the interrupt programs of this lab.

First interrupt demonstration: increment or decrement of display: If you want to justify this little program as more than just the simplest demonstration we could devise, then imagine that our goal is to keep track of the number of passengers on a platform with limited capacity. (Maybe it's the glass platform that is cantilevered over the Grand Canyon.) One turnstile produces a high-low transition when a person enters the platform; a second turnstile does the same when a person leaves the platform. The display shows the number on the platform (and, for simplicity, we will permit our display to show a *negative* number on the platform![2])

Interrupt 0 increments the value; Interrupt 1 decrements the value, and the main program simply loops, displaying the current number. We'll note the simple hardware, then the program.

Hardware: two pushbuttons, more-or-less debounced: We have learned that bounce on an edge-triggered input can cause mischief: it will look like multiple requests. You might be inclined to dismiss bounce in the present application for a mistaken reason. As our hypothetical confused designer said in Chapter 22W, "the bounce doesn't last long – just a few milliseconds; so don't worry about it." This was thoroughly wrong when applied to a controller that can respond within microseconds. We do need to debounce.

But it turns out that we can get away with a pseudo-debouncer that would not suffice if applied to a true edge-triggered input, like the clock of a flip-flop or counter. The simple *RC* shown in Fig. 23L.6 is a good-enough debouncer for the present case.

The 10k, $0.1\mu F$ *RC* provides the slowdown; the 4.7k is included to protect the '410 by limiting current from the charged capacitor on power-down.[3]

Why does a simple *RC* slowdown suffice? We can see why if we recall why a slowdown was *not*

---

[2] This won't happen often in life: only, perhaps, when someone parachutes in and then walks off.
[3] We borrowed this detail from the circuit of the SiLabs '410 development kit.

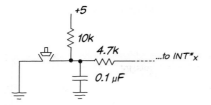

Figure 23L.6 Pseudo debouncer to feed interrupt request pins: just an *RC* slowdown circuit.

sufficient in the case of a true edge-triggered input. A slow edge caused trouble there because the edge-response was fast enough to cause indecision as the slow edge passed through the threshold region. A first response to the rising edge caused the flop output to switch; that event caused a power-supply disturbance that, in turn, caused the input stage to change its mind. And so on – reproducing some of the neurotic indecision that we saw in spectacular form when we teased a '311 comparator with a slow edge, back in Lab 8L (see Fig. 8L.3).

The controller's interrupt inputs, though described as *edge-sensitive*, are not truly so. Instead, the controller samples those inputs on successive cycles of the internal clock. The response to the interrupt request is not so quick or direct as to cause a power-supply disturbance and consequent indecision about whether the input has crossed the input threshold. So a simple *RC* slowdown circuit that eliminates large signal swings during bounce is sufficient here. The edge-triggered interrupt does not get confused by a slow edge, as a true edge-triggered input would.

Wire such a pseudo-debounced pushbutton to each of the two Interrupt request pins: INT0* and INT1*.

Code: INC or DEC within the ISRs: The program is odd; the MAIN loop simply displays the value of the A register, endlessly:

```
STUCK:  MOV DISPLAY, A ; show display--constant, till interrupt inc's it
        SJMP STUCK
```

All the action happens in the two tiny ISRs: one increments the value of the **A** register, the other decrements that value.

AoE (§14.3.8) warns against trying to do too much within an ISR. We think these ISRs pass that test handsomely: each one consists of one line of code, and then the RETI that returns to the main program from the ISR.

Here, as usual, are the register initializations. This time they are quite fussy. We won't reiterate the register loadings that appear in the "USUAL_SETUP" routine. As usual, feel free to ignore these bit-by-bit details. Instead, rely on the Configuration Wizard as we do.

Register	bit/byte-value	function
IT01CF	d7 (= IN1PL)	0 ⇒ INT1 active low
...	d6...d4 (= IN1SLx)	111b ⇒ assign INT1* to P0.7
...	d3 (= IN0PL)	0 ⇒ INT0 active low
...	d2...d0 (= IN0SLx)	101b ⇒ assign INT0 to P0.5
TCON	d2, d0 (= IT1, IT0)	11b ⇒ both interrupts edge-triggered
IE	d7, d2, d0 (= EA, EX1, EX0)	1 ⇒ enable interrupts: global, external 1, external 0
IP	d2 (= PX1)	1 ⇒ INT1 to high priority level

The Configuration Wizard can set up interrupts: The wizard eases this fussy work a great deal. The screens help to remind a user of the choices that must be made. Fig. 23L.7 shows the principal selections, including assignment of interrupts to particular pins. (Another menu screen, not shown, permits altering the "natural priority" among interrupt sources). We placed INT0* at P0.5, INT1* at P0.7. We did not alter natural priority.

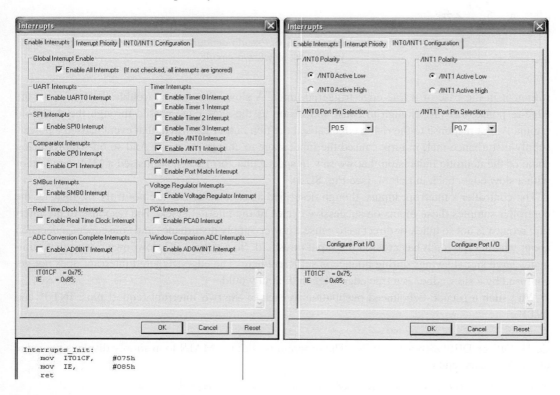

Figure 23L.7 Configuration Wizard helps set up interrupts too.

Strangely, the wizard seems to have omitted the issue of edge versus level sensing on the interrupts. *Edge-triggered* interrupts are useful. To get edge-triggered behavior, we write (as you can see in the full program listing on the book's website),

```
        SETB IT0    ; make INT0 Edge-sensitive (IT0 = TCON.0)
        SETB IT1    ; ditto for INT1
```

...but this time we'll configure by hand: We're a little chagrined to say that after showing the wonderful labor-saving wizard once again, in this case we prefer to set up interrupts by hand. We do this partly because the wizard forgot one option (edge-triggering), and more generally because we think it's easier to follow the code if it is done and explained (in a comment within the listing), one action at a time. But this is only a pedagogical point: we expect you will use the Wizard whenever you can; you can get the full program, `int_inc_dec.a51`, from the book's website.

Run the program:

Slow motion checkout: single-step: First try the program in single-step or multiple-step. You should see the program loop in the main "STUCK" loop until you press one of the interrupt pushbuttons. When you do press that button, hold it down for at least a couple of cycles of the multiple-stepping: a brief pulse on the interrupt will have no effect because these are not true edge-triggers, as we noted on page 932.

You should see execution move to one or the other of the ISRs in response to a press of a pushbutton. If you keep the button down, you should not see any further response to this interrupt request: in the important sense, these interrupts *are* edge-triggered, even though they do not respond to a brief pulse.

If you press the INT0* pushbutton while single-stepping through ISR1, does INT0* break into the ISR? It should not, even though INT0* does have priority over INT1*. This sort of priority – known by the strange name "natural" – determines only which interrupt request wins when both requests come at the same time.

Full speed: Run it at full-speed and try the two pushbuttons. As in single-step, one button-press should produce a single increment or decrement of the display value, thanks to the edge-triggered behavior of these interrupts.

Test the pseudo-debouncer: If you wonder whether the pseudo-debouncing works, try removing the slow-down capacitor from one of the switches. You should find, when the program is running at full speed, that instead of incrementing or decrementing, a button-press sometimes changes the display by several counts.

Try altering the natural priority of interrupts: Within the interrupt–initialization section appears one line that we have commented-out: SETB PX1. This line would have two effects:

- it would reverse the normal priority rank between interrupts 0 and 1;
- it would allow INT1 to break into an INT0 service routine, whereas the normal, natural priority scheme lets even a lower priority routine finish before the controller will respond to the higher-priority interrupt, as we mentioned above.

If you feel like watching these effects in action, remove the semicolon that comments out SETB PX1 and watch as you single-step (or multiple-step) the program. Does INT1* now break into the ISR for INT0* (let's call that ISR0)? If an INT0* request comes during ISR1, what happens after the return to Main from ISR1?[4]

23L.2.2 DAC

It would be nice to set the ADC and DAC ranges equal. We cannot. ADC input range is 0–2.2V (the value of $V_{\text{reference}}$); DAC "output compliance" range is smaller: 0–1.3V. We will set the DAC output range slightly below that limit: 0–1.2V.[5]

The DAC output is a *current*, not a voltage. The DAC *sources* this current, and the easiest scheme for converting this current to a voltage is just to pass it through a resistor to ground: see Fig. 23L.8. For a larger output swing we would need to add an op-amp.

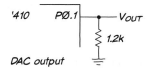

Figure 23L.8 DAC output is a current; R needed to convert it to a voltage.

We set the DAC full-scale current at 1mA (our old favorite) and let it flow through a 1.2k resistor to ground.

---

[4] Because we have made the interrupts edge-sensitive, the INT0* request will have been saved on a flip-flop and will be honored when the response to INT1* concludes.

[5] DAC "output voltage compliance" worst-case limit is $V_{\text{DD}} - 1.2\text{V} = 1.3\text{V}$ (C8051F410 datasheet, "IDAC Electrical Characteristics," Table 6.1, p. 75).

Justification: a choice, left- or right-: Like many DACs and ADCs, the DAC in Fig. 23L.9 allows us to choose whether to use *left-* or *right-* justification of its 12-bit result:

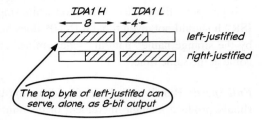

Figure 23L.9 DAC input, like ADC result, can be either left- or right-justified.

We have chosen to *left-justify* because this arrangement permits us to use the eight MSBs by themselves, treating DAC (and later, ADC) as if it were an 8-bit part. Ignoring the low four bits does not corrupt the high eight; it simply reduces the converters' resolution. *Left-justifying* the DAC input value makes it easy, in the exercise that follows (§23L.2.7), to change from 12-bit to 8-bit output.

23L.2.3 12-bit resolution? Yes and No

As we have said in Chapter 23N, DAC and ADC offer more resolution than we can handle in our breadboarded setup: noise normally buries the effect of the lowest two or three bits of the available twelve. But just to show off the 12-bit resolution, let's run the DAC in slow motion, watching its output with a DVM. The DVM's averaging effect will make even the LSB steps visible.

As usual, one is obliged to make many initialization choices in order to use this peripheral. We had to choose even what event should be used to update the DAC. We chose a *write* to the DAC register for this updating event; it seemed the most straightforward scheme. We rejected an update on any of four timer overflows and an update on an edge of an input line named CNVSTR.

23L.2.4 Setting up the DAC using the Configuration Wizard

Choosing the DAC as peripheral, and then DAC1 which we chose because we like its (rigid) assignment to P0.1 (leaving P0.0 free for other use), we get a menu screen, as usual: see Fig. 23L.10.

Choose

- enable DAC1,
- left-justify,
- 1mA output current[6]
- update on write to high byte,
- using the sub-menu, set the crossbar to "skip" P0.1, leaving it free for the DAC,
- . . . and set it as an analog pin.

Install a 1.2k resistor to ground. This will convert the 1mA (max) DAC0 current output to a voltage.

23L.2.5 DAC1 initializations from Configuration Wizard

The result will be a set of initialization commands, which we ask you to append to the incomplete program listed in §23L.2.6:

---

[6] . . . because of our longstanding love affair with the value ONE of course. It makes arithmetic so easy!

23L.2 SiLabs 4: Interrupt; DAC and ADC

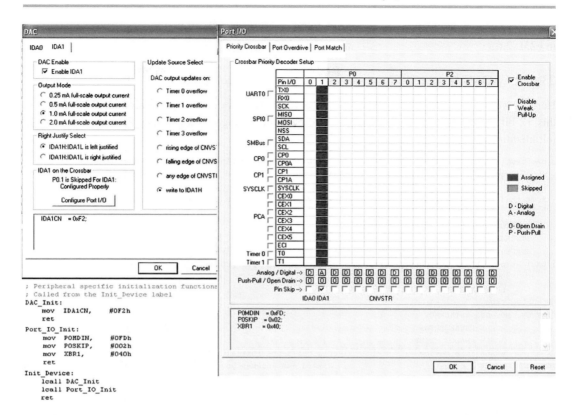

Figure 23L.10 Configuration Wizard sets up DAC; port "crossbar" configures DAC output pin as analog.

```
DAC_Init:
    mov  IDA1CN,   #0F2h
    ret

Port_IO_Init:
    ; P0.0 - Unassigned, Open-Drain, Digital
    ; P0.1 - Skipped,    Open-Drain, Analog
    ; P0.2 - Unassigned, Open-Drain, Digital
    ; P0.3 - Unassigned, Open-Drain, Digital
    ; P0.4 - Unassigned, Open-Drain, Digital
    ; P0.5 - Unassigned, Open-Drain, Digital
    ; P0.6 - Unassigned, Open-Drain, Digital
    ; P0.7 - Unassigned, Open-Drain, Digital

    mov  P0MDIN,   #0FDh
    mov  P0SKIP,   #002h
    mov  XBR1,     #040h
    ret

; Initialization function for device,
; Call Init_Device from your main program
Init_Device:
    lcall DAC_Init
    lcall Port_IO_Init
    ret
end
```

Append these initialization commands to the incomplete program called `dac_test_12_june15.a51` in §23L.2.6.

23L.2.6 DAC test program: a beginning for you to complete

Here we've written a loop that adds one to the 12-bit DAC value each time around. (The details of this 12-bit increment are spelled out after the program listing.) You can use the wizard to add the initializations.

```
; dac_test_12_june15.a51    12-bit resolution: apparent only in slow-motion, when watching with DVM
; changed to DAC1, and pin assignment noted in comment

; NOTE that this program is incomplete: it awaits your setup of DAC1, probably using Wizard

        $NOSYMBOLS ; keeps listing short
        $INCLUDE (C:\MICRO\8051\RAISON\INC\c8051f410.inc)
        STACKBOT EQU 080h   ; put stack at start of scratch indirectly-addressable block (80h and up)
; DAC1 at port P0.1, current converted to voltage by a 1k R to ground
        ORG 0h
        LJMP STARTUP
    ORG 080h
STARTUP: MOV SP, #STACKBOT-1
        ACALL USUAL_SETUP
        ACALL ANALOG_SETUP
        ACALL Init_Device           ; this calls the initializations generated by Configuration
                                    ;  Wizard and assumes you've pasted these in
; here's 12-bit ramp
        CLR A
        CLR C ; clear carry bit
        MOV IDA1L, A ; first-pass, to see voltage at zero, on DAC1
        MOV IDA1H, A

RAMP_12: MOV A, IDA1L ; recall low nybble
        ADD A, #10h ; increment low nybble, left-justified (updating carry)
        MOV IDA1L, A ; update low byte of DAC
        MOV A, IDA1H ; recall high byte
        ADDC A, #0 ; use carry out of low byte, if any, to update high byte
        MOV IDA1H, A
        SJMP RAMP_12

;------- INITIALIZATIONS
USUAL_SETUP: ANL   PCA0MD, #NOT(040h)     ; Disable the WDT, Clear Watchdog Enable bit
; Configure the Oscillator
        ORL   OSCICN, #04h      ; sysclk = 24.5 Mhz / 8
; Enable the Port I/O Crossbar
        MOV   XBR1, #40h        ; Enable Crossbar
        RET

ANALOG_SETUP: SETB P0.0 ; make sure latch is high
        RET

; NEXT SECTION IS INCOMPLETE: (here append the commands prescribed by the Configuration Wizard)-------
; You don't need to use the next four lines; the Wizard should generate these lines, and others as needed
;PORT_SETUP:
;; set up Vref to 2.2V (necessary for ADC, probably not for DAC)
;       MOV P0MDIN,    ; set up DAC1 pin for analog
;       MOV P0SKIP,    ; tell crossbar to skip DAC1 bit (P0.1)
;       MOV IDA1CN,    ; enable DAC1; update on write to high byte; left-justified; 1mA full-scale
;       RET
END
```

23L.2 SiLabs 4: Interrupt; DAC and ADC

Some details of the "12-bit increment": The 12-bit ramp loop extends its increment beyond one byte using the trick we saw in §21N.6: `ADDC A, #0`, incrementing if the CARRY bit is set. The *increment* of the low nybble may look odd, at first.

A preliminary point: we must not use INC A for any greater-than-byte operation because INC *affects no flags*. Thus INC could not be used for this 16-bit operation, which uses the Carry flag in its second step.[7]

A second point: given that we must use ADD rather than INC, why add 10h rather than 01h for an increment? Because the lowest nybble (that is, the lowest four bits) of the 12-bit value lives in the *high* nybble of the register, IDA1L. Adding 10h may look funny, but it does indeed simply increment the lowest nybble of the 12-bit value.

Trying the 12-bit ramp: Note that this program includes no delay. So don't expect to see anything coherent if you run it at full-speed. There are two ways to look for the tiny 12-bit LSB increments:

- single-step, to allow time for the DVM to settle and average out noise;
- plant a breakpoint somewhere in the loop (see details below).

Either method should show the increments of a single LSB.

Voltage value of an LSB: What LSB step size do you *observe*, for this 12-bit DAC? Given the full-scale value of 1.2V, what step size would you *expect*?[8]

Set up a watch window... You can watch the IDA1 value in either of two ways. The more laborious method, giving you better control, is to set up a watch window for particular registers (this lets you see only what you have chosen).

Watch window for particular registers: In Fig. 23L.11 we've begun the process of displaying as separate items the 2-byte input registers of the IDAC1. In the figure, we have selected the low byte. Fig. 23L.12 shows the results of setting up a *watch* of both bytes, IDA1H and IDA1L.

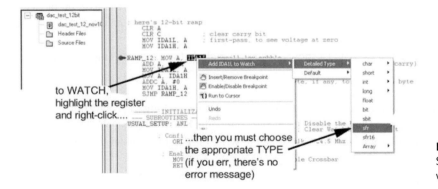

Figure 23L.11 Setting up a watch window.

---

[7] This kind of detail is buried in the description of the INC operation in the Philips "80C51 Programmer's Guide and Instruction Set," http://www.keil.com/dd/docs/datashts/philips/p51_pg.pdf.

[8] $1.2V/2^{12}$.

940   Lab Micro 4. Interrupts; ADC and DAC

...Add a breakpoint: A breakpoint – which halts execution when program flow reaches it – allows us to see the alteration of the IDAC1 input after each pass around the loop. Each time you click the green "Go" button, you can see the IDAOL value increment (in its peculiar way: it will grow by 10h, but that is a 1-bit increment for the lowest nybble of the DAC input register). In Fig. 23L.12 we placed both input registers in the Watch window.

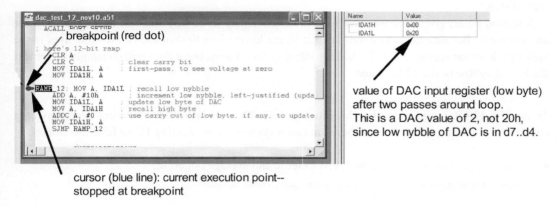

Figure 23L.12 Breakpoint can make it easy to watch effect of increments to DAC input.

We could display the DAC input registers by choosing View/DebugWindows/SFR's/IDAC instead of using a watch window.

Watch the DVM and at the same time you can note the size of the LSB step. Fig. 23L.12 shows a count of two going to the DAC (whose LSB lives at d4 of the IDA1L register). At this count we saw $V_{out} \approx 0.9\text{mV}$.

Incidentally, you need to ground the DVM as close as you can to the '410's ground: a few inches of breadboard ground line can drop several LSBs in voltage (we saw a drop of about 3mV, along the six-inch length of the ground line on the breadboard strip).

23L.2.7   8-bit resolution

Eight-bit resolution is more useful in our breadboarded circuit. A scope can display the individual steps of the DAC's 8-bit ramp, and we will from this point use both DAC and ADC as 8-bit devices. This we can do by ignoring the low four bits, using only the IDA1H register of the DAC and the equivalent high-byte register of the ADC.

At 8-bits, the scope image in Fig. 23L.13 shows, when we included some averaging, the LSB steps of about 5mV apiece.

Code: the RAMP loop, a task for you: This program is very simple, lacking the contortions required by the 12-bit increment. The ramp loop is as brief as the comments below suggest. We leave the code to you. Note that, for 8-bit control of the DAC, we need to write only to the register that holds the top eight of its 12 bits. This register is IDA1H.

Let's use the accumulator as scratch register where we can do the incrementing.

```
EIGHT_BIT:                   ; transfer current "ramp" value to DAC1
                             ; increment ramp value
                             ; waste a little time
                             ; keep doing this, forever
```

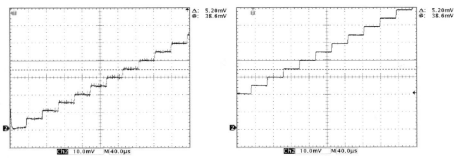

ramp steps, cleaned up by averaging... ...and still more averaging

Figure 23L.13 DAC ramp at 8-bit, averaged: LSB steps visible. (Right-hand figure's vertical shift is artifact of a changed trigger voltage.)

23L.2.8 Code: the full program – minus the RAMP, which we leave to you

The full program includes a delay so that we can run it at full-speed and watch the ramp on a scope. As usual, the initialization chores dwarf the code loop in fussiness.

```
; dac1_8bit_wizard_MT.a51    8-bit resolution

  $NOSYMBOLS      ; keeps listing short
$INCLUDE (C:\MICRO\8051\RAISON\INC\c8051f410.inc)
STACKBOT EQU 07Fh  ; put stack at start of scratch indirectly-addressable block (80h and up)
         ; DAC1 at P0.1

ORG 0h
LJMP STARTUP
ORG 080h

STARTUP: MOV SP, #STACKBOT
acall USUAL_SETUP
        acall Init_Device ; let the wizard do the work for us
        setb p0.1 ; make sure DAC pin is high

; here's 8-bit ramp
        CLR A
EIGHT_BIT:        _____   ; transfer current "ramp" value to DAC1 (high byte)
                  _____   ; increment ramp value
                  _____   ; waste a little time (use DELAYSHORT routine)
                  _____   ; keep doing this, forever

DELAYSHORT: PUSH ACC
MOV A, #10H
DJNZ ACC, $
POP ACC
RET
USUAL_SETUP:  ANL PCA0MD, #not(040h) ; disable WDT
      ret
;-----------------------------------
;- Generated Initialization File  --
;-----------------------------------
; Initialization function for device,
; Call Init_Device from your main program
Init_Device:
    lcall DAC_Init
    lcall Port_IO_Init
    ret
```

```
; Peripheral specific initialization functions,
; Called from the Init_Device label
DAC_Init:
    mov    IDA1CN,   #0F2h
    ret
Port_IO_Init:
    ; P0.0  -  Unassigned,  Open-Drain, Digital
    ; P0.1  -  Skipped,     Open-Drain, Analog
    ; P0.2  -  Unassigned,  Open-Drain, Digital
    ; P0.3  -  Unassigned,  Open-Drain, Digital
    mov    P0MDIN,   #0FDh
    mov    P0SKIP,   #002h
    mov    XBR1,     #040h
    ret
end
```

A DVM will show you the LSB value – the increment visible if you use multiple-step and the small delay. We saw about 8mV. A scope image is more exciting. You may have to limit bandwidth (hiding fuzz) in order to make the individual steps clear.

23L.2.9 ADC

Hardware: a signal source: In the initial ADC test, it's convenient to feed the ADC with a DC level. We can use a potentiometer, with its maximum value just a little beyond full-scale, and a moderate R_{Thevenin}. (What is the maximum R_{Thevenin} for this circuit? This question will warm your heart, no doubt, carrying you back to the start of this course!)[9]

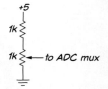

Figure 23L.14 Potentiometer can provide a DC source to feed the ADC.

Code: The task of setting up the ADC is similar to setting up the DAC: the many options call for many choices. Our first ADC test program displays the 8-bit ADC result on the LCD, and the program loop is as simple as most of our programs. The routine that picks up a sample from the ADC is this, as you saw in §23N.4.3:

```
GET_SAMPLE:   CLR CNVRT_START ;  low on AD0BUSY (to permit rising edge)
              SETB CNVRT_START ;  rising edge on AD0BUSY starts conversion
              JNB CNVRSN_DONE, $; hang here till conversion-done flag
              MOV SAMPLE_HI, ADC0H ; high byte
              RET
```

GET_SAMPLE includes a pair of commands that provide a rising edge on a bit that starts the ADC. We looked at the details of this code back in §23N.4.3.

23L.2.10 Set up ADC with Configuration Wizard

We described other ADC initialization process back in §23N.4.3. The *port* configuration also is more complicated for the ADC than for earlier programs. We'll leave the DAC setup as before, at P0.1, and will add an interrupt input, and will set up the voltage reference, as well. The crossbar makes both DAC and ADC pins analog (we'll soon be using both ADC and DAC in a program), and permits use of the voltage reference brought out at another pin, P1.2. This pin is made analog, and is brought out, even though the ADC uses it only on-chip. See Fig. 23L.15.

---

[9] In case you aren't instantly carried back to those happy days when you were younger, we will remind you that the maximum R_{Thevenin} comes when the pot slider is set to its topmost position. Then $R_{\text{Thevenin}} = (1\text{k} \parallel 1\text{k}) = 0.5\text{k}$. At every other setting R_{Thevenin} is lower.

23L.2 SiLabs 4: Interrupt; DAC and ADC

Table 23L.1

	byte-value	function
ADC0CN	d7	ADC0 enable: $1 \Rightarrow$ enable
...	d5	AD0INT: $1 \Rightarrow$ sample ready (this is a flag we test)
...	d4	AD0BUSY: rising edge starts conversion
...	d2	AD0LJST: $1 \Rightarrow$ Left Justify
...	d1, d0	selects event that starts conversion
...	00b	selects rising edge on AD0BUSY as event that starts conversion
ADC0MX	00h	ADC0 mux assignment $\Rightarrow$ to P0.0
ADC0TK	d3...d0	tracking mode
...	FBh	dual-mode (this is the default)
ADC0CF	d2, d1	number of conversions per start-convert
...	00b	$\Rightarrow$ one sample/start-convert
P0	d1, d0	both bits high, to permit use as analog pins
P0MDIN	0FCh	ADC & DAC pins set to analog (P0.0, P0.1)
P0SKIP	23h	skip INT0*, ADC and DAC, d1, d0
P1SKIP	04h	skip Vref pin
REF0CN	13h	enable Vref, set full-scale to 2.2V

Preliminary ADC test: potentiometer voltage on display: This ADC test amounts to demonstrating a digital voltmeter. The head of the program follows. The ADC initializations you can add with the wizard's help.

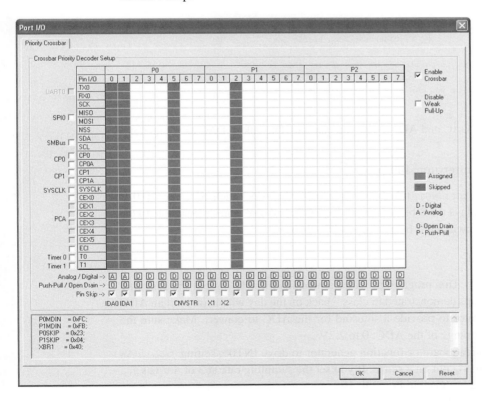

Figure 23L.15 Crossbar showing skipping and analog pins for use by ADC and DAC.

```
; adc_wizard_apr11.a51   ; ADC test, done with configuration wizard Jan 11: pins reassigned
; ADC as voltmeter (8 bits displayed on LCD), input on P0.0

        $NOSYMBOLS      ; keeps listing short, lest...
```

Lab Micro 4. Interrupts; ADC and DAC

```
$INCLUDE (C:\MICRO\8051\RAISON\INC\c8051f410.inc)  ; ...this line might produce huge list
                                                   ; of symbol definitions (all '51 registers)

$INCLUDE (C:\MICRO\8051\RAISON\INC\VECTORS320.INC) ; Tom's vectors definition file
STACKBOT EQU 080h    ; put stack at start of scratch indirectly-addressable block (80h and up)
DISPLAY_LO EQU P2
DISPLAY_HI EQU P1

; port use: ADC0 on P0.0

        ORG 0h
            LJMP STARTUP
        ORG 080h

STARTUP: MOV SP, #STACKBOT-1
            acall USUAL_SETUP
            acall SETUP_ANALOG
            acall Init_Device

START_ADC:  mov ADC0CN, #84h ; low on AD0BUSY
            mov ADC0CN, #94h ; high on AD0BUSY starts conversion
            jnb ADC0CN.5, $ ; hang here till conversion-done flag
            mov DISPLAY_HI, ADC0H
            mov DISPLAY_LO, ADC0L ; show ADC value on display
            sjmp START_ADC

;------- INITIALIZATIONS
USUAL_SETUP: anl   PCA0MD, #NOT(040h)        ; Disable the WDT.
        ret                                  ; Clear Watchdog Enable bit

SETUP_ANALOG: setb P0.0 ; set pin high, to permit use as analog pin (this for ADC)
            ret
;------- Wizard's INITIALIZATIONS (These you can fill in with the help of the Configuration Wizard)
; ...
```

Once you are satisfied that both DAC and ADC work you can let ADC talk to DAC as we do next.

23L.2.11 ADC to DAC upon interrupt: full program

The full program, `adc_dac_int_jan11.a51`, takes a sample from the ADC each time the '410 is interrupted. The program then displays the 8-bit sample on the LCD, and also sends the sample to the DAC. It is too long to type in, so download it from the book's website.

23L.2.12 Apply ADC and DAC

Watch ADC to DAC for sinusoid: This program provides a nice setup for trying out the *sampling* rules that we discussed – and perhaps demonstrated in class – back on the day when you built an ADC in Lab 18L. Use a function generator to provide a sinusoid for the ADC to convert. Make sure the sinusoid does not exceed the input range of the ADC: 0 to 2.2V.

Use a TTL square wave from breadboard or function generator to drive INT0*, setting f_{sample}. As a first test, watch analog in and analog out on a scope. If you set the sampling rate at 3 or 4 times f_{in}, you should see something like what's shown in Fig. 23L.16.

As you play with the relation between f_{in} and f_{sample} you can confirm, at least roughly, the teachings of Nyquist. For f_{sample} below $2 \times f_{in}$ you should be able to recognize *aliasing*. But the reconstructed waveform may be so weird that it is hard to recognize as a sinusoid at a strangely low frequency. The filter that we suggest next will help to make aliasing more obvious.

Filter the DAC output: A good low-pass "reconstruction filter" allows one to sample efficiently, with a sampling rate just a little above twice the maximum f_{in}. You may recall the rule that the lowest sampling artifact appears at $f_{sample} - f_{in-max}$. The good filter available to us, the Max294, attenuates to below -60dB (one part in a thousand) just 20% above $f_{cutoff} \approx f_{3dB}$. So, we can safely sample at $2.2 \times f_{in-max}$ or above. You're not likely to want to calculate this f_{sample} when, instead, you can just play with f_{sample} by adjusting a function-generator knob, while watching the results on a scope.

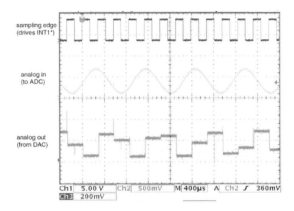

Figure 23L.16 Sinusoid to ADC, DAC output reconstructed from ADC samples.

We set out in Chapter 22N the details of the filter circuit and of the audio amplifier that permits listening to the reconstructed output. We will not repeat that detail here. You will need a *third* oscillator to drive the clock of the MAX294. If you don't have a function generator available, wire up a '555 oscillator to give yourself a square wave at a frequency adjustable from perhaps 5kHz to 1MHz. (See Lab 8L.)

...and listen to the DAC output: We hope you'll wire up the audio amp described in §23N.5, and listen to the output waveforms. The information delivered by your ears will supplement what scope and eyes can reveal.

23L.2.13 A task, if you're in the mood: modify waveforms

We propose in Chapter 22N some changes that the 8051 might apply to a sampled waveform rather than just spit it out at once to the DAC. A few lines of code could implement a *rectifier*, for example – full-wave or half-wave. Perhaps more fun is a digital-lowpass filter, done with weighted-averaging between a current sample and a running average. Certainly the filter is more fun to listen to.

As we admit in Chapter 22N, these exercises are essentially entertainments since the 8051 is indeed feeble as a signal processor. But the exercise will give you a little practice in coding, while delivering the reward of vivid results visible on a scope screen, and perhaps audible.

23S Supplementary Notes: Micro 4

23S.1 Using the RIDE assembler/compiler and simulator

23S.1.1 Installing RIDE

RIDE[1] is offered free for designs using code up to 8K. It comes in two modules: a framework called RIDE7 (at date of this writing) supporting a variety of microcontrollers; and a "module" for a particular processor. The particular module that we need is called RKit51, for the 8051.

You must register to get access to these two downloads. You will be sent a "serial key" that allows use of the "hobby" version of RKit51. After you install Ride7 and the "component" RKit51, you can use the program for a week without licensing. But you should proceed with the licensing process.

Under Help, click on License, choose "serial activation" and you will be able to enter the *serial key*. Click on "get activation key" and this second key code will be emailed to you. Enter that, and licensing is done.

Note, by the way, the *path* where RIDE is installed. You will need to specify this later to tell RIDE where to find your register equates files. On our machine, for example, the INC folder in which such equates for the Silabs controller live is here:

```
C:\Program Files\Raisonance\Ride\inc\51\Silabs
```

For the Dallas controller, our path is the same except that it ends with ...\51\Dallas.

Your installation may put your INC folder elsewhere. Make a note of this location, peculiar to your installation.

23S.1.2 Create a "project"

RIDE uses a "Project" folder to keep together your source file and your debug choices, which we will meet later in §23S.2.1.

Make a NEW project, under the PROJECT menu. The menu that appears, see Fig. 23S.1, invites you to decide where to put your project, what to call it, and which controller the code will run on. That last detail matters, although all the listed controllers are 8051 types, because particular recent versions of the 8051 add particular features and registers not defined for the original part. For the big-board, which uses a Dallas 8051, DS80C320 is the closest fit (Raisonance does not list the model that we use, 89C430). For SiLabs, choose C8051F410.

As you scan the available 8051s you may be impressed by the variety available. We chose the 8051, as we have said elsewhere, because it is the most widely sourced controller. RIDE lists 19 manufacturers. For one manufacturer, Silicon Laboratories, RIDE lists 244 variants – and we know

---

[1] Raisonance Integrated Design Environment. Raisonance is a French company.

23S.1 Using the RIDE assembler/compiler and simulator

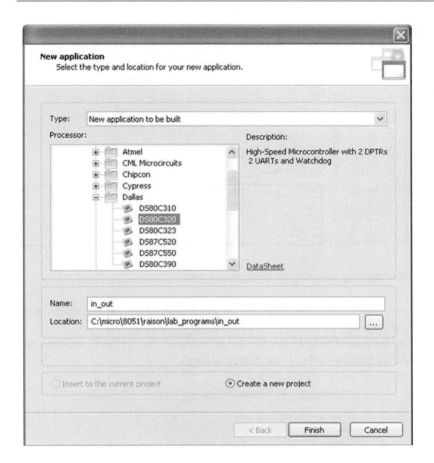

Figure 23S.1 Setting up a new project: you will choose location, name and controller variant.

that other 8051s are available, since RIDE's Dallas listing is not complete, omitting the controller that we use in our labs.

23S.1.3 Compose an assembly-language *source* file: (*.a51)

Write an assembly-language file – or, much easier – open one you've gotten from someone else so that you know the form is OK. You add it by choosing "Add item" under the "Project" menu, then browsing to find the file.

The in/out program of the big-board Lab 21L appears in the RIDE screenshot of Fig. 23S.2. The figure shows the full screen with the little program loop enlarged.

This program simply takes a byte from a keypad, then passes it to a display. You do not need to understand how the program operates at this point. For now, we are concerned only with the form of RIDE's listing. We will mention some peculiarities of this listing below.

Directives addressed to the *assembler* (not to the 8051): The source file includes several commands addressed to the assembler or compiler, not to the controller itself. Most of these are explained elsewhere in §20N.7. We will simply mention them here.

  NOSYMBOLS This line spares you the nuisance of generating a multi-page printout of DS80C320

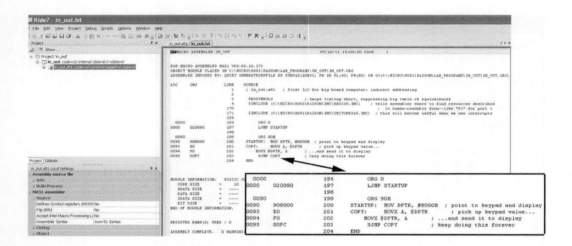

Figure 23S.2 Project window showing assembly-language source file.

register and bit names, each time you print the .lst file produced by the assembler, a file showing the machine language produced from your assembly language source.

Ride provides an alternative way to get the same result: in the "project" window, just choose NO for "display includes" (this is the default setting).

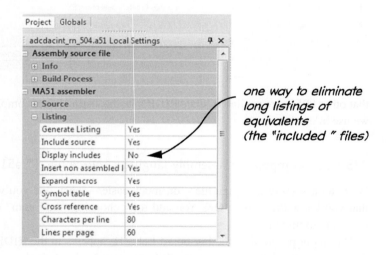

Figure 23S.3 One way of forestalling the display of excessive listings.

INCLUDE This line, described in Chapter 21N, is not needed in this very simple program. But we insert the line at the head of every program in order to cover those programs that do need it. The included file is listed in the top left window of the screenshot in Fig. 23S.4.

ORG This line, also described in Chapter 21N, tells the assembler the address at which to begin the code that follows this command.

Listing. Looking at the assembler's output: The micro labs include many program *listings* that show not just the source code (assembly language like "MOV A, DPTR," in Fig. 23S.2) but also the assembler's *output*: executable machine-language code (like the hexadecimal value "E0" in that listing).

23S.1 Using the RIDE assembler/compiler and simulator

Figure 23S.4 Ride window lists the included files.

The assembler's ability to translate human-intelligible assembly-language code like "MOV A..." into fussy machine code like "E0" is the tool's principal function. A human could look up each instruction to find the equivalent machine code – and we will ask you to do that, once or twice. But humans are not quick at this, and don't enjoy the work. When we enter programs into RAM by hand, as we often do in the big-board version of the labs, it is this hexadecimal machine code that we enter, using the keypad.

When working with the SiLabs device, and also late in the big-board labs, we don't concern ourselves with the machine code. Instead, we just use a laptop computer to send the executable code to the controller.

23S.1.4 The assembler's substitutions, permitting symbolic names

Sometimes an intermediate step lies between an assembly language command and its machine-code equivalent.

If we used built-in ports rather than external buses, for example, we might need to refer to those ports as "P2" and "P1."

The single line below does this and assumes that the keypad and display are wired to the two listed ports, P2 and P1 (respectively).

```
TRANSFER: MOV P1, P2   ; in one operation, copy byte from keypad to display
```

For this code, the REG320.INC file that was *included* is essential. It allows the assembler to make sense of "P1" and "P2."

The assembler looks up the addresses of these symbolic names and plugs those *addresses* into the executable code that it produces. Below is the "listing," showing the assembler's substitutions for P1 and P2, and the corresponding machine code. For "P1," the move destination, the assembler substitutes "144" (decimal); for "P2" it substitutes "160."

```
85A090           ...     TRANSFER:   MOV 144 , 160
```

And in the machine code listing in the left-hand column the assembler places – after the *op code*[2] 85 – the hexadecimal equivalent of those two decimal value: $A0_{16}$ for 160_{10}, 90_{16} for 144_{10}.

The screenshot in Fig. 23S.4 lists, in the top left window, the *included* REG320.INC, called a "dependency." Such included files appear in that window only after *assembly* (see §23S.1.7).

23S.1.5 Another assembler substitution using symbolic names

Even clearer than the port names "P1" and "P2" are descriptive names, as we noted in §21N.1.2. EQU permits this. The assembler can help us to define these. Once defined, these names would allow us to make the TRANSFER code almost self-explanatory. We could write it as

```
TRANSFER: MOV DISPLAY, KEYPAD
```

[2] "Operation code."

Below is a listing that shows the first stage of the two-step substitution that permits this way of referring to the ports: when we write "DISPLAY," the assembler does a text substitution, putting in "P1." You have read many of these details.

```
; byte_in_out_ports.a51
    $NOSYMBOLS ; keeps listing short
    $INCLUDE (C:\MICRO\8051\RAISON\INC\REG320.inc)
    KEYPAD EQU P2
    DISPLAY EQU P1
    ORG 0h
    SJMP TRANSFER
    ORG 020h
TRANSFER: MOV DISPLAY, KEYPAD
    SJMP TRANSFER
```

The translator then consults the *included* .INC file and translates "P1" to address A0 (hex) – as you have seen before in §23S.1.4.

```
85A090         178     TRANSFER:    MOV 144 , 160
```

If you are accustomed to high-level programming, you may not be impressed by this achievement. But you'll probably concede that it may help to make assembly code more readable.

23S.1.6 Associate the .A51 file with the project

This step RIDE calls "adding an Item..." Under the PROJECT menu, choose "Add Item" That will bring up a directory listing; work your way to where your source file lives, and click on it.

23S.1.7 Assemble the .A51 file

Under PROJECT click on BUILD. "Build" does two operations. First, it *compiles* (for source in C code) or *assembles* (for source in assembly-language). Then it *links* the assembled modules (sometimes there are several modules), assigning addresses in order to place code appropriately.

Catch an error: If the assembler finds an error in the code, it will display this in the window at the bottom of the screen. We have in Fig. 23S.5 altered the program of § 23S.1.3 on page 947 to inject an error. The assembler kindly flags our mistake.

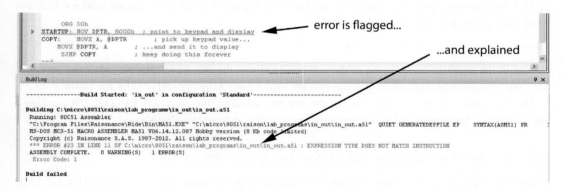

Figure 23S.5 Assembler flags errors for us (and underlines them in red).

A full *listing* of the assembler's output occasionally may help to illuminate an error by giving it some context.

23S.1.8 View the .lst file, if you like

Sometimes you may want to see a listing like those that we include in the micro labs, showing not just the assembly-language *source* code but also the addresses where this code is to be placed, and the executable machine code that the assembly language generated.

To see such a listing, use the VIEW menu and choose VIEW LISTING. If an error occurs, the error message will appear after the offending line.

```
   ...
    0030                    175            ORG 30H
                            176    STARTUP:  MOV DPTR, 8000H  ; point to keypad and display

*** ERROR #23 IN LINE 11 OF C:\micro\8051\raison\lab_programs\in_out\in_out.a51 :
EXPRESSION TYPE DOES NOT MATCH INSTRUCTION
    0030    E0              177    COPY:     MOVX A, @DPTR    ; pick up keypad value...
    0031    F0              178              MOVX @DPTR, A    ; ...and send it to display
    0032    8000    F       179              SJMP COPY        ; keep doing this forever
                            180            END
   ...
```

Our error here was to omit the "#" ahead of the "8000H." This pound sign is necessary – not only in order to do what we intend (something the assembler isn't smart enough to divine), but also just to form a legal instruction. Once you've assembled without producing errors, you can try your program in the simulator or "Debugger."

23S.2 Debugging

The assembler will warn you of syntax errors. But the RIDE *simulator* does much more: it will help you to judge whether your program does what you intended.

Under the *Debug* heading click on Start. This will open a window that looks much like the source file in the assembler but shows a blue diamond opposite the first executable line of code. You can *single-step* by pressing F7 or (equivalently) by using the Debug menu and choosing *Step_Into*. If you're curious about details, note that STEP comes in three flavors: "Step_Into" means "step through subroutines,"[3] whereas "Step_Over" means run at full speed through a subroutine, resuming single-step on return; "Step_Out" means exit subroutine (it immediately executes the subroutine's RETURN operation).

But single-stepping is not much fun until one takes advantage of the simulator's ability to show you the contents of 8051 registers. The simulator makes the controller *transparent*. In Fig. 23S.6 we have stepped through just three lines of the In/Out program, while we have chosen (from the left-hand pane) several items to display:

Main registers: you will always want to see this window, rich with information. It shows not only the principal registers but also what DPTR *points to*.
External ports: we have displayed locations 8000h through 8003h, the four locations dedicated to I/O on the big-board computer.

The bottom pane of the screenshot in Fig. 23S.6 shows a detailed *listing* of the program (as in §23S.1.8). One can see there, for example, that the line to be executed next is at address 34h.

---
[3] In C code what we call "subroutines" are called *functions*.

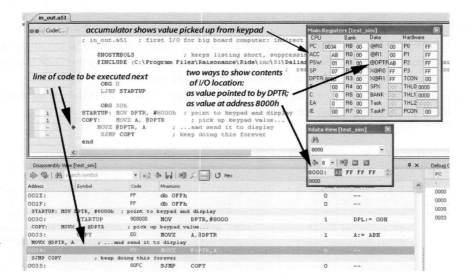

Figure 23S.6 Debugger window showing register values after a couple of program steps.

- The register "PC" (program counter) at the top of the Main Registers window shows the same value, as it should.

Other registers reflect what this simple program has accomplished in three steps:

- DPTR shows the value 8000h, a value loaded in the line labeled "STARTUP."
- ACC (the **A** or "accumulator" register) shows the value ABh, a value also shown at address 8000h in the Xdata window, and (equivalently) also in the "@DPTR" box, the location pointed to by DPTR.
- ACC shows this value because the line labeled "COPY" transferred the ABh from the I/O port to the **A** register (the same as "ACC")[4]

23S.2.1 Choosing what to VIEW during debugging

When you start the debugger, you will see in the left-hand pane, as in Fig. 23S.7, a set of options listed for viewing.

Some of these you will choose rarely; one you will choose always, as we have said: *Main Registers*. If your program does I/O using the external bus rather than the 8051's built-in ports, then you'll probably also want to watch "external memory": our I/O looks to the processor like "external memory." Under VIEW, select DATA DUMP, then XDATA VIEW. This will show you a window displaying the content of a block of addresses.

You can type in the address that interests you – say, 8000h, if you're watching an I/O operation on your little computer. Then you can also force a particular value into any location by clicking on the displayed value, typing in a new value, and hitting Return. Thus you can simulate an input from keypad, for example, or from an ADC.

You can of course also use Xdata to show *memory* data rather than I/O since the 8051 is agnostic on the I/O memory distinction that we humans like to make.

---

[4] As we have noted elsewhere (Chapter 22N), this nasty complication arises from the 8051's two distinct ways of addressing the **A** register. The annoying fact is that assemblers *require* that we use one name or the other, in distinct contexts (see Chapter 21N). We are required to write "MOV A, @DPTR," not using "ACC." But we are also required to write "PUSH ACC" in order to save the **A** register on the stack.

23S.2 Debugging

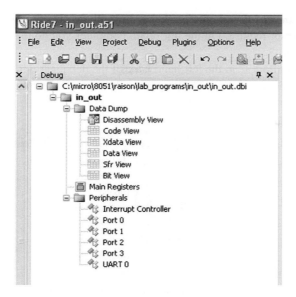

Figure 23S.7 Debugger offers a set of viewing options.

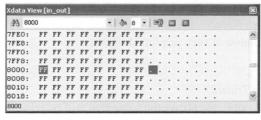

specify External Data (Xdata) address...

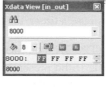

...then shrink window, to show what matters to you

Figure 23S.8 XDATA window shows port data.

WATCH window: This is a versatile tool that lets you set up a single window to display some detail important to you. To watch a signal or register, highlight it and right-click, then choose WATCH. Incidentally, WATCH works in RIDE exactly as it does in SiLabs IDE.

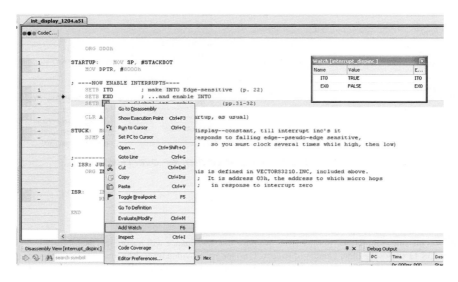

Figure 23S.9 WATCH window allows you to keep an eye on particular variables.

Supplementary Notes: Micro 4

Simulating interrupt: In order to simulate an *external interrupt* (that is, the sort initiated by driving an 8051 pin), you will need to simulate an interrupt-request. On the big-board computer you do this by providing a falling edge on bit 2 of built-in Port 3.

To simulate this interrupt request, double-click "Port 3" in the left-hand pane of the debugger window: see Fig. 23S.10. Then right-click P3.2 and choose Ground. The program will respond when you next *step*.

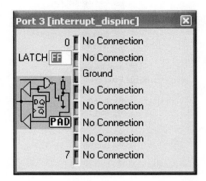

Figure 23S.10 To simulate interrupt request, ground P3.2 in Port 3 window.

Timers: When you begin to use a timer, the timer window is very helpful.[5] It shows:

- timer control register (T2CON: control);
- timer values (THL2, the 16-bit present value, for Timer2);
- "capture/reload" register (RCAP2) – for Timer2 only: this shows the value that will be reloaded when timer is in auto-reload mode;
- overflow flag (TF2, for Timer2);
- Run bit (TR2 for Timer2): set high to enable timer.

Figure 23S.11 shows Timer2's simulation window, before and after an overflow that caused a reload of the initial value.[6]

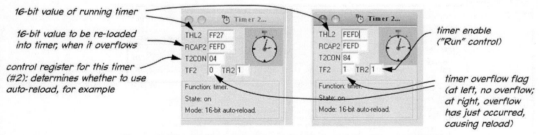

Figure 23S.11 Timer window (Timer2: more versatile than 0 or 1).

---

[5] The timer simulation was missing from RIDE7 in May 2015. We hope that Raisonance will restore this useful feature.
[6] The screenshot in Fig. 23S.11 shows the older version of Ride, 6.1, no longer provided by Raisonance. This version is available for download from the AoE website. We used this older version because the current Ride 7 does not include the Dallas timers in its simulation.

Serial port: A window can be dedicated to showing the serial port and its registers (Dallas 89C430 has two serial ports, SiLabs '410 has one). We'll not attempt to explain the serial port here; it will become interesting only at a time when you decide to use it, perhaps late in the term. Fig. 23S.12 is an excerpt from Ride's description of its simulation that shows a screenshot of the serial port simulation window.[7]

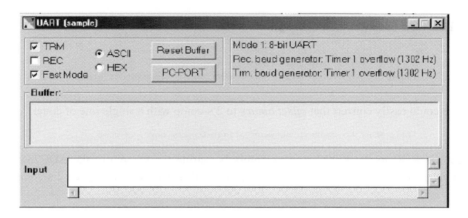

Figure 23S.12 Ride debugger window dedicated to the serial port.

As you would expect, the simulator can display a value sent from or received by the UART, and in receive mode REC allows display of a serial stream.

23S.2.2 Lots more features!

If you want to become expert in using RIDE, look at the thorough documentation provided in two PDF files whose links are posted on the book's website: RIDE1.PDF and RIDE2.PDF. Have fun!

23S.3 Waveform processing

You now have an ADC and DAC linked to your computer – or very soon you will. Once you've got these wired, it's fun to try altering waveforms: fun to see the result on a scope, and fun to hear the result if you put the output through an audio amplifier. Let's review some of those operations on waveforms.

A controller like the 8051 is not well adapted to signal processing. A Digital Signal Processor (DSP) is the device designed to do such a job. It includes fast, wide multipliers and accumulators, and usually speeds its processing with "pipelining." Nevertheless, it is useful and fun to try to compose some tiny signal-processing programs as you try out your ADC and DAC.

23S.3.1 Rectify

Half-wave: Half-wave is straightforward once we've determined the form of the ADC's digital output. The ADC as wired in Lab 23L is single supply, and speaks in *offset binary*: see Fig. 23S.13.

So, to implement a rectifier we can watch the MSB which indicates on which side of the midpoint the input voltage has fallen. In offset binary 0 indicates lower half, 1 upper half. 2's-complement inverts this MSB behavior.

---

[7] Image reproduced from Raisonance literature, "80C51 and 80C51XA Development Tools," Part 2, Debugging, §3.3.3.3.

956 Supplementary Notes: Micro 4

```
FFh   1111 1111   ←─── TOP ────→   7Fh   0111 1111
80h   1000 0000   ←─── MID ────→   00h   0000 0000
00h   0000 0000   ←─BOTTOM─→       80h   1000 0000
```

Figure 23S.13 Digital data formats. *offset binary*   *2's complement*

Half-wave code: Hence the code (for offset binary):

```
        jb acc.7, AS_IS   ; if MSB hi, leave it alone (it's already above midpoint)
        mov a, #80h       ; ...but if MSB low, force to midpoint
AS_IS:  ...               ; when program lands here, A holds value at midpoint or above
```

2's-comp form: We could easily convert that *offset binary* to 2's-comp with a single line of code:

```
        cpl acc.7         ; flip MSB, to convert between offset-binary and 2's-comp
```

Or we could get the ADC to speak in *2's-comp* simply by wiring the V_{SS} pin of the ADC7569 (on big-board computer) to -5V (the converter is that smart!). Either way, we could get a 2's-comp output.

Full-wave: A simple *INVERT* operation on *offset-binary* values does not work. Here is what happens when the signal comes in at full-scale, and then gets flipped (about its midpoint) in two alternative ways:

- CPL (flip every bit) versus
- NEG (2's-complement the sample: flip and add one[8]): see Fig. 23S.14. The NOT preserves zero.

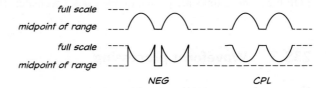

Figure 23S.14 2's-complement *invert* does notwork applied to offset-binary.

This difficulty suggests we need to be careful when we try a *full-wave rectify* operation. Again, how would our code differ in the two cases: if the ADC were talking 2's-comp versus offset-binary?[9]

23S.3.2 Filters

IIR filter: Chapter 23N suggest a scheme for a low-pass filter: average a new sample with a running average of all previous samples. This is called an *Infinite Impulse Response* filter because, using feedback, its response to an impulse never totally dies away. We'll see the contrasting case in a moment.

Preliminary question: does format matter? Will this work equally well with offset-binary and 2's-comp formats? Will our code depend on which we're using? (I always find this question difficult!) Fig. 23S.15 shows a block diagram of such a filter.

If we give old and new samples equal weight, as suggested in §23N.6, then we can calculate the

---

[8] There is no NEG operation in the 8051's repertoire of operations; the process would require two steps: CPL A then INC A.
[9] NEG works for 2's-comp, CPL works for offset-binary.

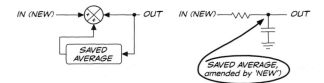

Figure 23S.15 IIR filter.

effective f_{3dB}. Here's the argument: imagine a step input, from 0 to 1V, at a time when the long-term average is zero. Watch what happens on a better-behaved filter.

You could easily implement either of these schemes on your lab microcomputer. In Chapter 23N you saw a scope image showing step-response waveforms like those in Fig. 23S.16.

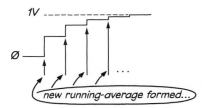

Figure 23S.16 Filter's response to step input.

If new and old are given equal weight, then each new sample takes the output halfway from where it started (after the previous sampling-and-averaging process) toward V_{in} (1V, in this case). (You may recall that these numbers happen to describe precisely what V_{cap} did on the '555.) Such a trip takes 0.7RC (as you may recall from '555 experience). Hence you could calculate f_{3dB}.

For example, if you sample at 8kHz, the time between samples is $1/(8\text{kHz}) = 0.12\text{ms} = 120\mu\text{s}$, and that, in turn, must be 0.7RC. Hence, $RC = (1/0.7) \times 120\mu\text{s} = 170\mu\text{s}$, and f_{3dB} for this filter ought to be $1/(2\pi \times 170\mu\text{s}) = 900\text{Hz}$.

To change f_{3dB} you can change the sampling rate (easy, but crude) or change the relative weights given to *new_sample* versus *running_average*. Changing those weights requires using the *multiply* command, as we mentioned in Chapter 23N. The multiply operation is slow – but maybe not too slow for modest audio signals: multiply takes 5 cycles, 20 clocks: about $2\mu\text{s}$: perfectly OK if we're sampling at 8kHz, as suggested above.

A DSP does this kind of operation nimbly, and could readily do such digital filtering; simplest and fastest would be a hard-wired digital filter, like those used in digital-audio playback DAC-chips (these are in the FIR form described just now).

23S.3.3 FIR filter

Finite Impulse Response (FIR) describes a filter that doesn't remember *all* past inputs (as an analog filter would do) but instead knows about a few samples – usually, but not always, *past* samples.[10] It lets those flow in and out of a "pipeline," and multiplies these by differing weights to form a weighted average that is the output.

For example, we could put out an *average* of newest sample and *one-ago sample*, as in Fig. 23S.17. This would be a not-very-conservative low-pass.

Or we could put out the *difference* between the newest sample and *one-ago sample*, as in Fig. 23S.18. This would be a differentiator or high-pass, of a very rudimentary sort. A longer pipeline would let us make a more refined filter.

---

[10] How can it know "future" samples? By letting us process or operate on a sample that is not the most recent, but rather is one surrounded by later as well as prior samples.

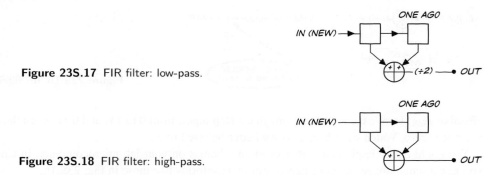

Figure 23S.17 FIR filter: low-pass.

Figure 23S.18 FIR filter: high-pass.

23S.3.4 Reverb

This is lots of fun to listen to and play with. If you use a microphone as input, you can put yourself in a pipe or a barrel or a hall or a canyon, as you vary echo time and the quickness of decay.

Reverb needs a "circular buffer" – a table in RAM that forms a pipeline as long as the echo time that we want. Each time a sample comes in, we combine it (add it) with an attenuated version of an old running-accumulation value – pulled from a place *DELAY* ago in the pipeline – and then we play this sum and store it away. If input falls silent, for example, we'll hear the old running-accumulation value gradually die away.

The circular buffer is no more than a table with moving pointer, and pointer reinitialization when it hits the end of the table. The algorithm that pulls out the "old" value also needs to be smart enough to "wrap around" the end of the table. Apart from that, it's straightforward.

24N Micro 5. Moving Pointers, Serial Buses

Contents

24N.1	**Moving pointers**	**959**
	24N.1.1 Table copy, in assembly language	960
	24N.1.2 Table copy, in C	961
	24N.1.3 An alternative to using DPTR (especially useful for Dallas version)	962
	24N.1.4 Store and playback code (big-board version)	963
24N.2	**DPTR can be useful for SiLabs '410, too: tables**	**964**
24N.3	**End tests in table eperations**	**964**
	24N.3.1 Test for end by counting transfers	964
	24N.3.2 Test for end by watching pointer value	965
24N.4	**Some serial buses**	**966**
	24N.4.1 UART/RS232	966
	24N.4.2 Universal Serial Bus (USB)	968
	24N.4.3 Serial Peripheral Interface (SPI)	969
	24N.4.4 SPI peripheral: a digital potentiometer	970
	24N.4.5 Other serial protocols…	973
24N.5	**Readings**	**974**

Why?

What we'd like to do: store and recover tables of digital data, including audio.

24N.1 Moving pointers

To this point, in the Dallas computer with its external buses, we have used the 8051's data pointer (DPTR) in almost all off-chip data transfers. That was just an awkward way that the 8051 works and we had to live with it. Today we will see ways to use DPTR that make it much more appealing: instead of an annoying piece of indirection, use of a data pointer turns out to be obviously useful as soon as the pointer begins to *move* within the program.

To appreciate the usefulness of *indirect addressing* – which is what DPTR imposes – let's imagine trying to copy a small table *without* the use of a pointer. This cannot be done with an 8051, but the pseudo-code is plausible, and can be implemented on most computers and microcontrollers.

Pseudocode to copy data without using a pointer: Let's suppose that the original data lives in a table that begins at RAM address 400h; the duplicate is to be written at address 800h.

```
pick up a sample at 400h
write it to 800h
pick up a sample at 401h
write it to 801h
pick up a sample at 402h
```

write it to 802h...

The process – if interpreted to mean a new line of code for each action – horrifies you no doubt: it cries out to be done by a loop, with addresses incremented within the loop. And that process requires the use of pointers (assuming that we don't try to write self-modifying code: a scary alternative).

24N.1.1 Table copy, in assembly language

So let's do it the reasonable way: use DPTR, and increment it each time through the loop:

```
STARTUP: mov dptr, #400h ; init pointer to start of original table
    movx a, @dptr ; pick up a sample...
    movx ; OOPS! Now what?  Need a second data pointer?
```

We started bravely – but now need a second pointer. How will we manage that? A couple of possibilities appear:

Table copy: one pointer, re-initialized We can reinitialize DPTR each time we need to shift from Original to Copy, and then back. But that code is pretty ugly and painful: here's what it might look like (it's approximately the way a C compiler does the job):

```
STARTUP: mov dptr, #0800h ; init pointer to start of Copy table

         mov r6, dph    ; save Copy dptr (a byte at a time:\\
         mov r7, dpl    ;   (high half, then low half)\\

         mov dptr, #0400h  ; init pointer to start of Original table\\
         mov r4, dph    ; save Original pointer\\
         mov r5, dpl

         mov r3, #10h   ; init copy counter for 16 transfers\\

AGAIN:   mov dph, r4    ; load Original pointer\\
         mov dpl, r5\\
         movx a, @dptr  ; pick up a sample...\\
         inc dptr       ; advance pointer\\
         mov r4, dph    ; save Original dptr (a byte at a time: \\
         mov r5, dpl    ;   (high half, then low half)\\

         mov dph, r6    ; load Copy pointer\\
         mov dpl, r7\\
         movx @dptr, a  ; save sample in Ccpy table\\
         inc dptr       ;   advance pointer\\
         mov r6, dph ; save copy pointer\\
         mov r7, dpl \\
         djnz r3, AGAIN ; keep copying, till 10h are done\\

DONE:    sjmp DONE      ; hang up here, when copying is finished \\
```

Table copy: using Dallas' second pointer A little better: use two pointers. The original 8051 has just one DPTR, but some more recent versions like our Dallas part have two, and this eases a copying task. That's the good news. The bad news is that the second data pointer is named DPTR – and it's up to the programmer to keep track of the single bit (the LSB of a register called "DPS" – data pointer select) that distinguishes the two. Here's what the code to copy 16 bytes looks like using the two DPTRs:

24N.1 Moving pointers

```
        START:
            MOV DPTR, #400h    ; init main pointer to start of original table ("source pointer")
            MOV DPS, #1        ; get other pointer
            MOV DPTR, #800h    ; init secondary pointer to start of duplicate table ("destination pointer")
            MOV DPS, #0        ; ...restore main pointer
            MOV R0, #010h      ; init sample counter
        COPY:   MOVX A, @DPTR  ; get a sample
            INC DPTR           ; advance source pointer
            MOV DPS, #1        ; get destination pointer
            MOVX @DPTR, A      ; store a sample in duplicate table
            INC DPTR           ; advance destination pointer
            MOV DPS, #0        ; ...restore source pointer
            DJNZ R0, COPY      ; till all are transferred, carry on
        STOP:   SJMP STOP      ; hang up here, when done
            END
```

In Chapter 24W we show several other ways of doing this task. But who wants to think that hard about saving pointers? Let's see what the C code to do the same task might look like.

24N.1.2 Table copy, in C

In C, the copying program calls for some rather fussy initializing declarations; but once that is done the transfer code is charmingly simple.

```
/* tblcopy_804.c */
// use compact notation to copy 16 bytes
        #include<Z:\MICRO\8051\RAISON\INC\REG320.H>
void main()
{
    xdata char *ORIG;      // declares 'type' of data is byte ("char"),
                           //    stored in external memory ("xdata")
                           //    location defined by a pointer
    xdata char *DUP;       // same setup, using a different pointer

    int i;     // just a counting index, used in FOR loop
    ORIG = 0x400; /* start address of table to be copied: this loads a pointer */
    DUP = 0x800; /* start address of table of copied data: loads another pointer */
    for (i=1; i<=16; i++)  // Just 16 transfers (tiny value, to ease debugging)
    {
      *(DUP++) = *(ORIG++); /* one-step transfer, with auto-increment of pointers,
                       though 8051 itself can't do this */
    }
}
```

Some details of this code call for explanation:

    `xdata char *ORIG;`

This line expresses several choices:
- ORIG is a pointer: that's what the "*" means;
- it points off-chip, to "external" data (that's expressed in the "x" of "xdata");
- the data is one byte wide: "char" ("character").

    `*(DUP++) = *(ORIG++);`

an extremely compact way to say what takes many lines of assembly code.
- the "=" sign implements the transfer – and, as you know, the "*" says that these are *pointers* used in the transfer;
- the "++" that follows DUP says "increment the pointer *after* its use."

Incidentally, the assembly code that the C compiler produces is not very elegant. It looks a little worse than what we showed on page 960 illustrating the one-pointer method:

```
ASSEMBLY LISTING OF GENERATED OBJECT CODE
            ; FUNCTION main (BEGIN)
          ; R4R5 is assigned to ORIG
                            ; SOURCE LINE \# 14\\
0000 7C04       MOV    R4,\#004H\\
0002 E4         CLR    A\\
0003 FD         MOV    R5,A\\
                            ; R6R7 is assigned to DUP\\
                            ; SOURCE LINE \# 15\\
0004 7E08       MOV    R6,\#008H\\
0006 FF         MOV    R7,A\\
                            ; SOURCE LINE \# 17 \\
0007 F500    R  MOV    i,A\\
0009 750001  R  MOV    i+01H,\#001H\\
000C            ?FOR1:\\
                            ; SOURCE LINE \# 19 \\
000C 8C83       MOV    DPH,R4\\
000E 8D82       MOV    DPL,R5\\
0010 0D         INC    R5\\
0011 ED         MOV    A,R5\\
0012 7001       JNZ    ?LAB5\\
0014 0C         INC    R4\\
0015            ?LAB5:\\
0015 E0         MOVX   A,@DPTR\\
0016 FA         MOV    R2,A\\
0017 8E83       MOV    DPH,R6\\
0019 8F82       MOV    DPL,R7\\
001B 0F         INC    R7\\
001C EF         MOV    A,R7\\
001D 7001       JNZ    ?LAB6\\
001F 0E         INC    R6\\
0020            ?LAB6:\\
0020 EA         MOV    A,R2\\
0021 F0         MOVX   @DPTR,A\\
                            ; SOURCE LINE \# 17\\
0022 0500    R  INC    i+01H\\
0024 E500    R  MOV    A,i+01H\\
0026 B411E3     CJNE   A,\#011H,?FOR1\\
                            ; SOURCE LINE \# 22\\
0029 22         RET \\
                            ; FUNCTION main (END)\\
```

But the beauty of using a higher-level language like C is that you don't much care how ugly the code is that the compiler generates, except when you're much concerned about speed. When you are so concerned, you can code a critical patch in assembly language.

24N.1.3 An alternative to using DPTR (especially useful for Dallas version)

Now back to assembly language. There is an alternative to using the DPTR for off-chip data transfers, and this alternative offers the great virtue that it leaves the DPTR free for other uses. (You know of this alternative if you looked through the note on addressing modes, §21S.1.4.) That freedom becomes important today when, for the first time, DPTR is likely to be busy writing or reading tables.

This alternative addressing mode is written as MOVX @R0, A for example. R1 is also available (but not other registers), and so is a transfer in the opposite direction of course.

In order to use this MOVX @Ri, A addressing mode, one first loads P2 with the high half of the 16-bit address. Fig. 24N.1 sketches the scheme. In our computer it is convenient to use this addressing mode for I/O devices so we load 80h into P2; R0 or R1 then provides the lower half of the I/O address. In the case shown here – which is taken from Lab 23L – those two addresses are 02h and 03h to complete the 16-bit addresses of the ADC and DAC. (§21S.1.4 spells out in greater detail how this MOVX @Ri, ... addressing mode works.)

Look at the MACRO ASSEMBLER ADCDACINT_RN_504 code on page 929 for an example using this addressing mode. It is used there to access ADC and DAC.

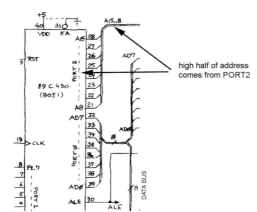

Figure 24N.1 Processor's PORT2 provides high half of address when using external buses.

24N.1.4 Store and playback code (big-board version)

Using this addressing mode to handle the converters makes it easy to fill a table from ADC or to read from a table to the DAC. Here is part of the storage scope program from big-board Lab 24L that fills a table with samples, then plays the table back to a DAC. The program is timed by an interrupt that sets the sampling rate.

```
; TABLEINT_Rn_408.A51  store ADC data in table, & play it out to DAC: TIMED BY INTERRUPT
.
.  ; details omitted, re interrupt enabling and use of interrupt flag
.
STARTUP: ACALL INITPTR ; INITIALIZE ptrs for ADC & DAC data, etc.
         MOV DPTR, #800h ; init main pointer to first writeable location
         ACALL OKINT ; go enable interrupts
FILLTABLE: SETB IE.7 ; re-enable interrupts
         JNB SAMPLEFLAG,FILLTABLE ; hang here till it's time to take a sample
         CLR SAMPLEFLAG ; clear the flag set by ISR
         CLR IE.7 ; disable interrupts till we've taken care of the task
         ACALL GETSAMPLE
         MOVX @DPTR, A ; store the sample
         INC DPTR
         ACALL COMP_PTR ; see if we've hit end address
         JC FILLTABLE ; ...if not, do it again (c ==> movx ptr < ref)
REINIT:  MOV DPTR, #800h ; init main pointer to first writeable location
PLAYBACK: SETB IE.7 ; re-enable interrupts
         JNB SAMPLEFLAG,PLAYBACK ; hang here till it's time to PLAY a sample
         CLR SAMPLEFLAG ; clear the flag set by ISR
         CLR IE.7 ; disable interrupts till we've taken care of the task
         MOVX A,@DPTR ; get the sample from table, not new from ADC
         MOVX @R0,A ; send sample out to DAC
         INC DPTR ; advance the table pointer
         ACALL COMP_PTR ; see if we've hit end address
```

```
        JC PLAYBACK      ; ...if not, do it again
        SJMP REINIT ; otherwise, reinit and play again
```

As you'll notice, the use of DPTR is notably straightforward: just one pointer used, and it increments until the table has been filled or entirely played back – at which point it is re-initialized. In the listing below, we omit some of the details, the better to show the simple use of DPTR:

```
        MOVX A,@DPTR ; get the sample from table, not new from ADC
        MOVX @R0,A ; send sample out to DAC
        INC DPTR ; advance the table pointer
        ACALL COMP_PTR ; see if we've hit end address
        JC PLAYBACK      ; ...if not, do it again
```

Thanks to the addressing of ADC with `MOVX @Ri,...`, the routine GETSAMPLE also is wonderfully simple:

```
GETSAMPLE:  MOVX A, @R1      ; Read ADC
    MOVX @R1, A    ; send anything--just a pulse to START pin of ADC
    RET
```

24N.2 DPTR can be useful for SiLabs '410, too: tables

To this point, the SiLabs '410 has not needed DPTR since it has not used MOVX operations. But it can use these to exploit its limited on-chip RAM (2K). This address space is sometimes called "MOVX RAM" because of the way that is addressed. We cannot build a large table, as the Dallas part can with its external 32K SRAM. But we can store 2K on the '410. In §24L.2 we store ADC samples in this MOVX RAM.

This memory could be applied, of course, to build a table of any sort of data. 2K might seem a more substantial memory were it used, for example, as a buffer for characters incoming from a keyboard. You may recall that we mentioned in Chapter 23N, the use of a "circular buffer" as a plausible use of *interrupt*: building a table of characters stored whenever a new character was ready and recalled whenever the processor was ready for such a value. The circular buffer could be useful for uncoupling the timing between keyboard entry (timed by the unpredictable rate at which a human types) and a running program.

In the program of §24L.2, Ram_Store, we apply the on-chip RAM to store ADC samples: we save 2K samples from the ADC, then replay them endlessly to the DAC, thus implementing what we call a storage scope. If you are ambitious, you may want to try modifying this playback program, converting it to a *reverb* or echo program by using the RAM as a circular buffer that combines a new input with an attenuated old entry. But chances are you're pressed for time, just now, and should return to this notion later.

24N.3 End tests in table eperations

24N.3.1 Test for end by counting transfers

For a small table using a counter is easiest. We counted down an 8-bit register in the little sample program of §24N.1.1:

```
...
  MOV R0, #010h       ; init sample counter
```

24N.3 End tests in table operations

```
COPY: MOVX A, @DPTR       ; get a sample
...
      DJNZ R0, COPY       ; till all are transferred, carry on
```

In the C code equivalent program, we used

```
for (i=1; i<=16; i++)  // Just 16 transfers (tiny value, to ease debugging)
```

If the number of iterations is not large, or if you think of the problem as doing a particular *number* of iterations, then using a counter makes sense. (In assembly language, counting beyond an 8-bit value requires a little more coding; in C there is no such penalty for increasing the count value.)

24N.3.2 Test for end by watching pointer value

But testing *address* may be a better way to check when a table operation is done. This we did in the storage scope program of §24N.1.4. In that program, we wrote

```
      ACALL COMP_PTR      ; see if we've hit end address.
```

That COMP_PTR routine is relatively complex, looking for an exact 16-bit match between the value of DPTR, the moving pointer, and the END address. You can look at that routine in the big-board Lab 24L if you like.

But more often you are likely to be content with an *8-bit* comparison as you do an end test (if you do it in assembly language; again in C you can be indifferent to the extra code called for by a 16-bit compare).

In the storage scope program for example, where the goal is to use all available RAM, we need only check the *high* byte of the moving pointer to know if it has passed the end of RAM (RAM space runs up to I/O space, which begins at 8000h). So if we advance the pointer before testing we can check for the value 80h in the high byte, DPH. When the pointer hits that value, stop.

Here is code for an 8-bit address compare. The end value is in register R6. This routine SETS the Carry flag as long as the pointer has not reached the end value.

```
; --POINTER COMPARE: this subroutine alters C flag
COMP_HI: XCH A,R5 ; quick way to save A (faster than PUSH ACC)
CLR C
MOV A, DPH
SUBB A, R6 ; see if ptr less than END: CY flag holds answer: true if ptr < END
XCH A,R5 ; recover old A value
RET
```

The preliminary `CLR C` is necessary, as you probably have picked up by now, because the 8051's sole *subtract* operation, SUBB, takes in the carry flag. SUB seems to be mis-spelled as SUBB. That second B indicates the *Borrow* flag (same as Carry) that the operation takes in. Hence the need to clear Carry before we do the subtraction that we are using as a sort of *signed* compare operation.

We could test for equality: If we were content to test for equality with the END value, we could use `CJNE A, #END_ADDRESS` rather than a *subtract* operation. But the code is not simpler:

```
; --POINTER COMPARE: this subroutine could control a software flag, which we'll call ALL_DONE
COMP_HI: XCH A,R5 ; quick way to save A (faster than PUSH ACC)
MOV A, DPH
CJNE A, R6, NOT_YET ; see if ptr less than END
SETB ALL_DONE         ; if hit limit, set soft flag to tell main program
SJMP EXIT
CLR ALL_DONE
EXIT: XCH A,R5 ; recover old A value
RET
```

Pointer value test in C: In C, address compare requires much less thought than in assembly language. The code is almost what you would write if you were describing the program flow to another human in a comment.

```
// Here is the table-fill portion: store one sample each time interrupted
            while (STORE_POINTER <= TABLE_END)
                {
```

The value TABLE_END, above, does the job of register R6 in the 8-bit assembly language versions of §24N.3.2. There is no difficulty (for the programmer) in using a 16-bit value for TABLE_END.

24N.4 Some serial buses

Our big-board lab computer speaks to its peripherals through its buses – which are of course *parallel* sets of wires – or through its built-in ports, each of which presents eight lines in parallel. The SiLabs controller also has used only *parallel* ports, to this point. Interfacing peripherals this way is appealingly straightforward.

> AoE §14.6.3

But we should be aware of another way to transfer data, a method that has taken over much interfacing, *serial* transmission. Serial is gaining in full-scale computers; but for use with microcontrollers it is still more appealing, because of controllers' chronic shortage of pins.[1] Serial buses have become fast enough so that even parts like FPGAs that might earlier have run many parallel lines now may include *serializer/deserializer* blocks ("SerDes").[2]

24N.4.1 UART/RS232

Let's sketch a few of the serial protocols. The oldest of the standard serial protocols is relatively complex, though slow: it is usually called RS232 (or, UART).[3] It is the protocol used by the now-obsolescent "COM" ports on computers. Most microcontrollers include an RS232 port; the 8051 does (your Dallas processor includes two, the '410 offers one). These ports are useful for communicating with a full-scale computer (a PC, usually); but are not much used for communication with *peripherals* (serial RAM, serial ADCs, real-time clocks, serial sensors, for example).

> AoE §14.6.3H

RS232 requires intelligence in the receiver because it uses just one line (plus ground), or two for bidirectional use.[4] The lack of any separate synchronizing signal – a clock or strobe – means first that the receiver must be set up to expect approximately the correct transmission bit rate; and second – and more subtly – that the receiver must also *synchronize* itself with the transmission in order to sample bit levels at appropriate times. A *start* bit kicks the receiver into life. The receiver then samples at a time that its clock tells it is the midpoint in each bit period. A *stop* bit terminates the byte transmission.[5]

The absence of a separate clock line is not unique to RS232. USB, Firewire, SATA, and PCIe share

---

[1] Another way to put this point is to say that people using controllers are chronically stingy: they could have lots of pins if they were willing to pay for them. But low cost is much of the appeal of microcontrollers. A 16-pin controller like, for example, the Microchip PIC24F04KA200 is very chatty – but only through its amazing *three* serial protocols: SPI, I²C and RS232/UART.

[2] See TI's discussion of alternative SerDes technologies, for example: http://www.ti.com/lit/ml/snla187/snla187.pdf.

[3] This acronym stands for a full name that is quite a mouthful: "Universal Asynchronous Receiver/Transmitter." Even RS232 has a history: "Recommended Standard #232," as AoE reminds us in §14.7.8.

[4] The standard connector for many years included *25 pins*, later reduced to 9 in recognition of the fact that people were not using all those signals.

[5] A student asked how a start bit can be distinguished from an ordinary data bit if no dead time is required between byte transmissions. It's a good question, to which the answer is that a loss of synchronization soon would be detectable because a stop bit would be lacking. At that point the receiver could ask for a re-transmission. If you wonder "couldn't a non-synchronized bit pattern by chance deliver what looked like a correct stop bit?" the answer is Yes – but this is a rare event so that soon, probably next byte, error should be detected. Error detection and re-transmission then will restore synchronization. See AoE, §14.7.8.

24N.4 Some serial buses

this characteristic. But the lack of clock does complicate the hardware needed to communicate, in contrast to the simple SPI for example (§24N.4.3).

If only seven bits of data are sent (enough for the standard ASCII character set) then one error-checking "PARITY" bit is usually included, to describe the transmitted data: a HIGH would indicate that the number of 1s transmitted was *odd*. More usually, a UART will send eight bits of data. So a typical UART transmission uses ten bits for seven or eight bits of data.

The top trace of Fig. 24N.2 shows an RS232 transmission from a computer to DS89C430 (which accepts downloaded code in this format). The second trace shows the controller's response, which it issues as soon as it has seen the initial ASCII[6] character for "Carriage Return"[7] (CR).

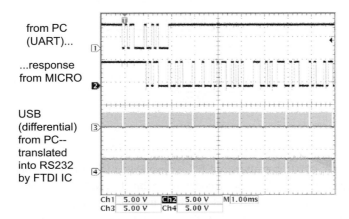

Figure 24N.2 Micro loader signal exchange, using UART and USB serial protocols.

If we scrutinize the bit stream on the top trace, see Fig. 24N.3, we can just make out the pattern – which ought to be CR followed by line-feed (LF). The hexadecimal values for these ASCII characters are 0D, 0A.

The micro's response is a longer declaration – the second line of Fig. 24N.2 announcing that this is a Dallas part speaking. On the screen of the computer linked to the microcontroller, Fig. 24N.4 shows the response.

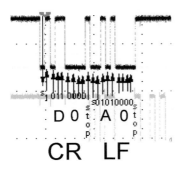

Figure 24N.3 Two RS232 characters decoded from the bit stream.

```
DS89C420 LOADER VERSION 1.0  COPYRIGHT (C) 2000 DALLAS SEMICONDUCTOR
>
```

Figure 24N.4 Controller's text response sent back to PC as RS232 serial stream of characters.

---

[6] ASCII: "American Standard Code for Information Interchange."
[7] How's that for an anachronism? "Carriage" refers to a typewriter's mechanism.

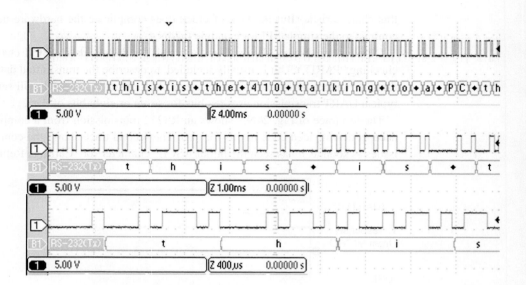

Figure 24N.5 A UART transmission at three sweep rates, decoded.

A SiLabs UART transmission from the '410 controller to a PC is displayed at increasing scope sweep rates in Fig. 24N.5. In this figure, the oscilloscope decodes the 8-bit data value (as ASCII characters). At the highest sweep rate (shown in the lowest of the three traces) you may be able to figure out the binary code that was sent. It is hard though. First, the scope has been set to invert the logic levels to make a graphical High correspond to a logic High. Second, the 8 bits of data are padded with two extra bits: a Start bit (graphically, a Low in the figure) and a Stop bit (graphically, a High).

The data is sent LSB first, so we are obliged to read the clusters of four bits "backwards" (MSB on the right). We have labeled the stream with hex values just above the bit stream in Fig. 24N.6.[8] The value for "t" is 74h, as it should be. See if you can make out the 68h for the character "h." This scope [9] finds the decoding easier than you and we do.

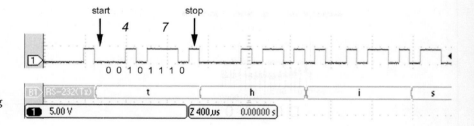

Figure 24N.6 UART detail, showing ASCII "74h" for character "t".

24N.4.2 Universal Serial Bus (USB)

AoE §14.6.3M

Because COM ports have disappeared from ordinary computers, the LCD display board that you use with the lab controller accepts USB, translating it to the RS232 form that the DS89C430 and the SiLabs '410 understand. USB is a much more complex protocol. The transmission gets its good noise immunity by using differential signalling in place of the old UART approach, which was simply to

---

[8] The ASCII code is only seven bits, but the eighth bit, the MSB, here is sent as zero.
[9] This is a Tektronix MSO2014, using a RS232 decoder module.

get noise immunity by use of very wide voltage swings (as wide as ±12V in the original specification, though the levels used by our RS232 translator are only standard 5V logic swings).

USB is not a true *bus*, in the sense that one cannot tie multiple devices to a shared set of USB lines; but with some additional electronics (a "hub" that fans from one to many, and many to one), one can achieve the equivalent result. USB ("full speed") achieves 12Mbits/s; the newer USB2 ("high speed") betters that rate 40-fold, to 480Mb/s, and USB3 ("SuperSpeed") specifies 5–10Gb/s, max.

Figure 24N.7 is a detail of a few of the bursts of USB activity that looks like no more than fuzz in Fig. 24N.2.

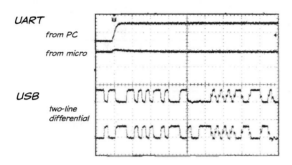

Figure 24N.7 Detail of USB differential signals.

The lower two lines show the differential signalling. The signals are complements – except for a couple of hiccups: one near the middle of the screen, one at far right. These are used to signal bus states such as "device reset."

USB can't stop talking: it is notably noisy, and this noise can be a nuisance if, for example, USB is feeding a circuit that includes analog functions.

24N.4.3 Serial Peripheral Interface (SPI)

AoE §14.7.1

Microcontroller peripherals need a serial interface simpler than RS232 and USB, and several protocols are available.[10] The one you will meet in the *digital potentiometer* exercise in §24N.4.4 and will demonstrate in Lab 24L, both Dallas and SiLabs versions, is probably both the simplest and the most widely used. It is called the Serial Peripheral Interface bus or protocol: SPI.

SPI was Motorola's serial bus; other manufacturers offered others. "I^2C" ("Inter-IC") from Phillips is more complex: it uses just three lines (clock, a data line in each direction, and ground), and allows multiple bus "masters." It forms a true *bus*. So, many devices can be tied to this pair of lines without the separate chip-select lines, one per peripheral, that SPI requires. See §24N.4.5.

SPI scheme: SPI sends a clock, and data that is to be sampled on one of the edges of the clock (exactly *which* edge is determined by the particular flavor or "mode" of SPI). Thus the scheme is "synchronous," in contrast to RS232. In addition, one must add one Chip_Select* line for each peripheral, since no *address* is transmitted. This requirement of a line per peripheral makes it not quite a true "bus." (But USB also is not a true bus, either; so we can't be too much shocked at SPI's inelegance.)

This synchronous scheme makes SPI very easy to transmit and to receive. The number of bits transmitted is flexible. The data is transmitted MSB first for the devices used in our labs (digital pot in Lab 24L, SPI RAM in Lab 25L.2). That bit order may be reversed in other applications. No start or stop bit, nor parity or checksum complicates the scheme (or protects it from error). The data

---

[10] RS232 is used by a few peripherals: we use such a *talker* chip in our lab, and RS232 liquid crystal displays (LCDs) are available.

rate depends, of course, on the hardware. On our Dallas 8051, where we "bit-banged" the SPI (that is, implemented it with code), the bit rate is about 120kHz as one can make out from the image in Fig. 24N.9 which shows two bytes being transmitted: 11h, then EEh. On the '410, which includes a hardware SPI port, the bit rate can go higher (to several MHz. See C8051F410 data sheet §23.5).

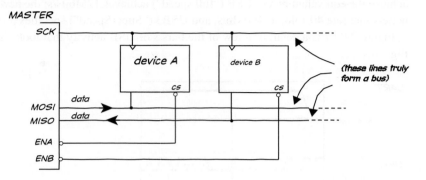

Figure 24N.8 SPI pseudo-bus.

SPI's virtue is that it asks so little of a peripheral: only a shift-register is needed. The Master device need only...

- select a device, by asserting its private ENABLE* line...
- put out a bit level on the transmit line (which goes by the ugly name MOSI ["Master Out Slave In"])...
- then provide a clock edge (some peripherals demand rising edge, others falling)
- meanwhile, the master can receive data on the MISO line ("Master In Slave Out"), and can do this even as data flows out (this is "full-duplex" communication)

The lab digital potentiometer ("digipot") that we suggest you might control with SPI takes an 8-bit command word followed by an 8-bit value. The command 11h tells the digipot to use the byte that follows (data) to set the potentiometer's position – in one of 256 possible locations. Fig.24N.9 shows one particular transmission to the digipot.

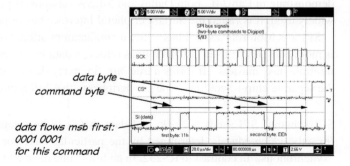

Figure 24N.9 SPI waveforms: two bytes sent to digital potentiometer. Signals are CLOCK, CHIP_SELECT* and DATA.

data flows msb first: 0001 0001 for this command

The 11h command appears near the left edge of the screen; the EEh data appears near the right edge. You will recognize the two "E" values as two cases of a long sequence of HIGH's followed by a LOW on the LSB.

24N.4.4 SPI peripheral: a digital potentiometer

We set forth here details that normally would appear in lab notes. We do this to avoid repetition because this SPI peripheral is available in both versions of today's lab.

We propose that you try the SPI bus by letting your 8051 control a *digital potentiometer*, using the three SPI lines. This digital potentiometer, the Microchip MCP41100, is a 100kΩ pot whose slider is moved from one end to the other, in 256 steps controlled by a byte sent to it.

We will use it to let the 8051 control the gain of an op-amp *inverting amplifier*. We use the inverting configuration to allow the circuit to attenuate or amplify. One possible application for such a circuit is to make a self-adjusting input to an ADC: the controller could reduce gain if input went to full-scale, increase gain if input fell below some minimum level. This would be a sort of automatic-gain-control, useful in some applications (such as recording audio). Fig. 24N.10 is a generic version of the circuit we propose that you build. In Lab 24L you will find the pinouts and suggested values.

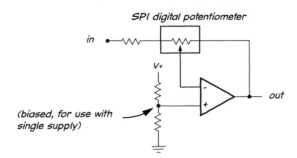

Figure 24N.10 MCP41100 digital pot used to adjust gain of amplifier.

An integrated version of such an amplifier – much souped-up – is available: the TI LMH6881. It is a very fast differential amplifier (to 2.5GHz – yes, gigahertz) with gain programmable through an SPI interface, in 0.25dB steps. Not needing that speed, we'll be happy if you'll putter along with a '358 when you build the circuit of Fig. 24N.10.

A possible use for this SPI-controlled amp: AGC: Your circuit could be used, as we have noted, as an *automatic gain control* in order to hold the output amplitude roughly constant as input amplitude varies. Such circuits (see Fig. 24N.11) are standard in the RF circuits of car radios (since signal strength varies widely as the car travels, into hollows, onto hilltops) and on speech recording devices (the speaker may shout or whisper, and the user doesn't want to be obliged to make continual adjustments of input level). A diode feeding a capacitor and resistor to ground will sense an average level.

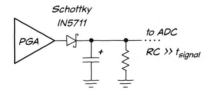

Figure 24N.11 Amplitude-averager: output could be fed to ADC to allow micro to adjust gain.

Figure 24N.12 shows a scope image of the amplifier output and the output of such an averaging circuit. This *average amplitude* voltage could go to the micro's ADC, and the ADC could then respond appropriately: crank up the gain when the output is very small, and vice versa.

The simplest micro program that might form such an AGC would simply increment or decrement the *gain* of the amp according to whether the average amplitude was above or below a set target. Such a program would be slow to adjust since it would need to wait for the averager to settle each time it had output a new gain setting. You could improve response by pairing fast "attack" response with slow release. Or you may invent a smarter scheme – such as a binary search for the best gain value.

Infrared remote (for control of television and other appliances): The *remote* that controls many household devices uses a scheme much like the UART's. The receiver and its remote need to share a transmission rate. A START pulse synchronizes the receiver, which then samples the bit level at regular intervals. We taught an 8051 controller to recognize ten keys of a Sony remote. The remote transmitter sends not quite binary levels, as a UART does, but instead bursts of 40kHz oscillations. A signal-conditioning circuit *integrates* the magnitude of these bursts (in a circuit much like the "averager" of Fig. 24N.11 and like any AM demodulator), and passes that result to a comparator. The comparator puts out the binary levels shown on the lower three traces of Fig. 24N.13.[11] The lower traces show three successive key pressings. The program that decodes the transmission first looks for the START* pulse, then samples four successive bits to get the codes for keys 0 through 9.

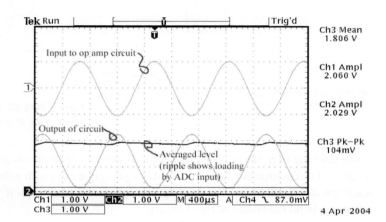

Figure 24N.12 Scope image showing amplitude-averaging level and its source.

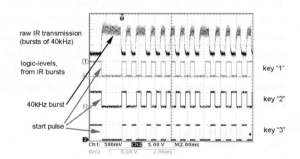

Figure 24N.13 Infrared remote signals: bursts of 40kHz, and conversion of this to logic levels.

Our IR remote receiver board is show in Fig. 24N.14. The two interrupts were used to provide an easy way to detect pulse widths.

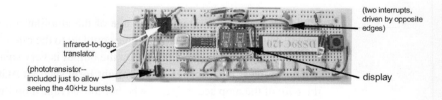

Figure 24N.14 IR remote receiver board using infrared-to-logic-level IC and microcontroller.

---

[11] The figure shows *three* transmitted codes; only the bottom trace matches the sequence of bursts shown in the top trace (in inverted form).

24N.4.5 Other serial protocols. . .

We will not try an exhaustive look at serial protocols; we leave that to AoE.

AoE §14.7.2

I-squared-C bus: I^2C is a fancier controller-to-peripheral scheme, as we have said.[12] It is a true bus, using two-lines (signal and clock, plus ground), and sends address as well as data. So, I^2C eliminates the need for SPI's individual CHIP_ENABLE lines. It also requires a good deal of cleverness from both sender and receiver. Device address must be extracted from the serial stream. It's neat – but elaborate to code.

Controllers from Phillips (the inventor) provide built-in I^2C, and so do many peripherals. Linking such hardware-equipped devices is easy; doing it all in code is not easy. In contrast, we did implement SPI in the big-board lab, with "bit banging." This required about 40 lines of code.

In the labs we do not use the elegant but complex I^2C protocol even with the SiLabs '410, which implements I^2C in hardware. Even given this assistance, we have steered away from I^2C in the labs, preferring the simplicity of SPI.

. . .and a wrapup of some serial protocols: Here we sketch some other leading serial formats.

AoE §14.7.3

One-wire: Really? *one* wire? Well, that's a bit of puffery: it's one wire plus ground; but still impressive. All devices are tied to one line that is normally high (and not-so-incidentally *powering* all the "slave" devices). A master device sends data and address information – encoded as longer or shorter low pulses. A peripheral that recognizes its address responds (pulling the line low; a capacitors keeps each part alive during the low pulse). You may have seen these around in stand-alone form: as mysterious stainless-steel buttons on walls: a night-watchman, making his rounds, touches the button with a device that reads the button's identifying code, proving that he passed that way (and when).

AoE §14.7.4

JTAG: Joint Test Action Group[13] is a serial scheme, originally invented to test printed-circuits by testing all the installed ICs. The circuits are daisy-chained, a test signal is routed through all of them, the result is read back. JTAG's use has been widely extended into both the programming of parts (as in the "pod" that programs our lab PALs) and in the debugging of circuits, such as microcontrollers (otherwise painfully opaque). A processor or controller, for example, can be designed with JTAG-readable internal registers so that it can be examined and even single-stepped while installed in a circuit.

Some SiLabs controllers – the ones in packages with many pins – use JTAG. Ours – the '410 – does not, preferring a protocol that does not sacrifice four of its precious pins.

AoE §14.7.14

Firewire: "IEEE 1394" and Thunderbolt: Firewire, introduced by Apple but now dropped from its computers in favor of Thunderbolt, was less widely used than USB; 1394a achieves same nominal speed as USB2, about 480Mb/s, but in practice is the faster of the two, coming closer than USB to using its full capacity. It also is fully duplex.

AoE §14.7.6

Ethernet: 10–100–1000Mb/s: This is what ties your PC to campus computers, for example – when you aren't using a wireless link.

---

[12] "Inter Integrated Circuit" Bus. For Philips' detailed description and promotion of this format, see
http://www.semiconductors.philips.com/acrobat/applicationnotes/AN10216_1.pdf.
[13] Named after the industry group that set up this standard.

A table listing some serial formats:

Format	speed (bits/second)
RS232	19.2k @ 50 feet (original spec; now up to perhaps 100k)
RS422 and RS 485	differential version: 100k @ 4000feet; 10M @50 feet
USB1	12M
USB2	480M
USB3.0	5G
USB3.1	10G
Thunderbolt	10G?
Firewire: 1394a	400M
Ethernet	10M to 1G

24N.5 Readings

AoE Chapter 14 ("Computers, Controllers and Data Links")
 §14.7: Serial Buses and Data Links
8051 Reference, 8052.com Tutorial (http://www.8052.com/tut8051)
 Timers: pp. 20–29;
 Timer interrupts: p. 35
 Priority among interrupts: pp. 36–37.
Dallas ... User's Guide: **Timers:** pp. 96–107
 Timer 2, especially pp. 99–105;
 Timer 2, auto-reload mode, particularly pp. 100–102;.

24L Lab Micro 5. Moving Pointers, Serial Buses

Overview

Today's lab offers several 8051 applications, and we assume that most people will not have time to try all of them. Here's what you might do today.

- Add loader hardware.
- **Storage Scope:** A chance to use your computer to store waveforms. You can download this program from the book's website. Rather than ask you to do coding, in this exercise we'd like you go directly to the pleasures of playing with sampling rate. Here, where you are constrained by the 30K available to you for storing samples, you will have an incentive to sample as thriftily as possible.
- **Serial bus:** The 8051 controls the gain of an amplifier using a 3-wire serial bus. You may be able to dream up a neat use for this. An automatic-gain-control circuit is one possible use.
- **Timers:** Then, a chance to try out two of the controller's timers in one or two programs:
  - The first program that we offer does nothing much: only increments the display each time the timer "times out" or rolls over. It demonstrates how the timer works, however – and you might decide to use this program to set the sampling rate for example. The ADC-to-DAC program lives already on your computer (see §23L.1.3: `adcdacint_rn_504.a51`); you'll need only to have the timer-program's ISR (interrupt service routine) call the ADC–DAC program. Then the pair of programs would sample at a rate set by the 8051 timer. This is tidier than the scheme we used last time – which dedicated an external oscillator to the task of setting the sampling rate.
  - The second program uses two timers. It is longer, and therefore more of a pain to key in if you are not using the downloader. One timer – the *slow* one – provides an interrupt that occurs a few times per second. The other timer – the *quick* one – sets the *width* of a HIGH pulse that goes out each time the *slow* timer rolls over. The quick timer's delay value is determined by the *keypad* value. Thus this is a keypad-controlled PWM setup, though running at an unusually slow rate.

    We propose that you use this pulse output to control the position of a *servomotor* of the sort used in radio-controlled models. We hope you'll consider it fun to see your computer make something *move*.

Lab 25L invites you to use the 8051 as a *standalone* device. It is the final prescribed lab; after that, we hope you may devise some project of your own – even if very modest. It's more fun to work on your own schemes, and it teaches you faster. We recognize that you may well have to skip parts of Lab 24L in order to have time to take off on your own. Feel free to do that.

24L.1 Data table; SPI bus; timers

Introduction

Today's programs are longer than those you have entered in the earlier micro labs. Therefore we start by proposing that you invest some time in getting your little computer to accept code loaded from a PC. We hope you will not have to invest a lot of time: when things go smoothly, this is a five-minute task. But bugs can slow things down of course.

Incidentally, you can get the source code for all these exercises from the book's website. You can use RIDE to make the downloadable file, *filename*.HEX, which you then can feed to your machine. You will find some executable files (".HEX" files) posted.

The first exercise, "storage scope," calls for almost no new hardware at all if you have already installed a switched-capacitor filter (MAX294), as we hope you did last time. You will need an amplifier in order to judge the effects of your minimized sampling and you may also have installed that last time. You'll also probably want to feed your computer from a microphone amplifier that you wired up back Lab 7L.

The second and third tasks, which probably you will treat as *alternatives* unless you are feeling very energetic, *do* call for a little hardware.

24L.1.1 Time to invoke the loader

As you connect your computer to the serial lines from the USB/LCD board, watch out for one potential trap: the serial lines are labeled by the names they carry at the *PC*:[1]

- DTR (Data Terminal Ready): this active-low signal drives your Glue PAL's LOADER* pin: a terminal that until now has been tied high.
- TXD (Transmit eXchange Data): this drives the micro's *RXD* pin (pin 10, also known as P3.0).
- RXD (Receive eXchange Data): this is driven by the micro's *TXD* pin (pin 11, also known as P3.1).

24L.1.2 A tiny program to test loading

It's a good idea to try loading something tiny and simple into your machine, as a first test. Here is such a program, which simply flips an LED On, then Off, at a few Hz. Here's the setup, which you wired in Lab 22L. With luck, you may find it still in place.

We have written it in two forms: in assembly-language, and in C. In either version, you may need to use RIDE to generate the .HEX file that the downloader accepts.

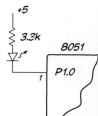

Figure 24L.1 Simple hardware to let 8051 toggle an LED.

Assembly-language bit-flip LED-blinker Here it is in assembly-language. This tiny program would run too fast to observe unless you single-stepped it.

```
; bit_flip_assy.a51 tiny test program to test Loading:
; toggles an LED at P1.0, at full speed: must be run in SINGLE-STEP
    $NOSYMBOLS      ; keeps listing short, lest...
    $INCLUDE (C:\MICRO\8051\RAISON\INC\REG320.INC)   ; ...this line should produce huge list
```

---

[1] This potential ambiguity results from the fact that both laptop computer and microcontroller can stand in role of sender, or "data terminal equipment" (DTE) in the jargon. Signals are defined from the point of view of DTE, so TxD means signal transmitted from DTE. In an unambiguous case the receiver would be a peripheral that was clearly not a DTE (instead, called "DCE;" Data Circuit-terminating Equipment). Laptop to microcontroller is not such a case.

24L.1 Data table; SPI bus; timers

```
    $INCLUDE (C:\MICRO\8051\RAISON\INC\VECTORS320.INC) ; of symbol definitions (all '51 registers)
ORG 0
    LJMP START
ORG 480h
START:  MOV 0C4h, #04h ; turn on ALE during ROM run (bit 2, ALEON, of PMR register, C4h)
FLIPIT: CPL P1.0 ; toggle the LED bit
SJMP FLIPIT ; ...forever
END
```

Assembly-language bit-flip LED-blinker with delay: The program invokes a DELAY routine that you wrote back in Lab 21L.

```
; bit_flip_assy.a51 tiny test program to test Loading:
;   toggles an LED at P1.0, at a few Hz
    $NOSYMBOLS ; keeps listing short, lest...
    $INCLUDE (C:\MICRO\8051\RAISON\INC\REG320.INC)   ; ...this line should produce huge list
        ORG 0
        LJMP START
        ORG 480h
START:  CLR A
            MOV R2, A ; a curious way to clear the delay registers,
            MOV R3, A ;  for maximum delay (since decrement comes before test)
FLIPIT:     CPL P1.0 ; toggle the LED bit
            LCALL DELAY ; this is a routine you wrote back in Lab 21L.  It lives at 100h
            SJMP FLIPIT ; ...forever

; -- -- -- -- in case you don't still have that delay routine:
                ORG 100h
DELAY:      PUSH PSW       ; save registers that'll get messed up
            PUSH ACC
            PUSH B
            MOV A,R2       ; get outer-loop delay value
INITINNER: MOV B,R3        ; initialize inner loop counter
INLOOP:     DJNZ B, INLOOP     ; count down inner loop, till inner hits zero
            DJNZ ACC,INITINNER ; ...then dec outer, and start inner again.
            POP B          ; restore saved registers
            POP ACC
            POP PSW
            RET            ; Now back to main program.
    END
```

C-language bit-flip LED-blinker

```c
/* delay_bitflip_408.c */
// flips a bit, with delay

    #include<Z:\MICRO\8051\RAISON\INC\REG320.H> // used to define "P1"
    // P1.0 drives LED


    sbit volatile OUT_BIT = P1^0;

    int n;      // this 16-bit value doesn't quite provide 0.5 sec
    void delay (void); // announce function that follows main

    void main (void)      // must be called "main," to satisfy setup.a51
    {
    while(1)
        {
        delay();
```

```
                OUT_BIT = ~OUT_BIT;
        }
}
void delay (void)
{
        for(n= 0; n< 0x8600; n++){}    // 2Hz delay value (250ms delay)
}
```

Compile this in RIDE (fixing the path to REG320.H to fit the machine you're using to run RIDE), and load the .hex file to the microcomputer (putting the code into external RAM: easier to debug if it's there).

24L.1.3 Storage scope

Once you have the Loader working, you can use it to send more interesting programs to your machine. The program listed here can fill available RAM You may want to begin by using a tiny table, say, two bytes long. Doing this would allow you to test it in single-step or in RIDE simulation.

When you are satisfied that the program works on your machine, change the END ADDRESS (the values in R7, R6) to 8000h. The table-filling will stop one byte before that at the highest available RAM address, 7FFFh, because the end-test occurs before the storing operation. 30K of the RAM's total 32K space is available, since the bottom 2K is write-protected. In that 2K of protected space lives the program itself, if running from RAM (in the full lab microcomputer). Alternatively, the program can run from ROM while using the external RAM for data storage.

After filling the RAM table, the program then plays out those samples to the DAC, endlessly. We call it "storage scope" because it can record waveforms. (Of course, you may now be so blasé and habituated to digital scopes that you assume that all scopes store waveforms! They didn't always, as you may recall if you used an analog scope earlier in this course.)

The rate at which the program takes in and plays out samples (the "sampling rate") is controlled by an external function generator whose TTL output drives Interrupt 0. The ISR simply set a software *flag* (a bit in the processor's RAM), while the MAIN program branches on that bit, hanging up till the bit is asserted. This is, we admit, a rather poll-like use of interrupt; but it's easy.

24L.1.4 Bonus Swedish translator

Just for fun, we've included a strange feature: a switch that allows you to play back the stored waveforms *in reverse*. The program senses the level at P3.4 to determine whether to play back the stored samples in the order received (a 1 at P3.4 gets this result) – or in reverse order (a 0 at P3.4). A slide switch connected to P3.4 gives you control. This direction switch is shown in Fig. 24L.2.

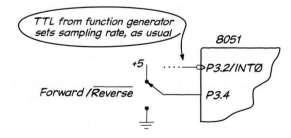

Figure 24L.2 A slide switch selects direction of playback (forward/reverse*).

The reverse playback feature we included to honor the memory of Alejandro Jenkins, a former student and teaching assistant who made a memorable "Swedish Translator." His "translator" was no more than his computer, with lots of memory to accommodate long speeches, and capable of playing back recordings in reverse. Alejandro's great achievement was not so much in building the machine as in becoming quite adept at speaking *backwards*. His parlor trick then – performed before the class, once he had made sure that no one in the room actually knew Swedish – was to speak backwards into the machine. His unintelligible speech then was rendered intelligible (though difficult, combining the effects of reverse-speech with a substantial Costa Rican accent). But it was impressive. Perhaps you can learn to speak in reverse. Perhaps not.

Once your Loader hardware is running, you can pump your storage scope code into the processor in a couple of seconds, and then spend your time trying it out. We have included a 16-bit compare-pointer routine in this program. This is more refined than necessary; an 8-bit comparison of the *high-order* byte in the address would be quite satisfactory. Still, you may be interested to see one more instance showing how an 8-bit processor can string together 8-bit operations to handle quantities larger than a byte.

You can download both source code and listing, if you like, from the book's website. The listing is useful in case you want to study details and perhaps change some by hand after loading the program. (We assume you are running code from RAM, not from the internal ROM). Changing the table length, for example, is easily done on your little computer: simply take the bus and alter the end values. Here are the two lines in the source code that do this (they take values defined at the program's head; these values produce a table of maximal length):

```
MOV R7, #END_HI   ; set reference address (end of table: last address below I/O space)
MOV R6, #END_LO   ; ...this is the low byte
```

The program *listing* shows that these values are loaded in at addresses 146h, 148h. To define a tiny 2-entry table for debugging you could replace the full-size table values with 08h for R7, 02h for R6. This table would begin at 800h, terminate at 801h. You can do this modification by hand much more quickly than you could do it by re-assembling: that process would require you to alter the source code, re-assemble it, and then re-load from the PC.

24L.1.5 SPI bus: a simple serial bus

The *synchronous* SPI scheme is very easy to transmit and to receive. SPI is described in §24N.4.3.

SPI peripheral: a digital potentiometer We propose that you try the SPI bus by letting your 8051 control a *digital potentiometer* with the three SPI lines. This *digital potentiometer*, the Microchip MCP41100, is a 100kΩ pot whose slider can be placed anywhere between one end and the other, in 256 steps controlled by a byte sent to it. The circuit is described in §24N.4.4. Fig. 24L.3 reiterates the circuit that we propose.

8051 code to send and receive SPI Here is some code that sends two bytes each time the program is interrupted (by INT0*). In addition, it reads serial data on an input line and stores those two input bytes. In the lab exercise described below, that input is a *dummy*, since the peripheral offers no output data for the 8051 to receive. But in other settings, the ability to receive and send at the same time can be useful.

Before sending the *pot setting* byte, the program first sends a *command* byte that tells the IC that the *next* byte is to be interpreted as *pot setting*. That command code happens to be 11h. (You saw this *11h* in the 2-byte sequence recorded in the scope image titled "SPI signals" in Fig. 24N.9.)

Because this code is pretty cryptic, we start with the *source* file – the ".a51" file, which is a bit easier

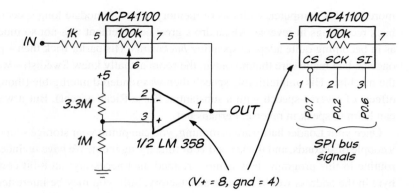

Figure 24L.3 MCP41100 digital pot used to adjust gain of amplifier.

to follow than the ".lst" file (especially because the latter loses symbols in several cases, translating them to bare numbers).

SPI code files: The source (`SPI_digipot_int_504.a51`) and list (`SPI_digipot_int_504.lst`) files for SPI code can be obtained from the book's website.

A possible use for this SPI-controlled amp: AGC Your circuit could be used, as we have remarked in Chapter 24N, as an *automatic gain control*, in order to hold the output amplitude roughly constant as input amplitude varied. If you're in the mood to try such a task, look back there for suggestions.

24L.1.6 Using the 8051's timer

This controller, like most, includes hardware features that are quite impressive – and also quite difficult to get a grip on because one needs to take care of a host of small and tedious details to take advantage of these features. The timers (*three* of them on the 8051, with differing characteristics) are perhaps the most extreme example of this painful wealth (the serial port perhaps competes for difficulty). You don't have time to learn the details of these timers so we are providing some code that makes a timer show one of its several abilities. Once you've convinced yourself that you can make the timer work, we hope you may find an application for it as you move on to your own schemes in the time that remains in this course.

The counter/timer[2] that we'll use is smarter than those on the earlier versions of the 8051: it is a 16-bit counter rather than 13; much more important, it knows how to reload itself automatically with a starting value programmed into it once it has counted to overflow. When it overflows it can either toggle an output pin, or interrupt the processor. We will use it for this latter purpose. As we've said, we never have very exciting ideas for a response to such an interrupt; so we'll just increment the display value. We'll leave to you the niftier applications for this timer.

A program to use counter 2 in auto-reload mode: The code is longish so download it (`TMR800.A51`) from the book's website. Notice that the symbol file provided by Dallas eases things a little: we can write `SETB ET2    ; timer-2 interrupt enable`, rather than look up the address and bit position of the timer-two enable bit. Nevertheless, the code is still pretty fussy: not transparent to read, not a great deal of fun to write. We hope the *results* please you.

---

[2] Both names are appropriate since this piece of hardware can be set up either to count edges on an input pin – in which case it is properly a "counter" – or to count processor clock cycles (slowed by a pre-scaler). In the latter case, where the clock rate is fixed, it is better called a "timer." Not that the name for the thing is a big deal.

24L.1 Data table; SPI bus; timers

Two timers used to set width of repeated pulses: The program `SERVOPULSE_512.a51`, which you can get from the book's website, uses Timer2 again, in *auto-reload* form, to keep interrupting a few times per second, each time the timer overflows. All Timer2 does in that event is to set a single bit – a "flag" bit in the micro's internal RAM. The MAIN program watches for that flag, and when the flag goes high, MAIN starts Timer0, loading it with a delay value determined by the *keypad*. Timer0 puts out a pulse at Port 3, a pulse whose width was set from the keypad. This program is a bit long, and heavy on tiresome initializations and quirky flag-clearings. Skip it, if you're impatient to get on to your own projects.

To give you the pleasure of seeing something *move* in response to control by your computer, we suggest you let the output pulse drive a so-called "servomotor" of the sort used in radio-control models; see Fig. 24L.4. This ingenious motor takes a *rotary position* that is determined by the width of the pulse fed to it. The motor has three leads: red, white and black. Red gets +5V, black goes to ground, white takes the logic-level pulse from your computer's P1.0.

Within the servomotor is a one-shot. Every time an input pulse arrives, the motor is driven one way or the other for the duration of the mismatch between the two pulses: *input* pulse versus *one-shot* pulse. Since the one-shot pulse width is adjusted by motor position (the motor turns a pot rigged as variable resistor), the one-shot pulse width adjusts to match input pulse width. If input outlasts one-shot, the motor runs in one direction during the mismatch time; if one-shot outlasts input, the motor is driven the other way during the mismatch. Full adjustment and consequent homing of the motor usually occurs over many pulse repetitions. The scheme is slow but neat, and very simple to drive.

Figure 24L.4 A hobby servomotor.

Timer2 sets the slow drumbeat – the repetition rate for the pulses: see Fig. 24L.5.

Figure 24L.5 The timer-program's output pulses repeat a lazy rate determined by Timer 2.

And a second timer, Timer0, whose pulse duration is loaded from the keypad, determines the duration of the pulse put out at each of Timer2's slow drumbeats. Fig. 24L.6 shows a scope image showing the variation in pulse widths achieved by the program over a range of keypad input values.

24L.1.7 Program listing: servomotor pulse using timers

Download `SERVOPULSE_512.a51` from the book's website. The program uses a single key value, from

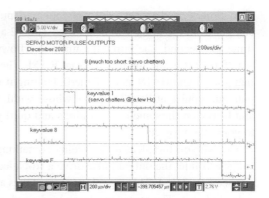

Figure 24L.6 Servo pulse widths corresponding to keypad inputs: set by Timer 0.

the keypad, to determine pulse width, and thus position of the servomotor. The range of key values that work before hitting the servo's limits, is 4 to F (both those values bump up against the motor's stops); 5 to E stop just short of those limits. The range of pulse widths is about 0.5–2ms. The pulses repeat at about 30Hz, with Timer 2 values shown.

In an effort to fine-tune the effect of key values on pulse width, we tacked in a use of *multiply*, backing off slightly from the range of pulse widths otherwise provided by the key value shifted-left-four-times. The multiplier value, 0DCh, keeps about 85% of that shifted value. Now maybe you can dream up your own application for this timer. If not, take a look at Lab 25L, where we invite you to take advantage of the 8051's ability to work as a self-sufficient *standalone* controller rather than as the brains of the computer that you have put together on the big board.

24L.1.8 SPI code listing

Download `SPI_digipot_int_504.a51`, and the listing file, `SPI_digipot_int_504.lst`, from the book's website.

24L.1.9 Two storage scope listings: in assembly and in C

Similarly, download the assembly code, `table_bidirectional_dec10_wi_ale.A51`, and its listing, `table_bidirectional_dec10_wi_lst.A51`; the C is `store_and_playback_bidirectional.c`. In the latter, you'll note several assembly-language-like C extensions used to specify port pins, an interrupt vector, and the use of external memory (`xdata char *STORE_POINTER`)[3].

24L.2 SiLabs 5: serial buses

24L.2.1 Port pin use in this lab

We need to use seven of the eight pins of PORT0 this time. You will need to disconnect any conflicting uses from earlier labs. We show in Fig. 24L.7 one signal, "MISO" (an SPI serial bus signal), though we do not use it in this lab. We use it next time in Lab 25L.2, but it seems to make sense to include the full SPI bus wiring at this stage.

P0.5 is put to two different uses in this lab. Be sure to disconnect the first use (INT0*) when you shift to the second (RXD).

---

[3] Presumably, `xdata` tells the compiler that using this pointer calls not for `mov` but for `movx`.

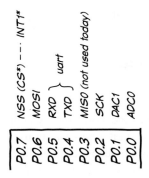

Figure 24L.7 PORT0 pin use in this lab.

24L.2.2   RAM data table

The '410 includes a modest amount of on-chip RAM: 2K. This is not enough to store much input from an ADC – the task we will ask of it today; it might be more appropriate as a buffer to store keyboard input for example. But to demonstrate use of the on-chip "MOVX RAM," we will use it here to store a couple of thousand samples from the ADC, and then play those back endlessly to the DAC. In this operation it will perform like a "storage scope" of small capacity.

This program uses GETSAMPLE from the ADC code you saw last time, and sends the ADC samples out to the DAC, as it did last time. The novelty today, is that for the first time we are using the 8051's "data pointer" (DPTR), a 16-bit register that you have read about in class notes, beginning with 22N, but may well have ignored, since it has not appeared in any lab program till now. DPTR defines the address used in an *off-chip* data transfer (with the Dallas computer on the other big-board branch of these micro labs, this is the normal way to read and write data; that computer uses *buses* and a relatively large off-chip RAM [32K]). For the '410, its "MOVX RAM" is not really *off-chip*, but, in addressing this memory, the controller treats it as if it were.

The routine STORE IT places one sample in RAM at an address defined by the contents of DPTR. It then advances the pointer and checks to see if the table is full:

```
STORE_IT:   MOV A, SAMPLE_HI ; if we haven't hit end of RAM, store the sample
            MOVX @DPTR, A ; store sample
            INC DPTR
            MOV A, DPH ; beyond end of RAM?
            CJNE A, #08h, OK
            SETB TABLEFULL ; if RAM's done, let the world know this
OK: RET
```

The way this code checks for "table full" is idiosyncratic: because the on-chip RAM resides at addresses 0 to 7FFh, we can detect "...full" by watching the *high* byte of DPTR (an 8-bit register named DPH: data pointer high). When an increment of DPTR takes DPH to 8h, the pointer has moved past available RAM, and it is time to stop.[4] The same scheme is used on *playback* from the table of stored data.

Code listing   The code listing for this program, adc_store_on_chip_apr15.a51, is available from the book's website. One other detail of it may call for explanation: in both *STOREIT* and *PLAYBACK* routines, we disable interrupts. We do that so that you can single-step the program even when it

---

[4] In this event, the DPTR "wraps around" to value 0, and further data writes would overwrite good data. But that's OK: the DPTR is tested after an increment and before its use, so the overwrite does not occur.

is getting interrupt signals at its usual high rate: you can watch the two routines in slow motion, untroubled by continual new responses to interrupt.

Serial buses

Controllers are always wishing they had more pins. The smaller the controller, the more urgency there is to the task of making efficient use of what pins are available. We have seen the pin-sharing used in the interface in §21L.2 adopted to avoid committing even two lines to the debugging interface. Serial buses sacrifice speed in some settings – in microcontroller applications, though not in full-scale computers, where serial links can outrun parallel – in the interest of pin-saving. At the extreme, one line can carry all information (assuming always that ground is defined as well). Dallas' "1-wire" interface is such a scheme. So is the traditional UART bus, often referred to as RS232. We will see the latter in this lab. But we will start with a serial bus that is less efficient but also easier to implement: Motorola's Serial Peripheral Interface (SPI) bus.

24L.2.3 Serial bus 1: SPI

SPI is not quite a true "bus," as you know from Chapter 24N, because it requires, along with the "bused" or *shared* lines, one line dedicated to each particular peripheral: a private *enable*. This is equivalent to providing an intercom buzzer line to each user in an office: the buzzer means "it's for you." The user who hears the buzzer picks up the phone on the shared line and uses it, listening, talking, or (most often) doing both. The bus is described in more detail in Chapter 24N.

Hardware: SPI controls digipot In the present example, SPI goes to a single peripheral, a digitally controlled potentiometer ("digipot"). The pot's slider can be placed anywhere along the fixed resistance (which in this case has the nominal value 100kΩ).[5]

Test setup The pinout of the digipot is shown in Fig. 24L.8. The SPI pin use on the '410 is fixed in hardware, so you will need to disconnect anything that conflicts. See this lab's pin use in Fig. 24L.7.

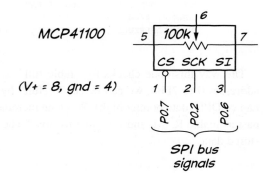

Figure 24L.8 SPI digipot wiring.

Code listing: The code listing for this program, which calls itself SPI_digipot_silabs_apr14.a51, is available from the book's website.

---
[5] 100k± a whopping 30%!

24L.2 SiLabs 5: serial buses

First test: just variable resistance: Initially, to test whether the digipot is working, let's just use a DVM to measure *resistance* between the slider and either end of the digipot. The program should allow the keypad value to set a resistance anywhere between 100k and a value close to zero.[6]

You can watch the SPI signals on a scope: NSS (P0.7), the chip-enable; SCK (P0.2); MOSI (P0.6).[7] See if you can make out the values of the bytes sent. The fixed *command* byte, 11h, may help you to get your time-bearings as you watch the effect of changing the *data* byte.

Register settings: port I/O: The Configuration Wizard sets things up as in Fig.24L.9.

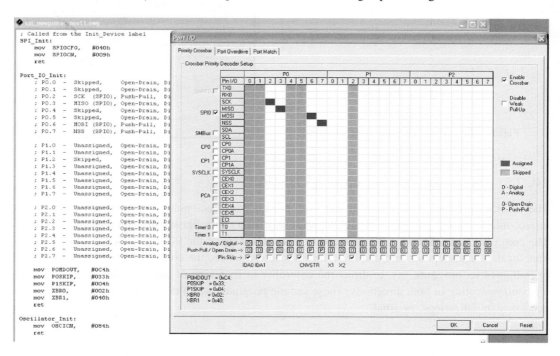

Figure 24L.9 SPI configuration using wizard.

Register settings: SPI choices: SPI obliges us to make several choices. Fortunately, the Configuration Wizard helps a lot. Fig. 24L.10 shows the choices we made.

24L.2.4 Apply the digipot

Once you are satisfied that the digipot is working, try making its behavior a bit more useful: use it to set the gain of an inverting amplifier. See Fig. 24L.8.

As suggested in Chapter 24N, adding an amplitude-averaging feature (just a diode and an *RC*) could allow one to make an automatic gain control. The '410 permits doing this also in digital form by use of its *window* function: the '410 can watch the ADC input and alert the program if the signal lies outside some allowed range. But the analog averager seems to allow an implementation with simpler software. Do what you find appealing.

---

[6] The residual resistance of the analog switch that routes the signals always remains; the minimum resistance is not zero ohms, but up to 100Ω (i.e. 100Ω, max.; 52Ω, typical).

[7] MOSI ("Master Out Slave In") is labeled "SI" in Fig. 24L.8.

Lab Micro 5. Moving Pointers, Serial Buses

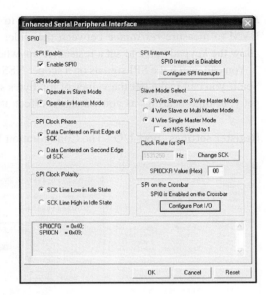

Figure 24L.10 SPI configuration choices.

24L.2.5 Serial bus 2: setting up an RS232 UART

The old RS232 "serial port" – now absent from most personal computers – is included on nearly all microcontrollers, and provides a straightforward way for a controller to talk with a bigger computer. The controller can report to a PC (perhaps to take advantage of the PC's more sophisticated capacities for data manipulation and display); the PC also can talk to the controller (perhaps to direct the operation of its program: say, to direct the controller in its taking of data). In this section we will try communicating in both directions. These sample programs won't do much – but will show some necessary initializations, as usual.

Hardware: serial link from PC to micro: A traditional RS232 link achieves good noise immunity by using very wide signal swings. A simplified RS232 can get by with TTL levels. We will use the latter, simplified scheme, taking advantage of the USB⇔RS232 translator that is included in the LCD display card. This converter uses the RS232 protocol, but at TTL levels. This is convenient, allowing us to omit the level-translator sometimes required.[8]

In fact, the LCD card includes *two* RS232⇔USB translators. One is a standalone IC (FTDI232R) which can be used as a general-purpose device and is required to talk to the Dallas 8051 (in the big-board labs). The other translator is included in a SiLabs controller on the LCD card: the same IC that does USB-to-C2 translation for the SiLabs IDE.

The '410's UART uses pins P0.4 and P0.5 (TX0 and RX0, respectively). This assignment is rigid. Connect those lines – which are named from the point of view of the *controller* to the RXD and TXD terminals on the LCD card. Those terminals are labeled from the viewpoint of the *PC* that is connected through the LCD card. Sorry – but that means that the '410's TX0 connects to the LCD's RXD, RX0 to the LCD board's TXD. Fig. 24L.11 shows what the link looked like on our breadboarded setup.

---

[8] Only "sometimes" required, because a simple clamp can be sufficient in place of a true level-translator like those described in AoE §12.10.4.

24L.2 SiLabs 5: serial buses

Figure 24L.11 Serial link using LCD board.

24L.2.6 SiLabs "terminal" program: serial interface

The easiest software for linking the laptop to micro is the SiLabs *Terminal* program. Before opening this program, load code into the '410; then *disconnect*. Then open the terminal program. Under *settings* make sure the baud rate is 9600; for the other characteristics, take the defaults (shown on the left side in Fig. 24L.12).

After you downloaded the program to the '410 and disconnected, the program began to run, talking endlessly on its serial port. When you click on *connect* in the terminal window, you should see the '410 sending its simple message, over and over (along with one character corresponding to the hex value input to the '410's PORT2, to which the keypad is attached). That keypad connection is not shown in Fig. 24L.11; the connection was shown in Fig. 21L.15,

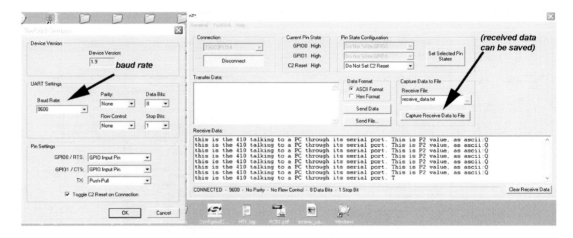

Figure 24L.12 SiLabs terminal screen: UART settings and main window as micro talks to PC through UART.

Serial interface, done in two directions

24L.2.7 Controller can talk to a PC

The program shown running in Fig. 24L.12 serves only to show that the controller can indeed talk to the world of bigger computers. The code listing is `serial_message_silabs_apr11.a51` on the book's website. The code recites some code listed in the "db" (define byte) line that follows the routine "GETSPEECH." How GETSPEECH works is a detail we tried to explain in §24L.2.8.

Register settings Here is a summary of the register initializations.

Register	bit/byte-value	function
SCON0	d7 (= S0MODE)	0 ⇒ 8-bit serial mode
...	d4 (=REN0)	1 ⇒ receiver enabled
...	d1 (= TI0)	Xmit interrupt flag: high when Xmit done; must be cleared
...	d0 (= RI0)	Receiver interrupt flag: high when Receive done; must be cleared
CKCON	d3 (= T1M)	0 ⇒ use SCA1, SCA0 to set timer 1 clock rate
...	d1, d0 (= SCA1, SCA0)	00 ⇒ controller clk = sys clk / 12
...	00h	sys clk / 12 (this is the reset value)
TMOD	d6 (= C/T1)	0 ⇒ timer 1 clock defined by T1M bit (CKCON.4)
...	d5, d4 (= T1M1, T1M0)	10b ⇒ 8-bit auto-reload
...	20h	ditto
TCON	d7 (= TF1)	1 ⇒ timer-1 overflow (clear by hand unless using interrupt)
...	d6 (TR1)	1 ⇒ timer 1 enable
...	40h	timer 1 enabled
XBR0	d0 (= UARTOE)	1 ⇒ route UART signals TX0, RX0 to P0.4, P0.5
...	01h	ditto
T1H	d7...d0 (= timer reload value)	96h ⇒ 9600 baud, when controller clk = sys/12

Figure 24L.13 shows the wizard screen indicating where these values come from if, as is likely, you don't want to check bit-by-bit in each register. We did insert one value by hand: the baud rate reload value, #096h.[9]

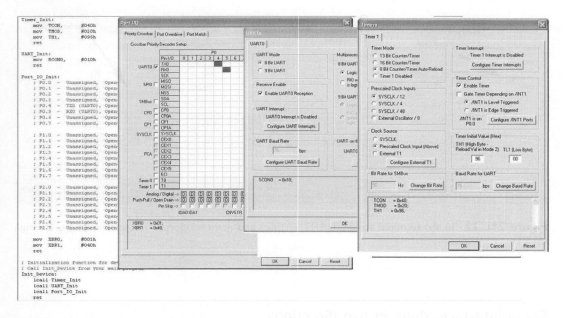

Figure 24L.13 UART configuration by wizard.

---

[9] The wizard wrongly prescribed a different value when we tried getting the reload value from a baud rate. We have not yet determined why.

24L.2.8 A detail of controller-to-PC program: how GETSPEECH gets the speech

In `serial_message_silabs_apr11.a51` – the program that has the controller talking endlessly to the PC, available on the book's website and illustrated in screenshots in Chapter 24N – we use a programming device we have not used elsewhere in these labs. The tiny subroutine GETSPEECH picks up ASCII characters to send, one at a time. And the novel code is MOVC, in "GETSPEECH: MOVC A, @A+PC." This line means "move code," and the source is an address formed by summing the PC (program counter) with A (accumulator). The destination is the accumulator, as you can see. We mentioned this addressing mode back in §21S.1.5.

GETSPEECH thus loads the accumulator, each time it is invoked, with the value stored at $PC+A$, where PC is the value of the PC at this point. In programming jargon, A serves as an *index* into a data table that begins at the value PC.

The PC value used is the address of the *next* instruction (here, RET).[10] So to get the very first character in the speech (that character is the "t" in "this is the '410..."), register **A** must hold the "index" value 1, so that $PC+A$ forms the address where the "t" is stored. That is why A is set up with the value "1" by the line

> START_OFFSET EQU 01h ; initial offset into speech table

This value, 1, is loaded into **A** in a two steps: first into R2, thence into **A**. And after use as the "index," **A** then serves as repository for the character picked up from the table.

This heavy use of the **A** register is characteristic of the 8051. Later processors, as we have said before, spread the labor among many registers; early processors including the 8051 made the accumulator a central stage for a great many operations: the results of all boolean and arithmetic operations, and the source and destination for off-chip *moves*, called "MOVX."

To make this program a little more fun – a little more like your *own* – we suggest that you change the message from that we have listed after "db...." That may let you better appreciate the power that the serial port now affords you: the power to let the controller report its results to a full-scale computer. How you use this power we leave to you, as usual. Some people may find it useful in a project at the end of this course.

24L.2.9 Bluetooth wireless link can carry serial signal

Once you have the serial program working – micro talking to PC – you can, if you like, route that signal through a Bluetooth radio. You can send it to a laptop, or (probably more exciting) you can send it to an Android cellphone. (You cannot send it to an iPhone, because Apple doesn't allow ordinary citizens like us to talk to their devices.)

Hardware The bluetooth module is a standalone device. It includes a microcontroller that translates UART serial signals to and from the bluetooth radio protocol. The module (made by Roving Networks) is mounted on a carrier (this done for us by Sparkfun).

Change power from 5V to 3V To use this as a wireless serial link, power it with 3.3V (*not* 5V) from the LCD card's SiLabs cable (the cable that terminates in a 5×2 connector). Fig. 24L.14 shows the way to provide that 3.3V supply.

---

[10] This definition of the "present" PC value you may recall from the discussion of subroutine CALL, where it is this address that is stored on the stack, and from JUMP calculations, where the PC value used is the address two *after* the address of the JUMP instruction itself.

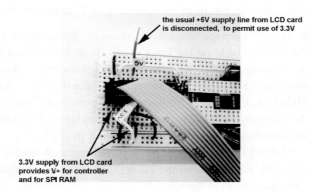

Figure 24L.14 3V supply for bluetooth and for the '410.

This 3V supply can power the '410 as well. (You will use this method in Lab 25L as well, where we meet another part that requires 3V: a serial RAM).

...and the bluetooth module Then feed the '410's Tx and Rx lines (though we're not now using Rx) to the module's Rx and Tx pins, respectively. Recall that each device names these signals from its own point of view; so when the '410 *talks* on its Tx line, the module *listens* on its Rx line.

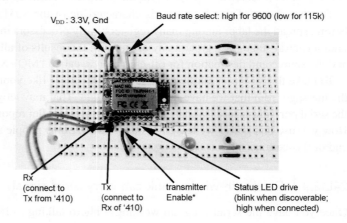

Figure 24L.15 Bluetooth module's wiring. (Bluetooth module shown is Sparkfun's Roving Networks WRL-12579.)

Connecting to a smartphone The procedure for connecting to an Android phone is pretty simple once your '410 is driving the bluetooth module with the endless serial message "this is the '410 talking...." The "status" LED of the module will blink at 1Hz.

- On the phone, turn on bluetooth.
- Scan for bluetooth devices.
- The module is likely to identify itself as "Firefly..."
- If you select that device, a window will open inviting you to provide a password. That password is "1234."
- After perhaps 30 seconds, the phone should say that it has "paired" with Firefly, and the *status* LED should cease blinking, going instead to a continuous ON.
- The phone screen should begin to fill with the '410's repeating message. If the excessive line length is bothersome, insert in the '410 program a carriage return partway through the message,.

Connecting to a laptop The laptop requires more hand-holding than the smartphone.

- Again you must pair and provide a password. Once you have *paired* the PC with Firefly, the PC will tell you what *COM ports* it will use for Bluetooth. Here is the process in more detail:
  - Click on bluetooth icon.
  - Right click "add a device";
  - ...this should show a set of bluetooth devices that Windows has detected;
  - ...click on Firefly, then Next;
  - enter device pairing code, 1234, then Next.
  - After perhaps a minute, the clock-like small circle icon at the foot of the small tower icon will disappear, indicating that the connection is achieved.
  - Now double-click on the tower icon.
  - "hardware" will show which COM port Windows has assigned to the device.
- Double-click on the PUTTY icon.
  - Choose SERIAL.
  - Set to the COM port that you discovered Windows had assigned to this device.
  - While in this window, set up text "translation" appropriately , to show the characters sent by the micro:
  - under "Window" section, then "translation," choose Latin 4 North Europe.
  - under "Session" /default settings /save. This will save both COM port assignment and text settings for next time.

24L.2.10 Controller can listen to a PC

The serial port also allows a PC to tell a controller how to behave. The combination of this capacity with that demonstrated in §24L.2.7 permits one to combine the familiar interface of a full-scale computer with the nimbleness of a dedicated controller. Again, `serial_receive_silabs.a51`, the program we offer on the book's website provides only the simplest demonstration. It will be up to you to put this capacity to work, if you choose to.

This program uses the same initializations needed in §24L.2.7, but the program loop is much simpler: the controller simply displays whatever character is sent to it by the PC. A useful version of this program would apply the received data to a more interesting purpose, steering the controller to this or that action.

This program's loop,

```
SHOW_ONE:   JNB RI0, $   ; await receipt of a byte from UART
            CLR RI0 ;  ...clear flag when it's asserted
            MOV DISPLAY, SBUF0 ; get received data
            SJMP SHOW_ONE ; go look for another character
```

...looks almost exactly like the loop of §24L.2.7 – except that it tests a flag that indicates a character has been *received* ("RIO") rather than *transmitted* ("TIO") as in the SENDIT loop of §24L.2.7.

Here is that earlier program, SENDIT which tests *TIO*.

```
SENDIT:  MOV SBUF0, A ; Put character in buffer (send it)--UART 1
LINGER: JNB TI0, LINGER ; wait here till told it's been sent
        CLR TI0
   RET
```

In order to make this sort of program really useful, rather than just a demonstration, the controller program probably should include a *parser* that could use the transmitted text to carry out this or that act. For example, the parser could check whether the incoming character is C or D. If C, then store the next byte for later use as the *command* word for the Digipot. If D, then store the byte that follows as the *data* word to be transmitted to the Digipot. Then transmit both, and go back to checking the UART input. Thus the PC might control the Digipot setting. Again, we have provided only the most skeletal code. Flesh it out, if its use interests you.

Using *Terminal* to talk to micro The SiLabs *Terminal* program, which can show what the pc *receives* from the micro – as in Fig. 24L.12, also can be used to write from pc to micro. Fig. 24L.16 shows the screen after four values have been sent from pc; a fifth has been typed but not yet sent. The values recorded in the screen shot are ASCII values "P123," values related to the name of our Harvard course.

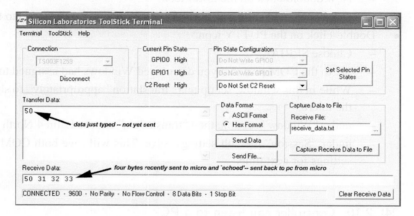

Figure 24L.16 SiLabs Terminal program permits writing from laptop to micro through serial port.

Code: controller listens to PC Download the code `serial_receive_silabs.a51` from the book's website.

24S Supplementary Note: Dallas Program Loader

24S.1 Dallas downloader

The Dallas 89C420/30/50 can have its onboard ROM loaded from a computer (let's call it a "PC") through its serial port.[1] The on-chip ROM includes a monitor program that normally is hidden. By asserting strange pin levels – $\overline{\text{PSEN}}$ along with RESET and $\overline{\text{EA}}$ – one can wake up this monitor program. Your GLUESTEP PAL does all of that when you assert its input pin named $\overline{\text{LOADER}}$. ($\overline{\text{LOADER}}$, until now, has been disasserted: tied high.)

Serial Load Hardware Configuration Figure 15-2

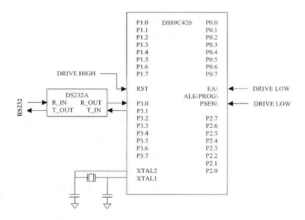

Figure 24S.1 Loader hardware connections: sketch, from Dallas datasheet.

24S.2 Hardware required

Almost no hardware addition is required, since your GLUEPAL is set up to exploit the Loader.

24S.2.1 USB-to-serial translator

The printed circuit that holds the LCD display also holds a USB-to-serial translator that allows a PC to talk to the Dallas 8051. Incidentally, it talks to the micro at TTL logic levels rather than the traditional wide RS232 serial levels. Three signals are used: TXD (transmit, from the PC), RXD (receive, from the point of view of the PC, again), and DTR (an enabling signal that we use to drive $\overline{\text{LOADER}}$). Incidentally, you can use this USB–UART link to allow a computer to communicate with your micro

---

[1] The *loader* program described in this note runs only under Windows. Dallas/Maxim provides no Macintosh or Linux version, to our knowledge.

for any purpose you like; not only to implement the loader function described in this note. Fig. 24S.2 shows the LCD/USB-translator board. The three relevant signal lines are indicated.

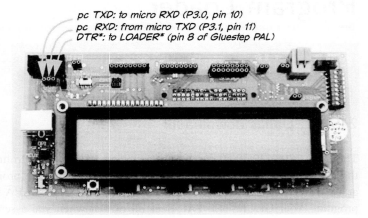

Figure 24S.2 USB to serial board: 3 lines from board to micro.

Note: The lines TXD and RXD are named as they are on the PC. They go to their complements on the micro. To say this more explicitly, ...

- TXD from the PC and LCD board connects to RXD on the micro;
- RXD from the PC and LCD board connects to TXD on the micro.

24S.2.2 Let the DTR Line drive $\overline{\text{LOADER}}$

We use DTR to determine whether to invoke the loader; DTR will be controlled by the loader program that runs on the PC. DTR (inverted by the translator) drives our GLUESTEP PAL's $\overline{\text{LOADER}}$ pin. $\overline{\text{LOADER}}$ evokes several GLUEPAL outputs in order to force the micro into the loading mode: the GLUEPAL asserts RESET51 and drives a low at PSEN* (turning on a 3-state at that output of the GLUE PAL); PSEN* is a line that till now has served only as an *output* from the controller. $\overline{\text{LOADER}}$ also evokes continuous clocking to the processor, necessary for making the device respond to the peculiar signals that wake up the built-in Loader/Monitor program.

Note: Don't forget to remove the wires that until now have *tied high* the $\overline{\text{LOADER}}$ pins on the gluestep PAL.

24S.3 Procedure to try the loader: two versions

Two loader programs are available. The current one offered by Dallas/Maxim is called MTK (Microcontroller Tool Kit). An older version, called LOADER420 uses an interface that we prefer. Dallas has dropped LOADER420, but it still is available on the web as of this writing.[2] We will treat LOADER420 first, then MTK.

Our most recent trials of the two loaders suggest that MTK is the more reliable. But you should read quickly through the LOADER420 discussion to see the operations that both versions permit. We have not written all this out in full for *both* loader programs. MTK can show you what LOADER420 can.

---

[2] We found it at Codeforge. You can also get it via our book's website.

24S.3.1 One loader program: LOADER420

After you have downloaded and installed LOADER 420, double click on the LOADER420 icon, and you'll get a window showing an "OPEN" button at the lower left, and a blank window. We need to initialize the serial port before going further.

Under PORTS/Select Port Settings choose a serial port: we like 3, and a baud rate: we like 9600. This is a *virtual* COM port, but Windows is willing to pretend that it is a traditional serial port (though Windows often gets confused as it does this; see §24S.5).

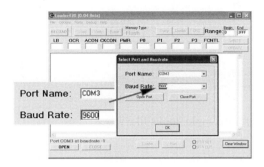

Figure 24S.3 LOADER420 serial port initialization.

Once the COM port is set up you can "OPEN" the COM port link, then click on "LOADER." You will notice that the screen now shows DTR "SET". DTR (active-low) drives the signal called $\overline{\text{LOADER}}$ on your GLUEPAL, and that signal tells the 8051 to wake up its monitor and loading program. On the LCD board, the DTR LED should be lit.

If everything works perfectly, you will get a display something like what's shown in Fig. 24S.4. This message comes from the DS89C420 and indicates both that it is properly linked to the PC and that the micro has received the combination of signals that starts up the LOADER program on the 89C420.[3]

We need to choose one more setup option, and then we can start to have some fun – downloading a program.

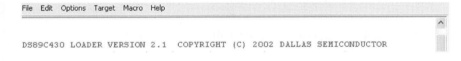

Figure 24S.4 Banner from micro indicates the connection worked. This is good news.

24S.3.2 Specify memory type: what to load

The 89C420, unlike the DS80C320 that you have been using, includes on-chip ROM (so-called "flash" type: reprogrammable, but non-volatile, much like the EEPROM you've been using on the PALs). Since we mean to load our programs not into that ROM but into the external RAM, we need to say this to the LOADER program. Under the Options/Memory type menu we choose External: see Fig. 24S.5.

24S.3.3 Try downloading a file from the PC

You need to use RIDE to generate a .hex file – a machine code file in Intel's format. You generate this file by clicking on the book icon to the right of the one you use to *assemble* the file. Copy that .hex file into the directory from which you are running the LOADER program.

---

[3] The screen shot does not show this reassuring message because it was taken from a PC not wired to the micro.

996 Supplementary Note: Dallas Program Loader

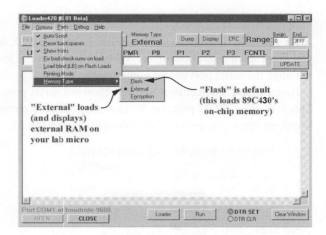

Figure 24S.5 We need to specify that the LOADER is to work on External memory.

The .hex file for the KEYSUM program of Lab 20L, for example, looks like this:

```
:02000000802E50
:0D0030009080007800E8F0F8E0280080F9EA
:00000001FF
```

The format is fussy – and the details probably don't interest you.[4]

24S.3.4 Procedure for shipping a file to the micro

Here are two ways to send a .hex file from PC to micro:

1. Option 1.
   - In the "File" pulldown menu, select "Load" This will allow you to select the .hex file for loading.
2. Option 2'
   - Issue either the "L" or "LX" command within the terminal entry window (L for on-chip memory, LX for external)
   - From the "File" pulldown menu, select, "Send File to Serial Port". This will again allow you to select the .hex file for loading.

Look at the downloaded code, using the PC: You can "verify" that the code in the micro's RAM does match the source file if you're very careful. Or (more entertaining), you can examine the contents of RAM to see if the miracle really has happened – see whether the code listed looks like the code in the source file (the hexadecimal entries, without the padding entries included in the .hex file itself).

To check what the code looks like in a particular block of your RAM, you specify a range of addresses, then click on DISPLAY.

---

[4] Just in case the fussy details *do* interest you, here's what the lines are telling us:

- the first character is a colon;
- each line begins with 2-digit data length (number of code or data bytes), then
- 4-digit start address; then
- 2-digit "record type" – 00 for code or data, 01 for execution address).
- The last byte in each line is a "checksum:" the 2's-complement of the eight-bit sum (that is, the sum "modulo 256") of all the bytes in the line, excluding the initial colon and the checksum itself.

Fanatics who want to know *more*, may want to look at http://www.keil.com/support/docs/1584.htm.

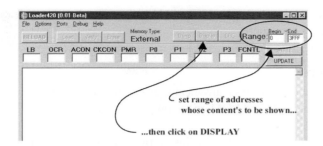

Figure 24S.6 Use DISPLAY to examine the contents of micro RAM.

In the present case, we're doing this just for fun – or to reassure ourselves. The process could be more useful: one could use it to pick up whatever code is in your RAM – say, after you had tinkered with it by hand – and then, on the PC one could use RIDE to *disassemble* the code into a tidy listing. (This is a trick we have not yet tried.)

Figure 24S.7 shows what the display looks like if we *display* the range 0 to 18h, after loading the tiny Lab 18L test program. You'll recognize the code.

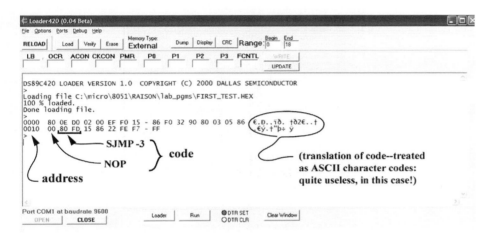

Figure 24S.7 Display of tiny test program of Lab 20L just after loading into micro's RAM.

24S.3.5 Running the micro again while tied to the PC

To allow you to run the micro's normal programs (rather than its internal monitor program that has been permitting the loading) we need to release DTR. Do that by clicking the "RUN" button at the bottom of the loader program screen. The program will not then take off, because it needs a new RESET and release – which you'll need to apply by hand. *Warning*: your micro may not work properly if you close the LOADER program and leave the serial line connected because the DTR line does not in that case simply float; it may drive DTR to a logic low, locking you in LOADER* mode at the micro. So leave the PC's LOADER420 program running, with RUN selected (DTR disasserted) in order to run your micro. If you disconnect the USB board from the micro, your computer will run as before: an internal pullup resistor on the PAL should disassert $\overline{\text{LOADER}}$ for you, but it is wiser to tie it high through a resistor of a few kΩ.

If you choose to display the range 0 to 10, for example (or if you type "DX 0 10"), the PC window should display the contents of the first 16 bytes in external RAM. You probably know what the first two or three bytes are anyway, since these hold the jump to the start of whatever program you worked with most recently: "80 xx" if you did a short jump; perhaps "02 xx xx" if you did a long jump.

If you're skeptical, you may want to take the bus to load a memorable pattern into low memory (such as "0123456..."), then invoke the loader once more.

24S.3.6 A second loader program: MTK

MTK (Microcontroller Tool Kit) is similar, but offers a less convenient user interface and lacks some features – such as the display of the contents of RAM and ROM on the programmed device.

Getting MTK: You can get MTK from the Dallas/Maxim site. The link carries you to an index where you'll want to choose the installer, MTK2_Install....

MTK: choose controller and set up serial port: As for LOADER, you'll need to do some initializations:

- under TARGET choose your controller;
- under OPTIONS/ConfigureSerialPort let's again choose COM3 and 9600

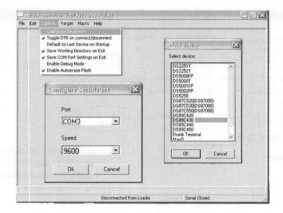

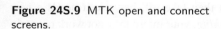
Figure 24S.8 MTK initializations: select processor and set up serial port.

Open and connect: As for the LOADER420, make the serial port then connect to Loader: see Fig. 24S.9.

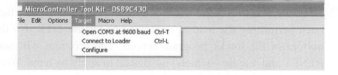

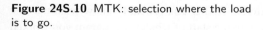
Figure 24S.9 MTK open and connect screens.

After a successful CONNECT, you will get the banner saying "LOADER420..." as in Fig. 24S.4. Again you will choose the load destination: RAM versus "FLASH" – the controller's internal ROM.

Figure 24S.10 MTK: selection where the load is to go.

24S.4 Debugging: LOADER420, in case you can't write to flash

Choose file, and send it: Then you'll need to select a file to load. It must be in the form of Intel .hex file.

Figure 24S.11 MTK: select file to load.

Then send it. When it has been loaded, you will want to *run* the program on your lab computer. To do that, you'll need to *disconnect* from LOADER.

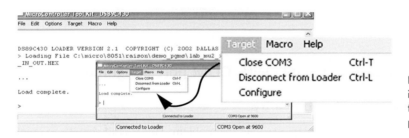

Figure 24S.12 When load is complete, *disconnect* from LOADER to run the program.

Disconnecting from the Loader *disasserts* DTR and the GLUEPAL's $\overline{\text{LOADER}}$ signal to which DTR is tied. The little computer reverts to its usual operation. You can run the program as usual, single-steppping or running full-speed.

24S.4 Debugging: LOADER420, in case you can't write to flash

You may sometimes get a cryptic error message saying "Cannot write 1's to 0's," on a write to *flash* memory. The remedy is strangely simple: under *options* choose "Load Blind to Flash." The error message and that option are shown in Fig. 24S.13.

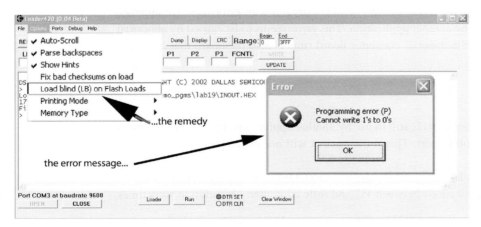

Figure 24S.13 An easy remedy if you get this *flash* error.

"Load Blind" means "omit two verification steps in the write process." Why this works we don't understand, not having understood the initial error. Fig. 24S.14 shows a successful write to flash after a failure. Between the two we selected the option "Load Blind to Flash."

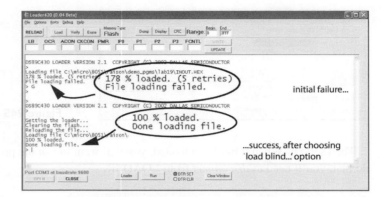

Figure 24S.14 Yes, the strange remedy works.

24S.5 Debugging in case of trouble with COM port assignments

24S.5.1 Easiest, primitive, remedy: worth a try

Sometimes simply changing USB ports at the laptop solves the problem. This is crude but it's so quick that you might try it if the LOADER tells you it cannot connect. Chances are that won't do it, in which case, read on.

24S.5.2 The problem

The PC talks to your little computer through an IC that translates USB signals to traditional serial signals. The PC runs a virtual serial port. Recent PCs have no real COM port connections to the outside world.[5] But they can emulate a COM port and, with some help from external hardware, can talk to a device that expects a traditional serial link. The PC emulates the standard serial port but sends the signals in USB form; a translator IC converts USB to serial – or vice versa when taking a message from the little computer to the PC.

Sometimes, when one tries to open the Loader420 or MTK program, an error message appears, saying that the COM port is in use by another application. This error seems to be caused by abnormal termination of the link with the USB-to-serial converter board (this is also the board that holds the LCD display).

24S.5.3 The remedy

Since the PC thinks COM3 (our usual port) is still in use by another application, the PC will automatically assign the USB board to another port. That reassignment will not match the setup within the LOADER420 or MTK program.

In the example below, Windows has assigned the device it to COM6. We need to force it back to COM3 in order to restore communications between PC and little computer. (A variation sometimes

---

[5] Until a few years ago, computers offered a so-called RS232 serial port connector: a DE9 9-pin connector carrying the wide-swinging TXD and RXD data lines, plus some control lines that sometimes but not always are used as well. Our scheme uses *one* of these other lines, DTR ("Data Terminal Ready," meaning that the computer is ready to receive data).

24S.5 Debugging in case of trouble with COM port assignments

occurs: PC assigns the board to COM3 but says, nevertheless, that the port is not available. In that event, we first must force the assignment *away* from COM3, then *back* again.)

First step: open Device Manager: Click on Start/Control Panel/System /Hardware/Device Manager. You'll see devices that include the PC's ports, including the COM ports (probably just one) – assuming the USB board still is connected to the computer.[6] In Fig. 24S.15, the single COM port has been assigned to Port 6. We must move it back to Port 3.

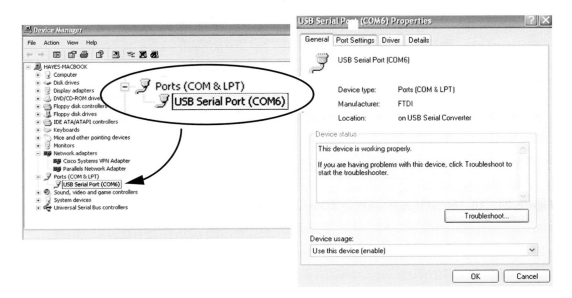

Figure 24S.15 First step in restoring COM3 connection.

Second step: force COM3 upon the PC, change the port setting: Select Port Settings, then Advanced, and you will see that the PC is using COM6. Scroll up from that port to select COM3. Close the window, going all the way back to the Device Manager window showing the COM port among many other devices. See Fig. 24S.16.

Final step: confirm that the reassignment to COM3 has succeeded: You may think you have failed because the device list still will show COM6 as in the left-hand side of Fig. 24S.17. But if you choose *Action/Scan for hardware changes*, after a few seconds the list will be updated to show COM3: the happy ending of the right-hand side.

---

[6] Recent PCs will not show a virtual COM port in Device Manager unless a serial device is present and attached to the USB port.

Supplementary Note: Dallas Program Loader

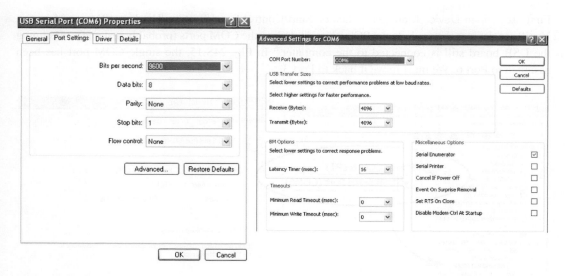

Figure 24S.16 Second step in restoring COM3 connection.

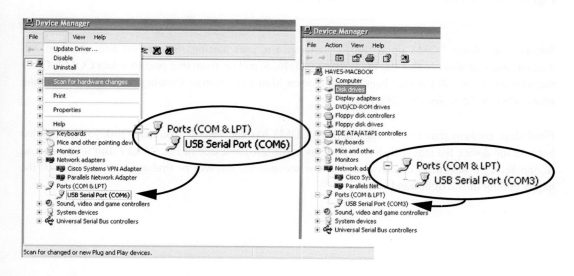

Figure 24S.17 Happy ending: COM3 again, at last.

24W Worked Example: Table Copy, Four Ways

24W.1 Several ways to copy a table

We've talked about several ways to copy data (a "table") from one location to another. Here we've printed out listings of *four* ways to do the job, for your viewing pleasure. We'll post these programs on the book site, too, in case you want to steal code from them.

Problem Write an assembly-language program that will copy 80h bytes from a *source* table beginning at address 100h to a *duplicate* table beginning at address 800h.

Solution (Use One Pointer)

```
; TABLECOPY_ONE_PTR_403.A51   TRY SINGLE DPTR FOR TABLE COPY 8/6/02, 4/03

        $NOSYMBOLS ; keeps listing short, lest...
        $INCLUDE (C:\MICRO\8051\RAISON\INC\REG320.INC)    ; ...this line should produce huge list
        $INCLUDE (C:\MICRO\8051\RAISON\INC\VECTORS320.INC) ;  of symbol defintions (all '51 registers)
        ORG 0
            AJMP START
        ORG 200h

START:MOV R6, #08 ; set copy table start address
        MOV R7, #00 ;  ...this is the low byte

        MOV DPTR, #100h ; init main pointer to source start
        MOV R0, #080h ; init sample counter

COPY:  MOVX A, @DPTR ; get a sample
        INC DPTR
        ACALL GETOTHER ; get the other pointer, and use it write
        DJNZ R0, COPY ; till all are transferred, carry on
STOP: SJMP STOP ; hang up here, when done

GETOTHER:  PUSH DPH                    ; save first pointer
            PUSH DPL
            MOV DPH,R6              ; recall second pointer
            MOV DPL,R7
            MOVX @DPTR, A ; store the sample
            INC DPTR
            MOV R7,DPL              ; save updated second pointer
            MOV R6,DPH
            POP DPL                 ; recall first pointer
            POP DPH
            RET
     END
```

Solution (Use Two Pointers)

```
; TABLECOPY_TWOPTRS_403.A51   TRY TWO DPTRS FOR TABLE COPY 4/22/03
        $NOSYMBOLS ; keeps listing short, lest...
```

1004 Worked Example: Table Copy, Four Ways

```
        $INCLUDE (C:\MICRO\8051\RAISON\INC\REG320.INC) ; ...this line should produce huge list
        $INCLUDE (C:\MICRO\8051\RAISON\INC\VECTORS320.INC) ; of symbol definitions (all '51 registers)

        ORG 0
          AJMP START
        ORG 160h
START:
        MOV DPTR, #100h ; init main pointer to start of original table ("source pointer")
        INC DPS ; get other pointer
        MOV DPTR, #800h ; secondary pointer to start of duplicate table ("destination pointer")
        DEC DPS ; ...restore main pointer
        MOV R0, #080h ; init sample counter

COPY:  MOVX A, @DPTR ; get a sample
        INC DPTR ; advance source pointer
        INC DPS ; get destination pointer
        MOVX @DPTR, A ; store a sample in duplicate table
        INC DPTR ; advance destination pointer
        DEC DPS ; ...restore source pointer
        DJNZ R0, COPY ; till all are transferred, carry on
STOP:  SJMP STOP ; hang up here, when done
END
```

Solution (Swap high-bytes) This works for small tables.

```
; tablecopy_DPH_swap_405.A51   copy 80h samples from source to second location;
; use Paul's DPH-swap method (4/28/05)

        $NOSYMBOLS ; keeps listing short, lest...
        $INCLUDE (C:\MICRO\8051\RAISON\INC\REG320.INC)  ; ...this line should produce huge list
        $INCLUDE (C:\MICRO\8051\RAISON\INC\VECTORS320.INC) ; of symbol defintions (all '51 registers)

        ORG 0
          LJMP START
        ORG 400h

START:MOV R6, #010h ; init transfer counter
        MOV DPL, #0h ; init low byte of both pointers (to 256 boundary)
        MOV R0,#01h ; init hi-byte of source pointer
        MOV R1,#08h ; ...and of copy pointer

COPY:  MOV DPH, R0 ; load source pointer (hi byte)
        MOVX A, @DPTR ; pick up a byte
        MOV DPH,R1 ; load copy pointer (hi byte)
        MOVX @DPTR,A ; store that byte
        INC DPL ; advance both pointers (lo byte shared)
        DJNZ R6, COPY ; all samples transferred? If not, do it again

STUCK:     SJMP STUCK ; when done, loop forever
    END
```

Solution (Use "Offset" from DPTR)

```
; TBLCOPY_OFFSET_403.A51  TABLE COPY USING OFFSET KLUGE

        $NOSYMBOLS ; keeps listing short, lest...
        $INCLUDE (C:\MICRO\8051\RAISON\INC\REG320.INC)  ; ...this line should produce huge list
        $INCLUDE (C:\MICRO\8051\RAISON\INC\VECTORS320.INC) ; of symbol defintions (all '51 registers)

        ORG 0
```

24W.1 Several ways to copy a table

```
        BLOCKSTART EQU 100h ; start address of source table
        BYTECOUNT EQU 6h ; arbitrary assumption, here: should reflect number of bytes in the
; block that is to be moved
        OFFSET EQU 07h    ; offset (high byte): diff betwe addresses of ORIG and DUPE tables
            AJMP START
        ORG 260h

START:      MOV R0, #BYTECOUNT ; init sample counter
            MOV DPTR, #BLOCKSTART ; init main pointer to start of original table
            MOV R1,#OFFSET ; set up offset constant (high half)

NUDGE:      MOVX A,@DPTR ; pick up byte to be transferred
            MOV R2,A ; save reg used as scratch in arithmetic
            MOV R3,DPH ; save main ptr, high half
            MOV A,DPH ; get ptr byte for manipulation
            ADD A, R1
            MOV DPH,A ; ...now points to DUPE table
            MOV A,R2 ; recover sample
            MOVX @DPTR,A ; ...and store it in DUPE table
            MOV DPH,R3 ; recover ORIG pointer
            INC DPTR ; ...and advance it

            DJNZ R0,NUDGE    ; continue, till all are transferred

STOP: SJMP STOP ; hang up here, when done
      END
```

25N Micro 6: Data Tables

Contents

25N.1	**Input and output devices for a microcontroller**	**1006**
	25N.1.1 Some sample INPUT interfacing examples	1006
25N.2	**Task for big-board users: standalone micro**	**1008**
25N.3	**Task for SiLabs users: off-chip RAM**	**1009**
	25N.3.1 The serial-access method	1009
	25N.3.2 An illustration: single-byte write and readback	1009
	25N.3.3 An illustration: multi-byte write and readback	1010
	25N.3.4 Code that uses the SPI RAM	1011

Why?

The class consists of two quite different projects for the two micro versions: for the SiLabs version, get data in and out of the controller using a few lines and a serial protocol. For the big-board, implement a design using a single-chip microcontroller (this is old hat for SiLabs people, of course).

25N.1 Input and output devices for a microcontroller

As you know, controllers can integrate many peripherals on-chip. The Dallas part includes a minimal set: timers and two serial ports, along with parallel data ports. The SiLabs '410 does better: ADC, DAC, PWM, several serial ports: SPI, I^2C, RS232, timers and a "real-time clock" that is aware of long-term timing (so, can implement long delays, or can provide timing markers for data).

But you'll sometimes need more, and routinely will need a few pushbuttons and an LED or two, at least. Fancy peripherals are likely to use one of the serial protocols – probably SPI or I^2C. But simpler devices don't warrant such heavy equipment. Here we offer examples of such simpler interfaces.

AoE offers a much richer set of examples in AoE §15.8. Figs. 15.20, 15.21 and 15.22 show about sixty parallel, SPI and I^2C peripheral examples including some sample part numbers.

25N.1.1 Some sample INPUT interfacing examples

Figure 25N.1 shows a few likely input devices. Some require some coding intelligence: the rotation sensors do, if they are to do more than "count ticks." The others are simpler. The "debouncer" uses just an *RC* slowdown, and this works because the "edge" sensor is not quite that; it senses the *level* on successive clock cycles. This slowness prevents comparator-like oscillations.

One more input device: keypad: Scanning a keypad matrix is not hard for a controller. Fig. 25N.2 shows the hardware for scanning a 4×4 switch matrix like that used in our lab keypads. The keypad connects a row and a column when a single button is pressed. The scanning algorithm therefore, is to

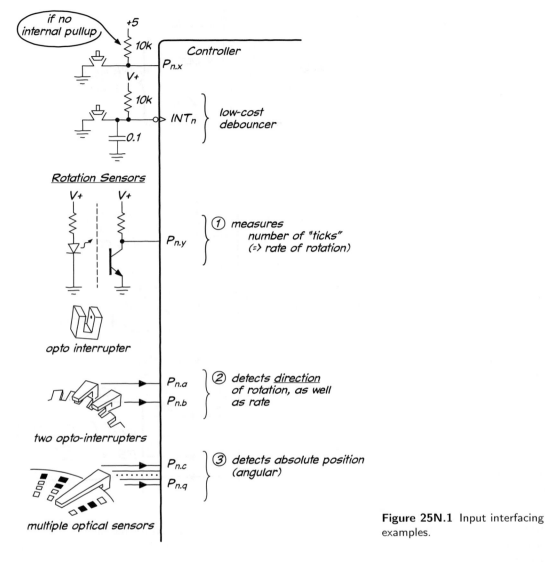

Figure 25N.1 Input interfacing examples.

continually assert just one-at-a-time of its four OUT lines, while sensing what comes in on the four IN.[1] If no input senses an assertion, no key is down. If one input senses an assertion, the program notes which OUT was driven and which IN sensed it. The pairing reveals which key was pressed and can steer the program to a table that delivers the key value. Debouncing can be implemented by putting in a delay after first sensing a press, then checking whether press persists.

Some sample OUTPUT interfacing examples: Some of the output examples in Fig. 25N.3 will be familiar. Others – particularly the Liquid Crystal Display (LCD) – probably are not.

The LCD includes a good deal of intelligence. The interface has been standardized and implements a set of commands, such as "clear" and "newline," as well as a built-in counter that places any new character next to the previous character. It is also clever enough to accept its 8-bit commands and

---

[1] The four inputs must be pulled up. Using the 8051, we relied on the built-in "weak pullup" resistors. But for a different microcontroller, pullup resistors of perhaps 10k could be required.

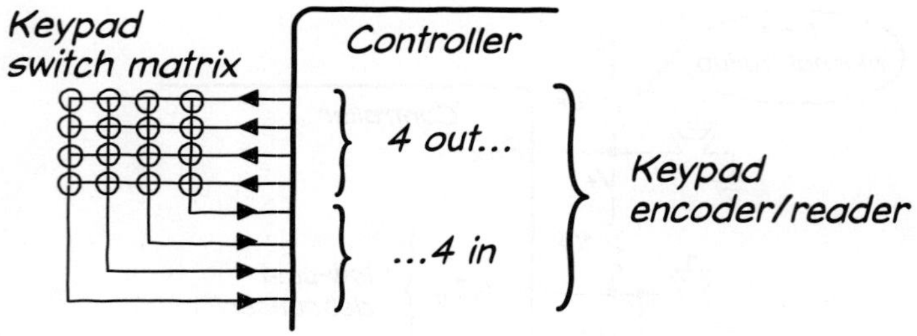

Figure 25N.2 Controller can scan a keypad (a matrix of switches). See AoE Fig. 15.8

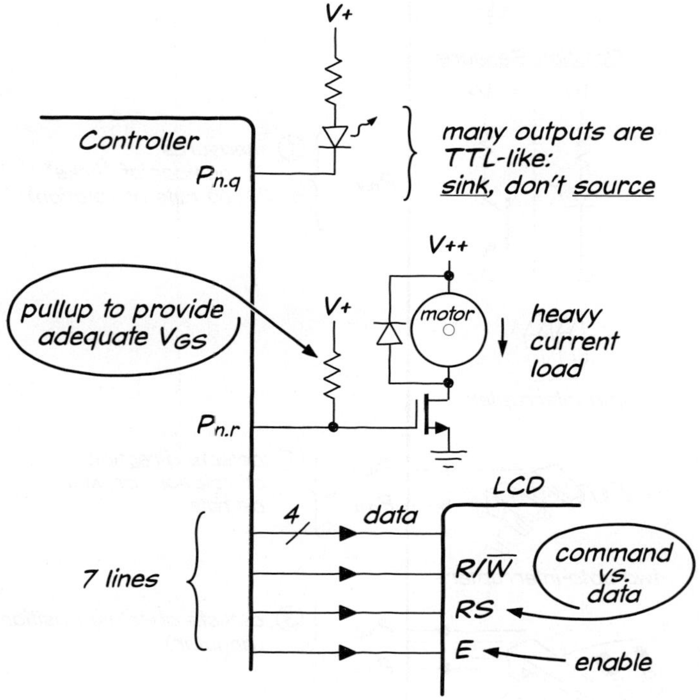

Figure 25N.3 Output interfacing examples.

characters 4-bits at a time, in order to limit the number of wires required for the interface.[2] If you consider seven lines too many (controller pins are indeed precious), you will find a smaller selection of LCDs that accept serial data: RS232, SPI or I^2C.[3]

25N.2 Task for big-board users: standalone micro

Today we ask SiLabs and Dallas people each to do something that the other group has been doing quite easily.

We ask that big-board people discover how easy it is to pump code into the 8051's on-board ROM. SiLabs people have been doing this all along of course. We don't want Dallas builders to leave the course thinking of a microcontroller as a circuit about 30 inches by 15 inches, with keypad attached.

---

[2] You can also eliminate one of the seven lines, R/W*, if the controller sets timing by relying on delays rather than by reading back a "busy" signal from the LCD.
[3] Newhaven Display International makes LCDs that accept all three: SPI, I^2C and RS232-serial; Adafruit provides an interface board that converts between SPI or I^2C and the standard LCD interface (using six pins of the LCD's standard 16; it uses six rather than the seven that we show in Fig. 25N.3 by omitting the R/W* line).

Now that you have had the satisfaction of building the big "transparent" computer with an 8051, we'd like you to enjoy how compact and simple a controller can be. We'd like you to watch it doing something with no external hardware attached, or almost none (it does need a crystal or ceramic resonator).

We hope that you had a chance to try the Loader program in Lab 24L. If not, you can try it now. We remind you of the two alternative ways to program the 8051, two methods described in Lab 25L:

- Program it in the big breadboard setup that you have been using all along. The GLUEPAL permits this loading, as you know.
- Put a little hardware on a single breadboard strip and program the 8051 there. Details appear in Lab 25L.

We hope that both big-board and SiLabs people will dream up some small task to do with this standalone machine. Almost anything is more fun when it's your own conception. You'll want to do more than blink an LED; but perhaps not much more. You may want to take an advance look at Chapter 26N which includes our list of gadgets that might be fun as peripherals – things like accelerometers, motors, a talker chip.

25N.3 Task for SiLabs users: off-chip RAM

In both versions of Lab 24L we built RAM data tables in a straightforward way using a moving *pointer* – the familiar DPTR – to store and to recover data, one byte at a time. This allowed the Dallas processor access to a 32K RAM, enough to allow many seconds of audio, for example. The SiLabs '410 controller looked a little sad with its mere 2K of on-chip RAM available for data storage.

But the '410 can make up for its lack of buses by using *serial* instead of parallel access to an *external* 32K RAM. The code that carries out this serial access is much more complicated than the code for the bused scheme. It is complicated even though the controller includes a hardware implementation of the SPI serial protocol.

25N.3.1 The serial-access method

The serial RAM available to the SiLabs controller has some brains within it, brains that allow it to communicate despite its paucity of pins: just eight. Like other SPI peripherals (including even the simple SPI Digipot mentioned in Lab 24L) the serial-access part needs to be told what to do with values fed to it. It therefore understands a small collection of *commands*.

The 32K SPI RAM that we use is capable of storing or reading a single byte at a specified address. But for building and reading tables it provides a neat alternative: it can store a sequence of bytes at successive addresses using its internal counter, automatically incremented upon each data write. The playback works the same way. Both multi-byte operations – write and read – require only an initial address specification. After that, the RAM is pretty-much autonomous.

25N.3.2 An illustration: single-byte write and readback

Before sending data we must tell the SPI RAM how we want it to handle the data that we present. Here, the "status write" command, 01h, tells the RAM "what's coming in the next byte will be a mode command." The byte that follows selects the mode: here, "single" byte, 01h.

Once we've told the RAM to take a single byte, we can issue a data write command (02h), followed by the address and data. Fig. 25N.4 shows just the beginning of the frame in which the data is written.

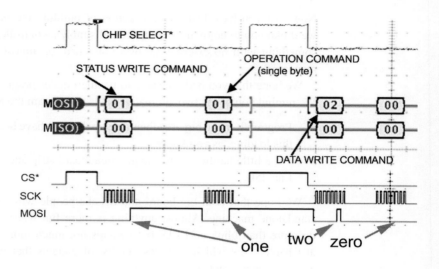

Figure 25N.4 First send the SPI RAM a command to tell it which of its "modes" to use.

Fig. 25N.5 shows the writing of a single byte to an arbitrary address, then its readback. The controller writes the data value AB into location 24 45, then reads it back, as you can see in the figure.

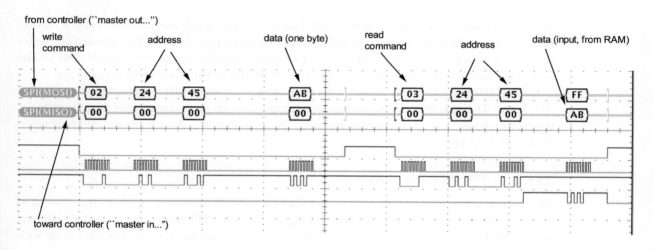

Figure 25N.5 SPI write and readback using Microchip 32K serial RAM.

The very first item in each frame – whose beginning is flagged by the assertion of CS* – is a *command*, as you would expect: *write* (02h), later *read* (03h).

25N.3.3 An illustration: multi-byte write and readback

When the goal is to store data in a table and then to read it back, the RAM's ability to handle a long sequence of data bytes is just what we need. In Fig. 25N.6 the controller sends three bytes of data (AAh, BBh, CCh) to three addresses beginning at the start of RAM (address zero).

The successive writes occur quickly, thanks to the automatic incrementing of address. In contrast to the single-byte action of §25N.3.2, here there is no need to specify the successive addresses, once the start address has been defined.

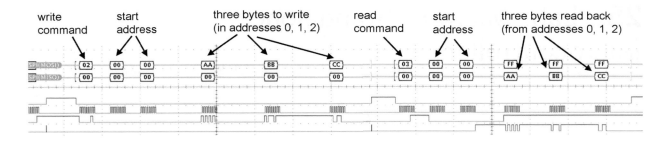

Figure 25N.6 SPI RAM takes and delivers three successive bytes, with only the start address specified.

25N.3.4 Code that uses the SPI RAM

The program below, which writes a series of bytes to the SPI RAM, is conceptually simple but full of fussy detail.

```
; now write  several bytes
WRITE_SEVERAL:   acall START_VALS  ; initialize sample counter
         acall INIT_ADDRESS ; set up start address

         mov R7, #WRITE_COM  ; START_OPERATION will use this to send appropriate command

         acall START_OPERATION ; assert CS*, send command
         acall SEND_ADDRESS
SEND_ONE: acall GET_SAMPLE ; get a byte from ADC
         acall SEND_DATA
         acall CHECK_END ; see if all done (this subroutine sets flag, ALL_DONE when table full)
         jnb ALL_DONE, SEND_ONE ; ...if not all done, send another
         clr All_DONE           ; if all done, clear the flag and proceed
         setb NSSMD0       ; disassert CS* at end of XmssN

; ...and read them back
READ_SEVERAL: acall START_VALS
         acall INIT_ADDRESS ; set up start address
         mov R7, #READ_COM ; now get the value back from RAM

         acall START_OPERATION ; assert CS*, send command
         acall SEND_ADDRESS
READ_ONE: acall GET_DATA
         mov DISPLAY, R3 ; show what RAM delivers (just for single-step debug)
         mov IDA0H, R3 ; ...and send to DAC

         acall CHECK_END ; see if all done
         jnb ALL_DONE, READ_ONE ; ...if not all done, recall another
         clr All_DONE           ; if all done, clear the flag and proceed
         setb NSSMD0       ; disassert CS* at end of reception


         sjmp $ ; this to test single transfer
         sjmp WRITE_SEVERAL ; ...this to do continual transfers
```

This listing omits the gritty detail of the several subroutines. The full program,

  `spi_ram_adc_dac_incomplete.a51`,

is available on the book's website. The word "incomplete" in the filename refers to the requirement that a user write the short block of code that implements an *end test* for the program's table write and read.

25L Lab: Micro 6: Standalone Microcontroller

Reminder: There's an easier way to use a microcontroller! Once again, we are saying – after you have invested a great many hours of hard work – "...of course, you'll never want to do it *this* way again!" We did this to you after you had learned to design with discrete transistors. We did it after you had done some building of logic with little 74xx ICs. Now we are doing it after having you wire up the big-board computer, mostly entering its code by hand. In the present case we mean to remind you, before you leave the course, of the normal way to use a microcontroller: program it using a computer, then let the controller work alone with little or no off-chip hardware. This the SiLabs users have been doing all along. (But they have missed the interesting hardware bugs that you have enjoyed.)

In this lab we suggest a skeletal application so that at least you will be able to play with the controller as a standalone computer on a single breadboard strip. You may want to do something more than this preliminary play – but that's up to you. You may, on the contrary, want to return to your Big Board home-made computer and use that to implement a project of your own devising.

If the pulse-width task described here in §25L.1.3 doesn't appeal to you, take a look at some other code that we're providing:

- LCD display (these displays will show successive characters when fed ASCII codes at a single port);
- keypad scan (produces a key value upon the press of a single key in a bare 4×4 keypad like the ones included in our lab keypads)

Neither of the above two programs is interesting by itself – but either might be useful as an interface in some application that you devise.

25L.1 Hardware alternatives: two ways to program the flash ROM

25L.1.1 The preferred scheme: use your bused computer to try loaded code

If your little computer will take loads and run them from RAM we recommend proceeding that way: load the code into the external RAM and run it as usual – with the useful facility of *single-step*. When you are satisfied with the program's behavior do a second download, this time into the 8051's on-chip ROM (so-called flash memory); pull the programmed part from the big board, and install it on a single strip, where it can run on its own. You can try running from ROM while the part is in your big circuit, if you now tie the EA* pin high (see Fig. 25L.1).

The GLUEPAL already takes care of the loading. All you need add to try running from internal ROM, is a switch that can tell the micro to take its code from internal memory, rather than to the external, buses. It's as simple as this:

Incidentally, code will run faster from internal memory than from external RAM – on average, about three times faster – since for many instructions it is able to reduce the normal 4-clock cycle to

25L.1 Hardware alternatives: two ways to program the flash ROM

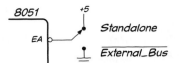

Figure 25L.1 A toggle switch will let you try running the micro from internal ROM.

one clock. This effect will not matter to you unless you happen to be running code that does timing in software.

25L.1.2 An alternative scheme: program the micro on a standalone board

About five minutes' wiring can produce a breadboard on which you can program the standalone controller without the use of your 'big board' computer. This works well,[1] but does not allow for debugging by watching bus activity as you single-step through code. A self-contained controller uses no external buses, and its opacity can be frustrating when a program doesn't do what you expect. Simulation in RIDE can make up in part for the black-box quality of a microcontroller.

It *is* possible to single-step the controller on your big-board setup, even when it is running code from its internal ROM. You can enable single-stepping by turning on the controller's ALEON bit (MOV 0C4h, #04h is the line of code that does the job). Doing this activates ALE, on which the GluePAL's single-step function relies for its timing. In the posted Standalone programs we have turned on ALE by setting that ALEON bit. Nevertheless, troubleshooting remains radically more difficult when one cannot see bus activity.

A circuit diagram: Figure 25L.2 shows the circuit. The serial connections are familiar; the use of DTR may not be. DTR (active low) here puts the micro into the peculiar state that makes it ready to receive a program load through its serial port.[2] DTR's assertion:

- drives RST high;
- drives PSEN* low – leading the micro to turn on its internal ROM and set up the serial port;
- drives EA* low, to permit the loading.

Disassertion of DTR releases all those lines, allowing the micro to run from its newly-loaded ROM code.

A circuit portrait: Figure 25L.3 shows what the standalone circuit looks like. This circuit is blind, deaf, and dumb, so it's not much use in itself. We hope, of course, that you will try adding some hardware to the programmed micro to make it do something satisfying.

The little tin can plugged into XTAL1 and XTAL2 is a crystal that makes use of the controller's on-chip oscillator.[3] It is smaller and less expensive than the oscillator used in the full computer. Or, you can clock the controller with the oscillator you have been using in the earlier labs.

---

[1] It is easier to make this scheme work than to load your full computer because this circuit sidesteps many possible circuit defects, such as the full micro's possible inability to write to RAM. That action was never called for in the micro labs to this point, and thus was not tested. So it is possible for an otherwise healthy lab computer to refuse to do a RAM download, though it may be able to download to the on-chip ROM.

[2] On the full big-board computer, your GLUEPAL drives these signals in the same way, in response to DTR. The gating in Fig. 25L.2 does one more useful task: it drives EA*. On the full computer, you are obliged to switch EA* by hand.

[3] More technically, the controller expects a fundamental-mode parallel-resonant AT-cut crystal. See DS89C430 datasheet, pin description section, p. 12.

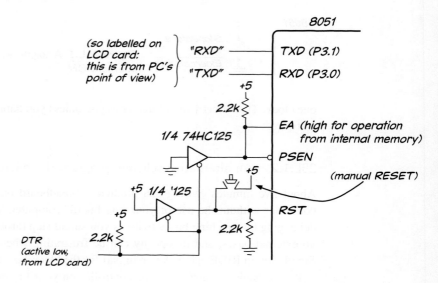

Figure 25L.2 Hardware that allows loading a standalone 8051.

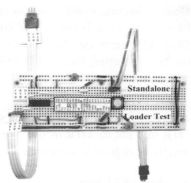

Figure 25L.3 A photo of breadboarded hardware that allows loading a standalone 8051.

LED to test your hardware: An LED driven from the corner pin, P1.0 will let you confirm that your loader hardware works (a 3.3k resistor will limit LED current). Load the controller with one of the BITFLIP-with-delay .HEX files available on the website. Disconnect the loader, and see if the LED blinks. If not, try a manual reset of the 8051. When you see the blinking, you can proceed to find a use for this loadable controller.

The remainder of this section suggests a little bit of I/O for the standalone, and some ready-made code that one might be able to put to entertaining use.

25L.1.3 Hardware: a circuit to measure pulse width

Figure 25L.4 shows a circuit that can be used to let the standalone micro measure and display period. The displays sketched are a hexadecimal LED type. You may prefer to use the LCD that you have been using in earlier micro labs.

25L.1.4 Code to use this hardware

The "hardware" is pretty generic of course, so it is capable of doing lots of tasks. It's well-adapted to counting, and those are the applications we have in mind. Later, we give two versions of a program to record (and display) the duration of a logic *low* input pulse: one version shows the duration in *decimal*

25L.1 Hardware alternatives: two ways to program the flash ROM

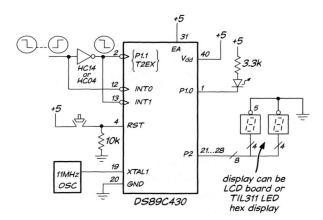

Figure 25L.4 DS89C420 wiring for standalone use on a single breadboard.

form, and is timed in software.[4] The other version records the duration in *binary* form, showing it in hexadecimal on the displays. That version uses the 8051's internal timers.

The decimal version is better if the pulse duration is offered primarily for display; the binary/hex version (which makes more efficient use of its counting bits) is better if the duration is measured not primarily for display, but rather for use in a larger program.

We have posted the source files (.a51), rather than the .lst files as in earlier labs, because we assume that you now will prefer to download code from a PC to your lab micro (and you will need such a facility if you want to burn code into the standalone controller; you will use RIDE to assemble and link the .A51 file, generating the .HEX file that you can ship to the controller). We also post the .HEX file, in case you want to get it ready-made.

Decimal duration program: The falling edge of the input pulse drives INT0*, whose ISR starts the software counter. The trailing edge of the input pulse drives INT1*, whose ISR stops the counting and displays the result. The range of pulse widths that this program displays (we're showing only the top byte of a 16-bit quantity) spans 100 counts, from about 0.3ms to 30ms, when running full-speed, as in its present form. When the count overflows, the Overflow LED warns you not to trust the displayed value.

25L.1.5 Binary duration program

This program too uses start and stop of the pulse to drive the respective interrupts, 0 and 1. The counting is done differently though here: INT0's ISR starts an internal 16-bit timer after clearing it; the end of the pulse has two effects: it triggers INT1, as in the other program, and INT1's ISR displays the resulting count; the end of the pulse also triggers a timer *capture* operation: the timer count is recorded in a 16-bit "capture" register. This trick probably gets us nothing in this program: we could just as well have read the timer registers themselves after stopping the timer. We just wanted to try the "capture" feature.

You can tinker with the rate at which the internal timer increments, to adjust for a convenient range of pulse width. With the counter running full-speed, the range of displayed values (again the top byte of a 16-bit count) is from $40\mu s$ to about 6ms.

---

[4] The program listed runs at full speed; to measure wider pulses without overflowing, insert a CALL to a DELAY routine.

25L.1.6 Some ways to use the programs

First, test the setup: If you load one of these two programs into a DS89C430, and plug that IC into a breadboard wired with a display, a function generator can provide a TTL square wave for the circuit to measure. Once you're satisfied that the device works, you'll probably begin to cast about to measure something more interesting than a the output of a function generator.

Capacitance meter (again?!): You may have done this before, and may be tired of the notion; if so, push on to an application of your own invention. If you're not yet bored, then you can use a '555 oscillator, powered from +5V and ground, to feed your duration meter with a square wave to measure. You may possibly recall, from Lab 8L, that the waveform's time *low* (approximately its half-period) is $0.7 \times RC$, where R is the resistor between the capacitor and the discharge pin and C is the capacitor that you install. You might choose an RC value that gives a reading that is close to full-scale. Then try plugging in smaller capacitors, to see if the count drops proportionately.

If you start with full-scale for a $0.33\mu F$ cap, for example, include a trimmer potentiometer (used as variable resistor) so that you can adjust the pulse duration to produce something close to a count of "33." See if $0.1\mu F$ produces a count close to "10," and so on. Does a second 0.33 cap, plugged in parallel, show the effect you expect?

If you have the energy, you can add a third digit to your display, increasing the range of your meter. You'll need to modify your code to pick up the high four bits of the bottom byte that measures pulse duration.

Reaction timer (again?!): Yes, you've seen this before – but done with dedicated hardware rather than with a micro. Again, skip this if it all looks too familiar.

If you let a debounced switch drive INT0* low, then another switch drive INT1* low, the micro can measure the time between these events: the circuit could serve as a *reaction timer*. But you'll need to slow the count of the decimal version: you need to increase the full-scale pulse range by about a factor of 10, to about 300ms.

IR remote reader (fancier): This one is more interesting. Once you can measure pulse width, you're just a few steps from a gadget to read the bit-sequences sent by an IR remote control. Fig. 25L.5 is a scope image showing three examples of codes sent by a Sony IR television remote. The raw signals from the remote that generate these logic levels are shown in Fig. 25L.7. The full code sequences are much longer than what we show. But what is shown would be enough to allow a program to distinguish at least the three codes here. (This figure is a detail of Fig. 24N.13.)

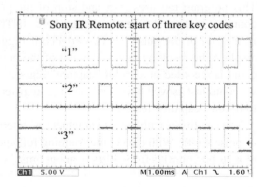

Figure 25L.5 Three key codes from Sony TV IR remote (detected by IR receiver IC).

25L.1 Hardware alternatives: two ways to program the flash ROM

IR translator IC: The remote puts out not quite the logic-level signals shown above but instead bursts of 40kHz oscillations. An integrated IR photodiode and detector circuit translates those bursts into logic levels. The part is shown is Fig. 25L.6.

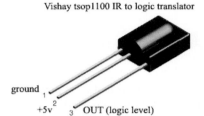

Figure 25L.6 IR translator.

Figure 25L.7 shows the ≈40kHz oscillation that the IR remote sends; this signal is integrated by a receiver IC, then sent to a comparator included on the IC to give a logic-level output as shown in the figure, where the bottom trace shows logic levels like those of Fig. 25L.5.

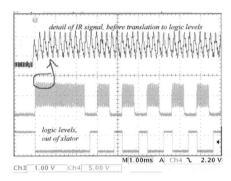

Figure 25L.7 Detail of IR remote transmission: bursts of 40kHz oscillations.

The IR remote signal is robust because it need only sense whether a burst of 40kHz is present; it needn't detect each transition as our audio project did back in Lab 13L.

A program strategy: Your program needs to detect the long START* pulse (2.5 ms), in order to get synchronized. To detect that, invoke a pulse-duration program, and compare the resulting count with a minimum-width reference. The Sony START* lasts for a count of about 6Dh by the binary "capture" program of §25L.1.5 when the timer runs at full speed. So set a minimum reference safely below that typical value.

Then, having found the START* pulse, detect the width of the next pulse. This next pulse is the first that carries code information: it will be either 600μs or double that. So make your *data* pulse reference something between. The timer shows 1Ah for 600μs, double that for the longer pulse.

When your program decides whether a pulse is Long or Short, it can branch to a patch of code that shifts a 1 or a 0 into a register. After doing this four times, you will have enough information to distinguish the keys 0 through 9.

A use for a remote reader? Every gadget these days has its remote, but you might enjoy controlling something across the room in the lab: making a motor spin forward or backward or stop? Selecting among several sound sources to feed a speaker? We admit it: we're groping. It's decoding the remote that's fun, rather than the experience of controlling something with it – for that experience is rather commonplace these days.

25L.1.7 Two 16-bit pulse-duration counting programs: one decimal output, one binary/hex

`pulse_measure_decimal_standalone_505.a51` and `pulse_measure_capture_standalone_505.a51` are posted on the book's website.

25L.2 SiLabs 6: SPI RAM

Port pin use in this lab: The wiring (see Fig. 25L.8) is essentially the same as that used in §24L.2. Again we use INT0* to time the program. INT0* replaces RXD that used P0.5 late in §24L.2.

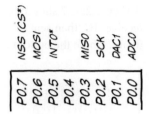

Figure 25L.8 PORT0 pin use in this lab: INT0* returns to P0.5.

25L.2.1 One last prescribed exercise: big RAM table

This exercise is brief: we propose that you download a longish piece of code that will allow your '410 access to a RAM big enough (at 32K bytes) to store some sound. Conceptually, this application is the same as one you met last time with the on-chip RAM. The differences are three-fold:

- the larger capacity of this RAM will let you have more fun: 32K will allow you to record a little speech, or sing a snatch of song, and hear it played back;
- this external RAM will introduce you to another serial-access peripheral (in addition to the SPI Digipot that you may have met last time);
- for the first time you will run into the need to take account of a device that cannot tolerate a 5V supply: the SPI RAM must be powered from 3.3V.

The 3.3V supply presents no difficulty for the '410 which is happy to run at that supply voltage. As you set up your circuit though, you must take care to avoid inadvertently powering or driving either '410 or RAM with 5V.

Power the '410 and RAM with 3.3V: You will need to *disconnect* the line that, until now, has used the USB power fed through the programming adapter to power the '410. That lead, from the top left corner pin of the cable, now just dangles; in its place, the 3.3V supply available on that cable's lower left corner now should power the '410 and the SPI RAM. This change is shown in the photo of Fig. 25L.9, and is sketched in the diagram of Fig. 25L.10.

Link '410 to SPI RAM: Four lines of the SPI bus – two data lines, one for each direction; a clock; a chip-select* – link controller to RAM. An interrupt input to the '410 comes from an external function generator: see Fig. 25L.10.

25L.2 SiLabs 6: SPI RAM

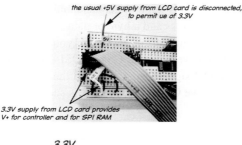

Figure 25L.9 3V supply replaces 5V, as supply from LCD card's cable, for both '410 and SPI RAM.

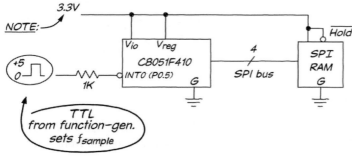

Figure 25L.10 Power the '410 and RAM with 3.3V. DISCONNECT the 5V used up to this point.

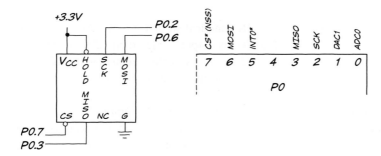

Figure 25L.11 Controller and RAM are linked with four SPI lines.

The 1k series resistor on the *interrupt* line – which sets the sampling rate – is included to limit current in case the '410 objects to a signal beyond its 3.3V supply.[5]

NOTE that the PORT 0 pin assignments shown in Fig. 25L.11 *differ* from those used in earlier labs. Please make the necessary changes. Particularly, don't forget that the DAC output at P0.1 is a *current* (0–1mA), and needs a resistor to ground (1.2k) to convert it to a voltage: see Fig. 25L.12.

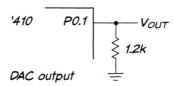

Figure 25L.12 DAC output is a current; *R* needed to convert it to a voltage.

25L.2.2 Trying out the SPI RAM

Watching the SPI signals: Some people will want to hurry on to using the RAM to record waveforms. But if you're curious to watch the SPI signals you may want to try a simpler program, one that sends just a single byte to RAM and reads it back. The program `spi_single_byte.a51` on the website does

---

[5] We think it does not so object at this open-drain port pin. See Datasheet at pp. 151–152: "Port I/Os on P0 are 5V tolerant over the operating range of VIO." But we're playing it safe.

this just once, so that a frozen screen image of the several signals is quite understandable. You have seen in Chapter 25N what to expect: a command followed by address and a byte of data.

If you watch the SPI in action, trigger on CS* (falling edge) in Normal mode (so you keep one, fixed image). If you have a four-channel scope, you can watch every signal: CS*, MOSI and MISO. (The latter two signals, as you recall, are the strangely named signals from and into the controller from the SPI RAM: "Master Out Slave In" and "Master In Slave Out.") Fig. 25L.13 is such a scope image, borrowed from Fig. 25N.4.

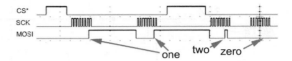

Figure 25L.13 SPI is just barely intelligible.

25L.2.3 More fun: try the 32K storage program

The fun, today, will lie in playing with storage and playback of audio. As usual, you will want the good reconstruction filter – the MAX294, 8-pole elliptic lowpass. Use a TTL output from a function generator to clock it. In Chapter 23N you saw the MAX294 circuit, and also a circuit for an audio amplifier to feed with the DAC output signal.

Today, for the first time, you have enough storage (32K samples) so that it is worthwhile to pull out the *microphone amplifier* that you built back in Lab 7L. Its output, centered on 1.25V, can be fed directly into the ADC input of the '410 (whose range is 0–2.2V; 1.25V is close enough to the midpoint of that range; if you're a perfectionist, you can move the quiescent voltage down to 1.1V by shifting the amplifier's bias at the non-inverting input).

The 32K samples will let you record many seconds of speech or song: a 1k samples/second rate is adequate for speech by a male voice. To get away with such stingy sampling, you will need the good reconstruction filter (the MAX294), of course. If you want to reproduce a song, your sampling should be less stingy.

If you can round them up, set up three function generators:

- One to provide a sinusoidal test input to feed the ADC (later, we expect you'll replace this with your mike amp).
- Another to provide a TTL *sampling* signal: this drives the '410's INT0*. We would again suggest a 1k series resistor, since the '410 is powered with 3.3V.
- A third (TTL) to drive the clock on the MAX294 filter. You'll recall that the '294 is a *switched-capacitor* filter: its f_{3dB} or f_{cutoff} is $f_{clock}/100$.

Figure 25L.14 is a reminder of the scheme. You'll want to start, no doubt, with a sinusoid from a function generator. But when you have played with sampling rate, seeing how stingy you can be with the help of your good filter, we hope you'll pull out the microphone amplifier and start talking to your machine (or singing, if you have courage).

25L.2.4 Complete the storage scope program

The program is *almost* complete. The SPI code makes this program long and fussy, and we don't want to stop you to make you study all that detail. Instead, to give you some sense of ownership over this program, we ask you to write just a routine that determines whether the table-filling is complete.

The goal is to use all available SPI RAM. Since it is a 32K part, its addresses run from zero up to

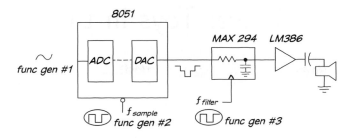

Figure 25L.14 Three function generators provide input signal, f_{sample} and f_{filter}.

$2^{15}=32$K. The address one beyond this limit – as you may recall from our many big-board examples that use the high half of address space for I/O – is the hexadecimal value 8000h.

This program uses a 16-bit *counter* to determine when storage is complete. Since the 8051 offers only one such register, DPTR, we take advantage of this register – but putting it to an unusual use: not to define an *address* as usual but simply to *count*. Here is the CHECK_END routine, waiting for you to implement it in assembly language:

```
;********************************************************************************
CHECK_END:                    ; save register(s) we'll mess up
                              ; increment 16-bit counter
                              ; beyond end of RAM?
                              ; If not, go get another a byte
                              ; ...if address 8000h, beyond RAM...
                              ; ...then set flag bit that main program tests
GO_AHEAD:                     ; restore register(s) we've messed up
                              ; and go back to main program
;********************************************************************************
```

CHECK_END is called after each *write* to RAM and after each *read*. Program initializations include a zeroing of DPTR before the *writes* begin, and then a reinitialization to zero before the *reads* begin.

In order to test the value of DPTR, you should consider the limitations on the compare operations available to you:

- the controller lacks a 16-bit compare operation;
- the available compare, an 8-bit operation, CJNE, is most versatile when operating on the accumulator or **A** register.

The flag bit that tells the main program that table-filling is complete is named "ALL_DONE." It has been defined at the head of the program, so your code can set a bit using that variable name. You needn't go look up where the bit lives; that, of course, is the beauty of descriptive signal names.

You can insert your patch of code by editing the file `spi_ram_adc_dac_incomplete.a51`, which is posted on the book website.

25L.3 Appendix: Program Listings

The following programs are available on the book's website:

Single-byte Write and Readback: `spi_single_byte.a51`.

32K RAM store and playback code: `spi_ram_adc_dac_incomplete.a51`.

Two versions of a pulse-width program:
  `pulse_measure_decimal_standalone_505.a51`.
  `pulse_measure_capture_standalone_505.a51`.

SPI RAM code to complete: `spi_ram_adc_dac_incomplete.a51`.

26N Project Possibilities: Toys in the Attic

Contents

26N.1	**One more microcontroller that may interest you**	**1023**
	26N.1.1 PSOC configuration	1024
26N.2	**Projects: an invitation and a caution**	**1025**
26N.3	**Some pretty projects**	**1025**
	26N.3.1 Laser character display	1025
	26N.3.2 Sound source detector	1026
	26N.3.3 Computer-driven Etch-a-Sketch	1027
	26N.3.4 A giant Etch-a-Sketch	1028
	26N.3.5 An attempted insect	1028
26N.4	**Some other memorable projects**	**1030**
	26N.4.1 X–Y scope displays	1030
	26N.4.2 Complications and improvements	1030
	26N.4.3 Size control; zoom	1032
	26N.4.4 True vector drawing: position-relative	1033
	26N.4.5 Animation	1034
	26N.4.6 Computed drawing	1035
	26N.4.7 Laser X–Y display	1035
	26N.4.8 Charmingly simple LED display	1036
	26N.4.9 Vehicles	1036
	26N.4.10 A neat application for the independent-drive steerable chassis	1037
	26N.4.11 A computer-steered toy car	1038
	26N.4.12 A perennial heartbreaker: inverted pendulum	1039
26N.5	**Games**	**1041**
	26N.5.1 Pacman on scope screen	1041
	26N.5.2 Asteroids on a scope screen	1042
	26N.5.3 Other great games	1043
26N.6	**Sensors, actuators, gadgets**	**1043**
	26N.6.1 A few good sources of sensors, actuators and other devices	1043
	26N.6.2 A source of small parts for mechanical linkages	1044
	26N.6.3 Mechanical: motors, etc.	1044
	26N.6.4 "Servo" motors	1044
	26N.6.5 Nitinol muscle wires	1045
	26N.6.6 Transducers	1046
	26N.6.7 Displays	1048
	26N.6.8 Interface devices	1048
26N.7	**Stepper motor drive**	**1049**
26N.8	**Project ideas**	**1051**
	26N.8.1 Nifty new and untested projects for your inspiration	1051
26N.9	**Two programs that could be useful: LCD, Keypad**	**1052**
	26N.9.1 LCD driver; keypad scanner	1052
26N.10	**And many examples are shown in AoE**	**1052**
26N.11	**Now go forth**	**1052**

26N.1 One more microcontroller that may interest you

We'll not have time to try these "PSOC" devices, but you may at some later time want to. These integrate a microcontroller (an 8051, in the instance shown below) with *analog* parts that can be configured by downloaded code. These parts, which Cypress Semiconductor calls, a bit grandiosely, "Programmable System on a Chip," begin to permit an analog equivalent of the programmability you are accustomed to seeing in PALs.

We include below a few pages from a PSOC data sheet, describing their integration of a rather standard controller with the novelty of a couple of *operational amplifiers*, as well as comparators (a feature that is not new: the SiLabs '410 includes one). It is not only the operational amplifier that is integrated; that would hardly be much of an achievement. It is also the configuration of the op-amp circuit that is to some extent programmable: the op-amp can be configured for various gains, or to form an I-to-V ("transimpedance") converter, or even a sample-and-hold.

We expect that this is just a modest beginning to a richer set of programmable analog functions. Already these parts have enjoyed commercial success.

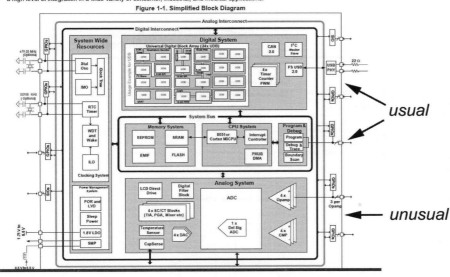

Figure 26N.1 PSOC block diagram: 8051 plus standard digital parts and converters – plus op-amps. (Figure used with permission of Cypress Semiconductor Corp.)

The analog parts are shown in a little more detail in the block diagram of Fig. 26N.2. Those "SC/CT" blocks mark *switched-capacitor* blocks[1] that implement the effective equivalent of *resistors* in the analog elements. They can be used to set amplifier gains, as in Fig. 26N.3, for example.

---

[1] CT, "continuous time," seems to be treated as equivalent rather than a contrast.

Project Possibilities: Toys in the Attic

Figure 26N.2 PSOC block diagram showing analog parts. (Figure used with permission of Cypress Semiconductor Corp.)

26N.1.1 PSOC configuration

Figure 26N.3 is an excerpt from the PSOC data sheet showing how op-amp circuit gains can be set. Only particular values are available.

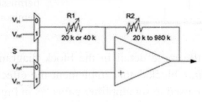
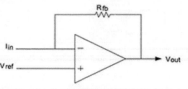

Figure 26N.3 PSOC: selected op-amp circuit gains set using digital codes. (Figure used with permission of Cypress Semiconductor Corp.)

26N.2 Projects: an invitation and a caution

In our own version of this course, only a minority of the busy students choose to do projects. But a project can be heaps of fun. To help you conceive of one, here is some information on gadgets and ideas that might inspire a project builder, along with sketches of some great projects of yester-term.

General advice:

- Try your ideas on someone else, early in the process. Someone with experience can steer you away from the impossibly grandiose; less often, she might tell you that your invention is too modest to be called a "project."
- Choose a project appropriate to the little computer, not a project that's better done on a full-scale computer. That means, among other points, keep the programming modest. It also may mean for example, *making a motor spin* rather than doing a computation and then displaying a number.
- Try to include some hardware additions along with your invented software. This point is partly a corollary of the preceding point, but also reflects a second aim: the project often works to help you firm up your grip on ideas, by letting you do some designing. It can serve as a sort of review. A broader review is better than a pure exercise in programming.
- Make your project *incremental* rather than all-or-nothing. For example, the people who built the computer-controlled RC car (§26N.4.11) started out proposing a game in which the computer would make one car chase a second car that was controlled by a human. We persuaded them to try this in stages:
    - first, see if the computer could control the car (pretty easy);
    - second, see if their ultrasonic ears could get significant information from the *delay* times between the car's ultrasonic squeak and the reception at the computer's three "ears;"
    - third, build the link between the little computer and the PC.
- You get the idea. In fact, these students were able to get most of their original scheme to work. But tasks nearly always take longer than one expects, and we want you to feel satisfied with a modest project, rather than frustrated by a grand one.

26N.3 Some pretty projects

Later in these chapter, beginning at §26N.4, we mention some memorable past projects from our course. Beginning at §26N.6 we suggest some hardware that we offer to builders – along with the much richer possibilities of hardware available from hobby and robotics sites. In this preliminary section, in hopes of exciting ambitions, we offer a glimpse of some of the more photogenic gizmos produced by the energetic minority who found time for a project.

26N.3.1 Laser character display

This student persevered despite our advice to try something less ambitious. What he proposed seemed much too complicated: to spin a set of differently inclined mirrors, letting a laser switch On and Off appropriately, to produce a projection of characters on a screen.

We gasped when he brought the thing in. We also felt uneasy for his beautiful machine, noting that it was mounted with *plasticene*. Even the aiming of the laser was accomplished by pushing the plasticene blob this way or that.

Fig. 26N.4 shows the prettily fabricated set of mirrors, each with its indexing toothpick. One other toothpick serves to index the start of a full rotation. These toothpicks break a light beam in two opto-interrupters read by the controller; these are the machine's only inputs. The set-screws above and below each mirror allowed hand-adjustment of each mirror's tilt. Software permitted still-finer tuning, to make up for small imperfections in the left–right positioning of each mirror. Fig. 26N.5 shows the machine in action. The text is illegible in this photo – but was perfectly legible in life. It spelled out a long message, beginning with the name of the course and finishing with the name of the author, Mr. Uscinski. He deserves to be remembered for this machine!

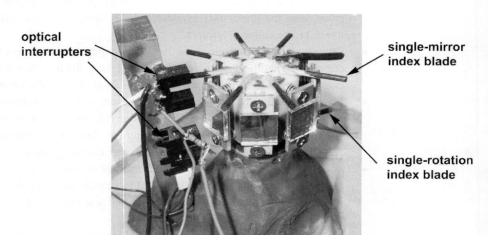

Figure 26N.4 Detail of spinning-mirrors structure in laser display.

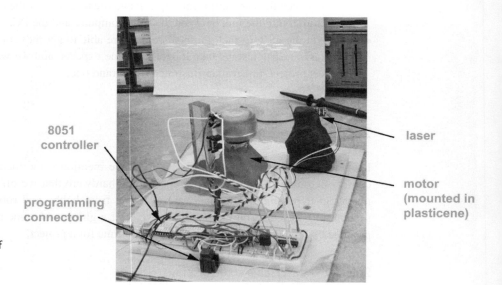

Figure 26N.5 Laser display; showing projection on a sheet of paper (incoherent image, in this photo).

26N.3.2 Sound source detector

This project, done with the big-board version of the computer rather than with a single-chip implementation as in the other cases, came together with surprising speed: in one long afternoon. It survives

26N.3 Some pretty projects

partly because its builders, Sian Kleindienst and Kristoffer Gauksheim, took the trouble to mount it on a big piece of foamcore.

It displays the compass heading of a loud sound (usually a handclap). The computer is fed three binary signals, taken from three microphone amplifiers and comparators. The microphones are separated by a few centimeters. The computer gets an approximate direction by noting simply the *order* in which the comparators fire. It gets a more detailed reading in one 60° pie-slice – implemented in just one segment of the circle because the student designers decided they should call it a day and move on to reviewing for exams. Their circuit gets that detailed reading by noting time delay among the three microphones.

Fig. 26N.6 shows the three microphone boards, and the computer board reading "320," the compass heading of a recent handclap. Fig. 26N.7 shows the circuit with a scope displaying the staggered arrival of the handclap that produced a particular compass reading.

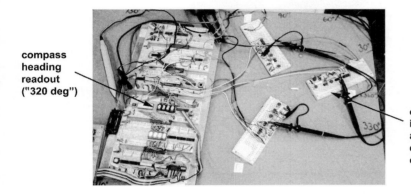

Figure 26N.6 Sound source detector project.

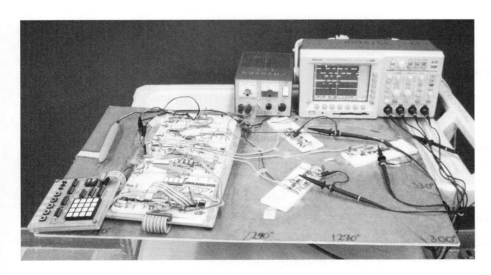

Figure 26N.7 Sound source detector showing scope display of microphone-comparator inputs.

26N.3.3 Computer-driven Etch-a-Sketch

This project implemented a notion less original than the ones above: let two stepper motors turn the knobs of a child's Etch-a-Sketch display. The mechanical part of the task was a pesky challenge, because the toy's knobs' resistance to turning is considerable, and the shafts wobble a bit. The gadget did, however, laboriously spell out "P 1 2 3." That doesn't sound like much of an achievement, perhaps.

But having witnessed the struggles that came before, as the designers implemented the stepper driver in software, including the refinement of half-step, we were thoroughly impressed when we saw our course number emerge on the screen (not shown in Fig. 26N.8, unfortunately).

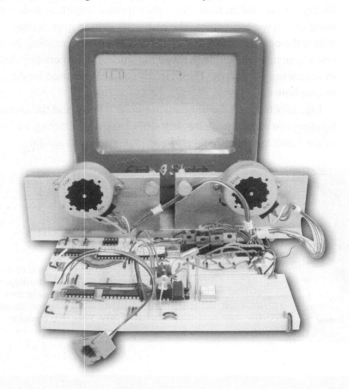

Figure 26N.8 Etch-a-sketch driven by stepper motors.

26N.3.4 A giant Etch-a-Sketch

A somewhat similar drawing machine was fashioned, again using two stepper motors, from a massive X–Y table that we rescued from the trash. The resolution of this machine is fantastically fine – but this project applied it to writing a short message in the manner of the toy etch-a-sketch.

A later student added a solenoid that permitted the pen to lift. It would be nice to see this excellent X–Y table used to rout copper for a printed circuit board layout. But the software for such a thing would be way out of line for the purposes of our students' projects.

26N.3.5 An attempted insect

Not all projects succeed, unsurprisingly. But failures after intelligent effort please us too. Fig. 26N.10 shows one: a nicely fabricated six-legged insect. Each leg is driven by a servo-motor (the sort that takes an angular position determined by the width of a pulse). This insect carries its battery, a flat Polaroid 6V type, in its belly. The tin can on top regulates that to 5V.

The designers discovered, when they set their machine to walking, that insects need knees – or some other scheme that lets a foot rise and fall from the ground. The creature of Fig. 26N.10 was unable to walk, as designed; it could only raise and lower its body, in the manner of a person salaaming. It could not proceed forward or backward.

The designers of the Radio Shack insect of Fig. 26N.11 understood that the creature needed to be

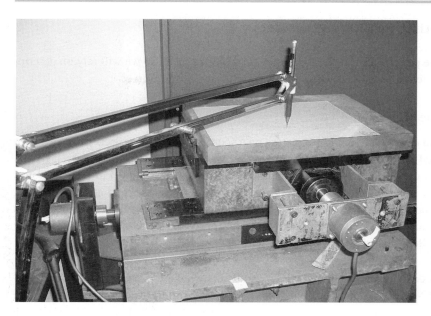

Figure 26N.9 Big X–Y table driven in the manner of the Etch-a-Sketch.

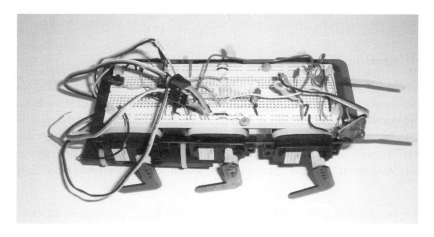

Figure 26N.10 A six-servo insect in need of further evolution.

eccentric leg mount allows middle leg to rise and fall, thus lifting front and rear legs from the ground as middle leg moves

Figure 26N.11 Insect legs better designed: these can lift from the ground as they reach forward.

able to lift some of its feet while they moved forward, then to lower them to the ground as they moved back.

The middle leg is an eccentric rod that periodically lifts the fore and rear legs from the ground. While lifted, these legs move forward, then they regain contact as they move backward relative to the insect's body. But if the creature of Fig. 26N.10 was not a successful insect, it still was a handsome one, and nicely constructed. Next time, no doubt, it will acquire knees.

26N.4 Some other memorable projects

Most of the memorable projects that we have seen do not look like much so we will rely on descriptions more than on photographs in this section.

26N.4.1 X–Y scope displays

This is the quickest and easiest way to draw a picture on a scope screen. The projects described used analog CRTs, not digital scopes with their LCD screens.

Using a CRT, if you use an *RC* slow-down circuit (basically, a low-pass filter) to join the points that your program puts out, you achieve a quick and easy "vector" display, good at drawing lines, including diagonals, whereas the more usual raster-scan (TV-like) display does lines laboriously with a succession of dots – and incidentally makes ugly lines whenever the slope approaches vertical or horizontal (an effect called "aliasing").

Hardware: You will need to give your computer a second DAC (*two* more if you implement the hardware "Zoom": see §26N.4.3). The complicating wrinkles suggested in the later subsections require a bit more hardware. Most of the work, however, lies in the programming, and in building the data tables your program is to send out. The payoff comes when you get to put whatever you choose on the scope screen.

Preliminary note: two DAC types: In the lab we have two sorts of DAC:

- "Microprocessor compatible," single supply: the AD7569 (A/D, DAC in one package); and the AD558, an 8-bit single-supply part (you used this DAC in Lab 17L, where it provided feedback for the successive-approximation A/D). Avoid the AD558 in any application where it is fed directly from the data bus, because of its unacceptably-long setup time.
- "Multiplying" ("MDAC"): this type offers an output that is the *product* of its digital input and an (analog-) input current. We exploit that characteristic in the Zoom feature described more fully in §26N.4.3.

If the hardware method appeals to you, you should decide at the outset whether you mean to zoom, in order to choose the appropriate DAC.

16×16 dot array hardware: For a 16×16 array of dots, two 4-bit DACs are sufficient. It is convenient to drive these from a single (8-bit) port, and to take advantage of the DAC's internal register. Fig. 26N.12 is the laziest scheme, which does not require disconnecting lines from your DAC–A/D. If you can tolerate an error of about 1 LSB in one of the outputs, this lazy scheme is good enough. (The Y value varies the four *low-order* bits fed to the X DAC. Note that these four bits are fed to the four *most* significant bits of the Y DAC.)

The program should fetch a byte at a time from successive locations in a table of X–Y display values, and put out those bytes. If you include a call to your delay subroutine, you will be able to tinker with the drawing rate.

26N.4.2 Complications and improvements

Connect the dots: As you learned in elementary school, connecting a few dots can give a coherent picture. So, with relatively little programming effort (few coordinates to list in your data table), you

26N.4 Some other memorable projects

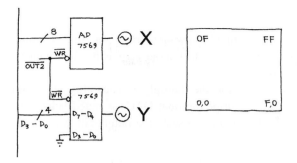

Figure 26N.12 Simple X–Y scope display; scope screen map.

can draw straight-line pictures. To connect the dots, add a low-pass filter at the output of each DAC. Try *RC* of a few microseconds (e.g., $R = 2.2k$, $C = 0.01$ μF). But you will have to experiment; the visual effect will vary with the drawing-update frequency, as well as with the filter's *RC*. The filter slows the movement of the output voltages, of course, so that the movement of the scope trace becomes visible. You will notice that this new scheme obliges you to pay attention to the sequence in which your program puts out these dots.

256×256 dot array: You can of course use two output ports to feed the full eight bits to each DAC. If you do this, you should arrange things so that the new information reaches the two DACs simultaneously. That requires use of an 8-bit register of D flip-flops. Here a 558, fed by the output of the D register, may be easier to wire than the 7569. We will leave to you the details of this hardware. (Use a 74HCT574 8-bit D register; enable the '574 3-states continuously.) In order to provide the same full-scale voltage from the '558 that you get from the '7569 (0 to 2.55V) you will need to adjust the '558's output gain. (See Lab 18L, and the '558's full data sheet.)

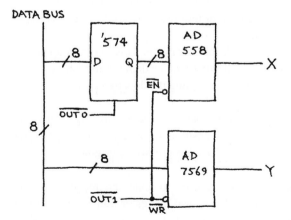

Figure 26N.13 256×256 X–Y display hardware: register added for simultaneous updates.

If you enter a large number of data points, you will begin to see flicker in the display, even if your program includes no delay routine. The screen needs to be refreshed about 30 times/second in order not to flicker annoyingly. You can calculate the number of dots you can get away with – or you can just try a big table and examine the resulting display for flicker.

26N.4.3 Size control; zoom

As we have suggested already, you can *zoom* in software. But if you want to make your programming task a little easier, and like the neat feature of a picture symmetrical about zero as its size changes, then consider the zoom hardware described just below.

Zooming, you will recall, is the operation that can exploit the multiplying capability of certain DACs. The most straightforward method would use a multiplying dual DAC, as noted below. An alternative scheme, detailed below, would use three DACs: one to scale the outputs of two others, those that determine X and Y values.

Zoom using dual MDACs:

- Analog Devices' DAC8229, a voltage-out device with a voltage-reference *IN* terminal that does the scaling. Well-suited to zooming X–Y because it offers two DACs in a package. Its X and Y outputs are the product of that voltage reference and the 8-bit digital input. This part requires split supplies, since its V_{out} is negative for a positive V_{ref} input.
- LTC7545A: similar: dual MDAC using single-supply – but again requiring split supply to provide the V_{ref} input, because once again V_{out} is negative for a positive V_{ref} input. This is a 12-bit part, but one can use it as an 8-bit part, driving only the eight MSBs.

Zoom using three DACs: Fig. 26N.14 details a 3-DAC circuit using MC1408:[2] Incidentally, in case you choose to modify your circuit, the '1408's reference current flows into the summing junction of an internal op amp; therefore, you can provide a *voltage* rather than *current* reference, if you prefer. In that case you simply feed the '1408's pin 14 through an appropriate series resistor (for example, 7.5k from a +15V supply).

In Fig. 26N.14, we feed a constant current into just the topmost of the three DACs – the "*Size*" DAC. The other two DACs are fed not a constant I_{ref} but, instead, the *Size* DAC's output current (mirrored by duplicate mirrors). Thus the computer can use the *Size* DAC to scale the *X–Y* outputs. If R_{scale} were 7.5k, for example, 2mA would be the maximum current out of X and Y DACs.

The mirror – a circuit we have tended to hide from you in this book – "bounces" the scaling DAC's *sunk* current from the positive supply in order to *source* current into the two current-reference inputs of the X and Y DACs.

Figure 26N.15 is a surprisingly subtle image drawn with this hardware. This pyramid appears to show cleverly gradated shading. In fact, the image shows nothing fancier than a diminishing square. The square was drawn as usual by defining four points, then connecting the dots by slowing each DAC output with an *RC*.

The apparent shading in Fig. 26N.15 results from the CRT beam's gradual slowing as it moves from source to destination. Since the movement of the dot's position slows exponentially (the rate approaching zero as the dot approaches its target), the trace grows progressively brighter.

Centering the zoomed image: A circuit refinement is proposed below to center the image on the CRT regardless of *size*. Without this circuit addition, a change of size also moves or translates the image: since X and Y are always negative, the visual effect will be as if a figure that grows were moving down and to the left on the screen, as well as toward you. If you prefer to make your figures "approach" head-on, then you should make the modification shown below: an op-amp is added to the output of each of the DAC's (X and Y); the op-amp allows you to center the coordinate output at 0V.

---

[2] You may prefer a newer single-supply multiplying DAC such as the AD7524; if you use one, you will need to modify the circuits we have shown below, which assume a '1408.

26N.4 Some other memorable projects

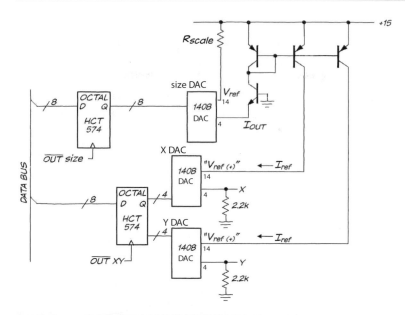

Figure 26N.14 Multiplying DAC can scale X–Y image.

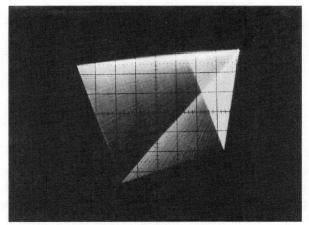

Figure 26N.15 Receding square drawn with scaled X–Y DACs.

In Fig. 26N.16's circuit, half the DAC input current is applied to the output op-amp's summing junction,[3] and thus is subtracted from the output, centering that voltage as the scaling current varies.[4]

26N.4.4 True vector drawing: position-relative

Your drawing table need not store absolute screen locations. Instead, it can store vectors: direction and length relative to present screen location.[5] The relative vectors suggested in Fig. 26N.17 are listed in the manner of compass headings: ESE = East–South–East. This scheme would work with a 4-bit direction specification (16 directions). The remaining four bits could define magnitude; that magnitude could be used with the multiply instruction to stretch the unit-length direction vector (approximate unit lengths will do!).

---

[3] The 1k resistors are included to equalize the sharing of current between the two DACs even in the presence of differing V_{offset} values that could put their two summing junctions at slightly different voltages.

[4] Thanks to D. Durlach for this nifty amendment.

[5] Thanks to Scott Lee for proposing and demonstrating this arrangement.

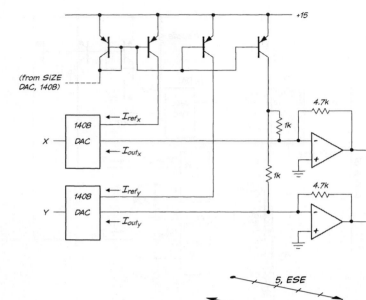

Figure 26N.16 Op-amp added to DAC to give V_{out} symmetrical about 0V.

Figure 26N.17 Drawing with true vectors: *relative* movement.

This way of defining a figure allows one to rotate it without much difficulty. Some rotations are shown in §26N.4.6. The programming is a good deal harder than for an absolute X–Y figure.

26N.4.5 Animation

If you enter the coordinates for two or more similar but different pictures in memory, and then draw these pictures in quick succession (changing the picture, say, every 10th of a second) then you can animate a drawing. Evidently, you will need a large table to animate even a simple figure, so start modestly. You will want to use the follow-the-dots scheme to minimize the number of data points required to draw a single image. Fig. 26N.18 is a crude example, showing first a hand-drawn sketch of a creature with two leg positions, then a scope image showing an implementation that was a little more ambitious. The creature seems to have morphed from perhaps horse to terrier, and the implementation apparently allowed the dog's head to bob, and its tail to wag.

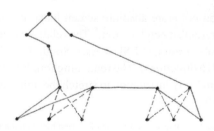

Figure 26N.18 Animation using vector drawing: a sketch and an implementation.

26N.4.6 Computed drawing

The tedious loading of values into memory demanded by the table-reading schemes we have suggested probably has made you yearn to turn over to the machine the job of determining what points it should draw. Of course you can do this. You could write a program that would draw a rectangle, say, by incrementing the X register for some steps while holding Y fixed, then incrementing the Y register while holding X fixed, then decrementing the X register..., and so on.

Two excessively experienced programmers have used the X–Y hardware to draw cubes, which they then could rotate about any of three axes.[6] Fig. 26N.19 showing multiple cubes is a multiple exposure showing the cube rotating over time. This sort of task requires too much programming sophistication – and too much code – for ordinary mortals.

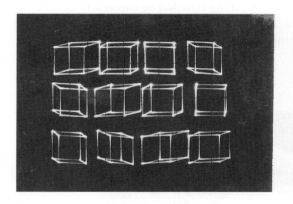
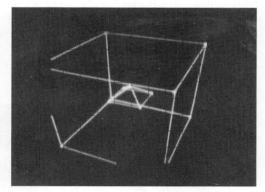

Figure 26N.19 Computed drawing: two versions of cube that can be rotated about its axes: Shumaker/Gingold.

26N.4.7 Laser X–Y display

A mirror that can be tilted on each of two orthogonal axes can steer a laser beam to project an image on a screen (or ceiling). The first students to try this built an admirable structure using two audio speakers, shown in Fig. 26N.20.

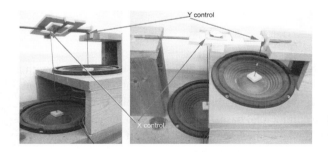

Figure 26N.20 X–Y laser mirror assembly – sadly oversized.

The massive mechanism of Fig. 26N.20 *worked*, sort of: the mirror did tilt in response to signals

---

[6] Grant Shumaker did this first (1986), and did it entirely in assembly language, entered by hand. This is a feat roughly comparable to climbing one of the middle-sized Himalayas without oxygen. David Gingold did it again, this time putting shapes within cubes; David used oxygen (in the form of a C compiler), as noted in §26N.5.2 in a discussion of David's remarkable asteroids game.

from two DACs; the projected laser beam thus could be placed here or there on the ceiling under program control. But as a way to draw an image it was a sad failure. The mechanism was much too massive and therefore much too slow.

A later student made the scheme work by radically reducing the scale of the mechanism. Instead of eight-inch woofers he used the tiny speakers in a set of headphones, and to these he attached a tiny shard of mirror.

26N.4.8 Charmingly simple LED display

You've seen displays made by spinning or oscillating sticks of LEDs, perhaps: some can be attached to a bicycle wheel; others attach to fan blades, or to an oscillating rod. A student wanted to make such a thing – and at first was stalled by the difficulty of getting electrical signals to a thing that was spinning.[7] To rig brushes for sliding contacts – one for each LED – seemed problematic. (One can buy such wiring harnesses, but we didn't have one at the time.) He wisely shied away from committing much effort to fabrication.

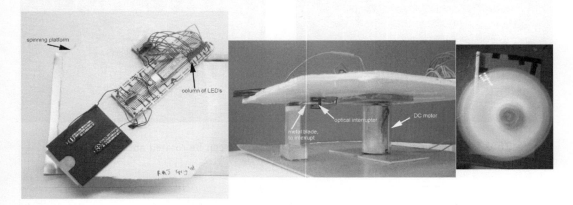

Figure 26N.21 Spinning display: computer and battery spin, along with display.

His solution was to spin *everything*: not just the display LEDs but also the computer and its power supply (a battery). His display didn't have much to say: it counted out the digits 0 through 9. But it *worked*, and that was gratifying to see.

26N.4.9 Vehicles

Tractors with independent drive for two wheels: Two motors mounted end to end, and each with a rubber-tired wheel on its shaft form a tractor that can be steered, in the manner of a bulldozer or tank, by driving the two wheels independently. It shows the useful ability to pivot in place.

The version on the right in Fig. 26N.22 is a commercial base. This model is no longer available, but similar chassis are.[8] Students have used it to make a line tracer and an odd sort of dancing machine that would learn what steps to dance by scanning a nearby radio-frequency "tag."

The home-built version on the left in Fig. 26N.22 was applied by a couple of ambitious students

---

[7] This was Ryan Jamiolkowski, 2008.
[8] Parallax offers a circular base, SEN-13285 (www.parallax.com). Pololu offers a similar round chassis plate with cutouts for wheels (Pololu 5" Robot Chassis RRC04A) (www.pololu.com). For either base one would need to select appropriate wheels and then improvise servo motor mounts.

26N.4 Some other memorable projects

Figure 26N.22 Tractor/tanks: steered by independent drive of the two rear wheels.

(Mike Pahre and Danny Vanderryn) to blunder its way through a maze. Its only sensor was its nose-bump switch, but this was enough to allow the tractor, after laboriously finding its way to the far side of a maze built with boxes and books, to make the return trip without going down any blind alleys. It cruised calmly out, avoiding all the dead-ends it had found on the way in. When the tractor had completed its graceful exit from the maze it then did a victory dance in the manner of a football player who has just scored a touchdown. We thought it was entitled to feel pleased with itself.

That was a challenging project, and putting together the tractor itself requires some machine-shop skills. Because the motors draw a lot of current, the tractor was fed through an umbilical cable, which carried logic signals as well as power for the motors.

The commercial tractor (really just a chassis), in contrast, uses low-power servo motors that have been modified for continuous rotation. See §26N.6.4. These motors run happily from a flat polaroid battery or a 3-cell AAA pack carried by the little vehicle. But lacking the stepper-motor drive, which allows easy repeatability of a travel path, this tractor would need some feedback and recording scheme to mimic the stepper-tractor's feat of maze mapping.

26N.4.10 A neat application for the independent-drive steerable chassis

Two students (Lucas Kocia and Ferrell Helbling) took advantage of the ready-made steerable chassis to make a neat car with a little bit of intelligence. The idea was simple and they implemented it in an afternoon: a car that is smart enough not to roll over the edge of a cliff (well, really the edge of a table).

They gave their car a long snout, and on it put an optical ranging device – a clever Sharp infrared module with range of 4–30cm[9] – pointing down. The ranger thus can sense when the snout protrudes over the edge of a table.

The ranging device uses *synchronous detection* to ignore ambient light. It applies a square wave Enable to its IR emitter. While the emitter is On, the integrator accumulates the reflected light plus ambient. When the emitter is Off, the receiver *inverts* its signal – ambient only – before feeding it to the integrator. Thus ambient light cancels itself.

The two students showed us this car flirting with disaster but always turning back from the cliff – and then startled us by revealing that they had done it with pure analog methods (apart from the arguably digital use of a '555 as a one-shot). Probably we looked disappointed to hear that they had

---

[9] Sharp GP2Y0A41SK0F, available from Sparkfun, which offers also a longer-range version.

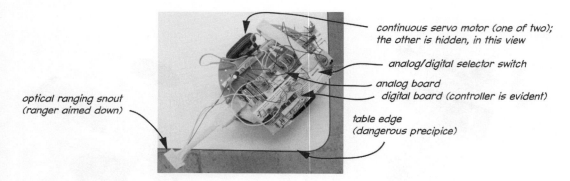

Figure 26N.23 Smart anti-suicide car.

not taken advantage of the microcontroller that they had met in the course. A day later they invited us to watch the car again do its suicide-avoidance routine. But this time they had done the task using the little SiLabs controller. A switch selected between the two modes, *analog* versus *digital*. We were charmed: what a neat way to demonstrate the wide range of skills that we hope students would have picked up by the end of this course.

26N.4.11 A computer-steered toy car

Two students (Jonathan Wolff and George Marcus) proposed a project that we thought wildly excessive: their little computer was to steer an *RC*-controlled car, chasing a car steered by a human. (We mentioned their ambitions in §26N.2.) "Please," we pleaded with them, "couldn't you start by seeing if the computer can just steer a car?" Reluctantly, they consented – looking as if they considered this a dull and trivial task. We had ruined their game.

Many days later, they demonstrated such a design, and it had turned out to be a very challenging task. The "chase me" game never arrived. Their design entailed several substantial tasks:

- The hardest part of their project lay in letting the computer locate the little car. Their solution was to put ultrasonic "squeakers" on the car, then to listen for the squeak with three ultrasonic detectors. Their computer would command a squeak;[10] the time delays to the three squeak detectors would, in principle, indicate the car's position, through triangulation.
- *Triangulation*?! Trigonometry on the little computer, in assembly language? They recognized that this sounded painful, and decided to turn this computation-heavy work over to a desktop PC. To that end, they built a little-computer-to-PC interface (a job that would have been a respectable project in itself). Little computer shipped the raw numbers – time delays; PC responded by shipping back coordinates of the car's location.
- Given these coordinates, the little computer now figured out how to get the car from where it was to where it ought to be.
- Finally, the little computer "drove" the car to its destination. To do this, it controlled switches that were designed to be controlled by a human pressing pushbuttons. Implementing such switches was not hard. But steering the car was. They had bought the cheapest car offered by Radio Shack, and this car could not be told simply "turn left." In order to turn left, it would have to turn the front wheels *right*, and then *back up*. Only then could it drive forward to "go left."

---

[10] Sad to say, under pressure of time, they sneaked a wire to the little car, rather than command the squeak through a radio signal.

26N.4 Some other memorable projects

Remarkably, they made all this work. Fig. 26N.24 shows the little car, now much battered, and also a snapshot of the two designers looking at their project with odd detachment. (Shouldn't they be dancing with excitement?) In the figure, one can make out two of the three ultrasonic detectors attached to breadboards placed on two stools.

Figure 26N.24 Little *RC* car. Triangulation located its ultrasonic squeak.

26N.4.12 A perennial heartbreaker: inverted pendulum

This is a problem routinely assigned in *controls* courses. It is also very difficult, done seat-of-the-pants in the style of this course.[11] People have come close to success, but real success has eluded everyone, so far.

Fearless pioneer tried two dimensions: The first person to try attacked the problem in a style that made success extra-unlikely, and yet made an impressive approach to doing the job. He had not studied controls and probably didn't know how hard a task he had set himself. If he had, would he have undertaken it in *two* rather than the usual *one* dimension?

He used simple and entirely analog methods to attempt to keep a pencil balanced on its point. The choice of a *pencil* rather than the more usual long rod made his job harder still. Nevertheless, his primitive machine was able to keep a pencil balanced on its point for a few seconds, by moving a little platform about on a smooth base.

The analog design shone an LED on the upright pencil while a pair of photodetectors watched to see if the pencil leaned away from the vertical. The photodetectors fed a differential amplifier, which drove a DC motor whose response tended to force the pencil upright. The motor did this by spinning a pulley on which a string was wound, and the string would draw the balancing-block one way or the other on the base platform.

Two such LED-photodetector-pair units took care of watching for X and Y lean. The negative feedback in this arrangement made the mechanism work quite well even though the construction was

---

[11] Two good examples of success appear on YouTube; if these are gone when you look, you'll find many more by searching for "inverted pendulum." See http://www.youtube.com/watch?v=MWJHcI7UcuE (stick on printer-like base); http://www.youtube.com/watch?v=d_8RYpHfRK4&feature=related (Lego Mindstorm car balances on its two wheels).

crude in the extreme: sewing thread, wooden pulleys, a composition-board base, and odds and ends of adhesive tape were its materials.[12]

The mechanism worked pretty well until the little platform reached one of the edges of the base. There, the pencil would fall. What the whole design lacked was any awareness of the position of the moving block upon the base – and the slowness of its response sets it up as vulnerable to instabilities like those you saw with the PID motor control. Later efforts were more sophisticated, but not more charming than Paul Titcomb's.

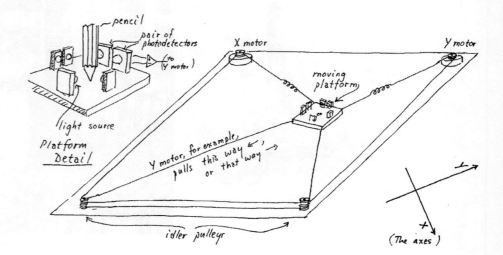

Figure 26N.25 A wonderful partial success: pencil balancer.

One-dimensional inverted pendulum: After this initial attempt, everyone tried the task in the more manageable form, in *one* dimension, using the guts of an old printer to move the base of a long stick. We began to collect discarded printers, and fitted one with a second potentiometer that indicated the *position* of the moving base in addition to the *tilt* information given by a first pot. This information provided a warning when adjustment range was close to finished.

The results have been mixed. One student whose computer control of this setup failed wrote a thorough and somewhat formula-rich report concluding that he had failed because the motor's response was too slow to allow it to catch the falling stick.[13] He proposed that the job could not be done till we got a heftier motor.

A term or so later, two students (Justin Albert and Partha Saha) came closer to success using purely analog methods. These students were sophisticated enough to be well aware of the inherent stability problems, and tried summing various sorts of error signal, making a PID loop that resembled the one

---

[12] This balancer was the work of the inspired tinkerer *extraordinaire*, Paul Titcomb. Another of Paul's amazing projects deserves mention here; it may inspire someone to a similar effort. Paul made a drawing-mimicker using his lab computer. Paul rigged a 2-jointed arm that held a pen. He placed a potentiometer at each joint, and as he guided the arm, drawing a picture by hand, he had the computer read the joint-potentiometers at regular time-intervals. Then, when he had finished a drawing, Paul would ask the computer to play back the motions it had sensed, by driving a pair of stepper motors on the joints of a similar arm also holding a pencil. The machine never quite worked: its drawings showed a rather severe palsy – probably an effect of its mechanical crudity and the fact that (unlike the pencil balancer) it operated *open loop* – without the error-forgiving magic of feedback. It was an impressive gadget nevertheless, and the arm and its last, shaky drawings languish at the back of our lab. They await someone who will perfect the scheme.

[13] Is it significant that this fellow was an MIT grad?

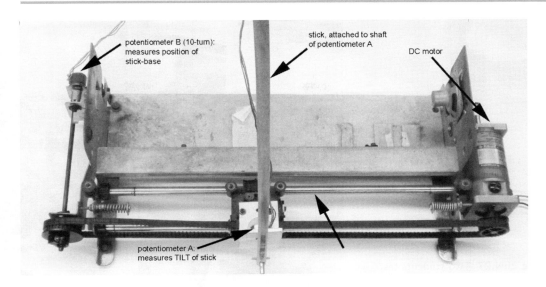

Figure 26N.26 Inverted pendulum hardware: the guts of an old dot-matrix printer drive.

you built in Lab 10L. Their circuit worked well enough so that we asked them to show it off to the new crop of students on opening day. In September, they pulled out the machine that had worked in May – and were unable to get it to work again. This is a sad tale. Is it an argument against analog circuits trimmed with multiple potentiometers – an argument in favor of digital methods?

A recent effort used a digital implementation of a PID loop. It showed admirable intelligence, shimmying the base in its efforts to keep the stick vertical. After a few seconds, it would lose the struggle, unable to provide sufficient corrections. It was clearly trying, though. The tantalizing *near* success of the machine is visible in the photos of Fig. 26N.27: first, with anxiety and apprehension, the designers launch the thing;[14] then they show their delight as the machine makes its frantic small adjustments, keeping the ruler upright for several seconds. We omit the sadder conclusion in which the ruler falls.

26N.5 Games

26N.5.1 Pacman on scope screen

This was insanely excessive: it required a home-made "video board" to refresh the scope screen; the processor was not quick enough to redraw the full screen, over and over. Instead, the designer set up RAM that was written initially with the screen layout, and then re-written rapidly using hardware counters.

The computer's display duties thus were reduced to writing only *changes*: Pacman's mouth opened and closed, and he moved, as monsters moved too.

We wish we could show it to you in action – but during one summer somebody pulled the battery plug, and we've been too lazy to re-enter the code so far. We could not bring ourselves, however, to destroy the hardware; so, maybe someday....

---

[14] These are Emily Russell and Lusann Wang (2012). Their collaborator, Tom Dimiduk, is not shown. They used an Arduino controller rather than the usual 8051, because Tom was familiar with the device.

Figure 26N.27 Two students test their brainchild. Their expressions show apprehension and hope.

here, pacman did some eating

Figure 26N.28 Two Pacman screenshots.

26N.5.2 Asteroids on a scope screen

Pacman used a raster-scanned display, with a mostly-fixed display defining the maze. Space Invaders used vector drawing like that described earlier in §26N.4.6.[15]

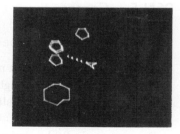

Figure 26N.29 Asteroids on a scope screen.

The game included the gravitational field of the "real" asteroids game. It was a marvel to watch

---

[15] This was David Gingold's project, so hugely ambitious that we have to note that he got credit for it in two courses; the other was one in computer animation.
He put together not only the code but also hardware that would draw lines of constant brightness, and having compiled the C code on a full-scale computer, he needed to pump it into his little home-made computer. This need arose on a weekend, so rather than ask us for help in finding a UART IC he wrote a serial interface in software – as if he needed a challenge to make his project a little harder. His serial link worked, as did the game.

(and this is another of those pieces of hardware that we keep around, unable to bring ourselves to tear it apart).

26N.5.3 Other great games

Other wonderful games have appeared, but we lack photos of these; we'll have to settle for words.

- **Stephane's *virtual-bubble* game.** Stephane Ryder figured out how to decode the signals from a Nintendo *Power-Glove* – a neat toy that can detect position and orientation of a glove, and also finger closings. He used these signals to let his computer implement a game. The computer would issue beeps at a rate that increased as the glove came nearer to the 3D coordinates of a "virtual-object" – just some arbitrary X,Y,Z point in the air. When the glove was at the target point, one could close the hand and then *move* the target point, depositing it wherever one opened the hand. Dazzling!
- **Nomeer & Clay's punch-out interface.** These students[16] replaced the keypad of a Nintendo boxing game with an array of photosensors set in a square that looks like an empty picture frame. Behind the frame they put a TV set showing the boxers. To punch the adversary the user punches air in the picture frame. The nine segments of the frame defined functions earlier assigned to keys on the keypad: uppercut, body-blow, etc. It was not a complicated project – but fun.
- **LED maze game.** Computer handled scope display, but the patterns displayed were stored as state machines in PALs. So, one could change the maze (a set of rooms: rectangles with passages to other rectangles) by swapping PALs. (Meredith Trauner & Robby Klein)
- **Tetris on an LED array.** (Kurt Shelton): 8051 running from flash. An awful lot of coding – but it did work.

26N.6 Sensors, actuators, gadgets

At this point in our university *course* we would list devices that we happened to have in the lab, hoping students might conceive new ways to use them. In this *book*, which should outlast any list of devices available at the time of writing, it seems more useful to steer a reader toward the sort of supplier that we have found useful. The Web can lead you to many other sources, of course.

26N.6.1 A few good sources of sensors, actuators and other devices

Here are some sites that we enjoy:

- **Adafruit** A really good site (www.adafruit.com), founded by a serious and thoughtful electronics geek who often integrates parts into helpful "breakout" boards that make the parts easy to wire. This aid is important because so many new parts are available in surface-mount packages only. The breakout boards permit breadboarding with such parts. Ada often provides code in addition to take advantage of programmable parts. At present, Ada leans toward Arduino interfaces and code.
- **Sparkfun** Similar to Adafruit, though more commercial (www.sparkfun.com). Sparkfun, too, often provides helpful breakout boards.
- **Parallax** Good for robotics parts such as chassis, and a good collection of sensors (www.parallax.com).

---

[16] Noah Helman, Sameer Bhalotra and Clay Scott.

Pololu For robotics; selection somewhat smaller than Sparkfun's (www.pololu.com).

All Electronics This is a surplus shop, so the selection is unpredictable and odd. But here you'll find strange
devices at low prices that may inspire invention. For example, "band brake for electric scooter."
(www.allelectronics.com).

eBay And don't forget to try eBay which sometimes, of course, offers great bargains.

26N.6.2 A source of small parts for mechanical linkages

Specialized parts like bearings, gearing, belt drives, and universal joints are available at Sparkfun (above) and also at both a specialized site and at Amazon's general "Industrial & Scientific Store."

Amazon Industrial & Scientific Store (www.amazon.com/industrial) Lacks the coziness of the Small Parts, Inc. that Amazon bought, but accesses a great variety of parts.

Stock Drive Products (http://www.sdp-si.com/) A smaller site, less overwhelming.

26N.6.3 Mechanical: motors, etc.

Stepper motors: The stepper's appeal is that it does not require use of *feedback* for precise control of positioning.

1. Two types: unipolar versus bipolar.
   - **Unipolar.** Unipolar is the easier configuration to drive, requiring only a transistor switch on each "phase" winding (typically, this is a switch to ground, as in Fig. 26N.31). The motor will provide six leads – the usual four leads for the opposite ends of the two windings, plus two more for the center tap of each winding. Sometimes the two center taps will be combined, reducing the number of leads to five.
   - **Bipolar.** These 4-lead motors omit the center tap, and thus require a pair of drivers that can both *source* and *sink* current. Such drivers are available in integrated form and are called "H-bridge." Though harder to interface, the bipolar is more efficient than the unipolar, whose windings keep half their length always idle.
2. Resolution and power. Typical stepper motors provide 200-steps-per-revolution and need substantial current 0.3–1A, at perhaps 10V.[17]
3. Small stepper motors. Smaller motors usually show coarser resolution – for example, Sparkfun offers a small motor with 48 steps per revolution.[18] The smallest that we have encountered is less than 1cm in diameter.
4. Stepper drivers. As noted in §26N.7, one can sidestep the work of writing code to drive a stepper motor by taking advantage of a driver IC.

26N.6.4 "Servo" motors

These pulse-width controlled motors lend themselves to computer control because pulse duration is not hard to regulate with a microcontroller. They come in two forms:

---

[17] For example, Adafruit bipolar NEMA-17 size – 200 steps/rev, 12V 350mA. Similar steppers cost less at All Electronics, but the selection there is small.
[18] Sparkfun ROB-10551.

1. Radio-control model type: position control.

   These are sold as actuators for use in *RC* models. (You may have read a description of these in Lab 24L.) They rotate through about 270°, holding a rotation position that is determined by the width (duration) of a logic pulse sent to the motor repeatedly. They are available in at least three standard sizes, use little power and provide substantial torque (the motor is small, but geared way down).

   Applications: useful to steer a car, e.g.; but strong enough for, say, positioning a small robotic arm.

2. Continuous-turn type: rotation rate control.

   These differ from the usual in using pulse width to determine not shaft *position* but rather rate and direction of rotation. A pulse of standard width (1.5ms) produces *no* rotation; a pulse of shorter or longer duration produces rotation in one direction or the other, at a rate roughly proportional to the difference from the standard pulse width. A 1ms pulse produces full speed in one direction, 2ms evokes full speed in the opposite direction. These are used in the car chassis described in §26N.4.9, one servo driving each of two wheels.

26N.6.5 Nitinol muscle wires

1. Muscle wires: tiny pullers.

   One company at least really does call these "muscle wires."[19] Their more serious name is shape memory or nitinol wire. They shorten (by 3–7%) with considerable force when heated if they've been stretched while in their cool, soft state.

   This slight shrinkage percentage implies that a long wire is needed to provide any substantial actuating "stroke." A small company called Miga Motors makes an ingenious stroke enhancer that essentially folds a longish wire. This scheme achieves a stroke of 0.3″ from a credit-card size circuit board. One such device is the MigaOne-15, shown in Fig. 26N.30.[20] It includes a contact that permits an intelligent driver circuit to sense full travel in order to shut off power, preventing damage from excessive heat.

Figure 26N.30 MigaOne amplifies stroke length for nitinol wire actuator.

---

[19] www.musclewires.com.
[20] This is neat but expensive at about $50. See migamotors.com.

Nitinol actuators call for very large currents, since the current's job is to heat the wire with I^2R power; but the power needn't be applied for long if a brief contraction is required. So a small battery can do the job, if duty cycle is very low, keeping the whole setup small.

2. Applications: tiny robots.

A couple of students made a six-legged bug using muscle-wire – and concluded that they would never touch muscle wire again: it gobbles current, doesn't pull very hard, and is pesky to build with (requiring tensioning adjustments). Still, you can't find an actuator lighter than the nitinol wire.

The wires can be used to mimic the behavior of tendons, making a crude hand with wiggleable fingers.[21]

26N.6.6 Transducers

Sound:

1. Ultrasonic send, receive "speaker"/microphone.

   These are small diaphragms tied to a piece of piezo-electric stuff, and are resonant at about 40kHz. Transmitter and receiver are either identical or very similar to one another.

   Applications: ranging (send a burst; listen for echo); communication (send data as sequence of bursts).

2. Integrated ultrasonic transducer and amp.

   This was the method Polaroid developed as rangefinder for its cameras. The method is mimicked in MaxBotix rangers.[22] The EZ1, for example, offers a sensing range from 0 to 255 inches, in 1″ increments. It provides an output in three forms: analog, pulse-width, and RS232 serial. These rangers are available in various sensitivities. Note an alternative ranging method that uses *infrared* light. This is mentioned in §26N.4.10. That IR device shows better resolution, but shorter range.

Infrared:

1. IR remote-to-logic signal translator.

   This was described in Chapter 24N. The translator converts each 40kHz burst put out by a typical IR remote into a single logic level. (In the example shown in Chapter 24N, a burst produces a logic Low; silence produces High.)

   Applications: a circuit with brains (your computer) can translate the serial stream of logic levels that an IR remote provides. A properly synchronized 8-bit shift-register, done with computer or with hardware, could, in principle, give you 256 options.

2. Passive differential IR motion detector.

   An "alarm" signal is asserted when the IR image of a room changes. 3.3V to 12V, logic-level output.[23]

Color discriminator: An IC array of detectors, each with a color filter. Two ingenious people paired this device with optical fibers in an effort to read resistor color codes. The resistor colors were badly defined and defeated them – but the machine was able to read the colors of a ribbon cable, much of the time. Pretty good. You may enjoy the challenge of improving on their machine to automate resistor sorting[24] For specifications, see the Parallax site.[25]

---

[21] A kit for such hand is offered by MuscleWires at www.musclewires.com.
[22] These are available from Sparkfun. See for example, MaxBotix LV-MaxSonar-EZ1.
[23] Sparkfun SEN-13285.
[24] Before you do this hard work, should we remind you that there is an easier way to determine the value of a resistor? Probably not.
[25] www.parallax.com/detail.asp?product_id=30054 (part is TAOS TCS230AMLN).

26N.6 Sensors, actuators, gadgets

Acceleration; rotation rate: Many accelerometers are offered by Adafruit, Parallax, and Sparkfun. You should check their sites for new models even niftier than the ones we have noticed. These companies typically offer a 'breakout board.

You probably want a range of just a few g.[26] With two axes, you can sense *tilt*.

Sparkfun offers a helpful discussion on their site of many alternative accelerometers.[27]

- Analog output, 3-axis ($\pm 3g$, 3V supply): ADXL335.
  Sometimes analog output is nice, letting you see on a scope screen a live image of the device outputs. The controller's ADC can read these values – but to read three axes with a single ADC requires use of an analog multiplexer. The SiLabs '410 includes such a multiplexer, allowing the single ADC to read all three axes in succession. Adafruit offers a breakout board for this accelerometer, a board that includes a 5V-to-3.3V regulator.
- Digital output: SPI or I^2C: ADXL 345 (selectable range, ± 2 to $16g$, 3V supply). SPI interface is not hard to handle, especially with SiLabs hardware implementation of SPI (see Lab 24L). The accelerometer outputs are not viewable on a scope screen, but their values can be displayed on a laptop's screen through the SiLabs IDE. And taking the inputs in digital form is tidier and easier than what's required if the accelerometer speaks analog. SPI obviates need for successive ADC operations that would include swapping ADC input pins.

Gyro: A small solid-state gyro gives an analog output proportional to angular acceleration. Here one axis may be enough, and analog output is more common than digital.

- 1-axis, analog output (3V supply): LY530AL (Sparkfun).
- 3-axis, I^2C and SPI output, breakout board permits 5V supply and 5V control signals: L3GD20H from STMicro (Adafruit).

Accelerometer and gyro combined: You can get accelerometer and gyro on one IC to make things more compact. For example MPU-6050 (Sparkfun). (Expensive, at $40.)

Surely lots of neat improved models will be coming along, so treat our list of devices as only a snapshot of what was available back in 2015.

Magnetic field: Two basic sorts are available: plain and fancy.

1. Plain: Just Hall-effect sensors. Linear output proportional to current: a breakout board by Sparkfun, ACS712. Simpler sensors give Yes/No detection. For example US1881 from Sparkfun.
2. Fancy: a digital compass. Adafruit provides a 5V-compatible breakout board using a Honeywell 3-axis
   HMC5883 IC. The circuit uses two orthogonal coils to sense variations in the orientation of the earth's magnetic field relative to each of the two coils. The 3-axis readings are delivered with I^2C.

Force: Small flexible sheets that put out a small voltage when flexed. For specifications see www.parallax.com/detail.asp?product_id=30056.

Speech generation: A two-IC board takes ASCII characters in serial form (UART) and does its best to pronounce the text phonetically. DEV-11711, Emic-2 chipset (Sparkfun; expensive, at $80).

If you're willing to work harder, you can put together two $25 ICs, a UART-text-to-allophone translator and a speech synthesizer IC, TTS256 and Speakjet (Sparkfun).

[26] One g, as may you know, is the acceleration due to gravity here on the surface of earth.
[27] Sparkfun's "Accelerometer Buying Guide," http://www.sparkfun.com/tutorials/167.

Speech recognition: This is a very hard task – rendered utterly easy by someone's IC and little printed-circuit board: EasyVR Shield 3.0 (obviously named with Arduino in mind), Sparkfun COM-13316. 28 user-independent commands, 32 user-definable.

Just getting it to work is nowhere near a project. The challenge is to think up an interesting application. The thing is small (about $1''$-by-$2''$), so it could, in principle, be mounted on a little vehicle (controlled by a microcontroller, probably – but not necessarily), permitting control of the vehicle by voice.

26N.6.7 Displays

1. liquid crystal (LCD).

   Nearly all use the industry-standard interface that looks like a single I/O location; the computer feeds the display successive ASCII codes (7 or 8 bits), and the display takes care of placing the characters properly. This requires six or seven I/O lines, as we noted in Chapter 25N, though an add-on board can convert this parallel interface to a serial protocol.[28]

   Applications: wordy output for many sorts of gadget. (These may not be very exciting since they only take our little computer a little way toward the very chatty LCDs that you're accustomed to on big computers.)

2. OLED bit-mapped display.

   These provide beautiful high-resolution images or text displays, usually quite small. Their interface, at the time of writing, remains difficult. Adafruit provides libraries of code to help. For example, Adafruit's 128-by-32 display.[29]

3. "Intelligent" LED display: 2-character or 4-character alphanumeric.

   These are small. Each character occupies a single I/O address, and wants to be fed ASCII. For example, Siemens PD2435. Handsome, but obsolescent. Try Octopart.com to find stock.

   Good for small, brief, bright verbal output.

4. Bar-graph.
   - LED: An array of LEDs. You may have seen similar displays used as VU meters on audio equipment. Simple ones call for wiring each LED. For example, Adafruit's KWL-R1025BB. Fancier ones use a serial interface (also at Adafruit).
   - Electrostatic: "E-Ink." The appeal of E-Ink displays is zero power-consumption after a level is written. The difficulty is their peculiar drive requirements: a brief pulse at $\pm 15V$ to turn a segment On or Off. But you might enjoy showing off your skill in making your own driver. The company takes the annoying line that their parts' drive demands are *confidential* (they ask a user to sign a non-disclosure agreement!). It's no secret though, and drive for a few segments is not hard to implement.

     A 14-segment bar graph is available: KWL-R1025BB (Digikey). So is a 3-character numeric display: SC004221 (Digikey).

26N.6.8 Interface devices

Output:

- Solid-state relay. These will turn On/Off a heavy AC load (motor, lamp, appliance) using a logic-level as input. For example, a small device for 8A load is Sparkfun's COM-10636. Input is optically coupled to the high-voltage parts, so your circuit remains isolated from the scary 120V supply.

---

[28] Adafruit provides this "back pack" board: I^{2}C/SPI Character LCD Backpack, PRODUCT ID: 292.

[29] Monochrome 128-by-32 SPI-interface OLED graphic display, PRODUCT ID: 661. These are available also from the usual distributors, including Digikey.

In and Out:

1. Radio serial link. A simple radio link could carry digital data: the simplest scheme would be to turn a transmitter ON, then OFF, then ON... The simplest transmitting device would be a fast logic-level oscillator gated ON or OFF (you could use a MOSFET to apply power or not). The simplest receiver would be an amplifier driving an "envelope detector" – a simple 1-diode AM detector like the one you met in Lab 3L. You might build the amplifier from scratch; or you may prefer to look for an IC amp. A comparator could then determine whether the envelope-detector showed that a burst of signal was coming in ("1") or not ("0").

   At its simplest, this burst/no-signal waveform could generate the logic-level pulse that controls the position of a *servo motor* (see above), steering a little car, say. If you were more ambitious, you could use this simple radio link to send serial data in one of the standard conventions – using conventional RS-232 (to a UART on the receiving end) or perhaps I^{2}C, mentioned just above.

   The easier way to proceed is to buy ready-made Send and Receive modules. See for example SparkFun's many "RF link" boards, some simple (such as transmitter WRL-10534 and receiver WRL-10532). These are "indiscriminate," so call for some intelligence on your part. Some integrate UART serial I/O.

   Still easier is to start with a board that converts between USB and radio, though the software needed is substantial. "Wixel" radio modules: WRL-10532 (Sparkfun). We have not tried these.

26N.7 Stepper motor drive

Generally: A stepper motor contains two coils surrounding a permanent-magnet rotor; the rotor looks like a gear, and its "teeth" like to line up with one or the other of the slightly offset coils, depending on how the current flows in the coils. A DC current through these coils holds the rotor fixed. A reversal of current in either coil moves the rotor one "step" clockwise or counter-clockwise (a coarse stepper may move tens of degrees in a step; a moderately fine stepper may move 1.875°: 200 steps/revolution). The sequence in which the currents are reversed (a gray code) determines the direction of rotation.[30] See Fig. 26N.32, where the signals to coils A and B are labeled D_1 and D_2.

Integrated stepper driver: Integrated stepper-motor driver chips can make driving a stepper extremely easy: the chip usually is just a bidirectional counter/shift-register, capable of sinking and sourcing more current than an ordinary logic gate. Fancier ICs, like the Allegro A3967SLB[31] can provide finer control: so-called "microstepping."[32] The A3967SLB permits 1/8 steps. This part also includes driver transistors capable of more than what one would expect in a modest 20-pin DIP: 750mA at up to 30V (though not both at the same time). This is not enough for a large stepper motor, but is impressive, indeed.

Such drivers reduce the controller's job to determining *direction* of rotation and rate of stepping (an edge on the IC pin initiates a single step).

The controller can drive the stepper: If you don't want to be so fancy, and so lazy, you can use a few pins of the controller to drive power transistors wired to the motor. Power MOSFETs will do

---

[30] You will find a helpful animation at http://www.pcbheaven.com/wikipages/How_Stepper_Motors_Work.
[31] This surface mount part is available in a handy breakout board from Sparkfun: ROB-10267, "Easy Driver...."
[32] Microstepping – which can be pushed to extremes such as 1/256 step – works by energizing *both* windings at graduated relative currents. The price one pays for this result is drastic reduction in torque so one should not rely on microstepping for extreme resolution. See http://www.micromo.com/microstepping-myths-and-realities.

the job: the IRLZ34 used in Lab 12L (55V, 30A max) would do.[33] Fig. 26N.31 shows the scheme. Integrated stepper-motor drivers normally include current-limiting. The home-brew scheme shown in Fig. 26N.31 lacks such limiting, so it is up to you to moderate the supply *voltage* that will not overheat the motor. It can get somewhat hot to the touch without damage.

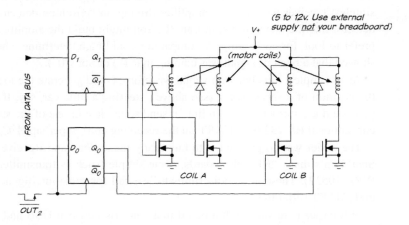

Figure 26N.31 Stepper driver hardware.

One way to generate such successive patterns is to load a value into a register and rotate that value, feeding two adjacent bits to the motor's two coil-drivers: see Fig. 26N.32.

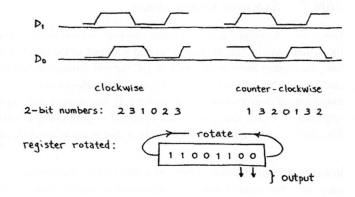

Figure 26N.32 Rotated register can produce the 2-bit drive pattern for stepper.

Using the stepper motor: You will dream up your own ways to use a motor, but here are a few uses students have tried.

Crane: Two motors, each with a spool on its shaft to wind cord, can rest on a tabletop with the load suspended from both cords. The two motors can work together to lift, move, then lower the load. This is easy to do crudely but would be challenging if one wanted to let the load move vertically, then horizontally, then vertically again. A small electromagnet could let the machine pick up and drop an iron load (a washer, perhaps). (A small spool of wire-wrap wire is handy for an instant home-made coil; put an iron bolt through its core). (One student, Dylan Jones, built such an electromagnet, hung it from a stepper-driven "winch," and mounted the whole thing on the tractor described below.)

---

[33] The snubber diodes shown in Fig. 26N.31 will protect the transistors from voltage spikes on transistor turn-off. These should be power diodes such as 1N400x (the "x" specifies max reverse voltage, not critical in this circuit). They should not be small-signal diodes like our usual 1N914.

Drawing machine: As we said back in §26N.3.3, two motors can drive the X and Y knobs of a child's sketching toy.

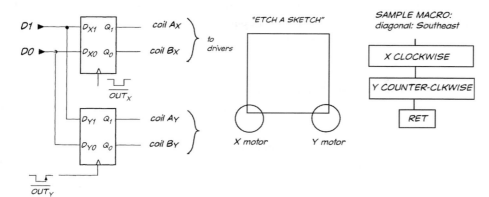

Figure 26N.33 Two stepper motors can drive an X–Y drawing toy.

26N.8 Project ideas

26N.8.1 Nifty new and untested projects for your inspiration

Speech recognition: You certainly don't want to try it the conventional, full-scale way: transforming to the frequency domain and then comparing against a template. That is a huge programming task, ill-suited to this little machine and our tools. (But see the cheap trick suggested on page 1048: use someone's IC that does the hard work for you.) We want a rough-and-ready method.

One strategy – used, for example, on an old Radio Shack chip that could distinguish "Stop" from "Go" for toys – is to look at just *zero-crossings*: in other words, find the dominant frequency (this method might detect the "Sssss" in "Stop"). Count zero-crossings in hardware or software, and check against a few range definitions: "$\geq$5kHz$\Longrightarrow$Sssss," and so on.

A little more refined: use a pair of steep filters, high-pass, low-pass, with little overlap; then compare amplitudes (time-averaged, we suppose) above and below this frequency boundary. Bell Labs experimented with this method in the early '50s, placing the boundary around 900Hz, and found they could distinguish the 10 digits pretty well.

Audio encryption: Here's an idea we saw proposed in a trade magazine: simple encryption of audio – say, for a cordless phone – by inverting the frequency spectrum of the speech.

Do this by exploiting a violation of Nyquist's rule: multiply the speech data (in digital form) by a large square wave at a frequency just slightly above the top signal frequency. The result is an inverted frequency spectrum (because of aliasing or folding). De-crypt by multiplying by the same frequency. The "multiplication" is simpler than it sounds: just flip the MSB of the data, at the multiplication rate (apply a 1, then a zero, then a 1... to an XOR with the MSB of data). This also can be done with analog methods.

Music effects: chorus, etc.? You may dimly recall a phase-shifter, back in Lab 7L, that could vary phase by varying a resistance to ground.[34] If you use a digipot like the one you met in Lab 24L you can make a computer-controlled phase shifter.

---

[34] This was the second of two phase-shifter designs.

An alternative way to get the same result might use a *multiplying DAC* – which can be described as a digitally-controlled attenuator; output is the product of the analog and digital inputs.

Combined with a summing circuit, in principle either shifter can make interesting rock-and-roll sounds – like those vaguely mentioned in Lab 7L. To hear a repertoire of such sounds, try http://www.harmony-central.com/Effects. This web page also includes circuits and helpful links.

26N.9 Two programs that could be useful: LCD,Keypad

26N.9.1 LCD driver; keypad scanner

These hardware interfaces are described in Chapter 24N. The code to implement these is posted on our book website, in case you want to use one of these devices in a project.

26N.10 And many examples are shown in AoE

We have suggested just a few possible peripherals for your controller as we said in Chapter 25N AoE suggests many more.

AoE §15.8

26N.11 Now go forth

We hope that you'll be moved to build a *project* now that you've made it through the book. But our larger wish of course is that you'll take what you've learned of electronics and apply it in many settings. And we hope too that you'll continue to find electronics intriguing and fun. The customer in Fig. 26N.34 goes a bit beyond our ideal: we do want your devices to be useful. But we admire his enthusiasm. Probably all of us geeks have in our basements some gizmos like the one he's buying – cool things that we just had to have.

Figure 26N.34 A model of enthusiasm for electronics – though maybe not for its utility.

A A Logic Compiler or HDL: Verilog

AoE §11.3.4B

Let's get started. As you know from your experience with other programming languages, the refinements are many, the options can be dazzling. But our goal in this course is modest: only to let you try out some of the capabilities of Verilog and PLDs. Later, when you undertake a more ambitious design, no doubt you will want to get yourself one or more Verilog reference texts and then explore the rich possibilities.[1]

A.1 The form of a Verilog file: design file

A Verilog file uses some rigid forms that are better shown than discussed. Examples are in Fig. A.1. When you write your first programs we assume that you will simply copy the forms that you see – and in early cases we will provide you with nearly-complete files, so that you can sidestep the fussy form requirements, simply entering the interesting stuff: the equations that describe the logic you intend. Here, for a first example, is a file that implements a two-input AND gate and a two-input OR gate:[2]

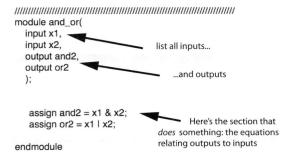

Figure A.1 Verilog source file for AND OR functions.

Here are the elements:

- the keyword "module" ...
- ...followed by filename[3], and then...
- ...the list of input and output variables, "(x1, x2, and2, or2)"
- ...and the concluding semicolon – required to end most lines.

---

[1] Here are two that we like: J. Bhasker, *Verilog HDL Synthesis* (1998): extremely concise, offering little discussion but at least one example of each topic treated; S. Palnitkar, *Verilog HDL* (2003): more discursive, and including fewer examples, but more detailed discussion of each.
[2] This is the form produced by Xilinx ISE 11.1. The form changes slightly, from revision to revision. The last ISE Design Suite revision is 14.7. Further revisions will not occur, as Xilinx moves to its "Vivado," which, however, does not support devices earlier than Series 7 FPGAs.
[3] This is the case for our simple designs. A file can, however, include multiple modules.

All these signals are of the default type, "wire." Soon (§A.3) we will meet the other important category, "reg."

Then, with these preliminary introductions of variables done, we can get down to business, which is done in the two "assign" lines. *Assign* is the keyword that defines the logic linking the output (for example, the first: "and2") with the relevant inputs (here, x1 and x2). The "&" means AND, "|" means OR. Assign is appropriate for combinational logic (though other methods are available). We will see soon that sequential circuits are better designed with a different form.

If one compiles this design, Xilinx's ISE will draw a schematic of the result.

In ISE 12, one can set up an easy default display of the "top level" schematic. We recommend doing that.

In a small design, it would be a nuisance always to have to specify signals of interest. But in a complex design it is useful to be able to look at only the network that produces a single output, or a few, among many. Here is the process that permits specifying which signals to display in the schematic.

- First, choose "top level ports" and add these.
- Then the signals themselves are shown in Fig. A.2; but not the linking logic.

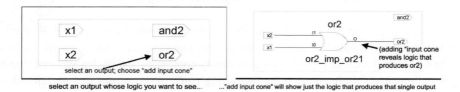

Figure A.2 Select signals to be shown on schematic.

- Then you can click on a particular output to see a "cone" that shows, see Fig. A.3, all the logic that produces that output (this is one way to get the schematic).

Figure A.3 Showing logic producing a particular output.

No surprises, here, but it is reassuring to see in Fig. A.4 that it looks like what you would expect.

A.2 Schematics can help one to debug

Looking at the schematic makes sense not just because we are beginners doing simple tasks. The schematic may help anyone to judge quickly whether the design compiled as expected. This can be true even for a design that is not simple. Xilinx's ISE permits one to look at the logic that generates just a piece of a complex design. One can look at what ISE calls the "input cone" for a single output, in a design with many outputs.

For example, Appendix B shows how displaying logic for a single output can tame a bewildering rat's nest of logic. The example there is the PAL that produces the many outputs of the "glue" PAL that helps to implement the big-board lab computer.

A.3 The form of a Verilog file: simulation testbench

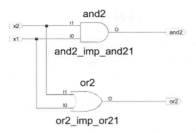

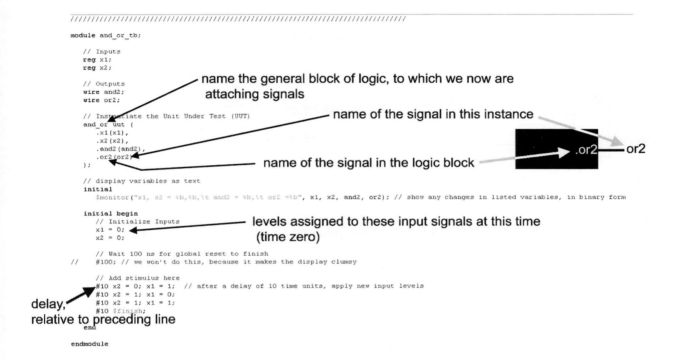

Figure A.4 Compiled AND–OR logic produces what one would expect: Verilog's schematic output.

Figure A.5 Verilog testbench.

A.3 The form of a Verilog file: simulation testbench

Once you have written a Verilog design file, usually you will want to simulate it to see if the design behaves as you expect. We do this for all the designs in this course. But we do that because our designs are not complex. There are respectable arguments *against* simulating. Some designers using FPGAs for complex designs find it quicker just to load the design into the hardware and try it. This used to be called the "burn and crash" method. But the perjorative sound of that name is not always appropriate.

Simulation files are difficult to write, and for a complex design often cannot be exhaustive: you cannot try every possible case. This is especially likely for sequential designs.

The simulation file, called a testbench,[4] is a Verilog text file that exercises a design. In the case of

---

[4] Strictly, Verilog calls such a test file "Test Fixture." Testbench is VHDL jargon, but we use it because it seems to be more widely used than test fixture, particularly in Xilinx literature.

an AND and an OR gate this is hardly necessary, but the extreme simplicity of the AND–OR logic should make the testbench easy to follow.

The testbench begins something like the design file, listing inputs and output variables, and stating their types – wire or reg. Just in case you were thinking you're beginning to understand the wire/reg distinction, note that the *inputs* in the simulation file are always of type *reg*, the outputs are *wire*. The wire/reg distinction is difficult. The type reg bears a ghostly relation to the hardware element, "register," but only ghostly. A variable of type *reg* has a value "retained in memory... until changed by a subsequent assignment."[5] That makes it sound like a collection of flip-flops, but it emphatically is not that, and does not imply use of flip-flops. Fig. A.5 shows a testbench for exercising the AND–OR logic.

Simulation results: The result, in Fig. A.6, of simulation is what one would expect. OR looks like an OR of x1, x2; AND looks like an AND.[6]

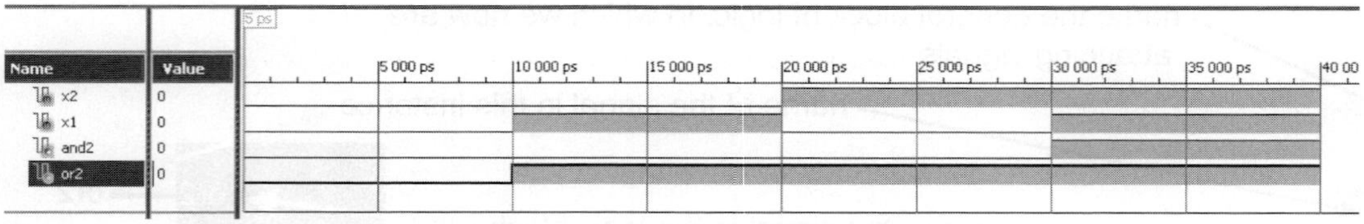

Figure A.6 Waveform output from Verilog testbench.

Instantiation: Some elements of the testbench require explanation. One is the "instantiation" block:

```
   // Instantiate the Unit Under Test (UUT)
   and_or uut (
       .x1(x1),
       .x2(x2),
       .and2(and2),
       .or2(or2)
   );
endmodule
```

Instantiate is a weird word for a less weird concept. It means that it takes the defined logic block (in this case, a block providing an AND and an OR gate) and uses it to process signals in a particular *instance*. The word is chosen to indicate that we are coming down from the abstraction of a high level design into an "instance" that make the design real.

The above code (.and2(and2)) indicates what we're doing in the present case: the mundane operation of linking the particular name in this instance ("and2") with the name of the variable coming out of the logic block (.and2).

But the fact that the two variables carry almost the same name – one from the design block, and another from the particular instance – obscures the fact that they are quite different, independent notions. To make that point, we've made a second testbench with sillier names. Fig. A.7 is a sketch that tries to make the point that the and_or logic block's inputs and outputs can be named as we please, when the block is invoked by a particular use or instantiation.

---

[5] J. Cavanagh, *Verilog HDL* (2007), p. 77.
[6] ISE 11 and 12 simulations show an annoying flaw: initially, it seems to have done nothing, showing constant levels. That's because ISE shows the waveforms in an extreme close-up zoom, and places the cursor at the extreme right of the window. To get intelligible waveforms, click on the "zoom to full view" icon (which, in our version, looks like a circular life preserver).

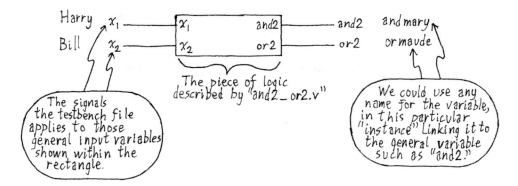

Figure A.7 Instantiation illustrated: the underlying logic block's signals can be assigned to any signal names in the particular "instance".

Here is the list of signals and the instantiation section of the silly names version:

```
// Inputs
reg harry;
reg bill;

// Outputs
wire andmary;
wire ormaude;

// Instantiate the Unit Under Test (UUT)
and_or uut (
    .x1(harry),
    .x2(bill),
    .and2(andmary),
    .or2(ormaude)
);
```

This testbench works just as well as the other. The simulation results in Fig. A.8 are the same except for the silly names.

Display of results in table form: The text display, in contrast to the waveform display done automatically by the Xilinx compiler, occurs because of the following lines of code:

```
// display variables as text
initial
    $monitor("x1, x2 = %b,%b, \t and2 = %b, \t or2 =%b", x1, x2, and2, or2);
// show any changes in listed variables, in binary form
```

`initial` is a keyword that says, "Do this once." `$monitor` displays the listed signals any time one of them changes. The `%b` is a "modulus" indicator: it says "use binary number format." `\t` inserts a tab, just to make the table easier to read. If the timing-diagram simulation result is clear to you, skip the tabular form.

Input "stimulus" values and delay times: At last we reach the core of the simulation, where we stimulate the design with specified levels of the input variables:

A Logic Compiler or HDL: Verilog

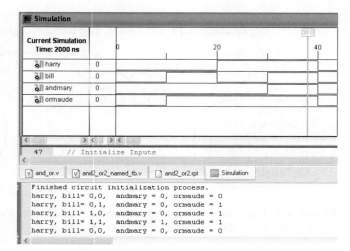

Figure A.8 Simulation results: same with silly-name "instantiation".

```
    initial begin
        // Initialize Inputs
        x1 = 0;
        x2 = 0;

        // Wait 100 ns for global reset to finish
//      #100; // we won't do this, because it makes the display clumsy

        // Add stimulus here
        #10 x2 = 0; x1 = 1; // after a delay of 10 time units, apply new input levels
        #10 x2 = 1; x1 = 0;
        #10 x2 = 1; x1 = 1;
        #10 $finish;
    end
```

The #10... values say "wait ten time units before applying the input levels that follow." The testbench determines what a time unit is (usually nanoseconds, but for present purposes is are of no importance since we are only looking at the results of the design without concern for timing). So the set of stimuli shown above provide changes at 10ns intervals. This timing shows on the waveform displays such as Fig. A.8. Again, the keyword "initial" says, "Go through this sequence once" at startup.

A.4 Self-checking testbench

It is possible, though laborious, to write a testbench that predicts results for each function, and compares these predictions against actual results of a design. We have done this for the "gluePAL" design that links the processor to its peripherals, in the "big board" version of the microcomputer labs later in this course. Here is an excerpt from the self-checking portion of the testbench: a truth table, and a "task" or function that refers to it.

```
// Truth tables list expected outputs (rightmost column), and invoke
//   particular functions (named ``Task") to compare this prediction against
//   the result of the logic in the source file

  // here's the  truth table for CTR_OE_bar: f(br, loader)
     check_CTR_OE_bar(0,0,1);
```

A.4 Self-checking testbench

```
        check_CTR_OE_bar(0,1,0);
        check_CTR_OE_bar(1,0,1);
        check_CTR_OE_bar(1,1,1);

// ...and the collection of "tasks" or functions invoked by truth tables

task check_CTR_OE_bar;
    input i_BR_bar;
    input i_LOADER_bar;
    input expect_CTR_OE_bar;

 begin
   #25; BR_bar = i_BR_bar; LOADER_bar = i_LOADER_bar;
   #25;
   if (CTR_OE_bar !== expect_CTR_OE_bar)
    begin
     $display("\t CTR_OE_bar ERROR: at time=%dns \t INPUTS: BR_bar=%b, LOADER_bar=%b,
      \t OUTPUT: CTR_OE_bar=%b, \t expected=%b", $time, BR_bar,LOADER_bar,
      CTR_OE_bar, expect_CTR_OE_bar);

     errors = errors + 1;
    end
 end
endtask
```

We will not try to explain this scheme fully, but will settle for the following sketch: the *truth table* for CTR_OE_bar shows two input levels and an expected output. The *task* named check_CTR_OE_bar watches for a mismatch between the truth table's output prediction and what the function CTR_OE_bar actually produces.

Any mismatch – where the actual and expected values don't match:

```
   CTR_OE_bar !== expect_CTR_OE_bar
```

evokes a display detailing the mismatch: what was expected and what was observed:

```
   if (CTR_OE_bar !== expect_CTR_OE_bar)
       begin
            $display(
```

This self-checking feature is useful to us in this course, where, if we ask students to design the *gluePAL*, the testbench warns the student of each case where the design does not produce the result we had hoped for.

Without error-flagging, a complex testbench is hard to use: In Fig. A.9 is the result of simulating the gluePAL logic: a set of waveforms that one would be hard-pressed to use for spotting logical errors.

But this testbench includes the self-checking feature that lists the one function that failed to match expectations. The instance is listed at the bottom of Fig. A.9 – too small to read. Fig. A.10 has it in more readable form.

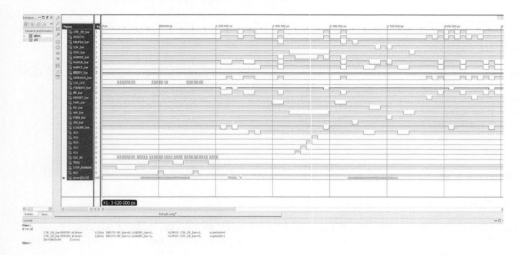

Figure A.9 Testbench waveforms alone can make spotting logical errors very difficult.

Figure A.10 Self-checking testbench picks out the single function that failed from the simulation plotted in Fig. A.9.

A.5 Flip-flops in Verilog

"always@(...)": test for change in a listed variable: We did a combinational task (AND and OR) using the keyword *assign* back in §A.1. In order to design a sequential circuit, one that uses flip-flops, another form is useful, with the strange keyword

$$\text{always@(VARIABLE)}.$$

The compiler then tests for a change in the value of VARIABLE, and proceeds when a change occurs, executing whatever follows that line. This form can be used to design combinational circuits, but a special form of it is *required* for design of edge-triggered flip-flop circuits.

"always@(posedge...)": edge-sensitive test for change: This flip-flop form is

$$\text{always@(posedge VARIABLE)}$$

for rising edge VARIABLE ("negedge" for falling edge). We will soon meet the complication that this form can indicate either *edge*-sensitive behavior or *level*-sensitive behavior. *Order*, in a listed set of variables, will determine which behavior applies to a particular variable.

As a first example here is about as simple a sequential circuit as is possible: a single D flip-flop, with asynchronous reset. On a rising edge of *clock*, q takes the value d; if reset_bar is asserted (low), q takes the value zero (this is an asynchronous reset, despite the word "negedge," below). Verilog is a rather big machine to invoke for the purpose of building this tiny thing, a single D flop, but a simple example is enough for a start.

```
module flop(d, clock, reset_bar, q);
    input d;
    input clock;
    input reset_bar;
    output q;

    wire d,clock,reset_bar;
    reg q;

    always @(negedge reset_bar or posedge clock)
       if (!reset_bar)    // the parens are mandatory, incidentally (this is an
                // asynchronous reset, despite the word ''negedge" -- see below)
          q <= 0;
       else
          q <= d;

endmodule
```

Even this simple design introduces some new Verilog conventions. Here are several.

- The type *reg* for the output is required for use with the always@ form. (We have specified "wire" for the other signals; this was not necessary, since "wire" is the default, a notion we relied upon in the AND–OR example, §A.1.) We noted, back in §A.3, that *reg* does not mean hardware "register" or "flip-flop." Instead, it means that a signal holds its value until overwritten.
- The "=" sign used with *assign* should be replaced, after "always@(posedge...)," with "<=" ("non-blocking" assignment). This seemingly superficial change is important and quite subtle. In §A.13 we dwell further on this "blocking" versus "non-blocking" distinction further. Feel free to avoid thinking about this subtlety; you can avoid thinking if you adopt the rule that you will use the *non-blocking* "<=" for flip-flops.[7]
- "if...else": the line that follows is executed when the "if" condition is fulfilled; it falls through to the "else" if the condition fails. ("if...else" sets may be nested.)
- *Edge-sensitive versus level-sensitive inputs*: a very subtle (and confusing point):
    - clock is an *edge-sensitive* input, whereas ...
    - reset_bar is *level-sensitive* (despite the word "negedge")

    Only the *last* signal listed among the "if...else ...else" statements is treated as *edge-sensitive* and as the *clocking* event;[8] the others are level-sensitive.[9]

The Verilog schematic in Fig. A.11, showing its implementation of the flop just described, confirms that only the clock is truly edge-sensitive.

And the simulation result is, of course, consistent with Verilog's design: reset* is *asynchronous*: Q clears as soon as reset* is asserted; Q does not wait for the clock edge: see Fig. A.12.

---

[7] In case you insist on thinking about what this distinction means: "<=," "non-blocking" assignment versus"=," blocking, here is a sketch. Roughly, the blocking forms will be executed in order (as in an ordinary computer program) whereas non-blocking lines are executed in parallel, all at the time specified. In the case above that means "if clock goes high or reset_bar goes low." Use of a "blocking" assignment after "always@(posedge...)" is permitted but not recommended because of potential timing problems. See Bhasker, p. 68.

[8] Bhasker, p. 78.

[9] Perhaps we are making too much of the word "edge," or are asking too much in expecting Verilog to use the term in its normal digital sense. To be kind to Verilog one might say that in that language "edge" means simply "change." In that sense, it is not wrong to say that the Verilog design responds to an "edge" on its reset* line: it responds to a *change* on that line.

A Logic Compiler or HDL: Verilog

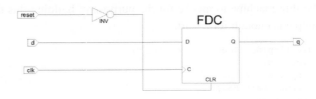

Figure A.11 Verilog schematic shows only clock is edge triggered.

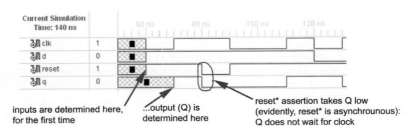

Figure A.12 Flip-flop simulation shows reset* is asynchronous.

inputs are determined here, for the first time

...output (Q) is determined here

reset* assertion takes Q low (evidently, reset* is asychrounous): Q does not wait for clock

Flip-flop with synchronous reset*: One could build a flop with a *synchronous* reset*. This is not something we have met, but Verilog permits it. The difference in the design is just that the "always@(...)" would mention only the clock, so that only this signal could cause the circuit to change state.

```
module flop(d, clock, reset_bar, q);
    input d;
    input clock;
    input reset_bar;
    output q;

    wire d,clock,reset_bar;
    reg q = 0;         // stating level initializes flop level for simulation

    always @(posedge clock)
      if (!reset_bar)  // this reset* is synchronous: it will not be
                       // applied until the rise of clock
        q <= 0;
      else
        q <= d;

endmodule
```

Verilog's schematic in Fig. A.13 shows what we would expect: the reset_bar signal can force the D input low, but any change of Q will await the next rising clock edge.

Simulation confirms that clearing awaits the next rising clock edge: see Fig. A.14.

No surprise here; but it's nice to find that we can choose either result – sync or async behavior.

A.5.1   Simulation of a sequential circuit must begin in a known state

Simulation can fail if your testbench fails to specify the circuit's initial state. In a combinational circuit, that will pose no problem, but it is a point that requires attention in a *sequential* circuit (a topic treated at §A.5 and after). If, for example, you test a divide-by-two circuit and fail to tell the simulator what is the circuit's initial state, the simulator will balk, giving you no results.

A.5 Flip-flops in Verilog

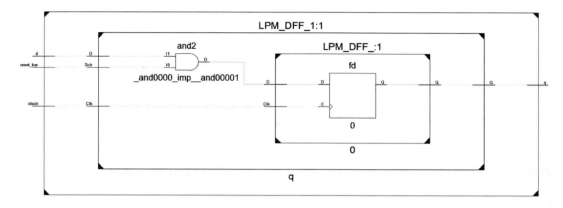

Figure A.13 Schematic of sync-clear* flip-flop, for contrast with the more usual async clear*.

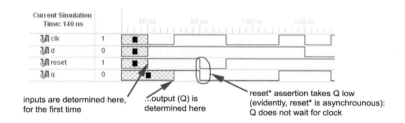

Figure A.14 Simulation shows sync-clear* takes effect only on the next clock.

One Can specify the initial state in the design file...: In Fig. A.15, a toggling flop circuit fails to simulate until we add a specification of the initial level of the flop output, in the design file (not in the testbench).

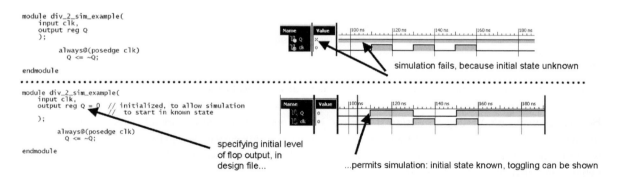

Figure A.15 Simulation of a sequential circuit requires initialization.

...but instead one should provide a hardware initialization: The previous method makes the simulation work, but has no effect on the implementation: the startup condition is unpredictable. So a hardware reset* should be added to the design:

```
module div_2_sim_example_better(
    input clk,
    input reset_bar,
    output reg Q
    );
```

```
always@(posedge clk or negedge reset_bar)
  if (!reset_bar)
    Q <= 0;
  else
    Q <= ~Q;
endmodule
```

This hardware reset* would take effect not only in simulation, but in the behavior of the programmed part.

A.6 Behavioral versus structural design description: easy versus hard

Using Verilog, one can design circuits, sequential or combinational, by detailing their innards, as if drawing gates and flops. Here is a tiny 2-bit synchronous counter designed by describing its structure. In a moment, we'll do this same job the easy way.

Structural way: detail the design ("hard"): With an XOR gate feeding the second flop we can make a T-flop, one that changes only when told to. We will tell it to change if Q_0 is high just before the clock edge (during set-up time). Here's some Verilog code for saying that.

```
module ctr_boolean_2bit(clk, reset_bar, Q);
  input clk;
  input reset_bar;
  output [1:0] Q;
    wire clk, reset_bar;
    reg Q;
    always@(posedge clk or negedge reset_bar)
      if (~reset_bar)
        Q[1:0] <= 2'b0;          // asynchronous reset
      else                        // these events occur on the rising edge of clk
        begin
          Q[0] <= ~Q[0]; // Q0 always toggles
          Q[1] <= Q[0] ^ Q[1]; // Q1 toggles if Q0 is high (XOR achieves that:
        end
endmodule
```

Verilog draws us the schematic, Fig. A.16, and it is what we expect to see.

Behavioral way: ask Verilog to design it ("Easy"): But one can also be remarkably lazy, leaving much of the design work to the compiler. It is almost embarrassingly easy to design the simple counter we just made at the higher "behavioral" level. We can describe how we would like the circuit to behave, and then let Verilog do the rest.

```
module two_bit_simplest_ctr(clk, reset_bar, count); // list the signals
    input clk; // ...say if they're in or out
    input reset_bar;
    output [1:0] count; // this is a 2-bit variable

      wire clk, reset_bar; // jargon used for inputs
      reg count = 0; // ...and outputs when using "always@(..."
                          // form (here initialized to zero)
      always@(posedge clk,negedge reset_bar)
                    // despite the name, reset_bar is not edge-sensitive
        if (~reset_bar)       // level-sensitive
          count <= 2'b00;
        else
```

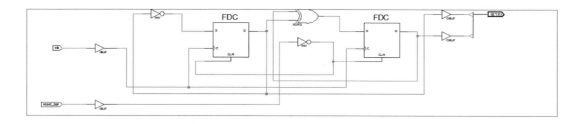

Figure A.16 2-bit counter designed with detailed Boolean description.

```
            count <= count +1;     // clock really IS edge-sensitive
endmodule
```

The resulting schematic is the same as before (the rectangle that feeds the second flop encompasses a single XOR gate, as one can confirm by exploring or "pushing into" that rectangle). You probably don't need to be persuaded that you are likely to use *behavioral* designs whenever you can.

A.7   Verilog allows hierarchical designs

When a design gets complicated, it's nice to be able to design a relatively simple "low level" module, like a D flip-flop with reset (in the example below); then invoke that block in a slightly more complex module (here, a flop that toggles each time it is clocked)...and so on, ultimately putting together a complex circuit that is a fairly simple collection of submodules.

This technique recalls our way of doing *modular* designs in the analog section of the course. We held to the "times-ten impedance rule" because following that rule permitted us to design module *A*, then hook it up to module *B* without having to re-analyze the behavior of *A* when so attached: we did not need to consider a complex supermodule called *AB*. Designing a module at a time helped keep our work of design and analysis simple. Verilog permits the equivalent method, which is illustrated below in an example very similar to one provided by Palnitkar in his book, *Verilog HDL* (pages 17ff).

The circuit to be designed is a 3-bit ripple counter. The example makes this out of D flops, each with its Q-bar output tied to its D input. Nothing clever here. But the simplicity of the example makes it useful. It shows each block made from lower-level sub-blocks. Specifically:

- The D flop is made "from scratch," invoking no simpler pieces.
- The toggling flop[10] is made by feeding the complement of Q back to D, on the D flop. Even the "complement" is generated by using a submodule called "NOT:" a Verilog "primitive."
- The 3-bit counter is formed from three of the toggling flops, the clock of each driven by the Q of the lower, preceding flop.

Here is what the modules look like.

D flop: This is getting familiar by now: clear on reset; D to Q on clock (this clock is falling-edge).

```
module D_FF (q, d, clk, reset);
output q;
input d, clk, reset;
reg q;
```

---

[10] This is not a T flop, but a simple divide-by-two: a flop that toggles each time it is clocked.

```
always @(posedge reset or negedge clk)
                   // the clock is made falling-edge, to produce an UP counter
if (reset)
q <= 1'b0;
else
q <= d;
endmodule
```

The toggle-every-time flop ("DIV2"): This is just a D flop with D fed Q-bar, so that the flop toggles each time clocked. (Note, as we said earlier, that this is simpler than a T flop.)

```
module DIV2_FF(q, clk, reset); // this is a new device built out of the module
                     // ``D_FF'', and also a built-in Verilog ``primitive''
                     // gate called ``not''--which does what you'd expect
output q;
input clk, reset;
wire d;
D_FF dff0(q, d, clk, reset); // instantiate D_FF. Call it dff0.
                     // The name "dff0" does not matter at all,
                     // and is never used elsewhere.
         not n1(d, q);     // not gate is a Verilog primitive:
                     // here, the built-in primitive is "not;"
                     // in this instance we give it the arbitrary name "n1".
                     // d is its output, q is its input.
                     // We are passing a signal named ``q'' to the "not" primitive,
                     // and taking back its output
                     // with a signal name, "d"
                     // In other words, "n1" inverts "q" applying its output
                     // to "d" of the D_FF.
                     // In case invoking "not" seems over-complicated,
                     // here is a more familiar way to get the same result:
                     //    assign d = !q;
endmodule
```

Even the job of *inversion* is done, here, by invoking a lower-level module; in this case it is the Verilog "primitive" named "not." Its output is listed first, its input second: "not n1(d, q)." This way of passing or defining parameters is discussed in §A.9.

The ripple counter: This *counter* takes three of the DIV2 flops, tying Q_n to Clk_{n+1}:

```
module ripple_carry_counter_3bit (q, clk, reset);
output [2:0] q;
input clk, reset;
// four DIV2_FF instances

DIV2_FF div2_0(q[0],clk, reset);
         // applies the named signals (clk and reset) to inputs of DIV2, and takes
         // DIV2's output, "q" in DIV2, and calls it "q[0]" in the top-level counter
DIV2_FF div2_1(q[1],q[0], reset);
         // ...similar -- but here's the rippling: clock for this DIV2 flop
         // is not "clk" but the output of the low-order flop: q[0]
DIV2_FF div2_2(q[2],q[1], reset);
endmodule
```

What's perhaps most striking in this design is the fact that the top-level module – what we call the "ripple counter" – which we might expect would be the most complex element, seems the simplest.

Here's a potentially confusing point: once again, the "instantiation" signal names – div2_0, div2_1, div2_2 – do not matter at all, and are never used elsewhere in this design. The signals actually used in the top level design are only those listed within the parens, such as (q[0],clk, reset). The 3-bit counter's outputs are just those three q[n] values.

Instantiation once again: The lines of the top-level module that assign signals to inputs and outputs of lower level modules may at first look unfamiliar:

```
// four DIV2_FF instances

        DIV2_FF div2_0(q[0],clk, reset);
            // applies the named signals (clk and reset) to inputs of DIV2,
AA      DIV2_FF div2_1(q[1],q[0], reset);
            // ...similar--but here's the rippling: clock for this DIV2 flop
            // is not "clk" but the output of the low-order flop: q[0]
```

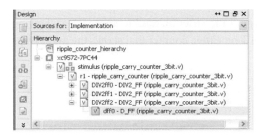

Figure A.17 Verilog modules, one built out of modules below its level.

And let's remind ourselves of the DIV2_FF's list of signals:

```
BB      module DIV2_FF(q, clk, reset);
```

Putting the two equations together – the lines marked above with AA and BB – we see that the particular instance, div2_1, applies the particular signal "reset" to the input of the same name in DIV2_FF. More interesting and illuminating is the application of q[0] to the DIV2_FF input named clk, and the assignment of an output signal name q[1] to the DIV2_FF output named q.

This looks more familiar if we recast this *instantiation* in the form that we have seen in testbench files.

The hierarchy: Figure A.17 shows ISE's representation of the hierarchy. The last of the listed DIV2 flops is shown expanded, so that the simple D_FF from which it is made can be seen.

Schematic of the modules:

Bottom level: DIV2_FF: The bottom-level module, D flop, is too familiar to need displaying. The DIV2_FF is pretty obvious; but maybe it's reassuring to see, in Fig. A.18, it drawn by ISE.

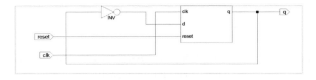

Figure A.18 DIV2_FF module: D flop feeding Q-bar to D.

A more familiar way of designing the always-toggle flops: Perhaps we pushed too far, going all the way back to re-designing the D flop and building a toggler from a D flop. If we skip that stage, the *always-toggle* flop looks more familiar:

```
always @(negedge clk or negedge reset)
  if (!reset)
    q <= 0;
  else
    q <= !q;
```

The ripple counter: Figure A.19 shows a set of three such DIV2s strung together to form the ripple counter. This schematic – showing just mysterious rectangles doing the DIV2 task – isn't very illuminating. But paired with an exploded view of one of those rectangles (the DIV2 of Fig. A.18) it makes sense. Again, the result is obvious; and again perhaps it is reassuring to see ISE producing what we would expect.

The schematic is easier to understand if we do *not* adopt the strict hierarchic approach – showing just high-level black boxes. Fig. A.20 gives a view of the 3-bit ripple counter that shows the underlying D flops.

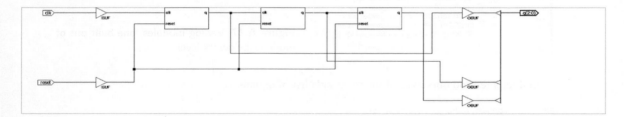

Figure A.19 Schematic of "top level" design: ripple counter stringing together DIV2 flops.

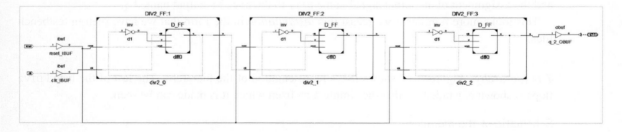

Figure A.20 Detailed schematic, showing innards of div-2 blocks within ripple counter.

A.8   A BCD counter

The simplest counter is *natural binary*. In Verilog, as you know, you can design an UP counter simply by writing, "... count <= count +1." But sometimes (when serving those ten-fingered creatures known as *humans*) it is better to use a divide-by-ten or "binary coded decimal" (BCD) counter.

A.8 A BCD counter

Such a counter is messier to code in Verilog, and the cascading of several stages calls for a hierarchical design like the one described for the ripple counter of §A.7.

A single-digit BCD counter (the lowest-level unit): A BCD counter looks like an ordinary counter (that is, natural-binary), except that it has to be told to roll over to zero after nine.

A subtler difference also is necessary, because this is to be a unit in a hierarchy. Because each counter is just one of several stages to be cascaded, the single-digit's Carry_Out must depend on not only the state of this digit but also on the state of all lesser digits. Therefore, Carry_Out depends also on Carry_In. (These signals use the shorter names "cin" and "cout," in the file just below.) The need to include Carry_In to permit cascading is a point treated again in Chapter 16N.

Code A.1 (Single-digit BCD)

```
module bcd_counter(
    input clk, cin, reset_bar,
    output reg [3:0] count,
                // "reg" because we'll use the "always @(" form ("non-blocking");
                //  "= 0" for clean startup in simulation
    output cout
    );
 assign c_out = cin & (count == 9);
                // the "cin" condition implements dependence on the lesser stages:
                //  all must be full, to generate a carry-out
 always @(posedge clk, negedge reset_bar)
  begin
    if (~reset_bar)
                count <= 0;         // async reset
                else                // here begin the transitions that depend on
                                    // the clock (that is, those that are synchronous)
    if (cin & (count == 9))
                                    // here's the line that makes this BCD rather
                                    // than the usual natural binary:
        count <= 0;
                                    // roll over only when this digit is full (nine)
                                    // and so are lesser digits, if any
        else
          if(cin)
            count <= count +1;
          else
            count <= count;
  end
endmodule
```

This time we have a hardware reset*, so we did not use the initializing device that we mentioned in §A.5.1. We did not write output reg [3:0] count = 0. We relied on the hardware reset* instead, following the advice we gave ourselves then.

Cascade several BCDs to make a larger counter: The BCD counter of §A.8 is designed – like the old MSI 4-bit IC counters mentioned in Chapter 16N (for example, the 74HC161) – to be cascaded so as to make a divide-by-one-hundred, divide-by-one-thousand, and so on. Another Verilog module can implement that cascading. Here, for example, is a divide-by-one-hundred made with two of the BCD modules.

Code A.2 (BCD divide-by-one-hundred)

```
module div_100_bcd(
    input clk,
```

```
   input cin,
   input reset_bar,
   output [7:0] count,
   output c_out
   );
bcd_counter units (clk, cin, reset_bar, count[3:0],c_out0);
             // 'instantiating' one div-10
bcd_counter tens (clk, c_out0, reset_bar, count[7:4],c_out);
             // ...this one same, except its cin is
             // cout of units (instantiated as "cout0")
endmodule
```

Again, as in the ripple counter, this "top-level module" is far simpler than the elements that it strings together.

A.9   Two alternative ways to instantiate a sub-module

Parameters set by *position*: The code that applies a sub-module named "bcd_counter" to make a particular stage, such as the units, is compact:

```
bcd_counter tens (clk, c_out0, reset_bar, count[7:4],c_out);
           // ...this one same, except its cin is
           // cout of units (instantiated as "c_out0")
```

That's good – but this compactness comes at a cost. The named signals `clk...`, `c_out0` are applied by the higher-level module (div_100_bcd) to the lower (bcd_counter) relying on *position* in a list. So, for example, c_out from the lower module is given the name c_out0. This works because c_out is last in the list within the module bcd_counter. If we got the order wrong, our design would go wrong. If, say, we thought that reset_bar came last in the listed signals of bcd_counter, we would be associating reset_bar with the sub-module's c_out. This association would make no sense.

Parameters set by *name*: A less compact but clearer way to associate signals with sub-module signals (that is, a clearer way to instantiate the sub-module) is to list the two, side-by-side, as we have seen done in testbenches.

```
bcd_counter units(
  .clk(clk),
  .cin(c_out0),
  .reset_bar(reset_bar),
  .count(count[7:4]),
  .c_out(c_out)
  );
```

Parameters set by name rather than by position in a list: the second BCD counter: Its cin, for example, is `c_out0 from the lesser stage`

If we use this scheme, we don't need to remember where in a list the bcd_counter signals were placed. The associations are explicit. For a simple module (or "function") – like, say, the NOT of page 1066, with its single input and single output – the ordered listing probably is quite manageable. When signals are more numerous, the explicit listing may be worthwhile.

A.10 State machines

A counter implemented with the conditional form, "case": For a simple counter, there is no tidier, easier design method than the "behavioral" description we saw earlier. But for odder machines that advance from one "state" to another (*state* defined as the levels of its flip-flops), other forms, such as nested "ifs" or "Case..." can be useful. "Finite state machine" is a fancy name for a circuit that steps through a sequence of states, steered by signals from the outside world. More usually, we call these circuits just "state machines." Counters are a special case of state machine – the simplest case. State machines are treated more fully in Chapter 17N.

How would you design such a thing? Nested "if...else" pairs can make complex branching structures, but where many conditions are to be tested, the "case" form is tidier. Here, for example, is a way to make an up/down counter conceived as a state machine. We can list each possible *present state* of the counter as a *case*, leading to one or another *next state*. Because you know an easier way to make a counter, and therefore aren't likely to find this method very appealing, we've given the counter a slight oddity – allowing it to step through only three states, and making the sequence a bit strange (we use gray code[11] rather than the usual "natural binary"). With these wrinkles we've tried to make our use of *state machine* form almost justifiable (soon we'll show an instance where this form surely is justified). We have also included two other novel elements, which we'll explain below: a *carry out* that switches meaning appropriately, according to direction of count, and *naming* of states in order to make the code more readable.

Code A.3 (Up/Down counter done with case statements: Version 1, asynchronous Carry_Out)

```
module up_down_state_machine(clk, startup_bar,up,c_out, state);
                    // state form to make a tiny counter
    input clk, startup_bar, up;
    output c_out; // carry out: asserted on zero, for down-count,
                    // on 11 for up-count
    output [1:0] state;
    wire clk,startup_bar,up, c_out;
    reg [1:0]state;
    parameter A = 'b00, B = 'b01, C = 'b11;  // let's make this a div-by-three
    assign c_out= ( (up & (state == C)) | (~up & (state == A)));
            // a Mealy output, since it depends on an input (up) as well as state
    always @(posedge clk, negedge startup_bar)
        begin
          if (~startup_bar)
              state <= A ;
          else
            case (state)
              A:
                if (up)
                    state <= B;
                else
                    state <= C;  // this reverses count direction
              B:
                if (up)
                    state <= C;
                else
                    state <= A;  // this reverses count direction
              C:
```

---

[11] Gray code allows only one bit to change, between successive states. This rule prevents false transient conditions. See AoE §10.1.3E.

```
            if (up)
               state <= A;
            else
               state <= B;    // this reverses count direction
      endcase
   end
endmodule
```

Two novelties, here:

parameter: Allows us to give intelligible names to states to make the code more readable.[12]

c_out is conditional: Assert `c_out` on *maximum* count (11) if counting UP, on *minimum* count (00) if counting DOWN. Nothing really new in this – but a feature we have not shown before in a counter.

Such a carry is asynchronous, so could glitch during a state transition in which flops didn't synchronize exactly, because their delays differed slightly. If `c_out` drives a *synchronous* input – such as a `carry_in` to a synchronous counter, such a glitch is harmless. If, in contrast, we use `c_out` as an overflow detector, say, using it to clock an overflow flop, the glitching can cause mischief.

Up/Down counter done with "case" statements: Version 2, synchronous Carry_Out: The up/down counter of §A.10 can be designed with a *synchronous* `c_out`. This should be glitch free: the carry, like the states, would change only shortly after each clock. Between clocks, like any synchronous function, it will be well-behaved: quiet.

The addition of `c_out` to the next state information makes the carry wait for the clock:[13] Then the level of `c_out` is specified for each state:

```
      case (state)
         A:
            if (up)
               begin
                  state <= B;
                  c_out <= 0;    // this is level we'll get after next clock
               end
```

And here is the full up/down counter, except for the start of file information, which we did not repeat because it looks like that for the async `c_out file` on page 1071.

Code A.4 (Up/down counter)

```
module up_down_state_machine_sync_carry(clk, startup_bar,up,c_out, state);
                                  // state form to make a tiny counter
// [INITIAL BLOCK OMITTED; same as for async carry-out]
      always @(posedge clk, negedge startup_bar)
      begin
            if (~startup_bar)    // this is the usual asynchronous reset function
               begin
                  state <= A ;
                  c_out <= 0;
```

---

[12] The scope of this definition is local to the module. One authority advocates using, in place of parameter, the *define* compiler directive. Define is not a part of Verilog and therefore is to be used outside any Verilog module (Monte Dalrymple, *Microprocessor Design Using Verilog HDL* (2012), p. 36.

[13] We also had to call `c_out` type "reg" at the head of the file:
```
wire clk,startup_bar,up;
reg [1:0]state;
reg c_out;
```

```
                            end
                        else
                          case (state)
                            A:
                              if (up)
                                begin
                                  state <= B;
                                  c_out <= 0;  // this is level we'll get after next clock
                                end
              else              // this is down counting
                                begin
                                  state <= C;
                                  c_out <= 0;
                                end
                            B:
                              if (up)
                                begin
                                  state <= C;
                                  c_out <= 1;    // try for sync carry (next state is max on up count)
                                end
              else            // this is down counting
                                begin
                                  state <= A;
                                  c_out <= 1;  // try for sync carry (next state is min, on down count)
                                end
                            C:
                              if (up)
                                begin
                                  state <= A;
                                  c_out <= 0;
                                end
                              else           // this is down counting
                                begin
                                  state <= B;
                                  c_out <= 0;
                                end
                          endcase
     end
```

A.11 An instance more appropriate to state form: a bus arbiter

The *case* form we just saw is tidy – but it is not a form appropriate to a counter. It worked all right for a tiny divide-by-three design: three cases, two branches for each. It would work very badly for, say, an 8-bit up/down counter: 256 cases, two branches for each. The state-machine form is made to order, however, for a machine that follows a pattern more complex than that of a counter. Here, we use state form to make a "bus arbiter" – a machine intended to let two users (probably two microprocessors) share one resource (perhaps a memory, perhaps a peripheral).

Figure A.21 is a block diagram of the thing: two requesters; the box can grant a request from either one.

The circuit is to enforce the sort of rule familiar from nursery school, a rule for sharing the sandbox shovel at recess:

1. don't keep the shovel if you're not using it;
2. don't grab it from your classmate if she's using it;
3. take turns.

A Logic Compiler or HDL: Verilog

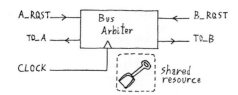

Figure A.21 Block digram of arbiter.

Flow diagram: Let's draw a flow diagram to say that. How many states do we need? Certainly we need a state in which we give the bus to A, and another state in which we give it to B. A subtler point, though, is the need for two more states, distinguished by *who's next*, in case of a new simultaneous pair of requests.

Figure A.22 is such a preliminary flow diagram showing only the states, given names that roughly describe the meaning of each state. The two left-hand states describe the waiting condition: no one has been given the bus. The GRANT_A or GRANT_B notation indicates an output or an action – the granting of the bus.

Transition rules: We can finish the flow diagram by adding notation showing what input conditions should take us from each state to another. In Fig. A.23 the arrows are labeled with the input conditions that lead to the transitions. A indicates a request from user A, whereas AB means both are requesting. This is a synchronous clocked circuit of course, so it is the *clock* that times the advance from any state to another. The circuit *output* is shown in the state in which it is asserted.[14]

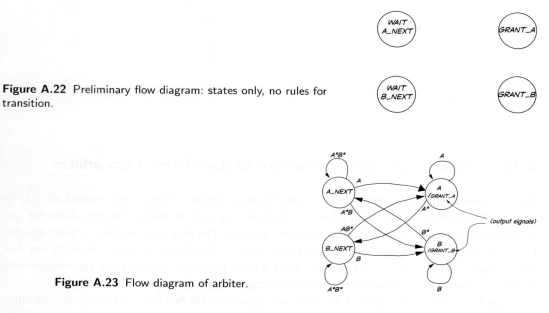

Figure A.22 Preliminary flow diagram: states only, no rules for transition.

Figure A.23 Flow diagram of arbiter.

This diagram makes a choice that wasn't required by the problem statement: the shovel never goes directly from user A to user B; there's always a nursery-school quiet time (lasting one clock period) between the grants. We may ask you to modify the design in some future exercise, eliminating this quiet time. The change would not be at all difficult to implement.

---

[14] This circuit's outputs depend on state only, not on state and present input. In state-machine jargon, that makes this a "Moore" machine rather than "Mealy," just in case you bump into these terms.

A.11 An instance more appropriate to state form: a bus arbiter

A.11.1 Simulation showing arbiter's behavior

Figure A.24 shows waveforms for the arbiter from the Verilog simulation. This may help to make the circuit's behavior less abstract.

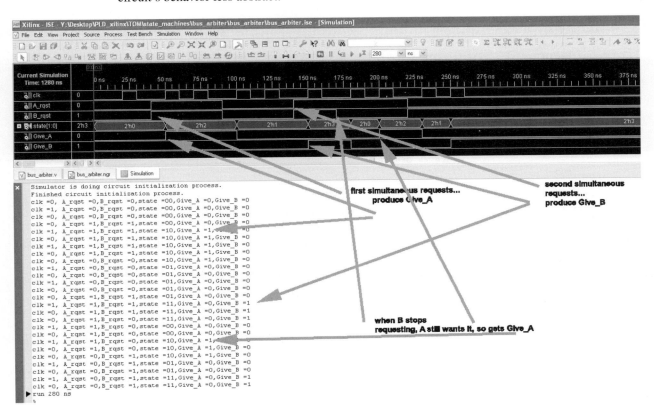

Figure A.24 Bus arbiter simulation.

Code A.5 (Bus arbiter)

```
module bus_arbiter(clk, startup, A_rqst, B_rqst, state, Give_A, Give_B);
    input clk;
    input startup;
    input A_rqst;
    input B_rqst;
    output [1:0] state;
    output Give_A;
    output Give_B;
    wire clk,startup,A_rqst,B_rqst;
    wire Give_A,Give_B; // wire for assign
    reg [1:0] state;
    parameter waitA = 'b00, waitB = 'b01, GrantA = 'b10, GrantB = 'b11;
            // give names to states, to make code more readable

    assign Give_A = (state == GrantA);
            // clumsy way to get my output, after Moore effort failed
    assign Give_B = (state == GrantB);
            // This is low-budget Moore (does not depend on any input)
    assign A_only = A_rqst & ~B_rqst;
```

```verilog
            assign B_only = B_rqst & ~A_rqst;
            assign no_rqst = ~B_rqst & ~A_rqst;
            always @(posedge clk, posedge startup)
            if (startup)
              state <= waitA;
            else
            case (state)
                  waitA:
                  begin
                        if (no_rqst) // hang here till A requests or B without A
                              state <= waitA;
                        else
                              if (A_rqst)
                                    state <= GrantA;
                              else
                                    if (B_only)
                                    state <= GrantB;
                  end
            waitB:
            begin
                  if (no_rqst) // hang here till A requests or B without A
                        state <= waitB;
                  else
                        if (B_rqst)
                              state <= GrantB;
                        else
                              if (A_only)
                                    state <= GrantA;
            end
            GrantA:
            begin
                  if (A_rqst) // hang here till A loses interest
                        state <= GrantA;
                  else
                        state <= waitB;
            end
            GrantB:
            begin
                  if (B_rqst) // hang here till B loses interest
                        state <= GrantB;
                  else
                        state <= waitA;
            end
      endcase
endmodule
```

The bus arbiter itself is not exciting – but we hope you feel the power you have with state-machine techniques to design a great variety of devices that might solve a problem quite elegantly.

A.12 Xilinx ISE offers to lead you by the hand

When you aren't in the mood to think, you can call up ISE's substantial library of ready-made designs. Many elements are included: gates, flops, multiplexers, counters, state machines – all in the form of "templates" showing the code (minus the usual head-of-program information).

Suppose, for example, that you've forgotten how to design a simple Up/Down counter. You can find a standard design template as follows from within the ISE:

- from Project Navigator, select **Edit>Language Templates.**
- under Verilog, choose **Synthesis Constructs.**
- ...expand the **Coding Examples**.

In the example below, we have then chosen

- **Counters,**
- **Binary**
- **Up/Down Counters**

and we then clicked on **Simple Counter** evoking the listing shown on the right in Fig. A.25.

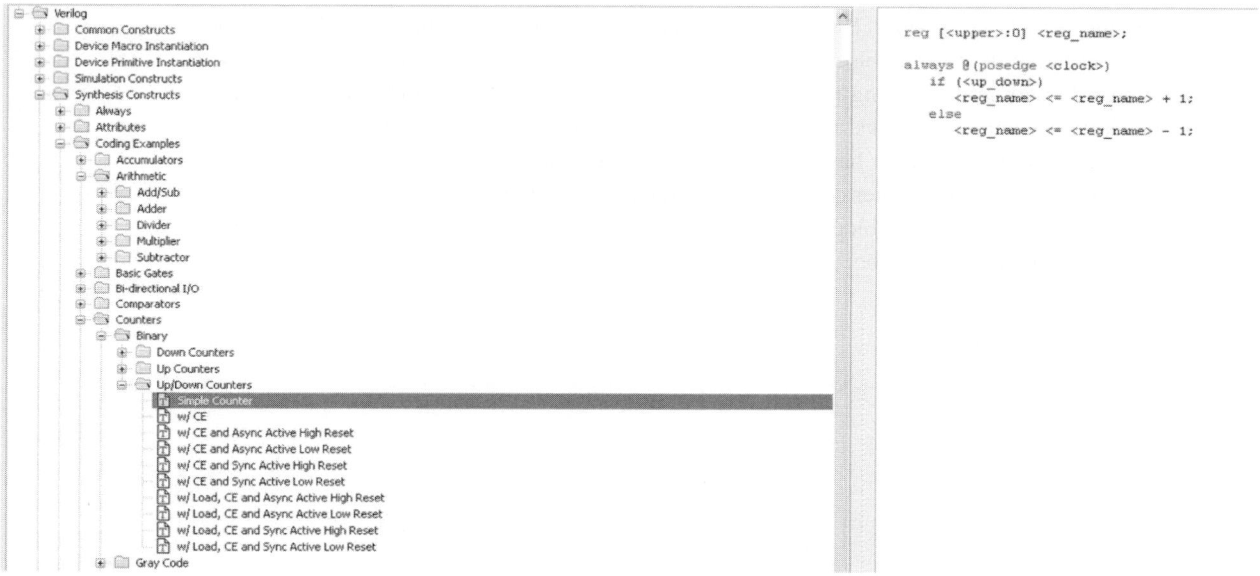

Figure A.25 Xilinx ISE includes a library of templates for standard logic blocks.

We can make this template less abstract by filling in the blanks so as to implement a particular design – say, an 8-bit up/down counter:

Code A.6 (8-it up/down counter)   First the usual header stuff, then

```
reg [7:0] count;
always @(posedge clock)
 begin
  if (up)
    count <= count + 1;
  else
    count <= count -1;
 end
```

A.13   Blocking versus non-blocking assignments

Here's a point you should feel free to ignore. This example is for those who are troubled by the odd distinction between these two sorts of "assignment," a distinction mentioned back in §A.5.

An example can make this abstract-seeming difference clear. Let's use Verilog to design a 3-bit shift-register.[15] Fig. A.26 shows what we intend to build. We will design this twice: once correctly – and once not, as if we had not digested the difference between *blocking* and *non-blocking* assignments.

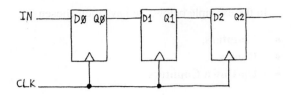

Figure A.26 3-bit shift-register: a design we intend.

Let's do it wrong: blocking version: Blocking assignments are executed in the order they appear within a sequential block. It is called blocking because it completes the assignment of left-hand side from right-hand side without permitting interruption by any other Verilog operation. As Cummings puts it, "A blocking assignment 'blocks' trailing assignments in the same *always* block from occurring until after the current assignment has been completed."[16] Blocking assignments behave as you probably are accustomed to in computer code: one line executes, then the next.

Let's design the shift-register with blocking assignments, letting IN generate Q_0 on the clock, Q_0 generate Q_1... and so on. Here's the code:

```
module shift_reg_blocking(
    input clk,
    input in,
    output reg q0,
    output reg q1,
    output reg q2
    );
  always@(posedge clk)
    begin
        q0 = in;
        q1 = q0;
        q2 = q1;
    end
endmodule
```

This code will compile properly. But it won't build what we intended. `always@(` takes action upon the rising edge of clk. The several assignments all are made in response to a single clock edge. So far, so good.

On a single clock then, the IN should sail right through to Q_2: *not* what we wanted. The schematic in Fig A.27 shows what Verilog built from this code:. It's not a shift register at all but a flop with three labels assigned to its single output.

Let's do it right: non-blocking version: When we use *non-blocking* assignments, on each clock edge the three flop outputs take the *old* levels that were present at their inputs just before the clock rose. As the MIT notes put this, "... all assignments [are] deferred until all right-hand sides have been evaluated (end of the simulation timestep)."[17]

---

[15] This example appears in the notes for MIT's 6.11, Spring 2004, and also in a paper by Clifford Cummings, "Nonblocking Assignments in Verilog Synthesis, Coding Styles that Kill,"
http://www.sunburst-design.com/papers/CummingsSNUG2000SJ_NBA. Bhasker treats the topic in his §2.18.

[16] Cummings, §3.

[17] MIT notes, p. 7. The use of the word "simulation" in what is a discussion not of simulation but of *synthesis* is odd – and reminds us of the discomfiting fact that Verilog was born as a simulation program. Its use for synthesis came later.

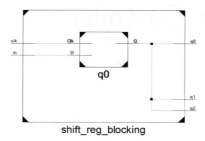

Figure A.27 Schematic of Verilog's implementation of blocking "3-bit shift-register" design.

Because each assignment takes in an *old*, pre-clock level, this design does, indeed build a shift-register, as the Verilog schematic in Fig. A.28 confirms.

```
module shift_reg_non_blocking(
  input clk,
  input in,
  output reg q0,
    output reg q1,
    output reg q2
);

  always@(posedge clk)
    begin
      q0 <= in;
      q1 <= q0;
      q2 <= q1;
    end

endmodule
```

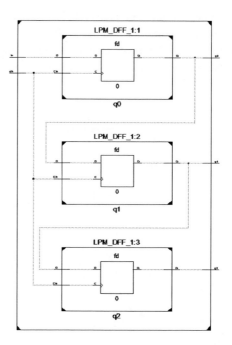

Figure A.28 Code and schematic of Verilog's implementation of non-blocking shift register. This works.

The example supports the point, which you probably believed anyway, that you should use *non-blocking* assignments for sequential designs.

B Using the Xilinx Logic Compiler

B.1 Xilinx, Verilog, and ABEL: an overview

As hardware description languages (HDLs) have evolved, Verilog and VHDL have largely displaced the older hardware compiler languages like ABEL. These newer languages, both widely used, look more like high-level computer programming languages. VHDL may be marginally more powerful, but is the harder to learn.[1] We will introduce you to the fundamentals of Verilog. We will not use VHDL or ABEL.

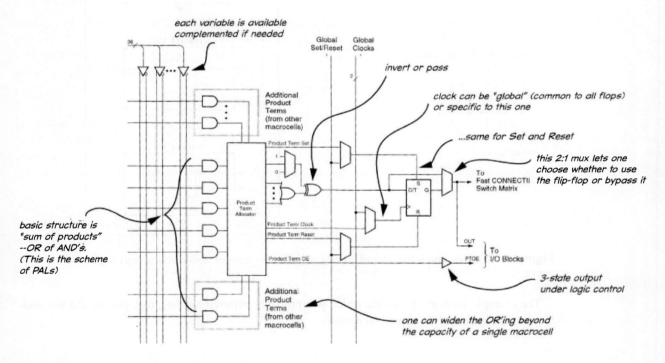

Figure B.1 Xilinx XL/XC95xx logic call (macrocell).

Whether the hardware in which your design is to be implemented is a CPLD (the smaller, more rigid structure) or an FPGA (the large and very versatile structure using *look-up tables* to map input to output), the procedure is fundamentally the same when using the Xilinx tools.

---

[1] If you happen to be interested in this strange dispute, see
http://www.cl.cam.ac.uk/users/mjcg/Verilog/Cooley-VHDL-Verilog.html.

B.1 Xilinx, Verilog, and ABEL: an overview

B.1.1 Download Xilinx's ISE code

Go to the Xilinx website, click on Downloads and take the current ISE (we took the current ISE 14.7 for Windows, the final version of ISE. Xilinx maintains ISE but is not offering further revisions, preferring Vivado which, however, will not compile PLDs).[2] Choose **webpack**, the free download. It's a big file (more than 3GB for our download; the installed size is above 8GB).

The downloaded files expand and among them you will find one called XSETUP.EXE. Though webpack is free it requires a license, which is issued once you have registered and which is valid for a year. When it expires you can get a new license. I guess Xilinx wants to keep your registration up to date. When installation is complete, you will find a *Project Navigator* icon on the desktop. Click on this.

B.1.2 Verilog process: making a source file

Define a project: name the project and place it: On the File menu, choose "new", and give your project a name. You may want to place your work somewhere other than in the main Xilinx directory, the default location.

Choose a device and language: The next menu will ask you to choose Device Family, Device and language ("Synthesis Tool"). We will choose one of the simplest, smallest parts, adequate for holding all the logic our lab computer needs: XC9500 CPLDs.[3] For the first part of this section we'll assume that it's Verilog that we mean to use. We will use the XC9572XL (a 3.3V part). Note that the simulator you want is the ISE, not ModelSim; see Fig. B.2.

Open or start a file: When you are using a template file that we have provided in this course, you will step past this "new source" window and instead will "add source" using the next window. But let's go through the steps we would use if starting from scratch. Click on new source; that will bring up a window as in Fig. B.3 where you choose "Verilog module" and name it.

Begin the source file: The compiler lets you start the new file by entering inputs and outputs (which can be one-bit values, as here, or multi-bit, as for a counter). The compiler then delivers a template file, with some of the fussy work already done. This template is shown at the center of Fig. B.4.

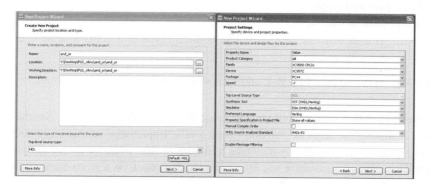

Figure B.2 Name a project, choose device and compiler.

The template file takes care of most of the boring stuff, including these details:

---

[2] If you want to use ABEL, as we do not, you must choose not the current version of the compiler (called ISE) but version 10.1, which includes the obsolescent compiler, ABEL.

[3] The smallest of these, the XC9536 holds 36 flip-flops and about 800 gates.

Using the Xilinx Logic Compiler

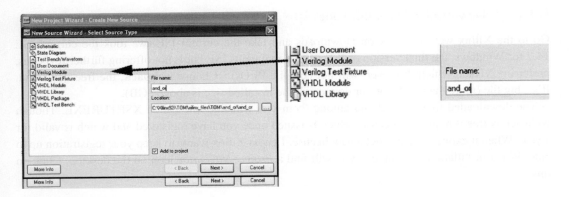

Figure B.3 Name the new Verilog module.

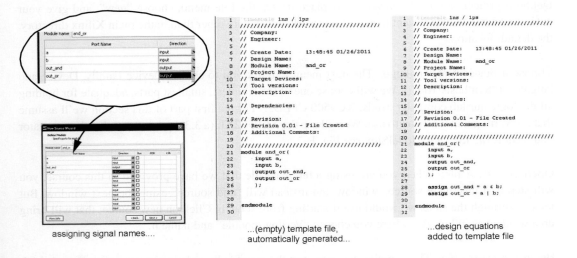

assigning signal names....   ...(empty) template file, automatically generated...   ...design equations added to template file

Figure B.4 List inputs and outputs; ISE will produce a template file; you then enter your design.

- lists all signals used as inputs and outputs;
- specifies whether each is input or output;
- classifies by type: "wire" (the default) or "reg" (the less common type that holds a value once given)

We then write the design – as on the right in Fig. B.4, adding it to the template.

Synthesize the file (compile it): Once we have written and saved the source file, we choose "Synthesize" under "Implement..." among the *Sources*. We don't need a full "Implement" at this stage, and won't until it is time to program a part. Synthesis may take a minute or two, even for a very simple file. When a stage of synthesis or implementation has been completed, a check on a green button will indicate completion, see Fig. B.5.

Look at schematics of the result if you like: There's not much reason to look at schematics usually – but it's fun: makes you believe the compiler has actually accomplished something. And sometimes a glance at the schematic can show that the compiler has not understood your intention. In the present

B.1 Xilinx, Verilog, and ABEL: an overview

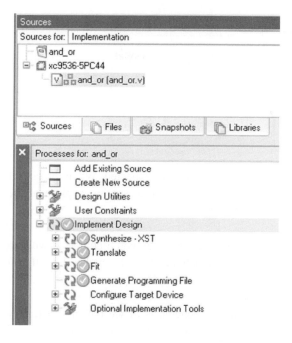

Figure B.5 Implement the design – or start with synthesis only, which is quicker.

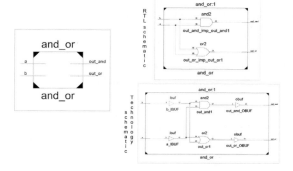

Figure B.6 Schematics: block, the RTL, finally "technology schematic," showing how it will be implemented.

case, where we're asking for AND and OR, there's not much room for confusion; the schematic can hardly hold surprises.

Two forms of schematic display: RTL versus technology schematic: The schematic is offered in two forms:

- RTL form (register transfer level – strange HDL jargon), a generic design, independent of the particular part that you are using; or
- technology schematic, a diagram that shows how the circuit will be constructed in the particular IC that you are using.

These alternative schematics are shown below in Fig. B.6. In the present case, the RTL is the simpler of the two. Sometimes the *technology* version is the more intelligible.[4]

---

[4] Occasionally, we have found the RTL schematic to be quite wrong: in a counter design it implemented an adder rather than counter in response to the line COUNT <= COUNT + 1. The technology schematic has not shown this error.

Using the Xilinx Logic Compiler

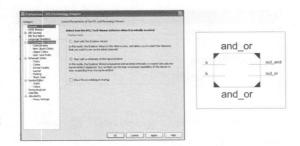

Figure B.7 For simple circuits let the schematic view show the entire circuit.

In Fig. B.6 you can see the schematics for the AND–OR logic look as expected. When the circuit is complex, a full schematic may not be helpful. Instead, you may want to see just a piece of the schematic, not to be overwhelmed. This option is illustrated next.

You can choose to see a partial schematic: The schematic viewer gives you a choice of what to diagram. This choice is visible the first time you ask for a schematic, and can be evoked later if you click on Edit/Preferences/RTL/Technology Viewers.

Full view: good for simple circuits: The simpler and handier scheme is to see the full design by default. This you will get if you choose the option "start with a schematic of the top-level block," as in Fig. B.7. Clicking on the "black box" will show the details of the design, as in Fig. B.6. A more complicated design may include further sub-layers, which can be exposed by clicking on each succeeding layer.

Partial view: good for complex circuits: If you want to see just part of your design drawn, then choose the other option in the viewer: select "start with explorer wizard," as in Fig. B.8. We have asked to see "top level ports" in the figure.

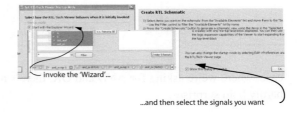

Figure B.8 Schematic viewer allows you to select signals you want to see diagrammed.

...and then select the signals you want

Clicking on "Create Schematic" brings up a black-box schematic (the box on the left in Fig. B.9), much like the one in Fig. B.7. But this one lets us select which signals to diagram. One can select an *input* signal, and diagram all that it drives (a so-called "output cone"). Or (more often useful) one can select an output signal and ask to "add input cone," as we have done in the right-hand image of Fig. B.9 (where we show the OR but not the AND function).

Figure B.9 Schematic view can show less than the entire design.

A more complex design for illustrating a reason to see less than full schematic: The AND OR design hardly motivates one to see less than the full schematic. Fig. B.10 is a more persuasive example: the "glue PAL" that you can use to link the large elements of the lab computer (if you build the "big board" version later in this course). The full schematic is intelligible but too much to see all at once.

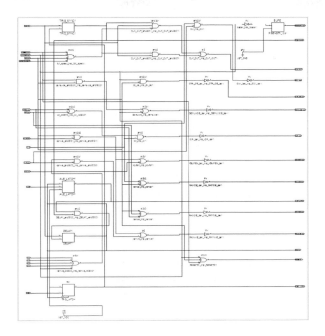

Figure B.10 Full schematic may show too much.

Figure B.11 shows how nice and clear a single function in this design can look if we ask to have only this single function drawn for us.

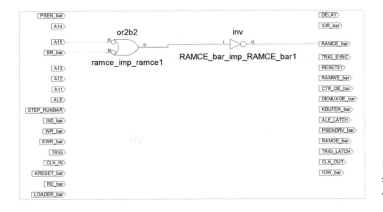

Figure B.11 Schematic of a small part of a design may be easier to understand.

B.1.3   Simulate the design

One can test the behavior of a design by writing another Verilog file called a *testbench*.[5] This file allows you to apply input stimuli to the design and see what output results. The simulator will produce a waveform plot and can additionally, if you like, show inputs and outputs in tabular form.

---

[5] Strictly, it should be called, in Verilog jargon, "Test Fixture;" "testbench" is a VHDL term, but widely used for both languages.

Using the Xilinx Logic Compiler

Process: add new source ("..._tb.v"); ISE makes a template file: We can use Project/New Source/Verilog Test Fixture to get ISE to make an empty template file for us, with most of the fussy work already done.

```verilog
////////////////////////////////////////////////////////////////
module and_or_tb;

    // Inputs
    reg a;
    reg b;

    // Outputs
    wire out_and;
    wire out_or;

    // Instantiate the Unit Under Test (UUT)
    and_or uut (
        .a(a),
        .b(b),
        .out_and(out_and),
    .out_or(out_or)
    );

    initial   // next line prints results in table form
            // (we'll also get a waveform simulation, by default)
    $monitor("(a, b) = %b,%b, out_and = %b, out_or = %b", a, b, out_and, out_or);

    initial begin
        // Initialize Inputs
        a = 0;
        b = 0;
        // Wait 100 ns for global reset to finish
        #100;
        // Add stimulus here
    end
endmodule
```

This template file does the work of associating this testbench with its design file:

```verilog
// Instantiate the Unit Under Test (UUT)
            and_or uut(
```

and "instantiates" the necessary signals – the inputs and outputs:

```verilog
        .a(a),
        .b(b),
        .out_and(out_and),
        .out_or(out_or)
    )
```

ISE's template also remembers what we might forget: that a simulation file requires that the inputs be of type *REG*, so as hold their values till overwritten with a new value, while the circuit outputs are of type *WIRE*.

Complete the testbench file...: In this testbench you specify input levels and when they are to change (the "#10," below, means, for example, "do this 10 time units after executing the preceding line").[6] Below, we have inserted the four possible input combinations for the AND–OR logic's two inputs, a and b.

```verilog
module and_or_tb;
    // Inputs
    reg a;
    reg b;
    // Outputs
    wire out_and;
    wire out_or;
    // Instantiate the Unit Under Test (UUT)
    and_or uut (
        .a(a),
        .b(b),
        .out_and(out_and),
        .out_or(out_or)
    );
    initial   // next line prints results in table form
              // (we'll also get a waveform simulation, by default)
    $monitor("(a, b) = %b,%b, out_and = %b, out_or = %b", a, b, out_and, out_or);
    initial begin
        // Initialize Inputs
        a = 0;
        b = 0;
        // Wait 100 ns for global reset to finish
        #100;
        // Add stimulus here
        #10 b = 1; // the "#10" defines delay relative to preceding line
        #10 b = 0;
            a = 1;
        #10 b = 1;
        #10 a = 0;
            b = 0;
    end
endmodule
```

ISE's simulator produces a waveform result automatically. We have added a "$monitor..." line to this testbench in order to get a textual output, as well. Sometimes this may be clearer than a waveform display, though the present case is so simple that a tabular output surely is not necessary.

...and simulate: Simulating this file soon gets us results like those shown in Fig. B.13's bottom right window: input and output waveforms. But the process that gets you this result is a bit convoluted.

B.1.4 Simulation procedure

Once you have a testbench defined and added to the project, you can run it. To do this, check the "Simulation" button rather than "Implementation" at top left of the screen shown in Fig. B.12.

Then click on "Simulate Behavioral Model." ISIM will open (at a leisurely pace) – and will show horizontal lines where you hoped to see waveforms (see top right of Fig. B.13). But don't lose heart. You can get ISIM to show you the waveforms (which by default it withholds, weirdly enough) by clicking on "Run All" under the Simulation menu. Finally, click on the lifesaver-like icon, and you should see the waveforms.

If you have set up a text display in your testbench you will see a tabular output along with the

---

[6] You can define "time units" how you like; in this file, a unit is defined (in a line at the very top of the file, not reproduced here) as a nanosecond.

Using the Xilinx Logic Compiler

Figure B.12 Simulation process: first select "Simulation" rather than "Implementation".

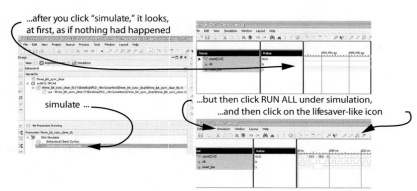

Figure B.13 Steps in simulation process.

default waveform display. Sometimes the result of a simulation is easier to see in its tabular form. It may be, for example, in the case the AND–OR shown in Fig. B.14. But that is a matter of taste.

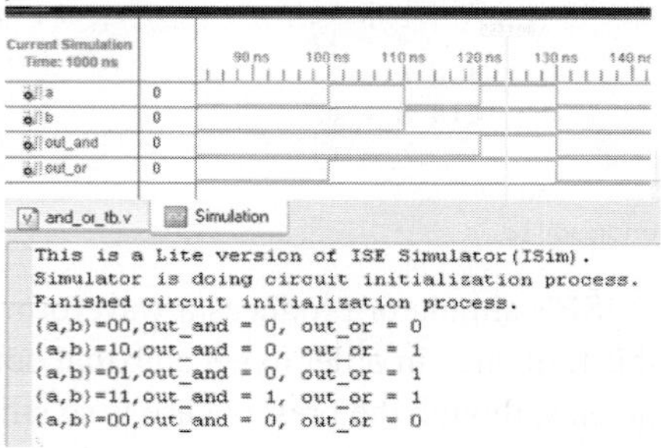

Figure B.14 Result of simulation, using textual testbench.

Self-checking testbench: For a complex design, it may be worthwhile to write a *self-checking* testbench that includes expected results and compares these against actual results. Such testbenches are painful to write. One is described in Appendix A

Verilog trouble: in case of weird errors, try cleanup: Occasionally, compilation errors corrupt files and the compiler than gives puzzling error messages such as "could not link simulation." If that occurs, try cleaning up project files: select the *Project* menu, then *Cleanup Project Files*.

C Transmission Lines

Contents

C.1	**A topic we have dodged till now**		**1089**
	C.1.1	The impedance rules we have applied so far, when A drives B	1089
C.2	**A new case: transmission line**		**1090**
	C.2.1	Considering frequency (of sinusoids)	1090
	C.2.2	Considering risetime	1091
C.3	**Reflections**		**1092**
	C.3.1	Pulses rather than sinusoids	1092
	C.3.2	Improper termination 1: end open	1092
	C.3.3	Improper termination 2: end shorted	1093
C.4	**But why do we care about reflections?**		**1094**
	C.4.1	Remedies: termination	1094
	C.4.2	Classic remedy: terminate the line in matched resistance	1094
	C.4.3	Series termination: include a driving resistor matched to the line's characteristic impedance	1095
C.5	**Transmission line effects for sinusoidal signals**		**1097**
	C.5.1	Familiar: low-pass behavior	1097
	C.5.2	Novel: transmission line effects	1097
	C.5.3	Partial impedance mismatch	1098

C.1 A topic we have dodged till now

Because, in our labs, we have not encountered the high frequencies that call for consideration of *transmission line* behaviors, you might leave this course unaware of this way of looking at impedance questions. You will need this way when you encounter frequencies significantly higher than the highest we have seen in this course. We have seen signals of only a few MHz, with perhaps a single exception: the fast parasitic oscillation of the discrete follower of Lab 9L. That frequency did come close to 100MHz for some of you. You clocked your microcontroller at 11MHz or 24.5MHz (the SiLabs version), but you were not obliged to do any handling of this clock.

C.1.1 The impedance rules we have applied so far, when A drives B

The most common case: voltage source at modest frequencies On the very first day of the course we began to drum away at a theme concerning the recurrent case in which circuit fragment A drives another fragment, B. We have said, over and over, that we like to make sure that A's Z_{out} is low relative to B's Z_{in}. And we elevated to the status of a *rule of thumb* the notion that this relation should be a ratio of 10 to 1:

$$Z_{\text{out\_A}} \leq \frac{1}{10}(Z_{\text{in\_B}}) \ .$$

This design rule remains valuable. Don't worry: we are not about to tell you that transmission line rules make this rule obsolete.

A less common case: current source at modest frequencies Now and then we have met signals that are *currents* rather than voltages. A *current source*, like the photodiode of Lab 6L (drawn in Fig. C.1 with a smaller R), has tastes opposite to those of a voltage source.

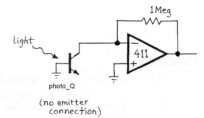

Figure C.1 Current source as signal calls for *low* R_{in}.

The photodiode in Fig. C.1 likes to drive a load whose R_{in} is *low*, and its favorite load is (of all things!) a short circuit. The op-amp current-to-voltage converter of Fig. C.1 (sometimes called by the fancy name "transimpedance amplifier") comes close to providing this ideal R_{in}. At this *virtual ground* R_{in} is very low; the golden rules would put it at *zero*.[1] (We doubt that you need to be reminded that R_{in} for this circuit is *not* 1M!)

C.2 A new case: transmission line

We have worked with coaxial cables throughout this course (we often call these BNC cables, referring to their connectors). They usually have worked for us quite transparently. Occasionally – if you found yourself using not a function generator but a circuit with a substantial R_{source} to drive a cable – you may have become aware of the cable's substantial capacitance (about 30pF per foot). That's as complicated as things have got so far.

But the behavior of the familiar coax cable becomes unfamiliar when we push on to frequencies higher than what we have been using in this course, while driving fairly long cables. (The effect is not limited to cables, but these are easiest to discuss because of their standardized characteristics).

Specifically, we get new results in cases that can be described in terms using either *frequency* or *time*. We can describe in two ways the cases where we must treat a signal path as a *transmission line*:

AoE §H.2

1. frequency: when the line length[2] is, to quote AoE, "a significant fraction...of the wavelength of the highest frequency";
2. time: when "the round-trip propagation time is a significant fraction of the signal risetimes."

C.2.1 Considering frequency (of sinusoids)

Let's try applying the first of these criteria – the one that speaks of *frequency* – to the signals we have met in this course. Let us see whether we can confirm that our experience makes sense: we noticed no transmission line effects.

---

[1] You're more sophisticated than that though. You recall that the actual $R_{in} = R_{feedback}/A$, where A is the open-loop gain. So, at a frequency where A was 10k, for example, R_{in} would be $1M/10k = 100\Omega$.

[2] Strictly, as AoE says, it is the "electrical length" that matters: electrical length takes into account the slowing of signal propagation in the cable dielectric. – and see §C.2.1.

C.2 A new case: transmission line

At what cable length should we begin to worry about transmission-line effects?[3] We are applying a rule of thumb that says we'll worry if cable length – the "electrical length," taking account of cable dielectric – is about *1/10* of the wavelength of the applied sinusoid.

Frequency	Wavelength, λ, (meters)	λ physical length (meters coax)	Xmssn-line threshold length (meters; cable = $\lambda/10$)
1M	300	200	20
3	100	66	6.6
10	30	20	2
30	10	6.6	0.66
100	3	2	0.2

The "physical length" of one wavelength in the *cable*, at a given frequency, is less than in free space because the signal travels more slowly in the dielectric. So, difficulties appear at shorter cable lengths than one might expect.

But the table makes clear how we managed to steer clear of transmission-line effects in this course: we used function generators most of which top out at 3MHz. And on the rare day when we set the generator to 3MHz we did not run sinusoids down a 7-meter cable.

C.2.2 Considering risetime

So much for sinusoids (except for one more glance in §C.5). But steep digital waveforms can raise transmission-line problems, even when the repetition rate of the waveform is low. The signals we met in our course did sometimes show the potentially troublesome combination of fast edges with fairly long lines. We did not worry about these effects because the result of these reflections was only to make our logic signals somewhat ugly, but not so ugly as to produce spurious crossings of logic thresholds. We admire digital circuitry for its ability to ignore a good deal of ugliness.

We did not drive long coaxial cables with logic signals, but we did drive as much as 12 inches of printed-circuit trace on the big-board computer, with its PC buses. Reflections did occur there, but caused no mischief because the signals were not very demanding. We sent only signals such as *data* and *address*, which had plenty of time to settle after a transition. You may recall that we did not send any *clock* signal down these bus lines.

We will look at such bused waveforms in §C.4.3. But first, let's note some rules of thumb that can relate *edge rates* (or "risetimes") to signal path lengths, so as to learn when we ought to worry about transmission-line issues.

Relating risetime to path length

The cause of trouble It's easiest to see what causes trouble by describing what sort of reflections do *not* cause trouble: those from short lines that return to the driving end while the driver is still in transition. The reflection may distort the edge somewhat, but typically is harmless. So where risetime is well over the round-trip signal propagation time, we can ignore transmission line effects.[4]

Rules of thumb that tell one when to worry can be formulated in several ways. Here is a very direct rule for estimating line length where transmission line effects should begin. The rule describes the length of a printed circuit trace whose reflection would arrive just at the completion of the transmitting edge.

Begin treating a line as transmission line at length...

---

[3] The table assumes signal speed is 2/3 the speed of light, about right for cable with solid polyethylene dielectric.
[4] See http://www.ultracad.com/mentor/transmission%20line%20critical%20length.pdf.

three inches $\times t_r$ (in nanoseconds)[5]

This rule is based on the argument that such a length allows a reflection to return during the transition time. A risetime of 1ns, for example, allows time for about 6 inches of travel. (Wave velocity on FR4 printed circuit is assumed to be about one half the free space velocity, which is approximately one foot per nanosecond.) A round-trip on a three-inch length of PC trace should take 1ns. Hence the rule. A risetime of 3.5ns (close to 74HC driver edge rates[6]) would put the transmission line "critical length" around 10 inches.

A less direct rule of thumb translates risetime to the highest frequency present in the waveform:

$$\text{Bandwidth (that is, max sine frequency)} \approx 0.35/t_r \text{ where } t_r \text{ is risetime}$$

So, for example, a t_r of 3.5ns translates to 100MHz. The table of §C.2.1 suggests, then, that for the 3.5ns edge, transmission line effects would begin at around 0.2m, or about 8 inches for coaxial cable or 6 inches for a PC trace. This rule is more conservative than the simpler risetime rule, "three inches $\times t_r$," which put the boundary at about ten inches. But the critical-length values from the two rules are not far apart.

C.3 Reflections

C.3.1 Pulses rather than sinusoids

The behaviors of pulses on transmission lines are easier to understand than the behavior of sinusoids, so let's start with these. We'll reach sinusoids later (§C.5). When everything is done right, as we'll see in §C.4.1, nothing remarkable happens when a pulse travels down even a long cable. The interesting – and troublesome – behaviors, called "reflections," result from improper *termination*. You may have seen equivalent effects demonstrated for waves on a taut rope.[7] We'll begin with the extremes of wrong terminations of a long cable to which we'll apply a pulse.

C.3.2 Improper termination 1: end open

If the end of the transmission line is left open, which in electrical terms is equivalent to driving a load whose impedance is much larger than the transmission line's impedance, a pulse is reflected at the end. The pulse is of the same polarity as the incident pulse (thus reaching double the level of the input pulse), and travels back to the input.[8] Here is an example. In Fig. C.2, the pulse is narrow enough so that the returning pulse arrives well after the extinction of the initial pulse. The two pulses at the driving end – the original and the reflection – lie about 32ns apart. This is about what we would predict for a 3m cable.[9]

---

[5] Rule proposed for FR4 printed circuit. http://www.ultracad.com/mentor/transmission%20line%20critical%20length.pdf. A nice informal explanation appears at http://www.ultracad.com/articles/criticallength.pdf

[6] See Philips user's guide: http://www.nxp.com/documents/user_manual/HCT_USER_GUIDE.pdf, indicating 4ns risetime for bus driver IC's, 6ns for ordinary 74HC. More recent HC parts may show faster edges.

[7] A good animation and film clip of pulses: http://www.animations.physics.unsw.edu.au/jw/waves_superposition_reflection.htm#reflections. Less good is Wikipedia's treatment: http://en.wikipedia.org/wiki/Pulse_%28physics%29.

[8] This "reflecting" behavior, like that of the case of *shorted* end, result from conservation of energy. For the open end, current is zero, for the shorted end voltage is zero. The induced reflection dissipates energy in the line and driving source impedance. See http://www.ti.com/lit/an/snla027b/snla027b.pdf.

[9] The round-trip length is 6m or 20 feet. Wave velocity in this coax is 0.66 what it is in free air, so about 0.66 feet/ns, and round-trip time therefore is about 30ns.

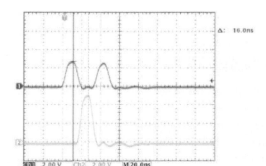

Figure C.2 Reflected pulse appears non-inverted one round-trip propagation time after the incident pulse when end is *open* (3m cable). (Scope settings: 2V/div; 20ns/div.)

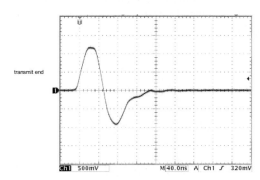

Figure C.3 Reflection from shorted end comes back *inverted* (3m cable). (Scope settings: 500mV/div; 40ns/div.)

C.3.3 Improper termination 2: end shorted

At the other extreme of mismatch, the far end is *shorted* rather than open. In this case the pulse is reflected, but returns *inverted*. Fig. C.3 shows the effect for a pulse sent down a 3 meter cable.

Step input, end shorted A curious variation on this circuit can be used to generate a narrow pulse from a step input. A first thought, if one considers what the transmit end would look like with the far end shorted, might be "You'll get nothing if $R_{\rm out}$ is 50Ω, since we're looking at a divider between 50Ω and 0Ω." That's almost correct: you *do* get nothing in the steady state. But it takes some time for the *transmit* end to discover that the end is shorted. Meanwhile, the voltage looks like the driving voltage (or, more exactly, like $V_{\rm source}/2$, since $R_{\rm out}$ forms a divider with the cable's characteristic impedance of 50Ω). Only when the reflected wave returns does the voltage drop to zero. The pulse width equals the round-trip delay time. In Fig. C.4, the pulse begins to fall perhaps 30–40ns after the rise to full amplitude. The calculated round-trip time is about that as we argued in §C.3.2.

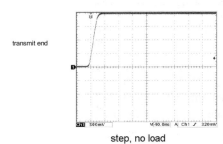

Figure C.4 Step input into shorted cable produces a narrow pulse (3.5m cable). (Scope settings: 500mV/div; 40ns/div.)

C.4 But why do we care about reflections?

One can make a plausible argument that these reflections may be harmless. Send a pulse down a long coaxial cable from a logic gate (low output impedance) to another logic gate (high input impedance). What happens?

This case, with a logic gate rather than function generator driving the cable, differs from the "...end open" case of §C.3.2 in that the signal source is not matched to the cable as the function generator was (50-ohm output impedance is built into the generator, as in your lab instruments). That 50Ω source impedance swallowed the reflected waveform, so only one reflection occurred. The low output impedance of the logic gate should, instead, produce a new reflection (and inversion), which travels anew, and will get reflected. In short, it looks as if things will get complicated and messy.

Indeed they do. Fig. C.5 shows the waveform into an unterminated 6-foot coax driven by a collection of 7HCT logic gates, paralleled, for extra current).[10] Ugly!

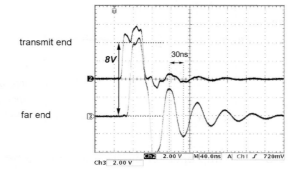

Figure C.5 74HCT logic driving 6-foot coaxial cable, unterminated. (Scope settings: 2V/div; 40ns/div.)

Not only is the waveform ugly, but the 8V peak at the far end of the cable could damage a 5V logic gate. The oscillations look, at first glance, like resonances that we have seen produced by stray L and C. You may have seen such junk if you forgot to attach a scope probe's ground lead, for example. But the oscillations are not that. Instead, they result from reflections that bounce off the far end, then off the transmitting end, and so on, banging back and forth. When the pulse reflects from the low-impedance at the transmitting end, it is *inverted*, and that would account for the negative-going excursion at the far end, a swing that does occur about one round-trip time (30ns) after the initial positive pulse reaches that far end. The very best evidence that this ugliness results from reflections is the fact that proper termination cleans things up (see §C.4.3, Fig. C.7).

C.4.1 Remedies: termination

AoE §H.5

The reflection effects described in §C.3.2 and §C.3.3 are intriguing. But they are, of course, undesirable – except for the short-pulse generator of Fig. C.4, which could be useful. For fast signals, proper *termination* will clean up signals that otherwise might be distorted by reflections.

C.4.2 Classic remedy: terminate the line in matched resistance

If we "terminate" the cable – that is, join signal line to ground or another voltage source through a resistor – with a resistance equal to the characteristic impedance of the cable, then the cable appears to continue forever. The wave or pulse never hits a discontinuity. If we terminate the 3.5m coaxial

---

[10] We used all eight gates of a 74HCT541, for nominal current of about 50mA, enough to drive a 50Ω termination to legal logic High.

C.4 But why do we care about reflections?

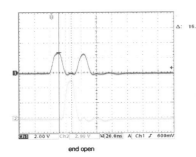
end open

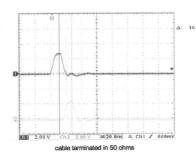
cable terminated in 50 ohms

Figure C.6 Open-ended versus terminated cable (10 feet): termination eliminates the reflection. (Scope settings: 2V/div; 20ns/div.)

cable that produced a double pulse back in Fig. C.2, no returning pulse occurs. In Fig. C.6 we show the two cases, unterminated and terminated.

The same remedy – a 50Ω resistor between the signal line and ground at the far end of the cable – cleans up the pulse sent by the logic gates of Fig. C.5.

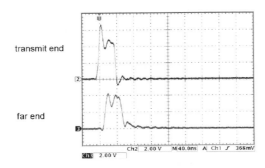
transmit end

far end

Figure C.7 50Ω termination cleans up 74HC-to-coax drive. (Scope settings: 2V/div; 40ns/div.)

This termination does clean things up – but it calls for a lot of current from the signal source: as much as 100mA, if the termination goes to ground and is driven to 5V.[11]

C.4.3 Series termination: include a driving resistor matched to the line's characteristic impedance

AoE §H.5

A less perfect but more practical protection is simply to install a series resistor on the transmitting end, matched to the cable's characteristic impedance. This is so standard on function generators that you may not even think of this as a "remedy." But that series resistance driving the cable helps a lot.

It does not prevent reflections from the far end (as we have seen in §C.3.2), but it does prevent a *second* reflection at the driving end, and prevents the later rebounds from both ends. It swallows up any reflected wave, preventing the bang-bang nonsense visible in Fig. C.5. It also attenuates the driving signal so that the level that reaches the far end of the line is full-size, not *double* size.

For logic gates, this simple scheme is preferable to the classic termination at the far end, which

Figure C.8 Series termination at the transmit end will swallow up reflections.

---

[11] The current would be less by half if we used a termination *divider* to one-half of the supply voltage, making its $R_{\text{Thevenin}} = 50\Omega$. This is a bit more trouble to wire. DC current could be eliminated if the termination were AC-coupled, using a blocking capacitor.

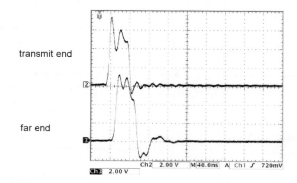

Figure C.9 47Ω series R at driving end of 6-foot coax cleans '541 logic drive pretty well. (Scope settings: 2V/div; 40ns/div.)

would call for large quiescent currents. Fig. C.9 shows the cleanup provided by insertion of a 47Ω resistor at the output of the '541 logic driver as it drives a 6-foot cable.

Driving PC board traces rather than coaxial cable For the similar case of driving printed circuit traces, inserting a series resistance at the driving end again helps a lot. Fig. C.10 shows the improvement where logic gates drive a 10-inch bus trace. The left-hand figure shows direct drive; the right-hand screen shows the cleanup provided by driving through 100Ω. The driver is 74HCT541.[12]

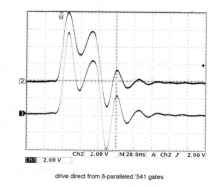 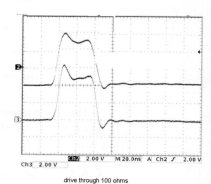

Figure C.10 Matched drive impedance cleans up 10-inch printed circuit drive. (Scope settings: 2V/div; 20ns/div.)

drive direct from 8-paralleled '541 gates   drive through 100 ohms

Use an inherently-terminated logic family LVDS, the differential digital logic you met back in Chapter 14N, solves the reflection problem elegantly. At first glance, the fast transitions of LVDS might seem troublesome – as low as about 0.25ns.[13] But a glance at the typical LVDS wiring in Fig. C.11 may reassure you.[14]

The characteristic impedance of the printed circuit traces is around 100Ω, and yes, that resistor at the receiver sure looks like a terminating resistor matched to the line's impedance. Indeed it is, and it is not an afterthought. The 100Ω is not just desirable as termination: it is also required in order to generate the signal at the receiver. The transmitter's current, sourced through one wire, sunk through the other, generates the *difference* signal of 350mV.

---

[12] The drive is again the set of eight gates paralleled for high current capability used for the coaxial cable conventionally terminated in Fig. C.7.
[13] http://www.ti.com/lit/an/slla014a/slla014a.pdf
[14] Compare National/TI LVDS User's Guide: http://www.ti.com/lit/ml/snla187/snla187.pdf, Fig. 1-1.

C.5 Transmission line effects for sinusoidal signals

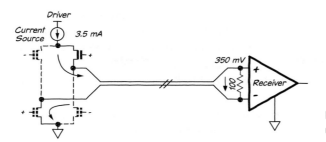

Figure C.11 LVDS transmitter and receiver wiring.

LVDS scope image Figure C.12 shows differential signals, small but pretty clean, as they arrive at the far end of a 10-inch printed circuit trace.

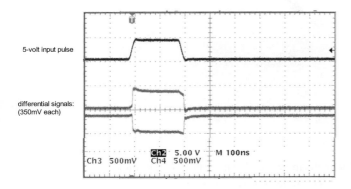

Figure C.12 LVDS inherently-terminated signal line produces pretty clean signals on 10in PC trace. (Scope settings: 5V/div (input), 500mV/div (diff signals); 100ns/div.)

C.5 Transmission line effects for sinusoidal signals

Where the signal applied to a transmission line is sinusoidal rather than a pulse, the effects are even farther out of the range of what we saw in lab. We did apply digital signals with fast edges, but never applied high frequency sinusoids. If we had, we would have found results surprisingly different from what we saw at lower frequencies.

C.5.1 Familiar: low-pass behavior

At lowish frequencies, a coaxial cable behaves like a low-valued capacitor to ground. If we use our function generators whose R_{out} is 50Ω to drive a long cable – say, the 3.5m cable we used in most of the pulse tests above – we usually see no attenuation. This is what we are accustomed to in lab.

But we would anticipate that at some high frequency we would begin to see a *low-pass* effect from the RC of 50Ω driving the cable's 30pF/foot of stray capacitance. If we calculate this effect for 3.5m of RG58 coax, we get an f_{3dB} of ≈9MHz. This relatively high frequency explains why we have not noticed any such effect in our labs: our function generators do not run that high – or if they do, we've not used them at such a frequency. And we certainly never drove 10 feet or so of coax at high frequencies.

C.5.2 Novel: transmission line effects

If we try driving this 3.5m coax with a sinusoid, pushing the frequency beyond what have tried in lab, we get a new behavior that results from transmission line effects. Specifically, as frequency climbs, approaching one where the cable length is 1/4 wavelength, amplitude at the transmitting end falls – looking superficially like the behavior of a low-pass. Amplitude goes right down to zero when the cable length is exactly a quarter wavelength. This occurs at about 16MHz. And if frequency climbs further, amplitude begins to rise, going all the way to *full* amplitude when cable length is 1/2 wavelength (32MHz). This behavior is brand new. What's going on? Reflections from the unterminated end of the cable are causing this result.

Figure C.13 shows a manually swept frequency that begins at 1MHz and moves up to about 110MHz. The scope is watching the *transmitting* end of the unterminated cable. The "nulls" are obvious, the first at about 16MHz.

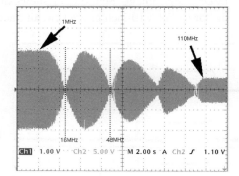

Figure C.13 Manually swept frequency shows several *nulls* where reflected waveform cancels the driving waveform. (Scope settings: 5V/div, 2s/div.)

When the cable length is 1/4 wavelength, the reflected waveform arrives back at the transmit end at a time when the driving signal has reached 1/2 its period. The reflected wave is then an inverted version of the driving signal. The two cancel. When the round-trip time is a full wavelength there is no interference, and again (at 32MHz) we see full amplitude.

Let's try the numbers to see if the *nulls* of Fig. C.13 make sense. Round-trip time for the 3.5m cable – about 11.5 feet – is the time to travel 23 feet. In the cable, where signal velocity is about 2/3 wave velocity in free space, 23 feet takes about 23×1.5ns $\approx$35ns. A null should appear when this time is a half period, making the full period around 70ns. That period corresponds to a null frequency of 1/70ns = 0.015gHz= 15MHz: about what we observed in Fig. C.13. Seems to make sense.

Since the cause of the problem is the same as the cause of the pulse reflections that we saw earlier, the remedies are the same as well: matching impedances of source and load to the cable.

C.5.3 Partial impedance mismatch

We have looked at only the extreme mismatch cases: far end open and far end shorted. Less extreme mismatches produce, as you might guess, *partial* reflections. These are also undesirable. We will not illustrate them here.

D Scope Advice

Contents

D.1	**What we don't intend to tell you**		**1099**
D.2	**What we'd like to tell you**		**1099**
	D.2.1	Triggering:	1099
	D.2.2	Aliasing on a digital scope	1103

D.1 What we don't intend to tell you

We don't believe we can tell you how to use a scope by writing a lot of words. You'll learn by trying it – and, if you're lucky, by having someone with experience look over your shoulder as you try. But there are a few points that may be worth writing out and trying to illustrate.

D.2 What we'd like to tell you

AoE Appendix O: The Oscilloscope

D.2.1 Triggering:

Triggering is the hard element in scope use. *Triggering* is jargon for the scope's mechanism for synchronizing its successive *sweeps* from left to right across the screen. This synchronization is necessary in order make the image coherent rather than a muddled overlaying of traces. On successive sweeps, the trace had better redraw the waveform in the same left-right position as the previous sweep. If it doesn't, the traces seem to wander horizontally, as in Fig. D.1. The visual effect is thoroughly annoying.

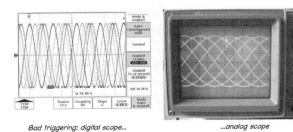

Figure D.1 Bad triggering produces muddled overlaying of images on scope screen.

Choose the appropriate signal to trigger on (reject all the duds) – starting with the most general issues...:

Don't start by adjusting LEVEL Novices often start by twiddling the trigger *LEVEL* knob hoping to see the waveform stabilize. That is the wrong way to begin. Figs. D.2 and D.3 illustrate this for analog and digital scopes, respectively.

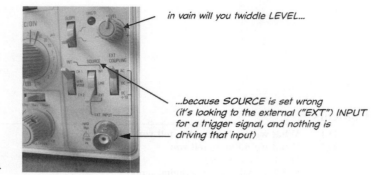

Figure D.2 Don't start by fussing with the LEVEL knob. First, set SOURCE appropriately.

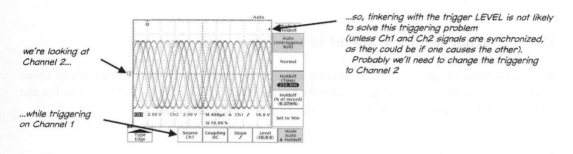

Figure D.3 LEVEL won't solve this problem either (digital scope version).

Instead, start with the most general issue: INT/LINE/EXT Before fine-tuning can work (as by adjustment of LEVEL) you must tell the scope which sort of signal to trigger on. The most general selections are those labeled "INT/LINE/EXT" on our scopes.
- INT means "internal," and refers (despite the strange name) to the scope's INPUT channels. Nearly always this is the correct choice
- LINE triggers the scope on the power supply of the scope. This is good for stabilizing any waveform synchronized with the 60Hz "line" voltage (the supply that comes out of a wall socket).[1] If you see low-frequency noise or wobble, and wonder if it's caused by the 60Hz noise that pervades indoor settings, LINE trigger usually answers that question: if LINE stabilizes the wobble, then, yes, the wobble is caused by power-supply noise. LINE, then, can be useful – but not often.
- EXT means "EXTERNAL," and takes its signal from a separate BNC input somewhere on the scope.[2] This is useful when the usual trigger source – one of the input channels (INT) – is marred by noise that can upset triggering.

...then choose the particular internal trigger source: You will nearly always trigger on an input channel rather than LINE or EXT. You will need to choose *which* channel – and there is a trap lurking, here.

---
[1] 50Hz in Europe.
[2] Usually on the front of older analog scopes; in Tektronix TDS3014 digital scopes, the connector is on the back.

Figure D.4 EXTERNAL trigger is useful when an input signal is muddled by noise.

Beware the evil "VERTICAL MODE": The analog scope shows two choices: Ch 1 and Ch 2. Between them – tempting the wishy-washy liberal – is a compromise: VERT MODE ("vertical," whatever that might mean).

Beware the evil VERTICAL MODE

Figure D.5 The evil VERTICAL mode tempts the unwary.

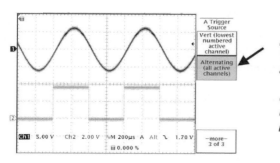

equivalent to analog scope's VERTICAL MODE-- and equally perilous

(Notice sine and square do not align: in fact, square wave's edge aligns with peak and trough of sine)

Figure D.6 VERTICAL mode (called "alternating..." on the digital scope) falsifies relative timing among channels.

This mode triggers on the two channels *independently* – and as a result *loses the relative phase or timing between the two channels.*

This mode is named differently on our digital scopes (TDS2014), and you would have to hunt through menus to find it – so it is very unlikely to lead you astray. The TDS2014 calls this triggering mode "Alternating" – not to be confused with the display mode of analog scopes that is called "ALT"![3]

But if you use this mode, with analog or digital scope, you will get a display that *falsely* displays relative timing of the input channels. Fig. D.6 for example, shows the function generator's MAIN output and SYNC output, falsely displayed by this mode, on the digital scope:

We know that the timing display in Fig. D.6 is false because the MAIN output (a sinusoid here) and the TRIG_SYNC output (a square wave) are locked in phase by the design of the function generator, the TRIG edges coinciding with peak and trough of the MAIN output. That relation is not shown here. The scope is misrepresenting the relative timing of the two signals.

Why, you ask, do scope designers provide this nasty trap? As you would expect, there is indeed a case for which VERT mode is useful. That is the very rare situation when signals on the several channels are not synchronized. In this case, VERT does give a stable display for all channels at the expense of losing information concerning relative timing. We have to be smart enough to remember that relative timing is lost, or VERT mode will mislead us. In the entire set of labs that we do in this course, you are unlikely to want this mode even once.

"AUTO" versus "NORM": Once you have told the scope which signal to use as trigger, you still face a choice, labeled "AUTO" versus "NORM." AUTO displays traces whether or not you have set the trigger LEVEL properly; NORM does not: it shows you nothing if you have not fed the scope a good trigger signal. Since a dark screen – or frozen screen, in the case of a digital scope, which holds the last image it caught – is even worse than a muddled one, we advise you to think of NORM as short for *ABNORMAL*. We use it only in exceptional cases.

Figure D.7 shows the contrasting results of AUTO and TRIG choices for a case where the LEVEL has been set too high (above the highest level of the Ch 2 waveform chosen as trigger source; the arrow shows this trigger level, in both left- and right-hand images). The left-hand image, using AUTO, shows a stuttering display; the NORM display shows us nothing. Usually we prefer something to nothing.

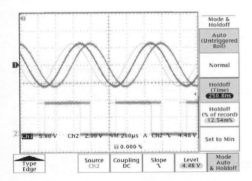

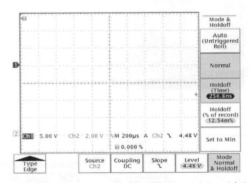

AUTO trigger, with LEVEL set too high...   ...NORMAL trigger, with LEVEL set too high

Figure D.7 AUTO and NORMAL trigger results when LEVEL is mistakenly set higher than the input waveform.

The case for NORM: NORM becomes useful (rather than just frustrating) when we want to display a waveform whose trigger event is infrequent. AUTO gets impatient and sweeps when it decides

---

[3] You probably recall the contrast between ALT and CHOP display modes for an analog scope: ALT draws one trace for a full sweep (say, Ch 1), then draws another trace (say, Ch 2). CHOP, in contrast, jumps the beam rapidly back and forth between the two channels on a single screen sweep.

D.2 What we'd like to tell you

enough time has passed; NORM waits patiently for the next occurrence of the trigger event. You will recognize the cases that call for NORM when you meet them. But expect these to be exceptional.

D.2.2 Aliasing on a digital scope

Digital scopes are pretty nearly irresistible, but they are vulnerable to one misbehavior that is quite baffling to someone who has not seen it before (and sometimes puzzling to someone who *has* seen it before, as we can attest). This is the effect called "aliasing," the generation of a false image when the scope does not sample the input waveform often enough.

The topic of sampling, and the artifacts it can introduce, is treated at some length in Chapters 18N and 18S. If you are reading this appendix early in the course, it is too early to work at a full understanding of sampling effects. We will settle here for showing you an example, and then adding a few words on why the aliasing looks as it does. If you return to this appendix after reading Chapter 18N we hope the points sketched here will look familiar.

Aliasing occurs when we violate the standard sampling rule (often referred to as Nyquist's or Shannon's sampling theorem). This rule says that one needs somewhat more than two samples per period of an input to get enough information about that signal. If one samples less often, one doesn't just miss some data. One gets *disinformation*: a pseudo signal called an "alias."

For example, a sinusoid at 5kHz should be sampled at more than 10kHz. Sampling creates a lowest frequency artifact at $f_{\text{sample}} - f_{\text{in}}$.[4] So a 5kHz signal sampled at 7.5kHz would create an artifact at 2.5KHz. In the example illustrated by Fig. D.8 we will see a similar effect.

Aliasing can produce a strange low-frequency image on a digital scope: To demonstrate this hazard, we fed a sinewave of just under 5kHz to a digital scope[5] that was sweeping rather slowly (200ms/division). Fig. D.8 is what the scope showed us.

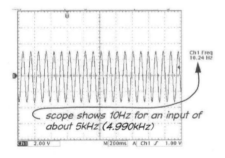
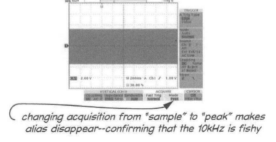

Figure D.8 A digital oscilloscope can show an unexpected low-frequency when sweeping too slowly to resolve the input signal.

Ten hertz, when we fed the scope about 5kHz (5kHz less 10Hz: 4,990Hz – a detail that turns out to be important)?[6] Is the scope crazy?

No, it's not crazy, it's just stuck with being a sampling machine. This scope collects 10,000 data points as it fills a screen. So, when sweeping slowly at 200ms/division, a trip across the scope screen takes 2 seconds. So the 10k data points must have been picked up (sampled) at 5kHz. This is much too low a rate to get an honest image of a signal coming in at close to 5kHz. The result is the spurious 10Hz

---

[4] More generally, sampling produces "images" at $n \times f_{\text{sample}} \pm f_{\text{in}}$. But usually it is only the image of lowest frequency that concerns us.
[5] A Tektronix TDS3014.
[6] The nearness of f_{in} to f_{sample} is what produces the striking low-frequency sinusoid in Fig. D.8, rather than just a puzzling messy display. But other input frequencies would have similar effects. An input close to any multiple of 5kHz evoke the same sort of strangeness.

signal that the scope shows us. (This alias occurs at the frequency $f_{\text{sample}} - f_{\text{in}} = 5\text{kHz} - 4.990\text{kHz} = 10\text{Hz}$.)

The right-hand scope image in Fig. D.8 shows what happens when (confronted by the weirdness of a 10Hz output) we change scope acquisition from "sample" (the usual setting) to "peak," which catches highest and lowest levels, acquiring samples at a high rate despite the slow sweep rate.

We don't want to scare you with this example of aliasing. You won't see this often and it isn't really likely to fool you (you're not likely to believe that 5kHz looks like 10Hz, despite what the scope says). But it's useful to have heard about the possibility of aliasing. When your digital scope suddenly talks nonsense – especially if it's surprising low-frequency nonsense – we hope you'll remember this possible cause.

E Parts List

The parts used in the labs that this book describes are listed below. It's a chore to order them all, so we intend to have someone "kit" the parts for us, in separate *analog* and *digital* collections. To find such kits please visit www.learningtheartofelectronics.com.

If you're unable to locate some part (and some on this list are bound to become hard to find), try the excellent parts search tool OCTOPART (octopart.com).

Part, generic	Description	Detail (pkg, etc.)	Mfgr	Part No., exact	Source	Particular Distributor's Part No.	Price (1) US$	Price (25) US$
LAMP								
#47	lamp	T3-1/4	Visual Communications Company	47	D	CM47-ND	1	0.65
#344	lamp	T3-1/4	JKL Components	344	M	344	1	0.85
Inductors								
10mH	Q = 150	radial	Murata	13R106C	D	811-2058-ND	0.59	0.58
100uH		radial	Panasonic	ELC-12D101E	M	67-ELC-12D101E	1.28	0.83
ferrite bead	length 4.3mm, inside diameter 1.5mm, 100MHz test frequency	cylinder	Kemet	B-20F-46	D	399-10820-ND	0.25	0.18
Transformer								
6.3 v, C.T.	1.2A		Stancor	P-6134	N	16P8886	16	15
heat sink								
TO-92	slip-on	TO-92	Aavid	575200B00000G	M	532-575200B00	0.68	0.63
Switches								
slide	SPDT	0.1" centers	E-switch	EG1218	D	EG1903-ND	0.58	0.55
snap action	SPDT	0.2" centers	C&K	ZMA00A150L04PC	D	CKN10157-ND	0.73	0.69
pushbuttons								
spst	square, flush mount	0.2" centers	Aspem	MJTP1212	D	679-2424-ND	0.36	0.35
spdt	square	0.1" centers	TE	KS12-R22CQD	D	CKN1595-ND		
DIP:								
4-position	SPST	0.1" centers	TE	1825057-3	D	450-1364-ND	0.81	0.67
8-position	SPST	0.1" centers	Alco	ADE0404	N	68K9398	0.81	0.75

Sources: A=Avnet Express, Arizona; ADI=Analog Devices Inc.; D=Digikey; J=Jameco; M=Mouser; N=Newark; PA is Proto Advantage (Ontario): PA/D means that Digikey carries the part; Q=Quest Components Inc., (California); R=Rochester Electronics (MA); S=Sparkfun; TTI=TTI Inc., (Texas).

Parts List

Part, generic	Description	Detail (pkg, etc.)		Mfgr	Part No., exact	Source	Particular Distributor's Part No.	Price (1) US$	Price (25) US$
Potentiometers									
trimpot	cermet, one turn	1/2W	1k	Bourns	3352T-1-102LF	D	3352T-102LF-ND	1.53	1.22
			10k		3352T-1-103LF	D	3352T-1-103LF-ND	1.53	1.22
			100k		3352T-1-104LF	D	3352T-1-104LF-ND	1.53	1.22
			1M		3352T-1-105LF	D	3352T-1-105LF-ND	1.53	1.22
motor-driven	rotary pot	4–6V, audio taper B3	100k	Alps	RK16812MG099	M	688-RK16812MG099	9.61	8.58
	slider pot	10V, audio taper 15A	10k	Top-up Industry	"100mm"	S	COM-10734	19.95	18.95
Microphone									
	electret, omi-direct.	0.1" centers		PUI Audio	AOM-6738P-R	D	668-1296-ND	1.65	1.18
DIODES									
silicon, ordinary									
1N914	silicon diode 0.3A fast recovery	DO-35		Fairchild	1N914BTR	D	1N914BCT-ND	0.1	0.06
1N4004	silicon diode, 1A	DO-41		Diodes, Inc.	1N4004-T	D	1N4004DICT-ND	0.13	0.11
zener									
1N5232	zener, 5.6V	1/2W		Vishay	1N5232B-TR	D	1N5232BVSCT-ND	0.19	0.15
Schottky									
1N5817	Schottky, 1 A.			Diodes, Inc.	1N5817T	D	1N5817DICT-ND	0.44	0.25
1N5711	Schottky, 15mA, low capacitance.			STM	1N5711	N	89K1806	0.86	0.5
LED									
HLMP-4700	Red, low current	5mm		Avago	HLMP-4700-C0002	D	516-2483-1-ND	0.6	0.4
C566C-AFS-CU0W0252	Yellow, bright	5mm		Cree	C566C-AFS-CU0W0252	D	C566C-AFS-CU0W0252CT-ND	0.18	0.17
HLMP-3950	Green, high efficiency	5mm		Avago	HLMP-3950	D	516-1347-ND	0.6	0.4
LNG995PFBW	blue, wide angle	5mm		Panasonic	LNG995PFBW	D	P468-ND	2.1	1.7
LTL-30EHJ	red/green, 3-lead	5mm		Lite-On	LTL-30EHJ	D	160-1057-ND	0.5	0.4
TSTS7100	IR LED, 9850nm, 10° angle	TO-18		Vishay	TSTS7100	D	TSTS7100-ND	3.45	2.42
TIL311	LED hex display with decoder	DIP, 14-pin		TI	TIL311	J	TIL311	19.95	19.95

Sources: A=Avnet Express, Arizona; ADI=Analog Devices Inc.; D=Digikey; J=Jameco; M=Mouser; N=Newark; PA is Proto Advantage (Ontario): PA/D means that Digikey carries the part; Q=Quest Components Inc., (California); R=Rochester Electronics (MA); S=Sparkfun; TTI=TTI Inc., (Texas).

Parts List

Part, generic	Description	Detail (pkg, etc.)	Mfgr	Part No., exact	Source	Particular Distributor's Part No.	Price (1) US$	Price (25) US$
TRANSISTORS								
BIPOLAR								
2N3904	NPN, small-signal	T0-92	Fairchild	2N3904TFR	D	2N3904D26ZCT-ND	0.21	0.18
2N3906	PNP, small-signal	T0-92	Fairchild	2N3906TFR	D	2N3906D26ZCT-ND	0.21	0.18
MJE3055T	power NPN (10A), min beta 20 @4A	TO220	ON	MJE3055TG	N	45J1504	0.97	0.75
MJE2955T	power PNP (10A), min beta 20 @4A	TO220	Fairchild	MJE2955TTU	D	MJE2955TTUFS-ND	0.64	0.5
Array								
CA3096	bipolar array, 3 NPN, 2 PNP	16 DIP	Intersil	CA3096CM96	R	CA3096CM96	0.88	0.81
HFA3096	bipolar array, 3 NPN, 2 PNP	16-SOIC	Intersil	HFA3096BZ	D	HFA3096BZ-ND	6.38	5.43
	SOIC-16 to DIP adapter	16 DIP	Proto Advantage*	PA0005	PD	PA0005	5.09	5.09
Optical								
BPV11	phototransistor, visible light	visible, 5mm	Vishay	BPV11	N	32C9138	0.68	0.64
QSD124	phototransistor, IR	IR, 5mm	Fairchild	QSD124	D	QSD124-ND	0.51	0.33
MOSFET								
BUK9509	power MOSFET, n-ch, logic-level Vgs	TO220	NXP	BUK9509-40B,127	D	568-5727-5-ND	1.6	1.28
IRLZ34	power MOSFET, n-ch, logic-level Vgs	TO220	IR	IRLZ34NPBF	N	63J7705	1.75	1.1
2N7000	NMOS	T092	Fairchild	2N7000TA	N	31Y5851	0.12	0.12
BS250P	PMOS	T092	Diodes,Inc	BS250P	D	BS250P-ND		
DG403	analog switch	16 DIP	Maxim	DG403CJ+	D	DG403CJ+-ND	8.43	7.47
Array								
CD4007	MOSFET array, 3 N-ch, 3 P-ch	14 DIP	TI	CD4007UBE	N	4C7958	0.48	0.37
JFET								
2N5485	JFET	T0-92	Central Solid State	2N5485	M	610-2N5485	1.19	0.98
1N5294	JFET, current source, 0.75 mA	DO-35		1N5294	N	79T5011	min. order 25	4.43

Sources: A=Avnet Express, Arizona; ADI=Analog Devices Inc.; D=Digikey; J=Jameco; M=Mouser; N=Newark; PA is Proto Advantage (Ontario): PA/D means that Digikey carries the part; Q=Quest Components Inc., (California); R=Rochester Electronics (MA); S=Sparkfun; TTI=TTI Inc., (Texas).

* Proto will procure, assemble & solder the IC to their carrier.
We like Proto for its ability to procure SMT parts and then solder these to their carriers.

Parts List

Part, generic	Description	Detail (pkg, etc.)	Mfgr	Part No., exact	Source	Particular Distributor's Part No.	Price (1) US$	Price (25) US$
SCR	SCR, 100V, 8A	TO220	ON	MCR218-4G	D	MCR218-4GOS-ND	1.29	1.03
OP-AMP								
LF411	op-amp, JFET input, 3MHz	8 DIP	TI	LF411CP	N	60K6163	1.69	1.35
LM741	op-amp, bipolar, 1,5MHz	8 DIP	TI	LM741CN	N	97K3586	0.73	0.58
LM358	op-amp, dual, single supply	8 DIP	TI	LM358P	D	296-1395-5-ND	0.49	0.33
LMC6482	op-amp, rail to rail, in & out, CMOS (16V max V+ to V-)	8 DIP	TI	LMC6482IN/NOPB	D	LMC6482IN/NOPB-ND	1.78	1.4
LMC662	op-amp, single supply, rail out, CMOS	8 DIP	TI	LMC662CN/NOPB	N	41K2696	1.69	1.32
comparators								
LM311	comparator	8 DIP	TI	LM311PE4	M	595-LM311PE4	0.61	0.47
TLC372CP	comparator, single supply	8 DIP	TI	TLC372CP	N	36K3644	1.14	0.9
voltage regulators								
LM317T	voltage reg, adjustable, 1.5A	TO220	TI	LM317HVT/NOPB	D	LM317HVT/NOPB-ND	2.42	1.93
LM385-2.5	voltage reference	TO-92	TI	LM385LPR-2-5	D	296-31451-1-ND	0.66	0.51
LM78L05	voltage reg, 3-term, 100mA	T0-92	Fairchild	LM78L05ACZ	D	LM78L05ACZFS-ND	0.43	0.25
LT1073	switching regulator, buck/boost	8 DIP	Linear	LT1073CN8#PBF	D	LT1073CN8#PBF-ND	5.53	3.67
oscillators								
TS555IN	timer/oscillator, CMOS	8 DIP	TI	TLC555CP	D	296-1857-5-ND	0.85	0.68
crystal oscillators								
11.0592MHz	"half size" metal can	8 DIP	ECS	ECS-2200B-110.5	D	XC268-ND	3.68	3.15
8MHz	"half size" metal can	8 DIP	CTS	MXO45HS-3C-8M0000	D	CTX743-ND	2.07	1.98
crystal								
11.0592MHz	0.2" centers	2-pin, welded	ECS	ECS-110.5-S-4X	D	X1064-ND	0.81	0.69
Filter								
MAX294	switched-capacitor 8-pole elliptic low-pass filter	8 DIP	Mx	MAX294CPA	D	MAX294CPA+-ND	5.66	5.44
audio Amps								
LM386N	power amplifier (single-supply; will drive 8-ohm speaker)	8 DIP	TI	LM386N-1/NOPB	D	LM386N-1/NOPB-ND	0.98	0.79
LM4667	switching amplifier	10 TSOP	TI	LM4667MM/NOPB	D	LM4667MM/NOPBCT-ND	1.2	0.96
	MSOP-10 to DIP adapter	10 DIP	Proto Advantage*	PA0027	PD	PA0027	10.99	10.99

Sources: A=Avnet Express, Arizona; ADI=Analog Devices Inc.; D=Digikey; J=Jameco; M=Mouser; N=Newark; PA is Proto Advantage (Ontario): PA/D means that Digikey carries the part; Q=Quest Components Inc., (California); R=Rochester Electronics (MA); S=Sparkfun; TTI=TTI Inc., (Texas).

* Proto will procure, assemble & solder the IC to their carrier.
We like Proto for its ability to procure SMT parts and then solder these to their carriers.

Parts List

Part, generic	Description	Detail (pkg, etc.)	Mfgr	Part No., exact	Source	Particular Distributor's Part No.	Price (1) US$	Price (25) US$
DIGITAL								
TTL (LS)								
74LS00	quad NAND	14 DIP	TI	SN74LS00N	N	60K6847	1.18	0.94
74LS503†	8-bit successive-approximation register	16 DIP	Fairchild	DM74LS503N	R		call	call
CMOS (HC)								
74HC00	quad NAND	14 DIP	TI	SN74HC00N	D	296-1563-5-ND	0.52	0.35
74HC02	quad NOR	14 DIP	TI	SN74HC02N	N	67K1077	0.49	0.33
74HC04	hex inverter	14 DIP	TI	SN74HC04N	D	296-1566-5-ND	0.56	0.43
74HC74	dual D flop	14 DIP	TI	SN74HC74N	D	296-1602-5-ND	0.52	0.35
74HC86	quad XOR	14 DIP	TI	SN74HC86N	D	296-8375-5-ND	0.42	0.33
74HC125	quad 3-state buffer	14 DIP	TI	SN74HC125N	D	296-1572-5-ND	0.45	0.35
74HC175	quad D register	16 DIP	TI	SN74HC175N	D	296-8257-5-ND	0.52	0.4
Counters								
CMOS: 74HC								
74HC160	4-bit BCD counter, jam clear	16 DIP (to be discontinued)	NXP	74HC160N,652	M	771-74HC160N	1.65	1.32
		16TSSOP	NXP	74HC160PW,118	D	568-8860-1-ND	0.62	0.48
	TSSOP 16 to DIP adapter	16 DIP	Proto Advantage*	PA0034	PD	PA0034	4.19	4.19
74HC161	4-bit binary counter, jam clear	16 DIP	STM	M74HC161B1R	D	497-1787-5-ND	0.96	0.76
74HC163	4-bit binary counter, sync clear	16 DIP	TI	CD74HC163E	D	296-33033-5-ND	0.88	0.7
74HC390	dual ripple decade counter	16 DIP	STM	M74HC390B1R	D	497-7370-5-ND	1.07	0.85
74HC393	dual ripple binary counter	16 DIP	TI	SN74HC393N	N	60K6838	0.48	0.37
74HC4040	12-bit ripple counter	16 DIP	TI	SN74HC4040N	D	296-8324-5-ND	0.71	0.54
PLL								
74HC4046	phase-locked loop	16 DIP	TI	CD74HC4046AE	D	296-9208-5-ND	0.66	0.53
CMOS: 74HCT								
74HCT14	hex Schmitt Trigger inverter	14 DIP	TI	SN74HCT14N	D	296-8394-5-ND	0.59	0.45
74HCT125	quad 3-state buffer	14 DIP	TI	SN74HCT125N	D	296-8386-5-ND	0.42	0.33
74HCT139	dual 2-to-4 decoder	16 DIP	TI	SN74HCT139N	D	296-8390-5-ND	0.63	0.51
74HCT541	octal 3-state buffer	20 DIP	TI	SN74HCT541N	D	296-1619-5-ND	0.68	0.52
74HCT573	octal transparent latch, 3-state	20 DIP	TI	SN74HCT573N	D	296-1621-5-ND	0.61	0.49
74HCT574	octal D register, 3-state	20 DIP	TI	SN74HCT574N	D	296-1623-5-ND	0.76	0.61

Sources: A=Avnet Express, Arizona; ADI=Analog Devices Inc.; D=Digikey; J=Jameco; M=Mouser; N=Newark; PA is Proto Advantage (Ontario): PA/D means that Digikey carries the part; Q=Quest Components Inc., (California); R=Rochester Electronics (MA); S=Sparkfun; TTI=TTI Inc., (Texas).

* Proto will procure, assemble & solder the IC to their carrier.
We like Proto for its ability to procure SMT parts and then solder these to their carriers.

† This part is obsolete, but available at several sources. If you can't find it, we offer a functional equivalent, made with a PAL.

Parts List

Part, generic	Description	Detail (pkg, etc.)	Mfgr	Part No., exact	Source	Particular Distributor's Part No.	Price (1) US$	Price (25) US$
PAL XC9572XL	44 TQFP-to-DIP adapter	44 TQFP	Xilinx	XC9572XL-10VQG44C PA0093*	D PD	122-1448-ND PA0093	2.86 10.49	2.86 10.49
Processor/ Controller C8051F410	controller, 50MHz	32 LQFP 32 DIP	Silicon Labs Proto Advantage*	C8051F410-GQ PA0091	D PD	336-1318-ND PA0091	6.02 10.99	5.8 10.99
DS89C430 or DS89C450	controller, 33MHz	40 DIP	Maxim/Dallas	DS89C430-MNG DS89C430-MNG+ DS89C430-MNL DS89C430-MNL+ or DS89C450: DS89C450-MNL+	R D R D D	DS89C430-MNG DS89C430-MNG+-ND DS89C430-MNL DS89C430-MNL+ DS89C450-MNL+-ND	19.54 21.31 15.42 21.31 28.29	17.13 18.69 14.31 18.69 24.81
RAM CY62256NLL-70PXC	32K×8 SRAM, low-power, 55ns. Wide pkg: better for affixing label Any equivalent is OK	28 DIP, 0.6" (wide)	Cypress	CY62256NLL-70PXC	N	48W3332	4.88	4.88
Converters								
DAC AD558JN	DAC, 8-bit, single supply, volt out	16 DIP	Analog Devices	AD558JN	D	AD558JN-ND	18.2	15.31
DAC08	multiplying DAC, parallel in, current out	16 DIP	Analog Devices	DAC08CP	D	DAC08CP-ND	3.22	2.59
AD7524JN	multiplying DAC, parallel in, current out	16 DIP	Analog Devices	AD7524JN	D	AD7524JN-ND	10.74	8.8
DAC8229	dual DAC, 8-bit, multiplying, volt out	20 SOIC 20 DIP	Analog Devices Proto Advantage*	DAC8229FSZ PA0008	ADI PD	DAC8229FSZ-ND PA0008	11.16 4.69	10.15 4.69
MAX5102	dual DAC, parallel in, volt out	16 TSSOP 16 DIP	Maxim Proto Advantage*	MAX5102BEUE+ PA0034	D PD	MAX5102BEUE+-ND PA0034	3.24 4.19	3.11 4.19
ADC AD7569	ADC, DAC, microprocessor-compatible	24 DIP	Analog Devices	AD7569JNZ	M	584-AD7569JNZ	20.42	17.36

Sources: A=Avnet Express, Arizona; ADI=Analog Devices Inc.; D=Digikey; J=Jameco; M=Mouser; N=Newark; PA is Proto Advantage (Ontario): PA/D means that Digikey carries the part; Q=Quest Components Inc., (California); R=Rochester Electronics (MA); S=Sparkfun; TTI=TTI Inc., (Texas).

* Proto will procure, assemble & solder the IC to their carrier.
We like Proto for its ability to procure SMT parts and then solder these to their carriers.

Parts List

Part, generic	Description	Detail (pkg, etc.)	Mfgr	Part No., exact	Source	Particular Distributor's Part No.	Price (1) US$	Price (25) US$
Hexadecimal display TIL311	hexadecimal display with decoder & latch	16 DIP	TI	TIL311	J	TIL311	19.95	17.95
Miscellaneous Peripherals	digital potentiometer, SPI, 100k	8 DIP	Microchip	MCP41100-I/P	D	MCP41100 MCP41100-I/P-ND	1.61	1.11
23K256-I/P	SPI RAM, 32K	8 DIP	Microchip	23K256-I/P	D	23K256-I/P-ND	1.34	1.08
MXD1210	battery-backup RAM power supply controller	8 DIP	Maxim	MXD1210CPA+	D	MXD1210CPA+-ND	9.08	6.29
	IR receiver module	3-pin, side view	Vishay	TSOP31240	D	TSOP31240-ND	1.45	0.98
5103310–1	header, 10-pin	5X2	TE Connectivity	5103310-1	D	A33179-ND	1.25	1.19
CEM-1205C	5V buzzer	0.3" pin spacing	CUI	CEM-1205C	D	102-1124-ND	2.51	2.08
PAL XC9572XL	72-cell PLD, 3.3V	44-PLCC	Xilinx	XC9572XL-10PCG44C	N	98K3382	8.91	8.91
Custom Parts LCD card	32-bit LCD display; USB⇔UART and USB⇔Silabs C2 translator		‡					
SAR (for ADC lab, D5)	functional equivalent to 74LS502, which now is obsolete and hard to find	24 DIP	‡					

							Price (1) US$	Price (100) US$
Capacitors ceramic, CK05	10pF 68pF 100pF 470pF	0.2"	AVX	CK05BX100K	M	581-CK05B100K	0.2	0.18
	0.01uF	axial, 630V	Cornell-Dubilier	150103K630BB	N	93B3223	0.62	0.57
polyester (mylar)	0.1uF	axial, 50V	Cornell-Dubilier	WMF05P1K-F	TTI	WMF05P1K-F	n.a.	1.14
	0.223uF	axial, 400V	Cornell-Dubilier	150224J400FE	N	91B1397	1.4	1.4
	1uF	axial, 35V	Kemet	T322B105K035AT7200	D	399-11374-1-ND	1.67	0.97
tantalum	4.7uF	axial, 20V	Vishay	173D475X9020VE3	N	06X1297	1.1	0.85
	15uF	axial, 20V	Vishay	173D156X9020XE3	N	06X1292	1.94	1.56
	100uF	axial, 20V	Kemet	T322F107K020AT	A	T322F107K020AT	9.53	6.47
OS-Con/ conductive polymer	100uF, low series resistance	axial, 16V	Panasonic	667-16SEPC100MW	M	16SEPC100MW	0.82	0.35

Sources: A=Avnet Express, Arizona; ADI=Analog Devices Inc.; D=Digikey; J=Jameco; M=Mouser; N=Newark; PA is Proto Advantage (Ontario): PA/D means that Digikey carries the part; Q=Quest Components Inc., (California); R=Rochester Electronics (MA); S=Sparkfun; TTI=TTI Inc., (Texas).

‡: go through www.learningtheartofelectronics.com

Parts List

Part, generic	Description	Detail (pkg, etc.)	Mfgr	Part No., exact	Source	Particular Distributor's Part No.	Price (1) US$	Price (100) US$
Resistors carbon composition, 1/4W	1, 10, 47, 100, 150, 220, 240, 270, 330, 390, 470, 560, 620, 680, 750, 820 1k, 1.5k, 2.0k, 2.2k, 2.7k, 3.3k, 4.7k, 5.6k, 6.2k, 6.8k, 7.5k, 8.2k, 10k, 12k, 15k, 20k, 22k, 33k, 47k, 56k, 68k, 82k, 100k, 120k, 150k, 220k, 330k, 470k, 560k 1M, 3.3M, 4.7M, 10M	axial	Allen-Bradley or SEI		Q		n.a	0.10 to 0.3
power resistor, wire-wound	10Ω, 25W	axial	TE-connectivity	1625971-5	D	A102130-ND	3.05	2.08

Sources: A=Avnet Express, Arizona; ADI=Analog Devices Inc.; D=Digikey; J=Jameco; M=Mouser; N=Newark; PA is Proto Advantage (Ontario): PA/D means that Digikey carries the part; Q=Quest Components Inc., (California); R=Rochester Electronics (MA); S=Sparkfun; TTI=TTI Inc., (Texas).

F The Big Picture

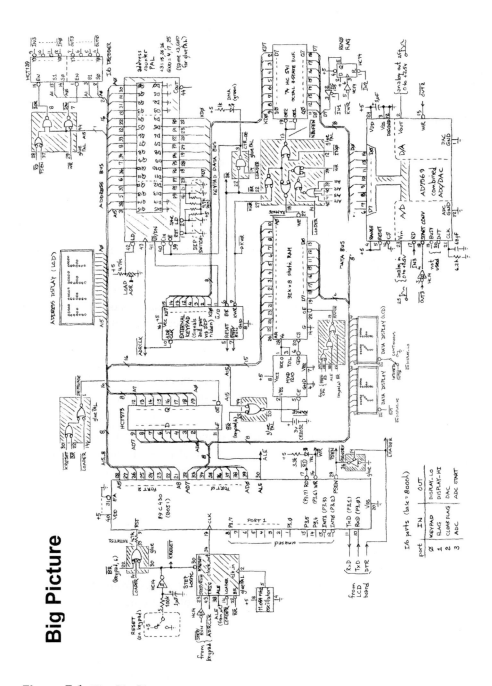

Figure F.1 The Big Picture.

G "Where Do I Go to Buy Electronic Goodies?"

Good question! Here are some hints, from our experiences.

I. Mail order and online

Digi-Key Corp (Thief River Falls, MN: 1–800–digikey). We used to say "*Get their catalog!!*" – but, sadly, they've abandoned paper, replacing it with an impressive search capability. You can get everything here, even in small quantities, with fast delivery. It's often worth designing with their webpage open in front of you. Online ordering and stock/price checking: www.digikey.com.

Mouser Electronics (www.mouser.com). Broad stocking distributor, with service comparable to Digi-Key's, and willingness to ship small quantities. Good selection of precision passives; and they are still printing a comprehensive paper catalog.

Newark Electronics + Farnell (1–800–2–newark; www.newark.com). Broadest stocking distributor, with service comparable to Digi-Key's and good selection of tools; paper catalog still in print.

"Stocking Distributors". These are the standard distribution channel for quantity buying; names like Allied (still publishing a paper catalog), Arrow, Avnet, FAI, Heilind, Insight, Pioneer, Wyle. Substantial minimum quantities – not generally useful for prototyping or small production.

Manufacturers' Direct. Many semiconductor manufacturers (Analog Devices, TI, Maxim,...) will not only send free samples with the slightest provocation, they will also sell in small quantities via credit card; check out Mini-Circuits for RF components, and Coilcraft for inductors, transformers, and RF filters.

Oddballs. Marlin P Jones, Jameco, B&D, Herbach & Rademan, Omnitron, ABRA, All Electronics; ephemeral collection of "surplus" stuff, some real bargains.

eBay (www.ebay.com). If you haven't been here, you've probably just arrived from Mars. LOTS of stuff, literally millions of items, an online auction. You can get plenty of electronic stuff, but CAVEAT BIGTIME EMPTOR. Feedback Forum helps.

Alibaba Small Pacific-rim companies that have stocks of obsolete components are easily found on Alibaba. You can place a quote request for a part number, and you'll get dozens of useful reasonably-priced suppliers.

II. Indexes and Locators

Octopart, FindChips, NetComponents (octopart.com, findchips.com, netcomponents.com). Give it a part number and it searches dozens of distributors, returning (sometimes unreliable) information on availability and pricing.

WhoMakesIt (www.whomakesit.com). A bit like the EEM catalog, helpful if you know the category of stuff you want, but not the manufacturer, etc.

Google (www.google.com). Our standard "portal," reads your mind and vectors you to the goodstuff. Helpful, sometimes, in finding parts and equipment manufacturers and vendors.

III. Local

Sometimes it's nice to shop in person; here are the sorts of places to go.

Radio Shack (www.radioshack.com). They call themselves "America's Electronic Supermarket"; we'd call them "America's Electronic Convenience Store." Their stores are everywhere, and they stock (pretty reliably) an idiosyncratic collection of parts and supplies, of uncertain quality or duration. However, in the changing marketplace of consumer electronics, their future path is unclear.

Electronics Flea Markets. Also known as "swap meets," perhaps somewhat in decline; two legendary meets are on opposite coasts: De Anza College (Cupertino, every second Saturday, March–October), and MIT (Cambridge, every third Sunday, April–October). What meets are three cultures (electronics, computers, hams); haggling is mandatory; caveat very emptor.

Electronics Surplus Supply Stores. These are incredibly cool! Several well-known haunts are Halted (www.halted.com; officially "HSC Electronics Supply"), in Santa Clara, Sacramento, and Santa Rosa; and Murphy's Surplus Warehouse (www.murphyjunk.bizland.com) in El Cajon.

IV. Miscellaneous

Obsolete ICs. Your best place to start is Rochester Electronics (www.rocelec.com), a wonderful place that apparently buys up inventories of ICs being discontinued. Jameco also has a lot of obsolete parts. Freetradezone has broker lists for obsolete parts. Also try Interfet (www.interfet.com), a manufacturer of small-signal FETs, including ones that the big guys have abandoned.

PC Board Manufacture. We like a place called Advanced Circuits, www.4pcb.com; you can get online quotes, and they do a good job and deliver pronto. Another inexpensive and fast PC house is Alberta Printed Circuits in Canada (www.apcircuits.com).

H Programs Available on Website

Here is a list of programs posted on the book's website. (Others may be added, from time to time.) We have not posted the very short programs of the early big-board labs.

Lab 20L

Silabs:

- `bitflip.a51`
- `bitflip_delay_inline.a51`

Lab 21L

Big-board:

- `16_sum_02.a51`
- `keysum_delay_inline_402.a51`
- `keysum_delayroutine_407.a51`

SiLabs:

- `bitflip_delay_subroutine.a51`
- `byte_in_out_ports.a51`
- `bitflip_if.a51`
- `byte_in_out.a51`
- `keysum_8bit.a51`
- `keysum_16bit.a51`

Lab 22L

Big-board:

- `getready_1208.a51`
- `intdsply_1204.a51`

SiLabs:

- `timer2_bitflip.a51`
- `pwm_by_wizard_nov10.a51`
- `comparator_oscillator_jan11.a51`
- `comparator_osc_display_jan11.a5`

Lab 23L

Big-board:

- `adcdacint_Rn_504.a51`
- `adcdacint_504.a51`
- `fullwave_401b.a51`
- `int_inc_dec.a51`

SiLabs:

- `adc_dac_int_jan11.a51`
- `dac_test_12_june15.a51` (incomplete)
- `dac1_8bit_wizard_MT.a51`
- `adc_wizard_apr11.a51`
- `adc_dac_int_jan11.a51`

Lab 24L

Big-board:

- `bit_flip_assy.a51`
- `tmr800.a51`
- `servopulse_512.a51`
- `spi_digipot_int_504.a51`
- `delay_bitflip_408.c`
- `table_bidirectional_dec10_wi_ale.a51`
- `table_bidirectional_dec10_wi_ale.lst`
- `store_and_playback_bidirectional.c`

SiLabs:

- `adc_store_on_chip_may15.a51`
- `spi_digipot_silabs_apr14.a51`
- `serial_message_silabs_apr11.a5`
- `serial_receive_silabs.a51`

Lab 25L

Big-board:

- `pulse_measure_decimal_standalone_505.a51`
- `pulse_measure_capture_standalone_505.a51`

SiLabs:

- `spi_ram_adc_dac_incomplete.a51`
- `spi_single_byte.a51`

Chapter 26N

Big-board:

- `lcd_4bit_407.a51`
- `keypad_encoder_606.a51`

Verilog code

Glue PAL files:

- `stepglue_qfp_oct15.v`
- `stepglue_qfp_oct15_tb.v`
- `stepglue_qfp_oct15.ucf`
- `stepglue_qfp_oct15.xise`

Micro counter files:

- `ctr_16bit_ud_3stt_10data.v`
- `ctr_16bit_ud_3stt_10data_tb.v`
- `ctr_16bit_ud_3stt_10data.ucf`
- `counter_micro_16bit_qfp_oct15.xise`

I   Equipment

I.1   Uses for This List

We know it's hard to know where to begin, in outfitting a home shop or a course lab. Our list, describing equipment that we have liked, may be helpful, at least as a starting point. Within few years after publication of this book, surely additional good alternatives will have appeared. So, we know that we are not saying the last word on this subject. Note, by the way, that our prices are what we find listed online. You may well get a better price, especially if you are buying for a university.

I.2   Oscilloscope

This is the most important piece of equipment in our lab, and the most expensive. For 30-odd years we have used mostly Tektronix scopes. We tried a scope made by HP (later called Agilent, now Keysight); we tried a Hameg and a LeCroy. We always returned to Tektronix. Most of our analog scopes were 2213s (2-channel, 60MHz); our digital scopes – which we continue to enjoy – were TDS3014s (4-channel, 100MHz).

Both these models now have gone out of production, and among the newer models that we have tried (including a Tek mixed-signal model, the MSO2014) the Rigol scopes seem to offer the best value: the MSO if you can afford it (digital channels occasionally are useful, and the 100MHz bandwidth is appealing, though far from essential).

I.2.1   Digital scopes

Rigol DS1054Z 50MHz, 4-channel. Really good value $399
Rigol MSO1104Z 100 MHz Mixed Signal Oscilloscope 4 "analog" channels,[1] 16 digital channels $997
Tektronix DPO2014B 100MHz oscilloscope, 4 channels $2074

I.2.2   Analog scopes

We have not tried any of these three. These are two of the last companies still manufacturing analog scopes:

B & K Precision 2160C 60MHz, 2-channels $1119
Instek 6051 50MHz, 2-channels $832
Instek GOS6103 100MHz, 2 channels, on-screen readout $1493

---

[1] These four "analog" channels are digitally-sampled channels, "analog" only in the sense that the display shows nearly continuous voltage variation, in contrast to the sixteen "digital" channels, which show only a binary display, High or Low.

You'll notice that these cost more than a very capable digital scope like the Rigols mentioned above, so you'll surely not be tempted to make a new analog oscilloscope your only scope. But it's worth recalling that discontinued scopes like the Tektronix 2213 are available *used*, and can cost very little (sometimes $200 or less). These can be especially appealing to a hobbyist.

I.3 Function generator

The evolution of function generators has not suited us. Most generators now use keypad or pushbutton frequency selection, which we find much less friendly than the knob of the analog generators that we have long used. The B & K model listed below comes close to the old style that we like.

B & K Precision Model 4017A, 10MHz, with linear/log sweep and digital readout $449

The newer, digital function generators offer capabilities not available from the old analog models, including arbitrary waveform synthesis. We simply don't use such capacities in our course, so we look for a simpler generator.

I.3.1 Doing without a dedicated function generator

Get a scope With function generator: A few scopes, including the Rigol MSO1104Z mentioned above, offer to include an optional function generator. The part number is Rigol MSO1104Z-S, and the 20MHz generator adds $232 to the price. Sharing a front panel with the oscilloscope is less convenient than using a separate generator – but the price is very appealing.

Rely on the powered breadboard's built-in function generator: A hobbyist planning to buy a powered breadboard like the PB-503 (see just below) could get by using the not-so-versatile function generator that is built into the PB-503. It lacks DC-offset variation, its waveforms are less perfect than those from a standalone generator, and its maximum frequency is only 100kHz.[2] But it could be adequate.

I.4 Powered breadboard

A unified breadboard with power supply, like the Global Specialties PB-503 that we have been using for decades, is very convenient. For a course, it seems nearly indispensable.

For an individual hobbyist, a cheaper alternative would be to buy a triple power supply (+5V, ±15V) and individual breadboard strips, or a set of three strips as in Global's PB-105 ($54). You would need to improvise other functions of the PB-503 ($433) that we find convenient:

- A simple function generator. We seldom use the one provided by the PB-503.[3] But when we want a second signal, it's handy.
- Debounced switches.
- Two potentiometers.
- LEDs.

None of those improvisations would be difficult.

---

[2] Another way to do without buying a function generator is to buy a fancier oscilloscope that includes a function generator. The Tektronix MDO3014 offers this (plus a spectrum analyzer), but at more than $4000 it probably is useful neither to a hobbyist nor to someone setting up an electronics lab.

[3] Most often, we use the built-in generator as a source of logic-level square waves. In Lab 5L we use it as a second sinewave source, where we feed two signals to a differential amplifier.

I.5 Meters, VOM and DVM

I.5.1 VOM (analog multimeter)

The $267 analog "Simpson meter" that we use only occasionally in our course (Simpson 260-8) is a luxury one can do without. We introduce it on Day 1 because we prefer that students know how to use the instrument (which requires more thought than a DVM does). But in the ordinary case, we all reach for a DVM, not a VOM. The VOM beats the DVM only in the exceptional instance where you would like a quick graphical indication of a slow trend – like the falling off of current in Lab 4L's transistor current source, as the circuit slips into *saturation*.

Hobbyists can happily skip the VOM, or could choose a small and inexpensive model like the Tenma 72-8170, $31 (Newark).

I.5.2 DVM (digital multimeter)

A great variety of DVMs is available. We need nothing fancy.

- **B & K Precision 2704C: Very simple** 3 1/2 digit DVM with transistor beta test, capacitance, frequency $66
  We like this very inexpensive device. Its frequency counter and transistor tester are nice features, not often found in an inexpensive meter.
- **Amprobe 37XR-A: Better, and more expensive** : 4 1/2 digit DVM: true RMS, capacitance, inductance, frequency. $152

I.6 Power supply

We have used only the first of these supplies – the expensive HP model. The Rigol's specifications are impressive.

- **Keysight (formerly HP) E3630A** triple power supply, 35W, to 20V @0.5A $670
- **Rigol DP832** triple output, 195W, to 30V @3A $450

I.7 Logic probe

- **Tenma 72-500** combined logic probe and pulser, detects 10ns pulse, switchable TTL vs. CMOS levels $27

I.8 Resistor substitution box

We like the one that offers fewer values, because it makes changing values easy. But occasionally the 1Ω resolution of the other is useful. For a course, perhaps many of the RS4000 and one of the RS5000.

- **Elenco RS4000** 24 values, 10Ω through 1MΩ $17 (Amazon)
- **Elenco RS5000** 1Ω increments, 0 to 11MΩ $24 (Robotshop.com)

I.9 PLD/FPGA programming pod

We use the Xilinx model DLC9G. $47 (Amazon)

USB links the pod to a computer. It provides leads to carry the JTAG signals to a PAL. We use a zero-insertion-force socket, wired to these leads plus ground and +5V. But these ZIF sockets (e.g., Aries 44-6551-10 at Digikey) are not available in single quantity. You could use an ordinary 44-position 0.6″-spacing socket, instead.

I.10 Hand tools

I.10.1 Pliers

We like small "chain nose" pliers. The these are 5 inches long and spring to their open position. We like the feel of the C & K, but we've moved to the less expensive Excelta.

Excelta 2644 smooth-surface chain nose or 2644D serrated-surface chain nose $17
C&K 3772D chain nose $52

I.10.2 Wire strippers

Ideal 45-125 22 to 30 AWG wire gauge $12

I.10.3 Screwdriver

Xcelite R3324 slotted head, 0.1" blade, 6.25" length $3.50

I.11 Wire

22 AWG solid hookup wire, insulated
- Alpha: 3051/1 BK005, solid 22 ga., l00-foot spool (8 colors; black color indicated here by "BK;" see other codes, e.g., "RD005" for red): $20 (Allied)
- Techedco 1000 foot spool (8 colors): $49 (Ebay)

Banana hookup wire banana double-ended, 18 inch. We cut most of these at midpoint and solder on solid leads, to make banana-to-wire cables, convenient for plugging into breadboards. We use red and black.
- Pomona: B-18-0 (black), B-18-2 (red) $5.22 @quantity 25 (Digikey)

J Pinouts

J.1 Analog

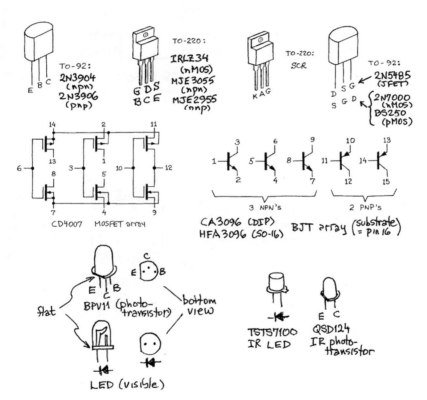

Figure J.1
Transistors.

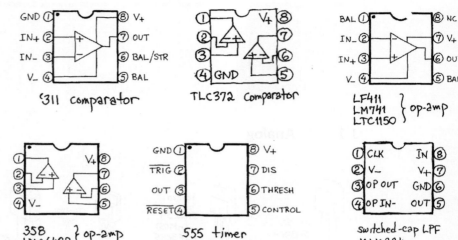

Figure J.2 Linear ICs: op-amps, comparators, etc.

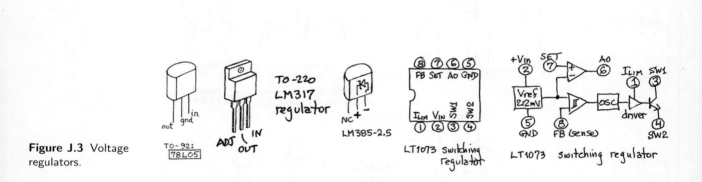

Figure J.3 Voltage regulators.

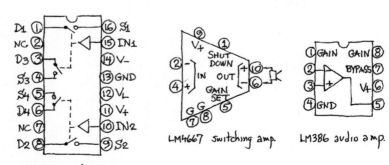

Figure J.4 Miscellaneous analog ICs.

J.2 Digital

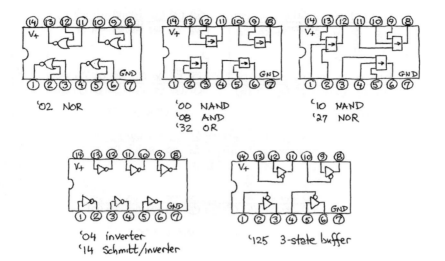

Figure J.5 Gates.

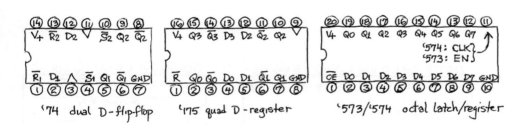

Figure J.6 Registers.

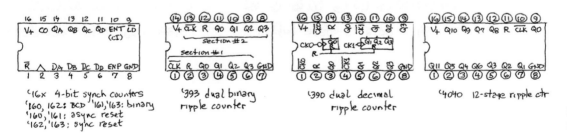

Figure J.7 Counters.

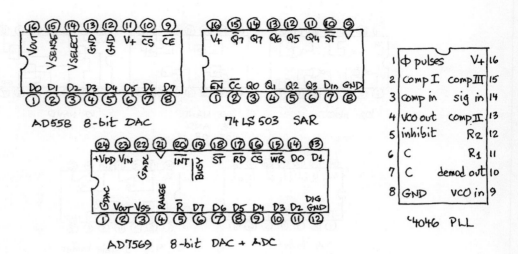

Figure J.8 Converters, PLL: ADC, DAC and SAR.

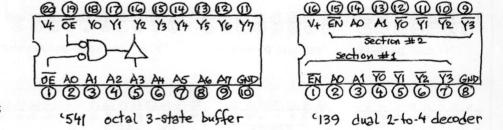

Figure J.9 Three-state buffer; decoder.

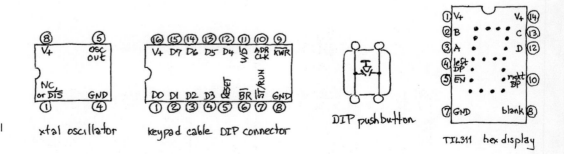

Figure J.10 Digital miscellany.

J.2 Digital

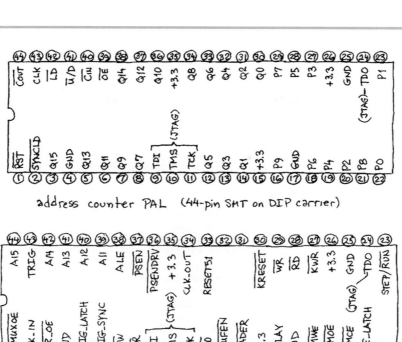

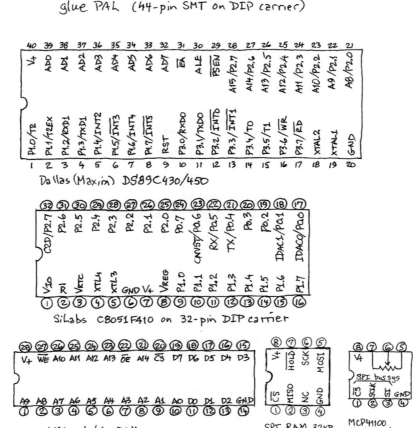

Figure J.11 Two PALs.

Figure J.12 Microcontrollers, RAM, SPI digipot.

Index

LC circuit, 113
$R:2R$ ladder, 697–698
RLC circuit, 113–118
    resonance, 114
R_{ON}
    analog switch (lab), 490
    MOSFET (lab), 489
R_{on}
    analog switch (lab), 490
$f_{crossover}$, 312
h_{fe}, 158
r_e
    ...deriving, 207
'311 (lab), 337
'555, 341–343
.INC file (assembler/compiler), 817
.lst file (RIDE assembler/compiler/simulator), 951
#344 lamp (lab), 343
#47 lamp (lab), 174
"WATCH" window
    RIDE assembler/compiler/simulator, 953
"active low", 522
"loop" (program): origin of term, 815
"saturation" (FET usage), 500
"sticky" flag, 600, 601
1N5294 current-limiting diode, 342
1N5294 current-limiting diode (lab), 213
1N5717 Schottky diode (lab), 135
2's-complement
    cheap trick to calculate, 816
2N3904 (lab), 169
2N3906 (lab), 271
4046 PLL, 717
74HC00 (lab):, 540
74HC14 (lab), 339
74HC175 (lab), 594
74HC4040 (lab), 730
74HC4046 (lab), 730
74HC574 (lab), 728
74HC74 (lab), 588
74HCT139 (lab), 835
74HCT573 (lab), 786
74LS00 (lab), 540
74LS503 (lab), 725
78L05 regulator (lab), 454
8051 controller, 765
8051 port, 770–771
8051

    pinout (lab), 786
    addressing modes
        bit operations, 862–864
        DPTR, 857–858
        examples (brief), 867–868
        indirect addressing, 860–861
        R_n, 859, 860
        multiple address spaces, 864–868
        off-chip operations, 857–860
        on-chip operations, 857
        portpins vs external buses (examples), 862
A in feedback: open-loop gain, 396
AB in feedback: loop gain, 397, 399
AC/DC (scope input select), 34
ACC
    vs "A" SiLabs (lab), 846
active filter, 354
    Bessel, 387
    Bessel (lab), 374
    Butterworth, 387
    Butterworth (lab), 374
    Chebyshev, 387
    Chebyshev (lab), 374
    transient response, 388
    VCVS, 354
active filter (lab), 373–374
active low
    ...in Verilog, 526, 571
    why prevalent, 529
active pullup gate output: CMOS (lab):, 545
active rectifier, 281
AD558 (lab), 724
AD7569 (lab), 926
ADC
    see "analog ↔ digital conversion", 701
ADC
    SiLabs (lab) , 942–944
ADC (lab), 725–729
ADC & DAC to micro
    modify waveforms
        lowpass filter, 923–925
        rectify: full-wave, 922
        rectify: half-wave, 923
    number formats, 922
    reconstruction filter, 917
    SiLabs
        left-justify, 917
ADDC (lab), 841, 854

Index

adder
  worked example, 556–559
address decoding, 675
address multiplexing (lab), 786
addressing modes
  see "8051: addressing modes", 857
ADx address/data lines (lab), 786
ALE (lab), 787
aliasing, 696, 735, 738–739
ALU
  design with gates, 563
  design with Verilog, 564
  worked example, 563
AM radio (lab), 135
ammeter (lab), 29
analog
  contrasted with digital, 513
analog ↔ digital conversion
  R:$2R$ ladder, 697–698
  ADC designs, 701–712
    ...compared, 701
    "flash" or parallel, 701
    binary search (SAR), 705–707
    delta–sigma, 706–712
    dual-slope, 703
    how delta–sigma high-passes quantization noise, 710–711
    how delta–sigma low-passes signal, 711
    oversampling, 709
    pipelining, 703
    tracking, 705
  aliasing, 714
  DAC designs, 697–701
    current summing R:$2R$, 697
    PWM, 700
    switched-capacitor, 699
    thermometer, 697
  dither, 714
  quantization error, 694
  resolution, 694
  sample and hold, 705
  sampling artifacts, 712–714
  sampling rate, 694–697
    aliasing, 696
analog switch
  application
    chopper (lab), 491
    flying capacitor (lab), 493
    sample-and-hold (lab), 492
    switched-capacitor filter (lab), 493–495
analog to digital conversion
  ADC designs
    single-slope, 616
AND as "if", 574
AND as Pass/Block* function, 573
assembler directive (RIDE assembler/compiler/simulator), 947
assembly language, 813
  assembler directive, 817
  bit versus byte operations, 775, 776, 864
  CALL subroutine, 826, 830
    stack, 826
  directives, 817
    INCLUDE, 817
  DPTR initialization, 822, 823
  EQU, 818
  extending operations beyond 8 bits, 830
  format, typical, 823
  INC, 817
  INCLUDE, 817
  indirect addressing, 823
  jump offsets, 815
  loop in slow motion, 818
  moving pointer, 959–961
  MOVX, 823
  ORG, 817
assertion level logic notation, 522–528
assign (in Verilog), 1054
audio amplifier (LM386), 920

B in feedback: fraction fed back, 396
ballast resistor., 469
bandpass filter (worked example), 103
battery backup (lab), 832
Bessel filter, 387
beta, 158
bias current, 288
bias divider, 179
biasing, 161
  details, 163
binary, 514
binary number codes, 518
binary numbers
  hexadecimal notation, 519
  signed, 518
  unsigned, 518
binary search (lab), 724
binary search ADC, 705–707
binary search display (lab), 727
bipolar transistor, 154
  four topologies, 167
  impedance at collector, 167
bipolar transistor (defined), 152
bit defined, 514
bit move
  using carry flag, 873
bit operations
  bit move, 873
  external bus, 869–870
  in C, 873
  port pins, 870–871
bit versus byte operations, 864
Black
  formalizing feedback, 249
Black, Harold
  memoir, 245
bleeder resistor, 143
blocking capacitor, 74
blocking vs non-blocking assignments, 1077–1079
Bluetooth, SiLabs (lab), 989–991
Boole, 520–521
bootstrap (op-amp input), 397

Index

BPV11 (lab), 899
BR* (lab), 786
breadboard (use of), 25
breadboard printed-circuit, microcomputer (lab), 620
breakpoint
    SiLabs (lab), 802
    SiLabs IDE, 806
BS250 (lab), 271
BS250P (lab), 453
BUK9509–40B (lab), 486
bullet timer (worked), 642
bus, 648
bus activity
    ISA/EISA, 768–770
Butterworth filter, 387

C code
    bit operations, 873
C language
    functions as subroutines, 829
c2 interface, SiLabs (lab), 793–796
C8051F410
    pin use, all labs (lab), 796
C8051F410 (lab), 792
CA3096 array (lab), 209
CA3600 (lab), 544
cache memory, 653
CALL
    see "subroutine", 826
    SiLabs (lab), 844
CALL (lab), 842
capacitance meter (lab), 749–750
capacitor
    blocking, 74
    decoupling, 75
    dynamic description, 52
    exponential RC charging), 53
    hydraulic analogy, 52
    polarized, 85
    reading value, 85
    static description, 52
    tolerance codes, 89
carrier frequency
    group audio project, 505
carry
    counter carry-out glitch (lab), 630
carry (counter), 611
    carry-out: synchronous, 612–613
carry flag
    for bit move, 873
CC* (lab), 726
CD standard, 745
CD4007 (lab), 544
CE*, 649
CEM-1205C lab), 899
Chebyshev filter, 387
chopper op-amp, 317
clamp, diode (lab)
    see "diode clamp", 136
clipping
    differential amplifier, 220

clock, 579
CMRR
    in op-amp (worked example), 240
CMRR, op-amp from array (lab), 218
combinational logic, 520
common emitter amplifier (lab), 174
common-emitter amplifier
    ...modest gain, 157
    bypassed-emitter, 197
    distortion, 194
        remedy: R_E, 195
common-emitter amplifier (lab), 172
common-mode amplifier (discrete)
    common-mode gain, 205
common-mode gain
    measuring (lab), 212
comparator, 320–326
    ...contrasted with op-amp, 320
    in SiLabs controller (lab), 897
    digital
        magnitude (design problem), 560–562
        worked example, 559
    hysteresis, 324–326
        AC, 326
        how much?, 325
        Woody Allen, 325
    Schmitt trigger, 324
comparator (lab), 337–338
compensation capacitor
    op-amp (lab), 218
compliance
    output voltage, 222
conditional branch, 874–875
counter, 583–586
    16-bit
        async load (lab), 622, 625
        reset, sync (lab), 622
        sync load (lab), 622
    16-bit (lab), 621–626
    application
        stopwatch (lab), 631–633
    Carry$_{in}$ & Carry$_{out}$, 611
    carry-out glitch (lab), 630
    cascading, 611
    load, 614
    ripple, 583
    ripple (lab), 591
    ripple versus synchronous, 606
    synchronous, 584–586
    synchronous (lab), 591–592
    synchronous clear, 613
counter (lab), 591–592
CPLD, 597
crossbar (lab), 798
crossbar switch
    SiLabs output port enable, 872
    SiLabs port enable, 775
crossover distortion
    op-amp remedy, 259
crossover frequency, 312
crystal oscillator (lab), 788

Index

CS*, 649
current limit
    op-amp output, 242
current mirror, 222
    as load for op-amp first stage, 240
current source
    op-amp, 257
    single-transistor, 155
current source (lab), 172
current-hogging, 469
current-to-voltage converter (op-amp), 258
D flip flop, 579
D flip-flop, 579
D flip-flop (lab), 588–589
DA (lab), 839, 853
DAC
    see "↔ digital conversion: DAC designs", 697
DAC
    SiLabs (lab), 935–942
DAC (lab), 724–725
DAC reconstruction filter, 917
Darlington
    switch, applied (worked example), 351
Darlington transistor pair, 349
debounce
    pseudo (lab), 932
    spst RC debouncer, 873
debouncer, 577, 582
debouncer (lab), 592–594
debug
    DAC kinks can show wiring errors (lab), 927
    folded pin, 671
    oscillation frequency as clue, 673
    power through I/O pin, 672
decibel, 64
decibel: two definitions, 41
decoder
    2-to-4, 553
decoder
    I/O (lab), 834
decoding (address and control), 675
decompensated op-amp, 384
decoupling capacitor, 75
decoupling capacitor (lab), 176
decoupling of power supply (lab), 176
decoupling power supplies (lab), 262
delay loop, Silabs (lab), 801
delta–sigma ADC, 706–712
deMorgan, 520
    theorem, 522
depletion mode, 498
DG403 (lab), 489
DG403 analog switch, 475
difference amplifier, 189
    op-amp version, 301
difference amplifier (discrete), 201–206
    ...evolves into op-amp, 206
    design, 204–206
    differential gain, 205
differential amplifier, 189

clipping, 220
op-amp version, 301
differential amplifier (discrete), 201–206
    ...evolves into op-amp, 206
    design, 204–206
    differential gain, 205
differential amplifier (lab), 209–213
differential gain
    measuring (lab), 213
differentiator
    op-amp, 300
    passive RC, 55
differentiator (lab), 79–80
Digikey, 463
digital
    alternatives to binary, 516
    binary, 514
    contrasted with analog, 513
    resolution, 515
    why, 514
digital logic
    minimizing, 569
Dilbert, 776
diode
    zener, 122
diode (lab) , 29
diode circuits, 118
diode clamp, 119
diode clamp (lab), 136–137
diode drop, 119
DIP package
    breadboarding, 210
directives, 817
discrete transistor (defined), 152
display board (lab), 627
distortion
    differential amplifier, 220–222
    symmetric, 221
dropout voltage (lab), 454
DS89C430 (lab), 785
DSP, 957
dual-slope ADC, 703
dynamic RAM, 653
dynamic resistance, 122
E12 resistor values, 38
EA* (lab), 786
Early effect, 224–227
    calculating R_{out}, 226
    emitter resistors as remedy, 227
    Wilson mirror, 227
Ebers–Moll, 190
    r_e, 193
Ebers–Moll view of transistor, 190
Ebers-Moll
    I_S, 192
    I versus V curve, 192
    reconciling with $I_C = \beta \times I_B$, 201
edge triggering
    advantages, 603
    how implemented, 605

Index

edge- versus level-sensitive input, 580
edge-triggering, 579
electret microphone (lab), 309
electronic justice, 196
emitter follower, 157
   ...as impedance changer, 159
   clipping, 164
   input impedance, 162
   push–pull, 165
emitter follower (lab), 170–172
EPROM, 652
EQU (lab), 850
exponential
   charge and discharge, 54
      estimating, 55
   function, 54

f_{3dB}
   placing (worked example), 101
FB73-110 ferrite bead (lab), 376
feedback
   ...versus open-loop, 250
   effect on R_{out}, 252
   examples without op-amps, 250
   generally, in electronics, 249
   positive, 320
   quantitative view
      A, 396
      B, 396
      benefit from phase shift, 399–400
      loop gain, 397, 399
   the golden rules, 251
feedthrough, analog switch (lab), 491
ferrite bead (lab), 376
FET, 152
   similarities to bipolar transistors, 466
   structure, 497
   virtues, 468
Field Effect Transistor, 466
filter
   software
      FIR, 957
      IIR, 956
filter, DAC reconstruction, 917
finite state machine, 655
FIR filter, 957
Firewire, 973
flag
   hardware (lab), 883
flag (hardware), 873, 880
flash ADC, 701
flip flop
   clock, 579
   D type, 579
   dangers
      slow clock edge, 607
   debouncer, 577, 582
   edge- versus level-sensitive input, 580
   edge-triggering, 579
   jam clear, 580
   master–slave, 579

   setup time, 586
   shift register, 586
   SR latch, 576
   T type
      built from D, 585
   transparent latch, 578
flip-flop, 575
   D, 579
   D type (lab), 588–589
   debouncer (lab), 592–594
   divide-by-two (lab), 589
   do-this/do-that (generalized), 601
   edge-sensitive "flag", 600
   one-shot, synchronous (lab), 595–596, 633
   pathology: slow clock edge (lab), 589, 590
   propagation delay (lab), 589
   setup time, 605
   shift-register (lab), 594–595
   SR, 599
   SR (lab), 588
   sync set, async clear, 600
   sync set, sync clear, 601
   T type, 584
   T type (lab), 590
float
   function generator common (lab), 211
floating logic inputs (lab), 541–543
FM demodulation
   PLL (lab), 732
FM demodulator
   group audio project, 506
FM using '555
   see "oscillator...", 342
folded pin, 671
follower (op-amp), 252
Ford, Henry, 520
Fourier (lab), 132
Fourier series, 117
   square wave, 117
FPGA, 526, 598
frequency compensation (op-amp), 380–384
frequency domain (lab), 81
FSM, 655
full-wave rectifier, 120
fuse
   duration of overload, 126
   slow-blow, 126

GAL, 597
gate array, 526
glitch (digital), 608
   asynchronous carry-out, 612
   redundant cover as remedy, 609
GLUEPAL
   Verilog code (lab), 803
   pinout (lab), 803
GLUEPAL (lab), 781–785
golden rules (feedback), 251
ground, 6
   scope probe (optimal), 789
   two senses, 20

ground loop, 391–392
ground noise, 391
grounded-emitter amplifier, 191, 193–195
Harvard Class computer, 771
hazard (digital), 608
HDL, 1053, 1080
heat sink
   gasket, 463
HFA3096 array (lab), 209
high-pass filter (lab), 82–83
hysteresis (lab), 338
I/O
   A register, 822
   8051, 821
   DPTR, 821
   using buses, 820, 821
      port address, 822
      waveforms, 824
   without buses, 821
I/O decoder (lab), 834
I/O decoding
   8051 "lazy", 818, 820
IC current source, 443
   JFET type, 443
   LT3092, 443
   REF200, 443
ICM7555 (lab), 341
IF
   bit test equivalents (lab), 849
IGBT, 470
IIR filter, 956
impedance
   RC circuit, 68
      worst-case, 68
   …in a direction, 164
   …output, 14
   capacitor, 59
impedance relations: a rule of thumb, 21
INA149 difference amplifier, 276
INCLUDE
   (RIDE assembler/compiler/simulator), 947
INCLUDE (assembler/compiler), 817
indirect addressing, 823, 860–861
inductor, 112
input protection clamp (lab), 542
instantiation (in Verilog), 1056–1058
integrator
   op-amp, 294–299
   passive *RC*, 57
integrator (lab), 80
interfacing
   ADC & DAC to micro
      big board, 912
      SiLabs, 915–916
   ADC &DAC to micro
      Big Board, 911
interfacing among logic families
   3V with 5V, 692
   CMOS can behave like TTL, 690
   low-voltage (<5V), 692–693

   TTL to CMOS (5V), 690–691
interfacing among logic families , 689–693
interrupt, 829, 906–912
   on timer overflow, SiLabs (lab), 891
   SiLabs (lab) , 932–935
   application examples, 906
   edge-sensitive, 909
   enabling, 909
   implementation, 907–909
   in C, 911
   latency, 907
   priority, 909–910
interrupt (lab) , 884–886
interrupt priority, SiLabs (lab), 935
inverting amplifier (op-amp), 254
IR remote (lab), 1016
IRLIB9343 (lab), 453
IRLZ34 (lab), 486
ISA/EISA bus, 766
ISA/EISA port, 767–770
ISR (lab), 886
JFET, 466, 497
   linear region, 499
JFET as Variable Resistance, 499
JTAG, 973
justification
   left/right, DAC, SiLabs (lab), 936
keypad (lab), 628
keypad scan, 1006
KiBi, 654
Kirchhoff's Laws, 9
KRESET* (lab), 786
latched display (lab), 833
latency (interrupt), 907
LCD
   interface, 1007
lead-lag filter, 720–721
LED
   bi-color (lab), 896
left-justify (ADC/DAC), 917
LF411 (lab), 262
line noise, 390–391
little r_e, 193
LM311 (lab), 337
LM317, 440
LM358 (lab), 308
LM385–2.5 (lab), 452
LM386, 920
LM386 (lab), 931
LM4667 (lab), 495
LM4667 switching amplifier, 471
LM50 temperature sensor, 348
LM723 (lab), 452
LM741 (lab), 262
LM78xx, 439
LMC6482 (lab), 453
LMC7555 (lab), 341
load (counter), 614
loader (Dallas micro), 993
   debugging

COM port trouble (windows), 1000–1001
    flash error, 999
  USB-to-serial translation , 993–994
loader program (lab), 976
LOADER* (lab), 786
loading, 14
logic families, 528
  TTL and CMOS circuitry, 528
logic gate
  active pullup, 533
  noise immunity
    differential transmission, 532
    hysteresis, 514, 531
  open drain/open collector, 533
  speed versus power consumption, 535
  three-state, 535
  threshold, 529
logic gates
  noise margin, 529
logic probe (lab), 537
loop gain, 397, 399
low-pass filter
  microcontroller version (lab), 931
low-pass filter (lab), 81–82
LT1073 regulator (lab), 458
LT1215 (lab), 378
LT3092, 443
LTC1150 (lab), 305
LVDS, 532

macrocell, 597, 598
Mao, Chairman, 901
masking (software), 875–880
master–slave flop, 579
MAX294, 917–919
MAX294 (lab), 931
MAX294 frequency response, 747
memory
  as data organization, 651
  RAM (lab), 661–663
meter movement (analog), 27
microcomputer
  ...as state machine, 760
  address multiplexing (8051), 772
  Big Board lab computer
    first test program, 815
    single-step logic, 773
  big-board lab computer
    first test program, 818
    single-step logic, 772
  bootstrap startup, 762
  bus activity (ISA/EISA), 768–770
  elements of a computer, 760
    ...minimal, 760
  Harvard Class vs Von Neumann, 771
  history
    microcontroller, 759
    microprocessor, 758
    pre-microprocessor, 757
  microcontroller
    ...contrasted with microprocessor, 763

  choosing a controller, 762, 765
  microprocessor
    ...contrasted with microcontroller, 763
  processor control signals, 765–767
    68000 (Motorola), 767
    8051, 770
    address, 767
    application examples, 767
    I/O vs memory, 767
    ISA/EISA bus, 766
  single-chip controller
    first test program, 774–776, 818
microcontroller
  ...contrasted with microprocessor, 763
  history, 759
  low-cost, 759
microprocessor
  ...contrasted with microcontroller, 763
  history, 758
Miller effect, 371, 470
MJE2955 (lab), 378
MJE3055 (lab), 174, 378
module (Verilog), 1053
MOSFET, 152, 497
  R_{ON}, 469
  analog switch, 474–485
    application: multiplexer, 477
    application: sample-and-hold, 478–485
    application: switched cap filter, 477
    charge injection, 479, 483
    CMOS, 474
    imperfections, 475
  body diode, 468, 474
  CMOS
    logic gate, 473
  IGBT, 470
  input capacitance, 470
  logic gate, 473
  power switch, 469
    application: switching amplifier, 471
    paralleling, 469
  symbols, 467
moving pointer, 959–966
  end tests, 964–966
  store and playback
    big-board version, 963
    BSiLabs version, 964
  table copy
    assembly language, 960
    in C, 961
MOVX at Rn, 962
MOVX at Rn (lab), 928
MOVX RAM, SiLabs(lab), 983
multimeter (lab), 27, 29
multiplexer
  decoder included, 551
  implemented with gates, 553
  implemented with three-states, 553
  implemented with transmission gates, 553
multiplexing, 550
  address (lab), 786

multiplexing
    LCD data (lab), 840
multiplier
    worked example, 563
MXD1210 (lab), 832
NAND latch (lab), 588
negative feedback
    in ordinary usage, 248–249
noise
    "line", 390–391
    ground, 393
    ground loop, 391–392
    ground noise, 391
    parasitic oscillation, 394–395
    power supply (lab), 176–177
    RF pickup, 390, 393
non-inverting amplifier (op-amp), 252
NOSYMBOLS
    (RIDE assembler/compiler/simulator), 947
Nyquist sampling theorem, 694
OE*, 649
offset binary, 922
offset current, 290
offset voltage (V_{OS}), 286
Ohm's Law
    why it holds, 7
Ohm's law, 5
one-shot, synchronous (lab), 595–596, 633
op-amp
    stability
        split feedback (worked example), 403
open-collector (lab), 538
operational amplifier
    . . . designed from discrete transistors, 206
    AC amplifier, 301
    active rectifier, 281
    amplifier, inverting (lab), 265
    amplifier, non-inverting (lab), 265
    amplifier, summing (lab), 266
    as comparator (lab), 336
    chopper, 317
    current source, 257
    current source (lab), 271–272
    current-to-voltage converter, 258
    current-to-voltage converter (lab), 269–271
    detailed schematic ('411), 395
    difference amplifier, 301
    differential amplifier, 301
    differentiator, 300
    differentiator (lab), 306–307
    feedback generalized, 259
    finite gain
        effect of phase shift, 402
    follower, 252
    follower (lab), 263
    frequency compensation, 380–384
        . . . imposes integration, 380–381
        uncompensated and decompensated op-amps, 384
    golden rules
        conditional application, 281
        when they apply, 256
    imperfections, 284
        balancing paths for I_{bias}, 288
        bias current (I_{bias}), 288
        DC, 284
        dynamic, 284
        gain rolloff (GBW product), 290
        input and output voltage range, 292
        noise, 292
        offset current (I_{offset}), 290
        offset voltage, 286
        output-current limit, 291
        slew rate, 290
    integrator, 294–299
    integrator (lab), 303–306
        offset trim, 304
        self-trimming op-amp, 305
    integrator tests op-amp, 298
    inverting amplifier, 254
    made from transistor array (lab), 214–219
    non-inverting amplifier, 252
    open-loop (lab), 263
    phase shifter (lab), 266–267
    photodetector (lab), 269–271
    power booster, 258
    push–pull buffer (lab), 268
    rail-to-rail, 293
    single-supply AC amplifier (lab), 308–309
    slew rate (lab), 308
    source of name, 251
    specifications, ordinary and premium, 294
    summing amplifier, 255
    transresistance amplifier, 258
    virtual ground, 255
OR as Set/Pass* function, 575
ORG
    (RIDE assembler/compiler/simulator), 947
ORG (assembler/compiler), 817
OS-CON capacitor (lab), 459
oscillator, 326–335
    applied as suntan alarm, SiLabs (lab), 899
    implemented with SiLabs comparator (lab), 897
    FM using '555 (lab), 342
    IC: '555, 329
    IC: recent, 330
    relaxation (IC inverter), 328
    relaxation (op-amp), 327
    sinusoid: Wien bridge, 331–334
    triangle (lab), 342
oscillator (lab)
    IC oscillator: '555, 341–343
    op-amp relaxation, 338
    PWM motor drive, 340
    RC: IC Schmitt trigger, 339–341
    sinusoid: Wien bridge, 343–344
oscilloscope
    probe, 109
oscilloscope (lab: first view), 32
    triggering, 32–33
oscilloscope advice
    triggering, 1099–1103

Index

"auto" vs "norm", 1102
vertical mode, 1101
oscilloscope probe
  compensation, 112
output current limit
  op-amp (lab), 264
output impedance, 14
overflow (worked), 644
oversampling, 709

PAL, 526, 597
parasitic oscillation, 394–395
  discrete follower, 367–371
  discrete follower (lab), 374–376
  general remedies, 372
  generally, 356
  op-amp
    capacitive load as problem, 362–365
    capacitive load: remedy 1, feed back less, 363
    capacitive load: remedy 2, move feedback, 364
    capacitive load: remedy 3, split feedback, 364
    capacitive load: remedy 4, specialized driver, 365
    frequency compensation as remedy, 359–360
    gain roll-off as remedy, 361
    generally, 356–359
    phase lag as cause, 358
  op-amps
    stability criteria, generally, 365–367
  operational amplifier (lab), 376–379
    remedy: snubber, 379
    remedy: split feedback, 379
  remedy: base resistor (lab), 376
  remedy: ferrite bead (lab), 376
parasitic oscillation (lab)
  comparator, 337
PCA, SiLabs (lab), 895
phase margin, 383
phase shift
  hiding from it, 60
  rules of thumb, 69
phase splitter, 189–190
phase splitter (worked), 182
phase-locked loop, 716–722
  applications, 720
  FM demodulation (lab), 732
  loop filter (lab), 730–731
  phase detector
    edge-sensitive, 717
    edge-sensitive compared to XOR, 717
    XOR, 716
    XOR (lab), 732
  stability, 720–722
    lead-lag filter, 720–721
phase-locked loop (lab), 729–733
phasor diagram, 70
PID
  'I: integral of error (lab), 432
  controller design, 410
  D: derivative of error
    calculating (lab), 429
  D: derivative of error (lab), 428

effect of derivative, 415–420
formal description, 411
I: integral of error (lab), 430
integral, 420
motor control loop, 408
P: proportional to error (lab), 427–428
phase margin, 413
sample applications, 408
setting D gain, 417
the problem, 407
PLD, 526, 597–1079
  structure, 598–599
PLL
  see "phase-locked loop", 716
POP
  SiLabs (lab), 846
positive feedback
  hysteresis, 324–326
  oscillators, 326–335
potentiometer, 12
  as variable resistor, 12
  construction, 13
power, 4, 8
  P = IV justified, 9
power dissipation
  variable resistor, 459
power supply
  filter capacitor, 124
  fuse, 125
  split, 120
  transformer current rating, 125
  transformer voltage, 123
  unregulated, 123–126
power, in resistor, 8
priority
  interrupt, natural (lab), 935
probe
  oscilloscope, 109
probe compensation, 112
program "loop: origin of term, 815
program microcontroller
  Dallas 89C430
    hardware (lab), 1013
  Dallas DS89C430 (lab), 1012
programmable logic array, 526
project
  examples
    etch-a-sketch, 1027
    game: asteroids on scope, 1042
    game: pacman on scope, 1041
    inverted pendulum, 1039
    laser character display, 1025–1026
    sound source detector, 1026
    spinning LED display, 1036
    vehicles, 1036–1039
    X–Y table, 1028
  inverted pendulum, 1041
projects
  hardware
    displays, 1048
    motors, 1044–1045

muscle wire, 1045
  stepper motor drive, 1049–1051
nitinol actuator, 1045
software
  LCD drive, keypad scan, 1052
transducers
  accelerometer, 1047
  color discriminator, 1046
  force, 1047
  gyro, 1047
  infrared, 1046
  magnetic field, 1047
  speech recognition, 1048
  ultrasonic, 1046
projects (general advice), 1025
PROM, 652
propagation delay (lab), 589
PSEN* (lab), 786
PSENDRV* (lab), 786
pseudo-static RAM, 653
PSOC (Cypress), 1023
pulse duration program (lab), 1018
pulse-width measuring program (lab), 1014
pulse-width modulation, 700
PUSH
  SiLabs (lab), 846
push–pull
  SiLabs port option, 871
push–pull follower
  …simple, 158
PWM
  controller hardware, SiLabs (lab), 894–897
  configuration, SiLabs (lab), 895
  see "pulse-width modulation", 700

Q, 115–116
Q (quality factor), 131–133
QAM, 516
Quality factor, 115
quiescent defined, 178

race (digital), 608
radio
  AM demodulation, 127, 130
  AM receiver, 126–130
radio (lab)
  see "AM radio", 135
rail-to-rail op-amp, 293
RAM
  block diagram, 649
  synchronous SRAM, 652
  types: static vs dynamic, 653
  vs ROM, 652
RAM (lab), 661–663
ramp waveform, 53
RC
  time constant, 55
RC circuit
  impedance, 68
    worst-case, 68
reactance, 59
  capacitor, 59

reaction timer
  controller version (lab), 1016
reaction timer (lab), 750
rectifier
  full-wave, 120
  half-wave, 119
rectifier, full-wave (lab), 134
rectifier, half-wave (lab), 133
redundant cover, 609
REF200, 443
REF200 current-source IC, 464
reflections, 1092
relaxation: see *oscillator*, 327
relay, 486
RESET vector, 815
RESET51 (lab), 786
resistance, 7
  dynamic, 7
  parallel, 10
  shortcuts, 11
resistor
  "E12" values, 37
  "ten percent" values, 37
  carbon composition, 35
  color code, 36
  fabrication, 7
  metal-film, 35
  power, 38
  tolerance, 36
resonance, 114
resonant *RLC* circuit, 113–118
resonant circuit (lab), 131–133
RETI, 910
RETI (lab), 886
reverb, 958
RIDE
  .LST file, 951
  "WATCH" window, 953
  assembler directive, 947
  simulation, 951–955
RIDE (assembler/compiler/simulator), 946–955
ringing, 117
ringing (lab), 133
ringing of LC circuit, 138
ripple counter, 583
  contrasted with synchronous, 606
ripple counter (lab), 591
RLC circuit
  ringing, 117
RLC (lab)
  see resonant circuit, 131
ROM
  varieties: EEPROM, 652
  varieties: EPROM, 652
  varieties: Flash, 652
  vs RAM, 652
Romeo and Juliet, 574
Romeo and Juliet , 690
RS232, 966
RS232, SiLabs (lab) , 986–989

Index

sample and hold, 705
sample-and-hold
   charge injection, 479, 483
   errors, 480
sampling
   aliasing, 735, 738–739
   analogy with amplitude modulation, 740–743
   artifacts of sampling, described in frequency domain, 739
   filtering artifacts, 735
   seen as multiplication, 742–743
sampling rate, 694
SAR
   see binary search ADC, 705
SAR (lab), 724
saturation (bipolar)
   effect upon R_{in}, 199
saturation voltage (discrete transistor) (lab), 175
Schmitt trigger, 324
Schmitt trigger (lab), 338
scope multiplexer (lab), 752
self-checking testbench, 1088
self-checking testbench (in Verilog), 1058–1060
sequential circuit, 575
SerDes, 966
serial buses, 966–974
   I^2C, 973
   …a comparison, 974
   "one-wire", 973
   Firewire and Thunderbolt, 973
   IR remote, 972
   JTAG, 973
   SerDes, 966
   several serial buses, compared, 973
   SPI, 969–971
   UART (RS232), 966
   USB, 968
serial link
   LCD, 1007
servomotor (lab), 981
SET*, RESET* (lab), 589
setup time, 586, 605
shadow RAM, 653
Shannon
   see Nyquist sampling theorem, 694
shift register, 586
shift-register (lab), 594–595
simulation
   RIDE assembler/compiler/simulator, 951–955
sinewave generator (lab), 751
single-breadboard counter lab, 620
single-slope ADC, 616
single-step
   SiLabs IDE, 806
slow clock edge (lab), 590
slow-blow fuse, 126, 143
SPI, 969–971
   SPI RAM, 1009
SPI bus (lab), 979
SPI RAM
   hardware (lab), 1018
SPI, SiLabs (lab), 984

split feedback (lab), 379
split feedback (worked example), 403
split power supply, 120
SR flip-flop (lab), 588
SR latch, 576
SR latch (lab), 588
stack, 826
   for short-term storage, generally, 827
state machine, 655–660
   (lab), 663–669
   RAM-based (lab), 665–669
state machine (in Verilog), 1071
static RAM, 653
STEP PAL
   schematic (lab), 803
   single-step (lab), 787
   Verilog code (lab), 804
STEP PAL (lab) , 787–790
stopwatch (lab), 631–633
storage scope program (lab), 978
subroutine, 826, 830
   equivalent to function in C, 829
   SiLabs (lab), 844
   simulation, 828
subroutine (lab), 842
summing circuit, 255
suntan alarm (lab), 899
sweeping frequencies (lab), 81
sweeping function generator frequencies, 93
   analog generator
      analog scope: timed display, 97
      analog scope: XY, 95–96
      digital scope, 97
   artifacts caused by fast sweep, 96, 98
   digital generator
      digital scope, 98
switch
   bipolar transistor, 166
switch debouncer, 577, 582
switch, bipolar transistor (lab), 174, 176
switched-capacitor DAC, 699
switching amplifier, 471
switching amplifier (lab), 495
symbols, descriptive (assembler/compiler), 818
SYNC_OUT (function generator output), 33
synchronizer
   on carry-out (worked), 646
synchronizer (lab), 595
synchronous
   meaning, 584
synchronous carry
   for overflow (worked), 645
synchronous carry-out, 612
synchronous circuit
   advantages, 603
synchronous counter, 584–586
   contrasted with ripple, 606
synchronous counter (lab), 591–592
synchronous SRAM, 652

T flip-flop, 590

Index

T network, 297
T-flop, 584
temperature effects
   current mirror, 223
temperature instability, 196–201
   remedy
      R_E, 197
      compensation with second transistor, 200
      explicit feedback, 199
temperature response (bipolar), 196
temperature stability
   diff-amp (worked example), 239
Terminal (SiLabs utility) (lab), 987, 992
test loop (lab), 791
testbench (in Verilog), 1055
thermal calculation: MOSFET (lab), 486
thermal runaway
   op-amp push–pull, 242
thermally conductive gasket, 463
Thevenin
   model, 15
   resistance, 15
      shortcut, 16
      shortcut: justifying, 15
   voltage, 15
Thevenin model (lab), 26–27
three-state, 535
   buffer keypad (lab), 662
   where needed, 648
three-state (lab), 545
three-state buffer (lab), 545
Thunderbolt, 973
time constant, 55
time constant (lab), 78–79
time-domain (lab), 78
timer
   code, SiLabs (lab), 891, 893
timer, 8051 (lab), 980–982
TIP110 Darlington transistor pair, 349
tolerance (resistor), 36
tolerance codes, capacitor, 89
tracking ADC, 705
transistor (bipolar)
   beta, 158–160
   Early effect, 224–227
      calculating R_{out}, 226
      emitter resistors as remedy, 227
      limit to amplifier gain, 230
      remedies, 227–230
      Wilson mirror, 227
   first model, 154
   symbol, 154
   temperature effects
      mirror, 223
transistor pinout (lab)
   by experiment, 169
transistor summary
   Early Effect, 235
   impedances, common-emitter amp, 234
   impedances, diff-amp, 234
   Miller Effect, 235

   switch, 235
transistor switch
   bipolar, 166
transistors (bipolar)
   . . . why they are hard, 189
   common-emitter amplifier
      bypassed-emitter, 197
   Ebers–Moll model, 190
   effect of saturation upon R_{in}, 199
   grounded-emitter amplifier, 193–195
   phase splitter, 189
   temperature instability, 196–201
transmission gate, 466
transmission lines, 23, 1089
   reflections, 1092
   series termination, 1095
transparent latch, 578
transresistance amplifier, 258
tri-state, 535
triangle waveform, 53
truth table, 520
TTL
   syncing output from function generator, 33
TTL (function generator output), 33
two's-complement, 518
   worked example, 555–556
two's-complement (lab), 841

UART, 966
uncompensated op-amp, 384
universal gate, 521
USB, 968

variable resistor, 12
VCO
   group audio project, 504, 505
VCVS (lab), 374
VCVS filter, 354, 386
VCVS filter (lab), 373
Verilog, 526, 527, 571, 1053–1079
   "always @" form, 1060–1062
   "case", 1072
   "parameter", 1072
   behavioral versus structural design, 1064, 1065
   blocking vs non-blocking assignments, 1077–1079
   case, 1071
   flip-flop, 1060–1062
   hierarchical design, 1065–1070
   instantiation, 1056–1058
   state machine, 1071
      bus arbiter example, 1075
   synchronous carry_out, 1072
   template files, 1076
   testbench, 1055
      initialize, 1062, 1069
      self-checking, 1058–1060
      tabular output, 1057
verilog
   schematic views, 1083
   simulation, 1085
   simulation procedure, 1087
   synthesize, 1082

Index

testbench, 1086
   self-checking, 1088
VHDL, 526
virtual ground, 255
volatility (of memory), 653
voltage divider, 11
voltage divider (lab), 26
voltage regulator
   317 (adjustable), 441
   78xx (fixed), 439
   crowbar (lab), 457
   crowbar circuit, 444
   current limit, 436
   dropout voltage, 437
   evolving a design, 434–437
   IC regulator
      adjustable, 440
      fixed, 439
   LM317 (adjustable), 440
   low-dropout regulator, 438
   stabilization, 436
   switching, 445–450
      configurations: boost, buck, invert, 446
      efficiency, 447
      inverting (lab), 461
      step-down ("buck") (lab), 460
      step-up ("boost") (lab), 458
      web applications for design, 449
      Williams, Jim, 450
   switching (lab), 457
   thermal calculation (lab), 454
   thermal design, 441
   three-terminal, adjustable (lab), 456
   three-terminal, fixed (lab), 454–456
   voltage reference IC, 435
   zener, 434
voltage, defined, 6
Von Neumann computer, 771
VP01 (lab), 271, 453

watch
   SiLabs IDE, 807
watchdog (microcontroller), 777
watchdog timer (lab), 798
waveform
   ramp, 53
   triangle, 53
WE*, 649
Wien bridge, 331
Wien bridge oscillator (lab), 343–344
Wilson mirror, 227
wired OR, 534
Woody Allen, 325
worst-case impedance, 68

XC9572, 1081
XC9572XL, 1081
Xilinx, 598
Xilinx ISE, 1081
XOR as Invert/Pass* function, 574

Z_C, 60
zener diode, 122